Advances in Intelligent Systems and Computing

Volume 935

The series "Advances in Intelligent Systems and Computing" contains publications on theory, applications, and design methods of Intelligent Systems and Intelligent Computing. Virtually all disciplines such as engineering, natural sciences, computer and information science, ICT, economics, business, e-commerce, environment, healthcare, life science are covered. The list of topics spans all the areas of modern intelligent systems and computing such as: computational intelligence, soft computing including neural networks, fuzzy systems, evolutionary computing and the fusion of these paradigms, social intelligence, ambient intelligence, computational neuroscience, artificial life, virtual worlds and society, cognitive science and systems, Perception and Vision, DNA and immune based systems, self-organizing and adaptive systems, e-Learning and teaching, human-centered and human-centric computing, recommender systems, intelligent control, robotics and mechatronics including human-machine teaming, knowledge-based paradigms, learning paradigms, machine ethics, intelligent data analysis, knowledge management, intelligent agents, intelligent decision making and support, intelligent network security, trust management, interactive entertainment, Web intelligence and multimedia.

The publications within "Advances in Intelligent Systems and Computing" are primarily proceedings of important conferences, symposia and congresses. They cover significant recent developments in the field, both of a foundational and applicable character. An important characteristic feature of the series is the short publication time and world-wide distribution. This permits a rapid and broad dissemination of research results.

** Indexing: The books of this series are submitted to ISI Proceedings, EI-Compendex, DBLP, SCOPUS, Google Scholar and Springerlink **

More information about this series at http://www.springer.com/series/11156

Sukhan Lee · Roslan Ismail ·
Hyunseung Choo
Editors

Proceedings of the 13th International Conference on Ubiquitous Information Management and Communication (IMCOM) 2019

 Springer

Editors
Sukhan Lee
Intelligent Systems Research Institute
Sungkyunkwan University
Suwon, Korea (Republic of)

Roslan Ismail
Software Engineering
Universiti Kuala Lumpur
Kuala Lumpur, Malaysia

Hyunseung Choo
College of Software
Sungkyunkwan University
Suwon, Korea (Republic of)

ISSN 2194-5357   ISSN 2194-5365   (electronic)
Advances in Intelligent Systems and Computing
ISBN 978-3-030-19062-0   ISBN 978-3-030-19063-7   (eBook)
https://doi.org/10.1007/978-3-030-19063-7

This Springer imprint is published by the registered company Springer Nature Switzerland AG
The registered company address is: Gewerbestrasse 11, 6330 Cham, Switzerland

Preface

A rapid and full scale of convergence among communication, computation, robotics, and artificial intelligence is underway with humans and things being evermore in integration and interaction while being increasingly embedded in ambient intelligence. This ongoing convergence, leading to the 4th Industrial Revolution, is driving a fundamental change in the way we live with a new generation of transportation, health care, education, manufacturing, business, personal service, security, and energy and environmental management. As we ride on the big wave of the 4th Industrial Revolution, deployment of intelligent functions and services over the Internet of things, cyber-physical systems, and robotics in distributed environments is expected rapidly rising in such a way that our society is transformed to being massively smart with physically and mentally augmented human beings within. Under this circumstance, information management and communication in ubiquitous and collective computing and communication environments become essential. Developing fundamental theories and advancing implementation tools to address the issues involved in information management and communication in the wake of the convergence among communication, computation, robotics, and artificial intelligence as described above is of critical value.

This volume represents the collection of papers presented at the 13th International Conference on Ubiquitous Information Management and Communication (IMCOM 2019), held in Phuket, Thailand, during January 4–6, 2019. Eighty-eight papers out of 228 papers submitted from all around the world are accepted for presentations. Eighty-eight contributions to this volume are organized into five parts: Part 1. Network Evolution, Part 2. Intelligent and Secure Network, Part 3. Image and Video Processing, Part 4. Information Technology and Society, and Part 5. Data Mining and Learning. To help readers understand better, a chapter summary written by the editors of the volume is included in each chapter as an introduction. It is the wish of the editors that readers find this volume informative and enjoyable.

The editors would like to express their appreciation to the members of the organizing and program committees of IMCOM 2019 for their hard work for the conference. The editors would also like to thank Springer-Verlag for undertaking the publication of this volume.

Sukhan Lee
Roslan Ismail
Hyunseung Choo

Contents

Image and Video Processing

Network Evolution

Part 1. Network Evolution

We live in an era of massive and high-speed interconnections with Internet of Everything under the framework of cyber physical systems. Things in homes, offices, hospitals, schools, shopping malls, factories, transportations, infrastructures, etc., are to be interconnected in and across their domains, including individuals, by wired or wireless networks into a holistic smart space in the coming years. Network evolution becomes mandatory due to the fact that, not only the scale of networking grows unprecedentedly, but also the requirements are highly diversified due to the heterogeneous nature of networking. For instance, a global networking is feasible with the availability of a proper infrastructure for internet, while a temporary local ad hoc networking is necessary without such an infrastructure available. The evolution of networking under a large scale of heterogeneous interconnections exposes a number of issues to handle: (1) minimizing delays between communication nodes and between networks, (2) dealing with efficiency in energy and resource utilization, (3) measuring and enhancing network throughputs, (4) monitoring fault and ensuring fault tolerance, and, last but not least, (5) providing network security.

This part is organized with sixteen papers that address the five issues described above. The sixteen papers are grouped into five sections based on the issues they deal with. In what follows, a brief summary of each group is presented by describing the key concepts of individual papers.

(1) **Minimizing Delays between Communication Nodes and between Networks:** The paper, entitled "Review of Internet Gateway Discovery Approaches in MANET," by Mufind Mukaz ebedon, Amna Saad, and Husna Osman addresses the problem related to the network delay in the Mobile ad hoc network (MANET). MANET form on the go network without any infrastructure, but for a mobile node which need to communicate outside the network, it needs a gateway. The current approaches for building gateway are limited in performance due to lack of stability. To improve stability, decrease overhead, and end to end packet delay, the combination of clustering approach and Internet gateway discovery approach has been performed. The paper, entitled "An Efficient Edge-Cloud Publish/Subscribe Model for Large-Scale IoT Applications," by Van-Nam Pham and Eui-Nam Huh addresses the delay in IoT related applications. IoT produces enormous data, which need to be analyzed and stored at best suited places. IoT devices often directly connect to their closest brokers to send and receive data, which requires an efficient broker overlay network for Pub/Sub systems to support wide geo-distribution IoT platforms. This paper proposes an efficient Edge-Cloud Pub/Sub model that coordinate brokers in the system for large-scale IoT application. It reduces network delay while delivering data for these applications. The paper, entitled "Split Point Selection based on Candidate Attachment Points for Mobility Management in SDN," by Bora Kim, Syed M. Raza, Rajesh Challa, Jongkwon Jang, and Hyunseung Choo address the handover delay in wireless sensor networks. Split point is switch on the flow path which reroute the traffic towards new

AP after handover. This paper presents a split point selection algorithm which having minimum weight by utilizing the list of candidate APs for next handover of Mobile Node (MN) to determine topologically most efficient split point in terms of handover delay.

(2) **Dealing with Efficiency in Energy and Resource Utilization:**
The paper, entitled "Energy-Efficient Computation Offloading with Multi-MEC Servers in 5G two-tier Heterogeneous Networks," by Luan N. T. Huynh, Quoc-Viet Pham, Quang D. Nguyen, Xuan-Qui Pham, VanDung Nguyen, and Eui-Nam Huh addresses the problems related to efficient energy utilization by Multi-Access Edge Computing in 5G networks. Multi-Access Edge Computing in 5G networks, enhances computation capabilities and power limitations of mobile devices by offloading computation task to the nearby MEC servers. However offloading increase network traffic and increase delay. To decrease the energy consumption as result of offloading, this paper propose random offloading search algorithm. The paper, entitled "Robust Bio-Inspired Routing Protocol in MAN-ETs using Ant Approach," by Arush Sharma and Dongsoo S. Kim proposes an energy efficient multi-path routing algorithm in MANET based on foraging nature of ant colonies. The MANETs suffers from instability during communication which requires efficient routing protocol. The paper, entitled "Energy Level-based Adaptive Backscatter and Active Communication in Energy-harvesting Network," by Kwanyoung Moon, Yunmin Kim, and Tae-Jin Lee address the issue related to energy optimization in Radio Frequency IDentification (RFID) system. RFID active tags and semi-passive tags have limited battery capacity and it affects overall network performance This paper improves network performance by performing data transmission and energy harvesting adaptively where the tags operate in the passive mode as well as the active mode for data transmission and perform selective energy harvesting mode to improve throughput and energy utilization ratio. The paper, entitled "Capacity Planning for Virtual Resource Management in Network Slicing," by Rajesh Challa, Syed M. Raza, Hyunseung Choo, and Siwon Kim addressed the capacity planning problem. To efficiently utilize the resources of the network, the service providers need to multiplex their physical infrastructure over multiple logical networks which is called network slicing. One key issue is, what is the minimum slicing criteria? To address this problem, this paper proposes a simple orchestration-centric discrete event simulator to develop and evaluate minimum slice capacity requirement.

(3) **Measuring and Enhancing Network Throughputs:**
The paper, entitled "On Evaluating a Network Throughput," by Alexey S. Rodionov calculates "throughput" of a network by arithmetical average of maximum flows between all pairs of nodes. This index shows how good a network is from the point of view of possibility of transferring flows through it. The paper, entitled "Frame Aggregation Control for High Throughput and Fairness in Densely Deployed WLANs," by Toshitaka Yagi1 and Tutomu Murase proposes a novel control method. This method controls the Frame Aggregation size and/or Contention Window size for achieving high throughput and fairness between multiple wireless LANs (WLANs). The paper, entitled "Novel Traffic

Classification Mechanism in Software Defined Networks with Experimental Analysis," by Youngkyoung Kim, Syed M. Raza, Van Vi Vo and Hyunseung Choo performs effective traffic classification using different IP addresses which are assigned to Mobile Node (MN) and improved performance of the network. The paper, entitled "Development of Sublayer Network State Inference Technology Based on Protocol State Dynamic Extraction for Improved Web Environment," by Yewon Oh and Keecheon Kim addresses the problems related to the web server environment by creating autonomously controlled protocols based on dynamic extraction in the sublayers. The paper, entitled "Flexible Deployment System based on IoT Standards using Containers in Smart Factory Environments," by Changyong Um, Jaehyeon Lee, and Jongpil Jeong build smart factory "The Internet of Things (IoT)" system through distribution of Doker Containers.

(4) **Monitoring Fault and Ensuring Fault Tolerance:**
The paper, entitled "Fault-Tolerant Topology Determination for IoT Network," by Ryuichi Takahashi, Mitsumasa Ota, and Yoshiaki Fukazawa proposes a method to generate a fault-tolerant routing topology by a genetic algorithm. The fault occurs when there are multiple communications on a gateway hence loss of data occurs in such cases. The paper, entitled "Design and Experimental Validation of SFC Monitoring Approach with Minimal Agent Deployment," by Jisoo Lee, Syed M. Raza, Rajesh Challa, Jaeyeop Jeong, and Hyunseung Choo presents an SFC monitoring approach for various critical functions like load balancing, fault management, and congestion avoidance that reduces the signaling cost by deploying Monitoring Agents in minimum number of SFs. The paper, entitled "A Disaster Resilient Local Communication System without Depending on the Internet Connectivity," by Arisa Tanaka, Taka Maeno, Mineo Takai, Yasunori Owada, and Masato Oguchi proposed an independent local communication network to avoid dependability on the global networks for communication in terms of telephone calls, e-mails, and social networking services (SNSs). This network act as an alternative in terms of the failure in global network due the damage as a result of disaster.

(5) **Network Security:**
The paper, entitled "ARP Request Trend Fitting for Detecting Malicious Activity in LAN," by Kai Matsufuji, Satoru Kobayashi, Hiroshi Esaki, and Hideya Ochiai handle the issue related to security in the Local Area Network. They detect the malicious network activity by ARP monitoring request trend.

A Disaster Resilient Local Communication System Without Depending on the Internet Connectivity

Arisa Tanaka[1](✉), Taka Maeno[2], Mineo Takai[3], Yasunori Owada[4],
and Masato Oguchi[1]

[1] Ochanomizu University, 2-1-1 Otsuka, Tokyo, Japan
arisa@ogl.is.ocha.ac.jp, oguchi@is.ocha.ac.jp
[2] Space-Time Engineering, 3-27-3 KandaSakumacho, Tokyo, Japan
tmaeno@spacetime-eng.com
[3] UCLA, 3803 Boelter Hall, Los Angeles, CA, USA
mineo@ieee.org
[4] NICT, 2-1-3 Katahira, Aoba-ku, Miyagi, Japan
yowada@nict.go.jp

Abstract. Currently, networks are essential to the lives of people and have spread widely as an important information infrastructure. Many people use networks as a way to contact others, for example, via telephone calls, e-mails, and social networking services (SNSs). Therefore, a communication failure, which can be caused by damage to network systems or network congestion when many people try to use the network at the same time in an emergency, may bring about anxiety or embarrassment during a disaster. Therefore, in this research, we aim at constructing an information communication system useful in large-scale disasters between subnets gathered in emergency time by using that environment. In particular, focusing on the chat system by P2P communication, we developed a communication infrastructure that can handle communications even without the Internet, and we developed an application to be placed on it. We also show the usefulness of the system.

Keywords: Disaster · P2P · NAT traversal · STUN/TURN

1 Introduction

Currently, networks are essential to the lives of people and have spread widely as an important information infrastructure. Many people use networks as a way to contact someone, for example, via telephone calls, e-mails, and SNSs. Therefore, a communication failure, which can be caused by damage to network systems or network congestion when many people try to use the network at the same time during an emergency, may bring about anxiety and embarrassment during a disaster. The impact is particularly great for Japan because it is a country with frequent earthquakes [1]. The main causes of communication failure are issues

© Springer Nature Switzerland AG 2019
S. Lee et al. (Eds.): IMCOM 2019, AISC 935, pp. 5–17, 2019.
https://doi.org/10.1007/978-3-030-19063-7_1

with the base stations and backbone networks [2,3]. If one cannot connect to the Internet because the base station or backbone networks become damaged or malfunction due to a disaster or if the bandwidth to the Internet becomes very narrow, it is impossible to share information over the network. There is also a desire to connect subnets at evacuation centers when they gather as on-the-spot needs. Thus, even when the Internet is disconnected, it is necessary to realize the configuration of internetworking between subnets gathered in emergency situations and a useful information communication system in case of a large-scale disaster. In particular, this paper focused on an Information-sharing system that enables terminals participating in wireless LAN to share data such as messages and files.

Normally, many existing networks used by individuals and small organizational units constitute private networks and are connected to the Internet via Network Address Translator (NAT) routers. Therefore, we propose the construction of a communication infrastructure that allows for NAT routers to autonomously construct and connect networks without going through the Internet and not making modifications to these private networks. At that time, we consider methods to realize Peer-to-Peer (P2P) communications such as telephone calls and messages among nodes of other private networks. As an example, we assume the subnets gathering at evacuation centers as on-the-spot needs, and private networks between emergency vehicles and private networks that are constructed temporarily among shelter centers at the time of a disaster are connected to each other in an ad hoc manner.

As a solution, since there are many isolated environments where terminals are connected to NAT routers at the time of a disaster, we propose the connection of the isolated private networks and information sharing by communication between terminals. For that reason, we build a communication infrastructure by creating a mechanism for NAT routers. The proposed infrastructure is considered the most useful in practice to apply a NAT traversal technology and to create a mechanism that allows for isolated private networks to be connected by a chat system (see Fig. 1).

2   Related Work

At present, much research has been completed on the NAT traversal problem that communication cannot be started from the outside of NAT to the internal network.

In [4], this is related to the problem that anyone of the conventional NAT-f (NAT-free protocol), which solves the NAT traversal problem by cooperation between the external node and the home gateway, can access internal nodes in the home network. Then, it has been shown that grouping in service units can simultaneously realize access control from external nodes and control of services provided by internal nodes.

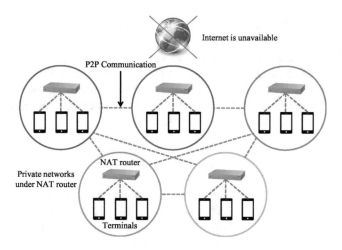

Fig. 1. Target figure

In [5], a scheme that realizes NAT traversal has been proposed by using a DNS server and a NAT router to cooperate with each other. The proposed method is called an NTS (NAT-Traversal Support) system. For this system, an NTS server with a remodeled DNS server, NTS router remodeling of the NAT router, and an NTS protocol as a protocol are used. Also, unlike NAT-f, which requires addition of terminal functions, this system can realize NAT traversal at the end-end without adding functions to general user terminals.

In [6], this makes it possible to exceed symmetric-type NAT by extending STUN that cannot originally correspond with the symmetric type NAT and implement a method to start TCP communication from outside the NAT environment.

Related research is used to solve the NAT traversal by preparing a server separately from the NAT router, and it is not NAT traversal between private networks but NAT traversal between the private network and the external network. For this reason, it is considered very useful especially in case of disaster to solve NAT traversal between private networks and creating a mechanism for the NAT router itself, which is the focus of this research.

End-to-end communication protocols are also related to our work. For example, Dual Stack Mobile IPv6 equips NAT detection function which support NAT traversal for end-to-end communication between nodes [7]. Host Identity Protocol (HIP) also provides a mechanism to realize NAT traversal so as to achieve end-to-end communication [8].

However, these protocols require complicated procedure, in which some servers and networks should be equipped to provide a service for NAT traversal. Basically, they are designed for mobile terminals in a normal situation, of which the conditions are completely different from an emergency case. We assume a large-scale disaster in our research work, and backbone networks and servers are unavailable mostly in such a case. Only a subnet connecting NAT routers and terminals exists. If such isolated subnets aggregate, each of them are a Wi-Fi subnet in most cases, and they would like to communicate with each other temporarily on the spot of disaster, we don't want change the setting of each terminal. Instead, it is desirable that NAT routers have a mechanism to connect with each other and provide NAT traversal function for their terminals so as to achieve end-to-end communication with terminals on other subnets. We focus on such a situation in this research work.

3   System Proposal

We propose a system that make it possible for terminals participating in the wireless LAN to share data such as messages and files in situations where the Internet cannot be used due during disasters and cannot make contact with other parties. In order to realize this system, we built a communication infrastructure and developed a usable application.

3.1   Communication Infrastructure

In particular, Wi-Fi AP and NAT routers (hereafter called Wi-Fi routers) can create local private Wi-Fi service individually, which is independent to the mobile network or backhaul network. This makes it possible to communicate through private networks; that is, a mobile ad hoc network (MANET) between Wi-Fi routers is constructed. Here, an MANET is an autonomous decentralized network that does not depend on base stations and fixed networks but rather on mobile terminals as constituent elements [9]. Despite the advantage of being able to configure a network without requiring existing facilities, this approach has not been used in practice. This is in part because many items using MANET were systems requiring advance preparation on the terminal side. Therefore, an ordinary MANET often constitutes a network between terminals, but this time, as mentioned above, we construct a network between Wi-Fi routers. In other words, by creating this mechanism on the Wi-Fi router instead of the terminal, we construct an MANET between Wi-Fi routers, which will enable communication between terminals using Wi-Fi routers without changing the terminal side.

However, since direct connection from the external network to a private network address is impossible, we use NAT traversal. We use the Session Traversal Utilities for NATs (STUN)/Traversal Using Relay NAT (TURN) server and signaling server for this NAT traversal. To specify the communication partner, the terminal finds both the STUN/TURN server and the signaling server ad hoc and realizes NAT traversal communication.

Subsequently, Web Real-Time Communication (WebRTC) is used for a system that performs P2P communication between terminals, which is a framework for realizing real-time communications such as video calls on the browser. Using WebRTC, voice chat, video chat, and file sharing among web browsers without plug-ins are available, which has mechanisms for P2P communication that communicate with the other party directly, realizing NAT traversal. Not all web browsers are supported, and they are limited to Chrome, Firefox, and a few other browsers. However, this method has many advantages. It does not require the installation of a dedicated application for terminals, and it can send large amounts of data at high speeds. DTLS is adopted for communication, and encryption is performed. This enables data sharing of messages and files among the terminals participating in the wireless LAN.

3.2 Application

As an application, we developed a chat system for the purpose of sharing data such as messages and files among terminals participating in wireless LAN by WebRTC. Figure 2 shows the actual screen, and Fig. 3 shows the actual administration screen, which is also assumed to be accessed from a personal computer. The rough usage method is as follows.

1. Connect to the wireless LAN from AP
2. Open the browser (Chrome or Firefox) and the site
3. Enter the user name, and start communication by pressing the Connect button
4. Enter the message; the communication partner connected in real time appears on the button such as "Send to name"; press the button of the person you want to communicate with and send
5. In the case of a file, press the connection button inside the frame in which the name of the communication partner is written, and wait for the success of the communication
6. When it succeeds, put the file you want to send, wait for approval from the other party, and send it if approved

3.3 Problems of the Proposed System

First, we explain the technology necessary for P2P communication on the Internet.

1. NAT Traversal
2. Knowing which service can be used by the communication partner
3. Resolving the partner ID and IP address
4. Determining whether the other party is online or offline

These four tasks are necessary, and some type of server functions are usually provided for them on the Internet. Determining how to perform these tasks between ad hoc connected networks in an environment where they are not connected to the Internet is a problem. At the present stage, only 1 has been addressed using the STUN/TURN server as outlined above, but 2, 3, and 4 remain open problems. For this reason, we assumed that all nodes can use common services and have a mechanism to resolve the ID and IP address of the other

Fig. 2. Information-sharing application

Fig. 3. Actual administration screen

party. An experiment was carried out while assuming both parties are online. To summarize, based on the experiments, we created two private networks using two Wi-Fi routers and confirmed the feasibility of the communication environment to share information using NAT traversal between them. We also proposed an application that works under such circumstances, and the experiments are described in Sect. 7.

4   Traditional Peer-to-Peer (P2P) Mechanism

P2P is a system that directly connects terminals on the network to each other as equal peers, which transmits and receives data (see Fig. 4). It is also a generic term for software and systems that communicate using such a scheme. In contrast to the client-server model (see Fig. 5), which prepares a specific server and exchanges information, a system that directly connects users and exchanges voice data and files has been put to practical use. As a result, communication between the terminals is enabled without going through a specific server. Configuration of P2P is useful in the case of a disaster for exchanging information. However, we need to modify a terminal communication function to realize Information-sharing application such as in Fig. 2, which is not realistic because terminals are owned by normal consumers in most cases. Therefore, we proposed a mechanism explained in the previous section, in which private networks under the Wi-Fi routers are connected. An MANET is constructed, on which terminals can communicate with each other in a traditional P2P manner.

Connect directly

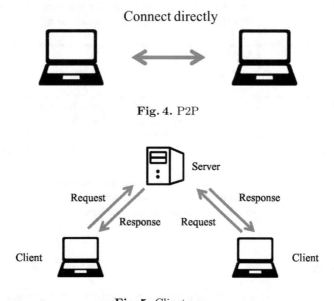

Fig. 4. P2P

Fig. 5. Client-server

5   NAT Traversal and WebRTC

5.1   NAT Traversal

Many terminals belong to a private network and are under NAT, and communication is common from a terminal to an external network. However, it is impossible to communicate directly from an external network to the terminal under NAT.

One method to solve this problem is called NAT traversal.

5.2   STUN and TURN

STUN is one method of NAT traversal. The communicating host makes a UDP connection to the STUN server and acquires the global IP address and port number assigned by the NAT. There are 4 types of NAT as follows: full cone NAT, address-restricted cone NAT, port-restricted cone NAT, and symmetric NAT (Table 1). The types of NAT that STUN can handle are limited to full cone NAT, address-restricted cone NAT and port-restricted cone NAT.

In Table 1, the lower an entry is in the table, the more restrictive the usage restrictions. In particular, the most severe symmetric NAT cannot be handled with STUN, but by using TURN for symmetric NAT, all types of NAT can be handled. However, since all communication is performed via the TURN server, it is no longer P2P communication, and the server is heavily loaded.

Table 1. Type of NAT

Type name	Restrictions of terminals on the Internet side that can access the port assigned by NAT	Severity of usage restrictions
Full cone	Any terminal	Gentle
Address-restricted cone	Terminals already accessed by terminals under NAT	Severe
Port-restricted cone	Only from the terminal and port number accessed by the terminal under NAT	More restrictive than address-restricted cone
Symmetric	Only when the communication source terminal and the communication destination terminal are one to one	Pretty severe

5.3   WebRTC

WebRTC is a mechanism for realizing real-time communication with a browser; interconnection between terminals can be achieved using P2P communication. Since UDP is used for the protocol, high-speed communication is possible. Signaling and interactive connectivity establishment (ICE) are required for P2P communication via WebRTC.

Signaling is a method for obtaining information such as the IP address information and the port number of a communication partner. ICE is a protocol for exchanging information to establish a session for NAT traversal communication.

5.4   NAT Traversal and P2P Communication

To perform P2P communication between terminals in a private network using WebRTC, NAT traversal is required. At this time, signaling and ICE mechanisms are used. For terminals to establish P2P sessions with each other, it is necessary to exchange information such as the IP address and connection port number of the communication partner by signaling. Thus, a signaling server that can be accessed from any terminal is required.

ICE is a summary of NAT traversal procedures such as STUN and TURN. It collects candidates that are likely to communicate, exchanges the candidate information with the opponent by signaling, and attempts to communicate with the other party. We illustrate P2P communication by STUN on an ordinary network in Fig. 6 and communication by TURN on an ordinary network in Fig. 7.

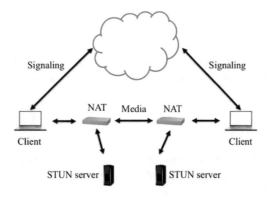

Fig. 6. P2P communication by STUN

6   Assumed Environment

Figure 8 shows the assumptions about the communication environment during a disaster.

At this time, when the private networks are connected ad hoc via Wi-Fi routers, it is supposed that Wi-Fi routers are used that have a mechanism that can uniquely detect the signaling server and STUN/TURN server autonomously and share information via P2P communication over NAT.

Even in situations where it is impossible to connect to the Internet in a disaster, by connecting to the Wi-Fi of this Wi-Fi router, it becomes possible for terminals connected to Wi-Fi to share information. Currently, we are building an environment to conduct experiments assuming such a communication environment. The experimental environment will be introduced in the next chapter.

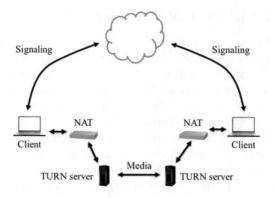

Fig. 7. Communication via TURN

7   Experiment Environment

In order to confirm whether the application actually moves while the Internet is shut down, the following experiment is conducted. The edge server is explained first.

7.1   Edge Server to Use

In this research work, an edge server is used to implement Wi-Fi router functions shown in Fig. 8. This can be regarded as an extension of Wi-Fi AP, which has a sort of server functions. The edge servers currently in use are as follows (see Fig. 9). Additionally, DTN technology is included because we will consider the development of new systems in the future.

– OS uses Debian GNU/Linux
– Signaling and STUN/TURN installed

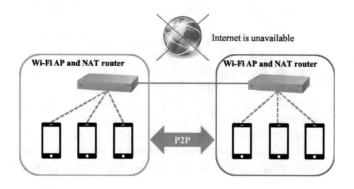

Fig. 8. Assumed environment

- A general-purpose platform that supports real system construction and system analysis/linkage
- Build a network autonomously
- Enable information synchronization between real systems by DTN technology

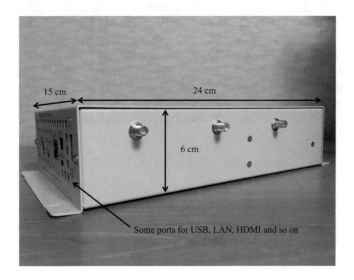

Fig. 9. Edge server to use

7.2 Experiment Environment

In this research, by using both WebRTC application and coturn, which is open-source software that can build STUN and TURN servers, we constructed a local experimental environment. The environment is shown in Fig. 10. In this environment, sharing information using NAT traversal in the local environment is experimentally examined. By doing this, we simulate experiments in the event of a large-scale disaster such as when the Internet is blocked.

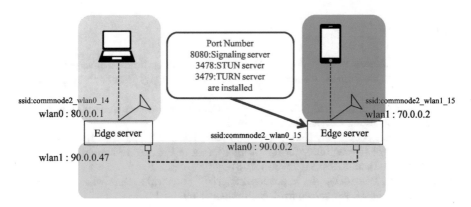

Fig. 10. Experiment environment

Hostapd was installed on each edge server equipped with wireless LAN and then operated as Wi-Fi routers. This created two different private networks.

For the Client, we installed one Android smartphone and one personal computer connected to different edge servers. Each IP address is acquired by DHCP.

As a server, the developed signaling server is constructed at port number 8080; coturn, which is the STUN and TURN server, is constructed at port number 3478 and 3479. The STUN/TURN server setting in the signaling server is set at port number 3478 and 3479 of the same server.

Since the two edge servers are connected by wireless LAN and are in the same network, the whole network is red, blue and green. Then, the server checks whether the clients can chat with each other. Through these, we confirm the usefulness of whether we can chat and share information without depending on the Internet.

8   Conclusion

We proposed and constructed a useful information communication system for how to respond when each subnet is gathered in the network when large-scale disasters occur. Among them, we devised a chat system between evacuees through NAT traversal and developed a communication infrastructure and application to address it.

Experiments simulating the environment are conducted in which the network has run out, and an appropriate environment is proposed. This confirms that the system is useful in an actual environment.

In the future, we will realize the scheduled experiment and evaluate it. In addition to a chat system, we aim to devise a system that is more useful in large-scale disasters.

Acknowledgement. This work is partially based on a collaborative research agreement between Ochanomizu University and the National Institute of Information and Communications Technology (NICT). Additionally, this work is partly supported by JST CREST JPMJCR1503, Japan.

References

1. Disaster Management in Japan: About the damage estimation and countermeasures of the earthquake immediately under the capital city (last report), December 2013
2. Nakamura, I.: Large scale disaster and communication network - I think about the Great East Japan Earthquake, July 2011
3. Kagawa, K., Kuno, Y., Tamura, H., Takada, H., Furutani, M., Minamikata, N.: Improvement of credibility for operation system in the case of large disaster. NTT DOCOMO Tech. J. **14**(4), 25–36 (2013)
4. Suzuki, H., Watanabe, A.: A proposal for NAT traversal system with communication group-based service control, September 2009

5. Miyazaki, Y., Suzuki, H., Akira, W.: Installation of the NAT traversal system independent of user terminals, July 2008
6. Kuroda, J., Nakayama, Y.: STUN-based connection sequence through symmetric NATs for TCP connections (2011)
7. Soliman, H.: Mobile IPv6 support for dual stack hosts and routers, RFC5555, IETF (2009)
8. Moskowitz, R., Heer, T., Jokela, P., Henderson, T.: Host Identity Protocol Version 2 (HIPv2), RFC7401, IETF (2015)
9. Mase, K.: Mobile ad hoc network, March 2002
10. WebRTC. https://webrtc.org/
11. Coturn. https://github.com/coturn/coturn
12. Dutton, S.: WebRTC in the real world: STUN, TURN and signaling, 4 November 2013
13. Real time communication with WebRTC. Google Developers
14. Motohashi, S., et al.: A proposal and study of an information sharing system on a Wi-Fi AP with server function in nomal times/disastrous situation. In: DEIM 2016, Hukuoka, pp. 1–5 (2016)

Flexible Deployment System Based on IoT Standards Using Containers in Smart Factory Environments

Changyong Um, Jaehyeong Lee, and Jongpil Jeong[✉]

Sungkyunkwan University, Suwon, Gyeonggi-do 16419, Republic of Korea
{e7217,objective,jpjeong}@skku.edu

Abstract. The Internet of Things in the Smart Factory is the most fundamental and important element for building the cyber physical system. The Internet of Things performs a key function that connects the physical and digital components of a manufacturing site and can be formed in a variety of structures. The system of smart factory needs to be designed by adopting IoT standards in order to have a reliable structure. System designers can adopt structures based on IoT standards to implement stable and reliable systems, but they must perform cumbersome tasks to comply with the standards. This paper proposes implementation of IoT system through deployment of docker containers, which is a virtualization technology, and ensures stability and flexible connectivity of rapid deployment and processes.

Keywords: Smart factory · Internet of Things ·
Cyber-Physical System · IoT Standards · Virtualization · Docker

1 Introduction

In the fourth Industrial Revolution, various technologies are exploding due to the development and fusion of related technologies, and emerging technologies are quickly applied to various industries. Since many countries regard the fourth industrial revolution as an important component of national competitiveness, they are investing in relevant laws and standards for the integration of various industries and technologies. The fourth industrial revolution is a widely known concept by Klaus Schwab [1], who claims that changes in speed, scope, depth and system impact are taking place based on Information Communication Technology (ICT) beyond the digital revolution. It refers to a new technological revolution that combines various technologies such as robots, the Internet of Things, artificial intelligence, big data, self-driving cars, 3D printing, and nanotechnology. A key element of innovation in manufacturing environments is Cyber-Physical System (CPS), which is based on Internet of Things (IoT), which closely connect between the digital layer and physical layer, and thus smart factory is established [2].

© Springer Nature Switzerland AG 2019
S. Lee et al. (Eds.): IMCOM 2019, AISC 935, pp. 18–33, 2019.
https://doi.org/10.1007/978-3-030-19063-7_2

Through IoT and Machine-to-Machine (M2M) communication, manufacturers prepared a framework for effective management of devices. The IoT system can be deployed in a variety of structures, and the system has a vertical type of solution depending on each environment. These solutions are deployed for specific industries, resulting in increased investment and operating costs with custom hardware and software requirements. International standards organizations have proposed oneM2M standards for linking independent systems. A system based on oneM2M standard provides structural stability and connectivity to other standards and devices. In a smart factory environment where CPS is based, connectivity to various devices and components is important. A stable and flexible CPS is implemented by establishing the system suitable for IoT standards [3–6].

Docker is open-source virtualization technology and uses container concepts in a lighter form than traditional virtualization. Docker containers use fewer resources than traditional hypervisor methods to provide developers with greater efficiency. Developers can build a common process-independent environment, assign functions, and adjust further as changes occur. Communicating an environment suitable for the process avoids environmental hazards and ensures stability. Because the type and usability of facilities in a manufacturing environment are limited compared to normal IoT device operating environments, the process of the system can be managed even if the number of Docker images is not large [7].

In the smart factory, CPS should have flexible connectivity to systems external environments as well as to inside the factory. Therefore, smart factory systems should adopt IoT standards-based structures to avoid isolation due to non-standard. For this system to be configured according to IoT standards, certain rules must be followed. Developers or managers face the challenge of having to deploy standard components of the system to these rules by the node and consider their suitability. Unlike other areas, manufacturing sites have structures that have a small number of device types on a large scale. In such an environment, building a system for each node and verifying its suitability results in unnecessary time-consuming waste. The oneM2M standard nodes have common components and individual components. Common components are basically deployed in equal containerization and individual components are deployed in containerization according to the function of each device. These characteristics of the oneM2M standard are appropriate for applying the Docker Container concept, which is process virtualization. In this paper, we propose to secure long-term consistency by reducing the initial time of devices and ensuring a common operating environment by virtualizing and distributing processes to nodes according to IoT standards. Also, the proposed system has strong connectivity to the external elements because it complies with the IoT standards.

In Sect. 2, we list research regarding virtualization and IoT standards, and Sect. 3 proposes to use process virtualization to configure IoT standards-based systems. Section 4 identifies the systems implemented and their operability; Sect. 5 describes the conclusions and future research tasks.

2   Related Work

2.1   The Smart Factory

Due to the rapid development of electronic, information and advanced manufacturing technologies, manufacturers' production methods are shifting from digital to intelligent foundations. A new generation of convergence of CPS and virtual reality technologies, artificial intelligence and big data has arrived. Innovations that create new value are increasingly reducing traditional manufacturing forms and introducing new architecture manufacturing businesses. As a result, intelligent manufacturing technology is one of the high-tech areas where industrial countries pay more attention [8–11]. Europe's 2020 strategy, Germany's industry 4.0 strategy and China's manufacturing 2025 were proposed. The innovation of intelligent manufacturing has sparked an explosion of interest in future manufacturing around the world [12,13].

2.2   Internet of Things

IoT is a worldwide infrastructure for information society and provides improved services by connecting objects together based on information and communication technologies that utilize advanced interoperability technologies. A large number of devices and information objects are connected through the network layer. The vision of IoT is to make smart changes in all fields such as smart devices, smart phones, smart cars, smart homes, smart cities, and smart factory. Using IoT allows real objects to be seamlessly integrated into global networks, allowing them to interact between physical objects and cause synergy effects through cyber layers. IoT is a basic networking infrastructure for configuring CPS. CPS is a system that is built and designed based on a complete integration of computing algorithms and physical components. CPS provides far greater functionality, adaptability, scalability, elasticity, safety, security and usability than today's simple embedded systems [14,15].

2.3   Cyber-Physical System

CPS is an integrated system that virtualizes the physical portion of a hardware function so that it can work closely with the digital layer. From an industrial standpoint, various objects are digitally connected to provide a basis for intelligent decision making. It also enables the implementation of sophisticated functions in manufacturing environments where a variety of components must be combined and stability assured. In production systems, CPS views are partially distinguished from existing automation pyramids and introduce functions in a more distributed manner than traditional hierarchical structures. The system will be highly fragmented and have the ability to organize responses organically to respond to unexpected events. It also dynamically resizes and reconfigures systems to meet a variety of business opportunities [16–19].

Through IoT, CPS can realize new patterns of interaction and business possibilities and implement more mutually organic manufacturing environments. The potential for CPS was due to the continued deployment of artificial intelligence, information and communication technologies and computing, communications and storage devices. CPS is not implemented in its own right; it includes Mergent System (MAS), Service Oriented Architecture (SOA), IoT, cloud computers, augmented reality (AR), big data, and M2M. There are still significant challenges related to safety, security and interoperability, and these visions must be addressed first before they can be realized in an industrial context [20].

2.4   Virtualization

Container virtualization, which is often used in Linux, differs from other virtualization technologies in terms of virtualization. This is a viable alternative to hypervisor-based virtualization, but it can work similarly. Traditional hypervisor-based virtualization technologies are commonly used for server virtualization and isolation, and virtual machines are utilized to virtualize various layers of the hierarchy, including the hardware layer. You can create independent virtual machines separate from the bottommost host and specify the corresponding hardware and applications. The Guset operating system is installed and functional for each VM instance [21–23].

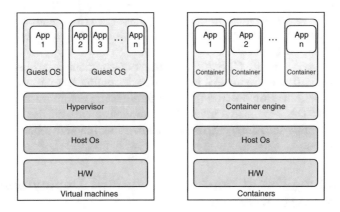

Fig. 1. The concepts of virtual machine and containers.

Figure 1 shows the formal differences between container-based virtualization technologies and virtual machine-based virtualization technologies. Container-based virtualization implements a process through a container engine, and a hypervisor-based virtual system implements a VM instance through a hypervisor. Container-based virtualization with OS-level process isolation does not require hardware and driver virtualization, unlike a hypervisor. In particular, the container shares the same OS kernel as the underlying host system. Thus providing independent characteristics of the virtual network interface, process space, and file system. Through these shared kernel capabilities, containers can reduce image volume and enable efficient virtualization systems [24, 25].

2.5 IoT Standard

Various vertical IoT systems are constructed in different ways. Each vertical system has its own characteristics and requirements because it has different ecosystems and different goals. The system for intelligent transportation has been specialized to deal with vehicle network and traffic information. On the other hand, the smart home system aims at using and transmitting the data of the connected device and external information. Smart Healthcare can check health status of users by using health data record according to body condition. The goal between the vertical systems is eventually in the transfer and recording of the data, which is not completely independent of each other. Therefore, it is not cost effective to implement the system separately for different purposes, and it is unreasonable for data to be transmitted to each other. Therefore, it is necessary to propose a vertical system for integration or to enable mutual transfer of data through a horizontal platform from the beginning [26, 27].

Main goal of oneM2M is to lead the unfragmented M2M service layer standard market by integrating M2M service layer standard activities of each other and different ecosystems and developing global specifications jointly. For this process, seven Standards Development Organizations (SDOs) interrupted duplicate M2M service layer work and pooled existing M2M documents. oneM2M refers to BroadbandForum (BBF) and Open Mobile Alliance (OMA) to take advantage of existing specifications [28].

Another way to build CPS in a factory environment is to use OPC as a core technology. It is an extensive information modeling framework for digitization and industrialization 4.0 and includes middleware communications technology. The OPC UA's extensive information modeling capabilities and semantic content-rich communications capabilities enable flexible and reliable delivery of data [29].

3 Container-Based oneM2M Standard Deployment System Architecture

3.1 oneM2M Standard Architecture

The common service layer of the oneM2M standard focuses on basic registration, messaging, or interoperability with other systems. OneM2M node consists of four functional nodes: Infrastructure node (IN), Middle node (MN), Application Service Node (ASN) and Application Dedicated Node (ADN). Each functional node can be composed of one or more Common Service Entities (CSEs) or Application Entities (AEs), CSE is a logical concept that runs within oneM2M service platform. This includes common oneM2M service functions, and AE represents servers, gateways, and devices. oneM2M defines three reference points (i.e., Mca, Mcc, and Mcn). The Mca reference point allows the use of services provided by AEs. The Mcc base point activates internal communication. Mcc' references are similar to Mcc, but provide an interface to another oneM2M system. In order to take advantage of features such as the device trigger service provided by the Third Generation Partnership Project (3GPP) network, a Men reference point exists between the CSE and the service entity in the base network. The oneM2M standard has a hierarchical resource structure and all services are provided through resources. Each resource can be used as a reference through a Uniform Resource Identifier (URI) through the MQTT, HTTP, and CoAP protocols. In addition, resources can be accessed through the CREATE, RETRIEVE, UPDATE, and DELETE (CRUD) commands commonly used in RESTful API architectures [30].

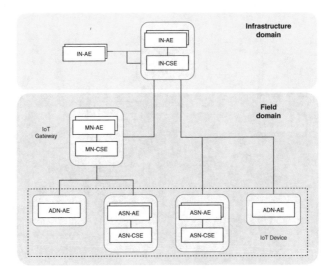

Fig. 2. oneM2M standard reference model.

Designing and building a service for each node and node according to the reference model shown in Fig. 2 is accompanied by an environmental consideration and customization to suit the system to be implemented. In a factory situation that is exposed to a variety of devices and environments, optimizations are usually made between the system and the equipment because different environments are typically built between the nodes. However, device and device functions do not change significantly in the manufacturing environment. Therefore, unnecessary waste occurs when a new environment is established to construct a system. Therefore, it is very useful to modularize the functions so that they can adapt to the environment of the equipment in the smart factory, and to share and use them as needed.

3.2 Container-Based oneM2M Standard Deployment System Architecture

Technology that makes it easy to implement repeated elements reduces the unnecessary tasks that comprise each node of an architecture. A key factor in the process of inefficient system deployment is the iterative deployment of the system environment and functions of each device. Because production is carried out only with specific facilities and functions in the manufacturing sector, similar facilities and functions are shared by the manufacturing sector. For example, companies that utilize injection equipment have an injection machine, a hopper, a counter, and a temperature controller. This is a common list of needs for many similar companies. It can be used to modularize peripheral components by function and environment and to build common manufacturing systems in factories around injection equipment.

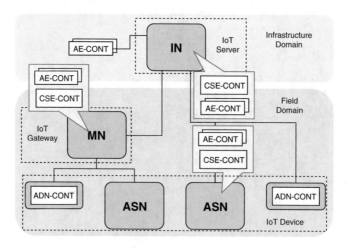

Fig. 3. Container-based oneM2M standard reference model.

Administrators can create and use multiple container images that are virtualized incorporating all the elements into each device type, but can easily vary their functionality by using the features of the oneM2M standard architecture. Figure 3 shows a container-based oneM2M reference model with processes isolated and virtualized. The CSE and AE section of each node are separately containerized and managed. Since CSE is common, containerize AE only according to function. The structure for constructing a flexible system is shown in Fig. 3 By containerizing only AE, various functional combinations are possible according to AE. For example, assume that node A uses only the functions of the temperature zone 1, 2, and the quantity coefficient and that node B performs only the lookup of temperature zone 2, 3. Node A then drives the temperature zone1 inquiry container, the temperature zone2 lookup container, and the quantity coefficient container. Node B drives the temperature zone 2 lookup container and the temperature zone 3 lookup container. Because each process is used in combination with containerization, the container image is created and managed as much as the function rather than creating a container image as the type of node.

Containerizing the device environment and functions can ensure operational stability by sharing the same environment and functions. Using virtualized modules in the host OS enables flexible process changes and management of the number of resources used in individual containers, thereby benefiting from efficient computing resources. Using the concept of a container to distribute images prevents changes in the environment or functions between devices. Containers are constructed based on images. The process of creating an image and the process required are specified in the area called Dockerfile. Dockerfile includes base images, image versions, network types, folders to mount, tasks to perform during build, open ports, and port assignment. Because images can be built based on Dockerfile, the environment and process of images are flexibly changed and created. Use Dockerfile to create a base image that meets the oneM2M standard and a new Dockerfile that matches the characteristics of each device based on the base image. If the administrator systematically manages Dockerfile, it is easy to restore and create images even if specific problems or additional changes occur.

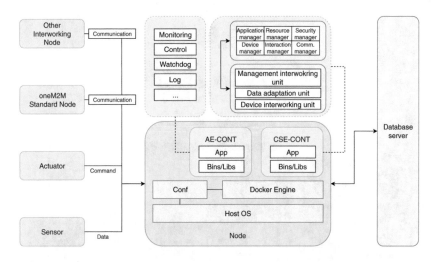

Fig. 4. Container-based oneM2M standard node model, containerized with common and individual elements and linked to external nodes

Figure 4 shows the internal components of oneM2M standard containers and the external environment associated with them. CSE containers include common service functions such as application and service layer management, communication management, group management, device management, security, and registration. AE container includes functions for specific applications and process operation. The CSE Container and AE Container are managed and executed through the Docker Engine and share the host system and network. Even if there are additional containers, each container operates separately. Generally, AE containers perform data collection and control functions for sensors and actuators. CSE containers interact with nodes that are oneM2M standard or interact with other standard nodes such as OCF and LWM2M. You can also activate a database as a container, but the repository itself is configured separately to maintain a stable database. Containers have great advantages in that they can easily change the execution environment and create multiple containers in the same environment. This feature is useful when administrators need to build flexible systems, but can be an unstable factor if they need to continuously manage and maintain data, such as databases. Therefore, database repositories

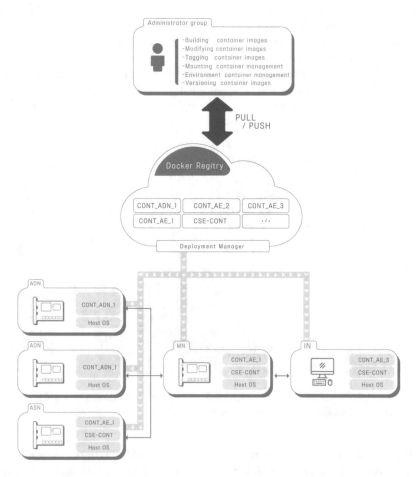

Fig. 5. The form in which each container image is deployed to systems that conform to the container-based oneM2M standard.

are considered stable systems that reside on separate database servers rather than on virtualized container types.

Figure 5 shows the structure of registering a container image in the Docker registry and distributing it to each node that complies with the OneM2M standard. To quickly use containers on a device, the container image must be stored in the registry. A group of system administrators in the enterprise infrastructure creates, registers, and manages the CSE container and the AE container suitable for each device in the registry. Although many images need to be created in other IoT environments, there are not many types of AE images required by the system because the manufacturing environment uses the same or similar facilities. When each node pools a container image built from the Docker Registry, it becomes a node that performs a function and role that meets the oneM2M standard. Each container is constructed based on oneM2M, ensuring the organic connectivity between container objects. The Administrators group only performs tasks such as creating, modifying, and deleting containers in the registry. Because the device uses the image in the registry, the manager's work is automatically reflected in the device.

4   Implementation

This chapter describes the implementation of the oneM2M standard using containers. Mobius was used as a platform to implement a system that complies with the oneM2M standard. Mobius is an open platform developed by the Korea Electronics Technology Institute (KETI) and aims to horizontally integrate IoT field. Mobius supports multiple services and a variety of devices, ensuring interoperability between different standards and platforms. Platforms provide configuration tools for each IN, MN, ASN, and ADN according to the reference model of the oneM2M standard, which are defined by Mobius, nCube:Rosemary, nCube:Lavender, nCube:Thyme. It provides monitoring web applications, mobile applications, and adduino configuration tools as well as reference model nodes. Mobius was considered suitable for containerizing systems as it is an IoT platform that complies with the oneM2M standard.

Table 1. Details of implementation system

	Application dedicated node	Infrastructure node
Name in the Mobius	nCube-Thyme	Mobius
Language	Nodejs, Python	Nodejs
Version	2.3.2	2.4
Device	Raspberry pi 3+	Desktop
Host OS	Raspbian	Ubuntu 16.04
Container image	Debian-stable	Ubuntu 16.04

This implementation was implemented in the simplest of the oneM2M reference models for ADN and IN structures because the purpose of the implementation is to ensure the possibility of containerizing common and individual components and operating on the oneM2M standard. Table 1 shows environments of the implementation system. The Mobius platform model for each reference model used nCube:Thyme and Mobius servers. Models are released on the Git hub, with versions 2.3.2 and 2.4 respectively. nCube:Thyme works containerized at Raspberry Pi 3+ and Mobius works containerized at the desktop. The base image of a container that operates in Raspberry is the Debian-standable version, and the base image of a container that operates on the desktop is Ubuntu 16.04. Docker files are created to build the image, and the contents of the Dockerfile define open ports, device mounts, folder mounts, and code modifications.

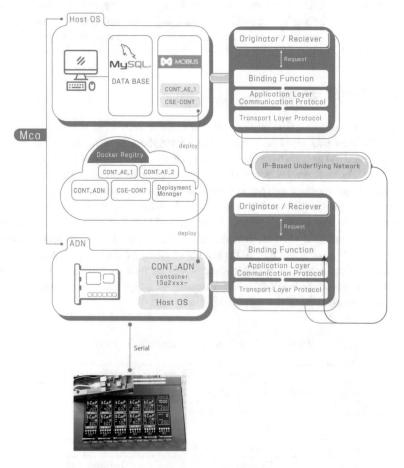

Fig. 6. Implementation of the IN-ADN reference model for container-based systems conforming to the oneM2M standard.

Figure 6 shows the architecture of the system that virtualizes and deploys the oneM2M standard node. Administrators build images and push them into the registry. In the implementation process, the Docker Hub was used as image registry.

Docker Hub is a public registry provided by the Docker Foundation. More than 1.5 million images are registered and widely used around the world. In order to pull images, images must be registered in the registry, so the ADN image, IN-CSE, and IN-AE images must be pushed to the registry first. The ADN container carries out the process of pulling data from the Fool-proof equipment from the Raspberry Pi and the process is accomplished through serial communication. The IN-CSE container is a common element on the Desktop that carries out processes such as device management, registration, and data management. IN-AE is an individual element on the Desktop that monitors the current time. In this implementation process, functional monitoring was targeted only because it was one of the issues that separated CSE and AE, which were divided into common and individual components, and operated containers.

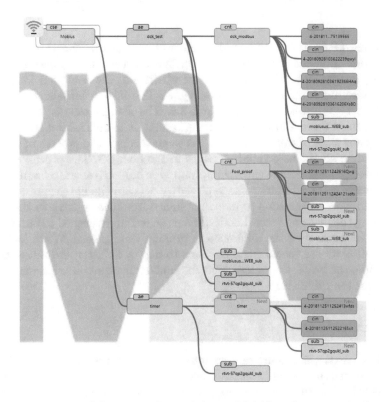

Fig. 7. Monitoring of the connectivity of the oneM2M-based containerized nodes.

Each node was deployed with a container and operated normally in compliance with the oneM2M standard. Connected nodes inside the platform can be viewed through the application provided by KETI and shown in Fig. 7.

The time required for the initial run was analysed when each node was initialized on the oneM2M basis. The minimum components to build a node were measured in time to be satisfied, including package updates, nodejs, python, Message Queuing Telemetry Transport (MQTT) broker and client, Git code updates. Table 2 shows a comparison of the time required to perform the initialization in three cases.

Table 2. Initial run time in three cases

Type	nCube-Thyme (sec)	Mobius (sec)
Initialize directly from host OS	210.850	250.341
Initialize when an image exists	3.702	3.890
Initialize when no image exists	102.157	111.233

In three cases, the initialization time was measured. The three cases are initialization immediately in the host OS, initialization in the presence of the container image, and initialization in the absence of the container image. The target nodes are Mobius and nCube-Thyme. The fastest initialization time is the second type, measured in 3,7 s and 3.9. The slowest case is the first type measured in 211 and 250 s. Based on the results, it is understood that the first type requires a lot of time because it has to go through the process of updating and installing all the associated packages in the host OS. The process seems to be executed in a short time since the second type only drives the container based on the image if the container image already exists. In the absence of an image, it takes up most of the time to move the image from the registry to the device. Comparing the first and second nodes, the more resources needed to initialize the process, the more time the difference in running time is seen. When looking at these results, we see that using containers to deploy is faster and more reliable. In addition, the larger the process and the smaller the container base image has, the greater the difference.

5   Conclusions

While IoT-based systems are rapidly spreading around the world and their fields are also diversifying, most industrial field systems still maintain traditional systems. This is because manufacturers are unsure of security, stability, and manageability when changing existing systems. Therefore, research is actively underway to containerize existing systems and maintain their functions and apply additional IoT systems. However, there is a lack of research to apply IoT standards based on containers.

This paper proposes a container-based industry system that complies with oneM2M standard that complies with IoT standards. The advantage of Docker registry is that it is easy to deploy and apply IoT standards-based systems and manage versions easily. Because same operating environment can be established on all nodes, single board computers are suitable for IoT environments that are actively used and it can reduce system errors that occur due to driving environment problems. Because the oneM2M standard model consists of common and individual elements, it is easy to isolate and virtualize processes and manage them. The implementation showed that the proposed system could be containerized and applied to the device and operated normally. It also confirmed that containerizing and deploying the process requires less time than installing and running the process directly on the node. These results show that it is possible to build a smart factory system by virtualizing processes based on IoT standards. In addition, this method is faster and more reliable than the method by which processes are installed directly in the hostOS.

We plan to study ways to use orchestration tools such as kubernetes and Docker swarm to better use container-based smart factory systems that operate as IoT standards in the future. We also plan to apply virtualization-based systems to various standards such as OCF and OPC UA.

Acknowledgement. This research was supported by the MSIT (Ministry of Science and ICT), Korea, under the ITRC (Information Technology Research Center) support program (IITP-2018-08-01417) supervised by the IITP (Institute for Information & communications Technology Promotion).

This research was supported by Basic Science Research Program through the National Research Foundation of Korea (NRF) funded by the Ministry of Education (NRF-2016R1D1A1B03933828).

References

1. Klaus, S.: The Forth Industrial Revolution. World Economic Forum (2011)
2. Lee, H.: Embedded system framework and its implementation for device-to-device intelligent communication of manufacturing IoT device considering smart factory. J. Korean Inst. Intell. Syst. **27**(5), 459–465 (2017)
3. Wu, G., Shilpa, T., Kerstin, J., Nageen, H., Kevin, D.J.: M2M: from mobile to embedded internet. IEEE Commun. Mag. **49**(4), 36–43 (2011)
4. Song, J., Andreas, K., Mischa, S., Piotr, S.: Connecting and managing M2M devices in the future internet. Mobile Netw. Appl. **19**(1), 4–17 (2014)
5. Husain, S., Athul, P., Andreas, K., Apostolos, P., JaeSeung, S.: Recent trends in standards related to the internet of things and machine-to-machine communications. J. Inf. Commun. Convergence Eng. **12**(4), 228–236 (2014)
6. Sheng, Z., Shusen, Y., Yifan, Y., Athanasios, V., Julie, M., Kin, L.: A survey on the IETF protocol suite for the internet of things: standards, challenges, and opportunities. IEEE Wirel. Commun. **20**(6), 91–98 (2013)
7. Xavier, M.G., Marcelo, V.N., Cesar, A.F.D.R.: A performance comparison of container-based virtualization systems for mapreduce clusters. In: 2014 22nd Euromicro International Conference on Parallel, Distributed and Network-Based Processing (PDP), pp. 299–306 (2014)

8. Brettel, M., Niklas, F., Michael, K., Marius, R.: How virtualization, decentralization and network building change the manufacturing landscape: an Industry 4.0 perspective. Int. J. Mech. Ind. Sci. Eng. **8**(1), 37–44 (2014)

9. Li, F., Jiafu, W., Ping, Z., Di, L., Daqiang, Z., Keliang, Z.: Usage-specific semantic integration for cyber-physical robot systems. ACM Trans. Embed. Comput. Syst. (TECS) **15**(3) (2016)

10. Wan, J., Daqiang, Z., Yantao, S., Kai, L., Caifeng, Z., Hu, C.: VCMIA: a novel architecture for integrating vehicular cyber-physical systems and mobile cloud computing. Mobile Netw. Appl. **19**(2), 153–160 (2014)

11. Wan, J., Hehua, Y., Di, L., Keliang, Z., Lu, Z.: Cyber-physical systems for optimal energy management scheme of autonomous electric vehicle. Comput. J. **56**(8), 947–956 (2013)

12. Lee, J., Behrad, B., Hung-An, K.: A cyber-physical systems architecture for industry 4.0-based manufacturing systems. Manuf. Lett. **3**, 18–23 (2015)

13. Zhou, J.: Intelligent manufacturing-main direction of "Made in China 2025". China Mech. Eng. **26**(17), 2273–2284 (2015)

14. Rose, K., Scott, E., Lyman, C.: The internet of things: an overview. In: The Internet Society (ISOC), pp. 1–50 (2015)

15. Stankovic, J.A.: Research directions for the internet of things. IEEE Internet Things J. **1**(1), 3–9 (2014)

16. Colombo, A.W., Stamatis, K.: Towards the factory of the future: a service-oriented cross-layer infrastructure. In: ICT Shaping the World: A Scientific View, vol. 65, p. 81 (2009)

17. Hellinger, A., Heinrich, S.: Cyber-Physical Systems. Driving force for innovation in mobility, health, energy and production. Acatech Position Paper, National Academy of Science and Engineering, p. 2 (2011)

18. Lin, S., Bradford, M., Jacques, D., Rajive, J., Paul, D., Amine, C., Reinier, T.: Industrial internet reference architecture. Industrial Internet Consortium (IIC), Technical report (2015)

19. Monostori, L.: Cyber-physical production systems: roots, expectations and R&D challenges. Procedia CIRP **17**, 9–13 (2015)

20. Leitao, P., Stamatis, K., Luis, R., Jay, L., Thomas, S., Armando, W.C.: Smart agents in industrial cyber-physical systems. Proc. IEEE **104**(5), 1086–1101 (2016)

21. Felter, W., Alexandre, F., Ram, R., Juan, R.: An updated performance comparison of virtual machines and linux containers. In: 2015 IEEE International Symposium on Performance Analysis of Systems and Software (ISPASS), pp. 171–172 (2015)

22. Morabito, R.: Virtualization on internet of things edge devices with container technologies: a performance evaluation. IEEE Access **5**, 8835–8850 (2017)

23. Morabito, R., Vittorio, C., Aaron, Y.D., Nicklas, B., Jorg, O.: Consolidate IoT edge computing with lightweight virtualization. IEEE Network **32**(1), 102–111 (2018)

24. Pérez, A., Germán, M., Miguel, C., Amanda, C.: Serverless computing for container-based architectures. Future Gener. Comput. Syst. **83**, 50–59 (2018)

25. Han-Chuan, H., Jiann-Liang, C., Abderrahim, B.: 5G virtualized multi-access edge computing platform for IoT applications. J. Netw. Comput. Appl. **115**, 94–102 (2018)

26. Jaewoo, K., Jaiyong, L., Jaeho, K., Jaeseok, Y.: M2M service platforms: survey, issues, and enabling technologies. IEEE Commun. Surv. Tutorials **16**(1), 61–76 (2014)

27. Martin, F., Apostolos, P., Anett, S., JaeSeung, S.: Horizontal M2M platforms boost vertical industry: effectiveness study for building energy management systems. In: 2014 IEEE World Forum on Internet of Things (WF-IoT), pp. 15–20 (2014)

28. Jorg, S., Guang, L., Philip, J., Francois, E., JaeSeung, S.: Toward a standardized common M2M service layer platform: introduction to oneM2M. IEEE Wirel. Commun. **21**(3), 20–26 (2014)
29. Markus, G., Stephan, H., Chris, I., Leon, U.: Information models in OPC UA and their advantages and disadvantages. In: 2017 22nd IEEE International Conference on Emerging Technologies and Factory Automation (ETFA), pp. 1–8 (2017)
30. Documents for oneM2M Architecture Information. https://www.etsi.org/deliver/etsits/118100118199/118101/02.10.0060/ts118101v021000p.pdf. Accessed 18 July 2018

On Evaluating a Network Throughput

Alexey S. Rodionov[1,2(✉)]

[1] ICM and MG SB RAS, Prospekt Lavrentieva, 6, Novosibirsk 630090, Russia
[2] Siberian State University of Telecommunications and Information Sciences,
Kirova str., 86, Novosibirsk 630102, Russia
alrod@sscc.ru
https://icmmg.nsc.ru/

Abstract. The task of obtaining and estimating maximum flows between all pairs of nodes in a given subset of a network's nodes is discussed. It is shown how significantly reduce number of operations by re-usage of some intermediate results while calculating maximum flows between several pairs of nodes. Obtaining all maximum flows allows calculate an arithmetical average of maximum flows between all pairs of nodes characterizes "throughput" of a network in a whole that may serve as the index of a citys road network quality, for example. Cumulative bounds are used for faster and strict estimation of maximum flows and their average.

Keywords: Graph · Maximum flow · Network throughput

1 Introduction

Finding maximum flow (MF) between nodes s and t in a graph is the classic task and is considered by a number of researchers [1–3,5,7]. We consider new task: obtaining an arithmetic mean of maximum flows between all pairs of nodes in a given subset of graph's nodes. This index shows how good a network is from the point of view of possibility of transferring flows through it in a whole. For example, it helps in estimating quality of transport infrastructure of a megalopolis. Let us use the following denotations:

$$
\begin{aligned}
G &\quad \text{undirected probabilistic network;} \\
N(G) &\quad \text{set of } n \text{ nodes of } G; \\
E(G) &\quad \text{set of } m \text{ edges of(G);} \\
S &\quad \text{subset of } s \text{ nodes under consideration;} \\
e_i\ (e_{ij}) &\quad i\text{-th edge or edge that connects } i\text{-th and } j\text{-th nodes, depending on context;} \\
F_{ij}(G) &\quad \text{maximum flow (MF) between nodes } i \text{ and } j \text{ in } G, \text{ simply } F_{ij} \text{ if } G \text{ is known;} \\
\bar{F}(G,S) &\quad \text{arithmetic mean of maximum flows between all pairs of nodes in } S; \\
Pth_{ij} &\quad \text{arbitrary path between nodes } i \text{ and } j; \\
t_{ij}\ (t(e_{ij})) &\quad \text{throughput (capacity) of the edge } e_{ij}; \\
Ngh(v) &\quad \text{set of nodes, adjacent to } v \text{ (neighborhood of } v\text{);}
\end{aligned}
$$

© Springer Nature Switzerland AG 2019
S. Lee et al. (Eds.): IMCOM 2019, AISC 935, pp. 34–41, 2019.
https://doi.org/10.1007/978-3-030-19063-7_3

$N_{u,v}$, $u \leq v$ set $\{u, \dots, v\}$.

It is obvious that

$$\bar{F}(G, S) = \frac{2}{s(s-1)} \sum_{i,j \in S, i < j} F_{ij}(G). \tag{1}$$

The problem is that when calculating all F_{ij} by turns, using known algorithms, a lot of operations repeats many times (for example, when calculating MFs in a long chain). In the current paper we discuss possible tricks for speeding up calculations and examine some special cases.

2   Cutnodes and Bridges

If there is a cutnode or a bridge in G (see Fig. 1), $S = S_1 \cup S_2$, $S_1 \cap S_2 = \emptyset$ and S_1 and S_2 belong to G_1 and G_2, correspondingly, for each u in S_1 and v in S_2, we have $F_{st} = \min\{F_{ux}, F_{xv}\}$ or $F_{st} = \min\{F_{ux}, t_{xy}, F_{yv}\}$, respectively. So we must obtain all F_{st} only after calculating all F_{ux} and F_{xv} (F_{yv}) thus saving a lot of operations.

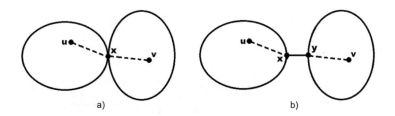

Fig. 1. Graphs with a cutnode (a) and bridge (b)

Partial case is existence of a dangling node.

Let us name a graph G without all pendant nodes as a kernel graph G', set of pendant nodes as D, $S^* = S \cap D$ and a set of nodes in G that are adjacent to nodes in S^* (if it is not empty) as K. MF from some pendant node $u \in S^*$ to a node $v \in V(G')$ (see Fig. 2) is a minimum of a MF from its adjacent node $x \in K$ to v and a throughput t_{ux} as it inevitably goes through x. So

$$F_{sv}(G) = \begin{cases} \min\{t_{ux}, F_{xv}(G \backslash e_{ux})\}, & \text{if } v \notin K; \\ t_{ux}, & \text{otherwise} \end{cases} \tag{2}$$

Now, we first must calculate all $F_{xy}(G')$, $x, y \in K$, that is equal to $F_{xy}(G)$, according to (1), and then we simply obtain all $F_{uy}(G)$, $F_{xv}(G)$, and $F_{uv}(G)$ for all nodes $u, v \in S^*$.

Second, we calculate all F_{xz} for $x \in K$ and $z \in V(G') \backslash K$, and then all F_{uz} for $u \in S^*$ may be obtained easily.

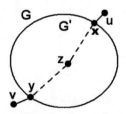

Fig. 2. Graph with a dangling node

3   Cases of Cycle and Chain

In a cycle and in a chain there are nodes (in a cycle – all), to which there are only two possible ways for incoming flow. It is possible to use this property for higher efficiency of algorithm.

3.1   Case of a Cycle

In the case of a n-node cycle, F_{uv} is obtained by considering two uniquely determined pathes between nodes u and v, so

$$F_{uv} = \min_{k \in N_{u,v}} t_k + \min_{k \in N_{1,n} \setminus N_{u,v-1}} t_k, \tag{3}$$

from which we easily obtain

$$\bar{F}(C_n) = \frac{2 \sum\limits_{i,j \in S,\ i<j} \left[\min\limits_{k \in N_{i,j-1}} t_k + \min\limits_{k \in N_{1,n} \setminus N_{i,j-1}} t_k \right]}{s(s-1)}, \tag{4}$$

and, if $s = n$, then

$$\bar{F}(C_n) = \frac{2 \sum\limits_{i=1}^{n-1} \sum\limits_{j=i+1}^{n} \left[\min\limits_{k \in N_{i,j-1}} t_k + \min\limits_{k \in N_{1,n} \setminus N_{i,j-1}} t_k \right]}{n(n-1)}. \tag{5}$$

3.2   Existence of a Chain

If there is a chain Ch in a graph G that connects nodes x and y (see Fig. 3), then for calculating MFs between its nodes, we close chain by a pseudo edge with a throughput equal to a MF between its ending nodes in G without this chain, and reliability equal to $R_{st}(G \setminus Ch)$. Now we can use previous results about a cycle for obtaining F_{ij} for all pairs of pivot nodes in a chain.

Let in the Fig. 3 $x = 1$, $y = k$ and inner nodes of the chain have numbers $2, ..., k-1$. Let $t_{i,i+1} = t_i$, $i = 1, \ldots, k-1$ also. After calculating $t_k = F_{1k}(G \setminus Ch)$ we easily obtain (let us consider the case $S = N(G)$)

$$\sum_{i=1}^{k-1} \sum_{j=i+1}^{k} F_{ij} = \sum_{i=1}^{n-1} \sum_{j=i+1}^{n} \left[\min_{k \in N_{i,j-1}} t_k + \min_{k \in N_{1,n} \setminus N_{i,j-1}} t_k \right]. \tag{6}$$

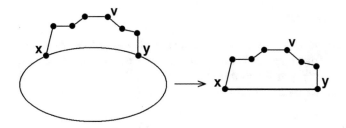

Fig. 3. Graph with a chain

Remark 1. If MFs from some node v to x and y in $G\backslash Ch$ goes through independent sets of edges, then all F_{vj} for $j \in Ch$ may be obtained by using these results also.

Remark 2. If no inner nodes of Ch are pivot ones, then chain is substituted to an edge with a throughput $t^* = \min\{t_i,\ i \in E(Ch)\}$, thus reducing graph's dimension and total number of operations.

4   Case of a Tree

In this case we have an unique path between any pair of nodes. As we need compute maximal flows between all pairs of nodes, we may gain from this property. First, we know MF between neighbour nodes: they are equal to throughputs of corresponding edges. Thus, if nodes s ant t are adjacent, then $F_{st} = t_{st}$.

For obtaining the remaining F_{ij}, depth-first traversal shows best results: for any intermediate node x, maximal flow through it between some node s in one of upper levels and some node t in lower one, is $F_{st} = min\{F_{ux}, F_{xv}\}$. As we use depth-first traversal, F_{ux} is known already, and thus for any node y that is successor of x we have

$$F_{sy} = min\{F_{ux}, t_{xy}\}. \tag{7}$$

By storing all MFs for pathes from the root to all other nodes, we provide easy computing of the remaining F_{ij} for nodes that belong to different branches that start from joint node x: $F_{ij} = min\{F_{ix}, F_{xj}\}$.

4.1   Case of a Chain

If a tree is a chain Ch with n nodes and $n-1$ edges $e_1, e_2, \ldots, e_{n-1}$ with throughputs $t_1, t_2, \ldots, t_{n-1}$, then we have

$$F_{ij} = min\{t_k,\ k = i, \ldots, j-1\}.$$

4.2   Case of a Star

Let a tree be a star S with $n + 1$ nodes and n edges e_i, with throughputs t_i, $i = 1, \ldots, n$, that connect center node 0 with rest nodes. In this case $F_{0i} = t_i$ and for $0 < i < j \leq n$ $F_{ij} = \min\{t_i, t_j\}$.

From this we easily obtain

$$\bar{F}(Ch) = \frac{2\left(\sum\limits_{i=1}^{n} t_i + \sum\limits_{i=1}^{n-1} \sum\limits_{j=i+1}^{n} \min\{t_i, t_j\}\right)}{n(n+1)}. \tag{8}$$

5   Cumulative Bounds of an MF

In most cases, obtaining the exact value or estimating some index of network quality is necessary for decision making about whether a network is good enough for its purpose. Thus, it is enough to know, if this index is less or exceeds some predefined threshold. In [8] authors propose the approach of cumulative updating bounds of a quality index for decision making on the example of all-terminal reliability.

In [6] we try to discuss this approach in general. We have shown that, if maximum (M) and minimum (m) values of some function F on finite set W are known, and we know exact values of this function for some k elements $w_i \in W_0 \subseteq W$, then current bounds for average value of F are

$$LB_k = \frac{\sum\limits_{w \in W_0} F(w) + m\left(|W| - |W_0|\right)}{|W|}; \tag{9}$$

$$UB_k = \frac{\sum\limits_{w \in W_0} F(w) + M\left(|W| - |W_0|\right)}{|W|}. \tag{10}$$

Obtaining value of F for some next w_{k+1} allows us to improve these bounds:

$$LB_{k+1} = LB_k + \left[F(w_{k+1}) - m\right]; UB_{k+1} = UB_k - \left[M - F(w_{k+1})\right]. \tag{11}$$

Let us examine bounds for a maximum flow. There is classical result, that for any pair of nodes v and u MF is equal to a minimum cut between these nodes [?]. Thus we can use *any* cut between these nodes as initial upper bound. Fastest way for finding a cut is using such, that cuts off v or u off the rest of G. So we use

$$UBF_0 = \min\left\{\sum\limits_{i \in Ngh(u)} t_{ui}, \sum\limits_{j \in Ngh(v)} t_{jv}\right\}$$

for this purpose. On the flip side, if nodes v and u are connected in G, then, obviously, MF cannot be less than minimum capacity of edges, that belongs to a path $PTh(u,v)$ between them, that was found during check of their connectivity, by breadth first search, for example. Thus

$$LBF_0 = \min_{e \in PTh(u,v)} t(e).$$

Now we transfer to estimation of $\bar{F}(G)$. In the simplest case, for example when we calculate all F_{ij} in parallel, we can assume for all of them that $M = UBF_0$ and $m = 0$ (no flow is possible). If we use considerations above, then initial bounds for $\bar{F}(G)$ are

$$LB = \bar{F}(G, S) = \min \left\{ (\min\{t_{ij},\ i, j \in S,\ i < j\},), \right.$$

$$UB = \frac{2\left(\sum_{i,j \in S,\ i<j} \min\left\{ \sum_{u \in Ngh(i)} t_{iu}, \sum_{v \in Ngh(j)} t_{vj} \right\} \right)}{s(s-1)} \left. \vphantom{\sum} \right\}. \tag{12}$$

Thus, it is possible to update cumulative bounds for $\bar{F}(G)$ after obtaining new value of some F_{ij} during executing, for example, the Ford-Falkerson algorithm [4] (low bound only), or after finding its final value (both bounds). As Ford-Falkerson algorithm takes many iterations for a large network, second approach gives more effective program. By the same reason we recommend not to update bounds after obtaining each F_{ij} in the case of existence of chains in network structure, but use partial sums of their values.

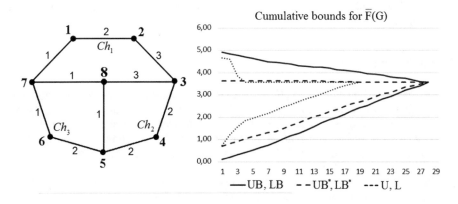

Fig. 4. Convergence of cumulative bounds for $\bar{F}(G)$

In the Fig. 4 bounds are shown for $\bar{F}(G)$ of the graph, presented in this figure in the case of $S = \{(i,j)\}$, $i = 1,\ldots,n-1$, $j = i+1,\ldots,n$ (exact value 3,571428). Bold lines correspond to

$$m = 0; M = \min\left\{\max\left(\sum_{i\in Ngh(u)} t_{ui}, \sum_{j\in Ngh(v)} t_{jv}\right), u,v \in N(G)\right\},$$

dash lines correspond to "smart" dynamic choice of m and M: they correspond to rest pairs of nodes only.

Dot lines show results of obtaining partial sums. There are three chains in the graph (7-1-2-3, 3-4-5, and 5-6-7) so, according to the Subsect. 3.2, we can obtain partial sums of MFs for some pairs of nodes (12 of them) using Ford-Falkerson algorithm 3 times only. We start from this partial sums and then consider rest MFs, thus obtaining result in 18 steps instead of 28.

6   Conclusion

In this paper we have proposed some methods that allow speeding up calculation of maximum flows between all pairs of nodes in a network with bi-directical links. Comparing with calculating them one-by-one, proposed methods allows re-using of some intermediate results thus significantly reducing number of operations. It is easy to see, that there are ample opportunities for parallelization in obtaining maximum flows with usage of the proposed methods or not. We can simultaneously calculate as many MFs as possible by sequential algorithms or calculate them one-by-one by parallel ones. Using average maximum flow as criteria for structural optimization of a network is multivariant problem and is the task of our further investigations.

References

1. Aldous, D.J., McDiarmid, C., Scott, A.: Uniform multicommodity flow through the complete graph with random edge-capacities. Oper. Res. Lett. **5**(37), 299–302 (2009)
2. Colin, M., Alex, S., Paul, W.: Uniform multicommodity flows in the hypercube with random edge-capacities. Random Struct. Algorithms **50**(3), 437–463 (2016). https://doi.org/10.1002/rsa.20672. https://onlinelibrary.wiley.com/doi/abs/10.1002/rsa.20672
3. Evans, J.: Maximum flow in probabilistic graphs - the discrete case. Networks **2**(6), 161–183 (1976). https://doi.org/10.1002/net.3230060208
4. Ford, L.R., Fulkerson, D.R.: Maximal flow through a network, pp. 243–248. Birkhäuser Boston, Boston (1987). https://doi.org/10.1007/978-0-8176-4842-8_15
5. Fujii, Y., Wadayama, T.: An analysis on minimum $s-t$ cut capacity of random graphs with specified degree distribution. In: 2013 IEEE International Symposium on Information Theory, pp. 2895–2899 (2013). https://doi.org/10.1109/ISIT.2013.6620755

6. Rodionov, A.: Cumulative estimated values of structural network's reliability indices and their usage. In: 2016 Dynamics of Systems, Mechanisms and Machines (Dynamics), pp. 1–4 (2016). https://doi.org/10.1109/Dynamics.2016.7819071
7. Wang, H., Fan, P., Cao, Z.: Analysis of maximum flow in random graphs for network coding. In: 2005 IEEE International Symposium on Microwave, Antenna, Propagation and EMC Technologies for Wireless Communications, vol. 2, pp. 1062–1065 (2005). https://doi.org/10.1109/MAPE.2005.1618103
8. Won, J.M., Karray, F.: Cumulative update of all-terminal reliability for faster feasibility decision. IEEE Trans. Reliab. **59**(3), 551–562 (2010). https://doi.org/10.1109/TR.2010.2055924

Frame Aggregation Control for High Throughput and Fairness in Densely Deployed WLANs

Toshitaka Yagi$^{(\boxtimes)}$ and Tutomu Murase$^{(\boxtimes)}$

Nagoya University, Nagoya, Aichi, Japan
yagi@net.itc.nagoya-u.ac.jp, tom@itc.nagoya-u.ac.jp

Abstract. To achieve high throughput and fairness between multiple wireless LANs (WLANs), we propose two novel control methods using frame aggregation of IEE802.11n/ac. WLANs which are densely deployed are suffered from radio interfere with each other. The proposed method controls *Frame Aggregation* size and/or *Contention Window* size. They are determined in each WLAN according to the interference intensity of each WLAN in order to maximize capture effect or to minimize error rate. Performance evaluation results showed that the proposed methods have 66% improved throughput than conventional method with keeping very high fairness.

Keywords: WLAN · Frame aggregation · Capture effect · IEEE802.11 · IEEE 802.11n/ac · Densely deployed WLAN · Interference · Throughput · Fairness

1 Introduction

Recently, the numbers of mobile wireless LANs (WLANs) that are composed of mobile APs or tethering terminals and terminals such as tablets and laptop PCs are increasing. Thanks to the penetration of free Wi-Fi spot services in places such as meeting room, cafes, and airports, we can access the Internet anytime, anywhere, and even on the go. As a result, however, many WLANs come close each other then they significantly interfere with each other. The influence of interference depends mainly on the number of terminals of WLANs and the distance between WLANs. This may decrease quality of services such as throughput in total of WLANs as well as cause unfairness in throughput in each WLAN. In order to increase throughput and fairly share the channel resources, methods to provide the user with desired resources.

Many previous researches showed that IEEE802.11e (EDCA) can flexibly control channel resources among terminals in a single WLAN. Many of them proposed changing the values of CW (Contention Window) or AIFS (Arbitration Inter Frame Space) for their objectives. However, the proposed methods could be are not effective in multiple WLANs case. This is because we have to consider that a terminal in a WLAN suffers from interference brought from outside of the WLAN as well as inside of the WLAN.

© Springer Nature Switzerland AG 2019
S. Lee et al. (Eds.): IMCOM 2019, AISC 935, pp. 42–53, 2019.
https://doi.org/10.1007/978-3-030-19063-7_4

Many other researches showed that frame aggregation technology implemented in IEEE 802.11n/ac is effective to improve throughput in a WLAN. In other words, we can change throughput by changing the number of frames aggregated by frame aggregation (referred to as frame aggregation size). Many of the researches pointed out appropriate frame aggregation control can achieve airtime fairness [7, 8] in a single WLAN. However, as far as we know, no researches have revealed control methods to provide desired fairness in throughput between multiple WLANs, especially with heterogeneous interference cases.

To cope with increasing total throughput and keeping desired fairness in throughput of each WLAN, we propose a novel control with combination of frame aggregation control and EDCA parameter control.

Our proposed control is composed of two methods for two different interference situations. The first method, SFF method, is a simple low-cost method. It is effective for multiple WLANs with homogeneous interference situations. The second method, FCCI method, is a rather complicated but accurate method. It takes into account the heterogeneous interference situations such as different distances between WLANs.

This paper is organized as follows. In Sect. 2, we describe related research, conventional methods, and their issues. In Sect. 3, we propose two frame aggregation methods, SFF and FCCI that realize both high fairness between WLANs and higher throughput than conventional. We describe how they achieve such throughput characteristics. In Sect. 4, we evaluate the effectiveness of the proposed methods. In Sect. 5, we summarize this paper.

2 Related Work

In this chapter, we describe related research. Then we clarify the conventional research on WLAN priority control and its issues.

2.1 Capture Effect

Capture Effect is one of the factors that greatly affect throughput in an environment where WLANs are close to each other. Generally, two or more frames are transmitted at the same time and collision occurs, the frame reception fails. But if the SINR of the frame is large, the reception of the frame is successful by capture effect. Therefore even in situations where multiple WLANs are in close and interfere with each other, if the distance between WLANs is larger than the distance between an AP and terminals, capture effect enables correct multiple packet reception.

2.2 Frame Aggregation

Frame aggregation is adopted by IEEE 802.11n. It is aimed at reducing overhead time such as backoff control and IFS by transmitting an arbitrary number of multiple data frames to one physical header. In this research we use A-MPDU. The A-MPDU aggregates frames on a MAC frame basis. Even when an MPDU frame error occurs, it is possible to retransmit only the MPDU frame (subframe) in which an error occurred by using the block ACK and the FCS (Frame Check Sequence) attached to each MPDU. The maximum value of the A-MPDU is set to 65535 bytes. In this paper, we assume that one frame data size is 1500 bytes and the maximum value of frame aggregation size is 40 frames. Frame aggregation remarkably reduces the overhead of the CSMA/CA mechanism. This is because the conventional 40 times the frame is transmitted at one transmission opportunity and the corresponding ACK is completed in one frame.

2.3   Related Researches and Issues

There are many researches about fairness among terminals in the WLAN. In [5] and [6], by adjusting the value of CW of terminals in the WLAN and making a difference in transmission opportunities among the terminals, fair throughput is achieved among the terminals. Similar to these methods, it is possible to attempt fairness between WLANs by adjusting the CW value of each WLAN. However, there are cases where the bandwidth could not be effectively utilized or the accuracy of fairness is lowered. This is because these conventional researches do not take into consideration capture effect. Capture effect is a phenomenon that can be normally received even if simultaneous transmission (collision) of frames occurs. The effect of capture effect on throughput in densely deployed WLANs is described in [1, 3, 4]. In [9, 10], research utilizing a Capture Effect realizes improvement of throughput and its effectiveness is shown.

Accordingly, to maintain fairness of each WLAN, a method of controlling CW while considering the capture effect can be assumed. However, this method could not effectively utilize the bandwidth. This is because, in order to achieve fairness, there are cases where it is necessary to reduce the transmission opportunities of WLANs that have good radio interference conditions. Therefore, they cannot take advantage of simultaneous transmission by capture effect. Moreover, fairness control which gives a difference in transmission opportunities such as priority control by CW has a problem that redundant time occurs at a node, not high priority. It reduces the total throughput.

To solve these problems, we employ IEEE 802.11n frame aggregation. In [11, 12], researches show effectiveness to throughput by using frame aggregation. Frame aggregation has been mainly used to maintain airtime fairness in environments where terminals communicating at multiple rates are mixed. In [7, 8], a high rate terminal transmits multiple frames at a time, thereby they secure the same communication time as the low rate terminal. Thus, they maintain airtime fairness. However, the fairness of throughput of each WLAN cannot be maintained even if guaranteeing airtime fairness. Therefore, the conventional airtime fairness control cannot achieve desired throughput fairness and the frame aggregation control aiming at realizing fairness (desired throughput ratio) by controlling the frame aggregation size for each WLAN is desired.

We solve the problems of the conventional frame aggregation method and propose novel methods. They achieve higher throughput and higher fairness among WLANs by both maximizing amount of capture effect according to either homogeneous or heterogeneous interference case and maximizing amount of effects of frame aggregation.

3 Frame Aggregation Size and Communication Opportunity Control

In this section, we describe two proposed methods using frame aggregation, then discuss throughput characteristics obtained by the proposed method.

3.1 SFF (Simple Frame Aggregation for Fairness) Method

The SFF method is an effective control method when each WLAN is in the same radio interference situation. To keep the throughput at an arbitrary ratio, we control frame aggregation size.

In the SFF method, each WLAN transmits frames by aggregating frames in the same way as the target throughput ratio. For example, if the throughput ratios are th1: th2: th3 among the three WLANs (WLAN 1, WLAN 2, WLAN 3), the ratio n1, n2, n3 of the respective frame aggregation size also becomes th1: th2: th3 accordingly. SFF assign the maximum frame aggregation size to the WLAN you want to maximize throughput.

We explain that the throughput ratio can be realized by the ratio of the frame aggregation size. When the transmission opportunities in each WLAN are equal and the interference conditions are also equal, the ratio of the amount of data received per unit time corresponds to the frame aggregation size. Therefore, the ratio of the frame aggregation size is the throughput ratio. Next, we explain the reason for assigning the maximum frame aggregation size. Overhead can be reduced by increasing frame aggregation size. Therefore, we can obtain the largest throughput by allocating the maximum frame aggregation size.

3.2 FCCI (Frame Aggregation and Communication Opportunity Control Considering Interferences) Method

The FCCI method is an effective method when the radio interference situation of each WLAN is different. When the interference situation of each WLAN is different, the transmission success rate is different. Therefore, by obtaining appropriate the transmission frequency and the frame aggregation size, we improve throughput while securing fairness. In the following, to explain the validity of the proposed method, we discuss the influence of the frame aggregation size and the transmission frequency on throughput between multiple wireless LANs.

Control of Frame Aggregation Size and Transmission Frequency Considering Interference

The throughput is obtained by the transmission frequency, the error probability, the frame aggregation size, and the backoff overhead. The error probability is determined by the simultaneous transmission probability and the interference amount determined by the positional relationship. In other words, it depends on the transmission frequency and the positional relationship which determine the simultaneous transmission probability. We describe these factors in order.

The characteristics of the frame aggregation size between multiple WLANs are as described in Sect. 3.1.

Next, we discuss the transmission frequency. By changing the transmission frequency, it is possible to change the error probability. The error probability is the probability that a frame becomes an error due to collision. Changing the transmission frequency changes the collision probability of the frame. When the transmission frequency is increased, collisions increase. Conversely, if the transmission frequency is reduced, the backoff overhead increases and lower the throughput. Further, the error probability also changes depending on the degree of the interference determined from the positional relationship. When the position is closed (when the interferences are large), the error probability of the frame increases. Conversely, if the position is far (when interferences are small), the error probability of the frame decreases. Therefore, in order to reduce the error probability, when the interference is large, it is necessary to suppress the transmission frequency of frames and avoid collision of frames. Conversely, when the interference is small and capture effects are effective, it is necessary to improve the transmission frequency of frames and aim for parallel transmission by collision of frames and improve throughput.

The backoff overhead can be reduced as the frame aggregation size increases and as the transmission frequency increases.

The throughput of each WLAN can be obtained from the error probability of the frame of each WLAN, the frame transmission frequency, the backoff overhead, and the frame aggregation size described above. By multiplying the above four factors, it is possible to find the number of frames received per unit time. The frame error probability is determined by the transmission frequency of each WLAN as described above.

In the proposed method, considering these basic characteristics, we control the throughput to the desired ratio and realize a high throughput. First, in order to achieve high throughput, the proposed method adjusts the transmission frequency. It maximizes the transmission frequency within the range where the error probability is the minimum degree. This avoids unnecessary errors (and retransmissions) and enables preferable parallel transmission, which leads to maximization of throughput. Further, in consideration of the error probability of the frame and the retransmission due to the error at this time, the frame aggregation size ratio is determined according to the desired throughput ratio. Further, the frame aggregation size is determined while avoiding the term with the largest frame aggregation size ratio exceed the maximum value of the frame aggregation size. As a result, fair and high throughput can be realized.

Frame Aggregation Size and Transmission Frequency

In this chapter, we discuss the case where the desired throughput ratio of each WLAN is equal. As described above, it is difficult to explicitly obtain the optimum value because throughput calculation requires complicated calculations based on information such as positional relationships in addition to frame aggregation size and transmission frequency control. Therefore, as a basic study, consider the case where the throughput considered to be the most general is controlled to the ratio 1: …: 1.

First, we describe the case where each WLAN is in the same radio interference situation. First, the transmission frequency is set making the frame error probability of each WLAN becomes the smallest. In other words, when simultaneous transmission by capture effect can be utilized, the proposed method increases the transmission frequency. When simultaneous transmission by capture effect cannot be utilized, the proposed method decreases the transmission frequency. Moreover, the proposed method increases the frame aggregation size of each WLAN to the maximum according to the target throughput ratio.

Next, we consider the case where each WLAN is in a different radio interference situation. Figure 1 is a diagram showing that three WLANs are in close. Since the WLAN 1 is distant from other WLANs 2 and 3, interference of WLAN1 is small. On the other hand, the WLAN 2 and 3 are close to each other and interference is large. The balloon in Fig. 1 is an example of frame transmission of each WLAN. The white squares in the balloon are the back off wait time and the red squares represent the frames. WLAN1 with small radio interference set high transmission frequency since simultaneous transmission by capture effect can be utilized. WLAN 2 and 3 with large interference does not overcome the benefits of the capture effect and the frame error probability rises. Therefore, to decrease the frame error probability of WLAN2 and 3, WLAN2 and 3 decreases the transmission frequency. After reducing the frame error probability as much as possible, we set the frame aggregation size considering the target ratio obtained. WLANs 2 and 3 with low transmission frequencies increase the frame aggregation size to achieve fair throughput. On the other hand, the WLAN 1 with a high transmission frequency set the frame aggregation size according to the frame aggregation size set by the WLANs 2 and 3 with the transmission frequency set low.

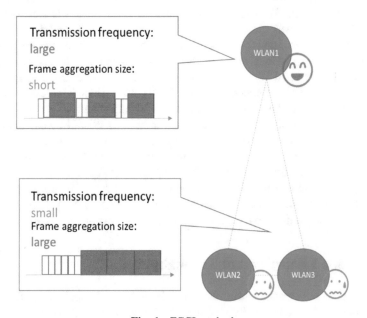

Fig. 1. FCCI method

Derivation of a concrete optimum frame aggregation size determination method is a future work. In this paper, we evaluate the extent to which the effectiveness of the proposed method can be obtained when we set the suboptimum frame aggregation size.

3.3 Throughput Characteristics of the Proposed Method

In this section, we describe the throughput characteristics of SFF and FCCI.

The SFF method achieves the throughput ratio of each WLAN by making the difference in the frame aggregation size. Since the throughput of each WLAN increases as the frame aggregation size increases, the throughput increases as much as the frame aggregation size increase. When the radio interference situation differs between WLANs, fairness in the SFF method is low. This is because the probability of successfully receiving frame in each WLAN is different. WLAN with good radio interference situation gets high throughput and fairness declines.

Next, we describe the throughput characteristics of the FCCI method with reference to Fig. 3. In Fig. 3, we describe the throughput characteristics in the case where the distance between the WLANs is increased while maintaining the distance ratio between WLANs. When the distance between WLANs is short, simultaneous transmission by capture effect couldn't occur. Therefore, each WLAN reduces the transmission frequency as much as possible to avoid frame collisions, and then increases the number of frame aggregation. The throughput increase by the amount of overhead reduced by frame concatenation. When the WLAN distance increases, the radio interference situation is improved in the WLAN 1, and the transmission frequency is increased and the capture effect is positively utilized. Therefore, in addition to reducing the overhead due to each frame aggregation, the throughput increase by utilizing the capture effect. Furthermore, as the distance between the WLANs increases, the radio interference situation of WLAN2 and WLAN3 is also improved. Therefore, the proposed method increases the transmission frequency of all WLANs. Then, since it increases the number of frame aggregation, the throughput increases by the overhead reduction of each WLAN and the use of simultaneous transmission of each WLAN by capture effect.

Since it is a prerequisite to maintain fairness, it is necessary to reduce the frame aggregation size. It may sacrifice throughput somewhat. For example, suppose that the radio interference situation of all WLANs is improved in Fig. 3. At that time, if all the WLANs increase the transmission frequency and the number of frame aggregation, the WLAN 1 with the best radio interference status obtains a very high throughput. Therefore, it is necessary to lower the throughput to suppress the frame aggregation size.

4 Evaluation Fairness and Throughput

In this section, we evaluate how the total throughput changes when the distance between the WLANs changes by the network simulator NS 3 using the basic model.

4.1   Two Models with Different Interference Degrees

We evaluated in the situation that WLANs are closely deployed. The number of WLANs is three.

First, we evaluated fairness of the SFF method. We showed how fairness changes depending on the difference in radio interference of each WLAN. An isosceles triangle was constructed as shown in Fig. 2. We change the ratio of the hypotenuse/base to change the interference between WLAN 1 and WLAN 2, 3. We investigated the relationship between the ratio of hypotenuse/base and fairness of throughput at that time.

Next, in order to show the effectiveness of the FCCI scheme, we place WLAN in an isosceles triangle type again as shown in Fig. 3. Then, we deform this triangle with similarly since we change the distance between the WLANs and the degree of interference.

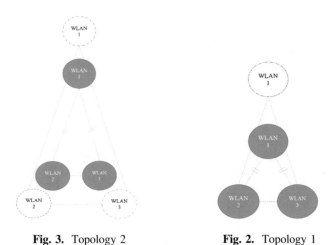

Fig. 3. Topology 2   **Fig. 2.** Topology 1

4.2   Effectiveness of the Proposed Method

First, we assume that each WLAN consists of one AP and one terminal in order to consider the simple model. Considering the communication by the mobile router and the tethering terminal, the distance between the AP and the terminal can be considered to be very close, we set the distance between the terminal and the AP to 0.3 m. The throughput ratio is controlled to be 1: 1: 1 so that WLAN can obtain equal throughput. In order to show the basic characteristics, we use bidirectional UDP saturated traffic. Other parameters of each simulation are summarized in Table 1. In this paper, we change the CW value to change to the transmission opportunity of each WLAN. At this time, we investigate the relationship between the WLAN distance and the throughput. For the isosceles triangle, the base side: hypotenuse was 1: 4.03 (the ratio of the base side to the height was 1: 4).

Under these circumstances, first, in order to evaluate the influence due only to the change in distance between each WLAN, the angle of the triangle is fixed. We fixed the shape of the triangle and evaluated the effective capacity while changing only the distance between WLANs. For simplicity, we set CWmin = CWmax. In the future, if CWmin and CWmax are independently controlled, high effect can be expected. Also, in order to manipulate the throughput to an arbitrary ratio, CW is more suitable than AIFS [5]. However, verification of the effect by adding AIFS control is also a future task. The conventional method to be compared is a method of realizing fairness by adjusting only CW (the frame aggregation size is 1). This is a method of using the CW value that can achieve a fair and high total throughput at each WLAN distance while taking the capture effect into account.

Table 1. Simulation parameter in ns3

Physical layer	IEEE802.11n
Frequency	2.4 GHz
Modulation method	Mcs7
Bandwidth	20 MHz
Traffic	UDP
Payload	1472 byte
A-MPDU maximum size	65535 byte
Simulation time	10.0 s
Carrier sense level	−82 dbm
Loss model	Log distance model
Path loss exponent	3.0
Fading model	Nakagami-Rice
Transmission power	10 dbm

This time, in order to obtain the suboptimal frame aggregation size and the transmission frequency showing high throughput, we evaluated throughput changing CW and the frame aggregation size at each distance. The values of CW were changed to 3, 4, 5, 6, 8, 10, 12, 15, and 20. First, when CW was 15, we confirmed that collision hardly occurred. Therefore, we selected many values smaller than 15. The number of frame aggregation size was changed to 2, 4, 6, 8, 10, 15, 20, 25, 30, and 40. At this time, the same parameters were used for WLAN 2 and WLAN 3. This is because in order to make the throughput ratio of each WLAN 1: 1: 1, it is appropriate to use the same parameters for WLAN 2 and WLAN 3 in the same radio interference situation. Three simulations were performed for each combination of parameters, and fairness and throughput were averaged to evaluate.

Fairness of SFF Method

Figure 4 shows the fairness of the SFF method when the hypotenuse/base side is increased as shown in Fig. 2 in order to increase the difference in the radio interference situation of each WLAN. When each WLAN is arranged in an equilateral triangle (hypotenuse/base = 1), fairness shows a high value of 0.98. In addition, when arranged in this equilateral triangle, the throughput was improved by 82% compared with the conventional method when the distance between WLANs was 4 m. However, when the hypotenuse/base becomes larger, that is, when the radio interference situation of each WLAN becomes larger, impairment of fairness becomes noticeable. When the hypotenuse/base = 2.8, the fairness drops to 0.957.

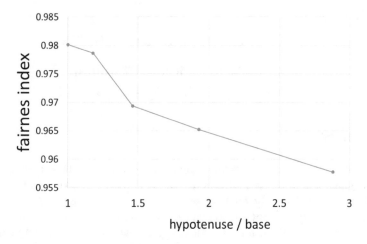

Fig. 4. Fairness of SFF method

Throughput of FCCI Method

Total throughput obtained by the proposed method is shown in Fig. 5 when the distance between WLANs is changed without changing the topology of WLAN 1, WLAN 2, and WLAN 3. Fairness was kept at 0.99 or more by each method. The horizontal axis is the distance between WLAN 2 and WLAN 3. When the distance between WLAN 2 and WLAN 3 is 2 m, the throughput of the proposed method is 56.3 Mbps, and the throughput is improved by 71% compared with the conventional method. At 4 m, the proposed method shows 61.2 Mbps which is as high as 66% compared with the conventional method. After that, the throughput of the proposed method rises to 93.7 Mbps around 10 m, which is expected to rise by 112% compared with the conventional method.

As shown above, the proposed method shows high throughput improvement in the environment where the distance between the WLAN 2 and the WLAN 3 is around 4 m in practice.

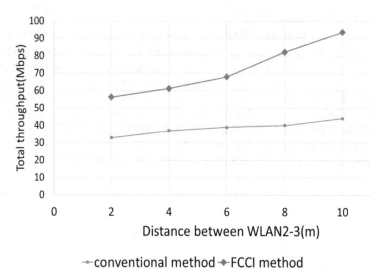

Fig. 5. Throughput of FCCI method

5 Summary

In this research, we proposed two frame aggregation control methods both to improve total throughput and to achieve fairness in densely deployed WLANs. SFF is a simple low-cost method and mainly for homogeneous interference cases. FCCI is a rather complicated but very efficient method and is for heterogeneous interference cases. In the heterogeneous interference case, frame aggregation sizes in IEEE802.11 and CW (contention window) values are carefully chosen in order to obtain both maximum capture effect gain and maximum frame aggregation gain.

Compared with the conventional method, simple CW control to keep low error rate, the proposed method showed 66% improvement than the conventional method in terms of throughput with keeping fairness among three WLAN with heterogeneous interference cases. We conclude that the proposed method is effective enough not only to homogeneous but heterogeneous interference cases.

Acknowledgments. This research is supported by Grant-in-Aid for Scientific Research (15H02688) (15H01684) (16H02817).

References

1. Isomura, M., Miyoshi, K., Murase, T., Oguchi, M., Segari, S., Baid, A., Seskar, I., Raychaudhuri, D.: Measurement and analysis on QoS of wireless LAN densely deployed with transmission rate control. In: Wireless Communications and Networking Conference, pp. 1865–1870. IEEE (2015)
2. Mangold, S., Choi, S., Hiertz, G.R., Klein, O., Walke, B.: Analysis of IEEE 802.11 e for QoS support in wireless LANs. IEEE Wirel. Commun. **10**(6), 40–50 (2003)

3. Kumatani, N., Isomura, M., Murase, T., Oguchi, M., Segari, S., Baid, A., Seskar, I., Raychaudhuri, D.: Context aware multi-rate control in densely deployed IEEE802. 11 WLAN for avoiding performance anomaly. In: 2016 IEEE 41st Conference, Local Computer Networks (LCN), pp. 363–370. IEEE (2016)
4. Lee, J., Kim, W., Lee, S.J., Jo, D., Ryu, J., Kwon, T., Choi, Y.: An experimental study on the capture effect in 802.11 a networks. In: Proceedings of the Second ACM International Workshop on Wireless Network Testbeds, Experimental Evaluation and Characterization, pp. 19–26. ACM (2007)
5. Banchs, A., Perez-Costa, X., Qiao, D.: Providing throughput guarantees in IEEE 802.11 e wireless LANs. In: Teletraffic Science and Engineering, pp. 1001–1010 (2003)
6. Banchs, A., Serrano, P., Oliver, H.: Proportional fair throughput allocation in multirate IEEE 802.11 e wireless LANs. Wirel. Netw. **13**(5), 649–662 (2007)
7. Abdul-Hadi, A.M., Tarasyuk, O., Gorbenko, A., Kharchenko, V., Hollstein, T.: Throughput estimation with regard to airtime consumption unfairness in mixed data rate wi-fi net-works. Communications **1**, 84–87 (2014)
8. Kim, M., Park, E.C., Choi, C.H.: Adaptive two-level frame aggregation for fairness and efficiency in IEEE 802.11 n wireless LANs. Mob. Inf. Syst. (2015)
9. Takahashi, K., Obata, H., Murase, T., Ishida, K.: Throughput improvement method exploiting capture effect in densely placed WLANs. In: IMCOM 2015, p. 36 (2015)
10. Kanda, M., Katto, J., Murase, T.: Enhancement of HCCA utilizing capture effect to support high QoS and DCF friendliness. In: 2016 13th IEEE Annual Consumer Communications & Networking Conference (CCNC), pp. 335–338. IEEE (2016)
11. Lin, Y., Wong, V.W.: WSN01-1: frame aggregation and optimal frame size adaptation for IEEE 802.11 n WLANs. In: 2006 Global Telecommunications Conference, GLOBECOM 2006, pp. 1–6. IEEE (2006)
12. Ginzburg, B., Kesselman, A.: Performance analysis of A-MPDU and A-MSDU aggregation in IEEE 802.11 n. In: 2007 Sarnoff Symposium, pp. 1–5. IEEE (2007)

Development of Sublayer Network State Inference Technology Based on Protocol State Dynamic Extraction for Improved Web Environment

Yewon Oh and Keecheon Kim[✉]

Konkuk University, Seoul, Korea
Iamoyw9857@gmail.com, kckim@konkuk.ac.kr

Abstract. A web server environment is the most commonly used core architecture. Web services may have limitations in their uses for various purposes by unspecified users. Moreover, formulated web services have made it difficult to change the environment according to usage patterns. Web-based networking analytics technology is considered to be a solution to this problem, and developing the technology requires research on inference of the network state in sublayers presentation, session, transport, network, data link, and physical layer in the seven layers of the open systems interconnection (OSI) model. The present study focuses on the sublayers of the application layer to make an inference of the network state of the sublayers. Specifically, individual attributes are assigned with thresholds to make an inference of the appropriate finite state for each attribute. Dynamic extraction of the protocol state as proposed by this study can improve the indicators of individual users' experiences. The problem of formulated web-based networking services can be resolved by creating autonomously controlled protocols based on dynamic extraction.

Keywords: Sublayer network state inference · Finite state machine (FSM) · Threshold inference

1 Introduction

Transport layer networking technology based on the user experience is formalized at the stage of web application implementation. In other words, users use formulated networking services instead of services that are specifically optimized for their use patterns and environment. To address this issue, the present study focuses on autonomous transfer networking technology. Generally, transport layer-level networking technology has the problem that it is difficult to make modifications due to its implementation in the kernel space. As a solution to the problem, this study proposes autonomous transfer networking enabled in the user space with minimum modification in the kernel space. The priority in the technology is to maximize the user experience in web applications by applying new requests and determining the user's experience level. As a result, all web-based network service users can use protocols with more accurate data packet flow and safely classify protocols that have not been predetermined.

© Springer Nature Switzerland AG 2019
S. Lee et al. (Eds.): IMCOM 2019, AISC 935, pp. 54–61, 2019.
https://doi.org/10.1007/978-3-030-19063-7_5

This study proposes a technique that incorporates the state of sublayers of the application layer, such as transport or session layers, in real time to use as a basis for decision making. The sublayers refer to the layers below the application layer in the open systems interconnection (OSI) model. They are any states of a protocol that are not considered in the application layer, including the congestion window and protocol state [1].

This study is organized as follows. Section 2 provides an overview of the literature as the background to this work. Section 3 briefly describes the technology pertaining to this study and its implementation. Section 4 presents the conclusions.

2   Previous Research

The purpose of the present study is to extract the required attributes dynamically from the various protocol states of the sublayers rather than the application layer in order to improve the user's experience. Relevant previous research areas for the present study include the Transmission Control Protocol (TCP) behavior inference tool, the finite state machine (FSM), and the Harris corner detection algorithm.

2.1   TCP Behavior Inference Tool (TBIT)

TBIT is a tool for characterizing the TCP behaviors of remote web servers [1]. TCP behaviors are often tested using TBIT in web servers because it is easy to perform tests on web servers because they respond to information requests with few prerequisites. Testing includes the TBIT test to measure the initial congestion window (ICW) value, which can have a significant impact on the performance of a web server, and six other tests to verify the operation in TCP. The TBIT test compares TCP congestion control. TCP congestion control algorithms reduce loss to the network and include Reno [2], Tahoe [3], and newReno [4].

TBIT may have limitations because it tests TCP only. While we cannot ignore TCP/IP, the most commonly used protocol in the web environment, we should also consider environments where different types of multiple protocols are used. Therefore, other protocols based on TCP will be considered in the future. Moreover, in contrast to previous research that focused on the transfer of certain packets to determine whether congestion control or SACK should be used, the present study proposes sublayer network state inference technology that requires extraction of significant output values for random packet transfer. The technology also requires a model for inference of the network communication status instead of validating intended packets along with the determination of status using congestion control.

2.2   Finite State Machine (FSM)

The FSM performs basic validations for situation-specific actions and easy management [6]. The FSM predefines individual states and makes transitions from the currentstate to another state according to the changing circumstances of the state. In other words, FSM is a tool for defining a behavior mode. As in TBIT, transmission and reception of SYN and ACK values are major factors for FSM in the TCP protocol. The present study needs to address the challenges in predefining SYN and ACK values because random packets are transferred.

Figure 1 shows the TCP protocol as Finite State Machine (FSM). The square box represents the state of the machine and the arrows represent the state transition. Passive open and active open in the process of converting and connecting the state. First, passive open is to specify the server to connect from a specific client or to wait for all clients to connect. Active open is the time when active clients send and connect SYN messages. As mentioned above, the Synchronize message initializes the connection, synchronizes the sequence number between the devices, and the Finish message notifies the device that the FIN pin is terminated in the TCP segment. Acknowledgment also acknowledges messages such as SYN and FIN messages [7].

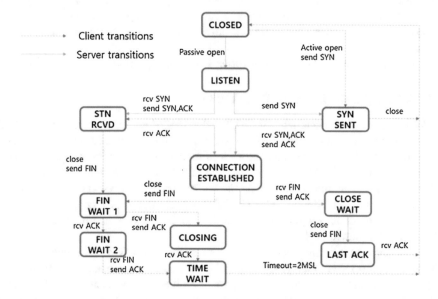

Fig. 1. TCP Finite State Machine

2.3   Harris Corner Algorithm

The Harris corner detector is the most common method for finding a key point in an image [8]. Figure 2 shows that a key point can be found by identifying the flat region, edge, and corner by slightly shifting a specific window in all directions, including vertically, horizontally, and diagonally.

To be specific, the corner value is determined by determining the direction and shift value of the window using the eigenvector and the sign of the resulting value. The corner derived from the algorithm is a significant state change and is used as a threshold.

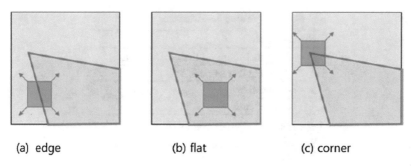

(a) edge   (b) flat   (c) corner

Fig. 2. Classification of state through window shift

3 Technical Scenario

TBIT, one of the aforementioned relevant research areas in the development of a sublayer network state inference technology for dynamic extraction of the protocol state, provides useful background knowledge for effective examination of the communication process.

The present section describes the method of implementation of the technology that best fits the study objectives.

The flow of the overall technology implementation can be summarized as follows.

- Test for attributes that can be specified when sending a packet
- Record specific attributes whose attribute value of the common attribute specified in the performed test was reversed
- Based on the above results, create n-dimensional axis of common properties
- Generate a graph of the values generated during the basic communication in the state of the axis
- Find the threshold value using the generated graph as the starting point using the Harris corner algorithm
- Generate FSM with threshold of common attribute
- Dynamic inference by comparing the attributes of the transport layer protocol coming in with the specified common attributes.

To extract various protocol states of sublayers, common attributes required for comparison of network protocols are first defined. The common attributes for this study are shown in Table 1.

The most basic of the 17 common attributes is assumed to duration. Other attributes are compared to duration to calculate appropriate thresholds for them [11].

In addition, to further infer the transport layer, additional attributes of the HTTP protocol belonging to the application layer are additionally examined and the network status reasoning in a more accurate web environment becomes possible by inferring the protocol of the transport layer.

Table 2 is a table arbitrarily selecting attributes required for network state reasoning among HTTP protocol attributes.

Table 1. Attributes

No.	Attribute
1	Inter packet arrival time (=Duration)
2	Total number of packet
3	Packet size
4	Host address
5	App ID
6	Duration
7	Payload size
8	Number of transferred bytes
9	CRC (Cyclic redundancy check) Error
10	CWR (Congestion Window Record)
11	TTL (Time to Live)
12	Initial advertised-window bytes
13	Congestion Loss Notification (ECN)
14	Advertised-wmdow bytes
15	Effective Bandwidth based upon entropy
16	Sequence number loss recovery tune
17	Flow table

Table 2. List of HTTP attributes

No.	HTTP packet	Attributes
1	HTTP request status	Number of HTTP requests
		Rate of HTTP requests
		Total data in headers
2	HTTP response status	Number of HTTP responses
		Rate of HTTP responses
		Number of responses with content length
3	HTTP request methods	GET
		POST
		PUT
		DELETE
4	Status code count	100–500
5	HTTP cookie	Hostname
		Connection_count
		Content_type
		Session

Important source code for parsing these attributes to improve the web environment can use the collector.getRequestStats() function to parse a variety of attributes, such as the number of HTTP responses and requests, status or total header size, when HTTP requests and responses (Table 3).

Table 3. Source code of HTTP attribute parsing

```
void printStatsSummary(HttpStatsCollector& collector)
{
    PRINT_STAT_HEADLINE("2. HTTP request");
    PRINT_STAT_LINE_INT("[1] Number of HTTP requests", collector.
    getRequestStats().numOfMessages, "Requests");
    PRINT_STAT_LINE_DOUBLE("[2] Rate of HTTP requests", collector.
    getRequestStats().messageRate.totalRate, "Requests/sec");
    PRINT_STAT_HEADLINE("3. HTTP response");
    PRINT_STAT_LINE_INT("[1]  Number  of  HTTP  responses",
    collector.getResponseStats().numOfMessages, "Responses");
    PRINT_STAT_LINE_DOUBLE("[2]  Rate  of  HTTP  responses",
    collector.getResponseStats().messageRate.totalRate, "Responses/sec");
}
```

The thresholds are calculated using the Harris corner detection algorithm described in Sect. 2.3. Based on the comparison, the corner is created, and the specified window value is modified in all directions; then the value is specified as the threshold for the attribute. Once the thresholds for all 17 common attributes are obtained, a test is configured using TBIT to store and transfer data, and the necessary state is configured based on the attributes' thresholds. As the values of the attributes approach their thresholds, the FSM is created based on the changes in the network state and it starts making inferences on the network states of sublayers.

In this process, it is important to study the TBIT, which is a technique to verify the status of the existing TCP protocol prior to inferring the state of the network. Therefore, we tested the packet loss rate of the existing TBIT and the improved network state reasoning technology, and the result is as shown in Fig. 3.

TBIT uses a packet capture engine that is basically provided by Linux. In the improved technology, it reduces the overhead of Linux by capturing packets through several threads. Zero-copy which does not copy memory to the kernel PF_RING library which has excellent performance is improved and used [12]. Because these packets collecting libraries are heavily influenced by the NIC, the NICs were tested in the same environment. Existing TBIT had an average 30% packet loss on 10 tests, but 10% of the improved technology showed 0% packet loss. Through this test, we can see that the improved technology research minimizes the loss of the packet and therefore can have better accuracy in reasoning the state.

Fig. 3. Packet loss rate comparison

4 Conclusion

As mentioned in the introduction, networking analytics technology based on web-based dynamic protocol extraction can provide users with an improved web environment. As the technology requires data on the user's experience, this study proposed technology for making inferences on the network states of the sublayers of the application layer. TBIT, a technology with a long history and that the present study heavily drew on, has constraints such as considering only the TCP environment and using predetermined congestion control, requiring improvements to be considered for current systems. The technology proposed in this study can perform more effective and precise dynamic extraction using thresholds than predetermined algorithm tests. This suggests that the proposed technology will make more accurate inferences than TBIT does. However, as thresholds alone can be insufficient for making highly accurate inferences, testing and inferences also need to create an FSM based on the final thresholds of common attributes.

This study suggests that the inference method is most suitable for inferring the network state of sublayers, and it allows dynamic extraction of random network protocols despite the use of FSM because of the use of threshold-based FSM rather than predetermined formulations.

Acknowledgment. This work was supported by Institute for Information & communications Technology Promotion (IITP) grant funded by the Korea government (MSIT) (2014-0-00547, Development of Core Technology for Autonomous Network Control and Management) and Next-Generation Information Computing Development Program through the National Research Foundation of Korea (NRF) funded by the Ministry of Science and ICT (No. NRF-2017M3C4A7083678).

References

1. Padhye, J., Floyd, S.: On inferring TCP behavior. In: Proceeding of SIGCOMM 2001, June 2001
2. Allman, M., Paxson, V., Stevens, W.: TCP Congestion Control, RFC2581, April 1999
3. Jacobson, V.: Congestion avoidance and control. Comput. Commun. Rev. **18**(4), 314–329 (1988)
4. Floyd, S., Henderson, T.: The New Reno Modification to TCP's Fast Recovery Algorithm, RFC 2582, April 1999
5. Li, Y., Peng, C., Yuan, Z., Li, J., Deng, H., Wan, T.: Mobile insight: extracting and analyzing cellular network information on smartphones. In: Proceedings of ACM MobiCom, New York City, NY, USA, October
6. Cheng, K.-T., Krishnakumar, A.S.: Automatic functional test generation using the extended finite state machine model. In: 1993 30th Conference on Design Automation. IEEE (1993)
7. Treurniet, J.R., Lefebvre, J.H.: A finite state machine model of TCP connections in the transport layer. Defence R&D Canada-Ottawa (2003)
8. Harris, C., Stephens, M.: A combined corner and edge detector. In: Alvey Vision Conference (1988)
9. Pei, Y., et al.: Effective image registration based on improved Harris corner detection. In: 2010 International Conference on Information Networking and Automation (ICINA), vol. 1. IEEE (2010)
10. Gogoi, P., et al.: Packet and flow based network intrusion dataset. In: International Conference on Contemporary Computing. Springer, Heidelberg (2012)
11. Kim, H., Kwon, D., Ju, H.: An inference method of stateless firewall policy considering attack detection threshold. J. Internet Comput. Serv. (JICS) **16**(2), 27–40 (2015)
12. Deri, L.: Improving passive packet capture: beyond device polling. In: Proceedings of SANE, vol. 2004 (2004)

Fault-Tolerant Topology Determination for IoT Network

Ryuichi Takahashi[1][✉], Mitsumasa Ota[2], and Yoshiaki Fukazawa[2]

[1] Ibaraki University, Hitachi, Ibaraki, Japan
ryuichi.takahashi.office@vc.ibaraki.ac.jp
[2] Waseda University, Shinjuku, Tokyo, Japan

Abstract. In the IoT, gateways are used as relay points so that resources deployed in an environment can communicate with each other. As the number communication paths connected to one gateway increases, the number of paths interrupted by a gateway failure increases. In an environment where real-time communication is required, the loss of data due to a gateway failure is unacceptable. Therefore, a method to determine the routing topology with fewer paths that are interrupted by a gateway failure is needed. In this paper, we propose a method to generate a fault-tolerant routing topology by a genetic algorithm. Our experimental results show that the proposed method can generate a topology where communication routes are not concentrated on one gateway in a realistic time.

Keywords: IoT network topology · Fault-tolerance ·
Genetic algorithm

1 Introduction

In an Internet of Things (IoT) environment, resources such as RFIDs, sensors, and actuators installed in various "things" communicate with and control each other [4]. IoT is rapidly spreading due to the miniaturization of power devices and the development of new communication technologies [15]. Currently, IoT environments of various scales from small scale IoT (called nano IoT) to global scale IoT are being deployed [8]. It is predicted that 50 billion devices will be connected on the Internet in 2020 [5].

Many devices are driven by batteries. Because battery capacity is limited, many resources exchange data, but not all resources can interconnect directly with geographically remote resources. In such case, a gateway is used because it allows IoT resources to be connected to traditional network [17,18]. A gateway is a device which becomes a relay point of resource communication.

In an IoT environment using gateways, it is necessary to determine not only the gateway that each resource uses but also the communication path between resources. The routing topology is usually a tree structure topology without a closed loop throughout the network to reduce the amount of data exchanged

© Springer Nature Switzerland AG 2019
S. Lee et al. (Eds.): IMCOM 2019, AISC 935, pp. 62–76, 2019.
https://doi.org/10.1007/978-3-030-19063-7_6

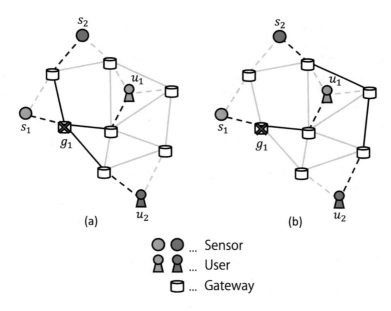

s_2

s_2

u_1

u_1

s_1

g_1

s_1

g_1

(a) u_2

(b) u_2

○ ● ... Sensor

🧍 🧍 ... User

🗄 ... Gateway

Fig. 1. Example of fault-tolerant topology

between gateways and to prevent circulation. Several studies have proposed methods to determine the routing topology between resources. However, many existing methods aim to minimize the communication cost by determining the most economical topology but do not consider the possibility of gateway failure. Herein a method is proposed to generate a high fault tolerant routing topology by consider the possibility of a gateway failure.

Figure 1 shows our motivating example where the same color indicates data obtained from the sensor is transmitted to a user via a gateway. In the figure, the data obtained from the sensor is transmitted to the user indicated by the same color via the gateway. There are two communications from s_1 to u_1 and from s_2 to u_2. Gray lines represent candidate communication paths, while the black lines represent actual communication paths. For the sake of clarity, only the part of the tree structure with the communication path is represented by black lines. In this environment, consider a case where a gateway failure occurs due to power outage or hardware crash. In topology (a), if a failure occurs in gateway g_1, both two communication paths are disconnected. On the other hand, in topology (b), communication is not concentrated on one gateway. Hence, even if a failure occurs in one gateway, the two communication paths are never disconnected at the same time.

Although the arrangement of gateways and resources are the same, the number of paths disconnected at the time of a gateway failure differs (Fig. 1). That is, the number of disconnected paths in a failure depends on the routing topology. A delay in information due to topology reconstruction is undesirable in applications requiring a real-time nature. Such situations require a fault-tolerant topology where fewer paths are disconnected in case of a gateway failure.

This paper aims to solve this problem by using a genetic algorithm. First, we convert the problem of resource allocation to the gateway into a minimum spanning tree problem. Next, we propose a method to generate a fault-tolerant topology by a genetic algorithm.

The rest of this paper is structured as follows. Section 2 introduces related works. Section 3 formally defines the problem dealt with in this paper, while a method to solve this problem is described in Sect. 4. In Sect. 5, we apply the proposed method to randomly generated networks. Finally, we conclude this paper in Sect. 6.

2   Related Work

2.1   Generating Routing Topology

This section introduces studies related to this paper. The following examines route determination methods for communication between resources.

Ma et al. [11] proposed a method to determine the communication paths between nodes in a wireless sensor network (WSN). By using a multi routing tree to construct the topology, fault tolerance is guaranteed while minimizing power consumption. However, their method assumes an environment that aggregates data from the sensor nodes to the sink nodes. In the sensor network of IoT, it is assumed that the deployed resources cooperate with each other rather than aggregate the data to one node. Hence, the network assumption differs from ours.

Ismail et al. [7] proposed a resource allocation method assuming a mobile network environment. Their method is similar to the resource allocation method that handles the gateway in this paper to determine the access point of each device. However, their method, a mobile device can connect to multiple access points simultaneously. This is different from the IoT environment in which communication is performed by selecting one gateway from the candidates.

Li et al. [10] researched a method to construct a routing tree for P2P network in distributed interactive applications by using a genetic algorithm. Their method aims to determine the route between two nodes in the environment and is similar to the problem dealt with in this paper. However, their method optimizes the communication speed, and does not consider fault tolerance.

Next, research on the IoT environment using a gateway is introduced. Karthikeya et al. [1] proposed a method to optimize the placement of the gateway. Their method, can minimize the introduction cost of the gateway into the environment by optimizing the gateway placement. Our method differs because the optimal communication path is generated in an environment where resources and gateways are already deployed.

Kim et al. [9] researched a method to solve the resource allocation problem to the gateway in the IoT environment. Their method uses a genetic algorithm to generate routing topologies. They strive to minimize the sum of the data transmission amount in the environment, but do not consider the impact of a gateway failure. Therefore, it is possible that a topology in which data transmission concentrates in one place may be obtained as a solution.

In addition, [2,3] can be cited as a method to select relay points in communication between resources. However none of the above methods consider a failure at a relay point.

Although several studies have investigated the generation method of the routing topology in a sensor network and an IoT environment using a gateway, no study has yet to propose a method to generate a routing topology with high fault tolerance in an IoT environment.

2.2   Genetic Algorithm

The Prüfer Code [14] is a well known way to express a spanning tree as an array of integers. However, it has been noted that the Prüfer Code has low locality and low heritability [6]. Due to these issues a genetic algorithm using the Prüfer Code has been shown to produce good results only for star-type trees [16]. Since the solution of the resource allocation problem does not always become a star-type topology, Prüfer Code is not suitable.

There are other coding schemes like Blob Code [13], Dandelion Code [12], etc. These improve the locality compared to the Prüfer Code and tend to provide better solutions. In this paper, we adopt the coding scheme of Kim et al. [9]. From the viewpoint of locality, this scheme has a higher locality than the abovementioned schemes. Furthermore, the code encoded by Minhyeop's scheme is easy to decode into a spanning tree.

3   IoT Environment

This section defines the environment dealt with in this paper. The environment consists of IoT resources, gateways and their connections. Within the environment, data is exchanged between mutually connected resources. The application determines which resource is used for data exchange.

A gateway is a device that relays communication between resources. Each gateway has a list of directly connectable gateways. At least one path must exist between gateways that exchange data.

Each resource communicates directly with one gateway to send data. Each resource has a list of gateways that it can communicate with directly and it must be able to communicate with at least one gateway.

This paper addresses how to determine the allocation gateway of each resource and to obtain the routing topology between gateways so that the data exchange between the resources does not concentrate on one gateway.

To solve this, we convert the resource allocation problem of the gateway into a minimum spanning tree problem according to Minhyeop's model [9]. Once resources and gateways are deployed in an environment, the routing paths among

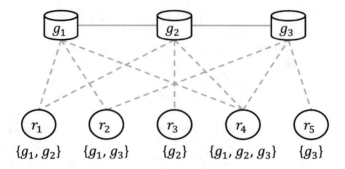

Fig. 2. Example of a gateway – resource connection graph

the gateways and the gateway – resource allocation are determined. Based on these connection relations, a gateway – resource connection graph is generated. The formal description of the gateway – resource connection graph G_C is shown as

$$G_C = (V, E)$$

$$V = V_r \cup V_g$$
$$V_r = \{r_n | n = ResourceID\}$$
$$V_g = \{g_n | n = GatewayID\}$$

$$E = E_r \cup E_g$$
$$E_r = \{(u, v) | u \in V_r, v \in V_g\}$$
$$E_g = \{(u, v) | v \in V_g, v \in V_g\}$$

G_C is a simple undirected graph consisting of a set of nodes V and a set of edges E. V consists of two node sets, V_r and V_g, V_r contains the resource nodes and V_g contains gateway nodes. E consists of two edge sets, E_r and E_g. E_r contains the edges expressing the connectable relation between the gateway and resource, while E_g contains the edges expressing the connectable relation among the gateways.

Figure 2 shows an example of a gateway – resource connection graph. Solid lines denote connectable relations among gateways. Dotted lines represent the connectable relation between resources and gateways. The list of gateways with which each resource can communicate is indicated in the parentheses under the resource. Determining the routing topology is equivalent to finding a spanning tree in this gateway – resource connection graph. We call this tree a connection tree.

The gateway – resource connection graph and resource pairs communicating in the environment $T = \{t = (r_i, r_j) | r_i, r_j \in V_r\}$ are treated as inputs. Here, the resource pair $t = (r_i, r_j)$ indicates that communication is performed between resources r_i and r_j. The output is a connection tree which represents a routing topology. The connection tree as the output requires that the communication path is not concentrated on one gateway.

4 Generating a Fault-Tolerant Topology

This section explains the method to obtain a fault-tolerant topology. Our method is an extension of Minhyeop's genetic algorithm [9]. Next-generation individuals are generated by a single-point crossover method and a mutation. To evaluate whether communication of topologies is not concentrated on one gateway, the fitness function must be defined. The encoding and the fitness function are described below.

4.1 Encoding of the Topology

In the genetic algorithm, each individual solution candidate is represented by a sequence. Converting a problem into a sequence is called encoding. In this section, we explain how the proposed method encodes the solutions.

Here, we adopt the encoding scheme proposed by [9]. In this scheme, each individual C is expressed as

$$C = Gene[R + G]$$

$$Gene[i] = \begin{cases} al_i & (i < R) \\ con_{(i-R)} & (i \geq R) \end{cases}$$

R represents the total number of resources. G represents the total number of gateways. al_i is the ID of the gateway to which resource r_i is allocated, while con_i is the ID of the gateway to which gateway g_i is connected. Each individual is expressed as connection tree by a sequence with the sum length of the number of resources in the environment and the number of gateways. Sequences are divided into the head and tail parts. The head part shows the specific gateway that each resource is allocated and the tail part shows the specific gateway that each gateway is connected. Figure 3 shows an example of this encoding.

The black lines represent a connection tree. In this encoding scheme, one topology may be expressed in plural ways. However, one topology can always be obtained for one code. Additionally, it is possible that this method generates a topology that is unable to communicate. For example, in Fig. 4 shows the case where a closed loop is formed instead of a spanning tree. In a closed loop, the route between two arbitrary points cannot be uniquely determined.

Figure 5 shows that the connection relationship between gateways is divided into two groups. A disconnection between the gateways results in one gateway being unreachable from a certain gateway. In other words, it is possible that a set of resources will be unable to communicate.

As these, the encoding method of Minhyeop may generate a code that does not form a tree. These codes are dealt with by evaluating with the fitness function described in the next section.

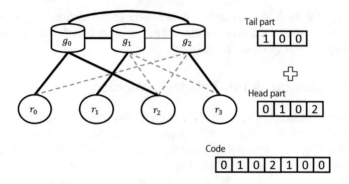

Fig. 3. Example of encoding

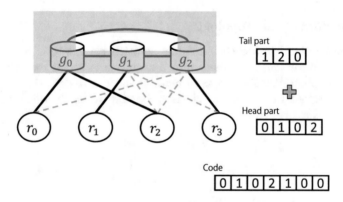

Fig. 4. Example of a code forming a closed loop

4.2   Fitness Function

The fitness function evaluates how much each individual matches the purpose. In the processes of crossovers and mutations, a better solution can be generated by leaving individuals with high fitness. The fitness function is defined so that the solutions without a concentration of communication to a specific gateway are highly evaluated.

As mentioned earlier, the adopted encoding scheme may not create a spanning tree in some cases. If a spanning tree is not formed, an unreachable gateway will exist and a gateway pair will cannot communicate with each other. Therefore, the fitness function is defined in two cases, that is, the case where the topology forms a spanning tree and the case where it does not form a spanning tree.

If the topology does not form a spanning tree, the function evaluates the fitness by using the number of gateways included in the largest connected component (maximum connected subgraph) in the entire environment. Since the spanning tree contains all nodes in the graph, the size of the connected component in the spanning tree takes the maximum value. Therefore, by leaving

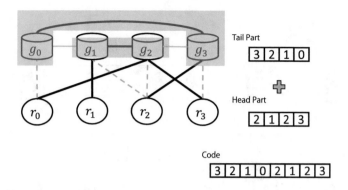

Fig. 5. Example of a code with a disconnection between gateways

individuals with a larger number of nodes included in the largest connected component, the individual approaches the spanning tree.

When the topology forms a spanning tree, the number of disconnected communication paths during a gateway failure is used to evaluate fitness. The number of disconnected communication paths when a gateway failure occurs is equal to the number of communication paths through that gateway. Therefore, we count the number of paths through each gateway.

We consider a fitness function that evaluates the communication concentration on each gateway. We use the average value of the communication path through the gateway. If the average value is small, the expected value of the number of disconnected communication paths during a gateway failure is small. Consequently, the fitness function is more highly evaluated as the average value of the number of communication paths decreases.

Considering only the average value of the number of communication paths through each gateway is insufficient.

Figure 6 represents two topologies that achieve a connection between three pairs of resources (r_1–r_5, r_2–r_4, r_3–r_6). The number in the circle of each gateway is the number of communication paths through that gateway. When calculating the average value, the same value of 1.33 is obtained for both topologies. If only the average value is considered, these topologies will be evaluated as the same. Because all three communication paths pass through gateway g_5 in Fig. 6 right, a failure of gateway g_5, disconnects all communications. On the other hand, in Fig. 6 left, the paths are not concentrated on one gateway. Thus, the left topology is more fault-tolerant as the paths are more dispersed. Hence, both the average and the dispersion of the communication path must be considered.

Let $pass(g)$ be the number of communication paths passing through each gateway g. Individuals with small average values and with small variance values of $pass(g)$ are evaluated highly. As a result, a topology with a minimal number of disconnected paths due to a gateway failure can be identified. The fitness function is defined as

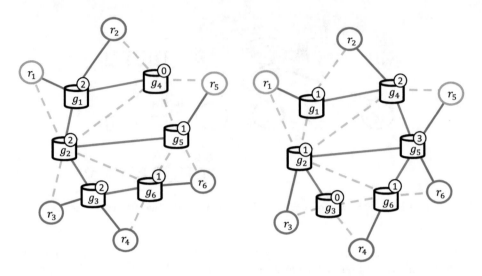

Fig. 6. Importance of the route distribution

$$f(c) = \begin{cases} size(c) & (spanning\ tree) \\ M - \left[aV(pass(g)) + \sum_{t_i \in T} l(path(c, t_i))\right] & (no\ spanning\ tree) \end{cases}$$

In individual c, $size(c)$ represents the number of gateways included in the largest connected component. $l(path(c, t_i))$ represents the path length for communication among resource pair t_i. $V(pass(g))$ is the variance of the number of communication paths through each gateway, and a is a weight for variance. By changing the value of a, the sum of the path lengths and the number of paths through each gateway can be adjusted. To prioritize a topology that does not concentrate the communication paths to one gateway over the route length, a is set to a large value.

As described above, the smaller the average value and the variance value are, the higher the fitness. However, forming a spanning tree must have a higher fitness than not forming a spanning tree. The fitness when not forming a spanning tree is the number of gateways included in the largest connected component and this value never exceeds the total number of gateways in the environment N_g. Therefore, by setting M to a value much larger than N_g, it is possible to prevent the fitness when a spanning tree is formed from being smaller than when a spanning tree is not formed.

To calculate the average value, the sum of the number of relay gateways of each transmission is divided by the total number of gateways in the environment. Since the total number of gateways in the environment N_g is constant, the calculation time can be reduced slightly by setting the value of a to $1/N_g$ beforehand.

5   Evaluation

A simulator was used to evaluate whether the proposed method can be provide a topology that does not concentrate the communication paths onto a specific gateway. For the experiment, we used a gateway – resource connection graph with randomly generated spanning trees and a list of resource pairs to communicate as the inputs. In this experiment, all resources in the environment communicate with one of the other resources.

In the first experiment, the execution times as the number of resources varied from 5 gateways to 20 gateways were measured. The results of our method were compared to the brute force method and the Minhyeop's method.

In the second experiment, the maximum value and the average value of the number of disconnected communication paths upon a gateway failure were measured. In the proposed method and Minhyeop's method, we measured 10 randomly generated gateway-resource connection graphs (20 gateways, 20 resources).

This experiments were measured using a laptop (Intel Core i5 CPU M560 2.67 GHz, 4.00 GB memory, Windows10 32 bit). We set the weighting parameter a of the variance in the fitness function to 3.0. Table 1 shows the other parameters of the genetic algorithm.

Table 1. Setting of the genetic algorithm

Number of individuals	100
Number of selections	10
Number of generations	50
Mutation rate	0.10

5.1   Experiment 1: Execution Time

Figures 7 and 8 show the measured execution time for 5 and 20 gateways, respectively. The y-axis of Fig. 7 is logarithmic axis. The execution time of the brute force method when the number of resources is 16 or more is very long, and a solution is not obtained. Except for the case where the number of resources is very small, the proposed method and the Minhyeop's method can find a solution in a short time (Fig. 7). The execution time of the brute force method exponentially increases as the number of nodes increases, and it is not practical.

In the case of 20 gateways, the execution time in the brute force method is very long and no solution is obtained. For methods using a genetic algorithm, the execution time linearly increases with number of resources (Fig. 8). Even in the case of 20 gateways and 100 resources, the execution time is less than 1 s. From these results, the genetic algorithm can provide a topology in realistic time. Hence, these methods have sufficient scalability.

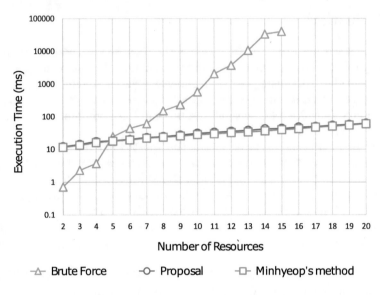

Fig. 7. Execution time (5 gateways)

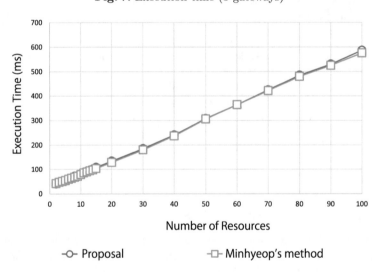

Fig. 8. Execution time (20 gateways)

In the proposed method, the variance is calculated by counting the number of communication paths through each gateway. For this process, the proposed method requires a longer execution time than the Minhyeop's method. However, the difference is very small, less than 10 ms. Hence, the increase in time cost to consider the fault tolerance using the proposed method is negligible.

Furthermore, we show convergence of fitness in the proposed method. Figure 9 shows the change in fitness when the proposed method is executed for five ran-

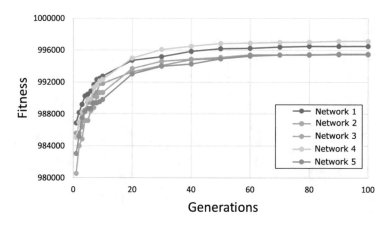

Fig. 9. Convergence of fitness in the proposed method

dom generated networks (20 gateways, 100 resources). In any networks, the fitness change is large up to 40 generations, and after 50 generations there is no big change in the fitness. The convergence time depends on the scale of the network, but if the network is about the same scale as the experimental environment, it is possible to discover the solution in less than 1 s in 50 generations.

5.2 Experiment 2: Fault-Tolerance and Trade-Off

Here we evaluate whether the proposed method can provide a fault-tolerant topology. The topology is identified by finding a resource allocation that does not concentrate communication paths on one gateway. The fault-tolerant topology is evaluated using two measurement results of the maximum value of the disconnected communication path length when a failure occurs in one gateway and the average length of the communication paths.

Figure 10 shows the maximum number of paths disconnected when a fault occurs. In almost all cases, the proposed path results in smaller number of disconnected paths. The reduction is about 40%. Thus, considering variance can prevent the concentration of communication paths. The topology obtained by the proposed method successfully reduces the number of disconnected communication paths due to a gateway fault.

The average value of the connection paths was evaluated in the same manner. Figure 11 shows the results. The communication path length is equal to the number of gateways through which each communication path passes. The results show that the proposed method tends to generate slightly longer communication paths in the topology. It means that the amount of data exchanged among the resources is larger than existing methods. However, the average path length increase is as small as about 10%, resulting in a suitable performance unless the amount of data transfer is the primary emphasis. Hence, the proposed method can greatly improve the fault tolerance in exchange for a slight increase in the communication path length.

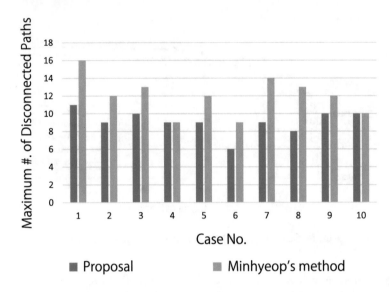

Fig. 10. Maximum number of disconnected paths

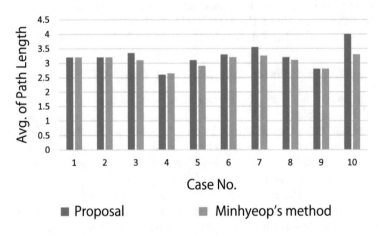

Fig. 11. Average length of the communication path

6   Conclusion

In an IoT environment using a gateway, it is necessary to determine the gateway that each resource communicates with. When communication paths are concentrated in one gateway, a gateway fault results in the disconnection of many paths.

In this paper, we propose a method to provide a fault-tolerant topology where fewer communication paths disconnect due to a fault using a genetic algorithm. Using an extension of Minhyeop's method, the individuals used in the genetic algorithm are encoded and the fitness function considering the variance of the number of communication paths through each gateway is defined. Experiments confirm that the proposed method can generate a fault-tolerant topology in a realistic time.

This method defines weight a to make the path variance and the path length compatible. Because it is generally difficult to quantify user preferences, a method to determine the weight a is necessary. The experiment conducted in this paper is a simulation experiment in randomly generated environments. In the future, we plan to reevaluate the effectiveness in an actual IoT environment.

References

1. Karthikeya, S.A., Kotagi, V., Siva Ram Murthy, C.: Leveraging solution-specific gateways for cost-effective and fault-tolerant IoT networking. In: 2016 IEEE Wireless Communications and Networking Conference, pp. 1–6 (2016). https://doi.org/10.1109/WCNC.2016.7564811
2. Karthikeya, S.A., Narayanan, R., Siva Ram Murthy, C.: Power-aware gateway connectivity in battery-powered dynamic IoT networks. Comput. Netw. **130**, 81–93 (2018). https://doi.org/10.1016/j.comnet.2017.11.001. http://www.sciencedirect.com/science/article/pii/S1389128617303936
3. Aoun, B., Boutaba, R., Iraqi, Y., Kenward, G.: Gateway placement optimization in wireless mesh networks with QoS constraints. IEEE J. Sel. Areas Commun. **24**(11), 2127–2136 (2006). https://doi.org/10.1109/JSAC.2006.881606
4. Atzori, L., Iera, A., Morabito, G.: The Internet of Things: a survey. Comput. Netw. **54**(15), 2787–2805 (2010). https://doi.org/10.1016/j.comnet.2010.05.010. http://www.sciencedirect.com/science/article/pii/S1389128610001568
5. Evans, D.: The Internet of Things: how the next evolution of the Internet is changing everything. Cisco Internet Business Solutions Group (IBSG) **1**, 1–11 (2011)
6. Gottlieb, J., Julstrom, B., Raidl, G., Rothlauf, F.: Prüfer numbers: a poor representation of spanning trees for evolutionary search. In: Proceedings of the Genetic and Evolutionary Computation Conference (GECCO 2001), pp. 343–350. Morgan Kaufmann Publishers, San Francisco (2001)
7. Ismail, M., Zhuang, W.: A distributed multi-service resource allocation algorithm in heterogeneous wireless access medium. IEEE J. Sel. Areas Commun. **30**(2), 425–432 (2012). https://doi.org/10.1109/JSAC.2012.120222
8. Kawamoto, Y., Nishiyama, H., Kato, N., Yoshimura, N., Yamamoto, S.: Internet of Things (IoT): present state and future prospects. IEICE Trans. Inf. Syst. **E97.D**, 2568–2575 (2014). https://doi.org/10.1587/transinf.2013THP0009
9. Kim, M., Ko, I.Y.: An efficient resource allocation approach based on a genetic algorithm for composite services in IoT environments. In: 2015 IEEE International Conference on Web Services, pp. 543–550 (2015). https://doi.org/10.1109/ICWS.2015.78
10. Li, Y., Yu, J., Tao, D.: Genetic algorithm for spanning tree construction in P2P distributed interactive applications. Neurocomputing **140**, 185–192 (2014). https://doi.org/10.1016/j.neucom.2014.02.035. http://www.sciencedirect.com/science/article/pii/S0925231214004421
11. Ma, G., Yang, Y., Xuesong, Q., Gao, Z., Li, H.: Fault-tolerant topology control for heterogeneous wireless sensor networks using multi-routing tree. In: 2017 IFIP/IEEE Symposium on Integrated Network and Service Management (IM), pp. 620–623 (2017). https://doi.org/10.23919/INM.2017.7987344

12. Paulden, T., Smith, D.: From the Dandelion Code to the Rainbow Code: a class of bijective spanning tree representations with linear complexity and bounded locality. IEEE Trans. Evol. Comput. **10**, 108–123 (2006). https://doi.org/10.1109/TEVC. 2006.871249
13. Picciotto, S.: How to encode a tree. University of California, San Diego (1999). https://books.google.co.jp/books?id=DXQ-AQAAIAAJ
14. Prüfer, H.: Neuer beweis eines satzes über permutationen. Arch. Math. Phys. **27**, 742–744 (1918). https://ci.nii.ac.jp/naid/10028221960/
15. Rawat, P., Singh, K.D., Bonnin, J.M.: Cognitive radio for M2M and Internet of Things: a survey. Comput. Commun. **94**, 1–29 (2016). https://doi.org/10. 1016/j.comcom.2016.07.012.   http://www.sciencedirect.com/science/article/pii/ S0140366416302699
16. Rothlauf, F., Goldberg, D.E.: Representations for Genetic and Evolutionary Algorithms. Physica-Verlag, Heidelberg (2002)
17. Singh, F., Kotagi, V., Siva Ram Murthy, C.: Parallel opportunistic routing in IoT networks. In: 2016 IEEE Wireless Communications and Networking Conference, pp. 1–6 (2016). https://doi.org/10.1109/WCNC.2016.7564825
18. Zhu, Q., Wang, R., Chen, Q., Liu, Y., Qin, W.: IoT gateway: bridgingwireless sensor networks into Internet of Things. In: 2010 IEEE/IFIP International Conference on Embedded and Ubiquitous Computing, pp. 347–352 (2010). https://doi.org/10. 1109/EUC.2010.58

Review of Internet Gateway Discovery Approaches in MANET

Mufind Mukaz ebedon$^{(\boxtimes)}$, Amna Saad, and Husna Osman

Universiti Kuala Lumpur, 1016 Jalan Sultan Ismail, 50250 Kuala Lumpur, Malaysia
ebedonmufind@gmail.com, {amna,husna}@unikl.edu.my

Abstract. Mobile ad hoc network (MANET) form a network on the go without the need of any infrastructure and centralize administration. For a mobile node to communicate with a node outside the MANET a gateway has to be put in place to translate one topology architecture (infrastructure) to another topology (uninfrastructure) to create a hybrid network. Different Internet discovery approaches have been created (proactive, reactive and hybrid) and a recent focus is on adaptive approach which adapts the time to live (TTL), Periodicity of Internet gateway advertisement and route path. Researcher in Cluster Internet discovery approach have shown a good performance in the MANET but less research has been done in this area. The discovery of Internet gateway still need further research due to the fact that MANET topology is not stable, different scenarios or environment will need a different approach for the discovery of the Internet gateway. Our contribution to this research is to combine the clustering approach and Internet gateway discovery approach, to make the Clusterhead as the mobility Internet gateway relay of the fixed Internet Gateway to decrease the Overhead, end to end packet delay. This paper review will analyze the pros and cons of different approaches (reactive, proactive, hybrid, adaptive and cluster Internet discovery approaches) in MANET. A comparison of these approaches will be done. A future vision of a new approach will be proposed.

Keywords: Integrated Internet-MANET · Hybrid networks ·
Clustering · Clusterhead · Gateway discovery

1 Introduction

Internet has changed the way people communicate, mobile devices have been a key element to the advance of the communication. People used social media to share their life with others at any geographical area they are located, all this is done by connecting devices to the Internet, however, this communication needs an infrastructure for mobile devices to communicate. In an area where it is difficult to put an infrastructure or there is not any coverage of any telecommunication Antenna, or it is expensive to install one, communication will be

© Springer Nature Switzerland AG 2019
S. Lee et al. (Eds.): IMCOM 2019, AISC 935, pp. 77–88, 2019.
https://doi.org/10.1007/978-3-030-19063-7_7

impossible. Mobile Ad hoc network (MANET) has been defined as a group of wireless mobile nodes dynamically forming a temporary network without the presence of any infrastructure or centralized administration. Each node in ad hoc act as a router and terminal mobile at the same time [1]. MANET can be a solution to the problem cited earlier, which can bring some benefits such as an extension of telecommunication coverage, reduce the cost of implementing an infrastructure, people can send message and access Internet application anywhere at any place they are located.

For mobile ad hoc to connect to the Internet a discovery mechanism has to be put in place to allow mobile node to share their message with other mobile node on the Internet. To be able to connect to the Internet, node has to discover and register with an Internet gateway which acts as a middle point between Ad hoc network and infrastructure network. Integrating MANET and Internet come with some difficulty, different approach was created to found the best suitable Internet Gateway discovery mechanism which can adapt to the dynamic changing of the topology and can give the best performance in term of packet delivery, Overhead, packet delay etc. Proactive, reactive approach, hybrid and adaptive approach was created.

In the proactive approach, the Internet Gateway broadcast packet advertisement periodically [2], which create an overhead in the network due to too much advertisement packet. Reactive approach does not send packet advertisement periodically, but mobile node sends a gateway solicitation to the Internet gateway [3]. The bad side of this approach is the creation of more end to end packet delay due to the route discovery process. Hybrid approach combines the proactive approach and the reactive approach to make sure that node which is near the Internet gateway will use proactive approach and those who are far from the Internet gateway advertisement will be used reactive approach [4,5], but the determination of an optimal proactive area will create more overhead if the area is large (large TTL value). A better approach is the one who adapt base on the environment and behavior of the mobile ad hoc network. Adaptive approach researcher come out with some mechanism to regulated the periodic advertisement from the Internet gateway, the coverage of the periodic advertisement base on the TTL and also some modification on gateway solicitation message to resolve the problem of overhead and end to end delay [6,7]. This approach will still create too much process for the Internet gateway to decide the right TTL and periodic advertisement, and this method make some assumption about the practical MANET which the environment change constantly.

Clustering has brought a lot of advantages in the MANET by grouping node, electing a Clusterhead which act as a coordinator of that group, by doing that the cluster decreases the routing table of each node, re-utilization of the frequency [8], but few research has been done in MANET on the discovery of Internet gateway in clustering [9]. The existing research do not show the overall performance of a real MANET discovery.

The current Internet discovery approaches in MANET give good performance in specific scenario and environment, that is why researcher improve the existing approach to come out with a better discovery mechanism but to achieve a better performance which can be implemented in any type of scenarios is very difficult, further research is still needed in this area to found the best suitable mechanism which can be adaptable in any scenario.

Our research contribution in this area is to mix the mobility of Internet gateway with a fixed gateway to get faster discovery and registration to the Internet gateway. To decrease the handover of mobility of the Internet gateway, we are going to use a clustering mechanism in which node will form a group and elect a clusterhead, and the clusterhead will be the Internet gateway relay. For clusterhead outside the fixed gateway advertisement coverage will send their solicitation packet to the nearest clusterheads which will reply back rather then sending the packet to the fixed gateway which can create an overhead in the network. Our approach will decrease the overhead due to the grouping of nodes in a cluster, the end to end delay will also be decreased by using the Clusterhead as a relay of Internet packet to the fixed Internet gateway (few hops). All mobile node in the group will only register to their Internet mobility Clusterhead, thus a decrease of handover and increase in the discovery, freshness of the route will be observed.

This review will be structured as; the second section will be talking about gateway discovery approaches, adaptive gateway discovery will be described in the third section, the fourth section will be focusing on Clustering discovery approach, the fifth section a comparison review of some research in Internet gateway will be analyzed, and finally we will conclude by giving our future alternative on Internet gateway discovery.

2   Gateway Discovery Approaches

Mobile node in MANET need an Internet gateway for them to communicate to the external world, the gateway has to understand the MANET network and the infrastructure network. Nodes need to found and register with a particular Internet gateway to be able to send a packet outside the MANET. Three approaches are used to discover the Internet gateway; Proactive approach, Reactive approach, and Hybrid approach.

2.1   Proactive Approach

In the proactive approach [6,10], the Internet Gateway broadcast packet advertisement periodically which is called Gateway advertisement (GWADV) containing all the information needed by the node to register to the Gateway. The packet contains a TTL field which determines the number of hops the packet has to be forwarded by receiving node. Each node decrease the TTL when forwarding the packet until the TTL reach zero, no further forwarding will be process and node outside the range will not receive the packet, consequently will not be able to connect to the Internet. The advantage with this approach is that mobile nodes have a fresh route to the Internet gateway, the disadvantage of it is too many packets are sent after x second witch can create an overhead in the network. This disadvantage lead to an exploration of others approach.

2.2   Reactive Approach

In this approach [6,11] the Internet Gateway does not send packet advertisement periodically. A node which wants a access to the Internet sends a gateway solicitation packet to the Gateway, when the gateway received the solicitation packet, it will reply with a Gateway advertisement message to the node in unicast mode. The advantage of this approach is that mobile nodes which are far from the Internet gateway coverage range can still access the Internet by sending solicitation message, which can also decrease the overhead in the network but will also increase the end to end delay in the sense that packet has to be forwarded by intermediate node until it reaches the Internet gateway.

2.3   Hybrid Approach

Hybrid approach combines those two methods so that any node can still access the Internet regardless of where is located in the network [12,13]. Figure 1 show an hybrid approach.

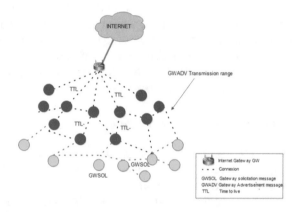

Fig. 1. Hybrid gateway in MANET

Hybrid approach solved one part of the problem by using proactive and reactive approach, note which is near the gateway used the proactive approach and those which are far from it use the reactive approach. This approach still has some drawback on the adjustment of the gateway periodic advertisement which can still create an overhead in the network and also too many gateway solicitation packets will create an end to end delay. Base on that further research has been focusing on an adaptive approach, which adapts to the condition of the network, further detail will be described in the next section.

3   Adaptive Discovery Approaches

Hybrid approach as mention earlier has some drawback on the time to send the advertisement packet, if the periodic time is high this will create an overhead in the network, if it is low node will not have a fresh route to the gateway and will start sending gateway solicitation message (GWSOL) to the Internet gateway [7]. The goal of the adaptive approach is to adapt base on the topology change. The time to live of a packet (TTL) will be increase or decrease base on the topology change and also the periodic advertisement will be adjusted each time the topology change. The adaptive approach used the proactive approach with some slice modification on TTL and GWADV, node outside the gateway coverage will still use the reactive approach.

Different technique for calculating the TTL and Gateway periodic solicitation message (GWSOL) has been proposed, here are the lists of some techniques which has been grouped on soft computing and nonsoft computing technique [14].

Non Soft computing technique focus on the adjustment of the TTL and gateway advertisement but not on the periodicity of the gateway message, maximum source coverage, maximum benefit coverage, Bidirectional link, Complete adaptive, Load adaptive access gateway, adaptive distributed discovery, and load aided adaptation. Soft computing focus on a choice of a better path to reach the internet gateway and also on Advertisement periodicity: Fuzzy with Genetic algorithm, Gateway Advertisement using FIS, Optimal route to the gateway, Adaptive way of Path Load Balancing, Fuzzy Logic based.

Adaptive mechanism for the discovery of Internet gateway have also some drawback some mechanism do not take in consideration the periodic advertisement and others the TTL. The non adjusting of these parameters will still create an overhead and end to end delay.

4   Clustering Discovery Approach

The process of dividing mobile node in MANET into virtual groups is called clustering and the virtual group is called a cluster. The formation of the virtual group is done according to some rules with different behaviors for nodes included in a group and those who are not in the group [15]. See an example of a cluster structure in Fig. 2. In Cluster, a mobile node can have different status and function, such as Clusterhead, clustergateway, cluster member. A Clusterhead is the coordinator of that group, its function is to forward a packet to others cluster. A clustergateway is a node which as a link to other clusters, its function is to forward a packet to other neighboring clusters. A cluster member is an ordinary node which is not a Clusterhead and does not have any links to other cluster [16].

Author which used this approach considers using the advantages of the cluster such as grouping node which decrease the overhead in the network due to packet restricted to go out of the group only by passing it to the Clusterhead.

In [17] they used DSDV for communication inside the cluster head gateway switch routing, DSDV make node have a table of all the node in the network for a fast packet send. The author used also a Cluster Gateway source routing (CGSR) for the communication in that cluster. The approach assumes that Clusterhead

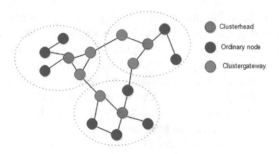

Fig. 2. An example of clustering in MANET

has to be near the foreign agent (FA) to be registered with it. If the Clusterhead cannot send the packet to the foreign agent, a mobile node can directly by itself send the packet. They used a combination of CGSR and Mobile IP for a bidirectional link. The drawback of this approach is to assume that Clusterhead are near the FA, in reality, node move in different direction, and using only proactive approach will not give a good result in a tense MANET.

In [18] they used Clusterhead to perform the discovery, IOADV was used for the communication between Clusterhead and Internet gateway. A reactive approach was also used for the discovery of Internet gateway at startup. For the formation of the cluster they used multimedoid, a comparison of Enhanced LEACH and multimedoid was done for the selection of a Clusterhead. No detail on Clusterhead communication to Internet gateway was given.

In [9] a formula has been created to compare the overhead in different researcher technique. It has been shown that clustering approach with Internet gateway has decreased the overhead compared to other techniques such as reactive, proactive and hybrid.

Few research has been done in clustering internet discovery in MANET, using proactive approach alone will not show the performance of a real environment when nodes are moving arbitrarily. More performance metrics is needed to give a better view of the existing proposal. The exchange of Clusterhead Internet solicitation message and fixed Internet gateway will create an overhead in the network. The idea is to decrease the flow of message and to make sure that node which is far from the fixed gateway in Cluster can still send a packet to the Internet without the MANET incurring more end to end delay.

This area needs more research in term of the choice of the formation of the cluster, protocol performance comparison, enhancement of routing protocol, and the discovery of Internet gateway, backup for Clusterhead if the Clusterhead is off due to some circumstances.

5   Related Works

This section will review some Internet Gateway Discovery approaches base on the impact of periodic advertisement on the MANET, the impact of having mobile and fixed internet gateway, the number of internet gateway, and lastly the approach used (proactive, reactive, hybrid, adaptive) see Table 1. This analysis will help us to determine if the increase of node in mobile ad hoc network will need more fixed or mobile Internet gateway to decrease the overhead. Is a certain periodic advertisement give a better result in decreasing or increasing the overhead and end to end delay?

In term of internet gateway periodic advertisement time varying or higher. The effect of varying the beacon intervals time on the total control overhead in [19] show initially decreases as the beacon interval increases and when it reaches 20 s the total control overhead remains constant, the approach used in this paper is proactive, as we know proactive approach needs an internet gateway to send periodic advertisement for node to update their route path and other information, a smaller periodic advertisement create more overhead but help node to update their route without any delay, a large periodic advertisement does not create an overhead but will not help node to have a current path to the internet gateway. In [20] they compared the performance of proactive, reactive and hybrid in term of periodic advertisement. In end to end delay metric with a periodic advertisement of 2 to 60 s with a higher speed of node, the proactive and hybrid show a higher average end to end compared to reactive approach, due to too much traffic node need a higher processing times, and that creates more interface queue, a reactive approach does not used a periodic approach that means it the better solution to used. In [21] they show that the periodic advertisement has a different result on a different model used Random Waypoint, CMM, and RPGM, a realistic model have to be put in place for a better performance. In [4] with 30 s of periodic advertisement, they proposed a hybrid approach which shows a better result in overhead and end to end delay.

Adaptive approaches are not using a fixed periodic advertisement, the advertisement is varying base on each author's algorithm. Most of the adaptive approach shows a better performance in term of overhead and end to end delay, due to the combination of proactive and reactive approach with a slice modification in term of periodic advertisement, area coverage base on TTL, node gateway solicitation. But some author did not compare their proposed algorithm with other author proposed adaptive algorithm to determine if their solution was better, they just compared their work to the so called conventional approach (the proactive, reactive and hybrid) like in [7,22].

In [10] they observed that when node moves at a higher speed the overhead in adaptive is a bit higher than the hybrid one due to the shorter broadcasting interval but the adaptive perform better in term of packet delivery ratio and decrease the packet transmission delay to the foreign agent. In [24] the performance is based on the handover if the existing connected gateway is not unavailable. The improved protocols compare to pure protocols shows a decrease in handover overhead for the modified mobile IP over AODV and OLSR compare to the original mobile IP over AODV. In [7] they proposed a mechanism where internet gateway advertisement message is sent only to some area where the link is expected to remain stable, they compare their approach with others authors

Table 1. Internet gateway review

This table provides a selected authors approaches in Internet Discovery							
Authors	Number of GW		Approach used				Internet GW_ADV
	Fixed	Mobile	Proactive	Reactive	Hybrid	Adaptive	
Sun et al. [19]	1,2		Yes				5 to 60 s
Lee et al. [4]	2				Yes		30 s
Ghassemian et al. [3]	4,12		Yes	Yes	Yes		10 s
Ruiz and Gómez-Skarmeta [22]	2					Yes	N/A
Bin et al. [10]	1,2					Yes	N/A
El-Moshrify et al. [20]	3		Yes	Yes	Yes		2 to 60 s
Kumar et al. [23]	2		Yes				N/A
Ratanchandani and Kravets [12]	2				Yes		10 s
Khan et al. [9]	2	4	Yes	Yes			10 s
Khurshid et al. [21]	2		Yes	Yes	Yes		10, 20, 30, 50 s
Lu et al. [24]		5				Yes	1 s
Iqbal and Kabir [25]	2				Yes		N/A
Iqbal et al. [26]	2		Yes				N/A
Majumder et al. [27]	2		Yes	Yes	Yes		N/A
Yuste et al. [7]	2					Yes	10 s
Saluja and Shrivvastava [5]	2		Yes	Yes	Yes		N/A

approaches in (RMD, ADD, GDSL, and REA) with different mobility model (RWP and TVCM), the proposed scheme shows a reduction in the packet loss rate, the end to end delay and the routing overhead, due to the ability to identify and utilizing stable route. In [6] they proposed on demand gateway broadcast scheme, where node send packet when they need to access the internet gateway. They compare they approach with a pure Proactive gateway. The performance shows a higher end to end delay due to the rebuilding of a route in reactive approach which is costly when node speed increase. In overhead, they found out that the increase of node speed create more overhead due to a constant data traffic to keeps the route update even the mobility is high. In both performances, the author approach has better performance.

In term of mobile as an internet gateway. In [21] compare the performance of mobile gateway (MG) with a speed of 2 ms and fixed gateway (FG) separately. They compare their own approach which integrates the mobile and fixed gateway (FGMG). In term of packet delivery ratio base on beacon interval, FGMG shows a better result, MG shows a bad result. In term of an end to end delay base on beacon interval FGMG shows a better result and FG show a bad result. The overhead is higher in comparison with FG or MG, combining fixed and mobile gateway shows a better result than using each method individually. As said earlier in [24] with 5 gateway which moves at 10 ms speed a decrease of overhead was observed for an Improvement of AODVV and OLSR compare with the original protocol. Having mobile gateway in MANET improve the performance, node choose the nearest gateway for a fresh route, short path and the lifetime between the node with the internet gateway is still constant, however using only mobile gateway is not a better solution, agent advertisement will flood the network and decrease the performance.

In term of higher number of fixed internet gateway, most of the researcher have been using 2 fixed internet gateway with 50 as the average of the number of mobile nodes in [3] they used 4 to 12 fixed gateway. They compare the effect of increasing the number of internet gateway (IGW) in a proactive and reactive approach. The packet delivery ratio shows a better performance for a limited number of internet gateways, for a signaling overhead, although higher numbers of IGW create more periodic internet gateway advertisement the signaling overhead is decreasing, due to the fact that more packets are transmitted successfully. The average packet delay is lower in proactive approach compare to a reactive approach. That is due to the fact that proactive discovery approach has a short path to the IGW and the route are often updated. In reactive discovery approach as long as the old path still available the traffic will be sent using the old path even if a new route with a short path is available. An increase in the number of fixed gateways show a better result, but periodic advertisement will be higher which can create overhead even if the node has a fresh route and send their packet successfully.

6   Research Gap

Proactive discovery approach depends on the frequent periodic advertisement. The high the periodic frequency, the performance decrease base on too much packet, the low the periodic frequency a high creation of end to end delay due to a fresh route to the Internet Gateway. Using only reactive discovery approach will create an end to end delay, due to a long path to reach the IGW. A solution is to use a hybrid approach which combines the proactive and reactive advantages. This approach has also some drawback in term of the interval beacon time used and also the time to live (TTL) of the advertisement see Sect. 2.

Adaptive approach has been created to solve hybrid problem, some researcher used mathematical solution to determine the TTL, some use message to trigger the Internet Gateway (IGW) to send advertisement, some used algorithm to make the IGW send advertisement only to stable link, some used algorithm to increase the TTL to cover the all MANET base on the behavior of the mobile node see Sect. 5, the drawback of these approaches is that a full coverage of the network cannot be achieved, the need for a sophisticated algorithm is still needed to make the IGW adjust it TTL, Periodic advertisement and to determine a stable link, which is also impossible in a MANET topology base on the arbitrary topology change.

Mobility Internet gateway alone will not give a good performance, a mixture of fixed and mobility Internet gateway is a good choice to use, but in a flat network, these will still create an overhead and end to end delay due to handover. So the solution is to used Clustering by using Clusterhead as the mobility gateway to have a better performance see Sect. 4. To our knowledge we are the first one who used a Clusterhead as a mobility Internet gateway relay, our approach will not just use the Clusterhead communication but will make Clusterhead an Internet Gateway, the Clusterhead will not broadcast any packet to the fixed Internet gateway to register each node which needs Internet Gateway but will have a duplicate image of a fixed gateway to reduce the discovery process and also to help neighboring Clusterhead which are far from the fixed Internet Gateway.

7   Conclusion

This review paper as shown that Majority of the researcher are using fixed Internet gateway for the discovery of Internet gateway in different scenarios, this will still create an overhead due to a long path in case of GWSOL message (reactive approach) and some path will be loaded with many packets. The mobility of Internet gateway has helped to solve this problem but will create too many handovers. Cluster in Internet gateway has proved that it can give a better performance compared to some other approaches, using it without making a Clusterhead as mobile gateway relay will not give a good result. For a better performance, our research is going to use clustering technique by using Clusterhead as a mobility Internet Gateway.

The benefit of our approach is to decrease the overhead due to the grouping of node in cluster, the end to end delay will also be decreased because only Clusterhead packet will be sent to the fixed gateway, the handover will not be seen because of the Clusterhead as the coordinator of the group, the response time of Internet gateway will be increased. Our undergoing research is on the choice of protocol to use in clustering discovery or the enhancement of it, choice of cluster formation algorithm, synchronization and communication with the fixed Internet Gateway from Clusterhead, those will be our future work to confirm or not our gaps.

References

1. Broch, J., Maltz, D.A., Johnson, D.B., Hu, Y.C., Jetcheva, J.: A performance comparison of multi-hop wireless ad hoc network routing protocols. In: Proceedings of the 4th Annual ACM/IEEE International Conference on Mobile Computing and Networking, pp. 85–97. ACM (1998)
2. Trivino-Cabrera, A., Casilari, E., González-Cañete, F.J.: An improved scheme for the integration of mobile ad hoc networks into the internet without dedicated gateways. In: 2006 11th International Workshop on Computer-Aided Modeling, Analysis and Design of Communication Links and Networks, pp. 16–21. IEEE (2006)

3. Ghassemian, M., Hofmann, P., Prehofer, C., Friderikos, V., Aghvami, H.: Performance analysis of internet gateway discovery protocols in ad hoc networks. In: 2004 IEEE Wireless Communications and Networking Conference, WCNC 2004. IEEE, vol. 1, pp. 120–125 (2004)

4. Lee, J., Kim, D., Garcia-Luna-Aceves, J., Choi, Y., Choi, J., Nam, S.: Hybrid gateway advertisement scheme for connecting mobile ad hoc networks to the internet. In: The 57th IEEE Semiannual Vehicular Technology Conference, VTC 2003-Spring, vol. 1, pp. 191–195. IEEE (2003)

5. Saluja, R.K., Shrivastava, R.: A scenario based approach for gateway discovery using MANET routing protocol. In: 2012 International Conference on Computer Communication and Informatics (ICCCI), pp. 1–6. IEEE (2012)

6. Xu, H., Ju, L., Guo, C., Jia, Z.: On-demand gateway broadcast scheme for connecting mobile ad hoc networks to the internet. In: 2014 International Conference on Smart Computing (SMARTCOMP), pp. 112–117. IEEE (2014)

7. Yuste, A.J., Trivino, A., Casilari, E., Trujillo, F.D.: Adaptive gateway discovery for mobile ad hoc networks based on the characterisation of the linklifetime. IET Commun. 5(15), 2241–2249 (2011)

8. Alinci, M., Spaho, E., Lala, A., Kolici, V.: Clustering algorithms in MANETs: a review. In: 2015 Ninth International Conference on Complex, Intelligent, and Software Intensive Systems (CISIS), pp. 330–335. IEEE (2015)

9. Khan, K.U.R., Ahmed, M.A., Reddy, A.V., Zaman, R.U.: A hybrid architecture for integrating mobile ad hoc network and the internet using fixed and mobile gateways. In: 1st IFIP Wireless Days, WD 2008, pp. 1–5. IEEE (2008)

10. Bin, S., Bingxin, S., Bo, L., Zhonggong, H., Li, Z.: Adaptive gateway discovery scheme for connecting mobile ad hoc networks to the internet. In: Proceedings of the 2005 International Conference on Wireless Communications, Networking and Mobile Computing, vol. 2, pp. 795–799. IEEE (2005)

11. Yuste, A.J., Trujillo, F.D., Trivino, A., Casilari, E., Diaz-Estrella, A.: An adaptive genetic fuzzy control gateway discovery to interconnect hybrid MANETs. In: IEEE Wireless Communications and Networking Conference, WCNC 2009, pp. 1–6. IEEE (2009)

12. Ratanchandani, P., Kravets, R.: A hybrid approach to internet connectivity for mobile ad hoc networks, vol. 3. IEEE (2003)

13. Zaman, R.U., Khan, K.U.R., Reddy, A.V.: A survey of adaptive gateway discovery mechanisms in heterogeneous networks. Int. J. Comput. Netw. Inf. Secur. 5(7), 34 (2013)

14. Mishra, R., Verma, P., Kumar, R.: Gateway discovery in MANET using machine learning and soft computing: a survey. In: 2017 International Conference on Innovations in Information, Embedded and Communication Systems (ICIIECS), pp. 1–6. IEEE (2017)

15. Anupama, M., Sathyanarayana, B.: Survey of cluster based routing protocols in mobile adhoc networks. Int. J. Comput. Theory Eng. 3(6), 806 (2011)

16. Yu, J.Y., Chong, P.H.J.: A survey of clustering schemes for mobile ad hoc networks. IEEE Commun. Surv. Tutorials 7(1), 32–48 (2005)

17. Khan, K.U.R., Zaman, R.U., Reddy, A.V.: A three-tier architecture for integrating mobile ad hoc network and the internet using a hierarchical integrated routing protocol. In: 2008 International Conference on Advanced Computer Theory and Engineering, pp. 518–522 (2008). https://doi.org/10.1109/ICACTE.2008.40

18. Manjula, S., et al.: Service discovery in clustered MANET and internet. In: International Conference on Computation System and Information Technology for Sustainable Solutions (CSITSS), pp. 204–209. IEEE (2016)

19. Sun, Y., Belding-Royer, E.M., Perkins, C.E.: Internet connectivity for ad hoc mobile networks. Int. J. Wireless Inf. Networks **9**(2), 75–88 (2002)
20. El-Moshrify, H., Mangoud, M., Rizk, M.: Gateway discovery in ad hoc on-demand distance vector (aodv) routing for internet connectivity. In: National Radio Science Conference, NRSC 2007, pp. 1–8. IEEE (2007)
21. Khurshid, H., Ghassemian, M., Aghvami, H.: Impact of applying realistic mobility models on the performance analysis of internet gateway discovery approaches for mobile ad hoc networks. In: International Conference on Innovations in Information Technology, IIT 2009, pp. 230–235. IEEE (2009)
22. Ruiz, P.M., Gómez-Skarmeta, A.F.: Enhanced internet connectivity for hybrid ad hoc networks through adaptive gateway discovery. In: 29th Annual IEEE International Conference on Local Computer Networks, pp. 370–377. IEEE (2004)
23. Kumar, R., Misra, M., Sarje, A.K.: An efficient gateway discovery in ad hoc networks for internet connectivity. In: International Conference on Conference on Computational Intelligence and Multimedia Applications, vol. 4, pp. 275–282. IEEE (2007)
24. Lu, B., Yu, B., Sun, B.: Adaptive discovery of internet gateways in mobile ad hoc networks: with mobile IP-based internet connectivity. In: 5th International Conference on Wireless Communications, Networking and Mobile Computing, WiCom 2009, pp. 1–5. IEEE (2009)
25. Iqbal, S.M.A., Monir, M.I., Osmani, S.M.R., Chowdhury, F.S., Chowdhury, A.K., Dhar, K.: A novel strategy to discover internet gateways in mobile ad hoc networks. In: 2010 13th International Conference on Computer and Information Technology (ICCIT), pp. 533–538. IEEE (2010)
26. Iqbal, S.M.A., Kabir, M.H.: Internet gateway discovery and selection scheme in mobile ad hoc network. In: 2011 14th International Conference on Computer and Information Technology (ICCIT), pp. 44–49. IEEE (2011)
27. Majumder, K., Ray, S., Sarkar, S.K.: Implementation and performance analysis of the gateway discovery approaches in the integrated MANET-internet scenario. In: 2011 IEEE 3rd International Conference on Communication Software and Networks (ICCSN), pp. 601–605. IEEE (2011)

ARP Request Trend Fitting for Detecting Malicious Activity in LAN

Kai Matsufuji[1]([✉]), Satoru Kobayashi[2], Hiroshi Esaki[1], and Hideya Ochiai[1]

[1] The University of Tokyo, Tokyo, Japan
mattun@hongo.wide.ad.jp, hiroshi@wide.ad.jp,
ochiai@elab.ic.i.u-tokyo.ac.jp
[2] National Institute of Informatics, Tokyo, Japan
sat@nii.ac.jp

Abstract. Security of local area networks (LANs) attract enormous attention these days. LANs are not safe even under a well-configured firewall, because malware can be easily delivered through network applications. Thus we need to take counter measures against malicious activities by focusing on each device itself. However, we cannot adopt many of existing methods to the devices which have limited resource capacity like IoT. Accordingly, we consider that the request pattern of address resolution protocol (ARP) may provide some indicators for finding such malicious activities without putting a burden on each device. In this paper, We propose a method to detect malicious network activity by ARP monitoring. The detection method is based on a fitting model of ARP request trend. We especially focus on the destination devices of ARP request in outlier detection with the model. We made an experiment with a data of monitored ARP requests in our network. We also discuss parameter tuning and validity of the model based on three notable requirements.

Keywords: Address resolution protocol · Local area network · Security

1 Introduction

Security of local area networks (LANs) is getting more and more serious these days. Malware makes some suspicious activities for expanding the malware itself, for searching sensitive data, or for overriding IoT application-layer protocols. Even if there is a firewall at the entrance of the network, malware can be easily delivered into LANs via phishing emails or malware infected smartphones through Wi-Fi. Thus we need to take counter measures against malicious activities in those cases by focusing on each device itself. However, it places a large load on the end-node devices to measure the network activity of them. It is difficult to make counter measures against malware, which works in the end-node devices, due to their resource capacity limitation. Some existing techniques [3] are proposed in recent days, but most of them are not reasonable in such as IoT

© Springer Nature Switzerland AG 2019
S. Lee et al. (Eds.): IMCOM 2019, AISC 935, pp. 89–96, 2019.
https://doi.org/10.1007/978-3-030-19063-7_8

networks. We propose based on address resolution protocol (ARP) to solve such problem. Because ARP request packets are broadcasted for finding their target MAC addresses in most cases, we can observe the packet in the other devices connected to the LAN. Therefore, we can observe the ARP request packets by adding a monitor machine to the LAN. This does not place an additional load to the end-node devices. Moreover, there is another advantage that we can introduce the ARP monitoring method without changing existing devices.

In this paper, we estimate whether malware invades into a LAN or not by observing the trend of ARP requests. When malware invades into a device, it tries the scanning attack to invade into other devices in the LAN or interrupts particular connections. The number of target hosts in a certain period may drastically increase or decrease in case of such malicious cases. Therefore, by observing the change of the number of target hosts about ARP request, we can judge whether malware invades into LAN or not in the period.

In this paper, we propose a fitting model which predicts the next situation of the number of destination hosts about ARP requests by using the number so far and judge whether the connection in the LAN is normal or not based on it.

2   Related Work

LAN security is paid more and more attention nowadays because many kinds of malware such as ransomware become a big threat in the modern society. However, we need to choose which method is the best when we apply it to capacity limited devices.

Kolbitsch et al. [3] proposed a malware detection approach which learns a program of malware and judges whether the malware invades or not. Their approach is implemented at the end host and makes the host free up some resources for the approach. However, we should not adopt such methods to the devices which have limited resource capacity like IoT.

In studies featuring LAN security, some of the studies focus on ARP just like our study.

Whyte et al. [5] record active systems within the network cell and devices they are trying to connect with by using ARP request and use them as an indicator for anomaly detection by counting ARP request which don't correspond to them. The study is related to ours in terms of anomaly detection using ARP request. However, it is difficult for their approach to deal with configuration changes because their method don't rewrite the record. Thus we need a method considering time variation and rewriting records.

Pandey [4], Jinhua et al. [2] and Hou et al. [1] study about LAN security focusing on ARP. However, these studies attach importance to taking measures against ARP spoofing, and it is difficult for us to adopt these studies to malware detection.

3   ARP Requests Classification Method

As mentioned in Sect. 1, we take measures against malicious activities by observing ARP request packets. We propose a fitting model to ARP request trend patterns. The fitting model predicts the trend of ARP requests in the next day from the past trends.

3.1   Requirements of Fitting Models

We list the requirements for fitting models in our study as follows:

Learn Automatically
    The model shouldn't require knowledge such as the list of malwares manually, otherwise we would update the list manually forever.
Follow Time Variation
    The model should expect the situation of the next day by learning the past trend. Because the system configuration in a LAN changes frequently, the model should be always updated.
Correspond to Amplitude
    The model should memorize drastic change experienced recently. It shouldn't be too sensitive to the trends. Otherwise, we can't adopt the model to the shaking trends.

Thus, we define $A_s(n)$, considering first and second items as mentioned before and memorizing the characteristic of $X_s(n)$ and make the variation smooth and $B_s(n)$, considering third item (mentioned again in Sect. 3.3).

3.2   The Extraction of the Degree of Destination

Our approach first makes a connection graph for a certain time duration of a network, i.e., device-to-device connections based on ARP request. We focus on the number of connected hosts in ARP (hereinafter, this is called the degree of destination). Our fitting model estimates the future trends learning from the past trends (Fig. 1).

Let S be the source that ARP request sends, and D_s be the list of IP addresses S had targeted at in a day. That is,

$$D_s = (IP_1, IP_2, ..., IP_k) \tag{1}$$

We define $X_s(n)$ as the degree of the target, more formally,

$$X_s(n) = k \tag{2}$$

Here, n represents the date.

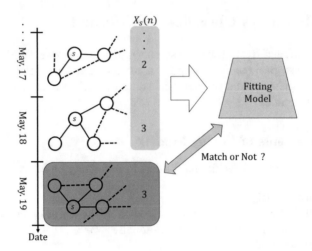

Fig. 1. Estimation the future trends of the degree of destination. The connection graphs are generated from the observations of broadcasted ARP requests.

3.3   Definitional Formulae of Fitting Model

The fitting model, we propose in this paper, gives an upper limit and a lower limit for possible changes. It expects that in the next time duration the degree of destination will be within those limits. We admit that, to actually show the fitness, we must introduce some fitness evaluation models, such that the gap between the upper and lower limits should be appropriate. However, this kind of evaluation model is dependent on the network operation, so we left this discussion open, in this paper.

In Sect. 3.1, we mentioned $A_s(n)$ and $B_s(n)$ considering three requirements. Further, we define an upper limit and a lower limit of the fitting model as $A_s(n) \pm \gamma B_s(n)$. By defining the two limits in this way, fitting model performs more efficiently meeting three requirements.

Figure 2 shows the example of the fitting model. $A_s(n) - \gamma B_s(n)$ and $A_s(n) + \gamma B_s(n)$ is the lower limit and the upper limit of the fitting model and when $X_s(n)$ doesn't lie between the boundaries, $X_s(n)$ is judged to be abnormal.

Now, when considering about definitional formulae of fitting model, we shouldn't define all cases in only one formula. This is because devices which always connect to somewhere every day are different from what occasionally connect to somewhere. In the former, the value of 0 in $X_s(n)$ should be judged as abnormal because as long as devices connected to somewhere, they keep on sending ARP request every certain period. In the latter, on the other hand, 0 in $X_s(n)$ is should be normal because they don't always need to send ARP request. Thus, we define formulae of fitting model distinguishing between the two cases.

Fitting Model About Permanently-Connecting Device. Many of the devices which always connect to somewhere (Permanently-Connecting devices)

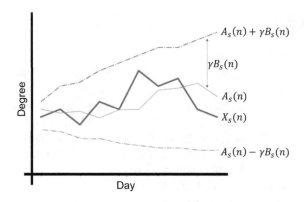

Fig. 2. An example of fitting model

connect some application servers (e.g., DNS, syslog, DHCP, Nagios and DB) during normal times, and it is assumed that the degree of destination approaches a constant value. Now we define the moving average about $X_s(n)$ $(A_s(n))$ as:

$$A_s(n+1) = \alpha A_s(n) + (1-\alpha)X_s(n), A_s(0) = 0 \ (0 < \alpha < 1) \tag{3}$$

and the variance $(Y_s(n))$ as:

$$Y_s(n) = \{X_s(n) - A_s(n)\}^2 \tag{4}$$

and the moving average about $Y_s(n)$ $(B_s(n))$ as:

$$B_s(n+1)^2 = \beta B_s(n)^2 + (1-\beta)Y_s(n)^2, B_s(0) = 0 \ (0 < \beta < 1) \tag{5}$$

Further, we define the determination condition to decide whether devices fit the model or not (C) as:

$$C : A_s(n) - \gamma B_s(n) < X_s(n) < A_s(n) + \gamma B_s(n) \ (\gamma > 0) \tag{6}$$

and we perform anomaly detection about S based on the inequality.

Fitting Model About Occasionally-Connecting Device. It is assumed that the degree of destination about devices which occasionally connect to somewhere basically exhibits the same behavior as mentioned in Sect. 3.3. However, we need to consider separately about when devices don't connect to anywhere, and the degree of destination evaluates to 0. In the above case, we redefine $A_s(n)$ and $B_s(n)$ as mentioned before as:

$$A_s(n+1) = \begin{cases} \alpha A_s(n) + (1-\alpha)X_s(n) & (X_s(n) > 0) \\ A_s(n) & (X_s(n) = 0) \end{cases} \tag{7}$$

$$B_s(n+1)^2 = \begin{cases} \beta B_s(n)^2 + (1-\beta)B_s(n)^2 & (X_s(n) > 0) \\ B_s(n)^2 & (X_s(n) = 0) \end{cases} \tag{8}$$

and C as:

$$C : X_s(n) = 0 \vee A_s(n) - \gamma B_s(n) < X_s(n) < A_s(n) + \gamma B_s(n) \tag{9}$$

4 Evaluation

4.1 Experiment Settings

To validate our approach, we collected ARP request packets from the network in our lab from April 17 to June 16, 2018. We investigated the degree of target per day by source IP address. Hereinafter, we call the upper and lower limits of the fitting model as mentioned in Sect. 3 as l_1 and l_2. In this experiment, we define γ as 2 ($l_1, l_2 : A_s(n) \pm 2B_s(n)$). Also, if over $1/4$ of measurement period about a device is 0, we call the device as Occasionally-Connecting device. Under these conditions, we describe l_1 and l_2 changing α and β and validate which parameter (α and β) fits the behavior of $X_s(n)$ the best.

We use first seven-days time-series (except days without observed communications) as a learning-period, which makes $A_s(n)$ stable enough for outlier detection.

4.2 Parameter Tuning

In this section, we discuss parameter tuning in the formulae of the fitting model. In other words, we consider how the fitting model works if the parameters are not tuned properly.

Figure 3 shows the transitions of $X_s(n), A_s(n), l_1$ and l_2 changing α and β to various values about three IP addresses in May, 2018. Now, IP_A and IP_B are Permanently-Connecting devices, and IP_C is Occasionally-Connecting device (not including the learning period).

In the case that both α and β is 0.99, $A_s(n)$ is almost constant in all three IP addresses in spite of the large change of $X_s(n)$. This is because $A_s(n)$ is fixed to data memorized in transient state because of too large α. Further, we can see that although $X_s(n)$ experienced large fluctuations, the fluctuations are not reflected in l_1 and l_2. This is because the amplitude memorized in transient state is reflected too strongly because of too large β.

On the other hand, we also evaluate the case which both α and β are 0.5. In $A_s(n)$, we can see that the value changes drastically compared with graphs as mentioned so far in all three IP addresses. This is because the value of last $X_s(n)$ is reflected too strongly because of too small α, and $A_s(n)$ doesn't memorize previous data. Further, we can see that unlike in 0.99, l_1 and l_2 change very intensely after $X_s(n)$ experienced drastic changes. This shows that the two value are too sensitive to the latest change because of too small β, and they can't predict correctly.

5 Discussion

As mentioned in Sect. 3, $A_s(n)$ and $B_s(n)$ in the formulae about fitting model meet the three requirements. Now, we list an example which had a good behavior as fitting model as a result of experiment (i.e., when both α and β is 0.9) and confirm that the fitting model in that case surely meets the three requirements (Fig. 3).

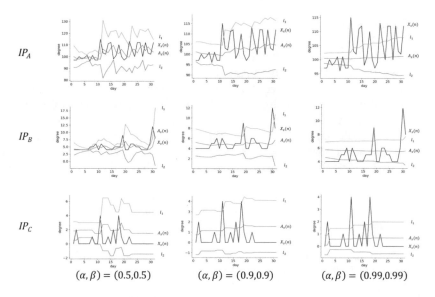

Fig. 3. The transitions of $X_s(n), A_s(n), l_1$ and l_2 changing α and β to various values about three IP addresses in May, 2018.

Learn Automatically

This condition is satisfied obviously because the fitting model doesn't require any information like the list of malware.

Follow Time Variation

It can be said that this condition is satisfied because l_1 and l_2 change in response to changes of $X_s(n)$.

Correspond to Amplitude

To judge whether this requirement is satisfied or not, we need to confirm drastic changes in Fig. 3. In all three IP addresses, when we observe a drastic change of $X_s(n)$ for the first time, the day's data is judged to be abnormal because the fitting model doesn't expect this change. On the other hand, when we observe them for the second time or later, the fitting model has memorized the drastic changes experienced in the previous several days. Therefore, we can see that in IP_A, large shakes occurring immediately after first shake is not detected as abnormal value, and in IP_B and IP_C, drastic changes occurring some time after the first drastic change is judged to be abnormal.

From the above, we derive the result that the fitting model we proposed has worked appropriately satisfying the three requirements if we choose the parameter properly.

In this paper, we have worked on our laboratories network. As a future work, we need to discuss other networks for confirming the reliability in our study. Also, we focused on the degree of destination in ARP request. In the future, we consider the different features such as the locality or frequency. Along with these future works, the achievement of this study will be more successful.

6   Conclusion

In this paper, we have proposed a fitting model of ARP request trend patterns. Firstly, we proposed three requirements for designing the fitting model. Secondly, we showed the formulae in accordance with the three requirements and defined an upper limit and a lower limit. Finally, we investigated and evaluated the transition of the fitting model changing parameters in the formula to various values. As a result, we concluded that the fitting model functioned meeting the three requirements when we choose the parameter properly. As a future work, we need to discuss other networks or consider the different features of the degree of destination.

References

1. Hou, X., Jiang, Z., Tian, X.: The detection and prevention for ARP spoofing based on snort. In: 2010 International Conference on Computer Application and System Modeling (ICCASM), vol. 5, p. V5-137. IEEE (2010)
2. Jinhua, G., Kejian, X.: ARP spoofing detection algorithm using ICMP protocol. In: 2013 International Conference on Computer Communication and Informatics (ICCCI), pp. 1–6. IEEE (2013)
3. Kolbitsch, C., Comparetti, P.M., Kruegel, C., Kirda, E., Zhou, X., Wang, X.: Effective and efficient malware detection at the end host. In: USENIX Security Symposium, vol. 4, pp. 351–366 (2009)
4. Pandey, P.: Prevention of ARP spoofing: a probe packet based technique. In: 2013 IEEE 3rd International Advance Computing Conference (IACC), pp. 147–153. IEEE (2013)
5. Whyte, D., Kranakis, E., van Oorschot, P.: ARP-based detection of scanning worms within an enterprise network. In: Proceedings of the Annual Computer Security Applications Conference (ACSAC) (2005)

Robust Bio-Inspired Routing Protocol in MANETs Using Ant Approach

Arush Sharma and Dongsoo S. Kim[(✉)]

Department of Electrical and Computer Engineering,
Indiana University Purdue University, Indianapolis, IN 46202, USA
{sharmaas, dskim}@iupui.edu

Abstract. This paper discusses about developing a routing protocol for mobile ad-hoc networks in a bio-inspired manner. Algorithms inspired by collective behavior of social insect colonies, bird flocking, honey bee dancing, etc., promise to be capable of catering to the challenges faced by tiny wireless sensor networks. We propose an energy efficient multi-path routing algorithm based on foraging nature of ant colonies. This paper considers many meta-heuristic factors to provide good robust paths from source node to destination node in a hope to overcome the challenges posed by resource-constrained sensors. The performance of proposed ant colony routing algorithm is compared against well-known AODV routing protocol.

Keywords: Ant Colony Optimization · Swarm Intelligence ·
Wireless Sensor Networks · Mobile ad-hoc networks

1 Introduction

Wireless Sensor Networks (WSN) consist of a large number of nodes equipped with sensing capabilities; communication interfaces which has limited memory and energy resources. WSN nodes are statically deployed over large areas. However, they can also be mobile interacting with the environment. WSNs have wide spectrum of applications which includes environmental sensing, health care, traffic control, tracking of wild life animals, etc. Usually individual sensor nodes send their data towards base station node (commonly known as sink node). Intermediate nodes perform relaying of sensed data towards the destination.

1.1 Design Challenges

Following are the design challenges faced by WSNs Routing Protocols.

1. Low Computational and memory requirements.
   Sensor nodes are equipped with a low end CPU and have limited memory. Therefore, it is mandatory that the routing algorithm has minimum overhead to make its execution feasible and effective.
2. Self-organization

© Springer Nature Switzerland AG 2019
S. Lee et al. (Eds.): IMCOM 2019, AISC 935, pp. 97–110, 2019.
https://doi.org/10.1007/978-3-030-19063-7_9

Wireless Sensor Network is expected to remain active for considerable period of time. Within that time period, few new nodes might be added to the network, while other nodes might die due to energy depletion or may become operational. A routing protocol therefore should be robust to such dynamic and unpredictable events. It should be empowered with self-organizing properties to let the network function as an autonomous system.

3. Energy Efficiency

Nodes are equipped with small non-rechargeable batteries. Therefore, the efficient battery usage of a sensor node is very important aspect to support the extended operational lifetime of network. The routing protocol is expected to forward the data packets across multiple paths, so that all nodes can deplete their battery source at a comparable rate. This results in load balance of network and increases the network lifetime.

4. Scalability

In WSN applications, thousands of nodes are generally deployed that have short communication ranges and high failure rates. Hence a routing protocol should be able to cope with the above challenges.

It is imperative to design such routing protocols which can cater to the aforementioned challenges and are robust and adaptive. Nature provides us examples of mobile, independently working agents which seamlessly work together to perform tasks efficiently, for example, flight of migratory birds, ant colony optimization, etc. Nature inspired algorithms also known as Swarm Intelligence (SI) are based on collective behavior of social insect colonies and other animal societies for solving different types of communication problems.

The remainder of paper is given as follows. At first general introduction and challenges are given which gives us the motivating factors to do research. Section 2 discusses about the related work done in the field of ant routing algorithms. Section 3 talks about the novel ideas implemented in ant routing algorithm. Section 4 does a comparison study against AODV routing protocol. Finally we conclude by giving a short summary and an outlook on future work.

1.2   Background

Ant algorithms are special class of SI algorithms, which consist of population of simple agents (ants) which interact locally and with their environment. The foraging behavior of ant colony inspires Ant Colony algorithm. Initially, ants randomly wander around the nest searching for food. When the food is found, they take it back to the colony and leave a trail of pheromones on the way. Other ants can sense the pheromone concentration and prefer to follow directions with higher pheromone density. Since shorter paths can be traversed faster, they will eventually outweigh the less optimal routes in terms of pheromone concentration. Additionally, pheromones evaporate over time, so ants are less likely to follow an older path which makes them search for newer paths simultaneously. In a case where an obstacle gets in their way, ants again initiates the route discovery process by randomly selecting the next hop until the ants converge on the paths with relatively higher concentration of pheromone.

2 Related Work

AntNet proposed by Di Caro and Dorigo [1] is a routing technique which is applied for best-effort IP networks. Optimizing the performance of entire network is its main aim according to the principles of Ant Colony Optimization, AntNet is based on a greedy stochastic policy, where each node maintains a routing table and an additional table containing statistics about the traffic distribution over the network. The routing table maintains for each destination and for each next hop a measure of the goodness of using the next hop to forward data packets to destination. These goodness measures, called pheromone variables, are normalized on the stochastic policy. This algorithm uses forward ants and backward ants to update the routing table. The forward ants use heuristic based on the routing table to move between a pair of given nodes and are used to collect information about the traffic distribution over the network. The backward ant stochastically follows the path of forward ants in reverse direction. At each node, the backward ant updates the routing table and the additional table which contains traffic statistics of the network.

The energy-efficient ant-based routing algorithm (EEABR) is a routing protocol for WSNs and extends AntNet proposed by Camilo et al. [2]. EEABR tries to minimize memory requirements as well as the overall energy consumption of the original AntNet algorithm. The ants retain information of only last two visited nodes because it takes into account the size of ant packet to update pheromone trail. In the typical ant-based algorithm, each ant carries the information of all the visited nodes. Then, in a network consisting of very large number of sensor nodes, the size of information would cause considerable energy to send ants through the network. Each node keeps the information of the received and sent ants in its memory. Each memory record contains the previous node, the forward node, the ant identification, and a timeout value. The transmission probability considers the artificial pheromone value and the remaining energy of the possible next hop.

Ladder Diffusion Algorithm proposed by Ho et al. [3] addresses the energy consumption and routing problem in WSNs. The algorithm tries to reduce the energy consumption and processing time to build the routing table and avoid the route loop. In this algorithm, the sink node broadcasts the ladder creating packet with the grade value of one. The grade value of one means that the sensor node receiving this ladder creating packet transmits data to the sink node requires only one hop. Then sensor nodes increments the grade value of ladder creating packet and broadcast the modified ladder-creating packet. A grade value of two means that the sensor node receiving this ladder-creating packet sends data to the sink node requires two hop counts. And this step repeats until all the sensor nodes get the ladder-creating packet. The ladder diffusion algorithm assures that the direction of data transfer always occurs from a high grade value to a low grade value, which means each relay is forwarded to the sink node since each sensor node records the grade value of relay nodes in the ladder table. The path decision is based on the estimated energy consumption of path and the pheromone.

Energy-Aware Ant Routing in Wireless Multi-Hop Networks proposed by Frey et al. [4] provides new mechanisms for estimating the fitness of a path and energy information dissemination thus enabling to prolong the network lifetime. The network lifetime is the time span a network can fulfill its service. Traditional Ant Routing Algorithm considers the pheromone value in its probabilistic routing decision process. This approach favours shortest paths over non-shortest paths which is not suitable for energy constrained networks. EARA extends the ant routing algorithm with an energy heuristic for determining the nodes residual energy and scheme for estimating a path's energy. EARA algorithm considers the residual energy of a node as an additional heuristic. Since the residual energy of a node changes over time, periodic Energy Ants are released for updating the energy values in the node's routing table. Periodic Energy Ants are sent occasionally as it can be a costly operation in terms of consumed energy.

Ant Colony and Load Balancing Optimizations for AODV Routing Protocol proposed by Abd Elmoniem et al. [5] discusses about improving the AODV routing protocol by taking the Ant Colony Optimization approach. Forward ant agents are sent as a part of route establishment request to find the route to destination. This route establishment phase is very much similar to Route Request (RREQ) phase of AODV routing protocol except for the fact that if the route to destination doesn't exist and there exist no neighbour, then the ant is broadcasted. Otherwise, if the active neighbour exists with highest pheromone, the forward ant is sent to that neighbour. In case of destination node receiving forward ant, backward ant is sent to the source node with a route to destination which comes under the part of Route establishment reply phase. The pheromone update policy is applied on the nodes receiving backward ants. Also it is applied differently depending upon whether the node is an source node or intermediate node or is destination node. Once the source node receives backward ant, the Data transmission phase begins. Each node receiving data packets forwards it to neighbor according to the pheromone values. Neighbor node having greater pheromone receives more data than those having less pheromone which leads to load balancing. If the route doesn't exist at all, a Route Error (RERR) packet is sent to the source node. If the routing table entry of the destination doesn't exist in source node, it deletes the route and again initiates the route discovery process.

Ant Colony Optimization for Routing and Load-Balancing: Survey and New Directions presented by Sim et al. [6] provides comparison of the approaches for solving the convergence problem in ACO algorithms. When the network reaches its equilibrium state, the already discovered optimal path is given more preference over other paths by the ants which leads to many problems such as congestion, reduction of probability for selecting other paths, network failure, etc. In order to mitigate this problem, some of the approaches include evaporation, aging, pheromone smoothing and limiting, privileged pheromone laying etc. Evaporation of pheromone is a technique to prevent the ants of favoring the older or stale paths which makes an ant to concurrently search for fresh paths. Aging refers to quantity of pheromone deposited by the ant. Older ant will deposit less pheromone compared to its young contemporary since they take more time in reaching destination. Limiting and Smoothing Pheromone refers to limiting the pheromone deposit by placing an upper bound which reduces preference of optimal paths over non-optimal paths In privileged pheromone laying, only certain ants are permitted to deposit extra pheromone. This makes the ant to converge to a solution by taking less time.

Ant-routing-algorithm (ARA) for mobile multi-hop ad-hoc networks- new features and results explored by Gunes et al. [7] is based on ant algorithms which makes it highly adaptive and efficient. The routing algorithm consists of three phases. Route Discovery Phase requires use of forward ant (FANT) and backward ant (BANT) control agents. FANT establishes the pheromone trail back to the source node. Similarly BANT establishes pheromone track back to the destination node. Node receiving FANT for the first time creates an entry in its routing table consisting of destination address which is the origin of FANT, next hop which is address of the previous node from which it received FANT and pheromone value which is computed based on the number of hops the FANT took to reach the node. The node forwards the FANT to its neighbors. Once the destination node receives FANT, it sends BANT back to the source node. Once the source nodes receives BANT from the destination node, the path is established and data packets can then be sent which comes under the Route Maintenance Phase. When data packets are relayed to destination by a node, it increases the pheromone value of the routing table entry. The last phase of ARA handles the routing failure caused by the mobility of node which are very common in MANETs. ARA assumes IEEE 802.11 on the MAC layer which enables routing algorithm to recognize the failure of route through a missing acknowledgement on the MAC layer. Node deactivates the link by setting the pheromone value to 0. The node then searches for an alternative link in its routing table. If there exist a route to destination in its routing table, it sends the packet via this path. If there exist multiple entries in the routing table, the node will not send any data packets. Instead it informs the source node which has to initiate the route discovery process again.

2.1 Working of Traditional Ant Colony Based Routing Algorithm

The basic ACO takes a reactive probabilistic approach of finding good robust paths between source and destination. The algorithm follows: At regular intervals, a foraging ant is launched with a mission to find a path to destination. This establishes backward pheromone trail from destination to source. When a node receives a foraging ant for the first time, it creates an entry in its routing table. The entry consists of destination address which is the source address of the foraging ant, next hop which implies the node from which it received the packet, and pheromone value. The foraging ant probabilistically selects the next hop to reach destination. Duplicate foraging ants are then removed by identifying their unique sequence ID. Once a forward ant reaches its destination, it triggers the flooding of backward ants which the destination will send to the source node. The backward ant establishes the pheromone trail from source to destination. After calculation of the selection probabilities, the node will forward the data packets to that neighbor node which has relatively higher selection probability among its neighbor set. The data packet is sent to the selected relayed node and is further relayed towards the destination node. The selected relay nodes increments their pheromone value by a specific amount. Like their natural counterpart, artificial pheromones decay over time. The evaporation process provides a negative feedback in the system which helps ant avoid the stale paths in the network. The evaporation of pheromones takes place constantly by decreasing the pheromone values. The procedure finishes once the data packet reaches the destination node. Nodes maintain the neighbor entity in its neighbor table by sending Hello Packet periodically to each other. Hello packet sending interval can be different according to different mobility scenarios.

If a node doesn't receive the neighbor's Hello packet after certain period of time, it then deletes neighbor information from its neighbor table.

3 Proposed Idea

This section discusses about various heuristic factors which are considered in the proposed routing algorithm which makes it novel compared to already existing routing algorithms for MANETs.

Pheromone and Repellent Pheromone: Ants in nature while travelling from their nest to food source make the routing decision when they reach intersection, i.e., when more than one path is available for their next hop. In such scenario, the probability to choose that path is more which has relatively higher concentration of pheromone compared to other available paths. After the robust path is established, the ants continue to use that path until they encounter some obstacle, for example, placing a stone or water coming from some source accidentally flowing through the path recently formed. In such case, ants no longer use that path and instead begin exploring new paths. Pheromones with repellent property are deposited by ants so that their contemporaries no longer use the earlier efficient path. This property if incorporated in algorithm would make delay less in the network as the nodes would be aware of failed paths in the network.

RSSI: In WSN, sensor nodes are aware of the proximity of their neighbors through RSSI. If the scenario is considered where the source node and destination node are placed very far apart such that the destination node barely comes under the transmission range of source node, both the nodes will receive packets with very low RSSI as power of received signal decreases with increase in distance. As a result the chances of packet drops are very high. But due to less number of hops between source and destination node, the destination node will experience less delay. Whereas if source node and destination node are connected such that there exist neighboring nodes through which the packets can be transferred in a multi hop fashion, then the nodes will receive packets with very high value of RSSI due to close vicinity with each other. Nevertheless, the delay experienced by the packet will be more as the packet would have travelled with more number of hops from source to destination.

Therefore, it is understood that extreme values of RSSI is not good for our proposed system. With the logic of RSSI explained above, it is vital that goodness of RSSI closely follows the Gaussian distribution [9] as shown in the figure below.

With reference to [12], if two nodes (source and destination) are placed such that the distance between them is less than the transmission radius R, the total expected hop count from source node to destination node is given by

$$\frac{d}{x}\left[\frac{1}{[p(x)(1-(1-p(x))^u)]} + \frac{u}{[(1-(1-p(x))^u)]}\right] \tag{1}$$

where x is the distance between two consecutive nodes. $p(x)$ is the probability of receiving packet and u is a constant and is selected as 1.

(x) is dependent on several measurements such as signal strengths, delay, etc. It is approximated by following equation:

$$P(x) = \begin{cases} 1 - \left(\left(\frac{x}{R}\right)^{2\beta}/2 \right), & x < R \\ ((2R - x)/R)^{2\beta}/2, & x \geq R \end{cases} \tag{2}$$

Where β is constant and is selected as 2.

In the experiment, R is selected such that the packet delivery ratio at destination node is 80%.

In order to make the Eq. 1 independent of particular distance d, it is optimized which is given as follows:

$$h(x, u, \beta, R) = \frac{R}{x} \left[\frac{1}{[p(x)(1 - (1 - p(x))^u)]} + \frac{u}{[(1 - (1 - p(x))^u)]} \right] \tag{3}$$

By plugging in the value of u in Eq. 3, it is simplified by following equation

$$h(x) = \frac{R(1 + p(x))}{x p^2(x)} \tag{4}$$

Taking the inverse of Eq. (4), Goodness of RSSI is achieved which follows the graph in Fig. 1. Hence

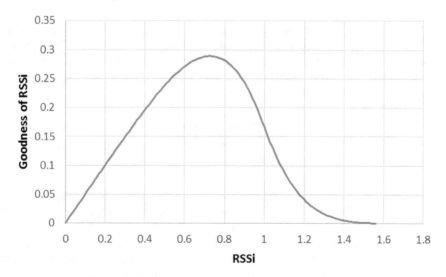

Fig. 1. Goodness of RSSI against varying range of RSSI

$$f(S) = \frac{1}{h(x)} = \frac{R(1+p(x))}{xp^2(x)} \tag{5}$$

Where f(S) is known as Goodness RSSI.

After the goodness of RSSI is calculated, the probabilistic formulae and goodness RSSI follows linear relationship.

In proposed routing algorithm, signal to noise ratio is computed by default at physical layer (with reference to OSI model). As the layers in OSI model are independent of each other, the Network layer (Layer 3) will not be aware of this parameter. Hence, it is imperative to add a packet tag at physical layer. The packet tag consists of signal to noise ratio information in dBm units. With respect to Network layer mechanism, when a node receives a hello packet from its neighbor, it scans for the packet tag so that it is aware of its neighbor's received strength. Each node then maintains a neighbor table where RSSI records are maintained against every neighbor. Also it is widely known that RSSI fluctuates too often even when nodes are static [10], the exponential moving weighted average approach chosen by us helps in smoothening the RSSI value.

Residual Energy: If the nodes among the discovered robust path are going to be used extensively for data packet transmission, their battery will deplete faster compared to nodes on non-efficient paths. This will result in creation of void nodes in the network which may lead to network partition. Inclusion of residual energy of node [11] in the routing decision will help in exploring paths other than already discovered robust paths. This technique will improve the lifetime of the network.

Proposed routing algorithm is an extended version of traditional ant colony based routing algorithm in which the main objective is to maximize the network lifetime. Traditional ant colony routing algorithm considers the pheromone value in its probabilistic routing decision process. This case is however not favorable for energy constrained networks. With additional heuristics discussed above, the extended probabilistic formula shall look like:

$$P_{n,d} = \frac{\left(\tau_{n,d}^i\right)^\alpha \left(h_{n,d}^i\right)^{-\beta} \left(E_n^i\right)^\delta \left(f(S)_n^i\right)^\gamma}{\sum_{j \in N_i} \left(\tau_{j,d}^i\right)^\alpha \left(h_{j,d}^i\right)^{-\beta} \left(E_{j,d}^i\right)^\delta \left(f(S)_{j,d}^i\right)^\gamma} \tag{6}$$

where n is the next hop selected by an ant to reach destination d from node i, $\tau_{n,d}^i$ is a pheromone value from neighbor n of node i to destination. h is the number of hops taken by an ant to reach destination from node i. N^i is the set of neighbors of node i. E is the remaining energy in the node, f(S) is the goodness RSSI which follows Gaussian distribution as shown in Fig. 1. α, β, δ, γ are the factors to adjust the relative importance of pheromone concentration, hops, residual energy, and goodness RSSI respectively.

As discussed in Sect. 2.1, nodes receiving the ants update the pheromone value in their routing table by depositing a constant value of pheromone in their routing table which acts as a positive feedback. As a result, an impulsive response is observed with regards to pheromone whenever a node receives an ant. Similar to the biological ants, the pheromone value is a function of time which means pheromone value decreases exponentially as the time progresses which makes it volatile.

4 Results

In this section, the experimental setup is discussed, followed by performance metrics and also simulation results.

4.1 Experimental Setup

The simulations are carried out using well known discrete event network simulator known as Network Simulator 3 (NS-3). In the simulation, the static nodes are deployed in a 10 × 10 grid with a distance of 50 m apart. As shown in Fig. 2, source and destination pair is indicated by blue color. Each node is configured with Friis Propagation Loss Model with respect to physical layer in which the transmission range of node can be varied by adjusting the receiver's sensitivity. Each node is installed with a battery of capacity 800 J. Approximately 4900 data packets are generated by the source node and are sent to the destination node. The Simulation runs for 500 s. With this experiment setup, the performance of Ant Colony Based Routing is compared against AODV Routing Protocol.

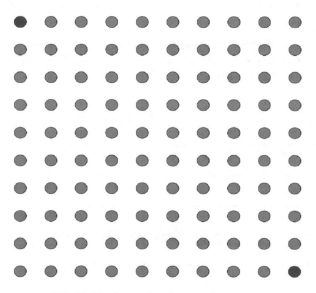

Fig. 2. Topology of static nodes in network

4.2   Performance Metrics

The following performance metrics are considered for the evaluation of Ant Colony Based Routing Algorithm and AODV Routing Protocol:

- Throughput

  The rate at which the data packets are delivered successfully by the network from a source to destination is known as throughput. It is also known as goodput. It is represented by bits or bytes per second. The throughput is affected by various factors such as limitation of physical medium, processing power, end user behavior etc. A higher throughput is what is desired in any communication based network.
- Mean Delay

  It is the time consumed by the data packet to reach the destination. The mean delay was calculated [8] by computing the ratio of sum of all end-to-end delays for all received packets of the flow to the total number of received packets.
- Network Lifetime

  It is defined as the time span a network can fulfill its service whereby source node and destination node communicate with each other by exchanging the data packets and other control packets.

4.3   Simulation Results and Analysis

The performance of both the routing protocols is analyzed under these three scenarios:

(a) Number of Hops

   In this scenario multiple source and destination pairs are selected at random which gives us an idea of the protocol performance against change in number of hops between source and destination (Fig. 3).

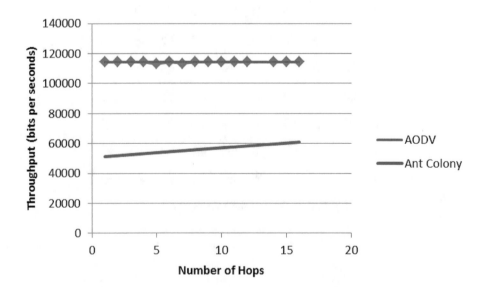

Fig. 3. Throughput versus number of hops

It is observed that AODV outperforms Ant Colony Based Routing Algorithm because Ant Colony Based Routing Algorithm has randomness property in terms of selecting the neighboring nodes. As a result, not all decisions would be favorable. Whereas in AODV, there are no multiple paths between source and destination which makes it more optimal in terms of hop counts (Fig. 4).

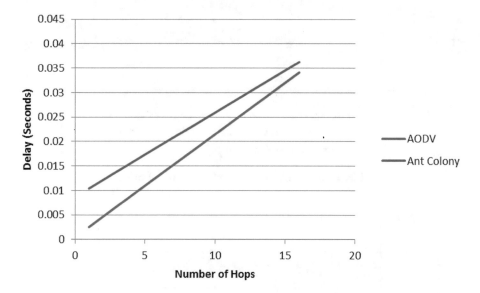

Fig. 4. Delay vs number of hops

As expected the increase in delay is caused by increase in number of hops between source and destination. In Ant Colony Algorithm, due to the property of randomness in path selection, its performance lacks in comparison with AODV.

(b) Background Traffic

Several source and destination node pairs are selected at random in the simulation which generates UDP traffic. The logic of using Background Traffic is to know how the algorithm performs in a real-life network environment where there is a disturbance along with regular traffic flow (Fig. 5).

Ant Colony Routing experiences less delay against AODV. This is due to the fact that whenever node faces congestion due to increase in background traffic, due to randomness property, the packet is received by node which has relatively less delay and is relayed to the destination node. Whereas in AODV, since there is no multipath routing, the delay will be more (Fig. 6).

In both the cases, due to increase in background traffic, decrease in throughput is expected. However Ant Colony Routing Algorithm experiences more packet drop ratio in comparison to AODV Routing Protocol which makes its throughput performance lesser than AODV (Fig. 7).

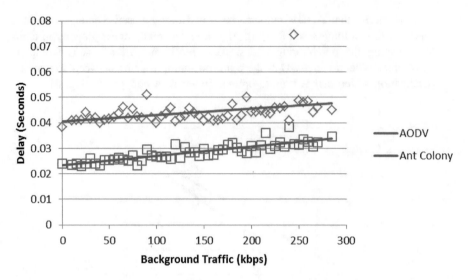

Fig. 5. Delay against increase in background traffic

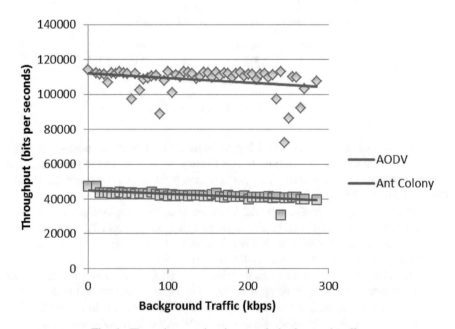

Fig. 6. Throughput against increase in background traffic

(c) Network Lifetime
Ant Colony Routing outperforms AODV in terms of Network Lifetime due to the
fact that energy based robust routing is implemented in layer 3. As a result, its
performance is better than AODV since AODV is not energy aware routing
protocol.

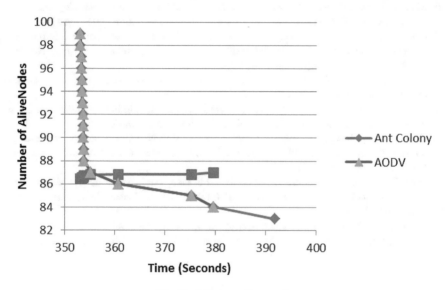

Fig. 7. Lifetime of network

5 Conclusion and Future Work

This work sought to develop a routing protocol for wireless sensor networks/MANETs
in NS-3. We analyzed the performance of network by considering the basic performance metrics such as throughput, delay, network lifetime and compared against
AODV routing protcol. In future we would include the repellent pheromone phase in
our routing protocol so that the nodes could detect the failure of its neighbor on the
IEEE 802.11 MAC layer. We will include various test scenarios in terms of mobility of
the network to test its robustness.

References

1. Di Caro, G., Dorigo, M.: AntNet: distributed stigmergetic control for communication
   networks. J. Artif. Intell. Res. (1998)
2. Camilo, T., Carreto, C., Silva, J., Boavida, F.: An energy-efficient ant-based routing
   algorithm for wireless sensor networks. In: Ant Colony Optimization and Swarm
   Intelligence, pp. 49–59 (2006)

3. Ho, J.-H., Shih, H.-C., Liao, B.-Y., Chu, S.-C.: A Ladder Diffusion Algorithm Using Ant Colony Optimization for Wireless Sensor Networks (2012)
4. Frey, M., Große, F., Gunes, M.: Energy-Aware Ant Routing in Wireless Multi-Hop Networks (2014)
5. Abd Elmoniem, A.M., Ibrahim, H.M., Mohamed, M.H., Hedar, A.-R.: Ant colony and load balancing optimizations for AODV routing protocol. Sensor Netw. Mob. Comput. (2011)
6. Sim, K.M., Sun, W.H.: Ant colony optimization for routing and load-balancing: survey and new directions. IEEE Trans. Syst. Man Cybern. Part A Syst. Hum. 33(5) (2003)
7. Gunes, M., Kahmer, M., Bouazizi, I.: Ant-routing-algorithm (ARA) for mobile multi-hop ad-hoc networks- new features and results. In: Proceedings of the Med-Hoc Net 2003 Workshop Mahdia, Tunisia 25–27 June 2003 (2003)
8. Carneiro, G., Fortuna, P., Ricardo, M.: FlowMonitor- a Network Monitoring Framework for the Network Simulator 3 (NS-3) (2009)
9. Zinonos, Z., Vassiliou, V., Christofides, T.: Radio Propagation in Industrial Wireless Sensor Network Envrionments: From Testbed to Simulation Evaluation (2012)
10. Chapre, Y., Mohapatra, P., Jha, S., Seneviratne, A.: Received signal strength indicator and its analysis in a typical WLAN system. In: 38th Annual IEEE Conference on Local Computer Networks (2013)
11. Jung, S.-G., Kang, B., Yeoum, S., Choo, H.: Trail-using ant behavior based energy-efficient routing protocol in wireless sensor networks. Int. J. Distrib. Sensor Netw. 2016, 9 (2016). Article ID 7350427
12. Kuruvila, J., Nayak, A., Stojmenovic, A.: Hop Count Optimal Position-Based Packet Routing Algorithms for Ad Hoc Wireless Networks With a Realistic Physical Layer (2004)

Energy Level-Based Adaptive Backscatter and Active Communication in Energy-Harvesting Network

Kwanyoung Moon, Yunmin Kim, and Tae-Jin Lee[✉]

College of Information and Communication Engineering, Sungkyunkwan University,
Suwon 16419, Korea
{mky709,kym0413,tjlee}@skku.edu

Abstract. In a Radio Frequency IDentification (RFID) system, active tags and semi-passive tags have limited battery capacity and it affects overall network performance. In this paper, we propose a new mechanism to perform two types of data transmission and energy harvesting adaptively based on the amount of residual energy of RFID tags in an energy-harvesting network compatible with the ISO/IEC 18000-7 standard. In the proposed method, tags operate the passive mode as well as the active mode for data transmission and perform selective energy harvesting mode to improve throughput and energy utilization ratio. The performance evaluation results show that the proposed method significantly improves throughput and energy utilization ratio compared to the existing.

Keywords: Energy harvesting · ISO/IEC 18000-7 standard · RFID

1 Introduction

Radio Frequency IDentification (RFID) is an automatic wireless data collection technology [1]. Recently, there are several researches about RFID sensor tags, an enhanced version of existing RFID tags. The RFID sensor tag can transmit sensing data by adding a computation function in the existing RFID tag [2].

RFID sensor tags are usually divided into semi-passive tags and active tags. The semi-passive tag uses internal battery to sense environment and it backscatters the signal of reader to transmit data. The active tag utilizes internal battery both for sensing and data transmission [3,4].

In the RFID system, the reader collects information of tags within the recognition range when there is periodic or necessary information. This information collection process is called tag collection, and the collision avoidance algorithm is used to prevent or reduce the collisions of tags when many tags perform transmission for tag collection. The typical collision avoidance algorithms are binary tree protocol, framed slotted ALOHA (FSA) and dynamic framed slotted ALOHA (DFSA) [5,6]. The collision avoidance algorithms are widely used in the standards of EPCglobal Class-1 Gen-2 and ISO/IEC 18000-7 [7,8].

© Springer Nature Switzerland AG 2019
S. Lee et al. (Eds.): IMCOM 2019, AISC 935, pp. 111–119, 2019.
https://doi.org/10.1007/978-3-030-19063-7_10

The ISO/IEC 18000-7 standard is a standard for supporting data transmission of active RFID tags, and defines a protocol for data transmission between the reader and the active RFID tag in 433 MHz frequency band [8]. The active RFID tags using the ISO/IEC 18000-7 standard have limited lifetime, confining the network performance due to limited energy storage capacity of tags. Recently, methods to increase the energy efficiency in the existing ISO/IEC 18000-7 standard has been studied in order to increase the operating time of such active RFID tags [9]. In order to supply sufficient energy to tags, we need to design a new energy harvesting method. In this paper, we propose a new selective energy harvesting method considering the amount of the remaining energy of tags, which is compatible with ISO/IEC 18000-7. Additionally, we propose a passive mode for data transmission as well as an active mode in order to minimize the energy harvesting time. By using the proposed method, it is possible to supply energy only to the tag in need of energy harvesting and save energy using the passive mode compared to the active mode. The proposed method is able to improve the throughput and the energy utilization ratio.

2   ISO/IEC 18000-7 Standard

The ISO/IEC 18000-7 standard [8] defines the method of information collection of active RFID tags. In the ISO/IEC 18000-7 standard the information collection process consists of Listen Period (LP) and Acknowledge Period (AP). Figure 1 shows the tag collection protocol of the ISO/IEC 18000-7 standard.

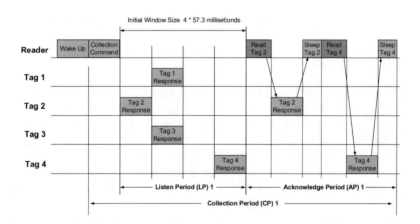

Fig. 1. ISO/IEC 18000-7 standard protocol.

Before the LP, the reader broadcasts a wake up command to wake up all tags within the communication range. All the tags switch to the ready state waiting for a command from the reader by completion of the wake up signal. Then the reader transmits a collection command message to induce the response of tags in the LP. The collection command message includes the value of window size and the maximum packet length. In the LP, the tags perform slotted ALOHA as the collision avoidance algorithm using the window size and the maximum packet length announced by the reader. Tags randomly select one of the slots in the LP and transmit response messages that have unique IDs of tags and information data. The state of a slot can be a successful slot if one tag is successfully transmitted in a slot, a collision slot if two or more tags are transmitted in a slot, and an idle slot if no tags transmit in a slot. The tags transmit response messages to the reader during the LP, and then the reader starts tag collection in the AP.

AP is used for additional data transmission of the successfully transmitted tags in the LP. The reader transmits the read message through the point-to-point communication to the tag. The tag receiving the read message performs additional data transmission. If the reader successfully receives the response message of a tag, it transmits the sleep message to the tag. After the tag receives the sleep message, the tag then switches to the sleep mode. The collection round continues until all the tags successfully transmit to the reader in the LP. The LP of the next collection round is calculated using the number of collision slots and idle slots of the current LP.

3   System Model

An energy-harvesting network consists of RFID sensor tags and one RFID reader. The RFID reader has functions of receiving data of the RFID tags and providing the RF energy signal to the tags. An RFID tag can operate as the active mode, passive mode, and energy harvesting mode depending on the amount of residual energy of the tag. The active mode uses the stored energy in the tag to perform data transmission. The passive mode performs backscatter communications for data transmission using the RF signal. In the energy harvesting mode, a tag receives energy from the RF energy signal by the reader. Figure 2 shows the operation modes of RFID tags and the RFID reader.

4   Proposed Scheme

Based on the ISO/IEC 18000-7 standard, we propose a new method to perform data transmission using the passive, active, and energy harvesting mode. Our proposed method consists of two steps. At the first step, tags determine the operation mode in the AP. In the second step, the tag performs operation based on the determined operation mode of the tag.

4.1   Operation Mode Selection Phase

The first step is to determine the operation mode of a tag based on the amount of residual energy of the tag. Before a tag transmits a M-response message in the

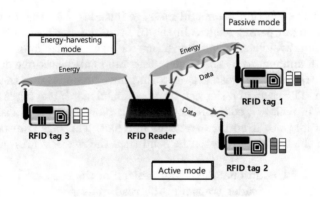

Fig. 2. Operation modes of RFID tags and RFID reader.

LP, the tag determines for additional data transmission and energy reception. Then the tag selects the operation mode in the LP according to the remaining amount of energy of the tag. If the energy level of the battery of the tag is lower than a transmission mode threshold, the tag selects the passive mode. Otherwise, the tag selects the active mode. Also, the tag determines whether to perform energy harvesting or not. When the energy level of the tag is less than an energy harvesting threshold, the tag selects the energy harvesting mode to receive energy. The tag includes the determined information in the M-response message and it transfers the M-response message to the reader. The proposed frame format of the M-response message in the LP is shown in Fig. 3. The M-response message includes the Transmission mode and the Energy-harvesting mode filed in the Tag status field.

4.2   Energy Harvesting and Data Transmission Phase

The second step is to perform the data transmission and energy harvesting for the tags. After the tags finish operation in the LP, they operate in the AP based on the determined operation mode. When the passive mode is used, the tag transmits the response message to the reader through backscatter communications. When the active mode is used, the tag transmits the response message using the stored energy of the tag. If the tag selects the energy harvesting mode, it harvests energy through the RF energy signal of the reader. The tags and the reader in the AP transmit the messages through point-to-point communications.

4.3   Operation of Proposed Method

Based on the ISO/IEC 18000-7 standard, the reader transmits a wake up signal and a collection command message to the tags. Then the tags perform slot competitions to transmit M-response messages in the LP. The tag that succeeds in

M-Response message

Protocol ID	Tag Status	Packet Length	Session ID	Tag Mfg ID	Tag Serial Number	Command Code	Response Data	CRC
1 Byte	2 Bytes	1 Byte	2 Bytes	2 Bytes	4 Bytes	1 Byte	N Bytes	2 Bytes

. . .	Alarm	Transmission mode	Energy-harvesting mode	Acknowledgement	. . .
. . .	1bit	1 bit	1 bit	1 bit	. . .

E-Read message

Protocol ID	Packet Options	Packet Length	Tag Mfg ID	Tag Serial Number	Session ID	Command Code	Command Arguments	CRC
1 Byte	1 Bytes	1 Byte	2 Bytes	4 Bytes	2 Bytes	1 Byte	1 Byte	2 Bytes

Energy-harvesting time
1Byte

Fig. 3. The frame formats of the M-response message in the LP and E-read message.

the slot competition determines the operation mode of the AP depending on the residual energy of the tag, and transmits the M-response message including the operation mode information and data. After the LP, the tags that successfully transmitted the M-response messages in the LP perform additional data transmission or energy harvesting in the AP. The operation of the tags in the AP is based on the determined mode of the tags in the LP.

Figure 4 shows the protocol of the proposed method. After the tags receive the wake up and the collection command messages from the reader, the tag 2 succeeds in the slot competition in the first slot of the LP. And, tag 2 determines the operation mode in the AP based on the amount of remaining energy. The amount of the residual energy of the tag 2 is smaller than the transmission mode threshold and greater than the energy harvesting threshold. So, the tag 2 selects the passive mode for additional data transmission in the AP. Next, the tag 2 transmits the M-response message in first slot of the LP. After the LP, the tag 2 receives the read message through point-to-point communications from the reader. Then the tag 2 performs backscatter communications for additional data transmission (passive mode). If the reader receives the response message from the tag 2, the reader sends a sleep message to the tag 2 and it switches to the sleep mode to save energy.

In the case of the tag 3, it succeeds in the slot competition in the second slot of the LP. In turn, tag 3 determines the passive mode and the energy harvesting mode since the tag 3 has insufficient amount of remaining energy. Then the tag 3 transmits the M-response message to the reader in the second slot of the LP. After the LP, the tag 3 receives the read message from the reader and transmits the response message to the reader by using the passive mode. The reader receives the response message from the tag 3 and then transmits the proposed E-read message for energy harvesting to the tag 3. The E-read message is defined in Fig. 3. The command code of E-read message uses a newly defined

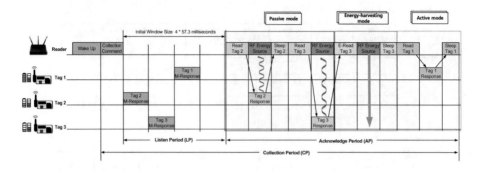

Fig. 4. An example of the proposed method.

command code 0×71 and the command argument of the e-read message contains the Energy-harvesting time field for the duration of energy harvesting in milliseconds. The tag 3 receives the E-read message and harvests energy from the reader. Then the tag switches to the sleep mode by receiving the sleep message. In the case of the tag 1, it determines only the active mode based on the amount of remaining energy of the tag 1 and then transmits the M-response message in the third slot of the LP. After the LP, the tag 1 receives the read message from the reader and transmits the additional response message by the active mode. Then the tag 1 switches to the sleep mode by receiving the sleep message.

5   Performance Evaluation

In this section, we evaluate the performance of throughput and energy utilization ratio. Our proposed method is compared with comparison method for varying numbers of tags. In the comparison method, data transmission of tags is performed based on the ISO/IEC 18000-7 standard and the tags perform slotted ALOHA for energy harvesting before the collection process. The parameters used in the simulation are shown in Table 1. We assume that tags collect data every data generation cycle with the data generation probability p. We perform simulation for low data generation probability case ($p = 0.1, 0.2$) and high data generation probability case ($p = 1$) to consider that all tags have not same data generation cycle.

Figure 5 shows the throughput of the proposed method and the comparison method. The throughput performance is calculated by the ratio between the total successfully transmitted data and total time. The throughput performance of the proposed method is greater than that of the comparison method because of efficient energy harvesting and the appropriate passive mode of the tags. In proposed method, tags have to wait to receive energy because these receive energy after finish data transmission in LP. However, the proposed method minimizes channel time for energy harvesting because it supplies energy only to the tags that need energy harvesting. The passive mode also reduces the number of energy harvesting times because energy is saved by backscattering compared to the active mode.

Table 1. Parameters

Parameter	Value
Duration of wake - up signal	2450 ms
Duration of collection command	5 ms
Duration of M-response/Response message	20.166 ms
Read/Sleep/E-read message in AP	5.898 ms
Energy-harvesting time	30.106 ms
Maximum battery capacity	10 mJ
Transmission power of tag	0.056 W
Reception power of tag	0.056 W
Data generation cycle	2000 ms

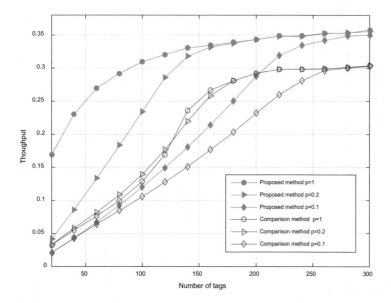

Fig. 5. Throughput for varying number of tags.

Figure 6 shows the performance of the energy utilization ratio, which is defined as the ratio between the saved energy and the used energy for data transmission. In our proposed method, the tags collect and save energy by using the energy harvesting mode and the passive mode. Thus, the saved energy of our proposed method is calculated by the sum of the supplied energy by using the energy harvesting mode and the saved energy by using the passive mode. The energy utilization ratio performance shows that the proposed method shows improved energy utilization ratio compared to the comparison method in most cases. In the comparison method, the tag performs energy harvesting through the slot competition. So the comparison method has a very low ratio between the actually used channel time for supplying energy and the allocated channel time for energy harvesting. However, the proposed method only supplies energy to the tags in need of energy harvesting without additional slot competitions for energy harvesting. Moreover, the proposed method exhibits high energy utilization ratio since the passive mode is used for data transmission even if energy harvesting is not performed.

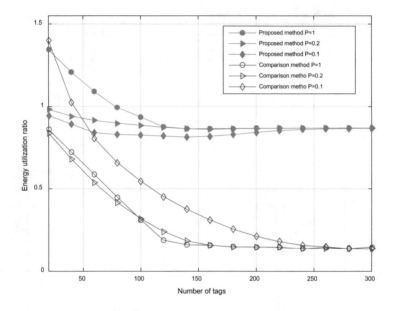

Fig. 6. Energy utilization ratio for varying number of tags.

6   Conclusion

In this paper we have propose a new energy harvesting and data transmission method considering the amount of remaining energy of the tags in an energy-harvesting network compatible with the ISO/IEC 18000-7 standard. The proposed method reduces the number of energy harvesting times and increases the energy harvesting efficiency by using the selective energy harvesting and the passive/active mode for data transmission. Thus the proposed method minimizes the throughput decrease caused by the energy harvesting. The performance evaluation shows that the proposed method has better performance in terms of throughput and the energy utilization ratio than the typical comparison method.

Acknowledgments. This work was supported by the National Research Foundation of Korea (NRF) grant funded by the Korea government (MSIT) (2014R1A5A1011478, NRF-2018R1A2B6009348) and Basic Science Research Program through NRF of Korea funded by MOE (NRF-2010-0020210).

References

1. Zhi-yuan, Z., He, R., Jie, T.: A method for optimizing the position of passive UHF RFID tags. In: Proceedings of IEEE International Conference on RFID-Technology and Applications, pp. 92–95, June 2010
2. Abdulhadi, A.E., Denidni, T.A.: Self-powered multi-port UHF RFID tag-based-sensor. IEEE J. Radio Freq. Identif. **1**(2), 115–123 (2017)
3. Kolarovszki, P., Kolarovszká, Z., Perakovic, D., Periša, M.: Laboratory testing of active and passive UHF RFID tags. Transp. Telecommun. **17**(2), 144–154 (2016)
4. Alsinglawi, B., Elkhodr, M., Vinh Nguyen, Q., Gunawardana, U., Maeder, A., Simoff, S.: RFID localisation for internet of things smart homes: a survey. Int. J. Comput. Netw. Commun. **9**(1), 81–99 (2017)
5. Zhi, B., Sainan, W., Yigang, H.: A novel anti-collision algorithm in RFID for Internet of Things. IEEE Access **6**, 45860–45874 (2018)
6. Jiang, A.: Collision avoidance anti-collision algorithm based on subset partition. In: Proceedings of International Conference on Natural Computation, Fuzzy Systems and Knowledge Discovery (ICNC-FSKD), pp. 2001–2008, August 2016
7. Solic, P., Radic, J., Rozic, N.: Early frame break policy for ALOHA-based RFID systems. IEEE Trans. Autom. Sci. Eng. **13**(2), 876–881 (2016)
8. ISO/IEC 18000-7: Information technology - Radio frequency identification for item management - Part 7: Parameters for active air interface communications at 433 MHz. ISO/IEC (2014)
9. Lee, C., Kim, D., Kim, J.: An energy efficient active RFID protocol to avoid over-hearing problem. IEEE Sens. J. **14**(1), 15–24 (2014)

Energy-Efficient Computation Offloading with Multi-MEC Servers in 5G Two-Tier Heterogeneous Networks

Luan N. T. Huynh[1], Quoc-Viet Pham[2], Quang D. Nguyen[1], Xuan-Qui Pham[1], VanDung Nguyen[1], and Eui-Nam Huh[1(✉)]

[1] Department of Computer Science and Engineering, Kyung Hee University, Yongin, Gyeonggi-do 17104, Republic of Korea
{luanhnt,quangnd,pxuanqui,ngvandung85,johnhuh}@khu.ac.kr
[2] ICT Convergence Center, Changwon National University, Changwon, Gyeongsangnam-do 51140, Republic of Korea
vietpq90@gmail.com

Abstract. Recently, Multi-Access Edge Computing (MEC), which has been emerged as a key technology in 5G networks, enhances computation capabilities and power limitations of mobile devices (MDs) by offloading computation task to the nearby MEC servers. However, offloading the computation tasks can increase network traffics and incur extra delays. Most existing approaches focus on the computation offloading with multi-user single-MEC scenarios to decrease energy consumption and latency of the MDs. Towards this goal, we investigate a computation offloading strategy for two-tier 5G heterogeneous networks integrated with multi-MEC. In addition, we propose a random offloading search algorithm, called ROSA, that rapidly achieve the minimized energy consumption of the system considering the computation offloading decision strategies. Simulation results show that our proposed algorithm based on offloading scheme outperforms other two schemes in terms of energy consumption.

Keywords: Computation offloading · Multi-Access Edge Computing · 5G · Heterogeneous · Resource allocation

1 Introduction

In accordance with the more population of mobile devices, such as smartphones, smartglasses, and smartwatches, many new applications are developed [1,6,7]. These applications such as face recognition, augmented reality, natural language processing, and online game require intensive computation and high power consumption. Nevertheless, mobile devices (MDs) have limited computing capabilities and battery capacity [1–3,8]. The conflict between the computation-intensive mobile applications and the resource-limited mobile devices is the challenges for the development of technologies. The Multi-Access Edge Computing (MEC) is proposed to address these challenges. MEC can provide cloud computing

© Springer Nature Switzerland AG 2019
S. Lee et al. (Eds.): IMCOM 2019, AISC 935, pp. 120–129, 2019.
https://doi.org/10.1007/978-3-030-19063-7_11

capabilities in close proximity to MDs to achieve high computing performance [3,7,11,12]. Accordingly, computation-intensive tasks of the mobile devices can be offloaded to the nearby resource-rich MEC servers, thus saving energy and extending the battery life [1,3].

With exponentially increasing data traffic, fifth-generation (5G), heterogeneous networks (HetNets), which consist of traditional macro base stations (MBSs) overlaid with small cell base stations (SBSs) (e.g., femtocells, picocells, and microcells), are expected to be an attractive technology to meet the increasing capacity requirements of future mobile networks. The MEC, as one key technology to achieve low-latency performance in 5G networks integrating MEC with MBS and SBS, is attracting lots of attentions [3,4].

In recent years, there are many research works on computation offloading for the MEC in 5G networks [3,4,12]. For instance, Zhang *et al.* in [12] considers a general multi-user in a heterogeneous network (HetNet). However, most existing works only focus on the computation offloading with multi-user single-server scenarios and a few approach studies on the computation offloading with multi-user multi-MEC [3,5]. The authors in [1] consider a system which consists of an MBS integrated with an MEC server and an SBS without the MEC server. Therefore, MD can offload to not only the MBS but also the SBS. Nevertheless, only the MBS has integrated the MEC server, so the computation tasks offloaded to the SBS are transmitted to the MBS, resulting in additional latency. A considered MEC-HerNet scenario is created by one macro base station and multi small cells base stations [11]. The authors [11] jointly optimize communication and computation resource allocation under the energy and latency requirements. But the increasing number of computation-intensive tasks can lead to the resource bottleneck of MEC servers.

In this article, we investigate a computation offloading strategy for two-tier 5G HetNets integrated with multi-MEC servers network to solve the problem in [11,12]. In addition, we propose a random offloading search algorithm, called ROSA, that rapidly achieve the minimized energy consumption of the system considering the computation offloading decision strategies that satisfies the latency requirements.

The remainder of this paper is organized as follows. In Sect. 2, we introduce the model of MEC HetNets. Then, we formulate the computation offloading problem in Sect. 3. The numerical results are presented in Sect. 4. Finally, we conclude this paper in Sect. 5.

2   System Model

The HetNets are consist of a set of small cell base stations connected with one macro base station. A considering two-tier 5G HetNets scenario is shown in Fig. 1. The SBSs are overlaid by the MBS. The SBSs, denoted by $\mathcal{M} = \{1, 2, ..., M\}$, are located closely to the MDs. The SBSs are associated to the MBS via the fiber links [3,11]. The MDs tend to access to these SBSs because SBS can carry lower network latency and higher data rates. SBSs and MBS are collocated with MEC servers in the order to provide the computing capabilities to MDs. We denote MBS and SBSs with MEC server as MBS-MEC and SBS-MEC, respectively.

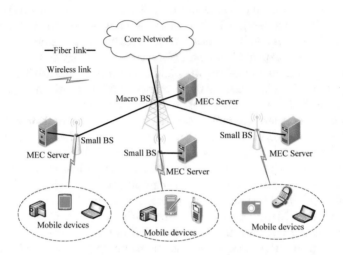

Fig. 1. Considered model of MEC HetNets

We denote the set of MDs as $\mathcal{U} = \{1, 2, ..., U\}$. The MD can be either computed its task locally or offloaded to SBS-MEC through the wireless links, or then continue to be offloaded to MBS-MEC through the fiber links. We assume that each task of the MD can be connected to only one SBS-MEC or MBS-MEC at a time instance. Each of the MDs has a computation task $T_i = \{d_i, w_i, \tau_i^{max}\}$, $i \in \mathcal{U}$, where d_i is input data size of the computation task (that is program codes, system settings, or input files), w_i is the number of CPU cycles required for completing the computation task T_i, τ_i^{max} is the maximum latency requirement.

2.1   Communication Model

Each MD can decide whether the computation task will be executed on the mobile device or offloaded to the MEC server. We denote $a_{i,m} \in \{0, 1, -1\}(i \in \mathcal{U}, m \in \mathcal{M})$ represents the offloading decision of task T_i. If the task T_i is executed remotely at SBS-MEC m, $a_{i,m}$ equals 1. On the other hand, if the task T_i is sent to the SBS and then be executed at MBS-MEC, $a_{i,m}$ equals -1. Otherwise, $a_{i,m}$ equals 0, that means it will be computed locally at MD i.

We apply orthogonal frequency division multiple access (OFDMA) for the uplink connection between MDs and BSs. In OFDMA, channels are allocated to MDs orthogonally, thus there is no inter-channel interference [8]. Channels allocated to MDs for computation offloading are all orthogonal, so interference can be ignored [10].

We can achieve the data transmission rate of MD i when it offloads the computation task T_i to the SBS-MEC m via the wireless access channel

$$r_{i,m} = W \log_2 \left(1 + \frac{p_i g_{i,m}}{\sigma^2} \right) \tag{1}$$

where W is the channel bandwidth, σ^2 is the Gaussian noise, p_i denotes the transmission power of MD i, $g_{i,m}$ denotes the channel gain between MD i and the SBS.

Latency time for transmitting T_i through wireless access and the committed fiber link can be given by

$$t_i^r = \sum_{m=1}^{M} F_{\{a_{i,m}=1\}} \frac{d_i}{r_{i,m}} + \sum_{m=1}^{M} F_{\{a_{i,m}=-1\}} \left(\frac{d_i}{r_{i,m}} + \frac{d_i}{\beta} \right) \tag{2}$$

In Eq. (2), $F_{\{z\}}$ is denotes as an indicator function. $F_{\{z\}} = 1$ if z is true and $F_{\{z\}} = 0$ otherwise. β is the data rate of the fiber link. The SBSs are associated to the MBS through the fiber links. We regard wired connection is shared with other communication infrastructures. Hence, we neglect the consumed energy of wired connection [12]. Besides, we ignore the latency and energy consumption for the MEC servers to send the results back to MD, because the data size after task calculating is much smaller than of the computation input data [2,8].

2.2 Local Execution Model

We define δ_i^l as the coefficient of the consumed energy for one CPU cycle of MD i and f_i^l as the computing capability of MD i. When computation task of MD i is executed locally, the processing time and energy consumption of computation task T_i can be given by

$$t_i^l = \frac{w_i}{f_i^l}, \tag{3}$$

and

$$e_i^l = w_i \cdot \delta_i^l. \tag{4}$$

2.3 Mobile Edge Execution Model

In this section, we classify into two schemes for offloading task: (1) through SBS-MEC and (2) through MBS-MEC.

Offloading to the SBS-MEC. We defines δ_m^s as the coefficient of energy consumption for per CPU cycle of SBS-MEC m. Let $f_{i,m}^{mb}$ denotes the number of computing capability allocated to the MD i by the SBS-MEC m.

The execution time of the task T_i on SBS-MEC m can be written as

$$t_i^s = \frac{w_i}{f_{i,m}^{mb}} \tag{5}$$

The total latency of task T_i on SBS-MEC consists of transmission time t_i^r and computing time can be obtained as follows

$$t_i^{sbs} = t_i^r + t_i^s \tag{6}$$

The transmission energy consumption of the task T_i from MD i to SBS-MEC m can be computed as

$$e_i^r = \sum_{m=1}^{M} F_{\{a_{i,m}=1\}} p_i \frac{d_i}{r_i} \tag{7}$$

The execution energy consumption of the task T_i can be as the following

$$e_i^s = w_i \cdot \delta_m^s \tag{8}$$

The total energy consumption of implementing the computing task T_i denoted by e_i^{sbs} consist of power consumption for transmitting denote e_i^r and power consumption for calculating e_i^s [9,12]. Hence, energy consumption e_i^{sbs} can be computed as

$$e_i^{sbs} = e_i^r + e_i^s \tag{9}$$

Offloading to the MBS-MEC. The tasks of the MDs can be transmitted to SBS through wireless links and then, it be executed in the MBS-MEC through the fiber link.

The execution time of the computation task on MBS-MEC can be given as

$$t_i^{mb} = \frac{w_i}{f_i^{mb}}, \tag{10}$$

where f_i^{mb} denotes as the computational ability that is allocated to MD i of the MBS-MEC, and δ^{mb} denotes as the coefficient of energy consumption for per CPU cycle of MBS-MEC.

The total latency of task T_i on MBS-MEC can be obtained as follows

$$t_i^{mbs} = t_i^r + t_i^{mb} \tag{11}$$

The transmission energy consumption of the task T_i from MD i to MBS-MEC is given by

$$e_i^r = \sum_{m=1}^{M} F_{\{a_{i,m}=-1\}} p_{i,m} \frac{d_i}{r_i}, \ \forall i \in \mathbf{U}, \forall m \in \mathbf{M} \tag{12}$$

The execution energy consumption on MBS-MEC to the computation task T_i can be computed as

$$e_i^{mbs} = w_i \cdot \delta^{mb} \tag{13}$$

The energy consumption of implementing the computation task T_i denote e_i^{sbs} composed of transmitting energy consumption denote e_i^r and computing energy consumption e^{mb} [9,12]. Hence, energy consumption e_i^{mbs} can be calculated as

$$e_i^{mbs} = e_i^r + e_i^{mb} \tag{14}$$

3   Problem Formulation

The computation offloading decision is computed to consider to choose the located execution for the task T_i, such as on the mobile device, or on the SBS-MEC, or on the MBS-MEC via SBS. We define $a = \{a_1, a_2, ..., a_i, ..., a_M\}$ as the offloading decision profile, where $a_i \in \{0, 1, -1\}$ is the offloading decision of MD i. Specifically, if $a_1 = \sum_{m=1}^{M} a_{i,m} = 0$ means the MD i accomplishs its computation task locally at the mobile device. In the case that $a_1 = \sum_{m=1}^{M} a_{i,m} = 1$ means the MD i chooses to offloads the task to the SBS-MEC. $a_1 = \sum_{m=1}^{M} a_{i,m} = -1$ means that the MD i decides to offloads the task to the MBS-MEC via SBS.

We study energy-efficient offloading scheme to minimizes the total energy consumption of the system considering the computation offloading decision strategies under MD task's the latency requirements. The total energy consumption consisting of the communication energy consumption and the execution energy consumption. Let t_i and e_i denote the latency and the total energy consumption of task T_i.

According to (4), (9), and (14), the total energy consumption of the computation task T_i can be given by

$$e_i = F_{\{a_i=0\}} \cdot e_i^l + F_{\{a_i=1\}} \cdot e_i^{sbs} + F_{\{a_i=-1\}} \cdot e_i^{mbs} \tag{15}$$

According to (3), (6), and (9), the total the latency for completing the computation task T_i can be given by

$$t_i = F_{\{a_i=0\}} \cdot t_i^l + F_{\{a_i=1\}} \cdot t_i^{sbs} + F_{\{a_i=-1\}} \cdot t_i^{mbs} \tag{16}$$

The optimization problem can be formulated as

$$
\begin{aligned}
& \underset{\{a_{i,m}\}}{\text{minimize}} && \sum_{i=1}^{U} e_i \\
& \text{subject to} && \text{C1: } t_i \leq \tau_i^{max}, \ \forall\, i \in \mathcal{U}, \\
& && \text{C2: } \sum_{m=1}^{M} |a_{i,m}| \leq 1, \quad \forall\, i \in \mathcal{U}, \\
& && \text{C3: } \sum_{i=1}^{U} f_{i,m}^s \leq f_m^{s,max}, \quad \forall\, i \in \mathcal{M}, \\
& && \text{C4: } \sum_{i=1}^{U} f_i^{mb} \leq f^{mb,max}, \\
& && \text{C5: } a_{i,m} \in \{0, 1, -1\}, \quad \forall\, i \in \mathcal{U}, \forall\, m \in \mathcal{M}.
\end{aligned}
\tag{17}
$$

In (17), the constraint C1 indicates that processing time makes sure the latency requirement of all MDs. The constraint C2 ensures only one option can be chosen for each MD i. Let $f_m^{s,max}$ and $f^{mb,max}$ denote the maximum computation capability of the SBS-MEC m and the MBS-MEC, respectively. The constraint C3 and C4 guarantee that the total resources of each MEC server selected to the computation tasks cannot exceed the computation capability. The constraint C5 denotes the binary offloading decisions.

The above optimization problem is NP-hard [2,12]. To solve the problem in (17), we find an optimal computation offloading decision, which has the minimum energy consumption of system that satisfies the latency requirements of the computation tasks and the limited resources of MEC servers.

We propose a random offloading search algorithm as an efficient approximated solution approach of the energy consumption minimization problem. The proposed algorithm called ROSA, consists of generate the environment for cloud computing and loop iterations to find the solution offloading decision with the minimum total energy consumption. The number of iterations depends on the number of MDs and the MEC servers.

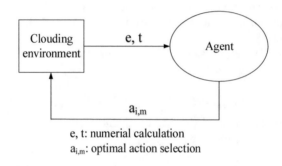

e, t: numerial calculation
$a_{i,m}$: optimal action selection

Fig. 2. Setup environment for cloud computing

Generate the environment for cloud computing shown in Fig. 2 with init and step function can be given by: (a) *Init* function: Store all of the environment characteristics. (b) *Step* function: Input: Action $a_{i,m}$ with i user number and m server number. Output: Set of the processing time t for task, the energy consumption e with corresponding action. On the other hand, random offloading search algorithm is implemented in *Agent*. The details of ROSA is discussed in Algorithm 1. The Algorithm 1 can find an solution to the problem.

4   Numerical Results

In our simulations, we consider the scenario with a set mobile devices in a range area with 3 SBS-MEC and an SBS-MEC. The transmission power is set to 100 mW. There are 20 mobile devices. The CPU computation ability of the mobile devices are 0.8 Ghz. The computation capability of each SBS-MEC and MBS-MEC are 4 GHz and 8 GHz, respectively. The uplink data rate in wired connection between SBS and MBS is 1 Gbps [3,12]. For the computation task,

---

Algorithm 1: Random offloading search algorithm (ROSA)

---

input : U, M, δ_i^l, δ^{mb}, δ_m^s, p_i, $f_m^{s,max}$, $f^{mb,max}$, σ, β

output: A solution computation offloading decision list and a minimal total
energy consumption

1. Generate the environment for cloud computing
2. Loop N Iterations

   Initialize matrix A with dimension $[U, M]$ with all 0

   Initialize min_energy with inf and $solution\_action$

   forall the *user i in U* **do**

   > Randomly select a server m in range $[1, M]$
   >
   > Randomly select action $a_{i,m}$ from $[-1, 0, 1]$
   >
   > Check the validity of action matrix A
   >
   > **if** *A is valid* **then**
   >
   > > Step with action A in environment
   > >
   > > Compare gained energy from environment to save smaller energy and
   > > update $solution\_action$
   >
   > **end**

   end

end

Return the $solution\_action$ and min_energy

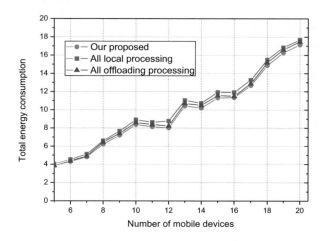

Fig. 3. Total energy consumption versus the number of devices

we assume that the maximum latency required for accomplishing the task is randomly assigned from 0.5 to 2 s, the number of CPU cycles required for accomplishing the task is set from 500 Megacycles to 1000 Megacycles. Each computation task has an input data size that is randomly distributed between 200 KB–600 KB.

We simulate our proposal compared with two following schemes:

(1) Local computing: all tasks are executed computations locally.
(2) Only offloading: all tasks always choose to offload on the MEC server.

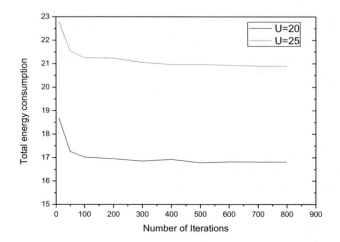

Fig. 4. Convergence of our proposed with different iterations

In Fig. 3 shows that the total energy consumption of our proposed is lower than that case of local computing and case of SBS offloading. Simulation results shows that our proposed algorithm based on the offloading scheme outperforms other two schemes in terms of energy consumption.

In Fig. 4, we have the convergence of our proposed algorithm with the iteration times. We consider number of mobile devices with U = 20 and U = 25. The total energy consumption decreases when the iteration times increase. This proves the impact of the iteration times to total energy consumption. The number of iterations also depends on the number of MDs. Our proposed algorithm can give an approximate solution to the problem.

5   Conclusions

In this article, we investigate the computation offloading scheme in two-tier 5G Hetnets. The major problem under consideration is a minimization of the energy consumptions including communication energy consumption and the execution energy consumption. Therefore, we propose a random offloading search algorithm, that rapidly achieve the minimized energy consumption of the system considering the computation offloading decision strategies. The simulation results demonstrate that our proposed algorithm outperforms other two schemes in terms of energy consumption. In future, we consider interference between other cell and design algorithm to the efficient solution as a genetic algorithm.

Acknowledgments. This research was supported by the MSIT (Ministry of Science and ICT), Korea, under the Grand Information Technology Research Center support program (IITP-2018-2015-0-00742) and the National Program for Excellence in SW (2017-0-00093), supervised by the IITP (Institute for Information & communications Technology Promotion).

References

1. Chen, X., Jiao, L., Li, W., Fu, X.: Efficient multi-user computation offloading for mobile-edge cloud computing. IEEE/ACM Trans. Netw. **24**(5), 2795–2808 (2016)
2. Guo, H., Liu, J.: Collaborative computation offloading for multiaccess edge computing over fiber–wireless networks. IEEE Trans. Veh. Technol. **67**(5), 4514–4526 (2018)
3. Guo, H., Liu, J., Zhang, J.: Efficient computation offloading for multi-access edge computing in 5G HetNets. In: 2018 IEEE International Conference on Communications (ICC), pp 1–6 (2018)
4. Hu, Y.C., Patel, M., Sabella, D., Sprecher, N., Young, V.: Mobile edge computing—a key technology towards 5G. ETSI White Pap. **11**(11), 1–16 (2015)
5. Lyu, X., Tian, H., Sengul, C., Zhang, P.: Multiuser joint task offloading and resource optimization in proximate clouds. IEEE Trans. Veh. Technol. **66**(4), 3435–3447 (2017)
6. Mach, P., Becvar, Z.: Mobile edge computing: a survey on architecture and computation offloading. IEEE Commun. Surv. Tutor. **19**(3), 1628–1656 (2017)
7. Mao, Y., You, C., Zhang, J., Huang, K., Letaief, K.B.: A survey on mobile edge computing: the communication perspective. IEEE Commun. Surv. Tutor. **19**(4), 2322–2358 (2017)
8. Pham, Q., Anh, T.L., Tran, N.H., Park, B.J., Hong, C.S.: Decentralized computation offloading and resource allocation for mobile-edge computing: a matching game approach. IEEE Access, 1 (2018)
9. Wang, C., Yu, F.R., Liang, C., Chen, Q., Tang, L.: Joint computation offloading and interference management in wireless cellular networks with mobile edge computing. IEEE Trans. Veh. Technol. **66**(8), 7432–7445 (2017)
10. Yu, Y., Zhang, J., Letaief, K.B.: Joint subcarrier and CPU time allocation for mobile edge computing. In: 2016 IEEE Global Communications Conference (GLOBECOM), pp. 1–6 (2016)
11. Zhang, J., Hu, X., Ning, Z., Ngai, E.C., Zhou, L., Wei, J., Cheng, J., Hu, B.: Energy-latency tradeoff for energy-aware offloading in mobile edge computing networks. IEEE Internet Things J. **5**(4), 2633–2645 (2018)
12. Zhang, K., Mao, Y., Leng, S., Zhao, Q., Li, L., Peng, X., Pan, L., Maharjan, S., Zhang, Y.: Energy-efficient offloading for mobile edge computing in 5G heterogeneous networks. IEEE Access **4**, 5896–5907 (2016)

An Efficient Edge-Cloud Publish/Subscribe Model for Large-Scale IoT Applications

Van-Nam Pham and Eui-Nam Huh[✉]

Department of Computer Science and Engineering, Kyung Hee University,
Yongin-si, Gyeonggi-do 17104, Republic of Korea
{nampv,johnhuh}@khu.ac.kr

Abstract. Internet of Things applications produce enormous amounts of data that need to be collected, analyzed, transferred, and stored at best-suited places. An interplay among Cloud Computing and Fog Computing proves to be a promising solution for the Internet of Things (IoT). Publish/Subscribe (Pub/Sub) systems have been widely deployed in many IoT projects. IoT devices often directly connect to their closest brokers to send and receive data. Building an efficient broker overlay network for Pub/Sub systems to support wide geo-distribution IoT platforms is a matter of immediate concern. In this paper, we propose an efficient Edge-Cloud Pub/Sub model that coordinate brokers in the system for large-scale IoT application. Our simulation results prove that our broker overlay network can support low delay data delivery for these applications in a high scalable manner.

Keywords: Fog Computing · Cloud Computing ·
Publish-subscribe systems · Broker overlay network · IoT applications

1 Introduction

The Internet of Things (IoT) is playing a pivotal role in collecting and exchanging data in recent years. IoT comprises an enormous number of interconnected physical devices/objects sensing or collecting data, interacting with their environment and communicating over geo-distributed networks to provide value-added services. IoT applications can be classified into five different categories: smart wearable, smart home, smart city, smart environment, and smart enterprise [1].

As a result, an unprecedented amount of data is collected and need to be stored, transferred, and analyzed at proper places. Cisco Global Cloud Index estimates that IoT applications will generate an extraordinary amount of data to the tune of nearly 850 zettabytes (ZB) by 2021. However, much of the data is not saved to the Cloud; just about 7.2 ZB of that content will be stored [2]. This is due to data are mostly generated close to the edge of the network, they will be

© Springer Nature Switzerland AG 2019
S. Lee et al. (Eds.): IMCOM 2019, AISC 935, pp. 130–140, 2019.
https://doi.org/10.1007/978-3-030-19063-7_12

processed by applications deployed at the edge to provide appropriate reactions at fast as possible, part of data and information extracted will be uploaded to the Cloud.

The wave of IoT deployments require specialized supports from the computing platform such as mobility support, geo-distribution, location awareness, and low latency; Fog Computing satisfies these requirements but traditional Cloud Computing does not [3]. The Fog can well support IoT applications by bringing computational power closer to the edge for producing analytics to provide the devices with quick reaction instructions. It also filters data to protect privacy, to enhance security and to reduce data sent to the Cloud (for further data mining, controlling, and long-term storage). Therefore, an interplay between the Cloud Computing and the Fog Computing paradigm is well suited for large-scale IoT applications.

Publish/Subscribe model provides a powerful paradigm for building large-scale distributed applications. Furthermore, Pub/Sub systems prove to be a useful, flexible data delivery model for IoT applications [4,5]. There are three main Pub/Sub architectures: Centralized, Peer-to-peer (P2P), and Broker Overlay [6,7]. First, in Centralized approach, the system uses a client/server model and relies upon a single matching engine to provide Pub/Sub services. This architecture is limited in scalability and some typical examples are CORBA Event Services [8] and JMS [9]. Obviously, this approach is not scalable to apply for large-scale IoT applications. Second, in Peer-to-peer approach, publishers and subscribers connected in P2P networks by using some kinds of distributed algorithms and coordinate to each other to match and route events [7,10]. They can be both publishers and subscribers in these systems to deal with high amount of topics or dynamic and diversified contents. However, high churn rate is a popular characteristic of P2P systems that might downgrade the systems. Finally, the other approach is using Broker overlay in which multiple Pub/Sub brokers are organized in a broker overlay network and clients will connect to some brokers to publish or subscribe to interested contents. Events are routed through overlay based on topic-based, content-based or typed-based routing [11–14].

In IoTivity [15], data are collected from IoT devices and ultimately stored in IoTivity Global Cloud using pub/sub brokers that might cause long delay and might not suit well with the above mentioned trend. Even though the Broker overlay approach could provide scalable, fault-tolerant, and cost-effective solutions for Pub/Sub services, it may cause high latency and low scalability for big IoT systems. A survey in [16] points out that topic-based Pub/Sub systems are widely used for messaging in IoT platforms. Subscribers express their concerns by subscribing to interested topic channels. When there is a new event on the subscribed topic, the system will notify and deliver the data to the subscribers in an appropriate and flexible manner. In reality, a lightweight data transport protocol such as Message Queue Telemetry Transport (MQTT) [17] is necessary for messaging among the devices in the IoT system. This often requires a message broker for each isolated network or area; it works well in LAN environment. However, in order to apply for large-scale IoT platforms, we still need a mechanism to link many of these brokers in an efficient and scalable way.

In this paper, we are going to propose an efficient Edge-Cloud pub/sub broker model for dynamic coordination and data delivery among brokers for large-scale IoT systems with high scalability and low latency properties. In the next section, we introduce the Edge-Cloud pub/sub model, explain its components and provide algorithms to bridge topics among brokers in the system. We provide simulation results in Sect. 3 to validate our claims; and we present our conclusion and future work in Sect. 4.

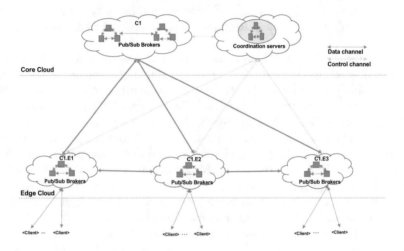

Fig. 1. The Edge-Cloud pub/sub model.

2   The Edge-Cloud Pub/Sub Model

Figure 1 shows the model of edge-cloud Pub/Sub broker system that supports large-scale IoT applications. There are three main components in the system: Pub/Sub brokers located at Edge Clouds or Core Cloud(s), clients that are IoT devices running Pub/Sub client applications, coordination servers located at core cloud to orchestrate all brokers of the whole system.

2.1   Interaction Among Components in the Model

At the bottom of the above model are Pub/Sub clients. These clients can be any IoT devices such as sensors, wearables or any standard devices like desktops, laptops, smartphones and tablets. They run Pub/Sub client applications to send and receive event notifications to/from the whole system by connecting to their nearest Pub/Sub brokers. In this paper, we assume that clients are statically configured or dynamically mapped to their closest Pub/Sub brokers by using techniques described in [18–20].

Pub/Sub brokers are strategically deployed at geographical locations in Edge Clouds or Core Cloud(s) to manage and exchange data among their clients. The number of Edge/Cloud Pub/Sub brokers can be scaled up or down adaptively depending on their number of clients and their workloads. As shown in Fig. 2, each broker has a "Bridging manager" module using the control channel to connect to one of coordination servers of the system. This integrated module interacts with the coordination server to help the broker knows which brokers publish/subscribe what topics by implementing our algorithm (described in Subsect. 2.2) and interacting with their coordinators in real time.

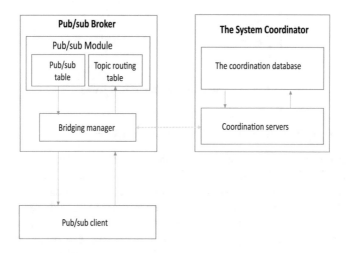

Fig. 2. The system model.

The system coordinator comprises a group of coordination servers that provides group communication functions in distributed environment. ZooKeeper [21] is a well-known tool for leader election, group membership, failure detection, and configuration management in distributed systems. Therefore, we can use an ensemble of main ZooKeeper servers located at the Core Cloud to coordinate among Pub/Sub brokers in the whole system. We design a ZooKeeper hierarchical namespace in the form of znodes to manage and coordinate among Pub/Sub brokers, see Fig. 3.

Each znode is a data object including its path and stored data that is read and written atomically. Brokers register listeners on involved znodes to watch for updates on the given data objects. Through this mechanism, brokers can coordinate to exchange information on topic publication and subscription. Interested readers can refer to [11, 21] for more information about this mechanism.

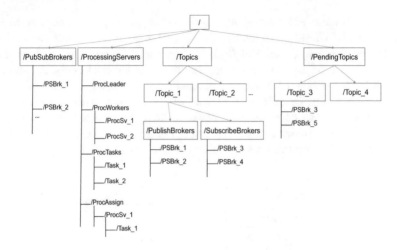

Fig. 3. The model's znode hierarchy.

2.2   Bridging Topics Among Pub/Sub Brokers

To support intensive volume of data transfer from many topics in large-scale IoT applications, the Pub/Sub system needs to provide mechanisms for coordination and dynamic adaptation among brokers. Moreover, an efficient overlay network of brokers that offers both scalability and low latency in data delivery over WAN scale systems is the utmost requirement. We devise protocols and algorithms to tackle these issues and Edge/Cloud Pub/Sub brokers are linked to play as a single message delivery domain. The main ideas are described as follows.

- The coordination database stores information about which brokers publish or subscribe what topics.
- Brokers inform their coordination servers about new topics published/subscribed from their clients.
- The coordinator informs its interested broker(s) about the new broker that has clients publish interested topics.
- The interested broker subscribes to the new publish broker on the topic to receive event notifications afterwards.

   To support many-to-many communication among clients, we need to link all brokers having clients publish or subscribe the same topic while avoiding routing loop in forwarding related event messages. We introduce a basic method and our proposed method to solve this issue as follows.

   An intuitive method is selecting one root publish broker per topic (ORPBT) as illustrated in Fig. 4. From this figure, supposes that broker BR1 is selected as the root publish broker for the topic TP1 due to it is the first known broker in the system has clients publishing events on the topic (we can use other methods to select the topic root when there many brokers publish the same topic). When there is a new publish broker on TP1, e.g. BR2, it has to publish related events

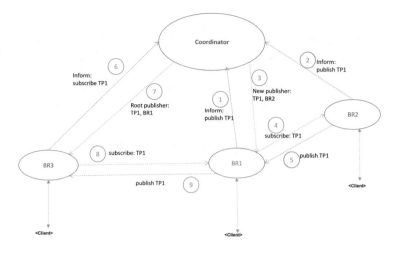

Fig. 4. Illustration of one root publish broker algorithm.

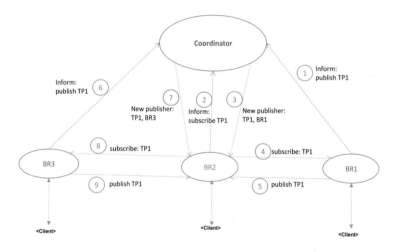

Fig. 5. Illustration of one hop from multiple publish brokers algorithm.

on this topic to the root (BR1). Interested brokers on the topic, e.g. BR3, just need to subscribe the common topic, TP1, at the root to receive all related event notifications of the topic.

We propose a better algorithm to link the common topic among brokers by direct bridging one hop from multiple publish brokers per topic (OHMPBT) as illustrated in Fig. 5.

As is clear from Fig. 5, BR2 was interested in TP1 and directly subscribed from BR1 at the beginning. When there is a new publish broker on TP1, e.g. BR3, the coordinator informs BR2 to subscribe TP1 on the new one. By doing this way, it just costs one hop among brokers to forward messages from sending clients to receiving clients. Nevertheless, each broker has to keep track of its bridging lists (topics publish/subscribe brokers) to avoid routing loop in the system: receiving brokers do not send back to the forwarding broker and refrain from re-forwarding messages to other brokers. We did many simulations to proof that our solution can help to build an efficient data delivery Pub/Sub broker overlay network for large-scale IoT applications as described in the next section.

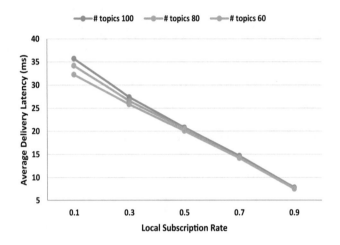

Fig. 6. Average delivery latency test.

3   Simulation Results and Discussion

This section presents our simulation scenarios and results we conducted to evaluate our algorithm, OHMPBT, through metrics such as the average delivery latency (ADL), the average forwarding traffic (AFT), and scalability of the broker overlay. Our simulations were written in Python programming language by using SimPy (a discrete event simulation tool for Python) simulation library [22]. We used about 2000 Starbucks locations for broker positions from [23]. We assume that the delay between two brokers is a random variable that is proportional to their geographical distances.

3.1   Average Delivery Latency Test for Different Local Subscription Rate

The scenario was designed as follows to evaluate ADL of the OHMPBT algorithm.

– 10 Pub/sub brokers are randomly selected;
– Number of publication lists per broker are 100, 80, 60: randomly selected from the 1000 topics pool;

- 1 publisher & 10 subscribers per topic;
- Local subscription rates are 0.9, 0.7, 0.5, 0.3, 0.1;
- Up to 50000 messages are collected.

The result of this simulation is shown in Fig. 6. From the figure we can see that when local subscription rate increases, the ADL decreases. This proves that using edge brokers to manage clients is efficient and it is suited with general IoT data transmission patterns. In addition, it is very appropriate with the trend: data generated locally are increasingly consumed locally.

3.2 Comparison Between ORPBT and OHMPBT Algorithms

We designed the following scenario to compare the two algorithms in terms of ADL, AFT, and to test the two algorithms' scalability. The results are shown in Figs. 7, 8, and 9.

- A pool of 2000 topics is randomly generated;
- Each Pub/Sub broker handles 100 topics: 1 publisher and 10 subscribers per topic;
- Each broker picks 100 topics from the pool randomly on behalf of its clients;
- The number of brokers increases from 10 to 100, 200, 300, 400, 500, and 600.

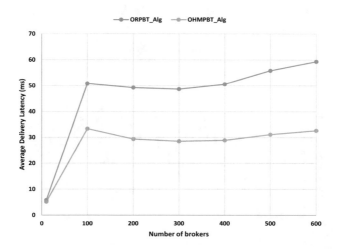

Fig. 7. Average delivery latency comparison.

As shown in Fig. 7, our algorithm, OHMPBT, is much better than the ORPBT algorithm in terms of ADL; two algorithms scaled well in case of increasing the number of brokers.

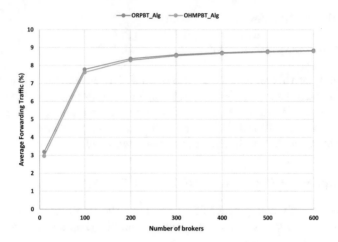

Fig. 8. Average forwarding traffic comparison.

The Fig. 8 shows that the AFT of the whole system between the two algorithms are similar. In terms of overall AFT of the system, both algorithms also scaled very well: when the number of brokers increased, the overall AFT did not increase linearly. However, when we examined the AFT of each broker as illustration in Fig. 9, our algorithm -OHMPBT- did show the much better result in terms of load balancing among Pub/Sub brokers in the system.

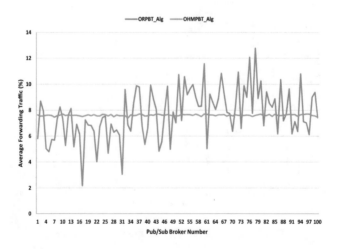

Fig. 9. Broker's average forwarding traffic comparison.

As clearly shown in Fig. 9, the AFT of each broker our algorithms just narrowly fluctuated around 7.8% while there were many brokers in the ORPBT algorithm experienced higher AFT (around 10% and above) due to they were selected as the root brokers. This could make ORPBT hard to scale well when the number of topics on each broker increases significantly.

4 Conclusions

In this paper, we proposed an efficient Edge-Cloud Pub/Sub model that can be deployed in WAN-scale distributed systems. Pub/sub brokers interact to each other with the help of coordination servers in finding, notifying, and subscribing to the publish brokers of interested topics. Clients in the system just need to connect to their nearest brokers to publish or subscribe their interested topics. Event notifications are delivered to the interested clients by just one hop between their host brokers. As proved in our simulation results, the broker overlay network created by our algorithm is low delay, high scalable and adaptive to clients' changing topic patterns. This makes our model very promising to apply for large-scale IoT applications that take advantages of publish/subscribe systems for data transmission.

This is our ongoing research; we have been doing more intensive simulations to evaluate our broker overlay performance in more large-scale setups. In addition, testing with real application data will be carried out. We will present these evaluation results in our future work.

Acknowledgment. This research was supported by the MSIT (Ministry of Science and ICT), Korea, under the Grand Information Technology Research Center support program (IITP-2018-2015-0-00742) and the National Program for Excellence in SW (2017-0-00093), supervised by the IITP (Institute for Information & communications Technology Promotion).

References

1. Perera, C., Liu, C.H., Jayawardena, S.: The emerging internet of things marketplace from an industrial perspective: a survey. IEEE Trans. Emerg. Top. Comput. **3**(4), 585–598 (2015)
2. CISCO, Cisco Global Cloud Index: Forecast and Methodology, 2016–2021 White Paper. https://www.cisco.com/c/en/us/solutions/collateral/service-provider/global-cloud-index-gci/white-paper-c11-738085.html
3. Bonomi, F., Milito, R., Natarajan, P., Zhu, J.: Fog computing: a platform for internet of things and analytics. In: Bessis, N., Dobre, C. (eds.) Big Data and Internet of Things: A Roadmap for Smart Environments. Studies in Computational Intelligence, vol. 546. Springer, Cham (2014)
4. Antonic, A., Roankovic, K., Marjanovic, M., Pripuic, K: A mobile crowdsensing ecosystem enabled by a cloud-based publish/subscribe middleware. In: 2014 International Conference on Future Internet of Things and Cloud, Barcelona, pp. 107–114 (2014)
5. Rowe, A., Berges, M.E., Bhatia, G., Goldman, E., Rajkumar, R., Garrett, J.H., Moura, J.M., Soibelman, L.: Sensor andrew: large-scale campus-wide sensing and actuation. IBM J. Res. Dev. **55**(1.2), 6:1–6:14 (2011)

6. Eugster, P.T., Felber, P.A., Guerraoui, R., Kermarrec, A.M.: The many faces of publish/subscribe. ACM Comput. Surv. (CSUR) **35**(2), 114–131 (2003)
7. Rahimian, F., Girdzijauskas, S., Payberah, A.H., Haridi, S.: Vitis: a gossip-based hybrid overlay for internet-scale publish/subscribe enabling rendezvous routing in unstructured overlay networks. In: 2011 IEEE International Parallel & Distributed Processing Symposium, pp. 746–757. IEEE, May 2011
8. Harrison, T.H., Levine, D.L., Schmidt, D.C.: The design and performance of a real-time CORBA event service. ACM SIGPLAN Not. **32**(10), 184–200 (1997)
9. Hapner, M., Burridge, R., Sharma, R., Fialli, J., Stout, K.: Java Message Service. Sun Microsystems Inc., Santa Clara (2002)
10. Baldoni, R., Beraldi, R., Quema, V., Querzoni, L., Tucci-Piergiovanni, S.: TERA: topic-based event routing for peer-to-peer architectures. In: Proceedings of the 2007 Inaugural International Conference on Distributed Event-based Systems, pp. 2–13. ACM, June 2007
11. An, K., Khare, S., Gokhale, A., Hakiri, A.: An autonomous and dynamic coordination and discovery service for wide-area peer-to-peer publish/subscribe: experience paper. In: Proceedings of the 11th ACM International Conference on Distributed and Event-Based Systems, pp. 239–248. ACM, June 2017
12. Baldoni, R., Beraldi, R., Querzoni, L., Virgillito, A.: Efficient publish/subscribe through a self-organizing broker overlay and its application to SIENA. Comput. J. **50**(4), 444–459 (2007)
13. Carzaniga, A., Rosenblum, D.S., Wolf, A.L.: Achieving scalability and expressiveness in an internet-scale event notification service. In: Proceedings of the Nineteenth Annual ACM Symposium on Principles of Distributed Computing, pp. 219–227. ACM, July 2000
14. Strom, R., Banavar, G., Chandra, T., Kaplan, M., Miller, K., Mukherjee, B., Sturman, D.C., Ward, M.: Gryphon: an information flow based approach to message brokering. arXiv preprint cs/9810019 (1998)
15. IoTivity software framework. https://www.iotivity.org
16. Menzel, T., Karowski, N., Happ, D., Handziski, V., Wolisz, A.: Social sensor cloud: an architecture meeting cloud-centric IoT platform requirements. In: 9th KuVS NGSDP Expert Talk on Next Generation Service Delivery Platforms (2014)
17. OASIS. MQTT Version 3.1.1 (2014). http://docs.oasis-open.org/mq./mq./v3.1.1/os/mq.-v3.1.1-os.html
18. Jafari, S.J., Naji, H.: GeoIP clustering: solving replica server placement problem in content delivery networks by clustering users according to their physical locations. In: 2013 5th Conference on Information and Knowledge Technology (IKT), pp. 502–507. IEEE, May 2013
19. Katz-Bassett, E., John, J.P., Krishnamurthy, A., Wetherall, D., Anderson, T., Chawathe, Y.: Towards IP geolocation using delay and topology measurements. In: Proceedings of the 6th ACM SIGCOMM Conference on Internet Measurement, pp. 71–84. ACM, October 2006
20. Eriksson, B., Barford, P., Maggs, B., Nowak, R.: Posit: an adaptive framework for lightweight IP geolocation. Computer Science Department, Boston University (2011)
21. Hunt, P., Konar, M., Junqueira, F.P., Reed, B.: ZooKeeper: wait-free coordination for internet-scale systems. In: USENIX Annual Technical Conference, vol. 8, no. 9, June 2010
22. SimPy. https://simpy.readthedocs.io/en/latest/
23. Starbucks shop locations. https://community.periscopedata.com/t/80fyna/starbucks-locations

Capacity Planning for Virtual Resource Management in Network Slicing

Rajesh Challa[1], Syed M. Raza[1], Hyunseung Choo[1(✉)], and Siwon Kim[2]

[1] College of Software, Sungkyunkwan University, Suwon, South Korea
{rajesh.c,s.moh.raza,choo}@skku.edu
[2] Sungkyunkwan University, Suwon, South Korea
siwon.kim59@gmail.com

Abstract. Network slicing is envisioned as one of the core enablers for 5G. It empowers service providers to multiplex their physical infrastructure over multiple logical networks. Slice orchestration performs lifecycle management and virtual resource management of a network slice. Slice capacity planning, one of the key responsibilities of slice orchestration faces two major challenges: (i) minimum slice capacity estimation for a given slice configuration, and (ii) lack of simple network slicing supported open source solutions due to which the research and development of slice resource orchestration activities are hit adversely. The feature-rich existing solutions have significantly high learning curve. This paper addresses these two issues. We present a probabilistic model for slice capacity planning and propose a simple orchestration-centric discrete event simulator to develop and evaluate minimum slice capacity requirement. The results taken on our custom simulator show around 96% of resource utilization, and have less than 5% relative error when compared with the theoretical model.

Keywords: Network slicing · Slice orchestration ·
Resource management · Event-driven model

1 Introduction

Network slicing (or simply, slicing) paradigm is widely accepted as one of the primary enablers for 5G [1,2]. Slice is a logically partitioned virtual network over the common physical infrastructure. The resources in a slice comprise of a collection of usecase (business) specific virtual network functions (VNFs) and virtual links [3]. However, slicing has brought in new challenges in the form of slice management and orchestration which include slice configuration, lifecycle management, and resource management. This article focuses on the slice resource management field of slice orchestration.

Slice capacity planning is one of the major challenges in resource management. Effective planning optimizes infrastructure cost, and satisfies the Service Level

© Springer Nature Switzerland AG 2019
S. Lee et al. (Eds.): IMCOM 2019, AISC 935, pp. 141–152, 2019.
https://doi.org/10.1007/978-3-030-19063-7_13

Agreement (SLA) driven resource demand with minimal rejection rate. An over-provisioning of resources to satisfy peak traffic demands may endup in lower average utilization, whereas addressing utilization drawback may result in higher rejections to avoid SLA violations. A major difference between cloud computing and slicing is the type of resources and its handling. Slice is an higher level abstraction over cloud resources and slice orchestration mainly focuses on virtual resource management unlike cloud that primarily manages physical embedding of these virtual resources. Such a separation of resource management follows the separation of concerns design principle, and results in efficient network management [4].

Telco operators are facing two key issues pertaining to slice capacity planning: (i) to determine best capacity required for a slice to meet a preset admission rate target, and (ii) to have a simplified resource management centric simulation platform which is easier to customize. Network slicing being emerging paradigm, relatively less related literature addresses these issues. Though there are existing open source that support network slicing, they have extremely steep learning curve [5–7]. Significant effort is required to understand these feature-rich solutions even to make a small feature addition. There is a need for a small and efficient solution, exclusively designed to research the resource management related algorithms and heuristics.

In this paper, we define the slice capacity in terms of service function chain virtual resources. We address the capacity planning problem in network slicing by proposing a probabilistic model based on $M/G/c/K$ queuing system. An efficient and fast converging hill-climbing algorithm computes the minimum capacity requirement of a slice to meet the operator-specific admission rate target. A novel custom discrete event simulator for slice resource management is presented, which is based on Observer and Publish-Subscribe design patterns. For empirical capacity planning, the flow profiling entity observes input flow requests and publishes these requests to the subscribed admission control heuristics. Our simulation results *(i)* estimates the slice capacity with respect to simulation results, and *(ii)* illustrates the impact of capacity size change and mean arrival rate on admission rate, and mean resource utilization.

Rest of the paper is organized as follows. The related work is examined in Sect. 2. Section 3 describes capacity planning probabilistic model based on $M/G/1$ queue system and discusses the evaluation method for given configuration. A novel discrete event simulator is described in Sect. 4 with detailed event control and management for capacity planning. The performance evaluation results of our capacity planning approach are shown in Sect. 5. Finally, Sect. 6 concludes the paper along with future direction.

2   Related Work

Resource management is one of the wide studied area across different domains, however, slicing is in its early stages and slice orchestration is a new research direction in current and near future [8–10]. We first present the literature related to slice orchestration from resource management perspective. Later, some relevant related work in cloud domain is discussed, as cloud is one of the enabling technologies for network slicing. We examine primarily from the capacity planning perspective.

Slice orchestration and management is one of the key focus areas in network slicing which is actively being pursued across standardization bodies as well as academia [8–10]. A list of recent slice orchestration articles is discussed in [8] along with a brief explanation of each article. The slice creation challenges with respect to capacity planning and resource management are enumerated in [9] which sets the future research direction in this field. An IETF draft is dedicated to mention the roles and responsibilities of an orchestrator from slice resource management perspective [10]. The cellular standardization body 3GPP has recorded the slice capacity planning related requirements in Release 15 [11–13].

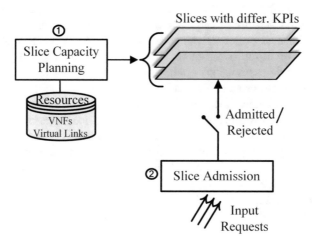

Fig. 1. Slice virtual resource management overview.

The capacity planning in network slicing and cloud computing is similar, hence we explore the contemporary study related to cloud capacity planning research activities and differentiate our work with the state-of-the-art work. Capacity planning in cloud infrastructure-as-a-service model is described in [14] where a probabilistic models based on G/GI/c/K and M/M/c/K queuing system are presented and validated it with the simulation results. The performance analysis for M/G/m/m+r queuing system is discussed in [15], which serves as a reference article for our paper. Another capacity planning article focuses on access networks and determines minimum fronthaul bandwidth required in 5G slices for a given infrastructure configuration [16]. In contrast, the probabilistic model in this article maps slice resource capacity planning to M/G/c/K system, propose a simplified discrete event simulator to assess the resource management algorithms, and presents a comparative evaluation between theoretical model and simulator results. For performance evaluation, we simplify the system to M/G/1/K as no closed-loop approximation for multi-server system is known which is easier to simulate.

3 Virtual Resource Management

Virtual resource management is one of the key subsystems in slice orchestration. It includes three phases namely, planning, admission control and network embedding. Embedding maps virtual resources to physical substrate and in this article. The resource management system illustrated in Fig. 1 has a pool of resources i.e., virtual links (VLs) and virtual network functions (VNFs) to be assigned to different types of slices (e.g., broadband, connected cars, Internet of Things and so on). Each slice serves user requests that need resources with specific Key Performance Indicators (KPIs) e.g., latency, bandwidth and error rate.

The planning phase projects an amount of resources required by a slice to serve an intended volume of requests without violating their respective SLAs. The second phase is policy-driven admission control that decides the acceptance/rejection of input requests to slices while respecting the capacity constraint. This article assumes an open loop system i.e., no feedback mechanism to predict the future resource demand.

The capacity of a slice is defined in terms of Service Function Chains (SFCs) as allocatable virtual resources. An SFC is a sequence of service-specific VNFs connected by VLs (links hereafter) which represents a service catered by the slice. Though every SFC comprises of multidimensional resources (e.g., CPU, bandwidth, storage, etc.), we assume only bandwidth resource as the slice capacity for brevity. Let $\mathcal{H} = \{1, 2, ..., H\}$ be the set of SFCs, each SFC $h \in \mathcal{H}$ has $l_1, l_2, ..., l_L$ links with individual link bandwidth $b_{l'}, l' = 1, ..., l_L$, and $I = \{1, 2, ..., I\}$ be the set of slices. The available capacity β_h of SFC h and that of a slice $C_i, i \in I$ are given by:

$$\beta_h = \min_{l'=1,...,l_L} b_{l'} \tag{1}$$

$$C_i = \sum_{h \in \mathcal{H}} \beta_h \tag{2}$$

As the duration of a flow request (flow hereafter) is unknown beforehand, the resource demand is measured and allocated for a minimum amount of time called *epoch*. Small epoch values are preferred as the larger values results in resource wastage after the flows departs. The allocated resources meet the user SLA and are continued to be held across epochs till the flow departs. The epochs are assumed to be of same length. We assume the arrival rate in each epoch to follow Poisson distribution and the per epoch flow duration to be a heavy-tailed Pareto distribution [17]. Both arrival rate and flow duration are independent and identically distributed random variables. If $|I|$ be the number of slices, each with finite available capacity C_i then the system can be modeled as $M/G/|I|/C_i$ queue.

It is difficult to find exact distribution functions for queuing model other than $M/M/./.$ as both the random variables are time and history dependent [12,13]. Therefore, we adopt an approximation approach for capacity planning based on steady state probabilities. Our queuing model driven capacity planning approach estimates the admission rates. Let λ be the mean arrival rate, $\mathbb{E}[D_i]$

Table 1. Summary of notation

Symbol	Definition
$h \in \mathcal{H}$	An SFC h in set of SFCs $\mathcal{H}$
I	Set of slices
$b_{l'}, l' = 1, ..., l_L$	Bandwidth of an individual link l'
β_h of SFC h	Available capacity of SFC h
$C_i, i \in I$	Available capacity of slice i
λ	Mean arrival rate at steady state
$\mathbb{E}[D_i]$	Mean flow duration at steady state
$\mathbb{E}[D_i^2]$	Second moment of $M/G/./.$ queue
$\mathbb{E}[F]$	Mean number of flows at steady state
V_λ^2	Squared coefficient of variation of arrival rate distribution
$V_{D_i}^2$	Squared coefficient of variation of flow duration distribution
ρ_i	Traffic intensity of slice i
$\mathcal{K}$	Maximum number of flow requests in the system at steady state
p_k	Blocking probability of $M/G/1/\mathcal{K}$ queue
Δ_i	Admission rate of slice i
r_{min}	Minimum resource demand
Δ_{min}	Minimum admission rate
ϵ	Capacity increment unit

be the mean flow duration at steady state and resources are allocated without violating the flow SLAs. We evaluate the minimum slice capacity required for certain admission rate target. Table 1 lists the notation and symbols used in this paper. The mean number of flows in the system $\mathbb{E}[F]$ is given by Pollaczek-Khinchin formula [18,19]:

$$\mathbb{E}[F] = \lambda \mathbb{E}[D_i] + \frac{\lambda^2 \mathbb{E}[D_i^2]}{2(1 - \lambda \mathbb{E}[D_i])} \tag{3}$$

where $\mathbb{E}[D_i^2]$ is the second moment of $M/G/|I|/C_i$ queue. If ρ_i represents the traffic intensity of a slice, $V_{D_i}^2$ is the squared coefficient of variation of the flow duration distribution, and then (3) can be simplified as

$$\mathbb{E}[F] = \sum_{i \in I} \left(\rho_i + \frac{1 + V_{D_i}^2}{2} \cdot \frac{\rho_i^2}{1 - \rho_i} \right) \tag{4}$$

where

$$\rho_i = \lambda \mathbb{E}[D_i] \tag{5}$$

There is no closed-loop approximation available for $M/G/c/K$, hence, we assume a single slice for modeling blocking probability (BP) (defined as the probability that an arriving flow request finds the system to be full). For multi-server scenario, we recommend readers to refer [18, 19]. Let $\mathcal{K}$ be the maximum number of flow requests in the system, p_k be the BP of $M/G/1/\mathcal{K}$ and V_λ^2 be the squared coefficient of variation of arrival rate distribution then $\mathcal{K}$, admission rate Δ_i of slice i and p_k are defined as

$$\Delta_i = 1 - p_k(\rho_i, \mathcal{K}) \tag{6}$$

$$\mathcal{K} = \frac{C_i}{r_{min}} \tag{7}$$

$$p_k = \frac{\rho_i^{\left(\sqrt{\rho_i}V_{D_i}^2 - \sqrt{\rho_i}+2\mathcal{K}\right)/\left(\sqrt{\rho_i}V_{D_i}^2 - \sqrt{\rho_i}+2\right)}(\rho_i - 1)}{\left(\rho_i^{2\left(\sqrt{\rho_i}V_{D_i}^2 - \sqrt{\rho}+1+\mathcal{K}\right)/\left(\sqrt{\rho_i}V_{D_i}^2 - \sqrt{\rho_i}+2\right)} - 1\right)} \tag{8}$$

$$p_0 = \frac{(\rho_i - 1)}{\left(\rho_i^{2\left(\sqrt{\rho_i}V_{D_i}^2 - \sqrt{\rho}+1+\mathcal{K}\right)/\left(\sqrt{\rho_i}V_{D_i}^2 - \sqrt{\rho_i}+2\right)} - 1\right)} \tag{9}$$

Equations (8) and (9) are based on the Smith's approximation of $M/G/1/\mathcal{K}$ queuing systems [18].

Algorithm 1. Slice capacity planning algorithm

Input: λ, $\mathbb{E}[D_i]$, $V_{D_i}^2$, V_λ^2, Δ_i, Δ_{min}, ϵ
Output: C_i (Capacity of slice i)
Initialization : $C_i = 0$; $\Delta_i = 0$

1 **Procedure** *SliceCapacity*
2   **while** $(\Delta_i < \Delta_{min})$ **do**
3         $C_i = C_i + \epsilon$
4         $\mathbb{E}[F] = \lambda\mathbb{E}[D_i] + \frac{\lambda^2\mathbb{E}[D_i^2]}{2(1-\lambda\mathbb{E}[D_i])}$
5         $\rho_i = \lambda.\mathbb{E}[D_i]$
6         $\mathcal{K} = \frac{C_i}{r_{min}}$
7         $\Delta_i = 1 - p_k(\rho_i, V_{D_i}^2, V_\lambda^2, \mathcal{K})$

A simple hill-climbing technique is implemented in Algorithm 1 to calculate the minimum slice capacity. As p_k is defined for $M/G/1$ queue, we perform the calculation assuming a single aggregate slice that combines all similar type slices, i.e., $i = 1$. The optimal capacity is iteratively computed by incrementing ϵ units per iteration and stops when the admission rates hit the target threshold (lines 2–3). The equations defined above for capacity C_i, mean number of flows $\mathbb{E}[F]$, traffic intensity ρ_i at slice i, maximum flows in the system $\mathcal{K}$ and BP p_k are used to evaluate the admission rate (lines 4–7). The time complexity of Algorithm 1 is dominated by $\mathcal{O}(n)$, where n denotes the number of iterations taken to reach the threshold. If we relax the assumption that all slices have same initial capacity then the time complexity is dominated by $\mathcal{O}(n|I|)$.

4   Simulation Model

We present a capacity planning simulation model in this section. Most of the open source software that supports network slicing are feature-rich and exhaustive [5–7], due to which the learning curve is too steep. Developers need to invest enduring efforts to synthesize required codeflow even to make a small simulation. Furthermore, expensive proprietary platforms lack the code customization flexibility to research novel heuristics. Our simulation model is significantly simple to research and develop different slice resource management algorithms and heuristics for capacity planning, admission control and resource allocation/assignment tasks.

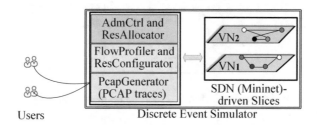

Fig. 2. Discrete event simulator for slice resource management.

The simulation model in Fig. 2 shows two major modules: *(i)* Mininet-based Software Defined Networking (SDN) controller that provides the network slices i.e., virtual networks [20]. Mininet topology is spawned via Python scripts where the nodes and links emulates VNFs and virtual links, respectively. The resources e.g., bandwidth can be assigned to each topology element which can become the basis for resource management simulations. *(ii)* The multithreaded software that simulates the layered asynchronous communication, resembling real world situations. The users generate flow requests that becomes the inputs for our event driven simulator developed using Python. Though the model can be used for various resource management tasks, we focus on the second module with capacity planning as the key objective in this article.

The input flows are sniffed by the PcapGenerator thread and generates the PCAP traces. The FlowProfiler/ResConfig thread is the core configuration entity that sets the essential capacity planning simulation parameters. The AdmCtrl/ResAlloc thread runs different admission control heuristics as per the slice orchestration policy. Finally, these flows are scheduled on the Mininet virtual networks (slices). The relevant flow entries generation in Mininet for flow scheduling are assumed to be existing. The communication between PcapGenerator and FlowProfiler/ResConfig follows Observer pattern, whereas Pub/Sub pattern driven communication exists between AdmCtrl/ResAlloc and FlowProfiler/ResConfig layers [REF].

A detailed functional view of the multithreaded software module is shown in Fig. 3. The solid lines represent synchronous communication such as

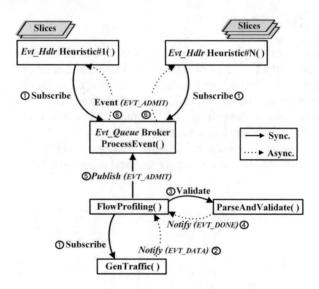

Fig. 3. Detailed functional view of resource management event simulation.

registration procedure i.e., subscription of the event listeners by observer `FlowProfiling()` and event handlers `Evt_Hdlr Heuristic#()`. The dotted lines depit asynchronous communication such as event posting. The subject function `GenTraffic()` captures flows periodically and store it as PCAP traces. This generated trace is then passed as `EVT_DATA` to `FlowProfiling()` through a notification event `Notify(EVT_xxx)`. The `FlowProfiling()` asynchronously get the PCAP trace parsed and validated. The one time initialization of simulation parameters is performed followed by populating the array of flows. Alternatively, flows can also be classified based on requested KPIs or traffic class if required. The array (of flows) is then published to a broker `Evt_Queue Broker()` that broadcasts the cloned array with `EVT_ADMIT` event to all the subscribed heuristics which are registered as the event handlers `Evt_HdlrHeuristic#()`.

The capacity planning procedure shown in Algorithm 1 is executed at `FlowProfiling()`. We examine $M/G/.$ queuing model simulated by random variables λ and $\mathbb{E}[D_i]$. One of the admission control heuristics event handler will assist in providing the admission rate Δ_i. Other event handlers can be utilized to apply empirical admission control heuristics for comparative analyses. This simulation model can be used for a continuous system too where flows arrive and depart periodically. The departure of flows are managed by a flow request specific timer callback that is registered once it is admitted into the slice.

5   Performance Evaluation

Our Python based custom simulator is running on Dell PowerEdge R730 server having 12 Cores/128 GB RAM. The Mininet and relevant SDN environment are also configured in the same virtual machine. Extensive simulations are done to verify our capacity planning probabilistic model and compare it against the empirically generated results. We assumed a continuous system where flow requests are arriving every epoch and are departing randomly. Artificial flows are generated and passed as input to the simulator. The M/G/1 as well as M/G/c systems were simulated to assess the performance.

Table 2. Simulation parameters.

Parameters	Values	Parameters	Values
Total flows	0.6 Million	λ	$300 - 700$
$\mathbb{E}[D_i]$	360 s	ϵ	50 units
Flow duration	Pareto distr.	Reso. demand	Uniform distr.
Δ_{min}	95%	r_{min}	10 Mbps
Simulation time	$5000 - 9000$ s	–	–

The list of simulation parameters and respective configurations are shown in Table 2. For evaluation purposes, we perform the capacity planning assuming single slice due to non-availability of a closed-loop blocking probability equation for multi-server M/G/./. queuing model. The impact of capacity planning is assessed across 2 metrics: Admission rate and mean resource utilization. The simulations were run for around 9000 s and assumed a batched arrival of input flows rather than one-by-one arrival or departure of flows with an intention to analyze the approximate behavior of our system. We define relative error e as the difference between observed $O[x]$ and estimated $M[x]$ values divided by the observed values. This ratio shows the error margin.

$$e = \frac{\left| O[x] - M[x] \right|}{O[x]} \tag{10}$$

The results in Fig. 4 shows the impact of capacity factor change on admission rate. The values for $V_{D_i}^2$ and V_λ^2 are taken from [19]. The capacity factor 1.0 implies 100%. Gradual increase of capacity by a factor of 0.1 (i.e., 10%) results in reducing blocking probability too. The result calculates the minimum capacity required for a slice to have 100% admission rate. The impact of mean arrival rate change on admission rate is depicted in Fig. 5 for a fixed capacity size. These admission rates are without violating any flow SLA constraint.

The mean resource utilization can be approximated from the results displayed in Fig. 6. The unit of Time slot (x-axis) is in minutes. The resources utilized for a fixed r_{min} by our custom simulator is almost 96 97% which is very close to the estimated values. The relative error e in all the cases is less than 5%, which validates our simulation results against well-known probabilistic model incorporated in this paper. A point to note is that in all the graphs, theoretical values are higher than the observed because random resource demand and mean service rate (bounded Pareto distributions) have higher variations in simulation environment.

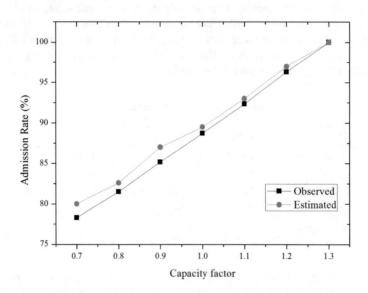

Fig. 4. Impact of capacity size change on admission rate for fixed mean arrival rate.

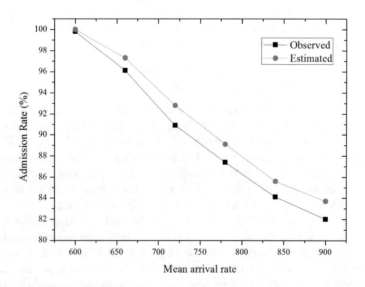

Fig. 5. Impact of mean arrival rate change on admission rate for fixed capacity factor.

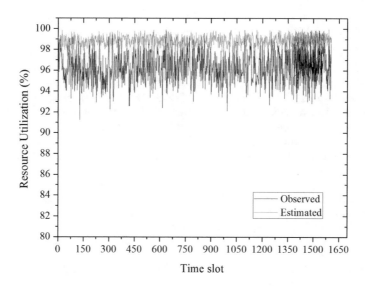

Fig. 6. Resource utilization comparison between theoretical and simulated (observed) results.

6   Conclusion

Network slicing offers an approach to multiplex the physical resources across multiple logical networks which are designed to serve a business usecase. Hence, key challenges is to minimize resource over-provisioning, rejection rate and SLA violations. The current open source software are too complex and have steep learning curve to carry out focused research on these resource orchestration issues. This paper presents probabilistic model for slice capacity planning based on $M/G/c/K$ queuing system to compute minimum slice capacity requirement. We further propose a simple discrete event simulator to develop and evaluate the resource orchestration algorithms and heuristics. Our simulation results estimate minimum capacity requirement, show admission rate and resource utilization values having less than 5% of relative error between observed and well-known theoretical results. In future, we would enhance our simulator to investigate slice admission control algorithms for a given capacity to maximize certain utility function e.g., revenue. We would like to integrate prediction model to compare the admission control trends.

Acknowledgments. This research was supported in part by Korean government (MSIT and MOE), under IITP (B0101-15-1366), G-ITRC (IITP-2017-2015-0-00742), and Human-plus research program (2018M3C1B8023550) supervised by the MSIT and MOE, Korea government.

References

1. Spyridon, V., Gkatzikis, L., Liakopoulos, N., Stiakogiannakis, I.N., Qi, M., Shi, L., Liu, L., Debbah, M., Paschos, G.S.: The algorithmic aspects of network slicing. IEEE Commun. Mag. **55**(8), 112–119 (2017)
2. Ordonez-Lucena, J., Ameigeiras, P., Lopez, D., Ramos-Munoz, J.J., Lorca, J., Folgueira, J.: Network slicing for 5G with SDN/NFV: concepts, architectures and challenges. arXiv preprint arXiv:1703.04676, May 2017
3. Bega, D., Gramaglia, M., Cano, C.J.B., Banchs, A., CostaPerez, X.: Toward the network of the future: from enabling technologies to 5G concepts. Trans. Emerg. Telecommun. Technol. **28**(8), 1–12 (2017)
4. Galis, A.: Key challenges - reliability, availability and serviceability in network slicing. In: IEEE Emerging Technology Reliability Roundtable, July 2017. http://cqr.committees.comsoc.org/files/2017/03/12-Alex_Galis.pdf
5. OpenDaylight Home. https://www.opendaylight.org
6. Open Network Operating System (ONOS) SDN Controller. https://www.opennetworking.org/onos
7. Floodlight OpenFlow Controller-Project Floodlight. http://www.projectfloodlight.org/floodlight
8. Al-Dulaimi, A., Ni, Q., Cao, J., Gatherer, A., Chih-Lin, I.: Orchestration of ultra-dense 5G networks. IEEE Commun. Mag. **56**(8), 68–69 (2018)
9. Kotulski, Z., Nowak, T.W., Sepczuk, M., Tunia, M., Artych, R., Bocianiak, K., Osko, T., Wary, J.-P.: Towards constructive approach to end-to-end slice isolation in 5G networks. EURASIP J. Inf. Secur. (2018)
10. IETF Draft: Network Slicing Management and Orchestration (2017). https://tools.ietf.org/html/draft-flinck-slicing-management-00
11. 3rd Generation Partnership Project (3GPP) TS 28.531 - Management and orchestration; Provisioning; Release 15, September 2018
12. 3GPP Management and orchestration; Generic management services TS 28.532 - Management and orchestration; Generic management services; Release 15, September 2018
13. Tovinger, T.: 3GPP SA5 chairman, management, orchestration and charging for 5G networks. http://www.3gpp.org/NEWS-EVENTS/3GPP-NEWS/1951-SA5_5G
14. Marcus, C., Menasc, D.A., Brasileiro, F.: Capacity planning for IaaS cloud providers offering multiple service classes. Future Gener. Comput. Syst. **77**, 97–111 (2017)
15. Khazaei, H., Mii, J., Mii, V.B.: Performance analysis of cloud computing centers using M/G/m/m+r queuing systems. IEEE Trans. Parallel Distrib. Syst. (TPDS) **14**(5), 936–943 (2012)
16. Halabian, H., Ashwood-Smith, P.: Capacity planning for 5G packet-based fronthaul. In: Wireless Communications and Networking Conference (WCNC), pp. 1–6, April 2018
17. Bega, D., Gramaglia, M., Banchs, A., Sciancalepore, V., Samdanis, K., Costa-Perez, X.: Optimising 5G infrastructure markets: the business of network slicing. In: INFOCOM 2017-IEEE Conference on Computer Communications, pp. 1–9, May 2017
18. MacGregor, S.J.: Optimal design and performance modelling of M/G/1/K queueing systems. Math. Comput. Model. **39**(9–10), 1049–1081 (2004)
19. Ward, W.: A diffusion approximation for the G/GI/n/m queue. Oper. Res. **52**(6), 922–941 (2004)
20. Mininet: An instant virtual network on your laptop. http://mininet.org/

Design and Experimental Validation of SFC Monitoring Approach with Minimal Agent Deployment

Jisoo Lee, Syed M. Raza, Rajesh Challa, Jaeyeop Jeong,
and Hyunseung Choo$^{(\boxtimes)}$

College of Software, Sungkyunkwan University, Suwon, South Korea
{jisoo49, s.moh.raza, rajesh.c, jaeyeop, choo}@skku.edu

Abstract. Service Function Chaining (SFC) is a new flexible network service deployment model to efficiently address the overwhelming increase in demand for new services. SFC consists of dynamically provisioned softwarized service functions (SFs), which are logically chained together to deliver a particular service. Software Defined Networking (SDN) simplifies the control and management of SFCs by centralizing the control plane, as it manages the SF links and controls the service flow traffic. Various critical functions like load balancing, fault management, and congestion avoidance in SFC are dependent on effective monitoring system. However, conventional monitoring approaches have high signaling cost due to the deployment of Monitoring Agents (MAs) in all SFs. In this paper, we present an SFC monitoring approach that reduces the signaling cost by deploying MAs in minimum number of SFs. We propose an SF selection algorithm that identifies the faulty SF using an optimized set of SFs to deploy the MAs. We conduct the testbed experiments to evaluate the effectiveness of our approach. The results show that our approach reduces the signaling cost by 59.2% compared with the conventional one. We further present the effect of various thresholds and data rates on the proposed SFC monitoring approach.

Keywords: Service Function Chaining (SFC) ·
Software Defined Networking (SDN) · SFC monitoring ·
Agent-based monitoring

1 Introduction

Network services offered to internet users are statically configured according to the network operator's specific requirements and policies [1]. These include security components such as firewall and intrusion detection/prevention system (IDS/IPS), network optimization tools such as load balancer (LB) and TCP optimizer, etc. Network service management is becoming increasingly complicated due to the exponential increase in traffic. In such scenario, softwarized Service Functions (SFs) provide a flexible and dynamic model for service provisioning and configuration [2]. These SFs are chained together in a specific order to realize a service and this is known as Service Function Chaining (SFC) [3]. Each traffic is assigned a service function chain, which is a sequence of network services (also called service functions) that the traffic needs to

© Springer Nature Switzerland AG 2019
S. Lee et al. (Eds.): IMCOM 2019, AISC 935, pp. 153–167, 2019.
https://doi.org/10.1007/978-3-030-19063-7_14

pass through before exiting the network. SFC enables the network operator to guarantee elastic service management.

SF is placed at a specific position in the Service Function Path (SFP), which is the traffic transmission path. Software Defined Networking (SDN) plays important role in SFC, as it provides the control through its control and data plane separation. An SDN controller in SDN based SFC (SD-SFC) [4] determines an ordered list of virtual and physical SFs in SFC for traffic steering [5]. The performance of the SFC depends on the performance of every related SF (i.e., if a single SF is faulty or overloaded, it will affect the entire SFC and may result in SLA violation). Therefore, in the SD-SFC, a monitoring system is required to dynamically change the SFP based on the performance of SF. In general, the monitoring approach is divided into two categories: agent-less monitoring and agent-based monitoring [6]. The agent-less monitoring approach gathers statistics for a particular traffic flow through an interface or open protocol. In the agent-based monitoring approach, an active probe called agent is deployed on the monitoring object to collect the monitoring information.

The agent-less approach is relatively lightweight and easy to implement because it only obtains flow information through polling. This approach focuses on flow bandwidth and network congestion, but fails to provide insight of SFs hosting server, which is usually a commodity server shared by multiple SFs. In contrast, the agent-based approach can measure and collect various information about memory, CPU, and network in real time. Therefore, in the SD-SFC where immediate monitoring and status awareness are required for dynamic SFC management, the agent-based approach is more advantageous. However, this approach must take into account the cost of signaling and deploying the agents, as normally Monitoring Agents (MAs) are deployed per SF.

In this paper, we propose agent based cost-efficient SFC monitoring approach through minimal agent deployment without compromising on monitoring level. The proposed overloaded SF selection algorithm identifies the possibility of SF overload in the SFC by taking the difference between the service rate of the ingress SF and the arrival rate of the egress SF, and comparing it with the predefined threshold. Location and identity of overloaded SF is then determined through recursive search. We conduct the experiments through testbed implementation to verify our analysis and evaluate the effectiveness of the proposed approach. We show that our proposed approach reduces the signaling cost by 59.2% compared to the conventional method that deploys agents in all SFs. Moreover, we analyze the effect of various thresholds and traffic rates on the total delay.

The rest of this paper consists of the following parts. Section 2 provides overview of agent-based Zabbix open source monitoring solution and the existing research in SDN environment. Section 3 presents the design of the proposed solution, its major components, and overloaded SF selection algorithm. Section 4 describes the testbed implementation for evaluating the performance of the proposed approach and analyses of the experimental results in terms of the signaling cost and total delay. Finally, Sect. 5 presents conclusions and future work.

2 Background

2.1 Open Source Monitoring Solution: Zabbix

Zabbix is a highly scalable open source monitoring solution. It can monitor and collect numerous real-time metrics from tens of thousands of servers, Virtual Machines (VMs), and network devices [7]. Zabbix supports monitoring at various layers such as Infrastructure-level, Cloud environment-level, VM-level, and Service-level. Infrastructure-level measures the status of resources that serve as infrastructure, for example, CPU, memory, and disk usage of Physical Machines (PMs). Cloud environment-level measures the state of the cloud environment and the number of virtual machines deployed in the cloud environment. VM-level measures the state of a VM, for example, computing resources, CPU, memory, and storage usage of VM. Service-level measures performance and status information for network services.

As shown in Fig. 1, Zabbix architecture is largely composed of Zabbix agent and Zabbix server [8]. Zabbix agents (i.e., active probes) are deployed on the monitoring objects. The agents measure the monitoring information from the objects and send it to the Zabbix server. The server is a central component that determines availability and integrity of object through analysis of information gathered after interaction with agent. It also alerts the administrator if there is a problem with the monitoring object's performance. All configuration information and data received from the agents are stored in Zabbix DB. Zabbix proxy is used to collect data on behalf of the server in order to reduce the number of CPU and memory I/O operations of the server. Zabbix front-end is a web-based interface that allows the administrator to easily access the Zabbix server. It provides real-time dashboard based on monitoring data. Zabbix offers event-based trigger function that evaluates the collected monitoring data and indicates the problem in the monitored objects. It also provides an action function that performs some operations as result of event-based trigger. The administrator can use these functions and quickly respond when the problem occurs with the monitored objects. In this paper, we implement the overloaded SF selection algorithm using Zabbix's trigger and action functions.

2.2 Related Work

The agent-less monitoring approach uses open protocols or interfaces like OpenFlow [9]. It is lightweight and easy to implement because of using SDN's flow statistics. However, it increases the overhead in the controller and affects the accuracy of the measurement [10]. OpenTM [11] intelligently selects a switch to obtain flow statistics using routing information obtained from the OpenFlow controller. It increases the communication cost of each flow according to the number of active flows. FlowSense [12] is a push-based approach to monitor performance change by receiving notifications that convey flow statistics on the switch. It cannot guarantee the accuracy of the monitoring data under certain flow conditions and meet real-time requirements.

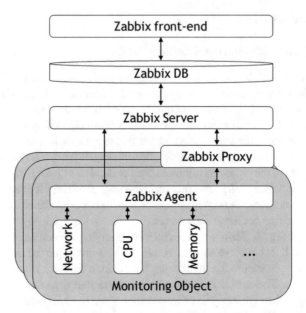

Fig. 1. Zabbix architecture

The agent-based monitoring approach enables collection and measurement of various information about memory, CPU, and network in real time. Therefore, in the SD-SFC environment, the agent-based approach is more advantageous as it provides warning and reporting through fine-grained performance monitoring. Traditional agent-based approaches such as Nagios [13], Opsview [14], and Zabbix [7] support various monitoring metrics. Since they perform monitoring without injecting additional traffic, they do not generate unnecessary network overhead [15]. However, these approaches have generally high signaling costs due to the deployment of the agents on all objects. To the best of our knowledge, none of the previous work gives the efficient monitoring solutions in terms of the signaling cost.

Zabbix does not limit the number of the monitoring objects and metrics. OpenBaton [16], an open source of Network Function Virtualization (NFV) Management and Orchestration (MANO) platform [17], uses Zabbix plugin based standard interface to receive notifications from the Zabbix Server. It uses the monitoring information to extend some other services (e.g., an Autoscaling Engine (ASE) and Fault Management System (FMS)) [18–20]. ASE compares the CPU idle time of the single SF with the threshold to automatically perform the scaling action [18], and FMS reroutes traffic to another path if there is a problem with a specific SF instance in the SFC [19]. X-FORCE [20] is an OpenBaton project which provides extensible framework for flexible management of SFCs in 5G networks. However, they have a high signaling cost because the MAs must be deployed in every SF running in a large SFC environment.

3 Proposed System

This section describes the design and interactions between major components of the proposed monitoring approach. The proposed system monitors the SD-SFC with minimal monitoring MA deployment to optimize the signaling cost. Although the proposed monitoring system reduces the number of deployed agents, no compromise is being made on the monitoring information, which is required to make critical decisions. The proposed approach is implemented in three steps: (1) Overload detection in the SFC, (2) overloaded SF identification, and (3) traffic rerouting to bypass overloaded SF.

3.1 System Design

Figure 2 shows the system design that integrates the monitoring system into the SD-SFC architecture. According to SFC internet-draft [21], the SD-SFC is consisted of SFC control plane and SFC data plane [20]. SFC control plane includes an SDN controller and the monitoring server with three modules: SF overload detection module, SF selection module, and Dynamic MA deployment module. The SFC control plan uses current state of the SF to configure the SFP according to service demand and manages Service Function Forwarder (SFF) policy. SFC data plane includes the SFC classifier, SFF, and SF.

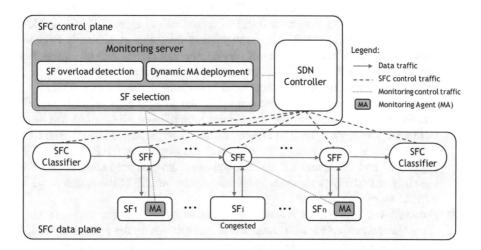

Fig. 2. System design

SDN Controller: In SD-SFC, it responsible for constructing SFPs according to each SFC by inserting flow rules into SFC data plane components (i.e., SFC classifier, SFF, and SF). If an SF encounters any problem or need to be scaled out, it updates the SFP that includes this SF. When the network traffic comes in, SFC classifier identifies each network service and assigns the information of the specific SFF and SFP. SFF transfers the traffic to the SF or other SFF based on the policies of SFP. SFs perform different middle box functionalities (e.g., Firewall, Deep Packet Inspection (DPI), and Network Address Translation (NAT)), and then return traffic to the previous SFF.

The SDN controller is informed about fault in SF by receiving an alarm from the monitoring server in SFC control plane. After the backup SF is instantiated and the failed SF is removed, the SDN controller updates the SFP by rerouting the traffic from faulty SF to backup SF and responses to the SFC classifier and SFF. The controller completes the SFP update process and notifies the monitoring server to remove the MA of the faulty SF.

Monitoring Server: It collects monitoring information from the MAs deployed in SFs in SFC and informs the SDN controller when a problem occurs. Generally, MAs are deployed in all SFs, which allow monitoring server to detect overloaded SF promptly. However, proposed approach in this paper argues that overloaded SF in SFC can be detected by initially deploying only two MAs at ingress and egress SFs, respectively. Ingress and egress MAs provide the service rate and arrival rate of ingress and egress SFs to monitoring server, respectively. The service rate is measured by the statistics of outgoing traffic per second and the arrival rate is measured by the statistics of incoming traffic per second. The monitoring server includes three modules to implement the overloaded SF selection algorithm.

(1) *SF overload detection module:* It determines whether there is an overloaded SF, which is the first step of the proposed approach. The service flow passes through the ordered SFs in turn according to SFP. The service rate/throughput of ingress SF is always same or greater than that of arrival rate of egress SF. If a problem occurs in any SF in the SFC, the throughput of this SF is significantly reduced. This module takes the difference of service rate and arrival rate of ingress and egress SFs, respectively. If the difference is greater than the threshold, it is determined that an overload has occurred at some SF in the current SFC.

(2) *SF selection module:* Once the overload in the SFC is detected, this module searches the overloaded SF. We use a binary search-based SF selection algorithm to find the overloaded SF. The number of search target SFs is halved with each search iteration, so the probability of finding the overloaded SF is doubled. The ingress SF and the egress SF are changed depending on the search range. The search is recursively performed, and finally, when one SF in the search range is left, the search is terminated.

(3) *Dynamic MA deployment module:* It dynamically deploys and removes MAs when the monitoring for additional SFs is required during the process of finding the overloaded SF. The operation is performed when the predefined trigger condition is satisfied, that is, when the problem occurs. The problem state has two cases. The first case is when an overload occurs in the SF where the MA is already deployed. In this case, since an overloaded SF is found, no additional MA deployment is required. The second case means that there is an overloaded SF between the ingress SF and the egress SF being monitored. In this case, we must deploy additional MAs. Based on the binary search-based SF selection algorithm, this module deploys the MAs in the SF located in the middle of the search range.

3.2   Overloaded SF Selection

This algorithm is proposed for detecting the overloaded SF by deploying the minimum number of agents in selected SFs. We exploit binary search that iteratively discards half of the search range for reducing complexity and performs until finding the overloaded SF. Since the binary search is performed on an ordered set, our algorithm applies it to the service rate set and arrival rate set sorted in descending order. In this algorithm, the next range of the iteration can be decided according to change in traffic rate. It aims to reduce the agent deployment cost and signaling cost through dynamic on demand agent deployment.

Algorithm 1. Overloaded SF selection algorithm

Input: μ_i, λ_i, $threshold$ α, S
Output: sf_x
    $Initialisation : s = 1$, $e = n$, $mid = 0$, $x = 0$,
    $A = \{sf_1, sf_n\}$
1: **function** SelectSF(s, e)
2: **if** $e - s > 1$ **then**
3:      **if** $|\mu_s - \lambda_e| > \alpha$ **then**
4:         $mid = \lfloor (s + e)/2 \rfloor$
5:         $A = A + \{sf_{mid}\}$
6:         **if** $|\mu_{mid} - \lambda_{mid}| > \alpha$ **then**
7:            $x = mid$
8:      **else**
9:         **if** $|\mu_s - \lambda_{mid}| > \alpha$ **then**
10:            **return** SelectSF(s, mid)
11:         **else**
12:            **return** SelectSF(mid, e)
13:         **end if**
14:      **end if**
15:    **end if**
16: **end if**
17: **if** $(1 < x < n)$ **return** sf_x
18: **else return** null

Algorithm 1 shows the pseudo code for overloaded SF selection algorithm for cost-efficient SFC monitoring approach. An SFC topology is denoted by $S = \{sf_i | i \in 1, 2, .., n\}$, where S is the set of SFs. Let $A = \{sf_k | k \leq n, sf_k \in S\}$ be a subset of S in which the monitoring agents are deployed. We first initialize the set A to have only sf_1 and sf_n elements. $M = \{\mu_i | i \in 1, 2, \ldots, n\}$ represents the set of service rates, where μ_i denotes throughput of sf_i. $\wedge = \{\lambda_i | i \in 1, 2, \ldots, n\}$ represents the set of arrival rates, where λ_i denotes per second arrival rate at sf_i. We use the service rate μ_s of the ingress SF $sf_s \in S$ and the arrival rate λ_e of the egress SF $sf_e \in S (e > s)$. If $\mu_s - \lambda_e$ is bigger than the congestion threshold (α), then it implies that some SF in the path is overloaded. The time complexity of SF selection algorithm is $O(\log n)$.

As shown in Algorithm 1, we use SF's service rate, arrival rate, and the threshold α as input. Also, the set S is used to include the SF elements in the set A. We get the overloaded SF (i.e., sf_x) as the output. The iterative procedure of this algorithm keeps track of the search boundaries with the two variables s and e. The s variable is the index value of first SF in the search range and is initialized with 1, and the e variable indicates the index value of last SF in the search range and it is initialized with n. The mid variable indicating the middle SF in search range and x variable indicating the over-loaded SF, are initialized with 0. The set A is initially included only two elements: sf_1 and sf_n. First, it is checked if the search range includes more than one SF, and it is done by taking the difference of e and s. If the difference is bigger than 1 the algorithm proceeds to the next step.

If there is no SF between sf_s and sf_e, the search has to terminated (line 2). If the difference between the service rate of sf_s and the arrival rate of sf_e exceeds the threshold, it is necessary to recalculate the mid variable. Also, sf_{mid} is included in set A and the agent is deployed (lines 3 to 5). If the difference between the arrival rate of sf_{mid} and the service rate of sf_{mid} exceeds the threshold, it means that the overload occurs at sf_{mid} (lines 6 to 7). Otherwise, the algorithm continues the binary search for the next range. If the difference between the service rate of sf_s and the arrival rate of sf_{mid} exceeds the threshold, the algorithm calls recursive function with s and mid values as parameters (lines 9 to 10). Otherwise, it calls the recursive function with mid and e values as parameters (line 12). These parameters determine the starting and ending points of next search range. If the overloaded SF exists, it is returned as sf_x, otherwise, the null value is returned (lines 17 to 18).

4   Performance Evaluation

4.1   Testbed Implementation

We implemented the experimental environment as shown in Fig. 3 to evaluate the performance of the proposed SFC monitoring approach. The test topology consists of one Linux server and eight Raspberry-Pi 3 boards which serve both as SFF and host for SF. All the Pi boards are connected in a serial topology through physical links, with host A and B at either end. Each Pi board runs Open VSwitch (OVS) 2.5.2 (i.e., an OpenFlow enabled switch) for controlling network flow, and Zabbix agent to measure monitoring information. Through direct link Pi boards communicate with Linux server which contains the open source monitoring solution Zabbix 3.4 and SDN control platform called ONOS [23] as an SDN controller. In order to minimize the processing delays we equip Linux server with 16G RAM and Intel (R) CPU 3.30 GHz 4 core processor. Flow traffic from host A to B is generated through Iperf [22].

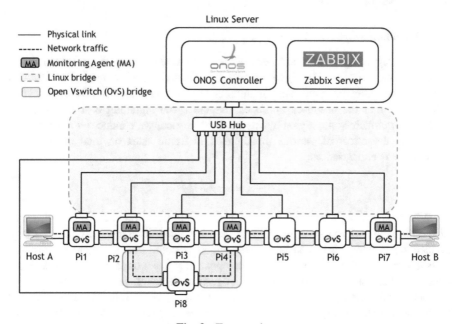

Fig. 3. Test topology

The experiment uses two bridges, Linux bridge for monitoring and OVS bridge for transporting network traffic. Linux provides kernel functionality that allows itself to act as a network bridge. The Linux bridge connects the Linux server's Ethernet interface with eight Pi board's Ethernet interfaces. This bridge enables interaction between the Zabbix server and the Zabbix agents. SDN controller adds the flow rules in OVS bridge of each Pi board for the flow traffic to route from host A to host B and back. Flow rules use IPs of the source Host A and the destination Host B to match and route the traffic. We configure hosts (i.e., SF hosts) information and items in the Zabbix server to activate the Zabbix agent and perform monitoring. The configured information consists of IP address and port of the Pi board where the Zabbix agent is installed. The items configure '*net.if.in [if]*', which is the statistics of arrival rate at the network interface *[if]*, and '*net.if.out [if]*', which is the throughput from outgoing network interface *[if]*.

We have implemented a scenario where the overload occurs in one of the seven SFs to demonstrate the reliability and cost-effectiveness of the proposed monitoring approach. When traffic is transmitted from Host A, the traffic passes through all the SFs in the SFC and is terminated at Host B. Therefore, if any one of the SFs get overloaded, traffic throughput decreases. We manipulate the output port bandwidth in one of the Pi boards to create an effect similar to network congestion. We have configured two types of triggers and actions in Zabbix server to implement SF selection algorithm. The first type of trigger performs an action to remove the agents placed in the SFs except the ingress and egress SF. The second type performs an action to deploy a new agent when there is an overloaded SF.

4.2   Results Analysis

We use the signaling cost for MAs deployment and total delay on the proposed system as performance metrics. The signaling cost is evaluated from the number of times the Zabbix server polls the agents and the number of active agents during the total delay τ. We confirm that the proposed approach has the lower signaling cost than the conventional approach of deploying agents in all SFs. Also, we conduct two experiments to observe the effect of various thresholds and traffic rates on total delay τ. The threshold α is expressed as:

$$\alpha = (|\mu_1 - \lambda_n|)/\beta \tag{3}$$

Where μ_1 and λ_n represent the service rate and arrival rate at ingress and egress SFs, before the congestion occurs. The threshold α is the difference between μ_1 and λ_n divided by β which is a constant value which is controlled to determine different threshold values. The optimal value of the constant β has a tradeoff between detection delay and false positive. The delay of congestion detection process is expressed as x which is affected by the bandwidth and threshold. The SF selection algorithm performs the search for $\log n$ times when there are n SFs, and y is defined as the computation time for the algorithm. Therefore, SF selection process has the delay of $y \times (\log n)$. The delay of congestion recovery process is defined as z. As a result, the total delay τ, consisting of congestion detection, overloaded SF selection, and congestion recovery process, is expressed as:

$$\tau = x + y \times (\log n) + z \tag{4}$$

We compare the signaling cost with the conventional monitoring approach through two indicators. First, Fig. 4 shows the impact of sampling rate at the number of control packets before the congestion. The x-axis represents the sampling rate, (i.e., number of times monitoring data is updated per second). The y-axis represents the number of packets (i.e., number of times the Zabbix server polls the data from the agents). Therefore, the increase in y value is proportional to the sampling rate regardless of the number of SFs. The proposed approach with only two agents has fewer packets than the conventional approach with seven agents. This result shows that the proposed approach reduces the control signaling cost by 71.4% compared to the signaling cost of the conventional approach. The result shown in Fig. 4 is only for one SFC, however, in real systems where hundreds of SFCs exist at any instance the impact of proposed monitoring approach will be much higher.

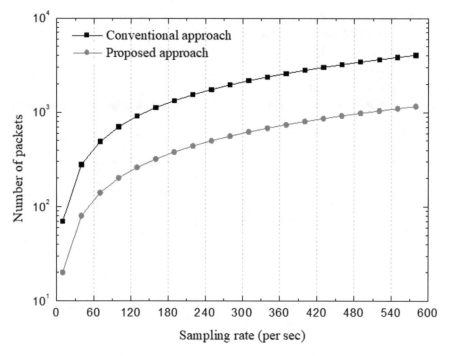

Fig. 4. The comparison of signaling cost against sampling rate

Figure 5 shows the comparison of number of active agents when $\alpha = 1.6$ Mbps, 2.5 Mbps, and 5 Mbps (i.e., $\beta = 6, 4,$ and 2) with the conventional approach. The x-axis represents total time including the entire process from congestion occurrence to recovery, and the y-axis represents the number of active agents. The proposed approach detects the congestion and deploys agents dynamically by using SF selection algorithm. If an overloaded SF is found, then the monitoring agents deployed at the SFs are removed. We could see that the lower the threshold, the faster deployment of the monitoring agent and the problem detection. For the given SFC of length seven, the maximum number of active agents in the proposed approach is five, which is still smaller than seven in the conventional approach. The result implies that the proposed monitoring approach reduces the signaling cost by 28.6% even when the agents are maximally deployed. On average, there is 59.2% reduction is signaling cost. As a result, it is worth noting that in the entire process, our proposed approach is more cost effective than the conventional approach.

Figure 6 shows the effect on total delay τ when $\alpha = 1.6$ Mbps, 2.5 Mbps, and 5 Mbps (i.e., $\beta = 6, 4,$ and 2). We transmit the traffic with the rate of 20 Mbps from Host A to B for 150 s (i.e., $\mu_1 = 20$ Mbps). After 30 s, we change the bandwidth of the third Pi board to 10 Mbps for creating the congestion (i.e., $\mu_3 = 10$ Mbps). We consider the time when the bandwidth returns to the initial value as the time when the SF is recovered. When α is 1.6 Mbps, we could see that the congestion recovery time is 75 s

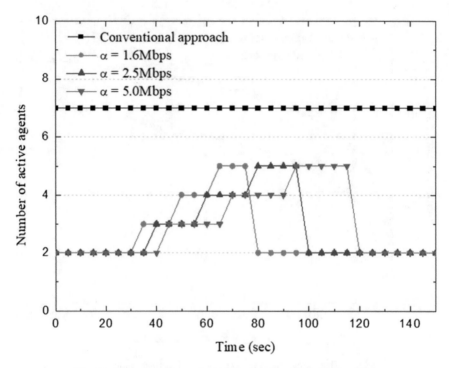

Fig. 5. The comparison of number of active agents against time

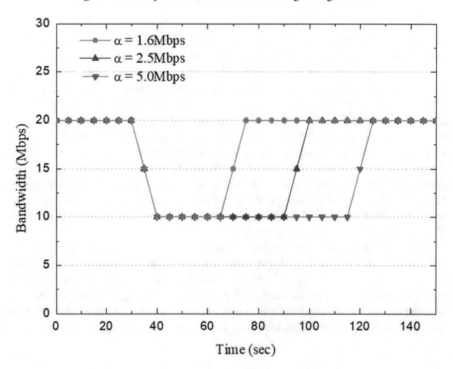

Fig. 6. The effects of varying thresholds on total delay τ

and τ is 45 s. When α is 2.5 Mbps, the congestion recovery time is 100 s. (i.e., 25 s increment) and τ is 70 s, and for α as 5 Mbps the congestion recovery time is 125 s (i.e., 25 s increment) and τ is 95 s. This variation in total delay is because congestion detection time goes up with increased in threshold values. When the threshold is too low, the delay is relatively short. However, there is a possibility that congestion in SF is misjudged because of small bandwidth fluctuations in network. Nevertheless, when the threshold is high, it is possible to determine the exact problem at the expense of longer delay.

Figure 7 shows the effect on total delay τ when $\mu_1 = 20$ Mbps, 40 Mbps, and 60 Mbps. As in the first experiment, we transmit the traffic for 150 s and change the bandwidth of the third Pi board to 10 Mbps (i.e., $\mu_3 = 10$ Mbps). Value of threshold is dependent on arrival rate, therefore, for different arrival rates we get values of α as: 2.5 Mbps, 7.5 Mbps, and 12.5 Mbps when the value of β is 4. When μ_1 is 20 Mbps, we could find that the congestion recovery time is 100 s and τ is 70 s. When μ_1 is 40 Mbps, the congestion recovery time is 115 s (i.e., 15 s increment) and τ is 85 s, and for μ_1 as 60 Mbps the congestion recovery time is 135 s (i.e., 20 s increment) and τ is 105 s. The difference in τ occurs because the threshold increases proportionately with the initial traffic rate (i.e., Congestion detection time increases). Therefore, an appropriate threshold needs to be selected according to the network environment with various rates.

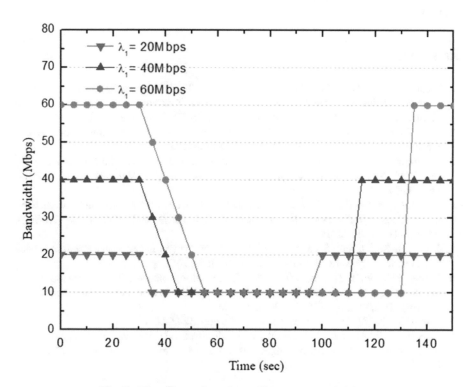

Fig. 7. The effects of varying traffic rates on total delay τ

5   Conclusion

In this paper, we propose a cost-effective SFC monitoring system with minimum agent deployment, which provides same level of monitoring as the conventional approach. We implemented the binary search based SF selection algorithm to reduce the control signaling overhead. We carried out the testbed implementation using open source Zabbix monitoring system and ONOS controller. Performance evaluation is performed in terms of the signaling cost and the total delay. The experimental results show that proposed monitoring system outperforms the conventional approach that deploys agents in all SFs. Comparing with the conventional approach, the signaling overhead is reduced by 59.2%. Furthermore, we verified the effect of different threshold and data rate values on the total delay. It has been identified that the appropriate value of threshold is dependent on tradeoff between delay and accuracy in congestion detection, and it must be selected as per operator policies. Our future work is also on determining the appropriate threshold range while considering various factors that effect it. We also plan to extend our algorithm and testbed for cases when there are multiple congested SFs at any given instance in the SFC.

Acknowledgment. This research was supported in part by PRCP (NRF-2010-0020210) through NRF, G-ITRC support program (IITP-2017-2015-0-00742), Information & communications Technology Promotion(IITP) grant funded by the Korea government(MSIT) (2014-3-00547), respectively.

References

1. Medhat, A.M., Pauls, M., Corici, M., Magedanz, T.: Resilient orchestration of service functions chains in a NFV environment. In: IEEE Conference on Network Function Virtualization and Software Defined Networks (NFV-SDN), pp. 7–12 (2016)
2. ONF White Paper. ONF TS-027: L4-L7 Service Function Chaining Solution Architecture. https://www.opennetworking.org/. Accessed 14 June 2015
3. Eichelberger, R.A.: SFC path tracer: a troubleshooting tool for service function chaining. In: IFIP/IEEE Symposium on Integrated Network and Service Management (IM) 2017, pp. 568–571 (2017)
4. Vu, A.-V., YoungHan, K.: An implementation of hierarchical service function chaining using OpenDaylight platform. In: NetSoft Conference and Workshops (NetSoft), pp. 411–416. IEEE (2019)
5. J. Lee, H. Ko, D. Suh, S. Jang, and S. Pack.: Overload and failure management in service function chaining. Network Softwarization (NetSoft), IEEE. pp. 1–5 (2017)
6. Wang, M.H., Wu, S.Y., Yen, L.H., Tseng, C.C.: PathMon: path-specific traffic monitoring in OpenFlow-enabled networks. In: Eighth International Conference on Ubiquitous and Future Networks (ICUFN), pp. 775–780. IEEE (2016)
7. Wikipedia (2018). Zabbix. https://en.wikipedia.org/wiki/Zabbix
8. Fatema, K., Emeakaroha, V.C., Healy, P.D., Morrison, J.P.: A survey of cloud monitoring tools: taxonomy, capabilities and objectives. J. Parallel Distrib. Comput. **74**(10), 2918–2933 (2014)

9. McKeown, N., Anderson, T., Balakrishnan, H., Parulkar, G., Peterson, L., Rexford, J., Shenker, S., Turner, J.: Openflow: enabling innovation in campus networks. ACM SIGCOMM Comput. Commun. Rev. **38**(2), 69–74 (2008)
10. Shokhin, Anatolii.: Network monitoring with Zabbix (2015)
11. Tootoonchian, A., Ghobadi, M., Ganjali, Y.: OpenTM: traffic matrix estimator for OpenFlow networks. In: International Conference on Passive and Active Network Measurement, pp. 201–210. Springer, Heidelberg (2010)
12. Yu, C., Lumezanu, C., Zhang, Y., Singh, V., Jiang, G., Madhyastha, H.V.: FlowSense: monitoring network utilization with zero measurement cost. In: International Conference on Passive and Active Network Measurement, pp. 31–41. Springer, Heidelberg (2013)
13. Barth, W.: Nagios: System and Network Monitoring (2018)
14. Križanić, J., Grgurić, A., Mošmondor, M., Lazarevski, P.: Load testing and performance monitoring tools in use with AJAX based web applications. In: MIPRO, Proceedings of the 33rd International Convention, IEEE, pp. 428–434 (2010)
15. Zaalouk, A., Khondoker, A., Marx, R., Bayarou, K.: OrchSec: an orchestrator-based architecture for enhancing network-security using network monitoring and SDN control functions. In: Network Operations and Management Symposium (NOMS), pp. 1–9 (2014)
16. Open Baton. http://www.openbaton.github.io/
17. ETSI NFV. Network Function Virtualization Management and Orchestration, ETSI GS NFV-MAN 001 V1.1.1 (2014-12). http://www.etsi.org/
18. Carella, G.A., Pauls, M., Grebe, L., Magedanz, T.: An extensible autoscaling engine (AE) for software-based network functions. In: IEEE Conference on Network Function Virtualization and Software Defined Networks (NFV-SDN), pp. 219–225 (2016)
19. Medhat, A.M., Carella, G.A., Pauls, M., Magedanz, T.: Orchestrating scalable service function chains in a NFV environment. In: IEEE Conference on Network Softwarization (NetSoft), pp. 1–5 (2017)
20. Medhat, A.M., Carella, G.A., Pauls, M., Magedanz, T.: Extensible framework for elastic orchestration of service function chains in 5G networks. In: Network Function Virtualization and Software Defined Networks (NFV-SDN), pp. 327–333 (2017)
21. Boucadair, M.: Service function chaining (SFC) control plane components and requirement draft-ietf-sfc-control-plane-06 (2016)
22. Tirumala, A., Qin, F., Dugan, J., Ferguson, J., Gibbs, K.: iPerf: the TCP/UDP bandwidth measurement tool (2015). http://iperf.sourceforge.net
23. Berde, P., Gerola, M., Hart, J., Higuchi, Y., Kobayshi, M.: ONOS: towards an open, distributed SDN OS. In: Proceedings of the Third Workshop on Hot Topics in Software Defined Networking, pp. 1–6. ACM (2014)

Split Point Selection Based on Candidate Attachment Points for Mobility Management in SDN

Bora Kim, Syed M. Raza, Rajesh Challa, Jongkwon Jang,
and Hyunseung Choo[✉]

Sungkyunkwan University, Seobu-ro, Suwon 2066, Korea
{bora007, s.moh.raza, rajesh.c, jangjk, choo}@skku.edu

Abstract. Seamless mobility is essential in future wireless networks to support multimedia rich realtime services. Software Defined Networking (SDN) is a new networking paradigm which can provide fine grained flow level mobility management in wireless networks through centralized controller, but suffers from intolerable handover delay. Split point approach is an effective way to reduce handover and end-to-end transmission delay in SDN wireless networks. The split point is a switch on the existing flow path from where traffic is rerouted towards new Attachment Point (AP) after handover. This paper presents a split point selection algorithm which utilizes a list of candidate APs for next handover of Mobile Node (MN) to determine topologically most efficient split point in terms of handover delay. Proposed algorithm calculates the weight of each switch in the MN-Corresponding Node (CN) path as the average distance (hop) between a switch and the candidate APs, and the switch with the minimum weight is selected as the split point. In addition to split point selection, this paper exploits the control and data plane separation provided by SDN to restore optimal path for a flow after the handover. The numerical analysis of the proposed scheme shows 9.6% to 13% total cost improvement in comparison to previous solution.

Keywords: Software Defined Networking (SDN) · IP mobility · Split point · Wireless network

1 Introduction

Advancements in wireless technologies and mobile devices have caused an exponential growth in the usage of mobile multimedia services [1]. Availability of more and more multimedia content and services increases the users Quality of Service (QoS) demands. A seamless mobility solution is a key to provide users with realtime connectivity QoS, but faces the challenges like intolerable handover and transmission delay. Software Defined Networking (SDN) is a new paradigm where control plane is logically centralized after separation from data plane [2, 3]. Centralized control plane in SDN provides the flexibility which is required to manage the user mobility with minimal handover and transmission delays.

© Springer Nature Switzerland AG 2019
S. Lee et al. (Eds.): IMCOM 2019, AISC 935, pp. 168–180, 2019.
https://doi.org/10.1007/978-3-030-19063-7_15

Centralized SDN controller assigns every Mobile Node (MN) an IPv6 Home Network Prefix (HNP) upon its registration in SDN enabled wireless network. As MN moves from one subnet to another in wireless network, the controller establishes new shortest path from MN to Corresponding Node (CN) of each active flow on MN at that moment and reassigns the same HNP. Split point based approaches have been proposed to further reduce the handover delays in SDN mobility management [4–6]. Instead of establishing a completely new shortest path with every handover, flow traffic is redirected from a switch (i.e., split point) in the current path towards the new Attachment Point (AP) of MN. Improvement in handover delay is highly dependent on efficient selection of split point, which is a challenging task as various factors like topological position, load, and centrality are involved.

One of the simplest criteria to select a split point for a flow is based on the load of the switches [4]. At the time of initial MN attachment, a switch in flow path with least load is selected as a split point for entire duration of the flow, and is not updated with MN movement. This selection mechanism may increase handover and transmission delays as MN moves away from the initial AP. Keeping the current path of a flow reduces the handover delay (due to less signaling at controller) [5], but may result in a suboptimal path. Hence, split point selection can be seen as a tradeoff between signaling cost (at the controller) and optimal path of a flow. This solution focuses on signaling overhead (i.e., handover delay) at the expense of transmission delay.

Service aware split point selection (SASS) solution focuses on the flow type while selecting a split point [6]. It categorizes flows into two types: (i) packet loss sensitive, and (ii) packet delay sensitive. Different weights are assigned to the flows based on their types, and a split point for each flow is selected based on its weight. Current flow path is updated to an optimal path (i.e., shortest path) only when optimal path length plus threshold is greater than current path length. None of the aforementioned solutions factors in the MN movement in their algorithms, whereas split point selection is topologically dependent on next AP of MN. Also, all mentioned solutions compromise on transmission delay, whereas minimal end-to-end latency is one of the major requirements of future wireless and mobile networks [7].

One of the simplest criteria to select a split point for a flow is based on the load of the switches [4]. At the time of initial MN attachment, a switch in flow path with least load is selected as a split point for entire duration of the flow, and is not updated with MN movement. This selection mechanism may increase handover and transmission delays as MN moves away from the initial AP. Keeping the current path of a flow reduces the handover delay (due to less signaling at controller) [5], but may result in a suboptimal path. Hence, split point selection can be seen as a tradeoff between signaling cost (at the controller) and optimal path of a flow. This solution focuses on signaling overhead (i.e., handover delay) at the expense of transmission delay.

In this paper, we propose a Candidate AP based Split point Selection (CASS) mobility solution for minimizing handover delay in SDN. It tackles the transmission delay by updating the flow path after the completion of handover. This way optimal path for a flow is ensured without compromising on the handover delay. Major contributions of this paper are as follows:

- Architecture design and component description for CASS mobility solution in SDN
- An algorithm which utilizes the information about next possible APs of MN to topologically select split point with minimum signaling cost
- Analytical model of signaling cost (i.e., handover delay) and packet delivery cost (i.e., transmission delay) for proposed solution
- Numerical analysis of proposed solution showing 9.6% to 13% total cost improvement over SASS

CASS architecture, operation and algorithm are described in Sect. 2, whereas analytical model and numerical results are presented under Sect. 3. Finally, Sect. 4 discusses concluding remarks and future work.

2   Candidate AP Based Split Point Selection (CASS) Mobility Solution

2.1   CASS Architecture

This paper utilizes the split point technique for efficient mobility management in SDN, where split point for a flow is selected based on candidate APs (candidates) for MN's next handover. APs whose transmission range overlaps with current AP of MN are labeled as candidates. Weight of each SDN enabled switch in current flow path is calculated using list of candidates, and switch with minimum weight is selected as split point. The rationale behind this proposed CASS approach is the possibility that any of the candidate is the next AP of MN. Optimal split point for a flow in terms of handover delay can be selected if the identity of next AP is given, which is not realistic. Hence, the proposed CASS approach facilitates a near optimal handover regardless of which AP MN next moves to from the list of candidates.

The SDN wireless network with CASS mobility solution is shown in Fig. 1, where SDN controller has an outband connection with SDN enabled switches and APs. MN initially attaches with AP1 in Fig. 1(a) and initiates a flow with CN1. Movement detection module in mobility control detects MN attachment with AP by receiving a control message from AP, and a path is installed and split point is calculated for a flow using path setup and CASS algorithm modules, respectively. The given topology manager module of SDN controller stores the topology information of the network. We have extended this module to store the list of candidates for each AP and CASS algorithm module retrieves it to calculate the split point for a flow. Information of each flow generated at each MN is stored in information sever, which includes current path of a flow and its selected split point.

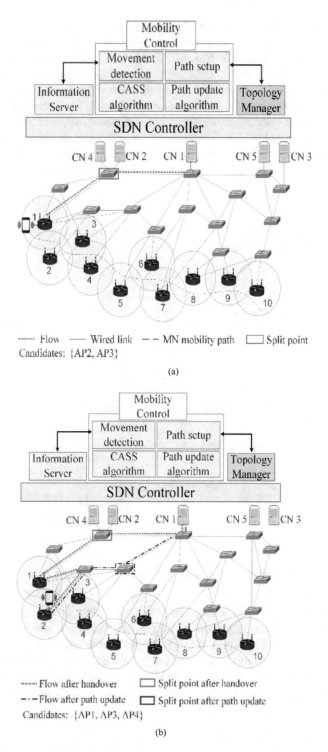

(a)

(b)

Fig. 1. Architecture of CASS mobility solution in SDN network: (a) Mobile node registration, (b) Mobile node handover.

As per MN mobility path shown in Fig. 1, MN next attaches to AP2, which is detected by movement detection module and triggers the path setup module. Path setup modules get the shortest path between AP2 and selected split point and install new path entries. Once the new path via split point is installed for a flow at MN after the handover, the path update module compares the current flow path (i.e., via split point) with shortest path in terms of hops. If the current flow path is not the shortest path, then new shortest path is installed, new split point is calculated, and traffic is switched on to new path, as shown in Fig. 1(b). Hence, the proposed CASS mobility solution reduces the handover latency through split point and provides optimal routing for minimum end-to-end latency at little cost of suboptimal routing for small period.

2.2 Operation

This subsection describes the control signaling procedure for first MN attachment, flow split point calculation, and handover process.

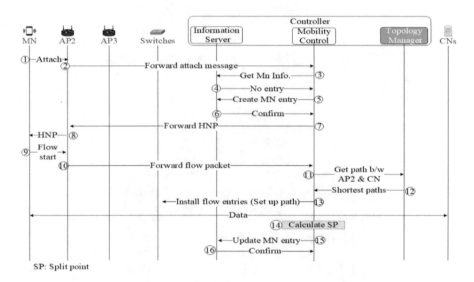

Fig. 2. Signaling call flow for MN registration and flow initiation.

As shown in Fig. 2, MN's first attachment with AP2 is forwarded to mobility control in SDN controller (step 1 and 2), where mobility control consists of movement detection, CASS algorithm, path setup and path update modules. Mobility control checks with information server for any existing entry of MN, and upon finding none it assigns and forwards HNP to MN after creating a new MN entry in information server (steps 3 to 8). Only after assignment of HNP, a flow can be initiated at MN. Upon flow

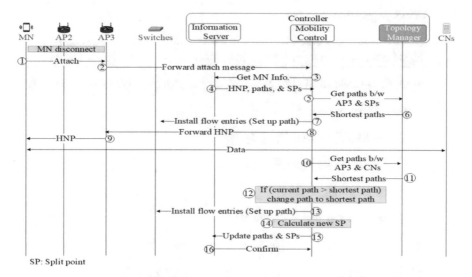

Fig. 3. Signaling call flow for MN handover.

arrival at AP2, the first packet is forwarded to the mobility control in SDN controller (steps 9 and 10). A shortest path is installed for the flow after communication with topology manager, and split point is calculated using CASS algorithm module (steps 11 to 15). Information regarding shortest path and split point of a flow are then updated in the information server.

Control signaling related to handover is shown in Fig. 3, where MN moves away from AP2 and decides to attach with AP3. This MN attachment is forwarded to SDN controller by AP3 (steps 1 and 2). Upon mobility control request, the information server provides current paths and split points of all flows which are active at MN (steps 3 and 4). Using this information, shortest paths between AP3 and split points is obtained and then installed through flow entries (steps 5 to 7). The flow traffic is routed on these new paths, after same HNP is forwarded to MN via AP3 (steps 8 and 9). Handover completes at step 9, however, the path update module gets the shortest path for all the flows and compares them with the newly installed flow paths which routes through the split points (steps 10 to 12). If the new path of the flow is greater than shortest path in terms of hops, only then shortest path is installed as a new path of the flow (step 13). Regardless of path update process, split point is recalculated for the new path of the flow (step 14). At the end of signaling process, all the new paths and split points of the flows are updated in the information server (steps 15 and 16).

2.3   CASS Algorithm

Algorithm I. Candidate AP based Split point Selection (CASS) algorithm

---

Input: P_i, Δ_i (candidate APs list), L_i, W_i //L_i, W_i: *load and weight of switches in* P_i

Output: $\overline{sw}_i$ (split point for flow i)

1:   **Procedure** *SplitPoint* $(P_i, \Delta_i, L_i, W_i)$

         /* Select split point for flow i */

2:   **Initialize** $\overline{sw}_i \leftarrow$ Null

3:   **for each** sw_j **do**   //Switch in P_i

4:         $w_j \leftarrow 0$ //Weight of switch $sw_j \in P_i$,

5:         **for each** a_{ik} **do**   //Candidate AP for flow i

6:               $w_j \leftarrow w_j +$ Num_of_hops (sw_j, a_{ik})

7:         $w_j \leftarrow w_j / |\Delta_i|$

8:         **if** $(\overline{sw}_i =$ Null) OR $(w_j < w_{\overline{sw}_i})$ **then**

9:               $\overline{sw}_i \leftarrow sw_j$

10:           **else if** $(w_j = w_{\overline{sw}_i})$ **then**

11:                 **if** $(l_j < l_{\overline{sw}_i})$ **then**

12:                       $\overline{sw}_i \leftarrow sw_j$

Let i be a flow and $S = \{sw_1, sw_2, \ldots, sw_N\}$ be the set of switches in a given network topology with respective weights $W = \{w_1, w_2, \ldots, w_N\}$. Assume, set $P_i = \{sw_1, sw_2, \ldots, sw_J\}, P_i \subset S$ denotes the current path of flow i, and set $L_i = \{l_1, l_2, \ldots, l_J\}, J \in P_i$ be the load of corresponding switches. The proposed CASS algorithm selects one of the switches in P_i as a split point for flow i. We assume that SDN controller has the load information of all switches through monitoring module, based on which it chooses a switch in case of a tie between the switch weights. If $A = \{1, 2, \ldots, X\}$ is the set of APs in the network then $\Delta_i = \{a_{ik}|k < X\}$ is the list of candidate APs for flow i. The sets P_i, L_i and Δ_i are the input for the CASS algorithm and are used to determine the split point $\overline{sw}_i \in P_i$.

CASS algorithm initializes the weights of all switches in P_i (lines 3 and 4). The weight w_j of switch $sw_j, j \in P_i$ is the average hop distance between itself and the candidates APs $a_{ik} \in \Delta_i$ (lines 5 to 7). The switch with minimum weight is then chosen as the split point n_i for flow i (lines 8 and 9). If two switches have same weights i.e., $w_j = w_{j'}|j \neq j'; w_j, w_{j'} \in P_i$ then the switch with lower load is selected (lines 10 to 12).

3 Performance Evaluation

We evaluate performance of the proposed CASS mobility solution in SDN based on the total cost and draw comparison against other solutions. In subsequent subsection we model the total cost for the proposed solution.

3.1 Analytical Modeling

Total cost (C_T) incurred by MN using the proposed CASS mobility solution is defined as the sum of total signaling cost (C_{SC}) and packet delivery cost (C_{DC}).

Signaling Cost: We assume that SDN network represents a single domain where each AP has a different subnet. Domain consists of total y subnets. If N is the number of subnet crossing performed by MN, then signaling cost is given by:

$$C_{SC} = E(N) \cdot E(U), \tag{1}$$

where $E(U)$ is the expected unit signaling cost for an MN. If F is the total number of flows at MN then number of flows at handover $\Gamma = \min\{F, \alpha F\}$ where $\alpha \in \mathbb{R}_{(0,1]}$ is the percentage of active flows at MN. Hence, $E(U)$ can be expressed as:

$$E(U) = \sum_{i=1}^{\Gamma} Q_i \cdot (H_i^s), \tag{2}$$

where Q_i is the signaling cost of flow i at MN, an H_i^sd is the distance function (number of hops) between the new AP $a \in A$ and split point $\overline{sw}_i \in S$. If t_f and t_s are the inter flow time and subnet sojourn time, respectively then, subnet crossing probability (P) per session is defined as [8]:

$$P = \Pr\left(t_f > t_s\right). \tag{3}$$

If Y is the random variable denoting number of subnet crossings, then its probability density function is given by:

$$\Pr(Y = y) = P_y \cdot (1 - P). \tag{4}$$

From (4) average number of subnet crossings are given as:

$$E(Y) = \sum_{y=0}^{\infty} y \cdot P_y \cdot (1 - P). \tag{5}$$

Let t_f and t_s follow the exponential distribution with rate λ_f and σ_s, respectively. Then, as per [8], $E(Y)$ is calculated as:

$$E(Y) = \frac{\sigma_s}{\lambda_f}. \tag{6}$$

Session-to-Mobility Ratio (SMR) [9] is defined as a ratio of session (flow) arrival rate (λ_f) to MN attachment period with an AP (σ_s), (i.e., $\frac{\lambda_f}{\sigma_s}$). From this $E(Y)$ can be express as $\frac{1}{SMR}$, and consequently signaling cost is defined as:

$$C_{SC} = \frac{1}{SMR} \sum_{i=1}^{\Gamma} Q_i \cdot H_i^s. \tag{7}$$

Packet Delivery Cost: CASS mobility solution performs the route optimization as shown in the Fig. 1. After handover, the packets belonging to flow i are routed from CN to MN on a path via $\overline{sw}_i$, and controller assumes it be a non-optimal path. Controller calculates new shortest path (optimal path) from CN to MN and compare it with the current path in terms of number of hops. In case, where current and new path are equal, controller declares the current path as optimal and performs no further operations. However, if new path is smaller, then controller installs the flow entries in the new path, directs the packets on it, and declares it the current path for i. Consequently, per flow packet delivery cost in CASS mobility solution can be expressed as:

$$R_i = \gamma_i \cdot G_i \cdot Z_i^n + (1 - \gamma_i) \cdot G_i \cdot Z_i^o, \tag{8}$$

where Z_i^n and Z_i^O express the packet delivery cost for flow i in case of non-optimal and optimal path, respectively. G_i is the average flow size during sojourn time in the AP, and γ_i is the ratio of packets belonging to flow i which route through non-optimal path before the optimal path is established. If d_i and δ_i denotes data rate and packet size of flow i, respectively, then we can express γ_i as:

$$\gamma_i = \begin{cases} \frac{d_i \cdot X_i \left(H_i^o, \Theta_S \right)}{\delta_i \cdot G_i}, & H_i^o < H_i^n, \\ 0, & otherwise, \end{cases} \tag{9}$$

where X_i is the time taken to establish the optimal path for a flow, which is a function of hops in the optimal path H_i^o and the processing time at the switches Θ_S. H_i^n is the number of hops in the non-optimal path.

Packet delivery cost includes the transmission delay through the network and processing delay at the network elements. Transmission delay is proportional to number of hops in the path, where processing delay is caused by flow table lookups and queuing. Therefore, Z_i^n and Z_i^o can be calculated as:

$$Z_i^n = H_i^n(\beta + \Theta_S) + \eta, \tag{10}$$

$$Z_i^o = H_i^o(\beta + \Theta_S) + \eta, \tag{11}$$

where β and η are the unit transmission cost over wired and wireless links, respectively. Therefore, the expected unit packet delivery cost $E(D)$ of MN can be give as:

$$E(D) = \sum_{i=1}^{\Gamma} R_i, \tag{12}$$

and total packet delivery cost (C_{DC}) is expressed as:

$$C_{DC} = \frac{1}{SMR} \sum_{i=1}^{\Gamma} R_i. \tag{13}$$

Table 1. Notation and values.

Parameter	Value	Parameter	Value
d_i	100–1000 Kbps	β	0.0075 ms
G_i	8,794–90,045 packets	η	15 ms
δ_i	1024 bytes	Θ_S	0.00228 ms

3.2 Numerical Analysis

We have used the test SDN enabled wireless network topology shown in Fig. 1 to perform comparative analysis between proposed CASS mobility solution and SASS. The test topology consists of 14 switches, 10 APs and 39 links, and MN starts with AP1 and moves on a predetermined path till AP10. APs are numbered in the order of MN attachment with them. Flows one and two initiate while MN is attached with AP1,

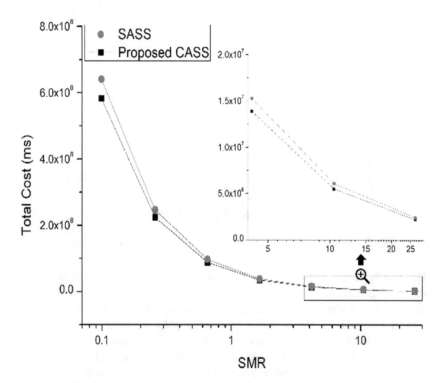

Fig. 4. Total cost comparison of CASS and SASS against SMR.

178   B. Kim et al.

flows three and four start during MN sojourn in AP4, and flow five is established when MN connects with AP8. Figure 1 shows the respective CNs for each flow, and all the flows continue till AP10. With each handover, SDN controller updates the path, recalculates the split point and ensures shortest path between MN and respective CNs. Table 1 shows the value of different parameters used in this analysis [10].

Handover frequency is an important factor in the performance evaluation of mobility solutions, which can be analyzed with varying values of SMR. Figure 4 compares the performance of proposed CASS solution with SASS for increasing SMR values, when network utilization is 50% and data rate of flows in 100 Kbps. CASS reduces the total cost by 9.6% compared to SASS under high (SMR>>1) as well as in low mobility scenario. This result signifies the role of CASS algorithm in reducing the handover and packet delivery cost, and highlights its agility under different mobility speeds.

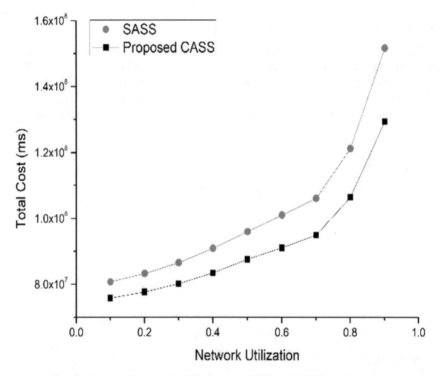

Fig. 5. Impact of network utilization on CASS and SASS performance.

The mobile traffic is expected to increase seven fold in coming few years [1]. Therefore, it is important to analysis the effect of network utilization on mobility solutions. Figure 5 shows that CASS achieves average 13% total cost improvement over SASS with varying network utilization, when SMR and flow data rates are 1 and 100 Kbps, respectively. This improvement is due to CASS feature of always restoring optimal path between MN and CN after the handover signaling. This result shows the robustness of the proposed solution under challenging traffic demands.

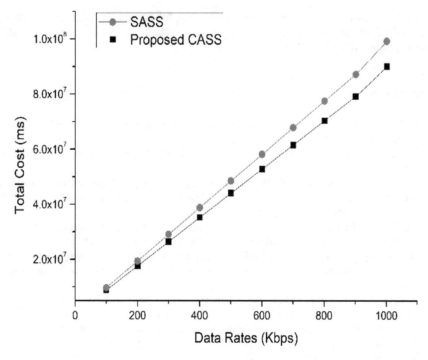

Fig. 6. Impact of data rates on total cost of CASS and SASS.

Bandwidth requirement of different flows at MN vary from few Kbps to Mbps depending on the flow service type. Total cost of a mobility solution depends on the flow bandwidth requirements. Figure 6 shows that CASS improves the total cost by 9.6% comparing to SASS, when SMR and network utilization are 1 and 0.5, respectively. This improvement is due to the reduced packet delivery cost which is achieved through the restoration of shortest path between MN and respective CNs. The result in Fig. 6 signifies that the CASS performs well under diverse service requirements.

4 Conclusion and Future Work

This paper presents a split point selection mechanism for SDN mobility management, based on the candidate APs for next MN handover. In addition to this, the proposed solution exploits the flexibility provided by SDN through separation of control and data plane to restore the optimal path for a flow after handover. The numerical results based on analytical model confirm that our proposed solution outperforms other split point based mobility solution in terms of total cost. We are working towards testing CASS mobility solution in real network through testbed implementation. In future, we plan to extend CASS by also factoring other variables in the decision.

Acknowledgments. This research was partly supported by PRCP (NRF-2010-0020210), Institute for IITP grant funded by the Korea government (MSIT) (2015-0-00547), and G-ITRC support program (IITP-2017-2015-0-00742), respectively.

References

1. Cisco, "Cisco visual networking index: Global mobile data traffic forecast update, 20162021 white paper," Cisco, Technical Report (2017)
2. ONF, "Openflow switch specification," Open Networking Foundation, TS 007 V1.3.1 (2012)
3. McKeown, N., Anderson, T., Balakrishnan, H., Parulkar, G., Peterson, L., Rexford, J., Shenker, S., Turner, J.: OpenFlow: enabling innovation in campus networks. SIGCOMM Comput. Commun. Rev. **38**(2) (2008)
4. Bradai, A., Benslimane, A., Singh, K.D.: Dynamic anchor points selection for mobility management in software defined networks. J. Netw. Comput. Appl. **57** (2015)
5. Shrivastava, P., Kataoka, K.: FastSplit: fast and dynamic IP mobility management in SDN. In: 26th International Telecommunication Networks and Applications Conference (ITNAC) (2016)
6. Kim, H., Jeon, S., Raza, S.M., Lee, J., Choo, H.: Service-aware split point selection for user-centric mobility enhancement in SDN. In: 12th International Conference on Ubiquitous Information Management and Communication (IMCOM), January 2018
7. El Hattachim, R., Erfanian, J.: "5G white paper", Next Generation Mobile Network, Technical Report (2015)
8. Pack, S., Nam, M., Kwon, T., Choi, Y.: An adaptive mobility anchor point selection scheme in hierarchical mobile IPv6 networks. Comput. Commun. **29**(16) (2006)
9. Makaya, C., Pierre, S.: An analytical framework for performance evaluation of ipv6-based mobility management protocols. IEEE Trans. Wirel. Commun. **7**(3) (2008)
10. Raza, S.M., Kim, D.S., Shin, D.R., Choo, H.: Leveraging proxy mobile IPv6 with software-defined networking. J. Commun. Netw. **18**(3) (2016)

Novel Traffic Classification Mechanism in Software Defined Networks with Experimental Analysis

Youngkyoung Kim, Syed M. Raza, Van Vi Vo,
and Hyunseung Choo$^{(\boxtimes)}$

College of Software, Sungkyunkwan University, Suwon, South Korea
{kimyk, s.moh.raza, vovanvi, choo}@skku.edu

Abstract. Recently, the usage of Internet through mobile devices has surpassed the access through wired network. It makes extremely difficult for the network operators to manage their network in resource efficient manner. Software Defined Networking (SDN), a key 5G enabling technology, makes network management easier for the network operators through logically centralized control plane. One of the benefits of SDN is specific treatment for different network traffic type, however it requires an efficient traffic classification solution. Comprehensive research has been done on deep packet inspection or machine learning based traffic classification methods which are unsuitable for delay intolerant networks due much overhead. This paper proposes a simple traffic classification mechanism using different IP addresses which are assigned to Mobile Node (MN). Different applications in MN uses one of the assigned IP addresses based on the type of network traffic they generate. Centralized SDN controller uses these IP addresses to classify the network traffic. This method is useful for various network functions such as load balancing, security and mobility. In this paper we take mobility as a use case and through emulation results establish that traffic classification significantly improves MN mobility performance.

Keywords: Software Defined Networking · Traffic classification · IP mobility

1   Introduction

Advancement of wireless access technologies has made Mobile Node (MN) the primary source of Internet access [1]. Which has caused the new services in communications, entertainment, commerce, and productivity to rapidly emerge. This makes it difficult for the network operator to efficiently manage the network traffic while satisfying the service requirements. In current traditional networks, all the traffic is treated under "one size fit all" approach which causes resource wastage and unfair treatment of services which require special attention. Software Defined Networking (SDN) can potentially solve this problem through traffic classification and by rendering specialized treatment towards different traffic types [2]. SDN separates the network control plane from the data plane and logically centralizes it in an entity called "SDN controller". The SDN controller not only control the network but also provides a perfect platform

© Springer Nature Switzerland AG 2019
S. Lee et al. (Eds.): IMCOM 2019, AISC 935, pp. 181–189, 2019.
https://doi.org/10.1007/978-3-030-19063-7_16

for efficient traffic classification which in turn makes bases for network functions like routing, load balancing, mobility, and security [3].

Currently, there are ongoing researches on various techniques for traffic classification. Most simple traffic classification mechanism is based on the port numbers, which was adopted by early stage protocols [3]. However, over the time more and more protocols and applications have been introduced and most of them use random or dynamic port numbers due to security reasons. Therefore, port based traffic classification technology is no longer useful. Recently, Machine Learning (ML) based techniques have been explored where the purpose is to classify different applications or cluster traffic flows into groups with similar patterns [4–6]. ML based flow classification mechanism classify traffic by extracting flow characteristics. Most of the proposed schemes use supervised learning training mechanism to extract simple flow features. Basic drawback of ML based traffic classification method is that it requires a large amount of sample data for efficient operation and accuracy.

One of the traffic classification challenges in SDN is the high control traffic between switches and the controller for the classification purposes. OpenState is a traffic classification architecture in SDN, which sends a minimum number of control packets to classification application on the controller [7]. It solves the problem of minimizing the interaction between the system's three main players, such as switches, SDN controllers and traffic classifiers, to improve the scalability of the approach and improve the overall system. While various traffic classification methods are being studied, there is yet no method to classify traffic completely. Most of the solutions are cumbersome and require high computation which makes them inappropriate for the real time traffic classification.

This paper proposes a simplistic approach of using different IPv6 addresses for traffic classification in SDN. The controller in proposed scheme assigns different Home Network Prefixes (HNPs) to MN, which are then changed to IPv6 addresses through stateless auto configuration method. Applications in MN then uses the IPv6 address appropriate for their traffic type, however, mechanism for this is not part of this paper. The controller then only checks the source IPv6 address of a flow to determine which traffic class it belongs to. To show the effectiveness and usability of the proposed approach, this paper also proposes IP mobility solution in SDN based on traffic classification, where delay intolerant flows are addressed first than delay tolerant flows. Emulation based performance evaluation of handover delay and signaling cost shows that the proposed mobility solution using traffic classification outperforms the traditional IP mobility mechanism in SDN.

2 Related Work

Traffic classification is significant for effective network management, and has been an important research topic for many years. Different network functionalities like load balancing, mobility and security are heavily dependent of traffic classification. In SDN networks, traffic classification is performed using various methods such as port number, ML, and OpenState.

There are two main types of traffic classification in traditional networks [3]. First, classical traffic classification based on traditional static TCP or UDP ports numbers.

The network devices must be layer four enabled to classify the traffic using transport layer port numbers, which makes networks devices computationally intense and expensive. However, in recent years, there has been a tendency in increasing number of applications to allocate a random port number or encrypt an IP layer to avoid spoofing, eavesdropping and identity theft. Second traffic classification solution requires a separate network device called Deep Packet Inspection (DPI). It is one of the most commonly used technologies, where it analyzes the inside of every data packet which requires a large amount of memory. Although the classification accuracy of DPI is high.

Recent studies have used ML technologies for traffic classification to overcome the disadvantages of aforementioned techniques [4–6]. There are various ML algorithms such as K-means clustering and regression which can classify or predict the data. ML based traffic classification requires centralized system, where traffic can be classified according to traffic pattern characteristics monitored from the network devices. In this regard SDN is the most suitable network environment even for ML based traffic classification. The problem with ML based solutions is that they are computationally intensive, cannot scale well and require a comprehensive training dataset for high classification accuracy.

OpenState is a research focused on developing a stateful data plane API for a SDN network [7]. Therefore, it can reduce the need to rely on a remote controller by using the state machines implemented inside the switch. It proposes an architecture in which the OpenState support switch sends the minimum number of packets to the traffic classifier on the controller in order to minimize the load on the classifier and improve the scalability of the approach. This approach reduces the load on the controller but makes the switches computation intensive. In this paper, the proposed traffic classification method keeps the computation at the controller, but minimizes it by only checking the IP address of the flow to determine its traffic type.

3 Traffic Classification in SDN

The proposed traffic classification mechanism uses IP addresses to determine different traffic classes. At the initial attachment of MN with the network, a default HNP is assigned to it by the controller, which is used to make the IPv6 address through stateless auto-configuration method. Here onwards in this paper IP address is used synonymously to HNP. This IP address is used by best effort traffic which does not require any specific treatment. Applications which have different and specific requirements request for a different IP address from kernel network stack. On receiving request from application, MN sends a request to the controller which is embedded in a Router Solicitation (RS) message. The controller assigns a different IP to MN through Router Advertisement (RA) message and maintains the Binding Cache Entry (BCE). Application traffic then uses the newly assigned IP address, and the controller uses this IP address to classify the traffic. Any other application in MN having similar requirements will also use this already assigned IP address, and MN does not need to send request to the controller. Subsequent subsections describe the proposed IP mobility management in SDN using the traffic classification mechanism in order to show its efficacy.

The proposed architecture consists of APs, OpenFlow enabled switches and SDN controller in the Mininet-WiFi [8] environment as shown in Fig. 1. The mobility management application on the controller consists of Management, BCE, Prefix Assignment, and Path Setup modules to provide mobility to different traffic (flow) types.

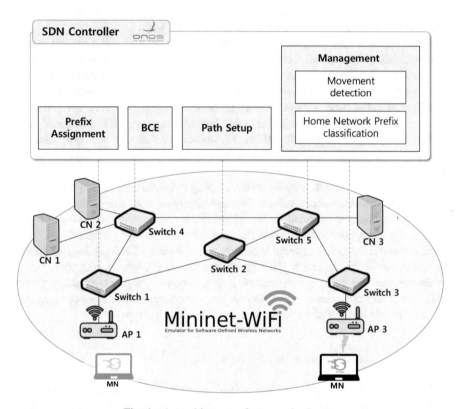

Fig. 1. An architecture of proposed scheme.

Management: This module consists of four functions; movement detection, HNP classification. When MN approaches the SDN domain, the controller must be able to grasp the connectivity of the MN. A request for an IP address belonging to a specific category in MN is received by this module through ICMPv6 RS message. Initially, RS message from MN is received by connected AP, which encapsulates it, and sends it to the controller. The controller parses the received message according to the MN ID and sends it to movement detection to match with the BCE. Network Prefix classification is used to distinguish flow types from MN.

Prefix Assignment: This module assigns the requested IP address type accordingly. In this paper we categorize IP addresses in only two types; real time and non-real time. To distinguish different flows according to the requirements, it classifies flows using the concepts presented in [9]. The IP for the non real time service is allocated as 2001: db8: 2222 :: / 64, and in case of real time service, 2002: db8: 2222 :: / 64 is assigned.

BCE: It stores the assigned IP address and information of the flow using that IP address. The state of MN can be grasped through such stored information. During handover, the mobile device informs the controller which IP addresses are currently in use by different flows.

Path Setup. This module determines whether to set the route according to the IP prefix address used by MN. During handover, this module recalculates and sets the path for the delay sensitive real time flow(s) using corresponding IP address type (type 2). However, defer the new path setting for flows which are using non-real time IP address type (type 1). After handling the handover of type 2 flow, new path is established for type 1 flow. Through this handover delay for real time services is reduced which helps in improving overall user experience.

The flow generated in the proposed solution are classified based on the IP addresses. Application in the MN determines which type of IP address it requires for the flow it will generate, and request it to the network stack. If the required type of IP address is already assigned to MN that application uses that IP address, otherwise MN requests the required type of IP address to the controller. The controller assigns the requested IP address to MN and makes the appropriate entry in the BCE. The operation procedure of the proposed scheme can be divided into two cases: MN registration and handover.

At the time of first registration of MN in the proposed system, MN sends a normal RS message which is forwarded to the controller by AP to which MN is attached. In response the controller assigns the default Prefix to MN which is for non-real time flows, and make an entry in BCE. If some application in MN needs to generate a real time flow, then MN requests a new Prefix to controller by sending a new RS message with request for real time prefix embedded in it. Upon receiving new request, the controller assigns an additional prefix for real time flows and updates the BCE accordingly. Now the non-real time and real time flows in MN uses different IP addresses based on different prefixes, and controller can conveniently distinguish the traffic type by just checking the IP address of the flow packets.

MN performs handover by moving from its current AP to new AP, and as a result of layers two attachment, new AP sends the message to the controller about attachment of MN. The controller then checks the BCE against MN ID and determines the active non-real time and real time flows at MN. The controller first update the path from new AP to CN for the real time flows and send RA message to MN with all the Prefixes currently assigned to MN. After this controller updates the path for the non-real time flows. This separation between non-real time and real time flows allows controller to send RA message much quicker which in turn reduces the disruption time for real time flows as they are delay intolerant.

4 Performance Evaluation

In this section we first describe the implementation of the proposed methods in the testbed and later show the performance results gathered through various experiments. The testbed environment of the proposed scheme is classified two parts: network and MN part. The network topology consists of 5 switches in the Mininet-WiFi environment with the ONOS controller as shown in Fig. 1. Mininet-WiFi can also be used to control and monitor MN movement. The MN uses two kind of services, and moves at 1 m/s from AP1 to AP3, which is the default speed. Access ranges for AP1 and AP3 are each 50 m.

In the second part, MN uses the packet manipulation tool Scapy [10] to add information to RS message. We can specify the contents of the reserved field or the code field in the RS message packet structure, or add a new option. The Reserved field is not used in RS messages and is set to 0 by default [11]. However, there is a problem that this field may cause compatibility problems with the field version change of the previous version. In the proposed scheme, we choose to put traffic information in the Code field. The Code field is set to 0 by default. However, traffic information is added 0001 or 0002 depending on the traffic type. The modified RS message packet is sent to the SDN controller in ICMPv6 format using the Scapy tool.

In this performance environment, the traffic generation tool analyzes packets generated using Iperf, ping and we can check by using Wireshark. There is one user using 3 kinds of service. One is real time service and the others are non real time services as shown in the Table 1. Bandwidth is differently applied to distinguish flows used in each MN.

Table 1. Performance evaluation settings.

Flow name	Flow type	Using prefix type	Bandwidth
flow1	Real time flow	Using prefix type 2	10 Mbps
flow2	Non real time flow	Using prefix type 1	7 Mbps
flow3	Non real time flow	Using prefix type 1	5 Mbps

Throughput comparison of the proposed scheme and the existing SDN based mobility solution is shown in Fig. 2. In existing mobility management based on SDN, the connection time for all flows after the handover is 4 s. On the other hand, since the proposed method handles mobility from a real time service with higher mobility requirements, it can be seen from the graph that real time flow resumes before the non real time flows. This plot shows that the proposed scheme improves the handover delay of real time flow by approximately 2 s. Handover delay values shown in these graphs include the MN disassociation time with AP1, layer two connection time with AP3 and proposed layer three mobility delay.

In handover delay, experimental environment is configured the same as the above experiment. Traffic generation tools in this environment use Iperf and ping, and analyze packets generated using Wireshark. But MN uses one real time service and one non

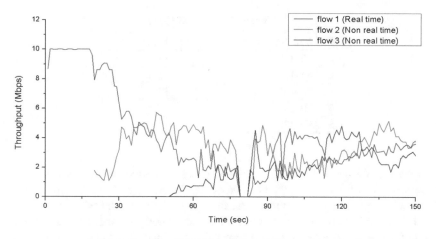

(a) Conventional SDN-based IP mobility management

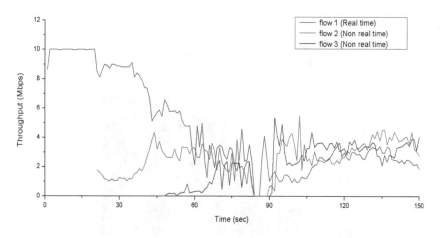

(b) Proposed mobility solution using traffic classification mechanism.

Fig. 2. Throughput comparison in testbed.

real-time service as shown in the Table 2. This evaluation is a comparison of the handover delay to the user speed. AP1 and AP3 are 100 m away from each other and change the user's speed by starting to move in Mininet-WiFi emulator environment. The user is moving from AP1 to AP3 and measures the handover delay for each user's walking speed [12] as shown in the following table. According to [12], the walking speed of 1 m/s is the speed of ordinary people and the speed of 3 m/s is the walking speed.

Table 2. User speed.

No	Mobility (start-end) time (sec)	Speed (m/s)
1	30–130	1
2	64–130	1.5
3	70–130	2
4	90–130	2.5
5	97–130	3

Figure 3 shows a graph of handover delay measured as the user step speed increases. It can be seen that the handover delay for flow1 and flow2 both increase as the user's stepping speed increases. As the speed increases, the time for the MN to attach to the AP decreases, but the overall handover delay increases because there are still unprocessed processes. Figure 3 shows that handover delay is smaller than flow 2 because flow 1 is processed from flow using real time service.

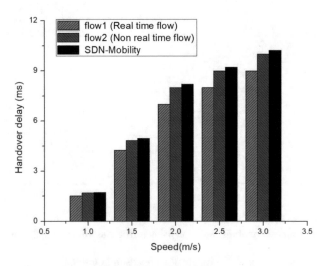

Fig. 3. Handover delay comparison with varying user speed.

5   Conclusion

In this paper, we propose a mechanism to transmit service information used in packets transmitted from MN to the network. The MN divides the traffic into real time services or non-real time dependencies according to mobility requirements, taking into account the method of informing the network about the IP type in the two operational procedures of the proposed method. Therefore, this paper focuses on the traffic classification mechanism for mobility and conducts performance evaluation experiments. This experiment proves its superiority by comparing the proposed schemes with existing SDN based mobility solutions that do not have a mechanism to classify traffic.

Acknowledgements. This research was partly supported by PRCP (NRF-2010-0020210), Institute for IITP grant funded by the Korea government (MSIT) (2015-0-00547), and G-ITRC support program (IITP-2017-2015-0-00742), respectively.

References

1. Yegin, A., Park, J., Kweon, K., Lee, J.: Terminal-centric distribution and orchestration of IP mobility for 5G networks. IEEE Commun. Mag. **52**(11), 86–92 (2014)
2. Kim, H., Feamster, N.: Improving network management with software defined net-working. IEEE Commun. Mag. **51**(2), 114–119 (2013)
3. Nguyen, T.T., Armitage, G.: A survey of techniques for internet traffic classification using machine learning. IEEE Commun. Surv. Tutorials **10**(4), 56–76 (2008)
4. Zander, S., Nguyen, T.T.T., Armitage, G.: Self-learning IP traffic classification based on statistical flow characteristics. In: PAM Conference, March 2005
5. Narayanan, V.A., Sureshkumar, V., Rajeswari, A.: Automatic traffic classification using machine learning algorithm for policy-based routing in UMTS–WLAN interwork-ing. AISC **324**, 305–312 (2014)
6. Reza, M., Sobouti, M.J., et al.: Network traffic classification using machine learning techniques over software defined networks. Int. J. Adv. Comput. Sci. Appl. (2017)
7. Bianco, A., et al.: On-the-fly traffic classification and control with a stateful SDN approach. In: IEEE International Conference on Communications (ICC) (2017)
8. Fontes, R.R., Afzal, S., Brito, S.H.B., Santos, M.A.S., Rothenberg, C.E.: Mininet-WiFi: emulating software-defined wireless networks. In: CNMS, November 2015
9. Kim, Y., Raza, S.M., Nguyen, D.T., Jeon, S., Choo, H.: Towards on-demand mobility management in SDN. In: IMCOM, January 2018
10. Scapy. https://scapy.net/
11. Neighbor Discovery for IP version 6 (IPv6). https://tools.ietf.org/html/rfc4861
12. Long III, L.L., Srinivasan, M.: Walking, running, and resting under time, distance, and average speed constraints: optimality of walk – run –rest mixtures. J. R. Soc. Interface (2013)

Intelligent and Secure Networks

Part 2: Intelligent and Secure Network

Intelligent network comprises of multiple concepts to improve network services towards a system that is less dependent to physical devices. This however, requires that the network is secure. Network security is vital to ensure that any data that pass through the communication channel cannot be read by any malicious party.

This part offers various improvements in the network intelligence and the ability to secure network. To start with, Deepika Pathinga Rajendiran and Melody Moh in "Adaptive hierarchical cache management for Cloud RAN and Multi-Access Edge Computing in 5G Networks" focuses on improving the runtime of Cloud Radio Access Network and Multi-access Edge Computing by adapting cache management. It's objective is to improve the cache hit rate, the network overhead and the access delay. In order to alight the cloud, Lin Yan, Dingsheng Wan, Qun Zhao and Yiyue Yang in "Research on implementation methods of Edge Computing in intelligent hydrology" simulate and analyse hydrological elements through edge computing to effectively reduce stress in the cloud.

In improving network performance, Aoi Yamamoto, Haruka Osanai, Akihiro Nakao, Shu Yamamoto, Saneyasu Yamaguchi, Takeshi Kamiyama, and Masato Oguchi in "Prediction of Traffic Congestion on Wired and Wireless Networks Using RNN" predict the fluctuation of network traffic using Recurrent Neural Network (RNN), which exhibits dynamic temporal behavior for a time sequence in Deep Learning classes. Aoi Yamamoto, Haruka Osanai, Akihiro Nakao, Shu Yamamoto, Saneyasu Yamaguchi, Takeshi Kamiyama, and Masato Oguchi in "Prediction of Traffic Congestion on Wired and Wireless Networks Using RNN" predict the fluctuation of network traffic using Recurrent Neural Network (RNN), which exhibits dynamic temporal behavior for a time sequence in Deep Learning classes. Haklin Kimm and Hanke Kimm in "Modeling and verification of starvation-free bitwise arbitration technique for Controller Area Network using SPIN Promela" on the other hand, introduces a formal model that solves starvation issues in the CAN arbitration protocol. In the same note, Tahira Mahboob, Young Rok Jung, and Min Young Chung in "Optimized routing in software defined networks – A reinforcement learning approach" proposed a routing mechanism for unicast routing in Software-defined Networks to minimise delay.

To ensure the improvement in security, Min Wei, Xu Yang, Jiuchao Mao and Keecheon Kim in "Secure framework and security mechanism for Edge Nodes in Industrial Internet" proposed a security framework and designs a security mechanism for edge nodes in the industrial internet.

Focusing on IoT, Supanat Limjitti, Hideya Ochiai, Hiroshi Esaki, and Kunwadee Sripanidkulchai in "IoT-VuLock: Locking IoT Device Vulnerability with enhanced network scans" proposed and developed a prototype of an architecture capable to locked remotely an IoT devices in order to prevent malware intrusion. Porapat Ongkanchana, Hiroshi Esaki, and Hideya Ochiai in "A rule-based algorithm of finding valid hosts for IoT device using its network traffic" focused on securing end-devices by creating a white list host resulted from filtered network traffic.

Focusing on the environment and power consumption, Le Yan, Jun Feng, and Tingting Hang in "Small watershed stream-flow forecasting based on LSTM" forecast the stream flow values over the next 6 hours. It has been used to predict river flow to forecast flood. Sunyoung Ahn, Young June Sah, and Sangwon Lee in "Moderating effects of spatial presence on the relationship between depth cues and user performance in virtual reality" studied on the depth cue effect on user psychology in virtual environment. Takashi Ito, Kenta Kamiya, Kenichi Takahashi, and Michimasa Inaba in "A method for identification of students' states using kinect" works on students learning state through gesture identification function in estimating their behaviour. And, Jun-Woo Cho, Song Kim, and Jae-Hyun Kim in "Multi-UAV Placements to Minimize Power Consumption in Urban Environment" optimised a series of Unmanned Aerial Vehicle placements in order to minimised it's power consumption.

Modeling and Verification of Starvation-Free Bitwise Arbitration Technique for Controller Area Network Using SPIN Promela

Haklin Kimm$^{(\boxtimes)}$ and Hanke Kimm

Computer Science Department, East Stroudsburg University of Pennsylvania,
East Stroudsburg, PA 18301, USA
hkimm@esu.edu, kimm.hanke@gmail.com

Abstract. The Controller Area Network (CAN) is a message based communication service working over high-speed serial bus systems, and mostly used in the automotive industries and real-time communication systems. The CAN bus connects several independent CAN modules and allows them to communicate and work together asynchronously and/or synchronously. All nodes can simultaneously transmit data to the CAN bus, and the collision of multiple messages on the bus is resolved by a bitwise arbitration technique that operates by assigning a node with a low-priority message to switch to a "listening" mode while a node with a high-priority message remains in a "transmitting" mode. This arbitration mechanism results in the starvation problem that lower-priority messages continuously lose arbitration to higher-priority ones. This starvation is seen as a critical section problem with priority scheduling, where a synchronization technique is required at the entry and exit of the bus. The techniques of non-starvation critical section with general semaphore and barrier synchronization are applied to enable the CAN bus to proceed without starvation. In this paper, we present a formal model of a starvation-free bitwise CAN arbitration protocol applying barrier synchronization and starvation-free mutual exclusion. Based on SPIN Promela, we provide that the proposed CAN starvation-free CAN bus model works correctly.

1 Introduction

CAN stands for Controller Area Network. The CAN bus is different from other bus systems that use one master processor interfaces with other sensors or slave processors while the CAN bus creates a network of devices, each of which on the network has a microcontroller or microprocessor that interfaces with other sensors or slave processors. The CAN bus does not have a master in control of the bus. Instead, each device operates independently. Unlike an Ethernet network which is location based, the CAN bus uses a message-based network. In a location-based network each device has a unique address while in a message-based network every device

© Springer Nature Switzerland AG 2019
S. Lee et al. (Eds.): IMCOM 2019, AISC 935, pp. 195–210, 2019.
https://doi.org/10.1007/978-3-030-19063-7_17

on the bus receives all network traffic, and the other unique aspect to the CAN bus is that unlike an Ethernet network, there are no collisions between messages from two different devices. The CAN bus employs a non-destructive arbitration scheme to determine which device transmits on the bus. Each CAN message has an arbitration field at the beginning. In case multiple devices are transmitting messages at the same time, the device that wins arbitration will send its data while the loosing devices wait until the transmitting device finishes. A device having the most dominant field wins arbitration. On the CAN bus a logic zero is dominant while a logic one is recessive. When multiple devices send a message the first part sent is the arbitration field. As each bit is transmitted all devices simultaneously listen to the bus. If a device transmits a recessive bit and hears a dominant bit on the bus then that device has lost arbitration and ceases transmission and continues to receive the remainder of the message sent by a winning device. Because the arbitration field comes before the data fields, in the end of arbitration only one device will transmit its data [4–9].

However, the aforementioned bitwise arbitration causes the lower priority CAN messages to keep losing their transmission turns at every collision of multiple messages on the CAN bus, which places the system with starvation problem as seen in operating systems, which is seen as priority scheduling algorithm that happens to make some low-priority processes waiting indefinitely. In a crowded CAN bus system, a steady stream of higher-priority messages blocks a low-priority message from occupying the CAN bus for transmission. One of solutions to this starvation problem is aging, which is a technique that increments the priority of CAN messages that keep losing their turns for the transmission [18–21]. There have been studies on resolving starvation problem on the CAN bus. Most studies on the CAN starvation problem have been approached as priority scheduling [11–16], not as barrier synchronization or starvation-free mutual exclusion.

The previous approaches based on the scheduling algorithm do not apply the bitwise arbitration properly, where all the CAN messages enter the bus simultaneously whenever the CAN bus is free, and the messages are compared bit by bit, leaving the dominant message on the bus. In previous studies, however, the candidate messages trying to occupy the CAN bus are compared first, and the dominant message is selected and placed on the CAN bus for transmission [12,14,16], and more than two nodes are not considered for arbitration simultaneously. Here we approach this CAN starvation problem, dealing multiple messages concurrently, while applying the bitwise arbitration correctly such as with bit-by-bit comparison. All the candidate messages enter the CAN bus simultaneously, but all the recessive messages leave the bus so that the dominant message can stay on the bus alone to transmit. In this paper we present a starvation-free bitwise arbitration with aging, but not solely based on the priority scheduling that considers two CAN messages at a time as seen in previous studies [12,14–16].

The remainder of this paper is organized as follows: Sect. 2 introduces Controller Area Network with its frame format of CAN 2.0 as well as arbitration method. Section 3 describes the previous studies related to CAN arbitration problem. Section 4 proposes and describes the starvation-free bitwise arbitration algorithm. Section 5 describes the SPIN program that has been developed to verify the proposed starvation-free algorithm, analyzes the program. Section 6 concludes the paper.

Field	S O F	Arbitration		Control				Data	CRC		ACK		EOF
	S O F	IDENTIFIER	R T R	I D E	r 0	DLC		Data	C_S	C_D	A_S	A_d	EOF
Length	1	11	1	1	1	4		0..64	15	1	1	1	7
Value	0	0...2031	0	0	0	0...8		x	X	1		1	127

Fig. 1. Frame format CAN 2.0

2 Controller Area Network

The CAN is known to be very suitable to any real time systems with its low cost and high reliability as a network. Furthermore, the amount of wiring between components or modules is drastically reduced by using a shared data bus, CAN, instead of hardwired point-to-point connections. In a CAN system, CAN controllers and other components are connected over a shared CAN bus so as to avoid the need of having the components with point-to-point connections that accrue a large amount of wiring, more complex electric circuits and noise, which result in a less effective and reliable system. However, the CAN bus is not well-suited for very fast data transmission such as multimedia applications but well-suited for soft real-time systems. Also, the CAN bus system is too sophisticated and expensive for applications with low data rates, in which only few parts of the system are involved in the transmission: sun roofs or heating systems [1–3, 5–9]. CAN standard frame format is shown in Fig. 1.

If two or more CAN controllers are transmitting at the same time then a dominant ('0') bit overrides a recessive ('1') bit and the value on the CAN bus will be in the dominant ('0') state. This mechanism is used to control access to the bus while comparing identifier fields. The CAN protocol calls for nodes to wait until a bus idle is detected before attempting to transmit again. If two or more nodes start to transmit at the same time, then by monitoring each bit on the bus, each node can determine if it is transmitting the highest priority message (with a numerically lower identifier) and should continue or if it should stop transmitting and wait for the next bus idle period before trying again. As the message identifiers are unique, a node transmitting the last bit of the identifier field, without detecting a '0' bit that it did not transmit, must be transmitting the highest priority message that was ready for transmission at the start of arbitration. This node then continues to transmit the remainder of its message, all other nodes having backed off. This arbitration scheme is equivalent to logical "and" operation. CAN protocol does not use a global time to control the bus. Instead each CAN controller uses its own clock. The CAN protocol therefore requires the nodes re-synchronize at each message transmission. Specifically, every node must synchronize with the leading edge of the start of frame bit caused by whichever node starts to transmit first in order to have bus arbitration work properly [5–9]. Four types of frames are used in the CAN protocol: Data frame, Remote frame to request the transmission of a data frame of the same frame identifier by a source, Error frame is used to destroy the frame, and Overload frame is used to provide for an extra delay between frames [1–3].

3   Previous Work

3.1   Distributed CAN System

With CAN, the functions of the distributed control systems perform well with more enhanced modularity and provide well-organized distributed controls. Under this setup the proposed distributed control system with CAN is able to perform and show the speedy synchronization moves while maintaining proper timing of the network in order to keep a flawless performance so that synchronization errors between components and robots are close to nil.

The developed program in our previous CAN work [10] provides general control of robots that communicate via CAN bus, and also presents the options to run a high-level coordinated task involving all robots with multiple moves where most movements are implemented in event-triggered control that provides a more efficient utilization of resources and a better serialization of tasks. With this, we are able to simulate the behavior of an assembly line or similar process that requires constraints on time and the need for individual activities to be performed in synchronous and/or asynchronous manners, or with/without coordinating robot moves. Our developed program works with the system commands that check the status of the distributed robots over CAN before implementing any move, since any fault or failure of a robot on our CAN system results in halting the whole CAN network and mandates a cold-start of the system again.

3.2   CAN Arbitration with Priority Scheduling

The previous CAN system had been implemented in two different ways: synchronous and asynchronous ways in order to see the accuracy and efficiency of the proposed CAN system. In synchronized move the robots of the system perform the same activities via control of the main microcontroller, which handles serial port initiation, robot arm initiation, macro job launching, interactive control via direct command, inverse kinematics calculation and control, monitoring and status querying and others. In asynchronous control, the microcontroller handles locally-connected robot I/O, and accepts the defined commands to initiate tasks. However, when implementing the system in asynchronous, we faced with unwanted situation of all the network of robots were stopped and in a deadlock situation so as to restart the whole network again. The CAN network faults of our proposed system were mostly caused by that low-priority CAN messages were dropped off because of failing to send over the CAN bus.

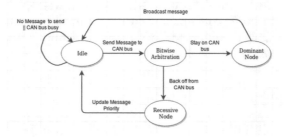

Fig. 2. State transition digram of bitwise arbitration mode

The low-priority CAN messages yielding to recessive mode are seen as a back-off protocol that has long been used to reduce contention on shared resources such as communication channels, and that has also been applied to improve the performance of shared memory algorithms in real systems [22]. One of the earlier work on resolving CAN arbitration problem related to starvation is applied by using a master node as shown in Fig. 2 [12]. The master node monitors CAN network all the time and remains silent as far as it receives messages of all the nodes but become active as it finds that some nodes are not contributing in the communication. The master node is in charge of changing the priorities of the low-priority nodes to higher priority so that the updated message can occupy the bus next round, broadcasting its message. This makes a CAN network in starvation-free mode. However in this approach, the master node should find out which low-priority messages have been backed off the CAN bus after the arbitration. It is not determinant in finding the messages that are backed off in this paper [13]. In [11], CAN arbitration model has been developed and implemented by using UPPAAL but the starvation-free CAN model has not been provided. The proposed model does not show the starvation problem as synchronization problem. In [15], the starvation-free CAN scheme has been developed and implemented using the Priority Inversion that takes 1's complement of the leftmost significant bit of a losing CAN message from the CAN arbitration. The proposed scheme works well for small number of messages in the CAN arbitration, and shows its correctness and implementation for one CAN message having lost its contention and tried the contention again after converting its leftmost bit to zero. In case, however, there are more than several messages on the bus to send their messages simultaneously, it is easily expected that there are more than one losing messages that participate in the next round of the CAN arbitration after resetting their left-most bit to zero in the proposed priority inversion scheme. Hence, more than several priority-inversion messages participate the CAN arbitration again to win the bus to send its message, so that the lowest-priority message lose to other priority-inversion messages. The priority-inversion scheme [15] and ID-rotation method [17] work well for the small number of messages and the arbitration between two messages, but do not work for large number of messages that participate again for the CAN arbitration after resetting their left-most bit to zero. In [16], CAN with flexible data rate (CAN FD) is applied in order to reduce the traffic congestion of current controller area network (CAN). In this paper we apply studies related to the synchronization with multiple processes and shared memory [22–25].

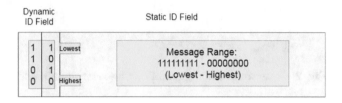

Fig. 3. Message ranges with dynamic and static IDs

In the following, we propose the CAN starvation problem as one of the synchronization problem, on which the well-known general semaphore algorithms [21, 24, 26] are updated to implement using SPIN Promela.

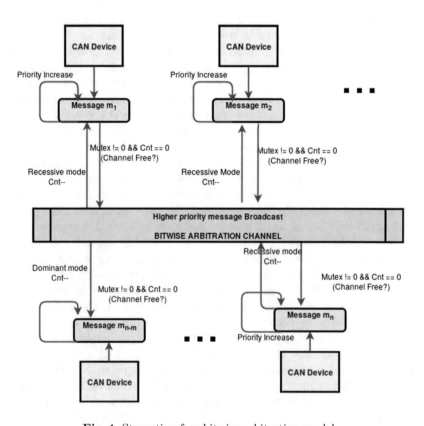

Fig. 4. Starvation-free bitwise arbitration model

4 Starvation-Free Bit-Wise Arbitration

4.1 Dynamic Message ID

The 11-bit message identifications are divided by 2-bit dynamic ID field with 9-bit static ID field as shown in Fig. 3. However, this dynamic-ID field can be flexible to be larger than 2 bits upon various network applications, and its size is lengthened more on CAN 2.0B Extended Frame where 29-bit identifiers are used. The messages using the static IDs start from the lowest number of 1023 with bit "111111111" end with the highest number of 0 with bit "000000000". With the 2-bit dynamic ID field, any recessive back-off messages can retry to occupy the CAN bus up to four times. Each 11-bit is initialed with the dynamic field ("11"), followed the static field with any number so that in the proposed system, none of the CAN messages can not use other values for its dynamic field unless it is recessive from losing in the CAN Arbitration. Each recessive message being caused by losing in the arbitration decrements its dynamic ID by one by one each round: $11 \rightarrow 10 \rightarrow 01 \rightarrow 00$. Thus each message in the proposed system is guaranteed to transmit its message in four tries.

4.2 Starvation-Free Model Correctness

As mentioned in the previous section, CAN arbitration problem, causing to starve low-priority messages but simply trying to resolve it as priority scheduling problem, does work exactly. The previous methods upon the starvation-free problem make all the messages to compare each other before allowing the dominant message occupy the CAN bus. In addition, the messages are not compared in bit by bit as CAN arbitration works, rather compared with 11-bit message identification numbers that can create up to 2048 different identifiers.

The CAN bus starvation problem has been as process scheduling in operating systems. In this paper, we consider the CAN bus as Critical Section since only one message occupy for broadcasting messages to all the CAN nodes. However, this CAN bus should open to any CAN nodes trying to send their messages simultaneously if the CAN bus is free at the time the CAN nodes check to see it open. This simultaneous access to the CAN bus leads the CAN Critical Section problem in the following changes.

- Property 1: Multiple nodes can access the CAN bus, Critical Section simultaneously.
- Property 2: Only one node stay on the Critical Section to enable for the transmission.
- Property 3: The CAN messages on the Critical Section run by bit-by-bit comparison.
- Property 4: Low-priority messages retire from the Critical Section if not dominant.
- Property 5: The identification number of the retired low-priority messages are decremented.

– Property 6: Any low-priority messages trying to access the CAN bus eventually enable to transmit their messages.

In operating systems [18–20], the critical section problem is defined such as no other process is allowed to run in its critical section when one process is running in its critical section. The critical section is considered a segment of code in which the process may be changing shared variables, writing to a file, and so on. A solution to the critical section problem demands to satisfy the following three conditions: Mutual exclusion, Progress, and Bounded Waiting.

The proposed properties are discussed to see whether to satisfy the three critical-section requirements. The first requirement of Mutual Exclusion says that if one process is running in its critical section, then no other processes are not allowed in their critical sections. The properties 1 and 2 enable the proposed algorithm to satisfy the Mutual Exclusion, but the CAN bus allows multiple messages to enter the critical section in which no other message runs for the transmission. Only one winning highest-priority message runs in the critical section, meaning that the winning on the CAN bus can transmit its own messages. The second requirement of Progress says that if the critical section is free and some processes wish to enter the critical section, then this process wanting to enter the critical section should not be delayed indefinitely. Only those processes not running in their remainder section can participate in deciding which one will enter the critical section. The properties 3 and 4 enable the Starvation-Free algorithm to satisfy the Progress. As mentioned above, only the messages in the critical section, say CAN bus, decide which message to keep staying and running in the critical section by making the lower-priority messages yield the CAN bus to higher-priority message. Finally only one node on the CAN bus running to broadcast its own message. This satisfy the second requirement that the nodes on the CAN bus decide which node stays in the critical section to run since these nodes are not in their remainder sections. The third requirement of Bounded Waiting says that no process should have to wait forever to enter the critical section if the process has made a request to enter. The properties 5 and 6 allow the algorithm to suffice the Bounded Waiting because any low-priority messages wishing to enter the CAN bus for transmitting their own messages

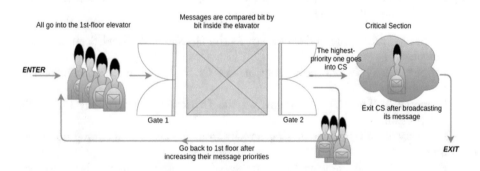

Fig. 5. Starvation-free mutual exclusion model

eventually enable to occupy the CAN bus because the property 5 allows the retired nodes to decrement its identification number of their corresponding fields, ended up increasing their message priorities.

4.3   Starvation-Free Mutual Exclusion with Semaphores

There are studies of critical section problem without starvation using weak semaphores. In weak-semaphore model, if a process performs a Wait operation and another process is waiting on that semaphore, then the first process is not allowed to immediately perform another Signal operation on that semaphore. Instead, one of the waiting processes must be given its turn. Hence, a single weak-semaphore would be in starvation situation. However, using few of them can guarantee starvation-free mutual exclusion [27]. In [26] Starvation-free critical section algorithm, shown in Algorithm 1, works with weak and split-binary semaphores. A system of semaphores $s_1+s_2+...+s_n$ forms a so-called split-binary semaphore if system preserves the invariant: $s_1+s_2+...+s_n+\#\{q \mid q \ in \ PV\} = 1$, where PV is a set of locations and q is a process [25].

The starvation-free algorithm of [26] is applied to explain Fig. 5. As seen in [26], each elevator door maps to a gate such as Gate1 to a elevator door on the first floor and Gate2 for the second floor. There will be no queue for waiting processes. This algorithm works with a batch of arrivals, which start to walk into the first-floor elevator and the door is closed when no more processes arrive. This Algorithm guarantees starvation-free mutual exclusion since the first elevator door is open again only after a whole set of processes in the second-floor elevator has been served.

4.4   Bit-Wise Arbitration Model Applying Non-starvation Mutual Exclusion with General Semaphores

The Starvation-Free Bitwise Arbitration can be seen as a synchronization problem such as starvation-free mutual exclusion shown in [26]. As seen in Fig. 5, our proposed starvation-free bitwise model can be explained as an elevator problem. CAN messages are arrived almost at the same time to enter an elevator on the first floor. When there are no more arriving CAN messages, the elevator goes to the second floor and lets the dominant highest-priority message in the elevator walks into CS, *Critical Section*. CAN bitwise arbitration is done inside the elevator while going from first floor to second. The remaining messages in the elevator, who could not be in CS, go down to the first floor by the stairs to enter the first-floor elevator again with other arriving new messages. While walking down the stairs, the priorities of remaining recessive messages are incremented.

Algorithm 1. Udding's Starvation-free Algorithm

```
1  byte gate1 = 1, gate2 = 0; //semaphores
2  byte cntGate1 = 0, cntGate2 = 0 ;
3  loop forever
4      Non Critical Section
5      wait(gate1);
6      cntGate1++;
7      signal(gate1); //First gate
8      cntGate2++;
9      cntGate1--;
10     if (cntGate1 > 0) →
11         signal(gate1)
12     else signal(gate2)
13     wait(gate2); //Second gate
14     cntGate2--;
15     Critical Section
16     if (cntGate2 > 0) →
17         signal(gate2);
18     else signal(gate1);
```

Algorithm 2. CAN Arbitration with Starvation-free Mutual Exclusion with General Semaphores

---

```
1  byte flagCAN = 1, busCAN = 1; //semaphores
2  byte cnt = 0, ent = #_of_processes;
3  loop forever
4      wait(flagCAN); //simulated wait
5      wait(busCAN);
6      cnt++;
7      if (cnt < ent) →
8          signal(flagCAN)
9      signal(busCAN);

10     Critical Section (CAN Bus Arbitration)

11     wait(busCAN); //simulated signal
12     cnt--;
13     if (cnt <= 0) →
14         signal(flagCAN);
15     else
16         wait(flagCAN);
17     signal(busCAN);
```

In the proposed Bit-wise Arbitration method shown in Algorithm 2, the CAN node with its message are defined as a process in lieu of the simplicity of describing the proposed synchronization problem. The processes in the algorithm share the following data structures: semaphore mutex, and int count. The semaphore mutex is initialized to 1, and count to 0. The mutex is used to ensure mutual exclusion of using CAN bus, the variable *count* is used to keep track of how may processes are currently on the bus before processing the Bit-wise Arbitration. The variable count is used by each lower-priority process yielding the CAN bus, say Critical Section, to higher-priority process. When the processes enter the first-floor simultaneously, each process increments the variable *count* by one; when the yielding processes exit the second-floor elevator, the variable *count* decrements by one as seen in Fig. 5. However, the semaphore mutex is changed to 0 by the first process that enters the critical section even though other processes enter the elevator simultaneously. Any process that enters the CAN bus do not change the mutex value if the variable count is greater than 0. Other processes outside the critical section have to wait if the semaphore mutex is 0. This kind of situation is seen in the reader-writer problem, where the first reader entering the critical section can change the semaphore, *write* to 0 and the last reader leaving the critical section can change the semaphore *write* to 1.

The proposed starvation-free technique shown in algorithm 2 provides a solution of the critical section problem that allows multiple processes, say CAN messages, in the critical section simultaneously but only the highest-priority message occupies the bus for broadcasting. The variable *ent* is initialized with the maximum number of processes to transmit messages on the CAN bus simultaneously. The Algorithm 2 starts in the following way. Since the semaphore busCAN is initialized to 1, the first CAN message attempting to enter a *simulatedwait* statements will succeed in executing s4: wait(busCAN). This process will promptly increment *cnt*, and implement a sequence of *simulatedwait* operations. This causes *cnt* to have a positive value. The first process and each subsequent processes, up to a total number of *ent*, are allowed to carry out s4 through s10. All the process are in CR simultaneously, where only the winner broadcasts its own message in CAN bus system. In exiting CR all the recessive messages including the winner in CR enter a *simulatedsignal* statements so that succeed in executing s11: wait(busCAN). This process will promptly decrement *cnt*, and a sequence of simulated signal operations from s12 thru s17. The last message in the *simulatedsignal* will cause *cnt* to have a zero so as to implement s14: signal(flagCAN). Then CAN bus is ready for the next round of bus arbitration again. This is shown clearly in Fig. 5.

In the proof, we will reduce the number of steps in the induction to three. The binary semaphore busCAN prevents the statements s5..9 from interleaving with the statements s11..17. S4 can interleave with statements s5..9 or s11..17, but cannot affect their execution, so the effect is the same as if all the statements s5..9 or s11..17 were executed before s4. The similar proof is shown well in [21,24,25]. The conjunction of the following formulas is invariant with the same notations, and the inductive proof of this general semaphore algorithm is also mentioned well in [21].

1. entering → (flagCAN = 0),
2. entering → (cnt >= 0),
3. #entering ≤ 1,
4. (flagCAN = 0) ∧ (¬ entering) → (cnt = 0),
5. (cnt ≥ ent) → (flagCAN = 1),
6. (flagCAN = 0) ∨ (flagCAN =1)

5   Starvation-Free Model Using SPIN

The proposed starvation-free algorithm provides a solution of the critical section problem that allows multiple processes, CAN messages, in the critical section, and its implementation is attached below. We illustrate this outcome with the SPIN Promela program shown in Algorithm 3.

```
stc369@stc369:/CAN-PROMELA/spin can5.pml
MSC: process 193 entered CAN bus
            with message 66 lowPri 193.
MSC: process 198 entered CAN bus
            with message 68 lowPri 193.
MSC: process 197 entered CAN bus
            with message 67 lowPri 193.
MSC: process 192 entered CAN bus
            with message 65 lowPri 192.
Winner pid 192 broadcasts its message 65
Recessive node with pid 194, message 66
Recessive node with pid 198, message 68
Recessive node with pid 197, message 67

MSC: process 198 entered CAN bus
            with message 69 lowPri 198.
MSC: process 132 entered CAN bus
            with message 67 lowPri 132.
MSC: process 129 entered CAN bus
            with message 66 lowPri 129.
MSC: process 196 entered CAN bus
            with message 65 lowPri 129.
Winner pid 129 broadcasts its message 66
Recessive node with pid 198, message 69
Recessive node with pid 196, message 65
Recessive node with pid 132, message 67
```

As seen in Algorithm 3, the variable *ent* is initialized with the maximum number of processes that can be in the critical section simultaneously. In this example *ent* is initialized with 4, so that at most 4 CAN messages are in for arbitration simultaneously. Statements s10..12 implement *signal(flag)* allow multiple

Algorithm 3. Implementation of Starvation-free Model

```
1  byte flag = 1, lock = 1;   //semaphores
2  byte cnt = 0;
3  byte ent = 4; //# of CAN messages for arbitration in this example
4  byte lowMsg, lowPri=255;
5  inline SendMSG() {
6        wait(flag);
7        wait(lock);
8        atomic { //each message for CAN arbitration
9            cnt++;
10           if
11           :: (cnt < ent) →
12             signal(flag);
13           if
14           :: (lowPri >= id) →
15             lowMsg = msg; lowPri = id;
16           :: else
17           fi
18        }
19       signal(lock);

20       if
21       :: (cnt == ent) →
22              the winner broadcasts in CR
23              lowMsg, lowPri = 255; //reset lowPri and lowMsg for next round;

24       wait(lock);
25       cnt--;
26       if (cnt <= 0) →
27         signal(flag);
28       else
29         wait(flag);
30       signal(lock);
31 }
32 proctype NodeCAN(byte id; byte msg) {
33       SendMSG();
34 }
35 init {
36       atomic {
37         run NodeCAN(0,65); run NodeCAN(1,66);
38         run NodeCAN(2,67); run NodeCAN(3,68);
39         }
40       /*updated messages with the higher priority
41         join the CAN bus again */
42 }
```

processes to enter the critical section that is immediately followed by the simulated arbitration for CAN bus shown in statements s20..23. Statements s24..30 simulate the *signal(flag)* operation that allow CAN bus to be available for the next round of message transmission.

As seen above, CAN messages have been created with 8 bits such that leftmost 2 bits are filled with 1 in order to provide dynamic priority, where message IDs are from 11000000 (=192) to 11111111 (=255), and dynamic IDs are from 11,10,01,00. As seen in the first round of CAN messages in Algorithm 3, the message 192 has been selected as winner from 4 messages that entered the CAN bus simultaneously, and the remaining IDs retired as recessive node. The 2 recessive nodes participated again in the second round of CAN bit arbitration: 129 (10000001) from 193 (11000001), and 132 (10000100) from 196 (11000100) after applying dynamic bit arbitration technique that replaces the leftmost two bits for the next round. Then this SPIN implementation enables low-priority IDs to transmit their messages.

6   Conclusion

In this paper we presented a new starvation-free model for CAN arbitration, where the low-priority messages yield the CAN bus to the higher-priority messages so that the low-priority keeps losing the arbitration not being able to transmit its message on time. Previous works [11–17] on this starvation problem were trying to resolve it as priority scheduling problem in which the CAN bus arbitration is executed just before entering the CAN bus so as to make one message access the CAN bus. The CAN bus arbitration allows all the messages to join the bus simultaneously and then selects the highest-priority message as dominant node that occupies the bus to broadcast its message to all the recessive nodes on the CAN network. Hence, this problem is interpreted as a non-starvation critical section model [25,26] not as priority scheduling. In our proposed model, we consider the CAN bus as critical section where multiple messages can access simultaneously for the arbitration, so that the multiple messages should access the critical section simultaneously once the CAN bus is free. This problem can be seen as a starvation-free critical section model where multiple processes are allowed to enter the critical section for the CAN bus arbitration. The well-known barrier technique and non-starvation critical section model applying general semaphore, that allow multiple process to join the critical section, have been applied to our proposed model, and its correctness has been provided. Furthermore, the proposed algorithm has been implemented by using SPIN Promela [28–30] to provide the model's correctness. The simulation of the proposed model using SPIN reveals that our starvation-free CAN model works well. At this time we are working further on a streamlined approach that the proposed model can be applied in other areas with ease.

References

1. Albert, A., Bosch, R.: Comparison of event-triggered and time-triggered concepts with regard to distributed control systems. Proc. Embed. World, 235–252 (2004)
2. CAN specification version 2.0. Robert Bosch GmbH, Stuttgart, Germany (1991)
3. Gil, J.A., et al.: A CAN architecture for an intelligent mobile robot. In: Proceedings of SICICA-97, pp. 65–70 (1997)
4. Fuhrer, T., et al.: Time-triggered Communication on CAN (Time-triggered CANTTCAN). In: Proceedings of ICC 2000, The Netherlands, Amsterdam (2000)
5. Hartwich F., et al.: Timing in the TTCAN network. In: Proceedings of 8th International CAN Conference, Las Vegas (2002)
6. Homepage of the organization CAN in Automation (CiA) (2004). http://www.can-cia.de
7. Rett, J.: Using the CANbus toolset software and the SELECONTROL MAS automation system, Control System Center, University of Sanderland (2001)
8. Magnenat S., et al.: ASEBA, an event-based middleware for distributed robot control. In: International Conference on Intelligent Robots and Systems (2007)
9. Davis, R.I., et al.: Controller Area Network (CAN) schedulability analysis: Refuted, revisited and revised. Real-Time Syst. **35**, 239–272 (2007)
10. Kimm, H., Kang, J.: Implementation of networked robot control system over controller area network. In: Proceedings of the Ninth IEEE International Conference on Ubiquitous Robots and Ambient Intelligence, Daejeon, Korea, 26–29 November 2012 (2012)
11. Pan, C., et al.: Modeling and verification of CAN bus with application Layer using UPPAL. Electron. Notes Theor. Comput. Sci. **309**, 31–49 (2014)
12. Murtaza, A., Khan, Z.: Starvation free controller area network using master node. In: Proceedings of IEEE 2nd International Conference on Electrical Engineering, Lahore, Pakistan, 25–26 March 2008 (2008)
13. MSC8122: Avoiding Arbitration Deadlock During Instruction Fetch, FreeScale Semiconductor, INC. (2008)
14. Lee, K, et al.: Starvation Prevention Scheme for a Fixed Priority Pci−Express Arbiter with Grant Counters Using Arbitration Pools. US Patent Application Publication, 14 January 2009
15. Lin, C.-M.: Analysis and modeling of a priority inversion scheme for starvation free controller area networks. IEICE Trans. Inf. Syst. **93−D**(6), 1504–1511 (2010)
16. Zago, G., de Freitas, E.: A quantitative performance study on CAN and CAN FD vehicular networks. IEEE Trans. Ind. Electron. **65**(5), 4413–4422 (2018)
17. Park, P., et al.: Performance evaluation of a method to improve fairness in in-vehicle non-destructive arbitration using ID rotation. KSII Trans. Internet Inf. Syst. **11**(10) (2017)
18. Silberschatz, A., et al.: Operating System Concepts, 9th edn. Wiley, Hoboken (2016)
19. Tanenbaum, A., Bos, H.: Modern Operating Systems, 4th edn. Pearson Prentice-Hall, Upper Saddle River (2015)
20. Stallings, W.: Operating Systems: Internals and Design Principles, 9th edn. Pearson Prentice-Hall, Upper Saddle River (2016)
21. Ben-Ari, M.: Principles of Concurrent and Distributed Programming, 2nd edn. Addison-Wesley, Boston (2006)
22. Ben-David, N., Belloch, G.: Analyzing contention and backoff in asynchronous share memory. In: Proceedings of PODC 2017, Washington, DC, USA, 25–27 July 2017 (2017)

23. Dalessandro, L., et al.: Transcational mutex locks. In: Proceeding of European Conference on Parallel Processing, Italy, 31 August–3 September 2010 (2010)
24. Trono, J., Taylor, W.: Further comments on "a correct and unrestrictive implementation of general semaphores". ACM SCIGOPS Oper. Syst. Rev. **34**(3), 5–10 (2000)
25. Hesselink, W.H., IJbema, M.: Starvation-free mutual exclusion with semaphores. Form. Asp. Comput. **25**(6), 947–969 (2013)
26. Udding, J.: Absence of individual starvation using weak semaphores. Inf. Process. Lett. **23**, 159–162 (1986)
27. Friedberg, S.A., Peterson, G.L.: An efficient solution to the mutual exclusion problem using weak semaphores. Inf. Process. Lett. **25**, 343–347 (1987)
28. Wikipedia page. https://en.wikipedia.org/wiki/Promela. Accessed 25 Oct 2017
29. Gerth, R.: Concise Promela reference (1997). http://spinroot.com/spin/Man/Quick.html. Accessed 25 Oct 2017
30. Ahrendt, W.: Lecture Slides: Introduction to Promela - in the course Software Engineering using Formal Methods, Chalmers University (2014). cse.chalmers.se/edu/year/2014/course/.../PROMELAIntroductionPS.pdf. Accessed 25 Oct 2017

Research on Implementation Methods of Edge Computing in Intelligent Hydrology

Lin Yan[1(⊠)], Dingsheng Wan[1], Qun Zhao[1], and Yiyue Yang[2]

[1] College of Computer and Information, Hohai University,
Nanjing, Jiangsu, China
linyanirene@hhu.edu.cn
[2] China Energy Engineering Group Jiangsu Power Design Institute,
Nanjing, Jiangsu, China
yangyiyue@qq.jspdi.com.cn

Abstract. In the hydrological big data scenarios, there are a large amount of data transmission and reception. The stressful pressure of data processing and analysis makes cloud computing confronted with a series of challenges. In order to overcome the challenges, edge computing emerges. An intelligent hydrological architecture based on edge computing is proposed in this paper. Making real-time processing and analysis of continuous data streams closer to the source of the objects or data can make full use of edge devices and network resources, reduce the amount of data transmission effectively and mitigate the pressure on the cloud side. The architecture has obvious advantages when dealing with tasks with high real-time requirements. Based on the Apache Edgent platform, this paper simulates the procedure of real-time processing and analyzing hydrological elements by the methods of edge computing. The experimental results reflect that edge computing applied in intelligent hydrology can effectively reduce the data transmission and relieve the pressure of data storage and analysis in the cloud.

Keywords: Edge computing · Intelligent hydrology · Apache Edgent · Real-time stream processing · Internet of Things

1 Introduction

Intelligent development has become increasingly dependent on the cloud. However, the cloud cannot solve all problems. The edge side of intelligent Internet faces the following challenges: insufficient bandwidth for the connection and analysis of massive data, steady increase of the demand of real-time experience, security and privacy protection and constant energy consumption [1]. Edge computing is an emerging technology that reduces transmission delay and bandwidth consumption by placing computing, storage, bandwidth, applications and other resources on the edge of the network [2]. Nowadays, edge computing has been applied into various industries, such as agriculture, transportation, electricity, medical care and home life, which provides a new idea for the construction of smart cities. A reference architecture of Internet of Things (IoT) for real-time applications in smart cities is proposed in article [3], which

© Springer Nature Switzerland AG 2019
S. Lee et al. (Eds.): IMCOM 2019, AISC 935, pp. 211–224, 2019.
https://doi.org/10.1007/978-3-030-19063-7_18

fully considers the prominent role of edge computing in real-time processing of current IoT systems. Reference [4] integrates vehicle networking and mobile edge computing to transmit delay-tolerant data and process data computing tasks. Reference [5] implements the off-line emergency communication based on edge computing of power Internet of Things, which ensures the intelligent, efficient and reliable operations of edge network in the state of disconnection. In article [6], Fog computing and edge computing are applied into the Medical Internet of Things (IOT) to monitor, analyze and transmit the blood of diabetic patients in real time. Considering the demand of smart home, reference [7] applies edge computing to smart home environment on the premise of saving computing resources and improving the utilization ratio of it. The hydrological construction in China is being transformed from "digital hydrology" to "intelligent hydrology". Nevertheless, the scenarios of edge computing applied in intelligent hydrology are rare.

In China, hydrological information collection system has been basically formed, with over 200,000 information collection points being established to gather water and rainfall conditions, engineering conditions and so on, and the rate of automatic collection has reached more than 80% [8]. The establishment of intelligent hydrological monitoring system will inevitably produce huge and complex sets of hydrological data, which will gradually present the features of big data: multi-source, multi-dimensional, large-scale and polymorphic [9]. At present, the development of hydrology and water conservancy informatization based on cloud platform have become more mature, but the architecture does not pay enough attention to the collection and real-time processing of raw data. It does not take the significant role of processing and analyzing data on edge devices and networks into fully consideration. In terms of actual data features and business scenarios in hydrology, applying edge computing services into intelligent hydrology to process and analyze data on the edge in real time can effectively reduce the amount of data transmission, improve the analysis efficiency, relieve the processing pressure of the cloud platform as well as solve the problems that the delay of business processing is high and the real-time requirements cannot be met.

2   Edge Computing and Apache Edgent

2.1   Edge Computing

Edge computing [10, 11] refers to the enabling technologies allowing computation to be performed at the edge of the network, on downstream data on behalf of cloud services and upstream data on behalf of IoT services. W. Shi defines "edge" as any computing and network resources along the path between data sources and cloud data centers, as shown in Fig. 1.

Figure 1 illustrates that the edge computing structure differs from the conventional cloud computing structure in that edge devices are not only data producers, but also data consumers. In addition to data collection and transmission, edge computing nodes can also process and analyze data. Edge and cloud collaborate to achieve two-way data transfer and task migration. Edge computing services involve computing offload, data storage, data processing, request distribution, service delivery, IoT management, as well as privacy protection, which has irreplaceable role in dealing with real-time data, short-cycle data and local decision-making scenarios. Migrating cloud computing tasks to the edge will be bound to reduce the data transmission, improve decision-making efficiency, and decrease the computing load of cloud.

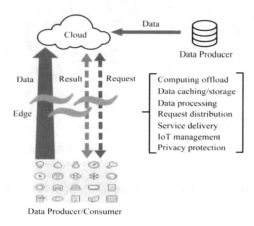

Fig. 1. Edge computing paradigm.

The core concept of edge computing is that "computing should be closer to the data source and the user" [12]. Meanwhile, it should reach the key requirements of industry digitalization in agile connection, real-time business, data optimization, intelligent application, security, privacy protection and so on [13].

2.2 Apache Edgent

Apache Edgent [14] is a programming model and micro-kernel style runtime that can be embedded in gateways and small footprint edge devices enabling local, real-time analytics on the continuous streams of data coming from equipment, vehicles, systems, appliances, devices and sensors of all kinds. Working in conjunction with centralized analytic systems, Apache Edgent provides efficient and timely analytics across the whole IoT ecosystem: from the center to the edge. Enabling streaming analytics on edge devices can reduce the amount of data which are transmitted and stored to analytics servers, the costs of communication and the load on data centers. Furthermore, edge analytics can decrease decision-making latency and enable device autonomy.

Edgent uses a functional API to compose processing pipelines with immediate per-tuple computation and provides a suite of connectors which greatly ease communication with a backend [14]. Edgent accelerates application development and pushes data analysis and machine learning to edge devices. Figure 2 shows the open IoT ecosystem of Apache Edgent, which is API driven, modular and can be used in conjunction with the analytics solutions such as Apache Kafka, Apache Spark and Apache Storm.

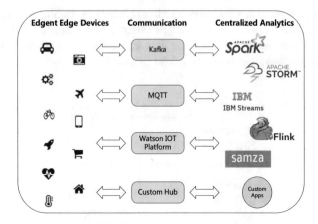

Fig. 2. An open IoT ecosystem of Apache Edgent.

3 Intelligent Hydrological Architecture Based on Edge Computing

Intelligent hydrology is mainly composed of intelligent hydrologic monitoring systems, intelligent hydrological data analysis systems and modern integrated application service platforms [15]. Its overall framework is reflected in four levels: perception layer, transmission layer, intelligence layer and application layer. The architecture proposed in this paper will improve the intelligence layer based on edge computing. The improved "intelligent hydrology" architecture is divided into five layers: data perception layer, edge computing services layer, network transport layer, cloud computing services layer and application layer. The intelligent hydrological architecture based on Apache Edgent is shown in Fig. 3.

The data perception layer uses devices or systems that sense, measure and transmit information anytime and anywhere to achieve a "more thorough perception" of the water environment. It includes the mechanism of the multi-source data collection, device control and communication [16]. Data such as water level, rainfall, water quality, weather, temperature, gate opening and so on are collected and transmitted through edge intelligent devices, such as various sensors and control equipment, and the intelligent control equipment in the edge can automatically adjust according to the feedback of the control system.

The edge computing services layer provides services on edge devices or networks. Services include six categories: data access, data preprocessing, data analysis, data storage and visualization, service delivery and strategy execution and cloud interaction. Data access is responsible for transmitting the collected raw data to the edge computing services layer through network interfaces, industrial serial ports or expansion interfaces. Data preprocessing uses real-time stream processing technologies to perform pre-processing on raw data, such as filtering, aggregation, semantic parsing and so on. Instantly processing the continuous data stream can not only filter the useless data, save storage resources, but also improve the efficiency of subsequent analysis. Data analysis is responsible for further data calculation and integration. It supports instant processing through real-time streams or centralized analysis based on the model library. Data storage and visualization include the storage of data after edge processing, the establishment and update of model library and the display of user interface. The visualization of edge services makes it easier for users to manage the tasks of the edge platform and monitor the execution status of all the tasks. Service delivery and strategy execution control the data transmission between the edge-edge and edge-cloud and control the response (such as instruction execution, device control, etc.) of related systems. Cloud interaction defines a variety of connectors that interact with the centralized analysis platforms and guarantees security and privacy during the data transmission. Meanwhile, it supports cloud computing offloading to the edge, extending the edge services. In addition to reduce the cost of data transmission and data storage, the introduction of the edge computing service layer also relieves pressures in cloud and improves decision-making efficiency.

The network transport layer is the basis for realizing the Internet of Everything, including LAN, WAN, mobile communication network, VPN and other wired or wireless networks.

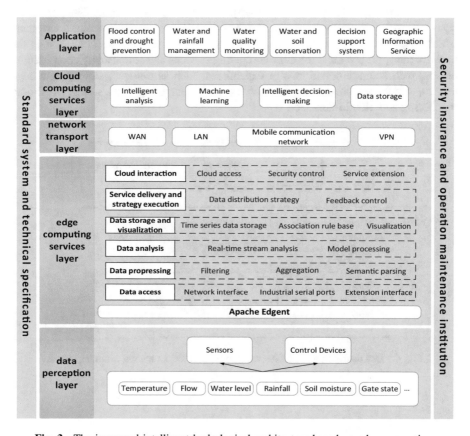

Fig. 3. The improved intelligent hydrological architecture based on edge computing.

The cloud computing services layer consists of a unified management of storage databases, virtual machine groups and network resources. It performs centralized computing and analysis of big data, and realizes unified scheduling and management of complex tasks, such as intelligent analysis, machine learning, intelligent decision-making and data storage.

The application layer is directly associated with the users and is responsible for processing the requests to or returns from the hydrological related business system. The end results and decisions are returned and displayed to users after invoking the interfaces in the service layer. This layer supports the integration of third-party services to provide better geographic information services, weather forecasting services and more. Users can access the business system in the application layer through browsers, desktop applications or mobile devices, which is convenient for workers to carry out works and make reasonable use of hydrological resources to achieve the harmony between the people and water.

In this paper, the edge computing services of the architecture are served by edge computing nodes (ECNs). ECN is a logical abstraction that contains multiple edge devices and the node distributed at different layers provides different services (as shown in Fig. 4). ECNs can be distributed across the data perception layer, the edge computing services layer and the network transport layer. The ECNs in the data perception layer are usually composed by monitoring terminals such as sensors and controllers, which usually provide real-time and lightweight computing services. The ECNs in the network transport layer and the edge computing services layer are usually composed by devices with computing and storage capabilities, such as routers, gateways and edge servers, which can provide services of complex computing, data storage, real-time monitoring and cloud interaction.

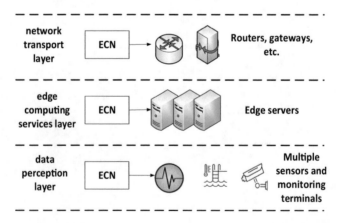

Fig. 4. The distribution of ECNs and the type of ECNs among different layers.

4 Experimental Analysis and Implementation

This paper refers to the historical data of three stations and Hongze lake located in the upper and middle reaches of the Huaihe River in China during 2017–2018. The historical data are respectively including four hydrological elements (water level, rainfall, evaporation and flow) of the Lutaizi station, the rainfall of the Runheji station, the flow of the Zhaopingtai Reservoir and three elements (water level, area, volume) of Hongze lake. We mimic 10,000 sets of real-time collected data as experimental data. The experimental data includes two types: normal values and outliers. The outliers are composed of null or empty values and values out of the normal range. The experiments simulate the process of collecting raw data by sensors and monitoring devices in the stations mentioned above on the Apache Edgent platform. The platform then performs real-time data preprocessing, data analysis, data monitoring and data storage on the collected data. In this paper, the following four experiments are aimed to verify the advantages and feasibility of edge computing applied in intelligent hydrology.

The experiments use two virtual machines to separately act as the edge real-time stream processing node (hereinafter called "stream node") and the edge server. The hardware environment of the stream node is CPU: Intel (R) Core (TM) i5-7200U CPU @ 2.50 GHz, memory: 2G, hard disk space: 80G; The software environment of the stream node is Microsoft Server 2008 operating system and the development language is Java. The deployment of the stream node in the actual application is closer to the data source so that the computing resources are insufficient and the configuration is relatively low. The CPU configuration, operating system and develop language of the edge server is the same as that of the stream node. The server's memory is 8G. The hard disk space of server is 100G and the database is MySQL 5.7.

4.1 Analysis of Experimental Results

Experiment 1. This experiment runs on the stream node. Apache Edgent supports data input at millisecond level. The experiment inputs a set of data per minute according to the actual situation of hydrological elements collection and then separately preprocesses and analyzes the corresponding hydrological elements of the above three stations by real-time streaming of Edgent to filter out the invalid raw data. The main logic of real-time stream processing in Apache Edgent is to aggregate data first, then filter out invalid data, and separate multi-source data at last. The corresponding filtering rules are adopted for different data streams. What's more, Edgent provides the topology for users to check the processing and results of the execution, which can clearly illustrate the data input and output of each stream processing monitoring node. The circle's size in the topology also reflects the amount of data through the node.

By executing the programs of Edgent to simulate the process of data acquisition and preprocessing in three stations, we obtain the following real-time flow topology diagram (as shown in Fig. 5). The topology is updated dynamically in real time. Clicking on the nodes in Fig. 5 can show the corresponding detailed processing information of the different nodes (as shown in Fig. 6), including the upstream and downstream nodes, the station to which the data belongs, the data type and the amount of input and output data.

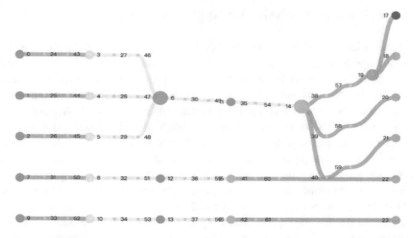

Fig. 5. The topology of the real-time data processing in three hydrological stations.

In Fig. 5, node 0, 1 and 2 respectively represent the raw input data of water level, evaporation, and rainfall elements in Lutaizi station. Node 7 represents the raw input data of rainfall in Runheji station. Node 9 is on behalf of the raw flow data of Zhaopingtai Reservoir. Node 6 is the union node, which aggregates the three hydrological elements of Lutaizi station. Node 11, 12 and 13 are the filter nodes that are responsible for filtering the different categories of data for the three stations by their corresponding rules. Node 14 separates the multi-source data of Lutaizi station after filtering. The 18, 20, 21 nodes respectively represent the valid data of water level, evaporation and rainfall after filtering in Lutaizi station. Similarly, Node 22 and 23 show the information of the valid data collected in Runheji station and Zhaopingtai Reservoir.

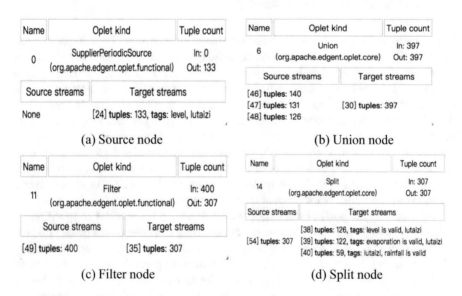

Fig. 6. The detailed information of different kinds of nodes in the topology.

Figure 6 shows the details of the main nodes. Figure 6(a) is the information of node 0, which is the source node of water level data in Lutaizi station. This kind of nodes only has data output but no input, because no upstream node. The node 0 in the (a) has generated 133 raw water level data in Lutaizi station. Figure 6(b) shows the detailed information of node 6, which is a union node, and the amount of input and output data of it is unchanged due to it only preforms aggregation operation on the data that cannot reduce the amount of data. The data that needs to be aggregated on node 6 comes from three nodes, named node 46, 47, and 48, and the amount of them are 140, 131, and 126 respectively. Figure 6(c) reflects the input and output information of the filter node, named node 11. The difference between the two represents the number of the invalid data. Node 14 in Fig. 6(d) is a split node, which identifies the categories of separated hydrological elements and the amount of valid data for each category. Each node in the topology can indicate the number of upstream and downstream nodes, which is convenient for locating data sources and destinations.

In summary, we can conclude from the experimental results that the edge computing services based on Apache Edgent can real-time filter and remove the invalid collected data close to the data sources. It will be bound to reduce the amount of data transmission on the networks effectively and improve the data processing efficiency of subsequent programs or projects. By monitoring the topologies of the data preprocessing on the edge, people can grasp the real-time situation of the data collection and make timely adjustments.

Experiment 2. This experiment runs on the server. The situation of data collection in the Experiment 1 is monitored in real time through the server-side edge computing services monitoring system (see Fig. 7).

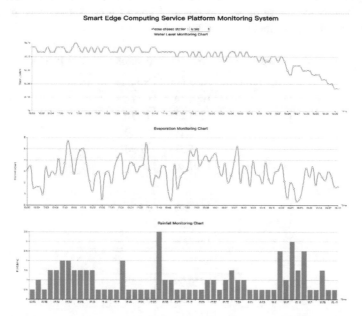

Fig. 7. The interface of the edge computing services monitoring system.

The monitoring system fetches the stream object on Edgent platform in real time, and displays the time points and values of the collected hydrological elements of each station. In Fig. 7, the acquisition situation of water level, evaporation and rainfall data in Lutaizi station are shown respectively from top to bottom. The system also supports to display the data of other stations by switching the selection in the top drop-down menu.

Experiment 3. This experiment runs both on the server and the stream node. The edge server trains the stored hydrological elements based on the defined models regularly. According to the results of the model training, the stream node performs model calculation on collected hydrological data by real-time streaming of Apache Edgent in order to achieve the real-time conversion of the data. The training and calculation of hydrological model can be deployed on edge nodes to realize on-demand service, which means that the results of model processing can be transmitted to the centralized analysis platform according to actual needs.

Taking the water level-area-volume model as an example, the model is trained on the edge server to fit and deduce the relationships among the water surface area, the volume of water and the water level synchronously observed at hydrological stations in different flood periods. The regression equations of water level-area-volume model [17] are as below:

$$V = a_0 + a_1 H + a_2 H^2 + \cdots + a_{n-1} H^{n-1} \tag{1}$$

$$A = b_0 + b_1 H + b_2 H^2 + \cdots + b_{n-1} H^{n-1} \tag{2}$$

In the equations, A is water surface area and V is the volume of water. Variable H is the corresponding water level.

Hongze lake, located in the middle and lower reaches of the main stream of the Huaihe River, is the fourth largest freshwater lake in China. Some of its hydrological characteristics data [18] are shown in Table 1.

Table 1. Data of water level, area and volume of Hongze lake.

Water level (m)	Area (km^2)	Volume ($10^8 m^3$)
11	1054.44	6.11
11.5	1188.29	12.70
12	1331.13	21.00
12.5	1375.32	27.90
12.83	1407.16	–
⋮		
13	1432.48	34.86
13.36	1482.39	–
13.5	1542.04	42.25
⋮		
14	1730.99	50.40
14.5	1990.36	59.68
15	2320.11	66.42

According to the above equations and data, the results of model training on the edge server are as following:

$$V = 0.0278H^3 - 0.7533H^2 + 20.58H - 166.1908 \qquad (3)$$

$$A = 38H^3 - 1402H^2 + 17560H - 72509 \qquad (4)$$

The water level-area relationship of Hongze lake illustrated in Fig. 8 is:

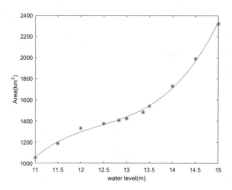

Fig. 8. The relationship between the water level and area of Hongze lake.

The water level-volume relationship of Hongze Lake illustrated in Fig. 9 is:

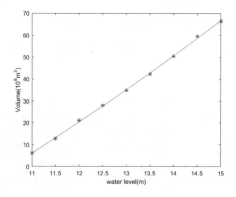

Fig. 9. The relationship between the water level and volume of Hongze lake.

When entering the flood periods, the water surface area and volume corresponding to the real-time water level are calculated on the edge side according to the above model, so that the impact of the flood occurrence can be quickly analyzed on the subsequent program.

This experiment simulates the real-time water level collection and transformation process of Hongze lake on the edge stream node during the flood period in August 2018. The water level-area-volume model transformation process is shown in Fig. 10. On the top of the figure is the data of water level collected in real-time. The second and third charts reflect the transformation results of the area and volume respectively.

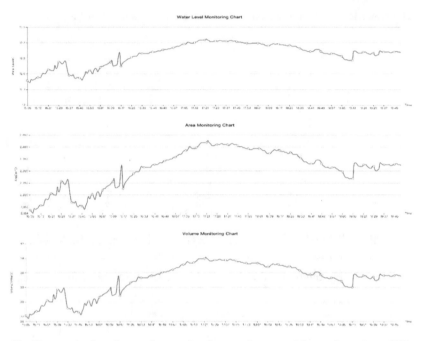

Fig. 10. The monitoring charts of water level-area-volume model transformation of Hongze lake.

The tasks of model analysis and processing are migrated from the centralized analysis platform to the edge. We train and update models regularly at the edge side near the data source and the different types of data transformed from the real-time collected data on the edge nodes are sent to the centralized analysis platform for further calculation, which can relieve the data processing pressure of centralized analysis platform, reduce data transmission effectively and improve data analysis efficiency.

Experiment 4. This experiment runs on the edge server. The edge servers are not only responsible for data monitoring and data analysis, but also the data storage. This experiment stores the hydrological data of the above three experiments on the edge server and checks the filtered data stored in the server to make up for the leak and make it into continuous time series data. Experiment 4 aims to illustrate that storing real-time collected data on edge servers can effectively reduce the data transmission on the network and data storage in the cloud. Some data processing and analysis tasks are executed on the edge side, which can realize local decision-making, meet the high real-time requirements of local tasks and improve the efficiency of analysis and decision-making.

In the actual process of hydrological data acquisition, video as a method of data acquisition will produce huge amount of data. If we transmit the video data to the cloud platform for centralized analysis, it is not a wise choice in terms of cost and efficiency. The analysis and feature extraction of video data realized on the edge side will greatly reduce the costs of transmission and storage, improve data analysis efficiency and reduce the pressure of centralized analysis platform. Edge computing will have great significance in the field of hydrological video data acquisition and analysis.

5 Conclusion and Future Work

In summary, the improved architecture proposed in this paper can be a guidance for the design and construction of intelligent hydrology. The edge computing services make full use of the edge devices and edge networks, which will accelerate the data computing and processing. Meanwhile, the real-time stream processing and cloud connection technologies implemented by Apache Edgent, an edge computing service platform used in this paper, provide the technical support for research and implementation of edge computing services in intelligent hydrology. Apache Edgent is still in the incubation and is not yet mature, but the edge computing is in constantly development, of which the functions will be more complete and powerful in the future. This paper will continue to do further research on the field of distributed edge computing and the collaborative theory between the edge and cloud.

Acknowledgement. This work has been partially supported by the National Key Research and Development Program of China (No. 2018YFC1508100) and the National Key Research and Development Program of China (No. 2018YFC0407900).

References

1. Tu, Y., Dong, Z., Yang, H.: Key technologies and application of edge computing. ZTE Commun. **15**(02), 26–34 (2017)
2. Hong, X., Yang, W.: Edge computing technology: development and countermeasures. Chin. Eng. Sci. **20**(02), 20–26 (2018)
3. Zhang, H., Xu, D., Li, J., Fu, G.: System reference architecture for Internet of Things for smart city real-time application. J. Beijing Inf. Sci. Technol. Univ. (Nat. Sci. Ed.) **32**(03), 7–12 (2017)
4. Li, M., Si, P., Sun, E., Zhang, Y.: Delay-tolerant data traffic based on connected vehicle network and mobile edge computing. J. Beijing Univ. Technol. **44**(04), 529–537 (2018)
5. Fu, D., Liu, D., Shi, G., Wang, A., Fan, C., et al.: Off-line emergency communication method of power Internet of Things based on edge computing. Telecommun. Sci. **34**(03), 183–191 (2018)

6. Klonoff, C.: Fog computing and edge computing architectures for processing data from diabetes devices connected to the medical Internet of Things. J. Diabetes Sci. Technol. **11**(4), 647–652 (2017)
7. Song, P., Li, C., Xu, L., Liang, X.: Edge computing system for smart home based on personal computer. Comput. Eng. **43**(11), 1–7 (2017)
8. Ministry of Water Resources Counselor Advisory Committee.: Analysis of the status of smart water conservancy and preliminary plan for its construction. China Water Conserv. **2018**(05), 1–4 (2018)
9. Ai, P., Yu, J., Ma, M.: Brief introduction of key technologies in intelligent hydrological monitoring system. Water Conserv. Inf. **2018**(1), 36–40+45 (2018)
10. Shi, W., Sun, H., Cao, J., et al.: Edge computing—an emerging computing model for the internet of everything era. Journey Comput. Res. Dev. **54**(5), 907–924 (2017)
11. Shi, W., Cao, J., Zhang, Q., et al.: Edge computing: version and challenges. IEEE Internet Things Journey **3**(5), 637–646 (2016)
12. Zhao, Z., Liu, F., Cai, Z., Xiao, N.: Edge computing: platforms, applications and challenges. Comput. Res. Dev. **55**(02), 327–337 (2018)
13. Edge computing reference architecture 2.0. automated expo, vol. 2018 (03), pp. 54–62(2018)
14. Apache Edgent Homepage. http://edgent.apache.org/. Accessed 20 Nov 2018
15. Xu, D.: Discussions on construction of smart hydrology. China Water Conserv. **2017**(19), 15–18 (2017)
16. Ye, F., Zhang, P., Xia, R., Gu, H., et al.: "Smart Chu River" system based on the new generation big data processing engine "Flink". In: The 6th Proceedings of China Water Conservancy Information Technology Forum, Shenzhen, pp. 37–47 (2018)
17. Wang, Z., Huo, L., Zhang, H., Du, Y.: Design of three- dimensional simulation of river flood. Geomat. Spat. Inf. Technol. **40**(05), 37–47 (2017)
18. Qi, X., Yang, L., Xia, B., Wang, Y.: Water level-capacity curve of Hongze lake based on remote sensing data. Adv. Sci. Technol. Water Resour. **37**(03), 77–83 (2017)

IoT-VuLock: Locking IoT Device Vulnerability with Enhanced Network Scans

Supanat Limjitti[1], Hideya Ochiai[2(✉)], Hiroshi Esaki[2],
and Kunwadee Sripanidkulchai[1]

[1] Department of Computer Engineering, Chulalongkorn University,
Bangkok, Thailand
supanat.l@student.chula.ac.th, kunwadee@cp.eng.chula.ac.th
[2] The University of Tokyo, Tokyo, Japan
ochiai@elab.ic.i.u-tokyo.ac.jp, hiroshi@wide.ad.jp

Abstract. With the ever increasing number of IoT devices, the threat of IoT malware is becoming more serious every single day. Remote login vulnerability, which is the focus of this paper, remains a crucial issue as system operators occasionally deploy IoT devices with default or well-known passwords for Telnet and SSH login. This paper proposes an architecture for IoT device vulnerability locking (IoT-VuLock), allowing these vulnerable devices to be locked remotely by network administrators which in turn reduces the risk of malware intrusion through Telnet or SSH. In this paper, a prototype of IoT-VuLock has been developed and tested. The experiment results indicate that IoT-VuLock is useful for network administrators to find and lock IoT devices with high risks of IoT malware intrusion.

Keywords: IoT · Remote login · Vulnerability · Telnet

1 Introduction

Due to the rapidly increasing number of Internet of Things (IoT) devices, malware threats are becoming more serious. Cyber-attacks based on IoT malware are currently growing at a rapid pace. For example, the DDos attack on Dyn DNS in October of 2016 is believed to be a result of an IoT malware – Mirai [1–3].

One of the weaknesses of IoT devices comes from its over-reliance on default settings including the login password. In order to allow configuration, sometimes remote access through Telnet or SSH ports is allowed with a generic default username and password pair causing a remote login vulnerability to remain in the network. These IoT devices are sometimes unmanaged or irresponsibly deployed.

This paper proposes an architecture of IoT device vulnerability locking (IoT-VuLock) in order to find and lock the vulnerabilities remaining in the network. An IoT-VuLock device is deployed in an intra-network by its administrator to (1) make a network scan similar to Nmap [4], (2) attempt log in with a well-known

© Springer Nature Switzerland AG 2019
S. Lee et al. (Eds.): IMCOM 2019, AISC 935, pp. 225–233, 2019.
https://doi.org/10.1007/978-3-030-19063-7_19

set of username and password pairs through open SSH and Telnet ports, and (3) change the password used to access the target device. Thus the IoT-VuLock device "locks" vulnerable IoT devices attached to the network.

Even if the network is operated behind a NAT or behind a firewall, the fact that IoT devices with default username and password can be easily intruded by malware must be kept in mind. This is because today's malware can easily be delivered into local-area networks through email attachments or Wi-Fi-enabled smartphones and laptops. IoT-VuLock will prevent malware from infecting IoT devices on the network by patching such vulnerabilities.

Having analyzed the behavior of Mirai malware, we were inspired by its scanning and brute-force login techniques. Our proposed model is designed to mimic parts of the behavior found on a number of Mirai variants to provide network administrators a tool to scan for and fix vulnerabilities in their private network.

Evaluation was done on two real-world networks to gather the performance numbers of IoT-VuLock. The locking feature of IoT-VuLock was also assessed on a network specifically designed for testing.

This paper is organized as follows: In Sect. 2, related work is addressed. In Sect. 3, the architecture for IoT device vulnerability locking is proposed. The evaluation is shown in Sect. 4, and the discussion is provided in Sect. 5. The paper is concluded in Sect. 6.

2   Related Work

Account locking is known as a common mechanism for protecting remote login services [8,10] against malicious login attempts. After several failed attempts, the server temporarily locks out further attempts in order to protect the account from the detected brute-force attack [5,6]. IoT-VuLock also locks accounts of IoT devices but in a different way. It first finds the vulnerable devices that can be logged into using weak username and password pairs and then fixes them with a stronger password for the corresponding username.

Approaches for detection and protection against brute-force attacks to remote login ports have been proposed. Satoh et al. [7] proposed flow-based detection against SSH dictionary attacks in the network. Alsaleh et al. [8] proposed a password guessing resistant protocol to protect the service from attacks made by a large botnet. IoT-VuLock also protects the remote login ports against such kind of attacks but in a different way. So our work can be used together with the mentioned work as a defense-in-depth.

The weakness of IoT devices also originates from its weak password management. Zhang-Kennedy et al. [9] and Florencio et al. [10] proposed new human-password management rules that are more effective and practical for today's Internet. Adapting those rules for IoT device management will probably protect the device itself. However, not everyone is strict, and vulnerable devices will still remain in the network.

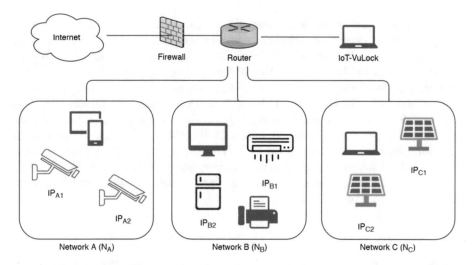

Fig. 1. IoT-VuLock Architecture. A network administrator deploys an IoT-VuLock device in their intra-network. It scans for vulnerable IoT devices, attempts to log in through SSH and Telnet, and updates weak passwords to stronger ones.

Patton et al. [11] also discusses the vulnerability of IoT devices in 2014 before the Mirai malware appeared. Their focus is on the exposure of IoT data (e.g., supervisory, control and data acquisition) while malware intrusion, which is our focus, is not discussed.

3   IoT Device Vulnerability Locking

3.1   Overview

Our proposed model is intended to be used by a network administrator in a private network. Inside this network, there may be a number of vulnerable Internet of Things (IoT) devices including those with open remote access ports e.g. SSH (TCP 22) and Telnet (TCP 23 and TCP 2323). Furthermore, they may have default or generic usernames and passwords. These ports can usually be blocked by a firewall at the entrance of the network. However, malware may intrude into the network through malicious email attachments or smartphones connected via Wi-Fi. Such malware can perform local network scans and infect vulnerable IoT devices.

This is where the proposed model comes in to prevent brute-force login attempts from attackers. In this architecture, we deploy an IoT-VuLock device whose process starts by scanning for devices or hosts with open ports and tries to log in with a predetermined set of username and password pairs. After successfully logging in, IoT-VuLock attempts to close open ports using access control lists and changes the password for the logged in user.

Let N be a set of IP addresses of an organization's intra-network. N is composed of a number of network segments which are denoted by N_A, N_B, N_C and so on. In this model, IoT devices in these networks each has specific IP address denoted by IP_{Xi} if $IP_{Xi} \in N_X$. For example, in Fig. 1, the HVAC is specified with IP_{B1} as it is in N_B. It is assumed that some IoT devices in the network are affected by the vulnerability described in the previous section. During the execution phase, $N = N_A \cup N_B \cup \cdots \cup N_M$ is given as an input for the model. Then a scan will commence on the listed networks. If open ports are found on hosts in these networks, attempts to log in will be made using a predetermined set of username and password pairs. If successful, IoT-VuLock attempts to perform actions as specified by the security level which will be discussed in the next sub-section.

3.2 Security Levels

For the model, four levels of security procedures are specified. This will determine the actions the model will take as follows:

Level 0 is a typical network mapping [4] that scans a specific network for connected hosts and their open ports. More formally, the model scans for all $IP_{Xi} \in N_X$ with a specified set of ports e.g. TCP 22, TCP 23, and TCP 2323.

Level 1 combines the network scanning activity with brute-force login attempts utilizing a list of common username and password pairs.

Level 2 scans the network, attempts to log in and if successful, tries to change the user password to a randomly generated string.

The network administrator or the security officer can select the security action level as they wish.

3.3 Login Procedures

The login procedure can be separated into two versions, one for SSH and another for Telnet.

Algorithm 1 describes the procedure for SSH login. It tries to log in to a host with well-known username and password pairs. Some SSH servers may simply disconnect the client, i.e., lock-out, if the client makes too many attempts with incorrect usernames and passwords. Our algorithm will deal with this using "timeout" which delays further attempts by a specific amount of time depending on the number of consecutive disconnections. We were inspired by the additive-increase/multiplicative-decrease feedback control algorithm used in TCP congestion control as shown in line 9 and 17.

Algorithm 2 describes the procedure for Telnet login. In the case of Telnet, the client first connects to the server with TCP and then pushes login username and password. The timeout behavior for Telnet is also the same as that of SSH.

The username and password pairs in the login credentials list (LoginCredList) can be obtained from publicly accessible sources such as the GitHub repository containing the Mirai malware source code [12].

Algorithm 1. SSH Login

LoginCredList is a list of well-known username and password pairs. **host** is the name of the target host or IP address.

```
1: i := 0
2: timeout := 0
3: sshloginsuccess := False
4: while i < len(LoginCredList) do
5:     (username, password) := LoginCredList[i]
6:     try:
7:         ssh.connect(host,22,username,password)
8:     catch AuthenticationException:
9:         timeout := timeout - 1
10:        if timeout <= 0 then
11:            timeout = 8
12:        end if
13:        i := i + 1
14:        continue
15:    catch RefusedConnectionException:
16:        sleep(timeout)
17:        timeout := timeout * 2
18:        continue
19:    catch GenericException:
20:        break
21:    sshloginsuccess = True
22:    break
23: end while
```

3.4 Overriding Passwords

After logging in, if the desired security action level is 2, the IoT-VuLock device tries to override the password of the target device, making note of both old and new passwords of the target device inside the security area of the IoT-VuLock device. The new password should be a randomly generated strong password, which cannot be easily broken by brute-force attacks. Depending on the target environment, the method or command for changing the password may be different. IoT-VuLock targets devices on the Linux platform with "passwd" command to push the stronger password.

4 Evaluation

We evaluated the scanning performance and the functions of overriding passwords with our prototype IoT-VuLock implementation on production and test networks. The prototype was written in Python, utilizing Paramiko library for

SSH connection and command execution, and TelnetLib for Telnet connection. The list of the username and password pairs was obtained from the Mirai source code hosted on github.com [12]. The number of unique username and password pairs is 62.

Algorithm 2. Telnet Login

LoginCredList is a list of well-known username and password pairs. **host** is the name of the target host or IP address.

1: **if** OpenPorts[host]['23'] **then**
2:     Telnet.connect(host,23)
3: **else if** OpenPorts[host]['2323'] **then**
4:     Telnet.connect(host,2323)
5: **end if**
6: i := 0
7: timeout := 0
8: telnetloginsuccess := False
9: **while** i < len(LoginCredList) **do**
10:    (username, password) := LoginCredList[i]
11:    **try:**
12:        Telnet.pushlogin(username,password)
13:    **catch** AuthenticationException:
14:        timeout := timeout - 1
15:        **if** timeout <= 0 **then**
16:            timeout := 8
17:        **end if**
18:        i := i + 1
19:        **continue**
20:    **catch** TelnetDisconnectedException:
21:        sleep(timeout)
22:        timeout := timeout * 2
23:        **continue**
24:    **catch** GenericException:
25:        **break**
26:    telnetloginsuccess = True
27:    **break**
28: **end while**

4.1   Scanning Performance

An evaluation of scanning performance was done on two university networks which we denote by A and B. Network A is a Linux user's network, and network B is a network for IoT building management, including light-control system, electric-power management system, heating, ventilation and air conditioning (HVAC) system. All of the devices of the login passwords were properly managed. The subnet of network A and B were both /25.

Table 1 shows the features of the two networks obtained by the scanning result at level 0. Table 2 shows the time consumed for SSH and Telnet login attempts.

Table 1. Number of open ports and scanning time (level 0)

Network	# of Hosts	Open TCP ports			Time [sec]
		22	23	2323	
A	51	40	3	0	166
B	43	21	5	0	39

Table 2. Time used to attempt logins (level 1)

Network	SSH [sec]	Telnet [sec]	Total [sec]
A	1713	1500	3213
B	890	347	1237

From these result, we can observe that level 0 scanning time cannot be simply estimated from the size of the network or the estimated number of open ports. This is due to varying network configurations on different networks. Furthermore, the time of login attempts in level 1 cannot simply be estimated especially in the case of Telnet. This is because the timeout values diverges from each other depending on the target service configurations. However, these scans were completed in realistic time. This indicates that IoT-VuLock can be used in practice.

4.2 Test of Password Overriding

This evaluation was done in a closed network with remote-login vulnerable devices. We prepared four Linux devices with different configurations in terms of their available ports and user accounts (username and password).

After executing our prototype at security level 2, it successfully changed the passwords for cases A, B, D, and E as shown in Table 3. It could not log into the device and could not change the password for case C, as the username and password pair was not included in the pre-defined username and password pairs list.

5 Discussion

IoT-VuLock can find devices with remote login vulnerability with level 0 and level 1 scanning and can fix such vulnerability by changing the password to a stronger one in level 2 operation. Level 0 operation is similar to the network scan provided by Nmap [4]. As an extension, we have proposed level 1 and 2. Users of IoT-VuLock must recognize that level 1 and 2 are very strong actions that should be used cautiously in the real networks. However, we have proposed these, due to the necessity to protect "unmanaged or irresponsibly deployed IoT devices" from malicious attackers.

IoT devices may be developed on many kinds of platforms, and Linux commands such as "passwd" are not always available. In this case, at security level 2 the action will be unsuccessful. Nevertheless, the list of devices with remote login vulnerabilities will be displayed to the network administrator as a level 1 action.

Table 3. Configuration for password overriding test (level 2)

Case	IP	Open ports	Username	Password
A	192.168.10.12	22	root	root
B	192.168.10.13	2323	root	default
C	192.168.10.14	22, 23	root	qwerty
D	192.168.10.14	22, 23	user	user
E	192.168.10.15	22	root	system

6 Conclusion

This paper proposed an architecture for IoT device vulnerability locking (IoT-VuLock) by enhancing network scans with login attempts, and proactively change weak passwords to stronger ones.

The evaluation result with our prototype implementation indicates that it can perform locking of vulnerable IoT devices in practical running time. Thus allowing network administrators to secure devices with high risks of infection.

References

1. Antonakakis, M., et al.: Understanding the mirai botnet. In: USENIX Security Symposium, pp. 1092–1110 (2017)
2. Kolias, C., et al.: DDoS in the IoT: mirai and other botnets. Computer **50**(7), 80–84 (2017)
3. Edwards, S., Profetis, I.: Hajime: Analysis of a decentralized internet worm for IoT devices. Rapidity Networks **16** (2016)
4. Lyon, G.F.: Nmap network scanning: the official Nmap project guide to network discovery and security scanning. Insecure (2009)
5. Abdou, A.R., Barrera, D., Van Oorschot, P.C.: What lies beneath? Analyzing automated SSH bruteforce attacks. In: International Conference on Passwords, pp. 72–91. Springer, Cham (2015)
6. Ramsbrock, D., Berthier, R., Cukier, M.: Profiling attacker behavior following SSH compromises. In: IEEE/IFIP DSN 2017, pp. 119–124 (2017)
7. Satoh, A., Nakamura, Y., Ikenaga, T.: SSH dictionary attack detection based on flow analysis. In: IEEE SAINT 2012, pp. 51–59 (2012)
8. Alsaleh, M., Mannan, M., Van Oorschot, P.C.: Revisiting defenses against large-scale online password guessing attacks. IEEE Trans. Dependable Secure Comput. **9**(1), 128–141 (2012)
9. Zhang-Kennedy, L., Chiasson, S., Van Oorschot, P.: Revisiting password rules: facilitating human management of passwords. In: IEEE APWG eCrime 2016, pp. 1–10 (2016)

10. Florencio, D., Herley, C., Van Oorschot, P.C.: An administrator's guide to internet password research. In: USENIX LISA 2014, pp. 35–52 (2014)
11. Patton, M., et al.: Uninvited connections: a study of vulnerable devices on the Internet of Things (IoT). In: IEEE JISIC 2014, pp. 232–235 (2014)
12. Mirai Source Code. https://github.com/jgamblin/Mirai-Source-Code. Accessed 30 Nov 2018

Adaptive Hierarchical Cache Management for Cloud RAN and Multi-access Edge Computing in 5G Networks

Deepika Pathinga Rajendiran and Melody Moh$^{(\boxtimes)}$

Department of Computer Science, San Jose State University, San Jose, CA, USA
{deepika.pathingarajendiran,melody.moh}@sjsu.edu

Abstract. Cloud Radio Access Networks (CRAN) and Multi-Access Edge Computing (MEC) are two vital technologies proposed for 5G mobile networks. CRAN utilizes the cloud model on top of the traditional RAN, thus provides scalability, flexibility, and improved resource utilization. MEC brings cloud computing services closer to the users to provide high bandwidth, low latency, real-time access and to improve the overall user experience. A cache may be included in both CRAN and MEC architectures to speed up time-critical communications as well as to provide preferential services. This work focuses on adaptive cache management of CRAN and MEC. First, a new hierarchical cache management algorithm, H-EXD-AHP (Hierarchical Exponential Decay and Analytical Hierarchy Process), is proposed to improve the existing EXD-AHP. Next, two adaptive versions of hierarchy-based cache management algorithms are described, with the goal to provide guaranteed-QoS. Finally, a third enhanced adaptive version is described to greatly speed up the run time. Performance evaluation of the new adaptive algorithms based on system parameters provided by Nokia Research are carried out with comparison with existing ones. Experimental results show that the new hierarchical H-EXD-AHP improves the performance in terms of increased cache hit rate as well as reduced network overhead and access delay. Furthermore, the three adaptive algorithms are effective in guaranteeing QoS and in reducing algorithm execution time. This work contributes significantly in realizing the support of 5G for IoT by enhancing CRAN and MEC performance and would have wide applications in other real-time systems that require efficient adaptive cache management with guaranteed QoS.

Keywords: 5G networks · Cache management ·
Cloud Radio Access Networks · Hierarchical cache management ·
Mobile networks · Multi-Access Edge Computing · Telecommunications

1 Introduction

The last decade has witnessed the rapid rise of intelligent mobile devices and Internet of Things (IoT) systems. Associating with this phenomenon is the increasing dependence of these technologies in everyday lives – including social, commercial, entertainment, education, healthcare, and government sectors. Fifth Generation (5G)

© Springer Nature Switzerland AG 2019
S. Lee et al. (Eds.): IMCOM 2019, AISC 935, pp. 234–253, 2019.
https://doi.org/10.1007/978-3-030-19063-7_20

mobile networks have been proposed to provide timely connectivity for these IoT and mobile devices in order to support the mounting services they deliver [21].

5G relies on many fast-growing technologies including Cloud Radio Access Networks (CRAN), Multi-Access Edge Computing (MEC), Millimeter Wave, Massive Multiple-Input Multiple-Output (MIMO), etc. Among them, CRAN and MEC utilizes cloud and virtualization models, making 5G systems flexible, scalable, and cost-effective. CRAN and MEC technologies will be described in a greater detail in Sect. 2.

Hierarchical memory systems usually make use of cache to reduce access delay for time-critical applications. Cache has been introduced in both CRAN and MEC to increase the speed of cellular network services as well as to improve resource utilization [7]. For MEC, caching on the edge (near the users) is even more effective. Instead of retrieving data from the Internet, it will be easier and quicker if there is content readily available as close to the users as possible [26].

The cost of these high-speed, carrier-grade cache resources is, however, extremely high, so their sizes are often limited. It is therefore necessary to efficiently manage these cache resources. In addition, as traffic load changes and the required Quality-of-Services (QoS) varies, it is critical to make the cache management adaptive. Furthermore, many applications in the 5G era demand some minimum QoS in terms of delay, throughput, or availability. It is thus highly desirable to provide guaranteed QoS in cache management schemes.

This work focuses on adaptive cache management to effectively utilize the limited cache resources in CRAN and in MEC. The major contributions may be summarized as follows:

Proposed H-EXD-AHP (Hierarchical Exponential Decay and Analytical Hierarchy Process), a hierarchical enhancement for an existing highly-effective cache management algorithm EXD-AHP [22].

Proposed AH (Adaptive Hierarchical) and IAH (Improved Adaptive Hierarchical); AH is an adaptive version for hierarchical cache management, and IAH is its improvement; both provide guaranteed QoS.

Proposed GGAH (Good Guess AH), a fast-convergence version for AH, which rapidly speeds up the execution time of the original AH.

AH, IAH, and GGAH have been applied to, and successfully demonstrated by both the newly proposed H-EXD-AHP and an existing hierarchical cache management algorithm H-PBPS (Probability Based Popularity Scoring) [11]. They both provide guaranteed QoS; GGAH in addition also swiftly speeds up the run time.

This paper is organized as follows: Sect. 2 discusses the background of CRAN and MEC architectures; it also presents major related works. Section 3 describes the most relevant existing cache management algorithms. Section 4 illustrates the preliminary results of the existing algorithms, which have motivated the design of new algorithms. Section 5 describes the proposed adaptive cache algorithms, including H-EXD-AHP, AH, IAH, and GGAH. Section 6 presents the performance evaluation of the proposed algorithms with comparison to existing ones. It is followed by the conclusion section that includes future directions. This work is a continuation of our research on cloud computing [8, 16], CRAN [10, 20], cache management of CRAN [11, 22–24], edge and fog computing [3, 5, 14], and IOT, mobile and 5G networks [18, 19, 21].

2 Background and Related Work

2.1 CRAN and MEC Architectures

Traditional Radio Access Networks (RAN) is a distributed architecture which consists of Long Term Evolution (LTE) Macro Base Station (MBS) (evolved Node B or eNodeB) and User Equipment (UE) [2]. Each Remote Radio Head (RRH) of eNodeB has its own Base Band Unit (BBU). They are connected to the core network or Evolved Packet Core (EPC) and the internet. CRAN is a centralized cloud architecture where BBUs are separated from their RRHs, virtualized and pooled together. The architecture of CRAN is shown in Fig. 1. Each BBU can be represented as a Virtual Machine (VM) and each VM has its own cache memory. The size of this cache memory involved in the BBU pool is limited, so it is important to manage this resource efficiently [24, 25].

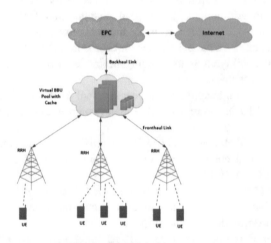

Fig. 1. CRAN architecture.

For each user, a User Equipment Context (UEC) record information is stored in the secondary cloud memory. This UEC information contains user's ID, state information of the current event or session, subscription details etc. So, instead of retrieving this information from the secondary cloud, it will be easier to access if it is stored in the BBU pool cache. CRAN aims for better scalability, flexibility and better resource utilization [24].

In MEC, cloud computing services are brought near the users to improve the user's experience [26]. These services include but not limited to content caching, task offloading, storage, and computation. The architecture of MEC is shown in Fig. 2. These servers can share its cached items with each other using the communications between MBS. To achieve communication between MBS, X2 interface can be used [7]. MEC aims to reduce latency and backhaul traffic flow of the networks.

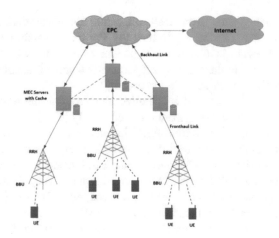

Fig. 2. MEC architecture.

CRAN and MEC technologies can be combined and form a new hybrid architecture as shown in Fig. 3. CRAN uses centralized BBU pool whereas MEC servers usually work with distributed MBS [12, 13]. In this hybrid architecture, MEC's request can be sent and received via MBS (RRH and BBU pool) [12]. For 5G networks, it can be either CRAN architecture or MEC architecture or combination of both architectures.

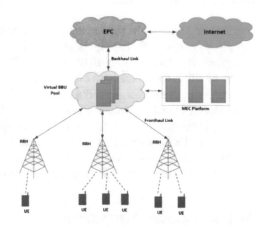

Fig. 3. Hybrid CRAN and MEC architecture.

2.2 Related Studies

Floratou et al. proposed adaptive Selective Least Recently Used – K (SLRU-K) and adaptive Exponential Decay (EXD) caching algorithms for Big SQL, in which they mentioned that their parameter K value and parameter a value for adaptive SLRU-K and adaptive EXD respectively has significant impact on changing workload, so they changed the parameter values dynamically according to the workload which resulted in better performance than existing algorithms [4]. For our research, we are using EXD for scoring the elements in the cache. If parameter a value is higher, more recent elements will be placed in the cache and if parameter a value is smaller, more frequent elements will be given importance in the cache [4].

Huang, Zhao, and Zhang used cooperative multicast caching mechanism in MEC between base stations to utilize the resources efficiently [9]. Cooperative means if the content is not in the small base station (SBS) instead of accessing it from MBS we can try to access from another SBS. Multicast means instead of serving multiple requests separately and storing it in each SBS, we can multicast the popular videos to all. Thereby saving storage space and decreasing energy consumption. They demonstrated that by caching in the edge, network latency can be reduced, and CHR can be increased [9].

Tran, Hajisami, and Pompili cached in a hierarchical manner [25]. Their research experiment showed us that caching in the edge (RRH) is better than caching in the BBU pool (cloud cache) since RRH is nearer to the users. Their performance metrics were CHR, latency, and backhaul traffic load [25].

Tsai and Moh adopted Exponential Decay [4] and Analytical Hierarchy Process [6, 17] came up with an algorithm called EXD-AHP scoring algorithm in which lowest score UEC is evicted to make space for highest score UEC [22, 24]. They have used four different levels of SLA users according to their mobility, basic and premium services. Using their algorithm, network traffic and cloud writes were reduced which increased their cache hit when compared to other existing algorithms [24]. The algorithm is described in greater detail in Sect. 5. We have enhanced this algorithm; i.e., H-EXD-AHP.

Kaur and Moh proposed new algorithms for cache management in 5G. In this, for scoring the UEC records, they have used Probability Based Popularity Scoring (PBPS). One such algorithm is Reverse Random Marking (RRM) with PBPS (RRM+PBPS) in which a certain percentage of UECs are marked after it reaches a threshold. If we want to evict the records, we can only evict from unmarked records. If all the records are marked, then increase the threshold and continue the same process. Another algorithm is that their PBPS scoring technique is combined with Hierarchy is called PBPS +Hierarchy (represented as H-PBPS in our paper) in which the cache is partitioned and allocated to users according to their service levels [11]. These algorithms which are simple in design were compared in terms of CHR, latency, and network traffic.

3   Existing Cache Management Algorithms

In this section, we are going to discuss some of the existing cache management algorithms for 5G mobile networks. Before that, the following flowchart gives a general idea of the structure of the algorithms. This flowchart shown in Fig. 4 is adapted from Tsai and Moh's work [24] and this flowchart is also used in Kaur and Moh's work [11]. A brief description of the flowchart follows.

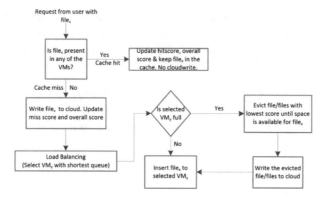

Fig. 4. Cache management flowchart.

In the existing papers, only UEC for CRAN is considered which is a small size record which is user's metadata. We have increased this record size and now it can be considered as user content files which can be used in MEC applications. Note that if we are using the algorithm or flowchart for BBU caching from CRAN then the "file" is a UEC record, and if we are using MEC server caching then it is file content. Each user with file or UEC request is incoming and if that file or UEC is already present in any of the Virtual Machine (VM) cache, then it is a cache hit and that file or UEC is updated in the cache. One must update the hit score and calculate the overall score according to the algorithm and there is no need to write in the cloud. On the other hand, if the requested file is not present in any of the VMs then it is a cache miss. One must write that file in the secondary cloud storage. After that, one must update the miss score and calculate the overall score according to the algorithm. While trying to add this file to any one of the VMs so that if it is requested in future again it will be present in the cache, we adopted S_{queue} LB algorithm [22, 23]. If the selected VM is full, evict files with the lowest score until space is available for this file. Write the evicted file or files to the cloud storage. After eviction when space is available, insert this file into VM [11, 24].

3.1   Least Frequently Used (LFU) Algorithm

LFU algorithm is used as a baseline comparison algorithm in the preliminary results section. LFU is a classic algorithm which keeps track of the number of times files are being accessed. When it's time of eviction, the least number of accessed or least frequently used files are evicted [15].

3.2   EXD-AHP Algorithm

Tsai and Moh proposed EXD-AHP algorithm which uses a scoring method to decide which files need to be present in the cache and which files need to be evicted from the cache [22, 24]. To calculate the scoring both EXD and AHP weights calculations are considered [4, 17]. For EXD, there is a parameter a in which we can tune it to keep the recently or frequently used files in the cache. Higher the parameter a value, the system keeps recently used files in the cache and smaller the parameter a value it leans towards the frequency of elements [4]. AHP weights are calculated by creating a matrix based on the SLA users [17]. If the file is accessed at time $ui1+\Delta u$ for the first time after $ui1$, then scoring is calculated as follows [24]:

$$S_i(ui1 + \Delta u) = S_i(ui1) * e^{-a\Delta u} + W_{AHP} \tag{1}$$

If the file didn't get requested at a time interval [$ui1$, ($ui1+\Delta u$)], then the score is calculated as follows [24]:

$$S_i(ui1 + \Delta u) = S_i(ui1) * e^{-a\Delta u} \tag{2}$$

Where $S_i(ui1)$ is the score (weight), $e^{-a\Delta u}$ is EXD calculation and W_{AHP} is AHP weight. From Eqs. (1) and (2), we can tell that the score of the file depends on both EXD and AHP weight calculations.

3.3   PBPS Cache Management Algorithms

Kaur and Moh proposed PBPS algorithms which also use a scoring technique to determine whether a file should be present in VM or not. This scoring is calculated using the Eq. (3).

$$\frac{hit\ score\ Li}{\#\ requests}(1 - \left(\frac{hit\ score\ Li}{\#\ requests}\right)^{\frac{miss\ score}{\#\ requests}}) \tag{3}$$

As we can see from Eq. (3), the score is calculated based on both cache hits and cache misses. At a given point an overall score can be calculated for a file which tells us how popular the file is based on hits and misses. The higher the score, the higher the popularity and less chance of eviction. This file will probably stay inside the cache because of its high PBPS score. The requests are constantly changing so the scores are continuously updated. The algorithm uses a rewarding system, in which the quantity of reward is varied according to the SLA users [11].

RRM with PBPS (RRM+PBPS) Algorithm. This algorithm proposed by Kaur and Moh uses PBPS scoring method and in addition to that, it uses RRM. When files are being requested and it is in the cache then it is a cache hit. If the file score exceeds a certain threshold it is marked. Then if we must evict some files to make room for new files, eviction process happens from unmarked files. If all the files are marked, then the threshold value M_t is marginally increased to unmark a certain number of files [11].

PBPS+Hierarchy (H-PBPS) Algorithm. This algorithm also uses the PBPS scoring method to calculate the addition and eviction of files in the cache. On top of this, the entire cache is divided, and each cache partition is dedicated to the users according to their SLA. This forms a hierarchy of users. We are giving preferential treatment, so the higher-SLA files will get the larger size of cache partition. The file addition or eviction happens in their allocated cache partition only [11].

4 Preliminary Results

4.1 Experiment Setup

For our experiment setup, we have used the CloudSim simulator [1]. This simulator is very popular and effective to cloud based applications. The following Table 1 shows the experiment parameter for our simulations [11, 24]; most have been provided by Nokia Lab researchers.

Table 1. Simulation parameters and values.

Parameter	Values
No. of host and of VM	1 and 4, respectively
VM cache sizes	0.75 GB & 2 GB
Arrival rate of files into the network	1400 files/sec
Total no. of users and of requests	25,000 and 420,000, respectively
File sizes	200 KB: fixed and distributed; 2000 KB: fixed and distributed; Distributed: uses the Normal (Gaussian) distribution
Network bandwidth	1 Gbps
QoS level: mobility; subscription	SLA 1: High Mobility; Premium SLA 2: Low Mobility; Premium SLA 3: High Mobility; Basic SLA 4: Low Mobility; Basic
EXD parameter a and LB algorithm	10^{-3} and S_{queue}
Analytical Hierarchical Process (AHP) weights	SLA1: 0.58; SLA2: 0.28; SLA3: 0.10; SLA4: 0.04
QoS level: PBPS hit reward	SLA1: 1; SLA2: 0.75; SLA3: 0.5; SLA4: 0.25

4.2 Preliminary Performance Evaluation

The preliminary performance evaluation section mainly serves the purpose of motivating the proposed new algorithms and the following performance evaluation on different file sizes, different cache distribution sizes are not evaluated in the existing papers. So, these new evaluation results from existing algorithms gave us a different perspective which inspired us to propose our algorithms.

Cache Hit Rate (CHR). CHR is an important performance evaluation metric for cache management problems. The following results use CHR for its performance measure. For different service levels L_i, CHR can be calculated using the following Eq. (4) [24]:

$$\text{CHR of } L_i = \frac{\text{Total number of Li cache hits}}{\textit{Total number of file requests from Li}} \tag{4}$$

Different File Sizes. In this, we used different file sizes for EXD-AHP and H-PBPS algorithms [11, 24]. Distributed means the requested files sizes are varied, Gaussian (Normal) distribution is used. In Table 2, different file sizes used for the simulations are displayed. Figure 5 comparing 200 KB and 2000 KB file sizes, the small file size gives the highest CHR; and comparing fixed and distributed file sizes, fixed file size gives the highest CHR. Note: EXD-AHP (1) means 200 KB fixed file size is used in EXD-AHP algorithm, H-PBPS (2) means 200 KB distributed file size is used in H-PBPS algorithm and so on. The EXD-AHP performs better than H-PBPS in all different settings.

Table 2. Different file sizes (cache size = 0.75 GB).

#	File size	Values
1.	200 KB fixed	CRAN
2.	200 KB distributed	MEC
3.	2000 KB fixed	MEC
4.	2000 KB distributed	MEC

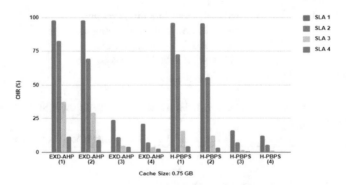

Fig. 5. Cache Hit Rate (CHR) (%) for different file sizes.

Different Cache Sizes. Table 3 displays the list of different cache sizes used for EXD-AHP and H-PBPS algorithms [11, 24]. CHR for different cache sizes is displayed in Fig. 6. Of course, cache size 2 GB gives better CHR than cache size 0.75 GB. In 0.75 GB cache size, EXD-AHP algorithm gives better CHR than H-PBPS algorithm. After this experiment, it is apparent that EXD-AHP performs better than H-PBPS.

Table 3. Different cache sizes.

#	File size	Cache size	Application
1.	200 KB fixed	0.75 GB	CRAN
2.	200 KB distributed	0.75 GB	MEC
3.	200 KB fixed	2 GB	CRAN
4.	200 KB distributed	2 GB	MEC

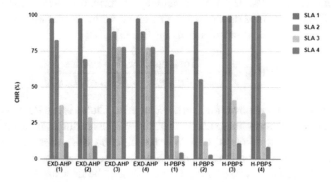

Fig. 6. Cache Hit Rate (CHR) (%) for different cache sizes.

Different Cache Management Algorithms. CHR for different existing cache manage-ment algorithms is evaluated in Fig. 7. For RRM+PBPS the threshold increase is 10%. Different cache size partitions for SLA levels are summarized in Table 4. For example, H-PBPS (2) means CD for SLA1 is 55%, SLA2 is 28% and SLA3 is 11% distributed in H-PBPS algorithm. We can see that if the cache size partition is changed, CHR is also changed accordingly. This motivates the design of H-EXD-AHP and of various AH's.

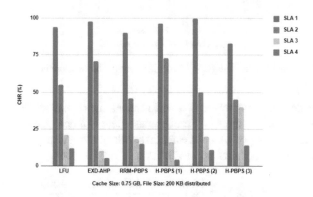

Fig. 7. Cache Hit Rate (CHR) (%) for different cache management algorithms.

Table 4. Different cache distribution (CD) for H-PBPS.

#	CD for SLA1	CD for SLA2	CD for SLA3
1.	70%	20%	8%
2.	55%	28%	11%
3.	47%	25%	20%

Cloud Write Rate (CWR). CWR is another performance metric for these cache algorithms. CWR is defined as the number of cloud writes by the number of requests. CWR for service level L_i is calculated using Eq. (5) [24]:

$$\text{CWR of } L_i = \frac{\text{Total number cloud writes}}{\textit{Total number of file arrivals in Li}} \tag{5}$$

Cloud write is needed whenever there is a cache miss. Cloud Write Rate and Cache Hit Rate are inversely proportional to each other. H-PBPS uses different cache distribution from Table 4. Figure 8 displays the CWR for different SLA users. Note that CWR varies according to the different cache distributions. Cache size also has a huge impact on CWR. As the cache size grows the CWR will also be decreased.

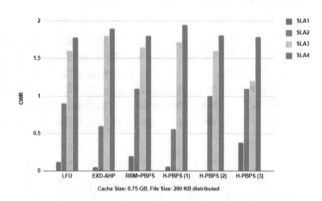

Cache Size: 0.75 GB, File Size: 200 KB distributed

Fig. 8. Cloud Write Rate (CWR) for different cache management algorithms.

Network Traffic. Network traffic is a performance metric calculated based on cloud writes. Because whenever there is a cache miss, there is a need for files to travel from cache to the cloud storage, which creates network traffic. Assuming each file size is 200 KB on average, it is calculated as follows in Eq. (6) [24]:

$$\text{Network Traffic of } L_i = \frac{\text{Number of Li cloudwrites} * 200 * 8 * 1000}{\textit{Simulation Time}} \tag{6}$$

Network traffic is shown in Fig. 9. This traffic result is not based on the different SLA, but the overall traffic result for different cache management algorithms. Note that CHR and network traffic are inversely proportional while CWR and network traffic are directly proportional. EXD-AHP algorithm has less network traffic overall. Different cache distribution of H-PBPS is used from Table 4. This shows that by changing the cache distribution the traffic can also be decreased. So, this also gave us the idea of applying the hierarchy cache partition to EXD-AHP and try to decrease the traffic even more.

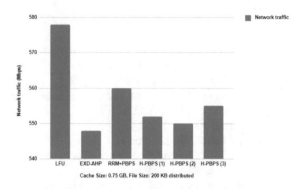

Fig. 9. Network traffic (Mbps) for different cache management algorithms.

5   Proposed Adaptive Cache Management Algorithms

The results presented in the previous section gave us the motivation to design the following proposed algorithms. For the first part, inspired by the change in cache distribution of H-PBPS, we applied the hierarchical part to EXD-AHP, called H-EXD-AHP. For the second part, we designed 3 algorithms that adaptively change the cache distribution to guarantee the QoS; more specifically, to meet the Minimum Guarantee of CHR for differentiated users according to their SLA. We applied these 3 algorithms to H-EXD-AHP and H-PBPS.

5.1   New: H-EXD-AHP Algorithm

Inspired by H-PBPS algorithm from the previous section, a hierarchical layer of cache allocation for differentiated SLA users according to their preferred treatment are applied in EXD-AHP [11]. The total cache is divided according to the type of SLA users. The H-EXD-AHP algorithm is as follows:

H-EXD-AHP Algorithm

```
1. For each file_x request with SLA L_i;
2.        Calculate new EXD-AHP score using equation (1);
3.        Update score of remaining files using
          equation (2);
4.        If  file_x is present in any one of the VMs;
          /* cache hit
5.            Update score of file_x;
6.            Return;
7.        Else; /* cache miss
8.            Write file_x L_i to the cloud;
9.            Select one of the VM using S_queue LB;
10.           If the selected VM has free space for file_i;
11.               Insert file_x with L_i;
12.               Return;
13.           Else if the VM is full;
14.               Find file_y which has the lowest score;
15.               Evict file_y from L_i;
16.               Write evicted file to the cloud;
17.           Else if file_x score is smaller than the
              lowest score;
18.               Evict lowest scored file from the cache;
19.               Write file_y to the cloud;
20.               Insert file_x L_i in the cache;
21.               Return;
```

5.2 New: Adaptive Hierarchy Cache Management Algorithms

AH Algorithm. This algorithm is an enhancement for both H-EXD-AHP and H-PBPS algorithms. In those algorithms, the cache size partitioned for SLA users is fixed and cannot be changed adaptively according to the runtime requirement. In the AH algorithm, the algorithm adaptively adjusts the Cache Distribution (CD) to improve CHR and to meet the Minimum Guarantee of CHR [11].

The first step is identical to either H-EXD-AHP scoring or H-PBPS scoring. For a given Traffic Distribution (TD), and with an initial Cache Distribution (CD), CHR is measured for each SLA. Minimum Guarantee of Cache Hit Rate is set for different SLA users. While any one of the SLA's CHR didn't meet the Minimum Guaranteed cache hit rate (MG) (CHR < MG), then, the surplus cache size is measured for each SLA whose CHR > MG. Next, the surpluses are arranged in increasing order. Then choose the highest surplus Li so that it can give some of its cache sizes to a deficit SLA. Choose the highest preferred SLA, whose CHR didn't meet the MG. Give X% (in our case 20%) of cache size from surplus CD to deficit CD. Remove X% (20%) of the surplus CD and update the new cache sizes. Do this until all SLA's MG is met. (The number of iterations to achieve the MG is also recorded and is used to compare the run time of all the AH's).

AH Algorithm

```
1.   Use H-EXD-AHP or H-PBPS algorithm;
2.       Measure CHRᵢ of each SLA Lᵢ for i = 1,2...n;
3.       While at least any one of the CHRᵢ <  MGᵢ for Lᵢ;
4.           Measure Surplusᵢ = CHRᵢ - MGᵢ for each Lᵢ
             where CHRᵢ > MGᵢ;
5.           Arrange Surplusᵢ in increasing order;
6.               Choose Lᵢ, which has the highest
                 Surplusᵢ;
7.               Choose Lⱼ, which has the highest SLA
                 preference AND CHRⱼ < MGⱼ for j = 1,2...n
                 and i≠j;
8.                   Set CDⱼ = CDⱼ + [CDᵢ * (X/100)];
9.                       Set CDᵢ = CDᵢ - [CDᵢ * (X/100)];
10.      GoTo Step 1;
11.      END While; \* when CHRᵢ >= MGᵢ for each Lᵢ
12.  Return;
```

Improved (I) AH Algorithm. This IAH algorithm is an improved version of AH algorithm, in this instead of borrowing from only one surplus SLA, we are going to borrow from top two highest surplus SLA and give it to deficit SLA users. From the first highest surplus we borrowed X% (in our case 20%) of its cache size and from second highest surplus we borrowed Y% (15%). Below is the IAH algorithm.

IAH Algorithm

```
1.   Use H-EXD-AHP or H-PBPS algorithm;
2.       Measure CHRᵢ of each SLA Lᵢ for i = 1,2...n;
3.       While at least any one of the CHRᵢ <  MGᵢ for Lᵢ;
4.           Measure Surplusᵢ = CHRᵢ - MGᵢ for each Lᵢ
             where CHRᵢ > MGᵢ;
5.           Arrange Surplusᵢ in increasing order;
6.               Choose Lᵢ, which has the highest
                 Surplusᵢ;
7.               Choose Lⱼ, which has the second highest
                 Surplusⱼ for j = 1,2...n and i≠j;
8.       Choose Lₖ, which has the highest SLA preference AND
         CHRₖ < MGₖ for k = 1,2...,n, k≠i and k≠j;
9.           Set CDₖ = CDₖ + [CDᵢ * (X/100)] +
             [CDⱼ * (Y/100)] where X > Y;
10.              Set CDᵢ = CDᵢ - [CDᵢ * (X/100)];
11.              Set CDⱼ = CDⱼ - [CDⱼ * (Y/100)];
12.      GoTo Step 1;
13.      END While; \* when CHRᵢ >= MGᵢ for each Lᵢ
14.  Return;
```

In this, we are Choosing the first and second highest surplus. Choose the highest preferred SLA Lj, whose CHR didn't meet the MG. Give X% (20%) of CDi AND Y% (15%) of CDj to CDk where X > Y. Remove X% (20%) of CDi from CDi. Remove Y % (15%) of CDj from CDj and update all cache sizes. End while all SLA's MG is met.

Good Guess AH (GGAH) Algorithm. The main purpose of GGAH is to speed up the convergence time in the above AH algorithms. In GGAH algorithm, instead of borrowing cache sizes from surplus SLAs, we try to meet the MG of CHR using the GG formula shown below. From previous algorithms, it is observed that for a given TD_i, CD_i is directly proportional to CHR_i for each L_i, that is, $CD_i = k_i * CHR_i$. Thus,

$$k_i = CD_i/CHR_i \tag{7}$$

To find Good Cache Distribution GCD_i, it should also be directly proportional to MG_i. So,

$$GCD_i = k_i * MG_i \tag{8}$$

To get the GCD_i value for each L_i, we first find the value of constant k_i for each L_i from Eq. (7) and substitute that k_i in Eq. (8). GGAH algorithm is as follows:

GGAH Algorithm

```
1. Use H-EXD-AHP or H-PBPS algorithm
2. Measure CHRᵢ of each SLA Lᵢ for i = 1,2...n;
3.      While at least any one of the CHRᵢ <  MGᵢ for Lᵢ;
4.           Calculate kᵢ using equation (7);
5.           Substitute kᵢ from equation (7) and calculate
             GCDᵢ using equation (8);
6.      If GCDᵢ of each Lᵢ sum is not equal to 100 then
        normalize;
7.      GoTo Step 1;
8.      END While; \* when CHRᵢ >= MGᵢ for each Lᵢ
9. Return;
```

5.3   New: Adaptive Hierarchy Cache Management Algorithms

The three versions of AH (namely, AH, IAH, and GGAH) are applied to H-EXD-AHP and H-PBPS. The algorithms are as follows:

Adaptive H-EXD-AHP Algorithms	Adaptive H-PBPS Algorithms
AH-EXD-AHP Algorithm	**AH-PBPS Algorithm**
1. Use H-EXD-AHP algorithm;	1. Use H-PBPS algorithm;
2. Use AH algorithm;	2. Use AH algorithm;
IAH-EXD-AHP Algorithm	**IAH-PBPS Algorithm**
1. Use H-EXD-AHP algorithm;	1. Use H-PBPS algorithm;
2. Use IAH algorithm;	2. Use IAH algorithm;
GGAH-EXD-AHP Algorithm	**GGAH-PBPS Algorithm**
1. Use H-EXD-AHP algorithm;	1. Use H-PBPS algorithm;
2. Use GGAH algorithm;	2. Use GGAH algorithm;

6 Performance Evaluation

In this section, the performance evaluation of proposed algorithms is discussed. First, H-EXD-AHP results are presented. Next, the 3 AH versions for H-EXD-AHP and H-PBPS are shown. The parameter values are the same as those shown on Table 1 unless otherwise specified.

6.1 Cache Hit Rate for H-EXD-AHP Algorithm

CHR for H-EXD-AHP is displayed and compared with H-PBPS algorithm in Fig. 10, using the cache partition shown in Table 4. Figure 10 uses 0.75 GB cache size and 200 KB distributed file size. Note that H-PBPS algorithms give good CHR for SLA 1 and SLA 2, but for SLA 3 and SLA 4 the CHR is less. Whereas H-EXD-AHP algorithm performs better because CHR is higher for all the SLA users when compared with H-PBPS algorithms. It is also clear that different cache partitions resulting in different performances.

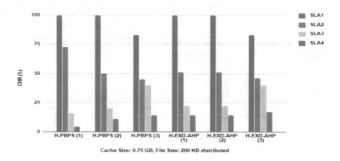

Fig. 10. Cache Hit Rate (CHR) (%) for H-EXD-AHP (Cache Size 0.75 GB, File Size 200 KB distributed).

6.2   Convergence Time for AH Algorithms

While all three AH versions guarantee QoS (in terms of CHR), this subsection evaluates the number of iterations each takes to achieve the Minimum Guarantee.

AH-EXD-AHP Results. Tables 5 and 6 show the number of iterations needed to reach MG for different adaptive cache algorithms. For Table 5, the traffic distribution is 25% for each SLA and MG is varied. For the result shown on Table 6, MG is 60% for SLA1, 50% for SLA2, 35% for SLA3 and 30% for SLA4 and traffic distribution is varied. Clearly from the results of the two tables, GGAH successfully speeds up the convergence time by requiring only 2 iterations, whereas IAH improves over AH in run time.

Table 5. Number of iterations for different MG (TD = 25%).

Minimum Guarantee of Cache Hit Rate (%)	95, 50, 20, 10	80, 45, 35, 15	70, 40, 30, 20	60, 50, 35, 30
AH-EXD-AHP	8	5	4	4
IAH-EXD-AHP	6	4	4	3
GGAH-EXD-AHP	2	2	2	2

Table 6. Number of iterations for different TD.

Traffic distribution (%)	30, 25, 15, 30	30, 20, 15, 35	25, 15, 10, 50	30, 20, 10, 40
AH-EXD-AHP	5	4	5	4
IAH-EXD-AHP	4	4	4	4
GGAH-EXD-AHP	2	2	2	2

AH-PBPS Results. Tables 7 and 8 are the corresponding results for AH-PBPS algorithms; they show the number of iterations needed to reach MG. For Table 7, the traffic distribution is 25% for each SLA and MG is varied. For Table 8, MG is 60% for SLA1, 50% for SLA2, 35% for SLA3 and 30% for SLA4 and traffic distribution is varied. Again, GGAH successfully speeds up the convergence time by requiring only 2 iterations, whereas IAH also improves over AH in run time.

Table 7. Number of iterations for different MG (TD = 25%).

Minimum Guarantee of Cache Hit Rate (%)	95, 50, 20, 10	80, 45, 35, 15	70, 40, 30, 20	60, 50, 35, 30
AH-PBPS	9	5	4	5
IAH-PBPS	7	4	3	3
GGAH-PBPS	2	2	2	2

Table 8. Number of iterations for different TD.

Traffic distribution (%)	30, 25, 15, 30	30, 20, 15, 35	25, 15, 10, 50	30, 20, 10, 40
AH-PBPS	6	4	6	4
IAH-PBPS	4	5	5	4
GGAH-PBPS	2	2	2	2

Furthermore, Comparing Table 7 with its counterpart in Table 5 (and Tables 8 and 6), we see that the first two AH-PBPS algorithms require more iterations than the corresponding AH-EXP-AHP ones except in the GGAH version. Thus, not only H-EXP-AHP algorithm not only performs better than H-PBPS (as illustrated in Sect. 6.1), its three adaptive versions also provide faster iteration than the ones of H-PBPS. It shows the AH-EXP-AHP is overall a more superior adaptive cache management algorithm than AH-PBPS.

7 Conclusion

This work has addressed the challenge of providing fast access and guaranteed QoS in the 5G era by proposing adaptive cache management algorithms for CRAN and MEC. While most existing cache management research has focused on improving cache hit rates, we design algorithms that not only provide high cache hits, but also adaptively adjust to incoming workload, and are able to guarantee minimum QoS for different SLA levels. First, preliminary results of several existing algorithms are illustrated to motivate new algorithm designs. Then, H-EXD-AHP, a hierarchical version is proposed for an existing high-performing algorithm, EXD-AHP. Next, two adaptive algorithms, AH and IAH, are described for hierarchical-based cache management, both guarantee the minimum QoS requested. Then, to improve the convergence time, GGAH is proposed, which first derives the proper cache distribution, so the algorithm can quickly meet the promised QoS. The three versions of AH are applied to and illustrated by both new H-EXD-AHP and an existing H-PBPS. Performance evaluation have shown that the three adaptive hierarchical algorithms have successfully guaranteed minimum QoS, among them GGAH is able to rapidly converge the algorithm execution to the desired QoS. Future work would apply the three AH algorithms to different cache management algorithms [4, 15, 25] and in other fog computing and edge computing settings [3, 5–7], and experiment the AH algorithms on real private and public cloud systems.

Acknowledgements. This work is supported in part by Nokia Foundation and by San Jose State University RSCA award.

References

1. Calheiros, R.N., Ranjan, R., Beloglazov, A., De Rose, C.A.F., Buyya, R.: CloudSim: a toolkit for modeling and simulation of cloud computing environments and evaluation of resource provisioning algorithms. Softw. Pract. Exp. **41**(1), 23–50 (2011)
2. Checko, A., et al.: Cloud RAN for mobile networks—a technology overview. IEEE Commun. Surv. Tutor. **17**(1), 405–426 (2015)
3. Choudharik, T., Moh, M., Moh, T.-S.: Prioritized task scheduling in fog computing. In: Proceedings of the ACM Annual Southeast Conference (ACMSE), Richmond, KY, March 2018
4. Floratou, A., et al.: Adaptive caching in big SQL using the HDFS cache. In: Proceedings of the Seventh ACM Symposium on Cloud Computing (SoCC 2016), New York, NY, USA, pp. 321–333 (2016)
5. Gao, L., Moh, M.: Joint computation offloading and prioritized scheduling in mobile edge computing. In: Proceedings of IEEE International Conference on High Performance Computing and Simulation, Orleans, France, July 2018
6. Gomes, A., Braun, T., Monteiro, E.: Enhanced caching strategies at the edge of LTE mobile networks. In: IFIP Networking Conference (IFIP Networking) and Workshops (2016)
7. Hou, T., Feng, G., Qin, S., Jiang, W.: Proactive content caching by exploiting transfer learning for mobile edge computing. In: GLOBECOM 2017 - 2017 IEEE Global Communications Conference, Singapore, pp. 1–6 (2017). https://doi.org/10.1109/glocom.2017.8254636
8. Huang, J., Wu, K., Moh, M.: Dynamic virtual machine migration algorithms using enhanced energy consumption model for green cloud datacenters. In: 2014 International Conference on High Performance Computing & Simulation (HPCS), pp. 902–910. IEEE (2014)
9. Huang, X., Zhao, Z., Zhang, H.: Cooperate caching with multicast for mobile edge computing in 5G networks. In: 2017 IEEE 85th Vehicular Technology Conference (VTC Spring), Sydney, NSW, pp. 1–6 (2017)
10. Karneyenka, U., Mohta, K., Moh, M.: Location and mobility aware resource management for 5G cloud radio access networks. In: 2017 International Conference on High Performance Computing and Simulation (HPCS), Genoa, Italy, July 2017
11. Kaur, G., Moh, M.: Cloud computing meets 5G networks: efficient cache management in cloud radio access networks. In: Proceedings of The 2018 Annual ACM Southeast Conference, Richmond, Kentucky, March 2018
12. Li, T., Magurawalage, C.S., Wang, K., Xu, K., Yang, K., Wang, H.: On efficient offloading control in cloud radio access network with mobile edge computing. In: 2017 IEEE 37th International Conference on Distributed Computing Systems (ICDCS), Atlanta, GA, pp. 2258–2263 (2017)
13. Lin, D., Hsu, Y., Wei, H.: A novel forwarding policy under cloud radio access network with mobile edge computing architecture. In: 2018 IEEE 2nd International Conference on Fog and Edge Computing (ICFEC), Washington, DC, pp. 1–9 (2018)
14. Moh, M., Raju, R.: Machine learning techniques for security of Internet of Things (IoT) and fog computing systems. In: Proceedings of IEEE International Conference on High Performance Computing and Simulation, Orleans, France, July 2018. (Invited Paper)
15. Podlipnig, S., Böszörmenyi, L.: A survey of web cache replacement strategies. ACM Comput. Surv. **35**(4), 374–398 (2003)
16. Reguri, V.R., Kogatam, S., Moh, M.: Energy efficient traffic-aware virtual machine migration in green cloud data centers. In: Proceedings of the Second IEEE International Conference on High Performance and Smart Computing, New York, April 2016

17. Saaty, R.: The analytic hierarchy process—what it is and how it is used. Math. Model. **9**(3–5), 161–176 (1987)
18. Sathyanarayana, S., Moh, M.: Joint route-server load balancing in software defined networks using ant colony optimization. In: Proceedings of the International Conference on High Performance Computing and Simulation (HPCS), Innsbruck, Austria, July 2016
19. Shahriari, B., Moh, M.: Intelligent mobile messaging for urban networks. In: Proceedings of the 12th IEEE International Conference on Wireless and Mobile Computing, Networking and Communications (WiMob), New York, 17–19 October 2016 (2016)
20. Shahriari, B., Moh, M., Moh, T.-S.: Generic online learning for partial visible dynamic environment with delayed feedback. In: Proceedings of the International Conference on High Performance Computing and Simulation (HPCS), Genoa, Italy, July 2017
21. Su, G., Moh, M.: Improving energy efficiency and scalability for IoT communications in 5G networks. In: Proceedings of 12th ACM International Conference on Ubiquitous Information Management and Communication (IMCOM), Langkawi, Malaysia, January 2018
22. Tsai, C., Moh, M.: Abstract: cache management and load balancing for 5G cloud radio access networks. In: ACM Symposium on Cloud Computing, Santa Clara, USA, September 2017
23. Tsai, C., Moh, M.: Load balancing in 5G cloud radio access networks supporting IoT communications for smart communities. In: 2017 IEEE International Symposium on Signal Processing and Information Technology (ISSPIT), Bilbao, pp. 259–264 (2017)
24. Tsai, C., Moh, M.: Cache management for 5G cloud radio access networks. In: Proceedings of ACM International Conference on Ubiquitous Information Management and Communication, Langkawi, Malaysia, January 2018
25. Tran, X.T., Hajisami, A., Pompili, D.: Cooperative hierarchical caching in 5G cloud radio access networks. IEEE Netw. **31**(4), 35–41 (2017)
26. Wang, S., Zhang, X., Zhang, Y., Wang, L., Yang, J., Wang, W.: A survey on mobile edge networks: convergence of computing, caching, and communications. IEEE Access **5**, 6757–6779 (2017)

Secure Framework and Security Mechanism for Edge Nodes in Industrial Internet

Min Wei[1(✉)], Xu Yang[1], Jiuchao Mao[1], and Keecheon Kim[2]

[1] Industrial IoT International S&T Cooperation Base,
Chongqing University of Posts and Telecommunications, Chongqing, China
weimin@cqupt.edu.cn, 1085418916@qq.com,
871133493@qq.com
[2] Department of Computer Science and Engineering, Konkuk University,
Seoul, Republic of Korea
kckim@konkuk.ac.kr

Abstract. The introduction of edge computing into the industrial internet brings many benefits. It can meet the demand of the real-time transmission and high reliability of industrial data. It is a challenge to ensure the data confidentiality between field nodes and edge nodes while meeting the requirements of high real-time in the industrial internet. This paper proposes a security framework and designs a security mechanism for edge nodes in the industrial internet. A test platform is implemented. The result shows that the proposed scheme is efficient, the data confidentiality in the industrial internet with edge side are effectively improved.

Keywords: Industrial internet · Edge nodes · Encryption

1 Introduction

The industrial internet is the integration and linking of big data, analytical tools and wireless networks with physical and industrial nodes, or otherwise applying meta-level networking functions, to distributed systems [1]. The industrial internet provides a method to integrate the machine sensors, middleware, software, cloud computing, and storage systems, improves operating efficiency and accelerates productivity [2].

A large amount of industrial data affects the bandwidth, energy and time in the process of transmission in the industrial internet [3–6]. Edge computing is performed at the edge near the industrial field, integrating network, computing, storage, and application core capabilities, and providing edge intelligent services. By introducing edge computing into the industrial internet and deploying edge nodes near the industrial field, the part computing function of the industrial cloud platform is moved from one or more central nodes to the other logical edge of the industrial internet. The new edge-to-cloud architecture provides the capabilities to enable customers to make more precise decisions in real-time.

Edge application services significantly decrease the volumes of data that must be moved, the consequent traffic, and the distance the data must travel, thereby reducing transmission costs, shrinking latency, and improving quality of service (QoS) [7]. The

© Springer Nature Switzerland AG 2019
S. Lee et al. (Eds.): IMCOM 2019, AISC 935, pp. 254–266, 2019.
https://doi.org/10.1007/978-3-030-19063-7_21

application services with high real-time requirements can be realized in local and edge side, which guarantees the real-time performance and reliability for data transmission.

Considering the edge nodes play the key roles in interconnecting the field network and the plant internet, as well as the cloud, the security issues became urgent [8]. Several previous studies have been conducted on this issue. Roman [9] conducted a security analysis on several common mobile edge paradigms and elaborated a general cooperative security protection system. Zuo [10] constructed an attribute encryption method with outsourced decryption in the fog computing. Louk [11] constructed a homomorphic encryption algorithm in a mobile cloud computing environment to provide data security protection for mobile users. Baharon [12] further proposed a lightweight homomorphic encryption algorithm for the computational efficiency problem in homomorphic encryption, which minimizes the encryption and key generation time while implementing the addition homomorphism and multiplication homomorphism. Liu [13] proposed a privacy protection authentication protocol SAPA based on shared rights and solved the privacy problem in cloud storage.

The paper focuses on the secure interact between the edge nodes and the industrial cloud, as well as the secure interact between the edge nodes and the field nodes. The paper is organized as follows. Section 2 presents a secure framework for the industrial internet with edge computing. In Sect. 3, we propose a security mechanism for edge nodes' north and south interaction. Test and result analyses are presented in Sect. 4. Finally, the conclusion is given in Sect. 5.

2 Security Framework for Industrial Internet with Edge Computing

A secure framework for the industrial internet with the edge computing is shown in Fig. 1. The interconnection between field node and the edge node may use the industrial Ethernet or industrial wireless, such as EPA [14], EtherCAT [15], WIA-PA [16] and WirelessHART [17]. The edge-cloud connection uses the backhaul network. The main entities in the framework include industrial cloud platform, security manager, edge node, filed node.

Industrial cloud platform is the place where the massive data will be analyzed and processed. Edge computing is near executing units and can collect high-value data required by the cloud. This edge capability supports the data analytics of cloud applications.

The security manager plays roles of distributing keys and devices authentication and managing the security mechanisms. In the proposal, the security mechanisms will be launched to ensure the entire network availability required by network entities in the industrial internet.

Edge nodes provide data processing capability including data analysis, processing, aggregation, privacy, and security, with bounded latency, adaptation, and agility. Edge node manages data, determines life cycle of data and creates value from data. Edge nodes may cooperate with multiple clouds. The edge node can execute much calculation, such as strategy execution, data encryption, and decryption.

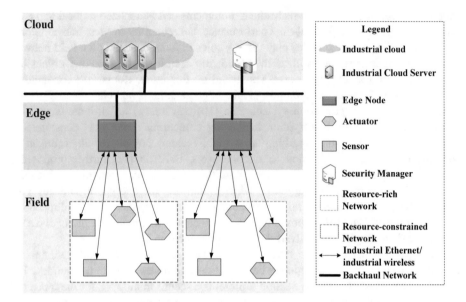

Fig. 1. Industrial internet security framework based on edge computing.

Field nodes can be classified resource-constrained field nodes and resource-rich nodes according to the resource capacity. The field nodes include sensor and actuator, which are expected to have secure communication with the edge nodes.

In order to guarantee the data security in south and north of the edge side, these secure mechanisms should be taken care according to the resources of the networks.

In the resource-constrained network, it is suggested that the field nodes use the lightweight encryption algorithm [18] to protect the data. The edge nodes decrypt the data and encrypt it, and make more precise decisions in real-time. Then the decisions may feed back to the field nodes.

In the resource-rich network, it is suggested that the field nodes use the fully homomorphic encryption algorithm [19] to protect the data, which is a form of encryption that allows computation on ciphertexts, generating an encrypted result which, when decrypted, matches the result of the operations as if they had been performed on the plaintext. The purpose of using homomorphic encryption in the area is to allow the edge side to do computation and make more precise decisions on encrypted data in real-time.

For the secure channel between edge nodes and the industrial cloud platform, it is suggested to use the fully homomorphic encryption.

The advantage of this framework includes that different security mechanism may be designed according to the classification of network. Moreover, resource-constrained network and resource-rich network connected with an edge node separately. It makes the framework more efficient.

3   Security Mechanism for Edge Nodes in Industrial Internet

3.1   General

In this section, security mechanisms in north and south for edge nodes is proposed, which consists of three parts: (1) key management mechanism and encryption algorithm, (2) security mechanism for the resource-constrained network, (3) and security mechanism for the resource-rich network.

3.2   Key Management Mechanism and Encryption Algorithm

Key Management Mechanism. The process of key management mechanism is shown in Fig. 2. The key management mechanism is designed to complete all the process, from the nodes joining the network to the key update.

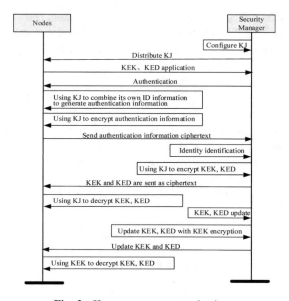

Fig. 2. Key management mechanism.

The nodes include field nodes and edge nodes. The process involves the following three keys: Joining key (KJ), Key encrypt key (KEK), Data encrypt key (KED).

The security manager configures KJ and distributes it to the nodes. Then it checks the nodes, if the nodes are legal, it will encrypt the KED and KEK by KJ and send it to the nodes. When the key updated, the security manager will encrypt the new key by KEK and send it to the nodes.

Encryption Algorithm. As discussed in Sect. 2, this paper considers two encryption algorithms for edge node, lightweight encryption algorithm, and fully homomorphic encryption algorithm.

The lightweight encryption algorithm is simple, so that the edge nodes may calculate and process in real-time. The hardware-based encryption algorithm and some lightweight encryption algorithm are welcome.

Fully homomorphic encryption algorithm has many applications on the internet, especially in the cloud area, but not many applications in the industrial internet. The advantage of fully homomorphic encryption is that the edge nodes can analyze and compute the ciphertext data without decrypting, and feedback the execution result to the field nodes directly.

This algorithm saves the time spent on encryption and decryption at the edge nodes and increases the security level of data protection. Thus, it can balance industrial data security and real-time. Therefore, this paper chooses the integer-based fully homomorphic encryption algorithm which has the lowest time complexity as the encryption algorithm.

The proposed fully homomorphic encryption algorithm using in the edge side is as follows:

KeyGen(λ): P is the private key, which was randomly chosen from the odd number of η bit, η is the length of the private key. $p \in [2\eta - 1, 2\eta)$; Randomly chosen τ couple integer r_i and q_i, we can acquire public key after calculating formula 1.

$$x_i = p\,q_i + 2r_i$$
$$0 \leq i \leq \tau, q \in (Z[0, 2^\gamma/p]), r_i \in (Z \cap (-2^\rho, 2^\rho)). \tag{1}$$

Then, x_i is rearranged, x_0 is an odd number and is the largest number of x_i. And $r_p(x_0)$ is even, and $r_p(x_0)$ is the remainder of r divide x_0. Finally, the public key $pk = \{x_0, x_1, \ldots, x_\tau\}$ can be acquired. The public key size is $\tau + 1$.

$E(pk, m \in \{0, 1\})$: A random subset $S' \subseteq \{1, 2 \ldots \tau\}$ can be selected, for any 1-bit plaintext m, the corresponding ciphertext can be calculated in formula 2.

$$c = \left[m + 2r + \sum\nolimits_{i \in S} x_i\right] x_0 \tag{2}$$

It represents the remainder of formula 3 divide x_0.

$$m + \delta + 2r + \sum\nolimits_{i \in S} x_i \tag{3}$$

Evaluate $(pk, C, c_1, c_2, C, c_t)$: C is a binary circuit with t input. There are t ciphertexts c_i. Ciphertext was added and multiplied in the circuit. Finally, an integer value c^* will be returned, and the returned value is satisfied formula 4 and 5.

$$Decrypt(sk, c^*) = C(m'_1, m'_2, \ldots, m'_t) \tag{4}$$

$$Decrypt(sk, c) : m' = (c \bmod p) \bmod 2 \tag{5}$$

The efficiency of the algorithm includes time efficiency and space efficiency. But with the continuous development of computer technology, the memory of existing computers is large enough to meet the vast majority of applications. Therefore, in the analysis of algorithm efficiency, the space efficiency will not be considered too much in this paper. Only the time efficiency of the algorithm will be analyzed. The time efficiency of the algorithm is determined by the computing complexity.

Because the computing complexity of fully homomorphic encryption algorithm based on the integer is the lowest, therefore, this paper chooses the integer-based fully homomorphic encryption algorithm as the encryption algorithm between the edge nodes and the industrial cloud platform. (Table 1)

Table 1. Symbols.

Symbol	Description
λ	Security parameter
p	Private key, the input of the algorithm
η	The length of the private key
τ	Number of public key samples
r_i, q_i, r	Random large prime number
γ	Bit length of the public key
xi	Bit length of noise
i	An integer greater than one
pk	The public key
ρ	Bit length of noise
m	Clear text
δ	Interference factor
S	Random subset
C	Binary circuit
c^*	Ciphertext, the output of the algorithm

3.3   Security Mechanism for Resource-Constrained Network

Overview. Considering the constrained-resource field nodes, it is suggested that the field nodes use the lightweight encryption algorithm to protect the data. The edge nodes may make more precise decisions in real-time after decryption. Then the decisions may feedback to the field nodes. In the north direction of the edge nodes, the communication resources between the edge nodes and the cloud platform are not limited, and a homomorphic encryption mechanism may be considered to ensure data security.

Security Mechanism of Southern Data. The lightweight encryption algorithm is used to protect the communication between the edge node and the field nodes (sensor and actuator). The security communication process is described in Fig. 3.

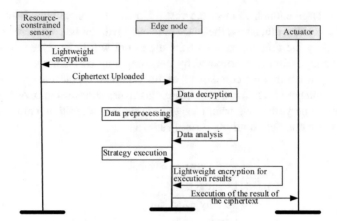

Fig. 3. Security communication process of the southern side for edge nodes with resource-constrained network.

The sensor encrypts the data and sends the data to the edge nodes with the lightweight encryption algorithm. The edge nodes decrypt the encrypted data and preprocess the data. The data may be analyzed according to the set mathematical model in the edge side. After the data analysis is completed, the decision will be prepared according to the established strategy. The result from edge node will be sent to the actuator after lightweight encrypted.

Security Mechanism of Northern Data. The edge nodes encrypt the decrypted data with the fully homomorphic encryption algorithm and send the ciphertext to the industrial cloud platform. The industrial cloud executes a homomorphic operation on the ciphertext data. Finally, the industrial cloud platform will calculate the ciphertext data and send it to the edge node. The data security communication process is described in Fig. 4.

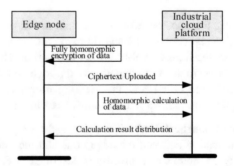

Fig. 4. Security communication process of the northern side for edge nodes with resource-constrained network.

3.4   Security Mechanism for the Resource-Rich Network

Overview. This paper proposes to use the fully homomorphic encryption mechanism in the south direction of the edge nodes considering enough resources of field nodes. The field nodes encrypt the data in full homomorphism, and the edge nodes do not need to decrypt the data and can process the data directly. In the north direction of the edge nodes, the communication resources between the edge nodes and the cloud platform are enough, and the homomorphic encryption mechanism may also be considered, and the process is same with 3.3.3.

Security Mechanism of Southern Data. In this paper, the fully homomorphic encryption algorithm is applied to the resource-rich nodes, and its data security communication process is shown in Fig. 5.

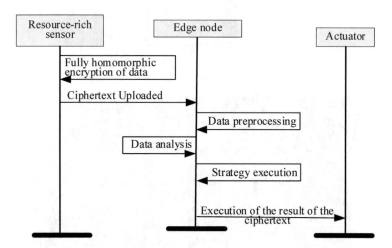

Fig. 5. Security communication process of southern data for edge nodes when nodes resource-rich.

The sensor encrypts the data and sends the data to the edge nodes with the fully homomorphic encryption algorithm. The edge nodes preprocess the data and don't need to decrypt it. The data may be analyzed according to the set mathematical model in the edge side. After the data analysis is completed, the decision will be prepared according to the established strategy. The result from edge node will be sent to the actuator directly.

4   Testing and Result Analysis

4.1   Experimental Platform Implementation

The experimental platform is built to test the security mechanism of the edge nodes both north and south. The test platform includes industrial cloud, edge nodes, and field nodes. As Fig. 6 shown, position 1 is Ali cloud server, which is implemented to play as the industrial cloud. Position 2 is the local interface to monitor the state of the network, which includes Wireshark tool and wireless sniffer tool. The wireless sniffer tool is used to capture packets sent by field nodes, and the Wireshark tool is used to capture packets between the edge nodes and the cloud. Position 3 is the resource-constrained sensor, position 4 is the resource-rich sensor, position 5 is edge node A, position 6 is edge node B. This platform is implemented by using WIA-PA nodes and its gateway which are developed by the previous work [20], where the WIA-PA gateway plays the role as an edge node.

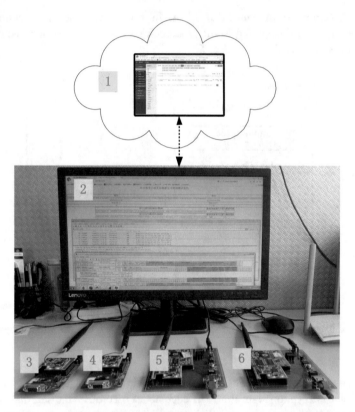

Fig. 6. Test platform for north-south security mechanism of edge nodes

4.2 Secure Communication Test Between the Edge Node and the Resource-Constrained Network

In order to verify the mechanism, the testing scenarios are implemented for secure communication between the edge node and the resource-constrained node. The data from the resource-constrained field node is sent to the edge node A, and the message is captured by the wireless sniffer tool, as shown in Fig. 7. The resource-constrained field nodes encrypt the data with the lightweight encryption algorithm-ITUbee [18] and the results is shown in Fig. 8.

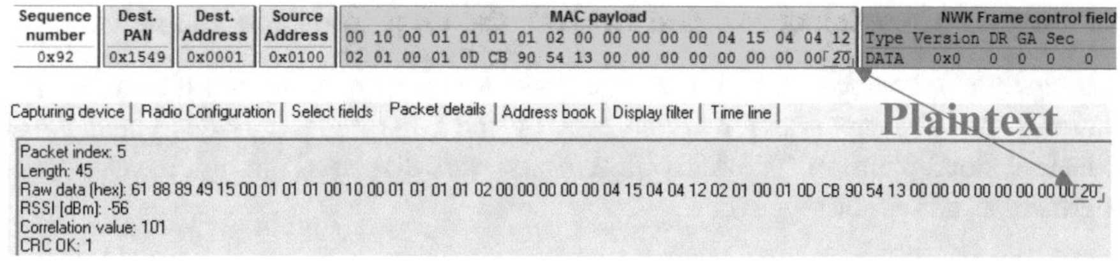

Sequence number	Dest. PAN	Dest. Address	Source Address	MAC payload	NWK Frame control field
0x92	0x1549	0x0001	0x0100	00 10 00 01 01 01 01 02 00 00 00 00 00 04 15 04 04 12 02 01 00 01 0D CB 90 54 13 00 00 00 00 00 00 00 00 20	Type Version DR GA Sec DATA 0x0 0 0 0 0

Capturing device | Radio Configuration | Select fields Packet details | Address book | Display filter | Time line | **Plaintext**

Packet index: 5
Length: 45
Raw data (hex): 61 88 89 49 15 00 01 01 01 00 10 00 01 01 01 01 02 00 00 00 00 00 00 04 15 04 04 12 02 01 00 01 0D CB 90 54 13 00 00 00 00 00 00 00 00 20
RSSI [dBm]: -56
Correlation value: 101
CRC OK: 1

Fig. 7. Plaintext of the resource-constrained field nodes

Dest. PAN	Dest. Address	Source Address	MAC payload	NWK Frame
0x1549	0x0100	0x0101	00 10 00 01 01 01 01 02 00 00 00 00 00 0A 1C 04 0A 19 02 01 00 01 0D 40 F9 97 12 00 00 00 00 00 00 00 00 E8 D0 99 B9 8A 3F EE 1F	Type Version DR G DATA 0x0 0 0

Capturing device | Radio Configuration | Select fields Packet details | Address book | Display filter | Time line | **Ciphertext**

Packet index: 13
Length: 52
Raw data (hex): 61 88 DF 49 15 01 00 00 01 00 10 00 01 01 01 01 02 00 00 00 00 00 02 1C 04 02 19 02 01 00 01 0D 3C 31 87 B3 00 00 00 00 00 00 00 00 E8 D0 99 B9 8A 3F EE 1F
RSSI [dBm]: -59
Correlation value: 104
CRC OK: 1

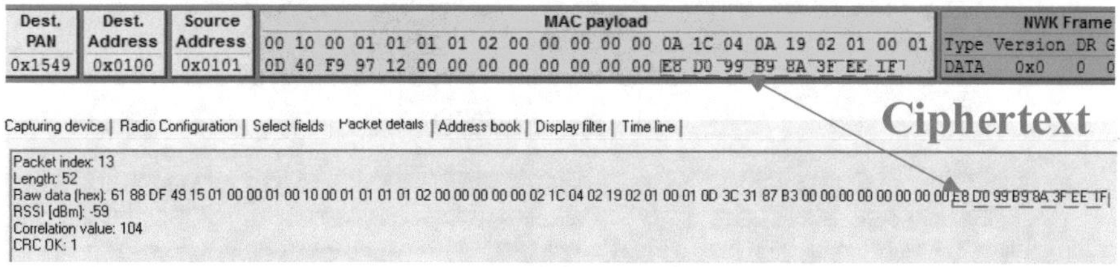

Fig. 8. Ciphertext of the resource-constrained field nodes

The results showed that the lightweight encryption algorithm used in this paper has been successfully applied to the resource-constrained field nodes.

4.3 Secure Communication Test Between the Edge Node and the Resource-Rich Network

In order to verify the mechanism, the testing scenarios are implemented for secure communication between the edge node and the resource-rich node. The data from the resource-rich field node is sent to the edge node B, and the message is captured by the wireless sniffer tool, as shown in Fig. 9. The results showed that the fully homomorphic encryption algorithm used in this paper has been successfully applied to the resource-rich field node.

```
▷ Transmission Control Protocol, Src Port: 60748, Dst Port: 43215, Seq: 1, Ack: 1, Len: 64
◢ Data (64 bytes)
      Data: 2176707487031197478373295533698056301151325559822...
      [Length: 64]
```

The Ciphertext which the resource-rich field node sent to edge node B

```
0000   30 9c 23 30 f2 e8 30 9c  23 30 f2 ff 08 00 45 00   0.#0..0. #0....E.
0010   00 68 19 e7 40 00 80 06  eb 8b 78 4f 02 01 78 4f   .h..@... ..x0..x0
0020   02 7e ed 4c a8 cf 89 63  ed 9f 05 5e 0a 9d 50 18   .~.L...c ...^..P.
0030   01 00 6a 8c 00 00 21 76  70 74 87 03 11 97 47 83   ..j...lv pt....G.
0040   73 29 55 33 69 80 56 30  11 51 32 55 98 22 84 04   s)U3i.V0 .Q2U."..
0050   07 14 21 85 82 22 82 38  81 24 13 18 88 23 90 08   ..!.."8 .$...#..
0060   42 70 11 41 14 38 03 79  08 11 30 80 55 63 80 91   Bp.A.8.y ..0.Uc..
0070   62 46 11 19 11 40                                   bF...@
```

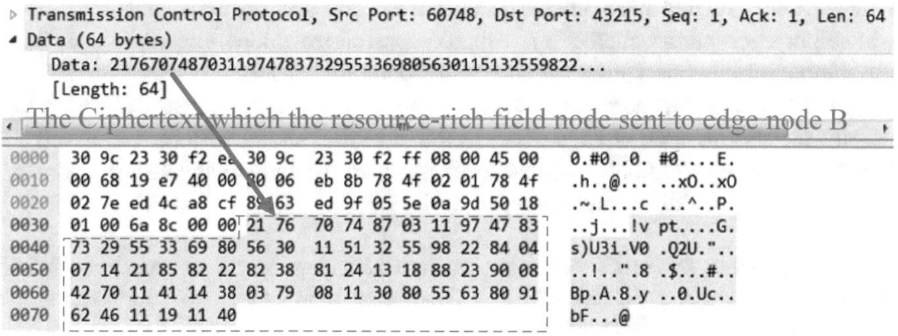

Fig. 9. The ciphertext which the resource-rich field node sent to the edge node B

4.4   Secure Communication Test Between the Edge Node and Cloud

The ciphertext 1 which sent by the edge node A is decrypted by the industrial cloud platform, and the ciphertext 2 which sent by the edge node B is decrypted by the industrial cloud platform. The decryption results were displayed through the test system interface, as shown in Fig. 10.

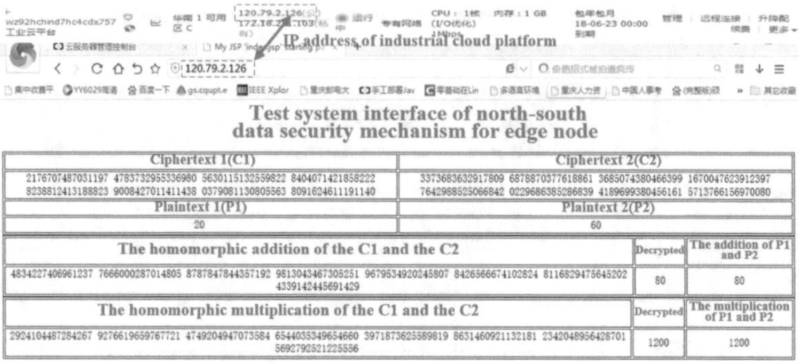

Fig. 10.   Test system interface of north-south security mechanism for edge node

The ciphertext 1 and the ciphertext 2 are homomorphic added by the industrial cloud platform and the result is decrypted by it. Then, the plaintexts are added, the result is shown in Fig. 10. The ciphertext 1 and the ciphertext 2 are homomorphic multiplied by the industrial cloud platform and the result is decrypted by it. Then, the plaintexts are multiplied, the result is shown in Fig. 10. The results are the same as the results we get in the security manager which is the fully homomorphic encryption algorithm test platform. Therefore, it is verified that the homomorphic calculation algorithm has been successfully applied in the industrial cloud.

5   Conclusion and Future Work

In order to balance the aspects of the security and the high real-time ability, the security framework and related mechanism are proposed. The security mechanism consists a lightweight encryption scheme for resource-constrained nodes and an integer-based fully homomorphic encryption scheme for resource-rich nodes.

In the future work, more research will focus on the security of the network which contains the coordination of edge nodes in the industrial internet. Moreover, in order to further validate the feasibility of this mechanism in the industrial internet, more key metrics will be considered and analyzed in the future.

Acknowledgments. This work was supported by Chongqing Education Committee Science and Technology Projects (KJ1600405) and Chongqing Basic and Frontier Research Project (cstc2017jcyjAX0235, cstc2015zdcy-ztzx40006).

References

1. Aazam, M., Zeadally, S., Harras, K.A.: Deploying fog computing in industrial Internet of Things and industry 4.0. IEEE Trans. Industr. Inf. **14**(10), 4674–4682 (2018)
2. Shan, Y.C., Lv, C., Cui, W.B.: A pilot study on the application of cloud CRM in industrial automation. In: Qi, E., Shen, J., Dou, R. (eds.) 20th International Conference on Industrial Engineering and Engineering Management 2013, vol. 1005, pp. 429–434. Atlantis Press, Zhuhai (2013)
3. Aldoghman, F., Chaczko, Z., Ajayan, A., Klempous, R.: A review on fog computing technology. In: IEEE International Conference on Systems, Budapest, vol. 1556, pp. 1525–1530. IEEE (2016)
4. Chiang, M., Zhang, T.: Fog and IoT: an overview of research opportunities. IEEE Internet Things J. **3**(6), 854–864 (2017)
5. Satyanarayanan, M.: The emergence of edge computing. Computer **50**(1), 30–39 (2017)
6. Sun, X., Ansari, N.: Edge IoT: mobile edge computing for the Internet of Things. IEEE Commun. Mag. **54**(12), 22–29 (2016)
7. Georgakopoulos, D., Jayaraman, P.P., Fazia, M., Villari, M., Ranjan, R.: Internet of Things and edge cloud computing roadmap for manufacturing. IEEE Cloud Comput. **3**(4), 66–73 (2016)
8. Usman, M., Abbas, N.: On the Application of IOT (Internet of Things) for securing industrial threats. In: 12th International Conference on Frontiers of Information Technology (FIT), Islamabad, pp. 37–40. IEEE (2015)
9. Roman, R., Lopez, J., Mambo, M.: Mobile edge computing, fog et al.: a survey and analysis of security threats and challenges. Future Gener. Comput. Syst. **78**(2), 680–698 (2016)
10. Zuo, C., Shao, J., Wei, G., Xie, M., Ji, M.: CCA-secure ABE with outsourced decryption for fog computing. Future Gener. Comput. Syst. **78**(2), 730–738 (2016)
11. Louk, M., Lim, H.: Homomorphic encryption in mobile multi cloud computing. In: International Conference on Information Networking, Siem Reap, pp 493–497. IEEE Computer Society (2013)
12. Baharon, M.R., Shi, Q., Llewellyn-Jones, D.: A new lightweight homomorphic encryption scheme for mobile cloud computing. In: IEEE International Conference on Computer & Information Technology, Liverpool, pp. 618–625. IEEE (2015)
13. Liu, H., Ning, H., Xiong, Q., Yang, L.T.: Shared authority based privacy-preserving authentication protocol in cloud computing. IEEE Trans. Parallel Distrib. Syst. **26**(1), 241–251 (2015)
14. Peng, D., Zhang, H., Zhang, K., Li, H., Xia, F.: Research of the embedded dynamic web monitoring system based on EPA protocol and ARM Linux. In: IEEE International Conference on Computer Science and Information Technology, Beijing, pp. 640–644. IEEE (2009)
15. Jiao, B., He, X.: Application of the real-time EtherCAT in steel plate loading and unloading system. In: Li, K., Xue, Y., Cui, S., Niu, Q. (eds.) Intelligent Computing in Smart Grid and Electrical Vehicles. ICSEE 2014. LSMS 2014. Communications in Computer and Information Science, vol 463, pp. 268–275. Springer, Heidelberg (2014)
16. Liang, W., Liu, S., Yang, Y., Li, S.: Research of adaptive frequency hopping technology in WIA-PA industrial wireless network. In: Wang, R., Xiao, F. (eds.) Advances in Wireless Sensor Networks. CWSN 2012. Communications in Computer and Information Science, vol. 334, pp. 248–262. Springer, Heidelberg (2012)
17. Saifullah, A., Xu, Y., Lu, C., Chen, Y.: Real-time scheduling for WirelessHART networks. In: 31st IEEE Real-Time Systems Symposium, San Diego, pp. 150–159. IEEE (2010)

18. Karakoç F., Demirci H., Harmancı A.E.: ITUbee: a software oriented lightweight block cipher. In: Avoine, G., Kara, O. (eds.) Lightweight Cryptography for Security and Privacy. LightSec 2013. Lecture Notes in Computer Science, vol 8162. Springer, Heidelberg (2013)
19. Van, M., Gentry, C., Halevi, S., Vaikuntanathan, V.: Fully homomorphic encryption over the integers. In: Gilbert, H. (ed.) Advances in Cryptology – EUROCRYPT 2010. EUROCRYPT 2010. Lecture Notes in Computer Science, vol 6110, pp. 24–43. Springer, Heidelberg (2010)
20. Wei, M., Kim, K.: An automatic test platform to verify the security functions for secure WIA-PA wireless sensor networks. Int. J. Distrib. Sens. Netw. **12**(11), 1–11 (2016)

Optimized Routing in Software Defined Networks – A Reinforcement Learning Approach

Tahira Mahboob, Young Rok Jung, and Min Young Chung[✉]

Department of Electrical and Computer Engineering, Sungkyunkwan University,
2066, Seobu-ro, Jangan-gu, Suwon-si, Gyeonggi-do 16419, Republic of Korea
{tahira,dudfhr2479,mychung}@skku.edu

Abstract. In this paper, we propose a reinforcement learning based
Q-learning routing mechanism for unicast routing in Software-defined
Networks (SDN). The main objective is to minimize the delay experienced by unicast traffic as it traverses the network. We consider unicast
traffic arriving with a Poisson arrival rate at each switch with exponentially distributed service times. We consider M/M/1 system for the forwarding tables in network switches (OpenFlow switches), and model the
delay function. Q-learning mechanism is adopted for dynamically updating the routing paths, based on the derived delay function. Efficacy of the
proposed routing algorithm has been evaluated using a component-based
framework i.e., OMNET++ simulator. The proposed routing scheme was
compared to the legacy shortest path routing mechanism (Dijkstra's algorithm). The proposed scheme effectively reduces the delay for unicast
traffic. We further proceed by exploiting system parameters, and observe
network behavior under the proposed scheme.

Keywords: SDN · Routing protocol · Reinforcement learning ·
Q-learning

1 Introduction

Massive influx in end-user devices and their corresponding bandwidth demands
has lead to implicit requirements for networks. Networks have to adapt with the
changing traffic and user patterns. Traditional networking solutions are inflexi-
ble, and are dependent on the off-the-shelf vendor equipment, offering a closed
set of services. Next generation paradigm such as Software-defined Networking
(SDN) requires scalable, flexible network architecture that responds nearly real
time to unpredictable resource demands while providing optimized services to
variety of users.

SDN is being leveraged across the globe as a networking solution for data
centers, enterprises, and service providers. It is able to provide flexibility in
the form of programmable infrastructure (control plane), data plane entities

© Springer Nature Switzerland AG 2019
S. Lee et al. (Eds.): IMCOM 2019, AISC 935, pp. 267–278, 2019.
https://doi.org/10.1007/978-3-030-19063-7_22

(switches) that are simple devices with match-action model easily programmable via controllers [1]. Also OpenFlow API, a simple vendor agnostic interface, and communication protocol managed by Open Networking Foundation (ONF) [2], supports multi-vendor network equipment.

Although SDN, and OpenFlow were initiated as an experiment in academia, it has gained popularity in past few years, and this trend is gaining momentum. Several enterprises such as Google, Yahoo, Facebook, Microsoft, Deutsche Telkom, and Verizon have adopted SDN. They are investing in ONF to support it's standardization, and adoption by the industry. However, the implementation of SDN technology for networks, telecom, and cloud service providers is an open research problem.

One of the fundamental challenge is an efficient routing mechanism that optimizes utilization of network resources, targets maximized network throughput and latency reductions. Several solutions were proposed by the industry, and research community. Huang, et al. [3], proposed an online unicasting, and multicasting routing algorithm that aimed to maximize the network throughput under resource constraints in SDN network. Authors in [4], proposed a low-complexity congestion-aware routing algorithm that asymptotically minimizes total network cost. Authors in [5] focused on utilizing traffic engineering principles, for adoption of SDN into existing networks.

Reinforcement learning techniques have been widely studied, and implemented for modeling routing in dynamically changing networks. Researchers in [6] propose the reinforcement learning mechanism for feedback routing in wireless sensor networks, and proposed reduction in the network overhead costs. Haeri [7] discussed applicability, and viability of implementing reinforcement learning technique for performance optimization in computer networks. The primary focus was to maximize performance of networks to achieve efficiency, reliability, and reduce system costs in conjunction with meeting user expectations.

Provisioning efficient services to users is critical, and is highly demanded. The forwarding elements (FEs) i.e., OpenFlow switches are resource constrained. Ternary Content Addressable Memory (TCAM) table containing forwarding rules have limited number of entries. Also, the links connecting the switches are bandwidth constrained. If not modeled proficiently, these constraints may emerge as a bottle neck in the scalability of SDN.

In this paper, we propose a machine learning based routing algorithm that targets reduction in latency in SDN network. Reinforcement learning approach has been adopted, based on Q-learning for handling routing decision by centralized SDN controller. The proposed technique exploits global view of the controller, aims to reduce delay experienced by unicast traffic in the network. We consider unicast routing requests as a Poisson process, and forwarding table in each switch as a queue. We use M/M/1 system to model delay of the unicast traffic at each switch, and measure transmission delay as the delay experienced by packets when they route from one switch to next. The proposed routing algorithm has been evaluated using OMENT++ network simulator, and results were compared with legacy routing protocol.

The rest of the paper is organized as follows. The system model is presented in Sect. 2, followed by a detailed discussion of the proposed routing algorithm in Sect. 3. We present evaluation of the proposed scheme in Sect. 4, and finally conclude the paper in Sect. 5.

2 Preliminaries

In this section we first present the system model, we then introduce the concept of routing in software defined networks. The basic mechanism involved in routing unicast traffic across SDN network has been discussed. We further elaborate the reinforcement learning algorithm, specifically the Q-learning algorithm in detail.

2.1 System Model

We consider a software-defined network (SDN) as $G = (N, E)$, as presented in Fig. 1, $\mathbb{N}$ is a set of vertices (OpenFlow-enabled switches), and $\mathbb{E}$ is a set of the edges (links connecting the switches). The OpenFlow-enabled switches are controlled by a logically centralized SDN controller, provided by networking frameworks such as Floodlight, Ryu, and NOX/POX. The controller communicates with traffic engineering and routing application using RESTful API as the North-Bound API. OpenFlow is used as the South-Bound API, used to communicate with OpenFlow enabled switches. Each switch $n \in \mathbb{N}$ has a forwarding table, the TCAM table having capacity of C_{max} rule entries. Routing in network is based on reinforcement learning, implemented as an application.

2.2 Routing in Software Defined Networks

Routing in SDN is implemented as an intelligent application implemented on OpenFlow architecture. Authors in [8] implemented routing by adapting traditional routing scheme for SDN environment. The programmable control plane in SDN simplifies network configuration, it helps policy enforcement, and facilitates evolution of network. The logically centralized SDN controller means physically distributed control planes (controllers), striving to achieve expected level of performance, scalability, and reliability. The forwarding plane (data plane) consists of routers and switches, responsible for forwarding network traffic. This traffic is controlled by the logically centralized SDN controller (control plane).

Each switch has a forwarding table containing list of rules with three major components: matching field, actions field, and a priority. The switch makes decisions regarding each flow based on instructions (rule entry) provided by the controller, where a flow is generically defined as a sequence of packets characterized by similarity. An incoming packet is matched with flow rules on the switch followed by corresponding action (i.e., drop packet or forward packet using port_ID/controller/group_ID or set address/port/VLAN_ID or set priority etc.). If the match does not exist, it is forwarded to the controller via Packet_In message by default in OpenFlow 1.0, 1.1, and 1.2 but for OpenFlow 1.3 and above switches, default rule has to be inserted by the controller. Controller further processes received packet by traffic engineering application running on it (e.g., install a new rule at the switch).

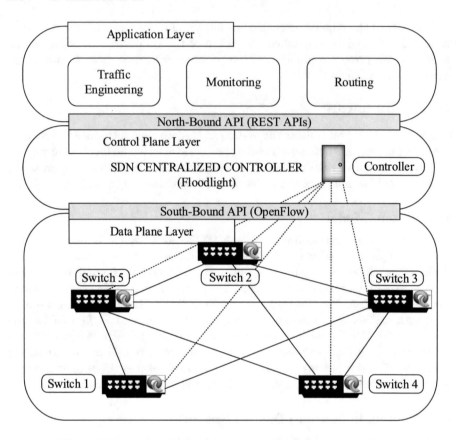

Fig. 1. System model for reinforcement learning routing in SDN.

2.3 Reinforcement Learning Approach: Q-Learning

Reinforcement learning algorithm is a type of machine learning algorithm in which agent receives a reward against current action in the following time step, evaluating the previous action. Q-learning, first proposed by Watkins [9], and the neural fitted Q-function proposed by Riedmiller [10], is a model free, off-policy reinforcement learning algorithm based on Markov decision process (MDP), but unlike MDP, learning is possible without prior knowledge of the environment.

In Q-learning, agent receives response from environment, it uses this response to set reward for previous action, assisting the algorithm in converging towards the goal. The state-to-state transition indicates traversal from one state to next, where action is selection of next optimal state. It is inherently grounded on Bellman's equation, $v(s) = \mathbb{E}[r_{t+1} + \lambda \cdot v(s_{t+1})|s_t = s]$, interpreted as value function implemented iteratively. This value function for Q-learning is based on action-value pair $Q^\pi(s, a)$ where $s \in \mathbb{S}$ (a set of states), and $a \in \mathbb{A}$ (a set of actions). Since, in a state we have multiple actions to choose from, we certainly want to execute the action that results in most value. The value is determined by reward, measured after action has been performed.

The action-value function for Q-learning can be expressed as

$$Q^{\pi}(s, a) = \mathbb{E}_{s'}[r + \gamma \cdot Q^{\pi}(s', a')|s, a], \tag{1}$$

where r is the reward observed for the current state, and $\gamma \in [0, 1)$ is the discount factor that determines the value of the future reward [9]. The optimal value function is defined as the maximum value of the action-value function under all strategies i.e.

$$Q^{*}(s, a) = \mathbb{E}_{s'}[r + \gamma \cdot \max_{a'}(Q^{*}(s', a'))|s, a], \tag{2}$$

where a' has the largest value. In order to obtain the updated Q value, the value based iterative technique updates the current Q value based on the original Q value, and the reward received [10], which can be represented as:

$$\begin{aligned} Q'(s_t, a_t) &= (1 - \alpha) \cdot Q(s_t, a_t) \\ &+ \alpha \cdot (r_{t+1} + \gamma \cdot \max_{a} Q(s_{t+1}, a)). \end{aligned} \tag{3}$$

However, this Q' value, is not directly assigned, rather updated using a method similar to gradient descent, depending on the α value (learning rate). One of the methods for assigning learning rate is to use a decaying function. Therefore, $Q^{\pi}(s_t, a_t)$ can converge to an optimal value by choosing an appropriate learning rate.

3 Proposed Scheme

In this section, we present the proposed scheme, detail delay estimation function, followed by details of the proposed routing mechanism. The main objective of the current study is to reduce delay experienced by unicast traffic in SDN network, every time a new unicast request is generated. Reinforcement learning approach based on Q-learning has been adopted. Traffic conditions of dynamic network may lead to updated routing paths based on delay experienced by traffic at each switch, and link in the network.

Without any loss of generality, we focus on delay function at switches (i.e., processing delay, queuing delay), and transmission delay at links. Propagation delay for network traffic has not been considered in current study. The traffic monitoring application collects network traffic statistics, uses it to measure delay. Weights at the links are updated if the current weight exceeds a predefined threshold, which may result in updated routes. $Q^{\pi}(s, a)$ is then utilized to compute optimal routing paths. New routes offering lesser delays may be computed, leading to updated forwarding rules (F_rs) at the switches. The proposed scheme is compared with the legacy shortest path routing (SP_routing), based on Dijkstra's algorithm.

3.1   Delay Estimation at Links and Switches

Reinforcement learning approach i.e., Q-learning has been adopted for routing updates in SDN network. Each open flow switch has an expensive and limited TCAM forwarding table entries ranging from 500 to 2000 rules, these costs increase proportionally with the number of rules stored in the flow tables, as studied by [11]. The incoming flows at the switch are matched with existing flow rules, and if no match is found, it is forwarded to the controller for appropriate processing. Matching ensures installation of only new rules (previously do not exist in the flow table). Efficient utilization of forwarding tables is a fundamental challenge for network designers, service providers, and operators. This observation motivates us to consider resource consumption at OpenFlow switches, and links for efficient utilization of available resources.

In this paper, we consider unicast packets arriving at each switch with Poisson process arrival rate of λ (packet/sec) having service rate μ [12]. Traffic from neighbouring switches arrives at a switch having arrival rates of $\lambda_1, \lambda_2, \lambda_3, ..., \lambda_e$, where $\lambda = \sum_{i=1}^{e} \lambda_i$. We consider M/M/1 system to model delay in SDN network, in terms of resource consumption at each switch, and link [13]. Also, it is to be noted that we exploit switch lookup time, and processing time of packets as parameters.

We consider a set $\mathbb{N} = \{1, 2, ..., M\}$, where M is total number of switches in the network, $\mathbb{X}_i$ is a set of switches a packet traverse before it reaches the destination, and $\mathbb{X}_i \subset \mathbb{N}$. We have

$$
\mathbb{X}_i = \begin{cases}
\mathbb{X}_1 = \{x_1\} & \text{for } x_1 \in \mathbb{N}, \\
\mathbb{X}_2 = \{x_1, x_2\} & \text{for } x_1, x_2 \in \mathbb{N} \\
\quad \vdots & \& \ x_1 \neq x_2, \\
\\
\mathbb{X}_n = \{x_1, x_2, ..., x_n\} & \text{for distinct } x_1, x_2, ..., x_n \in \mathbb{N} \\
& \& \ n < M.
\end{cases}
\tag{4}
$$

For Poisson distribution, delay experienced by a packet p_k at a switch k is

$$
delay_{p_k} = \frac{1}{|\mathbb{X}|} \left\{ \sum_{k=1}^{|\mathbb{X}|} \{ t_{xy_k} \cdot d_{que_k} \} \right\},
\tag{5}
$$

where $d_{que_k} = 1/(\mu_k - \lambda_k)$ is delay experienced by packets p_k at the switch k (i.e., the switching and the processing delay). We represent the transmission delay between two adjacent switches x and y as t_{xy}. Also, for simplicity we consider $t_{xy} = t_{yx}$.

The delay experienced by p_x packets, where $p_x \in \mathbb{P}$ and $\mathbb{P}$ is a set of packets arriving at a switch, can therefore be calculated as

$$
d_{total} = \frac{1}{|\mathbb{P}|} \left\{ \sum_{k=1}^{|\mathbb{P}|} \{ p_k \cdot delay_{p_k} \} \right\}.
\tag{6}
$$

3.2   Q-Learning Based Routing in SDN

The **Q** matrix is a matrix of size MxM, populated with routing information for unicast traffic, and entries are updated based on delay experienced by unicast packets as they travel from one switch to the next. This Q matrix is utilized to calculate optimal routing paths for unicast traffic.

The basic update mechanism of the $\mathbf{Q_{MxM}}$ is presented in Fig. 2. The agent receives response from the environment (e.g., in case of routing, the delay measurements), and sets the reward (i.e., link weights based on delay). Action is selection of next optimal switch to forward the packet (e.g., move the packet from switch x to switch y). The routing updates are handled dynamically using the Q-learning based routing update algorithm. The traffic monitoring module collects traffic statistics by sending probe packets from origin switch to destination switch. The MxM reward matrix $\mathbf{R_{MxM}}$ is populated with weights for each link using $w_k = \frac{1}{t_{xyk}}$, and the switches are assigned weights using $w_k = \frac{1}{d_{pk}}$, where $k = 1, 2, \ldots, N$. An example of weight assignment is illustrated in Fig. 3. A weight $w_{13} = w_{31} = 100$, is assigned to the links between switch one, and switch three. For simplicity, in our scheme we have used $t_{xy} = t_{yx}$, w_{13} is the outgoing link, and w_{31} is incoming link for switch 3.

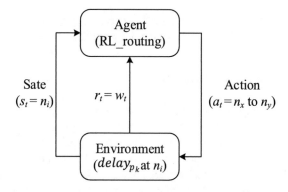

Fig. 2. Example of weight allocation.

Given an SDN network, having a finite set of switches $\mathbb{N}$, and links $\mathbb{E}$, the proposed routing algorithm is presented in Table 1. After initial assignment of weights at the initialization stage, weights assigned to each link are then updated in reward matrix if $w_k < w_{th}$ at next probe, where w_{th} is a pre-determined threshold. Here, probe is process of utilizing statistics fetched by the controller using RESTful API. The collected statistics are used to calculate weights for the links in the network using RL_routing module.

We, then update the Q matrix using

$$Q'(s_t, a_t) = (1 - \alpha) \cdot Q(s_t, a_t) \tag{7}$$
$$+ \alpha \cdot (r_{w_{t+1}} + \gamma \cdot \max_a Q(s_{t+1}, a)),$$

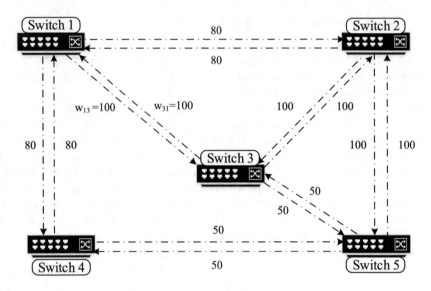

Fig. 3. Q-learning based routing in SDN.

and $r_{w_{t+1}}$ is used to update the reward matrix $\mathbf{R_{MxM}}$, where each entry is a weight corresponding to delay experienced by packets at that link.

4   Performance Evaluation

Evaluation of the proposed scheme has been done using the OMNET++ network simulator. We have tested the proposed routing algorithm i.e., RL_routing algorithm on a network of five switches. The simulation parameters used are presented in Table 2. The simulation was run for 400 s, and results were averaged over ten runs. The traffic statistics were collected for delay, and number switches visited by packets. The unicast traffic was generated from switch two, four and five (source switches), whereas switch one, and three were destination switches.

Performance of the proposed technique was compared with the legacy shortest path (SP_routing) algorithm i.e. Dijkstra's algorithm. Comparison of delay experienced by packets for SP_routing scheme, and the proposed RL_routing scheme is shown in Fig. 4, RL_routing decreases delay of packets compared to the legacy routing scheme. Therefore, delay experienced by packets as they travel through the network is decreased.

Change in number of switches visited by packets under the two routing schemes is presented in Fig. 5. Statistics were collected at destination switch three for packets originating from switches two, four, and five. Since RL_routing may lead to sub-optimal routes i.e., routes having longer paths with an increased hop count in order to avoid congested links. As depicted, the number of packets visiting single hop is greater in case of SP_routing since it chooses shortest

Table 1. RL_routing Algorithm.

Reinforcement Learning Routing (RL_routing) Algorithm

Initialization:
   Input: SDN network $G = (N, E)$, μ, λ, α, γ
   Output: $Q(s, a)$, F_r

1: **Initialize** $Q(s, a)$ as a zero matrix.
2: Calculate shortest path route at each switch n, $\forall\ n \in \mathbb{N}$, based on
   Dijkstra's algorithm (SP_routing) & store forwarding rules
   For (each episode or until learning stops):
3: **Probe** G for d_{p_k} at each switch n using (5)
      4: If (iternation = initial)
            Set $r_w \leftarrow$ (reward-values i.e. w_k from first probe)
      else
            Update r_w if ($w_k < w_{th}$)
5: **Update** $Q^\pi(s, a)$ using (7) and d_{total} using (6)
6: **Install** F_r at switches learned by 5
7: Go to step 3.
   End do
8: **return** $Q^*(s, a)$, F_r

Table 2. Simulation parameters.

Parameter	Value
Data rate	1: 100 kbps
	2: 200 kbps
Channel delay	Range (0.01 ms, 1 s)
Packet generation interval	1: random range (500 ms, 1500 ms)
	2: random range (250 ms, 750 ms)
Destination switch	1 and 3
Packet length	64 bytes

paths. However, there is an observed increase in the number of packets visiting two switches before reaching destination switch three in case of RL_routing, showing impact of updated routing paths having greater number of hops. Performance of the proposed RL_routing algorithm has further been evaluated by altering system parameters i.e. packet arrival rates, and link capacity.

Increasing the stated parameters further leads to a reduction in the delay as shown in sample run presented in Fig. 6. We doubled packet arrival rates, and link capacity at switches, and observed impact on delay. Delay is reduced to nearly a half of its previous value (approximately). This is observed since link

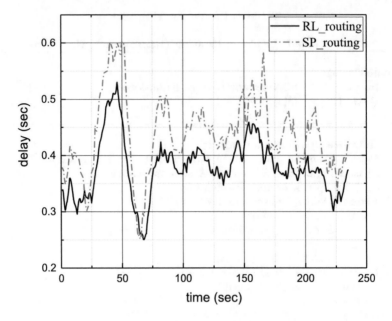

Fig. 4. Delay experienced by packets, measured at switch 3.

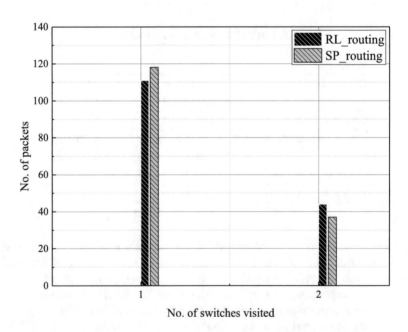

Fig. 5. No. of packets vs. visited number of switches 1, 2, 3.

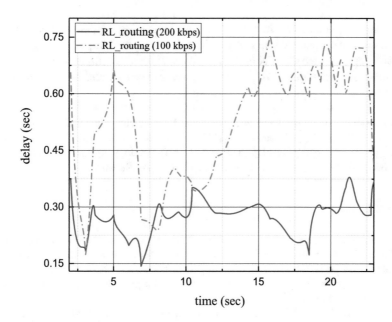

Fig. 6. Delay experienced by packets at different link capacities.

capacities are increased, number of packets transmitted by links increases resulting in reduced delay. During interval (t = 15 s to t = 25 s), delay for SP_routing
rises, and persists. This is observed, since dynamic traffic management is not
considered for legacy routing scheme, therefore leading to an increased delay.
However, in case of RL_routing scheme, rise in delay at time 11 s, and 22 s is
handled dynamically by routing updates, leading to reduction in delay.

5   Conclusions and Future Work

With dynamic, and growing traffic demands of today's network users, industry
is making efforts to handle diverse research challenges, and is proposing viable
solutions. Technologies such as SDN, virtualization, big data, and cloud services
have opened new research areas. In this paper, we propose a routing mechanism that adapts to network conditions, and aims at reducing delay experienced
by unicast traffic as it flows through the SDN network. Q-learning technique
RL_routing, has been proposed for routing updates in order to reduce latency
for unicast traffic. We evaluate the efficacy of RL_ routing algorithm using simulations. The results were compared with the shortest path routing algorithm, and
revealed a decrease in delay under the proposed scheme. The delay can further
be reduced by increasing the packets arrival rates (at switches), and increasing
the link capacities.

    This study is an attempt to excite the readers; exploring utilization of
machine learning techniques, and exploiting potential benefits for purpose of

performance optimization in networks. This is an ongoing research and in future we aim to implement variants of machine learning techniques in order to optimize the performance of SDN networks. We will focus on implementing reinforcement learning techniques for throughput maximization, and delay minimization problem.

Acknowledgment. This work was supported by Institute for Information & communications Technology Promotion (IITP) grant funded by the Korea government (MSIP) (B0190-18-2013, Development of Access Technology Agnostic Next-Generation Networking Technology for Wired-Wireless Converged Networks).

References

1. Kreutz, D., Ramos, F.M., Verissimo, P.E., Rothenberg, C.E., Azodolmolky, S., Uhlig, S.: Software-defined networking: a comprehensive survey. Proc. IEEE **103**(1), 14–76 (2015). https://doi.org/10.1109/JPROC.2014.2371999
2. Open Networking Foundation (ONF). https://www.opennetworking.org/. Accessed 16 Nov 2018
3. Huang, M., Liang, W., Xu, Z., Xu, W., Guo, S., Xu, Y.: Online unicasting and multicasting in software-defined networks. Comput. Netw. **132**, 26–39 (2018)
4. Shafiee, M., Ghaderi, J.: A simple congestion-aware algorithm for load balancing in datacenter networks. IEEE/ACM Trans. Netw. **25**(6), 3670–3682 (2017). https://doi.org/10.1109/TNET.2017.2751251
5. Agarwal, S., Kodialam, M., Lakshman, T.V.: Traffic engineering in software defined networks. In: 2013 IEEE Proceedings of International Conference on Computer Communications, IEEE INFOCOM, Turin, Italy, pp. 2211–2219. IEEE, April 2013
6. Forster, A., Murphy, A.L.: FROMS: feedback routing for optimizing multiple sinks in WSN with reinforcement learning. In: 3rd International Conference on Intelligent Sensors, Sensor Networks and Information, ISSNIP, Melbourne, Australia, pp. 371–376. IEEE, December 2007
7. Haeri, S.: Applications of reinforcement learning to routing and virtualization in computer networks. Doctoral dissertation, Applied Sciences: School of Engineering Science, Fraser University, Burnaby, Canada (2016)
8. Martins, J.L., Campos, N.: Short-sighted routing, or when less is more. IEEE Commun. Mag. **54**(10), 82–88 (2016)
9. Watkins, C.J.C.H.: Learning from delayed rewards. Doctoral dissertation, King's College, Cambridge (1989)
10. Riedmiller, M.: Neural fitted Q iteration–first experiences with a data efficient neural reinforcement learning method. In: European Conference on Machine Learning, pp. 317–328. Springer, Heidelberg, October 2005
11. Cohen, R., Lewin-Eytan, L., Naor, J.S., Raz, D.: On the effect of forwarding table size on SDN network utilization. In: 2014 IEEE Proceedings of International Conference on Computer Communications, IEEE INFOCOM, , Toronto, Canada, pp. 1734–1742. IEEE (2014)
12. Karlin, S.: A First Course in Stochastic Processes. Academic Press, Cambridge (2014)
13. Kleinrock, L.: Queueing Systems, Volume 2: Computer Applications. Wiley, New York (1976)

A Rule-Based Algorithm of Finding Valid Hosts for IoT Device Using Its Network Traffic

Porapat Ongkanchana$^{(\boxtimes)}$, Hiroshi Esaki, and Hideya Ochiai

The University of Tokyo, Tokyo, Japan
`pora@hongo.wide.ad.jp, hiroshi@wide.ad.jp, ochiai@elab.ic.i.u-tokyo.ac.jp`

Abstract. As the number of IoT devices continues to grow every day, we can say without a doubt that one day IoT devices will be used in every aspect of our life. However, devices without an appropriate security management can be easily exploited by adversaries and used for malicious reasons. This research focuses on a way to secure end-device and we have proposed a system which prevents malicious traffic from intruding into the system. Our target IoT devices are device which main functions are gathering data and transferring those data to its server. Our approach is to determine a set of rules based on network traffic, any hosts which not comply with the rules are considered invalid and all of the traffic related to it would be cut down by the switch. Hosts which followed the rule are acknowledged as secure and added to list called "White List". In our experiment, all of the hosts found in the white list are valid hosts, such as device's NTP, DNS, HTTP, DHCP servers. Experiment was only conducted on 2 devices; more experiment result is necessary before implementing into real environment.

Keywords: Security and privacy · Internet of Things · Sensor network · Packets analysis

1 Introduction

Nowadays, IoT devices are implemented everywhere, from something small like a security camera in your house or a heart rate monitor in a patient with heart disease to something sophisticated like sensor wireless networks used in weather stations. We are now only at the dawn of an IoT era, this technology would continue to grow and spread progressively and finally become a part of our society that we cannot live without. Our life has never been easier, but also never been more dangerous.

IoT devices, if handle unproperly, can be very venerable to an adversary. Imagine, your security camera leaking your photos from your own house, a false alarm of hurricane approaching from weather station or life-and-death situation like a fake pulse rate detected from a compromised heart monitor. Devices that designed to make our life easier are now tools used against us.

© Springer Nature Switzerland AG 2019
S. Lee et al. (Eds.): IMCOM 2019, AISC 935, pp. 279–287, 2019.
https://doi.org/10.1007/978-3-030-19063-7_23

Most of IoT devices are not easy to break in if managed by someone with appropriate security knowledge. Devices with a strong password and constant update are unlikely to be compromised. However, those are not an easy thing to ask for. Many users only consider what IoT device can do for them, not what it can do to them. Even when they start to notice the treat, without a proper knowledge, handling these devices is found to be difficult. In some case, setting a security measure is considered non-economical and burdensome. For example, in sensor network where a large number of devices are used and mostly located remotely far away from each other, setting a long-complicated password for all of the devices is quite problematic. Moreover, a constant software update for those devices would consume too much energy which lead to a regular battery replacement.

The motivation of this research is to find a way to help users secure their devices, a mean to alleviate the troublesome tasks of a large-scaled IoT devices network. In this paper, we have proposed a model which operate well with IoT device that gathers information then transfer those collected data to their corresponding servers. Our approach consisted of an intermediate switch filtering network traffics based on secured IP address list extracted from a rule-based algorithm.

2   Related Work

Several works have been done before in the field of IoT based network security. Jing et al. [1] has discussed various challenges related to security of Internet of Things. Shifa et al. [2] also mentioned numerous kinds of threats. Many researches were done to improve the security of IoT devices. Bull et al. [3] has discussed and demonstrated a potential of SDN based gateway while Sivaraman et al. [4] mentioned the benefit of dynamical traffic control provided by SDN. Gupta et al. [5] secure their home device, by implementing a firewall in their system. From previous works, we found that by implementing an intermediate filter in to our system, some level of security is achievable.

Many models were proposed to secure both end devices and their network. Naiket al. [6] concerned about the lack of authentication in IoT network which might lead to an unauthorized access and proposed a model using encrypted device's id as an authentication key. Liu et al. [7] designed a framework of Dynamic Defense Architecture separating defense mechanisms into 6 sections which connected and operate together.

Most approaches focus on secure end-to-end communication or the system but not the end-device itself. If end-device was compromised, malicious attacks can be conducted easily. Moreover, many systems require a sophisticated set up which might not simple enough for users to launch. Our proposed system was designed to require fewest change possible, so users can employ it without having to worry about cost and additional set up. System would identify the valid hosts and prevent device from communicated with unknown hosts securing the end device itself.

3   Designed Architecture

3.1   Proposed Model

Our main purpose is to create a system which can secure devices from adversaries with user's minimum effort. To achieve our goal, we designed the system to have an intermediate strictly filtering traffics out-going from and incoming to the device, while the intermediate should know by itself which traffics are safe and which are not.

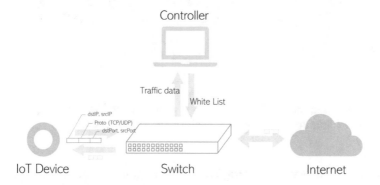

Fig. 1. Proposed Model: A smart switch filters traffics from and to the connected IoT device based on the White List constructed by controller.

Our proposed model, shown in Fig. 1, uses a SDN-switch as a median, filtering traffics between device and the internet. Our switch would only allow a set of specific IP addresses to communicate with the device. We called those IP addresses *"White list"*.

White list is constructed based on device's ordinary traffic data. Traffic is first captured at SDN-switch then passed to the controller PC. The controller PC will run a white list construction program determining the list of secure IP addresses then sends the list back to the switch. Switch starts working as a filter protecting the device.

3.2   White List Constructing Procedure

Rules. White list construction program uses rule-based algorithm to find secure IP addresses. Our rule is *"Only IP addresses to which the device initiates connection is considered a secured host"*.

Input Traffic Preparation. Before the traffic data is inputted into program, some preparation is necessary.

1. Only IP packets are examined.
2. Only packets related to IoT device are regarded.

3. Packets are divided into smaller groups of interval time τ.
4. In each group, packet is further divided in to smaller group based on port number. In this experiment, only IP packets with destination port or source port number less than 1024 (Well-Known port) are considered (Fig. 2).

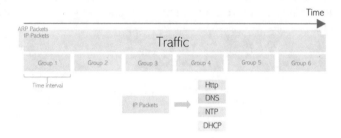

Fig. 2. Traffic Preparation: Only IP packets related to IoT device are picked up and divided into subgroups based on their timestamps and protocol.

White List Construction Program. The program would examine each group of traffic data chronologically. Anytime a packet with distinct IP address (never encountered before in this traffic group) is found, the program would investigate that packet's port number. If packet's destination port number is less than 1024, we consider that our device initiates the connection. On the other hand, if source port number is less than 1024 then, we consider otherwise. After all, the packets in a group are analyzed, we proceed to the next group (Algorithm 1).

The last step is to analyze data collected from each group. For each IP address, if in most groups (95%), the device is the one initiating the connection, then it is considered as a secure IP and added to the whitelist.

3.3   Judgement

All IP addresses in the white list is considered as safe and switch would let all connections go through. However, all the packets associated with IP addresses which are not allocated in the list are blocked.

4   Preliminary Experiment

At this moment, we only conducted our experiment of creating a whitelist. Switch has not been implemented into our system yet. Experiment model is shown in Fig. 3.

Algorithm 1. Extracting White List Algorithm

Input: Network Traffic
Output: The list of secure hosts or *White List*
1 **foreach** *Traffic group "TG"* **do**
2   **foreach** *Protocol "P"* **do**
3   **foreach** *IP* **do**
4   **if** *Host IP with P never found before within this TG* **then**
5   **if** *If packet Src Port is less than 1024* **then**
6   Mark **IP** as **secure** for this TG and P ;
7   Sgroup[P][IP]++ ;
8   **else if** *If packet Dst Port is less than 1024* **then**
9   Mark IP as **risk** for this TG and P ;
10   Wgroup[P][IP]++ ;
11   **else**
12   pass ;
13   **end**
14   **end**
15 **end**
16 **foreach** *Protocol "P"* **do**
17   **foreach** *IP* **do**
18   **if** *Ratio of **Sgroup[P][IP]** to **Sgroup[P][IP]** + **Wgroup[P][IP]*** $>$ *Threshold* **then**
19   add *IP* to *White List[p]* ;
20   **end**
21 **end**

Fig. 3. Experiment Model: We have setup a bridge between the device and the Internet to collect traffic for constructing its White List.

4.1   Experiment Method

2 IoT devices were tested in this experiment. These devices have an alike characteristic, they gather data then transmit collected data to server.

Device

1. **Device A:** Static IP address 203.178.135.96. Gathering electricity consumption and regularly transmit collected data to cloud server.
2. **Device B:** Dynamic IP address with mac address b0:12:66:01:01:05. Regularly communicates with its server.

Capturing the Traffic. First, we created a bridge connecting IoT device and internet on a PC. Then we captured all the traffic data which went through a bridge PC for 1–2 days using tcpdump command. Traffic data was saved in PCAP format.

Program. Program was implemented using Algorithm 1. Device's basic information was required in order for program to run: Mac address for device configured to use dynamic IP address and IP address if device has a static IP address. Program used device's network traffic as an input and output list of IP address it considered secure.

4.2   Result

Outputs of program are shown in Figs. 4 and 5 Some IP addresses are hidden because of security reason. However, after investigation, all of IP addresses listed from our program were secured hosts. IPA indicates a public IP address under management of Esaki-Ochiai labaratory, IPB is a public IP address allocated to The University of Tokyo.

1. *SGroup* : Number of traffic group which we believe our IoT device started the connection
2. *WGroup* : Number of traffic group which we believe host IP address started the connection
3. *Initiate*(%) : Proportion of SGroup to SGroup + WGroup. The higher Initiate (%) is, the higher probability that this IP address is secure.

Result of IoT Device A. Figure 4 shows a screen capture of program's result. Protocols necessary for IoT device to work properly were found within our list. After further examination (Table 1), we found that IP addresses extracted by program are the trusted hosts.

Table 1. Detailed information about white list IP addresses.

IP address	Description	Supervisor
172.104.105.31	ntp Server 1	Linode
129.250.35.251	ntp Server 2	NTT America
157.7.154.29	ntp Server 3	GMO Internet
160.16.75.242	ntp Server 4	SAKURA Internet
IPA1	Primary dns Server	Esaki Ochiai labaratory
IPA2	Secondary dns Server	Esaki Ochiai labaratory
IPB1	Tertiary dns Server	U-Tokyo
IPA3	DHCP Server	Esaki Ochiai labaratory
IPA4	Application Server	Esaki Ochiai labaratory

```
                SUMMARY
-------------------------------------
                ntp (123)
-------------------------------------
IP              Initiate(%)  SGroup WGroup
172.104.105.31  100.00       162    0
129.250.35.251  100.00       162    0
157.7.154.29    99.39        163    1
160.16.75.242   100.00       161    0
-------------------------------------
                domain (53)
-------------------------------------
IP              Initiate(%)  SGroup WGroup
IPA1            100.00       1109   0
IPA2            100.00       15     0
IPB1            100.00       13     0
-------------------------------------
                bootps (67)
-------------------------------------
IP              Initiate(%)  SGroup WGroup
IPA3            100.00       358    0
-------------------------------------
                http (80)
-------------------------------------
IP              Initiate(%)  SGroup WGroup
IPA4            99.78        1349   3
```

Fig. 4. Result of program when traffic of device A was used. Time interval for each group was set to 600 s and only IP addresses with Initate (%) higher than 95% was shown.

```
                    SUMMARY
---------------------------------------------
                    ntp (123)
---------------------------------------------
IP                Initiate(%)   SGroup WGroup
IPB2              100.00        519    0
133.243.238.243   100.00        520    0
---------------------------------------------
                    domain (53)
---------------------------------------------
IP                Initiate(%)   SGroup WGroup
IPA1              99.18         2537   21
---------------------------------------------
                    bootps (67)
---------------------------------------------
IP                Initiate(%)   SGroup WGroup
BC                100.00        4      0
---------------------------------------------
                    http (80)
---------------------------------------------
IP                Initiate(%)   SGroup WGroup
IPB3              99.65         2554   9
```

Fig. 5. Result of program when traffic of device B was used. Time interval for each group was set to 600 s and only IP addresses with Initate (%) higher than 95% was shown. BC means Boardcast IP or 255.255.255.255

Result of IoT Device B. Device B's result (Fig. 5) was similar to device A's. Core protocols were ntp, dns, bootps and http. IP addresses extracted from program were secure. However, DHCP server address was not detected. Instead, Boardcast IP address was listed as a secure IP.

5   Discussion

From experiment in Sect. 4, we found that our program can detect most of the hosts our IoT devices needed to operate. However, with this algorithm, when the device didn't have its own static IP address and had to retrieve its IP address from DHCP server, we couldn't detect IP address of DHCP server. The reason behind this was the way DHCP works, device would flood DHCPDISCOVER message to all hosts in LAN (255.255.255.255) to get its IP. So, every time the device tries to connect DHCP server, it will not send a message directly to server (Table 2).

Table 2. Detailed information about white list IP addresses.

IP address	Description	Supervisor
IPB2	ntp Server 1	U-Tokyo
133.243.238.243	ntp Server 2	NICT
IPA1	Primary dns Server	Esaki Ochiai labaratory
255.255.255.255	Boardcast	-
IPB3	Application Server	U-Tokyo

With this approach, it is difficult to extract IP address of DHCP server. We can add an exception to our rule when dealing with DHCP by looking deeper into every DHCP packets and interpret the meaning of each response. Another approach of solving this problem is to config our IoT devices with IPv6, the characteristic of IPv6 stateless autoconfiguration allows devices to perceive their IP without any helps from DHCP server.

For further experiment, we plan on making our program supporting more than one device at a time and get a result with in a reasonable time. Moreover, more types of device are necessary to test our approach. We may need to make an exception rule like in DHCP case. After all the tests, we plan on putting program into SDN switch and test in real environment.

6   Conclusion

In order for users to safely use IoT device, we have purposed a system which will prevent our devices from exposed to adversary. Our approach is to create a list of secure hosts. Note that some secured hosts might not be added to the list but even without those hosts, the system will work properly. Devices that gather data and initiate connection to their responding server are target group of this system. There is still a need to adjust program and test it on more devices before implementing it into a real system.

References

1. Jing, Q., Vasilakos, A., Wan, J., Lu, J., Qiu, D.: Security of the internet of things: perspectives and challenges. Wirel. Netw. **20**, 2481–2501 (2014)
2. Shifa, A., Asghar, M.N., Fleury, M.: Multimedia security perspectives in IoT. In: 2016 Sixth International Conference on Innovative Computing Technology (INTECH), pp. 550–555, August 2016
3. Bull, P., Austin, R., Popov, E., Sharma, M., Watson, R.: Flow based security for IoT devices using an SDN gateway. In: 2016 IEEE 4th International Conference on Future Internet of Things and Cloud (FiCloud), pp. 157–163, August 2016
4. Sivaraman, V., Gharakheili, H.H., Vishwanath, A., Boreli, R., Mehani, O.: Network-level security and privacy control for smart-home IoT devices. In: 2015 IEEE 11th International Conference on Wireless and Mobile Computing, Networking and Communications (WiMob), pp. 163–167, October 2015
5. Gupta, N., Naik, V., Sengupta, S.: A firewall for internet of things. In: 2017 9th International Conference on Communication Systems and Networks (COMSNETS), pp. 411–412, January 2017
6. Naik, S., Maral, V.: Cyber security—IoT. In: 2017 2nd IEEE International Conference on Recent Trends in Electronics, Information Communication Technology (RTEICT), pp. 764–767, May 2017
7. Liu, C., Zhang, Y., Li, Z., Zhang, J., Qin, H., Zeng, J.: Dynamic defense architecture for the security of the internet of things. In: 2015 11th International Conference on Computational Intelligence and Security (CIS), pp. 390–393, December 2015

Optimizing Color-Based Cooperative Caching in Telco-CDNs by Using Real Datasets

Anh-Tu Ngoc Tran[1]([✉]), Minh-Tri Nguyen[1], Thanh-Dang Diep[1],
Takuma Nakajima[2], and Nam Thoai[1]

[1] High Performance Computing Laboratory,
Faculty of Computer Science and Engineering,
Ho Chi Minh City University of Technology, VNUHCM,
Ho Chi Minh City, Vietnam
{51304672,nmtribk,dang,namthoai}@hcmut.edu.vn
[2] Graduate School of Information Systems,
The University of Electro-Communications, Chofu-shi, Tokyo, Japan
tnakajima@comp.is.uec.ac.jp

Abstract. Content Delivery Networks (CDNs) play a vital role in efficient content distribution inside the Internet to reduce network traffic and improve users' experience. Because CDNs are usually located outside Internet Service Providers (ISPs), they cannot decrease traffic inside ISPs and on peering links between ISPs and CDNs. Hence, there is a considerable need to deploy CDNs inside ISPs to tackle this issue. Additionally, these CDNs are called Telco-CDNs since we are in the control of ISPs. Traditional caching policies widely used in CDNs can be applied to Telco-CDNs. Nonetheless, the policies are inefficient in the context of Telco-CDNs in that network operators of CDNs have no knowledge of underlying network infrastructure while those of Telco-CDNs do. The fact leads to the emergence of caching algorithms in Telco-CDNs. Color-based caching strategy together with its routing algorithm is regarded as the most effective one with acceptable computation overhead. In principle, contents will be assigned to color tags and periodically re-colorized every interval in the approach. Since the previous study used a simulated dataset following gamma distribution, the characteristics of this dataset did not change every interval. For that reason, there was no experiment to verify the efficiency of the algorithm when the characteristics of the dataset varied. In this paper, we conduct numerous experiments to look the aspect over. The experimental results show that not only the color-based approach still remains prominently effective in comparison with Least Frequently Used (LFU) every interval, but also the strategy with periodical re-colorization outperforms the one without re-colorization. Moreover, the prior research only took account of users' interests in a global manner rather than geographically local regions, which is difficult to attain the optimizing traffic reduction. Thus, we also propose an iteration of the color-based caching strategy by making use of the insight of users' preference based on regional areas to optimize traffic. The experimental findings reveal that traffic can considerably be reduced

© Springer Nature Switzerland AG 2019
S. Lee et al. (Eds.): IMCOM 2019, AISC 935, pp. 288–305, 2019.
https://doi.org/10.1007/978-3-030-19063-7_24

for all local areas, especially by up to 27.3%. To sum up, the proposed extension of the color-based caching strategy surpasses the traditional color-based one in practice.

Keywords: Co-operative caching · Sub-optimal content placement · Hybrid caching · Routing algorithm · Dynamic content popularity

1 Introduction

Traffic on the Internet is getting more and more enormous due to Video-on-Demand (VoD) services. Although CDNs have helped to reduce such traffic by using caching servers, these servers are only placed outside of ISPs. Therefore, traffic on peering links between ISPs and CDNs cannot be reduced. It leads to the idea of placing CDNs inside ISPs; several ISPs have considered building CDNs for themselves, called Telco-CDNs or ISP-Operated CDNs. Owing to the global knowledge about network's physical properties, the network operators can readily deploy cache servers directly in their backbone network. In such a scenario, ISPs may apply several advanced techniques like hybrid caching and cooperative caching to improve network performance.

To design an effective caching strategy, the network operator should take many constraints into account such as network traffic, latency, the volume of content and so on. Since it is a challenging task, Nakajima et al. [10] limited the number of constraints and proposed a light-weight content distribution scheme for cooperative caching to map contents into cache servers within Telco-CDNs. The approach exploits the historical content accesses to calculate the separator ranks, which are used to colorize the contents. Based on the color tags, a cache server caches the contents carrying the same color as it. While comparing to the traditional caching LFU [5], the experimental results show that it achieves better performance, which is close to the sub-optimal outcome of the approach using the genetic algorithm [7]. The color-based routing proposed in the later study [11] also enhances the approach by further reducing 30% network traffic compared to the shortest path routing [4]. However, the authors only evaluate their color-based caching and its routing algorithm with the simulated dataset generated from the gamma distribution within a single colorizing interval. Notably, they calculate the separator ranks based on the content access log without considering the regional interests. It indicates that the approach may not optimize the network traffic for the specific geographic regions.

Since all evaluations in the previous studies [10,11] are not sufficient to demonstrate the effectiveness of the approach when being applied in practice, in this work, we mainly focus on exploiting the real trace of content accesses aggregated on a daily basis to evaluate the color-based approach and its variants thoroughly. In particular, we explore the characteristics of our dataset then utilize it to estimate the network traffic through 7 consecutive colorizing intervals. Moreover, as people living in different geographic regions may have different interests, we propose colorizing content based on regional content popularity to improve network traffic for each area.

The rest of this paper is organized as follows. Section 2 gives the background knowledge about Telco-CDNs, the color-based caching and its routing algorithm. Section 3 describes our optimization. Section 4 presents the experimental environment as well as explains the results conducted to assess the effectiveness of our improvement. Section 5 provides the overview of related work. Finally, Sect. 6 draws the conclusions and outlines the future work.

2   Background

2.1   Telco-CDNs

To serve the burgeoning VoD service, CDN is a fundamental technique referring to a geographically distributed group of servers which work together to provide fast delivery of internet content. At present, CDN is widely used by VoD service providers as it can speed up content delivery based on the geographic location of users, the origin servers, and the cache server, which makes it significantly reduces latency and improves the quality of services. Besides, most of the modern CDN techniques address improving the caching and routing algorithms. The goal is to not only minimize the network traffic but also increase the availability of contents. However, as the cache servers are usually located outside of the ISP network, these techniques cannot exploit the infrastructure of the ISP to thoroughly reduce network traffic. Without the knowledge of the underlying network, the service providers cannot obtain better traffic reduction even when placing their cache servers within the ISP's network. On the other hand, since Telco-CDN is organized and operated by the ISPs, it can efficiently distribute their cache servers across the backbone network and leverage the global knowledge of the infrastructure to improve the network traffic. Nonetheless, inefficient routing and content distribution may cause links congestion which would significantly degrade the performance [7].

2.2   Color-Based Caching

The color-based caching algorithm proposed by Nakajima et al. [10,11] was thoroughly recapitulated in the work of Tran et al. [12]. The algorithm itself in Telco-CDNs is a consolidation of both cooperative caching and hybrid caching. Cooperative caching is a method of aggregating many cache servers to extend space for storing. The two vital key factors to efficaciously utilize cache servers are content distribution and content duplication; the former targets to increase storage space, and the latter targets to duplicate popular contents, resulting in the increase in hit rates of cache servers. As we know transmission of contents to end users results in traffic in a network, and popularity of these contents quickly changes over an interval which can be an hour due to new popular contents, news or information from the social network like Facebook and Twitter [16,17]. In order to follow such changes, a hybrid caching strategy is proposed [18]. Storage space in a cache server is split into two components. The first one using LFU caching policy caches popular contents to increase hit rate, while the other using First In First Out (FIFO) caching algorithm caches recent contents to follow changes in content popularity.

Unfortunately, hybrid caching and cooperative caching are not compatible with each other because hybrid caching does not take the presence of other cache servers into consideration, which means that it may lead to replication of unpopular contents, causing ineffectively using cache storage. However, the color-based caching algorithm successfully coalesces advantages of both cooperative caching and hybrid caching by using colors. Cache servers and contents are assigned with color tags, and a content is only cached on a server if the color of this content matches the server's color. As a result, contents are disseminated to cache servers, which increases cache capacity. Furthermore, to increase hit rates of cache servers, multiple duplicates of popular contents are stored on different servers by attributing more than one color to them. The more popular a content is, the more colors in a tag it has. Figure 1 gives the general concept of how the color-based caching algorithm works. Additionally, the novel hybrid caching scheme proposed in the color-based caching algorithm can follow rapid changes in content popularity and work with cooperative caching. Storage space of a cache server is depicted in Fig. 2. A cache server has two separate spaces: the colorized one which only caches contents matching server's color and uses LFU [5] eviction policy, the other which stores any content mismatching server's color and uses modified LRU caching policy. Modified LFU [15] is adopted in this hybrid caching strategy rather than LFU because it yields better hit rates [3].

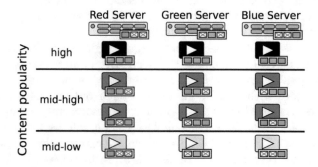

Fig. 1. Example of contents cached in three servers according to their color tags and popularities [10].

Server Colorization. A modified version of the well-known graph coloring algorithm of Welsh-Powell is used to assign colors to cache servers. In contrary to the original one, which is designed to colorize a graph using a minimal number of colors [14], the derivative one is modified to colorize servers with exact N colors and equally distribute all available colors to cache servers in the network.

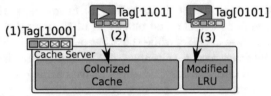

(1) Each server has a color tag
(2) The colorized area stores contents with matching tags
(3) The LRU area stores contents with unmatching tags

Fig. 2. Proposed LFU-LRU hybrid cache architecture [10].

For instance, presume that four colors, red, green, blue, yellow are used, the new version of this graph coloring algorithm will take these colors to colorize servers and the number of times red colors are used is approximately equal to the ones of green, blue and yellow. In details, the algorithm is described in [10].

Content Colorization. Unlike server colorization, in content colorization due to the disparity in the popularity of contents, they can have more than one color in their color tags. A color tag is represented by a bit vector of length N, in which 0 means no color and 1 means color; N is the number of colors used to colorize servers. As a result, a color tag can have no color or at most N colors. The more popular a content is, the more 1s its tag has. For example, with four colors red, green, blue, yellow (R, G, B, Y) used in the color-based caching algorithm, the popularity of contents is divided into five classes: the one with tags of all bit 0s, and other ones with one, two, three, four bit 1s. Table 1 delineates all possible tags corresponding with content popularity.

In order to assign color tags to contents, an origin server first gathers access logs from cache servers to calculate the popularity of all contents based on the number of requests, and sort them by popularity in descending order. Each content is assigned with a rank; the most popular one will have rank 1. After that, contents are distributed to popularity classes corresponding with their popularity. Contents in the same popularity class will be assigned tags in a cyclic fashion. Table 1 illustrates an example of content colorization with four colors. We can see that contents are classified into five classes by four numbers 22, 26, 47, 305. The first number specifies that contents that have rank from 1 to 22 will be classified into class with 4 colors, while the other numbers are for classifying other classes. That set of numbers is called separator ranks which are a set of content ranks to categorize contents to corresponding popularity classes.

The number of contents in each popular class significantly affects traffic in the network. With each set of separator ranks, traffic can be estimated without taking bandwidth and latency into consideration by the following formula:

$$T_{est} = \sum_I \sum_J \sum_K e_{ij} p_{ik} y_{ijk} \tag{1}$$

Table 1. Popularity classes and corresponding tags in a four-color case

Popularity class	Content rank	# of colors	Bit vector R	G	B	Y
High	1–22	4	1	1	1	1
Mid-high	23–26	3	1	1	1	0
			1	1	0	1
			1	0	1	1
			0	1	1	1
Middle	27–47	2	1	1	0	0
			1	0	1	0
			1	0	0	1
			0	1	1	0
			0	1	0	1
			0	0	1	1
Mid-low	48–305	1	1	0	0	0
			0	1	0	0
			0	0	1	0
			0	0	0	1
Mid-low	306-	0	0	0	0	0

where T_{est} is the total traffic, e_{ij} is a hop count for end-user i to fetch a content from cache server j, p_{ik} is the request probability that end-user i fetches content k, and y_{ijk} is a binary variable indicating whether cache server j storing content k is the nearest server from end-user i; I, J, K are the numbers of users, cache servers, and contents stored in origin servers, respectively [11]. The main goal is to find the optimal separator ranks that result in minimal network traffic, and satisfy two constraints. In case of four colors, the first one is

$$a \leq b \leq c \leq d \tag{2}$$

where a, b, c, d are separator ranks of popularity classes with four, three, two, one colors, respectively; it indicates that separator ranks of more popular classes are less than or equal to the ones of less popular classes. The second one

$$a * J * 100\% + (b - a) * J * 75\% + (c - b) * J * 50\%$$
$$+ (d - c) * J * 25\% = J * C$$

$$\Leftrightarrow a + b + c + d = 4 * C \tag{3}$$

describes the total cached contents in the network does not overflow the total storage capacity of all cache servers, where C is the number of contents a cache server can store. Expression $a * J * 100\%$, $(b - a) * J * 75\%$, $(c - b) * J * 50\%$, $(d - c) * J * 25\%$ are sizes of contents stored cache servers in each popularity class. Specifically, $a * J * 100\%$ shows that contents from rank 1 to a are supposed to be stored on 100% cache servers because these contents have four colors in their tags.

In general, for N colors, separator ranks stored in an array of N elements $S[0], S[1], ..., S[N - 1]$ need to satisfy the two following constraints

$$\begin{cases} S[0] \leq S[1] \leq ... \leq S[N - 1] \\ S[0] + S[1] + ... + S[N - 1] = N * C \end{cases}$$

The detailed algorithm to a sub-optimal solution of separator ranks is demonstrated in [11].

2.3   Color-Based Routing

In Telco-CDN, the routing algorithm is always a vital factor that can govern the traffic of a network. However, some common routing algorithms are showing several limitations. For example, the well-known Dijkstra algorithm [4], which is mostly used to find the shortest path in this domain, requires that each server much hold a large routing table. As a network's infrastructure changes, the network operator must update all routing tables even if it is a minor change. Meanwhile, some other network models, such as the models of Maggs and Sitaraman [8] and Wang et al. [13], utilize the hash function to distribute contents across the network. Cache servers are divided into small clusters, in which servers are adjacent and interoperable. It indicates that each server only interacts with the others in the same cluster and the origin server in the worst case. Although this approach would increase the cache capability as cooperative caching, it may cause congested links in some paths reaching to the servers that contain attractive contents.

To overcome these limitations, Nakajima et al. [11] continue to propose a color-based routing, which can leverage the color tags in the previous study [10] to diminish the network traffic and the routing cost. In particular, each cache server only holds two small routing tables. The first one is the Request Routing Table which stores all other colors (except its color) and its corresponding network interface IDs leading to the nearest caches servers carrying these color. This table is used to forward the request to the closest cache server which has the same color tag with the requested content when the cache server does not contain this content. In case of the requested content does not have any color tag, the request will be sent to the origin server by default. However, if any cache server in the path contains this content, it may send back the response before this request reaches the origin server. The second table is the Response Routing Table which stores the IP addresses corresponding network interface IDs. This table is used to send a response to the user when the cache server contains the requested content or receives it from another server.

Recall that all cache servers are colorized based on the modified version of Welsh-Powell algorithm [10]. It implies that the servers located near together as a cluster would have a different color. Hence, this routing algorithm not only increases the cache capability but also substantially reduces the average hops that a request must pass by to fetch a cached content from other servers when a cache miss occurs. Figure 3 shows a basic concept of the color-based routing. Unlike hash-based routing, an attractive content may carry many different color tags, which means it can be cached by different cache servers in a cluster. This feature ensures that the workload is balanced within the network, which reduces congested links significantly. The empirical results in the previous study show that the color-based routing can reduce more than 30% network traffic compared to the shortest path routing while applying with the color-based caching algorithm [11].

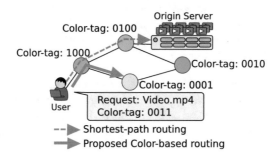

Fig. 3. Color-based routing example [11].

3   Optimization

In this section, we focus on describing how to optimize the color-based caching algorithm. Please refer to Table 2 for a summary of notations we use throughout the section.

Table 2. Notations

k	Content k
K	A set of contents
n	Region n
N	A set of regions in which Telco-CDNs are located
$P_G(k)$	The global popularity of content k
$P_{L_n}(k)$	The local popularity of content k in region n
T_{G_n}	Traffic of region n using global colorization
T_{L_n}	Traffic of region n using local colorization

In the work of Nakajima et al. [10], they assume that user preference is global. In other words, preference of users from different regions is identical.

Preference of users decides popularity of contents. Popularity of each content is calculated by gathering information from access logs of cache servers regardless of users' requests coming from any region. After that, the separator ranks algorithm described in [11] will take the popularity of contents as an input to find the best way of colorizing contents that results in minimal traffic in each region. This way of colorizing contents is called global colorization; The word "global" means that data popularity is considered global.

Nevertheless, people that come from different regions such as continents or countries have different interests, which means that their preference is local. Therefore, traffic can further be improved by colorizing contents in regions having different interests independently and locally. This way of colorizing contents is called local colorization. Contrary to global colorization, the popularity of data in each region is calculated by gathering information of requests coming within the region; Then, the information of contents' popularity of each region is used by the separator ranks algorithm to optimally colorize contents in a way that gives minimal traffic for this region.

Theorem. *If the separator ranks algorithm can find the optimal solution resulting in minimal traffic of a region by using contents' popularity of the region, then in that region, traffic using local colorization is always less than or equal to one using global colorization, no matter users' preference is local or global.*

Proof. The proof is by case analysis. Assume the separator ranks algorithm can find the optimal solution resulting in minimal traffic of a region by using contents' popularity of the region.

Case 1: Assume that users' preference is global. This means that $\forall k \in K$, $\forall n \in N$, $P_G(k) = P_{L_n}(k)$. Therefore, in region n there is no difference in traffic whether the separator ranks algorithm uses $P_G(k)$ or $P_{L_n}(k)$. In other words, $\forall n \in N$, $T_{L_n} = T_{G_n}$.

Case 2: Assume that users' preference is local. This means that $\forall k \in K$, $\forall n \in N$, $P_G(k) \neq P_{L_n}(k)$. While $P_{L_n}(k)$ exactly describes the contents' popularity of region n, $P_G(k)$ does not (1). Additionally, the separator ranks algorithm is previously assumed that it can find the optimal colorization of contents causing minimal traffic (2). From (1) and (2), we conclude that $\forall n \in N$, $T_{L_n} < T_{G_n}$. Therefore, the implication holds in all cases.   □

4   Evaluation

4.1   Experimental Environment

All of the experiments will be performed on an NTT-like network topology in Japan [2]; As shown in Fig. 4, it is composed of 55 cache servers and 1 origin server. Moreover, each cache server has a client that generates requests taken from a real dataset to simulate traffic. A set of requests comes from an access log of a website of Ho Chi Minh City University of Technology [1]; it was gathered in a week from November 7th, 2016 to November 13th, 2016, containing 208,478 requests for 2,817 distinct contents. In our study, the main criterion we concern is normalized traffic which is a ratio of traffic using a specific caching algorithm to one without cache servers.

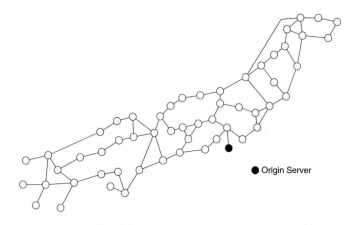

Fig. 4. NIT-like network.

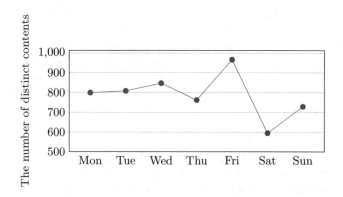

Fig. 5. The number of contents in a week.

In addition, as depicted in Fig. 5, the average quantity of different contents is about 800. On Friday, the number of contents requested reaches its peak, approximately 1,000 contents, while the lowest one is about 600. Figure 6 demonstrates users' activity in a week, they have a tendency to access our website more frequently in weekdays with a peak of about 35,000 requests than weekends; On Sunday, the number of requests is at its lowest with approximately 17,000 requests.

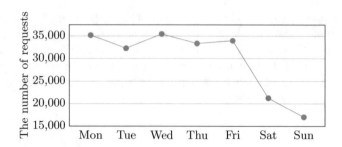

Fig. 6. The number of requests in a week.

4.2   Traffic Evaluation with Recolorization After Every Interval

In this experiment, caching algorithms: LFU, the color-based algorithm using 4 colors and shortest path routing (SPR) strategy, the color-based algorithm using 4 colors and color-based routing (CBR) strategy, the color-based algorithm using 8 colors and CBR strategy, are evaluated with the real dataset in 7 days.

The color-based caching algorithm is hybrid caching, as well as cooperative caching. The hybrid caching strategy is mainly designed to follow new popular contents, while rapid changes in access patterns are followed by periodically colorizing color tags of contents every interval. The hybrid cache ratio is appropriately adjusted according to how many new contents will appear to minimize the increase in traffic. Figure 7 shows the increase in traffic between Monday and Tuesday, and with 10% of the colorless-LRU and 90% of the colored-LFU area, the traffic increase between the 2 days is minimal. Since the hybrid caching strategy is not designed to change the ratio dynamically, optimal hybrid caching ratios for the other days cannot be changed on the run. Therefore, the optimal ratio for Monday and Tuesday will be used for other days. Moreover, the color-based algorithm in this experiment updates contents' tags every day to follow data popularity.

Figure 8 depicts normalized traffic of different caching algorithms with the real dataset. Except on Monday, LFU shows worse results than three other variants of the color-based algorithm. On Monday, the 4-color SPR gives the worse result than LFU because the solution found by separator ranks algorithm to colorize contents according to their popularity is sub-optimal. Specifically, the separator ranks of the 4-color SPR on Monday are [0 0 45 355], which means that there are no contents colorized with four or three different colors. Between the 4-color CBR and 8-color CBR, the results of 4-color CBR are sometimes larger or smaller than the ones of 8-color CBR. Furthermore, the normalized traffic on Sunday is higher than the others. It is because of the small number of requests requesting for the relatively big number of contents on Sunday. In general, when aggregating traffic in seven days, 4-color SPR, 4-color CBR, 8-color CBR achieve 9.7%, 23,2%, 21,1% smaller traffic than LFU, respectively (Fig. 9).

In the previous work [12], Tran et al. conducted a performance study of the color-based caching algorithm by using the real dataset. However, in that work,

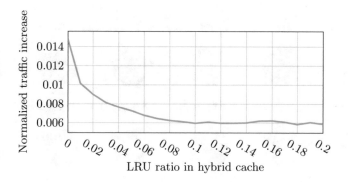

Fig. 7. Traffic increase between Monday and Tuesday.

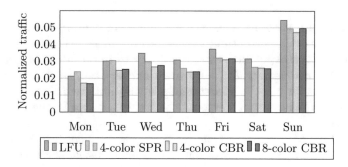

Fig. 8. Normalized traffic under different caching strategies.

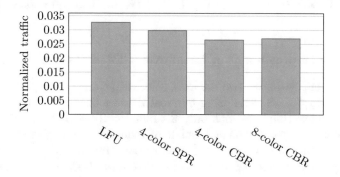

Fig. 9. Aggregate traffic in a week.

contents are not colorized after every interval, which results in larger traffic. The purpose of recolorization after every interval is to follow changes in content popularity. The following experiments compare color-based algorithms in case of recolorization and non-recolorization after every interval. These experiments evaluate traffic on 7th days in five alternate months from January to September instead of seven days like the previous experiment. The reason a day in every

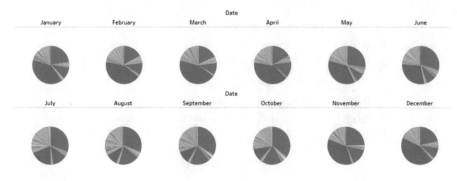

Fig. 10. Data changes.

alternate month is chosen to evaluate traffic instead of every day is because the content popularity in our dataset changes every month. Therefore, color-based algorithms with recolorization will be unlikely to improve traffic evaluated on seven days from November 7th, 2016 to November 13th, 2016. In Fig. 10, each circle represents for a set of contents in a month and each sector in the circle describes the popularity of a content. The bigger a sector is, the more popular a content is.

Figures 11, 12, 13 compare traffic of three variant color-based caching algorithms with and without recolorization. In every case, recolorization always yields better results than ones without recolorization. After aggregating the traffic of 7th days in five months, the traffic improvement of 4-color SPR, 4-color CBR, 8-color CBR with recolorization are 8.3%, 3.7%, 6.1% in comparison with non-recolorization.

4.3   Traffic Improvement of Local over Global Content Colorization

This experiment evaluates two different colorization strategies: global and local colorization, in 7 days. Analyzing our real dataset shows that users' requests come from many countries, but only 3 countries: Taiwan, U.S., Vietnam, are chosen because of the sufficient number of requests. Topologies of 3 countries are shown in Fig. 14. Every country is allocated a number of cache servers corresponding to the number of requests of that country. Specifically, Taiwan, U.S., and Vietnam have 4, 5, and 44 cache servers, respectively.

Figure 15 shows that traffic in 3 regions using local colorization is better than global colorization. Most of the users' requests come from Vietnam, so users' preference from Taiwan and U.S. does not have much influence on global preference. Therefore, applying local colorization to such regions definitely improves traffic within them. As depicted in Fig. 15, the traffic improvement in Taiwan and U.S. is 27.3% and 19.0%, respectively. Even though Vietnam users' preference is very similar to global preference, local colorization still improves about 6.2% traffic in Vietnam. It is because that local colorization just needs to store popular contents of Vietnam instead of storing all popular contents of three regions in global colorization.

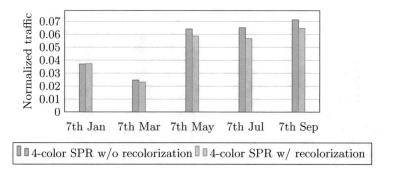

Fig. 11. Traffic of 4-color SPR with and without recolorization.

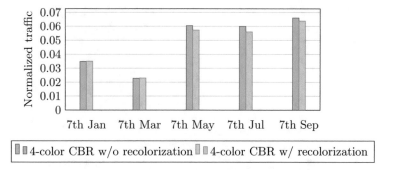

Fig. 12. Traffic of 4-color CBR with and without recolorization.

5   Related Work

Cooperative caching strategy has been considered in a wide area of CDN and Telco-CDN over the decade. One of the most important issues of this strategy is caching policy because it has a direct effect on the network performance. Many studies have aimed at designing an effective caching strategy to improve the performance of the network. For example, the hash-based caching in the Akamai's CDN model [8]. The network operator divides all cache servers into small clusters comprising adjacent servers. Contents are distributed within the cluster, in which cache servers would calculate the hash value of contents and compare them with server IDs to find the appropriate server to store this content. With a similar approach, the work of Li et al. [6] addresses the same problem but in a mesh network. Instead of using the hash function, Wang et al. [13] proposed a modulo function to distribute content replicas on k-closer surrogate servers, which addresses the problem in a ring network. In brief, although such approaches can increase the cache capacity by sharing the workload with adjacent cache servers, they always cause congestion on some paths reaching to the cache servers which store attractive contents. Moreover, without considering the underlying

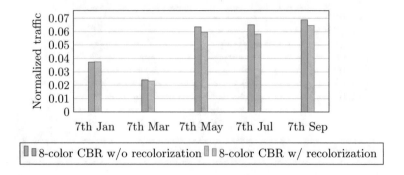

Fig. 13. Traffic of 8-color CBR with and without recolorization.

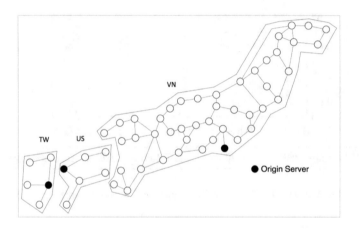

Fig. 14. Network topology of Taiwan, U.S., and Vietnam.

infrastructure, these approaches are unable to distribute contents effectively across the network.

In contrast, some heuristic approaches are applied in Telco-CDN, in which they can exploit the global knowledge of the whole network properties. For instance, Li and Simon [7] presented an efficient algorithm for content placement in Telco-CDN. The idea is to leverage a well-known meta-heuristic, namely Genetic Algorithm (GA), to predict user preference and find the sub-optimal location of contents within the network. Despite achieving an impressive performance, the approach seems impractical since its computational overhead is about 10 h while another study [17] revealed that the user access patterns change 20%–60% every hour.

The color-based caching and its routing algorithm were promoted in recent studies [10,11], whose experimental results show that they not only limit the computation time to a few seconds but also give a remarkable performance that is close to the results calculated by GA. This work is an extension of the previous

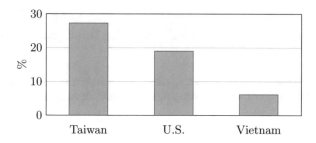

Fig. 15. Traffic improvement of three countries under local colored strategy over global colored strategy.

study [12], its goal is to further evaluate the color-based approach on the real dataset as well as improve the content distribution by colorizing them based on the regional popularity.

6   Conclusions and Future Work

As this work mainly focuses on verifying the feasibility of the color-based approach when being applied in practice, we not only explore the characteristic of our dataset and investigate how to leverage it on thoroughly evaluating the color-based approach but also improve the approach by colorizing contents based on their regional popularity.

In particular, we aggregate our user access log on a daily basis to have an overview of our dataset. The network traffic measurement in 7 consecutive days shows that the color-based caching strategy coupled with its routing algorithm usually achieves the best performance. Notably, the 4-color caching seems a bit better when running on this dataset, instead of the 8-color in the experiment of the previous study, which uses simulated data.

Besides, the content colorizing approach based on their regional popularity involves a notable improvement of network traffic. Our estimations prove that traffic reduction in Taiwan is 27.3%, in the U.S. is 19.0%. Meanwhile, network traffic in Vietnam only reduces 6.2% due to the similarity between global and regional interests in Vietnam.

However, the popularity of content is calculated based on the number of times that this content is accessed. It implies that the approach is utilizing the content popularity in the last interval. A recent study [9] shows that it is possible to predict the popularity of content in the subsequent interval by harnessing the rich source of information such as user access log, which would considerably improve the network traffic.

Moreover, although the color-based caching can achieve the performance which is close to the sub-optimal result calculated by the GA algorithm, and its computational overhead is limited to a few seconds, its separator rank (SR) algorithm has exponential complexity. Our measurement reveals that the computation time rises to a few hours or even days when the number of content is over 5000 contents. However, most computations in SR algorithm are independent of each other, simplifying and optimizing this algorithm is considerable work in the future.

Acknowledgments. This research was conducted within the project of Studying collaborative caching algorithms in content delivery network sponsored by TIS (IT Holding Group).

References

1. Website Ho Chi Minh City University of Technology. http://www.hcmut.edu.vn/. Accessed 27 Aug 2018
2. Arteta, A., Barán, B., Pinto, D.: Routing and wavelength assignment over WDM optical networks: a comparison between MOACOs and classical approaches. In: Proceedings of the 4th International IFIP/ACM Latin American Conference on Networking, pp. 53–63. ACM (2007)
3. De Vleeschauwer, D., Robinson, D.C.: Optimum caching strategies for a Telco CDN. Bell Labs Tech. J. **16**(2), 115–132 (2011)
4. Dijkstra, E.W.: A note on two problems in connexion with graphs. Numerische Mathematik **1**(1), 269–271 (1959)
5. Einziger, G., Friedman, R., Manes, B.: TinyLFU: a highly efficient cache admission policy. ACM Trans. Storage (TOS) **13**(4), 35 (2017)
6. Li, W., Li, Y., Wang, W., Xin, Y., Xu, Y.: A collaborative caching scheme with network clustering and hash-routing in CCN. In: 2016 IEEE 27th Annual International Symposium on Personal, Indoor, and Mobile Radio Communications (PIMRC), pp. 1–7. IEEE (2016)
7. Li, Z., Simon, G.: In a Telco-CDN, pushing content makes sense. IEEE Trans. Netw. Serv. Manag. **10**(3), 300–311 (2013)
8. Maggs, B.M., Sitaraman, R.K.: Algorithmic nuggets in content delivery. ACM SIGCOMM Comput. Commun. Rev. **45**(3), 52–66 (2015)
9. Meoni, M., Perego, R., Tonellotto, N.: Dataset popularity prediction for caching of CMS big data. J. Grid Comput. **16**(2), 211–228 (2018)
10. Nakajima, T., Yoshimi, M., Wu, C., Yoshinaga, T.: A light-weight content distribution scheme for cooperative caching in Telco-CDNs. In: 2016 Fourth International Symposium on Computing and Networking (CANDAR), pp. 126–132. IEEE (2016)
11. Nakajima, T., Yoshimi, M., Wu, C., Yoshinaga, T.: Color-based cooperative cache and its routing scheme for Telco-CDNs. IEICE Trans. Inf. Syst. **100**(12), 2847–2856 (2017)
12. Tran, A.T.N., Nguyen, M.T., Diep, T.D., Nakajima, T., Thoai, N.: A performance study of color-based cooperative caching in Telco-CDNs by using real datasets. In: Proceedings of the Ninth International Symposium on Information and Communication Technology. ACM (2018, to appear)
13. Wang, Z., Jiang, H., Sun, Y., Li, J., Liu, J., Dutkiewicz, E.: A k-coordinated decentralized replica placement algorithm for the ring-based CDN-P2P architecture. In: The IEEE symposium on Computers and Communications, pp. 811–816. IEEE (2010)
14. Welsh, D.J., Powell, M.B.: An upper bound for the chromatic number of a graph and its application to timetabling problems. Comput. J. **10**(1), 85–86 (1967)

15. Wong, W.A., Baer, J.L.: Modified LRU policies for improving second-level cache behavior. In: 2000 Proceedings of Sixth International Symposium on High-Performance Computer Architecture, HPCA-6, pp. 49–60. IEEE (2000)
16. Yin, H., Liu, X., Qiu, F., Xia, N., Lin, C., Zhang, H., Sekar, V., Min, G.: Inside the bird's nest: measurements of large-scale live VoD from the 2008 olympics. In: Proceedings of the 9th ACM SIGCOMM Conference on Internet Measurement, pp. 442–455. ACM (2009)
17. Yu, H., Zheng, D., Zhao, B.Y., Zheng, W.: Understanding user behavior in large-scale video-on-demand systems. In: ACM SIGOPS Operating Systems Review, vol. 40, pp. 333–344. ACM (2006)
18. Zhou, Y., Chen, L., Yang, C., Chiu, D.M., et al.: Video popularity dynamics and its implication for replication. IEEE Trans. Multimedia **17**(8), 1273–1285 (2015)

Multi-UAV Placements to Minimize Power Consumption in Urban Environment

Jun-Woo Cho, Song Kim, and Jae-Hyun Kim[✉]

Department of Electrical and Computer Engineering, Ajou University, Suwon, Korea
{cjw8945,ks5109,jkim}@ajou.ac.kr
http://winner.ajou.ac.kr

Abstract. In this paper, we find optimal multi-Unmanned Aerial Vehicle (UAV) placements with the low power consumption of UAV in an urban environment. To solve the NP-hard problem, we use non-hierarchical method. Performance analysis shows that average power consumption of UAV according to the number of UAVs and the relationship between power consumption and capacity fairness of UAV.

Keywords: UAV · Power consumption · Positioning · Optimization theory · Non-hierarchical

1 Introduction

In recent years, Unmanned Aerial Vehicle (UAV) has been studied as a mobile base station to provide continuous and reliable wireless services to users. In particular, UAV became a popular issue as 3GPP created a standard for Non-Terrestrial Networks (NTN) [1]. UAV mobile base stations can be used where it is difficult or impossible to build network infrastructures such as the mountainous area and the war-field [2].

However, in order to use a UAV mobile station in their areas, it is necessary to overcome some constraints. The power consumption associated with the UAV flight time is one of the key issues to consider because UAV has a limited power source such as fuel, a secondary battery, unlike existing base stations. Some studies have considered how to mount solar cells in UAVs, but they cannot guarantee the sustainability of the wireless services from UAV to users [3].

UAV placement optimization is one of the ways to minimize the power consumption of UAV. However, it is an NP-hard problem due to the pathloss for Air-to-Ground (A2G) which cannot be impossible to derive from mathematical deduction [4,5]. In particular, it should be to consider not only the pathloss but also the number of UAVs and the data rate of various wireless services, in order to find the three dimensional (3D) aerial position [6].

The main contribution of this paper is to derive of the optimal multi-UAV placement with the low power consumption of UAV in an urban environment.

© Springer Nature Switzerland AG 2019
S. Lee et al. (Eds.): IMCOM 2019, AISC 935, pp. 306–314, 2019.
https://doi.org/10.1007/978-3-030-19063-7_25

In order to find the optimal value for the NP-hard problem, we use the non-hierarchical method. As a result, we first show the relationship between the number of UAVs and average power consumption of UAV by changing UAV capacity. Next, we show the relationship between power consumption of UAV and capacity fairness of UAV during algorithm iteration.

2 System Model

We consider a system model where UAVs support wireless services for all ground users in an urban environment. U is the set of UAVs, and N is the set of ground users. We assume that the location of UAVs and the ground users are fixed. Let h_u be the altitude of UAV u ($u \in U$), $d_{u,n}$ and $\theta_{u,n}$ be the 3D distance and the elevation angle between a given UAV u and ground user n ($n \in N$), respectively [4].

We define a group of ground users that are served various data rate wireless services by UAV u as clusters, CT_u. UAVs can support wireless services to its clusters within their capacity boundary. C_u is the maximum capacity of UAV u, C_n is the data rate that a ground user n can support from UAV. In this system model, we assumed that the bandwidth, B, allocated by UAV to ground users in its cluster is constant. The system model is shown in Fig. 1.

Given this model, we can derive the multi-UAV placements where the power consumption of UAV is minimized. In order to do this, not only the number of UAVs required but also the UAV capacity should be considered.

2.1 Pathloss Model

Unlike terrestrial base stations, the location of UAV is modeled in three dimensions. So new models should be used instead of existing pathloss models. For deriving the power relationship between UAVs and their clusters, we use the pathloss model proposed in [7]. The LoS probability which is provided by the International Telecommunication Union (ITU-R) and [7] is given by [8]:

$$P_{LoS} = a \left(\frac{180}{\pi} \theta_{u,n} - 15 \right)^b ,$$

(1)

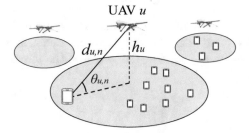

Fig. 1. System model

where, a and b are constant value which determined by environment and $\theta_{u,n}$ is the elevation angle which can be expressed as follow:

$$\theta_{u,n} = arcsin\left(\frac{h_u}{d_{u,n}}\right). \tag{2}$$

The pathloss can be derived:

$$PL_{u,n} = [P_{LoS}\mu_{LoS} + P_{NLoS}\mu_{NLoS}]\left(\frac{4\pi f}{C}\right)d_{u,n}^2. \tag{3}$$

where, μ_{LoS} and μ_{NLoS} are factors considered for LoS links and NLoS links, respectively. f is the central frequency and C is the speed of light. P_{NLoS} can be derived when the LoS signal and NLoS signal are the main influence on the pathloss [9]:

$$P_{NLoS} = 1 - P_{LoS}. \tag{4}$$

2.2   Optimization Model

Consider UAV u and a ground user n is in an urban area. The received data rate for user n is given by:

$$C_n = Blog_2\left\{1 + \frac{P_u}{PL_{u,n}I_n}\right\}, \tag{5}$$

where, P_u is the transmit power of UAV u, and I_n is the interference from all UAVs except UAV u [7]:

$$I_n = \alpha\sum_{v\neq u}^{U} P_v. \tag{6}$$

α is a constant between 0 and 1 to adjust the amount of interference.

If the ground users which are in a cluster CT_u are served in the coverage of UAV u, the required minimum power to satisfy the received data rate for each of ground users is given by:

$$P_u = \sum_{n\in CT_u} 2^{C_n/B}\,I_n\,PL_{u,n}. \tag{7}$$

The transmission power of UAV is affected by the distance between UAV and its associated ground user. Therefore, we assume that the UAVs have the same altitude in system model, and the height is the lowest value of the flight altitude. Finally, the optimization model, in which the power consumption of UAV is minimized, can be formulated to find x and y coordinates of the UAVs as [4]:

$$\{\boldsymbol{X}^*, \boldsymbol{Y}^*\} = argmin \sum_{u \in U} P_u. \tag{8}$$

$$s.t. \quad x_u \in \boldsymbol{X}^*, y_u \in \boldsymbol{Y}^*, \quad u = 1, ..., U \tag{9}$$

$$\sum_{u=1}^{U} \sum_{n \in CT_u} un = N, \tag{10}$$

$$\sum_{n \in CT_u} C_n < C_u. \tag{11}$$

3 Non-hierarchical Method

The UAV placement is NP-hard problem because it requires consideration of various variables such as pathloss model, user distribution, UAV movement, and so on. To solve this problem, some of the meta-heuristic algorithms are applied. Particle Swarm Optimization (PSO) is the most used algorithm in UAV placement. However, to solve our optimization problem, we need an algorithm to solve the clustering (classification) problem [10].

Non-hierarchical methods are algorithms that optimize a given criterion to divide the observed values into clusters. It is relatively simple than other clustering algorithms. The process of the algorithm is to measure the degree of similarity between observed values using criteria, perform classification, and then iterate the process of adjusting the cluster center [11].

3.1 Procedure of Finding Near-Optimal Multi-UAV Placements

Initialization. Non-hierarchical method must determine the number of clusters. This can be found by dividing the capacity of the UAV and the demand data rate of all users as follow [5]:

$$U = \left\lceil \frac{\sum_{n=1}^{N} C_n}{C_u} \right\rceil. \tag{12}$$

The number of clusters derived is the minimum required number of UAVs. If the number of UAVs is specified, the next step is to set the initial coordination of UAVs. The result of non-hierarchical method depends on the initial value of cluster center point. Therefore, it is general to set initial values of them using the heuristic algorithm. In this paper, we select U ground users randomly and select their coordinates as UAVs coordinates.

Clustering and Capacity Checking. After the initialization phase, we compute the matrix P by computing the power between the UAVs and all the ground users. The row of matrix P is the set of transmit powers of all UAVs from the

Table 1. Performance analysis environment

Parameter	Values
a, b	0.36, 0.21
c	3×10^8 m/s
μ_{LoS}, μ_{NLoS}	3 dB, 23 dB
f, B	2 GHz, 1 MHz
h_u	1 km
C_u	100 Mbps, 200 Mbps, 300 Mbps, 400 Mbps, 500 Mbps, 1 Gbps
C_g	100 Kbps, 1 Mbps, 10 Mbps
N	950 (100, 350, 500)

ground user, and the column is the set of transmit powers for all ground users from the UAV. After that, the minimum value of each row is found, and the column of the value is recorded in the matrix L, and then the clustering is performed based on this.

After clustering is complete, all of the UAVs capacity is calculated and recorded in matrix C. The size of the matrix C should be equal to the number of UAVs. If the largest value of the matrix elements exceeds the maximum capacity of the UAV, then δ_1 is counted. If δ_1 exceeds 8 in the iteration phase, the algorithm increases the number of UAVs and then proceeds to the initialization phase.

Centers of Clusters Renewal and Stopping Rule. The UAV placement update calculates the average value of the ground users coordinate value in the cluster. After that, the desired placements are found through iteration. If there is no further change value during the iteration, it is terminated.

4   Performance Analysis

To derive the near-optimal UAV placements, we perform the simulation using MATLAB. The performance analysis environment refers to [4,7,9]. We consider the ground users which consist of 50, 350 and 500 users with fixed data rates of 100 kbps, 1 Mbps, and 10 Mbps, respectively. Ground users are randomly distributed on the system model of size 1 Km × 1 Km. The other parameters are list in Table 1.

4.1   Results of the Number of UAVs and Average Power Consumption of UAV According to UAV Capacity

Figure 2 shows the required number of UAV and the average power of UAV as the UAV capacity change. The blue bar is the number of UAVs derived from (12), and the red bar is the number of UAVs derived after non-hierarchical

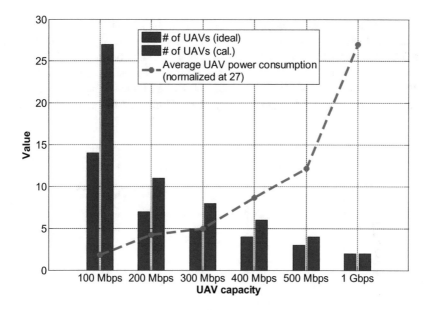

Fig. 2. Required number of UAVs and average power consumption of UAV

method operation. The difference between a blue bar and a red bar is larger with lower UAV capacity and decreases with higher UAV capacity. It is related to the distribution of ground users. If the ground users are uniformly distributed, the difference between the two bars will be small at any data rate. However, in the present simulation environment, the distribution of users is very random, and the deviation is large.

4.2 Results of Power Consumption vs. Capacity Fairness

Figure 3 show the change in power consumption and capacity fairness of UAV according to the algorithm iteration process. Capacity fairness is calculated using *Jain's Fairness Index* as follow: The green line is the normalized value of the average power consumption of UAV according to the UAV capacity. As the number of UAVs increases, the average power consumed of UAV decreases. UAV consumes more power for driving (such as propeller operation) than the power consumed for communication support. Therefore, it is essential for the network operator to set the appropriate number of UAVs to operate in the network according to the power consumption relation.

$$\Im(C_u) = \frac{\left(\sum_{u=1}^{U} C_u\right)^2}{U \sum_{u=1}^{U} C_u^2}. \tag{13}$$

As the number of iterations increases, the amount of power decreases in both graphs. It can also be seen that the variation in power consumption is reduced due to the solution convergence of the non-hierarchical method. However, capacity fairness does not increase even when power is lowered as shown in (b). It is caused by interference of the ground users located at the boundary of the cluster. As the capacity of the UAV decreases, the number of clusters that the UAV configures increases. As a result, the number of ground users existing at the cluster boundary increases, and the probability of receiving interference increases.

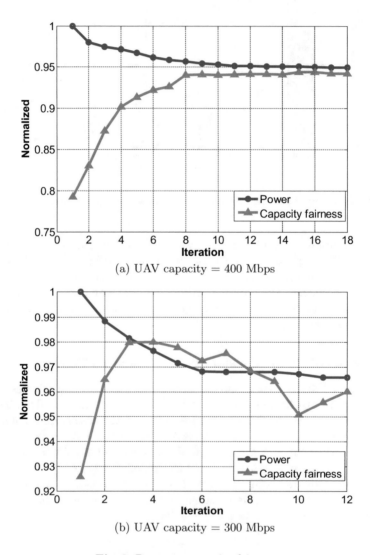

(a) UAV capacity = 400 Mbps

(b) UAV capacity = 300 Mbps

Fig. 3. Power vs. capacity fairness

5   Conclusion

In this paper, a near optimal solution for UAV placement that minimizes power consumption is derived considering the environment, where UAVs supporting various data rates are available for ground users. First, we have found the power relationships between UAVs and ground users, and perform optimization modeling to minimize power consumption. Moreover, used the non-hierarchical method to find the near-optimal value. As a result, we show the relationship between the number of UAVs and average power consumption of UAV, the relationship between power consumption of UAV and capacity fairness of UAV.

We do not consider the mobility of UAV and the mobility of the ground user. So we are going to do further research based on these considerations and plan to develop UAV's Medium Access Control (MAC) protocol.

Acknowledgments. This research was supported by the MSIT (Ministry of Science and ICT), Korea, under the ITRC (Information Technology Research Center) support program (2018-0-01424) supervised by the IITP (Institute for Information & communications Technology Promotion).

References

1. 3GPP: Study on New Radio (NR) to support non-terrestrial networks (Release 15); Technical report 3GPP TS 38.811 (2018)
2. Lee, D.H., Lee, K., Choi, H.J.: An iterative sort based resource scheduling for long distance airborne tactical communication in hub network. In: 9th International Conference on Ubiquitous Information Management and communication, pp. 11:1–11:4. ACM Press, New York (2015). https://doi.org/10.1145/2701126.2701201
3. Gupta, L., Jain, R., Vaszkun, G.: Survey of important issues in UAV communication networks. IEEE Commun. Surv. Tutor. **18**, 1123–1152 (2015). https://doi.org/10.1109/COMST.2015.2495297
4. Cho, J.W., Kim, J.H.: Performance comparison of heuristic algorithms for UAV deployment with low power consumption. In: International Conference on Information and Communication Technology Convergence, pp. 1067–1069. IEEE Press, New York (2018). https://doi.org/10.1109/ICTC.2018.8539485
5. Shi, W., Li, J., Xu, W., Zhou, H., Zhang, N., Zhang, S., Shen, X.: Multiple drone-cell deployment analyses and optimization in drone assisted radio access networks. IEEE Access **6**, 12518–12529 (2018). https://doi.org/10.1109/ACCESS.2018.2803788
6. Danoy, G., Brust, M.R., Bouvry, P.: Connectivity stability in autonomous multilevel UAV swarms for wide area monitoring. In: 5th ACM Symposium on Development and Analysis of Intelligent Vehicular Networks and Applications, pp. 1–8. ACM Press, New York (2015). https://doi.org/10.1145/2815347.2815351
7. Mozaffari, M., Saad, W., Bennis, M., Debbah, M.: Mobile Unmanned Aerial Vehicles (UAVs) for energy-efficient internet of things communication. IEEE Trans. Wirel. Commun. **16**, 7574–7589 (2017). https://doi.org/10.1109/TWC.2017.2751045

8. Al-Hourani, A., Kandeepan, S., Jamalipour, A.: Modeling air-to-ground path loss for low altitude platforms in urban environments. In: IEEE Global communication Conference, pp. 2898–2904. IEEE Press, New York (2014). https://doi.org/10.1109/GLOCOM.2014.7037248

9. Al-Hourani, A., Kandeepan, S., Lardner, S.: Optimal LAP altitude for maximum coverage. IEEE Wirel. Commun. Lett. **3**, 569–572 (2014). https://doi.org/10.1109/LWC.2014.2342736

10. Cho, J.W., Kim, J.H.: Multi-UAV positioning selecting scheme for minimizing power consumption using non-hierarchical cluster method. J. Korean Inst. Commun. Inf. Sci. **43**, 1263–1269 (2018). https://doi.org/10.7840/kics.2018.43.8.1263

11. Davidson, I., Ravi, S.S.: The complexity of non-hierarchical clustering with instance and cluster level constraints. Data Min. Knowl. Disc. **14**, 25–61 (2007). https://doi.org/10.1007/s10618-006-0053-7

Prediction of Traffic Congestion on Wired and Wireless Networks Using RNN

Aoi Yamamoto[1]($\boxtimes$), Haruka Osanai[1], Akihiro Nakao[2], Shu Yamamoto[2], Saneyasu Yamaguchi[3], Takeshi Kamiyama[4], and Masato Oguchi[1]

[1] Ochanomizu University, 2-1-1 Ohtuka, Bunkyo-ku, Tokyo 112-8610, Japan
{aoi,haruka_o}@ogl.is.ocha.ac.jp, oguchi@is.ocha.ac.jp
[2] University of Tokyo, 7-3-1 Hongo, Bunkyo-ku, Tokyo 113-8654, Japan
nakao@nakao-lab.org, shu@iii.u-tokyo.ac.jp
[3] Kogakuin University, 1-24-2 Nishi-Shinjuku, Shinjuku-ku, Tokyo 163-8677, Japan
sane@cc.kogakuin.ac.jp
[4] Kyushu University, 744 Motooka Nishi-ku, Fukuoka-shi, Fukuoka 819-0395, Japan
kami@f.ait.kyushu-u.ac.jp

Abstract. A large number of IoT devices are being used in the world. Data acquired by IoT devices are used as Big Data. These data sent from IoT devices to access points (AP) or base stations through wireless networks are stored in the cloud or data centers via wired networks such as back haul and backbone networks. The bandwidth of wireless or wired networks is oppressed by the transmission of a large amount of data. It causes network congestion, and the data cannot transmit and receive properly. To detect a sign of network congestion and prevent it is a very effective way to alleviate such congestion. However, it is difficult for us to predict a fluctuation of network traffic because it is caused by many factors and highly complex. Therefore, we try to predict it using Recurrent Neural Network (RNN), which exhibits dynamic temporal behavior for a time sequence in Deep Learning classes.

Keywords: Deep Learning · Congestion control · LSTM ·
Prediction of network traffic

1 Introduction

The age of Big Data has come about because of progress in technology and upgrading of the cloud environment. There are many kinds and forms of Big Data such as membership information and purchase history of consumers, text document obtained from e-mail and SNS, data of financial transaction and location information by GPS. It is difficult for conventional database systems to store and analyze this large amount of data. An analysis of valuable data concealed in Big Data by AI gives us useful knowledge. This knowledge is important to develop a growth strategy; thus, a lot of companies try to take advantage of it for marketing.

© Springer Nature Switzerland AG 2019
S. Lee et al. (Eds.): IMCOM 2019, AISC 935, pp. 315–328, 2019.
https://doi.org/10.1007/978-3-030-19063-7_26

IoT devices such as mobile phones, consumer electronics, cars, buildings and factories have increased rapidly in popularity, so various devices can access the Internet. In addition to the development of a network, IoT devices and sensors, which have been miniaturized and reduced cost, enable us to collect user information such as their attributes and activity efficiently. These data from IoT devices are stored in the cloud and used as Big Data.

Data collected by IoT devices are often time-series and real-time data; thus, an analysis of these data should be executed in a short period of time. Although these data tend to be analyzed after accumulation, it is sometimes too late when we detect an abnormality. Accordingly, demands for a real-time analysis is increasing. Therefore, data should be transmitted and processed rapidly in order to analyze them in real time. Nowadays, computers without advanced computing power can also analyze data by using cloud because of an improvement of high-speed networks.

Due to increasing demands of Big Data, a large amount of data is sent from real world to cyber world and dealt with. These data sent from IoT devices to access points (AP) or base stations through wireless network are stored in cloud or data centers via wired network such as back haul and backbone networks.

The bandwidth of wireless or wired networks is oppressed by transmission of a large amount of data. It causes network congestion and the data cannot be transmitted and received properly. Time-series data might be able to be predicted based on the past data, and packets causing network congestion are time-series data. Accordingly, we can possibly predict behavior of network traffic. There is a possibility that network congestion has a serious effect on the whole of a network system. To detect a signs of network congestion and prevent it is a very effective way to alleviate such congestion.

However, it is difficult to predict fluctuations of network traffic because they are caused by many factors and are highly complex. Network traffic congestion has never before been predicted effectively with Deep Learning as far as we know. Therefore, we try to predict network traffic congestion using one of Deep Learning techniques.

2   Related Work

When network congestion occurs, TCP/IP manages it by TCP congestion control. TCP congestion control is an improved algorithm based on slow start and congestion avoidance. It has been developed as TCP Tahoe, TCP Reno, and TCP NewReno [1,2]. In addition, improved TCP congestion control algorithms such as BIC TCP [3] and CUBIC TCP [4] is implemented in recent Linux releases. Moreover, Compound TCP [5] using a hybrid algorithm of loss base and delay base is implemented in Windows. Furthermore, Google has released TCP congestion control algorithm called TCP BBR based on Bandwidth Delay Product (BDP) in 2016, which has attracted notice [6].

At the same time, research on TCP congestion control suitable for wireless networks has also been conducted because a communication environment of wireless networks differs substantially from that of wired networks. WTCP, TCP Westwood, and TCP Westwood+ are known as typical TCP congestion control algorithms of wireless networks [7–9]. They are adapted for the property by which wireless network loses packets more often than does the wired network. Congestion control middleware developed in related works [10,11] aiming to control congestion among Android devices cooperatively. Each Android device notifies others of congestion window size indicating a number of segments that the android device sends from now. Then, they understand the transmission state of the surrounding devices. In addition, they automatically calculate an upper limit of congestion window size based on information from surrounding devices and revise it. This makes these devices share bandwidth, which they can use fairly. Therefore, it is possible to avoid accumulating ACK packets in wireless LAN AP.

In [12], congestion control middleware is improved. This study makes it possible to improve the transmission speed and fairness when many devices transmit data at the same time due to an adjustment of the timing that system works.

In [13], Rate-Adaptive TCP (RATCP) is developed. This is a TCP congestion control system in which the congestion window is adapted for bottleneck rate feedback. This compares RATCP and TCP under various network conditions in order to offer better feedback to TCP.

As stated above, TCP congestion control has been improved from a different perspective for many years. However, these are systems to control in response to an event such as packet loss due to network congestion. In contrast, if we can predict that a congestion is likely to occur, it is certain that we could realize very effective traffic control. Therefore, this research proposes predicting traffic behavior with Deep Learning.

Network traffic prediction has also been studied from various points of view. One of typical approaches is using statistical calculation model like autoregressive integrated moving average (ARIMA). Some research works employ Neural Network for the prediction of network traffic behavior [14–16]. Many methods of network traffic analysis and prediction are reviewed in [17]. However, most of them focus on the basic behavior of network and predict simple fluctuation of traffic, not a dynamic variation such as occurrence of congestion state. We try to predict a network traffic congestion in this research work.

3   Deep Learning Models

Deep Learning is an algorithm improved neural network and a way to implement machine learning. It is possible to learn a characteristic in more detail and improve a learning accuracy considerably. A neural network is an algorithm modeled after the neural circuit of brain. It has an input layer, a hidden layer and output layer. Multiple nodes in these layers are connected to edges, and each edge has a weight. This network calculates an error of output from an output layer and a correct data and propagates it from an output layer to an input layer. Then, it updates the weights of the edges and learns.

There are many kinds of neural networks such as those used for computer vision, time-series data, clustering and dimensionality reduction. The accuracy of the learned results changes which kind of neural network we use.

3.1 Recurrent Neural Network (RNN)

RNN is a model used to analyze and predict time-series data. It is able to memorize information of previous calculations, but in fact, it is not able to remember until two or three steps before.

3.2 Long Short-Term Memory (LSTM)

It is LSTM that solves this problem. LSTM is a model-improved RNN and a type of RNN. LSTM is realized by replacing units in the hidden layer of an RNN with a memory called an LSTM block and a block having three gates. The greatest feature of LSTM is that it can deal with long-term dependence, while RNN cannot.

In recent years, it has become possible to predict accurately using Deep Learning, for example, the learning of search logs when online users reserve flights with an airline company by RNN [18], making a proper e-mail reply with LSTM [19], making a model which returns an answer quickly that a questioner satisfies by Bi-directional LSTM [20]. Research using Deep Learning has achieved results in various fields, but studies on Deep Learning are not shown in the field of networks.

It is possible to predict the behavior of network traffic by LSTM because the packets of data dealt with in this study are time-series data, and we have to learn relatively long patterns.

4   An Overview of the Experiment

Network traffic is very complex, and its behavior is changed by many factors. Therefore, it is difficult to analyze and predict a complex state from the beginning. Thus, we begin to predict a simple case. This means that we focus on a purpose that we analyze rapid increases in traffic of wired and wireless networks, understand a sign that these traffics change and detect it before a network congestion occurs.

In experiments, we make a state that traffic increases rapidly by placing a load on the network which has no or almost no data transmission at the beginning. We try to analyze and predict these cases. After learning these congestion cases, we make a computer verify congestion traffic and see if the computer can recognize a trend for rapid increases in traffic before it really happens. The condition of network traffic congestion is different between cases of wired and wireless networks. Therefore, we choose the different input data suitable for each environment, respectively.

5   Experiment on Wired Network

5.1   Experiment Outline

In the experiment of this section, we use three computers. One is a server, and the others are clients. These machines transmit data to each other by iPerf. Client machines transmit UDP packets, and the server machine monitors this packet exchange (Fig. 1).

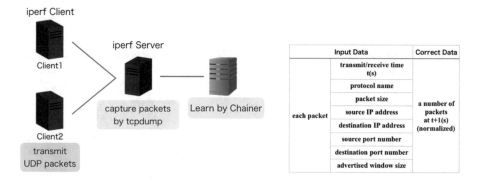

Fig. 1. Experiment environment   **Fig. 2.** Dataset

Figure 2 is a method of making input data and correct data for learning. Input data are eight kinds of feature values from each packet received between t–9 second to t seconds. Correct data are a normalized value from zero to one, which a number of packets send and receive at t+1 seconds.

We acquire three learning datasets and make learning models and evaluate them with respect to each dataset.

- Learning 1: Client 1 sends packets in a bandwidth of 10 MByte for 150 s. Client 2 sends packets in a bandwidth of 100 MByte for 60 s (Sect. 5.2).
- Learning 2: Client 1 sends packets in bandwidth of 10 MByte for 150 s. Client 2 sends packets in bandwidth of 100 MByte for 20 s, in bandwidth of 50 MByte for 20 s and in bandwidth of 100 MByte for 20 s continuously (Sect. 5.3).
- Learning 3: Client 1 sends packets in a bandwidth of 10 MByte for 150 s. Client 2 sends packets in bandwidth of 10 MByte for 10 s and send packets in bandwidth of 100 MByte for 60 s in 10 s (Sect. 5.4).

Table 1 is the performance of the computer used to generate traffic.

Table 1. The performance of the computer used to generate traffic

OS	Ubuntu 14.04.5LTS
CPU	Intel Xeon CPU E3-1270 V2 @ 3.50 GHz
Main memory	16 Gbyte

5.2   Wired Network: Learning 1

Learning data are packet data that client 1 sends packets in bandwidth of 10 MByte for 150 s. Client 2 sends packets in bandwidth of 100 MByte for 60 s.

Figure 3 shows a learning dataset and a prediction result that we input these learning data again into a learning model made by the data.

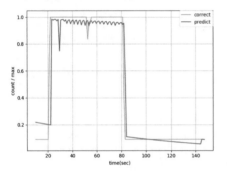

Fig. 3. Learning data and result of prediction of learning 1

Fig. 4. Test data 1 and result of prediction of learning 1

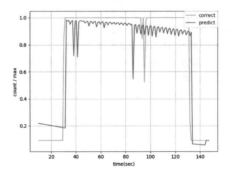

Fig. 5. Test data 2 and result of prediction of learning 1

The mean squared error is 0.01289. This figure and value show that this learning model succeeds in terms of learning characteristics of the learning data.

Figure 4 shows test data 1 and a result of the predicted output from the test data in the learning model. Test data 1 are data that are timed to start sending client 2's packets and are different from the learning data. The mean squared error is 0.02254, and this learning model can guess when a number of packets increases or decreases. This figure and value shows that the learning model can predict a traffic fluctuation even if the timing in which a number of packets increases or decreases is changed.

Figure 5 shows test data 2 and a result of the predicted output from the test data put into the learning model. Test data 2 are data that are timed to start sending client 2's packets and are changed from 60 s to 100 s. The mean squared error is 0.02487, and this learning model can guess when a number of packets increases or decreases. This figure and value shows that the learning model can predict a traffic fluctuation even if how long a time that client 2 sends packets is changed.

5.3   Wired Network: Learning 2

A learning data are packet data that client 1 sends in bandwidth of 10 MByte for 150 s, and client 2 sends packets in bandwidth of 100 MByte for 20 s, in bandwidth of 50 MByte for 20 s and in bandwidth of 100 MByte for 20 s continuously.

Figure 6 shows the learning data and a prediction result when we input these learning data again into a learning model made by the data.

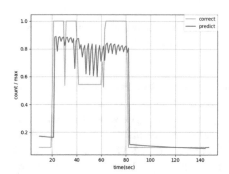

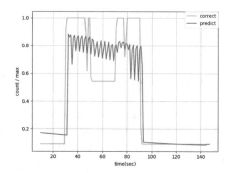

Fig. 6. Learning data and result of prediction of learning 2

Fig. 7. Test data and result of prediction of learning 2

The mean squared error is 0.02714. This learning model cannot predict a decrease in a number of packets between 40 s and 60 s precisely, but it is possible to presume that a number of packets between 20 s and 80 s is many. Therefore, this prediction model is sufficiently accuracy to predict network congestion.

Figure 7 shows test data and a result of predict output that the test data into the learning model. The test data are data that are timed to start sending client 2's packets and are differ from the learning data. The mean squared error is 0.03621. This model cannot guess that a number of packets decreases when it is used as input test data. However, it is able to predict a timing that a lot of packets are transmitted and received. Thus, it is no problem when we want to predict and deal with network congestion as well as inputting learning data.

5.4 Wired Network: Learning 3

Learning data are packets data that client 1 sends in bandwidth of 10 MByte for 150 s, and client 2 sends packets in bandwidth of 10 MByte for 10 s and send packets in bandwidth of 100 MByte for 60 s in 10 s.

Figure 8 is a graph showing the leaning data and a prediction result when we input these learning data again into a learning model made by the data.

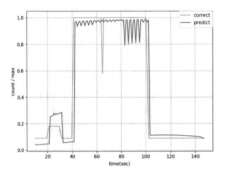

Fig. 8. Learning data and result of prediction of learning 3

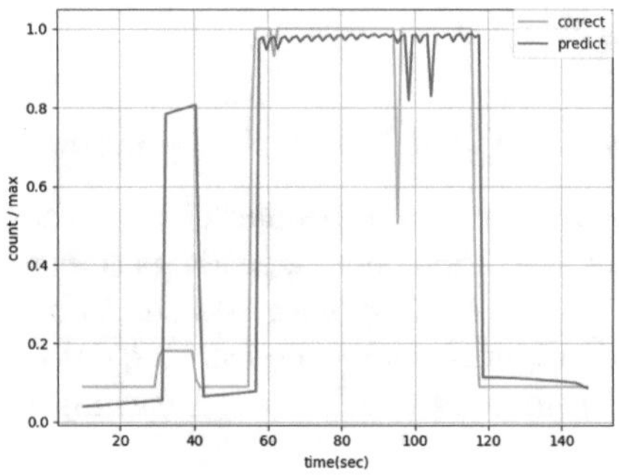

Fig. 9. Test data and result of prediction of learning 3

The mean squared error is 0.01932. This graphs and value show that this model succeeds in terms of the learning characteristics of the learning data.

Figure 9 shows test data and a result of predicted output when the test data were placed into the learning model. The test data are data that are timed to start sending client 2's packets and are different from the learning data, and a time that client 2 does not send packets is extended from 10 s to 15 s. The mean squared error is 0.04834. The error for 10 s that a number of packets increases the first time is large. However, this graph shows that the model can predict the timing of increasing and decreasing of packets and a fluctuation of a number of packets afterward.

6 Experiment on Wireless Network

In this chapter, communication is performed in a wireless environment, and traffic is generated. We explain learning and performance evaluation experiments using these data.

6.1 Investigation of Correct Data

First, we investigate the value of the correct data by observing the behavior of each parameter when Deep Learning. The correct data should be values indicating congestion. The correct data are decided by experiment.

We use a kernel monitoring tool developed by our previous work [22]. The Kernel Monitoring Tool is a system tool that can observe the behavior of parameters inside the Linux Kernel code, which is generally never observed. The Kernel Monitoring Tool can obtain logs from memory by inserting a monitor function and rebuilding kernel. Thereby, users can monitor parameters in real time, which include Congestion Window (CWND), RTT, and timing of errors. The experiment environment is shown in Fig. 10. Five Android terminals having the Kernel Monitoring Tool transmit packets, and we get TCP parameters on communication. In the interim, we measure RTT between PC and AP using a ping command from the PC.

Parameters are acquired for 60 s in order to observe the occurrence of congestion. For 0 to 19 s, no packets are transmitted from the Android terminal. We sent packets from 5 Android terminals to the server by iPerf all at once from 20 s. The congestion is generated by this extreme data communication in which all the Android terminals do not transmit or transmit packets. The three parameters in this experiment are graphed in Figs. 11, 12 and 13 for output.

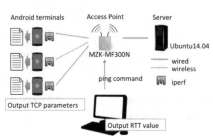

Fig. 10. Basic experimental environment

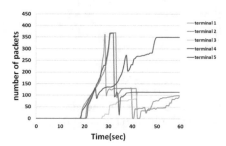

Fig. 11. Behavior of the CWND for Android devices

CWND Value for Android Device. Figure 11 shows the results obtained by measuring the behavior of the congestion window of the five Android terminals performs communication. CWND is a fixed value because of not transmitting packets up to 20 s. Each terminal is free to packets sent from near 20 s of being transmitting packets, and it can be seen that TCP controls each terminal. Twenty seconds after starting to send packets from the terminals, the TCP that detected congestion performs control by changing the CWND. The loss-based algorithm is a congestion control algorithm, which Android adopted, and it determines the congestion by detection of packet loss. As can be seen from the Fig. 11, the CWND increases exponentially until packet loss occurs by sending packets freely from each terminal. As a result, congestion occurs and TCP controls the traffic by changing the CWND.

RTT Value Between Android Terminal and Server. Figure 12 shows the results obtained by measuring the behavior of the RTT value between Android terminal and the server. The RTT value is gradually increased from near 20 s of being transmitting packets. Since the queue is not accumulated at about 20 s of being transmitting packets, the RTT value has not been significantly increased to process the packet appropriately. However, after 20 s, each terminal is sending a large number of packets. As a result, after 40 s, the RTT value seems to be rising sharply because packet processing cannot keep up.

RTT Value Between PC and AP. Figure 13 shows the results obtained by measuring behavior of the RTT value between PC and AP. Segments where the polygonal line disappears from the graph have not responded to the request and have timed out. This means that the AP is very busy, and congestion has occurred.

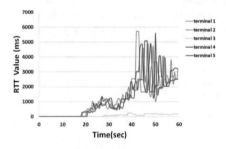 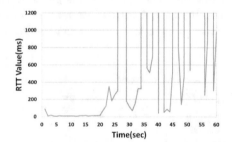

Fig. 12. Behavior of the RTT value between Android terminal and server

Fig. 13. Behavior of the RTT value between PC and AP

Decision of Correct Data. By experiment, RTT values between PC and AP measured using the ping command from the PC during data communication on Android terminal are used as correct learning data. This is because the RTT value between the PC and the AP also rises when congestion occurs from the Figs. 11 and 13, and it is considered that the rise of the RTT value represents the congestion of the AP.

6.2   Datasets Used in the Experiment

We explain the datasets used for Deep Learning. The correct data are RTT values between PC and AP according to Sect. 6.1. While measuring the RTT value between PC and AP, the request timed out several times. The measurement of the RTT value by the ping command results in a timeout if the response to the request is not more than 1000 ms. Therefore, the RTT value when timeout occurred was set to 1500 ms, and it was set as the correct data. Packets around the wireless LAN AP are used as input data. The device used is AirPcap [23] of Riverbed Company in the US. AirPcap USB-based adapters capture 802.11 wireless traffic for analysis by SteelCentral Packet Analyzer or Wireshark. Input data are five types of feature quantity derived from all packets captured by this device and two types of TCP parameters obtained from the Kernel Monitoring Tool aggregated every second. The details of input data are shown in Fig. 14.

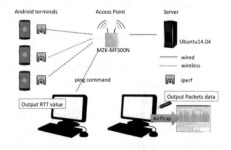

	Input Data	Correct Data
	time number of packets	
	average data amount	
	number of transmitters	
1 second	number of receiving devices	average RTT (PC-AP)
	average RTT value (Android terminal-Server)	
	average CWND	

Fig. 14. Dataset

Fig. 15. Experiment environment

6.3 Experiment Environment

Acquire data of Sect. 6.2 and conduct experiments. The experiment environment is shown Fig. 15. The Android terminals connected to one AP transmits the packet to the server. At the same time, the above data are acquired on the PC. After that, it processes the data into an appropriate data format and conducts Deep Learning using LSTM (Table 2).

Table 2. Specifications of devices

Android	Model number	Nexus S	Nexus 7 (2013)
	Firmware version	4.1.1	6.0.0
	CPU	1.0 GHz Cortex-A8	Quad-core 1.5 GHz Krait
	Memory (Internal)	16 GB 512 MB RAM	16 GB 2 GB RAM
	WLAN	Wi-Fi 802.11 b/g/n	Wi-Fi 802.11 a/b/g/n
Server	OS	Ubuntu 14.04 (64bit)/Linux 3.13.0	
	CPU	Intel(R) Core(TM)2 Quad CPU Q8400	
	Main memory	8.1 GiB	
AP	Model	MZK-MF300N (Planex)	
	Support format	IEEE 802.11 n/g/b	
	Channel	13	
	Frequency band	2.4 GHz (21412-2472 MHz)	

6.4 Wireless Network: Learning

This learning is a simpler experiment. One hundred thirty seconds of data is created using a total of 5 android terminals, four smartphones and one tablet terminal.

In 130 s, for 0 to 29 s, no packets are transmitted from the Android terminal. We started data communication from all Android terminals with iPerf in 30 s and continued to transmit packets for 70 s. After that, no packet is sent again for 30 s. The data set is extracted from the data captured in this experiment and uses learning data. Figure 16 is a graph of the prediction results. The RTT value, which is the correct answer data, is predicted using input data of t+9 s from t.

The orange line is the correct RTT value, and the green line is the predicted value.

We could not predict the exact value, but We can see that the timing at which the RTT value rises and the timing at which it falls were predictable. To verify the generalization ability of learning, we conduct experiments using test data. The test data are 120 s of newly created data. This time, we are sending packets for only 60 s of 20–80 s and not sending any packets at other times. Figure 17 is a graph of the prediction result that inputs the test data to the prediction model created earlier. Although accurate values cannot be predicted, the timing at which the RTT value increases and the timing at which it decreases can also be predicted with an error of a few seconds in the prediction by test data. Since the timing of sending packets and the time of the data set are also different from the learning data, it is considered that generalization ability is also present.

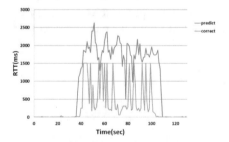
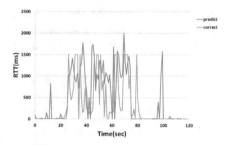

Fig. 16. Prediction result using learning data

Fig. 17. Prediction result using test data

7 Conclusions

In this study, we applied LSTM, a type of RNN, to the prediction of network traffic with the model of Deep Learning and evaluated performance in both wireless communication and wired communication.

In the experiment of wired connection, when learning and prediction are performed using data generated and captured using iPerf, it was found that traffic fluctuations can be predicted fairly accurately.

Although the value of the congestion window was not included as the feature value this time, by utilizing such values as feature value in the future, it is thought that prediction accuracy will be improved. For example, it is possible to predict the occurrence of congestion in advance.

In the experiment of wireless connection, since it was not possible to predict accurate values, we would like to make a more accurate prediction by adding the feature value of input data. In the experiment, in the case of extremely transmitting a packet, although it was not possible to predict to an accurate value, it was possible to predict the timing at which the RTT value increases and the timing at which it decreases. Since it gave a good result in experiments using test data, it is assumed that prediction is possible in this simple case.

Although we conducted experiments focusing on a simple case in which traffic increases rapidly for both wired and wireless connections, in the future, we will also conduct experiments assuming more realistic complex traffic and want to evaluate the performance. As soon as accurate prediction using data at the time of occurrence of congestion becomes possible, we want to control networks using Deep Learning.

References

1. Allman, M., Paxson, V., Blanton, E.: TCP Congestion Control, Internet RFC 5681, September 2009
2. Henderson, T., Floyd, S., Gurtov, A.: The NewReno modification to TCP's fast recovery algorithm. Technical report, IETF (2004)
3. Xu, L., Harfoush, K., Rhee, I.: Binary increase congestion control for fast long-distance networks. In: Proceedings of INFOCOM, March 2004
4. Ha, S., Rhee, I., Xu, L.: CUBIC: a new TCP-friendly high-speed TCP variant. http://netsrv.csc.ncsu.edu/export/cubic_a_new_tcp_2008.pdf
5. Tan, K., Song, J., Zhang, Q., Sridharan, M.: A compound TCP approach for high-speed and long-distance networks. In: Proceedings of INFOCOM, April 2006
6. Cardwell, N., Cheng, Y., Gunn, C.S., Yeganeh, S.H., Jacobson, V.: BBR: congestion-based congestion control. Queue **14**(5), 50 (2016). https://doi.org/10. 1145/3012426.3022184. 34 pages
7. Sinha, P., Nandagopal, T., Venkitaraman, N., Sivakumar, R., Bharghavan, V.: WTCP: a reliable transport protocol for wireless wide-area networks. Wirel. Netw. **8**(2/3), 301–316 (2002). Selected Papers from Mobicom 1999 Archive
8. Casetti, C., Geria, M., Mascolo, S., Sanadidi, M.Y., Wang, R.: TCP westwood: end-to-end congestion control for wired/wireless networks. Wirel. Netw. **8**(5), 467–479 (2002)
9. Grieco, L.A., Mascolo, S.: Performance evaluation and comparison of Westwood+, New Reno, and Vegas TCP congestion control. ACM SIGCOMM Comput. Commun. Rev. **34**(2), 25–38 (2004)
10. Hirai, H., Yamaguchi, S., Oguchi, M.: A proposal on cooperative transmission control middleware on a smartphone in a WLAN environment. In: Proceedings of IEEE WiMob 2013, pp. 701–717. IEEE (2013). http://ieeexplore.ieee.org/document/6673432/

11. Hayakawa, A., Yamaguchi, S., Oguchi, M.: Reducing the TCP ACK packet backlog at the WLAN access point. In: Proceedings of the 9th International Conference on Ubiquitous Information Management and Communication (IMCOM 2015), Article 37, 8 p. ACM, New York (2015). https://doi.org/10.1145/2701126.2701164

12. Shimada, A., Yamaguchi, S., Oguchi, M.: Performance improvement of TCP communication based on cooperative congestion control in Android terminals. In: Proceedings of the 12th International Conference on Ubiquitous Information Management and Communication (IMCOM 2018) (2018)

13. Karnik, A., Kumar, A.: Performance of TCP congestion control with explicit rate feedback. IEEE/ACM Trans. Netw. (TON) **13**(1), 108–120 (2005). https://ieeexplore.ieee.org/document/1402475/

14. Park, C., Woo, D.-M.: Prediction of network traffic by using dynamic bilinear recurrent neural network. IEEE (2009). Print ISBN 978-0-7695-3736-8

15. Chabaa, S., Zeroual, A., Antari, J.: Identification and prediction of internet traffic using artificial neural networks. Sci. Res. **2**, 147–155 (2010)

16. Junsong, W., Jiukun, W., Maohua, Z., Junjie, W.: Prediction of internet traffic based on Elman neural network. IEEE (2009). Print ISBN 978-1-4244-2722-2

17. Joshi, M., Hadi, T.H.: A review of network traffic analysis and prediction techniques. arXiv preprint arXiv:1507.05722 (2015)

18. Mottini, A., Acuna-Agost, R.: Deep choice model using pointer networks for airline itinerary prediction. In: Proceedings of the 23rd ACM SIGKDD International Conference on Knowledge Discovery and Data Mining, pp. 1575–1583 (2017). https://doi.org/10.1145/3097983.3098005

19. Kannan, A., Kurach, K., Ravi, S., Kaufmann, T., Tomkins, A., Miklos, B., Corrado, G., Lukács, L., Ganea, M., Young, P., Ramavajjala, V.: Smart reply: automated response suggestion for email. In: Proceedings of the 22nd ACM SIGKDD International Conference on Knowledge Discovery and Data Mining, pp. 955–964 (2016). https://doi.org/10.1145/2939672.2939801

20. Chen, Z., Gao, B., Zhang, H., Zhao, Z., Liu, H., Cai, D.: User personalized satisfaction prediction via multiple instance deep learning. In: Proceedings of the 26th International Conference on World Wide Web, pp. 907–915 (2017). https://doi.org/10.1145/3038912.3052599

21. Miki, K., Yamaguchi, S., Oguchi, M.: Kernel monitor of transport layer developed for Android working on mobile phone terminals. In: Proceedings of ICN 2011, pp. 297–302, January 2011

22. Miki, K., Yamaguchi, S., Oguchi, M.: Kernel monitor of transport layer developed for Android working on mobile phone terminals. In: Proceedings of the Tenth International Conference on Networks. ICN, pp. 297–302. https://doi.org/10.1109/WiMOB.2013.6673432

23. Riverbed Technology: Riverbed (1997). https://www.riverbed.com. Accessed 20 Sept 2018

Image and Video Processing

Summary of Part 3: Image and Video Processing

Hyunseung Choo, Duc-Tai Le, and Thien-Binh Dang

Image and video processing are an area of research that has seen huge growth in the recent past. Image processing is any form of signal processing for which the input is an image, such as photographs or frames of video; the output of image processing can be either an image or a set of characteristics or parameters related to the image. Most image-processing techniques involve treating the image as a two, and perhaps three, dimensional signal and applying standard signal-processing techniques to it. Video processing is a particular case of signal processing, where the input and output signals are video files or video streams. Video processing techniques are used in television sets, VCRs, DVDs, video codecs, video players and other devices. Image and video processing is everywhere around us, in smartphones, robotics, medicine, security systems, microscopy, remote sensing, video games, travel, shopping, environmental management and many other applications. The papers collected for this part present not only novel techniques but also diverse functionalities and services of image and video processing, as follows:

The paper, entitled "Moderating Effects of Spatial Presence on the Relationship Between Depth cues and User Performance in Virtual Reality" by Sunyoung Ahn, Young June Sah, and Sangwon Lee focus on creating a foundation to apply depth cue in virtual environment based on psychological understanding of users. The paper, entitled "A Method for Identification of Students' States Using Kinect" by Takashi Ito, Kenta Kamiya, Kenichi Takahashi, and Michimasa Inaba improve the performance of the previous work about identifying learning states of a student by adding learning data for generating identifiers and identifiers for the gestures. The paper, entitled "Micro-Expression Recognition Using Motion Magnification and Spatiotemporal Texture Map" by Shashank Shivaji Pawar, Melody Moh, and Teng-Sheng Moh propose a highly effective hybrid approach, which utilizes both Eulerian Video Magnification for motion magnification and Spatiotemporal Texture Map for feature extraction. The paper, entitled "Optimal non-uniformity correction for linear response and defective pixel removal of thermal imaging system" by Jae Hyun Lim, Jae Wook Jeon and Key Ho Kwon propose an optimal nonuniformity correction technique to reduces the number of bad pixels by considering thermal imaging system raw output characteristic. The paper, entitled "Azimuth Angle Detection with Single Shot MutilBox Detecting Model" by Quoc An Dang, Quan Mai Bao Nguyen, and Duc Dung Nguyen propose a method that enables detecting objects and estimating their pose simultaneously in a single model, without intermediate stages. The paper, entitled "An Alternative Deep Model for Turn-Based Conversations" by Van Quan Mai, Trieu Duong Le, and Duc Dung Nguyen propose a new model by extending the hierarchical recurrent encoder-decoder neural network for building a conversational dialogue system which provides natural, realistic and flexible interaction between human and machine based on large movie subtitles dataset. The paper, entitled "Tree Crown Detection and Delineation Using Digital Image Processing" by Zhafri Hariz Roslan, Ji Hong Kim, Roslan Ismail, and Robiah Hamzah conduct research on identification of tree species by using drone and the improved technique of image processing. The paper, entitled "Evaluation of Security and Usability of Individual Identification Using Image Preference" by Kotone Hoshina and Ryuya Uda propose a new personal identification method and system by selecting favorite images which are easy to be remembered and inputted. The paper, entitled "Learning Typical 3D Representation from a Single 2D Correspondence using 2D-3D Transformation Network" by Naeem Ul Islam and Sukhan Lee address the problem of typical 3D representation by transformation from its corresponding single 2D representation using Deep Autoencoder based Generate Adversarial Network termed as 2D-3D Transformation Network. The paper, entitled "Detection of Suspicious Person with Kinect by Action Coordinate" by Masaki Shiraishi and Ryuya Uda propose a method to judge a behavior patter of a person using Kinect in order to detect crimes at real-time even when skeleton coordinates are hidden by shielding objects. The paper, entitled "Invariant 3D Line Context Feature for Instance Matching" by Kyungsang Cho, Jaewoong Kim and Sukhan Lee presents a general method of identifying all the possible solutions or interpretations for an arbitrary pair of line segment sets by using invariant features associated with line segments. The paper, entitled "A Novel Performance Metric for the Video Stabilization Method" by Md Alamgir Hossain and Eui-Nam Huh present a new performance metric which can effectively measure the stability of a video based on the hypothesis is that in case of non-dynamic background, the background motion is zero without the foregrounds of a stable video when a video taken by a fixed camera.

Moderating Effects of Spatial Presence on the Relationship Between Depth Cues and User Performance in Virtual Reality

Sunyoung Ahn, Young June Sah, and Sangwon Lee[✉]

Department of Interaction Science, Sungkyunkwan University,
Seoul, South Korea
sunyoungain@gmail.com, somatosensor@gmail.com,
upcircle@skku.edu

Abstract. The user experience of Virtual Reality (VR) is a holistic experience related to internal and external aspects of users. Depth cues, which form depth perception, play an important role in enabling users not only to perceive the three-dimensional Virtual Environment (VE) from the two-dimensional surface but also to make interacts in the VE accurately. Spatial Presence (SP), a cognitive feeling of being there, addresses user's mental models related to the mediated environment. We assume that SP would have moderating effects on the relationship between depth cues and user performance of VR. We examine it through a user study. Considering the depth cue as an independent variable, participants are exposed to two experimental conditions: non-depth cue condition and depth cue provided condition. As moderator variables, Self-Location and Possible Action, a two-dimensional structure of spatial presence, are measured by a questionnaire. Task completion time is considered as a dependent variable. The results show that in high self-location situations, depth cues are not as useful, but they play an important role in low self-location situations.

Keywords: Virtual reality · Depth cues · Spatial presence · User performance

1 Introduction

Virtual Reality (VR) is a sensory-rich human-computer interface to create an effect of a three-dimensional world [1, 2]. A visual channel is a key sensory channel of the VR experience; how users see the mediated environment. Through visual representations, a form of visual stimulus, users obtain information about a physical space and an appearance of objects. Based on the obtained information, users perceive body positions related to the environment and make interactions with virtual objects [3].

In perceiving the Virtual Environment (VE), depth cues play an important role in the effective operation of cognitive processes; stimulus changing into information [e.g., 4]. Years of research on depth cues have shown that it influences not only depth perception but also sense of presence, and user performance. Hu et al. [5], have demonstrated that the shadow distribution, one of depth cues, allows users to have enhanced ability to

© Springer Nature Switzerland AG 2019
S. Lee et al. (Eds.): IMCOM 2019, AISC 935, pp. 333–340, 2019.
https://doi.org/10.1007/978-3-030-19063-7_27

judge contact between objects in the VE. Wann and Mon-Williams [6] have shown the effects of depth cues on recognizing three-dimensional features and making interactions in the VE. IJsselsteijn et al. [7] have found the effects of depth perception on increasing the sense of presence.

Depth cue has been a major research topic in VR, especially the link to the sense of presence. Presence has emerged in the psychological academia, which defines VR in terms of human experience, rather than technology hardware [8]. Presence, a sense of being there, has become sophisticated as a factor which affects user performance of VR [9, 10]. Spatial Presence (SP) is a subtype of presence which has emerged to reveal the cognitive mental process; what cognitive factors lead users to have a feeling of being there [11, 12]. In this study, SP is designated as a moderator.

The current study focuses on the depth cues, SP, and time performance to confirm the relationship in the VE. We assume that the sense of presence, a user's mental model, would moderate the effects of depth cues on user performance. It would be a primary step in identifying how cognitive mental process of presence affects the effects of sensory-dependent stimulus, depth cues, on user behavior, user performance.

2 Background

2.1 Spatial Presence (SP)

SP is an outcome of constructing a mental model of the self as being located in the VE [12]. SP focuses on the fact that people are conscious of the experience or feeling of being there. Researchers have investigated mental mechanisms that enable humans to feel presence – in other words, which cognitive factors or processes are involved in human perception and action, and how it leads to the feeling of presence [e.g., 13].

According to Vorderer et al. [13], SP is described by a two-dimensional structure: (1) Self-Location (SPSL) and (2) Possible Action (SPPA). SPSL is a core dimension which is the consciousness of being physically located within the mediated spatial environment. SPPA is a second dimension which refers to the perceived possible action in the given space based on mental representations of the mediated space. Based on the two dimensions, SP is formed through two steps: (1) Spatial Situation Model (SSM) and (2) Primary Egocentric Reference Frame (PERF) [11]. SSM is a precondition for SP to occur. On this level, users construct the mental model of the mediated environment which includes the space-related information. PERF is a second level which is the actual formation of SP. This level refers to the hypothesis-testing process to accept SSM as own egocentric viewpoint.

The study on SP has been conducted mainly in computer game research. Weibel and Wissmath [14] have shown that SP has a positive relationship with user performance. SP has been identified as a critical determinant of the effects of an electronic game on users in terms of game performance [15]. Since VR and computer games have a common aspect of providing an immersive experience, the effects of SP confirmed in computer game research is expected in VR research; however, there is a still lack of research on the effects of SP in VR research.

2.2   Depth Cues

Depth cues are one of the major factors which affect depth perception by providing information of distance and depth. Research on how humans perceive the depth information through visual cues in the natural environment has been extensively documented; however, naturally occurring depth cues may not occur equally in the VE. In other words, the perception of three-dimensional depth from the two-dimensional surface may not be same as the perception from the three-dimensional natural surface. Since all visual stimuli in the VE are created through the artificial process, designers are recommended to consider various cues that support more accurate depth judgments.

According to Sherman and Craig [3], four types of depth cues are typically considered in the VE: monoscopic depth cues, stereoscopic depth cues, motion depth cues, and physiological depth cues. Monoscopic depth cues refer to a relative depth cues [6] to enhance depth perception using empirically accumulated experiences in the natural world. It consists of eight components of static depth cues such as interpositions, linear perspectives, relative and known sizes, texture gradients, heights in the visual field, brightness, shadows, and aerial perspectives [3, 4, 16, 17] (see Table 1). These static cues are essential to providing information about the spatial scale of the VE when either stereoscopic depth cues or motion depth cues are corrupted.

Table 1. A description of monoscopic depth cues.

No.	Type	Description
1	Interposition	A depth order cue when the whole or part of one object occludes our view of another
2	Linear perspective	A depth order and distance cue due to the observance that parallel lines converge at a single vanishing point
3	Relative and known size	A distance cue due to the size difference between objects
4	Texture gradient	A distance cue due to the varying levels of visual clarity depending on the distance level
5	Height in the visual field	A distance cue based on the imagination that objects are lying on a horizontal surface below or above eye level
6	Brightness	A distance cue due to the difference of brightness between objects
7	Shadow	A size, location, and depth order cue between two objects
8	Aerial perspective	A distance cue due to the various saturation levels according to the distance level

3   Hypotheses

On the above, we studied the framework of SP and depth cues. SP is the feeling of being there and the mental model of one's body as a part of the VE. Depth cues refer to the spatial and depth information about the mediated environment. The study which reveals the holistic effects between information from the internal state of users such as SP and information from the external state of users such as sensory stimulus is essential that enable users to make rich, natural, and efficient interactions in VE. We expected the user's internal state, how much of a sensation of being there; how well mental representations are made related to the mediated environment, would control the effects of depth cues on user performance (see Fig. 1).

H1. SPSL moderates the effects of depth cues on the time performance.
H2. SPPA moderates the effects of depth cues on the time performance.

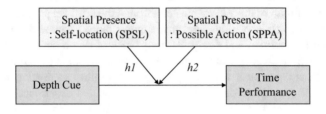

Fig. 1. Research model.

4   Method

4.1   Participants

Four females and four males between 23 and 27 years of age (mean age M = 25.2) participated in the study. They all had a normal or corrected-to-normal visual acuity. No sickness symptoms were observed during the experiment.

4.2   Stimuli

We designed the VE in two conditions (see Fig. 2). The condition 1 was called the "non-depth cue environment". No additional space and depth cues surrounded the scene (not even light). Only the impression of a horizon was induced. The condition 2 was called the "depth cue provided environment". Eight types of monoscopic depth cues were included to provide a rich depth cue condition (Table 1). In addition, there was a preliminary scene to learn how to operate the device; only three cubes were presented in the same distance.

Fig. 2. View of the first and second conditions.

4.3   Measurement

Independent Variable: Depth Cue. Eight types of monoscopic depth cues were chosen for controlling the viewing conditions of the VE (Table 1 and Fig. 2).

Moderator Variable: Spatial Presence. MEC Spatial Presence Questionnaire (MEC-SPQ) [13] was used to measure SP. SPSL consisted of 8 items (Cronbach's $\alpha = .92$), and SPPA consisted of 8 items (Cronbach's $\alpha = .91$).

Dependent Variable: Play Time. Task completion time of removing all ten cubes are tracked by the virtual reality software Unity to see user performance.

4.4   Experimental Design and Task

The experiment used a within-in subject design. All participant took part in both conditions. In one condition, the cubes presented all depth cues, and in the other condition, as many cues as possible were omitted. Ten cubes were displayed in the VE at each condition. Participants were asked to select cubes on an HMD in the order of how close they are to the user.

4.5   Procedure

Participants read an instruction material that included how to use the device and perform tasks. After reading the material, participants conducted the trial task to get used to make an interaction based on what they saw. Then the experiment started. When participants gazed at cubes, the pointer became active into clickable. Once participants clicked the cube in the order they felt closest, the cube was removed. When all of the cubes were removed, participants took the device off and filled out the questionnaire. This procedure continued for the next condition.

5   Results

The data of time performance and self-reporting were treated by PROCESS v3.1 [18]. Regression analyses on the additive multiple moderation model were conducted; the model was statistically supported (R-sq = .648, p = .038). The results of Ordinary Least Squares (OLS) regression shows that SPSL function as a moderator of the effects of depth cues on the time performance was statistically supported (Int_1 = 38.949 t (10) = 1.957, p = .079); H1 was supported. SPPA function as a moderator, however, was not statistically supported (Int_2 = –22.347, t(10) = -1.161, p = .273); H2 was not supported.

Table 2 describes conditional effects of the focal predictor at values of moderators. From the Table 2, the effect values of the 16th, 50th, and 84th percentiles increased positively; however, it was statistically effective only at the low level of SPSL. When the value of SPSL was held constant, the effects increased negatively depending on SPPA. The Fig. 3 shows that in the condition 1, SPSL significantly increased the play time negatively; the time performance was increased at the high level of SPSL, whereas in the condition 2, the overall time performance was high; however, unlike condition 1, it was more effective at the low level of SPSL.

Table 2. Results of conditional effects of the focal predictor at values of the moderators.

	SPSL	SPPA	Effect	se	t	p
16th	2.340	2.680	−23.343	8.252	−2.829	.018
	2.340	3.438	−40.271	16.139	−2.495	.032
	2.340	4.445	−62.786	34.280	−1.832	.097
50th	3.250	2.680	12.100	18.383	.658	.525
	3.250	3.438	−4.828	6.001	−.805	.440
	3.250	4.445	−27.343	17.124	−1.597	.141
84th	4.175	2.680	48.128	35.837	1.343	.209
	4.175	3.438	31.200	21.906	1.424	.185
	4.175	4.445	8.685	8.297	1.047	.320

Note. 16th, 50th 84th = the effect values of the 16th, 50th, 84th percentiles.

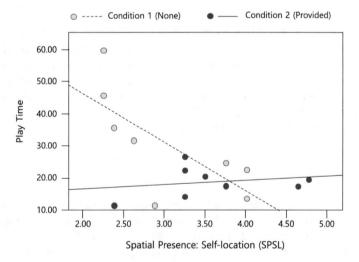

Fig. 3. A visualization of the relationship between SPSL and the play time under the two conditions.

6   Discussion and Conclusion

The present study focused on creating a foundation to apply depth cue in virtual environment based on psychological understanding of users. We expected spatial presence would control the effects of depth cues on user performance. To see the relations of the variables, we conducted a pilot test to see how each dimension of SP as moderators affects the effects of depth cues on the time performance.

The results indicated that spatial presence showed influence on the effects of depth cues on user performance. Particularly, Self-Location function as a moderator worked at the non-depth cue condition only. It implies that exclusively enhancing depth cues does not guarantee the high user performance. Performance degradation, however, may occur if depth cues are insufficient or improperly applied in the VE; and IJsselsteijn, de Ridder [7] have shown that naturalness of depth is not always preserved in the stimulus material. It implies that careful consideration is needed to construct depth cues in EV.

Possible Action refers to the perceived possibilities to act related to the mediated environment. Since Possible Action is strongly associated with making an interaction in the VE, we expected the level of Possible Action would affect the effects of depth cues on user performance. Unlike our expectations, the results show that Possible Action as a moderator had insufficient effects. It may mean that the consciousness of being physically located within the mediated spatial environment has effects on user performance without considering the Possible Action. It implies that to enhance user performance of VR, research on factors of forming the consciousness of being there (e.g., depth cues) is critical. On the other hands, there is a possibility that there may be errors in the experiment or measurement. Since the asked interaction to the participants during the task was performed in a static posture, it may be insufficient to measure the Possible Action. To address this issue, it is recommended to design the experimental task related to the dynamic action-based VR interaction, navigation and manipulation [3].

A limitation of this study is that the sample size was small to get clear outcomes. Despite the limitation, the results of this experiment provide insights that the exclusive enhancement of depth cues does not guarantee a significant improvement of user performance; however, depth cues have a significant effect when the VE provides an insufficient sense of presence to users. Thus, careful consideration on depth cues are needed. Further research is needed on the integration of individual depth cues to derive a method that can provide an optimal depth perception in the VE.

Acknowledgments. This work was supported by the National Research Foundation of Korea (NRF) grant funded by the Korea government (MSIP; Ministry of Science, ICT & Future Planning) (NRF-2017R1C 1B 1003650).

This research was supported by the MIST (Ministry of Science and ICT), Korea, under the National Program for Excellence in SW supervised by the IITP (Institute for Information & communications Technology Promotion) (2015-0-00914).

References

1. Burdea, G.C., Coiffet, P.: Virtual Reality Technology. Wiley, New Jersey (2003)
2. Bryson, S.: Virtual reality in scientific visualization. Commun. ACM **39**(5), 62–71 (1996)
3. Sherman, W.R., Craig, A.B.: Understanding Virtual Reality: Interface, Application, and Design. Elsevier, San Francisco (2002)
4. Cutting, J.E.: How the eye measures reality and virtual reality. Behav. Res. Methods Instrum. Comput. **29**(1), 27–36 (1997)
5. Hu, H.H., Gooch, A.A., Thompson, W.B., Smits, B.E., Rieser, J.J., Shirley, P.: Visual cues for imminent object contact in realistic virtual environment. In: Proceedings of the Conference on Visualization. IEEE Computer Society Press (2000)

6. Wann, J., Mon-Williams, M.: What does virtual reality NEED? Human factors issues in the design of three-dimensional computer environments. Int. J. Hum Comput Stud. **44**(6), 829–847 (1996)
7. IJsselsteijn, W., de Ridder, H., Hamberg, R., Bouwhuis, D., Freeman, J.: Perceived depth and the feeling of presence in 3DTV. Displays **18**(4), 207–214 (1998)
8. Biocca, F., Delaney, B.: Immersive virtual reality technology. Commun. Age Virtual Reality **15**, 32 (1995)
9. Barfield, W., Zeltzer, D., Sheridan, T., Slater, M.: Presence and performance within virtual environments. In: Virtual Environments and Advanced Interface Design, pp. 473–513 (1995)
10. Nash, E.B., Edwards, G.W., Thompson, J.A., Barfield, W.: A review of presence and performance in virtual environments. Int. J. Hum. Comput. Interact. **12**(1), 1–41 (2000)
11. Wirth, W., Hartmann, T., Böcking, S., Vorderer, P., Klimmt, C., Schramm, H., Saari, T., Laarni, J., Ravaja, N., Gouveia, F.R., Biocca, F.: A process model of the formation of spatial presence experiences. Media Psychol. **9**(3), 493–525 (2007)
12. Schubert, T.W.: A new conception of spatial presence: once again, with feeling. Commun. Theor. **19**(2), 161–187 (2009)
13. Vorderer, P., Wirth, W., Gouveia, F.R., Biocca, F., Saari, T., Jäncke, F., Böcking, S., Schramm, H., Gysbers, A., Hartmann, T., Klimmt, C.: MEC spatial presence questionnaire (MEC-SPQ): short documentation and instructions for application. Report to the European community, project presence: MEC (IST-2001-37661), 3 (2004)
14. Weibel, D., Wissmath, B.: Immersion in computer games: the role of spatial presence and flow. Int. J. Comput. Games Technol. **6** (2011)
15. Tamborini, R., Skalski, P.: The role of presence in the experience of electronic games. In: Playing Video Games: Motives, Responses, and Consequences, pp. 225–240 (2006)
16. Reichelt, S., Häussler, R., Fütterer, G., and Leister, N.: Depth cues in human visual perception and their realization in 3D displays. In: Three-Dimensional Imaging, Visualization, and Display 2010 and Display Technologies and Applications for Defense, Security, and Avionics IV. International Society for Optics and Photonics (2010)
17. Howard, I.P., Rogers, B.J.: Seeing in depth. Volume 2: Depth Perception. Canada: I. Porteous, Ontario (2002)
18. Hayes, A.F.: Introduction to Mediation, Moderation, and Conditional Process Analysis: A Regression-Based Approach. The Guilford Press, New York (2013)

A Method for Identification of Students' States Using Kinect

Takashi Ito, Kenta Kamiya, Kenichi Takahashi$^{(\boxtimes)}$,
and Michimasa Inaba

Hiroshima City University, Hiroshima 731-3194, Japan
takahasi@shiroshima-cu.ac.jp

Abstract. In study classes, teachers pay attention to students to find out whether they are following the study contents or not. A system for automatic identification of learning states of students would be useful for teachers to grasp learning states of all students. To develop the system to identify learning states, a method for identification of learning states of a student was proposed in previous work. In the method, a gesture identification function to estimate an action from information of a human skeleton and Computer Vision API which returns words and sentences representing the situation in an image of a student captured during learning are utilized. In this study, to make the situation closer to real situations, learning data to generate identifiers and identifiers for the gestures are added. In order to confirm the effectiveness of these improvements, the identification experiments in this laboratory are conducted.

Keywords: Kinect · Identification · E-learning

1 Introduction

In classrooms where education is provided, such as in schools, it is extremely difficult for the lecturer to keep track of the learning status of all of the students during the lecture. Lecturers are often occupied by writing on the blackboard, and often have their backs facing towards the students during lectures. Furthermore, lectures at universities are often held in large lecture halls, which makes it difficult for the lecturer to pay attention to the students sitting in the back of the classroom. Therefore, there is a need for systems that can identify the learning status of the students in place of the lecturer.

In a related research project, the researchers estimated the state of an audience based on video data of the audience members captured from an arbitrary location using convolutional neural networks [1]. In this research, the researchers extracted multiple audience members from the video of the lecture to estimate the state, and estimated whether the audience members were concentrating or not.

In this paper, we propose a method for identifying the learning state of a single student as a starting point for constructing a system that can identify the learning state of an audience of students. This system uses Kinect [2], available from Microsoft, to photograph the student. The Kinect includes a depth sensor as well as a camera, enabling it to collect a greater amount of information than that of a normal camera. In addition, a variety of Software Development Kits are also available for the Kinect, making it

© Springer Nature Switzerland AG 2019
S. Lee et al. (Eds.): IMCOM 2019, AISC 935, pp. 341–350, 2019.
https://doi.org/10.1007/978-3-030-19063-7_28

relatively easy to develop systems that use the features of Kinect [3, 4]. Our system uses the gesture recognition function, which is one of the functions offered by Kinect. The gesture recognition function enables users to obtain a confidence level that indicates the similarity of the motion of the photographed subject to each of the registered in a gesture database that is created in advance [5]. In addition, the proposed method also uses the Computer Vision API [6] available from Microsoft in order to improve the identification accuracy. The Computer Vision API analyzes information (states and objects) within the images that are sent to the API, and returns results. In the proposed method, we trim the image captured by the Kinect camera to include only the area around the student's hands, and send this image to the API. Then, we analyze the object that the student is holding and estimate the state of the student [7]. In order to verify the identification accuracy of this system, we conducted experiments on recognition of actual learning states with the cooperation of four students from our laboratory. Furthermore, we added training data representing more real postures of students.

2   Kinect for Windows V2

Kinect for Windows v2 (hereinafter referred to as Kinect) is a device developed by Microsoft for controlling video games through gestures and voice recognition [2]. The Kinect is shown in Fig. 1. The Kinect includes a color camera for capturing color images and a depth sensor that obtains depth information from projected infrared rays. The Kinect is able to handle a larger amount of data than a web camera. In order to handle those functions easily, Kinect for Windows SDK is provided. We can easily program functions for detecting human skeleton, acquiring face information such as eye directions, and so on.

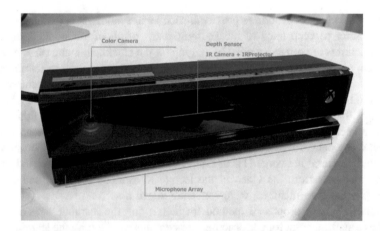

Fig. 1. A Kinect for Windows v2.

The Kinect includes a gesture recognition function that is capable of identifying the state of a human using classifiers. In addition, Microsoft provides the Visual Gesture

Builder to generate classifiers. Users can generate classifiers by registering arbitrary gestures from videos captured by the Kinect and having the Gesture Builder learn these gestures.

In this paper, we considered four different actions performed by the students, which consist of the actions of reading a book, taking notes, asking a question, and sleeping. Figure 2 shows examples of training data for classifiers. We generated a total of nine classifiers. The reason that the number of classifiers we used was larger than the number of actions was that we believe that a large amount of variation exists in how different individuals perform all of these actions, except for sleeping. Therefore, we generated multiple classifiers for these actions. We generated two classifiers for the action of reading a book, four classifiers for taking notes, and two classifiers for asking a question. In addition, we used the settings shown in Table 1 during the learning.

The classifier-based gesture recognition function can identify gestures within videos captured by Kinect in real time. The gesture recognition function returns the result in terms of a confidence level, which is a real number that expresses how similar the subject's actions are to each of the registered gestures.

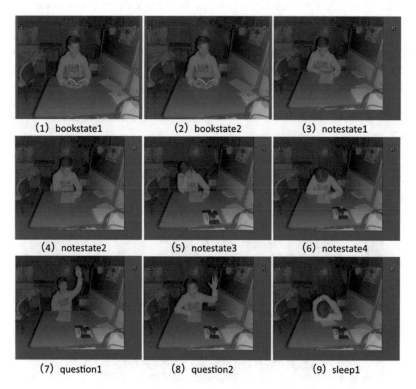

(1) bookstate1   (2) bookstate2   (3) notestate1

(4) notestate2   (5) notestate3   (6) notestate4

(7) question1   (8) question2   (9) sleep1

Fig. 2. Examples of training data for classifiers.

Table 1. Learning settings for Visual Gesture Builder.

	Taking notes	Other
Body side	Any	
Gesture type	Continuous	
Training settings		
Duplicate and mirror data during training	O	O
Use hands	O	X
Ignore left arm	X	X
Ignore right arm	X	X
Ignore lower body	O	O

3 Computer Vision API

At first, we tried identifying the state of a student using the gesture recognition function alone, but in order to improve the accuracy, we combined gesture recognition with the Computer Vision API provided by Microsoft. In the remainder of this paper, we refer to the Computer Vision API as simply API for convenience. The API was mainly used to identify the objects that students were holding. Therefore, instead of directly sending the image captured by Kinect to API, we first processed the image captured by Kinect by extracting the coordinates of the student's hands and trimming the image to include only the area around the student's hands. An example of a trimmed image is shown in Fig. 3. However, in the cases in which the student was sitting in a posture in which his/her hands were not in view, we directly used the image from Kinect.

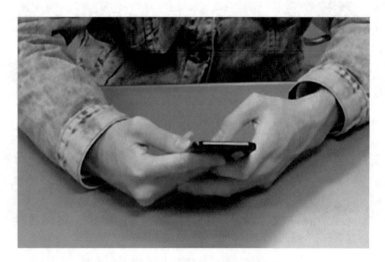

Fig. 3. An examples of a trimmed image.

API users can select one type from among seven types of analysis results provided by the API. For the proposed method, we selected "Description" and "Tags". The analysis results for the image shown in Fig. 3 are shown in Fig. 4. The Description option analyzes the image and returns the analysis results as an English sentence. In Fig. 4, the analysis has determined that the student is holding a cell phone. The Tags option analyzes the image and returns the results as English terms describing the action or object.

Description   { "type": 0, "captions": [{ "text": "a close up of a person holding a cell phone", "confidence": 0.3146926380632588 }] }

Tags   [{ "name": "person", "confidence": 0.9804659485816956 }, { "name": "indoor", "confidence": 0.9376060962677002 }]

Fig. 4. The analysis results for the image in Fig. 3.

The proposed method analyzes these two results and makes the determination that the student was using a smartphone if the API detects a "phone", was reading a book if the API detects "book" and "read", was taking notes if the API detects "pen" and "paper", and was sleeping if the API detects "sleep". The proposed method applies these results to the results of the gesture recognition function.

4   Method for Identifying States of Students

The proposed method for identifying student state uses the results of the gesture recognition function, which are obtained once every second, and the results from the API, which are available every 5 s, in order to display the results of the identified state of the student on a monitor every 10 s. The reason that the time interval for using the API results was set to 5 s was that attempting to make this interval even shorter would incur the risk of the results not being returned in time. The flowchart of the proposed method is shown in Fig. 5. First, Kinect is turned on, and the student is registered. After the student has been registered, *time*, *half_G*, and *result_G* are initialized to zero, and the identification begins. Here, *time* is a parameter that is used to control the algorithm such that the identification results are obtained every 1 s, *half_G* is an array that is used to store the results of the gesture recognition function for 5 s worth of data, and *result_G* is an array that is used to store the results for data in 10 s.

If the value of *time* is equal to zero, then the algorithm determines that the initial exception processing should be run, and the algorithm only trims and sends the image and only increments the parameter *time*. If the value of *time* is not equal to zero, then we acquire the gesture recognition results. The gesture recognition function provides the confidence level for each of the registered gestures. We select the gesture with the highest level of confidence, which we refer to as gesture i, as the result of the gesture recognition. We increment the value of $half\_G_i$ that corresponds to gesture i by +1.

In the case that the confidence levels for all of the gestures are all zero, we deem the state to be immeasurable, and do not increment any of *half_G*.

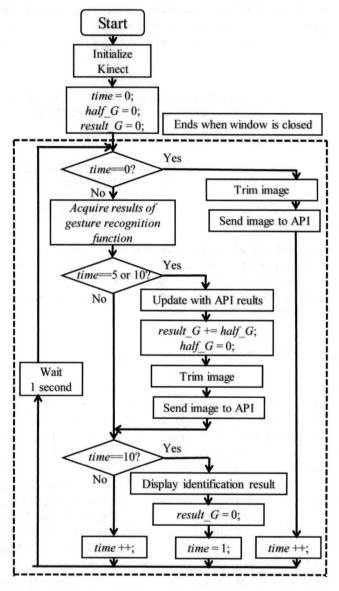

Fig. 5. The flowchart of the identification method.

Next, if time is equal to 5 or 10, then we use the API results to update *half_G*, in which the previous 5 s of data are stored. If result *i* from the API indicates that the student is reading a book, taking notes, or sleeping, then we update *half_Gi* using (1).

$$half\_Gi = (half\_Gi + 5.0)/2.0 \qquad (1)$$

If result i from the API indicates that the student is using a smartphone, then we update $half\_Gi$ using the following equation.

$$half\_Gi = 5.0 \qquad (2)$$

The reason that the update equation is different for the situation in which the student is using a smartphone is that the gesture recognition feature cannot recognize the action of using a smartphone, and $half\_Gi$ can only be updated through the API results. If the API result indicates that the API was unable to recognize the action, or the API result does not include any of the applicable terms, then the result is not updated. The updated value of $half\_G$ is added to $result\_G$, and $half\_G$ is reset to zero.

Lastly, if $time$ is equal to 10, then we display the identification results, and reset $result\_G$ to zero and $time$ to one. We display the gesture that corresponds to the highest value of $result\_G$. If $time$ is not equal to 10, then we increment $time$. Afterwards, we delay for one second, and repeat the process, starting from the condition that tests whether $time$ is equal to zero. In the proposed method, the process can be stopped anytime by closing the window.

5   Experiments

We conducted an experiment in which we instructed the cooperators to perform particular actions and used the system to analyze the states of the cooperators. The experimental cooperators consisted of four students from our laboratory. The cooperators were given instructions that were vague, such as "please take notes", rather than shown specific examples. The experimental cooperators were told to continue the action for 30 s. After the end of the experiment, we compared the actions of the experimental subjects with the identification results from the system using the recorded video in order to determine whether the system was able to successfully identify the states.

The experimental results for experimental cooperators A–D are shown in Table 2. The identification results for the case in which the subject was reading a book are labeled with the word "book", the results for the case in which the subject was taking notes are labeled with the word "notes", the results for the case in which the subject was asking a question are labeled with the word "question", and the results for the case in which the subject was using a smartphone are labeled with the word "smartphone". The reason that the denominator for each of the individual recognition results is three is that we made the experimental cooperators perform each of the actions for 30 s and acquired results every 10 s. The state which was recognized with the highest accuracy was the state of taking notes. For this state, four classifiers were provided to the gesture recognition function, which resulted in successful recognition in versatile situations.

However, it was difficult to differentiate this state from the state of reading a book. There were many cases in which reading a book was incorrectly identified as taking notes. The state which was recognized with the next highest accuracy was sleeping. Excluding cooperator C, the recognition accuracy was 8/9. The cause for these results for cooperator C was that cooperator C did not put his/her head down while sleeping, which created a situation that the classifier could not handle. In regards to the results for the experiment in which the cooperators were asking questions, although we assumed that the subjects would raise their hands while asking questions, experimental cooperators C and D did not interpret our instructions as we had expected, which resulted in the algorithm having difficulty in recognizing this state. Furthermore, although we were expecting to obtain the identification results for using a smartphone through API, the API was not able to analyze this state.

Table 2. Experimental results by the identification method.

	A	B	C	D	Accuracy
book	2/3	1/3	1/3	0/3	4/12
notes	3/3	2/3	3/3	1/3	9/12
question	3/3	1/3	0/3	0/3	4/12
sleeping	3/3	2/3	0/3	3/3	8/12
smartphone	0/3	3/3	0/3	0/3	3/12

Next, we added training data of five students to improve the accuracy of identification by let identifiers adapt to different body types and postures of students. In the added training data, new postures for "reading a book" and "sleeping". Figure 6 shows those images. As for "reading a book", we assumed that a student reads a book on a desk. We added images in which a student holds a book with his hands to read it. Additionally, we added two kinds of postures for "sleeping". We assumed that a student sleeps lying on his face in a desk. We added two kinds of images in which a student rests his elbows and leans on a chair, respectively.

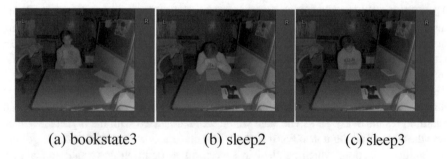

(a) bookstate3   (b) sleep2   (c) sleep3

Fig. 6. Examples of added training data

Table 3 lists experimental results for the training data including the added images. Comparing Table 2 and Table 3, we can see that the accuracy for "reading books," "taking notes," and "asking a question" is improved, while the accuracy for other two states is not improved.

Table 3. Experimental results using updated training data.

	A	B	C	D	Accuracy
book	3/3	1/3	0/3	2/3	6/12
notes	3/3	3/3	3/3	3/3	12/12
question	3/3	2/3	3/3	2/3	10/12
sleeping	3/3	0/3	0/3	2/3	5/12
smartphone	0/3	2/3	0/3	0/3	2/12

In order to improve the performance to identify "smartphone", we employed the API. However, the method didn't work well for identifying smartphones, and the improvement was not achieved. Thus, we recalculated the experimental results without using identification by the API. Table 4 lists the accuracy of identification without using the results by API. Table 4 doesn't include the accuracy for "smartphone", since identification of smartphones was not possible without the API. We can see from Table 4 that the accuracy for "book", "notes", and "question" is improved.

Table 4. Experimental results without using the API.

	A	B	C	D	Accuracy
book	3/3	3/3	0/3	3/3	9/12
notes	3/3	3/3	3/3	3/3	12/12
question	3/3	3/3	3/3	3/3	12/12
sleeping	3/3	0/3	0/3	3/3	6/12

6 Conclusions

In this research, we proposed a method for automatically identifying the state of students engaged in learning. The proposed method is composed of a gesture recognition function that uses Kinect and the Computer Vision API. We conducted identification experiments and showed that the proposed method achieves high accuracy in identifying the states that consist of taking notes and sleeping. In addition, we attempted to use the API to analyze the state based on an image of the area around the students' hands. However, this method failed to produce results that were as good as we had been expecting, and did not contribute to improvements in accuracy.

In the future, it is necessary to validate the analysis performance of the API. It is also future task to employ machine learning techniques for the improvement of the accuracy and compare the other ways of sustaining the concentration of learners.

Acknowledgments. This research was supported by Grant-in-Aid for Scientific Research JP16K00485.

References

1. Shimada, D., Yatomi, H.: Video-based estimation system, using convolutional neural networks for audiences' state in the classroom and discussion of its essential image features. In: Proceedings of Japan Society for Fuzzy Theory and Intelligent Informatics, no. 1, pp. 517–526 (2017)
2. Microsoft: Purchase an Xbox One Kinect Sensor – Microsoft Store Japan. https://www.microsoftstore.com/store/msjp/ja_JP/pdp/Xbox-One-Kinect-%E3%82%BB%E3%83%B3%E3%82%B5%E3%83%BC/productID.4601073900. Accessed 30 June 2017
3. Nakamura, K., Sugiura, T., Takada, T., Ueda, T.: KINECT for Windows SDK Programming – Kinect for Windows v2 Sensor Compatible Edition. Shuwa System, Tokyo (2015). Author, F., Author, S., Author, T.: Book title. 2nd edn. Publisher, Location (1999)
4. Microsoft: Kinect for windows product blog. https://blogs.msdn.microsoft.com/kinectforwindows/. Accessed 30 June 2017
5. Microsoft: Visual Gesture Builder (VGB). https://msdn.microsoft.com/en-us/library/dn785304.aspx. Accessed 30 June 2017
6. Microsoft: Computer Vision API – Image Processing and Image Analysis | Microsoft Azur. https://www.microsoft.com/cognitive-services/en-us/computer-vision-api. Accessed 30 June 2017
7. Ito, T., Ito, K., Takahashi, K., Inaba, M.: Experiments with Kinect for identification of a state of students. In: IEEE SMC Hiroshima Chapter Young Researchers' Workshop, pp. 26–29 (2017)

Micro-expression Recognition Using Motion Magnification and Spatiotemporal Texture Map

Shashank Shivaji Pawar, Melody Moh⬥, and Teng-Sheng Moh$^{(\boxtimes)}$⬥

Department of Computer Science, San José State University, San José, CA, USA
{shashankshivaji.pawar,melody.moh,teng.moh}@sjsu.edu

Abstract. Micro-expressions are short-lived, rapid facial expressions exhibited by individuals when they are in high stakes situations. Studying these micro-expressions is important as these cannot be controlled by an individual and hence offer a peek into what the individual is actually feeling and thinking as opposed to what he/she is trying to portray. The spotting and recognition of micro-expressions has vital applications in many fields including criminal investigation, psychotherapy, and education. However, due to their short-lived, rapid nature, spotting, recognizing and classifying micro-expressions is a major challenge. In this paper, a highly effective hybrid approach is proposed, which utilizes both Eulerian Video Magnification for motion magnification and Spatiotemporal Texture Map for feature extraction. The experiments are carried out on CASME II, the only spontaneous micro-expression dataset currently available. The proposed approach uses Support Vector Machines (SVM) classifier with kernels, and achieves an accuracy of 80%, viz. an increase by 5% comparing with the best existing method using SVM on CASME II. This work is a significant step for recognizing micro-expressions; the proposed method would be widely applicable to facial sentiment recognition in both images and videos.

Keywords: Micro-expressions (ME) · Motion magnification ·
Spontaneous datasets · Spotting · Recognition ·
Spatiotemporal texture map (STTM) · Eulerian Video Magnification (EVM) ·
Support Vector Machines (SVM)

1 Introduction

Human beings are a social species. We have evolved to express our feelings and emotions through a variety of ways such as facial expressions, body languages, and hand gestures. Of all these, facial expressions are the most common way of expressing one's emotions [19]. Even while expressing verbally, research has shown that 55% of the total information expressed is with facial expressions [14]. These facts illustrate the importance of facial expressions in analyzing an individual's psyche and the inner workings of his/her mind.

Facial expressions are categorized as macro-expressions and micro-expressions. Macro-expressions are what an individual can control using their facial muscles, and hence they can be manipulated by the individual to suit his/her needs. They typically

© Springer Nature Switzerland AG 2019
S. Lee et al. (Eds.): IMCOM 2019, AISC 935, pp. 351–369, 2019.
https://doi.org/10.1007/978-3-030-19063-7_29

last longer and are easier to detect and recognize. Micro-expressions, on the other hand, are involuntary and occur in high-stakes situations [2]. They also last for far shorter durations, typically 1/25 to 1/3 s [13, 30]. What micro-expressions worth studying is their ability to indicate the true thoughts and feelings of an individual, thus offering us a peek into the individual's mind. In short, we can ascertain if the words corroborate with one's true feelings. Yet, due to the innate short duration of these micro-expressions, human beings often find them extremely difficult to spot and recognize [3]. A study has shown that even experts have a chance (about 40%) worse than a coin toss of spotting a micro-expression [5].

Micro-expression spotting and recognition have applications in various fields, from criminal investigation, educational institutions, to psychological evaluation. Law enforcement agencies could use micro-expression recognition techniques to ascertain if an individual is telling the truth instead of polygraph tests which can be easily falsified [32, 33]. Individuals suffering from mental health issues such as autism [20] or schizophrenia [17] often cannot or do not want to express their emotions. Psychiatrists can then use micro-expression techniques to diagnose the patient not only from his verbal cues, but from his non-verbal cues as well. In educational settings, this technique can be used by educators to gauge the class's understanding of lectures.

While much research has been done in the macro-expression recognition paradigm, micro-expression recognition is still in its nascence. Spontaneous datasets such as SMIC [27] and CASME II [29] are recent and important additions to this field. In addition to being one of the few spontaneous micro-expression datasets, CASME II is also a video-based dataset. This allows extracting not only static but also dynamic information from the emotion sequence.

Existing works on micro-expression recognition using CASME II dataset had been able to achieve an accuracy of 63.41% [29] using Local Binary Pattern on Three Orthogonal Plane (LBP-TOP) for feature extraction [31] and Support Vector Machines (SVM) for classification; an accuracy of 75.30% [23, 24] by adding a preceding step to LBP-TOP while using SVM classification with a Radial Basis Function (RBF) kernel with leave-one-out cross-validation (LOOCV).

The main contribution of this paper may be summarized as follows:

1. Proposed an efficient process flow for micro-expression recognition that includes (a) preprocessing, (b) motion magnification using EVM or Eulerian Video Magnification [25], (c) feature extraction using spatiotemporal texture map (STTM) [18], and (d) classification using SVM with kernels.
2. Applied a highly effective motion magnification method, namely EVM or Eulerian Video Magnification [25], whose main functions include motion magnification and color amplification.
3. For feature extraction, instead of using the popular LBP-TOP, a 3D version of a Harris corner function [6] is used, which makes use of textural spatiotemporal variations in video sequences, called spatiotemporal texture map (STTM) [18]. It significantly reduces the overlapping features of three orthogonal planes created by LBP, thereby produces a less computationally-expensive feature space.

4. Achieved an accuracy of 80%, the highest among exiting works using SVM classification for micro-expression recognition with CASME II, the only spontaneous micro-expression dataset currently available.

This paper is organized as follows: Sect. 2 discusses the research that has been done in the field of macro and micro-expression spotting. In Sect. 3, we discuss the proposed approach in detail. Section 4 enlists the experiments conducted and their results, to validate the hypotheses presented in Sect. 3. We conclude in Sect. 5, while pointing out certain directions in which this approach could be moved forward.

2 Background and Related Works

2.1 Datasets

A dataset that is a representative sample of the problem space is a prerequisite to implement an automated approach for solving problems using machine learning. Facial expression datasets are majorly divided into posed and spontaneous datasets. Unavailability of spontaneous datasets has been one of the major roadblocks in micro-expression analysis. While several posed datasets such as JAFFE [11], CK [7], MMI [16] exist for facial expression analysis, spontaneous micro-expressions datasets are relatively rare. The main issue arises from the inherent nature of the micro-expressions. They are elicited in an individual only in high stakes situations. Ekman [2] suggests three ways to elicit micro-expressions in test subjects: (1) Asking the test subjects to lie about their reactions (2) Fabricating crime scenarios and (3) Asking the test subjects to lie about what they saw in videos. Despite these guidelines, a dataset that achieves micro-expressions elicitation is a recent addition to the field.

The literature shows a progressive trend in moving towards a dataset that captures micro-expressions in the true sense of the word. In [26], Wu et al. used the Cohn-Kanade (CK) dataset. This dataset consists of 374 sequences of 97 subjects posing micro-expressions using the guidelines given above. However, on analyzing the dataset, it is evident that the image sequences are very aggressive and lose the inherent subtlety of a micro-expressions. In [28] and [24] Li et al. and Wang et al. respectively, used a dataset called CASME II [29]. This dataset is by far the most accurate in capturing the true essence of micro-expressions as it doesn't additionally instruct the test subjects. It allows the test subjects to react naturally to the incident videos and captures their micro-expressions. This dataset contains 247 videos captured at 200 fps.

In [18], Kamarol et al. employed a combined approach which uses multiple datasets to compare and contrast the results. In addition to CK+ [10], an incremental update to the CK [7] dataset, it also used the CASME II dataset. The CK+ dataset contains a better sample of capturing the sample space than the CK dataset. It encompasses videos of test subjects from different ethnicities in a varied age group ranging from 18 to 50. The CK+ dataset consists of 304 labeled videos containing a total of 5521 frames.

The literature describes the various thought processes and approaches employed to capture micro-expression datasets [7, 10, 29]. An upward trend in terms of image quality and frame rate is observed in micro-expressions dataset capturing techniques.

Advancement in technology has also contributed to the increased quality in dataset preparation, as is evident from the 200 fps capture of CASME II dataset [29].

There is also a clear trend in improving the spontaneity of the datasets [7, 10, 29]. The initial datasets [7, 10] were posed and hence were not very descriptive of the micro-expression paradigm. However, the later datasets such as CASME II [29] have moved to a more spontaneous and natural expression elicitation techniques.

2.2   Preprocessing and Feature Extraction

Preprocessing is an important step in any form of image classification problem, as it levels the input space and makes the dataset as uniform as possible by removing unnecessary differences in illumination and other aspects. This enables us to analyze images for what they are without getting affected by local differences in the image quality. Once the images have been preprocessed, we need a mechanism to extract useful information from the images that can be quantified, convoluted and compared. The literature describes various approaches to extract these features.

Wu et al. [26] employed grey-scale conversion and detecting and cropping the face as preprocessing steps. The preprocessed images are 48 × 48 pixels in size. Wu et al. [26] used the algorithm described in [8] by Kienzle et al. to detect faces. Wu et al. [26] then utilized Gabor filters to extract features from the preprocessed images.

In contrast, Li et al. [28] did not employ any preprocessing to their images, as their technique for feature extraction thrives on a rich variety of the input image sequence. It employed the Kanade-Lucas-Tomasi algorithm [21] to detect faces. It then divided the image in 6 × 6 grid to extract its feature using Local Binary Pattern (LBP) [15] and Histogram of Optical Flow [9] techniques.

In [18] Kamarol et al. used an approach that is different from previous approaches of feature extraction, as it also utilized the dynamic information of the video sequences that is present in the time dimension. Kamarol et al. [18] first croped the facial area in the image sequences using the Viola-Jones Face detection algorithm [22]. The features were then extracted using a Gaussian kernel and a Gaussian weighting function with second-moment matrices of the video sequences.

2.3   Micro-expression Spotting and Recognition

Pinpointing where in an image sequence the micro-expression occurs is called spotting. Once the micro-expression has been spotted, it has to be recognized to be belonging to one of the pre-defined categories in a process known as the micro-expression recognition.

The literature differs on how it spots and recognizes micro-expressions. LBP [15] is a popular method for feature extraction from images. G. Zhao and M. Pietikainen in [31] also extended it to the 3D space to include the dynamic component in their ME spotting and recognition approach. Gabor wavelets proposed by Lyons et al. [12] is another popular approach for textural representation.

2.4 Classification

Classifying the feature vectors extracted in the previous stages of these algorithms is the next logical step. One common aspect that arises from the literature in this scenario is regarding classification.

A significant portion of the literature utilizes Support Vector Machines (SVM) to classify the features. Wu et al. [26] used the GentleBoost algorithm as a feature selector before using Linear SVM in an improved approach called as GentleSVM in 10-fold cross validation. Li. et al. [28] also utilized Linear SVM but with 5-fold cross validation. Kamarol et al. [18] utilized a multi-class SVM using one-against-one method. They used 2-fold cross validation and then employ Monte-Carlo simulation, repeating the process 100 times, till an average is achieved. Wang et al. [24] also used SVM with a RBF kernel using the LOOCV approach for classification

Other classifiers that have been used for recognizing micro-expressions include fusion of facial deformations using the transferable belief model [34], multi-conditional latent variable model [35], and deep networks [36]. In addition, Exponentiated gradient algorithms have been shown to be highly accurate [37].

3 Proposed Approach

We propose a hybrid approach, adopting, integrating, and improving the works done by Wang et al. [24], Wu et al. [25] and Kamarol et al. [18]. We use the EVM method described by Wu et al. [25] as a preprocessing step to amplify the micro-expressions, as done by Wang et al. [24]. We then provide these motion magnified images to the algorithm defined by Kamarol et al. [18] for feature extraction. We then use SVM with 10-fold cross validation with a linear kernel to classify the features into five expression categories.

The following steps provide an overview of the methodology that employed in this project to spot, recognize and categorize micro-expressions:

1. Capture live video feed of test subjects being interviewed. The questions asked in the interview will be such that they will elicit micro-expressions in the test subjects.
2. Crop input video using Viola-Jones facial detection method.
3. Motion magnification of input video using Butterworth filter.
4. The amount of magnification to be introduced is determined by a tunable parameter α. We arrive at an empirically derived value of α which maximizes accuracy.
5. Compute linear scale representation (L) of input video using Gaussian kernel.
6. Compute second moment matrix of L.
7. Compute μ using the second-moment matrices and a Gaussian weighting function.
8. Compute the $det(\mu)$ and $trace(\mu)$.
9. Compute the STTM.
10. The features are then classified using a multi-class SVM model with a linear kernel and 10-fold cross validation.

Figure 1 summarizes the process flow of the proposed approach; each step is described in detail in the subsequent subsections.

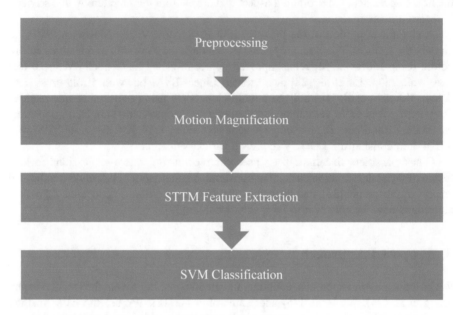

Fig. 1. Process flow

3.1 Preprocessing

In order to improve the chances of micro-expression recognition it is imperative to remove parts of the image sequence or videos that do not contribute to the end goal Features such as hair, the color of the subjects' clothing, the subjects' background are all potential causes for introduction of noise to the feature space. Hence, as a first step, we remove these from the image sequences.

We detect the face of the subject using the Viola-Jones face detection algorithm [22]. In the CASME II dataset, the position of the face is fairly constant and does not move much. Hence there is no need for facial alignment. After detecting the face, we crop the image and resize it an empirically derived resolution of 275×275. As we will see in Sect. 4, the resizing of the image to a particular resolution also affects the overall accuracy of the model.

After we have cropped the face, we then convert the image from RGB to grayscale. Information in the RGB range does not affect the overall accuracy of the model, but significantly reduces the computational complexity for feature extraction. Besides, as the feature extraction makes use of the textural information in the image sequences, keeping the image sequence in RGB color range is inconsequential.

3.2   Eulerian Video Magnification

The Eulerian Video Magnification (EVM) technique [25] is used to amplify the subtle emotions that are elicited in the subjects. Following is a brief overview of how the EVM works:

1. The input video sequence is first decomposed into various spatial frequency bands
2. A common temporal filter is then applied to the decomposed spatial frequency bands
3. The bands are then magnified by a tunable parameter α
4. These magnified bands are then added to the bands obtained after step 2 and are then collapsed to generate the final motion magnified video.

The temporal filters used: Ideal, Butterworth and Second Order Infinite Impulse Response (IIR) and the value of α all affect the overall accuracy of the proposed approach.

Wu et al. describe the relationship between the temporal processing and motion magnification in detail in [25]. We briefly illustrate this relationship: Given an image function $f(x)$, let $I(x, t)$ be the intensity of the input image at position x and at time t. As the image undergoes translational motion, we can express the intensities as a displacement function $\delta(t)$ such that:

$$I(x, t) = f(x + \delta(t)) \qquad \text{at time t}$$
$$I(x, 0) = f(x) \qquad \text{at time t} = 0$$

Assuming that the image can be approximated by a first-order Taylor series expansion, we rewrite the image at time t as

$$I(x, t) \approx f(x) + \delta(t) \frac{\partial f(x)}{\partial x} \tag{1}$$

Let $B(x, t)$ be the result of applying a temporal filter to the input signal at every position of x in Eq. (1) except at time $t = 0$ i.e. at $f(x)$. Thus, we have:

$$B(x, t) = \delta(t) \frac{\partial f(x)}{\partial x} \tag{2}$$

As specified above, we amplify the band-passed signal by a magnification factor α and then add it to the original input signal.

$$\widetilde{I}(x, t) = I(x, t) + \alpha B(x, t) \tag{3}$$

Combining Eqs. (1)–(3), we get,

$$\widetilde{I}(x,t) \approx f(x) + (1+\alpha)\delta(t)\frac{\partial f(x)}{\partial x}$$
$$\approx f(x) + (1+\alpha)\delta(t) \qquad (4)$$

Thus, we can see that the displacement in an image sequence $\delta(t)$ is magnified by a factor $(1+\alpha)$.

There are three main types of temporal bandpass filters, namely: Ideal, IIR & Butterworth filter. Figure 2a, 2b, 2c show the raw image sequence, the preprocessed image sequence, and the effect of EVM on preprocessed images, respectively. Note that the white highlighted parts in the image shown in Fig. 2c are hotspots for magnified motion as compared to the previous image in the sequence.

Fig. 2a. Raw image sequence

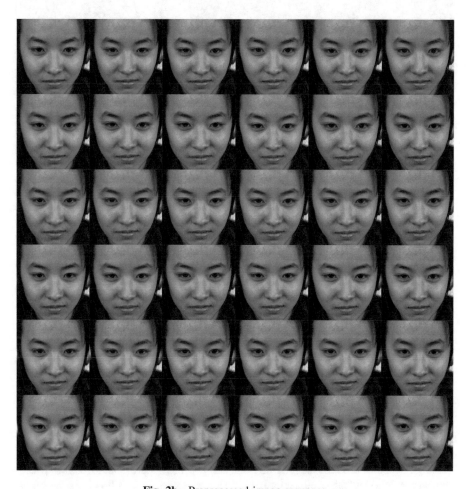

Fig. 2b. Preprocessed image sequence

Fig. 2c. Motion magnified image sequence

3.3   Feature Extraction Using Spatiotemporal Texture Map (STTM)

For feature extraction and representation, we adopt the algorithm proposed by Kamarol et al. [18]. It uses a 3-D Harris function to extract spatiotemporal information from an image sequence. A block based method is then employed to represent the feature space as a histogram. We shall take a closer look at both these aspects in the following subsections.

Feature Extraction. The feature extraction process may be described as follows; the corresponding flow diagram is shown in Fig. 3:

1. Crop the input image sequence
2. Compute linear scale representation (L) of image sequence using a Gaussian Kernel
3. Compute the convolution of a Gaussian weighting function with a second moment matrix composed of the temporal and spatial derivatives of the first-order derivatives (μ)

4. Calculate the trace and determinant of μ
5. Derive the STTM from $trace(\mu)$ and $\det(\mu)$

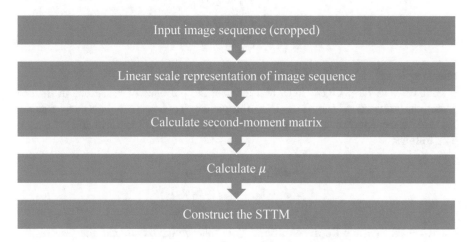

Fig. 3. STTM feature extraction process flow

The given image sequence f is converted into its linear scale-space representation by convoluting it with a three-dimensional Gaussian function.

$$L\left(\cdot; \sigma_l^2, \tau_l^2\right) = g\left(\cdot; \sigma_l^2, \tau_l^2\right) * f(\cdot) \tag{1}$$

where σ_l^2 is spatial variance of Gaussian kernel g and τ_l^2 is the temporal variance.

The spatial features of the input image sequence are represented by x and y which correspond to the respective axes of each image in the sequence while t indicates the temporal information on the time axis. The Gaussian kernel is defined as follows on these three axes.

$$g\left(x, y, t; \sigma_l^2, \tau_l^2\right) = \frac{\exp\left(-(x^2 + y^2)/2\sigma_l^2 - t^2/2\tau_l^2\right)}{\sqrt{(2\pi)^3 \sigma_l^4 \tau_l^2}} \tag{2}$$

To determine the changes in the spatiotemporal domain in the input image sequence f, we move a Gaussian window in various directions by small amounts [4, 24]. To determine the points at which these changes are evident, we apply convolution of a Gaussian weighting function $g\left(\cdot; \sigma_l^2, \tau_l^2\right)$ to a second-moment 3×3 matrix comprised of first order spatial and temporal derivatives.

$$\mu = g\left(\cdot; \sigma_l^2, \tau_l^2\right) * \begin{pmatrix} L_x^2 & L_x L_y & L_x L_t \\ L_x L_y & L_y^2 & L_y L_t \\ L_x L_t & L_y L_t & L_t^2 \end{pmatrix} \tag{3}$$

where L_x, L_y and L_t are first order derivatives defined as

$$L_x\left(\cdot; \sigma_i^2, \tau_i^2\right) = \partial_x(g * f),$$
$$L_y\left(\cdot; \sigma_i^2, \tau_i^2\right) = \partial_y(g * f),$$
$$L_t\left(\cdot; \sigma_i^2, \tau_i^2\right) = \partial_t(g * f)$$

where, $\sigma_i^2 = s\sigma_l^2$, $\tau_i^2 = s\tau_l^2$, and s is a constant.

Existence of large eigenvalues if μ indicate the presence of variations in the spatiotemporal domain of the given image sequence. To construct the Harris function (H) using these eigenvalues (λ_1, λ_2 and λ_3) we compute the determinant and the trace of μ as follows:

$$H = \det(\mu) - k\, trace^3(\mu)$$
$$= \lambda_1 \lambda_2 \lambda_3 - k(\lambda_1 + \lambda_2 + \lambda_3)^3$$

Block-Based Feature Representation. Now that we have extracted the features, we need a method to represent them such that the locality of the variation in the facial expressions in the image sequence is maintained. To do this, as shown in Fig. 4, we first divide each STTM into multiple blocks. Histograms could then be calculated for each block. This forms the feature vector of each of the image sequence. It has been found that the feature space is sparse, and a histogram computed using uniform width is not very effective at representing the features of the image sequence [18]. Hence, we follow the approach with non-uniform bin width, where the data in the lower range is has been allocated more bins. This is achieved by using the A-law compression before computing the histograms. This approach is able to capture subtler variations in the image sequence. Finally, the computed histograms of each block are then concatenated.

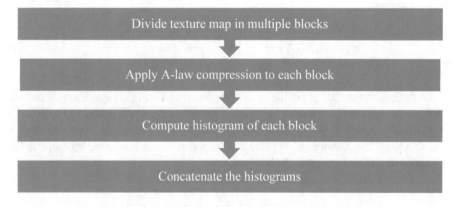

Fig. 4. Process flow for block-based feature representation

3.4   Classification

Analyzing the dataset, and the feature space generated by the STTM feature extraction approach, it is realized that the number of features are far more than the number of uniquely labelled data points. Hence, SVM is a very logical approach for classification. The CASME II dataset [29] has five major classes for the identified micro-expressions. We thus use the multi-class variant SVMs. We also use the one-against-one approach instead of the one-against-all. We employ multiple kernel shapes to find a hyper-plane that provides the maximum separation among all the classes.

4   Experiments and Results

4.1   Dataset

A comparison of various micro- and micro-expression datasets are illustrated in Table 1. As mentioned in Sect. 3, the CASME II dataset [29] is currently the only spontaneous dataset that is available for micro-expression analysis, and is therefore chosen for our experiments. The CASME II videos were shot using a 200 FPS camera, to capture the rapid and subtle characteristics of micro-expressions. High intensity and constant illumination apparatus were used to avoid flickering in illumination conditions that is caused due to inconsistent lighting in high FPS video capture.

Five expression categories were used: Happiness, Surprise, Disgust, Repression, and Others (includes emotions such as anger, sadness and tense etc.), in order to compare the results with the baseline [24]. We leave out videos labelled as Fear (note that there are only two videos labelled as Fear in the CASME II dataset).

Table 1. Comparison of various datasets

Datasets	CK	CK+	CASMEII	AFEW [1]
Type	Macro	Macro	Micro	NA
Resolution	640 × 480	640 × 480	280 × 340	NA
FPS	30	30	200	30
# of videos	374	304	247	601

4.2   Parameter Selection

To find out which parameters work the best for our proposed approach, we tested our hypotheses on smaller prototypes. This provided us with a guideline to work against, while saving the time involved in testing the entire dataset for each parameter and hypothesis. We created an equally supported sample dataset, which had 7 image sequences for each of the 5 expression categories. We conducted our experiments on this dataset, and then used the empirically derived parameters on the entire dataset of 246 videos. As our approach has four major components, we shall discuss the experiments conducted for each of the components in the following sections.

Preprocessing. To remove unnecessary regions of the image sequence such as hair, clothing etc. we first crop the image sequences. We first detected the subjects' face in the image sequences using the Viola-Jones face detection algorithm [22]. We then cropped the facial area from the image and resized it to 275×275 pixel size, and convert the cropped image to gray-scale to reduce the computational complexity needed during feature extraction. To identify if the cropping and resizing parameter selection affects the accuracy, we cropped the sample dataset to three resolutions 250×250, 275×275 and 300×300. We then proceeded with the feature extraction, representation and classification using the parameters given by Kamarol et al. [18]. We achieved the highest accuracy of 53.2% for a resolution of 275×275. Note that this is before EVM has been applied.

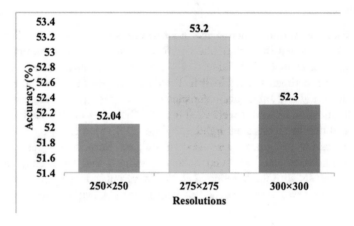

Fig. 5. Optimal value for cropping size

Figure 5 presents the effect of resolution size on accuracy, with other parameters that haven't been empirically derived to their optimal value. The results line up with our intuition as. It is possible that at a lower resolution, much of the inherent information in the image sequence is lost, which leads to a lower accuracy. On the other hand, a higher resolution of cropping leads to much noise in the image sequence, which hampers the performance. Hence preprocessing at an empirically derived right cropping size is instrumental building block of the pipeline.

Eulerian Video Magnification (EVM). There are two parameters that are of prime importance when considering EVM, namely; type of filter and amount of magnification (α). Wang et al. [24] achieved their highest accuracy using a second order IIR filter with $\alpha = 20$. To derive these parameters that work best for our approach, we first cropped and resized the images using the parameters derived in the previous section. We then magnified the motion in these image sequences using EVM. We varied α from 8 to 20 with a step size of two for IIR and Butterworth filters. We then extracted features using STTM parameters described in [18] and then applied Linear SVM for classifying them.

We observed that $\alpha = 10$ gives us the best accuracy of 71.43% for Butterworth filter, viz. in contrast to the findings of Wang et al. [24], where IIR performed better Butterworth filter at $\alpha = 10$. Furthermore, Wang et al. also noted that they have achieved the highest accuracy for $\alpha = 20$ for the IIR filter. Figure 6 illustrates the results of this experiment.

On observing the image sequences magnified at various α values we find that as α goes up, initially the motion is magnified. But at higher values of α the images got distorted; this results in very skewed images that do not represent the inherent subtleness of the micro-expressions. We concluded that the magnification at $\alpha = 10$ provides enough magnification for detection of micro-expressions without distorting the images.

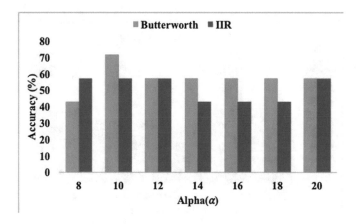

Fig. 6. Optimal value for type of filter and α

STTM Feature Extraction and Representation. For feature extraction and representation, we use the optimal parameters achieved by Kamarol et al. [18] for CASME II dataset. They are as follows:

$$\sigma_l^2 = 2$$
$$\tau_l^2 = 2$$
$$k = 0.04$$
$$A = 28$$
$$number\ of\ bins = 10$$
$$number\ of\ blocks = 9 \times 9$$
$$overlapping\ ratio = 30\%$$

Classification. Wang et al. have used leave-one-out cross-validation (LOOCV) for validating their method, and utilized various kernel shapes such a Linear, RBF and Polynomial for their SVMs. However, this method is very time consuming as it trains the model on n − 1 samples and then tests it on the remaining sample. This procedure is repeated for all the n samples in the dataset.

Instead we opted for K-Fold cross validation which is far more efficient than LOOCV. We preprocessed, extracted, and represented features using the parameters derived so far. Then, we applied SVM classification using one-against-one approach on various kernel shapes and various number of splits for K-Fold validation. We achieved the highest accuracy of 71.4% for linear and RBF kernels using 10-fold validation. Figure 7 illustrates the results.

The results of classification are in-line with the common understanding in machine learning. More training data leads to a better accuracy. The pipeline also makes the data linearly separable, as is evident from the choice of kernel shapes. The polynomial kernel we employed (degree = 5) performs the worst; it is interesting to note that the linear and the RBF kernels both perform equally better.

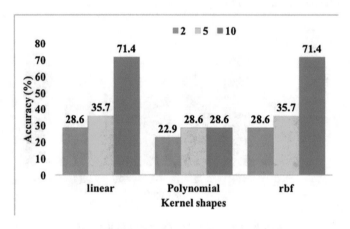

Fig. 7. Optimal values of kernel shape and K-Fold validation

Final Results on Complete Pipeline. As the best values of the parameters have been obtained, they are then applied in the corresponding phases of the proposed pipeline (as shown in Fig. 1) on the entire CASME II dataset of 246 videos. We repeat the classification 10,000 times and take the average; results are shown in Table 2. We achieve an accuracy of 80% with linear and RBF kernels, viz. about 5% higher than the baseline [24]. The results validated that the intermediate results achieved on prototypes were a step in the right direction. The maximum accuracy achieved is 80%.

Table 2. Classification report for emotion categories

	Precision	Recall	F1-score	Support
Disgust	–	–	–	2
Happiness	–	–	–	1
Others	0.8	1.0	0.89	20
Repression	–	–	–	1
Surprise	–	–	–	1
Avg./Total	0.64	0.8	0.71	25

The classification report in Table 2 reiterates the importance of a well-balanced and equally supported dataset. Expression category "Others" which has the highest support in a fold, gets the most precise prediction and contributes majorly to the overall accuracy of the approach. Note that an even higher accuracy might be achieved in an equally-supported dataset using stratified k-fold validation where each proportion of the expression categories is maintained across folds.

5 Conclusion and Future Work

A new approach using motion magnification and feature extraction has been proposed for recognizing micro-expressions. The hybrid method utilizes EVM for motion magnification, STTM for feature extraction, and SVM for classification. The complete process flow is described in detail. Comprehensive experiments have been carried out using 10-fold cross validation for SVM on CASME II, the only spontaneous micro-expression dataset currently available. With a linear kernel for SVM it has achieved the highest accuracy of 80%. This is 5% higher than the baseline [24], the best existing method using SVM. Future forays in this area would be to fine-tune the parameters used in STTM feature extraction and representation for possible improvement; applying multi-tier SVM classification to combat the problem of unequal support of micro-expression samples of each category, and exploring deep neural networks for the classification.

Acknowledgements. Melody Moh and Teng Moh are supported in part by SJSU RSCA awards.

References

1. Dhall, A.: Collecting large, richly annotated facial-expression databases from movies (2012)
2. Ekman, P.: Darwin, deception, and facial expression. In: Anonymous 2003. https://doi.org/10.1196/annals.1280.010
3. Ekman, P., O'Sullivan, M., Frank, M.G.: A few can catch a liar. Psychol. Sci. **10**(3), 263–266 (1999). https://doi.org/10.1111/1467-9280.00147
4. Förstner, W., Gülch, E.: A fast operator for detection and precise location of distinct points, corners and centres of circular features. In: Proceedings of ISPRS Intercommission Conference on Fast Processing of Photogrammetric Data (1987)
5. Frank, M., et al.: I see how you feel: training laypeople and professionals to recognize fleeting emotions. In: The Annual Meeting of the International Communication Association. Sheraton New York, New York City (2009)
6. Harris, C., Stephens, M.: A combined corner and edge detector. In: Alvey Vision Conference (1988)
7. Kanade, T., Cohn, J.F., Tian, Y.: Comprehensive database for facial expression analysis. In: Proceedings of the Fourth IEEE International Conference on Automatic Face and Gesture Recognition (2000)
8. Kienzle, W., et al.: Face detection - efficient and rank deficient, vol. 17 (2005)
9. Liu, C.: Beyond pixels: exploring new representations and applications for motion analysis (2009)

10. Lucey, P., et al.: The extended Cohn-Kanade dataset (CK+): a complete dataset for action unit and emotion-specified expression. In: IEEE Computer Society Conference on Computer Vision and Pattern Recognition Workshops (CVPRW) (2010)

11. Lyons, M.J., et al.: The Japanese Female Facial Expression (JAFFE) database. In: Proceedings of Third International Conference on Automatic Face and Gesture Recognition (1998)

12. Lyons, M., et al.: Coding facial expressions with Gabor wavelets. In: Proceedings Third IEEE International Conference on Automatic Face and Gesture Recognition (1998). https://doi.org/10.1109/afgr.1998.670949

13. Matsumoto, D., Hwang, H.S.: Evidence for training the ability to read microexpressions of emotion. Motiv. Emot. **35**(2), 181–191 (2011). https://doi.org/10.1007/s11031-011-9212-2

14. Mehrabian, A.: Communication without words. In: Communication Theory, pp. 193–200 (2008)

15. Ojala, T., Pietikainen, M., Maenpaa, T.: Multiresolution gray-scale and rotation invariant texture classification with local binary patterns. IEEE Trans. Pattern Anal. Mach. Intell. **24**(7), 971–987 (2002)

16. Pantic, M., et al.: Web-based database for facial expression analysis. In: IEEE International Conference on Multimedia and Expo, ICME 2005 (2005)

17. Russell, T.A., Chu, E., Phillips, M.L.: A pilot study to investigate the effectiveness of emotion recognition remediation in schizophrenia using the micro-expression training tool. Br. J. Clin. Psychol. **45**(4), 579–583 (2006)

18. Kamarol, S.K.A., et al.: Spatiotemporal feature extraction for facial expression recognition. IET Image Proc. **10**(7), 534–541 (2016). https://doi.org/10.1049/iet-ipr.2015.0519

19. Sandbach, G., et al.: Static and dynamic 3D facial expression recognition: a comprehensive survey. Image Vis. Comput. **30**(10), 683–697 (2012). https://doi.org/10.1016/j.imavis.2012.06.005. http://www.sciencedirect.com/science/article/pii/S0262885612000935

20. Schopler, E., Mesibov, G.B., Kunce, L.J.: Asperger syndrome or high-functioning autism? (1998)

21. Tomasi, C., Kanade, T.: Detection and tracking of point features (1991)

22. Viola, P., Jones, M.J.: Robust real-time face detection. Int. J. Comput. Vision **57**(2), 137–154 (2004)

23. Wang, Y., et al.: Efficient spatio-temporal local binary patterns for spontaneous facial micro-expression recognition. PLoS ONE **10**(5), e0124674 (2015)

24. Wang, Y., et al.: Effective recognition of facial micro-expressions with video motion magnification. Multimed. Tools Appl. **76**(20), 21665–21690 (2017). https://doi.org/10.1007/s11042-016-4079-6

25. Wu, H., et al.: Eulerian Video Magnification for Revealing Subtle Changes in the World. ACM Trans. Graph. **31**(4), 65:1–65:8 (2012). https://doi.org/10.1145/2185520.2185561. http://doi.acm.org/10.1145/2185520.2185561

26. Wu, Q., Shen, X., Fu, X.: The machine knows what you are hiding: an automatic micro-expression recognition system. In: Proceedings of the 4th International Conference on Affective Computing and Intelligent Interaction - Volume Part II (2011). http://dl.acm.org/citation.cfm?id=2062850.2062867

27. Li, X., et al.: A spontaneous micro-expression database: Inducement, collection and baseline. In: 10th IEEE International Conference and Workshops on Automatic Face and Gesture Recognition (FG) (2013). https://doi.org/10.1109/fg.2013.6553717

28. Li, X., et al.: Towards reading hidden emotions: a comparative study of spontaneous micro-expression spotting and recognition methods. IEEE Trans. Affect. Comput. **PP**(99), 1 (2017). https://doi.org/10.1109/taffc.2017.2667642

29. Yan, W., et al.: CASME II: an improved spontaneous micro-expression database and the baseline evaluation. PLoS ONE **9**(1), e86041 (2014)

30. Yan, W., et al.: How fast are the leaked facial expressions: the duration of micro-expressions. J. Nonverbal Behav. **37**(4), 217–230 (2013). https://doi.org/10.1007/s10919-013-0159-8

31. Zhao, G., Pietikainen, M.: Dynamic texture recognition using local binary patterns with an application to facial expressions. IEEE Trans. Pattern Anal. Mach. Intell. **29**(6), 915–928 (2007)

32. Matsumoto, D., Skinner and, L.G., Hwang, H.: Reading people: behavioral anomalies and investigative interviewing. The FBI Law Enforcement Bulletin (2015)

33. Handeyside, H.: Be careful with your face at airports (opinion) – CNN. CNN. https://www.cnn.com/2015/03/19/opinions/handeyside-tsa-spot-program/index.html. Accessed 5 May 2018

34. Hammal, Z., Couvreur, L., Caplier, A., Rombaut, M.: Facial expression classification: an approach based on the fusion of facial deformations using the transferable belief model. Int. J. Approximate Reasoning **46**(3), 542–567 (2007)

35. Eleftheriadis, S., Rudovic, O., Pantic, M.: Multi-conditional latent variable model for joint facial action unit detection. In: Proceedings of the IEEE International Conference on Computer Vision, pp. 3792–3800 (2015)

36. Liu, M., Li, S., Shan, S., Chen, X.: AU-inspired deep networks for facial expression feature learning. Neurocomputing **159**, 126–136 (2015)

37. Bartlett, P.L., Collins, M., Taskar, B., McAllester, D.A.: Exponentiated gradient algorithms for large-margin structured classification. In: Proceedings of the Neural Information Processing Systems (NIPS) (2004)

Optimal Non-uniformity Correction for Linear Response and Defective Pixel Removal of Thermal Imaging System

Jae Hyun Lim[1,2(✉)], Jae Wook Jeon[1], and Key Ho Kwon[1]

[1] Sungkyunkwan University, Suwon-si, Gyeonggi-do, Republic of Korea
bogyong00@skku.edu, jwjeon@yurim.skku.ac.kr,
khkwon@skku.ac.kr
[2] Hanwha Systems, Seongnam-si, Gyeonggi-do, Republic of Korea

Abstract. Most of thermal imaging system (TIS) has non-linear response characteristics of the s-curve due to the influence of its internal components. If these TIS raw outputs are calibrated using conventional two-point non-uniformity correction (NUC) without taking into account its response characteristics, the corrected output values cannot achieve a linear response characteristic in the dynamic range of the TIS. Furthermore, defective pixels can occur in the nonlinear response region in the TIS dynamic range. In order to solve this problem, we propose an optimal NUC technique considering TIS raw output characteristic. This proposed method can make it easier to develop many useful algorithms such as detection and tracking as well as image quality.

Keywords: Thermal image system · Infrared ·
Two-point non-uniformity correction · Optimal non-uniformity correction ·
Linearity response · Defective pixel · Dynamic range

1 Introduction

Thermal imaging system (TIS) was mainly used in military-oriented for applications requiring all-weather visibility, such as guided missile, weapon sight and so on. However TIS is expanding rapidly not only in medical industry to find influenza patients in public facilities but also in various industries for leakage gas, fire prevention, automotive and so on. In these TIS, Non-uniformity correction (NUC) algorithm is an essential signal processing process capable of correcting the non-uniform characteristics of the infrared sensor. NUC method has been developed into two types algorithms such as reference based calibration methods [1] and scene based calibration methods [2–4]. There are obvious advantages and disadvantages of two approaches [5] and many NUC algorithms are being developed based on both approaches. In particular, two-point NUC method, one of the referenced based calibration methods, is the most widely used and applied to many commercial products because of its ease of implementation [3] and cost-down of mass production. However, two-point NUC literally corrects the non-uniformity of the pixel of focal plane array (FPA) in TIS and cannot improve the linearity of the pixel output value within TIS dynamic range.

© Springer Nature Switzerland AG 2019
S. Lee et al. (Eds.): IMCOM 2019, AISC 935, pp. 370–382, 2019.
https://doi.org/10.1007/978-3-030-19063-7_30

Moreover, all pixel values corrected by two-point NUC in the nonlinear region of the TIS dynamic range are less uniform than in the linear region. This can result in defective pixels (or bad pixels) as the TIS operates for a longer period of time.

Therefore, we propose the optimal NUC method for constant response and defective pixel removal in the TIS dynamic range. In addition, we will show how to optimize the number of lookup tables efficiently for the optimal NUC method.

The rest of the paper is organized as follows: In Sect. 2, we describe the main causes of non-constant sensitivity of TIS. And in Sect. 3, we explained the idea of background for optimal NUC solution. In Sect. 4 we proposed several methods for optimal NUC solution and how to manage the multiple look-up tables for efficient memory operation. In Sect. 5, we compare optimal NUC results to conventional two-point NUC results and then show before and after image. Finally, Sect. 6 we offer conclusion and future research issues.

2   Main Factors of TIS Non-linear Response

Most of TIS is briefly composed of infrared sensor, electronic circuit, optic lens and housing part. Each part can affect the non-linear response of TIS directly or indirectly.

First, infrared sensors are the biggest cause of the nonlinear response characteristics of TIS, though there will be slight differences depending on FPA material and structure. Despite of improvements of infrared FPA manufacturing technology, detector is still the major problem for linear response. However, it is typically modeled as a linear equation.

$$V_{raw,\,ij} = G_{ij} \times \varnothing + V_{o,\,ij} \tag{1}$$

$V_{raw,\,ij}$ is pixel(i, j) raw output value from the detector, G_{ij} is sensitivity of each pixel, $\varnothing$ is infrared influx and $V_{o,ij}$ is each pixel offset value due to dark current [6]. i and j are pixel position in staring infrared sensor.

Second, electronic circuit can also affect the non-linear responses of TIS. This refers not only to readout integrated circuit (ROIC) inside the detector, such as capacitive transimpedance amplifier (CTIA), but also to components on PCB outside the detector, such as differential op amps (or general op amps), analog-to-digital converters and FPGA. This is because each component has unique electrical characteristics. For example, there are specifications of devices such as gain bandwidth products, integral nonlinearity (INL) and differential nonlinearity (DNL). These all of error sums in electronic components can affect non-linear response of TIS.

Third, optical lens is also one of the main factors. The magnitude of infrared incident radiation on the detector is related to lens design and specifications. The relationship is well known as follows [9].

$$\Phi(T) = \left[\tau_{eff} \bullet \int_{\lambda_1}^{\lambda_2} L(\lambda, T) d\lambda \right] A_i \bullet \Omega \,, where \, \Omega = \left[\frac{\pi cos^4 \theta}{4(F/\#)^2 + 1} \right] \tag{2}$$

Table 1. Parameters related to radiant power on the detector

Symbol	Name	Notes
Φ	Radiant flux	Incident power on detector from the source
τ_{eff}	Optical transmission	–
λ_1, λ_2	Wavelength interval	Detector spectral response
$L(\lambda, T)$	Spectral radiance @ temperature	Calculated from Plank's Law
A_i	Active area	Detector area
Ω	Solid angle	A portion of the area on the surface of a sphere per the square of the sphere radius
θ	–	Angle between the light ray and the normal to a point on the surface of a detector
$F/\#$	Optical f-number	EFL/Lens diameter

Finally, it is influenced by the physical housing that make up the entire TIS. It is very difficult to predict how the heat radiation generated by the internal TIS housing will be incident on the detector. The internal radiation problems are significantly related to TIS physical structure, material and electronic module power consumption (Table 1).

Because of these four factors, TIS output shape is very close to non-linear s-curve. Among them, the influence of the infrared detector has the greatest influence on the nonlinear response.

3   Background for Solution

3.1   Two Problems of Two-Point NUC Method

Two-point NUC is an essential pre-processing algorithm used in many industry TIS products. Each pixel of FPA has a unique gain (G_{ij}) and offset $(Offset_{ij})$ coefficients.

$$Corrected\ Output_{ij} = G_{ij} \times Raw_{ij} + Offset_{ij} \tag{3}$$

It starts from the assumption that the raw response characteristic of TIS is a perfect linear line. However, since raw output values of TIS usually form the s-curve, the corrected output values after two-point NUC do not have the intended values near the reference line. As a result, the corrected output values cannot be expected to have a linear response line, especially at both end regions within the dynamic range. In addition, these pixels can be potentially defective pixels, which can cause image non-uniformity and degrade image quality.

In summary, given two problems by two-point NUC method are as follows.

1. Non-linearity response characteristic after performing two-point NUC.
2. Defective pixels occurrence.

3.2   Limitations Due to Detector NETD

For TIS image uniformity, two additional points should be added to the two-point NUC reference points. There may be questions whether the additional 2 reference points are optimized for image uniformity, one of the most important performances of TIS. Theoretically, root-mean-square (rms, a quantitative index representing image uniformity) value of the NUC image data converges to zero as the reference point increases. However, it does not go below a certain threshold value as shown in the graph of Fig. 1 due to temporal pixel noise equivalent temperature difference (NETD) of the infrared detector.

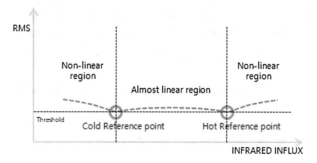

Fig. 1. TIS image uniformity over infrared incident radiation after performing two-point NUC

Even if more reference points are added to the middle or both end regions of the raw response s-curve, the rms value of the NUC image does not go below a certain threshold. In other words, the performance of the NUC algorithm is limited by the value of the temporal NETD of the infrared detector [10]. Therefore, many reference points excess of additional two points are useless in terms of image uniformity after carrying out NUC. It can also be confirmed by the detector NETD characteristic in Fig. 2. Note that spatial NETD is proportional to rms value of the NUC image. The shape of the rms value graph in Fig. 1 and the shape of the Spatial NETD graph in Fig. 2 are very similar.

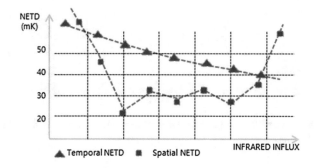

Fig. 2. Spatial and temporal NETD results by two-point NUC [10]

In addition to the TIS performance, additional two points are the most optimal points in terms of hardware constraints, mass productivity efficiency and cost reduction. These factors are very important for TIS commercialization.

4 Proposed Solution Section

4.1 Optimal Reference Line and Three Partitioning

For optimal performance, two additional reference points will be placed at both ends of the response curve when considering the raw response s-curve characteristics. Then find the optimal reference line. Key point is only how to find the optimal reference line compared to TIS raw response curve. The reference lines can be modeled as below.

$$Candidate\ reference\ lines = a_k \times \Phi + b_k \tag{4}$$

where k = 1, 2, 3••• and Ø is the infrared influx that could be obtained by adjusting the temperature of the homogeneous black body.

Optimal reference line can be determined by minimum difference areas between mean output data of all pixels and candidate line.

$$Error\ Sum = \int_{\lambda_{min}}^{\lambda_{max}} \left(\frac{\sum_{j=column} \sum_{i=row} \Phi(\lambda, T)_{i,j}}{No.\ of\ row \times No.\ of\ column} - reference\ line \right)^2 d\lambda \tag{5}$$

where Ø is the radiant influx, λ_{min} and λ_{max} are the detector spectral response.

If the area difference between the reference line and the average raw output s-curve is large, the image uniformity of the corrected output is degraded and fixed pattern noise (FPN) occurs in proportion to the difference area.

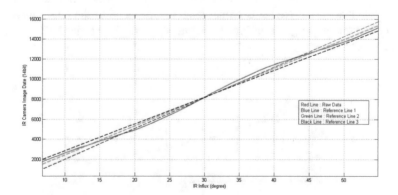

Fig. 3. Blue, green and black are 3 candidates reference lines. Green is the optimal reference line. y-axis: 14bit data, x-axis: black body temperature

In case of Fig. 3, there are the example of three candidate lines (blue, green, black) and one red line of the raw output that is the mean value lines of all pixels. Among three candidate lines, green line is the most optimal reference line than any other one because error sum is the smallest.

After determining the green line as the optimal reference line, divide the TIS raw output s-curve into three sections: middle linear region and both ends curved regions. Figure 4 shows the mean TIS raw data (red line) and the optimal reference line (green line) and partitioning as three segments.

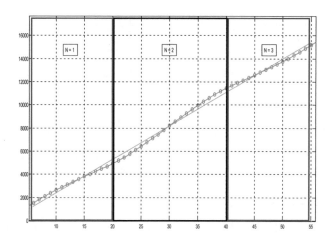

Fig. 4. Mean output line (red) and reference line (green) are divided into 3 parts. (N = 1, 2, 3) y-axis: 14bit data, x-axis: black body temperature

After three partitions within the dynamic range, conventional two-point NUC is carried out in each segment to calculate the gain and offset values and make three look-up tables. For each part of gain and offset values, those are calculated by

$$NUC\ Output_{ij} = Gain_{N,ij} \times Raw_{ij} + Offset_{N,ij}$$

$$Gain_{N,ij} = \frac{\overline{H_N} - \overline{C_N}}{H_{N,ij} - C_{N,ij}} \tag{6}$$

$$Offset_{N,ij} = \overline{C_N} - Gain_{N,ij} \times C_{N,ij}\ or\ Offset_{N,ij} = \overline{H_N} - Gain_{N,ij} \times H_{N,ij}$$

where i and j are pixel position in 2-Dimension of FPA (N = 1, 2, 3) Uppercase H and C stand for relatively hot and cold raw data over infrared influx. $\overline{H_N}$ and $\overline{C_N}$ are the reference values which are the mean of all pixels in FPA and $H_{N,ij}$, $C_{N,ij}$ is each individual pixel value in N-th segment.

Two-point NUC needs 2-reference points, but optimal NUC needs 4 reference points. H_1, C_1 is two points references in the first segment, and H_2, C_2 ($C_2 = H_1$ in 1st segment), H_3, C_3 ($C_3 = H_2$ in 2nd segment) is also two reference points respectively in 2nd and 3rd segment. Mid-region two H and C reference points are shared by the other regions (N = 1, N = 3).

Figure 5 shows the simulation results of the optimal NUC method. Each corrected line (three type green lines: solid and dash line in N = 1, 3 both ends regions and dash-dot line in N = 2, linear region) is almost consistent with the reference line (blue line). And the degree of the response linearity is improved compared to raw data (red line).

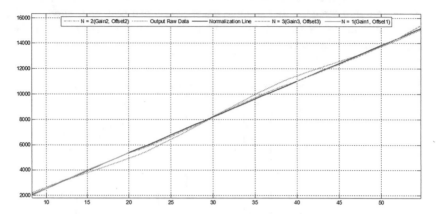

Fig. 5. Corrected output data in each segment (green) after multiplication of gain and addition of offset, y-axis: 14bit data, x-axis: black body temperature

4.2   Integrated Look-up Table

For optimal NUC method, the number of look-up tables has increased from one to three accordingly. Theoretically, FPGA has a longer memory access time than a single lookup table. In terms of hardware constraints and efficiency of memory bandwidth, three gain and offset coefficients for a single pixel must be successively placed in memory addresses. After interlacing each pixel gain and offset, we integrate the three look-up tables into single table (Fig. 6).

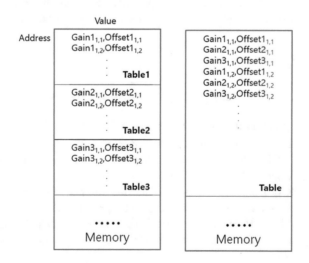

Fig. 6. General three tables 1, 2, 3 (left) are integrated into one table (right)

With single access to memory, FPGA takes three gain and offset sets corresponding to each single pixel from the memory at once, and then determines which gain and offset values are more appropriate depending on the dynamic range. This method improves TIS image quality and saves H/W resource at the same time. In other words, additional image processing algorithms can be applied to the remaining resources, or operating clock frequency can be decreased to reduce power consumption.

4.3   Proposed Optimal NUC Processing Flow Chart

Figure 7 flow chart is TIS conventional signal processing progress plus proposed optimal NUC processing. The difference compared with conventional two-point NUC is that there is a judgment statement about which value is applied to every single pixel. The summary of the procedure is as follows

1st step: Get raw data from TIS using black body.
2nd step: Determine the optimal reference line over the dynamic range.
3rd step: Three partitioning within the dynamic range.
4th step: Calculation gain and offset in each region.
5th step: Create an integrated look-up table.
6th step: Apply look-up table to NUC process.

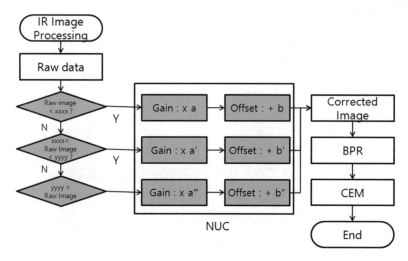

Fig. 7. White color flow chart is conventional IR image processing. Inserted green color flow chart is proposed optimal NUC processing. ▶ BPR: Bad Pixel Replacement ▶ CEM: Contrast Enhancement Method

In the above real time image signal process, TIS has almost the same response over the same infrared influx within its dynamic range.

4.4 Real Time IR Signal Processing Block Diagram

Figure 8 describes real time TIS signal processing flow between FPGA and memory in block diagram. In detector, infrared energy is converted to electrical analog signal. 14-bit digital data from ADC flows to the FPGA. In the FPGA, three gain and offset values corresponding to each pixel are obtained from the memory in advance. Gain and offset values are determined according to the input raw value range among three parts. The gain and offset values of each of the corresponding pixels are then multiplied and added to produce the corrected data value. Then BPR and CEM would be carried out.

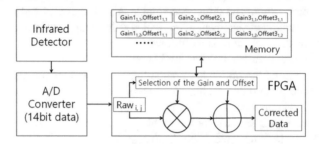

Fig. 8. Real time TIS signal processing between FPGA and memory

5 Performance Evaluation Section

The Fig. 9 shows the average values of the corrected signal outputs after performing the optimal NUC and two-point NUC, respectively.

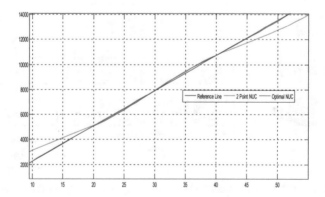

Fig. 9. Optimal NUC (blue) vs. two-point NUC (Red), y-axis: 14bit data, x-axis: black body temperature

Optimal NUC result (blue dash line) is almost the same graph line on the brown reference solid line, but two-point NUC result (red dash line) is inconsistent with both ends of the reference line. This result means that TIS raw data is originally the s-curve and 2 points NUC is insufficient to compensate both ends of s-curve line.

Table 2. Mean values of the gain and offset

	Two points NUC	Optimal NUC
Average gain	0.866	1.1978 (N = 1)
		0.866 (N = 2)
		1.2347 (N = 3)
Average offset	1104	–2696 (N = 1)
		1104 (N = 2)
		–710 (N = 3)
Sensitivity (data/degree)	220	280
	280	
	220	

We summarize the specific average values of gain and offset in optimal NUC method compared to conventional two-point NUC method in Table 2.

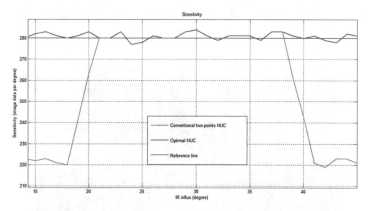

Fig. 10. This graph shows sensitivity (14bit data/degree): Two-point NUC (red), Optimal NUC (blue), Reference line (brown), y-axis: 14bit data, x-axis: black body temperature

We prove that the difference values of gain and offset in each segment make better linearity response than only one gain and one offset in Figs. 9 and 10. This is the same meaning that the sensitivity is almost constant within TIS dynamic range. In Fig. 10, blue line shows the constant output sensitivity (around 280 image data per degree) over infrared influx. Contrary, both edges of the red line have less sensitivity than mid-region sensitivity.

A TIS that requires the accurate temperature measurement of the target, the constant response characteristic makes algorithm development much easier and improves the system's temperature measurement accuracy. This is because it is easy to predict the output data for the incoming infrared radiation within the dynamic range.

This linear response characteristic reduces the occurrence of bad pixels during TIS operation. Bad pixels can be classified into two types: hard defect and soft defect. Hard defective pixels are predetermined by the detector manufacturer as bad pixels and can be known in advance by TIS designers before they are used in the system. Most defective pixels that occur during TIS operation are soft defective pixels. These bad pixels are determined by the pixel gain values in the look-up table and are usually replaced or interpolated by adjacent pixel values [7]. In the conventional two-point NUC method, the constant gain values of the pixel in the regions far from the reference line by two reference points does not reflect the raw output characteristic of the detector, and the probability of becoming a defective pixel is relatively high. Therefore, the pixel may become a bad pixel during TIS operation. This requires additional algorithms to find and correct bad pixels. Additional bad pixel resolution and correction methods require design complexity [8] and higher hardware performance. Contrary, the optimized NUC method decreases in the number of the defective pixels and does not require the additional defective pixel compensation algorithm compared to two-point NUC method.

Figures 11 and 12 shows the two corrected images with the same condition of the bad pixel determination (pixel gain coefficient value > 1.3 or < 0.7 is bad pixel). We can see that many bad pixels are removed in Fig. 12.

Fig. 11. Traditional two-point NUC method result: Bad Pixels (red circle)

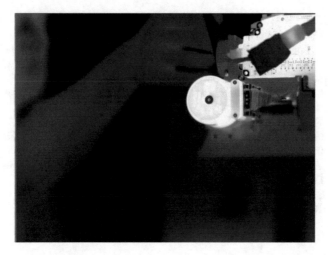

Fig. 12. Optimized NUC method result: Many bad pixels are removed.

6 Conclusion

In this paper, we present optimal NUC processing to insert into conventional two-point NUC. Our protocol has clear goals in terms of easy implementation, non-complexity and better performance. Considering characteristic of TIS output s-curve, 4 reference points are the most proper tradeoff. This method improves the linearity both ends of response s-curve and reduces the number of bad pixels accordingly. Three gain and offset look-up tables of the optimal NUC can mitigate H/W specification by integrating and operating into one table.

Considering the constant sensitivity and reduced bad pixels within the dynamic range, this work can be useful for precision applications such as temperature measurements imaging systems and search & tracking imaging systems. Especially in applications that require precision temperature measurement algorithms, it is easier to develop the algorithms because of their constant sensitivity.

References

1. Milton, A.F., Barone, F.R., Kruer, M.R.: Influence of nonuniformity on infrared focal plan array performance. SPIE Opt. Eng. **24**(5), 855–865 (1985)
2. Harris, J.G., Chiang, Y.M.: Nonuniformity correction of infrared image sequences using the constant-statistics constrain. IEEE Trans. Image Process. **8**, 1148–1151 (1999)
3. Torres, S.N., Pezoa, J.E., Hayat, M.M.: Scene-based nonuniformity correction for focal plane arrays by the method of the inverse covariance form. Appl. Optics **42**(29), 5872–5881 (2003)
4. Cao, S., et al.: Statistical non-uniformity correction for paddlebroom thermal infrared image. In: 2017 2nd International Conference on Multimedia and Image Processing (ICMIP), pp. 141–145. IEEE (2017)

5. Mooney, J.M., et al.: Responsivity nonuniformity limited performance of infrared staring cameras. Opt. Eng. **28**, 1151–1161 (1989)
6. Enke, L., Shangqian, L., Shimin, Y., Xiaoning, F.: Nonuniformity correction algorithms of IRFPA based on radiation source scaling. In: 2009 Fifth International Conference on Information Assurance and Security, vol. 1, pp. 317–321 (2009)
7. Isoz, W., Svensson, T., Renhorn, I.: Nonuniformity correction of infrared focal plane ar-rays. In: Proceedings of the SPIE, vol. 5783, pp. 949–960 (2005)
8. Celestre, R., Rosenberger, M., Notni, G.: A novel algorithm for bad pixel detection and correction to improve quality and stability of geometric measurements. J. Phys: Conf. Ser. **772**(1) (2016)
9. Mooney, J.M., et al.: Responsivity nonuniformity limited performance of infrared staring cameras. Opt. Eng. **28**(11), 281151 (1989)
10. Tissot, J.L., et al.: Uncooled microbolometer detector: recent developments at ULIS. Opto-Electron. Rev. **14**(1), 25–32 (2006)

Azimuth Angle Detection with Single Shot MultiBox Detecting Model

Quoc An Dang, Quan Mai Bao Nguyen, and Duc Dung Nguyen[✉]

HoChiMinh City University of Technology,
268 Ly Thuong Kiet, District 10, Ho Chi Minh City, Vietnam
dangquocan95@gmail.com, nguyenmbquan95@gmail.com, nddung@hcmut.edu.vn

Abstract. While detecting object becomes easier with deep models, estimating pose remains a challenging problem in modern vision research. In this work, we propose a method that enables detecting objects and estimating their pose simultaneously in a single model, without intermediate stages. Unlike some other approaches, we make the first attempt to hierarchically estimate objects pose using a deep network. We approach the problem by trying to mimic human perception in seeing objects, as we tend to estimate the object pose based on our experience and usually from coarse to fine fashion. We with enough resource to build a deeper model, our approach will be able to produce more accurate results for to complex tasks including object tracking, localization, and SLAM [2].

Keywords: Object detection · Pose estimation · Deep model

1 Introduction

Object recognition has played an important role in various applications in our modern life. From image understanding to applications in medical and robotic, it keeps raising many challenges in computer vision. With the goal of predicting object type and its position from a single image, this problem is the basic step to simulate a human vision. There are several models which can detect the objects in real time with high accuracy. The trend has changed in the last few years when researchers no longer prefer traditional approaches using designed features of computer vision. They have approached this problem using machine learning, in particular, neural networks for solving challenges in object detection. We are seeing more and more of deep networks that achieved really great results. You Only Look Once (YOLO) [11], for example, achieved 63.4% mAP while still maintaining a solid real-time performance (45 FPS) on the VOC2007 dataset. Another competitor, Single Shot Multibox Detector (SSD) [9], outperformed previous models by a large margin. The model achieved the state-of-the-art performance of 74.3% mAP at 59 FPS.

Q. A. Dang and Q. M. B. Nguyen—Contributed equally to this manuscript.

© Springer Nature Switzerland AG 2019
S. Lee et al. (Eds.): IMCOM 2019, AISC 935, pp. 383–395, 2019.
https://doi.org/10.1007/978-3-030-19063-7_31

We design our model based on one of these state-of-the-art models. The idea is that we can perform more than just object detection in the model by integrating an inference model inside the baseline model. We select SSD as our baseline model and extend it to detect the pose of the detected object. This is non-trivial due to the complexity of the object pose. As human cannot estimate object pose accurately, it is also hard to obtain the object pose at low error rate with deep neural networks. Even with state-of-the-art vision algorithms, it is unlikely that we can estimate the object pose without any prior to the object model. In order to reduce the complexity of this estimation, we focus on the estimation of the azimuth angle of the object. In this work, we introduce a new model named Multibox Angle that is capable of detecting objects class, their location, and their azimuth angle.

The problem we are trying to solve contain three smaller problems: location detection, class recognition, and azimuth angle estimation. We design a joint model to solve all three problems at the same time. This approach has been proved to give good results in [10,15]. With the aim of finding a powerful model which can predict numerous feature of images, we come up with an improved model based on [10].

2   Previous Works

Pose estimation is a classic problem in computer vision. However, it is still a challenging problem since many approaches failed to deliver good results. Initially, work on the topic focused primarily on instance alignment [7]. In those research, the model is trained on a specific instance. After training with enough images, a 3D model of the instance is constructed and then is matched with test images to produce viewpoint estimation. With the help of 3D sensors, compute the object pose is feasible [18]. It is, however, impossible to have a 3D model of every object we have, and matching algorithms will require a lot of computation to estimate the correct pose for the object. Therefore, some people try to detect a 3D pose from a 2D image without the model of every object [1]. The accuracy of this approach is still not enough for robot manipulation. That being said, the estimated pose is very useful for understanding the scene and make an appropriate move to continuously improve the pose estimation result.

Lately, numerous research gradually shifts to category-level to solve the problem. This introduces some challenges to be dealing with. The model must be capable of simultaneously handling different object appearances because of pose changes and large intra-class variations of the categories.

Tulsiani *et al.* [16] treat the problem as a classification problem and use Convolutional Neural Network to solve it. In this approach, the model includes a pipeline with of two networks. In the first stage, the image goes through Faster R-CNN [12] module to detect the class and bounding box of objects. These detected regions are cropped out and be pushed through the second network. This network is based on the pre-trained VGG model [14] with an additional fully-connected layer to implement the viewpoint detection. Their model is tested on the Pascal 3D+ dataset [17] and achieved state-of-the-art results.

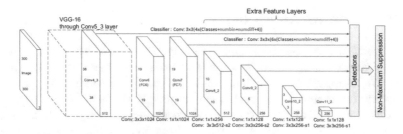

Fig. 1. SSD model with Azimuth detection. Our model adds some additional layers into existing detector blocks. The detector blocks now not only detect bounding box and classify object within that box but also show the orientation of the object. Red indicates our modification to the architecture of SSD [9]

Similarly, Su *et al.* [15] modify the pre-trained base network [8] with additional layers on top to cope with viewpoint estimation. They also use two-stage detection and pose estimation with R-CNN [6] detection. However, the authors manage to produce millions of synthetic images with their rendering pipeline based upon existing 3D model repositories. Altogether, their model surpasses [16] in viewpoint estimation.

Recently, Poirson *et al.* [10] take a different approach to the problem. Instead of using the two-stage approach which is slow and hard to optimize, they use only one network to cope with object detection and pose estimation. Their version modifies real-time, state-of-the-art detector [9] by adding an additional detector to estimate viewpoint. This additional detector classifies each default box into discrete viewpoints based on extracted features from the base network. As a result, their approach achieves comparable results with [16] on the Pascal 3D+ dataset but with 46 times speed-up.

3   Azimuth Angle of Object

In most of the recent researches, the authors always treat viewpoint as discrete values. In order to detect with continuous value, first, we have to understand the way viewpoints was treat as discrete values. The object viewpoint is determined based on the azimuth, which is one of the three numbers to determine a point in the spherical coordinate system. It was the angle of the object when we project it on a horizontal plane.

Based on the azimuth angle, the viewpoints are divided into discrete values. Those discrete values are actually the range of the azimuth angle of the object. For example, in a four viewpoints system, we named them as left, right, front, and back. In reality, they are four ranges of the azimuth angle of the object, and one viewpoint consists of 90°. Therefore in our approach, we tried to detect the real azimuth angle of the object.

4 Multibox Angle Model

In this work, we propose an improvement in Single Shot Multibox Detecting Model [9] for detecting object location and pose simultaneously using the join model. We address the continuous angle problem by breaking the prediction process into two stages. In the first stage, we predict the bin of angle. In another word, this process predicts the range in which the angle belongs to. In the second stage, we predict the different of the real angle against the bin. As a consequence, we add two more components to predict bin and the difference of the angle in that bin.

In order to predict the bin, we implement a classifier with a limited number of bins. And for predicting the offset of angle, we can either use a classifier or a regressor. It is hard to prove which one is better since the prediction relies a lot on how accurate we would like to estimate. When it comes to angle prediction, the human cannot give a very accurate answer either. Therefore, we decided to experiment on both approaches: using classifier and regressor for offset prediction.

In this work, we chose the SSD 300 as the baseline model. This means we have to scale the input image to 300×300. This scaling leads to another problem in predicting angle due to the change in the image ratio. To ensure that the aspect ratio of the image is maintained, we padded the input image before scaling.

Figure 1 shows our proposed model, which is based on the original SSD model [9]. We adopt the first few layers of the original model and add an additional classifier for estimating bin and azimuth angle inside a bin. For bin estimation, we treated it as a classification problem. And with azimuth angle prediction, we treated it as both classification and regression problem.

Loss Function

The training objective of our model is composed of three targets: bounding box localization, class classification, and angle estimation. Therefore, we propose the final loss function as the weighted sum of each target loss function. We utilize the approach presented in SSD [9], which categorize default bounding boxes into positive boxes and negative boxes. To classify each default box, a ground-truth box is matched against each default box based on IoU. A default box is said to be a positive box if it has the ratio IoU larger or equal to 0.5 and a negative one if otherwise. We can formulate the loss function for each target based on these two types of boxes. Let $x_{ij}^p = \{0, 1\}$ be an indicator for matching the i-th default box to the j-th groundtruth box of category p and N is the number of positive boxes. We adapt a bit the localization loss and classification loss from [9].

$$L_{loc}(x, l, g) = \sum_{i \in Pos}^{N} \sum_{m \in \{cx, cy, w, h\}} x_{ij}^{k} smooth_{L1}(l_i^m - \hat{g}_j^m) \tag{1}$$

$$\hat{g}_j^{cx} = (g_j^{cx} - d_i^{cx})/d_i^w$$

$$\hat{g}_j^{cy} = (g_j^{cy} - d_i^{cy})/d_i^h$$

$$\hat{g}_j^w = log(g_j^w/d_i^w)$$

$$\hat{g}_j^h = log(g_j^h/d_i^h)$$

$$L_{conf}(x, c) = - \sum_{i \in Pos}^{N} x_{ij}^p log(\hat{c}_i^p) - \sum_{i \in Neg} log(\hat{c}_i^0) \tag{2}$$

$$\text{where } \hat{c}_i^p = \frac{exp(c_i^p)}{\sum_p exp(c_i^p)}$$

To estimate the azimuth angle of an object, our model first predicts the bin the azimuth is in and then predicts the angle inside the bin. To accomplish this, there are two objectives needed to be optimized: bin prediction and inside-bin angle prediction. The loss function we used in the bin prediction is a softmax function over Q bin. Measuring the loss in negative boxes is unnecessary. Thus, we propose an alternative loss that only minimizes the angle loss of positive boxes.

$$L_{bin}(x, \hat{b}) = - \sum_{i \in Pos}^{N} x_{ij}^p log(\hat{b}_i^q) \tag{3}$$

$$\text{where } \hat{b}_i^q = \frac{exp(b_i^q)}{\sum_q^Q exp(b_i^q)}$$

There are two ways to predict the inside-bin angle: regression and classification. For classification, the loss function is a softmax function with R angles.

$$L_{angle}(x, \hat{a}) = - \sum_{i \in Pos}^{N} x_{ij}^p log(\hat{a}_i^r) \tag{4}$$

$$\text{where } \hat{a}_i^r = \frac{exp(a_i^r)}{\sum_r^R exp(a_i^r)}$$

For regression, the loss function is $smooth_{L1}$ loss between predicted angle and groundtruth angle.

$$L_{angle}(x, a^{pred}, a^{truth}) = \sum_{i \in Pos}^{N} x_{ij}^p smooth_{L1}(a_i^{pred} - a_i^{truth}) \tag{5}$$

The final loss function is the weighted sum of each target loss functions and normalized by N. Let α and β are weighs of localization loss and angle loss respectively. We set $\alpha = 1$ and $\beta = 0.8$ during our training.

$$L_{total} = \frac{1}{N}(L_{conf} + \alpha L_{loc} + \beta(L_{bin} + L_{angle})) \tag{6}$$

5   Training

Our model was implemented with Tenserflow [3] and trained on the PASCAL3D+ [17] dataset. This dataset contains annotations and images of 10 different objects. Furthermore, the images in this dataset are collected from Pascal [4] and Imagenet dataset [13]. Those two are both famous dataset for training object detection problems. However each object in PASCAL3D+ dataset only has around of 2000 images, this dataset is not large enough for training our model. Therefore we use data augmentation methods to generate more data for the dataset.

5.1   Data Augmentation

In object detection problems, there are various methods to generate more training images from an input image. For instance, using image rotation, we can get several images depending on the angle we choose. Or we can simply scale the image to different ratios. But we can not apply these methods in our work. The reason is that these image transformation may affect the result of predicting the azimuth angle. Therefore, we only crop the image with random positions. This augmentation step helps the network invariant to the location of the object in the scene.

As image transformation technique may affect the result of azimuth prediction while our model needs an input image with size 300×300. In order to preserve the image ratio, we padded image to ratio 1:1 before scaling image. After scaling, each input image is cropped with random positions to generate more training samples. The cropping process is performed under some conditions: the output image must have the original object, and the object in output image must not be cropped.

5.2   Training Strategy

After applying the data augmentation process, we split the new dataset into three subsets: training dataset, validation dataset, and test dataset with the ratio of 80%, 5% and 15% respectively. This resulted in the training dataset is only about 1600 images, quite a few for training a CNN.

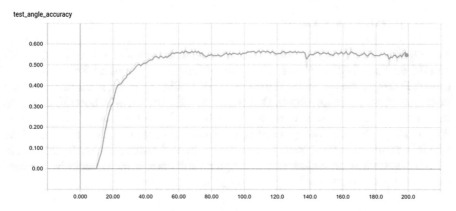

Fig. 2. Accuracy on testing data. At epoch 60th, the model achieves the highest accuracy and is stable at that accuracy.

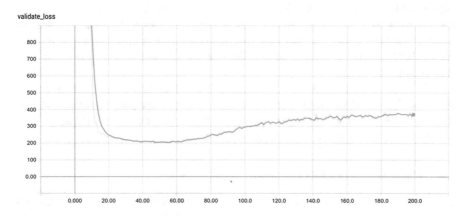

Fig. 3. Loss on validate data

In order to avoid overfit and underfit, we analyze the validation loss and test accuracy to select the optimal point. Using the Tensorboard [3] we draw two graphs. Figure 2 shows the accuracy of test data in the training process and Fig. 3 presents the loss on the validation data. The horizontal axis is the number of epochs and the vertical axis is the value at that epoch. One interesting observation is that at the epoch 50th, the loss on validation has stopped decreasing and begin to increase but the accuracy of test data keeps increasing. From epoch 60th forward, the accuracy is stable at 0.58.

6   Experiment Result

To evaluate the model, we ran several experiments for detection and angle estimation on Pascal 3D+ dataset [17]. We evaluate our model performance using the AVP metric proposed by *Xiang et al.* [17]. The AVP is an extension of AP metric used to evaluate object detection. AP [5] evaluates a bounding box to match groundtruth box if their $IoU >= 0.5$ and the box has a correct label. AVP extends that by adding another condition: the angle of the object must be within a range of some degree compared to ground truth's angle. In this work, all the results are presented with a range of 20°. However, we will also present our model performance at different ranges of degrees.

6.1   Base Network

One of the outstanding features of SSD [9] is that we can use a variety of base network models. In the model, we use VGG-16 [14] as our base network. This base model is used to extract features of input images. Therefore we can use pre-trained weights which was trained with a large and good dataset and attained state-of-the-art results. By this way, we can reduce our training cost and also utilize the greate feature of SSD model. We use pre-trained weights from Imagenet challenge [8].

VGG base layers are used as feature extractor their weights can be reused. However the additional layers from the SSD network are used for prediction, we can not reuse pre-trained weights for them. Therefore we retrain them with our additional layers to achieve the best results.

6.2   Classification or Regression

Our model uses a hierarchy structure to predict azimuth. The first layer is designed to predict the bin of angles and the second layer is responsible for a finer prediction in that bin. We use classification for the first layer and the second layer has two implementations: classification and regression. In this experiment, we train models with different num bin configure with two methods for the second layer. Table 1 shows the result of our experiment.

As we can see in Table 1, the results of the classification method are better than regression. Therefore the latter experiments were done with the classification method for the second layer. There is another observation on the bin number. With the increasing bin number, the accuracy of our model also increases.

Table 1. Accuracy of the two method for second layer with different bin number.

Method	4 bin	8 bin	18 bin	24 bin	36 bin
Classification	0.3216	0.3235	0.4062	0.4290	0.3473
Regression	0.1087	0.2960	0.3810	0.3482	0.3407

Table 2. Accuracy with different α, β configuration

Configuration	α	β	AVP
1	1.0	0.6	0.3533
2	1.0	1.0	0.3758
3	1.0	0.8	0.4290
4	0.8	0.8	0.3813
5	0.8	1.0	0.4028

However, there is a threshold on the number of bins, where the accuracy attains the maximum value.

6.3   Value of α and β

In the previous section, we introduced our loss function with hyperparameters α and β. We perform this experiment to choose the optimal α and β. Table 2 shows the result of our experiment. The best values for α and β are 1 and 0.8. This experiment also shows that prioritizing object detection optimization over azimuth detection gives a better result.

6.4   Multiple Objects

The previous experiment was performed with only one object. In this experiment, we trained our model with multiple objects. Table 3 presents the result of our experiment. As you can see, with the increasing of the number of objects, the accuracy of our model also increase. This observation contradicts our prediction. Because the capacity of our model is limited, therefore, increasing the object for prediction will decrease the accuracy of our model.

The interesting observation in this experiment is that the combination of similar shape object increases the accuracy. In model 4 and 5, we train an airplane, bus, car with another object. In model 4 we train bicycle and train for model 5. And the result in model 5 for airplane and bus is better than model 4.

From this result, we suggest a hypothesis that the extracted features may contain object axis. Which resulted in a larger training data for predicting azimuth from that axis, which give a better result. Figure 4 demonstrates some visual results from our experiments. It is obviously difficult to predict the angle when the main axis of the object does not appear explicitly on the image. This problem comes from the fact that detecting the reference coordinate of the plan where the object stand is non-trivial.

Another issue that we can observe from Fig. 4 is the unpredictable of object structure. It is no surprise that there are an unlimited number of variations in one object class. Even with the object model as prior, estimating the main axis will suffer extremely from the variations of object shape. Curves, abnormal edges, different surfaces can be the source of estimating error. Therefore, it is

392   Q. A. Dang et al.

Fig. 4. Prediction result. Images with green border are the correct results and red borders indicates wrong result.

Table 3. Accuracy with multiple Objects. The more objects we used to train our model, the more accuracy it gets. This shows a very promising results for future works with more training images [15] and more training objects.

Model	Aeroplane	Bus	Car	Train	Bicycle	mAVP
1	0.4290	-	-	-	-	0.4290
2	0.4284	0.5900	-	-	-	0.5092
3	0.4648	0.5873	0.6677	-	-	0.5733
4	0.4331	0.5625	0.6180	-	0.3597	0.4901
5	0.4518	0.5907	0.6079	0.6282	-	0.5696

better to accept the error in the estimation for this approach, as it is trying to give us what we really see from the scene.

6.5   Multiple Degrees Range

In order to evaluate our model accurately, we test our model with a different range of degree. Table 4 shows the result of our experiment. With the AVP of 0.13 at 0° and 0.32 at 5°. Our model archive quite good results at 0° and 5° with AVP 0.13 and 0.32 respectively.

Table 4. Accuracy with different degree range

Degree range	mAVP
0°	0.13
5°	0.32
10°	0.45
15°	0.51
20°	0.54
25°	0.57
30°	0.58
35°	0.59
40°	0.60
45°	0.61

7   Conclusion

In this work, we have proposed a new model to simultaneously detect objects and predict its angle based on the SSD model. The ability to detect objects and angles simultaneously opens the door to many possible applications. While the results are promising, there is still a lot of work to be done in other to bring this work to real-world applications. It also brings us closer to a stage where we can build a vision system at the human level.

References

1. Cheng, C., Chen, H., Lee, T., Lai, S., Tsai, Y.: Robust 3D object pose estimation from a single 2D image. In: 2011 Visual Communications and Image Processing (VCIP), pp. 1–4, November 2011
2. Durrant-Whyte, H., Bailey, T.: Simultaneous localization and mapping: part I. IEEE Robot. Autom. Mag. **13**(2), 99–110 (2006)
3. Abadi, M., Agarwal, A., Barham, P., Brevdo, E., Chen, Z., Citro, C., Corrado, G.S., Davis, A., Dean, J., Devin, M., Ghemawat, S., Goodfellow, I., Harp, A., Irving, G., Isard, M., Jia, Y., Jozefowicz, R., Kaiser, L., Kudlur, M., Levenberg, J., Mané, D., Monga, R., Moore, S., Murray, D., Olah, C., Schuster, M., Shlens, J., Steiner, B., Sutskever, I., Talwar, K., Tucker, P., Vanhoucke, V., Vasudevan, V., Viégas, F., Vinyals, O., Warden, P., Wattenberg, M., Wicke, M., Yu, Y., Zheng, X.: TensorFlow: Large-scale machine learning on heterogeneous systems. tensorflow.org (2015)
4. Everingham, M., Eslami, S.M.A., Van Gool, L., Williams, C.K.I., Winn, J., Zisserman, A.: The pascal visual object classes challenge: a retrospective. Int. J. Comput. Vis. **111**(1), 98–136 (2015)
5. Everingham, M., Van Gool, L., Williams, C.K., Winn, J., Zisserman, A.: The pascal visual object classes (voc) challenge. Int. J. Comput. Vis. **88**(2), 303–338 (2010)
6. Girshick, R.B., Donahue, J., Darrell, T., Malik, J.: Rich feature hierarchies for accurate object detection and semantic segmentation. CoRR (2013)
7. Huttenlocher, D.P., Ullman, S.: Recognizing solid objects by alignment with an image. Int. J. Comput. Vis. **5**(2), 195–212 (1990)
8. Krizhevsky, A., Sutskever, I., Hinton, G.E.: ImageNet classification with deep convolutional neural networks. In: Advances in Neural Information Processing Systems (2012)
9. Liu, W., Anguelov, D., Erhan, D., Szegedy, C., Reed, S.E., Fu, C.Y., Berg, A.C.: SSD: single shot multibox detector. CoRR (2015)
10. Poirson, P., Ammirato, P., Fu, C., Liu, W., Kosecka, J., Berg, A.C.: Fast single shot detection and pose estimation. CoRR abs/1609.05590 (2016)
11. Redmon, J., Divvala, S.K., Girshick, R.B., Farhadi, A.: You only look once: Unified, real-time object detection. CoRR (2015)
12. Ren, S., He, K., Girshick, R.B., Sun, J.: Faster R-CNN: towards real-time object detection with regional proposal networks. CoRR (2015)
13. Russakovsky, O., Deng, J., Su, H., Krause, J., Satheesh, S., Ma, S., Huang, Z., Karpathy, A., Khosla, A., Bernstein, M., Berg, A.C., Fei-Fei, L.: ImageNet large scale visual recognition challenge. Int. J. Comput. Vis. (IJCV) **115**(3), 211–252 (2015)
14. Simonyan, K., Zisserman, A.: Very deep convolutional networks for large-scale image recognition. CoRR (2014)
15. Su, H., Qi, C.R., Li, Y., Guibas, L.J.: Render for CNN: viewpoint estimation in images using CNNs trained with rendered 3D model views. CoRR, http://arxiv.org/abs/1505.05641 (2015)

16. Tulsiani, S., Malik, J.: Viewpoints and keypoints. CoRR, http://arxiv.org/abs/1411.6067 (2014)
17. Xiang, Y., Mottaghi, R., Savarese, S.: Beyond pascal: a benchmark for 3D object detection in the wild. In: 2014 IEEE Winter Conference on Applications of Computer Vision (WACV) (2014)
18. Zhu, M., Derpanis, K.G., Yang, Y., Brahmbhatt, S., Zhang, M., Phillips, C., Lecce, M., Daniilidis, K.: Single image 3D object detection and pose estimation for grasping. In: 2014 IEEE International Conference on Robotics and Automation (ICRA), pp. 3936–3943, May 2014

An Alternative Deep Model
for Turn-Based Conversations

Mai Van Quan, Trieu Duong Le, and Duc Dung Nguyen[✉]

HoChiMinh City University of Technology,
268 Ly Thuong Kiet, District 10, Ho Chi Minh City, Vietnam
`maivanquanbk@gmail.com`, `duongbk1602@gmail.com`, `nddung@hcmut.edu.vn`

Abstract. In this paper, we propose a new model for building a conversational dialogue system which provides natural, realistic and flexible interaction between human and machine based on large movie subtitles dataset. Our models are a generative model that is autonomously generated word-by-word, opening up the possibility of working on many different languages. To address this goal, we extend the hierarchical recurrent encoder-decoder neural network (HRED) for dealing with extracting features from long input turns and generating long output turns. Furthermore, we also apply an attention mechanism in order to attend to particular sentences on source side when predicting a turn response. The models are trained end-to-end without labeling data. The experiments show that our proposed model has improved 36% on BLUE score compared to the HRED model. It also shows many potential improvements in chatbot models.

Keywords: Chatbot · Conversation Model · Deep model

1 Introduction

A conversation is a process of exchanging information between two people through text and voice, and also gestures in form of images. In which one acts as the speaker, the other acts as the listener and then responds to the speaker. During the conservation, each person can exchange his role to the other and when one person finishes his speaking, his turn ends. Dialog systems simulate the process of conservation, which are also known as interactive conversational agents, virtual agents or chatbots, are used in a wide set of applications ranging from technical support services to virtual assistants, and entertainment.

Dialog systems can be classified into two groups: retrieval-based models and generative models. The retrieval-based model uses a set of predefined responses and then combines with heuristics to select the corresponding answer. On the other hand, the systems of the generative model are capable of generating the answer automatically by choosing word-by-word in the dictionary. The generative models have more challenging problems since it is very difficult to generate a

---

M. Van Quan and T. D. Le—Contributed equally to this manuscript.

S. Lee et al. (Eds.): IMCOM 2019, AISC 935, pp. 396–411, 2019.
https://doi.org/10.1007/978-3-030-19063-7_32

response to some questions, or some sentence that is not stored in the database. In this work, we target the generative model which can provide a response to an arbitrary sentence from the user.

One of the most general ways to build a dialogue system of the generative model is to use the Sequence-to-Sequence model [4]. The model consists of a two-tiered Encoder and Decoder, in which the Encoder extracts features from the input turn and emits it as a fixed vector. The Decoder uses this feature vector to generate a feedback. However, a conversation process involves multiple turns, while this model is only capable of receiving an input turn and generating a corresponding turn without any consideration from the previous turns. In order to generate a feedback appropriately to the context of the conversation, Hierarchical Recurrent Encoder-Decoder (HRED) model [13] uses a Context Encoder layer to store past information. However, when the input turn is too long, the Encoder cannot encode all the information required for decoding.

Our proposed model treats an input turn as multiple single sentences, then extracts characteristics from these single sentences independently at the Sentence Encoder layer. Finally, these features are then aggregated at the Turn Encoder layer to construct a feature vector of turn. As a result, important information will be fully extracted as an input turn becomes longer. In addition, at the Decoder layer, we use successive processing units, each will produce a response sentence based on a context vector and a vector produced by the previous unit. This helps the model generate a longer and more meaningful response than previous models. Moreover, we also propose to use an attention mechanism, which will improve extracting features process from input turn.

To training our models, we use in-house data which is collected from a lot of subtitle of films. We also evaluate our proposed model along with HRED model based on perplexity and BLUE score [8].

2   Related Work

Modeling conversations on micro-blogging websites with generative probabilistic models was first proposed by Ritter et al. [9]. They address the response generation problem as a translation problem, where a post needs to be translated into a response. Generating responses were found to be considerably more difficult than translating between languages, likely due to the wide range of plausible responses and lack of phase alignment between the post and the response.

The idea of using recurrent neural network framework for generating responses has been introduced by Shang et al. [11]. The authors focus on generating responses on micro-blogging websites. The framework has been extended by Sordoni et al. [12] to make the answer based on a set of three consecutive utterances rather than a pair of sentences.

It is uneasy to build such a large dataset for the dialog system due to the complexity and the context of human dialogs. Banchs et al. [2] have suggested that we can use movie scripts to build dialogue systems. Conditioned on one or more utterances, their model searches a database of movie scripts and retrieves

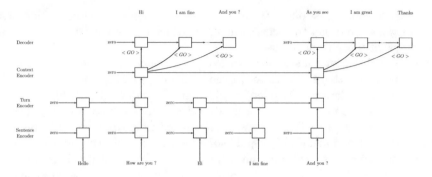

Fig. 1. Turn-Based Conversation Model Architecture

an appropriate response. It is interesting that the movie subtitles can be used to provide responses to out-of-domain questions. Ameixa et al. [1] has point this out and use an information retrieval system for this problem.

3   The Turn-Based Conversation Model

We proposed a new model to generate responses using four basic blocks: Sentence Encoder, Turn Encoder, Context Encoder, and Decoder.

The Sentence Encoder extracts and combines information in a stand-alone sentence that exists in a dialogue based on a sequence of sentences. We use this layer to encode a sentence of unknown length into a fixed-vector that describes core information of that sentence, i.e. the embedded context of the sentence.

Then, we stack the Turn Encoder on top of the Sentence Encoder. The input of this layer is fixed-vectors that are emitted from the Sentence Encoder layer. The output of Turn Encoder is a feature vector that captures all information when a person ends his turn.

We design the Context Encoder to capture general information for turns on whole dialogue, contains turns in the past. The Context Encoder's output depends on the information in the current turn and hidden context that this layer had accumulated from past turns.

Finally, we employ an RNN [3] (the Decoder) to decode all the information collected from the Context Encoder to emit a sequence of response sentence. We design the Decoder for generating one word each time and generate special tokens to identify when a turn or a sentence ends. Figure 1 provides an overview of this model.

3.1   Dialogue

A dialogue D contains T turns, $D = (T_1, T_2, \ldots, T_N)$ with N is the length of dialogue D. At turn T_i, a person could speak one or more sentences, at this point $T_i = (S_1^i, S_2^i, \ldots, S_{M_i}^i)$ where i is the order of turn in dialogue, $S_1^i, S_2^i, \ldots, S_{M_i}^i$

stand for sentences in turn T_i and M_i is the length of turn T_i. Each sentence j in turn i is constructed from words, so we have $S_j^i = (W_1^{i,j}, W_2^{i,j}, ..., W_{K_{i,j}}^{i,j})$ with $K_{i,j}$ is the length of sentence S_j^i. The proposed model is a probability model that uses to predict turn T_n when it knows $T_0, T_1, ..., T_{n-1}$. The model represents a probability $P(T_n | T_{n-1}, T_{n-2}, ..., T_1)$.

3.2   Sentence Encoder

The Sentence Encoder reads an input sentence S that belong to turn T and transforms it into a fixed-vector to describe information of that sentence.

At moment t, each word x_t will be processed by Sentence Encoder, then its extracted information will emit a state vector and this vector will be captured into a hidden state. This process continues until the last word is processed, and the last state vector becomes the final state vector. This vector holds information about the meaning of the sentence and rejects nonsensical information. The Eq. (1) will represent for state vector when a word is processed.

$$h_t = f(x_t, h_{t-1}) \tag{1}$$

where: $f(.)$ is a function, x_t is the word at position t in the sentence, and h_t is the state vector at position t. The initial state h_0 is a zero vector.

This layer extracts the context of the sentence and transforms the sentence of unknown length into a fixed-vector. In other words, the Sentence Encoder produces a context vector for each sentence so that we can utilize them for encoding the context of the whole paragraph.

3.3   Turn Encoder

We propose the Turn Encoder to get information from a turn. This means that the Turn Encoder connects extracted features from the Sentence Encoder layer to make another representation of the whole paragraph.

Given a turn $T = (S_t : t = 1, ..., M)$ has M sentences, each sentence S_t in turn T will be encoded by the Sentence Encoder. After the encode process finishes, we will have a sequence of final state vectors $H^S = (h_t^S : t = 1, ..., M)$ that stands for features from sentences $S_1, S_2, ..., S_M$.

The Turn Encoder uses a similar mechanism to the Sentence Encoder layer. It will continue to encode each vector in H^S at each time until the last vector h_M^S is encoded. At time step t, the network catches a hidden state h_t, and at point $t = M$, there will be a fixed vector representing the feature of a turn. Each hidden state at moment t is represented by (2).

$$h_t = f(h_t^S, h_{t-1}) \tag{2}$$

where $f(.)$ is an activation function, h_t^S is a feature vector extracted from the sentence S_t, and h_t is the state vector at moment t.

The Sentence Encoder and the Turn Encoder create an essential improvement that makes TBC model distinct from the HRED model:

- HRED model: the model treats a sequence of sentences in a turn as a long sentence. Therefore, it is really hard to extract features because it cannot remember everything in a long sentence and will get stuck at Vanishing/Exploding Gradients [14].
- TBC model: instead of treating a turn as a long sentence, TBC extracts features from sentences separately with the Sentence Encoder and then connects these information using the Turn Encoder. Hence, each layer does not have to process in a long sequence, this will reduce the loss of information too much when extracting information.

3.4   Context Encoder

In order to generate meaningful and natural responses as much as possible, the model should catch and extract all information about context while a person interacting with a dialogue system. The Context Encoder is designed for this purpose.

The Context Encoder operates similarly to the Sentence Encoder and Turn Encoder. However, the Context Encoder extracts features by collecting features from the current turn and previous turns that were captured in the past. Then, Context Encoder will be updated with this new information and its output will describe for the context of current dialogue. The memory of Context Encoder will be updated based on (3).

$$h_k = f(h_k^T, h_{k-1}) \tag{3}$$

where $f(.)$ is the activation function, h_k^T is the feature vector of turn T_k, and h_k is the state vector at moment t.

3.5   Decoder

The Decoder of the TBC model has the same architecture with the traditional Encoder-Decoder model. Three first layers extract and keep information from turns in the past by a state vector. Meanwhile, the Decoder layer uses this state vector with its own memory in order to come up with appropriate responses based on inputs context.

The Decoder is constructed by a sequence of processing units, each unit takes a responsibility to generate a single sentence. As a result, the number of units at the Decoder layer will equal with the number of sentences generated by the system. The input of each unit is a state vector from Encoder layers and the state vector of the previous unit.

A traditional decoder treats all responses in a turn as a long sentence, and therefore, it cannot actually know the way to start or end a sentence in a turn or the later sentence should depend on a part of the previous sentence in some way. By encoding each sentence separately and encode the whole text message using the encoded sentences, our proposed model can resolve the issues of previous models. The Decoder, thus, can generate sentences in the same fashion by generating contexts for a sequence of sentences and then extract that context to the sequence of words.

At processing unit k named D_k, model will decode sentence number k of turn t. D_k uses state vector c_t^C that gets from Context Encoder and a state vector h_{k-1}^t from adjacent previous unit with h_0^t is the initial state of Decoder layer and h_0^t is a zero vector. First, we will use h_{k-1}^t as the initial state $h_0^{t,k}$ of D_k. To start decoding, D_k will receive a start signal that is represented by a special token called <GO>. After that, the way we choose word number i at sentence k in turn t is shown in (4)

$$x_i^{t,k} = \arg\max_{x \in V} P(x|x_{i-1}^{t,k}, h_{i-1}^{t,k}, c_t^C) \tag{4}$$

where $x_i^{t,k}$ is the word at position i of the sentence k in turn t, $h_i^{t,k}$ is the state vector w.r.t. $x_i^{t,k}$, c_t^C is the final state vector from the Context Encoder at turn t, V is the dictionary that we constructed.

The probability could be calculated using (5)

$$\mathbf{P}_{\forall x \in V}(x|x_{i-1}^{t,k}, h_{i-1}^{t,k}, c_{t-1}^C) = g(x_{i-1}^{t,k}, h_{i-1}^{t,k}, c_{t-1}^C) \tag{5}$$

where $g(.)$ is a function represented in (6).

$$g(x_{i-1}^{t,k}, h_{i-1}^{t,k}, c_{t-1}^C) = z(W_D * f(x_{i-1}^{t,k}, h_{i-1}^{t,k}, c_{t-1}^C) + b_D) \tag{6}$$

where W_D, b_D are learning parameters of the Decoder layer, $z(.)$ is a function use to calculate the probability distribution, and $f(.)$ is the activation function.

The decoding process on a unit will stop when it generates a special token marking the end of a sentence, named "<EOS>" at moment i^*. After that, the state vector $h_{i^*}^{t,k}$ will be pass to next unit D_{k+1}. This process will continue until there is a unit that generates a token that represents for the end of turn called "<EOT>" in our case.

4   Turn-Based Conversation Model with Attention

In a conversation, a person response is based on not only the context of the dialogue but also the information that is given by that person. However, it is not true to say that all information in a turn is necessary because a person could speak as much as he can but to give a response, we only need to focus on some sentences. To solve this problem, we continue proposing an improved version of the TBC model, named the TBCA model (Turn-Based Conversation Model with Attention).

The TBCA model inherits architecture from the TBC model. This model still has 2 basic components: encoder and decoder. However, the decoder not only uses state vector that is also known as the context vector of the whole dialogue but also focus on extracting information from the Turn Encoder.

Based on the Attention Interfaces technique [5], the Decoder decides how much information from Turn Encoder could be used to give a response. In other

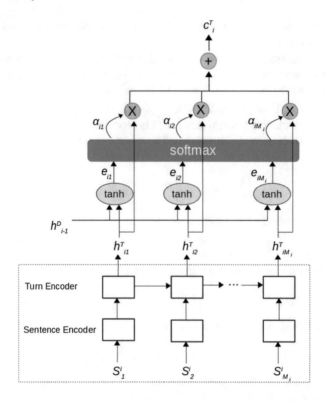

Fig. 2. Information from Turn Encoder is combined through Attention Interfaces

words, the Attention Interfaces technique allow keeping necessary information from a sequence of states, which is obtained from the Turn Encoder. After that, states will be combined into a context vector that describes for core information from the previous turn.

Different with state vector that is aggregated from Turn Encoder directly, this vector describes specific information from a sequence of state vectors returns from Turn Encoder with the importance of each state when they are used to give a response.

From i-th turn: $T_i = (S_j^i : j = 1, ..., M_i)$, after T_i is processed by Sentence Encoder and Turn Encoder, we will have a sequence of state vectors $H_i^T = (h_{ij}^T : j = 1, ..., M_i)$ that stores extracted information. Next, these states and a vector that is generated from previous decoder h_{i-1}^D is used as the input for Attention Interfaces.

Figure 2 demonstrates how to combine information from the Turn Encoder with Attention Interfaces. After each state in H_i^T is concentrated with h_{i-1}^D and go through activation function (tanh), the result will be passed through a softmax layer to emit a sequence of alpha factors. These alpha factors express how much the importance of each state that is extracted from Turn Encoder. Then, we will have a vector stand for a turn T_i

$$c_i^T = \sum_{j=1}^{M_i} \alpha_{ij} * h_{ij}^T \tag{7}$$

where $\alpha_{ij} = \dfrac{exp(e_{ij})}{\sum_{k=1}^{M_i} exp(e_{ik})}$, and $e_{ij} = W_a * tanh(h_{ij}^T \oplus h_{i-1}^D) + b_a$

Next step, this vector c_i^T will continue concentrating with context vector that is emitted from Context Encoder and use as the input for Decoder. So the Eq. (4) will be changed into (8).

$$x_i^{t,k} = \arg\max_{x \in V} P(x|x_{i-1}^{t,k}, h_{i-1}^{t,k}, c_{t-1}^*) \tag{8}$$

where c_{t-1}^* is calculated by the function $c_{t-1}^C \oplus c_{t-1}^T$.

The utilization of Attention Interfaces technique in the model will help increasing necessary information to give a response to the response will much more relate to context than the TBC model did before.

5   BLEU Metric

In this paper, we will use BLEU metric [8] - bilingual evaluation understudy, in order to evaluate how good our model is. BLEU metric is a method of automatic evaluation. This means that we will escape from the disadvantage of human evaluation: we can evaluate our model fastly, inexpensively, correlate highly with human evaluation, and we have a little marginal cost per run. The main idea of the BLEU metric compares how much different between candidate sentences and reference sentences.

BLEU score that has ranged from zero to one is computed by formula (9)

$$BLEU = BP * exp(\Sigma_{n=1}^N w_n \log p_n) \tag{9}$$

where BP is Brevity penalty, N is the maximum number of values we choose for n-gram, w_n is positive weight summing to one, and p_n is a modified n-gram precision on blocks of text.

BP is applied to the whole response with meaning that the response should best match with references. BP is calculated through 10

$$BP = \begin{cases} 1 & if\ r > c \\ e^{(1-r/c)} & if\ r \leq c \end{cases} \tag{10}$$

where r is the shortest length in reference corpuses, and c is the length of the candidate corpus.

Next, we will figure out how to calculate the modified n-gram precision on a sentence and then on a corpus. To compute this on a candidate sentence, we have four steps:

– Step 1: Count the maximum times of each word cluster that we could have based on n-gram occurs in any single reference sentence.

- Step 2: Clip the frequency of each word cluster by its maximum frequency in the reference sentence.

$$C_{clip} = min(Count, RC) \tag{11}$$

where $Count$ is the number of times that word cluster occurs in the candidate sentence, and RC is the maximum frequency of word cluster in the reference sentence.
- Step 3: Add this clipped frequency up.
- Step 4: Divide them by the total number of word clusters.

These steps could be shown through 12

$$p_n = \frac{\Sigma_{i=0}^{MC-1} C_{clip_i}}{MC} \tag{12}$$

where C_{clip_i} is C_{clip} of the word cluster i-th in the candidate sentence, MC is the maximum clusters of the candidate sentence that we could have based on n-gram we chose before. We calculate this value by getting the length of the candidate sentence divide by n-gram we chose and round that value down to its nearest integer.

Then, we could compute the modified n-gram precision on a corpus p_n by matching each candidate sentence with each reference sentence respectively. Then, we sum these C_{clip} and divide them with the total number of word clusters in corpus. The Eq. 13 shows us the way to calculate this value.

$$p_n = \frac{\Sigma_{C \in Candidates} \Sigma_{i=0}^{MC(C)-1} C_{clip_i}}{\Sigma_{C' \in Candidates} MC(C')} \tag{13}$$

where $MC(x)$ is maximum clusters of the candidate sentence x that we could have based on n-gram we chose before.

When we train our TBC and TBCA model, the model has to come up with responses from what it learned and we need to evaluate the quality of these using the loss function. There are many ways to find loss between what we expected and what our model gave us. In this case, we utilize Cross-Entropy [7] to measure distance or the difference between two independent probabilities, described in (15).

6 Model Hyperparameters

We use LSTM [10] with the size of the hidden layer is 1024 for the Encoder and Decoder layer in both TBC and TBCA model. Since LSTM can handle long-term dependency [6] in a sequence, it proves to be a suitable network for our model. The parameters of the model, however, need to be fine-tuned. The hyperparameters we discuss here are the batch size and the learning rate. For the tunning process, we use a sample of 10000 conversations that get randomly from our training set.

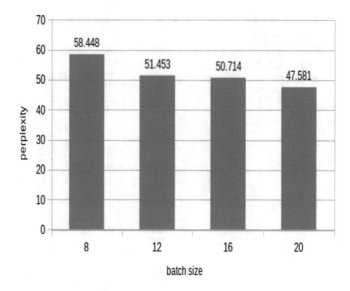

Fig. 3. Perplexity of TBC model after 1000 training steps with learning rate 0.03

6.1 Batch Size

First, we select the initial learning rate to be 0.03. Then, we simulate the training process with different batch sizes. Finally, based on perplexity after 1000 steps of feed forward and back propagation, we can choose a suitable batch size that has the smallest perplexity. Due to limited resources, the batch size is selected to be 8, 12, 16 and 20.

The result of this experiment is shown in Figs. 3 and 4. These results show that if we increase the batch size, the perplexity after 1000 training step decreased but not much. We decided that the batch size should be 20 for both TBC and TBCA training process.

6.2 Learning Rate

Next, we set up the experiment with fixed batch size 16. From there, train the model with different learning rates: 0.001, 0.003, 0.01, 0.03, 0.1, and 0.3. As shown in Figs. 5 and 6, the best learning rate we could choose for the TBC model was 0.1 and TBCA model was 0.3.

The purpose of tunning the learning rate is to speed up the convergence process. That said, as our model try to reach a local minimum, a large learning rate will make the model unable to converge. Therefore, we used another parameter called "learning rate decay factor". This parameter has to reduce the learning rate if our loss does not decrease after 1000 training steps. It is chosen to be 0.9.

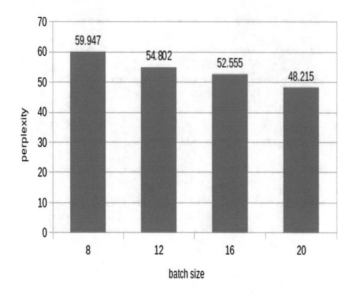

Fig. 4. Perplexity of TBCA model after 1000 training steps with learning rate 0.03

7   Experiments

We perform the experiments on the in-house dialogue dataset. The dataset consists of 3-turns dialogues that were extracted from movie scripts. In addition, all names and numbers are replaced with the <person> and <number> tokens respectively. The training set consists of 196000 dialogues, and the vocabulary size is 10000. The evaluation dataset has 25000 dialogues. We employ the SGD with momentum as our optimizer. The initial learning rate and momentum value are 0.03 and 0.9 respectively. The learning rate will decrease 1% each time when the losses in 3 last steps increased. One epoch of training is one pass on the training data. The order of training dialogues is randomly shuffled at the beginning of each epoch. The order of turns in the dialogue is however kept. Our vocabulary includes some special tokens that are described in Table 1.

We have trained three models: Turn-Based Conversation Model (TBC), Turn-Based Conversation Model with Attention (TBCA) which is applied Attention Interfaces technique [5] and HRED [13]. Each of these models is trained within 1 week. Then, we evaluate these models using two main metrics: Perplexity and BLEU. The results are recorded in Table 2.

Some observations from Table 2:

- Perplexity: the proposed models TBC/TBCA has higher perplexity because of its complexity. The implementation of the TBC/TBCA model is more complex than the HRED model. With more parameters, the training process of TBC/TBCA is hard to converge.

- BLEU score: the HRED model produces lower BLUE score than TBC/TBCA model since the dataset we use to compute BLEU score focuses on giving a response contains many sentences and 2 proposed models take care of linked sentences of a turn in dialogue while the HRED model does not.

We then apply these models in real-world conversations and results were shown in Table 3. We can see that the TBC/TBCA model tends to give responses

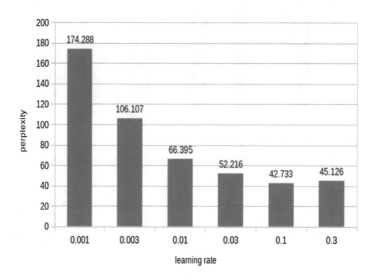

Fig. 5. Learning rate of TBC model after 1000 training steps with batch size 16

Table 1. Special token meanings

Token	Meaning
<GO>	Token is used to alert to start giving response
<PAD>	Standing for nonsensical token that is used to make sentence is filled
<EOS>	End of a sentence
<EOT>	End of a turn
_UNK	Standing for words that are out of vocabulary

Table 2. Evaluation result on models

Model	Perplexity	BLEU score
HRED	**6.4376**	0.2539
TBC	12.2453	0.3243
TBCA	12.7024	**0.3418**

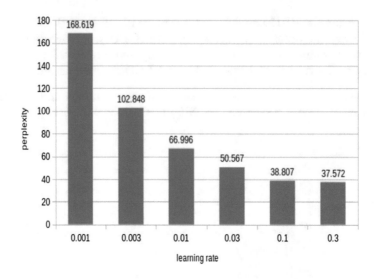

Fig. 6. Learning rate of TBCA model after 1000 training steps with batch size 16

that have long meaningful sentences, while the HRED model gives meaningful responses with only short sentences. It proves that we achieved our target: building a model could give sentences that have meaning based on each other in the same turn.

8   Conclusion

In this work, we have demonstrated that our Turn-Based Conversation Model can outperform the HRED model on the task of modeling utterances and speech acts. The TBC can produce a response that has many sentences and its learnability is much higher than the HRED model as demonstrated in our experiments. There is a huge potential for improving, such as embedding the sentence structures with attention into the model. The training time and performance can also be reduced by using an alternative approach, e.g. CNN models for NLP.

Appendix

9   Perplexity Metric

Perplexity is a measurement that is used to evaluate how well a probability model predicts a word at the moment t in our case. The smaller perplexity, the better the result. The Eq. (14) describes the perplexity:

$$Per = 2^{H(p,q)} = 2^{-\Sigma_x(p(x)*\log q(x))} \tag{14}$$

Table 3. Examples of outputs from the trained models

Input	Model response			Expected response
	HRED	TBC	TBCA	
How are you?	I am fine	I am fine	I am fine	I am fine
For what i am being paid, it is not out of line?	You have not even seen it	\<person\>, you are not a _UNK. you are not a _UNK. you are not a _UNK. you are not a _UNK. you are not a _UNK	I am not going to let you do that. i am not your brother	Who's hired you for this. so you are a hitman now. you could get busted
\<person\>, i know you do not love me now	I am sorry	I love you, \<person\>	I am not your wife. i want to be a _UNK	I have already married. take care of you words
Do not worry. i am gonna get you out of here	You are a _UNK	\<person\>, i am sorry	I am not going to leave you	Follow me
About what	I am not sure	I do not know	About the time you saw her last night	Everything

where $H(p, q)$ is the Cross-Entropy between probability distribution p and q, p is the probability distribution that we expect, and q is the probability distribution that our model learned.

To calculate the Cross-Entropy, we could use (15)

$$H(p^j, q^j) = -\Sigma_{i=1}^{C}(p_i^j * \log q_i^j) \tag{15}$$

where: p^j, q^j are the probability distributions p and q of word j-th, p_i^j is the probability that we would like to receive the word i-th on distribution p^j, and q_i^j is the probability that our model has for word i-th on distribution q^j. It means that if the larger probability at position i, the more chance the word i-th in the vocabulary is suitable for that position. C is the size of the vocabulary.

Here is a small example: assume that we will have a vocabulary with 3 word "\<EOS\>", "come", "home". With the expected output is "come home \<EOS\>", we will record response from first and second model into Tables 4 and 5 respectively. The losses from both models equal to 1/3. However, we can see that the second model is better than the first one because its probability is near our expected probability. Now, we apply (14) to calculate perplexity per each word on each model:

– First model:

$$Per_0^0 = 2^{H(p_0^0, q_0^0)} = 2^{-\Sigma_x(p_0^0(x)*\log q_0^0(x))}$$
$$= 2^{-(0*\log 0.3 + 0*\log 0.3 + 1*\log 0.4)} = 1.3176$$

$$Per_1^0 = 2^{H(p_1^0, q_1^0)} = 2^{-(0*\log 0.3 + 1*\log 0.4 + 0*\log 0.3)} = 1.3176$$
$$Per_2^0 = 2^{H(p_2^0, q_2^0)} = 2^{-(1*\log 0.1 + 0*\log 0.2 + 0*\log 0.7)} = 2$$

– Second model:

$$Per_0^1 = 2^{-(0*\log 0.1 + 0*\log 0.2 + 1*\log 0.7)} = 1.1133$$

$$Per_1^1 = 2^{-(0*\log 0.1 + 1*\log 0.7 + 0*\log 0.2)} = 1.1133$$

$$Per_2^1 = 2^{-(1*\log 0.3 + 0*\log 0.4 + 0*\log 0.3)} = 1.4368$$

Thus, we have the average perplexity:

– First model:

$$E(Per^0) = (Per_0^0 + Per_1^0 + Per_2^0)/3 = 1.5451$$

– Second model:

$$E(Per^1) = (Per_0^1 + Per_1^1 + Per_2^1)/3 = 1.2211$$

Since, we have $E(Per^1) < E(Per^0)$, the second model is probably better than the first model, just as what we expected.

Table 4. Response from first model

Probability of model	Expected probability	True or not
0.3 0.3 0.4	0 0 1 (come)	True
0.3 0.4 0.3	0 1 0 (home)	True
0.1 0.2 0.7	1 0 0 (<EOS>)	False

Table 5. Response from second model

Probability of model	Expected probability	True or not
0.1 0.2 0.7	0 0 1 (come)	True
0.1 0.7 0.2	0 1 0 (home)	True
0.3 0.4 0.3	1 0 0 (<EOS>)	False

Acknowledgment. This work is sponsored and supported by TIS INC. (Japan) under the collaboration between Hochiminh City University of Technology and TIS INC.

References

1. Ameixa, D., Coheur, L., Fialho, P., Quaresma, P.: Luke, I am your father: dealing with out-of-domain requests by using movies subtitles. In: Bickmore, T., Marsella, S., Sidner, C. (eds.) Intelligent Virtual Agents, pp. 13–21. Springer, Cham (2014)
2. Banchs, R.E.: Movie-DiC: a movie dialogue corpus for research and development. In: Proceedings of the 50th Annual Meeting of the Association for Computational Linguistics: Short Papers - Volume 2, ACL 2012, Stroudsburg, PA, USA, pp. 203–207. Association for Computational Linguistics (2012)
3. Britz, D.: Recurrent neural networks tutorial, part 1 - introduction to RNNs, September 2015. https://goo.gl/zfQtHi
4. Ilya, S., Oriol, V., V, L.Q.: Sequence to sequence learning with neural networks. In: Advances in Neural Information Processing Systems, pp. 3104–3112 (2014)
5. Mohandas, G.: Recurrent neural network (RNN) – part 4: attentional interfaces, November 2016. https://goo.gl/RjrwqW
6. Olah, C.: Understanding LSTM networks, August 2015. https://goo.gl/eyu2wT
7. Olah, C.: Visual information theory, October 2015. https://goo.gl/V2Up5C
8. Papineni, K., Roukos, S., Ward, T., Zhu, W.J.: BLEU: a method for automatic evaluation of machine translation. In: Proceedings of the 40th Annual Meeting on Association for Computational Linguistics (ACL), pp. 311–318, July 2002
9. Ritter, A., Cherry, C., Dolan, W.B.: Data-driven response generation in social media, January 2011
10. Sepp, H., Jürgen, S.: Long short-term memory. Neural Comput. **9**(8), 1735–1780 (1997)
11. Shang, L., Lu, Z., Li, H.: Neural responding machine for short-text conversation. In: Proceedings of the 53rd Annual Meeting of the Association for Computational Linguistics and the 7th International Joint Conference on Natural Language Processing (Volume 1: Long Papers), pp. 1577–1586. Association for Computational Linguistics (2015)
12. Sordoni, A., Bengio, Y., Vahabi, H., Lioma, C., Grue Simonsen, J., Nie, J.Y.: A hierarchical recurrent encoder-decoder for generative context-aware query suggestion. In: Proceedings of the 24th ACM International on Conference on Information and Knowledge Management, CIKM 2015, New York, NY, USA, pp. 553–562. ACM (2015)
13. Serban, I.V., Sordoni, A., Bengio, Y., Courville, A., Pineau, J.: Building end-to-end dialogue systems using generative hierarchical neural network models. In: Thirtieth AAAI Conference on Artificial Intelligence (2016)
14. Yoshua, B., Simard, P., Frasconi, P.: Learning long-term dependencies with gradient descent is difficult. IEEE Trans. Neural Netw. **5**, 157–166 (1994)

Tree Crown Detection and Delineation Using Digital Image Processing

Zhafri Hariz Roslan[1]([⊠]), Ji Hong Kim[1], Roslan Ismail[2],
and Robiah Hamzah[2]

[1] Department of Information and Communication,
Semyung University, Jecheon, Chungbuk, Korea
zhafriroslan@gmail.com, jhkim@semyung.ac.kr
[2] Department of Software Engineering,
University Kuala Lumpur (MIIT), Kuala Lumpur, Malaysia
{drroslan,robiah}@unikl.edu.my

Abstract. In this studies, we conduct research on identification of tree species prior to logging is one of the important new initiative to support the management of tropical forests. The use of drone and the improved technique of image processing will increase the potential of identifying tree species, particularly the commercial species, so that the value of timber in that area could be assessed. Only when the value of timber is found to be attractive for timber operation, the area could be opened for timber operation. Otherwise that area could be used for protection of biodiversity purposes. The drone will be used to capture the image of forests as the image data sets. Several algorithms are used for visual enhancement as well as the edge detection operator and morphological operation for segmentation and identification of crown image.

Keywords: Image processing · Quad-copter drone · Tree detection · Forestry

1 Introduction

The importance of managing tropical forest for sustainable basis has been acknowledged at global level, due to its important roles in global climate change, conservation of biodiversity and other socioeconomic development. Logging has been recognized as one of the most important economic activities of the tropical forests which generates a lot of revenues for countries development. Modern forest management requires efficient management of forest resources, not only for timber production but also for maintaining and monitoring the biodiversity of the environment.

Detection and classification of objects were and are still important fields for researchers in many areas of research including Remote Sensing and Photogrammetry [3]. Advances in technology allows new method applying new technology for tree detection and delineation. In the past decades, many researches are done on tree detection using satellite imagery and high-spatial imagery [1–3]. However, quad-copter drone has been shown to have better result in determining tree height and crown diameter [8]. In remote sensing, the Light Detection and Ranging (LiDAR) which is an active sensor can differentiate between man-made objects and natural grounds. Due to this, LiDAR has been actively used for tree detection and estimating the volume of trees [3].

© Springer Nature Switzerland AG 2019
S. Lee et al. (Eds.): IMCOM 2019, AISC 935, pp. 412–423, 2019.
https://doi.org/10.1007/978-3-030-19063-7_33

Another approach for tree detection is using image processing technique. Image processing technique is relatively new in tree detection. Because of this, there is a huge gap between remote sensing technique and image processing technique in terms of bench mark data set [3]. A few tree counting algorithms have been proposed over the past decades to detect natural vegetation. The commonly used algorithms are the valley finding algorithm, region growing algorithm, marker-watershed algorithm, and Object Based Image Analysis (OBIA) for tree crown delineation [1]. While there are abundance of researches using test data to detect between trees and buildings for both image processing and remote sensing technique, this paper concentrates on using dense forest images as the data set for timber production.

Remote sensing technique is usually implemented in tree crown delineation in dense forest due the advantages of spectral image produce by remote sensing technique [7, 8, 10]. Dense forest image tend to be more difficult to delineate due to the fact that the trees are too close to each rather than separated by a distance. Remote sensing excel in this area because it is capable to detect the position of the treetop hence perform tree counting more smoothly. However, remote sensing technique are not capable to find the tree crown edges. A research has been done by combining the spectral data with image processing technique [4]. Instead of using aerial image, a spectral image is used and the tree crown edges are determine using image processing technique. However, many assumption were made to compensate the lack of proving mechanism of the result produce in [4].

In this paper, the aerial image is used and converted to a multi-channel color image as to enhance the visual perception. The multi-channel image produce will resemble the spectral image as aerial image could not provide enough information for processing [5]. While most previous methods involve segmentation in RGB color format, it has proven to be fragile when the segmentation is done in visually complex environment. This paper will also focus on the trees that have a significantly higher height compared to the rest of the trees detected. This paper will use image processing technique to perform morphological opening-by-reconstruction for pre-processing, simple image segmentation, and followed by morphological operation for post-processing.

2 Methodology

2.1 Overview

The proposed method is based on the above flowchart. Our method consist of three stages. After image acquisition, image pre-processing will enhance the image and remove unnecessary noise in the image before moving on to the nest stage. The image segmentation stage will remove the background image (i.e. Ground floor and small trees). Furthermore, this stage is also responsible to segment each individual trees. The segmented image is smooths and extract the information in each tree crown in the image post-processing stage (Fig. 1).

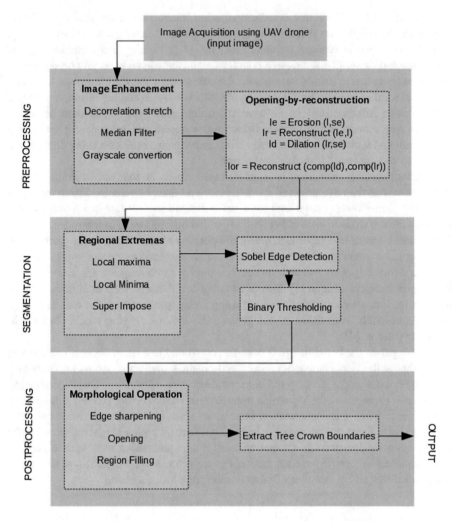

Fig. 1. Project workflow for the proposed method.

2.2 Image Acquisition

In this project, the DJI phantom drone is used as quad-copter drone. The drone possess a 1/2.3″ sensor and a focal length of 24 mm. Depending on the height of the drone when capturing the image, the area of the image can be estimated using the following formula;

$$\text{FOV} = \frac{\text{Sensor} \times \text{Working distance}}{\text{Focal length}} \tag{1}$$

The Field of Vision (FOV) is the area of the camera can capture while the working distance is the height between the drone and the object. The area of the image is an essential information to estimating the number of trees present in the image. The original RGB capture image can be seen in Fig. 2.

2.3   Image Pre-processing

Initially a decorrelation stretch is implemented to enhance the color separation and improve visual perception. The output of the decorrelation stretch can be seen in Fig. 3. Then a median filter to simultaneously remove noise and preserve the edges of the image. Opening-by-reconstruction is a combined of the morphological opening and the morphological reconstruction. Morphological opening is an erosion followed by a dilation. The shrinking by erosion and regrows by dilation depends on the shape of the structuring element to which the final image takes its shape. However, certain component of the targeted image will not be included in the final output image. Morphological reconstruction provide a solution due to the fact that the reconstruction grows in a flood-fill fashion which can include a complete connected components. The output of the opening-by-reconstruction can be seen in Fig. 4.

2.4   Image Segmentation

The opening-by-reconstruction provide a clear visual on the edges. Image segmentation is a process of separating the image into multiple segments. In order to have a proper segmentation of the targeted objects, firstly we separate the image into background and foreground. The background and foreground can be found using regional minima and maxima respectively. The background and foreground can be seen in Fig. 5. The background is depicted in black pixels (0) and foreground as white and gray color.

Edge detection: An edge can be stated as an observable difference in pixel values. This can be seen in the Fig. 5 as the borders of the tree crown. To correctly detect those edges, Sobel operator is applied to the image. The Sobel operator will find the horizontal and vertical edges in an image and produce a reasonably bright result. The Sobel operator uses the following filters and using kernel convolution to find the edges:

$$S(x) = \begin{bmatrix} -1 & 0 & 1 \\ -2 & 0 & 2 \\ -1 & 0 & 1 \end{bmatrix} \tag{2}$$

$$(y) = \begin{bmatrix} -1 & -2 & -1 \\ 0 & 0 & 0 \\ 1 & 2 & 1 \end{bmatrix} \tag{3}$$

	x-1	x	x+1
y+1	a	b	c
y	d	e	f
y-1	g	h	i

Color Values

$$e_x = \frac{(c-a)/2 + 2(f-d)/2 + (i-g)/2}{3} \tag{4}$$

$$e_y = \frac{(a-g)/2 + 2(b-h)/2 + (c-1)/2}{3} \tag{5}$$

$$|\nabla e| = \sqrt{e_x^2 + e_y^2} \tag{6}$$

The result of applying the Sobel operator can be seen in Fig. 6. Based on the Fig. 6, a morphological open is then apply to thicken the edge line. Then the regional extremas are superimpose to the edges of the trees. Superimposing the regional extremas to the edges allows to correctly identify which tree crown belong to which edges and delineation of the tree crown can be separated more which can be seen in Fig. 7.

The image is further segmented by using a global binary thresholding also known as the renown Otsu's method. Thresholding is one of the simplest image segmentation technique and it is useful to remove any unnecessary objects and noise present in the image. Based on the image we produce from the edge detection, the image is turned into a binary image from a gray scale by first choosing the threshold and then turning every pixel black or white according to whether its gray value is higher or lower than the threshold:

$$Pixel\ becomes \begin{cases} \text{White if its gray level is } > \text{ threshold} \\ \text{Black if its gray level is } < \text{ threshold} \end{cases}$$

2.5   Image Post-processing

The result for binary thresholding is presented in Fig. 8. Based from this image, image smoothing is done by using a morphological closing to sharpen the edges and image filling is also done afer that. Lastly, the individual tree crown borders are marked and extracted as shown in Fig. 9.

3   Result and Discussion

3.1   Simulation Result

The proposed method are implemented using MATLAB Image Processing Toolbox. The simulation are tested on 10 different test data with varying heights. All of the test data were obtained using the DJI Phantom 4 drone at various study sites.

Figure 2 shows the input image for which the proposed method is applied for which to detect and delineated tree crowns. The figure was taken at a height of 50 m. The decorrelation stretch is then implemented to the input image to enhance the color separation and improve the visual of the image for later processing. The decorrelation stretch output is shown in Fig. 3.

Fig. 2. Original RGB image.

Fig. 3. The decorrelation stretch to enhance the colour separation between each individual trees.

Next, the morphological opening-by-reconstruction is applied to the decorrelated image which gives the complete connected components of each tree crown. This in turn change the image to a grayscale image. Before applying the morphological operation, a median filter is used to reduce the noise present in the decorrelated image. The output of the morphological operation is shown in Fig. 4.

Fig. 4. Morphological opening-by-reconstruction.

Then, the regional maxima and minima are found separately first before combining into one image. This is to separate the background and the foreground of the image. The background image are considered to be the ground floor and short trees. Plus, the background is denoted as black blobs while the foreground is denoted as gray and white blobs as seen in Fig. 5.

Fig. 5. The background (black) and the foreground (gray and white).

After that, the edge of the individual tree crowns are found using Sobel edge detector. Since the test image comes from a densed forest site, using a first derivative edge detector will provide a reasonably good edges while also perform well under the presence of noise. The Sobel edge detector output is shown in Fig. 6.

Fig. 6. Sobel edge operator.

Once the edges are detected, a dilation operation is applied to thicken the edge lines for better processing. Then the regional extremas are superimpose to the edges as shown in Fig. 7. By implementing this, the edges of the tree crown are correctly matched according to the foreground while leaving the background. This step further improve the separation of the background and the foreground.

Fig. 7. Superimpose of the regional extremas with Sobel edge detection.

Then a binary thresholding operation using Otsu method is applied to the super-impose image to detect trees based on the threshold value. The output binary image is shown in Fig. 8. To further reduce any small objects, a binary morphological operation is applied to the binary image. Next, edges of the tree crown are smooth using a morphological closing operation and an image filling operation is applied to fill any holes in the blobs.

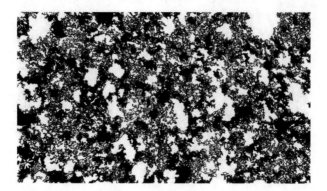

Fig. 8. Binary thresholding using Otsu method.

Lastly, all delineated tree crowns are detected and the number of tree crowns are recorded as shown in Fig. 9.

Fig. 9. Tree crown delineation.

3.2 Result Evaluation

This paper does not have a surveyed tree crown map to use as a reference data, however using the average tree density as proposed by [4] can give an estimated accuracy. Although the average tree density is different from one study site to another, but it does not vary greatly. Using the average tree density which is 960 trees/ha, an estimation number of trees based on the height of the drone when the image was taken that are from 50 m, 100 m, and 150 m. Using the formula (1) to find the area of the image, the estimated number of trees are 16, 11, and 6 trees for 150 m, 100 m, and 50 m respectively. Based on these numbers we can compute the accuracy of our method as shown in the table below:

The Table 1 shows that the proposed algorithm achieved a result of 67.68% accuracy. While the proposed algorithm perform optimally when the drone is at 100 m achieving 72.73% accuracy. Overall this result was expected due to the assumption that only trees with a relatively higher heights were prioritize in this paper. Nevertheless, the result is slightly lower as compared to the 75.6% from [4] and almost equal to [3]. The image data sets does plays a role as both [3] and [4] uses spectral images while [3] uses remote sensing technique and the later implement image processing technique.

Table 1. Result evaluation on the number of trees detected.

	Height of drone (meter)			Total number of trees
	50	100	150	
Estimated value	18	33	48	99
Experimental value	11	24	32	67
Accuracy(%)	61.11	72.73	66.67	67.68

3.3 Discussion

Although previous researches [3, 7, 8, 10] use remote sensing technique extensively, our proposed method provide a simple and adequately reasonable result. Rather than using spectral imagery which is costly, we propose using a pure RGB image and image processing technique. While other past research [1, 4, 6, 9] have used image processing technique, but the application of said research did not implemented on a dense forest image which is more prone to noise when processing.

As mention before, not all trees are taken into the tree counting as we only select trees that have a significant height to accommodate the tree logging production. Our estimated percentage on the tree counting accuracy was expected to be low as seen in the table above. While the shape of the tree crown also plays a role in the overall tree crown delineation as there is no reference data to confirm the accuracy of the shape detect using our proposed method.

Potential problems that deserve further research:

- **Limitation of the assumption.** The assumption that the tree crowns that have the highest gray value are assume to have the highest height. This assumption can be true depending the light condition during the test data was taken. If the test data was taken during the sun at the highest peak then this assumption might be possible. The same cannot be said during the evening when the light condition is declining. For future research, all of the test data should be taken in the afternoon for optimum light condition. Plus, other tools will need to be taken into consideration such as addition sensor to calculate the height of the trees for better result.
- **Overlapping trees.** Overlapping trees seems to a common trend in tree detection system. In our method as well. The overlapping trees effect the shape of the tree crown detected and this creates a false positive in the final result of our simulation. More research needs to be done to find the solution to this problem.

- **Tree crown boundaries.** The boundaries are sometime inconsistent as the drone fly higher. Some of the tree crowns which are separate but detects as one individual tree crown. However, a lower distance the tree crown can be detect separately. Shadows from the trees also affect the tree crown detection during the edge detection process. Several operation will be needed in the pre- processing stage to reduce the potential problems.

4   Conclusion and Future Work

In summary, considering the results it is obvious that the indices used in this paper are good for tree detection at a certain height. This method is related to choosing precise values for some parameters, which some are chosen manually. The algorithm used in this paper seems to performed quite well under the parameters set when experimenting with the data sets. Using the decorrelation stretch does improve the visual perception and the morphological opening-by-reconstruction is able capture the components of the tree crown without comprising any part of the object being detected. This paper does show that detection and delineation of tree crown can be achieved in densed forest image using quadcopter drone and image processing technique. As for future works, the accuracy of the tree crown needed to be evaluate as to prove that the tree crown detect are correctly detected. However, improvements in image segmentation either using more complex algorithm or changing some of the parameters can increase the accuracy of the result.

References

1. Hung, C., Bryson, M., Sukkarieh, S.: Vision-based Shadow-aided Tree Crown Detection and Classification Algorithm using Imagery from an Unmanned Airborne Vehicle. Australian Center for Field Robotics, The University of Sydney, Australia (2006)
2. Yang, L., Wu, X., Praun, E., Ma, X.: Tree Detection from Aerial Imagery. ACM GIS, Seattle (2009)
3. Talebi, S., Zarea, A., Sadeghian, S., Arefi, H.: Detection of tree crowns based on reclassi-fication using aerial images and lidar data. In: International Archives of the Photogrammetry, Remote Sensing and Spatial Information Sciences, vol. XL-1/W3. Tehran, Iran (2013)
4. Wang, L., Gong, P., Biging, G.: Individual tree-crown delineation and treetop detection in high-spatial-resolution aerial imagery. Photogram. Eng. Remote Sens. **70**(3) (2004)
5. Karimulla, S., Raja, A.: Tree crown delineation from high resolution satellite images. Indian J. Sci. Technol. **9**(S1) (2016)
6. Li, Z., Hayward, R., Zhang, J., Liu, Y., Walker, R.: IEEE (2009)
7. Cerutti, G., Tougne, L., Mille, J., Vacavant, A., Coquin, D.: Understanding leaves in natural images – a model-based approach for tree species identification. Comput. Vis. Image Underst. (2013)
8. Panagiotidis, D., Abdollahnejad, A., Surovy, P., Chiteculo, V.: Determining tree height and crown diameter from high-resolution UAV imagery. Int. J. Remote Sens. (2016)

9. Kumar, D., Padmaja, M.: A novel image processing technique for counting the number of tress in a satellite image. Eur. J. Appl. Eng. Sci. Res. **1**(4), 151–159 (2012)

10. Ullah, S., Dees, M., Datta, P., Adler, P., Koch, B.: Comparing airborne laser scanning and image-based point clouds by semi-matching and enhanced automatic terrain extraction to estimate forest timber volume. Forests **8**, 215 (2017). Molecular Diversity Preservation International and Multidisciplinary Digital Publishing Institute

11. Bulatov, D., Wayand, I., Schilling, H.: Automatic tree-crown detection in challenging scenarios. In: The International Archives of the Photogrammetry, Remote Sensing and Spatial Information Sciences, Volume XLI-B3 (2016)

Evaluation of Security and Usability of Individual Identification Using Image Preference

Kotone Hoshina[✉] and Ryuya Uda

School of Computer Science, Tokyo University of Technology,
1404-1, Katakuramachi, Hachioji, Tokyo 192-0982, Japan
c01152902a@edu.teu.ac.jp, uda@stf.teu.ac.jp

Abstract. Biometrics is one of the most attractive identification methods. However, any change makes the identification difficult such as injury. Moreover, if it is once stolen, the information for biometrics cannot be changed. On the other hand, password is easy to be forgotten and difficult to be memorized. Therefore, in this paper, we propose a new personal identification method and system by selecting favorite images which are easy to be remembered and inputted. Furthermore, we have also evaluated the security and the usability of the method.

Keywords: Individual identification · Image authentication · Usability

1 Introduction

Nowadays, biometrics identification becomes popular. One of the advantages is convenience since biometrics information is never lost and is never forgotten while identification card is sometimes missing and text password is gone out of our memories. Moreover, biometrics information is also easy to be inputted such as touching a device or seeing a camera while text password must be inputted by keyboard. However, injury may prevent a user from identification by biometrics and stolen information is never changed. Therefore, in this paper, we propose a personal identification method with images which are free from lost and forgotten and are easy to be inputted. We have also evaluated its security and usability.

There are three big categories in personal identification with images as secret information as follows;

1. drawmetric;
2. cued-recall;
3. recognition.

Methods in all of the categories are kinds of personal identification with corresponding knowledge and images are used for reducing of burden on remembering. Especially, recognition is thought the best way to reduce the burden on remembering. The reason is that it can be expected the following effects by interaction between presented images and memory of users since users can see the candidate images with correct images when they select the correct images;

© Springer Nature Switzerland AG 2019
S. Lee et al. (Eds.): IMCOM 2019, AISC 935, pp. 424–439, 2019.
https://doi.org/10.1007/978-3-030-19063-7_34

1. advantage of visual memory
2. possibility of recall the secret.

Therefore, in this research, we focus on the recognition of favorite images and propose a method and a system of personal identification with favorite images.

2   Related Researches

In this chapter, we describe existing personal identification technologies with images.

2.1   Evaluation for Security and Usability of Personal Identification System

Takada et al. evaluated security and usability of personal identification system named "Awase-E" in order to combine high usability as an advantage with improvement of unchangeable information as a disadvantage in biometrics [1]. "Awase-E" presents 1 set of 9 images from a group of images provided by a user. The user can register confidential information by performing four times to select 0 or 1 image from the 9 images.

Takada et al. evaluated possibility of long-term memorization and operation time of their system. In their experiments, they investigated the long-term memorization of Awase-E, PIN, password and random images for 4 weeks and 8 weeks with 45 examinees. As a result of the experiments, in the PIN and Awase-E, most of the examinees succeeded in remembering confidential information even after 8 weeks from the registration. In particular, in Awase-E, even in the experiments after 8 weeks, the success rate was 100%. On the other hand, in password and random images after 8 weeks, 25% of examinees failed to get the correct answer even when they tried three times. By the way, the average time taken to use Awase-E was 24.6 s. This is because the examinees were interested in the displayed images and watched them.

In addition, Takada et al. evaluated the security by speculation and extraction attacks which were considered to be attacks specific to this system. In the experiments of speculation attack, they adopted a speculation method in which the correct image was speculated with points in common by the examinees' friends and others. Also, in the experiments of extraction attack, they adopted a speculation method in which the correct image was speculated with six sets of images displayed on the system by strangers to the examinees. As a result of the experiments, the average of the number of successfully selected images in the selection of 4 images was 12 regardless whether the attackers were friends or others. However, when there was commonality in confidential information, it was found that there were cases where the correct images were selected by attackers.

As the experimental results above, it can be said that usability in Awase-E has been improved than that in the conventional technologies since users can store the images for a long time. On the other hand, the personal identification system with Awase-E has vulnerabilities in which attackers can speculate the correct image from the commonality of selected images and the position where the correct image is placed. The vulnerabilities are requested to be improved.

2.2 Secure Recognition-Based Image Authentication Without Changing Conditional Information

Mori et al. improved the security of the system for personal identification with images by Takada et al. and evaluated usability of a new proposed system [2]. The improved points were as follows. (1) Increase the number of candidates for images to be selected from 1 set of 9 images to 1 set of 25 images i.e. a total of 100 images. (2) Place the correct image in the specific area defined by the system. The first point improves security without changing confidential information of users. The second point suppresses the increase in the burden of operation due to the improvement in security. As a result of the implementation of the points, they adopted a system in which the correct images was selected after grabbing the images to put the correct image in the specific area, and there were four sets of 25 images. In addition, concentration of the place of the correct image in the specific area improved its usability.

Mori et al. evaluated the implemented system with 9 examinees and the evaluation subjects were the operation time and frustration value of the examinees. As a result of the experiment, the operation time was 16 s on average, and the operation time was shorter than that of the system by Takada et al. It shows that usability has improved. Also, the frustration value of the examinees was lower when the correct image was concentrated in the specific area. On the other hand, the correct images can be more easily speculated since the correct image is not randomly placed but concentrated in the specific area. Therefore, further measures against the vulnerability are required.

2.3 Image Authentication Method with Similar Image Against Shoulder Hack

Tsuya et al. proposed a personal identification method with similar images against shoulder hack, and evaluated its tolerance by experiments [3]. In their system, a user selects four correct images which are automatically generated as geometric patterns, and also selects eight dummy images correspond to each selected correct image. That is, there are four correct images and 32 dummy images. The dummy images are created by processing to invert the top and bottom, left and right, and distort one of the images. Four sets of nine images are displayed, and when the user succeeds in selecting all correct images from the four sets of images, it is regarded as a success of personal identification.

In order to investigate the tolerance of this system, Tsuya et al. conducted experiments simulating recording attacks. There were five examinees, and a screen of a phone tried to be taken by video three times from five different positions. After that, the five examinees watched the filmed video and tried logging in to the system three times. As a result of this experiment, none of the five examinees succeeded in hacking the system.

From this, it can be said that this system is resistant to shoulder hacks. On the other hand, usability of the system is lower than that of the systems by Takada and Tsuya. It is difficult for regular users to remember correct images since images used for the personal identification are similar geometric patterns. Moreover, it can be said that security is insufficient since it is possible to analyze the recorded images and find the correct images using a computer.

3 Proposed Method

3.1 Overview of Method

Based on the problems in the existing methods shown in Sect. 2, we propose a personal identification method in which non-geometric images are used as a measure against peeping and estimation by the third parties. Moreover, we try to improve usability of existing systems as follows.

In personal identification systems by image recognition, it is assumed that long-term memory of the correct answer image arbitrarily selected by a user works well. In this research, we classify images into categories and set the images related with user's favorite things as correct images in each category. We aim to let users memorize correct images for a long time by the relation with their favorite things.

The second point in the improvement is to prevent users' acquaintances from presuming the correct images. It is considered that the users' acquaintances can presume the correct image by analyzing the users' preference. Therefore, in order to prevent the analyzing, we set the number of image categories and number of images in a category to the number that the users' acquaintances can not presume the correct images. In the article by Takada et al., the number of images that the acquaintances of a user can not presume the correct images is three or more. Also, when the placement of the correct images is fixed or the correct images are gathered, it is considered that the third party can presume the correct image. In order to prevent the presumption, we construct a method in which all image groups that are candidates for correct images are arranged randomly. Also, in the article by Mori et al., it is considered that the fixed display of images of the system leads to vulnerability to peeping and recording attacks. Therefore, as a measure of the problem, several categories are randomly displayed from all image categories, and then, the user selects a correct image in one of the categories.

Finally, when images are displayed with a grid, not all the images in a category are displayed, so it takes time for users to recognize the correct images. In order to improve the display, we implemented a new user interface that images are arranged in a square panel and a user clicks or taps the correct image to select. By the new ideas mentioned above, we try to improve usability of existing methods.

3.2   Procedure of Method

We assume the proposal contents of Sect. 2.1 as a system for identifying individuals in the service on the website and implement it. The reason for implementing the service on the website is that it is estimated that personal identification on the web is the most popular and high usability and high security are required. The outline of the proposed method and the personal identification system implemented in this research is described in following subsections.

3.2.1   Registration and Operation of Users

In this system, images used for personal identification are divided into 8 categories, and 16 images are included in each category. Structures of image categories and composition of the images are shown in Fig. 1.

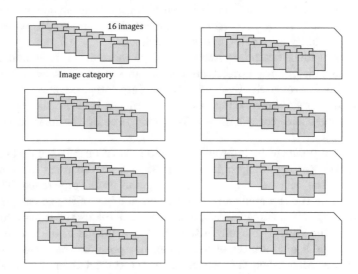

Fig. 1. Structures of image categories and composition of images

In the reference by Cowan [4], there was a description that some objects are stored in 4 or 3 chunks when a person memorizes them. Therefore, we set the number of categories to 8, and in order to make it easier for users to memorize pass images, we display 8 pass images in 4 rows at a time when the users set the pass images. At the setting, a user sets a user name as an arbitrary character string and a pass image. After that, the user selects 1 image that he/she likes from the categorized 16 images and sets it as a correct image. The user repeats this operation for 8 categories, and sets a total of 8 correct images. At this time, in order to prevent the third party from memorizing the order and the position of the images, the display order of the categories and the arrangement of the images in each category are randomized. At the identification phase of a user, 4 image categories are randomly displayed from 8 image categories, and the user selects a favorite image for each category. Also, arrange the images in each category are in random order at this time. When the user selects the correct image in all of the 4 categories, it is determined that the user is successfully identified. Further, when even one pass image in 4 categories selected by the user is different from the correct image, it is regarded as a failure of the user identification. In this case, the user is required to select the pass images again.

3.2.2 Image Category and Images in Category

In our system, users are required to select their favorite images. Therefore, it is necessary to categorize images and use images which a user can recognize whether they are his/her favorite or not. The rules for determining the category of images are described as follows.

1. Image categories must be categorized with common knowledge in a broad sense so that users will find their favorite categories.
2. All categories must be known by almost all users since our system should be used regardless of ages and occupation.
3. Category which is bound to a specific user must be avoided in order to prevent the leakage of personal information from images.

As an example of the item 1, we will give an example of creating a category "food" category instead of "bread". If "bread" is used, most of the images in this category will be similar, so it will be difficult for users to select images with their intuition. Next, as an example of item 2, we will give an example of creating "landscape" category instead of "historic building". If "historic building" is used, it is easier for specific users who have the knowledge of the building to judge each image whether they like or not. However, it becomes difficult for users who are lack of the knowledge to judge each image with their intuition. On the other hand, "landscape" can be recognized irrespective of the amount of knowledge, and users can judge each image with their intuition. Furthermore, as an example of item 3, we will give an example of creating a "clothing" category instead of "face". Human face images contain important personal information which can be used for personal identification, while the combination of clothes is never a factor for personal identification.

Also, the rules for selecting images in a category are described as follows.

1. Two or more images that can be recognized with the same name must not be included in one category e.g. two hamburger images in "food" category.
2. Images which contains personal information must not be used e.g. faces and documents.
3. Images with copyrighted works or portrait rights must not be used e.g. paintings.

Based on the above precautions, we decided 8 image categories as food, drink, animal, plant, landscape, furniture, clothing and vehicles in our method and system.

3.2.3   User Interface

In our system, we implemented a user interface with a method of selecting images by clicking or tapping in order to improve the usability of the existing methods. The operation procedure of displaying and selecting pass images is described as follows. First, 16 images in 1 category which is randomly selected from 8 image categories are displayed in a square panel. Next, a user selects a pass image. At this time, the image is double clicked or double tapped when the user set the pass image, or, the image is single clicked or single tapped when the user select the pass image. This operation prevents erroneous input on the registration and reduces the burden of the user. After the selection of the pass image is completed in one image category, images in the next category are displayed. When all of the images selected by the user are the correct images, the web page in a service is displayed. If there is a mistake in the image selection, the image category randomly appears again and the user must select the image again.

4   Implementation of System

4.1   Environment and Tools

In this research, we implemented the system of the method proposed in Sect. 3.1 as a web application for personal identification with images. We used XAMPP for Windows 7.2.6 and NetBeans 8.2 for the development environment, Apache 2.4.33 for a web server, MariaDB 10.1.33 for a database server, and PHP 7.2.6 for development language.

4.2   Pass Images

As described in the proposal in Sect. 3.2.2, images to be displayed on our system is determined in 8 categories, and 16 images which are appropriate to be classified in a category are stored in each category. In our system, we created eight image folders with the names of food, drink, animal, plant, landscape, furniture, clothing and vehicle.

The 16 images to be stored in a folder are decided as described in Sect. 3.2.2. In our system, images were collected from a copyright free image sharing site [5]. An example of created folders and 16 images are shown in Fig. 2.

4.3   Construction of Database

In our system, we used phpMyAdmin of XAMPP to construct a user information registration database on the local host. The name of the database was "system", the table name was "user", and the column was set as follows.

1. id: Set the data type to "int", autoincrement and primary key.
2. username: Set the data type to "varchar", the length to "20", and the collation sequence to "utf8_general_ci". In this implementation, student ID was set as "username".

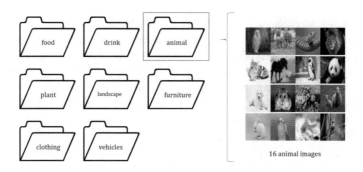

Fig. 2. Example of created folders and 16 images

3. p1: Set the data type to "varchar", the length to "255", and the collation sequence to "utf8_general_ci". The name of the selected pass image in the "food" folder is set to this column.
4. p2: Set the data type to "varchar", the length to "255", and the collation sequence to "utf8_general_ci". The name of the selected pass image in the "drink" folder is set to this column.
5. p3: Set the data type to "varchar", the length to "255", and the collation sequence to "utf8_general_ci". The name of the selected pass image in the "animal" folder is set to this column.
6. p4: Set the data type to "varchar", the length to "255", and the collation sequence to "utf8_general_ci". The name of the selected pass image in the "plant" folder is set to this column.
7. p5: Set the data type to "varchar", the length to "255", and the collation sequence to "utf8_general_ci". The name of the selected pass image in the "landscape" folder is set to this column.
8. p6: Set the data type to "varchar", the length to "255", and the collation sequence to "utf8_general_ci". The name of the selected pass image in the "furniture" folder is set to this column.
9. p7: Set the data type to "varchar", the length to "255", and the collation sequence to "utf8_general_ci". The name of the selected pass image in the "clothing" folder is set to this column.
10. p8: Set the data type to "varchar", the length to "255", and the collation sequence to "utf8_general_ci". The name of the selected pass image in the "vehicle" folder is set to this column.

4.4 Implementation of Web Application

We implemented the web application used for personal identification. We used NetBeans 8.2 for development environment and PHP for development language.

First, we created a new project on NetBeans and save it in the folder "htdocs" in the folder where we installed XAMPP. After the creation, we moved the 8 image folders described in Sect. 4.2 into the project folder. Outline of the Web page created in the project is described in the following subsections.

4.4.1   Implementation of Page for User Registration

In our system, we created a web page of "entry.php" as the user registration page. First, we implemented a text box to acquire the user name. Next, we stored the names of the 8 image folders in an array with character strings, and randomly selected a folder by the mt_rand() function. After that, we browsed the selected folder to obtain the names of 16 images in the folder, and store the names in an array as a character string. Finally, the 16 images were displayed with random order by selecting the name one by one from the array.

When one of the displayed 16 images is selected by clicking or tapping, the name of the selected image is acquired as a character string and the next image folder is displayed. After the selection of 8 images is completed, the user name and the names of the 8 selected images are posted to "check_entry.php" as described in Sect. 4.4.2. Layout and operation of "entry.php" is shown in Fig. 3.

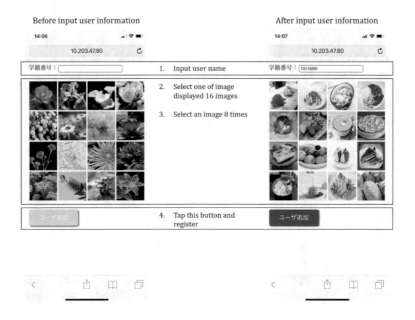

Fig. 3. Layout and operation of entry.php

4.4.2   Implementation of Page for User Registration Confirmation

In our system, we created a web page of "check_entry.php" as the user registration confirmation page. First, the posted user name from "entry.php" is displayed as described in Sect. 4.4.1. Also, referring to the folders acquired from the names of the 8 pass images, images in the folders are displayed. At this time, four of the pass images are displayed in one line according to the theory of human memory for objects as described in Sect. 3.2.1.

We also implemented "register" and "back" buttons. When "register" button is pushed, a text of "registration completed" is displayed and the user name and the name of the pass image are stored in the database in Sect. 4.3. When "back" button is pushed, "entry.php" is displayed and the user is requested to re-enter the user information. Layout of "entry.php" is shown in Fig. 4.

Fig. 4. Layout of check_entry.php

4.4.3 Implementation of Page for User Identification

In our system, we created a web page of "login.php" as the user identification page. First, we implemented a text box to acquire the user name. Next, we stored the names of the 8 image folders in an array with character strings, and randomly selected a folder by the mt_rand() function. After that, we browsed the selected folder to obtain the names of 16 images in the folder, and store the names in an array as a character string. Finally, the 16 images were displayed with random order by selecting the name one by one from the array.

When one of the displayed 16 images is selected by clicking or tapping, the name of the selected image is acquired as a character string and the next image folder is displayed. After the selection of 4 images is completed, the user name and the names of the 4 selected images are compared with the names in the database of Sect. 4.3. If the user name and all of the names of 4 pass images are matched, a text of "You have logged in successfully" is displayed. Also, if not matched, a text of "User name and images are not matched" is displayed in stead. Display and operation of "login.php" is shown in Fig. 5.

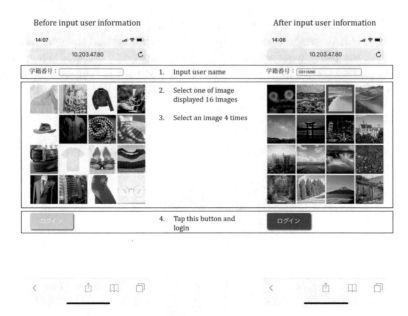

Fig. 5. Layout and operation of login.php

5   Evaluation

5.1   Evaluation of Usability

In this chapter, we describe a method to evaluate whether the usability of the personal identification system described in Sect. 3.1 is improved compared to existing methods. We implemented the personal identification system mentioned in Sect. 3 and evaluated it by experiments with examinees. There were 6 examinees who were students of our laboratory of the university. Also, all examinees were users of smartphones. The procedures of the experiments are described in the following subsections.

5.1.1   Experiments for Evaluation of the Proposed Method

First, we explain the experiment procedures of the personal identification system. The method for the proposed system is described in Sect. 3.2.1, and the operation of the interface is described in Sect. 3.2.3. When we explained the operation procedure to the examinees, we told them the following points; (1) Enter the user name in alphanumeric characters; (2) The number of registered pass images is 8; (3) The number of pass images used for identification is 4; (4) Operation for selecting a pass image; (5) To select favorite images at the registration of pass images. All of the examinees set their user names to their own student ID not to forget the user names. We intentionally did not tell them that layout of the pass images was randomly decided at each identification procedure.

Next, the examinees registered the user name and the pass images once by their smartphones. At this registration, we measured the time from the display of the first image category on the registration page to the display of the completion page of the registration.

Finally, the examinees did the identification procedure with their smartphones. The examinees inputted the registered user name and pass images, and succeeded the identification procedure with their smartphones until the completion page is displayed. At this identification, we measured the time from the display of the first image category on the identification page to the display of the completion page, and the number of times that the identification failed. The definition of one failure is that it is failed when the page for requesting input of the pass images again appears after completing the selection of the fourth pass image. Each examinee repeated the sequence of the procedures 30 times.

5.1.2   Experiments for Evaluation of the Existing Method

First, we explain the procedures of the experiments of the existing method. In the experiments, we used the existing method "S-IGA" proposed in Sect. 3 in the article [2]. The number of pass images was set to 4, and the number of images for candidates was 128. All of the images were the same as those in our system. When we explained the operation procedure to the examinees, we told them the following points; (1) Enter the user name in alphanumeric characters; (2) The number of registered pass images is 4; (3) Operation for selecting a pass image; (4) To select favorite images at the registration of pass images. All of the examinees set their user names to their own student ID not to forget the user names. We intentionally did not tell them that layout of the pass images of S-IGA was randomly decided at each identification procedure and all of the pass images except the first pass image were placed in the displayed screen.

Next, the examinees registered the user name and the pass images once to S-IGA by their smartphones. At this registration, the examinees selected a partial area of 5 rows times 4 columns images from the displayed 25 rows times 4 columns images by dragging the screen. After that, the examinees selected 4 pass images from the selected partial area. At this time, the order of selecting the images was also included in the secret information of users. We measured the time from the display of the selection of the partial area on the registration page to the display of the completion page of the registration.

Finally, the examinees did the identification procedure of S-IGA with their smartphones. The examinees inputted the registered user name and pass images, and succeeded the identification procedure with their smartphones until the completion page is displayed. At this identification, we measured the time from the display of the pass image selection on the identification page to the display of the completion page, and the number of times that the identification failed. The definition of one failure is that it is failed when the page for requesting input of the pass images again appears after completing the selection of the pass images. Each examinee repeated the sequence of the procedures 30 times.

5.1.3   Comparison of Proposed Method with Existing Method

We compared our method with the existing method in terms of "time taken to register pass images", "time to identification is successfully finished" and "number of failures in identification procedure" which are collected by the experiments described in Sects. 5.1.1 and 5.1.2. We compared the values of the average and the variance between our method and the existing method in order to evaluate the improvement of the usability of our system.

5.1.4   Results of the Experiment

We describe the results of the evaluation experiments described in Sects. 5.1.1, 5.1.2, and 5.1.3. The value of the average and the variance of the time taken to register a pass image are shown in the following Table 1.

Table 1. Comparison of time to register pass images.

	Our method	S-IGA
Average (sec)	74.83	72.50
Variance (sec)	1732.14	462.58

Next, the value of the average and the variance of the time to identification is successfully finished are shown in the following Table 2.

Table 2. Comparison of time to success identification.

	Our method	S-IGA
Average (sec)	15.04	22.18
Variance (sec)	13.61	161.69

Finally, the value of the average and the variance of the number of failures in identification procedure are shown in the following Table 3.

5.2   Evaluation of Recognition of Pass Images

In this section, we describe a method for evaluating usability whether the pass image is highly recognizable at selecting a pass image using the image categories and images described in Sect. 3.2.2. We implemented the personal identification system mentioned in Sect. 3 and evaluated by experiments with examinees. There were 30 examinees, all of whom are students of our university. All examinees for the experiments were completely different from those for experiments in Sect. 4.1 since we thought that influence by 30 times repeating operations must be eliminated. Also, all subjects were users of smartphones. The procedures for the experiments were as follows.

Table 3. Comparison of failure times individual identification.

	Our method	S-IGA
Average (times)	1.83	2.00
Variance (times)	2.81	13.00

First, we explain the procedures of the experiments in this evaluation of our system. When we explained the operation procedure to the examinees, we told them the following points; (1) Enter the user name in alphanumeric characters; (2) The number of registered pass images is 8; (3) Operation for selecting a pass image; (4) To select favorite images at the registration of pass images. All of the examinees set their user names to their own student ID not to forget the user names. We intentionally did not tell them that layout of the pass images was randomly decided at each identification procedure.

Next, the examinees registered the user name and the pass images by their smartphones. At this registration, the examinees inputted the user name and selected pass images in all of 8 image categories. We measured the number of times that the selection of the pass images was failed. At the selection, the order of the image categories and the arrangement of the pass images are random as described in Sect. 3.2.3.

After seven days when the examinees had registered the pass images, the examinees selected the pass images in all image categories. We measured the number of times that the selection of the pass images was failed. Also, 30 days after the registration, the examinees selected the pass images in all image categories and the number of times of the failures was measured. Finally, we evaluated the values of the average and the variance.

5.2.1 Results of the Experiment

We describe the results of the evaluation experiment described in Sect. 5.2. We measured the number of times of the failures immediately after and seven days after user registration, and the value of the average and the variance of them are shown in the following Table 4.

Table 4. The times the pass image was mistaken for each time.

	After 0 day	After 7 days
Average (times)	0.42	0.75
Variance (times)	0.74	0.69

5.3   Evaluation of Security

In this section, we evaluate the security against speculation and peeping to our system described in Sect. 3.1.

5.3.1   Evaluation of Security Against Speculation

In our method, it is assumed that the number of pass images that can be inferred by acquaintances of a user or the third party is one or two, based on the experimental results by Takada et al. [1].

When the number of successfully inferred pass images is one, the probability of successful impersonation by one trial on the system is as follows.

$$1/16^{(4-1)} = 1/4096$$

When the number of successfully inferred pass images is two, the probability of successful impersonation by one trial on the system is as follows.

$$1/16^{(4-2)} = 1/256$$

5.3.2   Evaluation of Security Against Peeping

In our method, it is assumed that the number of pass images that can be peeped and acquired by attackers is four. At this time, the probability that the image category which is used by an attacker contains four pass images is as follows.

$$1/{}_8C_4 = 1/70$$

Next, the probability that the image category which is used by an attacker contains three pass images and the attacker succeeds in login is as follows.

$$({}_4C_3 \times {}_1C_1/{}_8C_4)/16 = 1/280$$

Also, the probability that the image category which is used by an attacker contains two pass images and the attacker succeeds in login is as follows.

$$({}_4C_2 \times {}_2C_2/{}_8C_4)/16^2 = 3/8960$$

Finally, the probability that the image category which is used by an attacker contains one pass image and the attacker succeeds in login is as follows.

$$({}_4C_1 \times {}_3C_3/{}_8C_4)/16^3 = 1/71680$$

6   Summary

In recent years, various techniques for personal identification have attracted attention and related researches have been conducted. On the other hand, there are many disadvantages such as lack of usability and security. In this paper, we proposed and implemented a method to identify users by images with their favorite. We also evaluated the usability of the method by comparing with that of an existing method. In the future, we will consider the usability of personal identification systems and recognition of pass images based on the results of the experiments. Also, we will consider the security of the personal identification system of our proposed method compared with the method by Mori et al. [2].

Also, we found a problem of the proposed method that pass images of some users are easy to be speculated by attackers since individual preferences become public due to factors such as SNS. We are planning to seek out ways to make it difficult to speculate the individual preferences.

References

1. Takada, T., Onuki, T., Koike, H.: A user evaluation study about security and usability of Awase-E. IPSJ J. **47**(8), 2602–2612 (2006). (Japanese)
2. Mori, Y., Takada, T.: Stretchable image-grid authentication: additional decoy images and conditional layout of answer images realize secure recognition-based image authentication. IPSJ J. **57**(12), 2641–2653 (2016). (Japanese)
3. Tsuya, K., Ogura, K., Bista, B.B., Takada, T.: A proposal of authentication method using similar images with shoulder hack tolerance and user's momorability. In: The 80th National Convention of Information Processing Society of Japan, 3W-03, pp. 499–500 (2018). https://doi.org/10.11528/tsjc.2017.0_85. (Japanese)
4. Cowan, N.: The magical number 4 in short-term memory: a reconsideration of mental storage capacity. Behav. Brain Sci. **24**, 87–185 (2000)
5. Pixabay - Stunning Free Images. https://pixabay.com/

Learning Typical 3D Representation from a Single 2D Correspondence Using 2D-3D Transformation Network

Naeem Ul Islam$^{(\boxtimes)}$ and Sukhan Lee$^{(\boxtimes)}$

Intelligent Systems Research Institute,
College of Information and Communication Engineering,
Sungkyunkwan University, Suwon, South Korea
{naeem, lshl}@skku.edu

Abstract. Understanding 3D environment based on deep learning is the subject of interest in computer vision community due to its wide variety of applications. However, learning good 3D representation is a challenging problem due to many factors including high dimensionality of the data, modeling complexity in terms of symmetric objects, required computational costs and expected variations scale up by the order of 1.5 compared to learning 2D representations. In this paper we address the problem of typical 3D representation by transformation from its corresponding single 2D representation using Deep Autoencoder based Generate Adversarial Network termed as 2D-3D Transformation Network (2D-3D-TNET). The proposed model objective is based on traditional GAN loss along with the autoencoder loss, which allows the generator to effectively generate 3D objects corresponding to the input 2D images. Furthermore, instead of training the discriminator to discriminate real from generated image, we allow the discriminator to take both 2D image and 3D object as an input and learn to discriminate whether they are true correspondences or not. Thus, the discriminator learns and encodes the relationship between 3D objects and their corresponding projected 2D images. In addition, our model does not require labelled data for training and learns on unsupervised data. Experiments are conducted on ISRI_DB with real 3D daily-life objects as well as with the standard ModelNet40 dataset. The experimental results demonstrate that our model effectively transforms 2D images to their corresponding 3D representations and has the capability of learning rich relationship among them as compared to the traditional 3D Generative Adversarial Network (3D-GAN) and Deep Autoencoder (DAE).

Keywords: Deep autoencoder · GAN · 2D-3D Transformation Network

1 Introduction

Synthesizing realistic 3D objects from their corresponding projected 2D images is a challenging task. Although shadow and reflected patterns in the 2D images encodes information about the 3D geometry of the objects. However, without an exact reflectance model of the object and good estimation of the light direction, it is difficult to

© Springer Nature Switzerland AG 2019
S. Lee et al. (Eds.): IMCOM 2019, AISC 935, pp. 440–455, 2019.
https://doi.org/10.1007/978-3-030-19063-7_35

extract 3D surface geometry from 2D objects. Additionally, self-occluded regions are not present in the 2D images and are impossible to synthesize without prior knowledge about the objects.

In the past, researchers have made an effective progress in 3D mesh-based synthesis and modeling such as the work of [1–3]. Modeling and synthesizing 3D object using mesh is mostly task specific and cannot be generalized to the variation in the testing data.

With the advances in deep generative learning and the availability of large datasets such as 3DShapeNet [4, 5], there have been some inspiring efforts for synthesizing 3D objects using 3D voxels such as the work of [6–8]. The current deep generative model learning is different from part-based approaches as generative models learn the object representations rather than explicitly modeling the concept of parts or retrieve them from an object repository. However, learning 3D representation is a challenging problem due to many factors including the modeling complexity, required computational costs and expected variations scale up by order of 1.5 compared to learning 2D representations. Despite the promising results, performance of current generative architectures is limited especially when object is occluded.

Synthesizing 3D shapes from corresponding 2D images has many applications in computer vision and robotics. Considering the problem of robot grasping an object, synthesizing 3D objects from robot view enables the robot to have better scene understanding and increases the precision of successful grasping. In this paper, we demonstrate that using adversarial architecture for training a deep generative model leads to more realistic 3D synthesized objects. In our architecture we combine the merits of volumetric convolutional neural network [5, 7], deep autoencoder [25] and generative adversarial network [9, 10]. The generative adversarial is different from other heuristic models in a sense that it uses adversarial learning of discriminator to classify between real and synthesized objects. This learning rule pushes the network towards the generalized model which can avoid over-fitting while synthesizing objects. Secondly, the separate recognition and synthesis learning rule make the model more robust as compared to the single feature representation for both recognition and synthesis [6, 11]. Our model learns the relationship between 2D and 3D objects while synthesizing them using generator, and discriminating the pairs using discriminator. Our model has the ability to synthesize realistic shapes as well as learning the relationship between 2D image and the corresponding 3D representation.

Rest of the paper is organized as follows: Sect. 2 describes the related work of 3D representation of the objects based on geometric representation as well as generative approach. Section 3 discusses the proposed model for building the relationship between 2D images and its corresponding 3D representation. Section 4 describes the training procedure of the proposed 2D-3D-TNET. Section 5 shows the experimental analysis of our network whereas Sect. 6 concludes our work of 2D to 3D transformation.

2   Related Work

Most of the existing research for synthesizing 3D objects from its corresponding 2D images are based on two basic approaches: Geometric based approach and Generative based approach. First, we will discuss these approaches and then we will explain our proposed framework.

2.1   Geometry Based Models

Until recently, an effective progress has been made in View Synthesis problem using geometric reasoning. The shape of a particular object can be inferred by taking the intersection of projected silhouettes, given its views from known viewpoints using visual-hull techniques [12]. To estimate an accurate geometry of an object under fixed viewpoint and known lighting conditions, the photometric stereo algorithm has been proposed in [13]. Multiview stereo algorithms has been proposed in [14] which is used to reconstruct 3D object scenes when multiple view images are available. The multiple view images are then utilized to synthesize multi views of a scene. In [15] the authors use convolutional neural network to learn the relation between neighboring views, whereas the work in [16] uses two deep neural networks where they estimate homography by using one network and synthesizes the middle view images using the second network and then rectify the two view images. The single input view perspective approach has been used in [17], where they predict depth map and then synthesize novel view from the reconstructed 3D points transformation in the depth map. However, all these approaches are task specific and do not generalize well. Hence, fail under high variation with in the input data, whereas our proposed model generalizes well and is robust to the variations in the input data.

2.2   Generative Models

One of the first deep convolutional neural networks which is capable of synthesizing realistic images was proposed in [18]. However, such network requires viewpoint and color information as well as explicit factored representations of the object type which cannot be generalized well to unseen objects. Our model is based on GAN [9] for generating realistic 3D objects from corresponding 2D images. Recently [19] proposed a deep generative model of 3D shapes by using encoder along with GAN and with 3D convolutions. However, the model was trained using aligned 3D shape data. Our model solves the more challenging problem of effectively transforming 2D images to 3D shapes and learns the relationship among them using projection network. [20] shows 3D shape completion for simple shape when the views are available and also, they used 3D data as input to the model. [21] learns transformation from an input image to 3D object using multiple projections of the 3D shape from known viewpoints.

3 Model

Inspired by the recent advances in deep generative models, especially Generative Adversarial network [9] and deep autoencoders [25], we propose Deep Autoencoders and GAN based 2D-3D-TNET. Our proposed framework is shown in Fig. 1. The 2D-3D-TNET is based on GAN network. The conventional GAN network is composed of two sub networks: (1) Generator Network and (2) Discriminator Network. The Generator Network of the conventional GAN transforms the samples drawn from the noise distribution "z" to the data distribution with the aim to produce same distribution as the true distribution. The goal of the Discriminator Network is to distinguish between samples drawn from the true distribution and the samples generated by the generator and in return guide the generator to produce realistic samples from the noise distribution. The optimization is performed by jointly training both the generator and discriminator architectures. The loss function of discriminator in 3D-GAN is:

$$L_D = \log D(x) + \log(1 - D(G(z)))$$ (1)

This loss function is maximized by using stochastic gradient method. whereas the loss function of generator in 3D-GAN is:

$$L_G = \log(1 - D(G(z)))$$ (2)

The loss of generator is minimized using stochastic gradient method. The combined loss of 3D-GAN is:

$$L_{3D\_GAN} = L_G + L_D$$ (3)

3.1 Generator

The generator architecture in our proposed model is composed of deep autoencoder, where the input to the generator is 2D natural images of resolutions [128 × 128 × 3]. The autoencoder generator maps these inputs to [32 × 32 × 32] cube, which represents the corresponding 3D representation $G(x)$ of the input 2D images x in the 3D voxel space. The generator part of 2D-3D-TNET is composed of six convolution layers followed by three fully connected layers and six deconvolution layers. We use convolution filters of sizes [6-5-4-3-2-2] in the successive layers with stride of 2, followed by batch normalization and Leaky ReLU [22] activation function between each layer of encoder respectively. The number of filters are [64-196-512-512-2000] in the convolutional layers of encoder which is followed by three fully connected layers having dimensions of [1000-800-500] respectively. The decoder of the generator decodes the final fully connected layer output of the encoder to the voxel space having resolutions of [32 × 32 × 32]. In the decoder, we use deconvolution layers to upscale the one-dimensional encoded representation of the input space to the required cube size in the voxel space. The size of the filters and their dimensions are same as those of the encoder part of the generator. We also use batch normalization and same activation function in the decoder part of the network. There are no pooling layers in our architecture. Our generator loss function is based on the mean square error loss along with the conventional GAN loss. The mean square error uses to quantify the difference between the true distribution and the model distribution along with providing stability to the GAN learning.

$$L_{3D\_DAE\_G} = \gamma_g L_G + \gamma_1 L_2 \tag{4}$$

The first term on the right-hand side shows the conventional generator loss, whereas the last term shows mean square error loss.

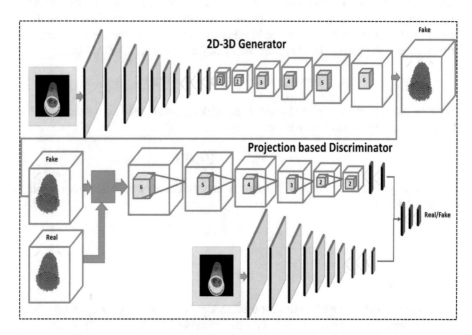

Fig. 1. 2D-3D Transformation Network (2D-3D-TNET): The upper part of the network shows generator which transform 2D input images to its corresponding 3D representation. The lower part of the network is discriminator which projects real or fake pair of 2D and its corresponding 3D representation to their corresponding latent representation for building association between them and for guiding the generator to generate realistic 3D representation. The real pair is based on real 2D and its corresponding 3D representation whereas the fake pair is based on real 2D and its corresponding 3D representation generated by the generator.

3.2   Discriminator

To guide the generator to produce realistic 3D transformation and to build the relationship among 2D images and their corresponding 3D objects, our discriminator architecture is composed of two projection networks: (1) 2D Projection Network which takes real 2D images as input and projects it to its corresponding one-dimensional latent representation. (2) 3D Projection Network projects the 3D object into its corresponding latent space representation. Input to the 3D Projection Network is either real or fake. In terms of real input, we use real 3D as input to the 3D Projection Network, whereas in terms of fake input, the input to the 3D Projection Network is the 3D output of the generator. The resolution of 3D input to the 3D Projection Network is [32 × 32 × 32] whereas the 2D input image resolution is [128 × 128 × 3]. The size of the filters used in each layer of both the projection networks are [6-5-4-3-2-2] respectively having stride of 2. Similarly, we also use batch normalization and Leaky ReLU in each layer of the projection networks. Now to build the relationship among 2D images and their corresponding 3D representations, we concatenate the projected latent representations of both the 2D and 3D projection networks into a single latent vector and build three layers fully connected architecture on top of it. The dimensions of fully connected network architecture are [500-800-500-10]. We use Leaky ReLU layer in between each layer as an activation, however, we do not use batch normalization in the fully connected network. The discriminator loss we use is:

$$L_D = \log D(x) + \log(1 - D(G(z))) \tag{5}$$

4   Training Details

The role of discriminator is to guide the generator to learn the data distribution properly, by discriminating between real data and fake data. In traditional GAN, the real data is the real input to the discriminator whereas the fake data is the fake input to the discriminator, which is transformed by the generator from the noise distribution to the data distribution. In our case, we consider the relation between 2D images and their corresponding 3D representations. We designed a network which provides projection from the 2D space to its correspondence 3D space by building the relationship between them. So, rather than taking noise as input to the generator, we provide 2D images as input to the generator which transforms it to its corresponding 3D representations. The discriminator input is either real pair or fake pair. In terms of the real pair, the discriminator is provided with the real 2D image and its corresponding 3D representation whereas in terms of the fake pair the input to the discriminator is the real 2D image and the corresponding 3D representation generated by the generator. Generative adversarial network map noise to the data distribution by the guidance of discriminator. However, to build relationship between 2D and 3D and to perform transformation from 2D to its correspondence 3D, we modified the architecture in an effective way where it builds relationship between 2D and 3D and project 2D to its correspondence 3D objects. Transformation from the lower dimension 2D space to its correspondence 3D object space is a challenging task for traditional GAN because the job of discriminator is to guide the generator by discriminating between real and fake samples, but what happens if the generator is provided with the samples which do not cover the whole input space, the discriminator stops guiding the generator to cover the whole input space causing the generative adversarial network to collapse. To avoid the problem of collapsing, we devised specific training methods where we replaced the generator by Deep autoencoder with L2 loss as an objective for the generator along with the traditional generator loss. The role of Deep autoencoder is to prevent the GAN from immature collapsing by training the generator parameters and in return provides a richer environment for fake samples. As GAN learning rule is based on game theory where generator tries to fool discriminator by producing realistic objects. The success of the training lies in the fact that both should learn with equal pace, however as the job of discriminator [9, 10] is very easy so it will converge very fast but the term $(1 - D(G(z)))$ in the cost function will push the generator away from the data distribution resulting in a noisy generator output. To avoid this problem, we force the discriminator to stop learning and wait for the generator to catch up the discriminator by using conditional learning rules in the discriminator. Deep autoencoder by itself provides effective generalized transformation from 2D images to its corresponding 3D objects if sufficient amount of training data is available. However, if the training data is sparse and having many variations then training of deep autoencoder becomes hard and fails in terms of generalization on the testing samples because of overfitting due to the exact one to one correspondence objective of the deep autoencoder. On the other hand, GAN provides generalization as its learning rule is based on discriminating real samples from fake. Although the GAN learning rules provides more general solution, however, this training of GAN is unstable and is very much dependent on the proper selection of the hyperparameters. In our proposed framework we combine both autoencoder objective and GAN objective. The autoencoder objective provides stability to the GAN which in turn produces more realistic representation of the input distributions.

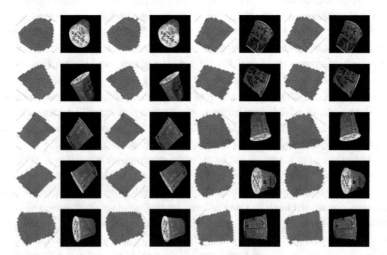

Fig. 2. Transformation results of the 2D Noodle Cups class to their corresponding 3D representations: the odd columns represent the generated 3D samples by the generator corresponding to its 2D representation in the even columns.

Fig. 3. Transformation results of the 2D Plastic Plate class to their corresponding 3D representations: the odd columns represent the generated 3D samples by the generator corresponding to its 2D representation in the even columns.

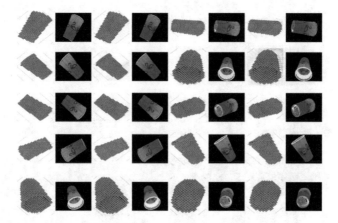

Fig. 4. Transformation results of the 2D Drinking Glass class to their corresponding 3D representations: the odd columns represent the generated 3D samples by the generator corresponding to its 2D representation in the even columns.

Learning rate also plays an important role in the successful training of the network. Very high learning rate will cause the network to overshoot from the desired optima and very small learning rate can keep the network lingering around local minima. Also, as the discriminator converges fast, so the discriminator learning rate plays role of slowing down its learning. In our experiments we use the learning rates of 0.0021 and

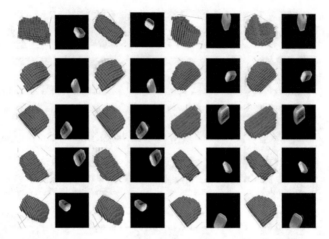

Fig. 5. Transformation results of the 2D Dental Floss class to their corresponding 3D representations: the odd columns represent the generated 3D samples by the generator corresponding to its 2D representation in the even columns.

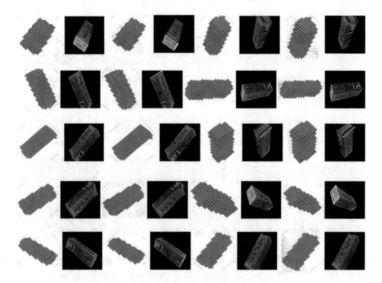

Fig. 6. Transformation results of the 2D Cookies Box class to their corresponding 3D representations: the odd columns represent the generated 3D samples by the generator corresponding to its 2D representation in the even columns.

0.0001 for the generator and discriminator respectively. Similarly, we use ADAM optimizer [23] for optimizing the parameters of the network with beta = 0.5. The total loss for the model is:

$$L_{3D\_DAE\_GAN} = \gamma_g L_G + \gamma_D L_D + \gamma_1 L_2 \tag{6}$$

where the first term on the right-hand side shows the loss of generator, second term shows the loss of discriminator and the last term defines the mean square error loss respectively.

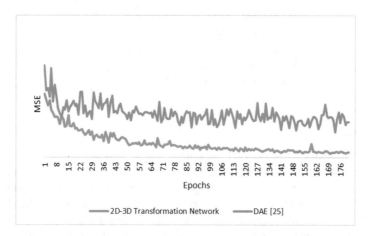

Fig. 7. Performance analysis on 500 multiclass testing samples for 180 epochs using deep autoencoder network and our proposed 2D-3D Transformation Network (2D-3D-TNET).

5   Experiments

5.1   Training Setup

The main purpose of our approach is to find the relationship between 2D images and its corresponding 3D representation along with the pose-based information. We use two different datasets: ISRI_DB [26] real dataset and ModelNet40 CAD dataset [5] for quantitatively evaluating transformation performance with a large-scale of 3D objects.

In the first experiment, we selected 6,000 training samples, and 500 testing samples, covering 5 different classes: Noodle Cup, Drinking Glass, Cookies Box, Dental Floss and Plastic Plate. Each sample consists of a 3D [32 × 32 × 32] voxel representation and a corresponding projected 2D representation having resolution of [128 × 128 × 3]. In order to create these training and testing samples, we placed each object on a rotating table and captured its point cloud from 16 different orientations using a structured light camera. we created a 3D model by registering the accumulated point clouds. we transformed the model using 1,300 random 6DOF poses. For each pose, we render projected RGB image of the object model in [128 × 128] 24-bit resolutions, and slice the point cloud into [32 × 32 × 32] binary voxels using octree representation [24]. Similarly, in the second experiment we used ModelNet40 [5] dataset. ModelNet40 having 40 classes with resolution of [$32 \times 32 \times 32$]. We created the corresponding 2D images by projecting the 3D voxel representation on (x,y) coordinates. We set the projected images to [128×128] 8-bit resolutions. We train our model with 180 epochs to learn the relationship between the input 2D images and corresponding 3D objects by following the specific procedure discussed above.

Table 1. Comparison of pixel wise error

Approaches	DAE [25]	2D-3D-TNET (**Ours**)
Noodle Cup	18.988	**18.829**
Drinking Glass	**18.756**	18.857
Cookies Box	13.526	**13.212**
Dental Floss	17.234	**16.720**
Plastic Plates	**11.735**	11.892

5.2   Results

We discuss the main findings of our experiments in the rest of this section. The input to our architecture is corresponding pairs of: (1) [$128 \times 128 \times 3$] 2D natural image, and (2) [$32 \times 32 \times 32$] 3D object in voxel space. After training deep autoencoder and our proposed 2D-3D-TNET for 180 epochs, we analyze the results qualitatively and quantitatively. In the first experiment, we have five different classes of objects. Class 1 contains different orientations of Noodles Cup. The results from the testing data of Noodles Cup are shown in Fig. 2, where the even columns represent 2D images of the corresponding class as input to the proposed 2D-3D-TNET. The 2D images are transformed to its corresponding 3D representation and the results of transformation is shown in the odd columns of the Fig. 2. The second class of the data is based on Plastic Plates as shown in Fig. 3, where the odd columns represents the generated 3D samples by the model from the corresponding 2D images in the even columns. Figure 4 shows the results of Drinking Glass class both in term of generated results in odd columns and their 2D real representation in the even columns. The Dental Floss and Cookies Box results are presented in Figs. 5 and 6 respectively, where the ground truth 2D samples are shown in even columns and the corresponding 3D transformation results are shown in odd columns.

From the experimental results of different classes, it shows that our proposed framework successfully transforms 2D images to its corresponding 3D representation. From the qualitative analysis of Dental Floss examples, as the object is very small and there are many variations in the data so both deep autoencoder and our model finds hard time to produce realistic output, but as compared to deep autoencoder, our model produces better results. Table 1 shows the comparative quantitative results based on our proposed model and deep autoencoder. The mean square error based on the transformed 3D samples and its corresponding ground truth samples is calculated based on each class separately as well as on the samples of all classes simultaneously. In terms of the single class analysis, it shows that pixel wise error between the generated samples and ground truth samples for Noodles Cup, Cookies Box and Dental Floss is small for our proposed model as compared to deep autoencoder. The variations in these classes are more and due to the generalization capability of our model, it produces better results. On the other hand, the variations in the Drinking Glass and Plastic Plates are small so deep autoencoder produce better results as compared to our proposed model. Figure 7 shows the testing samples error on one batch during each training epoch. The samples are picked randomly from the testing data during each epoch. The quantitative results show better performance of our proposed framework as compared to deep autoencoder. In terms of the second experiment where we used ModelNet40 for training and testing our proposed framework, we analyze the results based on two aspects: (1) we analyze our trained model on the testing samples of ModelNet40 in terms of 2D to 3D transformations. For this analysis we selected the random 2D testing images and transformed it to its corresponding 3D representation by using our trained 2D-3D-TNET. The 3D transformation results from its corresponding 2D images are shown in Fig. 8, where the 1st, 4th, 7th and 10th columns represents 2D input testing samples and the transformed 3D representations are shown in 3rd, 6th, 9th and 12th columns respectively. This shows effective typical 3D transformation of our proposed framework from its corresponding 2D images. To further verify the transformation of the framework, we analyze different views of the transformed 3D representation from its corresponding 2D image. This analysis gives insight into the learning capability of our proposed framework. For this analysis we picked 5 different class images from the testing samples including sofa, chair, monitor, airplane and mug and transformed these images to their corresponding 3D representations. The transformed 3D objects are then analyzed from different views where it shows robust transformation from 2D images. Figure 9 verify view-based analysis of the proposed framework.

5.3   Transformation Based on Similar Input

Different 3D objects having the same 2D representation, put constraint on the accurate transformation of 2D to its exact 3D representation due to the lack of available information. For example, the specific 2D images of table, desk, bookshelf and bathtub in Fig. 10 almost represents the same object in its 2D domain resulting in one to many relationships. However, transforming back from such 2D images to its corresponding exact 3D objects is an ill posed problem because of the limited available information in the 2D. The transformation network learns the intermediate 3D representation of these objects. The transformed 3D object shows the mean representation of all the samples as shown in the middle of Fig. 10.

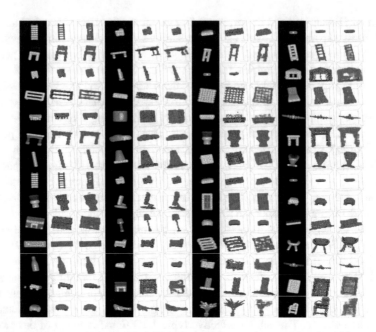

Fig. 8. Performance analysis on ModelNet40 multiclass testing samples for 180 epochs using the proposed 2D-3D Transformation Network (2D-3D-TNET). 1st, 4th, 7th and 10th column represents 2D input testing samples and the corresponding 3D ground truth 3D representations are shown in 2nd, 5th, 8th and 11th columns respectively whereas the transformed 3D representations are shown in 3rd, 6th, 9th and 12th columns respectively.

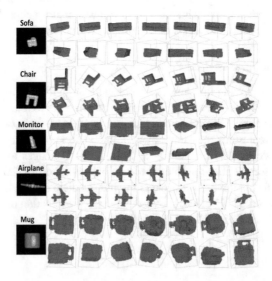

Fig. 9. Full 3D shape representation generated by the proposed framework from corresponding 2D image. First column represents 2D images and for each 2D image the corresponding two rows shows different view of the transformed 3D shapes for illustration purpose.

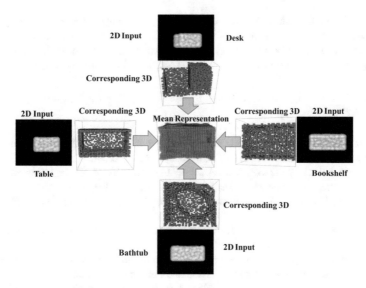

Fig. 10. Typical 3D transformation: the selected 2D images corresponds to different 3D objects and the transformation results from such 2D images represents the mean of all the samples as shown in middle. The edge samples are 2D input images whereas their corresponding 3D ground truth are next to the 2D images. The transformed 3D sample is in the middle.

6 Conclusion

In this paper we investigate the problem of typical 3D representation by transformation from its corresponding single 2D representation using 2D-3D Transformation Network (2D-3D-TNET). The proposed framework has the capability of generating realistic 3D objects along with the association between 2D and 3D representation. Our framework has two main parts: Generator and Discriminator. The generator part of our framework produces 2D to 3D transformation whereas the discriminator performs two tasks. First, it builds association between 2D and corresponding 3D by projecting 2D and 3D into its corresponding latent representation and associate them by using fully connected layers. Second, it guides the generator to produce realistic 2D to 3D transformation. The proposed framework offers advantages over conventional approaches in terms of generality and stability in training by applying conditional learning to the discriminator along with autoencoder objective function. Experiments indicate that the proposed framework works as intended while effectively transforming 2D to corresponding 3D representation. As a future work, we plan to handle the unique symmetric object representation as a general-purpose solution to various real-world domains.

Acknowledgement. Sukhan Lee proposed the concept of transformation of single 2D image to its corresponding typical 3D representation using 2D-3D Transformation Network, while Naeem Ul Islam implements the concept and carries out experimentation. This research was supported, in part by the "Robot Industry Fusion Core Technology Development Project" of KEIT (10048320), in part, by the "Project of e-Drive Train Platform Development for small and medium Commercial Electric Vehicles based on IoT Technology" of Korea Institute of Energy Technology Evaluation and Planning (KETEP) (20172010000420), sponsored by the Korea Ministry of Trade, Industry and Energy (MOTIE) and in part by the MSIP under the space technology development program (NRF-2016M1A3A9005563) supervised by the NRF.

References

1. Carlson, W.E.: An algorithm and data structure for 3D object synthesis using surface patch intersections. ACM SIGGRAPH Comput. Graph. **16**(3), 255–263 (1982)
2. Tangelder, J.W.H., Veltkamp, R.C.: A survey of content based 3D shape retrieval methods. In: Proceedings of Shape Modeling Applications. IEEE (2004)
3. Van Kaick, O., et al.: A survey on shape correspondence. Comput. Graph. Forum **30**(6) (2011)
4. Chang, A.X., et al.: Shapenet: An information-rich 3D model repository. arXiv preprint arXiv:1512.03012 (2015)
5. Wu, Z., et al.: 3D shapenets: a deep representation for volumetric shapes. In: Proceedings of the IEEE Conference on Computer Vision and Pattern Recognition (2015)
6. Girdhar, R., et al.: Learning a predictable and generative vector representation for objects. In: European Conference on Computer Vision. Springer, Cham (2016)
7. Qi, C.R., et al.: Volumetric and multi-view CNNs for object classification on 3D data. In: Proceedings of the IEEE Conference on Computer Vision and Pattern Recognition (2016)
8. Su, H., et al.: Multi-view convolutional neural networks for 3D shape recognition. In: Proceedings of the IEEE International Conference on Computer Vision (2015)
9. Goodfellow, I., et al.: Generative adversarial nets. In: Advances in Neural Information Processing Systems (2014)
10. Radford, A., Metz, L., Chintala, S.: Unsupervised representation learning with deep convolutional generative adversarial networks. arXiv preprint arXiv:1511.06434 (2015)
11. Sharma, A., Oliver, G., Fritz, M.: VConv-DAE: deep volumetric shape learning without object labels. In: European Conference on Computer Vision. Springer, Cham (2016)
12. Laurentini, A.: The visual hull concept for silhouette-based image understanding. IEEE Trans. Pattern Anal. Mach. Intell. **16**(2), 150–162 (1994)
13. Woodham, R.J.: Photometric method for determining surface orientation from multiple images. Opt. Eng. **19**(1), 191139 (1980)
14. Furukawa, Y., Hernández, C.: Multi-view stereo: a tutorial. Found. Trends® Comput. Graph. Vis. **9**(1-2), 1–148 (2015)
15. Flynn, J., et al.: Deepstereo: learning to predict new views from the world's imagery. In: Proceedings of the IEEE Conference on Computer Vision and Pattern Recognition (2016)
16. Ji, D., et al.: Deep view morphing. In: Computer Vision and Pattern Recognition (CVPR), vol. 2 (2017)
17. Garg, R., et al.: Unsupervised cnn for single view depth estimation: geometry to the rescue. In: European Conference on Computer Vision. Springer, Cham (2016)
18. Dosovitskiy, A., Tobias Springenberg, J., Brox, T.: Learning to generate chairs with convolutional neural networks. In: Proceedings of the IEEE Conference on Computer Vision and Pattern Recognition (2015)
19. Wu, J., et al.: Learning a probabilistic latent space of object shapes via 3D generative-adversarial modeling. In: Advances in Neural Information Processing Systems (2016)
20. Rezende, D.J., et al.: Unsupervised learning of 3D structure from images. In: Advances in Neural Information Processing Systems (2016)

21. Yan, X., et al.: Perspective transformer nets: learning single-view 3D object reconstruction without 3D supervision. In: Advances in Neural Information Processing Systems (2016)
22. Maas, A.L., Hannun, A.Y., Ng, A.Y.: Rectifier nonlinearities improve neural network acoustic models. In: Proceedings ICML, vol. 30, no. 1 (2013)
23. Kingma, D.P., Ba, J.: Adam: A method for stochastic optimization. arXiv preprint arXiv: 1412.6980 (2014)
24. Kim, J., et al.: Octree-based obstacle representation and registration for real-time. In: ICMIT 2007: Mechatronics, MEMS, and Smart Materials, vol. 6794. International Society for Optics and Photonics (2008)
25. Baldi, P.: Autoencoders, unsupervised learning, and deep architectures. In: Proceedings of ICML Workshop on Unsupervised and Transfer Learning (2012)
26. Intelligent Systems Research Institute Database of Household Objects (ISRI_DB). http://isrc. skku.ac.kr/DB3D/db.php

Detection of Suspicious Person with Kinect by Action Coordinate

Masaki Shiraishi[(⊠)] and Ryuya Uda

Tokyo University of Technology, 1404-1 Katakuramachi, Hachioji, Japan
g2117031c3@edu.teu.ac.jp, uda@stf.teu.ac.jp

Abstract. Surveillance cameras have been popular as measures against criminals. They are useful for deterrence of climes such as theft and murder, but they cannot prevent climes since real-time watching must be required for real-time detection of the climes. On the other hand, technologies of automatic recognition of people have been developed and the accuracy of image recognition by artificial intelligence has especially increased. That is, it is possible to detect crimes by installing artificial intelligence to surveillance cameras. However, a large amount of behavior data must be learned to determine a behavior pattern of a person by image recognition, and the data size is large since they are image data. Also, it is not confirmed whether a behavior pattern can be judged precisely when a part of the body is hidden by a shielding object such as another person. In this paper, we propose a method to judge a behavior patter of a person using Kinect in order to detect crimes at real-time. When a suspicious person is standing in front of the entrance door, there are cases where more than one person stands at the same time. In those cases, a part of the skeleton coordinates of the person who unlocks the door may be hidden, and it may be difficult to distinguish between behavior patterns. Therefore, in this research, we propose a method for detecting suspicious person even when skeleton coordinates are hidden by shielding objects. As a result of experiments which distinguish the picking operation from the unlocking operation by a key, the average of F-measure was 47%. We know that the score was quite poor to distinguish them. On the other hand, the average F-measure of some examinees showed more than 80%. We also mention about the value and the reason as consideration in this paper.

Keywords: Kinect · Position coordinate · Machine learning

1 Introduction

In recent years, cases such as theft and murder have been steadily occurring, and the number of occurrences is also increasing [1]. The number of placed surveillance cameras has been increasing in order to deter crimes or identify suspects after the incident. In addition, due to the development of computer technology such as image processing and artificial intelligence, suspicious people can be automatically detected and tracked not by real eyes but by computer eyes. There is a recent study by Mitsubishi Electric [2]. The company created a system in which suspicious people and socially vulnerable people in commercial facilities are detected with surveillance cameras using deep learning. Deep learning is an algorithm deepening the hierarchy of

© Springer Nature Switzerland AG 2019
S. Lee et al. (Eds.): IMCOM 2019, AISC 935, pp. 456–472, 2019.
https://doi.org/10.1007/978-3-030-19063-7_36

"neural network" which is a type of machine learning, and is an algorithm modeling neurons of the brain of an organism. In the research by Mitsubishi Electric, by defining things like sticks and baby carriages as attributes, videos are analyzed using deep learning and people who are taking suspicious behavior or who need help are classified.

Techniques for analyzing videos using deep learning can be dealt with by conventional techniques as long as systems are used in a limited small area. On the other hand, the purpose of our research is to set up a surveillance camera at entrance of a house to detect suspicious behaviors automatically. When deep learning is used to learn behavior or to update a learning model of behavior, it is necessary to collect and transmit a large amount of data from each surveillance camera to the center. When using video for learning behavior of people, it is not realistic to collect a large amount of data from each surveillance camera to the center via the Internet. Moreover, when deep learning is performed using images in a video, it is necessary to reduce the resolution of the images to reduce the size of data for learning by considering the learning speed, thereby lowering the accuracy for the classification of the behavior. Furthermore, when the video data leaks from the center, the privacy of people in the video is also infringed.

Therefore, in this research, we identify behavior of users by using Kinect which is a motion capture device capable of acquiring action coordinates, not image recognition. The action coordinates are smaller in data volume than video, and it is not difficult to concentrate data at the center via the Internet. Also, even if the information leaks out, the information contained in the data is only action information, so it is difficult to infringe the privacy of people.

There are many researches on behavior discrimination using Kinect and other sensor devices, but in many of these researches, skeleton coordinates of people are not hidden by shielding objects. Of course, there is a research in which skeleton coordinates are hidden in shields, but coordinates are only tracked and action patterns are not classified. Also, in these studies, since the positions of an examinee and a sensor device are fixed, highly accurate classification is not always possible when the installation place and environment are changed. Therefore, in this study, by using the convolutional neural network (CNN) which is one of the neural networks of deep learning, we classify behavior of people with high accuracy even when where a part of the skeleton coordinates of a person is hidden by shielding objects.

2 Related Researches

2.1 Behavior Classification of Kick and Punch by Kinect

In the research by Pang et al. they proposed a system to detect punching and kicking behavior using Kinect [3]. In their method, behavior is classified by joint position and velocity acquired from the joint coordinates. As a result of their experiments, it was able to detect punch with 95.83% accuracy and kick with 100% accuracy. However, in their method, machine learning is not used, and the behavior is defined by finely classifying the numerical values of the acquired joint coordinates, so it is necessary to define a new behavior with finely classifying new numerical values when the new behavior is added. Moreover, when adding new actions, it is necessary to confirm all of the definition of all existing actions. Furthermore, it is thought that it becomes difficult to define behavior when there are two or more similar behavior actions. Also, in their research, it is possible to compare the skeleton coordinates of a person since they are not hidden by obstacles. However, they do not mention the accuracy when some skeleton coordinates of the person are hidden in the shielding object.

2.2   Behavior Prediction by Skeleton Coordinates

In the research by Horiuchi et al. they propose a method that predicts the movement of a human body 0.5 s later using machine learning [4]. As a result of their experiments, the center of gravity of a human body can be predicted with the accuracy of 7.0 cm around when a person jumps. However, in their method, it seems impossible to predict when the user takes unexpected behavior. In other words, the next action can be predicted to some extent with a simple operation, but it may not work well for complicated actions or in actions when some skeleton coordinates of the person are hidden in the shielding object.

2.3   Detection of Disabled Person by Kinect

In the study of Dehbandi et al. they proposed a method to quantify the level of hand, arm and upper limb disorder by acquiring data using Kinect and machine learning [5].

In their method, a Wolf Motor Function Test, which is a test for healthy examinees to clinically verify upper limb function, was performed. In the test, healthy examinees emulated disability, and their motions were classified into healthy, mild disorder and moderate disorder. As a result of the experiment, the accuracy of healthy was 100%, that of mild disorder was 83.3%, that of moderate disorder was 91.7% and that of all items was 91.7%. The accuracy shows that they can be classified with high accuracy.

However, in their study, the movements were only classified in the difference between disabled and healthy people. Moreover, in the comparisons, Kinect was set at the same angle and the same distance against examinees. Furthermore, an examinee for learning and testing was the same person. Therefore, in their paper, it is not clear how movement between disabled person and healthy person is really different, and it is not applied to the detection of difference between suspicious person and a correct person. In the case of home surveillance cameras, there are cases where a person is not at the center of the picture, and the physique of people for the learning sample and that of criminal are different.

2.4   Motion Detection by Kinect

Yun et al. developed a data collection module using a Fresnel lens and an infrared sensor and proposed a method for detecting human motion [6]. In their experiments for their method, one data collection module was installed on the ceiling and two modules were also installed on two opposing walls of a corridor, and the acquired data were classified using a support vector machine to detect human motions. As a result of the experiments, the accuracy of the classification of the traveling direction, distance and speed was 94% or more. However, in their study, the motions are classified by traveling direction, distance and speed of an object, so it cannot classify behaviors that do not move like standing in front of the door.

2.5 Rate of Training Data and Test Data

In the research by Yamaguchi et al. intelligent data can be seamlessly processed even in distributed computing environments such as edge computing [7]. In evaluation of their study, 90% of the data set was used for training data and 10% is used for test data.

In the research by Kawamura et al. they propose an agent to classify the purpose of searching products as to whether the purpose of the search is for purchase or for collecting information when a user searches for a product [8]. In their study, 90% of the data set was used for learning data and the remaining 10% was used for test data.

Therefore, in our study, 90% of the data is used for learning and the remaining 10% is used for test of accuracy evaluation according to the rate of the research by Yamaguchi et al. and Kawamura et al. The test data is randomly extracted, replaced several times, and average and standard deviation are calculated.

3 Proposed Method

3.1 Overview of Proposed Method

In this paper, we propose a suspicious person detection method using Kinect. In our method, invasion of privacy as in the case of videos and images does not occur since Kinect acquires the skeleton coordinates of the human body by infrared rays. In addition, the surveillance cameras cost more than 20,000 yen, and the price becomes higher when the cameras can withstand security in real. On the other hand, Kinect can be bought at a low price around 15,000 yen. Moreover, Kinect can recognize a human body in three dimensions and can reproduce the joint information of the part hidden by obstacles, so it can be considered that the suspicious person can be detected by the recognition of actions of people.

There are various prevention measures against crimes at home. A suspicious person sometimes intrudes into the house from a window or a door. First of all, as a prevention measure against intrusion from windows, there is a method by attaching a film to a windowpane. It is inexpensive and it is possible to prevent the intrusion, so it is not necessary to protect the window by Kinect. On the other hand, in the case when a suspicious person intrudes a house from a door, a part of the skeleton coordinates of a person may be hidden when two or more people stand in front of the door.

Therefore, in this paper, by using Kinect, we propose a method to detect the suspicious person who is unlocking the door by picking even when a part of the skeleton coordinates of a person is hidden. First, the joint coordinates of human body are recognized and acquired as data. The action to unlock a door by the key and the action to pick are acquired as the data. The acquired data are shaped in a format, used for learning and test, and used for the evaluation of F-measure. A part of the skeleton coordinates of a person is shaped with zero padding in order to make the height of the person uniform.

3.2 Definition of Suspicious Behavior

First, it is necessary to define suspicious behavior in order to detect a suspicious person. In this paper, the suspicious behavior is defined to intrude a building such as a home or a facility by picking. Picking is to unlock the door with a wire-like special tool. Meanwhile, the other actions are to unlock the door using the key.

3.3 Installation Environment of Kinect

The installation environment of Kinect v2 is determined in the following two cases. We considered the environment was acquisition from diagonally behind and acquisition from the side by considering the position of the door since it was impossible to acquire the joint coordinates from the front of a person. The height of Kinect was 185 cm. Also, when placing a person in front of the door, the angle of Kinect was set diagonally to the right of the person, 30° to the door. The distance between the person and the Kinect was set to 300 cm. This is the best shooting position for Kinect, but the door prevents Kinect from setting at the distance in front of the person. In addition, it is necessary to watch hands of the person, so we set Kinect either sideways or diagonally behind. Also, the monitoring camera is generally hung from the ceiling, so we set Kinect at the height described above. If it was set at a very high position, it would be impossible to measure the joint positions of feet exactly, so it was 185 cm in this time. The details of the experiment environment are shown in Fig. 1.

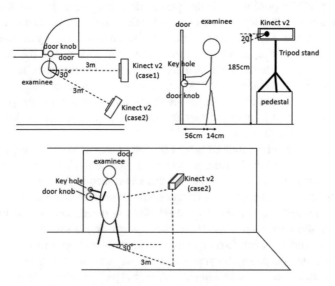

Fig. 1. Experiment environment of extracting action coordinates.

3.4   Acquisition of Joint Coordinates

The joint coordinates are acquired by skeleton tracking of Kinect. Kinect v2 can handle 25 joints of a human body, and each joint is stereoscopically expressed with X, Y and Z coordinate. In this research, we write the coordinates of 25 joints in one row, and X, Y and Z coordinate are delimited by commas. We call the row one frame and compile the frame with 40 sets as one sample.

3.5   Shaping of the Joint Coordinates Data

If the acquired data is used directly, personal information such as height which is contained in the data is sent to the data center. Therefore, we shape the data in a format in which all people are the same height in order to protect the privacy of people.

Also, we do it in order to decrease the amount of characteristics in height since not people but actions must be distinguished. In the data shaping, joint coordinates that are in contact with the door knob and the joint coordinates that are in contact with the floor are fixed, and other joint coordinates are changed. Also, in order to emulate the case where a part of the skeleton coordinates of a person is hidden, skeleton coordinates which are assumed to be hidden are set to zero. In this paper, we consider a case where only the skeleton coordinates of the palm and wrist can be acquired, so other skeleton coordinates are set to zero.

3.6   Joint Coordinate Data Generation

Joint coordinate samples are generated from samples of a pair of examinees to improve the F-measure of behavior classification. The samples are generated from the shaped data. Three samples are generated from two sample of a pair of examinees. In our research, we made the pairs with 10 examinees. The total number of the pairs is 45 in one experiment since samples of one examinee are used for test. When the samples of the 10th examinee are used for test, the pairs are made like that a pair of 1st and 2nd, 1st and 3rd, …, 8th and 7th, and 8th and 9th. The total number of generated samples is 27,000 from 2,000 original samples.

3.7   Network of Machine Learning

We evaluated the samples by convolutional neural network (CNN). In the CNN in our experiments, one-dimensional convolution was performed. The structure of this network is a convolution layer, a ReLU layer, a pooling layer, a convolution layer, a ReLU layer, a pooling layer, a fully-connected layer, a ReLU layer, a fully-connected layer, a ReLU layer, and a soft max layer. In addition, the joint coordinate data are three-axis coordinates of X, Y and Z coordinate, and 25 joint coordinates are used for the expression of a human body. One sample consists of 40 frames, and one frame consists of 25 joint coordinates values arranged in a line separated by commas. Therefore, the number of input units is one, and the number of units outputted by the soft max layer is two since it is binary classification. At this time, the batch size value to be used is chosen from two to the power of "n" between 32 and 512. The number of epochs is chosen from the value when F-measure is over 90%, and the value increases from 50 in increments of 50. The F-measure can be calculated as shown in Eq. (1).

$$\text{Fmeasure} = 2(P \cdot R)/(P + R) \qquad (1)$$

Also, those with a small standard deviation are used. When deciding the number of epochs and batch size, using 10-fold cross-validation method, 90% of the total data is used for training samples and the remaining 10% is used for test samples as described in Sect. 2.9. At this time, the average of F-measure in five times 10-fold cross-validation method is taken. In addition, when test samples consist of samples of one examinees and training samples consist of samples of the other examinees, each examinee in ten examinees is chosen for the test samples and F-measure is calculated in ten times.

3.8   Flow of Proposed Method

First of all, we confirm whether suspicious behavior for picking and unlocking action by key are classified or not and also confirm the batch size and the number of epochs whose average of F-measure exceeds 90%. In this paper, suspicious behavior is defined a rod-like movement and moves a rod up and down along a keyhole. Examinees are standing in front of a door. In case of suspicious action, examinees move a rod up and down along a keyhole with a right hand. In case of unlocking with a key, examinees insert the key in the keyhole with a right hand and unlock the door. In both cases, the left hand of examinees holds the door knob. Also, examinees stand in front of the door with a distance of 56 cm horizontally, and stand with their feet 14 cm apart slightly narrower than the shoulder width. This distance is decided as a distance that is easy to open the door. Then, examinees tuck their chin in and straighten themselves. Kinect is set 300 cm away from an examinee, and set based on the installation environment described in Sect. 3.3. When acquiring the behavior of an examinee by Kinect, before the acquisition of joint coordinates, the examinee turns in the direction in which Kinect is set. After the confirmation of starting the acquisition of joint coordinates, the examinee starts the actions. Action samples from turning in the direction of Kinect to starting the actions are acquired but thrown away.

The acquired data is shaped by scaling the coordinates to make all examinees have the same height since height must be eliminated from characteristics of examinees. By the shaping, height which is a kind of personal information of examinees is also eliminated so that the privacy of users is protected. The height is decided on the basis of the average height of the Japanese male in 2018 published by the Ministry of Internal Affairs and Communications [9]. Skeleton coordinates other than the palm and wrist are set to zero to reproduce the case where a part of the skeleton coordinates of a person is hidden by shielding object. The shaped text file data is put into two separate directories such as suspicious behavior and normal behavior. The average of F-measure is obtained five times by CNN using 10-fold cross-validation method. Then, we decide batch size and the number of epochs to be used in our experiments. The batch size is increased from 32 to 512 by double and double, and the number of epochs is incremented by 50 from 50 to the number whose F-measure exceeds 90%. At this time, we adopted the number when the standard deviation value is small. It should be noted that data of normal behavior is stored in the directory "0" and data of suspicious behavior is put in the directory "1".

Subsequently, in order to increase the number of examinees virtually and to increase the total amount of data, additional samples are generated based on the data from each pair of examinees. Then, data from one examinee are set to test samples, and the remaining data and the additional data are set to training samples. It should be noted that the data from an examinee for test samples are not used to generate the additional data. We evaluate the accuracy with the average of F-measure by ten times tests of each examinee.

4   Implementation

4.1   Skeleton Tracking

Acquisition of joint coordinates by skeleton tracking was performed in the following environment as described in Sect. 3.4. The OS was Windows 8.1, the IDE was Visual Studio 2015, the platform was x64, the OpenCV version was 3.1.0, and the programming language was C++.

Kinect used for skeleton tracking in this paper was the second generation Kinect v2. We acquired skeleton coordinates with Sensor, Source, Reader, Frame and Data in order [10]. First of all, Sensor defines the Sensor interface to handle Kinect, then Sensor is acquired and opened. Next, Source defines the Source interface for the Color frame, and then Source is acquired from the Sensor. Next, Reader defines the Reader interface for Color frame, and then Reader is opened from Source. Next, Frame and Data set the size of the Color image and the data, and then prepare the Open CV to handle the Color image. Also, the Frame interface for acquiring the Color image is defined, Frame is acquired from Reader, and the Color image is acquired from Frame. Then, the video is written in the video file. This video file is for checking whether the examinee is within the frame of Kinect. Color image is halved the size by cv::resize() since it is large in size to display a color image. Finally, Frame is released.

We explain the procedure of the acquisition of the joint coordinates of examinees. First, Frame interface for Body is defined and Frame is acquired from Reader. Then, preparation for the acquisition is completed by acquiring Body from Frame, and coordinates of each joint are acquired from an examinee.

Joint data is written in each text file for classification since Kinect v2 can acquire joint data up to six at the same time. Each sample is written in different file. One sample consists of 40 frames and one frame consists of one row with the X, Y and Z 25 joint coordinate separated by commas. The acquisition interval of joint coordinates for one row is 0.025 s in order to set the time for one sample to one second. It is necessary and sufficient time to acquire the behavior of examinees. The acquisition interval for one sample is 2 s.

When actually acquiring data, multiple samples are obtained at the same time in order to collect large amounts of data in a short time. First, when the program related to Kinect is run, a screen for entering the sample number and the number of samples to be acquired is displayed. After the entering the numerical values, the joint coordinates are acquired as data by Kinect. When data starts to be acquired, the examinee turns in a direction of Kinect, and then the examinee moves the action to the behavior of suspicious behavior or normal behavior. For this reason, samples immediately after starting data acquisition are not used for evaluation, and samples after the beginning of the action are used for it. We decided the time 5 s since it is necessary for 5 s at the longest to change the motion to the action. Here, the acquisition interval of samples is 2 s, and the time required to acquire one sample is about 1 s. Based on that, in our experiments, we set the number of samples taken at once to 25, and we repeated it twice to acquire 50 samples of data. Therefore, the number of acquisitions of joint data by Kinect is assumed to be 27, and the first two samples are not used for the evaluation, and the 25 samples after that are used.

4.2   Shaping of the Joint Coordinate Data

In this research, we shaped the data as described in Sect. 3.5. The data shaping was performed in the following environment. The OS was Windows 8.1, the IDE was NetBeans 8.0.1, and the programming language was Java.

First, two input and output arrays that store 3,000 values, which are values for one sample, are prepared. Also, an array for storing the number of rows for one sample was prepared. After that, text data are read, each row of the data is stored in one array, the values are separately retrieved by comma by comma using the split() method, and each value is stored in the array. After that, magnification for enlarging/reducing is found from the value of HEAD in order to set the height of all examinees to 170 cm. The magnification comes from the average of distance from the value of HEAD to the average value of FOOT_RIGHT and FOOT_LEFT. Finally, the height value is divided by 1.483041 in order to set the value to 170 cm since the height value is not 170 cm although the actual height calculated from examinees is 170 cm. After that, the height is scaled based on the obtained magnification, and stored in the output array. The scaled values are written in another text file as the same format as the original values. The data is shaped for 100 times while changing the text file name four times since there are 100 times 4 sets of data per examinee.

Specifically, joint coordinates that are in contact with the floor and joint coordinates that are in contact with the door knob are not changed during the shaping. That is, WRIST_RIGHT, HAND_RIGHT, WRIST_LEFT, HAND_LEFT, THUMB_RIGHT, HAND_TIP_RIGHT, THUMB_LEFT, HAND_TIP_LEFT, ANKLE_RIGHT, FOOT_RIGHT, ANKLE_LEFT and FOOT_LEFT are the same values as original values. We ignored the foot size since there was no correlation between height and foot size. When other joint coordinates are shaped, the magnification is obtained from HEAD, FOOT_RIGHT and FOOT_LEFT as mentioned above. In this shape, Y coordinate is changed since FOOT_RIGHT and FOOT_LEFT are in contact with the floor and hands are in contact with the door knob. The changed values of the coordinates are calculated with the magnification. Finally, they are obtained by adding the sum of the average of FOOT_RIGHT and FOOT_LEFT. ELROW_RIGHT and ELBOW_LEFT are calculated in the same way with HAND_RIGHT and HAND_LEFT.

Next, the hidden skeleton coordinates are set to zero by assuming that a part of the skeleton coordinates of a person is hidden by the shielding object. In this time, skeleton coordinates other than the coordinates of the palm and wrist are set to zero. That is, the eight joints of WRIST_RIGHT, HAND_RIGHT, WRIST_LEFT, HAND_LEFT, THUMB_RIGHT, HAND_TIP_RIGHT, THUMB_LEFT and HAND_TIP_LEFT are set to zero.

4.3   The Joint Coordinate Date Generation

A large amount of samples are generated from the shaped data of pairs of examinees as described in Sect. 3.6. The generation was performed in the following environment. The OS was Windows 8.1, the IDE was NetBeans 8.0.1, and the programming language was Java.

First of all, the average value of the same joint coordinate in the same row of each examinee is obtained. Next, the value of each examinee is subtracted from each value of joint coordinates of both examinees in a pair and divided by 2. Finally, the divided value is added to the value of joint coordinates of each examinee. The values of joint coordinates are used as generated values. Also, the average values of the value of the joint coordinates of both examinees in a pair are used as generated values.

4.4   Convolutional Neural Network

The structure of the CNN used in this study is shown in Fig. 2.

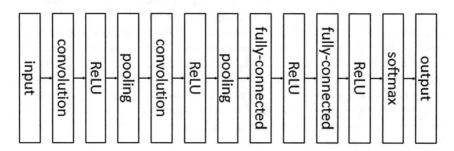

Fig. 2. Structure of CNN.

One-dimensional convolution is performed in this CNN and the network has 11 layers as described in Sect. 3.7. When a sample is inputted, it goes through a convolution layer, a ReLu layer, a pooling layer, a convolution layer, a ReLu layer, a pooling layer, a fully-connected layer, a ReLu layer, a fully-connected layer, a ReLu layer, and a soft max layer, and is outputted. More specifically, the number of input units is 1 for one sample. 25 joints times 3 coordinates are written in one row as one frame, and one sample consists of 40 frames. ConvolutionND of chainer.links is used for convolution. In the first layer of the convolution layer, the input channel is set to 1 since one sample is inputted, the output channel is set to 40 since the number of frames in one sample is 40, the filter size is set to 3, the stride is set to 3, the padding is set to 1. The second layer is the ReLu layer. The third layer is the average pooling layer, and the pooling size is set to 2. In the fourth layer of the convolution layer, the input channel is set to 40, the output channel is set to 20, the filter size is set to 3, the stride is set to 1, and the padding is set to 1. The fifth layer is ReLu. The sixth layer is the average pooling layer, and the pooling size is set to 2. The seventh layer is the fully-connected layer, the number of input units is set to 360, and the number of output units is set to 1,000 which is the number of numerical values for one sample. The eighth layer is the ReLu layer. The seventh layer is the fully-connected layer, the number of input units is set to 1,000, and the number of output units is set to 1,000. The tenth layer is the ReLu layer. The eleventh layer is the soft max layer, the number of input units is 1,000, and the number of output units is two since binary classification is performed. By the network with the layers above, we obtain the probability of the classification between normal behavior and suspicious behavior.

Next, the directory where the data to be stored for evaluation is set, and the list names of all the data are obtained from the directory. For test samples, one tenth of the total data is extracted from the data, and the list names are obtained. Then, the remaining list names are obtained for training samples. At this time, the training samples and the test samples are stored in float type 32 bit array. In addition, list names of training samples and test samples are stored in an int type 32 bit array. Both training samples and test samples are changed the dimension of the array to three dimensions of 1, 40 and 75 using reshape() method. Then, data sets are created with TupleDataset() method, and the samples are sequentially taken out with SerialIterator() method. Epoch, error function in learning or test, and F-measure are displayed during the operation. At the end of the learning, precision, recall, F-measure of each directory are displayed.

For the evaluation, if both the learning samples and the test samples contain samples of the same examinee, binary classification is carried out by taking the average F-measure by using 10-fold cross-validation method. The training samples are 90% of all samples, and the test samples are the remaining 10%. We assign numbers to the data, divide the data number by 10, and select the test samples by the remainder since it is necessary to extract test samples ten times so as not to overlap. On the other hand, when data of one examinee is included in the test samples and the remaining data is used for the learning samples, the average of the F-measure is obtained 10 times by changing examinees for the test samples.

5   Evaluation

5.1   Characteristic of Examinees

In this study, joint coordinate data was acquired from ten examinees. Height of the examinees is shown in Table 1.

Table 1. Height of examinees.

Examinee number	Body height [cm]
Examinee 1	169
Examinee 2	165
Examinee 3	163
Examinee 4	172
Examinee 5	161
Examinee 6	171
Examinee 7	170
Examinee 8	170
Examinee 9	170
Examinee 10	163

5.2 Decision of Epochs and Batch Sizes

In order to determine the number of epochs and the batch size to be used in the experiments, the average F-measure was calculated from the data of all examinees five times by using 10-fold cross-validation method. The number of samples per examinee was 200, and the total number of samples of all examinees was 2,000. In addition, samples from each examinee must be included both the learning samples and test samples. The number of test samples for picking was 100 and the number of test samples for normal unlocking was also 100. The number of epochs was increased from 50 by 50, the batch size was doubled from 32 to 1,800. We adopted the number of epochs and batch size when F-measure exceeded 90% and standard deviation was 5% or less. The results of F-measure when the number of epochs was fixed to 50 by changing the batch size are shown in Table 2.

Table 2. Results in 50 epochs and various batch size.

Epoch, batchsize		Precision	Recall	F-measure
Epoch 50, batchsize 32	Average	0.87	0.86	0.86
	Standard deviation	0.0732	0.0802	0.0828
Epoch 50, batchsize 64	Average	0.89	0.89	0.88
	Standard deviation	0.0402	0.0420	0.0410
Epoch 50, batchsize 128	Average	0.86	0.86	0.85
	Standard deviation	0.0364	0.0400	0.0387
Epoch 50, batchsize 256	Average	0.81	0.80	0.80
	Standard deviation	0.0337	0.0377	0.0400
Epoch 50, batchsize 512	Average	0.76	0.74	0.73
	Standard deviation	0.0347	0.0354	0.0382

The results of F-measure when the number of epochs was fixed to 100 by changing the batch size are shown in Table 3.

Table 3. Results in 100 epochs and various batch size.

Epoch, batchsize		Precision	Recall	F-measure
Epoch 50, batchsize 32	Average	0.92	0.92	0.92
	Standard deviation	0.0268	0.0248	0.0261
Epoch 50, batchsize 64	Average	0.92	0.92	0.92
	Standard deviation	0.0246	0.0271	0.0270
Epoch 50, batchsize 128	Average	0.92	0.92	0.92
	Standard deviation	0.0234	0.0247	0.0251
Epoch 50, batchsize 256	Average	0.90	0.90	0.90
	Standard deviation	0.342	0.0353	0.0350
Epoch 50, batchsize 512	Average	0.85	0.84	0.84
	Standard deviation	0.0281	0.0341	0.0342

In the results in Table 2, there was no case where the F-measure exceeds 90% in 50 epochs when changing the batch size. In the results in Table 3, there were cases where the F-measure exceeds 90% in 100 epochs when the batch size was 32, 64, 128 and 256. Among them, when the standard deviation was 5% or less, the batch size was 32, 64, 128, 256, and the standard deviation was 2.68%, 2.46%, 2.34% and 3.42%, respectively. Based on the above results, we decided that the batch size 128 at 100 epochs was used for our subsequent experiments since the standard deviation was the smallest when the average F-measure exceeded 90%.

5.3   F-measure Evaluation for Each Examinee

According to the decision in the previous section, we set 128 for the batch size and 100 for the number of epochs to our method as described in Sect. 3.8. We evaluated whether the action can be classified for each examinee by changing test samples of each examinee. The results of F-measure in behavior classification by changing test samples of each examinee are shown in Table 4. For each examinee, F-measure was obtained ten times, and the average F-measure and the standard deviation are shown.

Table 4. Results of test samples of each examinee.

Examinees number		Precision	Recall	F-measure
Examinee 1	Average	0.77	0.57	0.47
	Standard deviation	0.0239	0.0651	0.0961
Examinee 2	Average	0.68	0.56	0.49
	Standard deviation	0.0964	0.0466	0.0905
Examinee 3	Average	0.87	0.83	0.82
	Standard deviation	0.0464	0.0774	0.0844
Examinee 4	Average	0.31	0.33	0.29
	Standard deviation	0.129	0.0917	0.0681

(continued)

Table 4. (*continued*)

Examinees number		Precision	Recall	F-measure
Examinee 5	Average	0.55	0.55	0.54
	Standard deviation	0.059	0.0531	0.0595
Examinee 6	Average	0.89	0.87	0.87
	Standard deviation	0.0370	0.0463	0.0476
Examinee 7	Average	0.60	0.58	0.56
	Standard deviation	0.0736	0.0595	0.0860
Examinee 8	Average	0.64	0.64	0.63
	Standard deviation	0.0488	0.0467	0.0508
Examinee 9	Average	0.66	0.60	0.56
	Standard deviation	0.107	0.0653	0.0865
Examinee 10	Average	0.61	0.57	0.53
	Standard deviation	0.0559	0.0316	0.0489
All examinees	Average	0.77	0.57	0.47
	Standard deviation	0.0239	0.0651	0.0961

In the results in Table 4, the average F-measure of all examinees was 47%, that is, the actions could not be classified. The standard deviation was also 9.61% which was over 5%. Looking at each examinee, 29% was for one person, around 50% for 6, 63% for one person and over 80% for the remaining two. It indicates that it can be seen that the examinee 3 and the examinee 6 can be classified by the behavior patterns.

6   Consideration

In the results in Table 4, the average F-measure of all examinees is 47%, and the standard deviation is also 9.61%. Looking at each examinee, 29% was for one person, around 50% for 6, 63% for one person, 82% for examinee 3 and 87% for examinee 6. The 1-second action period in normal unlocking behavior for each examinee is shown in Table 5. The 1-second action period in picking unlocking behavior for each examinee is shown in Table 6.

Table 5. Action cycle in normal unlocking.

Examinees number	Number of time (time per second)					Standard deviation
	1	2	3	4	5	
Examinee 1	0.70	0.70	0.75	0.75	0.70	0.0245
Examinee 2	0.65	0.80	0.70	0.75	0.80	0.0583
Examinee 3	0.70	0.65	0.75	0.70	0.70	0.0316
Examinee 4	1.10	1.10	1.05	1.10	1.15	0.0316

(*continued*)

Table 5. (*continued*)

Examinees number	Number of time (time per second)					Standard deviation
	1	2	3	4	5	
Examinee 5	0.45	0.35	0.50	0.35	0.35	0.0632
Examinee 6	0.70	0.65	0.70	0.65	0.65	0.0245
Examinee 7	0.40	0.35	0.35	0.35	0.40	0.0245
Examinee 8	1.10	1.00	1.10	1.05	1.00	0.0447
Examinee 9	0.50	0.60	0.50	0.50	0.45	0.0490
Examinee 10	0.85	0.85	0.80	0.75	0.85	0.0400
All examinees	0.71					0.235

Table 6. Behavior cycle in unlocking behavior by picking

Examinees number	Number of time (time per second)					Standard deviation
	1	2	3	4	5	
Examinee 1	1.15	1.20	1.20	1.40	1.35	0.0245
Examinee 2	1.15	1.45	1.10	1.35	1.40	0.0583
Examinee 3	0.55	0.55	0.55	0.55	0.60	0.0316
Examinee 4	1.15	1.15	0.75	1.05	1.10	0.0316
Examinee 5	0.50	0.50	0.90	0.85	0.65	0.0632
Examinee 6	0.55	0.80	0.90	0.75	0.85	0.0245
Examinee 7	0.55	0.50	0.55	0.60	0.55	0.0245
Examinee 8	0.85	0.85	1.00	1.05	1.20	0.0447
Examinee 9	0.65	0.60	0.75	0.70	0.80	0.0490
Examinee 10	1.35	1.30	1.55	1.60	1.55	0.0400
All examinees	0.93					0.332

Examinee 3 and 6 with high F-measure in Table 5 mark similar values in Tables 5 and 6. The results indicate that examinees with high F-measure may appear when the number of examinees virtually increases by generating data with similar 1-second action period from the of the ten examinees. In other words, when an examinee with similar 1-second action period exists in training samples, the corresponding examinee in test samples will be classified with high F-measure. Or, when the number of examinees increases, the number of examinees with high F-measure will also increase.

7   Summary

Surveillance cameras are effective to deter theft and murder. However, they cannot prevent incidents in advance unless video is manually watching in real-time. Technologies of image recognition and artificial intelligence have made the surveillance cameras detect the incidents in real-time. However, the data size for learning has increased according to the increase of the size of video. Window grating and filming are also effective to prevent invasion from windows, and they are cheaper than the cost of Kinect. However, they cannot be applied to the entrance door since family members always go through the door. Therefore, we used Kinect to detect invasion from the entrance door. We obtained action coordinates rather than images for suspicious person detection by Kinect. Of course, there are existing researches in which human motions are classified using Kinect. However, in the researches, only big motions such as punch and kick are detected, or different motions of the same examinee are only classified. Moreover, in the researches, bodies of examinees are perfectly captured by Kinect from the best capturing position and there is no shielding object in front of the examinees. Therefore, we evaluated whether similar motions can be classified and whether the motions can be classified when a part of body is hidden by shielding object. Actually, we evaluated whether picking and normal unlocking was able to be classified when all skeleton coordinates except those of palm and wrist were set to zero.

Specifically, first, a total of 200 samples were acquired by Kinect. 100 samples of picking and 100 samples of normal unlocking were acquired from ten examinees. After that, the values of the joint coordinates were shaped by magnification based on the average height of a Japanese adult male, while the values of the joint coordinates in contact with the floor and the door knob were fixed. Then, the values of the joint coordinates except those of palm and wrist were set to zero. By using the data, we determined the number of epochs and the batch size used in the subsequent evaluation experiments by separating the data into learning samples and test samples and obtaining the average F-measure with 10-fold cross-validation method. The number of epochs and the batch size were selected when the average F-measure exceeded 90%. The number of epochs was determined to 100 and the batch size was determined to 128. Next, additional samples were generated based on the data of the ten examinees to increase the number of examinees virtually. We evaluated the data by setting test samples from data of one examinee and training samples from data of the other examinees. We used CNN and values of F-measure were outputted.

As a result, the average F-measure was 47%, the standard deviation was also 9.61%. On the other hand, some examinees showed high F-measure of more than 80%. We found that the examinees have an examinee whose cycle values of picking and normal unlocking were similar in training samples. We thought that it indicated that the more examinees were for learning the more examinees showed high F-measure would be. It is possible in our method since the size of joint coordinates data is quite smaller than that of video data to be sent to the center server. Also, joint coordinates of palm and wrist never leak the privacy of users. As future tasks of our research, we will virtually increase the examinees whose cycle values of picking and normal unlocking is various by linear interpolation.

References

1. Crime situation in 2018 and 2019 – Police department. https://www.npa.go.jp/toukei/seianki/h26-27hanzaizyousei.pdf. Accessed 22 Nov 2018
2. While refraining from the Olympic, suspicious person detection by AI. Mitsubishi Electric, you also support elderly people. http://style.nikkei.com/article/DGXMZO10563380S6A211 C1000000?channel=DF220420167276. Accessed 22 Nov 2018
3. Pang, J.M., Yap, V.V., Soh, C.S.: Human behavioral analytics system for video surveillance. In: 2014 IEEE International Conference on Control System, Computing and Engineering, pp. 23–28 (2014)
4. Horiuchi, Y., Shinoda, H., Makino, Y.: Computational Foresight: Forecasting Human Body Motion in Real-time for Reducing Delays in Interactive System. ISS (2017). (Japanese)
5. Dehbandi, B., Barachant, A., Harary, D., Long, J.D., Tsagaris, K.Z., Bumanlag, S.J., He, V., Putrino, D.: Using data from the microsoft kinect 2 to quantify upper limb behavior: a feasibility study. IEEE J. Biomed. Health Inform. **21**(5), 1386–1392 (2017)
6. Yun, J., Lee, S.S.: Human movement detection and idengification using pyroelectric infrared sensors. J. Sens. **14**(5), 8057–8081 (2014)
7. Yamaguchi, H., Yasumoto, K.: Smart Data Processing in Edge Computing. IEICE B Vol. J101-B, No.5, pp. 298–309 (2018). (Japanese)
8. Kawamura, T., Va, K., Nakagawa, H., Tahara, Y., Ohsuga, A.: Development of Web Agent to Extract Product Searching Intention according to Interaction Sequence. IEICE D, Vol. J94-D, No.11, pp. 1783–1790 (2011). (Japanese)
9. Statistics Japan. https://www.stat.go.jp/data/nenkan/65nenkan/zuhyou/y652402000.xls. Accessed 24 Nov 2018
10. Kinect for Windows v2 Isagoge – Serialization for C ++ Programmer. https://www.buildinsider.net/small/kinectv2cpp. Accessed 22 Nov 2018

Invariant 3D Line Context Feature for Instance Matching

Kyungsang Cho, Jaewoong Kim, and Sukhan Lee[✉]

Sungkyunkwan University, Suwon-si 16419, Republic of Korea
{mani0147, create, lshl}@skku.edu

Abstract. Conventional Approaches to line segment matching have shown their performances less satisfactory mainly since some of the features used for matching, such as the center and the starting/ending points of line segments, are not invariant. Furthermore, a pair of line segment sets to be matched may not have one to one correspondence, but each can be a subset of the other. This led to multiple solutions or interpretations in matching, where finding out all the possible solutions or interpretations out of an arbitrarily overlapping pair of line segment is of an issue. This paper presents a general method of identifying all the possible solutions or interpretations for an arbitrary pair of line segment sets by using invariant features associated with line segments. The invariant property of line segments comes from the orientation and location contexts of line segments that are defined based on infinite line representation of individual line segments. Simulation and experiment shown the effectiveness of the proposed method compared to conventional methods.

Keywords: 3D line matching · Directional/locational contexts ·
Multiple solutions or interpretations

1 Introduction and Problem Definition

In last decades, feature matching has always been proposed as a fundamental solution for applications such as 3D Reconstruction [1], 3D Registration [2, 3], Stereo SLAM [4, 5] and so on. Most of the existing solutions are based on local points or region features. These are not desirable for matching on poor environment which is non-textured environment. On the contrary, line segments provide relatively abundant even this situation. Line feature matching is indispensable for many applications and many different solutions have been proposed with successful results [6–8]. However, non-invariant characteristics that can be occurred in the line segment detection cause a problem of degrading the connectivity between a pair of line segments to be matched in two different scenes.

There are two main factors of deteriorating connectivity. First, locations of end points are not always invariant due to occlusion or noise. So effective point-to-point matching methods between end points are not applicable for line segment matching. Second, due to the producing fragmented line segments while detecting line segment, it causes problems that degrade geometric relationships. Second, line segments can be fragmented by shadow, occlusion or noise. It causes ambiguity when attempting geometrical approaches.

© Springer Nature Switzerland AG 2019
S. Lee et al. (Eds.): IMCOM 2019, AISC 935, pp. 473–485, 2019.
https://doi.org/10.1007/978-3-030-19063-7_37

This paper overcomes the above two factors and derives invariant properties: (1) Extend line segments infinitely to overcome positional errors and facilitate geometric access. (2) To create an invariant context regardless of the pose, every line segment has its own local coordinates and context. (3) Constructs a corresponding pair sets through a one-to-one comparison of all contexts. (4) Among the all interpretations, find an optimal pose which have most corresponding pairs.

In literature, there are many works based on used features. Rusu, et al. [9] use Fast point feature histograms (FPFH) for 3D registration. Guo et al. [10] use Rotational Projection Statistics (RoPS) for 3D object modeling. These features however tend to have high computational cost and are more suited for meshes rather than non-uniform point clouds. 3D registration based on line feature has been investigated for a long time. Most of the earlier works are either (1) Non-invariant to the definition of reference frame [11, 12]. (2) Approximate the solution [11]. (3) Only work under specific conditions [13]. It may be desirable to use line segments or planes to reduce computation while not losing 3D information.

In this paper, we propose a novel approach for multiple pose estimation based on 3D line segments that overcomes the problems of earlier works. Proposed approach is robust against uncertainties involved in extracting 3D line segments. This paper is organized as follows: Sect. 2 surveys recent works based on 3D line segments. Section 3 explains how we represent, select and refine extracted 3D line segments. Section 4 explains the process of context representation, matching them, estimating pose, and scoring candidate poses. Section 5 shows the experimental results using opened dataset. Section 6 discusses the results and future work.

1.1 Problem Definition

Extracted 3D lines are not invariant. They are partial segments that tend to have great deal of uncertainty as shown in Table 1. These uncertainties lead to mismatched correspondences or ambiguity interpretations. We introduce algorithms that overcome the

Table 1. Uncertainties of extracted 3D line segments.

Uncertainties in 3D lines segments	Challenges caused by uncertainties	Invariant methods used in proposed approach
Length variations	Length cannot be used as a base for matching	Extended and merged 3D lines
Location variations	Segments cannot be used directly for translation estimation	
Directional variations	Accommodate for matching direction tolerance	Use polar histogram with constant tolerance
Uncertainty in the direction sign	Incorrect estimation if direction is flipped	Consider both directions in matching
Symmetrical distribution of lines locations/directions	Produces many false matches	Prioritize poses with geometrical similarity by comparing each context*

* will be explained in the following sections.

non-invariant properties of the 3D line segment through representation of 3D line segments in Sects. 3 and 4. Furthermore, this paper introduces the 6DOF estimation method which is advantageous for 3D line matching.

2   Related Works

Parsi et al. [16] divide the problem of line matching into three categories with respect to the length of line segments: (1) Finite model, finite image. (2) Finite model, infinite image. (3) Infinite model, infinite image. They propose a novel approach to address single line segment matching in each case. In this paper, we are interested in case 3. Their proposed algorithm for case 3 arises when information about line segment lengths are either unavailable or ignored. Wang et al. [8] proposed a descriptor named Mean-Standard deviation Line Descriptor(MSLD) for line matching based on the appearance of the pixel support region. Meierhold and Schmich [15] present a method for estimating image orientation from line segments extracted from texture-less objects. They concluded that the accuracy of 3D lines is insufficient for image orientation. We believe that although the measured line segments do indeed have number of uncertainties, a set of selected lines can always infer enough information for a coarse rotation and translation estimation. Guan et al. [16] propose a 2-tuples based matching algorithm for the pose estimation of texture-less environments. They use the distance between the middle of the two line segments as an indicator to the distance between the 2-tuple lines and select 2-tuples with long lines and relatively close distance. They reject parallel 2-tuples as they produce intersecting point that is more sensitive to noise. Choi et al. [17] determine not to use Intensity-based edge detection for texture-less assembly objects. Intensity-based edge detection often gives too many edge pixels where only a few of them are useful compared to edges from depth discontinuities. They combined both point and line features for a coarse registration followed by ICP. Their 2-tuple descriptor consists of angle between lines and shortest distance between lines. Arth et al. [18] use LSD-based vertical lines of an image of surrounding buildings and façade to estimate the gravity vector of the camera. They align horizontal lines with horizontal vanishing point of façade 2D map of neighboring region obtained by smartphone's GPS, Compass, and Accelerometer from OpenStreetMap. A translation estimation process is used to complete the 6DOF camera pose estimation.

3   3D Line Representations and Preprocessing

3.1   3D Line Representations

We are interested in extracting 3D lines from captured 3D frames. There are many works in both 2D and 3D lines extraction. In this paper, we decided to use Kim's state-of-the-art Major line detection [19] algorithm for extracting 2D edges; and Lu's [20, 21] probabilistic approach for reconstructing 3D lines. In this paper, we represent each 3D line segment in two representation spaces.

Cartesian Space, Namely xyz Space

A 3D line segment is represented by 2 coordinates in xyz space representing its ends, as well as 2 additional coordinates in xyz space representing its extended infinitely ends.

Directional Vector Space, Namely ijk Space

A directional 3D line is represented by a point on the unity sphere in ijk space. This point is the end of a normalized vector started from the origin with the direction of the 3D line, i.e. for a line $\overrightarrow{AB}$, the ijk point is $|B - A|$. There is an inherit uncertainty in determining the sign of the direction of extracted line segment as it can be either $\overrightarrow{AB}$ or $\overrightarrow{BA}$. To allow this approach to be invariant to this uncertainty, we allow to using half-sphere by flipping vectors which are opposite side.

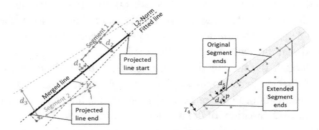

Fig. 1. Left: illustration of the method used for identifying and merging similar segments. Right: illustration of the method used to extend a line.

3.2 3D Line Preprocessing

True line may produce multiple extracted line segments which may have noisy directions along the true line. This effect happens due to poor illumination, occlusion, mildly contrasted edges, inaccuracy in localizing 2D edges, errors in the disparity map, missing/erroneous point cloud … etc. To provide more stable 3D lines to this algorithm, we are interested in estimating the original true 3D line from the extracted 3D line segments. We perform the following two preprocessing steps.

3D Line Segments Merging

Line segments $\overrightarrow{AB}$ and $\overrightarrow{CD}$ are said to be from the same line if: (1) The angle between the two segments $\gamma < T_1$. (2) The distance between the closest ends of the segments $d_1 < T_2$. (3) The shortest distance between each segment and the middle point of the other segment $d_2 \& d_3 < T_3$. Where T_i in this paper denotes to threshold i. Once two segments are identified as segments of the same line, we merge these segments by projecting the farthest ends of the segments onto L2-Norm fitted line, as shown in Fig. 1.

3D Line Segments Extension

A neighboring 3D point p is said to belong to a line segment and extends its ends if: (1) The shortest distance between the point and the line segment $d_4 < T_4$. (2) The shortest distance between the point and the closest extended segment end $d_5 < T_5$.

This algorithm extends 3D line segments if there are neighboring 3D points along its path as shown in Fig. 1. It's worth noting that extended lines are not necessarily physical true lines. They are however, consistent and invariant.

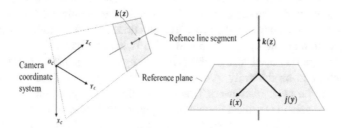

Fig. 2. Illustration of geometrical primitives for context.

4 Geometry Context and Matching

This section introduces representations of geometry context that contains angle/distance distributions of 3D line segments. We also propose a novel 3D line matching algorithm by comparing the geometry contexts. It can be approached in two ways that context representations, one is to describe whole segments in one context, and the other is to describe a context per a reference segment. In the former case, it may seem more appropriate because it reduces the hassle and is intuitive.

However, there are some difficulties to compare contexts. All factors of 6DOF should be considered unless strong assumptions or constraints are given. Also, this way depends on initial pose of two segment sets. Thereafter, the computation and the complexity will be increased Indispensably. In the latter case, the above difficulties can be overcome by using two assumptions: (1) matched pairs are always one-to-one corresponding; (2) The two reference segments that are the basis of the two contexts to be compared are corresponding truly. These two assumptions allow only two factors to be considered, unlike the former case, which had to consider the six elements. These two factors are indicators of rotation f_1 and location f_2 understandably. Specifically, RX, RY, and RZ are replaced by a rotation factor, and TX, TY, and TZ are replaced by a location factor, depending on geometry of the 3D line correspondences. These two factors are discussed in detail in Subsects. 4.1 and 4.2 respectively.

We classify geometry context into two categories: (1) directional context, which is to drive the rotation factor in ijk space; (2) locational context, which is to drive the location factor in xyz space. Even though they are different spaces, ijk and xyz share axial directions with each other. Let $k(z)$ be the axis of the reference line, and the axes of $i(x)$, $j(y)$ are $i(x) = C_Z \cdot k(z), j(y) = k(z) \cdot i(x)$, where C_Z is Z-axis of camera coordinate system. And there are primitives for describe context: (1) a plane through the origin of camera coordinate system and $k(z)$ is normal is reference plane. (2) the origin of xyz space is intersection point between the reference line segment and the reference plane. The primitives are shown in Fig. 2.

4.1   Directional Context

In Sect. 3, we have represented the direction of the 3D line segments as points on the unity sphere (ijk space). We grafted the horizontal coordinate system to explain the directional context. Each point has its own elevation and azimuth angle. Based on the second assumption, the corresponding pairs should have the same elevation angle and it is invariant. On the contrary, the azimuth angle does not have a unique value. In this subsection, obtaining the optimal azimuth value is the key issue.

Polar histogram is used to find the azimuth angle at which the contexts are most similar. Each point on the unity sphere is approximated to an ellipse with a Gaussian distribution on the polar histogram. This implies that the region considered for angle tolerance on the sphere is projected onto the i-j plane. Thickness of the ellipse will be in inverse proportion to distance between a point and origin. Also, Gaussian weight is assigned based on the distance from centroid of ellipse. Therefore, the Polar histogram is an index that describes the directional context and can be used to obtain the optimal azimuth angle or rotation factor f_1. These are shown in Fig. 3.

Through Polar histogram matching, corresponding pairs with the same direction can be listed. However, they are not one-to-one corresponding because excluded the locational information. Also, it is impossible that discrimination of parallel segments. And if there are parallel segments, it is impossible to decision.

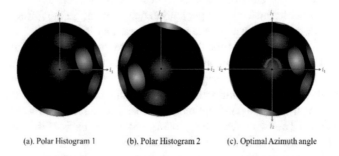

(a). Polar Histogram 1   (b). Polar Histogram 2   (c). Optimal Azimuth angle

Fig. 3. Two Polar histograms obtained in two different directional contexts are illustrated in (a) and (b). The optimal azimuth angle is illustrated in (c).

4.2   Locational Context

This section introduces the process of inferring the correspondence based on the positional relationship of 3D lines appearing in Cartesian space. The locational context is described using three projection planes. The three planes are perpendicular to the X, Y, and Z axes of the local coordinate system, respectively, and cross the origin. By projecting the corresponding candidates from the direction context to each of the three planes, they appear as 2D points and lines on the plane. Looking at the characteristics of the three planes mentioned above, the plane perpendicular to the Z axis (said as reference plane) is perpendicular to the reference line segment, while the other two planes (said as sub planes) contain the reference line segment. The line segment correspondences obtained by using rotation factor 1 should also match their geometric characteristics when projected onto three planes.

First, the decision is performed based on the elements projected onto the reference plane. In this plane, the reference line segment is projected to a point (said as reference point), and when comparing one pair of correspondence, it is possible to judge by comparing the distance and angle with this point. All line segments will be projected to the 2D line, except for lines parallel to the reference line segment. If projected onto a point, use the distance and angle to the reference point and, if projected on a line, do the same with the point in the shortest path.

Second, for sub-planes, the reference line segment will be projected to the 2D line. The corresponding candidate pairs projected on the plane should have the same angle and distance to the 2D line. The corresponding candidate pairs thus compared are left with only pairs having geometrically the same angle and position, and f_2 can be obtained based on the distribution of line segments projected on the sub-plane. Figures 4 and 5 shows the projection planes with projected elements respectively. And Fig. 6 shows the whole algorithm.

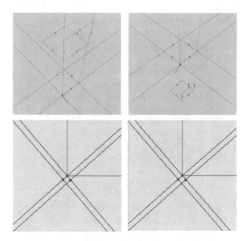

Fig. 4. Example for projection plane (first plane). The projected primitives are shape of points or lines on the plane.

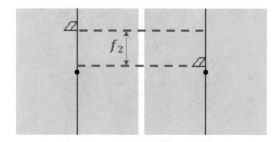

Fig. 5. Example for sub plane (perpendicular to X axis).

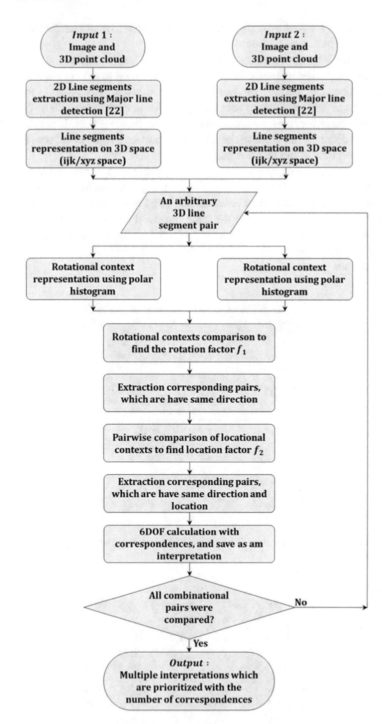

Fig. 6. The proposed whole algorithm.

5   Experimental Results

5.1   Experiment Using Augmented Noisy Lines

To analyze the effect of uncertainties of line extraction on this algorithm, we artificially added a Gaussian random noise to the location of each end points of each line to create an artificial new image. We then transform it with an arbitrary transformation matrix and used this algorithm to estimate that transformation. The following Table 2 shows the effect of noise on the estimated transformation.

Table 2. Transformation error with respect to the amount of artificial noise in line segments directions/locations.

STD (mm)	Rotation error (degrees)	Translation error (mm)
0	0.7	3
5	2.2	7
10	6.7	9
15	12.1	8
20	17.9	11
50	23.9	21

5.2   Experiment Using Stereo Dataset of Middlebury College

For statistical analysis of this algorithm reasoning accuracy; we hand-picked 4 scenes from Middlebury College's high-resolution stereo dataset [22], shown in Fig. 7 Each stereo scene contains camera calibration information and left/right scene's ground truth depth map.

The first interpretation estimated for all scenes up to 30% overlap of the two segment sets was the same as the ground truth. As the overlap ratio decreased, geometric ambiguity occurred.

Fig. 7. Matching results and ground truth depth images

5.3   Experiment Using RGB-D Data

In this experiment, we captured the point cloud of industrial object using 3D sensor. To analysis pose estimation accuracy and interpretation power, the RGB-D data were captured at varying angles. We placed the object on the turntable and measured the 6DOF error with ground truth and estimated multiple interpretations. Figure 8 shows the experiment environment.

The increase in angular difference naturally reduces the overlap region. We rotated the object up to 30° and analyzed the estimated interpretations. And Fig. 9 shows pose error graph. The two elements (circles and X) represent the true matching ratio according to the Overlap ratio. And the other three elements (diamond, square and triangle) are prioritized interpretations. Even if the angle difference increases, the first pose has little difference from the ground truth. And detail errors are shown in Table 3.

Fig. 8. Left: industrial object for experiment, right: experimental environment (industrial object placed on turntable).

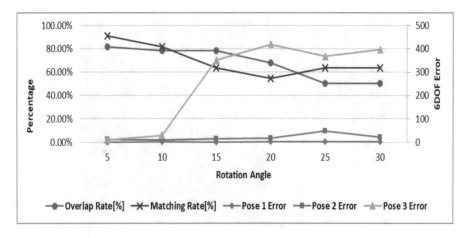

Fig. 9. 6DOF errors of interpretations, which are estimated per 5° rotated.

Table 3. Rotation and translation errors in detail about each interpretation.

Reference Image	Query Image	Rotation angle	1st Pose	2nd Pose	3rd Pose	Overlap
			Rotation Error[°]/ Translation Error[mm]/ Matching Rate			
		5°	1.07/ 1.05/ 90.9%	1.29/ 12.59/ 54.5%	0.85/ 10.81/ 45.4%	81.40%
		10°	1.96/ 1.86/ 81.8%	5.32/ 8.88/ 36%	2.45/ 29.06/ 27%	78.50%
		15°	1.05/ 1.51/ 63.6%	12.77/ 15.12/ 27.2%	19.82/ 350.58/ 27.2%	78.50%
		20°	0.12/ 2.65/ 54.5%	3.88/ 17.34/ 45.4%	58.91/ 418.23/ 45.4%	67.80%
		25°	2.50/ 3.27/ 63.6%	7.74/ 47.34/ 36%	59.52/ 367.42/ 36%	50.50%
		30°	1.64/ 1.77/ 63.6%	1.74/ 21.48/ 45.4%	3.4/ 395.58/ 45.4%	50.50%

6 Conclusions

In this paper we presented a novel algorithm for 3D line segments matching based on geometry context. We have conducted experiments on various situations such as indoor environments, poor text scenes interpretation and pose estimation of non-texture object. We have demonstrated the improvement of non-texture environment interpretations using proposed algorithm. However, performance of proposed approach is vulnerable to the absence of edges.

Future work revolves around improving the algorithm to be parameter-free as it adaptively determines the required thresholds based on the environment, used sensor, and object shape. We also aim to improve discrimination of true pose when ground truth is not provided. Extended line segments are having no distance information, but we are aiming to utilize that. Using the overlap or non-overlap ratio of the correspondences of each interpretation, the ground truth probability can be obtained.

Acknowledgements. Sukhan Lee proposed and guided the algorithm for contexts matching, whereas implementation and experimentation are carried out by Kyungsang Cho and Jaewoong Kim. This research was supported, in part, by the "Technology Innovation Program (or Industrial Strategic Technology Development Program, Report Number: 10048320)", sponsored by the Ministry of Trade, Industry & Energy (MOTIE), in part by the MSIP under the space technology development program (NRF-2016M1A3A9005563) supervised by the NRF and Basic Science Research Program through the National Research Foundation of Korea (NRF) funded by the Ministry of Education (2017R1A6A3A11036554). This project is also supported by the "Project of e-Drive Train Platform Development for small and medium Commercial Electric Vehicles based on IoT Technology" of Korea Institute of Energy Technology Evaluation & Planning (KETEP) (20172010000420), sponsored by the Korea Ministry of Trade, Industry & Energy (MOTIE).

References

1. Newcombe, R.A., Izadi, S., Hilliges, O., Molyneaux, D., Kim, D., Davison, A.J., Kohli, P., Shotton, J., Hodges, S., Fitzgibbon, A.: KinectFusion: Real-time dense surface mapping and tracking. In: Proceedings of the International Symposium on Mixed and Augmented Reality (ISMAR) (2011)
2. Zeng, A., Song, S., Niessner, M., Fisher, M., Xiao, J., Funkhouser, T.: 3Dmatch: Learning local geometric descriptors from RGB-D reconstructions, arXiv preprint arXiv:1603.08182 (2016)
3. Jaiswal, M., Xie, J., Sun, M.T.: 3D object modeling with a Kinect camera. In: 2014 Asia-Pacific Signal and Information Processing Association Annual Summit and Conference (APSIPA), Siem Reap, pp. 1–5 (2014)
4. Gomez-Ojeda, R., Zuñiga-Noël, D., Moreno, F.A., Scaramuzza, D., GonzalezJimenez, J.: PL-SLAM: a stereo SLAM system through the combination of points and line segments. arXiv: 1705.09479 (2017)
5. He, Y., Zhao, J., Guo, Y., He, W., Yuan, K.: PL-VIO: tightly-coupled monocular visual-inertial odometry using point and line features. Sensors **18**, 1159 (2018). https://doi.org/10.3390/s18041159
6. Zhang, L., Koch, R.: An efficient and robust line segment matching approach based on LBD descriptor and pairwise geometric consistency. JVCI **24**, 794–805 (2013)
7. Li, K., Yao, J.: Line segment matching and reconstruction via exploiting coplanar cues. ISPRS J. Photogrammetry Remote Sens. **125**, 33–49 (2017)
8. Wang, Z., Wu, F., Hu, Z.: MSLD: a robust descriptor for line matching, PR **42**, 941–953 (2009)
9. Rusu, R.B., Blodow, N., Beetz, M.: Fast point feature histograms (FPFH) for 3D registration. In: IEEE International Conference on Robotics and Automation, pp. 3212–3217 (2009)
10. Guo, Y., Sohel, F., Bennamoun, M., Wan, J., Lu, M.: An accurate and robust range image registration algorithm for 3D object modeling. IEEE Trans. Multimedia **16**(5), 1377–1390 (2014)
11. Daniilidis, K.: Hand-eye calibration using dual quaternions. Int. J. Robot. Res. **18**(3), 286–298 (1999)
12. Kamgar-Parsi, B.: An open problem in matching sets of 3D lines. In: IEEE Conference on Computer Vision and Pattern Recognition, vol. 1, pp. 651–656, December 2001

13. Guerra, C., Pascucci, V.: On matching sets of 3D segments. In: Conference on Vision Geometry, vol. 3811, pp. 157–167, July 1999
14. Kamgar-Parsi, B.: Algorithms for matching 3D line sets. In: IEEE Transactions on Pattern Analysis and Machine Intelligence, vol. 26, no. 5, pp. 582–593, May 2004
15. Meierhold, N., Schimch, A.: Referencing of images to laser scanning data using linear features extracted from digital images and range images. In: International Archives of Photogrammetry, Remote Sensing Spatial Information Science, vol. 38, 3/W8, pp. 164–170, September 2009
16. Guan, W., Wang, L., Mooser, J., You, S., Neumann, U.: Robust pose estimation in untextured environments for augmented reality applications. In: 8th IEEE International Symposium on Mixed and Augmented Reality ISMAR. Orlando, FL, pp. 191–192 (2009)
17. Choi, C., Taguchi, Y., Tuzel, O., Liu, M.Y., Ramalingam, S.: Voting-based pose estimation for robotic assembly using a 3D sensor. In: IEEE International Conference on Robotics and Automation (ICRA), Saint Paul, MN, pp. 1724–1731 (2012)
18. Arth, C., Pirchheim, C., Ventura, J., Schmalstieg, D., Lepetit, V.: Instant outdoor localization and SLAM initialization from 2.5D maps. IEEE Trans. Vis. Comput. Graph. 21(11), 1309–1318 (2015)
19. Kim, J., Lee, S.: Extracting major lines by recruiting zero-threshold canny edge links along sobel highlights. IEEE Sig. Process. Lett. 22(10), 1689–1692 (2015)
20. Lu, Z., Baek, S., Lee, S.: Robust 3D line extraction from stereo point clouds. In: IEEE Conference on Robotics, Automation and Mechatronics, Chengdu, pp. 1–5 (2008)
21. Nguyen, T.B., Sukhan, L.: Accurate 3D lines detection using stereo camera. In: IEEE International Symposium on Assembly and Manufacturing ISAM. Suwon, pp. 304–309 (2009)
22. Scharstein, D., Hirschmüller, H., Kitajima, Y., Krathwohl, G., Nešić, N., Wang, X., Westling, P.: High-resolution stereo datasets with subpixel-accurate ground truth. In: German Conference on Pattern Recognition (GCPR), September 2014

A Novel Performance Metric
for the Video Stabilization Method

Md Alamgir Hossain and Eui-Nam Huh[✉]

Kyung Hee University, Yongin-si 17104, Republic of Korea
johnhuh@khu.ac.kr
http://www.icnslab.net/#/home

Abstract. Video stabilization methods cannot compensate entire unwanted motions of an unstable video. In this paper, we present a new performance metric which can effectively measure the stability of a video. Our hypothesis is that in case of non dynamic background, the background motion is zero without the foregrounds of a stable video when a video taken by a fixed camera. Firstly, the foregrounds of a stabilized video are discarded, then, each background pixel motion is determined between the two consecutive background frames in two directions separately, afterwards, an average motion of each pixel is computed. These mean motions determine the stability of a video. The more unstable video will have more background motions. The background motion is a criterion to evaluate the stability of a video. The experimental results prove the efficacy of our proposed approach.

Keywords: Video stabilization performance metric ·
Video quality assessment · Mean Displacement Error

1 Introduction

The video stabilization (VS) is a very demanding task nowadays [1]. The quality of a video not only depends on the pixel quality but also relies on the stability of a video. The video registration has major applications in the motion analysis, areal video surveillance, autonomous vehicle navigation, model based compression, structure from motion, and mosaicking [1]. Microsoft [2–4], Facebook [5], and Adobe [6] have been doing much research in this area. The quantitative assessment of the VS method is a difficult task due to the lack of ground truth for undesired motions [7]. According to the Human Visual System (HVS) [8], a small amount of displacement is very difficult to detect.

Liu et al. [4,9] use distortion, cropping ratio, and stability criteria to measure the performance of the VS methods. The researchers mention that the stability corresponds to the high frequency components of a video. Zhang et al. [10] consider time complexity, accelerated components such as translation and rotation, distortion after video completion, and cropping ratio criteria due to lack of generally accepted evaluation metric. Niskanen et al. [8] apply divergence, jitter, and

© Springer Nature Switzerland AG 2019
S. Lee et al. (Eds.): IMCOM 2019, AISC 935, pp. 486–496, 2019.
https://doi.org/10.1007/978-3-030-19063-7_38

blur criteria to test VS accuracy, here, jitters represent a high frequency component in a video and the divergence estimates latency between consecutive frames. Besides, the researchers propose the Point Spread Function (PSF) to estimate the blur introduced by the video stabilization method. At first, the researchers find an intentional camera path by removing high frequency components up to 12 frequency components to get an intentional camera motion path (stable video). Then, they add some random noises to simulate an unstable video. After that, the synthetically unstable videos are processed by different VS algorithms. Finally, they calculate distance as a *Mean Square Error (MSE)* between the intentional camera motion path and the processed path (processed by VS algorithms). Zhang et al. [1] also add some known motions and some noises (Gaussian or Salt and Pepper noises) to each frame of a stable video separately to create a synthetically unstable video. Then, the unwanted motion estimation of the synthetically unstable videos is estimated by the different VS methods. Finally, they compute the error *(MSE)* between the given ground truth motion and the estimated motion of the motion estimation phase of different VS algorithms. The researchers only estimate the motion estimation error and they do not consider the transformation and accumulation errors of the VS methods.

The foregrounds in a video are one of the main cause of a video stability failure [11–13]. Most importantly, removing foreground after stabilization affects the accuracy of video stabilization [11–13]. Liu et.al [11] tell that rolling shutter distortion is a noise and it cannot be modeled accurately. Also, Grundmann et al. [14] mention that compensating rolling shutter effect [2,15,16] caused by Complementary Metal Oxide Semiconductor (CMOS) is a crucial task in video registration than a video captured by a Charge Coupled Device (CCD) camera. The VS approach cannot remove the wanted motions entirely that our proposed algorithm counts.

The sequence of video frame transformations in the video stabilization algorithms follow a background of a reference frame until the reference frame is changed. If a video is completely stable, no undesired motions will be remaining in the video except the foreground motion [17]. Motivated by the stabilization process of the video stabilization algorithms, in this paper, we use background motion which is key to judge the amount of instability of a video. At first, we remove the foregrounds from a stable video, then 2 directional (vertical and horizontal) displacements are determined between the consecutive background frames for each pixel separately. After that, the L^2 norm distance is estimated for each pixel. All the calculated L^2 norm distances are summed up between the two consecutive background frames and averaged out the summation. The average value is the mean displacement of a pixel between the two consecutive frames. The displacement is named as a *Mean Displacement Error (MDE)*. Finally, the total *Mean Displacement Errors (MDEs)* are averaged out based on the number of frames. The research has mainly two contributions: (1) The proposed method can quantitatively assess the stability of the VS method. (2) The proposed research evaluates the performance of the three up to date VS approaches. The proposed system model is depicted in Fig. 1.

In the rest of the paper, methods formulation, the simulated unstable video creation methods, and the pseudo code of our proposed approach are represented in Sect. 2. In Sect. 3, in addition to the dataset description and the comparison discussion, the section explains the results of the proposed method with the synthetically created videos and the real videos. Conclusions and future works of our proposed method are elaborated in Sect. 4.

2   Methodology

2.1   Metric Methodology

In this section, we describe our proposed performance metric model. The model consists of basic three steps: (1) *The Foreground Determination* (2) *The Background Separation* (3) *The Displacement Error*. Figure 1 displays the details of our proposed approach.

(1) **The Foreground Determination:** At first, the stabilized input video is converted into a grayscale $G_t(x,\ y)$ video. For the foreground determination, we use the SuBSENSE background subtraction algorithm [18], because the algorithm effectively models the dynamic background. For each background pixel, a set of N number of samples $BG_N(x,\ y)$ is stored. In case of the current frame $G_t(x,\ y)$, the foreground $FG_t(x,\ y)$ will be following,

$$FG_t(x,\ y) = \begin{cases} 1, & \{D(G_t(x,\ y),\ BG_n(x,\ y)) < R,\ \forall n\} < \#_m \\ 0, & \text{otherwise} \end{cases} \tag{1}$$

One and zero indicate the foreground and the background pixel respectively. $D(G_t(x,\ y),\ BG_n(x,\ y))$ is the distance between the background samples $BG_N(x,\ y)$ and the current sample $G_t(x,\ y)$, R represents distance threshold, $\#_m$ is the minimum number of matching.

(2) **The Background Separation:** The foreground $FG(x,\ y)$ is removed from the input frame $G(x,\ y)$ at time t and at time $t+1$, then, the background $BG(x,\ y)$ is as,

$$BG_t(x,\ y) = |G_t(x,\ y) - FG_t(x,\ y)| \tag{2}$$
$$BG_{t+1}(x,\ y) = |G_{t+1}(x,\ y) - FG_{t+1}(x,\ y)| \tag{3}$$

In case of the N number of image sequences $G_N(x,\ y)$, the N number of background samples $BG_N(x,\ y)$ is found.

(3) **The Displacement Error:** The vertical directional vector **u** and the horizontal directional vector **v** are estimated using the Farnebäcks dense optical flow algorithm [19]. A pixel $I(x,\ y,\ t)$ displaces the distance dx and dy in the next frame at time dt. According to the optical flow algorithm,

$$I(x,\ y,\ t) = I(x + dx,\ y + dy,\ t + dt)$$

Using the Taylor series of $I(x+dx,\ y+dy,\ t+dt)$, discarding the common term, and diving by dt, we get [20],

$$f_x u + f_y v + f_t$$

where, the image gradients fy and fx, and the displacements $\mathbf{v}$ and $\mathbf{u}$ will be,

$$f_x = \frac{\partial f}{\partial x}, \ f_y = \frac{\partial f}{\partial y}, \ u = \frac{dx}{dt}, \ v = \frac{dy}{dt}$$

The displacement vectors $\mathbf{v}$ and $\mathbf{u}$ are estimated for each pixel separately. We will get $M \times N$ numbers of $\mathbf{v}$ and $\mathbf{u}$ for the $M \times N$ pixels image frame. Then, the displacement errors are estimated as a Euclidean distance (L^2 norm) like the following,

$$||d_k||_2 = \sqrt{||u_i||^2 + ||v_j||^2}, \ k = M \times N \tag{4}$$

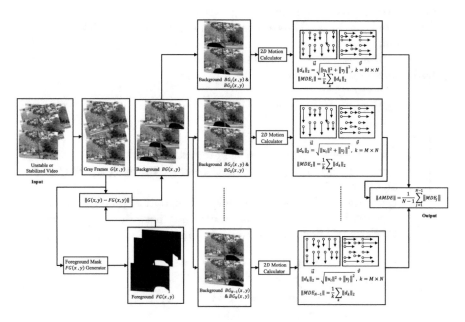

Fig. 1. A video stabilization performance metric model

The *Mean Displacement Error* of $||MDE_1||$ between the background frames $BG_1(x, y)$ and $BG_2(x, y)$, $||MDE_2||$ between the background frames $BG_2(x, y)$ and $BG_3(x, y)$, and $||MDE_{N-1}||$ between the background frames $BG_{N-1}(x, y)$ and $BG_N(x, y)$ are approximated as,

$$||MDE_1|| = \frac{1}{M \times N} \sqrt{\sum_{i=1}^{M} \sum_{j=1}^{N} ||u_i||^2 + ||v_j||^2} \tag{5}$$

$$||MDE_2|| = \frac{1}{M \times N} \sqrt{\sum_{i=1}^{M} \sum_{j=1}^{N} ||u_i||^2 + ||v_j||^2} \tag{6}$$

$$||MDE_{N-1}|| = \frac{1}{M \times N} \sqrt{\sum_{i=1}^{M} \sum_{j=1}^{N} ||u_i||^2 + ||v_j||^2} \tag{7}$$

Finally, we measure the *Average Mean Displacement Error (AMDE)* with the following Eq.,

$$||AMDE|| = ||MDE_1|| + ||MDE_2|| + \cdots + ||MDE_{N-1}||$$

$$||AMDE|| = \frac{1}{N-1} \sum_{j=1}^{N-1} ||MDE_{j-1}||$$

(8)

Algorithm 1. The pseudocode of the proposed model

Input : *Video was taken after stabilization*
Output: *MDE, AMDE*

1 read the *previousframe* from a video;
2 convert the *previousframe* into the grayscale;
3 remove foreground pixels from the *previousframe* using SubSENSE algorithm;
4 $amde \leftarrow 0$, $sum\_mde \leftarrow 0$, $count\_mde \leftarrow 0$;
5 **while** *video frame* $\neq 0$ **do**
6 | read the *nextframe*;
7 | **if** *nextframe empty* **then**
8 | | break;
9 | **else**
10 | | convert the *nextframe* into the grayscale;
11 | | remove foreground pixels from the *nextframe* using the SubSENSE algorithm ;
12 | | calculate the Gunnar Farnebäck's dense optical flow vectors u and v in the vertical and horizontal direction between the consecutive background *previousframe* and *nextframe* ;
13 | | $||d_k||_2 \leftarrow 0$, $sum\_d \leftarrow 0$, $mde \leftarrow 0$;
14 | | **for** $i \leftarrow 1$ **to** N **do**
15 | | | **for** $j \leftarrow 1$ **to** M **do**
16 | | | | $||d_k||_2 \leftarrow sqrt(u*u + v*v)$;
17 | | | | $sum\_d \leftarrow sum\_d + ||d_k||_2$;
18 | | | **end**
19 | | **end**
20 | | $mde \leftarrow sum\_d/(N*M)$;
21 | | $count\_mde + +$;
22 | | *print mde*;
23 | | $sum\_mde \leftarrow sum\_mde + mde$;
24 | | $previousframe \leftarrow nextframe$;
25 | **end**
26 **end**
27 $amde \leftarrow sum\_mde/count\_mde$;
28 *print amde*;

The unit of *MDE* and *AMDE* is pixel. The *MDE* is the mean displacement of each pixel per frame, and the *AMDE* represents an average of *MDE*. A zero value of *MDE* or *AMDE* means that the video is completely stable. A less value of *MDE* or *AMDE* is better for the stability of a video. The pseudocode of our proposed metric is shown in Algorithm 1.

2.2   Simulated Shaking Method

The simulated shaking videos are used to test our method accuracy. We take seven different images, then the different motions (translation, rotation, and scaling) are added to each frame to create a shaking video using the following translation and rotation (TR) and scaling, rotation, and translation (SRT) transformation. To create the videos, each pixel (x, y) of each frame is transformed by adding a translational vector $\langle h, k \rangle$, and rotated at an angle θ. The transformation will be,

$$\begin{bmatrix} x' \\ y' \\ 1 \end{bmatrix} = \begin{bmatrix} \cos\theta & -\sin\theta & h\cos\theta - k\sin\theta \\ \sin\theta & \cos\theta & h\sin\theta + k\cos\theta \\ 0 & 0 & 1 \end{bmatrix} \cdot \begin{bmatrix} x \\ y \\ 1 \end{bmatrix} \tag{9}$$

$$\begin{bmatrix} x' \\ y' \\ 1 \end{bmatrix} = \begin{bmatrix} s\cos\theta & -s\sin\theta & h \\ s\sin\theta & s\cos\theta & k \\ 0 & 0 & 1 \end{bmatrix} \cdot \begin{bmatrix} x \\ y \\ 1 \end{bmatrix} \tag{10}$$

3   Results and Discussion

To test a background motion of a stable video, we take the changedetection dataset in 2014 [21]. Each video in the changedetection dataset has more than thousand of frames. Besides, the seven different images are used to simulate the unstable videos. The video database of the five different categories (regular, quick rotation, crowd, parallax, and running) [22] is taken to estimate the stability of the videos which are stabilized by the three best methods named VirtualDub-Deshaker video stabilizer, YouTube stabilizer, and Adobe warp stabilizer.

3.1   Discussion

Our hypothesis is that non dynamic background motions will be zero if the videos are perfectly stable. Results from Table 1 show that the non dynamic background motions are approximately zero. Additionally, we observe the motions between the same video frames and we find *0.00019, 0.00051, 0.00034 MDEs*[1]. Besides, the background motions of the unstable and stabilized videos (stabilized by the VS method) of the same videos are experimented which is tabulated in Table 2 and shown in Fig. 2. To illustrate the difference between the stable and the unstable video, we depict the unstable and stabilized background motions frame by frame of a video named 2.avi (Fig. 3). Consequently, these results prove

---

[1] Mean Displacement Error.

our hypothesis. Therefore, background motions of a stable video are zero. In Tables 3, 4 and Figs. 4, 5 show the small differences between our added motions and the measured motions which are 0.126 $AMDE^2$ and 0.130 $AMDE$ respectively. The differences establish the evidence of our method robustness.

Afterwards, we compare the three methods (VirtualDub-Deshaker video stabilizer, YouTube stabilizer, and Adobe warp stabilizer) in respect of the stability of the stabilized videos. Firstly, the foregrounds are thrown away from the stabilized videos. Then, the stability of the videos is counted. The experiments are carried on the five different categories (regular, quick rotation, crowd, parallax, and running) videos separately. The regular, quick rotation, crowd, parallax, and running case videos are 10, 5, 14, 11, 10 in numbers. For simplicity, we only show the summary of the performance (Table 5, Fig. 6). The Adobe stabilizer and the Deshhaker stabilizer are equally success in two cases. The Deshhaker stabilizer is the best in the crowd and the running cases videos, whereas the Adobe stabilizer is finest in the quick rotation and running cases videos. The YouTube stabilizer gives the highest quality only in the case of regular videos. The Adobe stabilizer is slightly better than the Deshaker stabilizer, but, the Adobe stabilizer and the Deshaker stabilizer are almost equally efficacious.

Table 1. Background motion

Sl.	Video name	Only_BG Motion (AMDE)
1	Base/highway	0.0730
2	Base/office	0.0782
3	Base/pedestrians	0.0943
4	Shadow/busstation	0.0691
5	Shadow/backdoor	0.0816

Table 2. BG motion of the unstable and stabilized videos

Sl.	Video name	Unstable BG_Motion (AMDE)	Stable BG_Motion (AMDE)
1	Simple/0.avi	2.125	0.321
2	Simple/2.avi	2.423	0.207
3	Simple/4.avi	0.975	0.582
4	Simple/5.avi	1.451	0.255
5	Simple/8.avi	1.812	0.311
6	Crowd/1.avi	2.343	1.693
7	Crowd/2.avi	1.762	1.032
8	Crowd/4.avi	1.725	0.841
9	Crowd/14.avi	1.837	0.721
10	Parallax/5.avi	1.322	0.427
11	Parallax/13.avi	2.007	0.966
12	Parallax/17.avi	2.181	0.511
13	Running/3.avi	3.181	1.334
14	Running/15.avi	2.996	1.022
15	Running/16.avi	2.678	0.788

---

[2] Average Mean Displacement Error.

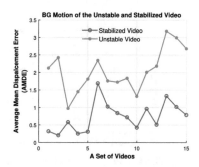

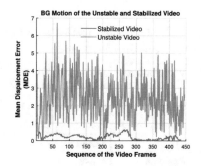

Fig. 2. Unstable and stabilized BG motions for a set of videos

Fig. 3. Frame by frame unstable and stabilized BG motions

Table 3. Accuracy for TR

Sl.	Video name	Ground truth (AMDE)	Our result (AMDE)	Diff. (AMDE)
1	leftTR	5.628	5.669	0.041
2	rightTR	5.628	5.687	0.061
3	lennaTR	3.330	3.122	0.208
4	ScheffaTR	6.546	6.418	0.128
5	sflowerTR	5.482	5.704	0.222
6	indoorTR	5.272	5.369	0.097
Average error = **0.126**				

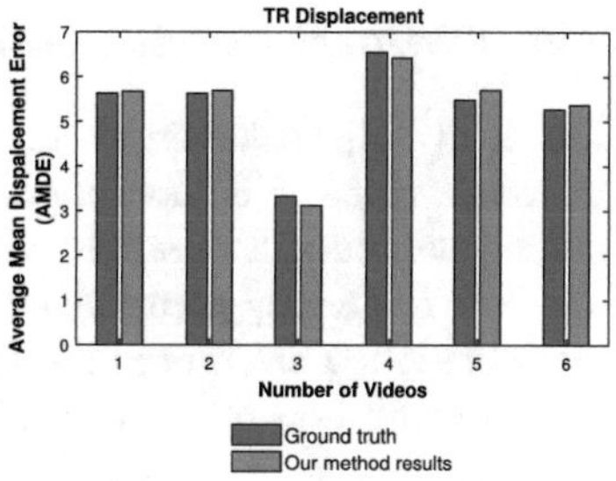

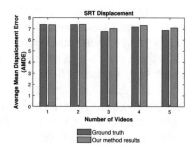

Fig. 4. Accuracy for TR

Table 4. Accuracy for SRT

Sl.	Video name	Ground truth (AMDE)	Our result (AMDE)	Diff. (AMDE)
1	leftSRT	7.388	7.368	0.020
2	rightSRT	7.388	7.396	0.008
3	flowerSRT	6.770	7.037	0.267
4	sflowerSRT	7.185	7.319	0.134
5	indoorTR	6.885	7.108	0.223
Average error = **0.130**				

Fig. 5. Accuracy for translation

Table 5. The three methods performance summary (without foreground)

Sl.	Video name	Deshaker (AMDE)	YouTube (AMDE)	Adobe (AMDE)
1	Regular	0.437	**0.259**	0.693
2	Quickrotation	2.105	1.926	**1.671**
3	Crowd	**1.233**	2.216	1.441
4	Parallax	1.782	1.790	**1.475**
5	Running	**1.555**	2.124	1.946

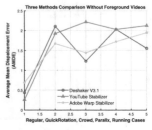

Fig. 6. The VS methods comparison for the real videos without foreground

3.2 Comparison with Other Methods

Liu et al. [4], Niskanen et al. [8], and Tanakian et al. [23], mentioned that an unstable video will have more high frequency components than the high frequency components a stable video. We know that every video has high frequency and low frequency components either stable or unstable. It is obvious that an unstable video has more high frequency components than a stable video, but we cannot count how much more high frequency components will have in an unstable video than a stable video. On the other hand, some other performance metrics [1,23] add some known motions to the stable videos to create the synthetically shaking videos, then the motions of the simulated unstable videos are estimated by each VS method separately, and errors are figured. The method that gives less error, the method is the best among them. However, the researchers only consider motion estimation phase error, whereas our proposed method measures total motion caused by the three phases of the VS method. The three phases of the VS method are motion estimation, frame transformation, and accumulation of the frame transformation [12]. Therefore, the errors come from the threes phase of the video stabilization process. Zhang et al. [10] mentioned in 2017 that there are no performance metrics that can estimate the stability of a video and they proposed a heuristic approach to figure the stability of a video. Some VS performance metrics [1,23–26] are synthetic motions based and some other methods [4,8,10,23] are the heuristic approaches, whereas we can directly compute the stability of the VS method, and we only use the synthetically shaking videos to test our method accuracy.

4 Conclusions

Our proposed performance metric can figure the stability of a video. We observe that the non dynamic background motions of the stable videos are approximately zero (Table 1). Besides, our method observes that the motions between the same video frames approach zero. Also, the background motion of a stabilized video (stabilized by the VS method) and the background motion of an unstable video have large differences (Table 2, Figs. 2, 3). Therefore, the background motion is a stability criterion. Our proposed model estimates the background motion after removing the foreground to test the stability of a video. We test our method accuracy in the case of the simulated shaking videos. The differences in Tables 3, 4 and Figs. 4, 5 between our given motions and the calculated motions by our proposed method are very small which prove our method robustness. Finally, the proposed performance metric successfully computes the performance of the three states of the art VS approaches in terms of the stability criterion of the video (Table 5, Fig. 6). Our proposed novel method is neither heuristic nor synthetic shaking video based, as other recent metrics do. In the future, we will add a distortion criterion and an unfilled image area criterion which are caused by video stabilization in addition to the stability criterion.

Acknowledgments. This work was supported by Korea Institute for Advancement of Technology (KIAT) grant funded by the Korea government (MSIT) (Tech Commercialization Supporting Business based on Research Institute-Academic Cooperation system).

References

1. Zhang, C., Chockalingam, P., Kumar, A., Burt, P., Lakshmikumar, A.: Qualitative assessment of video stabilization and mosaicking systems. In: IEEE Workshop on Applications of Computer Vision WACV 2008, pp. 1–6. IEEE (2008)
2. Baker, S., Bennett, E., Kang, S.B., Szeliski, R.: Removing rolling shutter wobble. In: 2010 IEEE Conference on Computer Vision and Pattern Recognition (CVPR), pp. 2392–2399. IEEE (2010)
3. Liu, S., Wang, Y., Yuan, L., Bu, J., Tan, P., Sun, J.: Video stabilization with a depth camera. In: 2012 IEEE Conference on Computer Vision and Pattern Recognition (CVPR), pp. 89–95. IEEE (2012)
4. Liu, S., Yuan, L., Tan, P., Sun, J.: Bundled camera paths for video stabilization. ACM Trans. Graph. (TOG) **32**(4), 78 (2013)
5. Kopf, J.: 360 video stabilization. ACM Trans. Graph. (TOG) **35**(6), 195 (2016)
6. Liu, F., Gleicher, M., Jin, H., Agarwala, A.: Content-preserving warps for 3D video stabilization. In: ACM Transactions on Graphics (TOG), vol. 28, p. 44. ACM (2009)
7. Marcenaro, L., Vernazza, G., Regazzoni, C.S.: Image stabilization algorithms for video-surveillance applications. In: Proceedings of 2001 International Conference on Image Processing, vol. 1, pp. 349–352. IEEE (2001)
8. Niskanen, M., Silvén, O., Tico, M.: Video stabilization performance assessment. In: 2006 IEEE International Conference on Multimedia and Expo, pp. 405–408. IEEE (2006)
9. Liu, S., Tan, P., Yuan, L., Sun, J., Zeng, B.: Meshflow: minimum latency online video stabilization. In: European Conference on Computer Vision, pp. 800–815. Springer (2016)
10. Zhang, L., Chen, X.Q., Kong, X.Y., Huang, H.: Geodesic video stabilization in transformation space. IEEE Trans. Image Process. **26**(5), 2219–2229 (2017)

11. Liu, F., Gleicher, M., Wang, J., Jin, H., Agarwala, A.: Subspace video stabilization. ACM Trans. Graph. (TOG) **30**(1), 4 (2011)
12. Matsushita, Y., Ofek, E., Ge, W., Tang, X., Shum, H.Y.: Full-frame video stabilization with motion inpainting. IEEE Trans. Pattern Anal. Mach. Intell. **28**(7), 1150–1163 (2006)
13. Safdarnejad, S.M., Atoum, Y., Liu, X.: Temporally robust global motion compensation by keypoint-based congealing. In: European Conference on Computer Vision, pp. 101–119. Springer (2016)
14. Grundmann, M., Kwatra, V., Castro, D., Essa, I.: Calibration-free rolling shutter removal. In: 2012 IEEE International Conference on Computational Photography (ICCP), pp. 1–8. IEEE (2012)
15. Forssén, P.E., Ringaby, E.: Rectifying rolling shutter video from hand-held devices. In: 2010 IEEE Conference on Computer Vision and Pattern Recognition (CVPR), pp. 507–514. IEEE (2010)
16. Karpenko, A., Jacobs, D., Baek, J., Levoy, M.: Digital video stabilization and rolling shutter correction using gyroscopes. CSTR **1**, 2 (2011)
17. Morimoto, C., Chellappa, R.: Evaluation of image stabilization algorithms. In: Proceedings of the 1998 IEEE International Conference on Acoustics, Speech and Signal Processing, vol. 5, pp. 2789–2792. IEEE (1998)
18. St-Charles, P.L., Bilodeau, G.A., Bergevin, R.: Subsense: a universal change detection method with local adaptive sensitivity. IEEE Trans. Image Process. **24**(1), 359–373 (2015)
19. Farnebäck, G.: Two-frame motion estimation based on polynomial expansion. In: Scandinavian Conference on Image Analysis, pp. 363–370. Springer (2003)
20. Opencv: Optical flow. https://docs.opencv.org/3.3.1/d7/d8b/tutorial_py_lucas_kanade.html. Accessed 03 Aug 2018
21. changedetection.net. http://www.changedetection.net/. Accessed 08 Aug 2018
22. Video database. http://liushuaicheng.org/SIGGRAPH2013/database.html. Accessed 21 Feb 2018
23. Tanakian, M., Rezaei, M., Mohanna, F.: Camera motion modeling for video stabilization performance assessment. In: 2011 7th Iranian Machine Vision and Image Processing (MVIP), pp. 1–4. IEEE (2011)
24. Wang, Z., Bovik, A.C., Sheikh, H.R., Simoncelli, E.P.: Image quality assessment: from error visibility to structural similarity. IEEE Trans. Image Process. **13**(4), 600–612 (2004)
25. Qu, H., Song, L., Xue, G.: Shaking video synthesis for video stabilization performance assessment. In: Visual Communications and Image Processing (VCIP), pp. 1–6. IEEE (2013)
26. Zhai, B., Zheng, J., Wang, Y., Zhang, C.: A multi-scale evaluation method for motion filtering in digital image stabilization. In: 2015 IEEE 27th International Conference on Tools with Artificial Intelligence (ICTAI), pp. 682–688. IEEE (2015)

Information Technology and Society

Part 4: Information Technology and Society

In the past few decades there has been a revolution in computing and communications, and all indications are that technological progress and use of information technology will continue at a rapid pace. Accompanying and supporting the dramatic increases in the power and use of new information technologies has been the declining cost of communications as a result of both technological improvements and increased competition. These advances present many significant opportunities but also pose major challenges. Today, innovations in information technology are having wide-ranging effects across numerous domains of society, and policymakers are acting on issues involving economic productivity, intellectual property rights, privacy protection, and affordability of and access to information. Several significant outcomes of the progress of information technology include the e-commerce, advanced information system for various factors, and the inclusion of intelligent system, the application of big data as well as the use of the embedded system is probably electronic commerce over the Internet, a new way of conducting business. Although these new development are seen only a few years ago, it may radically alter economic activities and the social environment. Already, it affects such large sectors as communications, finance and retail trade and are now expanding to areas such as education and health services. It implies the seamless application of information and communication technology including the entire value chain of a business that is conducted electronically.

Paper Presentation

This paper entitled "**A Framework of Skill Training: Is Skill in Training or in Learning?**" by Toyohide Watanabe discusses the skill training on the basis of characteristics of the embodiment and embodied knowledge in comparison with the knowledge learning with a view to investigating the procedural framework of skill training and establishing the research horizon in the educational/learning field. The project by Yasuyuki Maruyama and Toshiaki Miyazaki titled "**Smartphone Finder: Dedicated to Seeking Victims under Collapsed Buildings**" proposes a hand-held system, namely smartphone finder that is designed to locate a smartphone under a collapsed building effectively. They present in detail the basic mechanisms, system overview, and experimental results validating the smartphone finder and the proposed method. The paper entitled "**Preliminary Study of Haptic Media for Future Digital Textbooks**" by Noriyuki Iwane, Chunming Gao, Makoto Yoshida and Hajime Kishida presents a generic model explaining bodily knowledge acquisition and display with supportive haptic media for a future digital textbook system. Project titled "Effects of emotion-based color feedback on user' perceptions in diary context" by Jihye Han, Young June Sah, and Sangwon Lee suggests color feedback as a communication interface and investigates the effects of color feedback in a diary environment, specifically how color feedback on emotional words affect users' perceptions on social presence, attachment, enjoyment, and satisfaction. The paper entitled "**Mobile Pictogramming**" by Kazunari Ito proposes a smartphone version of Pictogramming and compared with other smartphone-enabled programming language environments and show advantages and merit of our proposed application.

Asad Masood Khattak, Salam Ismail Khanji, Wajahat Ali Khan project entitled "**Smart Meter Security: vulnerabilities, threat impacts, and countermeasures**" analyzes AMI from security perspectives; it discusses the possible vulnerabilities associated with different attack surfaces in the smart meter, their security and threat implications, and finally it recommends proper security controls and countermeasures. Project by Chen-Chi Hu, Hao-Xiang Wei, and Ming-Te Chi with the title of "**Shareow: A visualization tool for information discussion in social media**" - The first part focuses on propagation path visualization for a single post in social media. The hierarchical edge bundles method is adopted to optimize the layout and reduce visual clutter caused by excessive information. The second part focuses on the visualization for a summary of posts, which provides a tool for analyzing active users through their sharing and comments activities. Paper titled "**Does Crime Activity Report Reveal Regional Characteristics?**" by Tsunenori Mine, Sachio Hirokawa, and Takahiko Suzuki investigate whether or not the crime messages sent by e-mail can be further exploited as a valid source for analyzing the criminal characteristics of a region, i.e., whether or not they include the characteristics of the regional crime. Abdulaziz Aborujilah and Rasheed Mohammad Nassr, Mohd Nizam Husen and Nor Azlina Ali, AbdulAleem Al- Othman and Sultan Almotiri project titled "**A Conceptual Framework for Applying Telemedicine Mobile Applications in Treating Computer Games Addiction**" proposes a conceptual framework that highlights the capability of using telemedicine mobile applications (TMPs) and sentiment analysis method to assist psychologists, parents and games players to reduce and minimize the risks of digital gaming. we have carried out a preliminary investigation of school students' perspective of using TMPs in treating games addiction. Paper with the title "**NV-Cleaning: An Efficient Segment Cleaning Scheme for a Log-structured Filesystem with Hybrid Memory Architecture**" by Jonggyu Park and Young Ik Eom proposes NV-Cleaning, an efficient segment cleaning scheme for a log-structured file system with hybrid memory architecture.

Rasheed M. Nassr, Abdulaziz hadi Saleh, Hassan Dao, Md. Nazmus Saadat project with the title of "**Emotion-aware educational system: the lecturers and students perspectives in Malaysia**" - this paper surveyed lecturers and students from one Malaysian University, and the findings showed students and lecturers have high interest in consideration of emotions in the education process. This paper with the title "**Personal Identification by Human Motion Using Smartphone**" by Toshiki Furuya and Ryuya Uda recommends human motion is a kind of changeable biometric information even when biological information leaks out. In the authentication method, biometric information for authentication depends on an acceleration sensor and a gyro sensor. Hyohoon Ahn, Duc-Tai Le, Dung Tien Nguyen, and Hyunseung Choo project titled "**Real-Time Drone Formation Control for Group Display**" explains drones form group displays for entertainment and displaying application. The simulation shows that drone formation can display messages effectively. "**An approach to defense dictionary attack with message digest using image salt**" project by Sun-young Park and Keecheon Kim determine to increase the complexity of ciphertexts by postprocessing hash ciphertext. Yongwoo Oh, Taeyun Kim, and Hyoungshick Kim project titled

"An efficient software implementation of homomorphic encryption for addition and multiplication operations" propose a fast binary addition and multiplication algorithms to support various bit-wise operations.

Paper with the title "Pregnant Women's Condition and Awareness about Mood Swings: A Survey Study in Bangladesh" by Nusrat Jahan, Umme Salma Fariha, Musfika Rahman Ananna, and Amit Kumar Das presents findings from a three-month-long ethnography and online survey conducted in Bangladesh where the total number of the participant was 207. The analysis surfaces necessary care and cure of mood swing problem for pregnant women. Yongju Song and Young Ik Eom project title "HyPI: Reducing CPU Consumption of the I/O Completion Method in High-performance Storage Systems," discussed about exploit the non-volatile memory technologies, the I/O stack of operating systems needs to be revisited. Project with the title "Implicit Interaction Design in Public Installation Based on User's Uncon-scious Behaviors" by Hong YAN, QiuXia Li and Yun GUO aims to carry on the design of the seats in public areas, by the user's unconscious behavior of swing the seats unconsciously. It combines with interactive design that each user's behavior of swing the seats corresponds to music fragments played by different types of musical instruments, and finally all the fragments are aggregated to a music. YiLin Rong, Shizheng Zhou, Peng Cheng Fu and Hong Yan presented paper titled "Fish Swarm Simulation Fs Virtual Ocean Tourism" discussed the marine tourism as an example, the combination of virtual reality technology and tourism, through the establishment of the model of fish swarm, explores and studies the application and future development in VR marine tourism. Project title "TPP: Tradeo_ between Personalization and Pri-vacy" by Ubaid Ur Rehman and Sungyoung Lee considered the personalization aspect of recommendation, crowdsensing, and healthcare domains.

Sujune Lee, Jeongho Kim and Eunseok Lee presented a paper with the title "Dynamic Invariant Prioritization-based Fault Localization" proposed method by applying it to the Siemens project, which has been used as benchmark for various fault localization studies. Project title "Improving the Efficiency of Search-based Auto Program Repair by Adequate Modification Point" by Yoowon Jang, Quang-Ngoc Phung and Eunseok Lee proposed two techniques, namely selective validation and fine-grained fitness evaluation, to improve the effectiveness of generate-and-validate automatic program repair. Selective validation reduces the cost of validation by skipping the test execution of variants inadequate to repair the bugs. Yasmin Yahya and Roslan Ismail presented a paper on "Randomized Technique to Determine The New Seedlings for Simulation of Population Dynamic" discussed the use of inventory technique that creates the coordinate of trees, made the simulation for predicting future growth using individual trees, or tree mapping, possible. Paper title "A New Biometric Template Protection using Random Orthonormal Projection and Fuzzy Com-mitment" by Thi Ai Thao Nguyen, Tran Khanh Dang and Dinh Thanh Nguyen proposed a novel hybrid biometric template protection which takes benefits of both approaches while preventing their limitations. Masiath Mubassira and Amit Kumar Das paper title "The Impact of University Students' Smartphone Use and Academic Performance in Bangladesh: A quantitative study" intended to answer whether overuse of smartphones is hampering a student's performance and CGPA.

A Framework of Skill Training: Is Skill in Training or in Learning?

Toyohide Watanabe[✉]

Nagoya Industrial Science Research Institute,
#203, 2-1-16 Seimeiyama, Chikusa-Ku, Nagoya 464-0087, Japan
`watanabe@nagoya-u.jp`

Abstract. Currently the topics about embodiment or embodied knowledge are very popular. The interesting points are widely observed in the research subjects about robot development for the complementary of workers, nursing support for senior citizens, day-life assistance for physically handicapped persons, and so on. These research/development subjects focus on giving equivalent chances to all persons without any exceptions, with a view to solving various kinds of social problems, occurred by the rapid increasing of senior citizens and sharp decreasing of workers. Of course, the popularity for many sorts of human activities such as sports, cultures, etc. has been increasing year by year. In this paper, we discuss the skill training on the basis of characteristics of the embodiment and embodied knowledge in comparison with the knowledge learning with a view to investigating the procedural framework of skill training and establishing the research horizon in the educational/learning field.

Keywords: Knowledge learning · Skill training · Embodiment ·
Embodied knowledge · Knowledge transfer · Motion interpretation · Repeating

1 Introduction

Today, the topics about embodiment or embodied knowledge became interests of research/development subjects about various kinds of human behaviors and also have been investigated with respect to the complementary views of human activities [1]: robot development, nursing support, rehabilitation aid, training assistance, etc. These views focus observably on the embodiments for various kinds of human motions and/or their complex combinations, analyze the characteristics of the complex and concurrent behaviors with the related technologies, make the behaviors explicit as the mechanism based on the embodied knowledge, and contribute to the smart achievement of the behaviors partially or totally, corresponding to the objectives of individual research domains. Even in the educational/learning research field the embodiment and embodied knowledge are looked upon as one of interesting issues in promoting the growth of human ability, and are focused on as one of research topics to be investigated newly or attacked as the advanced personalities of human abilities [2].

The objectives in these researches/developments are not always the same, but different in general even if human behaviors were research focuses: a partial or total replacement of human works in the robot domain, help of human assistance in the

S. Lee et al. (Eds.): IMCOM 2019, AISC 935, pp. 501–516, 2019.
https://doi.org/10.1007/978-3-030-19063-7_39

nursing domain; full-automation or semi-automation in the industrial-production/ agricultural-growth domain, human skill-up support in the technical training domain, encouragement of human motions/operations in athletics, culture and artist domains, and so on. Even if the final objectives were different individually it is important in the researches/developments to grasp interpretatively human embodiment or embodied knowledge. In the educational/learning field, the issues about embodiment and/or embodied knowledge have been discussed as one of interesting topics with a view to growing up skills, which can be looked upon as excellent talents/abilities in individual domains, such as sports, cultures, games, music, arts, etc.

In this paper, we address the research issue about skill training: in particular, we make the architectural framework clear by analyzing the characteristics of behaviors observed in the skill training. In our investigation, the most important viewpoint is to discuss the framework and exercises of skill training, based on the embodiment or embodied knowledge, under our perspective that the skill training is different from the knowledge learning, contributed widely or historically to the educational/learning field. In our discussion, the fundamental mechanism of knowledge learning is naturally derived from the knowledge transfer scheme, which we had already proposed [3]. However, though this scheme is composed of three phases: *composition*, *acquisition*, and *understanding* (including *utilization*), we focus mainly on *acquisition* and *understanding* except for *composition*, and also regard *utilization* as one independent phase, separated from *understanding* specified in the original scheme. This is called as the knowledge learning mechanism, here.

On the other hand, as for the skill training we introduce a motion interpretation mechanism under the similar view that we got caught up institutionally when we designed the knowledge learning mechanism. In the skill training, based on the embodiment and embodied knowledge, we look upon the motion as the most important element: of course, in the knowledge learning the knowledge is the most important element. With respect to this motion interpretation mechanism, we propose a framework of skill training, which is composed of six stages: *basic* stage, *combined* stage, and *practical* stage as a main process, and *initial* stage, and *complementary* stage as a subordinate process.

2 Skill and Knowledge

It is useful to make our objective in this paper clear: we should assure the definite meanings for skill and knowledge individually before we distinguish directly the skill training from the knowledge learning. First, we refer to knowledge because many researches in the educational/learning field have been focusing on teaching actions and/or learning actions from a viewpoint of individual topics about the knowledge learning [4] and also the knowledge resources manipulated under these topics are defined as the learning contents or authoring texts, composed under course-ware or syllabus. Our knowledge transfer scheme as shown in Fig. 1 [3] can explain this situation sufficiently. At least, when we define the knowledge from a viewpoint of the knowledge learning, we can look upon the knowledge as a set of knowledge fragments with meaningful relationships for others as facts. On the other hand, the skill is in general defined as one of abilities which can been acquired by liberal arts or by training. Moreover, the ability must be always implemented as "action": this means that the skill is represented by "action" and that the action can be achieved by various means or methods.

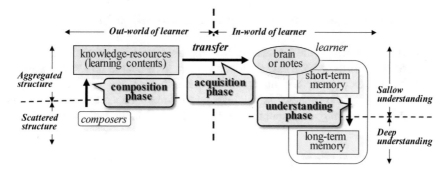

Fig. 1. Knowledge transfer scheme

Table 1. Types of skills

Type	Subtype	Examples	Notion
Embodiment-based skill	Sport-oriented skill	Skating, football-playing swimming, sumo, wrestling, …	Motor skill
	Culture-oriented skill	Calligraphy, trumpet-playing gaming, go-playing, …	—
Cognition-based skill	—	Reading/writing/hearing ability, computing ability, communication, …	Cognitive skill

Thus, we can look upon the skill as a special/smart ability to achieve a technical goal. This means that the skill is not always dependent on the embodiment or the embodied knowledge, but may be independent of the embodiment or the embodied knowledge. For example, the "wisdom" person with high-performance computation ability, cost-effective planning ability, logical conversation ability, etc., who is superior in comparison with many average persons, may be said a skilled person. These abilities are so called cognitive skills, in general [5]; and they are not related to the embodiment, but are derived from the elementary attainments. Namely, we can find out two types of skills: the embodiment-based ability and the cognition-based ability, as shown in Table 1. At least, we can define the ability, which enforces to achieve the desirable/finally-attainable goal by technical means or elementary attainments, as the skill, and classify into two classes by depending on the embodiment or the cognition. The embodiment-based skills are generated through human motions observed in many sports, while cognition-based skills are created from reactions in human activity such as many competitions, computer-games, social events, etc. Of course, skills are technical abilities and thus will be possibly stepwise achieved. Our discussion focuses on only the embodiment-based skills because our interesting viewpoint is the embodiment or the embodied knowledge.

3   Frameworks of Skill Training and Knowledge Learning

First of all, we focus on the knowledge learning and survey the features because of making the difference points clear in our comparison between the skill training and the knowledge learning, and then under such a coordination we address the topical viewpoints such as the framework, the features of domain-dependent skills, etc. and discuss the skill training, corresponding to the features of individual domains.

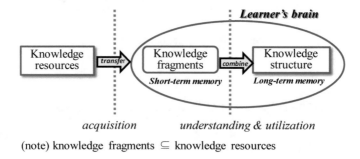

(note) knowledge fragments ⊆ knowledge resources

Fig. 2. Knowledge learning mechanism

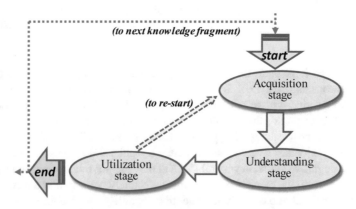

Fig. 3. Framework of knowledge learning

In the knowledge learning, the fundamental mechanism is directly derived from the knowledge transfer scheme in Fig. 1 [3], and is shown in Fig. 2. In this knowledge learning mechanism, the knowledge resources in the out-world of learners is transferred by the learner's acquisition action into his/her brain as the knowledge fragments in the short-term memory, and then edited/reorganized as the knowledge structure held in the long-term memory. Of course, the knowledge resources are already composed as the learning contents in the course-ware. Figure 3 is a framework of knowledge learning, modelled interpretatively from the knowledge learning mechanism. The supporting request in the knowledge learning process is to compose the procedure, which can perform the acquisition stage, the understanding stage, and the utilization stage effectively and/or effectually. Thus, the knowledge learning is specified as the issue to be investigated on the successive process of three stage:

- In the acquisition stage, how many knowledge fragments can the learner transfer?
- In the understanding stage, how can the learner combine the newly transferred knowledge fragments with the existing knowledge structure meaningfully?
- In the utilization stage, how can the learner solve problems, occurred or asked externally/internally, with their own knowledge?

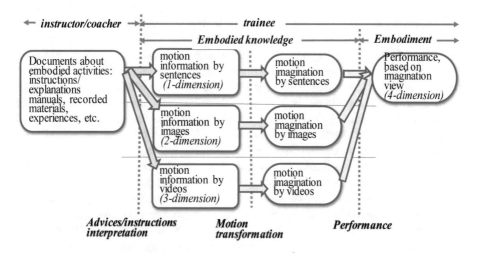

Fig. 4. Motion interpretation mechanism

Ordinarily, this process is performed on every knowledge fragment: if this process could not finish successfully, the acquisition stage for the same knowledge fragment may be restarted.

Here, we have a question "Can we explain human activity in the skill training similarly with this framework of knowledge learning?" For example, we consider the speed skating. Of course, in the skating sports there are various kinds of skating categories: the figure skating, the ice hockey, and the short-track skating, in addition to the speed skating. Their skating shoes are distinguished at glance among them as one of the most interesting features, and also the skating forms, the skating uniforms, the skating rules, the skating places, etc. are different individually, like the ordinal motion-oriented features. However, even if the skating shoes were different, it is common to first stand up straightly on the ice and then walk over the ice, leg by leg, as the initial stage. This initial stage is always requested commonly to all beginners in skating sports, and they must perform these motions repeatedly until their motions are completely achieved. If not so, they cannot try the next skating motions as the following process. If they tried the next stage without their own completions, they will be not able well to slide over the ice.

This situation cannot be specified by the framework of knowledge learning: the knowledge cannot become a clue basis to understand any motions, but may become any starter to enforce the motion achievement. The motion related to the embodiment is not an activity, dependent fully on intelligence or knowledge working. Even if the person acquired the knowledge about the speed skating and understood them completely, it is not sure that he/she can always skate well. This difference is generated as one gap between what the knowledge fragments are facts (declarative representations) or rules (procedural representations) and have explicit meanings in themselves, and what the instructions or explanations in the coaching documents about the embodiment are ambiguous and tacit representations for human motions/movements.

We show the motion interpretation mechanism in Fig. 4, illustrated conceptually against the knowledge learning mechanism in Fig. 2. Documents about embodied activities are corresponded to knowledge resources in Fig. 2, but are even explained case by case with the experiences or motion-specific interpretations and/or by various kinds of representation means/methods/media. Really, the trainees cannot easily perform the motions/movements without helps of instructors or coaches. This is because all of these documents cannot be absolutely described as 4-dimensional sequences of motions or movements, point by point. So, the knowledge fragments are meaningful specifications defined by themselves, but the coaching documents about embodiment are not rigid/explicit representations for interpretative explanations/instructions. Thus, the motion interpretation mechanism is looked upon as the mapping function from the embodied documents to the trainees' performances through three steps such as the advices/instructions interpretation procedure, the motion transformation procedure, and the performance procedure. The advices/instructions interpretation procedure is mainly controlled and executed by instructors/coaches. Of course, this motion interpretation mechanism is not only controlled by the forward processing, but also is done by the backward processing [6]. This procedure is important that instructors/coaches should take a play to interpret practically the already-edited instruction manuals/documents with their own experiences or performance theories.

The motion transformation procedure and the performance procedure are executed by the trainee in his/her-self: in the motion transformation step, the trainee is to understand the motion with instructions/advices, interpreted by coaches/instructors, and make up the corresponded imagination of the understood motion. In the performance procedure, the trainee is to execute the imagined motion practically as his/her embodied action. As we can grasp in Fig. 4, the fundamental functionality is not to select parts of instruction/advice contents in the existing manuals/documents and move them into the brain of trainee as the knowledge learning mechanism stepwise, but to interpret the instruction/advice contents as for his/her embodied knowledge and his/her experiences and perform the interpreted motion with his/her own embodied ability. The knowledge learning mechanism moves and understands knowledge resources, while the motion interpretation mechanism interprets the action contents and performs them as the realized motions.

Here, we compare Fig. 4 with Fig. 2. In Fig. 2, the learner catches directly the knowledge fragments, organized definitely as the knowledge resources, while in Fig. 4 the trainee cannot catch up directly with the instructions/advices about various motions: in this case, even if the trainee could understand the instructions/advices as the meanings about motions it is not effective if the trainee cannot perform the motion as his/her behavior. This is because in the skill training the action is not to acquire as the information about motion. Thus, this difference is important and must be taken into consideration carefully. In Fig. 4, the performance procedure from the embodied knowledge to the embodiment (or the embodied movement) is never observed in the knowledge learning: this performance procedure is repeated until the trainee can master the movement even if it were not complete. If not so, the trainee cannot execute the next motion absolutely. For example, consider the skating exercises. The first step is that the trainee should stand up straightly on the ice, but should not slide on the ice. If the trainee cannot stand up straightly on the ice, he/she cannot walk over the ice, one by one, as the next step. We can consider another simple example. In the driving, the first step is to do steering well: when the trainee cannot do the steering operation suc- cessfully, he/she cannot drive on the curve road.

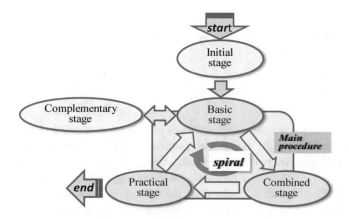

Fig. 5. Framework of skill training

We can show the stepwise stages in the embodied movement: the initial stage and the complementary stage as the subordinate process; and the basic stage, the combined stage, and the practical stage as the main process. Figure 5 is a framework of skill training, illustrated similarly with respect to that of the knowledge learning. These stages distinguish the motions to be performed stepwise because in each stage the motion to be achieved is differently pre-set and the pre-set motions must be always completed before the next motion. For example, we can list up the typical motions in the speed skating, corresponded respectively in each stage as follows:

(1) initial stage: the most pre-requisite movement on the ice:

  − stand up straightly on ice;
  − walk one by one over ice;

(2) basic stage: the basic movement in skating:

  − weight movement;
  − sliding;
  − stopping;
  − crossing;

(3) combined stage: the movement in speed skating:

  − straight-line gliding;
  − curve-gliding (cornering);

(4) practical stage: the technical movement in speed skating competition;

  − taking course;
  − start dash;
  − positioning;
  − pacing;
  − strategy under various conditions such as ice situation, weather situation, course exchanging timing, etc.

(5) complementary stage: growing up of embodied ability:

  − physical training;
  − land training;
  − pseudo forming and skating on land;
  − mental training.

Table 2 arranges the embodied activities of calligrapher, trumpet player in brass-band, and football player under our framework of skill training, based on the embodiment, in addition to skaters in speed skating. At least, we can similarly grasp the stepwise process in the embodiment-based skill training when we adapt our framework to these activities. Moreover, each embodied activity in Table 2 is characterized by a kind of activity, necessary tools, and evaluation means: in the embodied activities their own particular tools must be always prepared together with the executive embodiment and also various kinds of achievement criterion are used to evaluate the effects of motions in each stage. Of course, the means to evaluate the achievement degree is useful as absolute or relative judgement method.

4 Repeating Control Between Skill Training and Knowledge Learning

Here, consider again the training in the speed skating. In the initial stage, the primal instructions are:

(1) Stand up straightly on the ice;
(2) Walk over the ice one by one.

What does the instruction "Stand up straightly on the ice" indicate as the concrete exercise? Even if beginners could understand rightly its instruction as the sentence or command, can they perform it exactly? The answer is that many of them cannot stand up straightly on the ice: they cannot become "upright posture" as the result because they move their bodies, heads, legs, hands, etc. by their own unnecessary forces so as to keep their own balances stable as possible as they can.

To stand up straightly on the ice is:

(1) The metal blades, attached to the skating shoes, must keep the right angle for the ice plane so as to hold all of the skater's body weight;
(2) The center of gravity must be put on the metal blade to keep skater's balance stable;
(3) The skater's eyes must look straightly into the distance, but not look down.

Similarly, to walk over the ice one by one is:

(1) The center of gravity must be put on only one blade, but not on the right and left blades at the same time;
(2) The blade, in which the center of gravity is held, must be mutually exchanged with the walking actions;
(3) The posture of skater must be straightly kept;
(4) Also, the skater's eyes must look straightly into the distance, but not look down;
(5) In this case, the most important behavior is that the skater should pay a caution to move his/her center of gravity, but not slide on the ice.

In these two instructions, to slide over the ice plane is never requested. To put the center of gravity on only one metal blade is important even at any time. These instruction sequences for two different initial motions are not always easily performed for beginners, because in addition that these representations are too abstract, individual motions cannot be achieved independently but must be done cooperatively at the same time.

Also, the movement is too dependent on the characteristics of individual motions, attended inherently to the private experience history. Therefore, their interpretations for instructions are derived from their experiences, and must be challenged person by person.

Table 2. Features of embodied activities

	Skater in speed skating	Calligrapher	Trumpet player in brass-band	Football player
Activity	Personal		Organizational (or group)	
Tools	Skating shoes; grindstones, sharpening table	Writing brush paper, inkstones, ink, paperweight, mat	Trumpet, musical score	Spike; ball, goal box
Initial stage	Stand up on ice, walk over ice	Holding brush, raising and lowering of brush, carrying way of brush	Holding trumpet, using mouthpiece, sounding, breather	Kicking way of ball, catching way of ball, passing
Basic stage	Weight movement, sliding, stopping, crossing	Character writing	Scale practice, tonguing, intensity sound practice	Dribbling, shooting, trapping
Combined stage	Straight line gliding, curve gliding	String writing	Rendition	Pass-working, goal-shooting
Practical stage	Taking course, start dash positioning, pacing	Spatial arrangement	Joint performance, playing in concert	Formation situational awareness
Complementary stage	Physical training, mental training	Configuration boosters	Kansei force training	Physical strength cognitive training
Evaluation	Time (absolute), order (relative)	Subjective examination	Subjective examination	Score

We showed individually the frameworks of the knowledge learning in Fig. 3 and the skill training in Fig. 5, so as to make the difference between these activity processes explicit:

(1) In the knowledge learning, the process deals with many knowledge fragments, and arranges the newly acquired fragments systematically into the existing fragments under the understanding degree or ability of learner. This process is completely dependent on the existing structure of public knowledge resources;

(2) In the skill training, the process is to design imaginatively the mapping functions from some instructions to the corresponding interpreted/transformed motions with personal experiences, and then to perform the embodied actions with many repeated exercises under the mapping functions. The process is too strongly dependent on the personal parameters of mapping functions as the experience and training of trainee.

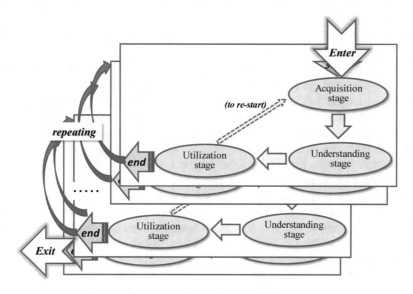

Fig. 6. Enhanced framework of knowledge learning

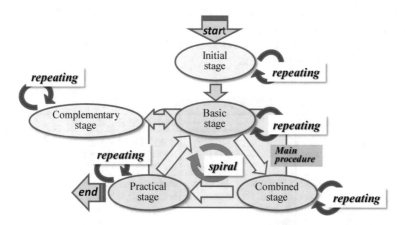

Fig. 7. Enhanced framework of skill training

Here, to make the difference between the knowledge learning and the skill training clear from a viewpoint of the repeating control, we improve Figs. 2 and 5, individually: Figs. 6 and 7 are respectively modified from their original frameworks. When we compare Fig. 7 with Fig. 5, the important feature in the skill training focuses on the adjustment procedure for motion performance with a view to implementing the mapping functions, corresponded to the successive motions. While, that in the knowledge learning controls the organization procedure for managing well-structure and consistent-relationships among knowledge fragments with a view to maintaining the understood knowledge.

In the skill training, the stages have successive relationships mutually: the basic stage is always located after the initial stage: the weight movement cannot be absolutely attained if the trainee could not stand up straightly on the ice plane or if he/she could not walk over the ice plane. This is because the weight movement is the most primitive and important action, and the trainee can go ahead without any sliding motions if the trainee could move smoothly the weight from left/right blade to right/left blade one by one. This motion will be attained stepwise as the result that the trainee has been performing repeatedly day by day, but not once. In the basic stage, the trainee is permitted to slide with the skating shoes/blades mutually: the newly instructed "sliding" makes the movement actions more speedily on the ice plane. Also, the trainee will be instructed how to use the both left-side/right-side edges of each blade, cooperated with the action "weight movement". In the combined stage, the actions to be possibly performed after many repeated actions in the basic stage are successively instructed as a combined action "sliding on ice" are cooperated movements with other parts of body such as hands, a head, legs, etc. as well as the forming posture [7]. Namely, the skating is composed of many simple/complex movements, as a unified parallel behavior. In addition, the trainee will be taught about the role of "lock", which arranges the blade sharp so as to be able to go ahead without any decreasing the sliding speed together with cornering techniques. At the last practical stage, various techniques for sliding in the competition are instructed so as to manage the lap time under the well-planed strategy: the start dash, the cornering, the positioning for competitors, etc. The trainee must catch up with instructor's advices and perform repeatedly so that he/she should feel that the motion became perfect.

Table 3. Categories of embodied activities

	Culture-culture-oriented			Sport (Athletics)-oriented		
	1 (1-to-n)	1-to-1	n (n-to-n)	1 (1-to-n)	1-to-1	n (n-to-n)
	Individual match	Competition	Team competition	Individual match	Competition	Team competition
Examples	Calligraphy, tea ceremony, flower arrangement, painting, ...	Go, Chess, othello, Japanese chess, ...	Orchestra, chorus, ...	Skating, skiing, golf, swimming, athletics, ...	Sumo, kendo, tennis, wrestling, ...	Baseball, football, set gymnastics, synchro swimming, ...
Judgement	Grading	Victory or defeat	Grading	Rank, record, grading	Victory or defeat	Victory or defeat, grading

These stages can be observed similarly in the calligrapher, the trumpet player in brass-band and the football player. So, the skill training is the repeating process to map the actions to the behaviors. On the other hand, the knowledge learning is the repeating procedure of knowledge learning mechanism for individual knowledge fragments. In addition, the complementary stage is set to grow up the fundamental physical-stretch: for example, in the speed skating training, the land conditioning exercises, the forming exercises, etc. are motions in this complementary stage.

In the skating training, the critical knowledge related to the skating technics is used as the instruction form. This instruction about how to do individual motions in the skating activity is very tacit and ambiguous: therefore, the understanding is different person by person; and also the adaptation means for persons are interpretively varied. The human embodiments are often inherently generated from personal features: the interpretation adaptable to one person is not always to be sure for other person; and the interpretation applied at the time t_0 is not constantly sure to be successful means at the time t_1. The skill must be manipulated under the dynamical situation, while the meanings of knowledge is stable and invariant in many cases over all persons and through all time periods. The embodied knowledge instructed in the skill training is generally so-called tacit knowledge, which is different from the explicit knowledge to be transferred in the knowledge learning. Therefore, the interpretation, transformation and performance observed in individual mapping steps are more or less different by trainees, even if the instructor indicated with the same tone, the same phase, the same timing, the same period, etc. for all trainees. In addition, the fact that the embodiment attached inherently to each trainee is not the same, has been growing up trainees one by one until today, and are also dependent on their own personal experience histories. Thus, in the skill training the instructions/advices generated originally from the coacher or instructional books are interpreted as various kinds of performances, like the interpretation mechanism in Fig. 4. While, in the knowledge learning the knowledge fragments preset in the learning course-ware are moved smoothly to learner's brain and materials: of course, a part of the knowledge fragments may be not received on some learners because of the learner's missed actions.

Next, we arrange the features of culture-oriented activities and sport-oriented activities, derived categorically from Table 2. Table 3 shows them with respect to organic-based activity or personal-based activity. In spite of both culture-oriented activity and sport-oriented activity, or both organic-based activity and personal-based activity, the judgement/measurement criteria for the attained skills or performed results are always defined as the check-points or training goals: scores, records, ranks, victory or defeat, etc.

The skill is not generated from the learning for managing knowledge on brain, based on intelligence, but derived from the training for performing behaviors on body, based on the embodiment.

5 Effects in Skill Training

We can know the training results with the degrees of goal achievement, as well as the test/examination results in the knowledge learning. This measurement for the degrees of goal achievement becomes possible in the practical stages as the final stage. It takes a long period to reach the practical stage from the basic stage (or the initial stage) because the trainee takes many repeated plays to master the actions in each stage until his/her behavior is naturally fit to the motion under the good balance. This is different from the knowledge learning: in the knowledge learning the examination can be applied every cycle from the acquisition stage to the utilization stage. Therefore, in the skill training it is better to apply the judgement for individual motions and give tacit advices or ambiguous instructions by the coachers or instructors. In this case, the advices or instructions generated from the judgements are strongly dependent on the subjective-specific observations/estimations. This is because the instruction procedure is directly dependent on the motion interpretation mechanism.

Here, we consider how to observe or know the degrees of attained skills. The skill is a technical ability represented as the embodied activity and the degree may become clear as a product generated by the trainee or with behaviors performed by the activity. In general, the result of skill training is grasped possibly [6]:

(1) Technically/talentedly mastered ability (implicit result)
(2) Product, generated by performance (explicit result)
(3) Score, recorded by performance (explicit result).

Namely, we can recognize the effects in the skill training through two means: one is a mastered ability of the trainee, as (1); and another is a product or a result score, generated/recorded by the trainee, as (2) and (3). (1) is a totally estimated result and is not observable directly as the explicit state. In this case, the ability level is arbitrarily assigned on the basis of predefined criteria: the first, the second, ...; upper, lower; and so on. The work positions, sorts of jobs, roles in projects, etc. pre-set in the companies or public organizations are dependent on such similar policies.

On the other hand, (2) is the activity to create some products such as pictures, photographs, arts, etc. as visible works; and this type can often be observed in culture-oriented activities. Similarly, (3) is represented with accountable values, the positioned order, or the victory or defeat, and so on; and this type can be observed in the sport-oriented activities.

As a performance result in the skill training, the ability in (1), the product in (2) and the score in (3) are standard. As you see, the ability is different from the product and the score: un-visible versus visible, or implicit vs. explicit. However, the product in (2) and the score in (3) are attained in the practical stage in Fig. 7: when we perform each motion in the stage such as the initial stage, the basic stage, and the combined stage, how to judge the completion? In general, we cannot find out the best means for this judgement. So, this is almost the same problem as the ability in (1). This is absolutely dependent on the subjective judgement of coacher or instructor with many experiences.

6 Support in Skill Training

When we compare the skill training with the knowledge learning, we can catch up the difference directly if we refer to Figs. 6 and 7 comparatively. In the knowledge learning, the progress is to execute repeatedly the successive knowledge cycle of acquisition, understanding and utilization for many knowledge fragments, while in the skill straining the progress is to repeatedly perform individual motions in each step, preassigned in the definite stage, so that the trainee should master embodied actions. In the skill training, the number of instances about the motions is not so many, but a few. However, their contents are too ambiguous to be performed as embodied actions: many explanations suitable individually to each trainee are needed, corresponded to the trainee's inherent behaviors. This is because the embodied actions is not on knowledge, but for motion.

Table 4. Knowledge learning and skill training

	Knowledge learning	Skill training
Subjective	Brain activity	Body (hands, legs, etc.) activity
Mechanism	Logical	Physical, situational
Processing	Sequential	Parallel (with timing, balance, etc.)
Process	Acquisition stage, understanding stage, utilization stage	Basic stage, combined stage, practical stage; initial stage, complementary stage
Contents	Explicit knowledge resources	Tacit/implicit advices/instructions
Goal	Comprehension (understanding)	Mastery (experience)
Evaluation and results	Quality; examination, question and answer; creation (discovery, invention)	Quantity and quality; product, record, order, victory or defeat; experience (work, demonstration)

In the knowledge learning the explanation functions for knowledge fragments or their relationships among knowledge fragments take important roles to understand the meanings of knowledge fragments, in addition to Q&A functionality, simulation as dynamic example or for behavior's observation, etc. While, in the skill training, the functionality for grasping and performing the motion comparatively under the gap recognition between the really observed motions and the ideal motions, customized privately for the trainee, is useful with help of instructor's/coacher's advices/instructions. In many cases, even if the trainee could understand the meanings of advices/instructions as the embodied knowledge it is not easy to perform the understood actions and achieve it sufficiently so as to make the progress of the next motions successive. In the skill training, our support viewpoints focus on the physical behaviors, and design the functionality or environment in the training, program and stepwise exercises, under our framework of skill training. How to support "repeating" phase is a key point.

7 Conclusion

In this paper, we discussed the features of skill training, based on the embodiment in comparison with that of knowledge learning with a view to making the framework of skill training clear, and characterizing the skill training in part of educational/learning fields. As a result, we proposed a framework of skill training, composed of two different composite procedures: the basic stage, the combined stage and the practical stage as the main process; and the initial stage and the complementary stage as the subordinate process. In comparison with the framework of knowledge learning, driven naturally from our knowledge learning mechanism, its essential feature is the concept "repeating".

In the knowledge learning, the repeating of knowledge cycle from the acquisition stage to the utilization stage through the understanding stage is meaningful to transfer many knowledge fragments effectually and successively. On the other hand, in the skill training the repeating control, achieved independently or successfully in individual stages such as the initial stage, the basic stage, the combined stage, the practical stage, and the complementary stage, takes important roles in each stage. Also, this repeating control is effective to perform embodied actions, transformed interpretatively, but not to understand embodied knowledge for the motion. We may be able to arrange the characteristics, as shown in Table 4, with a view to making the difference between the skill training and the knowledge learning comparatively clear.

As our next research progress, it is necessary to investigate the support method or the supporting functionality on the basis of the difference between the skill training and the knowledge learning, mechanically.

References

1. Bernstein, N.: The Co-ordination and Regulation of Movements. Pergamon Press, New York (1967)
2. Fujinami, T.: Embodied knowledge analyzed through information technology and its relevance with industry. J. IEICE **100**(11), 306–311 (2017). (in Japanese)
3. Watanabe, T.: Learning support specification, based on viewpoint of knowledge management. In: Proceedings of E-Learn 2012 (AACE), pp. 1596–1605 (2012)
4. Watanabe, T.: Research view shift for supporting learning action from teaching action, In: Proceedings of KES/IIMSS 2017, pp. 534–543 (2017)
5. Karni, A., Meyer, G., Rey-Hipolito, C., Jezzard, P., Adams, M.M., Turner, R., Ungerleider, L.G.: The acquisition of skilled motor performance – fast and slow experience-driven changes in primary motor cortex. Proc. Natl. Acad. Sci. USA **95**, 861–868 (1998)
6. Wulf, G., Shea, C., Lewthwaite, R.: Motor skill learning and performance: a re view of influential factors. Med. Educ. **44**, 75–84 (2010)
7. Kanemaru, N., Watanabe, H., Taga, G.: Increasing selectivity of interlimb coordination during spontaneous movements in 2- to 4- month-old infants. Exp. Brain Res. **218**(1), 49–61 (2012)

Smartphone Finder: Dedicated to Seeking Victims Under Collapsed Buildings

Yasuyuki Maruyama and Toshiaki Miyazaki[(✉)]

Graduate School of Computer Science and Engineering,
The University of Aizu, Aizuwakamatsu, Fukushima, Japan
{m5221147,miyazaki}@u-aizu.ac.jp

Abstract. It is critical to discover the victims buried under collapsed buildings urgently after a natural disaster. In recent years, people are habituated to carry smartphones always along with them. The smartphones are provided with a feature to transmit the Wi-Fi signals called "probe requests" to connect with access points. This fact motivated us to develop a novel method that can estimate the locations of the buried victims using the Wi-Fi signals. In this paper, we propose a hand-held system, namely smartphone finder that is designed to locate a smartphone under a collapsed building effectively. We present in detail the basic mechanisms, system overview, and experimental results validating the smartphone finder and the proposed method.

Keywords: Wi-Fi probe request · Localization · Disaster ·
Smartphone · Tablet · RSSI

1 Introduction

Natural disasters such as torrential rains and earthquakes frequently occur worldwide [1]. When a building collapses in a disaster, it is critical that the rescue workers discover the buried victims on an emergency basis. Historically, smart aids such as fiber scope [2] and rescue robots [3, 4] have been developed and implemented in rescue operations successfully; however, these aids are applicable only to a very limited target area despite their ease of portability or mobility.

In recent years, it is a common sight that people always carry smartphones along with them. The number of smartphone users in the world has been rising steadily, and it is forecasted to reach a whopping 2.87 billion, which is 1.24 times higher compared to 2017 [5]. The fact that a smartphone sends out a Wi-Fi signal called "probe requests" to the access points (APs) is motivating in this context as this feature could be effectively used to help the rescue workers estimate the locations of the victims in a disaster site such as a collapsed building. In this paper, we propose a smartphone finder that enables a rescue team to discover the buried victims quickly.

Localization methods using radio waves and their received signal strength indicator (RSSI) values have been proposed by various studies [6–10]. Zanella [6] introduced a method that considers the variability and accuracy of RSSI. Takashima et al. [7] proposed a method based on maximum likelihood (ML) that estimates the position of the target device probabilistically using the relationship between RSSI and distance;

© Springer Nature Switzerland AG 2019
S. Lee et al. (Eds.): IMCOM 2019, AISC 935, pp. 517–528, 2019.
https://doi.org/10.1007/978-3-030-19063-7_40

however, because it needs pre-measurement of the environment, it is not applicable to our present study. A similar method [8] that proposes a position estimation using ML does not require pre-measurements; however, it is found to be impracticable in our context, as it requires a sophisticated computer to perform its complex computations. Thus, it is evident that the smartphone finder should be realized using a hand-held PC or a tablet. In [9] and [10], the log-normal shadowing model is used for localization; this model that is widely used due to its simplicity and practicability estimates both the position of a target device and the propagation parameters using simple computations. Therefore, we propose to adopt this model in our method. Furthermore, the original works usually do not consider realistic conditions such as radio wave propagation and obstacles of different types in the disaster-hit area. Similarly, they presume availability of unmanned aerial vehicles for execution. These are a few significant factors to be considered while adopting the localization methods from previous studies.

In this study, we assume a situation whereby the victims are buried under the collapsed buildings, and the rescue workers move around the disaster-hit area to locate the victims. Figure 1 illustrates this scenario to locate a victim buried together with his smartphone using the smartphone finder that can capture Wi-Fi probe-request signals sent from the smartphone at regular time intervals. Then, when the received signals are found adequate, the finder starts to estimate the position of the smartphone and to course-correct it eventually using the newly captured signals. The estimated positions are displayed on the screen of the finder enabling the rescue worker to reach to the smartphone and the victim.

Rescue worker
and
smartphone finder

Buried victim with
smartphone

Fig. 1. A scenario to find a buried victim with a smartphone using the smartphone finder

The main contributions of this paper are as follows:

- Development of packet sniffer: It enables us to collect and analyze the Wi-Fi signals from smartphones. It is functionally simple as it extracts only the information necessary for localization.
- Development of a novel localization method suitable for the disaster environment: It is a modified version of the log-normal shadowing model offering high accuracy using simple computations.
- Development of a unique graphical user interface (GUI): To capture the estimation results easily, three types of GUI were developed.

2 System Overview

Figure 2 shows the system overview of the smartphone finder. It consists of a simplified packet sniffer, a localization function, and a unique set of GUIs that are realized as software components in a personal computer (PC). A network interface card (NIC) is attached to the PC to capture the Wi-Fi signals. Similarly, a GPS receiver attached to the PC helps to know the location of the smartphone finder.

The following sub-sections discuss in detail the three software components mentioned above.

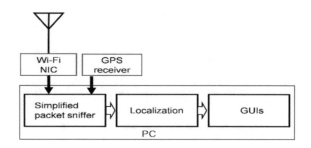

Fig. 2. A system overview of the smartphone finder

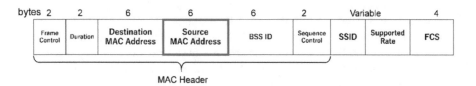

Fig. 3. The Wi-Fi probe-request frame format

2.1 Simplified Packet Sniffer

The sniffer captures the MAC address from the Wi-Fi probe-request signal whose complete frame is shown in Fig. 3 and logs it along with the RSSI value and its acquisition time. In addition, it logs the location information of the smartphone finder captured by the GPS receiver. Compared to the conventional packet sniffers such as Wireshark [11], our sniffer extracts only the MAC address from the probe request frame, owing to which it is simple and light. Figure 4 shows an example of the data captured by our packet sniffer.

520   Y. Maruyama and T. Miyazaki

Most smartphones can use both 5-GHz and 2.4-GHz band Wi-Fi; however, we use 2.4-GHz band Wi-Fi for localization, because the radio signal of lower frequency is better for propagation. The path loss of radio wave signal in free space is often represented as follows [12]:

$$Path\,loss\,(dB) = 32.5 + 20\log F + \log d,$$

where F is frequency [GHz] and d is distance [m]. As this formula indicates, the 2.4-GHz band is better than the 5-GHz band with regard to path loss.

Location of observation point	MAC address	RSSI	Time stamp
[001] N 37 525375 - E 139 937880 /	MAC address B4 CE F6 09 A4 41 /	RSSI -67 /	Time Sun Nov 6 16 12 49 2016
[002] N 37 525375 - E 139 937880 /	MAC address 9A DD 01 14 C1 D0 /	RSSI -59 /	Time Sun Nov 6 16 12 49 2016
[003] N 37 525375 - E 139 937880 /	MAC address B4 CE F6 09 A4 41 /	RSSI -71 /	Time Sun Nov 6 16 12 51 2016
[004] N 37 525375 - E 139 937880 /	MAC address B4 CE F6 09 A4 41 /	RSSI -69 /	Time Sun Nov 6 16 12 51 2016
[005] N 37 525375 - E 139 937880 /	MAC address C2 92 23 C6 69 59 /	RSSI -63 /	Time Sun Nov 6 16 12 53 2016
[006] N 37 525377 - E 139 937892 /	MAC address 6A 11 79 6B D9 3D /	RSSI -59 /	Time Sun Nov 6 16 12 57 2016
[007] N 37 525377 - E 139 937892 /	MAC address B4 CE F6 09 A4 41 /	RSSI -61 /	Time Sun Nov 6 16 12 59 2016
[008] N 37 525377 - E 139 937892 /	MAC address B4 CE F6 09 A4 41 /	RSSI -59 /	Time Sun Nov 6 16 12 59 2016
[009] N 37 525377 - E 139 937893 /	MAC address A4 B8 05 90 FC D8 /	RSSI -63 /	Time Sun Nov 6 16 13 01 2016
[010] N 37 525378 - E 139 937895 /	MAC address B4 CE F6 09 A4 41 /	RSSI -69 /	Time Sun Nov 6 16 13 01 2016

Fig. 4. An example of captured packets

To confirm the actual path loss, we conducted some preliminary experiments. Figure 5 shows the experimental results of RSSI plotted against the distance between sniffer and smartphone in an open space environment. In this experiment, the sniffer was carried by a person, while the target smartphone was placed on a table with a height of 1 m above the ground. For each distance, we measured the RSSI values for 1 min. In the figure, the small dots indicate the observed RSSI values, whereas the large dots indicate the average of the observed values at each distance. We failed to obtain the RSSI values in some cases owing to situational factors, and the number of values was found to vary with distance; however, the experimental results confirmed the satisfactory operation of our sniffer.

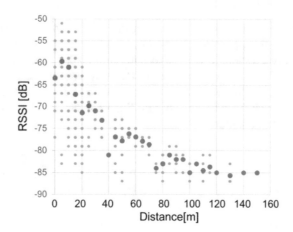

Fig. 5. Relationship between RSSI and distance in an open space environment

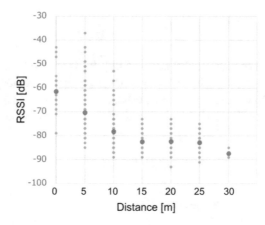

Fig. 6. Relationship between RSSI and distance in an adverse environment

Figure 6 shows the experimental results obtained for an adverse environment analogous to a disaster. The target smartphone was placed on the ground and covered with obstacles. Compared to Fig. 5 (open space environment), both the number of probe requests acquired and the maximum reachable distance decreased, and the variation in RSSI increased. This observation implies that we cannot use the obtained RSSI values directly to estimate the distance between the smartphone finder and the target smartphone. This aspect is discussed further in the following section on localization method.

There are two types of Wi-Fi antenna: directional and non-directional. Although the directional antenna is powerful in its specific directions, it has high implementation cost as the location estimation needs a geomagnetic sensor apart from extra calculations. Moreover, as we were keen on a small and portable system, we preferred to use a non-directional antenna for our sniffer.

2.2 Proposed Localization Method

Because the received RSSI values were found to be unstable as discussed above, we introduced a new localization method that is especially suitable for such unstable values.

The Log-Normal Shadowing Model

We suppose that a target smartphone is located at $(\hat{x}, \hat{y})$ and the RSSI value P_i of radio waves from the terminal is observed at (x_i, y_i). Let d_i denote the distance between $(\hat{x}, \hat{y})$ and (x_i, y_i). In the log-normal shadowing model [9, 10], the RSSI value P_i is theoretically calculated as follows:

$$\hat{P}_i = \hat{P}_{d_0} - 10\hat{n} \log_{10}\left(\frac{d_i}{d_0}\right),\tag{1}$$

$$d_i = \sqrt{(x_i - \hat{x})^2 + (y_i - \hat{y})^2}, \tag{2}$$

where d_0 is a reference distance that equals 1 [m], $\hat{P}_{d_0}$ is an RSSI value observed at d_0, and $\hat{n}$ is a path loss exponent. In the conventional localization method, we calculate the least squares as below using the observed value P_i and the theoretical value $\hat{P}_i$ obtained from Eq. (1) for θ, which includes the location of target smartphone and propagation parameters:

$$\min_{\theta} \sum_i \left(P_i - \hat{P}_i \right)^2 \tag{3}$$

$$\text{Subject to : } \theta = \left\{ (\hat{x}, \hat{y}), \hat{P}_{d_0}, \hat{n} \right\}; \tag{4}$$

Weighted Least Squares Method
In this method, we consider the "reliability" of the measured radio waves (RSSI values). The radio waves have higher reliability as their RSSI values become larger, since the variation in RSSI values increases with distance. Considering this fact, an appropriate weight is introduced to each term in the conventional least squares method, and the resultant weighted least squares method as follows is solved.

$$\min_{\theta} \sum_i \frac{P_i}{P_{max}} \left(P_i - \hat{P}_i \right)^2 \tag{5}$$

$$\text{Subject to : } \theta = \left\{ (\hat{x}, \hat{y}), \hat{P}_{d_0}, \hat{n} \right\}; \tag{6}$$

$$2 \le \hat{n} \le 4; \tag{7}$$

$$-50 < \hat{P}_{d_0} < -48; \tag{8}$$

where P_{max} is the largest value among the observed RSSI values. Since the estimation error is affected by feasible region of each parameter, we try to minimize it by providing additional constraints (7) and (8) whereby the rigid numbers were decided based on the data obtained in preliminary experiments. Using this method, we estimate both the location of the target smartphone and propagation parameters with high accuracy concurrently.

2.3   Proposed Graphical User Interface

The GUI of the smartphone finder is critical to achieve rapid rescue of the victims for which we designed it. A snapshot of the developed GUI is shown in Fig. 7 whereby the blue arrow indicates the direction of movement of the smartphone finder, whereas the black arrow indicates the direction of the estimated position of a target smartphone. In addition, each small dot shows the past trajectory of the smartphone finder, and its color represents the observed RSSI value at that point. The color changes from a light color to red depending on the RSSI value. By checking this kind of information displayed on the screen of the smartphone finder, the rescue worker can navigate continuously and quickly to the victim expected to be near the target smartphone.

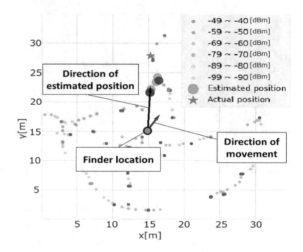

Fig. 7. A snapshot of the first GUI for navigating the user to the target smartphone

In a real-time disaster site, there would be many people with smartphones. Therefore, it is critical that multiple smartphone locations are displayed to facilitate appropriate assessment of the situation prevailing at the disaster-hit area. For this purpose, we developed two types of GUI, the snapshots of which are shown in Figs. 8 and 9. In the GUI of Fig. 8, the site area is divided into 10-m meshes, and the number of estimated smartphones is displayed in the corresponding mesh. The color of the mesh is changed depending on the number of smartphones in it. The dark red color indicates that many smartphones exist in the meshed area. By using this GUI, the rescue worker can recognize the specific area that needs priority quickly.

Another GUI considering multiple smartphones is shown in Fig. 9. In this figure, the red marker indicates the current location of smartphone finder, whereas the blue marker indicates the estimated position of the smartphone. If the rescue worker selects the estimated position, the smartphone is locked on, and a blue navigation line is displayed between the target smartphone and the finder. Subsequently, by following the navigation line, the worker can find the smartphone together with the victim easily. In addition, the number "5" on the green circle in this snapshot indicates the number of estimated smartphones in that area. If the circle is clicked, detailed information would be displayed on the screen. This GUI is simple as it helps recognize the estimated position of the smartphone easily by way of a simple online display of an appropriate information. This is an improved version of GUI shown in Fig. 7, as it is found to be relatively more effective particularly for the cases where many smartphones exist in the search area.

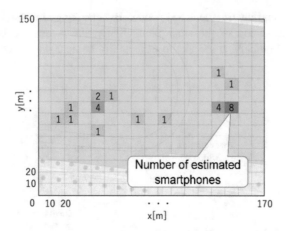

Fig. 8. A snapshot of the second GUI for displaying multiple smartphones

In all, we proposed three different GUIs enabling the user to switch from one GUI to another easily depending on the situation. To realize the GUIs, we used "Open-StreetMap" [13], JavaScript language, open-source JavaScript library (namely "Leaf-let") [14] for interactive maps, and several other techniques. A large-scale disaster often damages the infrastructure such as internet and telecommunication network. In such a case, we cannot use an internet-based map service like Google map [15]. Thus, we introduced OpenStreetMap that supports off-line map services when installed in a local map server, and the smartphone finder can display a map any time on each of the GUIs explained above without an internet connection.

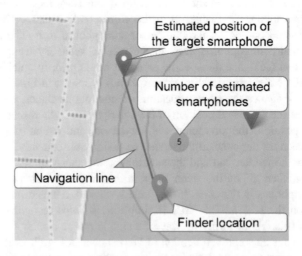

Fig. 9. A snapshot of the third GUI for displaying the target and multiple smartphones

2.4   Prototype System

We developed a prototype system of the smartphone finder, a snapshot of which is shown in Fig. 10. We used a small, lightweight Linux PC (ASUS VivoBook E200H, OS: Ubuntu 16.04) to realize the functions of the simplified packet sniffer, localization, and unique GUIs. In addition, it equips an external network interface card (NIC) (WLI-UC-GNME) for packet capturing. The NIC has a non-directional antenna and supports a monitor mode, which is needed to handle the raw packets received from the application layer in order to realize the sniffer function. Moreover, a GPS receiver (Globalsat BU 353 S4) was connected to the PC using USB interface to capture the location of the smartphone finder.

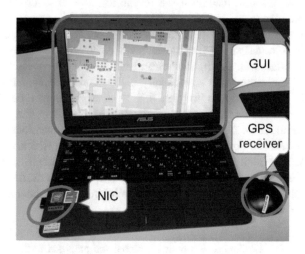

Fig. 10. Prototype system of the smartphone finder

The simplified packet sniffer function was developed using C language and "libpcap" library [16] to capture and analyze the Wi-Fi probe requests. The sniffer extracts only those information necessary for the localization function that was developed using Python language and a mathematics package, namely, "SciPy" [17] for estimating the locations of the target smartphones. The location estimation is performed for each smartphone one by one using the captured data. A smartphone is identified using the captured MAC address. In this way, the location estimation can be achieved for multiple smartphones by applying the proposed localization method to each smartphone. The unique GUIs were realized using Python and JavaScript languages.

3 Performance Evaluation

Performance evaluation was conducted using the prototype system. The experimental conditions are as follows:

1. Place of the experiments: open space in our university campus (outdoor).
2. Number of trials: seven times. (272 to 559 packets were acquired in one trial.)
3. The smartphone is placed on the ground and covered with obstacles (Clothes, Cardboard, or Iron).

A person randomly walked around the smartphone for 10 min with the smartphone finder and captured Wi-Fi probe requests. For comparison purposes, we also evaluated a conventional localization method based on the log-normal shadowing model using the same acquired data.

Table 1 compares our proposed method with a conventional localization method based on the log-normal shadowing model with regard to the average and standard deviation of the values of the estimation error. Evidently, the proposed method is found to perform the localization with significantly high accuracy compared to the conventional method. The result also implies that a smartphone can be discovered by searching in an area with a radius of approximately 13 m (7.6 m + 5.0 m) around the estimated position. Table 2 compares the proposed and conventional methods with regard to the estimation error for each type of obstacle with which the target smartphone was covered. In all cases, the estimation error of our proposed method is relatively smaller, and in particular, our method substantially outperformed in the case of iron. This fact clearly establishes that our localization method is suitable for tracing the smartphones under a collapsed building.

Table 1. Comparison of conventional and proposed method

	Averaged error (m)	Standard deviation (m)
Conventional method	36.1	41.6
Proposed method	7.6	5.0

Table 2. Comparison of conventional method and proposed method for each obstacle

Obstacle	Conventional method (m)	Proposed method (m)
Clothes	6.5	5.7
Cardboard	47.5	9.9
Two cardboards	34.4	12.5
Iron	114.6	12.8
None	1.2	1.1

Figure 11 plots an estimation result whereby the star mark (★), the large circle (●), and the small dots indicate the actual position of the smartphone, the estimated position, and the past trajectory of the smartphone finder, respectively. This result too proves the ability of our smartphone finder in locating the target smartphones satisfactorily.

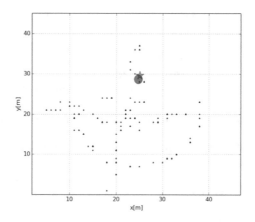

Fig. 11. Plots of the experimental results

4 Conclusion

In this paper, we proposed a smartphone finder to rescue the victims buried together with their smartphones under the collapsed buildings in a disaster site. To capture and analyze the Wi-Fi signals transmitted from the smartphones, we developed a novel packet sniffer that is relatively lighter compared to the general purpose sniffers owing to its special design to extract only those information necessary for the location estimation. Then, to estimate the location of the target smartphone with high accuracy, we proposed a new method called weighted least squares method that considers the reliability of the obtained RSSI values. The experimental results show that the estimation accuracy of our proposed method is 7.6 m, which is significantly better compared to a conventional method. Finally, we developed three unique types of graphical user interface to help the rescue workers search the smartphones easily. By just moving around a disaster-hit area with the proposed finder, the workers can find the victims with smartphones easily even if they are buried under the collapsed buildings. We firmly believe that our smartphone finder can contribute to the rescue operations related to natural disasters, because it has become a common practice for the people to carry the smartphones always with them.

Collaboration among the multiple smartphone finders by way of sharing the collected Wi-Fi packet data and the estimated location of the target smartphone is one of our future works in order to accelerate the speed of the rescue operations in locating the target smartphone.

Acknowledgments. This work was supported by Cross-ministerial Strategic Innovation Promotion Program (SIP), "Enhancement of societal resiliency against natural disasters." (Funding agency: JST).

References

1. Statista: Natural disasters – statistics & facts. https://www.statista.com/topics/2155/natural-disasters/. Accessed 18 Aug 2018
2. Hatazaki, K., Konyo, M., Isaki, K., Tadokoro, S., Takemura, F.: Active scope camera for urban search and rescue. In: 2007 IEEE/RSJ International Conference on Intelligent Robots and Systems, San Diego, pp. 2596–2602 (2007)
3. Liu, Y., Nejat, G.: Robotic urban search and rescue: a survey from the control perspective. J. Intell. Rob. Syst. **72**(2), 147–165 (2013)
4. Shah, B., Choset, H.: Survey on urban search and rescue robots. J. Robot. Soc. Jpn **22**(5), 582–586 (2004)
5. Statista: Number of smartphone users worldwide from 2014 to 2020 (in billions). https://www.statista.com/statistics/330695/number-of-smartphone-users-worldwide/. Accessed 18 Aug 2018
6. Zenella, A.: Best practice in RSS measurements and ranging. IEEE Commun. Surv. Tutor. **18**(4), 2662–2686 (2016)
7. Takashima, M., Zhao, D., Yanagihara, K., Fukui, K., Fukunaga, S., Hara, S., Kitayama, K.-I.: Location estimation using received signal power and maximum likelihood estimation method in wireless sensor networks. Electron. Commun. Jpn **90**, 62–72 (2007)
8. Zemek, M., Hara, S.: A novel target localization without a prior knowledge on channel model parameters. IEICE Tech. Rep. **107**(81), 13–18 (2007)
9. Sundqvist, J., Ekskog, J., Dil, B.J., Gustafsson, F., Tordenlid, J., Petterstedt, M.: Feasibility study on smartphone localization using mobile anchors in search and rescue operations. In: 19th International Conference on Information Fusion (FUSION), Heidelberg, pp. 1448–1453 (2016)
10. Dil, B.J., Havinga, P.J.M.: RSS-based self-adaptive localization in dynamic environments. In: 3rd IEEE International Conference on the Internet of Things, pp. 55–62. Wuki (2012)
11. Wireshark (2018). https://www.wireshark.org/. Accessed 18 Aug 2018
12. Gast, M.: 802.11 Wireless Networks: The Definitive Guide, 2nd edn. O'Reilly Media Inc., Sebastopol (2006)
13. OpenStreetMap: A free wiki world map. https://www.openstreetmap.org/. Accessed 14 Aug 2018
14. Leaflet: A open source JavaScript library for interactive maps. https://leafletjs.com/. Accessed 18 Aug 2018
15. Google map: A web mapping service developed by Google. https://www.google.com/maps. Accessed 18 Aug 2018
16. Libpcap: A portable C/C++ library for network traffic capture. https://www.tcpdump.org/. Accessed 31 Jan 2018
17. SciPy: Open source scientific tools for Python. http://www.scipy.org/. Accessed 18 Aug 2018

Preliminary Study of Haptic Media for Future Digital Textbooks

Noriyuki Iwane[1]([✉]), Chunming Gao[2], Makoto Yoshida[3], and Hajime Kishida[4]

[1] Hiroshima City University, Hiroshima, Japan
iwane@hiroshima-cu.ac.jp
[2] University of Washington, Tacoma, USA
[3] Okayama University of Science, Okayama, Japan
[4] Hokkaido Information University, Ebetsu, Hokkaido, Japan

Abstract. A digital textbook is capable to convey knowledge through a variety of learning media. Nowadays a digital textbook is still mainly consisted of verbal explanations which are traditionally called expository text. The conveying of knowledge in the format of expository text can be supported by various media such as graphs, figures, audios, and videos embedded in a digital textbook. These conventional learning media focus on auditory and visual human senses. Therefore, they are insufficient to support the learning of subjective bodily information regarding feeling of touch and sense of force. In this case, adoption of haptic media can potentially fill in the gap. Up to now there is no digital textbook system which is equipped with such supporting media. In a foreseeable future, a digital textbook system can be attached with haptic devices to pass on subjective information such as feeling of touch through human senses other than auditory and visual senses. To make that a reality, it is necessary to explore the potentials of these media to convey subjective bodily information. In this paper, we first present a generic model explaining bodily knowledge acquisition and display with the new media for a future digital textbook system. We then present a preliminary case study exploring possibilities of using a haptic device to convey a potter's knowledge, specifically the potter's skills in forming clay cups with a trowel. We introduce our initial experiments and discuss future work in using haptic media to support learning of expository text.

Keywords: Haptic interface · Sense of force · Bodily knowledge · Expository text · Knowledge transfer · Pottery

1 Introduction

A digital textbook is an educational tool and capable of conveying knowledge through various types of media such as text, figures, graphs, audios, and videos attached as digital data embedded in the contents. In most cases, a digital textbook is effective in conveying objective information by using expository text and visual demonstrations. However, when passing on subjective knowledge such as feeling of touch or sense of force using traditional media, it will be a different situation. We may assume that we can understand other people's bodily knowledge through their explanations, especially

S. Lee et al. (Eds.): IMCOM 2019, AISC 935, pp. 529–538, 2019.
https://doi.org/10.1007/978-3-030-19063-7_41

with the assistance of additional supportive media such as audios and videos. The fact is that most likely it is still uncertain what we obtained is accurately the same as what to be conveyed, because the knowledge to be conveyed in this case is subjective and they will result differently on the human sensors or receptors.

For example, if we obtained knowledge through haptic perception then it will be difficult to explain accurately using verbal language because of its subjective characteristics. Even if using audio or visual media to support the explanation, it will be insufficient to convey the actual feeling of touch and sense of force. In this case, it is reasonable to assume that if a tailored educational haptic medium could let the learners feel the force then it would help them to understand the correspondent expository text more intuitively. It might be feasible to pass on subjective information such as feeling of touch or sense of force using haptic media. However, up to now there is no digital textbook system that has been equipped with such supporting media.

Therefore, it is necessary to first investigate the influence of haptic media on users' comprehensibility of expository text as a preliminary study. On the other hand, there are commercial hardware and software systems which have implemented haptic tools in robotics and video game applications. However, the cost of such systems is usually too high for the purpose of supplementing a digital textbook system. For the purpose of personal educational use, it is also necessary to investigate potential designs and implementations which are inexpensive. In this paper, we present our preliminary study of a haptic medium for future digital textbook systems. As a case study, we aimed to convey pottery knowledge using a haptic medium along with traditional tools. The pottery knowledge to be conveyed is regarding bodily experience during forming clay cups with a trowel on a potter's wheel. More specifically, the knowledge is about under what kind of circumstances a clay forming will fail, and how it feels exactly when the clay collapses.

The rest of the paper is organized as follow: In Sect. 2, we discussed related work and presented our proposed generic model for bodily knowledge acquisition and display using a haptic medium, which can be integrated along with other tools such as video display into a digital textbook system. Section 3 introduced our case study model which targeted conveying pottery knowledge using a supportive haptic medium. Section 4 explained the preliminary experiment with the case study model and discussed the results. Finally, in Sect. 5 we will conclude and discuss the future work.

2   The Generic Model

2.1   Background

Haptic media have been under extensive study in literature and there have been numerous applications adopting haptic media [1]. A haptic component in multimedia systems is becoming an important feature to enable creative human interactions with the systems. Authors in [1] also demonstrated an application case for communication interfaces in a virtual environment. With the assistance of haptic sensors and force feedback devices people can experience feeling of touch or sense of force without being in the actual environment.

Haptic media can be adopted for educational purpose and used to support self-directed learning. There have been studies in the educational domain from theory to practice [2–5]. In practice, haptic media have been developed as haptic guidance to learn skills [6, 7] and used to support virtual lessons [8]. Effective methods have been explored to transfer bodily knowledge and skills involved in haptic experiences such as carpentry skills [9], pottery skills [8, 10–15], and so on. On the other hand, it has been in high demand to effectively train new learners to preserve these skills and expertise. How to support the effective training becomes one important research topic [16, 17]. In our previous related work [18, 19], we have studied using haptic devices to support verbal explanations in the teaching of pottery skills. We designed a system which can obtain data demonstrating pottery processing. Specifically, the force data was obtained by a force feedback device and later displayed using a haptic tool. Users can experience the pottery processing by holding the haptic tool and feel its movement. In real world, categorized by learning styles, there are two ways for skill transfer and learning, i.e., active learning and passive learning. While [6, 7] demonstrated possibilities of using haptic guidance for active learning, our approach to design the learning haptic interface has aimed for passive learning.

Our approach to support learning of expository text using haptic media has focused on the transfer of bodily knowledge such as sense of force passively. This will be desirable for a digital textbook system to integrate with a new learning medium. When conveying knowledge using a digital textbook, we may expect users to be able to understand the knowledge in the textbook. It might be hard for users to really apply the knowledge if it is involved with dexterity skills. For bodily knowledge such as craftsmanship skills, it will be unreasonable to expect users to apply them successfully without enough active practice. However, there have been supportive passive learning methods in introductory bodily knowledge transfer, which are effective for a novice learner to obtain the skills in a more hands-on experience. For example [6], an instructor of penmanship could stand behind a student to guide with the appropriate movement and apply appropriate force on a writing brush. While the student holds the writing brush with a hand, the instructor places his/her hand on the student's hand and teaches how to write characters by demonstrating his/her writing skill. In the process, the student holds the brush softly and feels the sense of force through the writing brush. This type of support is usually used at early stage of skill learning and it is effective for training used along with verbal instructions. We can imagine that this kind supportive learning will be similarly effective for a future digital textbook system to be equipped with a haptic demonstration tool. There is no doubt that further practices of such skills are necessary for a novice user to fully grasp the knowledge and skill. Assuming the contents of an expository text digital textbook is for introductory purpose, the focus of designing and studying a model for passive transfer of bodily knowledge is reasonable.

2.2 A Generic Model for Future Digital Textbook Systems

In this section, we introduce the concept of a future digital textbook system which is attached with supportive media. Figure 1 shows a generic framework explaining how bodily knowledge can be acquired for a digital textbook and how the bodily knowledge can be displayed using a digital textbook system.

When an experienced person manipulates an object, that person's skill and experience become the source of the knowledge. In the knowledge acquisition stage, that person can generate knowledge by verbally explaining the process and his/her feelings. The verbal explanation becomes the expository text which forms the main content of a digital textbook. Other than using verbal explanations, the experienced person's skills can all be represented by other media. In this model, we demonstrated usage of video and haptic media. The visual information and haptic information of bodily knowledge will be acquired separately while concurrently, so that they can be displayed by using a video display and a force display respectively. In the platform, the visual information of the process of manipulating the object is obtained by video recording. This video recording will be embedded in the digital textbook which can be played alone by the user. Additionally, in the knowledge acquisition process, the measurement of force is also recorded by using a force feedback device to obtain a sequence of force measurement. This force data will be embedded along with the video data into the digital textbook.

Therefore, a digital textbook contains the data for the expository text, video recording, and force data. A digital textbook system will have a display for the user to read the text and watch the video. Additionally, a haptic device will be required to be attached to the digital textbook system to display the sequence of force data. We propose that the sequence of data should be displayed along with the video display, during which the synchronization of those displays will be critical to let the user experience the process accurately.

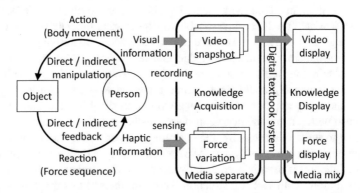

Fig. 1. Bodily knowledge acquisition and display for a digital textbook system

3   Supportive Knowledge Acquisition Using Haptic Devices

3.1   Acquire a Potter's Knowledge

We intended to examine the proposed generic model by conducting a case study in pottery learning. In the previous work, we had built an experimental system to acquire bodily knowledge and skills of a potter, of which the design was presented in [18]. We had also successfully acquired the knowledge from a professional potter using the proposed system [19]. In this section we briefly introduce how the system works.

Figure 2 shows our idea to measure force while forming clay and to store the force sequences as digital data. A digital force gauge is attached to a stand which extends to the clay forming area. At the head of the gauge, a real pottery trowel is fixed. The working surface of the trowel contacts the inside of a clay cup on a turntable. In our study, the turntable is a real electric pottery wheel. When the working surface of the trowel slides up inside the clay, the head of the force gauge will be pushed upward, hence the measurement of force can be taken. For the purpose of preliminary study, a professional potter took part in the work. The potter explained how the clay forming would fail by demonstrating at a real pottery studio using certain ceramic clay. In acquiring the potter's knowledge, we transcribed the potter's explanation as verbal knowledge. In addition, we recorded the scenes with a video camera and recorded the force sequence using the digital force gauge. In our case study, the force sequence and clay forming video compose the complementary bodily knowledge.

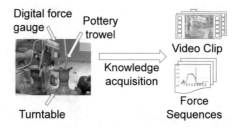

Digital force gauge   Pottery trowel

Video Clip

Knowledge acquisition

Force Sequences

Turntable

Fig. 2. Haptic data acquisition for pottery learning

Figure 3 shows a sequence of force data and their correspondent snapshots at collapsing scenes. Slow-Collapse represents a scene when the slow moving of the trowel caused the collapse of the clay. The verbal information is "When it is too slow of moving the trowel upward with regular rotation of the potter's wheel, the force will work at a certain place for too long and the clay will become too thin, then the clay will collapse." Force-Collapse represents a scene when there was strong force applied to the trowel and the clay failed. The verbal information is "When the force applied by the

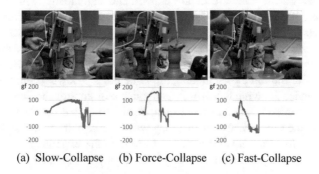

(a) Slow-Collapse   (b) Force-Collapse   (c) Fast-Collapse

Fig. 3. Force data sequence and correspondent video snapshots

trowel from inside of a clay cup is over the limit, the clay becomes too thin and will collapse." Fast-Collapse represents a scene when there is fast moving of the trowel. The verbal information is "When the trowel moves too fast and the force applied to the inside of the clay will be irregular, the rotating clay will tend to be out of center, and then the clay becomes non-uniform in thickness and it will collapse."

4   Experiment with the Haptic Display

4.1   Method

The research work of the preliminary study is to evaluate the effectiveness of the proposed haptic system along with video presentation to support learning of pottery expository text. We want to answer the research question: Will the media make the text more comprehensible when learning a potter's knowledge? If it is viable, how can we make it more comprehensible by using the proposed system?

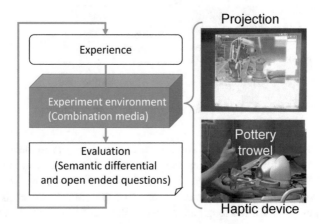

Fig. 4. Experiment structure

Figure 4 shows the structure of the experiment. Participants will experience the pottery learning process using the system and answer evaluation questions. In order to compare effects of the different media, we experimented with the two subsystems. One is the visual display subsystem and the other is the haptic display subsystem. The former presents a video clip by a projector and the latter presents a force sequence via a haptic device. For the force presentation, the same pottery trowel is attached to a force presentation device (Novint Falcon). A computer will control the force presentation device using recorded sequential force data. Participants will passively feel the generated force on the trowel. It means that they will not move the trowel actively but just hold it softly to feel the movement of the trowel. In the experiment, video clips are embedded in a power point slide and they can be played on demand.

There are two tasks. In task 1, the participants read the three pieces of verbal information (Slow-Collapse/SK, Force-Collapse/AK, and Fast-Collapse/FK) and watch the correspondent video clips twice. They then evaluate their understanding of the verbal information. In Task 2, they read the verbal information and watch the correspondent video clips along with experiencing haptic feelings twice. Then they evaluate their understanding of the verbal and bodily knowledge.

Evaluation items consists of 7-point scale multiple-choice questions based on semantic differential method (SD). The word pairs are "comprehensible-incomprehensible", "comfortable-uncomfortable", "intuitive-counterintuitive", "clear-unclear", "necessary-unnecessary", "rough-smooth", and "synchronous-asynchronous". The "comprehensible-incomprehensible" is for evaluating verbal information and the others are for bodily knowledge or haptic feeling. After completing the two tasks, participants also answered a questionnaire with open-ended questions, including: "What is a helpful feature that can support your understanding of the verbal explanations in the experimental system?", "What is a feature which is unnecessary to support your learning of the verbal explanations in the experimental system?", "What feature can be improved to help your understanding of the verbal explanations?", and so on.

4.2 Results and Discussions

Twelve students participated in the experiment. The left-hand side table in Fig. 5 shows the change of comprehensibility after experiencing the haptic devices. An upward arrow means that the experience of haptic feeling might improve the understanding of verbal explanation, i.e. an expository text. A right arrow means that there was no change, and the downward arrow means decreasing. There were 9 arrows showing up, 2 arrows showing down, and others showing no change. There is no downward arrow in SK (Slow-Collapse). Students "A" and "G" have no downward arrows. The right-hand side graph in Fig. 5 shows SD profiles of these two students. The center vertical line means neutral evaluations between respective word pairs. The top-most word pair corresponds to verbal knowledge and the others correspond to bodily knowledge evaluations. These profiles showed coincident responses. In the questionnaire, Student "A" stated that the haptic device was helpful because the system demonstrated the pressure to the hand in the process. He also stated that the system needs to improve in presenting the movements with more details. Student "G" stated both the video clip and the haptic device have helped understanding the verbal explanations.

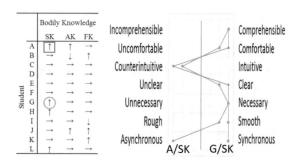

Fig. 5. Change of comprehensibility to expository text and the SD graph examples

On the other hand, Student "B" and student "I" showed opposite responses as described in Fig. 6. Student "B" claimed that the movement reproduced by the system movement was difficult to relate with the verbal explanations. Furthermore, he also stated that it was difficult to feel the differences in the movements. However, he suggested it might be helpful to demonstrate a sequence of forces of successful clay formation. On the contrary, Student "I" stated that it was very helpful to present feeling of the clay collapsing. He also suggested he would want to experience the successful forming pattern. These results indicate that there are personal differences in feeling of almost identical haptic experiences and it is hard to analyze this kind of experiments.

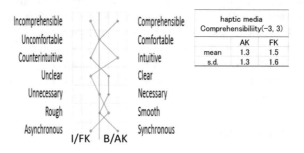

Fig. 6. SD graph for Student "B" and Student "I"

We have noticed that Students "A" and "G" indicated that haptic information of sense of force was more counterintuitive to understand the expository text in the Slow-Collapse experience evaluation. The interpretation of this response was difficult because we would assume that haptic information will be intuitive. This might have been due to the side effects of uncomfortableness or asynchronous presentations to the learners. When active learners read a text, they try to comprehend the descriptive knowledge using additional media such as figures and graphs. These additional representations together support comprehending the text both logically and intuitively. In our experiments, some students experienced the asynchronous presentations between the video and the haptic devices, which would inevitably cause confusion, and their SD graphs would be different. Interestingly, after taking the other two experiences with sense of force, these two students indicated it was more intuitive to understand the counterintuitive explanations.

5   Conclusions

In order to make verbal explanation about haptic knowledge more comprehensible, we proposed to design a supportive haptic medium which can be integrated into future digital textbook systems. In this paper, one of our main contributions is to present a generic model for the research and for the future development of a digital textbook with haptic media support. We also presented our preliminary study of using pottery learning as a special case to explore the possibility of adopting a haptic media to support learning bodily knowledge. The semantic differential method indicates that a haptic media might change users' comprehensibility of expository text regarding bodily knowledge. There remain the possibility of a haptic media. However, the responses were different among the limited number of participants. As future work, we will conduct experiments with more participants so as to conduct meaningful statistical analysis. And also we will explore the possibility of a haptic media support in the other bodily knowledge like to strike something, to shave, to polish, and so on. Our future work will also include exploring the opportunities to integrate an active learning haptic medium into a digital textbook system.

Acknowledgement. This work is supported by JSPS KAKENHI Grant Number 15K12174.

References

1. Saddik, A., Orozco, M., Eid, M., Cha, J.: Haptics: general principles. In: Haptics Technologies. Springer Series on Touch and Haptic Systems. Springer, Heidelberg (2011)
2. Martina-Ortiz, M., Schneider, O., MacLean, K.E., Okamura, A.M., Blikstein, P.: The haptic bridge: towards a theory of haptic-supported learning. In: Proceedings of the 16th International Conference on Interaction Design and Children, pp. 51–60 (2017)
3. Liu, L., Vernica, R., Hassan, T., Venkata, N.D., Lei, Y., Fan, J., Liu, J., Simske, S.J., Wu, S.: METIS: a multi-faceted hybrid book learning platform. In: Proceedings of ACM Symposium on Document Engineering, pp. 31–34 (2016)
4. Alam, K.M., Rahman, A.S., Saddik, A.E.: Mobile haptic e-book system to support 3D immersive reading in ubiquitous environments. ACM Trans. Multimedia Comput. Commun. Appl. **9**(4), 1–20 (2013)
5. Seim, C., Chandler, J., DesPortes, K., Dhingra, S., Park, M., Starner, T.: Passive haptic learning of Braille typing. In: Proceedings of the 2014 ACM International Symposium on Wearable Computers, pp. 111–118 (2014)
6. Teranishi, A., Mulumba, T., Karafotias, G., Alja'am, J.M., Eid, M.: Effects of full/partial haptic guidance on handwriting skills development. In: IEEE World Haptics 2017, Munich, Germany, 6–9 June 2017
7. Feygin, D., Keehner, M., Tendick, F.: Haptic guidance: experimental evaluation of a haptic training method for a perceptual motor skill. In: Proceedings of 10th Symposium on Haptic Interfaces for Virtual Environment and Teleoperator Systems, HAPTICS 2002, pp. 40–47 (2002)
8. Henmi, K., Yoshikawa, T.: Virtual lesson and its application to virtual calligraphy system. In: International Conference on Robotics & Automation Leuven, Belgium (1998)
9. Jose, J., Unnikrishnan, R., Marshall, D., Rao, R.B.: Haptics enhanced multi-tool virtual interfaces for training carpentry skills. In: International Conference on Robotics & Automation For Humanitarian Applications, India (2016)
10. Avila Mireles, E.J., De Santis, D., Morasso, P., Zenzeri, J.: Transferring knowledge during dyadic interaction: the role of the expert in the learning process. In: 38th Annual International IEEE EMBS Conference, Orlando, pp. 2149–2152. IEEE (2016)
11. Manitsaris, S., Glushkova, A., Bevilacqua, F., Moutarde, F.: Capture, modeling, and recognition of expert technical gestures in wheel-throwing art of pottery. J. Comput. Cult. Heritage **7**(2), 1–15 (2014)

12. Vinayak, Lee, K., Ramani, K., Jasti, R.: zPots: a virtual pottery experience through spatial interactions using the leap motion device. In: Proceedings of Conference on Human Factors in Computing Systems, Toronto, pp. 371–374. ACM (2014)
13. Arango, J.S.M., Neiria, C.C.: POTEL: low cost realtime virtual pottery maker simulator. In: Proceedings of Virtual Reality International Conference (2016)
14. Chiang, P.Y., Chang, H.Y., Chang, Y.J.: PotteryGo: a virtual pottery making training system. IEEE Comput. Graphics Appl. **38**(2), 74–88 (2018)
15. Ziat, M., Konieczky, C., Kakas, B.: Throwing of a ceramic cylindrical vessel: how height is affected by sensory deprivation. In: IEEE Haptics Symposium, pp. 529–530 (2014)
16. Park, C.H., Yoo, J.W., Howard, A.M.: Transfer of skills between human operators through haptic training with robot coordination. In: Proceedings of the International Conference on Robotics and Automation, Anchorage, pp. 229–235. IEEE (2010)
17. Avila Mireles, E.J., Zenzeri, J., Squeri, V., Morasso, P., De Santis, D.: Skill learning and skill transfer mediated by cooperative haptic interaction. IEEE Trans. Neural Syst. Rehabil. Eng. **25**(7), 832–843 (2017)
18. Iwane, N., Gao, C., Yoshida, M.: Building an experimental system to examine a method for supporting verbal explanation. In: Proceedings of 13th International Conference on Remote Engineering and Virtual Instrumentation, Madrid, pp. 341–343 (2016)
19. Iwane, N., Gao, C., Yoshida, M., Kishida, H.: A study on haptic media to support verbal explanations. In: International Conference on Interactive Learning (2016)

Effects of Emotion-Based Color Feedback on User' Perceptions in Diary Context

Jihye Han, Young June Sah, and Sangwon Lee[✉]

Department of Interaction Science, Sungkyunkwan University, Seoul, Korea
ripe-persimmon@naver.com, sahyoungjune@gmail.com,
upcircle@skku.edu

Abstract. With social network service, people have become share their emotions with others broader, and the demands of the private channel have decreased. Since the private channel has a limitation that there is no opponent of social communication, people could not share their emotions effectively. In order to supplement this part of the social conversation, we apply the communication interface to the diary of the personal channel. This study suggests color feedback as a communication interface and investigates the effects of color feedback in a diary environment, specifically how color feedback on emotional words affect users' perceptions on social presence, attachment, enjoyment, and satisfaction. To address this issue, we performed a between-subjects experiment (N = 15) with three levels of diary interfaces for color feedback: (1) emotion-based color feedback, (2) random color feedback, and (3) without color feedback (control condition). Results show that existing color feedback (i.e., emotion-based and random color feedback) was more effective than without color feedback on social presence and enjoyment; however, results of emotion-based color feedback and random color feedback had not any difference. These findings demonstrate the effects of color feedback and suggest several implications of the experiment which helps in future research.

Keywords: Social communication · Private channel · Diary interface ·
Color feedback · Emotions

1 Introduction

With the advancement of computer technology, people have become able to share their emotions through social network service (SNS). People do not write their recent status and feelings in the diary anymore; instead they put on Facebook or Twitter [4]. As a result, the demand of diary has decreased, and people have filled their social needs by sharing emotions in SNS [3, 4]. The private diary has the advantage of being able to tell the contents of the private area honestly, but there is no opponent of "social communication." In order to supplement this part of the social communication, we apply the communication interface to the digital diary channel.

Since color has been used visual communication tool in several studies [12, 13], this study applies to color as a feedback tool in the diary interface. Based on the chameleon effect theory which refers to the tendency to adopt the nonverbal behavior of interaction partners [6], we allow colors of the interface to mimic a user's emotions.

© Springer Nature Switzerland AG 2019
S. Lee et al. (Eds.): IMCOM 2019, AISC 935, pp. 539–546, 2019.
https://doi.org/10.1007/978-3-030-19063-7_42

We expect that participants would be more social present, attachable, enjoyable, and satisfactory by diary interface with color feedback than diary without color feedback interface, in a specific of color feedback, emotion-based color feedback would be more effective than random color feedback.

2 Background

2.1 Color and Emotion

The perception of color is essential to human's visual experience and plays an essential role in the visual channel [1]. This makes color a compelling visual cue for persuasive communication [12, 13]. Given the above, research on color affectivity has raised in many aspects, and therefore the emotional response to color has been investigated in multiple disciplines [17, 18].

Thus, we could apply colors to emotions. The interaction interface, which response to emotions in colors, allows the user to be more immersed in their emotions and feelings that the interface empathizes with their emotions as they do with people.

2.2 Social Presence, Attachment, Enjoyment, and Satisfaction

For users' perceptions, we have referred social presence, attachment, enjoyment, and satisfaction. At first, social presence has been argued as a factor of the awareness of existence in a series of interactions [15]. It also has defined the psychological sense of being together with intelligence that simulated other's mind [5, 10]. In terms of giving and receiving feedback communication, people can feel the social presence by feedbacks [4].

Second, regarding consumer products, Schifferstein and Zwartkruis-Pelgrim [8] have defined attachment as the power of emotional ties that consumers experience in their products. Empirical evidence suggests that when someone mimics the emotional expressions of another, and it facilitates affiliation, empathy, and prosocial behavior [6, 9]. Given that, the emotional mimicking of feedbacks can lead to increase interpersonal closeness and attachment.

Furthermore, user satisfaction plays an important role in product evaluation. According to Lee et al. [11], the more people have signs of compassion for other people's feelings, the more satisfied they are. Therefore, emotional feedbacks could also increase the satisfaction of products.

Lastly, enjoyment could be the same relationship with empathy signals. We applied that empathy signals as color feedback.

3 Hypotheses

In this paper, we present two different hypotheses which examine by the experiment:

Hypothesis 1. Users are more likely to perceive a diary interface with color feedback to be (a) social present, (b) attachable, (c) enjoyable, (d) satisfactory than a diary interface without color feedback.

Hypothesis 2. Users are more likely to perceive a diary interface with emotion-based color feedback to be (a) social present, (b) attachable, (c) enjoyable, (d) satisfactory than a diary interface with random color feedback.

4   Method

4.1   Participants

Participants included 15 graduate students (10 female, 5 male) who volunteered for the experiment were analyzed. The recruitment conditions were people who can see color (except for color weakness or color blindness) and who do not have difficulty in typing on computer keyboard. They ranged in age from 20 to 30 years, with a mean of 25.7 years (SD = 2.74).

4.2   Experimental Manipulations

In the experiment, all participants used a digital diary and received color feedbacks. The experimental diary detected the entered word and returned color feedback as highlighting the emotional word. Since we experimented with Koreans, the use of diaries was all conducted in Korean.

In the emotion-based color feedback condition, participants were received color feedback that matches the emotions of the scenario. Color stimulations of emotion-based feedback were formed of deep red, vivid red, brilliant blue-violet, brilliant blue, and vivid yellow which is related to the suggested scenario's emotions (Fig. 1). Also, in the random color feedback condition, participants were received achromatic color feedback (i.e., gray colors) which did not match the emotions of the scenario (Fig. 2). The participants in the without color feedback condition were not given any color feedback.

보았다. 그녀는 내가 꿈꾸던 이상형과 너무 닮아있었고 눈을 뗄 수가 없었다. 한 순간도 놓치기 싫어 그녀가 눈치를 채고도 남을 정도로 계속 쳐다보았고, 눈이 마주쳤다. 잠시 정신을 차리고 다시 수업에 열중하는 척을 하였지만 내 마음은 이미 그녀로 가득찼고 수업이 끝나기 전에 내 마음을 전달해야되나 끝나고 해야되나 계속 고민을 하다가 인생은 한방이지라는 친구의 말에 당장 실행하였다. 마침 오늘의 일을 예상이라도 한 듯이 주머니에 사랑해라고 적힌 풍선껌 쪽지가 있었다. 쪽지에 내 전화번호와 수줍은 마음을 담은 이모티콘을 그려 전달하였다. 그녀의 표정은 나의 예상처럼 기다렸다는듯이 쪽지를 덥석 받아채며 수줍게 웃으며 입을 가렸다. 입은 가렸지만 사슴같은 눈망울은 또 다시 나의 마음을 흔들어놨다. 어느덧 수업이 마치고 갑자기 부끄러워 도망치듯 친구들과 함께 교실을 뛰쳐나왔다. 설레는 마음으로 문자가 오길 기다렸고 얼마 지나지 않아 답장이 왔다. 끝.

Fig. 1. An example of emotion-based color feedback

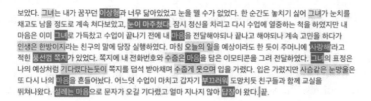

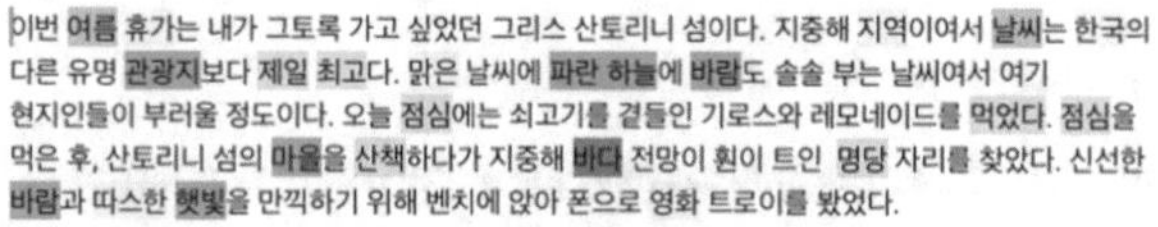

이번 여름 휴가는 내가 그토록 가고 싶었던 그리스 산토리니 섬이다. 지중해 지역이여서 날씨는 한국의 다른 유명 관광지보다 제일 최고다. 맑은 날씨에 파란 하늘에 바람도 솔솔 부는 날씨여서 여기 현지인들이 부러울 정도이다. 오늘 점심에는 쇠고기를 곁들인 기로스와 레모네이드를 먹었다. 점심을 먹은 후, 산토리니 섬의 마을을 산책하다가 지중해 바다 전망이 훤이 트인 명당 자리를 찾았다. 신선한 바람과 따스한 햇빛을 만끽하기 위해 벤치에 앉아 폰으로 영화 트로이를 봤다.

Fig. 2. An example of random color feedback

4.3   Writing Task and Measures

Participants were randomly assigned to one of the three conditions, and they are asked in writing tasks according to a hypothetical scenario (i.e., One's first love scenario). The writing task was processed with the web document application Zoho Docs.

Hypothetical Scenario. The prepared scenario was being intended to extract high valence and high arousal of emotions. In more detail, the scenario was related to joy, pride, love, happy-for, hope, and gratitude. The following is the one's first love scenario.

> *"Today is the new semester, and you go into liberal arts class. Oh my God. Look at the student sitting next to you! You have a crush on her at first sight. She is the ideal type you've dreamed of love for a long time. You can't control the smile of your mouth. All the attention is focused on the student sitting next to you. You glance at her seat, and finally, make eye contact with her. At that time, the class is over, and you talk to her with courage."*

After the writing task, participants filled out questionnaires about their perceptions on color feedbacks which suggested by a diary when they wrote the diary.

In the questionnaires, participants evaluated a social presence, ranging from 1 = "Not at all" to 10 = "Absolutely." Also, participants rated their level of agreement with each statement on attachment and enjoyment, ranging from 1 = "strongly disagree" to 7 = "strongly agree." For the satisfaction, questions were evaluated on a 5-point scale, with 1 = "Not at all" and 5 = "Very much."

Social Presence. Social presence was measured by seven items [11]: "How much did you feel as if you were interacting with an intelligent being?," "How much did you feel as if you were accompanied with an intelligent being?," "How much did you feel as if you were alone?," "How much attention did you pat to it?," "How much did you feel involved with it?," "How much did you feel as if it was responding to you?," and "How much did you feel as if you and the diary were communicating to each other?"

Attachment. The attachment was measured with 12 items [14]: "This diary has no special meaning to me," "This diary is very dear to me," "I have a bond with this diary," "This diary does not move me," "I am very attached to this diary," and "I feel emotionally connected to the diary."

Enjoyment. Enjoyment of communicating with the diary was measured by four items [7, 8]: "Interacting with the diary was exciting," "I enjoyed interacting with the diary," "Interacting with the diary was interesting," and "Interacting with the diary was fun."

Satisfaction. Satisfaction in the diary was measured by two items [4]: "After sharing this message, to what extent do you feel satisfied with the outcome?" and "To what extent do you think your goal for this message has been achieved?"

4.4   Procedure and Experimental Design

When participants arrived at the laboratory, an experimenter instructed them on how to use the diary and the process of the experiment. Prior to the experiment, participants in emotion-based color feedback condition were told that they were going to meet a diary specialized in empathy, the response to your emotions. In the random color feedback condition, participants were told that they were going to meet a diary specialized in responding to your behaviors, and in the without color feedback condition, participants were told that they were going to meet a diary. Next, participants of three conditions were given a hypothetical scenario and wrote a diary for about thirty minutes. After the writing task, participants completed a post-experiment questionnaire.

The study was conducted on a between-subjects experiment with color feedback conditions representing emotion-based color feedback, random color feedback, and without color feedback (Fig. 3). Participants were randomly assigned to one of the three conditions, and they all engaged in writing tasks according to a hypothetical scenario (i.e., One's first love scenario).

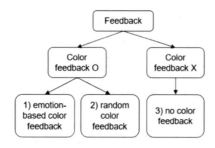

Fig. 3. Three color feedback conditions

5 Results

5.1 Descriptive Statistics

The descriptive statistics of social presence, attachment, enjoyment, and satisfaction on emotion-based color feedback, random color feedback, and without color feedback group were described Table 1.

Table 1. Descriptive statistics for measured variables.

Measured variable	Mean (SD)		
	Emotion-based color feedback	Random color feedback	No color feedback
Social presence	6.97 (1.12)	7.40 (0.69)	4.31 (1.15)
Attachment	5.68 (0.64)	5.20 (0.97)	4.84 (0.48)
Enjoyment	5.70 (1.30)	6.10 (0.91)	3.40 (1.18)
Satisfaction	4.80 (1.35)	4.80 (0.97)	3.00 (2.00)

5.2 Hypothesis Testing

To examine whether there are differences in users' ratings of the perceptions of social presence, attachment, enjoyment, and satisfaction of the diary interface depending on emotional feedback, one-way analysis of variance (ANOVA) was conducted on evaluation constructs.

The results of an ANOVA revealed significant color feedback differences on the construct of social presence, $F(2, 12) = 13.683$, $p < .01$; and enjoyment, $F(2, 12) = 8.115$, $p < .01$. Emotion-based, random, and without color feedback group reported similar levels of the attachment, $F(2, 12) = 1.686$, $p = .226$ and satisfaction, $F(2, 12) = 2.391$, $p = .134$. Thus, social presence and enjoyment showed significant differences depending on three levels of color feedback, and attachment and satisfaction did not show any significant difference.

As a post-hoc test, Tukey HSD indicated that the mean score of "social presence" on the emotion-based color feedback condition ($M = 6.97$, $SD = 1.12$) and random color feedback condition ($M = 7.40$, $SD = 0.69$) was significantly different than without color feedback condition ($M = 4.31$, $SD = 1.15$). However, the emotion-based color feedback condition ($M = 6.97$, $SD = 1.12$) did not significantly differ from the random color feedback conditions.

For the post-hoc analysis of "enjoyment" revealed that the emotion-based color feedback condition ($M = 6.97$, $SD = 1.12$) and random color feedback condition ($M = 7.40$, $SD = 0.69$) was significantly different than without color feedback condition ($M = 4.31$, $SD = 1.15$). However, the emotion-based color feedback condition ($M = 6.97$, $SD = 1.12$) did not significantly differ from the random color feedback conditions, the same results as "social presence."

Taken together, results of both "social presence" and "enjoyment" suggest that color feedback condition have effects on evaluations of the experimental diary than non-color feedback, while emotion-based and random color feedback conditions were not significantly different from each other. Therefore, in social presence and enjoyment, the model of hypothesis 1 was partially supported, and hypothesis 2 was not supported.

6 Discussion and Conclusion

Based on social presence, attachment, enjoyment, and satisfaction, we proposed three levels of color feedback of diary interface. Results of ANOVA and post-hoc analysis demonstrated the effects of the color feedback on user perceptions.

Depending on the presence of color feedback (i.e., color feedback and without color feedback), user' social presence and enjoyment about a diary had a different. In other words, when the diary interface presented color feedback, the participants felt more social present and enjoyable.

Besides, (1) color feedback condition showed a noticeable effect on social presence and satisfaction as compared with the condition without color feedback, while (2) among the color feedback conditions, the results of emotional color feedback and random color feedback conditions were not significantly different on social presence and satisfaction.

It was hypothesized that users are more likely to recognize a diary interface with color feedback as (a) social present, (b) attachable, (c) enjoyable, and (d) satisfying than a diary interface without color feedback (H1). The hypothesis was confirmed on two factors which are social presence and enjoyment, and this finding is in accordance with previous research, for instance Sas et al. [16], who have reported that feedback left on an individual's wall can enhance a positive experience in Facebook. It implies that existing the color feedback leads user' perceptions on social presence and enjoyment of the diary.

The second hypothesis assumed that users are more likely to recognize a diary interface with emotional-based color feedback as (a) social present, (b) attachable, (c) enjoyable, and (d) satisfying than a diary interface with random color feedback (H2). The hypothesis was not confirmed even on the social presence and enjoyment, and this is uncontradictory to Bazarova et al. [4], who reported that people experience greater positive effect after receiving replies to emotional posts. This is also inconsistent with our expectation that emotion-based color feedback has a positive assessment of diary than random color feedback because they are looking at users' emotions more.

A possible limitation of the experiment may be that hypothetical scenario failed to be considered participants' personal experience thus they could not draw experimental emotions. As such, they may not be representative of high valence and high arousal emotions we have been trying to draw. Another limitation is in the size and gender ratio of the sample. Ideally, group differences in color feedback conditions would be analyzed based on a larger sample; however, the experiment assigned five subjects per condition. We also note that the subject's gender may affect the intensity of emotions from the scenario and color feedbacks [2].

Future research should look into the emotional feedback effect more closely via all dimensions of emotions. In other words, our scenario does not provide other emotions; (a) high valence and low arousal, (b) low valence and high arousal, and (c) low valence and low arousal. They only suggest a scenario related to high valence and high arousal. In this regard, future research could refer to four types of emotional dimensions. Findings of the study will help future research to further the development of feedbacks of "diary interface" for social communication.

Acknowledgments. This work was supported by the National Research Foundation of Korea (NRF) grant funded by the Korea government (MSIP; Ministry of Science, ICT & Future Planning) (NRF-2017R1C 1B 1003650). This research was supported by the MIST (Ministry of Science and ICT), Korea, under the National Program for Excellence in SW supervised by the IITP (Institute for Information & communications Technology Promotion) (2015-0-00914).

References

1. Adams, F.M., Osgood, C.E.: A cross-cultural study of the affective meanings of color. J. Cross Cult. Psychol. **4**(2), 135–156 (1973)
2. Baylor, A.L.: The impact of pedagogical agent image on affective outcomes. In: International Conference on Intelligent User Interfaces, San Diego, CA, p. 29 (2005)

3. Bazarova, N.N., Taft, J.G., Choi, Y.H., Cosley, D.: Managing impressions and relationships on Facebook: self-presentational and relational concerns revealed through the analysis of language style. J. Lang. Soc. Psychol. **32**(2), 121–141 (2013)
4. Bazarova, N.N., Choi, Y.H., Schwanda Sosik, V., Cosley, D., Whitlock, J.: Social sharing of emotions on Facebook: channel differences, satisfaction, and replies. In: Proceedings of the 18th ACM Conference on Computer Supported Cooperative Work & Social Computing, pp. 154–164. ACM (2015)
5. Biocca, F., Harms, C., Burgoon, J.K.: Toward a more robust theory and measure of social presence: review and suggested criteria. Presence Teleoperators Virtual Environ. **12**(5), 456–480 (2003)
6. Chartrand, T.L., Bargh, J.A.: The chameleon effect: the perception–behavior link and social interaction. J. Pers. Soc. Psychol. **76**(6), 893 (1999)
7. Kim, K.J., Park, E., Sundar, S.S.: Caregiving role in human–robot interaction: a study of the mediating effects of perceived benefit and social presence. Comput. Hum. Behav. **29**(4), 1799–1806 (2013)
8. Koufaris, M.: Applying the technology acceptance model and flow theory to online consumer behavior. Inf. Syst. Res. **13**(2), 205–223 (2002)
9. Lakin, J.L., Jefferis, V.E., Cheng, C.M., Chartrand, T.L.: The chameleon effect as social glue: evidence for the evolutionary significance of nonconscious mimicry. J. Nonverbal Behav. **27**(3), 145–162 (2003)
10. Lee, K.M.: Presence, explicated. Commun. Theory **14**(1), 27–50 (2004)
11. Lee, K.M., Peng, W., Jin, S.A., Yan, C.: Can robots manifest personality?: an empirical test of personality recognition, social responses, and social presence in human–robot interaction. J. Commun. **56**(4), 754–772 (2006)
12. Miller, E.G., Kahn, B.E.: Shades of meaning: the effects of color and flavor names on purchase intentions. English (Philadelphia, Pennsylvania, USA: University of Pennsylvania, The Wharton School, Ph.D. dissertation). Psychology, terminology (2002)
13. Miller, E.G., Kahn, B.E.: Shades of meaning: the effect of color and flavor names on consumer choice. J. Consum. Res. **32**(1), 86–92 (2005)
14. Mugge, R., Schifferstein, H.N., Schoormans, J.P.: Product attachment and satisfaction: understanding consumers' post-purchase behavior. J. Consum. Mark. **27**(3), 271–282 (2010)
15. Richardson, J., Swan, K.: Examining social presence in online courses in relation to students' perceived learning and satisfaction (2003)
16. Sas, C., Dix, A., Hart, J., Su, R.: Dramaturgical capitalization of positive emotions: the answer for Facebook success? In: Proceedings of the 23rd British HCI Group Annual Conference on People and Computers: Celebrating People and Technology, pp. 120–129. British Computer Society (2009)
17. Suk, H.J., Irtel, H.: Emotional response to simple color stimuli. Kansei Eng. Int. **7**(2), 181–188 (2008)
18. Suk, H.J., Irtel, H.: Emotional response to color across media. Color Res. Appl. **35**(1), 64–77 (2010)

Mobile Pictogramming

Kazunari Ito[(✉)]

Aoyama Gakuin University, 5-10-1 Fuchinobe, Chuou-ku, Sagamihara, Japan
kaz@si.aoyama.ac.jp

Abstract. We have been developing a new programming learning environment called "Pictogramming". Pictogramming is coined from two words, "pictogram" and "programming". The basis of this application is using a human shaped pictogram (e.g., "human pictogram"). Pictograms have the effect of reminding oneself of some person or some object because of an abstract representation. The Human pictogram is associated with syntonic learning. This application moves these parts using a command sequence called the "Pictogram Animation Command". The application can draw movement history (similar to turtle graphics) called "Pictogram graphics". The combination of both functions enables learning using small steps in a short period of time using a small command set. This paper proposes a smartphone version of Pictogramming. We compared with other smartphone enabled programming language environments and show advantages and merit of our proposed application.

Keywords: Pictogram · Programming · Syntonic learning · Mobile learning

1 Introduction

We have been developing a new programming learning environment called "pictogramming" [1]. This application can be accessed at http://pictogramming.org/. Pictogramming is coined from two words, "pictogram" and "programming".

A pictogram is a graphical symbol used to understand a semantic concept based on the meaning of its shape. Pictograms are widely used in various fields such as counseling, safety, and facilities.

Pictograms are widely used and are designed to provide information regarding a human's action or status. So ISO Appendix 3864 includes guidelines on depicting a human shaped pictogram (e.g. a "human pictogram"). Figure 1 shows standardized shape of human pictogram from front and side direction.

The pictogram has been researched in various fields [2–4]. Pictograms are said to remind oneself of some person or some object due to a high level of abstraction present in pictograms. As for the well-known "emergency exit" pictogram, design effort focused on a person identifying with the human pictogram running towards the exit.

© Springer Nature Switzerland AG 2019
S. Lee et al. (Eds.): IMCOM 2019, AISC 935, pp. 547–553, 2019.
https://doi.org/10.1007/978-3-030-19063-7_43

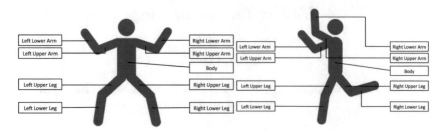

Fig. 1. Front and side direction of the human pictogram

2   Mobile Pictogramming

2.1   The Reason to Focus on a Human Pictogram

Papert, who developed LOGO [5], found great importance in the feature that children could execute LOGO's commands by pretending to be a turtle using their own bodies. This was called "syntonic learning" [6]. The book [6] mentions the following.

1. Body syntonic: It is strongly associated with children's senses and knowledge of their bodies.
2. Ego syntonic: It is consistent with the self-consciousness of children, as a human with intention, purpose, desire, likes, and dislikes.
3. Cultural syntonic: It is linked to one's activities firmly and positively rooted in one's culture.

The human pictogram recalls one's body, can represent one's ego, and created and illustrated pictograms may represent one's culture. So, the human pictogram fits with these three types of syntonic learning [7]. Such an object-first strategy for teaching introductory programming language has researched [8], but is never focused to human shaped pictogram.

2.2   PC Version of Pictogramming

Figure 2 shows a screenshot with PC browser. The application consists of three areas —"the human pictogram display panel" (left side of screen), the top right panel is "the program code description area", and "command assist buttons" to assist inputting the commands is arranged in the bottom area.

A large human shaped pictogram is displayed in the human pictogram display panel. The Panel can display the human pictogram in the front direction or the side direction, as defined by ISO 3864 Appendix. It consists of nine parts, body and head (considered a single part), two upper arms, two lower arms, two upper legs and two lower legs. Operations to the human pictogram are inputted and defined in the program code description area (upper right in Fig. 2). The input string to change states (positions) follows a format that separates opcode and arguments with blanks:

```
opcode arg1 arg2 ...
```

Opcode names, arguments, and each process can be easily predicted from one's experiences or existing knowledge. Moreover, the command and movement of the human pictogram are associated with fine granularity.

Fig. 2. Screenshot of PC version

2.3 Code Example and How to Input Code

The commands are classified into three types. First is a command to change the shape of the human pictogram, the "Pictogram Animation Command". The second type of command is equivalent to the turtle graphics, the "Pictogram Graphic Command". Not only "Pictogram Animation Command", we also implemented the function of turtle graphics to further stimulate syntonic learning, and we expect a synergistic effect by using both types. Figure 3 shows a sample code, which contains both "Pictogram Animation" and "Pictogram Graphics".

Line 1 changes the scale of the human pictogram. Lines 2 to 8 are in accordance with the LOGO programmer; this draws a rectangle that represents the victor's platform. The "M(ove)" command shown in line 10 means move 50 px in the positive x-axis direction and 200 px in the negative y-axis to get up on the box. The "R(otate)W (ait)" command means rotate a part of the body over some seconds and the next command is not executed until the movement is complete. For example, "RW LUA −120 1" shown in line 12 means "rotate the Left Upper Arm (LUA) 120° clockwise for 1 s". Lines 13 to 19 represent waving the left hand three times at a probability of 50%.

```
01:  SC 0.3
02:  PEN DOWN
03:  // Draw victor's platform
04:  REPEAT 4
05:  FD 100
06:  RT 90
07:  END
08:  PEN UP
09:  // Get up on the victor's platform
10:  M 50 -200
11:  // Raise the left arm diagonally
12:  RW LUA -120 1
13:  // Wave hands 3 times
14:  IF [rand(1,6) >= 4]
15:  REPEAT 3
16:  RW LLA -60 0.3
17:  RW LLA 60 0.3
18:  END
19:  END
```

Fig. 3. Sample code

The code are generated by following 3 ways:

(1) Input commands to code description area directly.
(2) By pressing a command assist button.
(3) By dragging the parts of the human pictogram, and rotate these about a fulcrum, the "R arg1 arg2 0" format instruction is automatically added at the cursor position of the program code description area. Here R means R(otate) command, arg1 is label of body parts, arg2 is rotate angle and last 0 means rotates instantly (0 s). Similarly, when right dragging, the entire human pictogram body translates and the "M arg1 arg2 0" format command is automatically added.

2.4   Smartphone Version of Pictogramming

The Pictogramming PC version application has implemented using HTML5, CSS, and JavaScript. The JavaScript library "processing.js", which can execute the Processing format program, is used for displaying the human pictogram. This application is compatible with typical browsers, and does not require installing any other applications. So we implements smartphone version by converting the layout and modify from PC version of application. The left of Fig. 4 is a screenshot of smartphone version. The screen width is rather short, so "the human pictogram display panel", "the program code description area", and "the command assist buttons" are arranged vertically (Fig. 4 right).

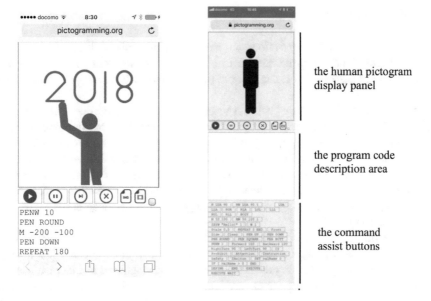

Fig. 4. Screenshot of smartphone version

We define two modes to realize programming process. One is "execute mode", which is top half of application web page. That mode displays "the program code description area", some operation buttons and top part of "the program code description area". Other one is "coding mode", which is lower of application web page. That mode displays some operation buttons, "the program code description area" and "the command assist buttons". The screenshot of these two modes are shown in Fig. 5.

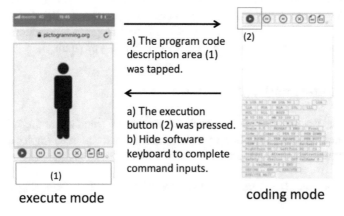

Fig. 5. Screen transition

3 Discussion

There have been developing various programming languages for beginner.

Alice allows novice programmers to be able to generate virtual worlds by controlling 3D objects [8]. But it can not run on smartphone, because it implemented by Java.

Scratch [9] and ScratchJr [10] is a visual programing environment developed by MIT Media Laboratory. Scratch programming is accomplished by dropping blocks of code and is fit as an introductory programming course because no specific knowledge about syntax is needed, and no syntax errors occur. But Scratch 2.0 works on only PC, and ScratchJr works on only iPad. Scratch 3.0 is planning to integrate with ScratchJr, and implements by HTML and JavaScript. But it would be difficult to create on the smartphone, because dropping blocks based programming language needs lot of space to illustrate blocks.

Pocketcode [11] is also a visual programming environment by dropping block based. It focus on working on smartphone, so lot of smartphone specific functions are equipped, for example, sensing data, real-time image recognition of inner camera, and so on. But it needs some steps to choose blocks and set blocks, it is not easy to switch between execute mode and coding mode than Pictogramming smartphone version. Pictogramming adopts simple text area to input code, and length of each command is not so long in Pictogramming. This is one reason why Pictogramming is also fit for smartphone version.

Okamoto stated the importance of a visual manifestation regarding the operation of the output to enhance the conceptual understanding of programming [12], where the evaluation efficiency is based on four statuses: visibility, epicritic, predictability, and independence, which she proposed as guidelines for creating programming teaching materials. A pictogram is a graphic symbol with high visibility and high epicritic sensation. An animated human pictogram (i.e., a human's motion) can used to represent the output of a program where it is strongly related to an operator's motion(s), and thus the output has a high level of predictability. Moreover, high independence is achieved if the language specification satisfies the fine-grain connections between a human pictogram's motions and the statements within a program.

4 Conclusions and Future Work

In this paper, we proposed the smartphone version of "pictogramming". We plan to perform quantitative and qualitative evaluations. And we plan to implement contents sharing platform to share programs between one's PC and smartphone.

Acknowledgements. This work was supported by JSPS KAKENHI Grant Number 16K01125 and 17K01115.

References

1. Kazunari, I.: Pictogramming—programming learning environment using human pictograms. In: IEEE Global Engineering Education Conference (EDUCON 2018), pp. 134–141 (2018)
2. Lafuente, A.S., Acevedo, R.J., Fernández, C.A., López, F.F., Maldonado, B.S.: An optimization on pictogram identification for the road-sign recognition task using SVMs. Comput. Vis. Image Underst. **114**(3), 373–383 (2010)
3. Federica, C., Eugenio, C.: Comprehension of safety pictograms affixed to agricultural machinery: a survey of users. J. Saf. Res. **55**, 151–158 (2015)
4. Yumiko, M., Toshiyuki, T., Toru, I.: Patterns in pictogram communication. In: Proceedings of the 2009 International Workshop on Intercultural Collaboration (IWIC 2009), pp. 277–280. ACM (2009)
5. Papert, S.: A learning environment for children. In: Seidel, R.J., Rubin, M. (eds.) Computers and Communication: Implications for Education, pp. 271–278. Academic Press, New York (1977)
6. Papert, S.: Mindstorms, Children, Computers, and Powerful Ideas. Basic Books, Inc., New Yorks (1980)
7. Kazunari, I.: Pictogramming—Learning environment using human pictograms based on constructionism. In: Constructionism 2018, pp. 592–599 (2018)
8. Stephen, C., Wanda, D., Randy, P.: Teaching objects-first in introductory computer science. In: Proceedings of the 34th SIGCSE Technical Symposium on Computer Science Education (SIGCSE 2003), pp. 191–195. ACM (2003)
9. Mitchel, R., et al.: Scratch: programming for all. Commun. ACM **52**(11), 60–67 (2009)
10. Louise, P.F., Brian, S., Elizabeth, R.K., Marina, U.B., Paula, B., Mitchel, R.: Designing ScratchJr: support for early childhood learning through computer programming. In: Proceedings of the 12th International Conference on Interaction Design and Children (IDC 2013), pp. 1–10. ACM (2013)
11. Kirshan, K.L., Matthias, M., Christian, S., Wolfgang, S., Bernadette, S.: Rock bottom, the world, the sky: Catrobat, an extremely large-scaling and long-term visual coding project relying purely on smartphones. In: Constructionism 2018, pp. 104–119 (2018)
12. Masako, O., Masayuki, M., Naoto, Y., Hajime, K.: Development and assessment of learning materials for computer programming focusing on "visual manifestation". Educ. Technol. Res. **37**, 51–60 (2014)

Smart Meter Security: Vulnerabilities, Threat Impacts, and Countermeasures

Asad Masood Khattak[1(✉)] [iD], Salam Ismail Khanji[1],
and Wajahat Ali Khan[2]

[1] College of Technological Innovation,
Zayed University, 144534 Abu Dhabi, UAE
Asad.Khattak@zu.ac.ae, salamkhanji@gmail.com
[2] Department of Computer Science and Engineering,
Kyung Hee University, Seoul, South Korea
wajahat.alikhan@oslab.khu.ac.kr

Abstract. Advanced Metering Infrastructure (AMI) is the aggregation of smart meters, communications networks, and data management systems that are tailored to meet the efficient integration of renewable energy resources. The more complex features and soundless functionalities the AMI is enhanced with, the more cyber security concerns are raised and must be taken into consideration. It is imperative to assure consumer's privacy and security to guarantee the proliferation of rolling out smart metering infrastructure. This research paper analyzes AMI from security perspectives; it discusses the possible vulnerabilities associated with different attack surfaces in the smart meter, their security and threat implications, and finally it recommends proper security controls and countermeasures. The research findings draw the foundation upon which robust security by design approach is geared for the deployment of the AMI in the future.

Keywords: Smart meter · AMI · Vulnerability · Smart grid

1 Introduction

One of the major challenges of the electrical system is the lack of demand response; it gives the consumers the opportunity to shift their electrical usage during peak period in response to some form of incentives [9]. However, the technological evolution through deploying the smart meters has been boosting up the demand side management and demand response concepts. Smart meters are the crucial component of the AMI where they offer two-way communication between consumers and the provider utilities. It empowers consumers to play a significant role in controlling their energy consumption. Consequently, consumers can produce energy to reduce their usage by selling it back to the provider utility as a response to some forms of incentives [1].

The real-time measures offered by the smart meter creates huge amount of data that can be easily communicated to consumers through web in-home displays (IHDs). Though soundless features of the smart meter that consumers may benefit from; its security issues and threats are of major concerns that must be addressed. Without careful preparation, consumers may not be willing to make use of the tremendous

© Springer Nature Switzerland AG 2019
S. Lee et al. (Eds.): IMCOM 2019, AISC 935, pp. 554–562, 2019.
https://doi.org/10.1007/978-3-030-19063-7_44

features offered by smart meters. Preparations must encompass its feasibility, invest-ment outlay, and the necessity to maintain an adequate privacy level [1]. On December 2015, Sandworm, the Russian hacker group, attacked the Ukrainian power grid [2]. The attack targeted substations at three electrical utilities resulting blackout affecting over 225,000 customers.

This research paper highlights possible vulnerabilities and security concerns in the AMI utilized in the energy sector (power grid). It addresses possible attack surfaces that are likely vulnerable to confidentiality, availability, and integrity exploitations. The research also highlights the built-in security mechanisms and does recommend proper security controls and countermeasures. Our findings are of vital importance to draw the foundation upon which security by design approach is aligned with the future deployment of the AMI.

The reset of this paper is organized as follows: Sect. 2 explains briefly the structure of the smart mater and AMI. Security issues and potential security attacks are addressed in Sect. 3, while proper security controls and countermeasures are recommended in Sect. 4. Section 5 concludes the paper with future research directions.

2 Advanced Metering Infrastructure: Structure Overview

AMI is an integrated system of smart meters (smart devices), communication networks, and data management systems. It offers utility providers with enormous data through the two-way communication including real-time measures, voltage measurement, load control, power outage detection, and many more. This valuable information can be communicated to third party for further analysis as well. Moreover, it gives the utility providers to remotely control, connect, disconnect, and even to configure the electricity service [10]. On the other hand, consumers may access their data through plenty of Internet-based applications to control their power usage, pay bills, and to sell back electricity on different peak timings of the day. This infrastructure efficiently integrates the renewable energy generated in the consumer's premises back to the smart grid [3].

Smart meter collects the metering data and transmits it through fixed network as in power line communications, fixed radio frequency, or the broadband over the power line (BPL) to be received by the AMI host system. Then, the data is sent to the data management systems for storage and further analysis which in turn send it to the service providers or the utilities [4].

2.1 Smart Devices

The AMI is composed of different types of smart devices that might differ whether the user is from the consumer side or from the provider side. For the clients or consumers; they are equipped with smart meter in their premises to collect the metering data and send it to the service provider. It also accepts commands from both consumer and service provider and acts accordingly. Other smart devices are the IHDs that help consumers to be aware of their energy consumption and to control it through mobile applications [11]. On the other hand, the service providers are usually equipped with load controlling devices (thermostats) to regulate the user consumption based on different criteria. Other sensors can be used to collect other forms of data that smart meter might collect to further control the energy consumption as in humidity, light, temperature, and motion detection sensors. In general, smart meters can be characterized in: net metering, the ability to communicate with other intelligent devices as in sensors, demand response commands, and time-based pricing. Other security features might include data encryption and energy theft detection [12].

2.2   Networking

A reliable communication network is a valuable component in the AMI which it is responsible for the transmission of the metering data between smart meters and other forms of smart devices to the service providers and vice versa. Consequently, it must be robust enough to transfer huge amount of data, control the access, preserve the confidentiality of the data, maintain authenticity and precision of the transferred data, and to be cost effective as well. Based on the implementation requirements; the communication network might be mesh or point-to-multipoint topology [5]. In mesh networking, neighboring smart meters are daisy-chained to forward, receive, and store transmitted data until the data is received by the data collector as depicted in Fig. 1. Typically, the mesh network operates in the unlicensed industrial, scientific, and medical (ISM) band of the RF spectrum.

However, in point-to-multipoint network topology as depicted in Fig. 2; smart meter can communicate with one data collector while data collectors can communicate with multiple meters. It typically uses the licensed RF band which has less noise compared to the unlicensed ISM band. Consequently, the two-way communication is enabled between data collectors and meters at substantial distances.

There are different types of network architectures for the realization of the AMI communication as in WiMax, ZigBee, Cellular, or power line carrier. In general, the network architecture must support the acquisition of data over Home Area Network (HAN), communication over Wide Area Network (WAN), display energy consumption via the IHDs, and managing the bi-directional communication between consumers and the service provider to facilitate account billings and consumer's awareness [13].

2.3   Data Management System

Data management systems (DMS) comprises of several subsystems that have different implementations as in consumer's interaction, power grid optimization, and service provider utility management. For instance, consumers can interact using a customized website to monitor their electrical consumption and to control it as well. Other utilities might access the consumption data for further refinement and analysis. Whereas service provider can offer better quality of power, outage control, and to adjust the power to meet their consumers' needs [6]. Moreover, DMS is also responsible for data editing, validation, and estimation to maintain complete accurate flow of traffic between consumers to data centers. Other benefits might rise of deploying the DMS in the smart grid is the analytical data available for data mining techniques and to build accurate information for decision making.

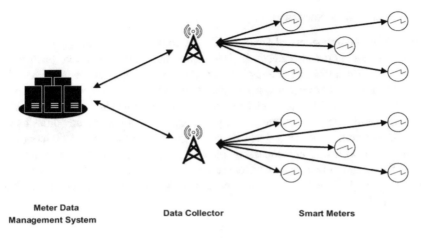

**Meter Data
Management System**   **Data Collector**   **Smart Meters**

Fig. 1. Point-to-multipoint network topology

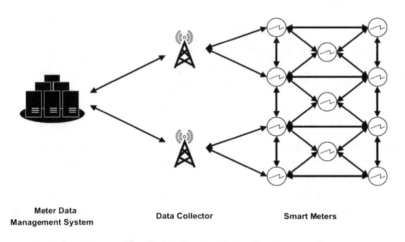

**Meter Data
Management System**   **Data Collector**   **Smart Meters**

Fig. 2. Mesh network topology

3   AMI Security Issues

According to the architecture of the AMI described in the previous section, we may identify the critical components that must adhere to specific security standards for the sake of maintaining consumer's privacy. As the information being transmitted, stored, or processed will go through the network infrastructure, data collectors, and smart meters; it is imperative to maintain its security so that to guarantee the integrity, authenticity, confidentiality, and availability of information. In this section, we will discuss possible attacks on each main components of the AMI; smart meter, data collector, and communication network.

3.1 Smart Meter Security

The main five components of smart meter are: control unit, metrology system, smart meter collector, home area network (HAN), and the optical interface. Each of these components has different targeting attacks which would impact the AMI security in general (Table 1). For instance, the control unit and metrology system are vulnerable against hardware and firmware reverse engineering. This would enable the attacker to steal information and to escalate privileges to exploit possible vulnerabilities as in the side channel attacks. Another form of attack is the modification of the control unit circuit board with one encompassing parasitic electronic components which would enable the attacker to remotely control the smart meter. Hence, the service might not be available for legitimate users and nevertheless, their data could be stolen or manipulated.

Table 1. AMI main components and subcomponents

Smart meter	Data collector	Communication network
Control unite	Control unite	Smart meter collector (RF)
Metrology system	Smart meter collector	HAN (ZigBee)
Smart meter collector	GPRS receiver	Optical port
HAN	USB interface	
Optical interface	Ethernet interface	
	UART interface	

Smart meter collector is a radio system for the communication between the smart meter and the data collector in the AMI. It sends real-time consumption readings to the data collector and in return it receives commands from the data collector as in firmware update and other configuration data. Possible attack is the interception that targets the proprietary design, as well as the data. Therefore, disruption of the power grid, power theft, and denial of power might take place.

On the other hand, HAN is responsible for the transmission of real-time consumption readings from the smart meter to other devices located in the consumer's premises. The major attack is the interception which would cause inconvenience for the consumer as well as data theft and denial of data. Finally, the optical interface is designated to the technician for the installation and configuration of the smart meter. Any interception or firmware attack might result in a massive denial of power and disruption of the grid.

3.2 Data Collector Security

Data collector is the heart of the AMI infrastructure. Since it connects to multiple smart meters; its security is vital as it would impact large portion of consumers. As depicted in Table 1; the data collector's control unit and smart meter collector are the same as the one in the smart meter. However; attacking the main unit or the smart meter collector in the data collector has a dramatic impact on the metering operations and denial of power. Moreover, the modification on its hardware would leverage unprotected interface to install malicious software to control the unit and thus, infecting all smart meters connected to it.

Table 2. AMI possible attacks

Attack	Attack surface	Scope of impact
Eavesdropping	Remotely through WAN or power line	Difficult to detect and expose consumer's privacy
Denial of Service (DoS)	Remotely through WAN	The whole or part of the grid including the AMI
Packet injection	Through WAN	False billings for both consumers and service provider
Remote connect/disconnect	Through WAN	Ranges from single premises to the whole grid
Firmware manipulation	Physical access to the smart meter. Remotely upgrade through WAN via the gateway	Affect the metrological and non-metrological part of the smart meter; can affect single to multiple gateways
Man-in-the-middle	Inside local metrological network (LMN) or WAN	False measurements for one gateway in case inside the LMN or all meters connected to the gateway in case via WAN

The GPS receiver is responsible for the synchronization between all the AMI components data along with its timestamp. Interception attacks that would lead to manipulation would indeed affect the precise timing of the infrastructure configuration, message transmission, and consumer's billing.

The USB interface has the same level of threat severity as the optical interface in the smart meter, but in addition it enables the attacker to install any malware easily through the USB flash drive. Consequently; interrupting the data collector and all connected smart meters as well. Meanwhile, modifications on the Ethernet interface or the UART interface via inserting malicious device as the Raspberry Pi would infect the backhaul communication function with the meter data management system. This would result in data theft, power theft, denial of power, and interrupting the grid.

3.3 Communication Network Security

There are two main routes for the communication; one between smart meters and data collectors and the other one is between smart meter and the devices within the consumer's premises. Typically, RF and ZigBee standards are deployed in the smart meter collector and HAN in the communication network respectively [7]. Intercepting any of these standards would expose consumer's information and manipulate its integrity. Other serious attacks might include man-in-the-middle attack, DoS, data theft, power theft, and grid interruption as well.

Table 2 summarizes the possible attacks that would take place on the AMI along with its scope of impact and the attack surface as well.

4 Security Controls and Countermeasures

It is imperative to consider security by design of the AMI in order to address all possible attacks and to prevent them accordingly. In the previous section, we have highlighted several security risks that the AMI is vulnerable to. And hence, proper security countermeasures must be addressed to maintain the availability and to enhance the efficiency of the AMI:

4.1 Encryption

Communication between different nodes of the AMI must be encrypted. Dependency on the TLS protocol is no longer efficient enough to guarantee data confidentiality. Hence, it is vital to integrate other encryption mechanisms as in RSA at both application and transport layers as well. In [7], authors proposed to use the physical unclonable function (PUF) through which it offers strong hardware authentication and efficient key management to assure the integrity of message exchange in the AMI. Another approach is to deploy message authentication codes (MAC) to protect the integrity of the AMI readings.

Another encryption scheme is proposed in [8] where aggregated data at the gateway is encrypted using homomorphic encryption which allows selective calculations on the cipher text to produce the encrypted results. And in return, when decrypted; it matches the results of processes performed on the plaintext.

4.2 Authentication Mechanism

Accurate verification of the source of the data is critical for consumer to authenticate the AMI she/he is communicating with, and for the AMI to validate the authenticity of the legitimate consumers as well. For instance, the consumer is supplied with a password to generate public-private key to build the authentication between the AMI and the power utility. Then, any new appliance asks to join the AMI for the first time is challenged to authenticate itself and to set up the communication channel to control the appliance from within the AMI. Finally, the newly joined appliance, AMI and power utility can decide whether to use the same public-private key used before or to generate a new one for each session to manage the flow of traffic between them.

4.3 Availability Mechanism

The availability of the AMI can be severely affected due to several vulnerabilities as in network jamming and packet flooding. A robust AMI can be designed to go through defined channels of alternative frequencies if the default channel is unavailable for specific period. Moreover, filtering network traffic would exclude ping requests that overwhelm the network and make it unavailable for legitimate users. Other systematic approach might help in preventing the DoS is through allowing the ARP cache static so that suspicious ARP cannot update its content malicious IP/MAC payloads. Moreover, preventing high speed traffic from reaching the kernel of the meter would significantly overcome the DoS attacks [8].

4.4   Jamming-Prevention Mechanism

In [13], authors proposed an algorithm by which the AMI would move entirely through a sequence of predefined channels so that to mitigate the jamming problem. Consequently, AMI's nodes are hard-coded with this algorithm to move through a common random sequence of channels in case the default channel was reachable for a specific period. Another scheme is suggested in [14] where a frequency agility interference avoidance algorithm utilizes energy detection and active scan to conduct a smart communication frequency channel selection. Their scheme was evaluated using ZigBee and WiFi coexistence in the context of packet error rate (PER).

5   Conclusion

Embracing state of art technologies to automate our daily life actions is still an immature field and a debatable issue. In this paper, we have investigated the advanced metering infrastructure from security perspective being the most critical matter to consider. We have evaluated the AMI attack surface and vulnerabilities associated with each and recommended proper security requirements respectively. The paper findings suggested the early security by design approach to be adopted to build more robust AMI to guarantee consumer's privacy.

Acknowledgement. This research work was supported by Zayed University Cluster Research Award # R18038.

References

1. Pepermans, G.: Valuing smart meters. Energy Econ. **45**, 280–294 (2014)
2. Finkle, J.: U.S. firm blames Russian "Sandworm" hackers for Ukraine outage, Reuters, 7 January 2016. http://www.mailonsunday.co.uk/wires/reuters/article-3389651/U-S-firm-blames-Russian-Sandworm-hackers-Ukraine-outage.html. Accessed Aug 2017
3. Yan, Y., Qian, Y., Sharif, H., Tipper, D.: A survey of smart grid communication infrastructures: motivation, requirements and challenges. IEEE Commun. Surv. Tutorials **15** (1), 5–20 (2013)
4. National Energy Technology Laboratory for the U.S. Department of Energy Advanced metering infrastructure, NETL modern grid strategy (2008). https://www.smartgrid.gov/files/NIST_SG_Interop_Report_Postcommentperiod_version_200808.pdf. Accessed Oct 2017
5. Hansen, A., Staggs, J., Shenoi, S.: Security analysis of an advanced metering infrastructure. Int. J. Crit. Infrastruct. Prot. **18**, 3–19 (2017)
6. Mohassel, R.R., Fung, A., Mohammadi, F., Raahemifar, K.: A survey on advanced metering infrastructure. Int. J. Electr. Power Energy Syst. **63**, 473–484 (2014)

7. Nabeel, M., Ding, X., Seo, S.H., Bertino, E.: Scalable end-to-end security for advanced metering infrastructures. Inf. Syst. **53**, 213–223 (2015)
8. Shuaib, K., Trabelsi, Z., Abed-Hafez, M., Gaouda, A., Alahmad, M.: Resiliency of smart power meters to common security attacks. Procedia Comput. Sci. **52**, 145–152 (2015)
9. Leiva, J., Palacios, A., Aguado, J.A.: Smart metering trends, implications and necessities: a policy review. Renew. Sustain. Energy Rev. **55**, 227–233 (2016)
10. Barnicoat, G., Danson, M.: The ageing population and smart metering: a field study of householders' attitudes and behaviours towards energy use in Scotland. Energy Res. Soc. Sci. **9**, 107–115 (2015)
11. Schultz, P.W., Estrada, M., Schmitt, J., Sokoloski, R., Silva-Send, N.: Using in-home displays to provide smart meter feedback about household electricity consumption: a randomized control trial comparing kilowatts, cost, and social norms. Energy **90**, 351–358 (2015)
12. Ueno, T., Inada, R., Saeki, O., Tsuji, K.: Effectiveness of an energy-consumption information system for residential buildings. Appl. Energy **83**(8), 868–883 (2006)
13. Aravinthan, V., Namboodiri, V., Sunku, S., Jewell, W.: Wireless AMI application and security for controlled home area networks. In: 2011 IEEE Power and Energy Society General Meeting, pp. 1–8. IEEE, July 2011
14. Yi, P., Iwayemi, A., Zhou, C.: Frequency agility in a ZigBee network for smart grid application. In: 2010 Innovative Smart Grid Technologies (ISGT), pp. 1–6. IEEE, January 2010

Shareflow: A Visualization Tool for Information Diffusion in Social Media

Chen-Chi Hu, Hao-Xiang Wei, and Ming-Te Chi[✉]

National Chengchi University, Taipei, Taiwan
mtchi@cs.nccu.edu.tw

Abstract. The popularity of social network has rapidly increased for many years. Social network continues to provide great functions to its user like sharing or diffusing the information on the social network. There are some interesting issues to the journalists are embedded inside the information diffusion. Therefore, we are encouraged to propose a visualization tool, "Shareflow", which can be used to observe information sharing activities in the social network and to explore opinion leaders and the propagation path caused by the information diffusion. Our approach consists of two parts. The first part focuses on propagation path visualization for a single post in social media. The hierarchical edge bundles method is adopted to optimize the layout and reduce visual clutter caused by excessive information. The second part focuses on the visualization for a summary of posts, which provides a tool for analyzing active users through their sharing and comments activities. A real-time interactive interface that demonstrates concurrently the breadth and depth of information diffusion is also provided.

Keywords: Opinion leader · Edge bundling · Information diffusion

1 Introduction

In recent years, social media has become popular and ubiquitous, people build connections to a wider audience by sharing their feelings and thoughts on this platform. The trends in the social media change rapidly, and social media may be suitable for people to receive information, understanding the actual process of information diffusion, however, the identification of hot topics and key persons is still a challenge. This study collected and analyzed activities in social media and provided a visualization tool that can help in understanding the propagation of topics. An overview of topics in specific periods was also provided.

In social media, people can interact with other users by message or the other expressions. In Facebook, the user can choose the like button, leave a comment, and share the messages with his friends. These activities show different degrees of agreement to a message. The number of likes and replies accumulate directly in the original message, although user diffuses his thoughts to the other friends. However, a user shows approval of a message by his stronger willingness and share

© Springer Nature Switzerland AG 2019
S. Lee et al. (Eds.): IMCOM 2019, AISC 935, pp. 563–581, 2019.
https://doi.org/10.1007/978-3-030-19063-7_45

the message to his friend by the "share" behavior. If his friend receives the shared message, his friend can also share the message with his other friends. In the diffusing process, the opinion leader can draw more people to share information when he shares a message. This chain reaction can be expressed as a node-link diagram to represent the relationship between the opinion leader and the other users and link these relationships in the information diffusion process.

For multiple posts in a given period, we have two approaches. The first approach pertains to the visualization of the diffusion path from a single post, and the second approach pertains to the overview of multiple posts. First, a node-link diagram is used to present the structure of the social network. The node represents the social network user, and the link describes the relationship between the users. However, it can cause visual clutter when the links are highly intersected. Hence, we adopt the edge bundling method to tie the adjacent edges into bundles to facilitate the flow path in the information diffusion. In addition to solving the resolution limitation for large data, we convert the node-link diagram into a radical hierarchical graph. These interactive nodes allow the user to highlight the diffusion flow for a specific node by hovering the mouse over it. Second, in addition to the timeline attribute of multiple posts. We build the hierarchical relationship among multiple posts by finding the active users. By switching to the timeline/active user mode, a user can easily explore the significant date/user between the posts, and trace the diffusion path of a specific post to find more interesting information.

The contributions of this study are summarized as follows:

- A clear and compact visualization of the information diffusion path for a single post, which can emphasize who the opinion leader is through a hierarchical visualization.
- A visual summary that can identify the active users and highlight the common relationship between multiple posts is created.
- An interactive social network exploration tool that combines the above two visualizations is formulated.

In Sect. 2, we introduce related work, including social network visualization, graph simplification, and edge bundling. Section 3 provides the main framework and related processes, such as data crawler, data analysis, data integration, local information diffusion path, and active user visualization. Section 4 presents the results and analysis, including case studies and user evaluations. Section 5 concludes this visualization and provides suggestions for future work.

2   Related Work

2.1   Social Network Visualization

Various types of social media platforms have emerged in recent years, including Facebook, Google+, Twitter, Plurk, and other well-known social media services. Google+ proposes the Ripples [17] that show users how posts are shared

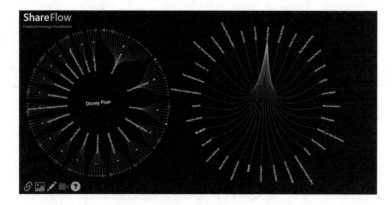

Fig. 1. ShareFlow. Left radial shows the visualization of a post in the "Disney Pixar" fanpage, and uses edge bundling to search for the active users. The right radial shows the dominant opinion leader and his followers, and it shows the diffusion path from the opinion leader into multiple hierarchies.

on Google+ and describe some constraints and possibilities for design. Their visualization layout mix of node-link and circular treemap that allows users to determine whom the important opinion leaders are based on the patterns of sharing from user interactions. This technique is useful in learning about the article trends and the influence of some leaders in the time sequence. Ho et al. defines a set of methods for measuring information dissemination on Plurk [8], and consider the quantification of a persons capability to disseminate ideas, measure the extent of propagation of a concept, and visualize information propagation in a microblog. The inference tree of Plurpagation shows the relationship between the central user and its followers, the dynamic visualization of information flow within Google Map based on the geographical scope, and the information dissemination of specific topics in the microblog. Cao et al. proposed Whisper [1] to trace the information diffusion of retweets on Twitter and visualize the sentiments in real time. Whisper retrieves information from Twitter APIs, traces events from users interest, presents information diffusion scenario of events in a timeline. Tracing from a regional perspective, and analyzes the emotional responses of communities on a given topic to identify temporal and spatial patterns for a burst event.

Online social networks have the feature in the spread of information ranging from popular topic detection to information diffusion. Guille et al. [7] present a representative method dealing with information diffusion and propose the taxonomy that summarizes the state-of-the-art. Ren et al. proposed WeiboEvents [14], a visual analytic system for analyzing events in Weibo. The system of the Retweet Tree View transforms each Weibo message into a node and connects the nodes according to the relationship between information propagation. The system interface provides an intuitive and powerful retweet tree visualization that inspires the creativity of users. OpinionFlow [20] is an interactive visualization combining with a Sankey diagram and a tailored density map, and they also provide an information diffusion model to characterize the propagation of opinions among users on social media. D-Map [3] is a visualization which could explore and analysis of social behaviors during information diffusion and propagation on social media through a map metaphor. On the branch, E-Map [2] is a visual analytics approach using map-like visualization tools for multi-facet analysis of social media data on the development of significant events, it confirms the capacities of event evolution with E-map. Sun et al. proposed SocialWave [16], a dynamic social gravity model to quantify the dynamic spatial interaction between the users on the social media in information diffusion. SocialWave utilized a new spatial visualization which integrates a node-link diagram into a circular cartogram and enables in-depth, multi-faceted analysis of the information diffusion on social media.

2.2   Graphic Simplification

When the node-link diagram shows a large number of closely interconnected nodes, these nodes can easily overlap and their edges become intertwined, thereby increasing the complexity of the node-link diagram and resulting in visual clutter. Hence, to solve this problem, the visual clutter can be reduced by changing the position of the nodes, reducing the number of nodes and edges, or combining the interaction [18,19]. Edgelens provides an interactive tool that reduces the edge overlap by curving graph edges away interactively without changing the node positions and preserving the relationship between the nodes and edges. The force-directed placement [5] is a vertice-edge layout algorithm based on physical simulation. This algorithm simulates the effect of spring expansion and contraction, adding the elements of attraction and mutual repulsion to the structure of the graph while maintaining a certain distance between each node. This method can reduce visual clutter without reducing data nodes. However, because some of the original data nodes have geographical significance, changing the location of the node may result in the loss of some information.

Flow map [12] uses hierarchical clustering given a set of nodes, positions, and flow data between the nodes, maintains the relative position between the nodes, and minimizes edge crossings and distorted node positions. It also uses binary hierarchical clustering to merge the edges with the same target and adjacent ones using spatial information by binary hierarchical clustering for edge routing. Edge routing is the mechanism used to avoid the edge of intersection through the intelligent distortion of positions. Many studies deal only with the edge of the bundle to achieve graphics simplification; similarly, the present study also uses edge bundling for single post-visualization.

2.3   Edge Bundling

Edge bundling binds the adjacent edges and reduces the overlap of the edges as much as possible without reducing the number of nodes and edges. This method solves the problem of visual clutter while retaining the details of the data and highlighting relevant information. Hierarchical Edge Bundles (HEB) is a classic algorithm for edge bundles [9]. For hierarchical data, the adjacency edges are bundled and each adjacency edge is bent and then modeled as a B-spline curve toward the polyline defined by the path via the inclusion edges from one node to another. This method visualizes implicit adjacency edges between parent nodes that result from the explicit adjacency edges between their respective child nodes. The HEB algorithm is applicable only to hierarchical structures, and hence, Geometry-Based Edge Bundling (GBEB) proposes an edge bundling framework without hierarchical structures [4]. Under GBEB, the average vector for each edge is first computed and a control mesh generated. Then, the control mesh is merged with other control meshes positioned adjacent to the direction of the line. Through these merged control meshes, and with the intersection of the control mesh to obtain the control point from the original graph, the forced and controlled meshes that intersect at the edge can only pass from the control point to achieve the effect of edge bundles.

Because HEB can only input the hierarchical data and GBEB needs to be bundled before generating a control mesh, Force-Directed Edge Bundling (FDEB), which is a self-organizing edge-bundling method [10] that treats edges as flexible springs, is proposed. The edges attract each other when both edges meet certain conditions, and then bundle together. The multilevel agglomerative edge bundling solves the issue of scalability of the bundles [6], and allows the processing of high variation in the curvature. The control mesh can be constructed hierarchically according to the location of the node. The proposed method can bundle hundreds of thousands of edges in seconds, and one million edges in a few minutes. GBEB and FDEB algorithms ignore the directionality of the edges, the weight, and graph connectivity. Selassie et al. proposed the divided edge bundling [15] as a method for tackling high-level directional edge patterns such as bundling edges in the direction of the same node, weighing the edge, preserving the directionality of the original data, and showing the overall information flow. Lhuillier et al. [11] propose a description of tasks that bundling edges and provide a wide set of example applications of bundling techniques.

3   Methods

We introduce the system flowchart of Shareflow in Sect. 3.1. The Sect. 3.2 introduces the characteristics of Facebook and discusses a tool for crawling data, data integration, and data analysis. The Sect. 3.3 introduces hierarchical edge bundling technology. The Sect. 3.4 introduces the visualization method of a single post diffuse path and the layout results. The Sect. 3.5 introduces the fanpage visualized according to three different purposes of the visualization method. The Sect. 3.6 introduces the interactive mechanism of Shareflow to assist the user conduct a deeper exploration of the information.

3.1   Overview

The system has three main components. The first part involves gathering public information from the Graph API and crawling the data by Node.js. According to the source of diffusion, data are then transformed into a node-link diagram

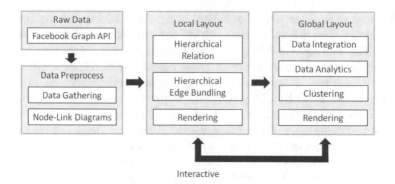

Fig. 2. System overview.

and the prototype built up in hierarchical visualization. The second and third parts use the D3.js library to implement the graphical interfaces and interactive mechanisms. The second part focuses on a single post visualization, by identifying the user who belongs to the opinion leader, by using hierarchical edge bundling for visualization. The method solves the visual clutter because of the too much interconnected and diffused paths. The opinion leaders here are defined as the user who creates a new discussion group or shares an issue from other opinion leader and also has a relative volume of users in discussion. The third part provides a global view to visualize the posts and active users in a fanpage. The system identified the active users by collecting and analyzing the multiple posts. The Rotating Cluster Layout [13] adopted to show all posts in the fanpage, the active users during the period and their concerned posts, and conclude on the trends of posts and issues which active users have discussed in the fanpage. The active user is the user who active response on the issue by replies or leaves comments on a post from the opinion leaders over a few times, we defined these users as active users.

3.2   Data Crawling, Integration, and Analysis

Data Crawling. A tool was developed for crawling post information from a Facebook fanpage. This tool provides an interface for the user to enter a fanpage ID and a period. After obtaining the users authorization, the tool can be used to crawl data by Facebook Graph API and display the post information published in the query fanpage. Figure 3 shows the query results, which includes post type, post creation time, post ID, post content, like counts, share counts, and comment counts. The data structure of our crawling tool is derived from MongoDB, it is a document schema. Every attributes store its data in a document, and a document is a post in our case.

Many studies have presented the interaction between users on the social network using the node-link diagram [8,14,17]. However, a simple node-link diagram can only show a limited relationship. Take the share visualization, for example,

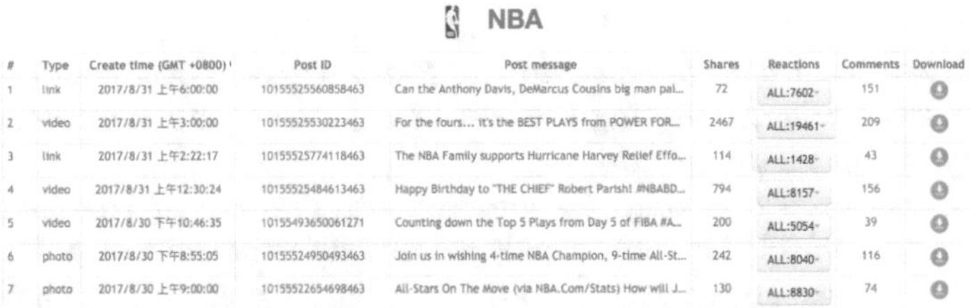

NBA

#	Type	Create time (GMT +0800)	Post ID	Post message	Shares	Reactions	Comments	Download
1	link	2017/8/31 上午6:00:00	10155525560858463	Can the Anthony Davis, DeMarcus Cousins big man pai...	72	ALL:7602	151	
2	video	2017/8/31 上午3:00:00	10155529530223463	For the fours... It's the BEST PLAYS from POWER FOR...	2467	ALL:19461	209	
3	link	2017/8/31 上午2:22:17	10155525774118463	The NBA Family supports Hurricane Harvey Relief Effo...	114	ALL:1428	43	
4	video	2017/8/31 上午12:30:24	10155525484613463	Happy Birthday to "THE CHIEF" Robert Parish! #NBABD...	794	ALL:8157	156	
5	video	2017/8/30 下午10:46:35	10155493650061271	Counting down the Top 5 Plays from Day 5 of FIBA #A...	200	ALL:5054	39	
6	photo	2017/8/30 下午8:55:05	10155524950493463	Join us in wishing 4-time NBA Champion, 9-time All-St...	242	ALL:8040	116	
7	photo	2017/8/30 上午9:00:00	10155522654698463	All-Stars On The Move (via NBA.Com/Stats) How will J...	130	ALL:8830	74	

Fig. 3. The crawling tool for a social media.

a node-link diagram can easily show the sharing relationship between the nodes, but without further design, showing the hierarchy relationship is difficult. The hierarchical edge bundling technology [9] is based on the hierarchy level to bundle edge and to show the hierarchy of the results. To obtain the hierarchical data for the following diffused path visualization, the raw data have to be integrated and the hierarchy of the sharing pattern built up after the crawling stage.

Data Integration and Analysis. The crawler tool uses the traditional table view to present post information in a fanpage. In addition to visualizing a single post, this study further aimed to visualize multiple posts in a fanpage. The tool for collecting posts in this period after the user submits the fanpage ID and query interval was revised. We stored multiple posts and data into the JSON format to serve as input for the following process. We built the tree hierarchy for collected data to provide a clear visualization layout. Because the number of users involved in the query period was too large to be displayed, we filtered the top 1% active users for visualization. The tree will root in the query fanpage as the first level, active users are in the second layer, and the posts, which are related to the active users are in the third level (leaf nodes).

3.3   Hierarchical Edge Bundling

The paths of the information diffusion can be built by active users and their followers. However, the challenge is how to reduce visual clutter caused by the highly interconnected graph. Hence, we adopted the hierarchical edge bundling method [9] to bundle the adjacent edges and solve this problem. The method is described briefly. The path of the two nodes is determined by the control points according to the distance between two nodes in the hierarchy. Assuming that P_0 and P_4 are related, a straight line is connected between P_0 and P_4, P_0 is extended to P_4, and the line corresponds to the hierarchical tree. Finding the minimum common ancestor of the path according to the two nodes, the nodes that the path passes through are the control nodes. The control nodes of this example are P_0, P_1, P_2, P_3, P_4, and the minimum common ancestor control node is P_2 from the starting point, and the control nodes are the control polygon. The control polygon affects the curvature of the spline curve connected to the two nodes, and the final spline curve is the edge of the visualization.

The spline model is rendered using the method of cubic B-spline, and cubic B-spline has a local control effect to the Bezier curve. Local control modifies the local curve without affecting the rest of the curve, and the computational complexity of this method is low.

Hierarchical edge bundling provides two ways to adjust the spline curve. The first is to control the control polygon through the control point P_i from the hierarchical tree structure. The adjusted control point Pi' is shown in the following Eq. 1:

$$Pi' = \beta \cdot P_i + (1 - \beta)(P_0 + \frac{i}{N-1}(P_{N-1} - P_0)) \tag{1}$$

where N denotes the number of control nodes, i denotes the index value of each control node, and β denotes the bundling intensity, $\beta \in [0,1]$. The user can determine the degree of modification for the control polygon by the value of β, and the control polygon will have a new spline curve.

The second method is to adjust every spline point $S(t)$ from each new curve and the new spline point $S'(t)$ as in the following Eq. 2:

$$S'(t) = \beta \cdot S(t) + (1 - \beta)(P_0 + t(P_{N-1} - P_0)) \tag{2}$$

where t is the spline curve parameter, $t \in [0,1]$, that determines the effects from spline point to spline curve by t. Hierarchical edge bundles adjust the spline curve using the two above methods, and thus affect the layout of the edge bundles. The higher the β value is, the stronger the bundle edge. The value of β in the visualization of the diffuse path from a single post is 0.7.

3.4   Diffuse Path Visualization with a Single Post

The diffusion graph is dominated by a node-link diagram, where the nodes represent users who diffuse the posts, and the links represent the direct diffusion relationship between the users. The nodes are placed on the contours of the circle to maximize display space. The nodes are arranged in the clockwise direction starting from the top of the circle. Based on the principle of hierarchical bundling, we classify active users into several groups according to the source of the sharing from and set a higher priority to the larger number of the source. The order of users in the same group is also prioritized in a clockwise order. In this layout, the nodes on the circle are not all sorted according to the time points of the users diffused activities but rather according to the result of the grouping, because the layout is hierarchical-oriented to edge bundles.

Compared with the tree structure, a circle layout can use space more efficiently and present the maximal sub-nodes in the same space. While the tree structure presents numerous sub-nodes, the size of the screen is a limitation, as such, some parts of the sub-nodes which cannot be displayed altogether. A circle layout design allows users to see the whole structure.

For the hierarchical edge bundling, the input data must be a hierarchical JSON file. Preprocessing the diffusion relationship of the posts, the name of the source in front of each follower was added, following the format of "the opinion leader - the follower", and the diffusion of users and opinion leaders as nodes was defined. The nodes and edges were colored according to their classification, the diffusion paths by different opinion leaders were emphasized, and the hierarchical relationship was used to process the hierarchical edge bundles.

To enhance the difference between direct information diffusion and indirect information diffusion, each opinion leader was assigned a different color. According to the classification of opinion leaders, the edges were drawn in different colors. As such, a user can visually see the color of the posts and identify whether there exists an indirect information diffusion. Moreover, this allows the users to easily see the number of opinion leaders. Through edge coloring, the difference of the users between different groups was emphasized, and the percentage axis assisted in exploring the power opinion leaders. In addition to the color emphasizing the importance of opinion leaders, we used other elements, such as the name of nodes and the percentage axis of opinion leaders, to provide the detailed information about the active users.

The information flow is another key data in understanding the information diffusion in a social network. In this study, the line width and the line color show the directional attribute, the width of the line from wide to narrow indicates the flow from a single opinion leader to his follower. Depending on the level of depth of the sharing that occurred, the color of the line denotes from white to dark blue. Base on the color gradient, a user can observe the development sequence of the graph and the participated order of the important opinion leaders in the process.

3.5 Fanpage Overview Visualization

An overview of the posts in a fanpage to help a user understand the information flow is provided below. The Rotating Cluster Layout [13] layout was adopted based on the preprocessed hierarchical data in Sect. 3.2. Each layer is connected by edges to indicate the relevance of the related information. We provide three modes in the overview layout: timeline, active sharer, and active comment. The active comment is decided by the comment has the relative volume of users express consent on it. The query fanpage is shown in the center of the circle, as the first layer in the hierarchy. The external nodes in the circle are posts in the third layer. The main difference between each mode is the node type in the second layer.

In the timeline mode, the node type in the second layer is the publish date of a post. The layout arranges the node by date in the clockwise order and starts from 12 o'clock. The post along with the clockwise direction also matches the sequence that the post was published to indicate the time sequence of the data. The radius of the node denotes the number of diffusion activities (ex: share or comments). Moreover, we use the node color to indicate the message type of the post (ex: link, photo, feel, and video).

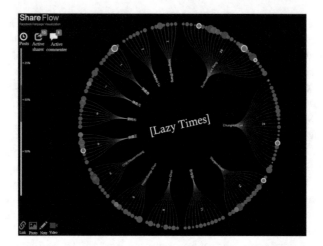

Fig. 4. The comments visualization by active commenters in "Lazy Times". The outermost nodes are denoted as different share types of the posts: Link (blue), Photo (red), Note (green), and Video (purple). The active sharers are denoted in the inner circle, and the colors are defined as the share ratio in the same shared information: less than 25% (orange), less than 50% (blue), and more than 50% (white).

The other two modes are related to the active user (sharer and commenter). In this study, the data on active users was used as the medium to establish the relationship between the multiple posts. However, considering the popular fanpages with a huge number of fans, the number of posts during the period of diffusion and the number of diffusions by each user may exceed the space of the layout. A threshold was identified to filter in the significantly active user. For example, we selected the top 1% active users in the "Lazy Times" fanpage as shown in Fig. 4. The text in the middle of the ring represents the active users, and the node along the line that extends to the outside of the ring indicates the source of the post diffused by the active users.

The active users are categorized by the quartile of the number of active users in the bar chart on the left side in Fig. 4. The active users are divided into three categories: The first tier is the top 25% of the active users (Q1), the second tier is 25% to 50% of the active users (Q2), and the third tier is 50% to 100% of the active users (Q3 + Q4). The bar chart is colored orange, light blue, and white to show the differences among the categories, and through the interactive mechanism of tooltip box to obtain the preference of the post issues which were diffused from these active users.

3.6   Interaction Mechanism

To present a simple and clear layout, we hid the detailed information of the post and active user and designed an interactive mechanism to display them on demand. First, we integrated the fanpage overview and the diffuse path in the above section. When a user clicks a post node in the fan page overview, then the diffused path of the post will be juxtaposed on the right view as shown in Fig. 5. We also highlighted the selected post node in orange color. In the active sharer/comment mode, we can also find multiple orange post nodes when the post is interacted by multiple active users. This design will help a user to discover the relationship between active users. Second, when a user mouse over a post/user node, the tooltip box will pop up to display the detailed information. Thus he can get the title, message, and comments of a post node. These will help users explore the popular posts and the hot posts which active users concern are concerned with.

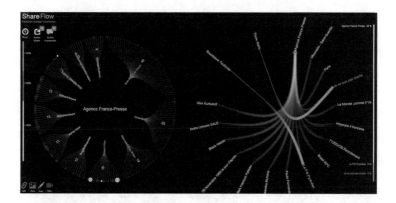

Fig. 5. Left view shows the overview for a fanpage, and the right view shows the diffuse path from the selected post (orange node).

4   Results and Discussion

4.1   Implementation Environment

The main programming language used in this study was JavaScript, and we used D3.js library for web-based visualization. In the experimental environment, the operating system is Window 7 64-bit, Intel®Core^{TM} i5-4430 CPU@3.00 GHz, 16 GB memory, and NVIDIA GeForce GTX 670 GPU.

4.2   Case Study

In this section, we discuss domestic and international fanpages separately. First, we demonstrate the effectiveness of the visualization on the active user and popular posts in a fanpage called "Lazy Times". Second, a comparison of an event in Paris in November 2015, was used. Using different news media fanpages, we can observe the issues that caught the attention of active users in various fanpages and explore noteworthy users. The related fanpage posts who actively participated in the discussion.

Single Fanpage Observation. Lazy Times is a fanpage created by a journalist who loves sharing information, thus information diffusion can be observed by following the sharing activities. Figure 6 show the results from the data collected from August 1, 2015, to September 15, 2015. A total of 973 users shared the posts and had 13 active users. Moreover, 477 users who left comments and 5 can be considered as active commenters. We chose a post (denote an external node in orange color in the left radical) to display the diffusion path visualized by edge bundling layout with color classification in the left. We can observe from the post that (denote in blue, green, and yellow) continued share and post. The top two active users ("Lazy Time" and green node) receive over 80% shares by their followers.

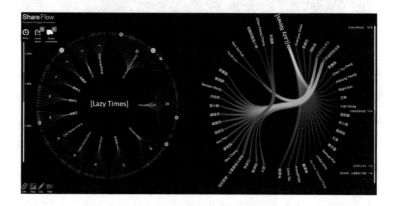

Fig. 6. Visualization of using ShareFlow.

Single Event Observation. Considering a news event occurred in Paris as an example, Shareflow can visualize the comment activities from the fanpage. The event occurred on November 13, 2015, and our crawling date was set between 13 and 16. However, because of the enormous number of posts published by CNN and BBC, even one percent of the total data cannot be displayed in whole with the current layout space. CNN and BBC change the crawling date was changed from 13 to 15. We analyzed the comment information from five well-known foreign news media fanpage for the period as shown in Fig. 7. We can estimate the scalability and impact of the five fanpages through the overall appearance of the visualization. In order to avoid visual clutter, we have set a maximum of display nodes around 2,500 in our experiment. In Fig. 7, in CNN fanpage, there are 225 active commenters in totally 2,517 posts.

Table 1. Questionnaires in user study.

Part 1: The readability of the layouts   Each nodes denote as a post and its area is proportional to the number of shares.   Q1: What is the number of shares of the post which has the most shares?
Part 2: The readability of the user's issue   The visualization tool shows the active commenters of a fanpage and each nodes denote the posts commented by the active commenters.   Q2: Which issue preference of the posts that the active commenter contributed to?
Part 3: The hierarchy of the diffusion path   Q3: What is the most share depth of the specific post?   (e.g. B shared from A; and C shared from B, then the most share depth is 2.)
Part 4: The overall evaluation of the diffusion path with different diffusion paths   Q4: Compared with the three visualization, evaluate the overall aesthetics, information flow, the hierarchy, and the opinion leader?

4.3 User Study

In this study, our target users are News related practitioners and the fanpage owners. News related practitioners study the social event and try to discover the important information inside these events. And the fanpage owners would focus on every situation on his/her fanpage. Therefore, the common demand for these target users is that they all focus on the information diffusion on the social network.

We evaluate our layout design and compared it with the Cluster Dendrogram and the Treemap (shown in Fig. 8). We also conducted a web-based questionnaire that allows the users to compare the differences between the layouts by different active sharers, active commenters and the diffusion path. We also added a different style on the comparison of the visualization of a single post diffuse path and recorded the cost of the users response time separately. This user study allowed a user to try each operation from three different layouts and respond to the questionnaire. This user study contained four parts, with the first two parts focused mainly on the overview visualization for a single fanpage. The last two parts concentrated on the evaluation of diffuse path visualization from a single post.

The user study is three steps. First of all, we describe our research purpose and the follow-up test. The second step, we introduce the initial of our system and some operations inside this visualization tool. The source of pre-processed data and the schematic layout with the visual encoding. At the last step, users

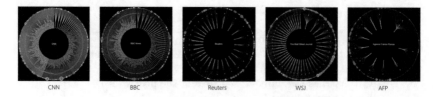

Fig. 7. Active commenter visualization of international news media in ShareFlow. Below each graph can be found, the name of fanpage, the time interval of data collection, the total number of the commenters, and the number of active users.

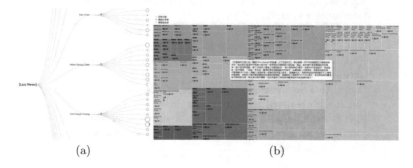

(a)   (b)

Fig. 8. Comparable visualizations in pilot study. (a) Cluster Dendrogram. (b) Treemap.

actually operate our system. And in this step, it can be subdivided into four steps: the readability of the layouts, the readability of the user's issue, the hierarchy of the diffusion path, and the overall evaluation of the diffusion path with different diffusion paths.

In total, 30 participants, (20 males and 10 females) participated in the study. Regarding the participants field of research, 23 majored in Computer Science, two majored in Journalism, and one majored in both areas, and four participants were not from these two majors. The participants ages ranged from 21 to 31 years old. We wanted to know the differences between the different layouts and tested the participants if they can interpret the visualizations correctly, and considering their time costs. Our task allowed users to find the most iconic characteristics target in the diagram, enabling users to find the clue through interactive operators. In this part, all 30 participants were correct, and the only difference was the time costs in the operation process.

The second part pertains to the readability of the layouts. The user is asked to find a specific user in the graph, observe which post the user is involved in, and estimate the issue preference of the user. The results can be found in Fig. 10, and show the readability of the participants is relatively consistent. Our layout has the minor advantage than the other two methods on the readability of the issues because the users can identify the same category of issues from our layout, while the other two methods still identify non-single categories of issues. However, we found the

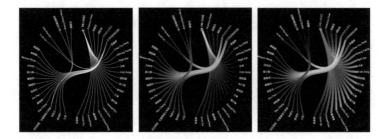

Fig. 9. Color design in ShareFlow. (a) Original visualization, the edge has no color and the two widths have no discernible difference. Edge transparent: 1. (b) Monochrome gradient visualization. Color of edge changes with the gradient (white to blue). The width of edge change with direction (wide to narrow). Edge transparent: 0.7. (c) Multi-color gradient visualization Color of the edge change with gradient and layer (red→ yellow→ green→ blue). Width of edge change with direction (wide to narrow). Edge transparent: 0.7.

differences in the time cost when the users operated the layouts. These data were processed by the TRIMMEAN function. First to remove the top 5% and the bottom 5% extreme data and get the average and the standard deviation of the data as Fig. 11. In Q1, the cost time of Treemap is obviously lower than the other two visualizations. The analysis shows that Treemap has the advantage of visualizing some important data characteristics. In Q2, our layout costs lower time than the other two. It means that the user spent less time to know the issue preference of a user.

The third part pertains to a test on a hierarchical diffuse path, where is your visualization by three different color styles as Fig. 9. The basic style is to use the initial hierarchical edge binding method, and no change in the color, width, and transparency of the edge can be observed. The monochrome gradient style and the multi-color gradient style are based on the depth of the diffuse hierarchy, and the line color and width will change with the number of hierarchies. Moreover, our tool also can highlight the influence of each opinion leaders with different gradient and color style from the different share behavior of each opinion leaders in our study case. Some results are shown in Fig. 12. The three kinds of style can successfully present over 70% of the primary hierarchy. However, the basic style has more opinions on the disadvantage of the hierarchy than the other two layouts using color gradient.

The last part evaluated the three different styles of diffusion path visualization. At first, the results of the overall aesthetics are shown in Fig. 13. The color gradient assisted in identifying the different hierarchies. The hierarchies come from opinion leaders, and so the users can still identify the different opinion leaders through the different colors. We concluded from the results that changing the color gradient and the width of the line assisted in the presentation of the direction of the graphics and hierarchy.

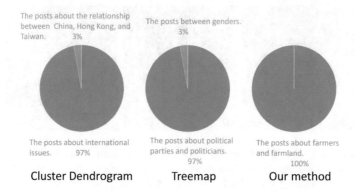

Fig. 10. The readability of the posts from users by three different visualizations.

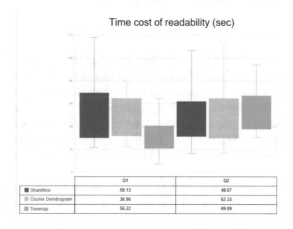

Fig. 11. The time cost from the readability of the posts that users find out the preference issue by three different visualizations.

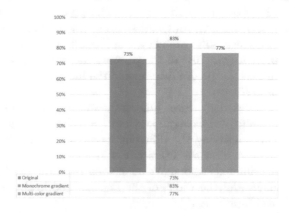

Fig. 12. Hierarchies of the diffuse path with three different color layout.

4.4   Limitation

The current visualization tools can only present a limited amount of data. If the data size is too large, the browser cannot render the data visualization effectively. Among the possible reasons are the length of the query interval is too long, a large number of posts on the fanpage and users interact with each other on the fanpage too frequently. In our experiment on the CNN fanpage, the maximum

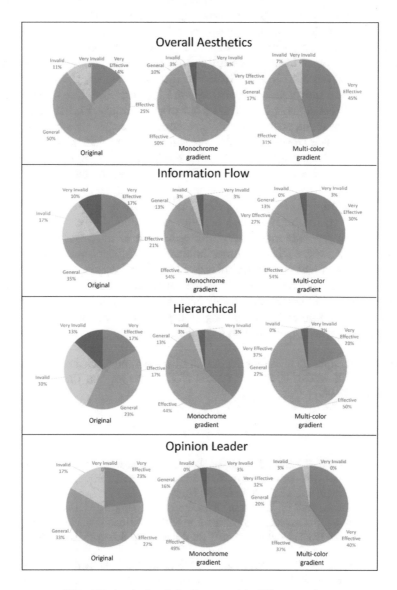

Fig. 13. Analysis of the layout with different colors.

number of graphs that can be displayed was about 2,500 data nodes (the number of post nodes in the external ring), the time interval is five days, with a total of 225 active commenters and 2,517 posts.

5 Conclusion and Future Work

This paper aimed to analyze user interactions using posts in social media and proposed a visualization tool for exploring the information diffusion, opinion leader, and hot topics. The fanpage overview layout provides three modes: time-line, active sharer, and active commenter. An analysis could be done by integrating multiple posts from the Graph API to analyzing and defining the active sharers and active commenters. After the filtering process, the tool visualizes the active groups and popular topics users are interested in. Users could explore posts with indirect sharing from the layout of the fanpage, and then visualize a single post. For a single post, we establish an opinion-leader-centered hierarchical structure to discover the relationship between the sharer and the opinion leader. We adopted the hierarchical edge bundling method, color design, and interactive mechanism to build a clear and compact visualization for information sharing and diffusion.

This study aims to help users view information with breadth and depth. Breadth refers to the active social network users and the relevant hot topics, while depth refers to the information diffusion path of the posts that help users explore the key person who promotes information diffusion and the rest of the diffusion. The scope of this study included the identification of the active users and the issues they are interested in by analyzing the frequency of social network activities (share and comment) from a single fanpage. We want to extend the scope to multiple fanpages that belong to a particular area and the proportion of activities of typical users and their preferable issues.

Acknokledgement. This work is supported by the National Science Council, Taiwan under MOST 106-2221-E-004-010-MY2 and NCCU Aim for Top University Research Project 107H121-08.

References

1. Cao, N., Lin, Y.R., Sun, X., Lazer, D., Liu, S., Qu, H.: Whisper: tracing the spatiotemporal process of information diffusion in real time. IEEE Trans. Visual Comput. Graphics **18**(12), 2649–2658 (2012)
2. Chen, S., Chen, S., Lin, L., Yuan, X., Liang, J., Zhang, X.: E-Map: a visual analytics approach for exploring significant event evolutions in social media. In: Proceedings of the IEEE Conference on Visual Analytics Science&Technology (VAST) (2017)
3. Chen, S., Chen, S., Wang, Z., Liang, J., Yuan, X., Cao, N., Wu, Y.: D-Map: visual analysis of ego-centric information diffusion patterns in social media. In: IEEE Conference on Visual Analytics Science and Technology (VAST), pp. 41–50. IEEE (2016)

4. Cui, W., Zhou, H., Qu, H., Wong, P.C., Li, X.: Geometry-based edge clustering for graph visualization. IEEE Trans. Visual Comput. Graphics **14**(6), 1277–1284 (2008)
5. Fruchterman, T.M., Reingold, E.M.: Graph drawing by force-directed placement. Softw. Pract. Experience **21**(11), 1129–1164 (1991)
6. Gansner, E.R., Hu, Y., North, S., Scheidegger, C.: Multilevel agglomerative edge bundling for visualizing large graphs. In: Pacific Visualization Symposium (PacificVis), pp. 187–194. IEEE (2011)
7. Guille, A., Hacid, H., Favre, C., Zighed, D.A.: Information diffusion in online social networks: a survey. ACM SIGMOD Rec. **42**(2), 17–28 (2013)
8. Ho, C.T., Li, C.T., Lin, S.D.: Modeling and visualizing information propagation in a micro-blogging platform. In: International Conference on Advances in Social Networks Analysis and Mining (ASONAM), pp. 328–335. IEEE (2011)
9. Holten, D.: Hierarchical edge bundles: visualization of adjacency relations in hierarchical data. IEEE Trans. Visual Comput. Graphics **12**(5), 741–748 (2006)
10. Holten, D., Van Wijk, J.J.: Force-directed edge bundling for graph visualization. In: Computer Graphics Forum, vol. 28, pp. 983–990. Wiley Online Library (2009)
11. Lhuillier, A., Hurter, C., Telea, A.: State of the art in edge and trail bundling techniques. In: Computer Graphics Forum, vol. 36, pp. 619–645. Wiley Online Library (2017)
12. Phan, D., Xiao, L., Yeh, R., Hanrahan, P.: Flow map layout. In: IEEE Symposium on Information Visualization, INFOVIS 2005, pp. 219–224. IEEE (2005)
13. Reingold, E.M., Tilford, J.S.: Tidier drawings of trees. IEEE Trans. Software Eng. **2**, 223–228 (1981)
14. Ren, D., Zhang, X., Wang, Z., Li, J., Yuan, X.: WeiboEvents: a crowd sourcing Weibo visual analytic system. In: IEEE Pacific Visualization Symposium (PacificVis), pp. 330–334. IEEE (2014)
15. Selassie, D., Heller, B., Heer, J.: Divided edge bundling for directional network data. IEEE Trans. Visual Comput. Graphics **17**(12), 2354–2363 (2011)
16. Sun, G., Tang, T., Peng, T.Q., Liang, R., Wu, Y.: SocialWave: visual analysis of spatio-temporal diffusion of information on social media. ACM Trans. Intell. Syst. Technol. (TIST) **9**(2), 15 (2017)
17. Viégas, F., Wattenberg, M., Hebert, J., Borggaard, G., Cichowlas, A., Feinberg, J., Orwant, J., Wren, C.: Google+ ripples: a native visualization of information flow. In: Proceedings of the 22nd International Conference on World Wide Web, pp. 1389–1398. ACM (2013)
18. Wong, N., Carpendale, S.: Interactive poster: using edge plucking for interactive graph exploration. In: Poster in the IEEE Symposium on Information Visualization (2005)
19. Wong, N., Carpendale, S., Greenberg, S.: Edgelens: an interactive method for managing edge congestion in graphs. In: IEEE Symposium on Information Visualization, INFOVIS 2003, pp. 51–58. IEEE (2003)
20. Wu, Y., Liu, S., Yan, K., Liu, M., Wu, F.: OpinionFlow: visual analysis of opinion diffusion on social media. IEEE Trans. Visual Comput. Graphics **20**(12), 1763–1772 (2014)

Does Crime Activity Report Reveal Regional Characteristics?

Tsunenori Mine[1]([✉]), Sachio Hirokawa[2], and Takahiko Suzuki[2]

[1] Department of Advanced Information Technology,
Faculty of Information Systems and Electrical Engineering, Kyushu University,
744 Motooka, Nishi-ku, Fukuoka 819-0395, Japan
`mine@ait.kyushu-u.ac.jp`
[2] Research Institute for Information Technology, Kyushu University,
744 Motooka, Nishi-ku, Fukuoka 819-0395, Japan
`{hirokawa,suzuki}@cc.kyushu-u.ac.jp`

Abstract. Crime is one of the most important social problems for administrative region. Ascertaining the detailed characteristics of crime and preparing countermeasures are important to keep community life safe and secure. A lot of studies using crime data and geographical data have been carried out with a view to crime prevention. These studies include analyzing geographical features of crime, mapping crime-related information and crime hotspots on the map, predicting crime rate and so on. In addition, police stations have recently begun emailing notifications regarding crime to citizens to help them avoid crime. The e-mail messages include rich information about regional crime; they are actively used by services providing guidance to people in how to avoid crime. These services map the messages onto regional maps using the location information in the messages and show the relations between the locations and crime on the map. In addition, some services send alarms to their users when the GPS information of the users indicates that they are passing by the places where crime has occurred. However, these services only use the location and crime information extracted from the messages. Thus, we cannot say the messages have been fully used to clarify characteristics of regional crime. Therefore, in this paper, we investigate whether or not the crime messages sent by e-mail can be further exploited as a valid source for analyzing the criminal characteristics of a region, i.e., whether or not they include the characteristics of regional crime. To this end, in this research, we conducted experiments to make clear whether or not the crime messages sent by e-mail can help to distinguish regions. Experimental results illustrate that the contents of e-mail crime messages helped to distinguish regions having greater than or equal to 100 reports, with an average F-measure of about 90.3%, while only using the names of the areas where crime has occurred cannot match that F-measure.

Keywords: Crime report analysis · Regional characteristics ·
Classification · Machine learning · Text mining

© Springer Nature Switzerland AG 2019
S. Lee et al. (Eds.): IMCOM 2019, AISC 935, pp. 582–598, 2019.
https://doi.org/10.1007/978-3-030-19063-7_46

1   Introduction

Crime is one of the most important social problems. A lot of studies have been conducted to understand crime characteristics, to distinguish crime hotspots and to infer regional crime rate with a view to preventing crime activities from occurring. These studies have examined regional situations where crime occurred and have analyzed them to predict crime hotspots considering demographic information and the presence of facilities or other points of interest (POIs) around the hotspots, such as parks, parking lots, schools in the region [7,17]. Wang et al. [19] used POI information such as area names, 10 major categories defined by FourSquare[1] and locations of facilities in the region, and taxi flow data showing how people commute the city, for crime prediction.

These studies employ geographical information analysis methods, which are useful for grasping the relationships between regional characteristics and the characteristics of crime, and for predicting the occurrence of crime in similar areas, but may risk losing specific characteristics of regional crime and its concrete situations when summarizing the data as geographical information. For example, Yamamoto [21] described the current situations of sex-related crime in Fukuoka prefecture, Japan in detail and discussed enlightenment activities for avoiding such crime. However, he summarized regional crime from the ward or block level to the city level based on geographical information.

And now, police stations have recently been notifying citizens about crime by e-mail so that they can take measures to avoid it. The e-mail messages are rich in information about regional crime, such as the characteristics of criminals and victims, the methods employed in the crimes, and the location of crime hotspots. To help people avoid crime, smartphone app-based warning services such as "Moly"[2] and "Digi Police"[3] extract locations of crime from e-mail messages, map them onto a map and send out a caution message to app users if they are near one of the locations.

However, the messages have not yet been thoroughly exploited to clarify the characteristics of regional crime. In this paper, we investigate whether or not the contents of e-mail crime notifications can be used to analyze the characteristics of crime in the region, i.e., whether or not they include characteristics of regional crime. To this end, we use crime reports provided by Fukuoka prefecture police stations[4] because the prefecture includes both big cities and small towns. The characteristics of Fukuoka prefecture enable us to analyze a variety of regional characteristics of crime, and the crime rate for Fukuoka prefecture places it among the five prefectures in Japan with the highest rate. In this paper, we try to answer the following research questions:

RQ1: Can we distinguish a region from others on the basis of crime type name occurring in the crime reports? i.e. does the region have its specific crime type names?

---

[1] https://ja.foursquare.com/dev/overview/venues.

[2] http://ascii.jp/elem/000/001/493/1493358/.

[3] http://www.appbank.net/2016/03/05/iphone-application/1175390.php.

[4] http://www.police.pref.fukuoka.jp/fukkei-mail/mailmg.html.

584 T. Mine et al.

RQ2: Can we distinguish a region from others on the basis of area names in the crime reports? i.e. are there strong relationships between the region and the names of areas where crime occurred?

RQ3: Can other specific words in a crime report distinguish a region from others? i.e. does the report include some characteristics related to the region alone?

If the answer to RQ1 is yes, the types of crime are characteristics of a region, and what we have to do is to analyze the crime types related to the region. If the answer to RQ2 is yes, we can distinguish a region solely on the basis of the set of area names in the message. Finally, if the answer to RQ3 is yes, crime reports can be used as a good resource for distinguishing regions because they include information expressing the distinct characteristics of regional crime.

Meanwhile, since the answers to these questions may depend on the particular region in question, we should make clear which regions can be characterized by which type of information. To answer the research questions above, we conducted experiments to clarify whether or not crime reports can help to distinguish regions. The experimental results indicate that the crime reports helped to distinguish regions where the number of reported crimes was greater than or equal to 100, with an average F-measure of about 90%. On the other hand, using only information about the types of crime or about the names of the areas where crime occurred, in the e-mail messages could not achieve the same F-measure.

In what follows, Sect. 2 discusses the relevant crime research literature and makes clear the position of this research; Sect. 3 describes crime reports sent out by police stations and shows some of the results of statistical analysis; Sect. 4 describes the results of experiments in distinguishing regions by crime reports and discusses whether crime reports are a good resource for analyzing the characteristics of regional crime. Finally we conclude the paper and discuss our future work.

2 Related Work

Research on geographical crime prediction has been actively conducted mainly in Europe and the United States [14]. For crime prediction, it is important to utilize data regarding the types of crime and the locations where the crime occurred [8].

Kylen et al. [11] showed that regional regularity influences crime occurrence patterns not only in urban areas but also in rural areas. Buczak et al. [3] extracted rules which indicate the relationships between regional characteristics and crime, from a UCI data set on "community and crime." [15] Various kinds of data are used to increase accuracy in predicting crime hotspots. Gerber [6] used the contents of regional messages on Twitter to predict criminal hotspots. Wang et al. [19] and Belesiotis et al. [2] used the relative locations where crime occurred, points of interest data and regional traffic volumes. Wang et al. [20] finely divided the areas of focus and predicted the occurrence of crime from the appearance of similar crime in neighboring areas. Takahashi et al. [18] extracted crime related posts from Twitter data. Amemiya and Shimada [1] conducted crime prevention research considering regional characteristics. They considered the categorization of districts, focusing on the space composition, and conducted an experiment on 40 areas in Ichikawa city, Chiba prefecture, Japan. They carried out a questionnaire survey on the anxiety residents feel about crime and which places make them feel uneasy. After mapping the survey results, they analyzed the need for crime prevention measures according to the crime types revealed by the survey.

To develop geographical crime prediction methods in Japan, Oyama [14] first systematized methods that have been developed in the Europe and the United States; he applied them to the data of a certain region in Fukuoka, Japan, and evaluated them by comparison. In this way, he proposed a method effective for predicting crime occurrence in Japan.

In contrast to the above work, this paper analyzed crime reports sent out by police stations to ascertain the crime characteristics of the regions. Specifically, we applied Support Vector Machine (SVM) to the crime reports to distinguish regions. As a result, we confirmed that regions issuing a large number of crime reports greater than or equal to 100 can be distinguished with an accuracy (F-measure) of about 90%.

3   Public Criminal Reports

We analyzed 4741 criminal reports issued by Fukuoka Prefectural Police Head-quarters (police stations for short) from 2014 to 2018[5]. As of 2018-06-21, there were 4741 reports. Note that 108 of them were corrections of earlier reports. The numbers of reports per year from 2014 to 2018 are 136, 470, 1564, 1724 and 847, respectively. These numbers denote the number of reports in a year, not the number of occurrences of crime in a year.

3.1   Format of Criminal Reports

Each report consists of six items. Table 1 illustrates an example of the contents of a report.

The first item "status" shows the situation of crime, "occurred" and/or "arrested." The second item "location" presents a place name specifying the city-ward-block where the crime occurred. The third item "crime" denotes the type of crime, which does not only consist of short words representing the type of crime, but also a word followed by "case," "crime case," "attempted," "arrested" or "solved." The fourth item "date" gives the date when the crime occurred, which consists of three attributes: month, day, and time of the day. The fifth item "reception date and time" are the date and time when the report was issued. The sixth item is the body of the report consisting of two parts: the first part consists of multiple sentences describing the crime situation; the second part also consists of multiple sentences, each beginning with an "@" symbol, which

---

[5] https://ckan.open-governmentdata.org/dataset/fukkeimail (As of Jun 21, 2018).

Table 1. Report items

No.	Item	Example
1	Status	Occurred, arrested
2	Location	Hakata-ward Sumiyoshi 2-chome
3	Crime	Attempted snatch
4	Date	Around 11 pm on October 6
5	Reception date and time	2014-10-07 08:55:06
6	The content of the email	Around 11 PM on October 6, a woman walking on Sumiyoshi 2-chome, Hakata-ku, realized that her handbag was about to be snatched. The culprits were two men who were riding on a moped bike, wearing a red hoodie and a full face helmet @In order not to become a victim of such a crime, carry bags on the side furthest from the roadway. @Do not walk while using a cell phone @Let's pay attention to what's happening around us by looking over our shoulders sometimes

presents specific advice from the police. In this paper, we used only the fifth and sixth items to build classification models. From the fifth item, we extracted four attributes: year (y), month (m), day (d) and hour (h) and put the symbol of each attribute on the attribute word, e.g. y:2014, m:10, d:07, and h:08. From the sixth item, we only used the first part. Since each sentence in the second part begins with the special symbol "@," we could easily separate them from the first part. So, the advisory messages from the police were excluded in analyzing the reports. In addition, we excluded from the first part any words exactly or partially included in the place name in the second item.

3.2 Extraction of Region Names

The second item "location" represents a region described in the municipal structure of city-ward-block or town. In this paper, for each report, we assigned three attributes as the names of region, i.e., a city-name, a ward-name and a block/town-name. As a result, 2205 place names were extracted in total. The number of reports corresponding to each location (place name) is shown in Fig. 1, confirming to what is known as Zipf's distribution.

From the place names, we selected 99 regions with 10 or more reports. Table 2 shows the number of region names for cities, wards and towns.

Table 3 shows the top 10 regions of cities, wards and towns for the occurrence of crime. Note that Fukuoka-city, the biggest city in Fukuoka prefecture, is not included in the table. Region names that start with East-ward or West-ward should be regions in Fukuoka-city. However, there is no explicit mention of the city. So, we assigned only West-ward or East-ward as their region name for

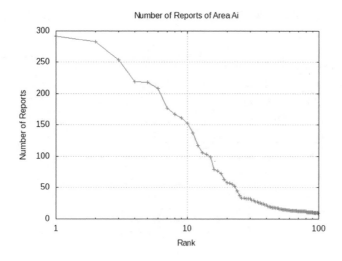

Fig. 1. Distribution of number of reports

Table 2. Number of region names

Region type	# of regions
City	40
Ward	27
Town	2138

those reports. We did not assign Fukuoka-city to them. As a result, there are only 32 reports whose region is Fukuoka-city, and Fukuoka-city does not appear in Table 3.

3.3 Extraction of Crime Types

The descriptions of crime types in the third item shown in Table 1 are not standardized. For example, there are a variety of expressions mentioning molester, such as molester case, molester incident, molester crime, Fukuoka prefecture anti-nuisance prevention ordinance violation (molester), molester case in Dazaifu-city, molester information in Dazaifu-city, arrest of suspects of a molesting case, molester case in Central-ward solved, molester suspect, and so on.

However, it turned out that there are only 688 sentences for the third items of 4741 reports to describe the crime types. So, we manually labeled the reports with 52 types of crimes. Table 4 shows the 35 crime types that occurred in two or more reports.

Table 3. Top 10 region names (city, ward, town)

City		Ward		Town	
254	Kurume	383	Hakata	32	Kawasaki town
137	Kazuga	292	South	30	Kasuya town
117	Ohnojo	283	East	24	Ijiri South-ward
106	Itoshima	219	Sawara	24	Yoshizuka Hakata-ward
103	Ohmuta	218	West Yahata	22	Chikuzen
99	Dazaifu	208	North Kokura	22	Hakata-Station-South Hakata-ward
79	Munakata	177	West	20	Sumiyoshi Hakata-ward
77	Chikushino	167	Central	19	Chiyo Hakata-ward
73	Tagawa	161	Johnan	18	Sasaguri town
58	Asakura	153	South Kokura	18	Nanakuma Jonan-ward

3.4   Regional Statistic Analysis

We conducted statistical analyses to see if crime types and area names can be useful information for distinguishing regions. We first calculated the probability of occurrence of a crime for each region as the ratio of the number of reports of the crime in the region to the total number of reports of the crime. Figure 2 shows the performance of the discrimination of regions by crime type.

As the number of reports corresponding to each crime type increases, the performance of the discrimination of regions by that crime type decreased. For a crime type whose occurrence frequency in reports is five or more, the rate is only about 0.06, which means it is difficult to distinguish regions by crime type.

Next, we calculated the occurrence probability of an area name for each region as the ratio of the number of reports of the area name in the region compared with the total number of reports of the area name. Figure 3 shows the performance of the discrimination of regions by the area names which appear in reports, but are not included even partially in the 2205 place names mentioned in Sect. 3.2.

As shown in Fig. 3, area names are more effective than crime types for distinguishing regions. For the area names which appear in two or more reports, the performance of the discrimination of regions is more than 0.82. As the number of reports corresponding to each area name increases, the performance of the discrimination of regions by the area name decreased as well.

However, for area names which appear in ten or more reports, the performance of the discrimination of regions is still more than 0.43. So, we can expect that using a set of area names appearing in reports compared to using each of them individually may be useful for building a classification model of regions.

Table 4. Crime type and number of reports

# of reports	Crime type
1231	Obscenity
940	Talking to
530	Molester
379	Stalking
314	Telephone fraud
235	Suspicious person
199	Snatch
134	Suspicious phone
124	Assault
113	Peeking
75	Peeping photo
52	Monkey
48	Vehicle burglary
48	Robbery
30	Fraud
21	Braking into a residence
21	Fictitious claim
20	Correction
19	Knife
18	Stealing
16	Dog
7	Suspicious fire
7	Murder
5	Malicious commercial code
4	Wild boar
3	Exposure
3	Suspicious document
3	Suspicious vehicle
3	Injury
3	Sneak thief
3	Vessel damage
2	Bite
2	Theft
2	Body contact
2	House invasion

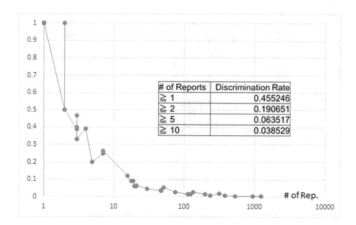

Fig. 2. Performance of discrimination of regions by crime names

4   Classification of Regions by Criminal Reports

4.1   Generation of Classification Model

We applied the morphological analysis tool MeCab[6] to 4741 reports and vector-ized them as hot-BoW (Bag of Words) vectors, where the value of a component of the vector is either 1 or 0 according to whether the word appears in each report or not. The words appearing in the reports mainly represent criminal types, criminals, victims, criminal situations, and the names of areas in regions.

When converting the reports to input vectors, we built three types of vectors: (a) the area name vector, (b) the report word vector without area names, and (c) the report vector with area names. Please note that the three vectors do not include any words exactly or even partially matched with a target region name. We used MeCab to generate such a list of the exceptional words.

By comparing the performance of classification models built by using the three vectors, we can evaluate the effect of components in each vector. To build the area name vector, we manually extracted the words representing area names in regions of the reports. The number of the area name words extracted is 714 and the number of target regions, each of which appears in 10 or more reports, is 99. When building the other two vectors, we added the reception date and time of the report to the words in the report. Those attributes are represented with tags that distinguish them from ordinary words. For example, the report in Table 1 contains the tagged words y:2015, m:01 and h:08. We can also use words denoting region names or crime types appearing in the reports as tagged words such as l:Hakata-ward, l:Sumiyoshi-Hakata-ward, l:2nd-block-Sumiyoshi-Hakata-ward, X:snatch, where tag "l" denotes an areal location and "X" denotes a crime type.

---

[6] http://taku910.github.io/mecab/.

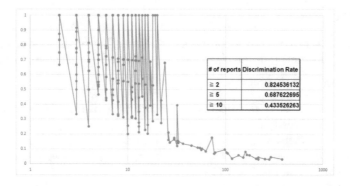

# of reports	Discrimination Rate
≧ 2	0.824536132
≧ 5	0.687622695
≧ 10	0.433526263

Fig. 3. Performance of discrimination of regions by area name

Please note again we did not use any region names or constituent parts of those names that appear in the reports when constructing the classification models. We only used them when constructing conceptual maps [4] that represent the characteristics of regional crime, which will be described later.

We built classification models to distinguish each region based on Support Vector Machine (SVM). We applied svm^{perf} [10][7] with the default parameter $C = 20$. The classification models we built are roughly divided into two: a model without Word Feature Selection (WFS) and a model with WFS. To build the model with WFS, we employed a SVM with WFS method proposed by Sakai and Hirokawa [16] (SVM+FS for short) because SVM+FS showed the best performance in our previous study [13]. We will explain SVM+FS later. Each model is further divided into three types according to the three types of vectors mentioned earlier.

We used F-measure defined by Eq. (3) for the evaluation of the model. TP (True Positive), FP (False Positive), FN (False Negative) and TN (True Negative) in the equations are the number of matched conditions as shown in Table 5.

Table 5. Confusion matrix

		Observed	
		True	False
Predicted	Positive	TP	FP
	Negative	FN	TN

$$precision = \frac{TP}{TP + FP} \tag{1}$$

$$recall = \frac{TP}{TP + FN} \tag{2}$$

$$F - measure = \frac{2 * precsion * recall}{precision + recall} \tag{3}$$

---

[7] http://www.cs.cornell.edu/people/tj/svm_light/svm_perf.html.

4.2   SVM+FS

SVM+FS carries out word feature selection according to the *svm-score* of each word in the documents to be classified. The *svm-score* is calculated as follows:

Calculation of *svm-score*

1. Let D be a set of N documents, which are classified into M classes, where $M = 2$ in this research.
2. When a document $d_i \in D$ $(1 \le i \le N)$ includes m distinct words, SVM+FS produces m one-word documents from d_i, where each one-word document $d_{i,j}$ only includes one word $w_{i,j}$ $(1 \le j \le m)$.
3. SVM+FS converts each document $d_{i,j}$ to a word vector $v(d_{i,j}) = \{v_1, \ldots, v_k, \ldots, v_n\}$, where n is the total number of distinct words in D. $v_k = 1$ if $w_{i,j}$ corresponds to the k_{th} element in the vector, and $v_k = 0$ otherwise.
4. For each target class, SVM+FS assigns $d_{i,j}$ a positive flag if d_i belongs to the target class, otherwise assigns a negative flag.
5. SVM+FS builds a model using SVM from a set of word vectors produced in step (3), and obtains a score of each word calculated by the SVM model. The score is called *svm-score*.

Using the *svm-score* mentioned above, SVM+FS is carried out in the following three steps.

SVM+FS

1. SVM+FS selects top K positive and top K negative words based on the *svm-score* of the words.
2. SVM+FS converts documents into input vectors only using the 2 K positive and negative words selected in step (1).
   - An input vector converted from a document is a hot-bow vector as mentioned in step (3) of the *svm-score* calculation procedure, i.e., if the input vector's element corresponds to a word in the 2 K words, the value of the element is 1, otherwise 0.
   - If a document belongs to a target class, the input vector converted from the document is assigned a positive class label, otherwise, the input vector is assigned a negative class label as mentioned in step (4) of the *svm-score* calculation procedure.
3. SVM+FS builds a classification model using SVM from the input vectors produced in step (2).

In addition to the above basic model building, in this paper, we performed oversampling to increase the number of positive examples as follows:

OverSampling for SVM+FS

1. Randomly select N pairs of documents belonging to a target class, where N = 10,000 in this paper.

2. Convert each pair of documents to an input vector as mentioned in the previous step.
3. Add the input vectors generated by the oversampling method to training data as positive class data.

5   Experimental Results

We carried out five-fold cross-validation, repeated it twice and took the average so that we could roughly check the classification performance of the models with and without WFS.

Figures 4 and 5 illustrate the experimental results of comparing classification methods with and without WFS. We used three types of vectors as mentioned in Sect. 4.1. As shown in Figs. 4 and 5, the effect of the area name vector is bigger than that of the report word vector without the area names, and the effect of the report word vector with the area name features is the best. The results are similar to the results shown in Sect. 3.4.

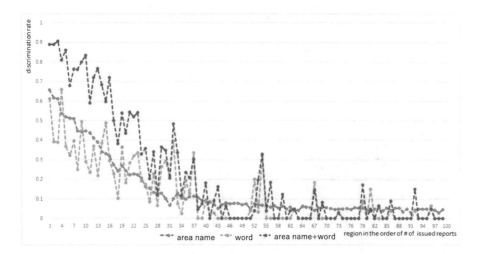

Fig. 4. Discrimination performance (F-measure) by area names, words related to crime and both the area names and the words without word feature selection

Figure 6 illustrates the effect of WFS when comparing the two models with and without WFS. The results shown in Fig. 6 pick up the F-measure values of the two models with and without WFS when using report words with area name features. As shown in Fig. 6, the classification model built by SVM+FS clearly outperforms the model built without WFS. Table 6 shows the classification performance of the two models when varying the number of reports where either area names or other words appeared. Although the area names are useful to discriminate regions, the classification performance even after applying WFS is only 0.711, where the number of reports in which the area names appeared is greater than or equal to 100. This means that the area name words are good features, but only using the area names is not enough to distinguish regions; the characteristics of regional crime are not only presented by the area names, but also by other expressions related to crime in the reports. Actually, when applying WFS to the reports whose features are all report words including area names, we obtained the best classification performance. In particular, when the number of reports where the feature words appeared is greater than or equal to 100, the average F-measure reaches 0.903.

Table 6. Average discrimination performance (F-measure) of regions with and without word feature selection

# of reports	$\geq$100		$\geq$50		$\geq$30	
Features	ALL	WFS	ALL	WFS	ALL	WFS
Area name	0.481	**0.711**	0.394	**0.616**	0.320	**0.526**
Report words wo. area name	0.392	**0.517**	0.341	**0.474**	0.310	**0.460**
Report words w. area name	0.770	**0.903**	0.670	**0.838**	0.578	**0.777**

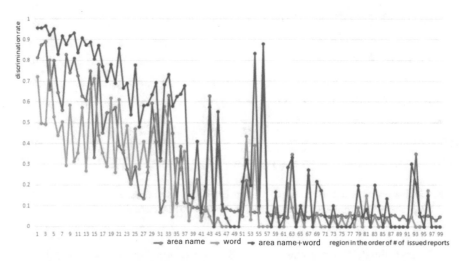

Fig. 5. Discrimination performance (F-measure) by area names, words related to crime and both the area names and the words with word feature selection

5.1   Discussion

In this paper, we analyzed the crime reports sent out by police stations. We examined whether this information can be used as a source for analyzing the characteristics of regional crime. We tried to answer the three research questions: RQ1, RQ2 and RQ3 mentioned in Sect. 1.

For the first research question (RQ1), we confirmed that the crime type information is not enough to achieve high discrimination performance in classifying regions as shown in Sect. 3.4. In fact, the average performance in distinguishing regions was very low, at about 0.06 when the number of reports where the crime type appeared is greater than or equal to five. We can say that it is difficult to distinguish regions only by using crime types. As a result, the answer to the RQ1 is "difficult since its classification accuracy is quite low."

On the other hand, for the second research question (RQ2), we confirmed that the average discrimination performance for regions when using area names is 0.771 when the number of reports from regions is greater than or equal to 100. As a result, the answer to the RQ2 is "useful for distinguishing regions, but not sufficiently."

For the third research question (RQ3), the average discrimination performance of regions was improved by using all words that appeared in the reports, and the average F-measure was 0.903 when the number of reports from regions was greater than or equal to 100. From this, the answer to RQ3 is "Yes," and we can conclude that police criminal report data is sufficiently useful as a resource for analyzing the characteristics of regional crime.

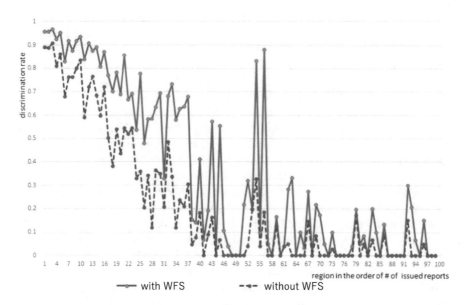

Fig. 6. Comparison of discrimination performance (F-measure) with and without word feature selection

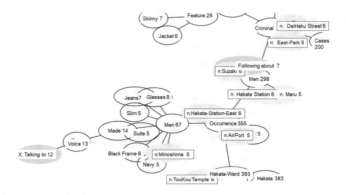

Fig. 7. Conceptual map for Hakata-ward (partial)

To confirm the results in an intuitive way, we show part of a conceptual map constructed from report words related to Hakata-Ward in Fig. 7. We can see that some areas in the Hakata-Ward region, which have similar crime-type characteristics, appear near each other on the map. This means that crime reports sent out by police stations can play an important role in analyzing the characteristics of regional crime. Furthermore, this mapping indicates that we need to make use of any feature words appearing in the reports, as well as the area names.

6   Conclusion

With a view to facilitating the prevention of crime, which is one of the most important social problems for the regions, in the research reported here, we analyzed 4741 crime reports sent out by police stations and investigated whether or not the contents of these e-mail crime messages can be turned into a valid source for analyzing the criminal characteristics of a region, i.e., whether or not they include differential characteristics of regional crime.

We hypothesized that, if the crime reports can be used as good resources representing regional characteristics, it would be possible to distinguish regions by using the reports. To test this hypothesis, we conducted experiments in distinguishing the regions on the basis of the reports. Experimental results illustrate that the classification success rate of regions increased as the number of relevant region reports increased. In particular, the average F-measure achieved in distinguishing regions is about 0.903 when the number of reports from each region is greater than or equal to 100. These results support the initial hypothesis. Thus, we can say that crime reports are good resources for analyzing regional crime characteristics.

Much work on the topics addressed in this study remains to be done. Since we only employed SVM+FS to build classification models for regions with a high classification performance, we should seek to obtain better methods by employing other machine learning techniques, in particular, Gradient Boosting Decision Trees [5, 22] and Deep Neural Networks [9, 12].

We also need to develop a method of automatically extracting important features from the crime reports such as area names, crime types, and expressions regarding the criminals and victims. Using the extracted feature words, we would like to develop methods for analyzing and tracking the temporal and geospatial changes in crime so as to build a method for predicting regional crime.

We will report on this research in the near future.

Acknowledgement. This work was partially supported by JSPS KAKENHI Grant No. JP15H05708, JP16H02926, JP17H01843, and JP18K18656.

References

1. Amemiya, M., Shimada, T.: The relationship between residents' fear of crime and spatial compositions in urban areas: a basic study of community-based urban planning for crime prevention. J. City Plan. Inst. Japan **44**(3) (2009). (in Japanese)
2. Belesiotis, A., Papadakis, G., Skoutas, D.: Analyzing and predicting spatial crime distribution using crowdsourced and open data. ACM Trans. Spat. Algorithms Syst. (TSAS) **3**(4), 12 (2018)
3. Buczak, A.L., Gifford, C.M.: Fuzzy association rule mining for community crime pattern discovery. In: ACM SIGKDD Workshop on Intelligence and Security Informatics, p. 2. ACM (2010)
4. Flanagan, B., Yin, C., Inokuchi, Y., Hirokawa, S.: Supporting interpersonal communication using mind maps. J. Inf. Syst. Educ. **12**(1), 13–18 (2013)
5. Friedman, J.H.: Stochastic gradient boosting. Comput. Stat. Data Anal. **38**(4), 367–378 (2002)
6. Gerber, M.S.: Predicting crime using twitter and kernel density estimation. Decis. Support Syst. **61**, 115–125 (2014)
7. Graif, C., Sampson, R.J.: Spatial heterogeneity in the effects of immigration and diversity on neighborhood homicide rates. Homicide Stud. **13**(3), 242–260 (2009)
8. Harries, K.D., et al.: Mapping crime: principle and practice. Technical report, US Department of Justice, Office of Justice Programs, National Institute of Justice, Crime Mapping Research Center (1999). https://www.ncjrs.gov/pdffiles1/nij/178919.pdf
9. Hochreiter, S., Schmidhuber, J.: Long short-term memory. Neural Comput. **9**(8), 1735–1780 (1997)
10. Joachims, T.: A support vector method for multivariate performance measures. In: Proceedings of the 22nd International Conference on Machine Learning, pp. 377–384. ACM (2005)
11. Kaylen, M.T., Pridemore, W.A.: Social disorganization and crime in rural communities: the first direct test of the systemic model. Br. J. Criminol. **53**(5), 905–923 (2013). https://academic.oup.com/bjc/article/53/5/905/337849
12. LeCun, Y., Bengio, Y., Hinton, G.: Deep learning. Nature **521**(7553), 436 (2015)
13. Lin, Y., Yamaguchi, K., Mine, T., Hirokawa, S.: Is SVM+FS better to satisfy decision by majority ? In: The 3rd International Conference on Soft Computing and Data Mining 2018 (SCDM 2018), pp. 261–271 (2018)
14. Oyama, T.: Development of geographic crime prediction method in Japan (in Japanese). Technical report, The Nikkoso Research Foundation for Safe Society (2017)

15. Redmond, M.: Communities and crime data set. http://archive.ics.uci.edu/ml/datasets/Communities+and+Crime
16. Sakai, T., Hirokawa, S.: Feature words that classify problem sentence in scientific article. In: the 14th International Conference on Information Integration and Web-based Applications & Services, pp. 360–367. ACM (2012)
17. Shihadeh, E.S., Winters, L.: Church, place, and crime: latinos and homicide in new destinations. Sociol. Inq. **80**(4), 628–649 (2010). https://onlinelibrary.wiley.com/doi/abs/10.1111/j.1475-682X.2010.00355.x
18. Takahashi, S., Kikuchi, H., Ochiai, K., Fukazawa, Y.: Exraction of criminal related posts from microblogs based on rarity and influence. J. Inf. Process. **58**(8), 1376–1386 (2017). (in Japanese)
19. Wang, H., Kifer, D., Graif, C., Li, Z.: Crime rate inference with big data. In: Proceedings of the 22nd ACM SIGKDD International Conference on Knowledge Discovery and Data Mining, pp. 635–644. ACM (2016). https://dl.acm.org/citation.cfm?id=2939736
20. Wang, K., Cai, Z., Zhu, P., Cui, P., Zhu, H., Li, Y.: Adopting data interpretation on mining fine-grained near-repeat patterns in crimes. J. Forensic Legal Med. **55**, 76–86 (2018). https://www.sciencedirect.com/science/article/pii/S1752928X18300313
21. Yamamoto, K.: Current situations of sex-related crime in Fukuoka prefecture and its enlightenment activities (in Japanese). Technical report (2017). http://www.police.pref.fukuoka.jp/data/open/cnt/3/42504/1/04yamamoto.pdf
22. Zhang, Y., Haghani, A.: A gradient boosting method to improve travel time prediction. Transp. Res. Part C Emerg. Technol. **58**, 308–324 (2015)

A Conceptual Framework for Applying Telemedicine Mobile Applications in Treating Computer Games Addiction

Abdulaziz Aborujilah[1](✉), Rasheed Mohammad Nassr[1],
Mohd Nizam Husen[1], Nor Azlina Ali[1], AbdulAleem Al-Othman[1],
and Sultan Almotiri[2]

[1] University Kuala Lumpur, 50250 Kuala Lumpur, Malaysia
{Abdualzizsaleh,rasheed,azlinaali,
abdulaleem}@unikl.edu.my
[2] Umm Al-Qura University, Makkah 21514, Saudi Arabia
shmotiri@uqu.edu.sa

Abstract. The growth of digital games development and usage has raised many issues in education, entertainment, psychology, and other fields. The number of people becoming addicted to digital games is growing rapidly. This has led researchers to propose solutions to mitigate the harmful consequences of digital games addiction. However, there is a lack of studies that define a comprehensive framework to monitor and treat people suffering from digital game addiction. This paper proposes a conceptual framework that highlights the capability of using telemedicine mobile applications (TMPs) and sentiment analysis method to assist psychologists, parents and games players to reduce and minimise the risks of digital gaming. we have carried out a preliminary investigation of school students' perspective of using TMPs in treating games addiction. The results show that both males and female participants are interested in using TMPs in addiction treatment with slightly different levels of motivation.

1 Introduction

Digital games are used in many applications in education, entrainment, brine training, and health. Digital games are defined as *"games played through an electronic format such as a computer, console or digital phone"* [6]. Computer games might include some harmful features. For example, violent contents could have unwanted results on human behaviour [1]. The sale of digital games has increased rapidly in the cyber world. For example, the Counter-Strike franchise has sold over 60 million copies [4]. Many of the mass shooting incidents have been widely attributed to people who are publicly known for playing violence games. For example, the Munich shooter in 2016 who was known for playing Counter-Strike [5].

Over the past few years, researchers have been looking for ways to reduce the risk of digital games addiction. Public interest in computer games violence led to the formation of the Entertainment Software Rating Board in the USA to monitor the sales of video games [2, 3]. There is a public health concern that violent games could have a

© Springer Nature Switzerland AG 2019
S. Lee et al. (Eds.): IMCOM 2019, AISC 935, pp. 599–609, 2019.
https://doi.org/10.1007/978-3-030-19063-7_47

direct harmful influence on children's mental health since they easily learn violent or aggressive behaviour [7].

E-Service is common in many different aspects of modern life such as home services payment management, tourist, airline, education, eHealth and more services. Several examples of eHealth services are adopted in the health sector and addiction treatment. For example, interactive voice recognition (IVR) technology which is interactive telephone-based technology has been used to advise people on the merits of better health [10]. Also, eHealth services were tested for drinking addition treatment [11].

Mobile-based mental health treatment is an eHealth service that is growing in importance in the mental health sector. Almost everybody frequently uses mobile phone applications. In contrast, web-based online treatment requires a stand-alone PC that is not always available [29]. Little attention has been paid to exploring how TMPs might be used in digital games addiction treatment. Hence, this study seeks to offer a solution that could assist in monitoring and treating those who are addicted to digital games. Also, it investigates the initial acceptance and intention of school students to use TMPs. This study focuses on school students as most digital games addicts start playing games at school age.

The first section of this paper will examine the recent literature on digital games, its influence on childhood and the methods that are used to treat the kids' addiction to digital games. The second section is concerned with the methodology used in this study. The third section presents the findings of the research. The paper ends with a conclusion and directions for future work.

2   Related Studies

Digital games players feel enjoyment and achievement in such games [16]. Thus they spend long times playing [18, 19]. However, up to 89% of teenagers' games include violent content [20–22]. The games with violent contents have a high potential to influence the violent or aggressive behaviours of youth [6]. This has led to concerns regarding the influence of violent games on players' beliefs about real aggression [6]. Those who become familiar with violent games are more likely accepting of violence in reality and become less empathic [6].

Mobile mental health applications are another promising solution for treating mental disorders. For example, Shakshuki [31] research the extent to which young adults in Germany are motivated to use mobile mental health applications. They applied the extended version of the technology acceptance model (TAM). The results showed that mobile applications are easy to use but young adults were questioning their effectiveness in treating mental disorders. The concerns about personal privacy were also raised. The authors suggested promoting these applications as a tool to facilitate mental treatment anytime and anywhere.

Mobile technology can play a vital role in treating people who are addicted to digital games. For example, TMPs is a software solution used to exchange the health information between the patients and the health service providers [6] and can assist in connecting patient with their therapist. TMPs render patients under constant medical care and give them a sense of comfort as they are 'close' to their doctors [9]. It also gives patients proper treatment when they need it. As face to face treatment and meeting counsellors is not always viable and cost-effective, TMPs extend the health functions and offers patients immediate health care services [6]. In recent years, there has been more interest in using TMPs in health and addiction treatment and recovery [6].

Strengthening Treatment Access and Retention State Initiative (STAR-SI) has developed software in the USA for those interested in using TMPs in addiction treatment. It is a useful tool for patients who need to be connected to his/her doctor over extended periods. Web-based telemedicine services provide the patients with open access to several web-based platforms and apps for substance use disorders (SUDs) assessment according to their preferred time [12, 13]. Such apps have been found to be more effective in treat alcohol addicts than medical conditions methods. It helps them reduce their weekly consumption [14, 15]. Despite the merits of eHealth technology, telemedicine services are not widely known in addiction treatment sectors. For example, Molfenter [6] conducted a project between February 2013 and June 2014 to investigate the adoption of telemedicine services in addiction treatment or to view how purchasers of addiction treatment services are ready to use telemedicine apps for addiction treatment. The project measured people's intention perceived facilitators and barriers to using telemedicine modalities such as web-based screening, telephone-based care, videoconferencing, web-based treatment, mobile smartphone applications (apps), and the virtual world. They found that purchasers are more interested in using video conferencing and smartphone mobile devices. Purchasers expressed interest to be treated through video conferencing and smartphone mobile devices.

van et al. [29] developed a conceptual model to assess patient trust in telemedicine service and validated it using the Patient Trust Assessment Tool (PATAT). A survey instrument was used to quantitatively assess the measure the participants' trust in this treatment service through the following factors: trust in the care organisation, care professional, treatment, technology, and telemedicine service. The scale also measures a holistic vision of trust in the telemedicine service. 795 patients completed the survey.

It measured their opinions on using telemedicine service in anticoagulation treatment. Partial Least Squares Structural Equation Modelling (PLS-SEM) was used in data analysis. The results showed that participants' trust in healthcare professionals and treatment have limited influence on overall trust. In contrast, their trust in the technology has a significant impact on their overall trust. While their trust in the care organisation has no impact on overall trust.

Most of the studies in this field focus on either proposing conceptual models for digital games adductors treatment or on designing and developing some TMPs prototypes. Very little work has been done to integrate TPMs and sentiment analysis methods for tracking and treating people who are addicted to digital games.

3   Proposed Framework

Researchers to date have tended to focus on proposing a conceptual model for digital games addictions. For example, Kongkarn et al. [16] suggested a framework entitled "A framework for applying an intelligent agent to monitor, interpret, and report the risk of online computer game addiction in children and early adolescents in Thailand".

This framework aims to help parents, clinicians and experts in Thailand to track kid's behaviour during playing games. The framework consists of an intelligent agent which tracks the actual interaction between the kids and computer games. It is also responsible for monitoring mouse clicking, device task changing, keyboard clicking to recreate children's pattern of playing games and store it in the knowledge base (KB). Some of the suggestions are generated based on player habits.

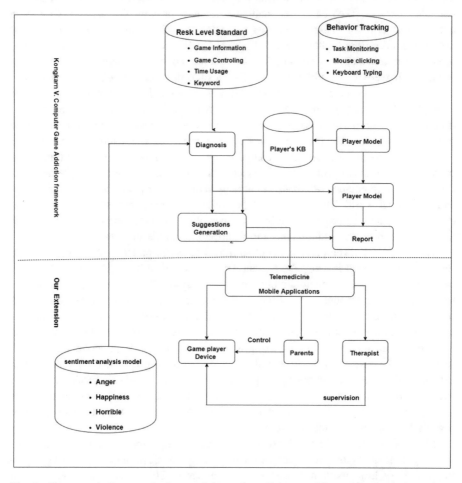

Fig. 1. Conceptual framework for applying telemedicine mobile applications in treating computer games addiction

In this era, a huge amount of information about people's daily activities, feelings and opinions are posted on social media. This information could be a primary source to track people's habits and attitudes toward different types of addiction such as drug, alcohol and digital games. For example, Raja et al. [34] proposed a framework to use sentiment analysis of social media in the prohibited use of pharmaceutical drug abuse. It consists of five layers: data collection, data transformation, sematic conversion, data analysis and interpretation and rule-based interpretation. Another study by MacLean et al. [35] implemented the Conditional Random Field (CRF) classification method to monitor prescription drug abuse behaviour in online health forums. The sentiment and feeling of people who misuse prescription drug are analysed to identify the level of their addiction.

Based on the study of Kongkarn and the other studies that employee the sentiment analysis in treating people who are addicted drug, alcohol and digital games, also the studies that suggest using telemedicine mobile application in in the health care services.

We propose adding the sentiment analysis and telemedicine mobile application to Kongkarn's framework. The sentiment analysis will be able to analysis people opining about the digital games. The games players used to post their views and comments on the games they play at the online games stores. These posts are very beneficial source of information that can be used to classify the games severity based on the predefined risk levels such as anger, happiness, horrible and violence levels. Moreover, Tele-medicine mobile applications can play impotent rule of associating the games players with their parents and psychologist doctors or therapist.

Figure 1 show conceptual framework for applying telemedicine mobile applications in treating computer games addiction.

4 Method

This study adopted a qualitative approach. An extended questionnaire was design based on [31] with some modifications.

4.1 Measurement Tool

The final questionnaire includes ten items in addition to three questions about the demographic information of participants. Then, questions 1 to 7 related to the partic-ipants' perception of usefulness and ease of TMPs and privacy level of TMPs. Questions 8, 9 and 10 concerned the attention of participants to use TPMs. The details can be found in Appendix A.

4.2 Data Collection

The participants were called to answers the questionnaire through online and paper-based formats. 92 participants answered the questions comprising 45 males and 47 females within three ages categories, 9–13 years, 14–17 years, and 18 years and above. The participants' feedback was collected from December 16, 2017, to February 16, 2017. The qualitative statistical model was applied using SPSS.

4.3 Results

The participants' characteristics are shown in Table 1. This study focused on males and female students aged between 9 and 18 who play computer games with different levels of addiction.

Table 1. Demographic information of the participants

Variable	Frequency	Percent (%)	Variable	Frequency	Percent (%)
Gender			CGP		
Male	45	48.90%	A	15	16.30
Female	47	51.10%	A-	23	25.00
Age			B	25	27.17
9–13	18	19.56	B-	19	20.65
14–17	22	23.91	Less than B-	10	10.86
18	52	56.52			

Normality distribution of the sample was between (0.009 and 0.04) which means that the sample population is normally distributed. Tables 2 and 3 shows the participants responses to the questions based on a Likert Scale [32] (Table 4).

Table 2. Respondents answer frequencies result for male's participants

	Question	Strongly disagree	Disagree	Neutral	Agree	Strongly agree
Q1	TMP are useful to provide information to those who has psychological problems caused by computer games	3 06.00%	3 06.00%	23 51.11%	14 31.11%	2 04.00%
Q2	TMP are an improvement to the services it supersedes	2 04.44%	7 15.55%	15 33.33%	16 35.55%	5 11.11%
Q3	Using TMP would not disclose my personal information	4 08.88	3 06.66%	23 51.11%	12 26.66%	3 06.66%
Q4	Telemedicine application are helpful	0 00.00%	3 06.66%	16 35.55%	16 35.55%	10 22.22%
Q5	I find TMP are easy to use	2 04.44%	6 13.33	14 31.11%	14 31.11%	9 20.00%
Q6	I find it easy to get the benefits from a TMP	5 11.11%	3 06.66%	22 48.88%	15 33.33%	0 00.00%
Q7	I find mobile mental health to be useful to improve my life in general	3 06.66%	8 17.77%	12 26.66%	20 44.44%	2 04.44%
Q8	I intend to check the availability of a suited TMP	4 08.88%	16 35.55%	16 35.55%	6 13.33%	3 06.66%
Q9	I feel confident to work through all interventions that the application provides me	8 17.77%	7 15.55%	16 35.55%	10 22.22%	4 08.88%
Q10	I intend to use a TMP	4 08.88%	5 11.11%	18 40.00%	13 28.88%	5 11.11%

5 Discussion

The results of the study show the readiness of male and female participants to use telemedicine applications as the communication method to deal with problems caused by digital games. There is a contrast between male and females in this issue. For example, for Q1 (TMPs are useful to provide information to those who has psychological problems caused by computer games) 51.11% of males are not interested (neutral) in using TMPs while 51.06% of females are interested (agree). This could be because females are more attracted to using new technology relative to males. In Q2 (TMPs are an improvement to the services it supersedes), 40.42% of females agree that TMPs are an improvement to the services they supersede, whereas the majority of males are divided between agree 35.55% and neutral 33.33% which means they are less attracted to using this technology. In Q3 (Using TMPS would not disclose my personal information), it seems that 51.11% of males and 51.06% of females are not concerned (neutral) with how TMPs will affect their privacy. This could occur due to the widespread usage of communication applications. Q4 (Telemedicine applications are helpful) shows that females 51.06% are more encouraged to use telemedicine applications than males 35.55%, while in Q5 (I find TMPs are easy to use) the majority of males feel neutral 31.11% and agree 31.11% that telemedicine applications are easy to use applications.

Similarly, females feel neutral 34.04% and agree 38.29% that telemedicine applications are easy to use. This could be because the modern generation is highly familiar with using mobile applications. In Q6 (I find it easy to get the benefits from a TMPs), 48.88% of males seem more enthusiastic using TMPs whereas 34.04% of female are more unsure of its benefits. In Q7 (I find mobile mental health to be useful to improve my life in general), both males and females share the same perspective that mental mobile applications will enhance their mental health in general.

In Q8 (I intend to check the availability of a suited TMPs), it seems that females 42.55% are more interested in finding available TMPs relative to males 35.55%. This is because most of the females are more curious about finding and using new things. In Q9 (I feel confident to work through all interventions that the application provides me), both males 35.55% and females 38.29% tend not to show clear confidence (neutral) in the intervention the TMPS could provide to them. This is normal because this kind of medical communication techniques is still new. In Q10 (I intend to use a TMPs), both males 44.44% and females 42.55% have similar intentions of use TMPs.

An independent-samples t-test was conducted to compare male and female responses. It was found that there is a significant difference in the scores for males in Q1 (M = 3.0000, SD = .63246) and females (M = 3.0625, SD = .72658) conditions, p = .029. Also, independent-samples t-test showed that there is a significant difference in the scores for males in Q4 (M = 3.7174, SD = .91075) and females (M = 3.7083, SD = .68287) conditions, p = .029.

The summary of results also shows that both males and female are pleased to use telemedicine applications. However, females are more attached and enthusiastic to use and experience TMPs as shown in their responses to Q1 and Q4.

Table 3. Respondents answer frequencies result for females' participants

	Question	Strongly disagree	Disagree	Neutral	Agree	Strongly agree
Q1	TMP are useful to provide information to those who has psychological problems caused by computer games	0 00.0%	1 02.00%	17 36.17%	24 51.06%	5 10.63%
Q2	TMP are an improvement to the services it supersedes	2 04.25%	6 12.76%	15 31.91%	19 40.42%	5 10.63%
Q3	Using TMP would not disclose my personal information	7 14.89%	5 10.63%	24 51.06%	7 14.89%	4 08.51%
Q4	Telemedicine application are helpful	2 04.25%	2 04.25%	13 27.65%	24 51.06%	6 12.76%
Q5	I find TMP are easy to use	2 04.25%	8 17.02%	16 34.04%	18 38.29%	3 06.38%
Q6	I find it easy to get the benefits from a TMP	2 04.25%	10 21.27%	16 34.04%	13 27.65%	6 12.76%
Q7	I find mobile mental health to be useful to improve my life in general	4 08.51%	6 12.76%	12 25.53%	20 42.55%	7 14.89%
Q8	I intend to check the availability of a suited TMP	6 12.76%	9 19.14%	20 42.55%	8 17.02%	4 08.51%
Q9	I feel confident to work through all interventions that the application provides me	7 14.89%	5 10.63%	18 38.29%	12 25.53%	7 14.89%
Q10	I intend to use a TMP	5 10.63	9 19.14%	8 17.02%	23 48.93%	2 04.25%

Table 4. T-test result of Q1 and Q4

		F	Sig.	t	df	Sig. (2-tailed)	95% Confidence interval of the difference		
							Lower		Upper
Q1.	Equal variances assumed	7.867	0.006	1.504	92	0.136	−1.74937		12.66604
Q4.	Equal variances assumed	4.922	0.029	0.055	92	0.956	0.16557	−0.31978	0.33790

6 Conclusion

The purpose of the current study was to examine the existing ICT based frameworks used in treating digital games addiction. The framework proposed by Kongkarn [6] consisted of the most important components required in digital games addiction treatment. This paper suggested an extension of this framework by adding TMPs and sentiment analysis items.

To examine the trend of school students to using telemedicine mobile applications in computer games addiction treatment, a questionnaire with ten questions was distributed to participants of different ages. Frequency analysis showed that females are slightly more attracted to using this technology relative to males. However, the t-test shows that both genders are having similar intentions to use TPMs.

Overall, this study strengthens the idea that TPMs might be used to treat school students who are addicted to computer games. The insights gained from this study may assist in knowing how TPMs could be adapted to the computer games addiction healthcare sector. The major limitation of this study is that it not explores the main factors in implementing the proposed framework in the real environment. Also, this study has not explored a sufficient sample of school students. Additional uncontrolled factors such as level of familiarity of participants with using TPMs might affect the results. A natural progression of this work is to analyse how to integrate TPMs and sentiments analysis methods in the digital games industry and parental control applications.

Appendix A

Q1.	TMPS are useful to provide information to those who has psychological problems caused by computer games
Q2.	TMPS are an improvement to the services it supersedes
Q3.	Using TMPS would not disclose my personal information
Q4.	I feel confident to work through all interventions that the application provides me
Q5.	I intend to check the availability of a suited TMPS
Q6.	Telemedicine application are helpful
Q7.	I find TMPS are easy to use
Q8.	I find it easy to get the benefits from a TMPS
Q9.	I find mobile mental health to be useful to improve my life in general
Q10.	I intend to use a TMPS

References

1. Burešová, I., Steinhausel, A., Havigerová, J.M.: Computer gaming and risk behaviour in adolescence: a pilot study. Procedia Soc. Behav. Sc. **69**, 247–255 (2012)
2. Crossley, R.: Mortal Kombat: violent game that changed video games industry. BBC News 2 June 2014 (2014). http://www.bbc.com/news/technology-27620071
3. De Camp, W.: Impersonal agencies of communication: comparing the effects of video games and other risk factors on violence. Psychol. Popular Media Cult. **4**, 296–304 (2015). https://doi.org/10.1037/ppm0000037
4. Steam Spy. Search results: Counter-Strike. https://steamspy.com/search.php?s=counter-strike. Accessed 13 Aug 2016
5. Reuters. Munich gunman, a fan of violent video games, rampage killers, had planned attack for a year. CNBC, 24 July 2016 (2016). http://www.cnbc.com/2016/07/24/munich-gunman-a-fan-of-violent-video-games-rampage-killers-hadplanned-attack-for-a-year.html

6. Ferguson, C.J., et al.: Digital poison? Three studies examining the influence of violent video games on youth. Comput. Hum. Behav. **50**, 399–410 (2015)
7. Hall, R., Day, T., Hall, R.: A plea for caution: violent video games, the supreme court, and the role of science. Mayo Clin. Proc. **86**(4), 315–321 (2011)
8. Molfenter, T., et al.: Trends in telemedicine use in addiction treatment. Addict. Sci. Clin. Pract. **10**(1), 14 (2015)
9. Wang, J., Wang, Y., Wei, C., Yao, N.A., Yuan, A., Shan, Y., et al.: Smartphone interventions for long-term health management of chronic diseases: an integrative review. Telemed. J. E Health **20**(6), 570–583 (2014)
10. Perrine, M.W., Mundt, J.C., Searles, J.S., Lester, L.S.: Validation of daily self-reported alcohol consumption using interactive voice response (IVR) technology. J. Stud. Alcohol **56**(5), 487–490 (1995)
11. Rose, G.L., Skelly, J.M., Badger, G.J., Ferraro, T.A., Helzer, J.E.: Efficacy of automated telephone continuing care following outpatient therapy for alcohol dependence. Addict. Behav. **41**, 223–231 (2015)
12. Hester, R.K., Delaney, H.D., Campbell, W., Handmaker, N.: A web application for moderation training: initial results of a randomized clinical trial. J. Subst. Abuse Treat. **37**(3), 266–276 (2009)
13. Moore, B.A., Fazzino, T., Garnet, B., Cutter, C.J., Barry, D.T.: Computer-based interventions for drug use disorders: a systematic review. J. Subst. Abuse Treat. **40**(3), 215–223 (2011)
14. Khadjesari, Z., Murray, E., Hewitt, C., Hartley, S., Godfrey, C.: Can stand-alone computer-based interventions reduce alcohol consumption? A systematic review. Addiction **106**(2), 267–282 (2011)
15. Gainsbury, S., Blaszczynski, A.: A systematic review of Internet-based therapy for the treatment of addictions. Clin. Psychol. Rev. **31**(3), 490–498 (2011)
16. Kongkarn, V., Sukree, S.: A framework for applying an intelligent agent to monitor, interpret, and report risk of online computer game addiction in children and early adolescents in Thailand. In: 2012 IEEE-EMBS International Conference on Biomedical and Health Informatics (BHI). IEEE (2012)
17. Wu, J.H., Wang, S.C., Tsai, H.H.: Falling in love with online games: the uses and gratifications perspective. Comput. Hum. Behav. **26**, 1862–1871 (2010)
18. Gentile, D.A., Lynch, P.L., Linder, J.R., Walsh, D.A.: The effects of violent video game habits on adolescent hostility, aggressive behaviors, and school performance (2004)
19. Gentile, D.A., Walsh, D.A.: A normative study of family media habits. Appl. Dev. Psychol. **23**, 157–178 (2002)
20. Carnagey, N.L., Anderson, C.A.: The effects of reward and punishment in violent video games on aggressive affect, cognition, and behavior. Psychol. Sci. **16**, 882–889 (2005)
21. Smith, S.L., Lachlan, K., Tamborini, R.: Popular video games: quantifying the presentation of violence and its context. J. Broadcast. Electron. Media **47**, 58–76 (2003)
22. Wallenius, M., Punamäki, R.-L.: Digital game violence and direct aggression in adolescence: a longitudinal study of the roles of sex, age, and parent-child communication. J. Appl. Dev. Psychol. **29**, 286–294 (2008)
23. Ferguson, C.J.: Violent video games and the supreme court: Lessons for the scientific community in the wake of brown v. Entertainment merchants association. Am. Psychol. **68**, 57 (2013)
24. Balci, K., Salah, A.A.: Automatic analysis and identification of verbal aggression and abusive behaviours for online social games. Comput. Hum. Behav. **53**, 517–526 (2015)
25. Gotterbarn, D.: The ethics of video games: Mayhem, death, and the training of the next generation. Inf. Syst. Front. **12**, 369–377 (2010)

26. Gotterbarn, D., Moor, J.: Virtual decisions: Video game ethics, just consequentialism, and ethics on the fly. ACM SIGCAS Comput. Soc. **39**, 27–42 (2009)
27. Kade, D.: Ethics of virtual reality applications in computer game production. Philosophies **1**(1), 73–86 (2015)
28. Molfenter, T.: Trends in telemedicine use in addiction treatment (2015)
29. van Velsen, L., Tabak, M., Hermens, H.: Measuring patient trust in telemedicine services: Development of a survey instrument and its validation for an anticoagulation web-service. Int. J. Med. Inf. **97**, 52–58 (2017)
30. Shakshuki, E.: The 4th international conference on current and future trends of information and communication technologies in healthcare. Procedia Comput. Sci. **37**, 252 (2014)
31. Hsiao, C.-H., Chen, M.-C., Tang, K.-Y.: Investigating the success factors for the acceptance of mobile healthcare technology. In: PACIS (2013)
32. Becker, D.: Acceptance of mobile mental health treatment applications. Procedia Comput. Sci. **98**, 220–227 (2016)
33. Nemoto, T., Beglar, D.: Likert-scale questionnaires. In: JALT 2013 Conference Proceedings (2014)
34. Raja, B.S., et al.: Semantics enabled role based sentiment analysis for drug abuse on social media: a framework. In: 2016 IEEE Symposium on Computer Applications & Industrial Electronics (ISCAIE). IEEE (2016)
35. MacLean, D., Gupta, S., Lembke, A., Manning, C., Heer, J.: Forum77: an analysis of an online health forum dedicated to addiction recovery. In: Proceedings of the 18th ACM Conference on Computer Supported Cooperative Work & Social Computing, pp. 1511–1526. ACM, February 2015

NV-Cleaning: An Efficient Segment Cleaning Scheme for a Log-Structured Filesystem with Hybrid Memory Architecture

Jonggyu Park and Young Ik Eom$^{(\boxtimes)}$

Sungkyunkwan University, Suwon 16419, South Korea
{jonggyu,yieom}@skku.edu

Abstract. This paper proposes NV-Cleaning, an efficient segment cleaning scheme for a log-structured filesystem with hybrid memory architecture. During segment cleaning, NV-Cleaning copies valid data to NVRAM rather than to the underlying slow storage. By doing so, NV-Cleaning quickly finishes segment cleaning and resume I/O of user processes faster than the conventional scheme. Our experimental results show that NV-Cleaning outperforms the conventional scheme in terms of both throughput and latency.

Keywords: Log-structured filesystem · Flash memory · Segment cleaning

1 Introduction

In the last decade, flash-based storage, such as SSD (Solid State Drive), has replaced conventional storage devices in various fields, from mobile systems to server systems, due to its low power consumption and high performance. However, flash-based storage has several well-known critical problems [1]. First, since it is not permitted to overwrite data on flash memory, flash-based storage should collect invalid space to reclaim free space, which is called garbage collection (GC). Second, flash memory has a limited number of erase cycles, and thus we should avoid intensive writes to the same location. The internal software layer in flash memory, FTL, performs such functions to solve the problems of flash-based storage devices.

There has been a substantial amount of research [1, 2] on how to exploit flash memory in the filesystem layer. For example, log-structured filesystems (LFS) [1–3] have gracefully addressed the aforementioned problems of flash memory by adopting log-structuring. Log-structuring transforms host I/Os into sequential ones, in order to eliminate random writes, which are harmful to flash memory [1]. Besides, log-structuring eliminates redundant writes of journaling filesystems by prohibiting over-writing, thereby reducing the number of writes in the system. For these reasons, a wide variety of fields adopt LFS on flash-based storage devices.

However, LFS has a critical drawback which is incurred by log-structuring. When LFS handles updates, it invalidates the previous data and writes the new data into a new space, because it does not allow overwriting. Therefore, LFS needs to collect the invalidated data, in order to reclaim free space. We refer to this as segment cleaning in this paper to distinguish it from GC of flash memory. LFS performs

© Springer Nature Switzerland AG 2019
S. Lee et al. (Eds.): IMCOM 2019, AISC 935, pp. 610–617, 2019.
https://doi.org/10.1007/978-3-030-19063-7_48

segment cleaning in a unit of a segment which consists of multiple blocks [4]. During the segment cleaning process, LFS selects a victim segment and copies valid blocks from the victim segment to a new segment. Unfortunately, LFS suspends I/O of user processes in the meantime. Therefore, segment cleaning can degrade I/O performance of user processes, and also diminish the lifetime of the underlying flash storage due to additional writes.

In this paper, we propose NV-Cleaning, an efficient segment cleaning scheme for a log-structured filesystem with hybrid memory architecture that comprises a combination of DRAM and NVRAM. During the segment cleaning process, NV-Cleaning copies valid blocks to NVRAM rather than to the underlying storage, and thus it quickly resumes handling I/Os of user processes. In addition, the valid blocks in NVRAM can absorb upcoming I/Os to the blocks, resulting in a decrease of the amount of I/Os. As a result, NV-Cleaning improves I/O performance of user processes while increasing the lifetime of the underlying flash memory. The current version of this work focuses on copying valid blocks to NVRAM.

2 Background

2.1 Log-Structured Filesystems and Segment Cleaning

LFS [2, 3] divides storage space into multiple segments, each of which consists of multiple blocks. LFS is designed to issue only sequential writes for both data and metadata in an append-only way without overwriting existing valid data. Therefore, LFS maintains filesystem consistency with low overheads, and provides high I/O performance on flash memory which is vulnerable to random writes.

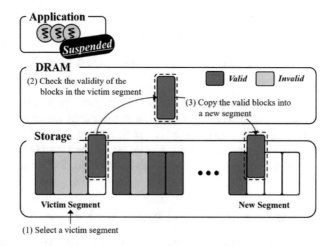

Fig. 1. Conventional segment cleaning process.

However, when LFS updates existing data, it invalidates the data and stores new data at the next block of what the filesystem lastly allocated. Since LFS does not overwrite existing data, it should collect the invalidated blocks to convert them into free blocks. LFS performs this process in a unit of a segment. We refer to this as segment cleaning. Figure 1 shows how LFS conducts segment cleaning. First, LFS selects a victim segment by calculating the number of valid blocks in each segment. Second, LFS checks whether each block in the victim segment is valid or not. Third, if the block is valid, LFS copies the block into a new segment. Otherwise, LFS skips the block and checks the validity of the next block. Finally, LFS sets the victim segment as a free segment after LFS examines all the blocks in the victim segment.

Unfortunately, the conventional segment cleaning can degrade I/O performance of user processes. When there are not enough free segments to serve the I/O of user processes, LFS suspends the I/O and performs segment cleaning to reclaim free space. We conducted a motivation experiment to measure the performance drop caused by segment cleaning. In the motivation experiment, we performed 10 GB of sequential writes under high space utilization, while increasing the space utilization from 90% to 97%. Here, the space utilization includes both valid and invalid space, excluding free space. The details of the experimental setup are explained in Sect. 4.

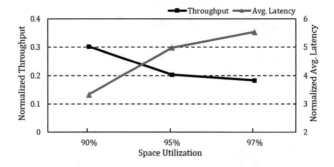

Fig. 2. Performance degradation incurred by segment cleaning.

Figure 2 shows the normalized throughput and latency. They are relative to the results obtained on a clean SSD (baseline). The sequential write throughput shows only 0.18% of baseline when the space utilization is 97%, because of frequent segment cleaning. For the same reason, the average latency is also increased by up to 5.5 times at most. Consequently, segment cleaning severely degrades I/O performance under high utilization.

2.2   A Hybrid Memory Architecture of DRAM and NVRAM

A new type of memory, NVRAM, has been introduced with fascinating characteristics, such as low latency, byte-addressability, and non-volatility as shown in Table 1. A considerable amount of research has been conducted to exploit the characteristics of NVRAM. Some of the research converges on a hybrid memory architecture where

DRAM and NVRAM are collaboratively utilized in the same system. For example, Oh et al. [6] adopts NVRAM unified with DRAM and utilizes it as a logging device for SQLite in mobile systems. Also, Zhong et al. [7] proposes an efficient swapping scheme for a hybrid memory system of NVRAM and DRAM in mobile systems. Unlike DRAM-only or NVRAM-only systems, hybrid memory systems can benefit from both DRAM and NVRAM. In this paper, we utilize some portion of NVRAM as a buffer for valid blocks, which is used by segment cleaning (Fig. 3).

Table 1. The characteristics of memories [5].

Technology	DRAM	MRAM	PCM	NAND
Non-volatility	No	Yes	Yes	Yes
Endurance	10^{15}	10^{12}	10^{8}	10^{3}
Write Time	10 ns	$\sim$10 ns	$\sim$75 ns	10 µs
Read Time	10 ns	10 ns	20 ns	25 µs

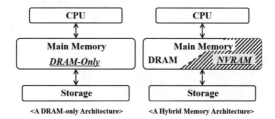

Fig. 3. A hybrid memory architecture.

3   NV-Cleaning

As introduced in the previous section, frequent segment cleaning can cause a severe performance drop when LFS has a lack of free space, since it suspends I/Os generated by user processes. In this paper, we present NV-Cleaning, an efficient segment cleaning scheme for a log-structured filesystem with hybrid memory architecture. During segment cleaning, NV-Cleaning copies valid blocks to NVRAM rather than the underlying storage, in order to improve I/O performance under high space utilization.

Figure 4 shows how NV-Cleaning works in detail. First, as with the conventional segment cleaning, NV-Cleaning chooses a victim segment considering the number of valid blocks in each segment. Second, it checks the validity of each block in the victim segment. Third, NV-Cleaning copies the valid blocks in the victim segment to NVRAM rather than to the underlying storage. Since NVRAM has less write time than storage devices, NV-Cleaning can finish segment cleaning faster than the conventional cleaning, thereby instantly resuming I/O of user processes. The valid blocks in NVRAM are batched and flushed to the underlying storage at idle time. Batching valid blocks into a single I/O can also reduce the number of I/Os and accelerate the flushing operation.

When upcoming I/Os try to access data of the valid blocks in NVRAM, NV-Cleaning can handle the I/Os in NVRAM without transferring the I/Os to the storage. In that case, NV-cleaning can decrease the number of I/Os towards the storage. Additionally, when the valid blocks are invalidated before being flushed to the storage, NV-Cleaning erases the blocks in NVRAM without writing the blocks in the storage. Currently, we implemented the scheme that flushes victim blocks to NVRAM rather than storage, by ignoring the copy operation and injecting delays considering MRAM write time of Table 1.

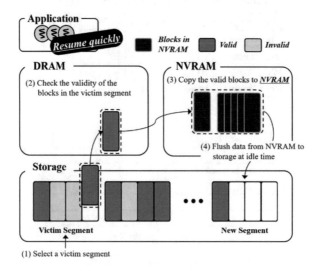

Fig. 4. The segment cleaning mechanism of NV-Cleaning.

4 Evaluation

4.1 Evaluation Setup

We implemented NV-Cleaning in the F2FS filesystem, which is the most popular log-structured filesystem in recent years. F2FS provides two different logging modes, LFS and SSR. In our experiments, we mounted the filesystem with LFS mode to generate segment cleaning. To evaluate our scheme, we performed experiments on a single computer equipped with Intel i7-3770 K, 8 GB DRAM, and Samsung SSD 850 Pro 128 GB. We selected Linux 4.15.0 kernel to compare our scheme with the state-of-the-art segment cleaning. In order to focus on the segment cleaning of LFS, we ensured that SSD-level GC does not occur during experiments through intentionally setting aside abundant free space in the SSD. To achieve this, we formatted a 128 GB SSD and used a partition of 50 GB (Table 2).

Table 2. Experimental environment.

Hardware	
Processor	Intel i7-3770 K
Memory	8 GB
Storage	128 GB Samsung 850 Pro SSD
Software	
Operating System	Ubuntu 14.04
Kernel Version	Linux Kernel 4.15.0
Benchmark	IOzone [8]

To measure the performance difference depending on the amount of free space, we compared NV-Cleaning with the conventional segment cleaning (F2FS-LFS mode without NVRAM), while gradually decreasing free space. For our experiments, we created dummy files to consume 50% of the partitioned storage and filled 40%, 45%, and 47% of the remained storage with valid and invalid data. We mixed valid and invalid data within a segment to increase the overheads of segment cleaning. To achieve this, we simultaneously created four files with a request size of 4 KB using four threads of IOzone and deleted two of them. In this environment, we performed sequential write workloads, and compared the throughput and latency of NV-Cleaning with those of the conventional segment cleaning.

4.2 Sequential Write Throughput

In the experiments for sequential write throughput, each scheme performs a sequential write workload of 10 GB with a request size of 128 KB using IOzone, while varying the free space from 90% to 97%. Figure 5 shows that NV-Cleaning outperforms the conventional segment cleaning by 39% on average and 41% at most. The conventional scheme copies valid blocks to the underlying storage during segment cleaning. Therefore, I/Os from the workloads are suspended until all of the I/Os related to

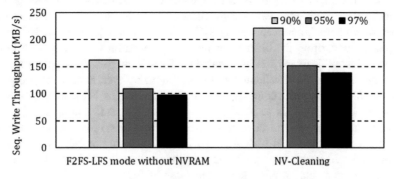

Fig. 5. Comparison between the conventional scheme and NV-Cleaning in terms of the throughput of sequential write workloads.

segment cleaning are transferred to the storage. On the other hand, NV-Cleaning copies valid data to NVRAM rather than storage. Consequently, NV-Cleaning can resume processing I/Os of the workloads faster than the conventional scheme.

4.3 Sequential Write Latency

We also measured the average latencies of the previous experiments. Figure 6 shows NV-Cleaning reduces the average latency by up to 30%, compared to the conventional scheme. On average, NV-Cleaning shows 28% lower average latency than the conventional scheme. The results are originated from the fact that NV-Cleaning copies valid blocks to NVRAM rather than storage, during segment cleaning.

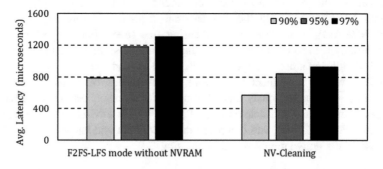

Fig. 6. Comparison between the conventional scheme and NV-Cleaning in terms of the average latency of sequential write workloads.

5   Discussion

There has been prior research [9] on how to reduce the overheads of segment cleaning. For example, SSR of F2FS overwrites invalidated blocks without segment cleaning. It eliminates the need for segment cleaning, thereby improving I/O performance under high space utilization. However, SSR also has several drawbacks. First, SSR generates random writes, which aggravate the overheads of GC inside SSD via increasing the number of data to be moved during GC. Second, when invalid blocks are non-contiguous, overwriting the blocks can incur fragmentation which degrades I/O performance. Since SSR overwrites invalidated blocks without permutating their locations, the fragmentation of the invalidated blocks still remains even after overwriting them.

In the future, we will complete the implementation of NV-Cleaning including handling of upcoming I/Os to valid blocks in NVRAM and designing the recovery process. In addition, we will develop our scheme by devising an efficient data management for valid blocks in NVRAM. To further evaluate the effectiveness of NV-Cleaning, we will perform supplementary experiments using real-world benchmarks, and compare the results of NV-Cleaning to those of SSR.

6 Conclusion

This paper presents NV-Cleaning, an efficient segment cleaning scheme for a log-structured filesystem with hybrid memory architecture. NV-Cleaning copies valid blocks to NVRAM rather than the underlying storage, thereby improving I/O performance under low free space. Our experimental results show that NV-Cleaning outperforms the conventional scheme in terms of both throughput and latency. In the future, we will complete the implementation of NV-Cleaning and perform further experiments with real-world workloads on various types of computing systems.

Acknowledgements. This research was supported by the MSIT (Ministry of Science and ICT), Korea, under the SW Starlab support program (IITP-2015-0-00284) supervised by the IITP (Institute for Information & communications Technology Promotion) and Basic Science Research Program through the National Research Foundation of Korea (NRF) funded by the Ministry of Science and ICT (NRF-2017R1A2B3004660).

References

1. Min, C., Kim, K., Cho, H., Lee, S.-W., Eom, Y.I.: SFS: random write considered harmful in solid state drives. In: USENIX Conference on File and Storage Technologies, pp. 139–154. USENIX, San Jose (2012)
2. Lee, C., Sim, D., Hwang, J., Cho, S.: F2FS: a new file system for flash storage. In: USENIX Conference on File and Storage Technologies, pp. 273–286. USENIX, Santa Clara (2015)
3. Rosenblum, M., Ousterhour, J.K.: The design and implementation of a log-structured file system. ACM Trans. Comput. Syst. **10**(1), 26–52 (1992)
4. Park, D., Cheon, S., Won, Y.: Suspend-aware segment cleaning in log-structured file system. In: USENIX Workshop on Hot Topics in Storage and File Systems, pp. 1–5. USENIX, Santa Clara (2015)
5. Analysts Weigh In On Persistent Memory. https://www.snia.org/sites/default/files/PM-Summit/2018/presentations/14_PM_Summit_18_Analysts_Session_Oros_Final_Post_UPDATED_R2.pdf
6. Oh, G., Kim, S., Lee, S.-W., Moon, B.: SQLite optimization with phase change memory for mobile applications. VLDB Endowment **8**(12), 1454–1465 (2015)
7. Zhong, K., Liu, D., Long, L., Ren, J., Li, Y., Sha, E.H.: Building NVRAM-aware swapping through code migration in mobile devices. IEEE Trans. Parallel Distrib. Syst. **28**(11), 3089–3099 (2017)
8. IOzone Filesystem Benchmark. http://www.iozone.org
9. Oh, Y., Kim, E., Choi, J., Lee, D., Noh, S.H.: Optimizations of LFS with slack space recycling and lazy indirect block update. In: Annual Haifa Experimental Systems Conference, pp. 1–9. ACM, Haifa (2010)

Emotion-Aware Educational System: The Lecturers and Students Perspectives in Malaysia

Rasheed M. Nassr[✉], Abdulaziz Hadi Saleh, Hassan Dao, and Md. Nazmus Saadat

University of Kula Lumpur (UniKL), Kuala Lumpur, Malaysia
{rasheed,Abdulazizsaleh,hassan,mdnazmus}@unikl.edu.my

Abstract. Emotion-aware educational system may feature the online education system in the near future. However, the current studies discussed more the technical implementation and how important to consider emotion in education. Receiving an input from lecturers and students may enrich the knowledge of the developers of emotion aware systems. This paper surveyed lecturers and students from one Malaysian University, and the findings showed students and lecturers have high interest in consideration of emotions in education process. However they raised many challenges such as to what extent lecturers should consider emotions when engaged with students? Do students provide enough input particularly in blended learning system where students prefer meeting lecturers face-to-face? It has been noticed that lecturers were motivating students to engage online and students show lack of self-motivation to engage independently. Eventually, lecturers were concerned about what types of emotion extraction/recognition tools should be considered? For instance, facial recognition and sound tone analysis require student to have visual/audio interaction with the system, as well as they are expensive and complicated to be implemented. Lectures proposed that statistical procedures and artificial intelligence techniques should be used to understand better the emotional patterns. Lecturers consider that utilizing the emotion analysis for mouse movement and keystroke while student are doing quizzes, assignments, tests, and exams will provide more findings than analyzing only textual communication with lecturers.

Keywords: Education · Virtual learning environment · Emotion aware · Student

1 Introduction

Students in face-to-face education have the opportunity to express their emotions to their lecturers using words, verbal tone, body language, and/or facial expression. In such ways, lecturers can consider those emotions and adjust the education process for improved student engagement and performance [1]. However, such an opportunity is not available for online students, as they are distanced physically from their lecturers. In such a situation, different tools can be used to capture students' emotions such as the

---

The original version of this chapter was revised: A reference to an earlier chapter which was omitted has been added. The correction to this chapter is available at https://doi.org/10.1007/978-3-030-19063-7_89

S. Lee et al. (Eds.): IMCOM 2019, AISC 935, pp. 618–628, 2019.
https://doi.org/10.1007/978-3-030-19063-7_49

analysis of textual communication between the lecturer and students [2], analysis of keystrokes, analysis of mouse movements, facial recognition or voice analysis.

Many proposals have been presented to develop an emotionally aware online learning system. However, before that, it was seen to consider the feedback of lecturers and students is important, as they are the main stakeholders of such systems. This paper sought the perspectives of lecturers and students in one of Malaysian university regarding proposing upgrade to their current Virtual Learning Education (VLE) system to be capable to recognize student's emotional status. In addition to identifying the most dominant students' emotions, the feedback from lecturers and students will help developing a prototype to capture those emotions. Eventually, this paper proposed a prototype to capture, analyze, store, and predict emotions. The proposed prototype will be based on previous studies and the feedbacks from the lecturers and students.

2   Literature Review

Emotion is strongly associated to academic achievement [3–7]. Commonly students communicate with the lecturer asking questions, complaining, or seeking advice. This happens over time, during the course of an entire module, and possibly a programme of study. Considering emotions in these interchanged messages and the way they use keyboards and mouse might help develop an emotional profile for students, which could "help identify not only whether the student is manifesting specific distress or emotional engagement at a particular moment in time, but also monitor the situation over time, and identify changes in patterns" [38, p. 16].

Recently, studies in emotion awareness in learning situations have concentrated on issues that involve capturing the sentiments and emotional states included in textual information. So that, the embedded opinions and emotions could play a key role first, in decision-making processes [8]; inspecting the influence of academic emotions (enjoyment, anxiety, pride, anger, hope, shame/fault, relief, boredom, hopelessness) on students' ways of thinking [9]. Secondly, entrenching emotional awareness into e-learning environments "ecologically," by evading introducing obtrusiveness or invasiveness in the learning process [10, 11]. Third, recognising patterns of emotional behaviour by witnessing motor-behavioural activity (facial expressions, voice intonation, mouse movements, log files, sentiment analysis, etc.) [12–14]. Fourth, using affect grid to measure emotions in software requirements engineering [15].

Kim et al. [16] employed motivation, emotion, and learning strategies as predictors of student achievement. They also found that emotions (boredom, enjoyment, and anger) significantly forecast students' achievement in a self-paced online mathematics course [17, 18]. Many studies have investigated the emotional status of students and its relationship with their performance and knowledge-acquiring [17]. They focused on developing proposals based on the control-value theory, as it is commonly used to describe students' emotions, performance, and achievements [9]. For instance, Pekrun et al. [9] proposed an instrument (The Achievement Emotions Questionnaire (AEQ)) to measure the student's emotion and its relationship with achievement. The AEQ with 24 scales is supposed to measure the enjoyment, hope, pride, relief, anger, anxiety, shame, hopelessness, and boredom. AEQ developers assumed that students' emotions could arise during class while studying, and when taking tests and exams. However, this tool is manual, and students need to report their emotions. Consequently, it is difficult to maintain emotional records in this way as the lecturers need to remind students to report their emotions regularly. Second, AEQ more effectively measures students' emotions when they have learning achievements, whereas students develop emotions while studying. Students may explicitly and implicitly reveal emotions when communicate with lecturers via text. Moreover, student's use of the mouse and keyboard in Virtual Learning Education (VLE), if captured, could show student's emotion. This approach is more accurate as students implicitly reveal emotions compared to reporting emotions explicitly, which could lead students to record ideal emotions rather than real ones [19].

Another project is "EmotionsOnto", which is concerned with students' emotions to reduce their intention to drop out. It was developed by Gil et al. [20]. Even though the aim of reducing dropout rates among students is good, yet, there is a need for comprehensive solutions for study-related issues by capturing and analyzing emotions. In other words, there is a need for generic emotion capturing and analyzing modules that can be utilized for many issues of studying such as dropout intention, low performance, stress with exams etc. Moreover, other studies such as Rodriguez et al. [2], Arguedas et al. [10], Arguedas et al. [17], Arguedas et al. [21], Arguedas et al. [22] rely on textual communication with lecturers as a source of emotion. Even though it is the cheapest, it may not be enough as many aspects such as analysis of student's usage of mouse/keystroke while interact with VLE may reveal more emotions than text.

The comprehensive emotion analyzing model found in Lang [23] comprises three systems: Subjective or verbal information, where individuals describe their emotions. The second system is behavioral extracted from the facial and postural expressions and speech. The third system is psychophysiological answers from the heart rate, galvanic skin response (GSR) and electroencephalographic response. However, maintaining all of these systems will be costly and impractical [19]. A more cost- effective proposal was developed by Arguedas et al. [21], which utilizes only textual students' feedback in online learning environments and ontology to quantify emotions for a better educational process. However, their proposal relies on the text from students, which might not be enough as it requires students to be aware of and express their emotions, which might not be easy. On the other hand, Arguedas et al. [21] took steps towards quantifying emotions and utilise them in the education process.

This study seeks a feasible approach entailing minimum costs but might not be as accurate compared to heart rate and galvanic skin responses. However, further analysis and utilizing AI could increase the accuracy of the proposed model. Therefore, the focus is on verbal information, in addition to analyzing mouse movements and keystrokes. In contrast to other studies that focused on emotional extraction to enhance students' performance and achievement, this study sees emotion extraction to enhance student- lecturer engagement and to motivate students to engage more with their studies will be more considerable. Therefore, the feedback from lecturers and students is necessary to figure out whether education context in Malaysia is in need to emotion-aware VLE system or lecturers and students have different perspective.

3 Method

3.1 Procedure

An announcement in VLE learning system of the intended computer science campus call students to participate in this open- ended survey on voluntary basis. Regarding the lecturers, from an academic email list, emails were sent to lecturers. While this non-probability sampling method does not guarantee the representativeness of the sample collected, however, it is the most suitable method to reach a wide variety of participants from all departments of the intended campus. Anonymous data were collected from the participants through an online survey. Content analysis was used to analyse the answers, which are collected from participants. Due to the nature of this study, where there is no specific theory to be followed; the conventional approach was used to analyse the answers. The conventional approach has the advantage that information is collected directly from participants without imposing preconceived categories or theories [24].

3.2 Participants

The participants were 30 students from four departments. There were 11 males and 19 females. There are 90% undergraduate, and 9% postgraduate. 40% of undergraduate are in their 4th year. All those students are familiar with e-learning system (VLE). On other hand, 18 lecturers participated, 60% of them are male, and 40% are female. In terms of academic title, 2 professors, 4 associate prof. and 12 assistant professors. 40% of them has above 10 years of teaching experience. Others have teaching experience between 5– 10 years. However, all those lecturers are specialized in computer science and information systems domains (software engineering, human interaction, computer engineering, electrical engineering, computer science, IT, and knowledge management).

4 Data Analysis and Discussion

Regarding the question "Considering students' emotion and mood in teaching process will improve students' capacity of learning", all lecturers-except one was neutral-strongly supported that students are emotionally driven when it comes to study. Similarly, students (90%) reported that their emotions and mood influence their study. Students think such system increase their engagement. When it comes to the most emotional status they see in their students, professors and lecturers (50%) reported frustration was the often negative emotions among their students. On other hand, regarding the positive emotion, the happy students were often been seen by lecturers. Students reported happy as the top positive emotion they feel during study, and boredom is the top negative feeling they experience, in contrast to lecturers' feedback, which considers the frustration as the top negative emotion among students. The frustration was the second negative emotion among students.

The frustration and other negative emotions may lead students to drop the course, regarding that, lecturers show disagreement with it- only 20% of them support that. The disagreed lecturers point out that students struggle first to cope up with the course, otherwise "possibly if the negative feelings are highly intense, otherwise most just tough it out and eventually things get better as they gradually understand the course" – one lecturer reported. Other aspects that may enforce students to not drop the course, yet, they are frustrated, is that "….because they can't afford to lose financially when not graduating on time. Students often push themselves till the end of semester, which would make it worse for their study"-one lecturer reported; or it could be their own personal and family problem as another lecturer reported. Overall, the lecturers, and based on their experience, reported that they encounter few cases of students drop out the course due to frustration, but they found more frustrated students who perform less "When it comes to dropping out, only few students may stop studying due to being frustrated, but the majority will continue ineffectively".

The feedback of lecturers regarding integrating emotion-detection with virtual learning environment systems. Few of the lecturers (22%) were neutral, others either agree or strongly agree. Moreover, all participants (except one was neutral) seeing integrating emotion-detection with VLE will improve the engagement of lecturers and students, as lecturers realize more regarding emotional status. Similarly, students supported having

VLE that emotion-aware. The majority of lecturers is seeing the VLE with emotional-detection is a competitive feature, only, one lecturer were against that consideration. Students have similar perspective of seeing emotion-aware VLE as a competitive feature for their university. However, the obstacle that could obstruct integrating emotion-detection particularly textual-based is that students provide insufficient text to be analysed and emotions retrieve. This is the opinion of 60% of lecturers. Actually, this is true. Seeing the statistics for one year for many courses provided in VLE (Fig. 1) showed that the overwhelming activities are not related to exchange message with lecturers, posting comments or threads in the forum.

This reveals that students' overall time in VLE for activities related to assignments, quiz, downloading files, browsing folders etc. Then, this make analyzing keystrokes and mouse movements surpass analyzing text. More evidence found, Fig. 2 presents the activities of lecturers and students. It could be seen that students' activities are triggered by lecturers' activities. Once lecturer increases activities in VLE, students' activities reach their top. Otherwise, students become idle. This indicates student's coming in to VLE often when lecturers post new assignment, quiz, or test. Those activities include considerable mouse movements and keystroke. Similarly, Students-50% of the participants- reported their agreement that students provide less text in VLE system. This might be explained as that the major of students prefer face-to-face delivering emotions rather than using VLE system. Lecturers reported students' tendency to deliver their emotional status in face-to-face style, referring to textual communication is an alternative when students have difficulty accessing lecturers. This show underutilization for the current VLE implementation-as one lecturer reported.

Regarding whether textual communication carry the emotional status of the students, lecturers-participants- reported that many times students explicitly show their emotions, however, 30% of lecturers were neutral. On other hand, the majority of students- participants- 80% reported they put their emotions in their words. Then, it could be said that students occasionally communicate textually with lecturers through VLE, but, they deliver their emotions in their words. Lecturers have some agreement regarding the insufficiency of textual communication to reveal students' emotions (Fig. 1 support that), due to that, they support to include other tools such as keystroke and mouse movement analysis to detect students' emotions. They see that students' interaction with VLE system reveal more than their words. Students share with the lecturers the same perspective. Moreover, lecturers discourage utilizing camera and sound with VLE to extract emotions from facial expressions and sound tone, as they think it involve many ethical considerations and may complicate VLE systems.

The lecturers and students showed positive intention to involve effectively in VLE integrated with emotion detection, whenever it is imposed. Lecturers and students showed supporting adopting emotion-aware VLE. Lecturers think their feedback to and engagement with students will be more effective, when VLE provides enough statistics regarding students' emotional status- providing simple statistics such how much time student and lecturer spend on VLE were been criticized [1]. Additionally, lecturers supported sharing the emotional status generated by VLE with students, as they (lecturers and students) consider that students become more conscious of their situation, and may prompt them to have behavioral change. This gets along with findings and recommendations of Arguedas et al. [10].

On other hand, the side-effect of emotion-aware VLE is that "Although VLE is important to some extent, but it's also tricky and allows for more emotion effects on student performance. It's a good feeling to the student to feel that his or her emotions is considered but it drives to weak resistance of these emotions and less self- control"- one lecturer reported. However, students against that. Students think this will increase their engagement and performance. This argument might be need further studying. Another challenge, students provide insufficient textual input in the VLE system for quality assessment. Then it may better to ask students to provide reflection for every class [1] as regular practice to bring students to engage.

5   Emotion-Aware VLE

Emotionally aware applications are receiving significant attention as a way to improve the user experience [7, 20]. Considering students' emotions and letting students be aware to their emotions helps increase their engagement and performance [2, 10]. Thus, projects collecting emotion-related data are increasing. Some initiatives perform social networks sentiment analysis or tracking students' engagement to decrease Massive Online Open Courses (MOOCs) dropout rates [20]. The project proposed by Gil et al. [20] introduced utilizing emotions to enhance the quality of education. Even though the project concerns students' dropout rate, there is a need for capturing, analyzing and utilizing emotions. Awareness of students' emotions by lecturers and teachers make their attitude and feedback more effective and timely [10, 17]. Then it could explicitly reveal the important to have dashboard summarizes students' behavior and attitude for better student's engagement [1, 25].

Moreover, discovering Malaysian students' emotions may reveal special features that characterize Malaysian students in terms of the type of emotion and its link with certain activity, which in turn helps to tailor the education process for Malaysian students. Integrating emotions in e-learning systems broadens their possibilities, allowing dynamic recommendation of activities according to the student emotions, as well as emotion-based content adaptation, among others [2]. As have been told no 'one size fits all', there is a need to design a system that consider the specific characteristics of each establishments as "differing educational philosophies, staff and student familiarity with the technologies, staff and student educational backgrounds, course design, the teacher's conception of learning and teaching, the level and discipline of the course, institutional policies and processes" [1].

Current VLE implementation provides pure statistical data regarding time spent and activities which was seen insufficient as reported by Aljohani et al. [1] in many similar systems; which also supported by the participated lecturers who seeing utilizing some techniques that extract emotional and attitude-related data from student interaction with VLE will be beneficial for lecturers. An overview of the proposed emotional profile framework is presented in Fig. 3. The proposed system was based on the conceptual model of Feidakis [18]. Four lecturers with various educational, and technological background have given their feedback regarding the best way to learn from some current implementations/proposals that have been reviewed in this study such as [26–36].

They were in agree to include emotion-extraction tools that are based on text, keystroke, and mouse movement, they think those tools are easy to integrate and the cost approaching zero. However, including facial recognition and sound tone analysis have been discouraged. Those techniques (facial recognition, and sound tone analysis) will be little expensive. Moreover, they require students to have visualized/audio interactions with VLE, which currently is not activated. The prediction of emotions using statistical analysis and artificial intelligence (IA) were encouraged, the lecturers see that may compensated the insufficient input from students and provide future insightful, assisting lecturers for suitable engagement with students. Moreover;

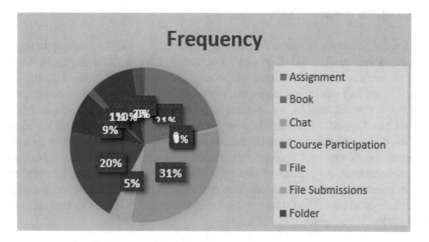

Fig. 1. Activities of students in VLE in one year

traditional statistical information presentation has been considered as a weakness of previous studies [25], due to that, utilizing advanced statistical techniques and artificial intelligence were the major concern in this study.

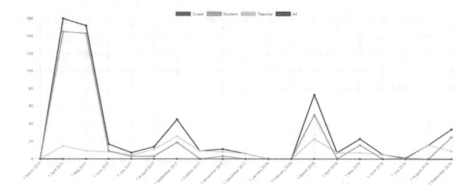

Fig. 2. Students' activity pattern in VLE triggered by lecturer's activity in VLE

In overall, the system shows that there is an emotion capture component that automatically scans text exchanged between the student and the lecturer. The second component is capturing students' keystroke/mouse emotional pattern [37]. The results of the two components will be stored in the database. These emotions can be visualized and analyzed [1] using statistical analysis for a deeper understanding of students' emotions and to predict them in similar situations in the future. Vieira et al. [25] found previous studies mainly utilize traditional statistical visualization techniques which are

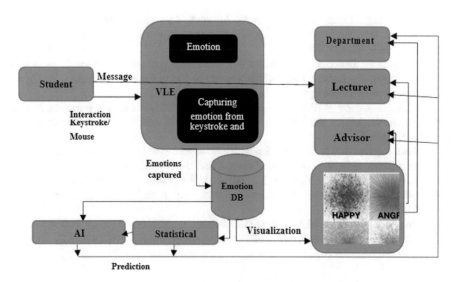

Fig. 3. Lecturers' perspective for emotion capturing and analysis

not enough, as student's emotions prediction is preferred rather than presented. This proposal also tracks students' emotions for the long term for better pedagogical design. Statistical procedures may help identifying the relationship between students' emotions and timeline of the learning process of the students (early time of the course, during assignment preparation and submission, during quiz period, during final exams, and lab projects) in order to predict the emotions during those circumstances. This is significant for better management of emotions. Additionally, the proposed system may aim to assist lecturers and management to understand students' emotions and effectively utilize them to improve students' engagement. Considering emotions effectively has been proven to enhance the quality of learning [10].

6   Conclusion and Future Work

This study has evaluated the significant of student's emotions consideration in education process from perspective of lecturers and students. The agreement regarding consideration emotions in education was undeniable, however, to what extent, was the argument. Moreover, developing emotion aware system is seen as an acceptable idea, however, there are challenges related to have enough interaction between the system and students that is sufficient to provide credible results. Another challenge is the complexity of the system, as including facial recognition-as an example, require installation cameras and visual interaction with system, which is seen by lecturers not available sufficiently. The limitations of this study are presented by limit number of participants, covering one institution, and lack of statistical data.

References

1. Aljohani, N.R., Daud, A., Abbasi, R.A., Alowibdi, J.S., Basheri, M., Aslam, M.A.: An integrated framework for course adapted student learning analytics dashboard. Comput. Hum. Behav. **92**, 679–690 (2018)
2. Rodriguez, P., Ortigosa, A., Carro, R.M.: Extracting emotions from texts in e-learning environments. Presented at the 2012 Sixth International Conference on Complex, Intelligent, and Software Intensive Systems (2012)
3. Gil-Olarte Márquez, P., Palomera Martín, R., Brackett, M.: Relating emotional intelligence to social competence and academic achievement in high school students. Psicothema **18**, 118–123 (2006)
4. Nelson, R., Benner, G., Lane, K., Smith, B.: Academic achievement of K-12 students with emotional and behavioral disorders. Except. Child. **71**, 59–73 (2004)
5. Parker, J., Creque, R., Barnhart, D., Harris, J.I., Majeski, S.A., Wood, L.: Academic achievement in high school: does emotional intelligence matter? Pers. Individ. Differ. **37**(7), 1321–1330 (2004)
6. Reyes, M.R., Brackett, M.A., Rivers, S.E., White, M., Salovey, P.: Classroom emotional climate, student engagement, and academic achievement. J. Educ. Psychol. **104**(3), 700–712 (2012)

7. Feidakis, M., Daradoumis, T., Caballé, S., Conesa, J.: Measuring the impact of emotion awareness on e-learning situations. Presented at the 2013 Seventh International Conference on Complex, Intelligent, and Software Intensive Systems (2013)

8. Loia, V., Senatore, S.: A Fuzzy-oriented sentic analysis to capture the human emotion in Web-based content. Knowl. Based Syst. **58**, 75–85 (2014)

9. Pekrun, R., Goetz, T., Frenzel, A.C., Barchfeld, P., Perry, R.P.: Measuring emotions in students' learning and performance: the Achievement Emotions Questionnaire (AEQ). Contemp. Educ. Psychol. **36**, 36–48 (2011)

10. Arguedas, M., Daradoumis, T., Xhafa, F.: Analyzing how emotion awareness influences students' motivation, engagement, self-regulation and learning outcome. Educ. Technol. Soc. **19**(2), 87–103 (2016)

11. Feidakis, M., Caballé, S., Daradoumis, T., Gañán, D., Conesa, J.: Providing emotion awareness and affective feedback to virtualized collaborative learning scenarios. Int. J. Contin. Eng. Educ. Life-Long Learn. (IJCEELL) **24**(2), 141–167 (2014)

12. Heylen, D., Nijholt, A., op den Akker, R.: Affect in tutoring dialogues. Appl. Artif. Intell. **19** (3–4), 287–311 (2005)

13. Davis, H.A., DiStefano, C., Schutz, P.A.: Identifying patterns of appraising tests in first-year college students: implications for anxiety and emotion regulation during test taking. J. Educ. Psychol. **100**(4), 942–960 (2008)

14. Bahreini, K., Nadolski, R., Westera, W.: Toward multimodal emotion recognition in e-learning environments. Interact. Learn. Environ. **24**(3), 590–605 (2016)

15. Colomo-Palacios, R., Casado-Lumbreras, C., Soto-Acosta, P., García-Crespo, A.: Using the affect grid to measure emotions in software requirements engineering. J. Univers. Comput. Sci. **17**(9), 1281–1298 (2011)

16. Kim, C., Park, S.W., Cozart, J.: Affective and motivational factors of learning in online mathematics courses. Br. J. Educ. Technol. **45**, 171–185 (2014)

17. Arguedas, M., Daradoumis, T., Xhafa, F.: Analyzing the effects of emotion management on time and selfmanagement in computer-based learning. Comput. Hum. Behav. **63**, 517–529 (2016)

18. Feidakis, M.: A review of emotion-aware systems for e-learning in virtual environments. In: Formative Assessment, Learning Data Analytics and Gamification. Elsevier (2016)

19. Feidakis, M., Daradoumis, T., Caballé, S.: Emotion measurement in intelligent tutoring systems: what, when and how to measure. Presented at the 2011 Third International Conference on Intelligent Networking and Collaborative Systems (2011)

20. Gil, R., Virgili-Gomá, J., García, R., Mason, C.: Emotions ontology for collaborative modelling and learning of emotional responses. Comput. Hum. Behav. **51**, 610–617 (2015)

21. Arguedas, M., Xhafa, F., Daradoumis, T., Caballe, S.: An ontology about emotion awareness and affective feedback in elearning. Presented at the 2015 International Conference on Intelligent Networking and Collaborative Systems (2015)

22. Arguedas, M., Daradoumis, T., Xhafa, F.: Towards an emotion labeling model to detect emotions in educational discourse. Presented at the 8th International Conference on Complex, Intelligent and Software Intensive Systems (CISIS-2014), Los Alamitos, CA (2014)

23. Lang, P.J.: A bio-informational theory of emotional imagery. Psychophysiology **16**, 495–512 (1979)

24. Hsieh, H.-F., Shannon, S.E.: Three approaches to qualitative content analysis. Qual. Health Res. **15**(9), 1277–1289 (2005)

25. Vieira, C., Parsons, P., Byrd, V.: Visual learning analytics of educational data: a systematic literature review and research agenda. Comput. Educ. **122**, 119–135 (2018)

26. Che-Cheng, L., Chiung-Hui, C.: Correlation between course tracking variables and academic performance in blended online courses. Presented at the IEEE 13th International Conference on Advanced Learning Technologies (ICALT), Hong Kong (2013)

27. Roberge, D., Rojas, A., Baker, R.: Does the length of time off-task matter? Presented at the 2nd International Conference on Learning Analytics and Knowledge, Vancouver, BC, Canada (2012)

28. Laur, E.J.M., Baron, J.D., Devireddy, M., Sundararaju, V., Jayaprakash, S.M.: Mining academic data to improve college student retention: an open source perspective. Presented at the 2nd International Conference on Learning Analytics and Knowledge, Vancouver, BC, Canada (2012)

29. Pistilli, M.D., Arnold, K.E.: In practice: purdue signals: mining real-time academic data to enhance student success. About Campus 15(3), 22–24 (2010)

30. Romero, C., López, M.-I., Luna, J.-M., Ventura, S.: Predicting students' final performance from participation in on-line discussion forums. Comput. Educ. 60, 458–472 (2013)

31. Bakharia, A., Dawson, S.: SNAPP: a bird's-eye view of temporal participant interaction. Presented at the 1st International Conference on Learning Analytics and Knowledge, Banff, Alberta, Canada (2011)

32. Ferguson, R., Shum, S.B.: Social learning analytics: five approaches. Presented at the 2nd International Conference on Learning Analytics and Knowledge, Vancouver, BC, Canada (2012)

33. Shum, S.B., Ferguson, R.: Social learning analytics. J. Educ. Technol. Soc. 15(3), 3–26 (2012)

34. Gómez Aguilar, D.A., García-Peñalvo, F.J., Therón, R.: Tap into visual analysis of the customization of grouping of activities in eLearning. Presented at the 1st International Conference on Technological Ecosystem for Enhancing Multiculturality, Salamanca, Spain (2013)

35. Rosen, D., Miagkikh, V., Suthers, D.: Social and semantic network analysis of chat logs. Presented at the 1st International Conference on Learning Analytics and Knowledge, Banff, Alberta, Canada (2011)

36. Softic, S., Taraghi, B., Ebner, M., Vocht, L., Mannens, E., Walle, R.: Monitoring learning activities in PLE using semantic modelling of learner behaviour. Hum. Factors Comput. Inform. 7946, 74–90 (2013)

37. Shikder, R., Rahaman, S., Afroze, F., Al Islam, A.B.M.: Keystroke/mouse usage based emotion detection and user identification. Presented at the 2017 International Conference on Networking, Systems and Security (NSysS), Dhaka, Bangladesh (2017)

38. Alharbi, L., Grasso, F., Jimmieson, P.: An experiment with an off-the-shelf tool to identify emotions in students' self-reported accounts. In: Symposium on Emotion Modelling and Detection in Social Media and Online Interaction, pp. 16–21. AISB, Liverpool (2018)

Personal Identification by Human Motion Using Smartphone

Toshiki Furuya[1(✉)] and Ryuya Uda[2]

[1] Graduate School of Bionics, Computer and Media Sciences, Tokyo University of Technology, 1404-1 Katakuramachi, Hachioji, Tokyo 192-0982, Japan
g211801510@edu.teu.ac.jp
[2] School of Computer Science, Tokyo University of Technology,
1404-1 Katakuramachi, Hachioji, Tokyo 192-0982, Japan
uda@stf.teu.ac.jp

Abstract. There are two types of personal authentication methods, that is, biometric authentication method and password authentication method. Although biometric authentication has an advantage that security is higher than password authentication, there is also a disadvantage that biometric information can not be replaced. On the other hand, an authentication method using human motion has been proposed in an existing research. Human motion is a kind of changeable biometric information even when biological information leaks out. In the authentication method, biometric information for authentication depends on an acceleration sensor and a gyro sensor. DP matching is used for the comparison of users in the research. However, false acceptance rate is 10.71% and false rejection rate is 56.90% in a single action. There is no way to use this method for personal identification. In our research, we also obtained data from acceleration sensors and gyroscopes which are standard equipments on usual smartphones. We used convolutional neural networks for the comparison of individuals instead of DP matching. As a result of personal identification by convolutional neural network with data of "circle", "rectangle" and "check" which are acquired from examinees, individuals can be identified with a rate of more than 88.5% on average in multi-level classification and 78.2% on average in binary classification.

Keywords: Personal identification · Smartphone · Human motion · Convolutional neural network

1 Introduction

In recent years, password authentication method has been used for personal identification by the development of information technology. Currently, PIN (Personal Identification Number) method of inputting an arbitrary number of four to six digits is used as a mainstream password authentication. However, in the PIN method, the password is broken with a probability of 1/10000 to 1/1000000, or stolen by a shoulder hack which is to see passwords over the shoulder. Also, in

S. Lee et al. (Eds.): IMCOM 2019, AISC 935, pp. 629–645, 2019.
https://doi.org/10.1007/978-3-030-19063-7_50

a password system that combines arbitrary alphanumeric characters, when the password leaks out from one site, the damage also expands to other accounts since the same password may be used for other sites. As a countermeasure of the disadvantage, biometric authentication methods are studied as an alternative to the password authentication method. Biometric authentication method is free from the management of keys since it uses information which is given by nature such as fingerprints, handwriting, iris etc. The information makes impersonation difficult. In fingerprint and iris authentication, a person is identified by different fingerprints or iris patterns. The accuracy is high since the patterns are completely different among people. On the other hand, in handwriting authentication, a person is identified by handwriting and writing speeds so that the accuracy is not so high. However, unchangeable information is used in fingerprint and iris authentication. Fingerprint can be stolen by residual fingerprints or photographs and iris information can also be stolen by photographs. Handwriting can also be stolen from the remaining handwriting but the script can be changed. On the other hand, handwriting authentication requires a device to acquire the writing. As described above, the biometric authentication methods currently studied have a disadvantage that biometric information cannot be changed or a special device is additionally required. Therefore, in this research, we propose a personal identification method by aerial signature using smartphones which are generally prevalent. It can be said that the smartphone is a general device since the penetration rate of smartphones in Japan is 71.8% [1] and the rate in the world is over 50% [2]. An acceleration sensor, a gyro sensor, a luminance sensor etc. are mounted on the smartphone as standards. In this research, we use the acceleration sensor and the gyro sensor for the aerial signature. In our method, an aerial signature as biometric information can be changed and no additional special device is required so that there is no disadvantage which appears in existing biometric authentication methods.

2   Related Researches

2.1   Personal Identification Studies by Human Action

Umemoto et al. propose an authentication method of "aerial signature" as a kind of biometric authentication methods which is free from the problems of leakage and privacy [3]. In their research, acceleration is acquired by a device that fits in a palm and normalized values of the acceleration are used as the data set. Users are identified by the variance of the acceleration per unit time. However, the problem in this research is that the acceleration sensor built in the device is affected by gravity. As a result, when a user performs an action, the user has to attach a hand on a desk closely, so that the movement of the user must be restricted. On the other hand, we use the values of the acceleration sensor excluding the gravitational acceleration and the values of the gyro sensor to eliminate the restriction of the behavior of users for identification. In addition, they doubt that the variance of the acceleration is never influenced by the proficiency of behavior so that they used DP matching, which is a statistical method, by using the variance of the acceleration as an index of proficiency levels. However, because we need to find new indicators to replace the variance, we use CNN, a type of deep learning, as a new index in this research.

Centeno et al. propose a method for personal identification based on human behavior patterns [4]. They acquired the data of "movement", "orientation", "touch", "gesture" and "pausality" by accelerometer, gyroscope and magnetometer sensors while examinees were in the state of "reading", "writing" and "map navigation" and the state of "sitting" and "walking", and then the combinations of these two kinds of the states such as "reading-sitting", "reading-walking", "writing-sitting", "writing-walking", "navigating-sitting" and "navigating-walking". As a result of personal identification of 60 examinees with the acquired data by Siames Convolutional neural network and one class support vector machine, the identification rate has been improved to 97.8%. The result indicates that the motion of human is characteristic. However, the accuracy is still insufficient for an authentication method since the false acceptance rate has improved only from 7.3% to 3%.

2.2   Personal Identification by Human Motion Using Smartphone

In order to increase the password strength, Hamano et al. propose an authentication method combining a single stroke of the brush and a single action as a gesture authentication using a gyro sensor that is standard device in a smartphone in addition to an acceleration sensor [5]. As a result of DP matching with a combination of a single stroke and a single action, the false acceptance rate increased to 30.71% although the rate was 10.71% by single stroke only. On the other hand, in our study, we believe that we can improve the false acceptance rate by keeping the false rejection rate low by using CNN instead of DP matching.

Izuta et al. evaluated the resistance of impersonation by imitating the motion of taking out a phone from a pocket of pants and lifting it from a desk, assuming the use of the mobile terminal in usual life [6]. In the evaluation of the experiments with five methods for taking out a phone from a pocket of pants, the minimum FAR was 0.00 and the maximum FAR was 0.10. Also, in the evaluation of the experiments with five methods for lifting a phone from a desk, the minimum FAR was 0.00 and the maximum FAR was 0.52. Their results indicate that characteristics of human motions was enough different for personal identification. However, in their study, a pressure sensor must be mounted on a smartphone. The sensor is not mounted on usual smartphones so that additional equipment is required. Also, the disadvantage is that the accuracy of the identification decreases when a person changes pants to different one.

Kasahara et al. propose an authentication method by walk with acceleration data obtained from the acceleration sensor of a mobile terminal on walking [7]. They propose to use LPC cepstrum used in voice processing as feature quantity of the acceleration information during walking. They showed the effectiveness of the method by quantitative evaluation of vector quantization by comparing with the feature quantity based on the statistic values used in existing researches.

When LPC cepstrum is used as a feature quantity, the effectiveness in personal authentication was proved since Euclidean distances between target - target and target - others were clearly divided by comparing with the features in existing researches. The disadvantage of their method is that it does not deal with other than flatland walking. We also evaluated our method on flatland since we knew that the acquisition of acceleration values on other than flatland was difficult.

3   Proposed Scheme

In the existing researches, there are problems that the acceleration sensor is affected by gravity of the area where the user can perform the operation is limited to the place beside the desk, and FRR 56.90% and FAR 10.71% in DP matching. Therefore, in our research, we eliminate motion restrictions on users by using acceleration sensors and gyroscopes on smartphones. We also identify users by CNN which is a kind of deep learning algorithms instead of DP matching which is a statistical method. We acquire data of the inclination and speed of the smartphone using the acceleration sensor and the gyro sensor by JavaScript on HTML, and identify users by CNN on a desktop computer in order that the performance of the classification does not depend on users' smartphone.

3.1   Acquisition of Inclination and Acceleration of Smartphone

In our method, inclination and acceleration of smartphone are acquired from a web page created using Apache, HTML and JavaScript. Parameters which are acquired in our method are as follows.

- X axis value of acceleration sensor $[m/s^2]$
- Y axis value of acceleration sensor $[m/s^2]$
- Z axis value of acceleration sensor $[m/s^2]$
- X axis value of gyro sensor $[m/s^2]$
- Y axis value of gyro sensor $[m/s^2]$
- Z axis value of gyro sensor $[m/s^2]$

These parameters are acquired during writing "aerial signature". The definition of "aerial signature" in this paper is to write an arbitrary letter or character in the air by a smartphone as a pen. We acquire the value of the acceleration sensor of a smartphone and the value of the gyro sensor at intervals of 1/4 second, and store the values in "access.log" which is the access log file for Apache. The acceleration sensor is used for acquiring how fast the smartphone is moving in the X axis, Y axis, and Z axis directions, and the gyro sensor is also used for acquiring how much the smartphone is inclined in the X axis, Y axis, and Z axis directions.

3.2   Extraction of Motion Section

The values of the inclination and acceleration of a smartphone are saved in the access log for Apache at intervals of $1/4$ s from when the start button on the web page is pressed to when the finish button on the web page is pressed. We extract the values of the inclination and acceleration from the log file to a CSV file since HTTP requests and JavaScript loading messages are also stored in the access log in addition to the data for our method.

3.3   Normalization of Data Length

In this study, we use CNN which is a type of machine learning and the length of all input samples must be the same. On the other hand, the data length of the values of the inclination and acceleration of a smartphone is different not only among users but also in the same user since time length is different from motion by motion. Therefore, we normalize the data length as preprocessing for the classification of the data by CNN. In our method, normalization by zero padding is performed as a data length normalization method. The reason of adopting the zero padding is that the difference of the length between users is large while dispersion of the length in the same user is small. In this point, we chose the zero padding since we thought the length is one of the characteristics in aerial signature. Specifically, "0" is added to the end of data until the data length becomes the same. Zero calculation erases the characteristics for convolution and max pooling layers so that the acquired data is only used for personal identification without influence of additional data.

3.4   Personal Identification by CNN

We identify users with two types of classifications by CNN. One is multilevel classification to find a user among users. In this classification, it is outputted that who did it among users. The other is binary classification to judge a user. In this classification, it is outputted whether a user is correct user or not. The structure of CNN used in our method consists of 10 layers in which three convolution 2D layers and two average pooling layers are alternately combined, and then three linear layers and two maxout layers are alternately combined.

4   Implementation

4.1   Implementation of Acquisition of Inclination and Acceleration of Smartphone

We will explain the implementation of acquisition of the inclination and the acceleration of a smartphone and the method of storing the data as shown in Sect. 3.1. In this research, we set up a measurement page combining HTML and JavaScript in htdocs of Apache, and obtain the inclination and the acceleration data by accessing measurement page from a web browser of a smartphone.

The inclination of the smartphone can be notified by "deviceorientation" event. The values of X, Y and Z axes at the inclination are acquired from "event.gamma", "event.beta" and "event.alpha". When the change of the values of X, Y and Z axes are detected by window.addEventListener('deviceorientation', function(event){}), the values are acquired.

The acceleration of the smartphone can be notified by "devicemotion" event. The values of X, Y and Z axes at the acceleration are acquired from "acc.x", "acc.y" and "acc.z". When the change of the values of X, Y and Z axes are detected by window.addEventListener('devicemotion', function(event){}), the values are acquired.

To save the values of inclination and acceleration data in the access log, ajax and "XMLHttpRequest()" object are used in HTTP communication between the web server and the web browser. The Web browser sends the data to the server by "XMLHttpRequest.open (method, url+"data")". In this research, we fix the method to Get method and also fix the url to IP address of the server.

The data to be transmitted by "XMLHttpRequest()" are expressed in one line with comma separated values of the X, Y, and Z axis in the inclination and acceleration. The data are transmitted to access.log of Apache at 250 ms intervals by setting the setInterval function as "setInterval(function, 250)". The data acquisition starts when the examinees push the "start measurement" button on screen, and the push can be found by window.addEventLister("onclick", function(), once:true). Also, the acquisition ends when the examinees push the "finish measurement" button on the screen, and the push can be found by window.addEventLister("onclick", function(window.location.href="http://192.168.1.128/b/exp.php";).

4.2   Extraction of Data from Log

The inclination and the acceleration data are stored in "access.log" of Apache as follows.

192.168.1.113– [19/Jul/2018:18:40:13 + 0900] "GET ?0.02215, 0.00095, 0.08149, 11.891776738804058, 47.79822171025447, 357.2549861916015, %HTTP1.1" 200 221

The meaning of the data is as follows.

IP address of the smartphone – [time of data acquisition] "Get method /? Value of X axis of inclination, Value of Y axis of inclination, Value of Z axis of inclination, Value of X axis of acceleration, Value of Y axis of acceleration, Value of Z axis of acceleration, %HTTP/1.1" 200 221 The data necessary for personal identification are those three values of the inclination and those three values of the acceleration so that we the data between "?" and "%" are extracted. The extracted data are written to another file as CSV file.

4.3   Data Length Normalization and Labeling

We explain the implementation of zero padding shown in Sect. 3.3. In zero padding, the number of lines per file of the CSV file is counted. After the decision of the maximum number of the lines, new lines with comma separated '0' are added until the number of lines reaches to the maximum number. When the data length normalization is finished, files of an examinee are stored in a specific folder examinee by examinee. Each of the folders is labeled with a different number.

4.4 Implementation of CNN

We explain the implementation of CNN shown in Sect. 3.4. In our study, CNN is implemented using Python2 system and chainer. The values of the functions and arguments used in each layer are mentioned in the following sub-sections.

Convolution Layer. Convolution 2D is used in the Convolution layer. Convolution2D(1,2,1) in the Convolution layer as the first layer shows the number of input channels is "1" in the first argument, the number of output channels is "2" in the second argument, and the number of filter size is "1" in the third argument. The number of input channels "1" means that the construction of input data is N rows M columns 1 channel. Convolution2D(1,1,1) is used in the Convolution layer of the third layer and the fifth layer. In these layers, the number of input channels is "1" in the first argument, the number of output channels is "1" in the second argument, and the number of filter size is "1" in the third argument.

Pooling Layer. In the Pooling layers as the second and the fourth layers, max_pooling_2d(self,l(x),2) is used. The values obtained from the Convolution layer are set as input values of the Pooling layer. Max pooling is done with a general size of poolwindows "2".

Fully Connected Layer. Linear is used in fully connected layer. In a Linear layer as the seventh layer, Linear(X,1000) is used. The size of the input vector is set to X in the first argument, and the size of the output vector is set to 1000 in the second argument. The first parameter X means that the values of the parameter changes depending on the size of input data so that X changes to the appropriate numerical value when an input sample changes. In a Linear layer as the ninth layer, Linear(1000,72) is used. The size of the input vector is set to 1000 in the first argument, and the size of the output vector is set to 72 in the second argument.

Maxout Layer. In the Maxout layer, maxout is used. In the Maxout layer as the sixth layer, maxout(self.l3(h2),1) is used. The return value of the function "self.l3" in which the result of Convolution layer of the fifth layer is inputted is set in the first argument, and poolsize is set to 1 in the second argument. In the Maxout layers as the eighth and tenth layers, the return value of the function "self.l3" in which the result of Linear layer is inputted is set in the first argument, and poolsize is set to 1 in the second argument.

Structure of CNN. The structure of CNN implemented using the above layers is shown in the Fig. 1. As shown in the figure, we use a convolution neural network composed of Convolution layers as layers 1, 3 and 5, Pooling layers as layers 2 and 4, Maxout layers as layers 6, 8 and 10, and linear layers as layers 7 and 9. In the configuration of our method, hyperparameters are set with the default values as follows; the number of epochs 50, batch size 5 and the number of units 1000.

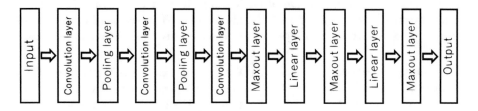

Fig. 1. Data flow and preservation

5   Evaluation

In the existing study by Hamano et al., DP matching is used for comparing the similarity between data. However, values of inclination and acceleration of the same user cannot be stably acquired although DP matching is a comparison method based on the similarity between two data. Therefore, in this research, we use CNN which is a kind of machine learning instead of DP matching which is a statistical data comparison method. In CNN, even if whole data are not similar, similarity in some parts of the data can be found. In our method, acceleration and angular velocity are acquired in aerial signature, and the acquired values are classified with binary and multilevel classification by CNN. F-measure is used for the evaluation.

5.1   Conditions for Data Acquisition

In our method, there are some conditions as follows in the acquisition method mentioned in Sect. 3.1.

Examinees must stand with straightening their back and with their feet shoulder-width apart on the flatland. There is no obstacle in 1.0 m horizontally from a user.

Examinees hold a smartphone on their dominant hand.

In aerial signature, examinees hold a smartphone in front of their chest. At that time, the screen of the smartphone turns to the examinees, and the examinees push the "start measurement" button. Next, the examinees do an arbitrary aerial signature operation. Finally, the examinees bring back the smartphone in front of their chest again after finishing the signature operation, and push "finish measurement" button on the screen. We call the sequence of the motions "the sequence in the aerial signature operation", and we count it one sequence.

In the drawing of "circle", examinees must draw a circle clockwise from the bottom of the circle with stretching their elbow. In the drawing of "rectangle", examinees must draw a rectangle clockwise from the bottom left with stretching their elbow. In the drawing of "checkmark", examinees must draw a checkmark in front of their chest.

Under the above conditions, 150 samples per character per examinee are acquired. The characters are "circle", "rectangle" and "checkmark" in the experiments in this paper. The reason of the number of 150 is that we want to have 100 samples per character per examinee at least and we also consider the load of examinees.

5.2 Evaluation in Binary and Multilevel Classification

We evaluate our method in binary and multilevel classification. In both of the classification methods, the average value and the standard deviation of f1-score are evaluated. In binary classification, one examinee is tested whether the examinee is classified in the examinee or not. In multilevel classification, one examinee is identified among all examinees.

6   Results of Evaluation

In this section, we evaluate the data described in Sect. 3.1 by CNN described in Sect. 3.4. The main indicator of the evaluation is f-measure in binary and multilevel classification. We also use the values of precision and recall. The target characters for the evaluation are "circle", "rectangle" and "checkmark" as mentioned in Sect. 4. Examinees are junior and senior fourteen students of Ryuya Uda Laboratory at Tokyo University of Technology.

6.1   Information of Examinees and Data

The information of the examinees is listed in Table 1.

In the Table 1, the height of the examinees was 163 cm to 182 cm, and the average height was 168.9 cm. Each examinee drew 150 samples per character so that 450 samples were acquired from one examinee since they drew three characters "circle", "rectangle" and "checkmark". In this experiment, 70% of whole samples were used as training samples, and the remaining 30% were used as test samples. Total number of the training samples from fourteen examinees was 1470 and that of the test samples was 630 as shown in Table 2.

The number of frames in one aerial signature is shown in Table 3. One frame means a set of data from sensors at a moment.

In Table 3, the maximum number is shown as "Max", the minimum number is shown as "Min" and the average value is shown as "Ave". The largest number in "circle" was 22 by examinee A and F, and the smallest number was 7 by examinee M. The largest number in "rectangle" was 31 by examinee A, and the smallest number was 7 by examinee L. The largest number in "checkmark" was 24 by examinee A, and the smallest number was 8 by examinee B and K.

Table 1. Information of examinees

Examinees	Smartphone model	Height (cm)
A	infober	165
B	Xperia XZ	165
C	Xperia X Compact	172
D	iPhone7	172
E	iPhone7	170
F	iPhone6s	171
G	iPhone6	182
H	Xperia X performance	164
I	iPhone6s	165
J	iPhone7	170
K	Xperia Z4	163
L	iPhone7 PLUS	172
M	Android S1	171
N	iPhone6	163

Table 2. Number of training samples

Training data	1470
Test data	630

Table 3. Number of frames in one aerial signature

Examinees	circle			rectangle			checkmark		
	Max	Ave	Min	Max	Ave	Min	Max	Ave	Min
A	22	17.28	14	31	25.3	21	24	19.59	17
B	12	9.71	8	17	13.75	11	12	9.71	8
C	15	13.10	12	16	12.99	11	15	13.08	9
D	16	13.43	11	17	13.43	11	16	13.43	11
E	19	16.36	13	26	20.31	16	20	16.43	13
F	22	16.50	12	20	15.94	12	21	17.28	14
G	14	11.8	11	16	13.92	12	15	12.31	11
H	17	13.03	10	16	13.47	11	19	13.13	10
I	15	11.64	9	16	12.18	10	14	11.68	9
J	19	15.47	13	16	14.16	12	17	15.47	13
K	15	10.07	8	15	10.10	8	14	10.42	8
L	14	10.56	8	14	9.95	7	14	10.56	9
M	14	9.76	7	16	12.38	10	14	12.51	11
N	17	13.63	11	21	17.58	14	20	15.66	12

6.2 Multilevel Classification Result

The average and standard deviation of precision, recall and f1-score in 5 times multilevel classification are in Tables 4, 5 and 6. The average and standard deviation of precision, recall and f1-score in 5 times binary classification are in Tables 7, 8 and 9 In those tables, "Ave" means the average and "SD" means the standard deviation. The bottom row in the tables shows the average of all examinees in each item.

Table 4. Multilevel classification result of aerial signature "circle"

Examinees	Precision		Recall		F1-score	
	Ave	SD	Ave	SD	Ave	SD
A	0.966	0.026	0.974	0.027	0.97	0.025
B	0.942	0.050	0.944	0.026	0.942	0.014
C	0.952	0.026	0.924	0.023	0.938	0.019
D	0.87	0.076	0.896	0.040	0.88	0.048
E	0.858	0.122	0.868	0.081	0.854	0.041
F	0.938	0.039	0.85	0.076	0.89	0.029
G	0.948	0.040	0.948	0.038	0.946	0.016
H	0.88	0.053	0.948	0.019	0.914	0.023
I	0.922	0.049	0.898	0.016	0.906	0.027
J	0.942	0.0349	0.944	0.047	0.942	0.019
K	0.918	0.008	0.858	0.021	0.886	0.015
L	0.826	0.033	0.846	0.043	0.834	0.020
M	0.882	0.030	0.922	0.022	0.902	0.016
N	0.958	0.037	0.94	0.028	0.948	0.019
ave	0.914	0.045	0.911	0.036	0.910	0.024

As a result of multilevel classification in "circle" by CNN in Table 4, the maximum value of precision was 0.966, the minimum was 0.870, and the average was 0.910. The maximum value of recall was 0.974, the minimum was 0.834, and the average was 0.910. The maximum value of the f1-score was 0.970, the minimum value was 0.834, and the average was 0.910. In addition, the average of standard deviation in each item was 0.045, 0.036 and 0.024. The highest three examinees were A, N and G in the sequence and the lowest three examinees were L, E and D in the sequence.

As a result of multilevel classification in "rectangle" by CNN in Table 5, the maximum value of precision was 0.962, the minimum was 0.716, and the average was 0.890. The maximum value of recall was 0.992, the minimum was 0.742, and the average was 0.882. The maximum value of the f1-score was 0.992, the minimum value was 0.754, and the average was 0.880. In addition, the average

Table 5. Multilevel classification result of aerial signature "rectangle"

Examinees	Precision		Recall		F1-score	
	Ave	SD	Ave	SD	Ave	SD
A	0.984	0.017	0.978	0.018	0.98	0.01
B	0.868	0.043	0.914	0.034	0.888	0.033
C	0.89	0.0480	0.918	0.029	0.914	0.030
D	0.886	0.030	0.742	0.097	0.802	0.063
E	0.934	0.053	0.952	0.033	0.94	0.021
F	0.814	0.095	0.742	0.088	0.77	0.054
G	0.716	0.107	0.81	0.062	0.754	0.033
H	0.912	0.071	0.816	0.083	0.858	0.051
I	0.924	0.048	0.912	0.095	0.912	0.033
J	0.86	0.056	0.822	0.040	0.838	0.025
K	0.914	0.029	0.94	0.031	0.926	0.022
L	0.87	0.107	0.886	0.061	0.872	0.034
M	0.914	0.061	0.96	0.035	0.938	0.032
N	0.992	0.011	0.992	0.011	0.992	0.008
ave	0.891	0.055	0.882	0.051	0.885	0.032

Table 6. Multilevel classification result of aerial signature "checkmark"

Examinees	Precision		Recall		F1-score	
	Ave	SD	Ave	SD	Ave	SD
A	0.986	0.019	0.922	0.036	0.95	0.021
B	0.904	0.038	0.974	0.028	0.936	0.015
C	0.914	0.058	0.902	0.070	0.906	0.013
D	0.898	0.090	0.81	0.092	0.846	0.025
E	0.926	0.036	0.974	0.030	0.95	0.027
F	0.952	0.027	0.962	0.025	0.956	0.005
G	0.966	0.030	0.964	0.036	0.966	0.021
H	0.886	0.068	0.85	0.075	0.862	0.026
I	0.962	0.035	0.966	0.026	0.962	0.023
J	0.906	0.063	0.974	0.037	0.934	0.018
K	0.988	0.011	0.928	0.035	0.956	0.015
L	0.888	0.080	0.89	0.047	0.888	0.022
M	0.92	0.039	0.97	0.01	0.944	0.021
N	1.00	0	0.976	0.037	0.986	0.019
ave	0.935	0.042	0.933	0.041	0.932	0.019

of standard deviation in each item was 0.055, 0.051 and 0.032. The highest three examinees were N, A and E in the sequence and the lowest three examinees were G, F and D in the sequence.

As a result of multilevel classification in "checkmark" by CNN in Table 6, the maximum value of precision was 1.000, the minimum was 0.886, and the average was 0.935. The maximum value of recall was 0.976, the minimum was 0.810, and the average was 0.933. The maximum value of the f1-score was 0.986, the minimum value was 0.846, and the average was 0.932. In addition, the average of standard deviation in each item was 0.042, 0.041 and 0.019. The highest three examinees were N, G and I in the sequence and the lowest three examinees were D, H and L in the sequence.

In Tables 4, 5 and 6, all of the precision, recall and f1-score of all aerial signatures were 0.880 or more in average. Moreover, all of the standard deviation were 0.083 or less in average.

6.3   Binary Classification Result

As a result of binary classification in "circle" by CNN in Table 7, the maximum value of precision was 0.948, the minimum was 0.588, and the average was 0.850. The maximum value of recall was 0.958, the minimum was 0.570, and the average was 0.760. The maximum value of the f1-score was 0.950, the minimum value was 0.216, and the average was 0.780. In addition, the average of standard deviation

Table 7. Binary classification result of aerial signature "circle"

Examinees	Precision		Recall		F1-score	
	Ave	SD	Ave	SD	Ave	SD
A	0.948	0.048	0.956	0.023	0.95	0.029
B	0.918	0.041	0.846	0.040	0.88	0.014
C	0.934	0.030	0.928	0.054	0.932	0.015
D	0.76	0.085	0.726	0.070	0.726	0.070
E	0.832	0.070	0.566	0.100	0.666	0.065
F	0.786	0.171	0.822	0.120	0.782	0.073
G	0.952	0.019	0.944	0.035	0.948	0.018
H	0.896	0.068	0.798	0.094	0.836	0.027
I	0.902	0.091	0.796	0.072	0.844	0.071
J	0.93	0.076	0.84	0.203	0.866	0.118
K	0.838	0.076	0.79	0.097	0.806	0.035
L	0.588	0.218	0.138	0.079	0.216	0.113
M	0.702	0.061	0.57	0.112	0.624	0.066
N	0.94	0.061	0.92	0.07	0.924	0.030
ave	0.852	0.080	0.76	0.084	0.79	0.053

Table 8. Binary classification result of aerial signature "rectangle"

Examinees	Precision		Recall		F1-score	
	Ave	SD	Ave	SD	Ave	SD
A	0.984	0.017	0.966	0.019	0.974	0.017
B	0.644	0.065	0.526	0.194	0.56	0.120
C	0.908	0.028	0.904	0.034	0.906	0.018
D	0.898	0.054	0.618	0.027	0.73	0.036
E	0.948	0.008	0.93	0.0245	0.94	0.01
F	0.794	0.100	0.604	0.127	0.67	0.084
G	0.828	0.134	0.598	0.135	0.678	0.079
H	0.79	0.032	0.644	0.051	0.71	0.033
I	0.81	0.113	0.654	0.086	0.716	0.050
J	0.762	0.065	0.766	0.101	0.758	0.048
K	0.922	0.069	0.828	0.081	0.866	0.029
L	0.798	0.082	0.756	0.055	0.772	0.028
M	0.78	0.049	0.622	0.132	0.686	0.086
N	0.976	0.018	0.988	0.011	0.982	0.013
ave	0.846	0.060	0.743	0.077	0.782	0.047

Table 9. Binary classification result of aerial signature "checkmark"

Examinees	Precision		Recall		F1-score	
	Ave	SD	Ave	SD	Ave	SD
A	0.962	0.043	0.914	0.080	0.934	0.036
B	0.886	0.071	0.924	0.026	0.904	0.042
C	0.898	0.043	0.856	0.054	0.874	0.030
D	0.744	0.093	0.582	0.152	0.634	0.063
E	0.93	0.097	0.862	0.070	0.89	0.051
F	0.97	0.052	0.926	0.036	0.948	0.023
G	0.926	0.035	0.946	0.033	0.936	0.011
H	0.886	0.072	0.774	0.112	0.816	0.036
I	0.836	0.061	0.942	0.053	0.882	0.018
J	0.95	0.069	0.85	0.043	0.898	0.027
K	0.9	0.062	0.966	0.030	0.93	0.021
L	0.8	0.110	0.59	0.122	0.666	0.051
M	0.836	0.102	0.858	0.114	0.838	0.052
N	1.00	0	0.992	0.011	0.996	0.005
ave	0.895	0.065	0.856	0.067	0.868	0.033

in each item was 0.055, 0.051 and 0.032. The highest three examinees were A, G and C in the sequence and the lowest three examinees were L, M and E in the sequence.

As a result of binary classification in "rectangle" by CNN in Table 8, the maximum value of precision was 0.984, the minimum was 0.644, and the average was 0.845. The maximum value of recall was 0.986, the minimum was 0.526, and the average was 0.743. The maximum value of the f1-score was 0.974, the minimum value was 0.560, and the average was 0.782. In addition, the average of standard deviation in each item was 0.060, 0.077 and 0.047. The highest three examinees were N, A and E in the sequence and the lowest three examinees were F, G and M in the sequence.

As a result of binary classification in "checkmark" by CNN in Table 9, the maximum value of precision was 1.000, the minimum was 0.744, and the average was 0.895. The maximum value of recall was 0.992, the minimum was 0.590, and the average was 0.856. The maximum value of the f1-score was 0.996, the minimum value was 0.666, and the average was 0.868. In addition, the average of standard deviation in each item was 0.065, 0.067 and 0.033. The highest three examinees were N, F and G in the sequence and the lowest three examinees were D, L and H in the sequence.

In Tables 7, 8 and 9, all of the precision, recall and f1-score of all aerial signatures were 0.740 or more in average. Moreover, all of the standard deviation were 0.083 or less in average.

7 Consideration

In this section, we consider the result of the personal identification from the F-measure in Sect. 6. As a result of binary and multilevel classification of "circle", "rectangle", and "checkmark" in Tables 4 and 7 of Sect. 6, examinees were identified with a probability of 78% or more in both classification methods. However, in multilevel classification, f1-score of an examinee was 75.4% which was 13% lower than the average, and in binary classification, that of an examinee was 56.4% lower than the average. We think the reason why f1-score was not good is that some examinees did not push "start measurement" and "finish measurement" button appropriately since we saw it in the experiment. Moreover, some examinees did not push "finish measurement" button on the same timing after finishing writing aerial signature although they were able to move their hand in the same speed. We think that these two factors lead to a decrease of f1-score.

Also, we found that the data length of examinee A with high f1-score was quite longer than that of other examinees. On the other hand, the data length of examinee L, F and D with low f1-score was almost the same. It means that the data length is one of the characteristics in the classification. However, the data length of examinee N with high f1-score was almost the same as most of the examinees. The f1-score was 92% and it means that the classification depends not only on the data length.

8 Summary

Biometric authentication methods such as fingerprints and iris authentication have an advantage that the requirement for users is not so hard in authentication information management since physical features are never forgotten and lost. On the other hand, there is a problem that the authentication information is not easy to be changed. Therefore, we proposed a personal identification method by aerial signature using a smartphone in this research. In our method, characteristics of handwriting as behavioral feature are used and they can be changed.

In the experiments, inclination and acceleration data on aerial signature were acquired using acceleration and a gyro sensors mounted on a smartphone, and the acquired data were classified by convolutional neural network. It was possible to identify examinees with the average of 85.2% for multilevel classification "circle", the average of 89.0% for "rectangle", and the average of 93.5% for "checkmark" as shown in Sect. 6. It was also possible to identify examinees with the average of 85.2% for binary classification "circle", the average of 84.6% for "rectangle", and the average of 89.5% for "checkmark". Those results indicate that personal identification by aerial signature is possible, and the change of registration characters is also possible since the rate of all of the three characters were about 85%. However, some of the rate marked considerably lower than the average value depending on examinees. We think that the low rates came from the timing of pushing buttons. In future, we will eliminate the buttons and set the start and the finish without the buttons. Actually, we will acquire the data when the sum of the square of the value of angular velocity in X, Y and Z axes is greater than the predefined threshold.

Furthermore, the number of similar data sets was small in our experiments since there were only fourteen examinees. We think the accuracy would decrease according to the increase of the number of examinees, especially in multilevel classification. Therefore, it is necessary to evaluate the accuracy again considering the influence the number of examinees.

References

1. Ministry of Internal Affairs and Communications. http://www.soumu.go.jp/english/index.html
2. AUN Consulting, Inc.: 40 countries in the world, main OS/model share situation. https://www.auncon.co.jp/corporate/2017/0317.html
3. Umemoto, K., Nishigaki, M.: A study on the user authentication using human motion. DICOMO2007, pp. 1338–1346 (2007). (Japanese)
4. Centeno, M.P., Guan, Y., Moorsel, A.: Mobile based continuous authentication using deep features. In: Proceedings of the 2nd International Workshop on Embedded and Mobile Deep Learning, EMDL 2018, pp. 19–24. ACM (2018). https://doi.org/10.1145/3212725.3212732
5. Hamano, M., Arai, I.: Proposal of authentication method for smartphone with accelerometer/gyro sensor. IPSJ SIG Technical report 2014-UBI-41, No. 17, pp. 1–8 (2014). (Japanese)

6. Izuta, R., Murao, K., Terada, T., Iso, T., Inamura, H., Tsukamoto, M.: Evaluation of spoofing tolerance of screen unlocking method based on eject operation of mobile phone. IPSJ SIG Technical report Vol.2017-UBI-55, No. 4, pp. 1–8 (2017) (Japanese)
7. Kasahara, H., Ichino, M., Yoshii, H., Tsurumaru, K., Katto, J., Komatsu, N.: A Study on the personal authentication using sensor data of mobile terminal. In: Proceedings of Biometrics Workshop BioX, pp. 45–50. The Institute Of Electronics, Information and Communication Engineers (2012). (Japanese)

HyPI: Reducing CPU Consumption of the I/O Completion Method in High-Performance Storage Systems

Yongju Song and Young Ik Eom[(✉)]

Department of Electrical and Computer Engineering, Sungkyunkwan University,
Suwon 16419, South Korea
{yongju,yieom}@skku.edu

Abstract. As non-volatile memory technologies being matured, the performance of the state-of-the-art storage device has been improved considerably. To fully exploit the non-volatile memory technologies, the I/O stack of operating systems needs to be revisited. There have been several works to optimize the way that I/O requests are transferred to the storage device and completed. One of them is the polling-based I/O completion method which can improve the I/O performance. However, it has a problem of using all CPU resources for I/O handling. Through analyzing the I/O completion methods, we propose an enhanced scheme, called HyPI, which consumes fewer CPU resources along with the reasonable performance. Our experimental results show that HyPI achieves 87.98% lower CPU consumption than that of the polling-based I/O completion method with a negligible performance drop.

Keywords: High-performance storage · I/O completion method · Storage system

1 Introduction

As new non-volatile memory technologies are developed recently, the state-of-the-art solid-state drives (SSDs), which is based on 3D XPoint memory [1] or Z-NAND [2], are being released. To support the potential of these new memory technologies sufficiently, the host controller interface (e.g., NVM Express) and the system I/O bus (e.g., PCI express) also have evolved accordingly. In practice, the high performance storage devices such as Intel Optane SSD 900P can achieve up to 550,000 IOPS for random read workload, while the conventional SSD such as Samsung 850 PRO can provide just 100,000 IOPS for random read workload. A data-intensive systems such as cloud servers and data center servers employ the high-performance storage devices widely for their performance requirements [3].

Meanwhile, it has begun to carefully revisit the conventional I/O stack of operating systems as the high-performance storage devices are released. This is because the traditional I/O stack has focused on the low-performance storage

S. Lee et al. (Eds.): IMCOM 2019, AISC 935, pp. 646–653, 2019.
https://doi.org/10.1007/978-3-030-19063-7_51

devices such as hard-disk drive (HDD). The Linux I/O stack which can be optimized for high-performance storage devices can be divided into 4 components: page cache, file system, block layer, and I/O completion methods. In this paper, we concentrate on the I/O completion methods among them.

I/O completion methods refer to the way that submitted I/O requests are transferred to the storage device and completed. There has been several works to improve the I/O completion methods to exploit the potential of the high-performance storage devices. Yang finds out that I/O completion using interrupts is not suitable for high performance storage devices and suggests the polling-based I/O service [4]. Also, Shin implements the dynamic polling intervàls technique to optimize the polling-based I/O completion method [5]. Besides, the Linux kernel supports three I/O completion methods such as interrupt-based I/O service, polling-based I/O service, and hybrid polling-based I/O service. A user can select one of them via `sysfs` interface. These three I/O completion methods have different characteristics. First, the interrupt-based policy consumes fewer CPU resources, but it shows the worst performance among them. The polling-based policy can achieve the best performance, whereas it consumes the CPU resources significantly. Lastly, the hybrid polling-based policy is similar to the combination of the interrupt and polling-based policy, while providing high performance and low CPU consumption. Meanwhile, even though the size of I/O request can affect the effectiveness of each I/O completion method, the aforementioned existing schemes do not consider it.

In this paper, we analyze the characteristics of each I/O completion method and find out what is the most appropriate way depending on the I/O size. Based on this observation, we propose a novel I/O completion method called **HyPI**, which serves low CPU consumption and reasonable performance. The experimental results show that HyPI achieves 87.98% lower CPU consumption compared to that of the conventional I/O completion method.

2   Analyzing I/O Completion Methods

In this section, we explain the basic I/O completion methods and perform experiments to measure IOPS, latency, and CPU consumption of each policy according to the size of I/O request.

Interrupt-Based I/O Service. Figure 1(a) illustrates an I/O completion process when using interrupts. In this I/O service, the operating system asynchronously deals with all the I/O requests. During the command execution in the storage device, the process which submitted I/O request gets to sleep until the I/O completion message is returned. At this time, CPU scheduler can delegate other tasks on the CPU or the CPU is allowed to go to low power state. Once the storage device finishes the I/O operation, CPU can resume processing of the interrupted task.

Polling-Based I/O Service. Unlike the interrupt-based I/O service, the operating system continuously checks the storage device whether its I/O request is

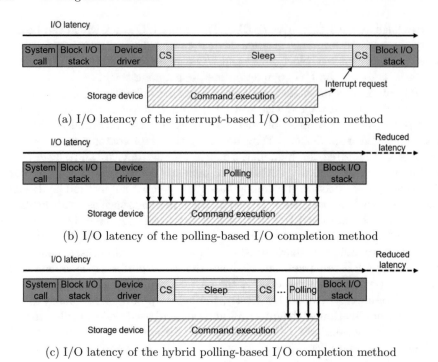

(a) I/O latency of the interrupt-based I/O completion method

(b) I/O latency of the polling-based I/O completion method

(c) I/O latency of the hybrid polling-based I/O completion method

Fig. 1. Overview of each I/O completion method

completed or not. As shown in Fig. 1(b), this I/O service does not trigger a context switch or a sleep operation. Thus, if the storage device is fast enough, it can be highly effective for improving I/O latency and throughput. However, this policy consumes CPU resources significantly for polling the storage device.

Hybrid Polling-Based I/O Service. To remedy excessive CPU consumption of polling-based I/O service, the Linux 4.10 added technical support for hybrid polling-based I/O service [6]. As illustrated in Fig. 1(c), this policy goes to sleep for a while at the beginning of the I/O service, and then polls.

To determine efficient period of sleep, the kernel classifies submitted I/O requests into 16 buckets according to their I/O size. Then, the period of sleep is calculated as a half of the mean service time (by default) of the past I/O requests corresponding to each bucket. With this I/O service, the Linux kernel can deliver low I/O latency while consuming less CPU resources compared to the polling-based I/O service.

Overall Performance Comparison. As the size of I/O request gets smaller, polling policy shows the best performance among the three I/O completion methods, as shown in Fig. 2(a) and (b). In case of 4 KB I/O request, the performance of the polling policy can be increased by up to 1.27×, compared with the interrupt policy. Meanwhile, we can see that the performance of hybrid polling policy also is similar to that of the polling policy. However, the performance gap between the interrupt and polling policy decreases gradually, as the size of the I/O request gets larger. In particular, this performance gap is just 4.5%, when a process submits 128 KB I/O request.

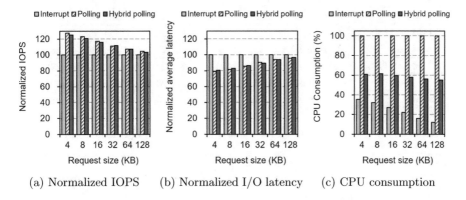

(a) Normalized IOPS   (b) Normalized I/O latency   (c) CPU consumption

Fig. 2. Performance characteristics of each I/O completion method

CPU Consumption Comparison. As inferred in Fig. 2(c), in all of the cases, the interrupt policy consumes the fewest CPU resource than other policies. In contrast, the polling policy spends almost 100% of CPU cycles. This notable result comes from the fact that, as the size of I/O request gets larger, the CPU consumption of the interrupt policy is diminished rapidly, whereas that of the hybrid polling policy drops slightly. This is because the hybrid polling policy polls the storage device for I/O completion after a predetermined threshold interval.

To summarize, we can characterize the above three I/O completion methods depending on the size of the I/O request. When the process submits a small I/O request, the hybrid polling policy is most effective. Otherwise, the efficiency of interrupt policy is acceptable. In case of the polling policy, it shows the best performance, but its performance does not differ quite from that of the hybrid-polling based policy and has the disadvantage of consuming too much CPU resources.

3   Design and Implementation

In this work, we try to achieve the combined benefits of interrupt and hybrid polling-based I/O completion method. In other words, our goal is providing high performance, while consuming fewer CPU resources. The experimental results in Sect. 2 show that hybrid polling-based policy can be highly effective for small-size I/O requests and the interrupt policy is suitable for large-size I/O requests. Therefore, we decide to choose adequate I/O completion method according to the pattern on the size of I/O request.

Table 1. I/O latency (μs) when employing each I/O completion method

I/O completion method	Size of I/O requests (KB)									
	[4,6)	[6,8)	[8,12)	[12,16)	[16,23)	[23,32)	[32,46)	[46,64)	[64,91)	[91,128)
Poll	8.9	10.3	10.6	12.7	14.3	18.1	21.2	27.6	34.6	46.3
Hybrid poll (10%)	9.1	10.5	10.9	12.8	14.5	18.1	21.3	28.0	34.7	46.5
Hybrid poll (30%)	9.2	10.5	10.8	12.9	14.6	18.2	21.3	27.9	35.1	46.8
Hybrid poll (50%)	**9.3**	**10.5**	**10.8**	13.6	14.5	18.1	21.1	27.5	34.5	46.6
Hybrid poll (70%)	11.5	12.0	12.1	**13.7**	**15.0**	**18.2**	**21.2**	27.8	34.6	47.3
Hybrid poll (90%)	11.7	13.0	13.6	15.5	17.0	20.5	23.4	30.2	37.3	49.0
Interrupt	11.3	12.8	13.0	15.1	16.9	20.5	23.8	**30.1**	**37.3**	**49.1**

Table 2. CPU consumption (%) when employing each I/O completion method

I/O Completion method	Size of I/O requests (KB)									
	[4,6)	[6,8)	[8,12)	[12,16)	[16,23)	[23,32)	[32,46)	[46,64)	[64,91)	[91,128)
Poll	100	100	100	100	100	100	100	100	100	100
Hybrid poll (10%)	81.5	83.4	84.2	86.5	88.0	89.1	90.2	90.4	90.5	90.4
Hybrid poll (30%)	75.5	75.4	75.4	75.0	74.7	74.3	74.4	74.0	74.1	73.2
Hybrid poll (50%)	**61.2**	**61.5**	**61.0**	60.4	59.7	58.8	58.4	57.0	56.4	55.8
Hybrid poll (70%)	50.4	49.6	49.2	**45.7**	**43.5**	**43.7**	**42.7**	40.9	39.9	39.90
Hybrid poll (90%)	44.2	41.7	41.2	37.6	35.2	31.4	28.9	25.8	23.2	21.5
Interrupt	34.2	32.2	30.7	28.0	26.4	22.5	22.2	**18.2**	**15.6**	**13.3**

First of all, we need to examine the hybrid polling policy in detail. This I/O completion policy performs contiguously polling after a certain time threshold (sleep), as described in the Fig. 1. By default, the Linux kernel configures this time threshold as 50% of mean service time for an I/O of the given size.

First of all, we need to examine the hybrid polling policy in detail. This I/O completion policy performs polling after a certain time threshold (sleep), as described in the Sect. 2. By default, the Linux kernel configures this time threshold as 50% of the mean service time for the I/Os of the given size. But, if the Linux collects service time of all the I/Os and calculates the mean of them to determine the time threshold, it will be certainly not effective. Therefore, the Linux kernel manages the submitted I/O requests by separating them into 16 buckets depending to the I/O size. In each bucket, there are only I/O requests which have similar size, thereby convincing that average service time can be used meaningfully. The Linux kernel calculates the value of bucket as follows:

$$bucket = \begin{cases} 2 * (\log_2(size) - 9) & Read\ request \\ 2 * (\log_2(size) - 9) + 1 & Write\ request \end{cases}$$

For example, if a process submits a read request whose size is from 4 KB (4096B) to 6 KB (6144B), the value of the bucket is calculated as 6. By using these parameters, the hybrid polling policy can deliver low I/O latency while consuming fewer CPU resources, compared to the polling policy.

Meanwhile, hybrid polling policy can be similar to interrupt policy or polling policy, depending on the period of sleep. To improve the effectiveness of HyPI, we need various versions of the hybrid polling policy which have different periods of sleep. So, we adjust the time threshold from 10% to 90% by modifying the kernel source code.

Tables 1 and 2 show the I/O latency and CPU consumption when employing different I/O completion methods. By comparing these I/O completion methods, we can see that polling policy generally shows the best performance while interrupt policy provides low performance. Also, the hybrid polling policy gets worse as the time threshold increases because the period of sleep gets longer before polling the storage device. In order to choose optimal I/O completion method for each I/O size, we decided to select the scheme that consumes the fewest CPU resources, while keeping the performance similar to that of the basic hybrid polling method. To select candidates, we consider I/O completion methods whose performance gap with basic hybrid polling policy is less than 10%. Consequently, the highlighted numbers in Table 1 are chosen. For example, in the case of the bucket for 12 KB–16 KB I/Os, the hybrid polling policy whose time threshold is 70% is considered as the optimal one. For larger-size I/Os than 46 KB, there is no significant performance gap regardless of the type of I/O completion methods, and thus we can choose the interrupt policy as the most appropriate way.

4 Evaluation

4.1 Experimental Setup

All experiments were carried out on a workstation that has two Intel Xeon E5-2620 processors (2.10 GHz, 8-core). As a high-performance storage, we employed an Intel Optane SSD 900P, which makes use of 3D XPoint memory technology [1,7]. To conduct experiments, we used a Flexible I/O Tester (FIO v3.05) [8] and set O_DIRECT flag to eliminate the disturbance of the page cache. In order to deal with synchronous I/O requests, the I/O engine of FIO is configured pvsync2, which enables the use of the polling policy.

4.2 Evaluation Results

Workload with I/Os of Same Size. To verify the effectiveness of HyPI, we conducted our evaluation by changing the size of the I/O request ranging from 4 KB to 128 KB. As depicted in Fig. 3(a), the latency difference between HyPI and hybrid-polling policy is less than 1.19%, on average. When large-size I/Os are submitted, HyPI has a slight performance drop. However, this decline is only 3.4% and 4.0% for 64 KB and 128 KB I/Os, respectively, and thus we consider this flaw is negligible. In the mean time, HyPI achieves 43.14% lower CPU consumption compared to that of the conventional hybrid polling policy as shown in Fig. 3(b). The reason is that HyPI completes I/O requests similarly to the hybrid polling policy for small-size I/Os, whereas performs in the same way as interrupt policy for the large-size I/Os. Experimental results show that HyPI can bring an advantage of low CPU consumption, and this benefit becomes greater with larger-size I/Os.

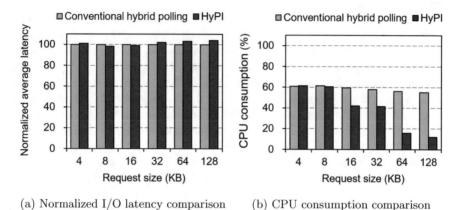

(a) Normalized I/O latency comparison   (b) CPU consumption comparison

Fig. 3. Experimental results of HyPI for a workload with same size I/Os

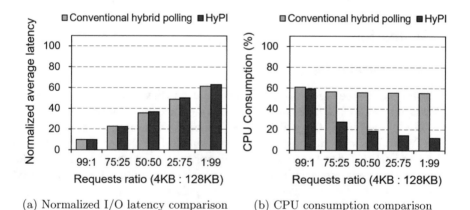

(a) Normalized I/O latency comparison   (b) CPU consumption comparison

Fig. 4. Experimental results of HyPI for a workload with mixed I/Os of different sizes

Workload With Mixed I/Os of Different Sizes. We also performed our evaluation for a workload with mixed I/Os of different sizes. The workload of our experiments consists of two I/O sizes: 4 KB and 128 KB. To clearly show the effectiveness of HyPI, we changed the ratio of 4 KB and 128 KB I/Os ranging from 1:99 to 99:1. Figure 4(a) shows that there is almost no performance penalty on HyPI. Figure 4(b) compares the CPU consumption of HyPI and the conventional hybrid polling policy. We found that, when employing the conventional hybrid polling policy, CPU consumption does not decrease significantly despite a drop in average I/O requests size. However, in the case of HyPI, CPU consumption is significantly diminished as the ratio of large I/O increases. Practically, the CPU consumption of conventional hybrid polling policy decreases from 60.8% to 55.2% for 99:1 to 1:99 (4 KB:128 KB) whereas that of HyPI is dropped from 59.23% to 12.17%.

5   Conclusion

We first analyzed three I/O completion methods which are supported by the Linux kernel. Based on this analysis, we proposed a new scheme, called HyPI (Hybrid Polling and Interrupt), which combines both interrupt and hybrid polling-based I/O completion methods. HyPI consumes fewer CPU resources along with the reasonable performance. Our experimental results show that HyPI achieves 87.98% lower CPU consumption than polling-based I/O completion method, while its performance is comparable to the conventional method. We believe that the benefit of HyPI will be greater for larger I/Os.

Acknowledgements. This research was supported by Next-Generation Information Computing Development Program through the National Research Foundation of Korea(NRF) funded by the Ministry of Science, ICT (No. NRF-2015M3C4A7065696).

References

1. 3D XPoint: Speed at What Cost? https://www.flashmemorysummit.com/English/Collaterals/Proceedings/2017/20170808_FR12_Handy.pdf
2. Samsung Z-SSD. https://www.samsung.com/semiconductor/ssd/z-ssd/
3. Koh, S., Lee, C., Kwon, M., Jung, M.: Exploring system challenges of ultra-low latency solid state drives. In: USENIX Workshop on Hot Topics in Storage and File Systems, pp. 1–7. USENIX, Boston (2018)
4. Yang, J., Minturn, D.B., Hady, F.: When poll is better than interrupt. In: USENIX Conference of File and Storage Technologies, pp. 1–7. USENIX, San Jose (2012)
5. Shin, D.I., Yu, Y.J., Kim, H.S., Choi, J.W., Jung, Y., Yoem, H.Y.: Dynamic interval polling and pipelined post I/O processing for low-latency storage class memory. In: USENIX Workshop on Hot Topics in Storage and File Systems, pp. 1–5. USENIX, San Jose (2013)
6. Past and Present of the Linux NVMe driver. https://www.snia.org/sites/default/files/SDC/2017/presentations/NVMe/Hellwig_Christoph_Past_and_Present_of_the_Linux_NVMe_Driver.pdf
7. Analysts Weigh In on Persistent Memory. https://www.snia.org/sites/default/files/PM-Summit/2018/presentations/14_PM_Summit_18_Analysts_Session_Oros_Final_Post_UPDATED_R2.pdf
8. Flexible I/O Tester (FIO). https://github.com/axboe/fio

Implicit Interaction Design in Public Installation Based on User's Unconscious Behaviors

Hong Yan[✉], QiuXia Li, and Yun Guo

Hainan University, Haikou, China
yanhong@hainu.edu.cn, Qiu_Xia_Li@hotmail.com

Abstract. With the widespread popularity of smartphones, communication between people in public places has decreased. This article aims to carry on the design of the seats in public areas, by the user's unconscious behavior of swing the seats unconsciously, combining with interactive design that each user's behavior of swing the seats corresponds to music fragments played by different types of musical instruments, and finally all the fragments are aggregated to a music. Experiments show that the interactive device designed by us not only provides a new mode of communication, but also strengthens the connection between people, and creates a more harmonious and exciting public space.

Keywords: Unconscious behavior · Public installation · Interaction design

1 Introduction

In a highly developed modern society, functional design results in the absence of emotional elements. Such as in the public places, many people become "phubbing" (phone snub), relying on mobile phones or the Internet to dispose of inner loneliness and emotional communication. The high pace of life and long urban commuter routes have objectively fragmented private time. In Fig. 1, no matter in what kind of occasions, such as walking, eating, waiting for a bus, taking a car, etc., whenever you are free, you will take out the mobile phone, keep your eyes on the screen and play games, twitter, Facebook and so on.

On the other hand, installation is the most common form of art that appears in public areas and public view. American sociologist John Nasby indicates that there is a need for high compensatory emotions everywhere. The more high-tech our society it is, the more we hope to create a high-emotional environment. As a product of human thinking and a socio-economic cultural activity, installation shows the people's demand for social essence and spiritual culture. This study aims to integrate unconscious behavior into the design of the art installation. Through the interaction and communication between users with unconscious behaviors, existing communication defamiliarization and social difficulties can be improved. In the product experience process, people can interact with the surrounding people, no longer ignore each other and create a harmonious environment to get along with each other.

S. Lee et al. (Eds.): IMCOM 2019, AISC 935, pp. 654–663, 2019.
https://doi.org/10.1007/978-3-030-19063-7_52

Fig. 1. People becomes "phubbing" in public place.

The "unconscious" behavior means a kind of subconscious behavior that has not been subjectively analyzed and judged. Unconscious behavior often occurs in people's daily life. For example, one wants to blow up the plastic bubble when seeing it; one wants to tear off the agnail on the finger when seeing or touching it; one will look back when hearing someone say "Hi!" in the back; and shaking leg, straining the head when one is nervous, looking upward when one is recalling, and so on.

Fig. 2. Unconscious behavior of swing the seat.

Our study integrates the user's behavior of swing the seats unconsciously as shown in Fig. 2, and through the unconscious behavior opens the way to play music fragment. Combining with the interactive design that each user's behavior of swing the seats corresponds to music fragments played by different types of musical instruments, and finally all the fragments are aggregated to a song as shown in Fig. 3. In our study, the Kinect 2.0 is used to analyze the user's positions. According recognized the angle and rate of rotation of the user's body; it achieves the effect of controlling the volume and rhythm of the musical instruments. It enables users to strengthen communication with others by music, instead of focusing on their own electronic devices.

Fig. 3. The framework of musical communications based on user's unconscious behaviors.

2 Related Works

Referring to the "unconscious" design, the work of Japanese designer Naoto Fukazawa is deeply rooted in the hearts of people. His design applies unconscious behaviors to the interaction functions, such as wall-mounted CD player. People know that this is the switch of the CD player but didn't realize why they knew it.

Unconscious behavior is not only reflected by physical touch, but also by the smell. The smell keeps the impression in memory longest. The project "Over the Horizon" in the Shanghai Disneyland brings smell experience into interaction experience. For example, when broadcast to the African savannah, the device emits the smell as same as fresh grass, giving the audience an immersive sensory experience.

At the same time, many excellent installations have emerged with the advancement of technology. These installation arts also incorporate interactive design. For example, the renovation of the New York Science Museum presented the largest interactive projection. The presentation shows a complete virtual ecosystem where exhibitors can understand the nature while transforming the virtual ecosystem. The interactive device has transformed the audience into an indispensable part of the design, and its inter-activity has become more and more important.

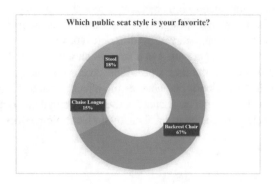

Fig. 4. The favorite public seat style.

3   Investigations of Public Seat

The investigation conducted in the rest area of the public place. 100 participants among the 16–45 years old conducted an online questionnaire survey. The main questions directed at the opinions and suggestions of the existing public seats. Based on this investigation, the research direction is established and combined with the interactive design rationally.

The results show that 67% of participants prefer to a seat with a backrest (as shown in Fig. 4). This result reflects a truth that there is usually a bench without a backrest, and the user cannot let the waist and back to have a rest.

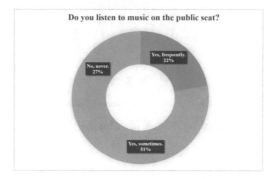

Fig. 5. The rate of listen to music on public seat.

According to questionnaires, 78% of the people listened to music while resting on the seat (as shown in Fig. 5). If there were strangers sitting next to them, 84% would feel embarrassed (as shown in Fig. 6). Besides, mostly were communicated with electronic devices, and rarely communicated with strangers around. The results reflect that many people have social discomfort and social fears in public. Therefore, our crucial issue is not only how to reduce the communication cost among people, but also how to ease the cautious situation, for enhancing communication between people and creating a harmonious environment.

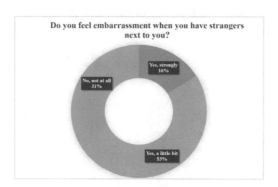

Fig. 6. The rate of embarrassment feeling when someone next to you.

In addition, according to survey statistics, 68% of users would unconsciously sway slightly when they are resting on a rotating seat (Fig. 7). This is one kind of unconscious behavior, and the important entry point of our study, which plays an important role in the innovation of the interactive design ideas with this unconscious behavior. Therefore, space can be remodeled in the rest area, making the waiting time less boring, and music can improve communication between people.

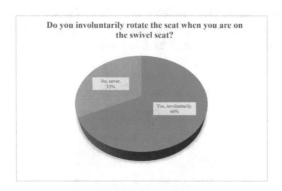

Fig. 7. The rate of swing the seat.

4   Interaction Design

This study uses "Processing" to program, through Kinect 2.0 for motion recognition, combined with the dynamic and static properties of human motion, constructing motion recognition feature vectors for human shoulders, elbows, knees, etc. The coordinate set $X = \{X_0, X_1, \ldots, X_i, \ldots X_n\}$, $Y = \{Y_1, Y_2, \ldots, Y_i, \ldots Y_n\}$, $0 \leq i \leq n$, respectively.

n is the total number of joints, and m is the total number of somatosensory recognitions. In order to facilitate the calculation, we only consider the rotation arc of the plane, regardless of the rotation arc of the three-dimensional space. The coordinate analysis is performed on the moving position trajectory of the joint, for example, the value of the arc of the rotation becomes smaller and the volume becomes smaller. Oppositely, the volume of the arc of the rotation becomes larger, and the corresponding volume becomes larger.

To calculate the rotation angle of the i-th joint, assume that the center of the circle is $(x_{i,t}, y_{i,t})$ and the radius is r_i. Then this circle can be expressed as:

$$\left(x_{i,j} - x_i\right)^2 + \left(y_{i,j} - y_i\right)^2 = r_i^2 \tag{1}$$

When time t, the i-th joint coordinates $(x_{i,t}, y_{i,t}), 2 \leq t \leq m$. **then**

$$(x_{i,t} - x_i)^2 + (y_{i,t} - y_i)^2 = r_i^2 \tag{2}$$

$$(x_{i,t-1} - x_i)^2 + (y_{i,t-1} - y_i)^2 = r_i^2 \tag{3}$$

$$(x_{i,t-2} - x_i)^2 + (y_{i,t-2} - y_i)^2 = r_i^2 \tag{4}$$

if

$$\left| \begin{matrix} (x_{i,t} - x_{i,t-1}) & (y_{i,t} - y_{i,t-1}) \\ (x_{i,t} - x_{i,t-2}) & (y_{i,t} - y_{i,t-2}) \end{matrix} \right| \neq 0 \tag{5}$$

then

$$A = 2 * (x_{i,t} - x_{i,t-1}) \tag{6}$$

$$B = 2 * (y_{i,t} - y_{i,t-1}) \tag{7}$$

$$C = 2 * (x_{i,t-1} - x_{i,t-2}) \tag{8}$$

$$D = 2 * (y_{i,t-1} - y_{i,t-2}) \tag{9}$$

$$E = x_{i,t}^2 + y_{i,t}^2 - x_{i,t-1}^2 - y_{i,t-1}^2 \tag{10}$$

$$F = x_{i,t-1}^2 + y_{i,t-1}^2 - x_{i,t-2}^2 - y_{i,t-2}^2 \tag{11}$$

also

$$x_i = \frac{D * E - B * F}{A * D - B * C} \tag{12}$$

$$y_i = \frac{A * F - C * E}{A * D - B * C} \tag{13}$$

$$r_i = \sqrt{(x_{i,t} - x_i)^2 + (y_{i,t} - y_i)^2} \tag{14}$$

The angle between the j-th coordinate and the previous coordinate is obtained:

$$\theta_i = \left| \arctan\left(\frac{y_{i,j-1} - y_i}{x_{i,j-1} - x_i}\right) - \arctan\left(\frac{y_{i,j} - y_i}{x_{i,j} - x_i}\right) \right| \tag{15}$$

So we can calculate the volume *VOL* and play speed *SPD* of music fragment:

$$VOL = \sum_{i=0}^{n} \gamma * \theta_i \tag{16}$$

$$SPD = \sum_{i=0}^{n} \delta * \theta_i \tag{17}$$

5   Appearance and Material

In Fig. 8, the seat we designed is to mimic the appearance of vinyl records. The user sits on the seat likes that the stylus is placed on the record. Moreover, the user is rotating seat likes that player is switching a vinyl record. Based on our "unconscious" interaction concept, this design reduces the time of understanding the product and achieve a better interactive experience.

Fig. 8. Our installation design.

An ergonomic is used in our design. The curvature of the back of the human body is set to 106° depending on the curvature of the human spine during the relaxation. The spine and pelvis can be cushioned in a relatively comfortable position. At the same time, a lumbar pillow is made on the back of the chair, so that the waist is supported and protected. The shoulders are naturally drooping and bent over the edge of the armrests. The shoulders and elbows are relaxed. The height of the backrest is the ankle can make the angle between the thigh and the calf at 30°, which allows the user to relax the leg muscles better.

6 Experimental Evaluations

In order to evaluate the effectiveness of the proposed method, we conducted two experiments. We selected common public seat and the proposed method without music interaction as baselines. As the evaluation criteria, we employ mean user evaluations. It represents a deviation that shows how much dispersion of satisfactions from 1 (lowest) to 5 (highest).

In the experiment, 20 participants (do not know each other) are assigned to 5 groups. The results by each method are shown in Fig. 9. The average satisfaction of common public seat is 2.15, standard deviation is 0.81; the average satisfaction of proposed method without music interaction is 3.2, standard deviation is 1.05; the average satisfaction of proposed method is 4.05, standard deviation is 0.88. In addition, the satisfaction of proposal method was proved higher and has a relatively good effect.

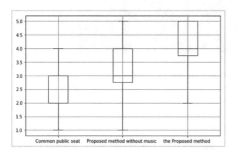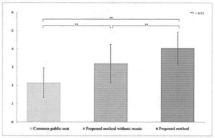

Fig. 9. User's satisfaction results by each method.

Subsequently, we divided the 20 participants into 2 groups, 10 in group A and 10 in group B. Let each participant evaluate a 1–5 rating on the question which is "Do you interested in strangers next to you?". Group A is answered before trying the proposed method, and Group B is answered after trying the proposed method. The average evaluations of Group A is 2.20, standard deviation is 0.91; the average evaluations of Group A is 3.50, standard deviation is 0.88 as shown in Fig. 10.

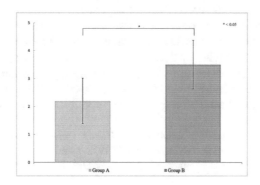

Fig. 10. The results of different groups.

However, the experimental result was advanced in a closed environment relatively. In reality world, there are a noisy environment and the deviation of some behaviors by different users. This problem would be deeply considered in the next step of this research.

Fig. 11. Usage scenarios.

7 Conclusion

This paper injects the "unconscious" design concept into the interactive device, which not only enriches the public space design form but also allows users to have a richer sensory experience in the "unconscious," making the design more intuitive and life more interesting. The combination of interaction design and installation art has become a new medium for information presentation. It not only changes the presentation form of installation art but also has vitality and new definition. More importantly, this enables users not only to get information by relying on the vision but also to have a more intuitive understanding and operation through other senses.

Our design combines interactive design and installation art to rationalize the usage and creative innovation of the rest area of public space. Moreover, it integrates music elements, so that users can carry out multi-sensory and immersive interaction through unconscious behavior in the process of rest experience.

Our design is suitable to be placed in the rest area of parks and shopping malls. Instead of ordinary seats, it can also be placed in the entertainment and leisure areas in the office building or shopping mall as shown in Fig. 11. It brings person the ease and pleasure when busy and exhausting. People can not only have touchable interaction with chairs, but also enhance communication with others. This kind of design can relieve the tired and awkward atmosphere, and make waiting time less boring.

Acknowledgements. This work is supported by "Hainan University Education and Teaching Reform Research Project (Project Number: HDJY1978)" and "Hainan University Research Initiation Fund Project (Project Number: KYQD(SK)1709)".

References

1. Lombard, M.: Direct responses to people on the screen: television and personal space. Commun. Rep. **22**, 288–324 (1995)
2. Lombard, M., Ditton, T.: At the heart of it all: the concept of presence. J. Comput. Mediat. Commun. (1997)
3. Bailey, J., Bailenson, J.N., Won, A.S., Flora, J., Armel, K.C.: Presence and memory: immersive virtual reality effects on cued recall. In: The International Society for Presence Research Annual Conference (2012)
4. Atterer, R., Wnuk, M., Schmidt, A.: Knowing the user's every move: user activity tracking for website usability evaluation and implicit interaction. In: Proceedings of the International World Wide Web Conference WWW06, pp. 203–212 (2006)
5. Davis, K., Owusu, E., Marcenaro, L., Feijs, L., Regazzoni, C., Hu, J.: Evaluating human activity-based ambient lighting displays for effective peripheral communication. In: Proceedings of the 11th EAI International Conference on Body Area Networks, pp. 148–154 (2016)
6. Barabasi, A.: The origin of bursts and heavy tails in human dynamics. Nature **435**(7039), 207–211 (2005)
7. Steinicke, F.: Being Really Virtual: Immersive Natives and the Future of Virtual Reality (2016)
8. Cummings, J.J., Bailenson, J.N.: How immersive is enough? A meta-analysis of the effect of immersive technology on user presence. Media Psychol. **19**, 272–309 (2016)
9. Kamil, M.J.M., Abidin, S.Z.: Unconscious human behavior at visceral level of emotional design. Soc. Behav. Sci. **105**, 149–161 (2013)
10. Goel, S., Hofman, J., Sirer, M.: Who does what on the web: studying web browsing behavior at scale. In: Proceedings of the 26th Conference on Artificial Intelligence AAAI (2012)
11. Jain, A., Bolle, R., Pankanti, S.: Biometrics: Personal Identification in Networked Society. Kluwer Academic Publishers, London (1999)
12. Kade, D.: Ethics of virtual reality applications in computer game production. Philosophies (2015)
13. Hua, M., Fei, Q.: The value of unconscious behavior on interaction design. In: IEEE International Conference on Computer-Aided (2010)

Fish Swarm Simulation Fs Virtual Ocean Tourism

YiLin Rong[1], Shizheng Zhou[2], Peng Cheng Fu[2], and Hong Yan[1(✉)]

[1] Design School, Hainan University, Haikou, China
yilinrong@hotmail.com, yanhong@hainu.edu.cn
[2] State Key Laboratory of Marine Resource Utilization in South China Sea,
Hainan University, Haikou, China
xpzzzsz@hotmail.com, pcfu@hainu.edu.cn

Abstract. As a key technology for the marine visualization, fish swarm simulation plays an important role in the development of marine tourism resources, marine ecotourism and environmental protection, and marine public welfare services. However, the traditional fish swarm simulation methods don't consider the viewer for marine visualization. The lack of interaction between the fish and the viewer, resulting in viewer cannot reach a deeper marine tourism experience. In this paper, we take marine tourism as an example, the combination of virtual reality technology and tourism, through the establishment of the model of fish swarm, explores and studies the application and future development in VR marine tourism.

Keywords: Fish swarm simulation · Virtual reality ·
Ocean tourism · Interaction design

1 Introduction

Today, the global marine economy is booming, and all countries have turned the sight to the ocean and began to formulate marine development strategies and plans. Including the marine tourism consumer market is vast, and its economic and social benefits are enormous. Ocean tourism is playing an increasingly important role in leading new consumer fashion, forming new consumption hotspots, expanding new investment fields, and creating new economic growth points. From the perspective of the development of global tourism, marine tourism is a new challenge. At present, Ocean Tourism is enriching the high-end marine tourism products of the South China Sea by developing marine tourism projects such as coral reef sightseeing, scientific investigation, underwater diving, marine exploration, sports, entertainment, luxury cruises, and expanding the development of tourism.

At the same time, the emergence of VR tourism breaks the traditional industry model and brings to the marine tourism. VR ocean tourism experience model allows visitors to experience some new destinations, allowing visitors to explore more fun in marine tourism. However, the current VR ocean tourism display content is still single, and there is a lack of interaction with users, resulting in users unable to reach a deeper travel experience. Through a variety of interactive ways to bring visitors fresh, exciting,

© Springer Nature Switzerland AG 2019
S. Lee et al. (Eds.): IMCOM 2019, AISC 935, pp. 664–671, 2019.
https://doi.org/10.1007/978-3-030-19063-7_53

fun, real experience, break the space and time constraints, solve the "pain point" is precisely what VR tourism needs to understand and explore.

Fish swarm simulation is the most important part of virtual ocean tourism. Because the fish swarm is a self-organizing group, there is no fixed leadership. Observational studies on self-organizing groups, including birds, insects, and fish stocks, indicate that individuals in a group may have an orderly coordinated state of motion while following simple behavioral rules. For example, to avoid collisions, individuals must maintain a minimum distance each other and similar ones in the group attract each other or repel. The influence and importance of these natural factors in-group behavior are worth exploring and studying. To this end, scientists have conducted a large number of experiments and observations to record the respective position of fish in the long-term swimming movement, such as the Aberdeen Marine Institute in the United Kingdom built a circular sink, the long-sleeved squid group (Scomber japonicus) conduct continuous observations to understand the swarm structure. However, the traditional fish swarm simulation algorithm does not consider the role of the user, so the lack of interaction between the fish and the user, resulting in users cannot reach a deeper travel experience.

This study considers the contact between the fish swarm and the user while studying the artificial fish swarm simulation. This study will adopt the concept of individual behaviors IBMs (Individual-Based Models) model, treating individuals in the group as independent units, following the exclusion-parallel-attraction rules in the Reynolds model, and considering collision avoidance as the primary premise, including evading fish. A collision between a group and a user. On this basis, this study improves the fish swarm algorithm to make the fish swarm proactively approach the user, so that the user is present and gets a better user experience.

2   Related Research

2.1   VR+Tourism

The current common areas of VR technology include medical, military aerospace and games, as well as used in real estate development, interior design, and rail transit. In recent years, many VR research attempts applied in the field of tourism. Through VR, they can dispel their fears and uneasiness about the unknown factors of the destination before the trip.

In 2016, Thorpe Park and Alton Towers announced the launch of the virtual reality roller coaster experience and the virtual reality ghost train experience. In the same year, eLong, an online travel service provider, released the hotel VR experience video, introducing VR technology in the user experience, providing users with the service of "not living in the prophet, immersive" when choosing a hotel. In the tourism development of Xi'an, China, through the consideration of the consumption characteristics, tourism preferences and market composition of tourists. VR technology was used to design the virtual tourism system of Xi'an city scenic area, a three-dimensional topographic landscape map and virtual panoramic roaming system.

The application of VR technology in tourism is mainly focused on VR immersive interactive experience, advertising, and marketing used to stimulate potential tourists to start traveling. However, for ocean tourism, this application is still in the initial stage of trial.

2.2 Swarm Algorithm Research

The study of swarm intelligence originated from the study of the group behavior of social insects (such as ants, bees, etc.). The existing research on group intelligence mostly starts from a group behavior with a large number of individuals, extracts models from their group behaviors, and establishes some rules for these behaviors, thus proposing algorithms to solve the reality. At present, there are ant optimization algorithms, ant-clustering algorithms, and particle swarm optimization algorithms.

Ant Optimization Algorithm (ACO). During the foraging process, the ant individual leaves a pheromone in the path that its passes and the subsequent ant individual determine the path by sensing the concentration of the pheromone. Since the pheromone volatilizes over time, the pheromone concentration on the relatively short track is also relatively large. Therefore, ants can find a shorter path for food in this way, and provide a new method for solving various optimization problems.

Particle Swarm Optimization Algorithm (PSO). PSO, proposed by James Kennedy and R. C. Eberhart in 1995, is a heuristic search algorithm based on swarm optimization, which is derived from the simulation of bird swarm and fish foraging behavior. First, a bunch of random particles (random solutions) is initialized, and then the optimal solution is found by iteration. In each iteration, the particles update their speed and position by tracking two extremums (single extremum and global extremum), applying optimization problems in functions, neural network training.

2.3 Artificial Fish Simulation

In view of the formation, structure, and behavior of fish schools, researchers have proposed some relevant theories and models from different angles. Some researchers have focused on studying the interactions between neighbors in fish schools, how to cooperate to escape danger and prey. Partridge proposes that from a local perspective, the fish visually perceives the movement of neighboring other fish and changes its movement accordingly based on this information. Niwa sees the formation and structure of the fish as an interactive particle system, using the Langevin equation to describe individual fish. Simon Hubbard believes that fish is an interactive and self-organizing particle, and that individual fish are governed by two forces (one that mimics the movement of other fish in the neighborhood and the other from external environmental factors). Breder defines the fish group as a specific state of motion of the fish. In the fish group, each fish moves in the same direction and at a uniform speed. He believes that the factors in which the individuals in the fish are separated from each other are: the distance is greater than the critical. The value is expressed as attractive; if the distance is less than the critical value, it is expressed as the repulsive force. Steven studied the effects of fish swarm on fish behavior and individual fish interactions.

Furthermore, some researchers have improved the artificial fish swarm algorithm from different aspects and proposed some improved artificial fish swarm algorithms. For example, Park uses an adaptive adjustment parameter to improve the algorithm and propose an improved artificial. Fish group algorithm; Cho introduces a chaotic system with sign ability and other characteristics into artificial fish swarm algorithm, and proposes a chaotic artificial fish swarm algorithm; Fu proposes an improved fish swarm algorithm with escape behavior and improved fish with reproductive ability. Group algorithm and artificial fish behavior algorithm based on multiple operators; Zhang uses the optimal individual retention strategy to improve the foraging behavior, at the same time, improves the clustering behavior, rear-end behavior in the algorithm, and "shrinks" the search domain. An improved artificial fish swarm algorithm is proposed. These improved algorithms have improved the efficiency of the basic artificial fish swarm algorithm to a certain extent, but in the context of marine tourism, the human factors in the existing artificial fish swarm algorithm have not been well solved.

3   Solutions

Fish naturally gather in the swimming process, which is also a vivid example of the formation of a living habit formed to ensure the survival of the group and avoid the hazard. Reyno lds believes that birds and fish. The formation of the cluster does not require a leader, only each bird or each fish should follow some local interaction rules, and then the cluster phenomenon emerges as a whole model from the individual interaction of the individual. Reyno lds defined three rules:

(1) Separation rules: try to avoid overcrowding with neighboring partners
(2) Alignment rules: try to be consistent with the average direction of neighboring partners
(3) Cohesion rules: try to move towards the center of the neighboring partner.

First, the collision avoidance between individuals is set as the primary condition, and the distance between individuals must be kept no less than the NND (Nearest Neighbor Distance). When the neighbor fish approaches, the corresponding behavioral responses, and actions are taken according to the distance. Secondly, the fish body reaction mode is defined according to the division of fish visual ability range. According to the division of fish visual ability range r, the fish body reaction mode is defined. In the model, the fish visual ability range r is divided into 4 distances:

(1) r_s: is a strong rejection distance. According to the principle of avoiding the collision, the distance is equivalent to the minimum distance NND, and the neighboring fish j is in the exclusion zone of the fish i, and the fish i will quickly leave away from the neighboring fish j direction;
(2) r_z: According to the fish's habit of maintaining and following the target movement, the distance between the parallel movements of the parallel movements is established, and the neighboring fish j is in the parallel labeling area of the fish i, and the fish i will advance at the speed of the label. The direction is unchanged;

(3) r_a: To attract each other to close the distance, which is within the visual range of the fish individual, and tends to approach each other due to the aggregation behavior of the individual fish. That is when the neighbor fish j is in the attraction area of the fish i, the fish i will advance toward the neighbor fish j;

(4) r_l: for the non-reactive distance, the neighbor fish j has been outside the scope of the visual ability of the fish i, and does not affect the fish i. Create a model of fish movement in a rectangular two-dimensional space.

In the model, the fish visual ability range r is divided into 4 distances, where: r_s is a strong repulsive distance. According to the principle of collision avoidance, the distance is equivalent to the minimum distance NND, when the neighboring fish is in the exclusion zone of the fish, and the fish will quickly leave in the direction away from the neighboring fish; r_z established according to the habit of the fish to maintain and follow the target movement. The labeling distance of parallel movements, the neighbor fish, is in the parallel labeling area of the fish, the fish will advance at the labeling speed v_{vol}, the direction is unchanged. We found that the swimming direction in a group tends to be random.

Our research adds user factors based on the Reynolds' model. We made the following definitions:

Natural Attraction User U to Fish F

When the fish swarm is in a state of non-frightening, the fish begins to surround the user, and the radius around it is $r_d = dis(u, f_i)$. u represents the current coordinates of the user, and f_i represents the coordinates of the i-th fish.

(1) The radius of the fish around the user gradually increases with time. After time t, the radius around the user is $r_u = r_d - \alpha * t$. α is the weight coefficient;

(2) When the user and the fish reach a strong repulsive distance, the forward direction of the fish becomes mirror refraction, that is, the fish becomes horrified, and the duration is Δt.

Frightening Rule for Fish Swarm

When the distance between the user and the fish is less than the strong repulsive distance, that is $r_s > r_u$, the duration of the panic state is $\Delta t' = (1 + \beta * |r_s - r_u|) * \Delta t$. β is the weight coefficient.

We tried to establish a fish swarm simulation based on the proposed method (see Fig. 1). In the future works, we try to explore the influence of the maximum visual ability range on the fish swarm structure, random coefficients are used to the mathematical model to maintain a certain degree of chaotic state, to avoid the abnormal order in the running process. This simulation is conducted in the absence of external predators to interfere with the stimulus, only when the individuals in the group interact with each other. In addition, external environment such as light, temperature can be considered for the structure changes of the fish.

Fig. 1. Fish swarm simulation for tourism.

4   Conclusion

In the era of modern information technology, tourism needs to adapt to the intensive information knowledge, pursue high technology, and realize the personalized information collection and processing of the tourism industry, to meet the growing tourism demand of people. In the application research of virtual tourism, the frequency of tourism transactions can be greatly improved. In the process of shortening the virtual tour of intermediate links, people can use fewer rest gaps to experience endless fun in virtual tourism. At the same time, virtual tourism can also meet the needs of various tourism groups. Tourism under VR technology can make some special tourists enjoy the same tourism experience, such as some disabled people, non-age-oriented tourism groups and some high-end tourism or preference. Sporty tourists. Secondly, it can also meet the needs of sustainable development of tourism resources. With the continuous advancement of society and the vigorous development of tourism, the sustainable development of environmental protection resources has become the most important thing at present. In the face of tourism resources, we should protect. It is a historical and cultural heritage or a natural ecological heritage. Through the application of VR technology, a large number of tourists experience the virtual tourism experience, thus achieving the sustainable development of tourism resources.

The three meanings of this study: First, cost savings. Generally, due to the limitations of equipment, site, and funding, many ocean simulation experiments cannot be performed, and with VR technology, users can get the same experience as real experiments. The second is to avoid risks. Using VR technology for virtual experiments can circumvent the risks of real-world operations, and you can safely do various dangerous experiments in a virtual environment. The third is to break the limits of time and space. With VR technology, it is possible to replace the long-term work required in reality within a short time, and the efficiency is greatly improved. In order to make marine tourism not limited to traditional forms, to create a new, international and rich cultural tourism industry, in order to make tourists more convenient to understand and feel the tourism culture of the South China Sea, to make the tourist experience more rich and diverse, to create an international marine tourism culture market. South China Sea Ocean Tourism requires a more innovative approach to visitors and promotion. The reference to VR technology has brought new vitality to the traditional tourism industry.

Acknowledgements. This work is supported by "Hainan University Education and Teaching Reform Research Project (Project Number: HDJY1978)" and "Hainan University Research Initiation Fund Project (Project Number: KYQD(SK)1709)".

References

1. Kato, M., Mori, Y., Sakaguchi, M.: VR batting training system for timing. J-STAGE (2017)
2. Tutzauer, P., Becker, S., Haala, N.: Perceptual rules for building enhancements in 3D virtual worlds. De Gruyter (2017)
3. Takada, R., Onishi, T., Kawata, S., Iwata, H.: The development of the Immersive 3D-VR training system for improving sports vision in spike receive. J-STAGE (2017)
4. Thomas, L.M., Glowacki, D.R.: Seeing and feeling in VR: bodily perception in the gaps between layered realities. Taylor J. (2018)
5. Durgin, F.H., Li, Z.: Controlled interaction: strategies for using virtual reality to study perception. PubMed (2010)
6. Crane, L.: Control a real space robot with VR. Elsevier J. (2018)
7. Steuer, J.: Defining virtual reality: dimensions determining telepresence. J. Commun. **42**, 73–93 (1992)
8. Bowman, D.A., Koller, D., Hodges, L.F.: Travel in immersive virtual environments: an evaluation of viewpoint motion control techniques. In: IEEE Virtual Reality Conference (1997)
9. Wei, S.G., Hao, Y., Sun, F., Yang, Z.Y., Zhu, Y.: The research of 3-dimension ocean environment virtual reality technology based on true environment (2006)
10. Guttentag, D.: Virtual reality: applications and implications for tourism. Tour. Manag. **31**, 637–651 (2010)
11. Berger, H., Dittenbach, M., Merkl, D., Bogdanovych, A., Simeon, S.J., Sierra, C.: Opening new dimensions for e-Tourism. Virtual Real. **11**, 75–87 (2007)
12. Gale, T.: Urban beaches, virtual worlds and 'the end of tourism'. Mobilities **4**, 119–138 (2009)
13. Parrinello, G.L.: The technological body in tourism research and praxis. Int. Sociol. **16**, 205–219 (2001)
14. Seidel, I., Gartne, M., Froschauer, J., Berger, H., Merkl, D.: Towards a holistic methodology for engineering 3D virtual world applications. International Conference on Information Society (2010)
15. Sevrani, K., Elmazi, L.: ICT and the changing landscape of tourism distribution- a new dimension of tourism in the global conditions. Revista de Turism: Studii si Cercetari in Turism **6**, 22–29 (2008)

16. Chen, S.E.: Quick time VR: an image-based approach to virtual environment navigation. In: International Conference on Computer Graphics and Interactive Techniques (1995)
17. Zyda, M.: From visual simulation to virtual reality to games. IEEE Comput. (2005)
18. Levis, P., Culler, D.E.: Maté: a tiny virtual machine for sensor networks. In: Architectural Support for Programming Languages and Operating Systems (2002)
19. Abdulrahman, A., Hailes, S.: Supporting trust in virtual communities. In: Hawaii International Conference on System Sciences (2000)
20. Najafipour, A.A., Heidari, M., Foroozanfar, M.H.: Describing the virtual reality and virtual tourist community: applications and implications for tourism industry. Kuwait Chapter Arab. J. Bus. Manag. Rev. **33**, 1–12 (2014)

TPP: Tradeoff Between Personalization and Privacy

Ubaid Ur Rehman and Sungyoung Lee[(✉)]

Kyung Hee University, Global Campus, Yongin-si, Republic of Korea
ubaid.rehman@khu.ac.kr, sylee@oslab.khu.ac.kr

Abstract. Modern technology relies on personalization due to its appealing services. It suggests the most relevant information to the users. Beside it several benefits, there may be some privacy leakage due to the personalization. As it analyzes and collects the user behavior's data, and generates a personalized decision. In this paper, we have considered the personalization aspect of recommendation, crowdsensing, and healthcare domains. We have identified the state-of-the-art research, specifically emphasizing on the personalization and privacy aspect. Also, we have conducted a survey, in order to identify the literacy of personalization and privacy. Moreover, we have discussed the attacks that exploit the vulnerability of personalization.

Keywords: Personalization · Crowdsensing · Recommendation · Healthcare · Privacy

1 Introduction

With the emergence of technology, online social networking platforms have become an important aspect of every individual. Several social media platforms are available, such as Facebook, Twitter, and Flickr, which can be used for networking, microblogging, and site tagging respectively. These platforms facilitate users in different degree of interaction, which includes resource sharing, chatting, online gaming, and other services. Due to the direct involvement of users, a large quantity of social information gets generated that provide assistance in many situations, such as user feedback regarding a social concern or product helps in retrieving an accurate information. In order to understand the preferences and behavior of a user, these quality social information are used by different technologies to provide a personalized suggestion.

Personalization helps the user to retrieve the information efficiently based on their preferences. The most common example is the recommender system, which generates a personalized recommendation based on the user browsing behavior. It has been applied in a variety of domains such as online multimedia, news, shopping, social media, and tourism. According to [24], 90% of marketer has deployed personalization strategies, due to the high impact on economy. Beside the recommender system, the crowdsensing and healthcare domains also use the

© Springer Nature Switzerland AG 2019
S. Lee et al. (Eds.): IMCOM 2019, AISC 935, pp. 672–681, 2019.
https://doi.org/10.1007/978-3-030-19063-7_54

personalization aspect to facilitate the users. Crowdsensing utilizes mobile phone sensors to collect user data, instead of deploying a sensor network. The sensing is classified into personal and community, which monitor the individual and group of users respectively. Currently, most of the healthcare applications use this concept of crowdsensing to monitor patient vitals, behaviors, and emotion. Based on these constraints, the application recommends a diagnosis, treatment, or follow-up plan for the patient.

Despite the fact of several benefits, privacy is considered a secondary requirement in these domains. Most of the applications focused on efficient, reliable, and accurate decision making/recommendation. According to Jeong et al. [25], users are interested in personalization, but they are also curious about how the service provider handles their personal data. Therefore, in the case of a recommender system, a small hint may spoil a big surprise. Suppose, in a husband-wife relationship, the husband bought an expensive birthday gift for his wife using a shared online shopping account. The purpose was to surprise her, but when his wife login to the online shopping account, the system will provide a similar recommendation based on the recent purchase, which will spoil the surprise. Similarly, in crowdsensing and health domain, mobile phone's sensors and computation are used, where the data leakage regarding a specific diagnosis may lead to a serious consequence. Most of the users desired high-quality personalization that required more personal data, but the users are unwilling to share their personally identifiable information. Therefore, we have emphasized on the tradeoff between personalization and user privacy, targeting recommender system, crowdsensing, and healthcare domains. In this research study, we have focused on the following research questions (RQs):

RQ1: How personalization is deployed in recommender system, crowdsensing, and healthcare domains?
RQ2: How personalization will affect the privacy?
RQ3: What will be the tradeoff between personalization and privacy?

The rest of the paper is classified as Sect. 2 describes the state-of-the-art techniques that use the personally identified information. Section 3 represents the survey result regarding the data privacy awareness in the community. The attacks that can occur due to personalization is discussed in Sect. 4. Finally, Sect. 5 will summarize the conclusion and future work.

2   Related Work

The personalization has been evolved and attracted great attention of the research community. We have covered the literature of recommender system, crowdsensing, and healthcare domains. Based on our literature review, we have selected the most relevant papers that cover the aspect of personalization.

2.1   Recommender System

Most of the personalized recommender systems, applications, engines, and frameworks have been proposed in the last few years. This mostly endorses the user to use their social network's account to avail the services. The goal is to collect and analyze the social network's activity and then generate a personalized recommendation accordingly. We have classified the recommendation systems into *(a)* Everyday Items, *(b)* Like-minded People, and *(c)* Article Suggestion.

(a) Everyday Items: In everyday items recommendation, the user is considered as an independent entity and suggested with a few random items, which leads to a cold start problem. In order to tackle the cold start problem, many solutions have been proposed [9]. A little interaction or social network activity analysis leads to an effective and personalized recommendation [10]. Similarly, Guy et al. in [12] described the relationship between the user and the items using the proposed ranking function.

(b) Like-Minded People: The social network platforms compared the user profile contents and recommend based on the similarity index. Groh et al. [14] used the neighborhood information from the social network and analyzed statistically using a collaborative filtering method. A unified framework for user recommendation was proposed in [15]. In [13], the authors designed an approach that collects information from different sources and generates a recommendation. Moreover, [17] uses the content and collaborative based approaches to generate recommendation and evaluate user profiles. Wang et al. designed an approach that measures the similarity and generates a recommendation based on an inferred network tags [18].

(c) Article Suggestion: Article suggestion is based on the topic and tag recommendation, which helps the user to choose the right tag. A survey of tag-based recommendation along with the evaluation was proposed in [19]. In [20], the tag was assigned automatically after analyzing the content of the article. Approaches for tagging based on correlation, rankboost, and neighbor voting graph was proposed in [21], [22], and [23] respectively.

2.2   Crowdsensing

Crowdsensing helps in reducing the overhead of data collection and processing by using mobile phones. Farkas et al. designed an application that provides information about public transport using participatory sensing [26]. The application crowdsourced data collection and feedback for visualizing the actual position of the vehicle. An end-to-end participatory noise mapping system titled as Ear-Phone, was proposed in [27] that uses compressive sensing for noise map recovery and outsource the environmental data collection. In [28], the authors develop an Adverse Drug Reactions (ADR) system that collects the ADR report through questions. Then apply machine learning algorithms to automate data collection procedures and efficiently track the adverse events. Hu et al. designed SmartRoad sensing system that collects data from in-vehicle smartphones GPS sensor using participatory sensing and detects traffic regulators, traffic lights, and stop signs [29]. Similarly, Wang et al. developed an application that shares the user's location along with vehicle speed in real-time [30]. The application uses participatory sensing for identifying the traffic condition and user location.

2.3   Healthcare

With the emergence of wearable and smart technology, there is a rapid growth in personalized healthcare management. A personalized wellness service recommendation system was proposed in [31], which monitors and quantifies the user activities using mobile phone sensors. In [1], the authors presented cyber-physical recommendation system that monitors user enjoyment while playing exergames. The system learns from user behavior and recommends similar games. Moreover, Dharia et al. proposed PRO-Fit framework [2], which monitors user activity and timetable using accelerometer and user's calendar respectively. Based on this information recommend a workout activity.

3   Personalization and Privacy Literacy

As from the literature, we have identified that most of the research work proposed different algorithms, frameworks, and applications, which collects the user's identifiable information along with the user's activities and behaviors. Based on this information, a personalized suggestion is generated. According to the best of our knowledge, none of the research work in the specified scope (recommender system, crowdsensing, and healthcare) has considered the effect of personalization on privacy. However, few of the studies [4,5,11,16] have developed a personalized recommendation application by integrating functional interactions and user's privacy preferences. In [3], the authors have described the tradeoff between personalization and privacy, targeting online social networks only. Therefore, the purpose of our research is to create awareness among the community, because these applications not only monitor user behavior using mobile phone sensors, but also access sensitive resources such as storage, camera, microphone, and contacts. Any malicious attempt may lead to serious consequences.

In order to understand the level of personalization and privacy literacy, we have created a questionnaire using Google Forms and share the links with the students using the university mailing list. The responses were collected anonymously and allowed to use for research purpose only. Table 1 shows the demographic information of the participant. A total of 196 students has completed the survey, which includes 103 males and 93 females in the young (16 to 22), early adult (22 to 35), and middle adult (35 to 50) groups. These participants are enrolled in undergraduate, graduate, and post-graduate studies. They belong to diverse nationalities,

Table 1. Demographic details of participants

		Participants
Gender	Male	103
	Female	93
Age	16 to 22	117
	22 to 35	66
	35 to 50	13
Education	Undergraduate	121
	Graduate	35
	Post Graduate	13
Nationality	South Korea	124
	China	2
	Nepal	1
	India	8
	Vietnam	9
	Ecuador	5
	Pakistan	33
	Egypt	1
	Yemen	3
	Bangladesh	10

which includes South Korea (124), China (02), Nepal (01), India (08), Vietnam (09), Ecuador (05), Pakistan (33), Yemen (03), and Bangladesh (10).

Each participant has answered eleven questions based on their understanding. The statistic of the survey is presented in Table 2. The survey questions were classified into personalization (5), privacy (5), and willingness (1). The results show some interesting facts, almost all the participants know about personalization and privacy. Most of the participants were satisfied with the high-quality personalization, but some do not want to share their personal data with the application. According to their reviews, they have no choice because they need to get the permission in order to use the application. Moreover, participants were also very concern about their privacy over the internet and they believe that it may be exploited by the attacker. But still, most of the participants share their post or comments without any privacy preference. According to them, it is just a post or comment, which will be useless for the attacker. However, all the participants show their interest to learn about the privacy attack that may cause by the personalization aspect. Figure 1 shows the graphical representation of the overall statistical result.

Table 2. Personalization and privacy questionnaire

Categorization	Survey questions	Description	Yes	No
Personalization	SQ1	Do you know about personalization?	193	3
	SQ2	Do you like the personalization aspect used in different websites and mobile phone application?	189	7
	SQ3	Do you like personalized recommendation feature of different applications?	146	50
	SQ4	Do you allow the third party application to use your personal data for recommendation?	110	89
	SQ5	Do you feel comfortable, if your personal data helps in generating a high quality personalized recommendation?	170	26
Privacy	SQ6	Do you know about privacy?	196	0
	SQ7	Do you really concern about your privacy over the internet?	177	19
	SQ8	Do you believe that activity monitoring and tracking along with personal information may lead to certain privacy threats?	152	44
	SQ9	Do you think it would be a privacy threat, if your mobile phone exchange the information (location, driving speed) with the nearby mobile phones for a specific task?	124	72
	SQ10	Do you share your posts or comments on the social media without any privacy preference?	181	15
Willingness	SQ11	Do you like to be aware of privacy attacks that may cause by personalization aspect?	196	0

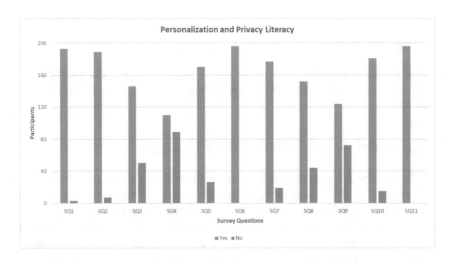

Fig. 1. Statistical representation of personalization and privacy survey results

4   Exploiting Personalization Vulnerabilities

Modern service based on personalization has appealed the users due to its numerous benefits. On the other hand, it also provides a new attacking surface for the attacker. In this section, we have discussed some of the attacks that exploit the personalization aspect and caused serious damage.

4.1   Pollution Attack

Pollution attack is categorized as the most effective attack for personalization services, as it allows third-party applications to modify the customize content and affect a user's choice set. In [6], Xing et al. used pollution attack and exploit the personalization aspect of YouTube, Google, and Amazon.

4.2   Data Poisoning

In data poisoning, the attacker recruited a group of malicious user to submit malicious data, which degrade the efficiency of a crowdsensing system. In [7], the authors have focused on the two types of data poisoning. (a) Availability Attack, (b) Target Attack.

(a) Availability Attack: The purpose of availability attack is to engage the malicious worker to increase the chances of error in the crowdsensing systems.

(b) Target Attack: The attacker tries to skew the victim to a specific answer by poisoning the sensory data. If the victim is changed to the target answer, the attack gets to succeed.

4.3   Whaling Phishing

Whaling phishing exploits the user to reveal personal or organization information [8]. It is a special type of phishing that targets a user having sensitive information.

5   Conclusion and Future Work

In the last few years, personalization has become very popular due to its value-added services. Besides these services, it has several privacy issues, as it collects the user information based on online surfing behavior or using mobile phone sensors. The collected data have sensitive information that leads to several privacy issues. The purpose of this research is to create an awareness among the community regarding their privacy. According to the best of our knowledge, this is the first study that considered the effect of personalization on privacy, emphasizing on the emerging domains of crowdsensing and healthcare along with the recommender system. Currently, we are working on designing an algorithm that will be used by the recommender system, crowdsensing, and healthcare domains, in order to ensure personalization and privacy. Also, our design mechanism will provide prevention against the existing attacks, such as pollution, data poisoning, and whaling phishing attacks.

Acknowledgments. This research was supported by an Institute for Information & Communications Technology Promotion (IITP) grant funded by the Korean government (MSIT) (No. 2017-0-00655). This work was supported by the MSIT (Ministry of Science and ICT), Korea, under the ITRC (Information Technology Research Center) support program (IITP-2017-0-01629) supervised by the IITP (Institute for Information & communications Technology Promotion) and NRF- 2016K1A3A7A03951968.

References

1. Agu, E., Claypool, M.: Cypress: a cyber-physical recommender system to discover smartphone exergame enjoyment. In: Proceedings of the ACM Workshop on Engendering Health with Recommender Systems (2016)
2. Dharia, S., Jain, V., Patel, J., Vora, J., Chawla, S., Eirinaki, M.: PRO-Fit: a personalized fitness assistant framework. In: SEKE, pp. 386–389 (2016)
3. Weinberger, M., Bouhnik, D.: Place determinants for the personalization-privacy tradeoff among students. Issues Informing Sci. Inf. Technol. **15**, 079–095 (2018)
4. Katragadda, B., Sharife, S.M.: Supporting privacy protection in personalized web search. J. Sci. Technol. (JST) **2**, 17–21 (2017)
5. Divekar, M.J., Patil, D.R.: Enabling personalized search over encrypted outsourced data with efficiency improvement. Int. J. **3** (2018)
6. Xing, X., Meng, W., Doozan, D., Snoeren, A.C., Feamster, N., Lee, W.: Take this personally: pollution attacks on personalized services. In: USENIX Security Symposium, pp. 671–686 (2013)
7. Miao, C., Li, Q., Xiao, H., Jiang, W., Huai, M., Su, L.: Towards data poisoning attacks in crowd sensing systems. In: Proceedings of the Eighteenth ACM International Symposium on Mobile Ad Hoc Networking and Computing, pp. 111–120. ACM (2018)
8. Bansal, G.: Got Phished! Role of Top Management Support in Creating Phishing Safe Organizations (2018)
9. Sedhain, S., Sanner, S., Braziunas, D., Xie, L., Christensen, J.: Social collaborative filtering for cold-start recommendations. In: Proceedings of the 8th ACM Conference on Recommender systems, pp. 345–348. ACM (2014)
10. Sedhain, S., Sanner, S., Xie, L., Kidd, R., Tran, K.-N., Christen, P.: Social affinity filtering: recommendation through fine-grained analysis of user interactions and activities. In: Proceedings of the First ACM Conference on Online Social Networks, pp. 51–62. ACM (2013)
11. Gardner, Z., Leibovici, D., Basiri, A., Foody, G.: Trading-off location accuracy and service quality: privacy concerns and user profiles. In: 2017 International Conference on Localization and GNSS (ICL-GNSS), pp. 1–5. IEEE (2017)
12. Guy, I., Zwerdling, N., Ronen, I., Carmel, D., Uziel, E.: Social media recommendation based on people and tags. In: Proceedings of the 33rd International ACM SIGIR Conference on Research and Development in Information Retrieval, pp. 194–201. ACM (2010)

13. Chen, J., Geyer, W., Dugan, C., Muller, M., Guy, I.: Make new friends, but keep the old: recommending people on social networking sites. In: Proceedings of the SIGCHI Conference on Human Factors in Computing Systems, pp. 201–210. ACM (2009)

14. Groh, G., Ehmig, C.: Recommendations in taste related domains: collaborative filtering vs. social filtering. In: Proceedings of the 2007 International ACM Conference on Supporting Group Work. pp. 127–136. ACM (2007)

15. Symeonidis, P., Nanopoulos, A., Manolopoulos, Y.: A unified framework for providing recommendations in social tagging systems based on ternary semantic analysis. IEEE Trans. Knowl. Data Eng. **22**, 179–192 (2010)

16. Bhamidipati, S., Fawaz, N., Kveton, B., Zhang, A.: PriView: personalized media consumption meets privacy against inference attacks. IEEE Softw. **32**, 53–59 (2015)

17. Hannon, J., McCarthy, K., Smyth, B.: Finding useful users on twitter: twittomender the followee recommender. In: European Conference on Information Retrieval, pp. 784–787. Springer (2011)

18. Wang, X., Liu, H., Fan, W.: Connecting users with similar interests via tag network inference. In: Proceedings of the 20th ACM International Conference on Information and Knowledge Management, pp. 1019–1024. ACM (2011)

19. Jüschke, R., Marinho, L., Hotho, A., Schmidt-Thieme, L., Stumme, G.: Tag recommendations in social bookmarking systems: AI Commun. **21**, 231–247 (2008)

20. Mishne, G.: Autotag: a collaborative approach to automated tag assignment for weblog posts. In: Proceedings of the 15th International Conference on World Wide Web, pp. 953–954. ACM (2006)

21. Krestel, R., Fankhauser, P., Nejdl, W.: Latent dirichlet allocation for tag recommendation. In: Proceedings of the Third ACM Conference on Recommender Systems, pp. 61–68. ACM (2009)

22. Wu, L., Yang, L., Yu, N., Hua, X.-S.: Learning to tag. In: Proceedings of the 18th International Conference on World Wide Web, pp. 361–370. ACM (2009)

23. Freund, Y., Iyer, R., Schapire, R.E., Singer, Y.: An efficient boosting algorithm for combining preferences. J. Mach. Learn. Res. **4**, 933–969 (2003)

24. Personalization Trends. https://www.emarketer.com/Report/Personalization-Retail-Latest-Trends-Challenges/2002008

25. Jeong, Y., Kim, Y.: Privacy concerns on social networking sites: Interplay among posting types, content, and audiences. Comput. Hum. Behav. **69**, 302–310 (2017)

26. Farkas, K., Nagy, A.Z., Tomás, T., Szabó, R.: Participatory sensing based real-time public transport information service. In: 2014 IEEE International Conference on Pervasive Computing and Communications Workshops (PERCOM Workshops), pp. 141–144. IEEE (2014)

27. Rana, R.K., Chou, C.T., Kanhere, S.S., Bulusu, N., Hu, W.: Ear-phone: an end-to-end participatory urban noise mapping system. In: Proceedings of the 9th ACM/IEEE International Conference on Information Processing in Sensor Networks, pp. 105–116. ACM (2010)

28. Chen, C., Huang, Y., Liu, Y., Liu, C., Meng, L., Sun, Y., Bian, K., Huang, X., Jiao, B.: Interactive crowdsourcing to spontaneous reporting of adverse drug reactions. In: 2014 IEEE International Conference on Communications (ICC), pp. 4275–4280. IEEE (2014)

29. Hu, S., Su, L., Liu, H., Wang, H., Abdelzaher, T.F.: Smartroad: smartphone-based crowd sensing for traffic regulator detection and identification. ACM Trans. Sens. Netw. (TOSN) **11**(4), 55 (2015)

30. Wang, C., Liu, H., Wright, K.L., Krishnamachari, B., Annavaram, M.: A privacy mechanism for mobile-based urban traffic monitoring. Pervasive Mobile Comput. **20**, 1–12 (2015)
31. Afzal, M., Ali, S.I., Ali, R., Hussain, M., Ali, T., Khan, W.A., Amin, M.B., Kang, B.H., Lee, S.: Personalization of wellness recommendations using contextual interpretation. Expert Syst. Appl. **96**, 506–521 (2018)

Dynamic Invariant Prioritization-Based Fault Localization

Sujune Lee[1], Jeongho Kim[1], and Eunseok Lee[2(✉)]

[1] Department of Electrical and Computer Engineering,
Sungkyunkwan University, Seoul, South Korea
{hoakw, jeonghodot}@skku.edu
[2] School of Software, Sungkyunkwan University, Seoul, South Korea
leees@skku.edu

Abstract. The Differences in a dynamic invariant provides an important clue to analyzing and locating software faults. However, generating a dynamic invariant is costly and generates noise that might be not related to the fault. In this paper, we propose a new technique called Dynamic Invariant Prioritization-based Fault Localization (DIPFL). This technique reduces the noise by prioritizing the variable by observing the change in the value of the invariant that is extracted based on the number of test cases. The DIPFL identifies the location of a fault by re-measuring the weight of the list of suspiciousness score ranking of the spectrum-based fault localization based on the priority data. We evaluated the proposed method by applying it to the Siemens project, which has been used as benchmark for various fault localization studies. Through the 118 versions, we verified the performance enhancement of 53.39%. Also, when evaluated using the EXAM score metrics, DIPFL improved performance by up to 14.7% over Naish2 when running 5% of the source code.

Keywords: Software debugging · Fault localization · Dynamic invariant · Prioritization

1 Introduction

During the last decade, there has been constant research on software debugging. Each task can be divided into fault localization, fault fixing, and regression testing. In particular, Fault localization (FL) is an important task in the debugging process [2]. If the correct faults cannot be found at this step, then fault fixing and regression testing are not effective because they are being performed for something that is not related to the fault. Therefore, several studies focused on improving the accuracy of FL. One is a study using a dynamic invariant [1–3]. This is an FL study in which the information of variables stored during the program execution is extracted to locate the fault. Given that the information of this variable is concrete data on the variables used in the software and the assigned values, it can be effective in identifying the location of the fault [2, 4–6]. The concept is to locate the fault using the difference between the invariant in the failure test case and the invariant in the successful test case for one variable. The difference in this invariant is used as a clue to locating the fault [1]. However, generating a dynamic

© Springer Nature Switzerland AG 2019
S. Lee et al. (Eds.): IMCOM 2019, AISC 935, pp. 682–693, 2019.
https://doi.org/10.1007/978-3-030-19063-7_55

invariant can be time consuming. Generating the invariant for an entire program is very expensive and generates a large amount of data that are unrelated to the fault. FDDI [1] reduces cost by reducing the number of test cases and functions in source code. Despite these efforts, unnecessary noise still remains.

Thus, we propose DIPFL, a dynamic invariant-based FL technique that can improve FL performance by prioritizing invariants associated with faults. DIPFL generates a invariant priority list by assigning high priority to the invariant related to the fault statistically. Then, it identifies the location of a fault by re-measuring the weight of the list of suspiciousness score ranking of the spectrum-based fault localization(SBFL) based on the priority data.

The rest of the paper is organized as follows: Sect. 2 provides background knowledge. In Sect. 3, we propose a new FL technique. Section 4 shows the results of the experiment. Section 5 discusses threats to validity. The conclusion and future works are presented in Sect. 6.

2 Introduction

In this section, we describe related research and motivation for this paper. First, we describe SBFL. Then, we describe a dynamic invariant and demonstrate an example of its execution. Finally, we describe the motivation

2.1 Spectrum-Based Fault Localization

Spectrum-based fault localization(SBFL) [8, 14] is one of the most popular techniques for FL. It is lightweight and has a high performance in FL [7]. This technique uses the statement coverage to determine the location of faulty statement. For one statement, when the result of the test case is a pass, it is denoted as N_{CS} if it covers the corresponding statement and N_{US} if it does not cover the statement. Conversely, when the result of the test case is a fail, then it is denoted by N_{CF} if it covers the corresponding statement, and N_{UF} if it does not cover the statement. The SBFL calculates the suspicious score of the statement based on the coverage mentioned above. SBFL has a number of suspicious score calculation formulas [7]. In this paper, DIPFL is compared with Tarantula [8] and Naish2 [9] to evaluate its performance.

$$\text{Tarantula} = \frac{\frac{N_{CF}}{N_{CF} + N_{UF}}}{\frac{N_{CF}}{N_{CF} + N_{UF}} + \frac{N_{CS}}{N_{CS} + N_{US}}} \tag{1}$$

$$\text{Naish2} = N_{CF} + \frac{N_{CS}}{N_{CS} + N_{US} + 1} \tag{2}$$

Each statement has a suspicious score calculated according to the calculation formula. The SBFL generates a statement ranking list in the order of the highest suspicious score, and the developer confirms from the highest ranking until a fault is found.

2.2   Dynamic Invariant

The dynamic invariant is the actual value of the variables when the program is executed. This data, obtained through execution of the program, can grasp the characteristics of the source code and the intention of the developer [2]. Recently, FL techniques using a dynamic invariant have been studied [1–3, 6, 10, 11]. The dynamic invariant generated in the successful test case represents value that causes the program to operate correctly. If the invariant extracted from the failure test case is outside the range of the results of the successful test case, this is a violation of the correct value and may be the result of a fault. Figure 1 shows an example of FL using a dynamic invariant.

```
main(int argc, char *argv[])
1:  int x = atoi(argv[1]);
2:  int y = atoi(argv[2]);
3:  int z = atoi(argv[3]);              Dynamic invariant of failed test case
4:  int a, b, result;                   x one of { 2, 3 }   x = 2, y = 3, z = 5  , fail
5:  if (y < z)                          y one of { 3, 4 }   x = 3, y = 4, z = 5  , fail
                                        z one of { 3, 4 }
6:      a = x + y;                      a == -1
7:  else                                b one of { 6, 12 }
8:      a = x - y;                      result one of { 5, 11 }
9:      if(z < 6) // if(z < 5) error
10:     b = x * y;                      Dynamic invariant of passed test case
11: else                               x one of { 1, 5 }   x = 1, y = 1, z = 4  , pass
12:     b = x * y + 3;                  y one of { 1, 6 }   x = 5, y = 6, z = 8  , pass
                                        z one of { 4, 8 }
13: result = a + b;                     a one of { -1, 2 }
14: print("result = %d", result);       b one of { 1, 33 }
                                        result one of { 3, 32 }
```

Fig. 1.   An example of the difference in the dynamic invariant.

In Fig. 1, the left box is source code, and the right box is data that can be obtained by using a dynamic invariant extraction tool such as the Daikon Tool [12, 15], which is the result of dividing the failure and successful test cases. The location of the fault is in line 9. Successful test cases and failure test cases can be used to generate a dynamic invariant to check the range of values of each variable. Then the differences for each variable are analyzed. In a successful test case, the z variable has only five values, however, the z variable in the failure test case has three and four values. Therefore, the z variable is selected as the difference between the dynamic invariant that occurred between the successful test case and the failure test case. Then, we can find line 9 that includes the z variable. As shown in this example, the data can be used as a clue in determining the fault.

2.3   Motivation

The difference in the dynamic invariant is a clue to locating faults. However, the computational cost to generate this is large. To solve these problems [1], the time cost is reduced by adding preprocessing procedures such as decreasing the number of functions and test cases. Despite the preprocessing procedure, a lot of unnecessary dynamic invariants are still generated. It is difficult to distinguish between the fault-related data between them. In fact, the exact cause of the fault is unknown, and several dynamic invariants may occur depending on the type and number of test cases. In the Fig. 1, all variables are extracted as the difference of the identities. However, except for the z variable, the remaining variables are noise, which is data that is not related to the fault. These are unnecessary data for identifying the location of the fault. Therefore, there is a need for a technique to reduce the noise while leaving the difference of the dynamic invariant related to the fault.

3   Proposal Approach

In this section, we describe the proposed technique DIPFL. First, we describe the overall process flow diagram. Then, each step is described in detail.

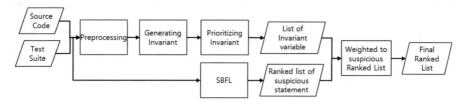

Fig. 2. Overall process flow diagram of DIPFL

3.1   Overview

Figure 2 show a process flow diagram of the DIPFL technology. The input data of DIPFL is source code and test suite. The first step is preprocessing. This step reduces the number of functions and test cases. The second step is generating the dynamic invariant. The third step is invariant prioritization. This step prioritizes the variables related to faults in the generated invariant. The fourth step is SBFL. This step creates a suspicious ranked list. The final step combines the results returned in steps 3 and 4 to generate the final ranked list.

3.2   Preprocessing

The preprocessing procedure reduces the number of functions and test cases. Firstly, it reduces the number of functions. DIPFL select function that always covered in failure test case. As a result, non-fault-related functions are excluded from the dynamic invariant generation. Secondly, it reduces the number of test cases. The failure test case fails because it covers faulty statement. When comparing the coverage of a test case with that of a failure test case, a test case with a low degree of similarity will include coverage that is not related to the cause of the failure. As a result, the invariant that is not related to the fault is extracted and this becomes noise. Therefore, DIPFL randomly selects a failure test case. Then, it selects a test case that is similar to the coverage of the selected test case to reduce the number of test cases.

3.3   Generating Invariant

Generating an invariant is the step of generating dynamic invariant. We used Daikon tool, which dynamic invariant extraction tool. To obtain the difference between the successful test case and failure test case, pass and fail sets are classified and generated. The next task is to analyze the difference between successful test cases and failure test cases.

3.4   Invariant Prioritization

Many invariants are generated in the generating-invariant phase. It is difficult to distinguish which data is fault-related data. In the paper, invariants are prioritized using the change frequency of the value of the invariant. Figure 3 shows part of the example code for the schedule in the Siemens project. The faults are located on lines 8 and 9. Figure 4 shows the result of the dynamic invariant generated by the Daikon tool. The first column is the number of failed test cases used to generate the dynamic invariant. The second column is the name of the function that generated the dynamic invariant. The third column is the name of the variable that was created. The fourth and fifth columns are the actual values of the dynamic invariant generated by the failure test case and successful test case, respectively. The number of successful test case is always fixed.

In the Unblock_process function, the variable count was changed twice in five failure test case increments. However, in the new_ele function, the variable new_num was changed in all five failure test case increments. That is, the variable count has little change in the value of the variable even though the number of test cases increases. This variable is a direct relationship to the fault. The variable new_num is a non-related-fault variable, and many changes occur depending on the characteristics of the source code. Therefore, a variable with a low rate of change of variable value is given a high weight, while a variable with a high rate of change is given a low weight value. In this way, all invariants are weighted, and a list of invariants is created. This is used to identify the location of the fault.

3.5   Weighted to Suspicious Ranked List

As a final step, the final suspicious ranking list is generated by combining the suspicious ranking lists, which are the result of the SBFL, and the list of invariants. In the list of invariants, the suspicious score of the statement, including the corresponding variable, is weighted to create the final suspicious score. The formula used is as follows:

$$Score_N = Score_o + (Score_{Max} - Score_o) \times \frac{value_{Inv}}{value_{max}} \tag{3}$$

$Score_O$ is the original suspicious score of the statement and $Score_{Max}$ is the highest suspicious score in the corresponding version. $value_{inv}$ is the weight value of the invariants in the list of invariants and $value_{Max}$ is the weight value of the highest invariant in the corresponding version. The final suspicious score calculation method gives a higher score when $value_{Inv}$ is higher. Furthermore, when the suspicious score of the faulty statement is lower in the ranked list of SBFL, a calculation formula is derived through $Score_{Max} - Score_O$ to give a high score.

Line	Source Code
1	void unblock_process(float ratio)
2	{
3	int count;
4	int n;
5	Ele *proc;
6	int prio;
7	if(block_queue) {
8	count=block_queue->mem_count + 1; // error -> have to remove +1
9	n = (int) (count*ratio); // error -> original code : n=(int)(count*ratio+1);
10	proc = find_nth(block_queue, n);
11	if (proc) {
12	block_queue = del_ele(block_queue, proc);
13	prio = proc->priority;
14	prio_queue[prio] = append_ele(prio_queue[prio], proc);
15	}
16	}
17	}
18	Ele* new_ele(int new_num)
19	{
20	Ele *ele;
21	ele =(Ele *)malloc(sizeof(Ele));
22	ele->next = NULL;
23	ele->prev = NULL;
24	ele->val = new_num;
25	return ele;
26	}

Fig. 3. Code example from the Siemens project schedule.

Dynamic Invariant				
# of TC	Function Name	Variable	Fail TC	Pass TC
10	unblock_process	count	count<=5	count<=11
20	unblock_process	count	count<=6	count<=11
30	unblock_process	count	count<=10	count<=11
40	unblock_process	count	count<=10	count<=11
full	unblock_process	count	count<=10	count<=11
10	new_ele	new_num	new_num<=23	new_num<=31
20	new_ele	new_num	new_num<=27	new_num<=31
30	new_ele	new_num	new_num<=30	new_num<=31
40	new_ele	new_num	new_num<=34	new_num<=31
full	new_ele	new_num	new_num<=35	new_num<=31

Fig. 4. Depending on the number of failed test cases, the generated dynamic invariant.

4   Experiment

In this section, we evaluate the performance of DIPFL. First, we describe our experimental project. Then, we describe the evaluation metrics. Finally, we analyze the experimental results. The experimental results are evaluated through two questions:

- RQ.1: Has influence of the invariant that are not related to the fault decreased?
- RQ.2: Has the effectiveness of fault localization improved?

4.1   Experiment Project

In this paper, we selected the project provided by SIR(software artifact infrastructure repository) [12] and conducted the experiment. SIR is a popular project in software testing and has many versions and test cases. We mainly used the Seimens projects, which are often used in FL research papers [7]. Each project is named using C language-based programs, such as Replace, Tcas, Tot_info, Print_tokens, Print_tokens2, Schedule, Schedule2. A brief description of each project can be found in Table 1.

Table 1. Test project descriptions.

Project name	Faulty version	LoC	Brief description
Replace	30	563	Patten recognition
Tcas	36	173	Altitude separation
Tot_info	19	406	Information measure
Print_token	5	565	Lexical analyzer
Print_token2	10	510	Lexical analyzer
Schedule	9	412	Priority scheduler
Schedule2	9	307	Priority scheduler

4.2   Evaluation Metrics

The evaluation metric used in this paper is the EXAM score, which is the percentage of the statement that must be executed before the faulty statement is executed. In FL, if a small number of statements are executed and more faulty statements are found, then this is evaluated as a more efficient technique. Another evaluation metric is the Best EXAM score and the Worst EXAM score. Problems that occur in SBFL are those that have the same suspicious scores and are ranked in the same order. Among the statements enclosed in the same order if the first faulty statement is found, it is called Best. If the last faulty statement is found, it is called Worst. There are many ways to evaluate Best and Worst EXAM score.

In this paper, Best and Worst EXAM scores are evaluated separately. There are omission errors in the fault version. Omission errors are difficult to identify because the fault statement is omitted. In such cases, the statement surrounding the faulty statement is recognized as an error-containing statement [13]. Finally, the invariant prioritization was evaluated in the same way as the EXAM Score. In the list of invariants, when we checked the variable from the top rank, we calculated the percentage of when we could identify the variables associated with the faults.

4.3   Result and Analysis

We describe the experimental results by answering the two questions described above. For RQ1, the results of invariant prioritization are shown in Table 2. The columns are the same as those in Table 1 of Sect. 4.1, and the last column is the average percentage of invariants associated with faults for each project total version.

Table 2. Result table of invariant prioritization.

Project name	Faulty version	LoC	Location of faulty invariant (%)
Replace	30	563	38.57
Tcas	36	173	43.91
Tot_info	19	406	46.39
Print_token	5	565	30.06
Print_token2	10	510	34.18
Schedule	9	412	12.98
Schedule2	9	307	31.35
Total	118		33.92

In other words, schedule project means that the invariants associated with the fault is located at an average higher than 12.98%. therefore, in the invariant prioritization results, DIPFL ranks the data related to faults in the top 33.92% on average. This means that about 70% of the noise can be reduced. Thus, the weight of the invariant that is not associated with the fault is reduced, and the weight of the invariant associated with the fault increases. In conclusion, DIPFL can assign a higher weight to the faulty statement.

For RQ2, we show the result of the EXAM score in Fig. 5. The x axis is the execution percentage of the source code, and the y axis is the number of the version that found the fault. The graph shows the performance of Tarantula, Naish2, and DIPFL, and consists of Replace, Tcas, Tot_info, Print_tokens, Print_tokens2, Schedule, and Schedule2. In Fig. 5b, when running only 5% of the source code, DIPFL, Naish2, and Tarantula can detect faults of 58.3%, 44.4%, and 25%, respectively. DIPFL improved 13.9% than Naish2. In Fig. 5c, we show the results of 42.1%, 26.3%, and 52.6%. DIPFL improved 15.8% than Naish2. In Fig. 5g, we show the results of 55.5%, 22.2%, and 11.1%. DIPFL improved 33.3% than Naish2. When the same number of codes were executed, DIPFL could identify more faults.

Therefore, the DIPFL technique is more efficient than the conventional Tarantula and Naish2. In Fig. 5f, we show the results of 88.8%, 88.8%, and 100%. In Fig. 5c, we show the results of 60%, 70%, and 60%. In Fig. 5d, we show the result of 80%, 80%, and 40%. In Fig. 5a, we show results of 88.6%, 88.6%, and 66.6%. Print_tokens and Print_tokens2 are cases in which the performance drops or does not change. In this case, the number of projects is relatively small, and the types of faults do not include variables. An example would be the printf statement. The statement for a simple sentence output does not contain variables and cannot extract a dynamic invariant. However, the number of versions with this fault is small, and the fault can be identified by weighting the result of the SBFL. Replace shows the results that are not significantly different from the existing technique when running 5% of the source code. However, when running 10% of the source code, DIPFL, Naish2, and Tarantula identified 93.3%, 90%, and 80% faults, respectively. Finally, Fig. 5h show result of total version of projects. When running only 5% of the source code, DIPFL, Naish2 and Tarantula can detect faults of 66.1%, 57.6%, and 48.3%, respectively. DIPFL performs better than Naish2 and Tarantula in the code examined range from 0% to 40%.

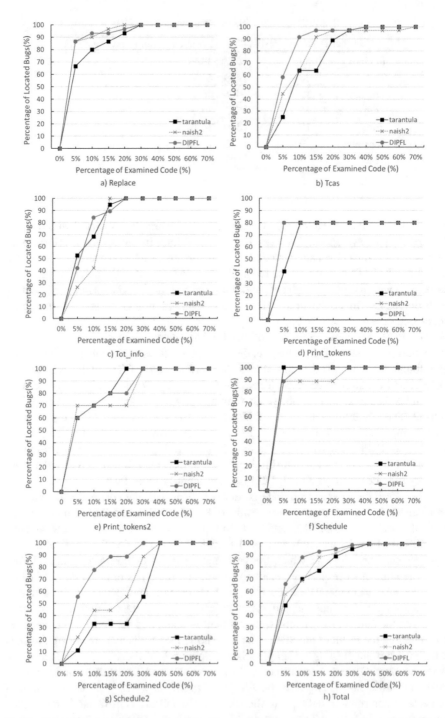

Fig. 5. EXAM score performance graph for Siemens projects (the order of each graph is replace, Tcas, Tot_info, Print_tokens, Print_tokens2, Schedule, Schedule2, and total).

In addition, Tables 3 and 4 show the percentages of the Best and Worst performance charts for each project. The first column is the name of the project. The second column is the percentage of the performance-enhanced version. The third column is the percentage of the version with no change in performance. The fourth column is the percentage of the degraded version.

Table 3. Results table of best EXAM metric

Project name	Positive (%)	Normal (%)	Negative (%)
Replace	50.00	20.00	30.00
Tcas	55.60	19.40	25.00
Tot_info	73.68	0.00	26.32
Print_token	40.00	0.00	60.00
Print_token2	20.00	20.00	60.00
Schedule	33.33	22.22	44.44
Schedule2	77.78	0.00	22.22
Total	53.39	14.41	32.20

For Replace, Tcas, Tot_info, and Schedule2, more than 50% of the versions improved. Conversely, the remainder can see results of less than 50%. This is because there are faulty statements that do not contain variables. This is the same as the above example using the printf statement. However, because the number of versions of this project is small, the overall results are not significantly affected. In conclusion the DIPFL performed better than the Naish2.

Table 4. Result table of worst EXAM metric

Project Name	Positive (%)	Normal (%)	Negative. (%)
Replace	60.00	7.00	33.00
Tcas	75.00	6.00	19.44
Tot_info	73.68	0.00	26.32
Print_token	40.00	0.00	60.00
Print_token2	20.00	10.00	70.00
Schedule	33.33	0.00	66.67
Schedule2	77.78	0.00	22.22
Total	61.86	4.24	34.00

5   Threats to Validity

We briefly introduce threats to the validity. DIPFL is an SBFL technique a using dynamic invariant, the value observed for the variable during execution. However, there are faulty statements that do not include variables. In this case, as a dynamic invariant can-not be generated, improvement of SBFL performance is limited. However, these types of faults are few in number, and there is no problem in fault location identification with the basic SBFL performance.

Also, we experimented with the Siemens project, which is often used in fault localization. Although it is a popular and often used project, their size is small. Therefore, experiments a larger scale are necessary and future research will be done. Finally, there is an influence on the preprocessing procedure. The Daikon tool may return different results depending on which test case is used. In other words, the size of the dynamic invariant that can be generated depends on how the test case is selected. Therefore, the generated dynamic invariant may have no data associated with the faults. Nonetheless, the reason it is used is likely to be generated in most case, and because this data is a great clue to finding faults, the dynamic invariant is used as highly-utilized data.

6   Conclusion and Future Works

In summary, we have proposed DIPFL, an efficient FL technique, to reduce the noise during generation of dynamic invariants. In order to reduce noise, which is a problem of invariants, we prioritized the invariant related to faults among them generated. In future research, there is a test case selection issue as the dynamic invariant may depend on what the input value is. This affects the output of invariant prioritization. Because the prioritization step is based on the invariant that is generated. Furthermore, as the project we tested is small, we must verify their effectiveness on a large scale. In the future, this research will be applied to a large project and further research will be conducted to improve the generation of dynamic invariants, including test case selection issues.

Acknowledgments. This research supported in part by the Technology and Basic Science Research Program through the NRF funded by the Ministry of Education (2016R1D1A1B03934610), MIST, Korea, under the National Program for Excellence in SW supervised by the IITP (2015-0-00914), and Next-Generation Information Computing Development Program through the National Research Foundation of Korea (NRF) funded by the Ministry of Science, ICT (2017M3C4A7068179).

References

1. Wang, X., Liu, Y.: Fault localization using disparities of dynamic invariantas. J. Syst. Softw. **122**, 144–154 (2016)
2. Sahoo, S.K., Criswell, J., Geigle, C., Adve, V.: Using likely invariants for automated software fault localization. In: Proceedings of ACM International Conference on Architectural Support Programming Languages and Operating systems, pp. 139–152 (2013)
3. Le, T.-D.B., Lo, D., Le Goues, C., Grunske, L.: A learning-to-rank based fault localization approach using likely invariants. In: Proceeding of ACM International Symposium on Software Testing and Analysis, pp. 177–188 (2016)
4. Ernst, M.D., Cockrell, J., Griswold, W.G., Notkin, D.: Dynamically discovering likely program invariants to support program evolution. In: Proceedings of IEEE Transactions on Software Engineering, pp. 99–123 (2001)

5. Hangal, S., Lam, M.S.: Tracking down software bugs using automatic anomaly detection. In: Proceedings of ACM International conference on Software Engineering, pp. 291–301 (2002)
6. Pytlik B., Renieris M., S. Krishnamurthi, Reiss S. P.: Automated fault localization using potential invariants. In: Proceeding International Workshop Automated Debugging, pp. 273–276 (2002)
7. Wong, W.E., Gao, R., Li, Y., Abreu, R., Wotawa, F.: A survey of software fault localization. In: Proceedings of IEEE Transactions on Software Engineering, pp. 707–740 (2016)
8. Jones, J.A., Harrold, M.J.: Empirical evaluation of the tarantula automatic fault-localization technique. In: Proceedings of ACM International Conference on Automated Software Engineering, pp. 273–282 (2005)
9. Naish, L., Lee, H.J., Ramamohanarao, K.: A model for spectra-based software diagnosis. In: Proceedings of ACM Transactions on Software Engineering (2011)
10. Wang, R., Ding, Z., Gui, N., Liu, Y.: Detecting bugs of concurrent programs with program invariants. In: Proceedings of IEEE Transactions on Reliability, pp. 425–439 (2017)
11. Abreu, R., Gonzalez, A., Zoeteweij, P., Gemund van, A.J.: Automatic software fault localization using generic program invariants. In: Proceeding of ACM Symposium on Applied Computing, pp. 712–717 (2008)
12. SIR Homepage. http://sir.unl.edu/portal/index.php
13. Zhang, X., Tallam, S., Gupta, N.: Towards locating execution omission errors. In: Proceedings of ACM Conference on Programming Language Design and Implementation, pp. 415–424 (2007)
14. Lee, J., Kim, J., Lee, E.: Enhanced fault localization by weighting test case with multiple faults. In: The International Conference on Software Engineering Research and Practice, pp. 116–122 (2016)
15. Ernst, M.D., Perkins, J.H., Guo, P.J., McCamant, S., Pacheco, C., Tschantz, M.S., Xiao, C.: The Daikon system for dynamic detection of likely invariants. Sci. Comput. Program. **69**, 35–45 (2007)

Improving the Efficiency of Search-Based Auto Program Repair by Adequate Modification Point

Yoowon Jang[1], Quang-Ngoc Phung[2], and Eunseok Lee[3(✉)]

[1] Department of Software Platform,
Sungkyunkwan University, Seoul, Republic of Korea
jangyoowon@skku.edu
[2] Department of Electrical and Computer Engineering,
Sungkyunkwan University, Seoul, Republic of Korea
ngocpq@skku.edu
[3] School of Software, Sungkyunkwan University, Seoul, Republic of Korea
leees@skku.edu

Abstract. In recent years Automated Program Repair (APR), especially generate-and-validate program repair techniques, has shown the possibility of fixing bugs in software automatically. However, current APR techniques still take long running time for repairing a bug. The dominant contributor to the runtime of generate-and-validate program repair techniques is the validation, which runs the regression test on all generated variants.

In this paper, we proposed two techniques, namely selective validation and fine-grained fitness evaluation, to improve the effectiveness of generate-and-validate automatic program repair. Selective validation reduces the cost of validation by skipping the test execution of variants inadequate to repair the bugs. The fine-grained fitness function evaluates variants based on both test cases execution result and the impact of modified points to original failing test cases. It increases the differential between variants, thus, makes the search more effective. We implement the proposed techniques in an automatic program repair tool, called AdqFix. We evaluate the AdqFix on bugs in the Defects4j benchmarks. The experimental result shows that the AdqFix reduces about 72% of executed test cases and saved about 49% of the overall running time while repaired three more bugs compared to the jGenProg, a state-of-the-art generate-and-validate automatic program repair.

Keywords: Automated program repair · Genetic programmer · Validation · Fitness function

1 Introduction

Recently the size of the software is increased, and their complexity is also growing. Followed by these, software debugging is getting more time-consuming and expensive process. Recent studies showed that the cost of debugging process is about 50% of the cost of the whole development process [1, 2].

S. Lee et al. (Eds.): IMCOM 2019, AISC 935, pp. 694–710, 2019.
https://doi.org/10.1007/978-3-030-19063-7_56

One of the factors contributing to the expensive cost of debugging is bug fixing [3]. Bug fixing is costly and tedious works because developers traditionally have to fix the software on manual. Recently researchers have focused on an automatic way to fix the program, namely Automated Program Repair. Given a buggy program and a test suit, which represents program's correctness, Automated Program Repair tries to find a patch (changes to the buggy program code) that can make the program pass all provided test cases, including the original failing ones. There are two main classes of Automated Program Repair: generate-and-validate techniques and semantics-driven techniques. Generate-and-validate approaches find program repair by executing the iterative process consisting of the generation and the validation activity. A set of change operators modifies the original program, and these modified programs are added to the candidate patches on generate process. Then they check the correctness of candidate patches based on the result of running test cases on the validation process while Semantics-driven approaches encode the program formally as an analytical procedure whose output is a fix [3].

GenProg is a general generate-and-validate technique for automated program repair. It is based on the fundamentals of Genetic Programming evolving of a set of changes to the original program [4–7]. At every iteration, new program variants, changed with certain modification points from the original program, are generated by modifying suspected statements randomly based on a suspected statements list of fault localization. Then compile the generated variant and run the compiled programs with all test cases to identify their correctness using fitness function. These generation and validation are repeated until finding a patch which makes all test cases pass. Although automatic program repair shows the possibility to repaired program without any human efforts, it is still time-consuming and may fail to find any valid patches.

There are two causes for this long running time. The first reason is expensive of the patch validation activities (fitness computation) [19, 33]. The patch validation consists of applying each candidate patch, compiling the program variant to obtain an executable program, and then running the program on provided test cases to determine their correctness. Since both the number of candidate patches and the number of test cases in the regression test suite are usually large. Reducing the validation cost can result in a significant improvement for the cost of the whole process.

Moreover, the variant generating process selects modification points at random, they often generate many inadequate variants which modify the original program at a set of modification points that has no effects to at least one failing test case. Program variants which are generated by these inadequate modifications undoubtedly fail the regression test. Thus, validating these program variants has incurred redundant validations

The second cause is ineffective of the search strategies which is encoded in the fitness function of the genetic algorithm. The genetic algorithm based on the fitness score to evolve better variants into the next generation. However, the canonical fitness function of GenProg fails to indeed distinguish between variants of varying quality, because it provides the same fitness score to all variants that pass the same proportion of test cases. This coarse-grained fitness function is inefficient to guide the search for promising regions in the search spaces.

In this paper, we introduce a new automated program repair tool AdqFix based on genetic programming, which can minimize these problems. AdqFix uses a selective validation strategy to reduce patch validation cost and avoid redundant validation. Selective validation skips the test execution of patches that inadequate to repair the bugs. We determine whether a patch is an inadequate variant or not by analyzing the impact of its change operators to all failing test cases. Specifically, we consider a variant is an inadequate variant if no modifications are hitting at least one failing test case. In order to make the fitness function more fine-grained, AdqFix considers not only the number of test cases that a program variant passed but also the characteristic of the change operators that have been applied. Considering two program variants that have the same number of failed test cases, the new fitness function can make different fitness score between these two program variants. Using minimal hitting set of failing test cases we can get the degree of adequate modifications for each variant. So we can distinguish the different variants and then it can have more adequate variants at the population.

We evaluate AdqFix on the DEFECTS4J benchmark [9]. We compare AdqFix with jGenprog in both terms of repairability and execution time. jGenProg is an implementation of GenProg, a state-of-the-art automatic repair tool for Java programs. The experimental results show that AdqFix outperforms jGenProg in both terms of repairability and execution time. More specifically, the AdqFix reduce about 72% number of test cases ran and saved about 49% of the overall running time while patched three more bugs compared to the jGenProg.

The remainder of this paper is organized as follows. Section 2 introduces key concepts that are necessary to understand our idea and presents a set of related work that improved jGenprog's algorithm. Section 3 presents the main contribution of this paper, AdqFix. Section 4 presents and analyzes our experiments. In Sect. 5 we discuss opportunities for threats to validity and future work. We conclude our main contribution of this paper and the experimental results.

2   Related Work

In the last a few years, several Automated Program Repair (APR) tools have been proposed using different techniques. The most representative of Automated Program Repair is Genprog [3–7], guided by genetic programming. Genprog tries to find the valid patch through evolving by many iterations. After fault localization, Genprog generates the patch candidates using atomic changes (mutations) and validates them by fitness function from running test suite (regression test). Genprog iterates these two steps, generation of patch candidates and validating them, until finding the valid patch. Starting from this, researchers had proposed different extended versions of Genprog.

Martinez et al. [10, 11] presents jGenprog, an implementation of Genprog for Java (Genprog works for C code). Their target to repair program is the Java program, and they present the results of running jGenprog for Java on the bugs of Defects4j [10–12]. Defects4j is a benchmark set for APR tools and consists of 357 real-world Java bugs by Just et al. [9]. Furthermore, there are other APR tools for considering the effectiveness of Genprog. On validating step of Genprog, it takes a long time to recompile and reinstall a program. Qi et al. [14, 15] presents that weak recompilation for a program can be used to run the test cases without recompiling the program entirely. Schulte et al. [16, 17] used Assembly code and Java byte code in order to recompile the changed function. These studies have shown that there is an expensive cost issue of Genprog. And many researchers are trying to address this issue. There are several ways to try to solve this issue. We focused on the test suite for validation (regression test) and fitness evaluation using fitness function.

2.1 Running the Test Suite on Validation

Large test suites make the automated repairing slow because they have to run all test cases for each variant which is generated as candidate patches [18–20]. To address this issue, Qi et al. [18] proposed RSRepair based on random search without any evolutions. RSRepair generates one single candidate patch and validates this at the same time. If it does fail at least one test case, they discard this candidate patch (variant) and then iterate these two processes until they find the valid patch. In order to optimize, RSRepair prioritizes the test cases. The prioritization strategy is that RSRepair executes one of the test cases first which could detect the candidate patch to be discarded [21].

Fast et al. [22] studied that reducing the test cases for validation could enhance the efficiency of APR techniques. Qi et al. [23] demonstrated that prioritizing the test cases could reduce the cost of regression test. Fast et al. [22] also investigated how to improve the efficiency of APR by selecting the subset of test suites. They evaluated the time cost of each candidate patch running the subset of the test suite instead of the whole test suites. There are reported to reduce the time of Genprog by 81%. Moreover, Walcott et al. [24] proposed the technique to select the test suite by random sampling. Li et al. [25] empirically evaluated the test suite prioritization using greedy algorithms.

However, all of them did not consider the repairing problem that only running a partial set of the test suite can induce APRs the repair fault by dropping the functionality that excluded test case may indicate from Qi et al. [26]. It is harmful to lose the specific functionality of the original program by selecting or sampling the partial test cases. We need to consider the efficiency of APR without undermining the functionality of the program. So we tried to approach to remove the variants which are likely to be discarded at validation process. Remaining all set of test suites, we can decrease the total number of running test cases by removing the unnecessary variants for evolution.

None of these works are concerned with evaluating each variant before running the test suite. Most of the researches to improve the efficiency of APR techniques considered treating the number of test cases. However, it could lose some functionality of the program by running a partial test suite. Our proposed technique evaluates each variant by checking how many the changes of variant hit failing test cases before regression test without removing any test cases. It can identify the state of each variant if the variant is related to the fault.

2.2 Fitness Function for Evolution

Fitness function determines the evolving strategy to select the variants which have the high probability of becoming the valid patch. The canonical fitness function of Genprog evaluates each variant by the sum of failed test cases. Through many evolutions, a variant finally becomes a valid patch when all test cases passed. However, many variants are having same fitness score with this canonical fitness function because the two variants usually could get the same number of failed test cases. This canonical fitness function generated this noise to select variant but Jin [27] showed the impact of the noise from fitness function in different contexts. So we need some studies to resolve this problem.

Fast et al. [22] proposed a dynamic predicate-based fitness function to enhance canonical fitness evaluation. They use test suite sampling and dynamic predicates to design their fitness function for resolving the noise. However, test suite sampling does not ensure the functionality of the original program and an overhead appear because they have to record a dynamic predicate to collect this information. Souza et al. [8] studied a new fitness function based on source code checkpoints to reduce the noise. They collect the intermediate program state called checkpoints throughout test case executions to distinguish the variants. However, they have to instrument source code to collect runtime data, and there are overheads of collecting and using the checkpoints.

Our proposed approach is possible to resolve the noise of canonical fitness function without collecting other resources to avoid many overheads. We take advantage of the coverage information of failing test cases from the fault localization. The resource already existed, so we do not have to collect new resources.

3 Proposal Approach

This section presents our automatic program repair approach, named AdqFix (Adequate Fix). AdqFix is based on genetic programming with considering adequate modifications for better evolution.

3.1 Overview

We propose an AdqFix approach, which works on patch generation and validation using useful information of the impact of its change operators to all failing test cases. There are four main phases for AdqFix: (1) Getting test results and coverage information of the buggy program with running all test suite, (2) Finding suspicious statement ranking from fault localization, (3) Computing adequate modification points, and (4) Patch generation and validation applying adequate modification points (see Fig. 1).

First, we start with running all test suite for the buggy program. By running all test suite, we get the test results and coverage information.

Second, we find the suspicious statement ranking from the coverage information. Coverage information includes which test cases are passed or failed. So we use coverage information statistically in order to find the faulty statements. Each statement could know the number of passed and failed test cases covered and calculate the suspicious score focused on the number of failed test cases. So we find the suspicious statement ranking, and we refer to this work as the fault localization.

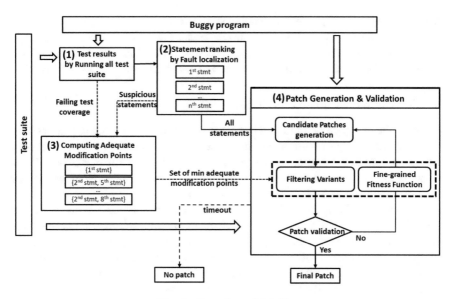

Fig. 1. Overview of AdqFix.

At the third phase, AdqFix computes adequate modification points using the coverage of failing test case and suspicious statement ranking. We need a standard to evaluate the candidate patch (variant) for our validation. We focus on the coverage of failing test cases because that coverage revealed the bugs on the program. As the standard, we make a set of statements that hit all failing test cases for each set. Even the statements that have consisted of the set are from the suspicious statement ranking. The set of adequate modification points must be minimal form. The reason why do not consider all statements of the buggy program and must keep minimal foam will be explained in Sect. 3.2.

Finally, AdqFix generates candidate patches and validates them with adequate modification points. Selecting modification point randomly from all statements, we generate the candidate patches (variants) using mutation operators. However, all variants are not essential for validation. Some of the variants are utterly irrelevant to chance to be selected to the next-generation. After generating candidate patches, AdqFix filters the candidate patches which include only inadequate modifications. Before running all test cases as the validation process, AdqFix can skip the validation of these variants by selective validation. And then we run all test cases for each variant which includes at least one adequate modification. We calculate the fitness score per each variant from a new fine-grained fitness function. A new fine-grained fitness function includes both the number of failed test cases from canonical fitness function and the degree of hitting the set of minimal adequate modification points. Considering the degree of hitting the set of minimal adequate modification points, AdqFix can distinguish the variants. We represent the way to find the score about the degree of adequate modification points in details on the Sect. 3.3.2. From the new fitness function, we can get the elaborately ranked variant list and the variants with higher fitness score are selected for next generation. We work at this stage iteratively until finding the valid patch. If the time is out, we stop Adqfix without the valid patch.

3.2 Computing Adequate Modification Points

We collect the coverage of all failing test cases as they already do running all test cases given as input a buggy program and its test suite. We also collect the statements that have the suspicious score greater than zero from suspicious statement ranking. We find the intersection of two collected sets. With these intersection statements, we compute all set of statements as hitting all failing test cases.

We must consider whether the sets are minimal. If the failing test cases are overlapped from the two statements, the two statements are not be composed of one hitting set. If we do not prevent overlapping, there are too many minimal sets of adequate modification from redundant sets so the expressiveness of adequate modification point worsens. It is the same reason why we use the statements which have the suspicious score. After considering the minimal form, we can make the set of minimal adequate modification points by computing hitting all failing test cases.

3.3 Patch Validation

3.3.1 Filtering Variants for Validation

In order to filter the variants which would be eventually discarded after the validation step, we should predict which variants are not essential for repair. For prediction, AdqFix considers whether the variants generated by inadequate modifications.

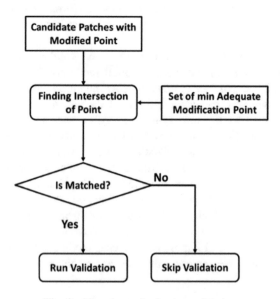

Fig. 2. Flowchart of selective validation.

We firstly identify the modification point of each variant generated. With the minimal adequate modification points collected in the previous step, we could check each variant if they have no any adequate modifications. If all modifications of the variant do not hit any set of minimal adequate modification points, AdqFix skips this variant for validation (see Fig. 2).

3.3.2 Fine-Grained Fitness Function

Let v be a variant for validation. A new Fine-grained Fitness Function of AdqFix calculates the Fitness Score (FS) based on two factors: test case valuation score $T(v)$, and adequate modification score $A(v)$, as shown in the formula (1). Where, the $T(v)$ represent the quality of variant v concerning program's correctness, while $A(v)$ evaluate the quality of variant v in term of the quality of modified points.

$$FS(v) = T(v) - A(v) \tag{1}$$

Depending on whether the variant v is adequate or not, the test case validation score $T(v)$ of v is computed in different ways. Specifically, if v is an adequate patch, the T(v) is computed as similar to the original canonical fitness function, which required to compile and run the program variant on all test cases. Otherwise, in case of variant v is an inadequate patch, the T(v) is evaluated to the size of the test suite (e.g., the worst fitness score of the original canonical fitness function) without required to run any test cases. Formula (2) represents how to evaluate the test case validation score of a variant.

$$T(v) = \begin{cases} \#\,FailedTC, & (if\ v\ is\ adequate) \\ |testsuite|, & (if\ v\ is\ inadequate) \end{cases} \tag{2}$$

The computation of adequate modification score of a variant v is shown in the formula (3).

$$A(v) = \begin{cases} 0.75 + 0.24 * \frac{1}{|v|}, & (if\ v\ is\ adequate) \\ 0.5 + 0.24 * \frac{1}{|v|} * \frac{|SS(v)|}{|AMP|}, & (if\ v\ is\ over\ adequate). \\ 0.25 + 0.24 * \frac{1}{|v|} * \frac{|BS(v)|}{|AMP|}, & (if\ v\ is\ inadequate) \end{cases} \tag{3}$$

Where

- $|v|$ is the number of modified statements in v,
- $|AMP|$ is the total number of adequate modification points, obtained from the Adequate Modification Points computation phase,
- $|SS(v)|$ is the number of adequate modification points that subsumed by the set of modified statements in v,
- $|BS(v)|$ is the number of adequate modification points that subsume the set of modified statements in v.

This formula guarantee that $(0 \leq A(v) < 1)$. Thus, the maximal effect of the A(v) factor is always smaller than the effect of one test cases in the T(v) factor. As the result, the test case validation score is always more important than the adequate modification factor in the new fitness function.

There is one more condition that the over adequate means modifications of the variant include one of the minimal Adequate Modification Points (AMP) set but include more statements except the one of minimal set. $size(v)$ is the number of modification points of the variant. The best adequate modification is when it is identical to one element in the AMP set, the smaller number of modification point is better. The totalNumberOfAMPs is the total number of the set of minimal adequate modification points.

For example, if the variant is inadequate, $A(v) = 0.25 + 0.24 * 1/size(v)*$ besubsumed(v)/totalNumberOfAMPs, and the $T(v)$ score is the size of test suite. Then subtract the $A(v)$ score from the $T(v)$ score. The variant with lower $T(v)$ score is the higher fitness variant. Through this formula, we can distinguish the variants more elaborately considering the degree of hitting AMP set. It works on the new fine-grained fitness function.

4   Evaluation

This section describes our evaluation and experimental results. Section 4.1 introduces the three research questions for evaluation. Section 4.2 shows the experimental environments including the subject program and experimental setup. On the remained sections we can see the experimental results addressing the research questions respectively.

4.1   Research Questions

Our experimental evaluation tries to answer the following three research questions:

- RQ1: How effectively does AdqFix fix the real-world bugs comparing with jGenprog?
- RQ2: Whether can AdqFix search a valid patch with lowering the number of evaluations compared to jGenprog?
- RQ3: Does AdqFix find a valid patch much faster than jGenprog?

RQ1 asks how well AdqFix runs on the real-world bugs in terms of the number of repairs and patches. We also identify the variant id when the first valid patch was found. From these aspects, we compare the effectiveness of AdqFix with jGenprog. To answer RQ1, we run AdqFix and jGenprog technique on the DEFECTS4J benchmark set. Both techniques generated and validated the candidate patches repeatedly within the time or generation budget. Then we collected the valid patches which pass all test suite. The valid patch includes the specific log information of repairing process. We can identify which bugs are repaired and the number of patches per bug. It is obvious that finding more repairs on all bug is better. Regarding getting more valid patches per bug, it has the benefit of maximizing the candidate of repair patterns. By collecting the valid patches, we can identify the variant id of the first patch. The bigger number of variant id means that the variant was old one.

There are some variants with the expensive cost of regression tests as running all test cases for each variant. Given the problem, RQ2 asks whether AdqFix can find the first valid patch with the smaller number of Test Cases Executed (TCE). To answer RQ2, we inserted the code which can count the number of TCE on the validation step. We identified the number of TCE until finding the first valid patch for each bug from both AdqFix and jGenprog.

RQ3 asks if AdqFix required smaller time for generating the first valid patch compared to jGenprog. By collecting the valid patch per each bug, we could identify the time to generate the valid patch. We compared the time to generate the first patch between AdqFix and jGenprog.

4.2 Experimental Environments

We selected the subject Java programs used on jGenprog as the experimental benchmark datasets for program repair, DEFECTS4J [9]. We ran AdqFix on this benchmark set that was made in order to use controlled experiments in software debugging and testing. They come with real-world bugs existing in history versions and have been widely selected by recent studies on APR [9, 32]. We use DEFECT4 J of version 1.0.1 and exclude the Closure Compiler project because it does not use JUnit test. We use the four projects with the total of 224 real-world bugs for experiments: JFreeChart, Commons Lang, Commons Math, and Joda-Time (see Table 1).

Table 1. Subjects for evaluation.

Subject	#Bugs	KLOC	Test KLOC	#Test cases
JFree Chart	26	96	50	2,205
Commons Lang	65	22	6	2,245
Commons Math	106	85	19	3,602
Joda-Time	27	28	53	4,130
Total	224	231	128	12,182

We implemented AdqFix by extending the ASTOR, automatic program repair framework for Java programs [10–12]. We use the GZoltar [30] as fault localization tool with the Ochiai [28] algorithm to find the suspicious statement ranking. All of our experiments are run on Ubuntu virtual machine with Intel Core i5-7600 CPU @3.50 GHz and 30 GB physical memory. We set the time limit to 180 min and set the maximum generation (iteration) to 10000 generations for each bug. It means that AdqFix tries to find the valid patch within 3 h or 10000 generations. We also set the fault localization threshold as 0.1 suspicious score. It considers only suspicious statements with more than 0.1 suspicious scores for randomly generated variants. In order to reduce the randomness in the result, we ran the experiment three times, each time with different seeds, and reported the average result over three experiment runs.

4.3   RQ1: Effectiveness of AdqFix on Real Bugs

AdqFix generates the valid patches for 22 bugs among all the 224 bugs while jGenprog finds the valid patches for 19 bugs. AdqFix finds three more repairs on Chart 14, Chart 19, and Math 64 bug. Table 2 shows the details of these 22. The variant Id on the 4th column of Table 2 means that the number of generated variants until finding the first valid patch. The 5th column #Patch means the number of generated valid patches within the time limit or the maximum generation. The #TCE on the 6th column is the number of test cases executed during validation. It represents the total number of test cases run during every iteration of the first valid patch. The 8th Column the Reducing ratio of #TCE is computed as subtraction the #TCE of jGenprog from the #TCE of AdqFix and then dividing by the #TCE of jGenprog. The 7th column Time (m) means that the time to find the first valid patch. They are organized by the average of values respectively from the different result of seeds. Especially at this RQ1, we focus on which bugs are repaired, Variant Id and the number of generated patches. We handle the other columns at next research questions.

We compare the effectiveness of AdqFix with jGenprog on the 19 bugs that are commonly repaired by these two techniques. The nine bugs are on Chart project and the other ten bugs are on the Math project. AdqFix had less variant Id number of the first patch on 11 bugs and needed more variant to find the first patch on six bugs compared with jGenprog. The other two bugs in Chart 1 and Chart 13 had the same variant Id. We could see that AdqFix generate fewer variants on about 42% of bugs (11/19). However, AdqFix generated more number of valid patches on only seven bugs and less number of valid patches on eight bugs. The other four bugs in Chart 1, 5, 12 and 18 had the same number of patches generated. AdqFix showed very similar results with jGenprog. AdqFix did not make many patches for maximizing the repair patterns. Overall, the Math project had worse results than the Chart project in terms of AdqFix. Even on Math 70 AdqFix generates about six times more variants than jGenprog, while on Chart 3 only about 1.26 times. Only two bugs on Chart were shown as AdqFix needed more variants but four bugs on Math.

Table 2. The experimental results of both AdqFix and jGenprog on DEFECTS4J.

Project	Bug Id	Approach	Variant Id	#Patch	#TCE	Time (m)	Reducing ratio of #TCE (%)
Chart	1	AdqFix	233.33	86.67	5562.00	63.67	44.24
		jGenprog	233.33	86.67	9974.00	64.00	
Chart	3	AdqFix	816.00	33.00	14799.33	102.00	12.98
		jGenprog	650.33	38.67	17006.33	78.67	
Chart	5	AdqFix	534.33	2.00	7007.00	44.00	34.32
		jGenprog	484.33	2.00	10668.00	36.00	
Chart	7	AdqFix	1145.00	5.67	9078.00	126.33	54.26
		jGenprog	3114.33	2.67	19849.00	479.67	
Chart	12	AdqFix	715.67	1.00	4875.67	130.00	44.93
		jGenprog	897.00	1.00	8854.00	164.33	

(continued)

Table 2. (*continued*)

Project	Bug Id	Approach	Variant Id	#Patch	#TCE	Time (m)	Reducing ratio of #TCE (%)
Chart	13	AdqFix	85.33	476.33	3783.67	14.00	48.77
		jGenprog	85.33	424.67	7385.67	127.67	
Chart	14	AdqFix	3501.0	11.00	33970.00	1359.0	x
		jGenprog	x	x	x	x	
Chart	15	AdqFix	1384.00	15.00	5451.00	696.00	35.05
		jGenprog	2753.00	10.00	8392.50	444.50	
Chart	18	AdqFix	2928.00	2.00	56256.00	337.00	22.15
		jGenprog	3493.00	2.00	72264.00	417.00	
Chart	19	AdqFix	5874.67	1.33	44083.00	2336.33	x
		jGenprog	x	x	x	x	
Chart	26	AdqFix	454.00	10.33	16421.00	249.67	31.08
		jGenprog	470.50	8.50	23826.00	278.00	
Math	60	AdqFix	578.00	1.50	5935.00	1395.50	87.14
		jGenprog	782.67	1.67	46138.00	2413.00	
Math	64	AdqFix	2952.00	1.00	31200.00	5991.00	x
		jGenprog	x	x	x	x	
Math	70	AdqFix	1004.67	2.33	3124.00	6.00	38.79
		jGenprog	152.67	2.00	5104.00	11.67	
Math	71	AdqFix	258.00	1.50	3546.50	2424.50	99.30
		jGenprog	329.00	1.00	507813.00	7935.00	
Math	73	AdqFix	241.00	17.33	2886.00	6.33	37.15
		jGenprog	99.67	16.67	4592.00	11.67	
Math	78	AdqFix	5596.00	1.00	186051.00	782.00	60.11
		jGenprog	7647.67	2.33	466433.67	2412.33	
Math	80	AdqFix	35.67	11.37	2390.00	105.00	56.69
		jGenprog	152.33	15.00	5518.00	343.67	
Math	81	AdqFix	186.00	12.33	4195.33	896.33	12.80
		jGenprog	78.00	15.67	4811.00	358.00	
Math	82	AdqFix	218.33	11.33	2924.00	319.67	57.81
		jGenprog	653.67	16.33	6930.67	419.00	
Math	85	AdqFix	154.33	23.33	2537.67	352.00	39.33
		jGenprog	56.67	34.33	4182.67	86.00	
Math	95	AdqFix	78.00	11.00	1466.00	170.33	51.82
		jGenprog	179.67	14.33	3043.00	138.67	

We could find more repairs on three bugs because AdqFix would shorten the path for generating the valid patch. Within the same constraints, AdqFix can arrive at the valid patch by removing redundant validation and differentiating the variants.

4.4   RQ2: Number of TCE of AdqFix

As the number of TCE (#TCE) shown in the 6th column of Table 2, the number of TCE is less on AdqFix than jGenprog for all 19 bugs. The 8th Column, the Reducing ratio of #TCE, shows that the ratio of skipping TCE to generate the first patch ranges from 12.8 to 99.3%. There is no any negative ratio value, so it could be concluded that AdqFix succeeded in reducing the number of TCE for each bug. For example, jGenprog had to execute 507813 test cases to find the first valid patch, but AdqFix only needed about 3547 test cases to run on Math 81 as the biggest reducing ratio, 99.3%. Comparing the two projects, the number of reducing ratio with 50% or more was one on the Chart project but six on the Math project. On the Math project, we skip the test cases more. Considering the sum of TCE from each approach, we can derive that the total reducing ratio of TCE is about 72.56%.

We can see the effectiveness of filtering test cases on AdqFix. We skip the running test cases for inadequate variants per generation so we need the smaller number of test cases executed for finding the valid patch.

4.5   RQ3: Efficiency of AdqFix

On Table 2, we compare the length of time consumed between AdqFix and jGenprog for each bug. The 9th column time (m) shows that there are different results for each bug. On the 13 bugs (6 Chart bugs and 7 Math bugs), AdqFix is much faster in generating the first valid patch than jGenprog. While, on the six bugs (3 Chart bugs and 3 Math bugs), AdqFix generated the valid patch slowly than jGenprog. Drawing on the reduction of the TCE previously mentioned, the 13 bugs were possible to find the first patch faster.

However, the six bugs are taking more times: Chart 1, Chart 5, Chart 15, Time 81, Time 85, and Time 95. Because of computing the adequate modification point set and calculating the fitness score with fine-grained fitness function, AdqFix has some overheads. Most of these bugs take much time by overheads because they would ignore the inadequate variants from fine-grained fitness function which are actually close to the valid patch except the Chart 15 bug. From the Table 2, we can see that the other five bugs have the bigger number of variant Id, which means the path to find the valid patch is longer. However, in case of the Chart 15 bug, AdqFix has the smaller number of variant Id so we can know that computing the adequate modification point set incurs the significant overheads.

However, AdqFix was still better than jGenprog about 58% of bugs (13/19). The time to find the first valid patch for total project was about 16219 min for jGenprog while it was about 8220 min on AdqFix. In Fig. 3, we can see that the time cost has reduced for AdqFix. AdqFix saved 7998.5 min (about 133 h) to repair the 19 bugs on DEFECTS4J. It can improve the efficiency of jGenprog, the original technique.

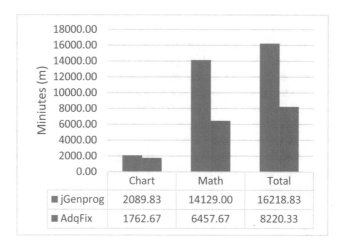

Fig. 3. Time cost of the first valid patch on both jGenprog and AdqFix.

5 Discussions

5.1 Threats to Validity

One potential threat to validity is the small number of different seeds for evaluation. In order to identify the performance of Genetic programming, we should run with many different seeds. According to the value of seed, the pattern of results would be different. However, our experiment uses only three different seeds. Even though we can check the effectiveness and efficiency from the number of seeds, we have to run AdqFix with more number of different seeds for generalizing the performance. We do not entirely match with the original result from jGenprog on DEFECTS4J cause of running on only three different seeds. Running with more different seeds, we expect to get the more close result from original and then can compare to AdqFix with more accurate results.

The noise of AdqFix is the other threat to validity. We can see the considerable size of the overheads on Chart 3, 5, 15 and Math 81, 85, 95. AdqFix on those bugs finds the longer path to find the valid patch by our fitness function. Our fitness function considers the adequate modification points from the coverage of failing test cases. If there are many sets of adequate modification points, computing the sets would take much time. We can decrease the noise by reducing the size of set or change to the ranked list.

We do not guarantee the correctness of the valid patches from AdqFix. We do not consider the correctness of the patch but only passing all test suite. In recent studies, it has been important to find more correct patches. In order to consider the correctness, we have to change the way to generate candidate patches, but we have a contribution on only validation step.

5.2   Future Work

In the future, we plan to run on the DEFECTS4J with more different seeds even on the other benchmark data set, IntroClassJava [29, 31]. The core methodology of AdqFix is selective validation, so if any APR techniques have evolution or validation stage, our proposed approach can be applied on other APR techniques. We will try to apply and verify on other evolution APR techniques. We also plan to apply the other resources for adequate modification points. Considering not only the coverage of failing test cases of the original buggy program but also the prioritization strategy for the coverage, we will try to alleviate the noise problem. We can make a particular strategy for the variant generation using historical information of the valid patches from AdqFix in order to address the correct patch.

6   Conclusion

The cost of validating patches is dominating the overall cost of search-based automatic program repair. In this paper, we present a search-based APR technique, AdqFix which has main contribution to the validation step. AdqFix identifies adequate modification points that can affect result of all failing test cases. Based on this adequate modification points, AdqFix can selectively skip the test execution of program variants. Moreover, we designed a fine-grained fitness function to find better evolving path for valid patches. We evaluated AdqFix on the DEFECTS4J benchmark. The results show that AdqFix outperforms jGenprog, a state-of-the-art APR technique in both repairability and running time. Specifically jGenprog repaired 19 bugs while AdqFix repaired 22 bugs including 19 bugs of jGenprog and three new bugs. AdqFix reduced the number of test cases executed for all common 19 bugs, with the reducing ratio about 72.56%. Drawing upon the reduction of test cases executed, AdqFix saved about 133 h. By resolving the limitation, we can improve the efficiency of the APR technique.

Acknowledgments. I would like to acknowledge M. Martinez et al. for their work on ASTOR, jGenprog. I thank Q. Phung for help on implementation of AdqFix. This research supported in part by the Technology and Basic Science Research Program through the NRF funded by the Ministry of Education (2016R1D1A1B03934610), MIST, Korea, under the National Program for Excellence in SW supervised by the IITP (2015-0-00914), and Next-Generation Information Computing Development Program through the National Research Foundation of Korea (NRF) funded by the Ministry of Science, ICT (2017M3C4A7068179).

References

1. Britton, T., Jeng, L., Carver, G., Cheak, P.: Reversible debugging software -quantify the time and cost saved using reversible debuggers (2013)
2. Undo Software: Increasing software development productivity with reversible debugging, Technical report white paper (2014)
3. Gazzola, L., Micucci, D., Mariani, L.: Automatic software repair: a survey. In: Proceedings of IEEE Transactions on Software Engineering, p. 1 (2017)

4. Weimer, W., Nguyen, T., Goues, C.L., Forrest, S.: Automatically finding patches using genetic programming. In: Proceedings of IEEE the International Conference on Software Engineering, pp. 364–374 (2009)
5. Forrest, S., Nguyen, T., Weimer, W., Goues, C.L.: A genetic programming approach to automated software repair. In: Proceedings of ACM the Annual Conference on Genetic and Evolutionary Computation, pp. 947–954 (2009)
6. Weimer, W., Forrest, S., Goues, C.L., Nguyen, T.: Automatic program repair with evolutionary computation. Commun. ACM Mag. **53**, 109–116 (2010)
7. Goues, C.L.: Automatic program repair using genetic programming. Ph.D. dissertation, pp. 364–374 (2013)
8. de Souza, E.F., Goues, C.L., Camilo-Junior, C.G.: A novel fitness function for automated program repair based on source code checkpoints. In: GECCO 2018 (2018)
9. Just, R., Jalali, D., Ernst, M.D.: Defects4J: a database of existing faults to enable controlled testing studies for Java programs. In: Proceedings of ACM the International Symposium on Software Testing and Analysis, pp. 437–440 (2014)
10. Martinez, M., Monperrus, M.: Astor: Exploring the Design Space of Generate-and-Validate Program Repair beyond GenProg. Cornell University Library (2018)
11. Martinez, M., Monperrus, M.: ASTOR: Evolutionary Automatic Software Repair for Java Matias. Cornell University Library (2014)
12. Martinez, M., Monperrus, M.: ASTOR: a program repair library for java. In: Proceedings of ACM the International Symposium on Software Testing and Analysis, pp. 441–444 (2016)
13. Le Goues, C., Dewey-Vogt, M., Forrest, S., Weimer, W.: A systematic study of automated program repair: fixing 55 out of 105 bugs for $8 each. In: Software Engineering (ICSE), 34th International Conference, pp. 3–13 (2012)
14. Qi, Y., Mao, X., Lei, Y.: Making automatic repair for largescale programs more efficient using weak recompilation. In: Proceedings of IEEE the International Conference on Software Maintenance, pp. 254–263 (2012)
15. Qi, Y., Mao, X., Dai, Z., Qi, Y.: Efficient automatic program repair using function-based part-execution. In: Proceedings of IEEE the International Conference on Software Engineering and Service Science, pp. 235–238 (2013)
16. Schulte, E., DiLorenzo, J., Weimer, W., Forrest, S.: Automated repair of binary and assembly programs for cooperating embedded devices. In: Proceedings of ACM the International Conference on Architectural Support for Programming Languages and Operating Systems, pp. 317–328 (2013)
17. Schulte, E., Forrest, S., Weimer, W.: Automated program repair through the evolution of assembly code. In: Proceedings of ACM the International Conference on Automated Software Engineering, pp. 313–316 (2010)
18. Qi, Y., Mao, X., Lei, Y., Dai, Z., Wang, C.: The strength of random search on automated program repair. In: Proceedings of ACM the International Conference on Software Engineering, pp. 254–265 (2014)
19. Goues, C.L., Forrest, S., Weimer, W.: Current challenges in automatic software repair. Software Qual. J. **21**, 421–443 (2013)
20. Kong, X., Zhang, L., Wong, W.E., Li, B.: Experience report: How do techniques, programs, and tests impact automated program repair? In: Proceedings of IEEE the International Symposium on Software Reliability Engineering (2015)
21. Rothermel, G., Untch, R.H., Chu, C., Harrold, M.J.: Prioritizing test cases for regression testing. In: Proceedings of IEEE Transactions on Software Engineering, pp. 929–948 (2001)
22. Fast, E., Goues, C.L., Forrest, S., Weimer, W.: Designing better fitness functions for automated program repair. In: Proceedings of ACM the Conference on Genetic and Evolutionary Computation, pp. 965–972 (2010)

23. Qi, Y., Mao, X., Lei, Y.: Efficient automated program repair through fault-recorded testing prioritization. In: Proceedings of IEEE the International Conference on Software Maintenance, pp. 180–189 (2013)
24. Walcott, K., Soffa, M., Kapfhammer, G., Roos, R.: TimeAware test suite prioritization. In: Proceedings of ACM the International Symposium on Software Testing and Analysis (2006)
25. Wappler, S., Wegener, J.: Evolutionary unit testing of object-oriented software using strongly-typed genetic programming. In: Proceedings of ACM Conference on Genetic and Evolutionary Computation, pp. 1925–1932 (2006)
26. Qi, Z., Long, F., Achour, S., Rinard, M.: An analysis of patch plausibility and correctness for generate-and-validate patch generation systems. In: Proceedings of ACM the International Symposium on Software Testing and Analysis, pp. 24–36 (2015)
27. Jin, Y.: A comprehensive survey of fitness approximation in evolutionary computation. Soft. Comput. 9(1), 3–12 (2005)
28. Abreu, R., Zoeteweij, P., van Gemund, A.J.C.: On the accuracy of spectrum-based fault localization. In: Proceedings of IEEE the Testing: Academic and Industrial Conference Practice and Research Techniques, pp. 89–98 (2007)
29. Goues, C.L., Holtschulte, N., Smith, E.K., Brun, Y., Devanbu, P.: The ManyBugs and IntroClass benchmarks for automated repair of C programs. In: Proceedings of IEEE Transactions on Software Engineering, pp. 1236–1256 (2015)
30. Gzoltar Homepage (2017). http://www.gzoltar.com
31. Durieux, T., Monperrus, M.: IntroClassJava: a benchmark of 297 small and buggy java programs. Ph.D. Dissertation. Universite Lille 1 (2016)
32. Le, X.B.D., Lo, D., Goues, C.L.: History driven program repair. In: Proceedings of IEEE 23rd International Conference on Software Analysis, Evolution, and Reengineering (SANER), pp. 213–224 (2016)
33. Weimer, W., Fry, Z.P., Forrest, S.: Leveraging program equivalence for adaptive program repair: Models and first results. In: Automated Software Engineering (ASE), IEEE/ACM 28th International Conference on. Piscataway, pp. 356–366 (2013)

Randomized Technique to Determine the New Seedlings for Simulation of Population Dynamic

Yasmin Yahya[(✉)] and Roslan Ismail

Universiti Kuala Lumpur, 1016 Jalan Sultan Ismail, 50250 Kuala Lumpur, Malaysia
{yasmin,drroslan}@unikl.edu.my

Abstract. The use of inventory technique that creates the coordinate of trees, made the simulation for predicting future growth using individual trees, or tree mapping, possible. The techniques predict the growth of trees, as well as predicting the growth of saplings and seedlings. In this case, not only the growth of tree is projected, trees that are going to die, new trees and seedlings with their locations, are also projected. The locations (coordinate) of the new seedlings will be identified based on the potential mother trees. As the result of this approach, the distribution of trees could be predicted in future. The system, however, made available various formulas to be applied in this simulation. Hence, this approach is not only applied for inputs to the decision making during logging operation, but also could improve the management of forest resources. That includes the needs to report the status of forest resources, including the distribution of its trees and species.

Keywords: Recruitment · Random number · Mother tree · Simulation

1   Introduction

The current and conventional technique of forest growth projection is based on empirical approach of statistical analysis where aggregated group values of volume and number of trees species and their respective sizes are used. The use of inventory technique that creates the coordinate of trees, made the simulation for predicting future growth using individual trees, or tree mapping, possible. The techniques predict the growth of trees, as well as predicting the growth of saplings and seedlings. In this case, not only the growth of tree is projected, trees that are going to die, new trees and seedlings with their locations, are also projected. Many countries in the tropics are adopting the concept of sustainable forest management in their management system. In this system, the timber that are going to be harvested, in terms of species and sizes, will also need to identify on its ability to regrowth at specific period of time. In many tropical countries, the selective felling based on 30 years rotation period is adopted [8, 10, 13]. In this case, the trees selected to be felled should be decided based upon the ability of the

© Springer Nature Switzerland AG 2019
S. Lee et al. (Eds.): IMCOM 2019, AISC 935, pp. 711–722, 2019.
https://doi.org/10.1007/978-3-030-19063-7_57

forest to regrow and to be harvested 30 years from the date of logging. However, in current technology technique, it is unable to produce the distribution of trees for the next cycle of logging. The inventory technique that currently practice is applying the systematic sampling of cluster 50 m by 20 m. This system can only obtain the mean of analyzed parameters such as tree volume and number of trees. This information is used to make decision for cutting system for that particular forest area. The system; however, is not able to produce:

1. The location of each tree for the forest area
2. The potential trees that are going to die
3. Seedlings, saplings and their locations
4. The distribution of the trees, tree species and tree size of the forest area
5. The direction of tree felling before logging
6. The prediction of damage trees based on the suggested logging limit.

A number of functions have been used to estimate tree survival, including the Linear [2, 28], Weibull [11, 28], Exponential [28] and Gompertz function [25]. The Logistic function has been used mostly for estimating individual tree survival [4, 6, 7, 9, 14, 17, 19, 22, 24, 26, 29, 30, 33, 36, 37]. Most the more recent survival models have been concentrating on the individual tree level [15]. One reason is that the single tree level seems to allow more specific estimates in uneven-aged, species-rich forests. Most of the seedlings died within six months although there were more than eleven thousand seedlings per hectare. The seedlings are hardly to survive more than five years [5].

This study focusing on developing individual-based simulation model for forest growth in a tropical rain forest. The simulation models will consider some improvements from the previous simulation models. With the introduction of geographic information system (GIS) to forest management, there have been a lot of improvement with regard the forest management [33]. The uncertainty randomized techniques are used to select the potential mother trees and to locate the new coordinate for the survived seedlings. The data are stored in the database which will be applied to integrate with the growth simulation system. The main gap is how to put to coordinates for the new seedlings.

2   Methodology

2.1   New Recruitment

Natural regeneration forms an essential component of selection harvesting systems used in rain forest management, and long-term yield forecasts must take account of the presence and amount of this regeneration [27]. Unlike mortality, recruitment rates even in homogeneous populations cannot be estimated in a simple way [31]. In this study, the process of new recruitment is based on the ecological approach where the process started with selecting the mother tree until creating the new coordinates. The basic equation for population in natural forest is divided into three categories which are tree, sapling and seedling.

$$P = P_{tree} + P_{sapling} + P_{seedling} \qquad (1)$$

$$P = \{x_1, x_2, ..., x_n\} \tag{2}$$

$$C = \{l_1, l_2, ..., l_n\} \tag{3}$$

where

$$P = \text{Population of trees (in the forest)} \tag{4}$$

$$C = \text{Location of the trees} \tag{5}$$

Potential mother trees are trees that are fruiting, and diameter breast height is greater than 30 cm. It is therefore subset of the total population of the trees.

$$P_{PotentialMother} \subset P \tag{6}$$

$$P_{PotentialMother} = \{P_{m1}, P_{m2}, ..., P_{mn}\} \tag{7}$$

$$P_{Mother} = \{m_1, m_2, ..., m_n\} \tag{8}$$

$$P_{Mother} \subset P_{PotentialMother} \tag{9}$$

The corresponding location of the selected mother trees are as follows:

$$C_{mother} = \{l_{mother1}, l_{mother2}, ..., l_{mothern}\} \tag{10}$$

There are two questions to be solved at this level:

1. Which trees among the potential mother trees that are going to be fruiting in that year. These issues will again arise the following year as well as throughout the duration of the simulation. Of course, there are no specific guide to predict the future mother trees. In that case to use random number to choose trees that are going to be mother trees during that period. The trees to be selected therefore will vary from year to year according to this approach.
2. The number of seeds and its corresponded locations. In this case it is well understood that only a few seeds will survive for the first year and the number is very few. In this study, we are using the number to be in the range of 5 to 8 seeds per each mother tree. As the result of this situation, there is a question such as the location of the seeds. It is well understood that the fruits fall around the mother trees. In this case we assume around the diameter of 10 m of their respected mother trees.

The actual mother trees that will produce fruit and produce seeds are also subset of the potential mother trees. In normal cases, the percentage of the trees that are fruiting is about 20% – 30% of the potential mother trees. So, the actual trees that are going to produce fruits and then seeds are a subset of the potential mother trees. There are three essential steps involved in generating new seedlings which are listed below.

STEP 1: Mother Tree

1. Establish the potential mother trees.
2. The mother trees should be trees at least with the diameter greater than 30 cm.

STEP 2: Creating New Coordinates

1. Number of seedlings created should be based on the number of seedlings that could be survived at the end of growing period.
2. The coordinate of the seedlings should be closer to the mother trees.
3. Hence, they will also share the percentage of shading as their mother trees.

STEP 3: Seedling, Sapling and Tree

1. For sapling to become a tree with diameter more than 5 cm, it will take maximum 5 years.
2. For seedling, the time to move to new category, i.e. sapling, is also 5 years.
3. Figure 1 illustrates the stages or cycle of the tree from seed, seedling, sapling and become an adult tree.

2.2   Generate New Location for Seedlings

The new seedlings will be given their new location (coordinates). Before locating the new coordinates to the potential survival seedlings, the first step is to check the potential mother trees. The system will do checking for mother trees which fulfill the criteria of diameter greater than 30 cm. Expert from forest management recommended that only 20% to 30% of the mother trees will produce seed. The mother trees will produce a lot of seeds such as 100 seeds but only around ten percent will survive [1]. Around seven to ten survival seedlings were chosen to be given the new coordinates. The distance of the new coordinates is around 1 m to 3 m from their mother trees. New tree characteristics need to be generated; such as tree number, tree diameter, new location and species name. The diameter of the new tree should be less than 3 cm; the height is less than 1 m and the species group are based on mother tree's group.

The survived new seedlings will grow around the mother tree (Fig. 4), which their coordinates are determined by random number. The details of the procedure to generate new coordinates are elaborated below. To generate new coordinate:

STEP 1: Potential mother trees

1. Determine number of potential mother trees.
2. N_1 = Potential mother tree.

STEP 2: How many of mother trees

1. Obtain the random number for percentage of mother trees.
2. N_2 = percentage of potential mother tree will be 20%–30%.

STEP 3: random number to select mother trees

1. Random number for trees to be mother trees are based on the number selected from N_1 with N_2 percentage.

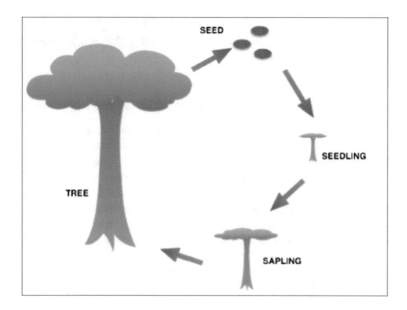

Fig. 1. The cycle of the tree.

STEP 4: Use random number to select seedlings

1. For each mother tree, random number for seedling to be produced is between five to seven seedlings.
2. S_1 = number of seedlings (5–7 seedlings).

STEP 5: Determine the coordinate for seedlings

1. For each seedling, determine the coordinate. The location of the coordinates is shown in Fig. 2.
2. The distance of the first seedling from mother tree should be five to eight meters.
3. Then the following seedlings are placed in one meter which is break into ten locations.

STEP 6: Example of calculating the coordinate of seedling

1. Suppose the coordinate of mother tree is (X, Y).
2. X can be right or left from mother tree and Y can be up or down from mother tree.
3. The distance of the seedling from mother tree = d_2; d_2 = random (1, dd_1 × 10).
4. New coordinate for x-axis = X_{seed1}; X_{seed1} = X + (d_2/10). If X-coordinate change, Y-coordinate remain, so the new coordinate for seed1 = [(X + d_2/10), Y].

5. For example, if coordinate of the mother tree is (3, 4); random number for d_2 is 7, then, coordinate for X_{seed1} should be $3 + 0.7 = 3.7$. Therefore, the location of seed1is (3.7, 4).
6. For each cycle, the seedling

$$C_{yi} = \{S_1, ..., S_{n1}\}\, C_{y2} = \{S_1, ..., S_{n2}\} ... C_{yn} = \{S_1, ..., S_{nn}\} \qquad (11)$$

The S_1 in Cycle 1 will be linked to its mother tree. The identity for that seedling is represented by $S_1 = S_{TnYcSn}$ where T is the tree number, Y is the year when it was created, and S is the seedling number (Refer to Table 1, third column).

The tree status will be updated yearly. The system will do the checking of the tree status whether it is seedling, sapling or tree. The whole process of recruiting new trees and checking their status is presented in Fig. 3. Three important elements for locating new tree recruitment that cover in this study are location and species, mother tree and number of seedlings.

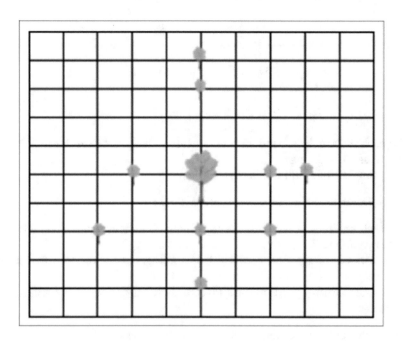

Fig. 2. Example of distribution of new tree recruitment.

3 Implementation and Result

The location of the new seedlings should be around their mothers' trees. Therefore, the first step in this procedure is to select the potential mother trees. In this study, potential mother tree is defined as tree with diameter 30 cm and above. Theoretically, not all mother trees will produce seed. According to forestry expert, only 20% to 30% of mother trees will produce seed. The system will select all trees with diameter at least 30 cm, then random number will be applied to choose the number of mother tree that will produce seed (20% to 30% of the total potential mother trees). The third step is to decide how many seeds will be produced by each of the mother tree. Mother tree will produce thousands of seeds but only 7 to 10 seeds will survive. The system will use random number to select the best number of survived seedlings which is between 7 and 10. These random numbers (for mother trees and seedlings) will keep changing in every simulation cycle due to the influence of forest dynamics. The fourth step is to locate and assign the coordinates to these new seedlings. The system will make division of 10000 small squares from 100 m by 100 m area around the mother's tree. Then, the location of the seedling will be randomly placed at any point in the small squares. The system will calculate the coordinate of the seedling $(U_x and U_y)$ based on the mother's tree coordinate. The calculation of the new coordinate is described in Sect. 2.2 under Step 6. The seedlings' species name will follow their mother trees species. The output information will be stored in database includes the year of simulation, year in system, seed number, species name, coordinate of the seedling and status of the seedling. The algorithm for this module is displayed in Fig. 3 and the example of output is shown in Table 1.

Table 1. Example of tree end status after one cycle simulation

Year	Year In System	Seed_Num	Species	SUx	SUy	Status
2	1	T1Y1SI	BOPR	−60489	−4992	ALIVE
2	1	T1Y1S2	BOPR	60540	5025	ALIVE
2	1	T1Y1S3	BOPR	60506	−4981	ALIVE
2	1	T1Y1S4	BOPR	60506	5029	ALIVE
2	1	T1Y1S5	BOPR	60514	−5000	ALIVE
2	1	T1Y1S6	BOPR	−60489	5022	ALIVE
2	1	T4Y1SI	UNKN	60512	−4981	ALIVE
2	1	T4Y1S2	UNKN	60524	5072	ALIVE
2	1	T4Y1S3	UNKN	60546	−5016	ALIVE
2	1	T4Y1S4	UNKN	60507	−5017	ALIVE
2	1	T7Y1SI	PHDK	60350	−4898	ALIVE
2	1	T7Y1S2	PHDK	−60263	−4896	ALIVE
2	1	T7Y1S3	PHDK	60337	−4888	ALIVE
2	1	T7Y1S4	PHDK	−60287	−4874	ALIVE
2	1	T7Y1S5	PHDK	60307	−4896	ALIVE
2	1	T7Y1S6	PHDK	60324	−4879	ALIVE

The results of the simulation in master databases (Fig. 5) are transferred to excel spreadsheet to generate further analysis. The spreadsheet has capability to manipulate and store huge amount of data as well as to illustrate the graphs

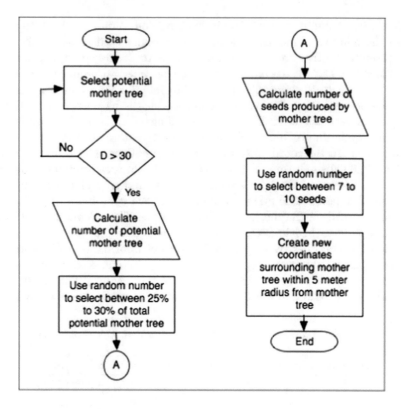

Fig. 3. Algorithm to locate new seedlings.

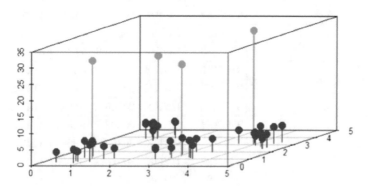

Fig. 4. Example of distribution of new tree recruitment.

and charts in meaningful manner. In addition, we could display results regarding number of trees, seedlings, saplings and total volume for commercial and non-commercial species.

This study focuses on the development of forest growth simulation system based on individual tree approach. Each tree will be simulated on its future growth based on its surrounding trees' neighbors and its ability to compete for sunlight to grow. One of major advances in the tropical forest simulation is the ability to use the individual trees, rather than the stockings. In this aspect, the simulation is capable to identify not only the rate of growth of the individual trees, but also the identification of trees to die and location of seedlings. Besides, the locations (coordinate) of the new seedlings will be identified based on the potential mother trees. As the result of this approach, the distribution of trees could be predicted in future. Obviously, there would be a lot of advantages when the locations of trees are known.

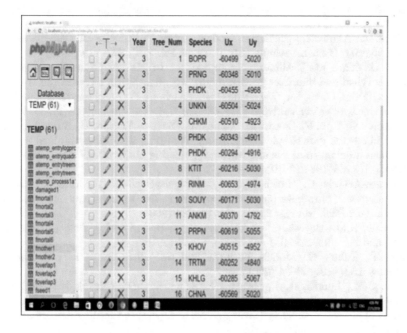

Fig. 5. Algorithm to locate new seedlings.

4 Conclusion

It is a great challenge to discover the toleration between simplicity and complexity in modeling and simulating forest growth in general and particularly in generating new location for new seedling. In this study, we showed how our new methods achieve this toleration by combining uncertainty with randomized technique in tree recruitment. The simulations are individual-tree based with the display of tree locations. We compared our system with various earlier systems which using different concepts. We found that our new simulation technique is far superior in a way of, first, recruiting new trees which are not yet implemented in any previous growth simulation system and second, the method of creating new coordinates by using random number for tree recruitment surrounding their mothers' trees was totally noble and impressive.

Acknowledgement. Our thanks to Universiti Kuala Lumpur (UniKL) for giving us opportunity to carry out this research and for their financial support.

References

1. Ashton, M.S., Hooper, E.R., Singhakumara, B., Ediriweera, S.: Regeneration recruitment and survival in an Asian tropical rain forest: implications for sustainable management. Ecosphere **9**(2) (2018)
2. Belote, R.T., Larson, A.J., Dietz, M.S.: Tree survival scales to community-level effects following mixed-severity fire in a mixed-conifer forest. For. Ecol. Manag. **353**, 221–231 (2015)
3. Buba, T.: Allometric prediction models of growth variables of Daniella oliveri in the Nigerian Guinea Savanna. Afr. J. Plant Sci. **7**(6), 213–218 (2013)
4. Cao, Q., Strub, M.: Evaluation of four methods to estimate parameters of an annual tree survival and diameter growth model. For. Sci. **54**(6), 617–624 (2008)
5. Chang-Yang, C.-H., Lu, C.-L., Sun, I.-F., Hsieh, C.-F.: Long-term seedling dynamics of tree species in a subtropical rain forest, Taiwan. Taiwania **58**(1), 35–43 (2013)
6. Crecente-Campo, F., Soares, P., Tomé, M., Diéguez-Aranda, U.: Modelling annual individual-tree growth and mortality of scots pine with data obtained at irregular measurement intervals and containing missing observations. For. Ecol. Manag. **260**(11), 1965–1974 (2010)
7. Diéguez-Aranda, U., Burkhart, H.E., Rodríguez-Soalleiro, R.: Modeling dominant height growth of radiata pine (Pinus radiata D. Don) plantations in north-western Spain. For. Ecol. Manag. **215**(1–3), 271–284 (2005)
8. Dorazio, R.M.: Bayesian data analysis in population ecology: motivations, methods, and benefits. Popul. Ecol. **58**(1), 31–44 (2016)
9. Eid, T., Tuhus, E.: Models for individual tree mortality in norway. For. Ecol. Manag. **154**(1–2), 69–84 (2001)
10. Free, C.M., Landis, R.M., Grogan, J., Schulze, M.D., Lentini, M., Dünisch, O.: Management implications of long-term tree growth and mortality rates: a modeling study of big-leaf mahogany (Swietenia macrophylla) in the Brazilian Amazon. For. Ecol. Manag. **330**, 46–54 (2014)
11. Fu, W., Simonoff, J.S.: Survival trees for interval-censored survival data. Stat. Med. **36**(30), 4831–4842 (2017)
12. Grogan, J., Landis, R.M., Free, C.M., Schulze, M.D., Lentini, M., Ashton, M.S.: Big-leaf mahogany Swietenia macrophylla population dynamics and implications for sustainable management. J. Appl. Ecol. **51**(3), 664–674 (2014)
13. Hartig, F., Calabrese, J.M., Reineking, B., Wiegand, T., Huth, A.: Statistical inference for stochastic simulation models-theory and application. Ecol. Lett. **14**(8), 816–827 (2011)
14. Holzwarth, F., Kahl, A., Bauhus, J., Wirth, C.: Many ways to die–partitioning tree mortality dynamics in a near-natural mixed deciduous forest. J. Ecol. **101**(1), 220–230 (2013)
15. Hülsmann, L., Bugmann, H., Brang, P.: How to predict tree death from inventory data—lessons from a systematic assessment of european tree mortality models. Can. J. For. Res. **47**(7), 890–900 (2017)
16. Indrajaya, Y., van der Werf, E., Weikard, H.-P., Mohren, F., van Ierland, E.C.: The potential of REDD+ for carbon sequestration in tropical forests: supply curves for carbon storage for Kalimantan, Indonesia. For. Policy Econ. **71**, 1–10 (2016)

17. Jutras, S., Hökkä, H., Alenius, V., Salminen, H., et al.: Modeling mortality of individual trees in drained peatland sites in Finland (2003)
18. Katz, R.W.: Techniques for estimating uncertainty in climate change scenarios and impact studies. Clim. Res. **20**(2), 167–185 (2002)
19. Lines, E.R., Coomes, D.A., Purves, D.W.: Influences of forest structure, climate and species composition on tree mortality across the eastern US. PLoS One **5**(10), e13212 (2010)
20. Mäkelä, A., Hari, P.: Stand growth model based on carbon uptake and allocation in individual trees. Ecol. Model. **33**(2–4), 205–229 (1986)
21. Mäkelä, J.M., Maso, M.D., Pirjola, L., Keronen, P., Laakso, L., Kulmala, M., Laaksonen, A.: Characteristics of the atmospheric particle formation events observed at a borel forest site in southern Finland. Boreal Environ. Res. **5**(4), 299–313 (2000)
22. Misir, M., Misir, N., Yavuz, H.: Modeling individual tree mortality for crimean pine plantations. J. Environ. Biol. **28**(2), 167 (2007)
23. Monserud, R.A., Sterba, H.: Modeling individual tree mortality for Austrian forest species. For. Ecol. Manag. **113**(2–3), 109–123 (1999)
24. Palahí, M., Pukkala, T.: Optimising the management of Scots pine (Pinus sylvestris L.) stands in Spain based on individual-tree models. Ann. For. Sci. **60**(2), 105–114 (2003)
25. Promislow, D.E.L., Tatar, M., Pletcher, S., Carey, J.R.: Below-threshold mortality: implications for studies in evolution, ecology and demography. J. Evol. Biol. **12**(2), 314–328 (1999)
26. Ruiz-Benito, P., Lines, E.R., Gómez-Aparicio, L., Zavala, M.A., Coomes, D.A.: Patterns and drivers of tree mortality in iberian forests: climatic effects are modified by competition. PLoS one **8**(2), e56843 (2013)
27. Sheil, D., May, R.M.: Mortality and recruitment rate evaluations in heterogeneous tropical forests. J. Ecol., 91–100 (1996)
28. Sims, A., Kiviste, A., Hordo, M., Laarmann, D., von Gadow, K.: Estimating tree survival: a study based on the Estonian Forest Research Plots Network. Annales Botanici Fennici **46**, 336–352 (2009). JSTOR
29. Uzoh, F.C.C., Mori, S.R.: Applying survival analysis to managed even-aged stands of ponderosa pine for assessment of tree mortality in the western United States. For. Ecol. Manag. **285**, 101–122 (2012)
30. Vanclay, F., Lawrence, G.: The environmental imperative: eco-social concerns for Australian agriculture. Central Queensland University (1995)
31. Vanclay, J.K.: Modelling forest growth and yield: applications to mixed tropical forests. School of Environmental Science and Management Papers, p. 537 (1994)
32. Woods, A., Coates, K.D.: Are biotic disturbance agents challenging basic tenets of growth and yield and sustainable forest management? Forestry **86**(5), 543–554 (2013)
33. Yahya, Y., Ismail, R.: Computer simulation of tree mapping approach to project the future growth of forest. In: Proceedings of the 11th International Conference on Ubiquitous Information Management and Communication, pp. 54. ACM (2017)
34. Yahya, Y., Ismail, R., Vanna, S., Saret, K.: Using data mining techniques for predicting individual tree mortality in tropical rain forest: logistic regression and decision trees approach. In: Proceedings of the 8th International Conference on Ubiquitous Information Management and Communication, p. 91. ACM (2014)
35. Yang, Y., Monserud, R.A., Huang, S.: An evaluation of diagnostic tests and their roles in validating forest biometric models. Can. J. For. Res. **34**(3), 619–629 (2004)

36. Yang, Y., Titus, S.J., Huang, S.: Modeling individual tree mortality for white spruce in Alberta. Ecol. Model. **163**(3), 209–222 (2003)
37. Yao, X., Titus, S.J., MacDonald, S.E.: A generalized logistic model of individual tree mortality for aspen, white spruce, and lodgepole pine in Alberta mixedwood forests. Can. J. For. Res. **31**(2), 283–291 (2001)

A New Biometric Template Protection Using Random Orthonormal Projection and Fuzzy Commitment

Thi Ai Thao Nguyen[✉], Tran Khanh Dang, and Dinh Thanh Nguyen

Faculty of Computer Science and Engineering, Ho Chi Minh University
of Technology, VNU-HCMU, Ho Chi Minh City, Vietnam
{thaonguyen, khanh, dinhthanh}@hcmut.edu.vn

Abstract. Biometric template protection is one of most essential parts in putting a biometric-based authentication system into practice. There have been many researches proposing different solutions to secure biometric templates of users. They can be categorized into two approaches: feature transformation and biometric cryptosystem. However, no one single template protection approach can satisfy all the requirements of a secure biometric-based authentication system. In this work, we will propose a novel hybrid biometric template protection which takes benefits of both approaches while preventing their limitations. The experiments demonstrate that the performance of the system can be maintained with the support of a new random orthonormal project technique, which reduces the computational complexity while preserving the accuracy. Meanwhile, the security of biometric templates is guaranteed by employing fuzzy commitment protocol.

Keywords: Biometric template protection · Fuzzy commitment ·
Orthonormal matric · Discriminability

1 Introduction

Biometric based authentication systems have been applied more and more in our daily life. Nowadays, we can access to our personal devices by providing to sensors our fingerprint, face, voice, iris, … We also do the same thing in order to access to thousands of online services. There is no need to remember or carry anything like password, token, … to prove ourselves to systems. It is obviously so convenience. However, benefits always come along with challenges. These challenges are related to the security of biometrics. As we all know, human has a limited number of biometric traits; therefore, we cannot change our biometrics frequently like password once we suspect that the templates are revealed [1]. Moreover, the fact that people may register thousands of online services makes them use the same biometrics to sign in some services. That leads to the cross-matching attacks when attackers follow user's biometric template cross the online services in order to track their activities. Another concern relates to the natural set-backs of biometrics. The fact that biometrics reflects a specific individual means it contains sensitive information (e.g. medical conditions)

© Springer Nature Switzerland AG 2019
S. Lee et al. (Eds.): IMCOM 2019, AISC 935, pp. 723–733, 2019.
https://doi.org/10.1007/978-3-030-19063-7_58

which users do not want attacker or even the server storing users' authentication data to discover. Last but not least, the network security needs to be discussed when user's private information is transmitted over insecure network [2]. As a result, biometric template protection is apparently indispensable part in every biometric system applications. To solve the security issues, the commonly approach is to store the transformed biometric template instead of the original one. According to [1], there are three requirements that a biometric template protection technique should possess.

1. Cancelability (Revocability + Diversity): it should be straightforward to revoke a compromised template and reissue a new one based on the same biometric data. In addition, the scheme should not generate the same transformed templates of an individual for different applications.
2. Security: An original biometric template must be computationally hard to recover from the secure template. This property guarantees that an adversary does not have the ability to create a physical spoof of the biometric trait from a stolen template.
3. Performance: the biometric template protection scheme should not degrade the recognition performance of the biometric system. This requirement can refer to the discriminability of the original biometric template which should be preserved after transformed. It also means transformed templates from the same user should have high similarity in the transformed space. And the ones from different users should be dissimilar after transformed.

To meet as many requirements as possible, many techniques have been proposed which can be classified into two main approaches: the feature transform approach and the biometric cryptosystem approach.

In the first approach, biometric templates are transformed using a function defined by a user-specific factor such as a key, a password, or a random string… The goal of this approach is to provide diversity and unlinkability by using different transforming functions for different applications involving the same set of users. The intrinsic strength of biometric characteristics should not be reduced applying transforms (represented by FAR – False Accept Rate) while on the other hand transforms should be tolerant to intra-class variation (represented by FRR – False Reject Rate). According to [3], two main categories of this approach are distinguished:

1. Salting: This transformation is invertible to large extent, therefore, the factor (also called an authentication key) has to be kept in secret. Thanks to the key, the cancelability requirement is guaranteed, and it also results in low false accept rate. However, the main drawback is also lied on this key. If the key is compromised, the original biometric template can be revealed. Some examples of salting approach have been proposed in previous works [4–6].
2. Non-invertible transform: This approach applies a noninvertible transformation function to protect the original biometric template. Noninvertible transformation refers to a one-way function which is "easy to compute" but "hard to invert". The key to generate this function can be public. Even if an adversary knows the key and the transformed template, it is computationally hard to recover the original biometric template. For this point, this approach provides better security than the salting. The cancelability is also easily achieved by changing the key which

generates the transformation function. However, the main drawback is the tradeoff between discriminability and non-invertibility of the transformation function. It is difficult to design a transformation function which satisfies both the discriminability and non-invertibility properties as the same time. In addition, each function is usually suitable for each kind of biometric feature. In [7, 8], the authors presented some proposals for generating cancellable fingerprint templates. In [9], the robust hash function was designed to transform face templates...

The biometric cryptosystems were originally developed for securing a cryptographic key using biometric data or directly generating a cryptographic key from biometric data. However, it has also been used in biometric template protection. In this approach, some helper data which is public is stored in the database. While the helper data does not reveal any significant information about the original biometric template, it is needed during matching to extract a cryptographic key from the query biometric data. According to [3], two main categories of this approach are distinguished: key-binding and key generation

1. Key-binding: In this approach, the original biometric template is bound with a key within a cryptographic framework. The result of this combination is stored in the database as helper data. This helper data does not reveal much information about key or the biometric template. However, the huge setback of this approach is the lack of the cancelability property because it is clearly not designed for this. In addition, the matching process depends on the error correction scheme applied. This can possibly lead to a reduction in the matching accuracy. So many techniques belonging to this approach have been proposed, e.g. fuzzy commitment scheme in [10], fuzzy vault scheme in [11] and its enhanced version for fingerprint, face and iris in [12–14], shielding functions in [15], ...
2. Key generation: In this approach, the authentication key is directly generated from biometric template. Therefore, there is no need to be concerned about the key security. However, it usually suffer from low discriminability which can be assessed in term of *key stability* and *key entropy*. Key stability refers to the extent to which the key generated from the biometric data is repeatable, and key entropy relates to the number of possible keys that can be generated. For example, if a scheme generates the same key regardless of the input template, it has high key stability but zero key entropy leading to high false accept rate. Otherwise, if a scheme generates different keys for different templates of the same user, the scheme has high entropy but no stability leading to high false reject rate. Therefore, the limitation of this approach is it is difficult to generate key with high stability and entropy. Moreover, the key generation seems to lose the cancelability property. Some techniques belonging to this approach have been proposed in [16–20] for fingerprint, face, ...

From this view, it can be seen that no one single template protection approach can simultaneously satisfy three both requirements (cancelability, security, and performance). On this account, some recent studies tend to integrate the advantages of both approaches while avoiding their limitations. Hybrid approach is the combination two or more methods to create a single template protection scheme. Very recently, in 2018, the combination of secure sketch and ANN (Artificial Neural Network) was proposed [21].

The ANN with high noisy tolerance capacity can not only enhance the recognition by learning the distinct features, but also assure the revocable and non-invertible properties for the transformed template. In addition, the secure sketch's construction can reduce the false rejection rate significantly due to its error correction ability. The fuzzy vault was combined with periodic function based transformation in [22], or with the non-invertible transformation to conduct a secure online authentication in [23]. The homomorphic cryptosystem was employed in fuzzy commitment scheme to achieve the blind authentication in [24]. Another combination approach was introduced in [25]. In this work, we are going to integrate the random orthonormal project and ideal of fuzzy commitment to guarantee the security for user's biometric template.

2 Background

2.1 Random Orthonormal Projection

Random Orthonormal Projection (ROP) is a technique that utilizes an orthonormal matrix to project a set of points into other space while preserving the distances between points. It was firstly presented as a secure transform for biometric templates in [4], and its remarkable feature is the ability to meet the revocability requirement. When considered as a standalone biometric template protection, it can be classified into a salting approach by the categorization of Jain [3]. As discussed in Sect. 1, the salting approach has low security because the secure template can be inverted if the key is revealed. Therefore, in [26] the authors proposed an additional module for ROP, which is considered as quantization module (Discriminability Preserving), to make it noninvertible. In this work, we try another way to ensure the secure template stored in database cannot be inverted.

The main idea of random projection is to create k orthonormal vectors size of l (l is also the size of the feature vector extracted from an original biometric template and usually $l = k$). To generate these k orthonormal vectors, k pseudo random vectors are created first and then an orthonormalization process like Gram-Schmidt process is employed to transform these random vectors into the orthonormal ones. This process succeeds if and only if the input vectors are linearly independent. However, this condition cannot always be guaranteed when a set of random vectors are generated. In summary, generating an orthonormal matrix of size $k \times k$ from Gram-Schmidt process may be a critical problem when applied on constraint computationally devices like PDA and handheld devices.

Another method to effectively deliver orthonormal matrix was introduced in [27]. The main idea of this new method is based on the fact that a small size orthonormal matrix can be generated without applying Gram-Schmidt process. This matrix is presented as below:

$$I_\theta = \begin{bmatrix} \cos\theta & \sin\theta \\ -\sin\theta & \cos\theta \end{bmatrix} (\forall \theta \in [0, 2\pi])$$

In the same way, we can create an orthonormal matrix A of size $2n \times 2n$ owns a diagonal which is a set of n orthonormal matrices of size 2×2. The other entries of A are zeros. We present the example of matrix A of size $2n \times 2n$ as shown in the formula below where the values $\{\theta_1, \theta_2, \ldots, \theta_n\}$ are the random numbers in the range $[0 : 2\pi]$

$$A = \begin{bmatrix} I_{\theta 1} & 0 & \cdots & 0 \\ 0 & I_{\theta 2} & \cdots & 0 \\ \vdots & \vdots & \ddots & \vdots \\ 0 & 0 & \cdots & I_{\theta n} \end{bmatrix}$$

$$A = \begin{bmatrix} \cos\theta_1 & \sin\theta_1 & \cdots & \cdots & 0 \\ -\sin\theta_1 & \cos\theta_1 & \cdots & \cdots & 0 \\ \vdots & \vdots & \ddots & \ddots & \vdots \\ \vdots & \vdots & \ddots & \ddots & \vdots \\ 0 & 0 & \cdots & \cos\theta_n & \sin\theta_n \\ 0 & 0 & \cdots & -\sin\theta_n & \cos\theta_n \end{bmatrix} \quad (\forall \theta_i \in [0, 2\pi])$$

In result, given the biometric feature vector x of size 2n, orthonormal random matrix A of size $2n \times 2n$, random vector b of size 2n, we have the transformation

$$y = Ax + b$$

Whenever, users have a thought that their secure biometric templates are compromised, these transformed templates can be replaced by the new ones which can be easily constructed by choosing a new set of values $\{\theta_1, \theta_2, \ldots, \theta_n\}$ and vector b.

By using this technique to produce the orthonormal matrix, there is no need for a complex process such as Gram-Schmidt. Beside its effectiveness in computational complexity, it can also improve the security while guaranteeing intra-class variation.

2.2 Fuzzy Commitment

Fuzzy commitment scheme belongs to the key-binding approach in biometric cryptosystem [10]. It is the combination between Error Correcting Codes (ECC) and cryptography. To understand how fuzzy commitment scheme works, we have to learn about ECC. Formally speaking, ECC plays a central role in the fuzzy commitment scheme. ECC checks and corrects the corrupted messages if they contains a certain number of errors which this ECC can afford to check.

In fuzzy commitment scheme, a biometric data is treated as a corrupted codeword. During enrollment phase, a user registers biometric template B_T, randomly chooses a key K (considered as a codeword). Helper Data Extraction module computes the helper data HD of biometric template B_T and key K. During authentication phase, a biometric query B (B and B_T can be quite different because of noise) and the helper data HD are put into the Recover module to extract the key K'. If the difference between B and B_T is smaller than the error correction capability of the ECC employed in this fuzzy commitment scheme, the Recover module can recover exactly the same key. The extracted key K', then, is checked with the enrollment key K to decide whether the biometric query is valid or not. This process is demonstrated in Fig. 1.

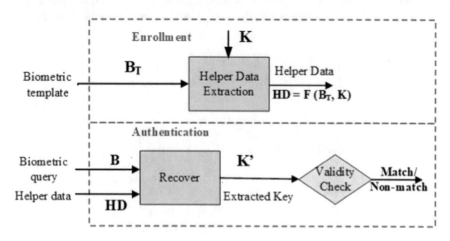

Fig. 1. Fuzzy commitment scheme.

3   Proposal Scheme

The main idea of this hybrid scheme is to secure a non-invertible cryptographic key from cancelable templates. The proposal assumes that the extracted feature of a biometric template can be presented by a vector in continuous domain. Figure 2 depicts the main idea of the proposal scheme.

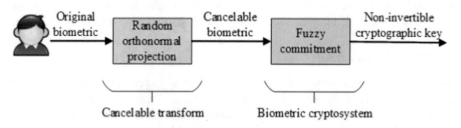

Fig. 2. The main idea of proposal scheme.

3.1   Enrollment Phase

In the enrollment phase, the user creates a random number K_M which is used to generate a random orthonormal matrix M (as described in Sect. 2.1). This K_M is stored on the user's device to regenerate the matrix M for authentication phase. If user wants to change the transformed template, he/she just has to create new value of number K_M. The feature extraction module extracts a feature vector B from the input biometric data. Then, B is combined with matrix M to produce the cancelable version B_{TC}. The user also creates a random key K. The fuzzy commitment scheme is applied with the inputs (cancelable biometric B_{TC} and the random key K). The output of this process is the helper data HD which is stored in the database after that. User also stores the hash version of K into the database for the authentication phase. The enrollment phase is demonstrated in Fig. 3.

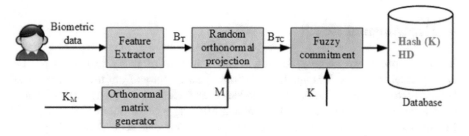

Fig. 3. Enrollment phase.

3.2   Authentication Phase

In this phase, user provides his/her query biometric data to feature extractor module. It results in the feature vector B (note that with the same user, B and B_T cannot be the same because of noise). Meanwhile, user retrieves the K_M in his/her device, and put it into the orthonormal matrix generator module to regenerate the matrix M. Then, B and M are put into the random orthonormal projection module to create another version of B, which is B_C. Since the orthonormal projection provides the discriminability, if B_T and B are extracted from the same person, their corresponding cancelable versions B_{TC} and B_C must be similar. In this case, the fuzzy commitment is absolutely able to recover the same key, which the user provided in enrollment phase, from the cancelable biometric B_C and the helper data HD retrieved from the database. We call the result of this

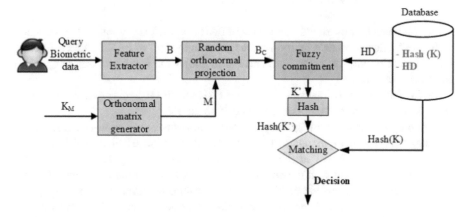

Fig. 4. Authentication phase.

process K'. The hash version of K' is compared with *hash(K)* stored in the database. If they are matched, the user is authenticated; otherwise, the authentication process rejects the access of this user. In this scheme, only *hash(K)* and *HD* are stored. These two parameters cannot recover or invert to the cancelable biometric data or even the original biometric data. Therefore, we can affirm that this combination provides the non-invertible transform. The whole process of authentication phase is illustrated in Fig. 4.

4   Experimental Result

In this proposal, we employ the set of facial images of many people to get the experimental result. PCA is applied to extract feature vectors from user's faces. Each feature vector has size of 200, and the value of each element is in range $[0, 1]$. PCA is trained under the training data set containing 50 images of 10 South East Asians, 22 Middle and West Asians, and 18 Europeans. The testing data set includes 153 people; each has 20 different facial expression. The first image of each user is registered in the enrollment phase, and the others are used to be tested. The accuracy of the biometric authentication system is evaluated through these error rates: FAR, FRR, EER. The meanings of these rates are explained below:

- FAR, also known as False Acceptance Rate, accepts an entrance when a visitor is invalid. This shows probability of the imposter logging in and succeeding.
- FRR, also known as False Reject Rate, rejects an entrance when a visitor is valid. This shows probability of the visitor logging in & getting rejected
- EER, also known as Equal Error Rate, is intersection of FAR & FRR, at which FAR equals FRR.

In any biometric based authentication system, we need to decide the parameter called threshold. If match score is greater than this threshold, the person who sends the request will be authenticated, otherwise, the system will refuse the request. The value of threshold t is calculated by the formula

$$t_i = 0.1 + 0.01 \times i, \quad for\ i \in \mathbb{N},\ and\ i \in [0, 49]$$

For each value of threshold, the values of FAR and FRR are calculated by statistics.

Figure 5 shows the recognition accuracy results in term of FAR and FRR in cases of no biometric template protection, and Fig. 6 shows the result in cases of applying our hybrid scheme.

In the first case illustrated in Fig. 5, the FRR and FAR intersect at the threshold $t \approx 0.22$. At this intersection, the error rate is about 9%.

In the hybrid scheme, integrating random orthonormal projection and fuzzy commitment, which is demonstrated in Fig. 6, the intersection of FRR and FAR (also known as EER) all values at 9%. This figure proves the proposed hybrid scheme delivers a positive result with the probability of correct recognition of around 91% (the threshold value also depends on the quantizing value; however, the result stays the same; in this experimental result, the quantizing value stands at 200). Hence, it is pertinent that the recognition performance of our hybrid scheme is competitive with the

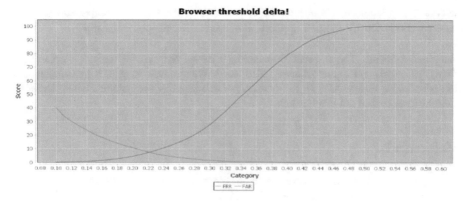

Fig. 5. FAR and FRR with no template protection.

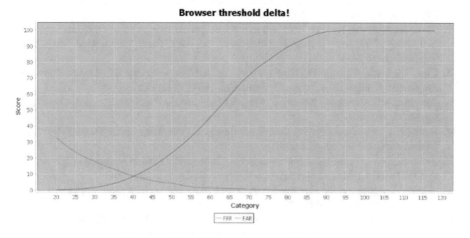

Fig. 6. FAR and FRR with our hybrid scheme.

non-template protection ones. In summary, the combination of orthonormal matrix and fuzzy commitment in biometric-based authentication system is absolutely feasible and can be put into practice.

5 Conclusions

In this paper, we have presented an efficient and innovative hybrid biometric template protection scheme which takes benefits of random orthonormal projection and fuzzy commitment while preventing their limitations. The hybrid scheme satisfies all requirements of: cancelability (revocability and diversity), security, and good performance. By using the lightweight approach for generating orthonormal matrix instead of the traditional Gram-Schmidt in random orthonormal projection, the computational complexity is reduced significantly. The experimental result also demonstrates that the accuracy of recognition is around 91% which stays the same in comparison with the case of none security method applied. That makes the proposal feasible.

Acknowledgments. This research is funded by Vietnam National University - Ho Chi Minh City (VNUHCM) under grant number C2018-20-13. We also want to show a great appreciation to each member of DSTAR Lab for their enthusiastic supports and helpful advices during the time we have carried out this research.

References

1. Rathgeb, C., Uhl, A.: A survey on biometric cryptosystems and cancelable biometrics. EURASIP J. Inf. Secur. **2011**(1), 1–25 (2011)
2. Maneesh, U., et al.: Blind authentication: a secure crypto-biometric verification protocol. Trans. Inf. Forensics Secur. IEEE **5**(2), 255–268 (2010)
3. Jain, A.K., Nandakumar, K., Nagar, A.: Biometric template security. EURASIP J. Adv. Signal Process. **2008**, 1–17 (2008)
4. Teoh, A., Goh, A., Ngo, D.: Random multispace quantization as an analytic mechanism for biohashing of biometric and random identity inputs. IEEE Trans. Pattern Anal. Mach. Intell. **28**, 1892–1901 (2006)
5. Chin, C., Teoh, A., Ngo, D.: High security Iris verification system based on random secret integration. Comput. Vis. Image Underst. **102**, 169–177 (2006)
6. Savvides, M., Kumar, B., Khosla, P.: Cancelable biometric filters for face recognition, vol.3, pp. 922–925 (2004)
7. Ratha, N.K., et al.: Generating cancelable fingerprint templates. IEEE Trans. Pattern Anal. Mach. Intell. **29**(4), 561–572 (2007)
8. Teoh, A., Ngo, D., Goh, A.: Biohashing: two factor authentication featuring fingerprint data and tokenised random number. Pattern Recogn. **37**, 2245–2255 (2004)
9. Sutcu, Y., Sencar, H.T, Memon, N.: A secure biometric authentication scheme based on robust hashing, pp. 111–116 (2005)
10. Juels, A. Wattenberg, M.: A fuzzy commitment scheme. In: Proceedings of the 6th ACM Conference on Computer and Communications Security, pp. 28–36. ACM, Kent Ridge Digital Labs, Singapore (1999)
11. Juels, A., Sudan, M.: A fuzzy vault scheme. In: Proceedings IEEE International Symposium on Information Theory (2002)
12. Lee, Y.J., et al.: Biometric Key Binding: Fuzzy Vault Based on Iris Images, pp. 800–808 (2007)
13. Feng, Y.C., Yuen, P.C.: Protecting face biometric data on smartcard with reed-solomon code, p. 29 (2006)
14. Uludag, U., Jain, A.: Securing fingerprint template: fuzzy vault with helper data (2006)
15. Tuyls, P., et al.: Practical Biometric Authentication with Template Protection, vol. 3546, pp. 436–446 (2005)
16. Dodis, Y., et al.: Fuzzy extractors: how to generate strong keys from biometrics and other noisy data. SIAM J. Comput. **38**(1), 97–139 (2008)
17. Li, Q., Chang, E.-C.: Robust, short and sensitive authentication tags using secure sketch, vol. 2006, pp. 56–61 (2006)
18. Sutcu, Y., Li, Q., Memon, N.: Protecting biometric templates with sketch: theory and practice, vol. 2, pp. 503–512 (2007)

19. Yang, W., Hu, J., Wang, S.: A delaunay triangle-based fuzzy extractor for fingerprint authentication, pp. 66–70 (2012)
20. Sutcu, Y., Li, Q., Memon, N.: Design and analysis of fuzzy extractors for faces, vol. 7306 (2009)
21. Dang, T.K., Huynh, V.Q.P., Truong, Q.H.: A hybrid template protection approach using secure sketch and ann for strong biometric key generation with revocability guarantee. Int. Arab J. Inf. Technol. (IAJIT) 15(2), 331–340 (2018)
22. Dang, T.K., et al.: A Combination of Fuzzy Vault and Periodic Transformation for Cancelable Biometric Template. IET biometrics, The Institution of Engineering and Technology, United Kingdom, vol. 5, pp. 229–235 (2016)
23. Lifang, W. Songlong, Y.: A face based fuzzy vault scheme for secure online authentication, In: Second International Symposium on Data, Privacy and E-Commerce (ISDPE), pp. 45–49 (2010)
24. Failla, P., Sutcu, Y., Barni, M.: eSketch: a privacy-preserving fuzzy commitment scheme for authentication using encrypted biometrics. In: Proceedings of the 12th ACM Workshop on Multimedia and Security, pp. 241–246. ACM, Roma (2010)
25. Iovane, G., et al.: An encryption approach using information fusion techniques involving prime numbers and face biometrics. IEEE Trans. Sustain. Comput. 1 (2018)
26. Feng, Y.C., Yuen, P.C., Jain, A.K.: A hybrid approach for generating secure and discriminating face template, vol. 5, pp. 103–117 (2010)
27. Hisham, A.-A., Harin, S., Sabah, J.: A lightweight approach for biometric template protection. In: Proceedings of SPIE (2009)

The Impact of University Students' Smartphone Use and Academic Performance in Bangladesh: A Quantitative Study

Masiath Mubassira and Amit Kumar Das[✉]

East West University, Dhaka, Bangladesh
masiathmubassira@gmail.com, amit.csedu@gmail.com

Abstract. Smartphone is the most desirable gadget among the young generation for a decade. Greater availability of Wifi connection and mobile data made it easier to access the Internet on smartphones. Students' performance may deteriorate due to excessive usage of smartphones. It is a concerned issue because proper education is vital for a promising career. However, the effects of smartphones on education and study habit are not well developed in HCI. This research paper intended to answer whether overuse of smartphones is hampering a student's performance and CGPA. An app is being installed in the smartphones of 237 students from some universities in Dhaka, Bangladesh. This app recorded their activities on their smartphones for two months. This paper also conducted a two-month-long online survey (n = 244) and also took interview of some students. This paper suggested a prototype for minimizing overuse of smartphones and increasing productivity among students.

Keywords: Smartphone · Impact · CGPA · University · Survey · Bangladesh

1 Introduction

The tremendous growth in availability of global smartphone access is initiating people from previously deprived of digital technology to gain the benefits given by the Internet and smart technology. Smartphones with its extensive computing capabilities and high-speed access to the Internet have reached its zenith throughout the globe. Smartphone industry is now one of the most potent industries in the world, and with that, the competition between various brands are getting more extreme than ever [1]. Today smartphones satisfy their users by fulfilling their needs for a communications device, a microphone, a handheld audio recorder, a digital camera, weather forecasting, a web browser, a satellite navigation system, a compass, a note creator, etc. Smartphones can also assist in gaining knowledge via any web browser or any other applications.

© Springer Nature Switzerland AG 2019
S. Lee et al. (Eds.): IMCOM 2019, AISC 935, pp. 734–748, 2019.
https://doi.org/10.1007/978-3-030-19063-7_59

Physical things such as books, lecture notes were the primary medium of study for the students in the last century. However, after the evolution of the Internet and computing technology, students are gaining knowledge via electronic devices. The introduction to the tablet technology created a shift in the way students learn: the devices provide media-rich, interactive and enthusiastic new environments [2].

However, it has also been used widely by youngsters as a source of global communication via many social networking applications such as Facebook, Instagram, Snapchat, Twitter, Viber, Whatsapp, etc. Ownership of smartphones has become a nearly omnipresent element of teen life: 95% of teenagers reported that they have a smartphone or access to one. 45% of teens said that they are online on a near-constant basis [3]. Students adore spending their valuable time clicking photos and uploading these on the social application. They also love to chat, make video calls, update status or upload stories on those networking apps. Subconsciously they are spending lots of time on the virtual world.

The impact of computing technology and smartphones on the overall academic performance of University students are not well known. There has been very little research on this sector of HCI. *So our first research question is to examine the relationship between the University students' academic performance and usage of smartphones.*

The intensity of the impact of smartphones and mobile applications usage behavior of juniors and seniors of universities may not be identical. Thus, there is a gap in knowledge on whether smartphones are creating a more significant impact on one group than the other. *So our second research question is to find and compare the average usage behavior of smartphones for both juniors and seniors and also to find its impact with respect to CGPA.*

This paper did a two-month-long online survey on students from East West University, University of Dhaka, United International University, North South University, and American International University Bangladesh in Dhaka, Bangladesh. This paper also took interview of 40 students from East West University. It also took real-time data from the app which was installed on the students' smartphones for two months. It used Cumulative Grade Points Average (CGPA) as the standard scale for measuring academic performance.

The next section did some background study on this topic. Later, it discussed the methods used to find the correlation between usage of smartphones and study behavior. Later on, it provided the comments from the interview and also the results and analysis of the survey data and the app data. Finally, it suggested a prototype for an interactive app for minimizing extensive usage of smartphones and increasing time spent on education.

2   Background Study

The Internet brought a revolutionary change and helped the students to enhance their academic knowledge. However, it is reported that Internet hindered academic performance. This work tried to find the association between academic performance and Internet usage of some students studying in a public University of Malaysia. It conducted a cross-sectional study of 186 fourth-year medical students. It recorded either 'yes' or 'no' responses via the Internet Addiction Diagnostic Questionnaire Instrument which contained eight questions. Students who answered 'yes' to 5 or more questions were declared dependent Internet users. There was no significant difference between CGPA of dependent and non-dependent Internet users [4].

This research paper investigated the impact of usage of the Internet on the academic performance of some students of tertiary institutions in Nigeria. A survey research design was used to collect data by using a questionnaire. It reported that the survey participants used the Internet from the Cyber Cafe. It discovered that the Internet is helping students to gain academic knowledge and it is also helping them to get better prepared for their examinations. It also suggested that lecturers should give more assignments for increasing the students' habit of using the Internet [5].

In this paper, the author claimed that WeChat is one of the most popular social networking apps in China and Hong Kong. This research paper mainly investigated the relationship between academic performance and learning interests of students who used WeChat and who did not. Only Accounting students were considered for this study. The students were continuously assessed for three months, and after that, the students were given a questionnaire for giving their comments and feedback on the WeChat learning scheme. The study found that students who used WeChat performed better than the students who did not use WeChat learning scheme [6].

This paper argued that Facebook was responsible for students' widespread attention from the study and thus it created a negative impact on academic performance. However, it claimed that the previous studies were mainly done with survey responses. So the study was insufficient, and so it used automatic logging and experience sampling to investigate the context of Facebook use. It used a data collection tool named ROSE which stands for Research Tool for Online Social Environments.

It collected user activities, usage statistics and conducted experience sampling (ES) surveys. Two versions of ROSE, a Chrome version, and a Safari version were used to collect data. It concluded that though past studies suggested a negative impact on academic performance, the difference between frequent and non-frequent users was nuanced. It also said that low GPA students spent more time in each Facebook use [7].

This paper described that people use networking technology to seek information as well as to share information. Learning in the context of social media has become an integral part of the experience in college. It suggested that how Personal Learning Environment (PLE) can assist both formal and informal learning and boost self-regulated learning in higher education. It described a pedagogical framework which is used by college students to demonstrate how to use social media to create PLEs and to support self-regulated learning. This paper concluded that not all students have the knowledge to customize social media and create PLEs and have self-regulated learning as they desired. So this paper suggested the pedagogical framework for enhancing complex personal knowledge management skills which would help them to create and manage PLEs [8].

3 Method

This paper conducted a two-month-long study in Dhaka, Bangladesh to analyze the students' usage of smartphones and study behavior. We conducted semi-structured interviews with the students of various departments in East West University. The students willingly participated in the study. In total, we performed 40 interviews with the students. The interviews were all voluntary, and the duration of each interview was approximately 2 min long. The interviews were all recorded by using an audio recorder named voice recorder app after taking permission from the students. The interview was taken in Bengali, and it asked questions about the time spent on various apps and also on the study, the most used apps and their opinions about whether they think smartphones are creating an impact on their study behavior and CGPA.

In addition to the semi-structured interview, we also conducted an online survey. The survey form was created by using the Google form. The survey form was written in both English and Bengali. It took some quantitative as well as qualitative data. It asked questions and demanded opinions just like the way the interview was taken. Although most of the questions had checkboxes and multiple choice questions, some questions were open-end questions which asked opinions from the participants. In total there were 11 questions on this survey form. The survey form was publicized via Facebook and Messenger. The survey form was shared with hundreds of students. However, only 244 students willingly responded to the online survey and face-to-face survey. The online survey form was open from 18th July 2018.

An app was also installed in the smartphones of some students from University of Dhaka, Bangladesh University of Engineering and Technology, Jagannath University, Rajshahi University, United International University, Daffodil University, East West University and University of Asia Pacific. The app collected data for two months. This app monitored and recorded the names of all the apps used by the students and also the length of time for which those apps were used. So this procedure gave a thorough and accurate study of how the students were using their smartphones.

4 Findings from the Methods

For an exhaustive study, we collected data which were analyzed by three methods as mentioned previously: (a) interview, (b) online survey, and (c) an app collecting real-time data which is being installed in the smartphones. This paper discussed and analyzed all these three methods for finding the correlation between the performance of students and the usage of smartphones.

4.1 Interview Results

In total, 40 participants were willingly interviewed by the first author. The interview was around 2 min long. All the interviewees are studying at East West University, Dhaka, Bangladesh. The participants belong to many departments such as Bachelors of Business Administration (BBA), Pharmacy, Economics, English, Electronics and Communication Engineering (ECE), Electronics and Telecommunications Engineering (ETE), Law and Information Studies and Library Management (LISM).

We got to know about their opinions, and they expressed their thoughts and ideas, and also shared their using habits of smartphones with us. The smartphone is not only affecting study time. As a participant said:

"I only study around 1.5 h daily, but I love to chat in Whatsapp the whole day. I use smartphones even when I am studying. My phone is always connected to the Wifi in my home. So whenever any notifications come on my phone, I lose concentration from my study. Indeed, smartphones are having a negative impact on me. I even sleep late at night because I use social networking apps for chatting."- (Farzana Alam, 6th Semester, ECE Department)

Another participant also discussed the problems she is facing due to the overuse of smartphones. The student told us:

"Since I study in the English Department, I have to access the Internet a lot to gain knowledge. However, I also use Facebook for an unlimited amount of time every day. After finishing my classes at my university, I go home and do chatting and messaging from 9 pm to 4 am. So it is hampering my studies, and also it is harmful to my eyesight." - (Fahria Parvez, 9th semester, English Department)

She also concluded that she is getting addicted to smartphones. As she also said:

"Whenever any notifications pop out on my smartphone during my study time, I think that I will only check the notification for 1 min and then come back to study. However, that does not happen as I keep using it for 1 h. So I keep losing my attention from my study."

Another participant also shared his addiction to smartphones with us. The student discussed:

"I spent a lot of time on chatting and messaging. I always need to keep my phone in my hand. Otherwise, I feel very restless. I believe if I used smartphone less, my overall CGPA would have been better." - (Tushar Khan, 8th semester, BBA Department)

Another student from the English Department expressed her weakness towards her phone by saying:

"I know it is affecting my study, but I cannot get rid of my smartphone. I try to keep my phone away whenever I have an exam the day after. However, I cannot

stay away from my cell phone for long. It is creating a bad effect on me." - *(Fazilatun Nesha Tanvi, 5th semester, English Department)*

A few participants even said they do not think smartphones are having a negative impact on their study. Rather, it is helping them in various ways. A student who is also a businessman explained:

"Since I do export and import business with China, I use WeChat app for communication because they cannot use any other apps such as Facebook. I use my phone during my study to find the meaning of an unknown word. Otherwise, I do not use my phone since my mother taught me not to use the phone unnecessarily during my study time. So it is my habit not to use phone unnecessarily while I sit to study." - (Zubair Islam, 7th semester, BBA Department)

We asked the participants about any solution on how to lower the excess usage of smartphones. One suggested that:

"After the establishment of Wifi, the Internet has become so cheap and affordable that we are way too much dependent on smartphones. Rather than wasting time on Instagram and Facebook, we should give time to ourselves. Even the children can also gain access to these social networking sites, and the government of Bangladesh should take care of this." - (Shanta Zerin, 8th semester, BBA Department)

Another concerned participant also shared her views with us saying:

"We should positively use smartphones. We should use these for a limited amount of time. We should distribute our work in time slots such as one slot is for study and another slot is for using smartphones and social networking apps." - (Momo Tayeba, 3rd semester, English Department)

4.2 Survey Results

In total 244 students from several public and private universities willingly participated in the online survey and face-to-face survey. Students from East West University, North South University, Bangladesh University of Professionals, Khulna University, Bangladesh University of Engineering and Technology, Institute of Business Administration, American International University Bangladesh, etc. completed the online survey. We collected 193 survey data via an online survey, and 51 survey data was collected using hard copies of the survey form. We only considered average usage per day of Facebook and apps for chatting or messaging for the survey. We considered first-year and second-year students as juniors and third-year and fourth-year students as seniors for the second research question.

(A) The following analysis is based on research question 1:

From Fig. 1 we can see that the kind of app which is used mostly by the students is social networking apps. The second most used app is the Google apps, and the least used app is the Microsoft apps. Thus, we can conclude that the Bangladeshi students mainly use smartphones for social networking and also for fun and enjoyment purpose.

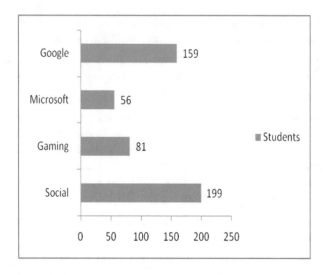

Fig. 1. Apps used by the University students

We asked the students whether the usage of smartphones is creating a negative impact on the study. From Fig. 2, we can see that 38.40% of students agreed, and they believe it is creating a negative impact. However, 32.30% students said no and the rest, that is, 29.30% of students were unsure about it. So we can see that the majority believed that the usage of mobile applications hinders academic performance.

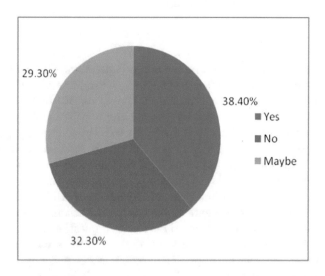

Fig. 2. Percentages showing students' opinions on the negative impact on the study

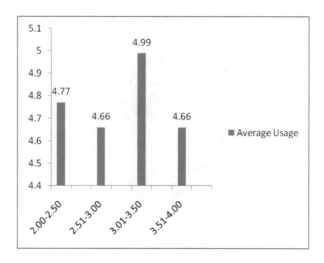

Fig. 3. Average usage per day of smartphones and the students' corresponding ranges of CGPA

From Fig. 3 we can see that 3.01–3.50 CGPA has the highest average usage of smartphones. 2.00–2.50 CGPA has the second highest average usage of smartphones. It is higher than that of ranges 2.51–3.00 and 3.51–4.00 CGPA. So we can say that the lowest range of CGPA 2.00–2.50 occurred when the average usage of smartphones per day is higher. However, 3.01–3.50 CGPA is showing an irregular pattern compared to the rest of the ranges of CGPA.

(B) The following analysis is based on research question 2:

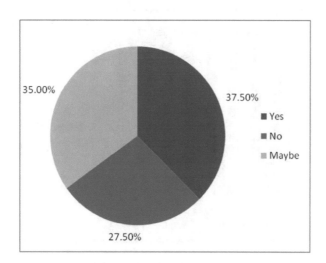

Fig. 4. Percentages showing juniors' opinion on the negative impact on the study

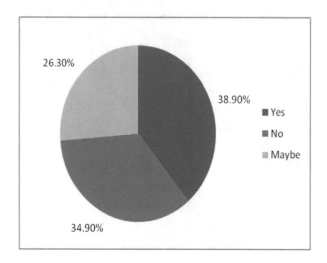

Fig. 5. Percentages showing seniors' opinion on the negative impact on the study

From Fig. 4, we can see that the majority of juniors agreed that mobile applications hinder their study. However, more juniors are unsure about the impact than that of juniors who believe it is not creating any bad impact on their study.

From Fig. 5, we can visualize that maximum number of seniors, that is, 38.90% believe smartphones scatter their attention towards their study. 34.90% said they do not believe it is right and the rest are undecided. We can conclude that a greater percentage of seniors believe that smartphones are not creating

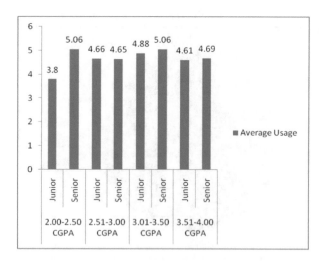

Fig. 6. Average usage (hours) per day of smartphones for juniors and seniors and their corresponding ranges of CGPA.

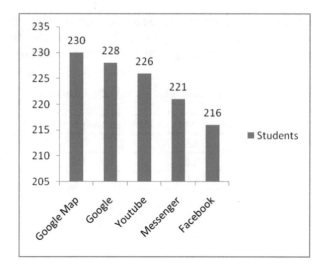

Fig. 7. Top 5 most used mobile applications

a bad impact compared to that of juniors. Also more juniors are unsure than seniors about the negative impact on study.

From Fig. 6, we can see that for ranges 2.00–2.50, 3.01–3.50, and 3.51–4.00 CGPA the seniors' average usage per day of smartphones is higher than that of juniors. So we can conclude that seniors and juniors have the same range of CGPA although seniors are using smartphones more than the juniors. For range 2.51–3.00 CGPA the average usage per day for both juniors and seniors

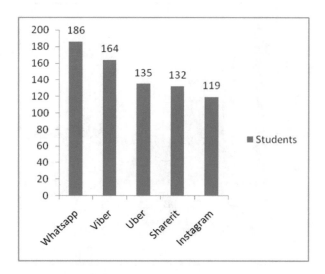

Fig. 8. The next top 5 used mobile applications.

is approximately the same. Thus we can conclude that smartphones are having a more significant negative impact on juniors than seniors because seniors can maintain the same CGPA range by spending more time on smartphones.

Table 1. Average usage (in hours) of smartphones

Total number of students	Average usage (hours) 6 p.m. to 12 a.m.
132	4.69
42	3.82
36	2.67
19	1.38
8	0.76

4.3   App-Based Results

An app was installed in some participants' smartphones which monitored them for two months. In total, 237 students of different public and private universities participated in this study. Their age range was from 18 to 25 years. We only considered Android users in this study. The app collected real-time data by recording the total number of minutes spent by the participants in all mobile applications. It also kept a record of all the mobile applications used by the participants.

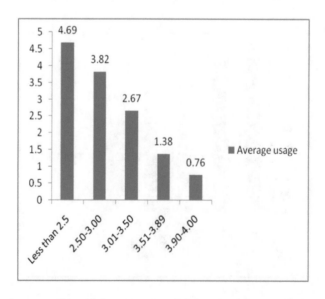

Fig. 9. Average usage (hours) from 6 p.m. to 12 a.m. and corresponding CGPA range

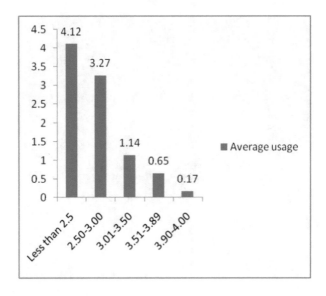

Fig. 10. Average usage (hours) from 12 a.m. to 8 a.m. and corresponding CGPA range

(A) The following analysis is based on research question 1:

From Fig. 7, we can see the top 5 most used apps by the students. Google Map is the most popular of all. So we can see that it is quite a different result than that of the online survey. From online survey results, we saw that social networking apps were the most popular and the second most popular were Google apps. So we can conclude that the students' views and their actual usage of mobile

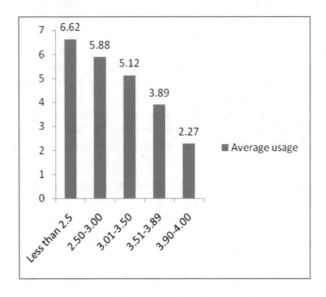

Fig. 11. Average usage from 8 a.m. to 6 p.m. and corresponding CGPA range

Fig. 12. Home user interface

applications differ slightly. In the top 5, there are only two social networking apps present which is Facebook and Messenger. The top 3 apps belong to Google apps. From Fig. 8, we can see the next five apps used by the students. Among the five apps, 3 are social networking apps Whatsapp, Viber, and Instagram. Among the top 10 apps mostly used, five apps were social networking apps. Thus, we can conclude that social networking apps are very much popular among the Bangladeshi students. From Table 1 we can see that majority of students, that is 132, has the highest average usage per day of mobile applications from 6 p.m. to 12 a.m. Evening time is the peak time for studying in Bangladesh perspective. However, a maximum number of students uses mobile applications in this time range. From Fig. 9, we can observe that there is a negative correlation between CGPA ranges and average usage of smartphones. As CGPA increases, average usage of smartphones decreases. This is the peak time for studying at home. Thus students who use phones for a minimal time have higher CGPA than the rest. From Fig. 10, we can observe that there is also a negative correlation between

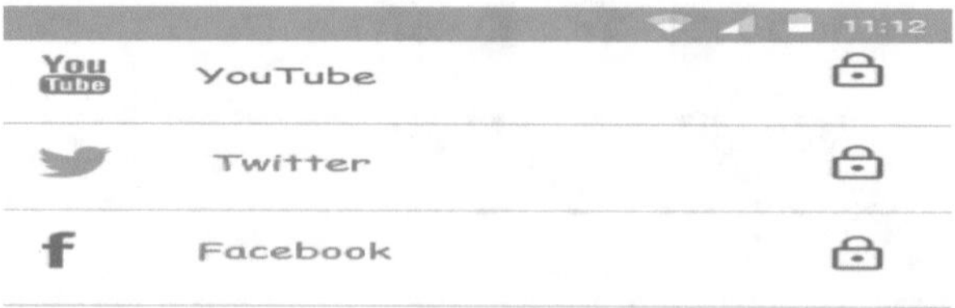

Fig. 13. App locking user interface

CGPA ranges and average usage of smartphones. As CGPA is ascending, the average usage of smartphones is descending. This is the peak time for sleeping. Thus students who are using smartphones instead of sleeping are not getting enough rest. As a result, it is hampering academic performance.

From Fig. 11, we can see that there is also a negative correlation between ranges of CGPA and average usage of mobile applications. As CGPA is increasing, the average usage of smartphones is decreasing. University mainly has classes in this period. Since students who are spending more time on phones than on attending lectures with rapt attention are having lower CGPA than the rest of the students.

5   Prototype

We designed the user interface by using proto.io tool. We tried to suggest a prototype which may reduce excessive usage of mobile applications and improve academic performance. Excessive usage can be reduced if the apps which are mostly used are locked after a certain length of time.

Figure 12 has an option named 'All Apps' which will show the user all the available apps in the mobile. It also has an option named 'View Statistics' by which the user can view his previous usage statistics.

Figure 13 will show all the apps available in the smartphone and has the option to lock any of the apps individually. This will greatly reduce overuse of smartphones since students may lock the apps which distract them the most during study hours.

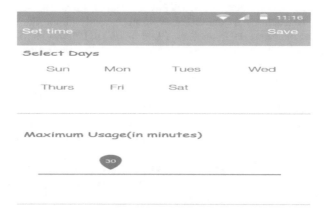

Fig. 14. Maximum usage and day's selection user interface

Figure 14 will give the opportunity to select any of the days from Sunday to Saturday and also to give input of maximum usage for each day. After it crossed the maximum usage limit, it will lock the app automatically. Thus, students cannot use more than the time set per day and thus spend more productive time on tasks which will assist them to develop their knowledge and skills.

6   Conclusion

This paper aimed at finding any correlation between smartphone usage and academic performance of university students in Bangladesh. It successfully found a negative correlation between CGPA and usage of mobile applications by using the data collected via the installed app in the students' smartphones. As the intensity of smartphones' usage increases, CGPA of students starts decreasing. The app thoroughly monitored and provided data for three-time slots. The survey results also suggested that the students believed that smartphones are creating a negative impact on their study. The survey results also suggested that senior students can maintain the same CGPA range as the juniors by using smartphones more than the juniors. We also found that Google and social networking apps are the most used apps by the students in Bangladesh. However, due to lack of time and also due to the lack of students' willingness to complete the survey form, we could not get around 300 survey data. Also, the survey data showed some anomalies which may be due to the input of some wrong data into the online survey form. Nonetheless, we got very accurate real-time data via the installed app in the students' smartphones and were able to examine the relationship accurately.

References

1. Anh, H.N.: Smartphone industry: the new era of competition and strategy. Bachelor thesis (2016)
2. Montrieux, H., Vanderlinde, R., Schellens, T., Marez, L.D.: Teaching and learning with mobile technology: a qualitative explorative study about the introduction of tablet devices in secondary education. PLoS ONE **10**(12), e0144008 (2015)
3. http://www.pewinternet.org/2018/05/31/teens-social-media-technology-2018/. Accessed 12 Aug 2018
4. Siraj, H.H., Salam, A., Hasan, N.A.B., Jin, T.H., Roslan, R.B., Othman, M.N.B.: Internet usage and academic performance: a study in a Malaysian public university. Int. Med. J. **22**(2), 83–86 (2015)
5. Ivwighreghweta, O., Igere, M.A.: Impact of the internet on academic performance of students in tertiary institutions in Nigeria. J. Inf. Knowl. Manage. **5**(2), 47–56 (2014)
6. Ng, K.K., Luk, C.H., Lam, W.M.: The impact of social mobile application on students' learning interest and academic performance in Hong Kong's sub-degree education. In: 2016 International Symposium on Educational Technology (ISET), Beijing, pp. 18–22 (2016)
7. Wang, Y., Mark, G.: The context of college students' Facebook use and academic performance: an empirical study. In: CHI 2018 Proceedings of the 2018 CHI Conference on Human Factors in Computing Systems, Montreal QC, Canada (2018)
8. Wang, Y., Mark, G.: Personal learning environments, social media, and self-regulated learning: a natural formula for connecting formal and informal learning. Internet High. Educ. **15**(1) (2012)

Pregnant Women's Condition and Awareness About Mood Swings: A Survey Study in Bangladesh

Nusrat Jahan, Umme Salma Fariha, Musfika Rahman Ananna,
and Amit Kumar Das[⊠]

East West University, Dhaka, Bangladesh
jahannira@gmail.com, salmafariha13@gmail.com,
musfika.ananna2015@gmail.com, amit.csedu@gmail.com

Abstract. Mood swing is a widespread problem during pregnancy. However, in the developing country like Bangladesh where 38% populations are women and their family never seek for any help during pregnancy, even they do not become aware of this complication which can be harmful to fetal. They cannot imagine it may cause miscarriage, and even a mother is suffering from a severe mood swing can deliver an abnormal baby. This paper presents findings from a three-month-long ethnography and online survey conducted in Bangladesh where the total number of the participant was 207. Our analysis surfaces necessary care and cure of mood swing problem for pregnant women. We also proposed an application using which a woman can get initial suggestion and also can enjoy an environment she prefers to make her happy for instance.

Keywords: Pregnancy · Mood swing · Bangladesh · Women · Fetal

1 Introduction

Pregnancy is a precious feeling for every woman. Every woman wishes to give birth to her child. The average pregnancy lasts around 37 weeks to 40 weeks which is full of physical and emotional changes. This mood swings percentage increase massively during pregnancy because of the changes in metabolism, stress, tension, fatigue and hormones named 'Estrogen' and 'Progesterone'. However, mostly during the first trimester between 6–10 weeks and then again in the third trimester, women experience this problem. The changes in hormone levels affect the level of neurotransmitters, where the neurotransmitter is the brain chemical which can regulate mood [1]. Moreover, 14%–23% of women are experiencing this problem in the whole world [13].

Psychologist defines 'mood swing during pregnancy' as 'a sharp and inexplicable change in mood during pregnancy for hormonal change' or, 'extreme or rapid change in mood. Such a mood swing can play a positive part in promoting problem-solving and in producing flexible planning. However, when a mood

© Springer Nature Switzerland AG 2019
S. Lee et al. (Eds.): IMCOM 2019, AISC 935, pp. 749–760, 2019.
https://doi.org/10.1007/978-3-030-19063-7_60

swing is also so strong that they are disruptive, they may be the main part of a bipolar disorder'. This is so normal, but sometimes in extreme case, it is treated as one of the risk factors which can causes miscarriage or, maternal death [6].

Women in developing countries like Bangladesh where women of the house treated as a sad part of the family and society may face more depression, stress and mental difficulties. Here, women may never realize, they are going through a 'mood swing' for their pregnancy. Women may never feel any need to share this issue with friends and family or, doctor. They do not know it can be harmful to their child in their womb if it reaches in extreme mood. Especially changes about appetite, changes in sleep, lack of energy, feeling sad, hopeless or, worthless, frequently weeping with no reasons, loss of interest in doing regular work was the common problem [10]. The women who live with their in-laws may don't have enough food or preferable food if she wants so, may don't have enough sleep, medical facilities though she needs it, she may not have an opportunity to take rest, or, may have to hear or bear something very unpleasant during her pregnancy. According to doctors, all these reasons may lead to some serious problem like miscarriage, preterm, low birth weight [3].

We want to provide them an environment which may release them from the difficulty of their life and can enjoy something that they wish to, will help them to give birth to a healthy baby and to avoid those risk factors. So, for doing something new, we might gather more knowledge about it. So, we talk to them and try to find out what they are going through during their pregnancy and what they need to stay happy, relax, refresh all the time. It is not possible to change the mentality or attitude of the people surrounding those women, but we may do something for them in which they can enjoy some moment as they want to spend.

2   Related Work

In this decade, some research took place on the mental health of pregnant women where it is concluded as financial concerns. These were strongly and independently associated with the mental health, and it is a factor such as working status, education, and family structure. From some of them, we inspired to do something new [7]. The diversity between and within ethnic groups need to take into consideration individual social, migration and economic circumstances [4]. Our work builds upon this concept of the researches.

Existing many types of research analyze the reasons behind poor mental health during pregnancy, when, why and what kind of psychological change happens and what could be the solution. Some of them directly talk to the patient, some of them take interview of psychologist, gynecologist to do so [13]. In those extreme case, the contextual factors can be the cultural and religious factors, physical and psychological readiness, time pressure, economic factors, family history, social support, local service provision, fear like fear of becoming ill, fear of medication causing harm to the baby, fear of being a lousy parent [10]. However, as a developing country, Bangladesh has entirely a very different environment than the western world. Here the relationship between the pregnant women and other family member is altogether different. In the rural areas, women even don't have the right to share their every feeling and also believe in prejudice in this era. So, that women even don't know what is 'mood swing,' do women can have a mood! In other words, they may face almost all the contextual factors.

Some study took place in the last some years, which shows that how women valued apps. The researchers reveal the importance of using digital information for establishing and maintaining social connections and intimate relationships with other mothers [12]. With the increasing development of Bangladesh here, the people are getting more conscious about the risk factors of pregnancy as a result now 30% of women taking treatment due to different pregnancy-related difficulties. Women of Bangladesh are getting more familiar to the internet world. In this decade, the internet is now 300,000 monthly 'pregnancy' searches on Google BD and 10,000,000(est.) female internet users in Bangladesh. Women are 62% more active on Facebook than before, they are joining different types of pregnancy-related groups or community and sharing their problems [5]. They are also using android apps related to the risk factors of pregnancy. 'Aponjon Pregnancy-Shogorbha' is a free health and fitness categorized application which was launched by the Government of Bangladesh where the estimated number of the app downloads range between 10000 and 50000 as per Google play store.

Regardless of how important the care of a pregnant woman who is passing through mood swing and what should the family member do for that, should be known by all. Our application is going to help the women who have no opportunity to have family support during this hard time by giving them a touch of their excellent environment for some instance which make her happy and hopefully will reduce the percentage of mentally abnormal childbirth.

3   Survey and Analysis

We conducted a three-month ethnographic study in some places by visiting some hospitals and houses from Dhaka, Cumilla, Lakshmipur, Narayanganj, and Munshiganj of Bangladesh to study the psychological states of the pregnant women during different phases of pregnancy. In the semi-structured interviews, the number of the participants were 186 who willingly participated in the study. We tried to cover women from different socioeconomic classes (low income is below 10,000 Taka per month, high income is above 30,000 Taka per month, middle income is between 10,000 to 30,000 Taka per month), and also who are educated or, uneducated and lives in rural, town and city areas and most importantly we covered women aged 17 to 45 years. We asked them some questions which can reveal their mental condition during that time like did they ever experience any miscarriage, how their family members were supporting them, any problem related to food habit, sleeping routine, financial problems, the baby in their womb was expected or not, with who, where and how they like to pass their days.

We also conducted an online survey where the participants were the members of different Facebook groups related to pregnancy, and it was an anonymous online survey which was seven days long. That survey form was in two formats Bengali and English. We asked the same questions to them too. Moreover, it was about 21 women who shared their data with us.

3.1   Urban (City and Town) Women

Increased motorization, mechanization related activities lack of civic amenities like park and other recreational facilities, worsened air quality makes the urban women physically unfit. In another case, if they conceive at the early of their working carrier, then they feel insecure about their future, and if they suffer from extreme depression, in some case they may harm themselves or their baby too. We met with one woman in Holy Family Red Crescent Hospital who was admitted there for menstrual problems. She was having the problem of several time bleeding. So, her doctor suggested her to get admitted in the hospital so that she can observe her on a regular basis. The city women who are highly educated with substantial financial support, mostly working women, even some of them are an entrepreneur. She was saying

– *"I am troubling from bleeding problem frequently from the beginning of my pregnancy, and my doctor suggested me to inject a medicine named 'HPC' one per week until my delivery and she may suggest me to stop taking this one any time if I get well but before this I will do this as I'm already experiencing a miscarriage".* (27 years old, BSc completed city woman, Dhaka).

Another was saying,

– *"I got married too late for completing my study and now I have experienced five miscarriage due to chromosomal complications, and this is my sixth time I become pregnant, and I do not know am I ever be able to be a mom or, not".* (40 years old woman, Dhaka).

Moreover, in case of middle class educated women some of them get married before completing their higher studies and conceive too early are also become depressed in the worry about their future, and stay upset by seeing their dreams are broken. Moreover, others who want to support their husband financially by doing job or business may have to leave their job in most of the cases due to their upcoming child's health and care, and in some case, they are in fear in that after maternity leave the company may not receive her again. Such as woman we met in another hospital who was saying

– *"I do not know, will it ever be possible for me to complete my study or not?"* (26 years old woman, Dhaka).

She became stressed and crabbed after she became pregnant and behaved very rude to her family members and the relations were being worse day by day. Moreover, some of them who lives with their in-laws or already having children or as a working woman, may not get an opportunity to think about their own or upcoming baby. One of this kind of woman was saying,

– *"Sometimes I daunt and feel lonely too."* (22 years old woman, Dhaka).

One was saying,

– *"Due to the shortage of time it is not possible for me to know or think something new or needed, as I have to handle all my housework in a single hand."* (31 years old, housewife, Dhaka).

In case of the lower class women of Dhaka city where a maximum of them lives in slums and are uneducated or less educated, have to handle all the house chores, manage expenditure in a single hand or, as a partner of her husband by doing work on garments factory or as a maid. When we asked her about sickness, then she said,

– *"I feel back pain, headache and feel irritated most of the time."* (27 years old, maid, Dhaka).

3.2 Rural Women

The village women who are treated as the most neglected person in their family may never think about themselves and may never get any extra care during their pregnancy mostly. Though the scenario is changing day by day as people are getting connected to the technological world and getting educated more than before, so they have an opportunity to know the right things now. The mass media plays a vital role, and the government of Bangladesh has taken many steps to increases awareness about pregnancy care which is being very active. Nowadays as using the Android phone is being more available and they are getting connected to the internet world so the use of app may get famous to them. There is some pregnancy Bengali app which makes it easier to use it for the women who are less educated or at least can read Bengali. Although we had to face some difficulties like, we have stayed some much unknown places outside of Dhaka, have encountered so many challenges to get permission from the hospital's authorities and so on to collect data.

The practical data collection method helps us to know the thinking of the pregnant women and their family in Bangladesh with a very close view.

3.3 Data Analysis

The data we got from our survey, evaluated and analyzed by us to find the appropriate feature for our application [11]. We found 83% pregnancy was unplanned and the rest of the 17% was planned. We also tried to find out the complications, the women of Bangladesh go through during pregnancy. Moreover, we got nine victims of Gestational Diabetes, 8 of Stomach Ache, 7 of Anemia, 5 of Asthma, Malnutrition, Hypertension, Premature rupture of amniotic membrane, Bleeding/Spotting, 4 of back pain, 3 of Overweight, many of them have vomiting problem. Many women face another complications also. So, this is important to alert them for which complication they need to consult with a doctor as soon as possible to save their baby from any harm.

Here in Fig. 1 most of the pregnant women become tensed, crabbed after the 32th week of pregnancy. As they are exceeding their delivery date, so it is obvious to happen to them. They remain happy, stable at their 28th week.

Duration of Pregneancy(week) by Emotional condition

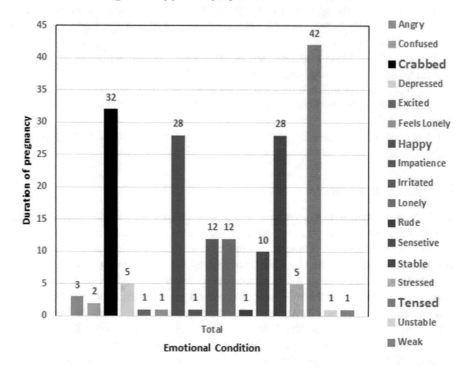

Fig. 1. Emotional condition of the pregnant women

In Fig. 2(a) we can see among all the participant 59% women have behavioral change during their pregnancy. Where 41% of women have no change, or they cannot understand that. Where in Fig. 2(b) among all the participant 78% of women get mental support from their family during pregnancy.

Although most of them told us they never feel any psychological change due to their pregnancy, among the women who noticed anger, irritation, caprice, crabbiness due to their pregnancy was prevalent for all the three trimester. When we talk to their family member, some of them told us something about the patient's mood swing. Like, one of the children who came with her mother to a hospital told us,

– "*Ammu forget things frequently, but never can feel that. Moreover, she is becoming caprice day by day.*"

However, when we asked them

– "*Did you ever discuss this issue with the doctor?*",

They answered,

– "*No! This is not any serious issue.*"

This is the way people accept any psychological change during pregnancy.

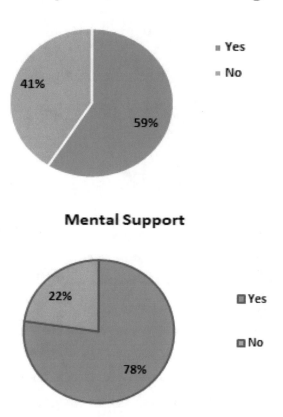

Fig. 2. (a) The pregnant women go through behavioral change due to pregnancy, Bangladesh during different periods of their pregnancy, (b) The pregnant women of Bangladesh get mental support during pregnancy.

In Fig. 3(a) among all of our participant, 71% of women (including the village/illiterate women) were familiar with the Internet world, and only 29% had no idea about it.

Moreover, in Fig. 3(b) where it shows the demand of feature what they think, these should be included in a pregnancy-related app. Although, 123 pregnant women among 180 don't have any suggestion about the mobile application. The main features they suggested are a consultation of a doctor during an emergency, a reminder of their daily chores, know the baby growths and foods value, cautions before and during the pregnancy, any exercise (all first knowledge) and so on.

The result of the data analysis portion gives us the main feature of our application as output. For example, when we analyzed the collected data about the suggestion said by the participant then most of them demand an option through which they can consult with a doctor on an emergency, they require a

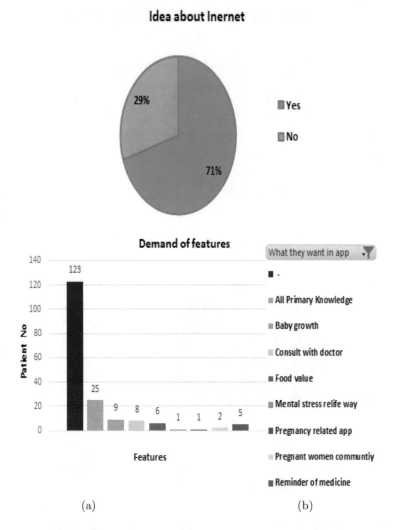

(a)   (b)

Fig. 3. (a) The idea about internet world among the pregnant women of Bangladesh, (b) The demand of features from the pregnant women of Bangladesh.

reminder system which will remind them about their essential chores, the baby picks up time, etc.

4   Proposed Application

Based on analyzed survey data, we have introduced an application for the pregnant women of Bangladesh. In that application, we tried to meet up all the necessities of upcoming mothers which they have desired to have or to get the services during pregnancy. We develop an application named "Matritto."

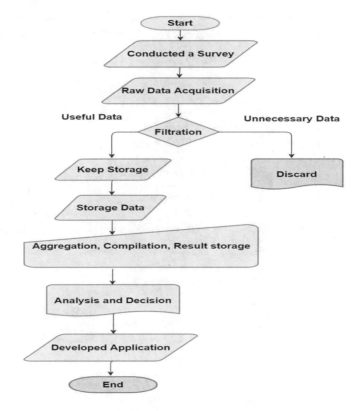

Fig. 4. The flow-chart of the work.

According to the rules of software engineering, we need to design the system before developing it. So, we create a use-case diagram to follow that rules which makes our work easier. Any user can take a quick and clear overview of the system by this diagram and Fig. 5 represents that diagram. Where Fig. 4 shows the flow of the whole work.

According to the design, we develop our application. Through *Due Date Prediction* option, user can know her delivery date. Through the *Advice of Doctor,* option user can view four more options named *what to do, what shouldn't do, risk signs of pregnancy and doctor's contacts.* Each option will provide relevant information which a doctor also suggests a pregnant woman. A more important part of our app is we are giving some trusted doctor's contact which will help them in their emergency condition.

The user may want to know how is her baby developing in her womb. Using the *Development of Baby* option, she can know about her baby growth development through reading and also week by week pregnancy videos. The user may forget what the doctor suggested to follow regarding foods. *Through suggestion for food* option, she can recall that. The food suggestion sector divided into

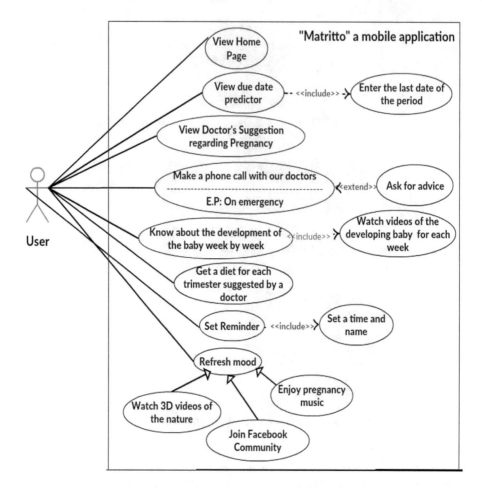

Fig. 5. The use-case diagram for the system.

three parts: *first trimester, second trimester* and *third trimester*. There has a suggestion of food for pregnant women which is good for baby and mother both.

Through *Alarm* option, the user will input their alarm name and set the alarm time. When the set alarm time and the current time are the same, then the alarm will ring and show the notification. The alarm notification will work even after the application close. Through *My World* option, the user will find more three options which are the *place, music,* and *community*. Through the *place* option, pregnant women can visualize three different 3-dimensional natural views. The videos will help a pregnant woman who desires to roam in nature. In music option, she can hear smooth music which is suitable for unborn baby and mother. In the *community* option, she will be able to connect a pregnancy-related Facebook group where pregnant women share their problems, views, feelings, and knowledge about other women condition during their pregnancy.

The interactive part of our application is *Emergency* option, through which user can communicate with our doctors. When any pregnant women face any problems, they need to consult a doctor but can't go at that time. They can use an emergency option and can directly call a doctor. In this way, it can primarily help a woman from any severe problems.

For solving their mood swing problem, we created a 3-dimensional video play list where she can visit at any time when her mood wants. During pregnancy, many women told us that they want to go out but because of their condition can not go out. So, in this way, they can view hills, seas, sunsets, mountains. We also created a smooth music play list which also suggests the doctors. According to the doctor, those smooth music helps in the development of the baby brain in their mother's womb. As hearing that music, a mom can feel relief during pregnancy especially. They also can connect with a Facebook community group if they have an account. So, *My world* option, will help the user to stay calm during their depressed period.

5   Conclusion

The sections above present a qualitative analysis of our ethnographic findings and key observations from our online and physical survey in addition to developing a vibrant, field-level understanding, regarding the implementation of the mood swing problems on the pregnant women. Our ethnography has demonstrated what kind of mood swings occur, and how it can be prevented. Our participants expressed their feelings, complications what they face and suggestion to develop our app. These findings help us to understand a pregnant woman's mood swings and develop a pregnancy-related mobile application. However, before synthesizing our findings into a set of key takeaways, we want to acknowledge that there are some limitations to our study. Our research only reveals a subset of the challenges encountered in some part of the country, Bangladesh. The pregnant woman's and the families that were studied were chosen based on convenience and participant availability. Hence, the findings of our study should not be generalized over the entire country. Also, the participants in our online survey represent only a small portion of the Bangladeshi population. Many pregnant women do not use the internet during her pregnancy. Many don't feel any need for a pregnancy-related application. We design some of our app works in the app simulator which should be done in our app but we could not. When we develop our app we planned to do virtual reality to give them a real scenario of their desirable places but we unable to do that exact way. We thought of connecting the database with our app so that we can find the other desires of pregnant women through our app. Despite these limitations, our research offers several key insights and takeaways that will be beneficial for the HCI community at large.

References

1. Harms, R.W., et al.: Mayo Clinic Guide to a Healthy Pregnancy. Part 3
2. Tyrlik, M., Konecny, S., Kukla, L.: Predictors of pregnancy-related emotions. J. Clin. Med. Res. **5**(2), 112–120 (2013)
3. Satyanarayana, V.A., Lukose, A., Srinivasan, K.: Maternal mental health in pregnancy and child behavior. Indian J. Psychiatry **53**(4), 351–361 (2011)
4. Prady, S.L., et al.: Psychological distress during pregnancy in a multi-ethnic community: findings from the born in Bradford cohort study. PLoS ONE **8**(4), e60693 (2013). Ed. Renato Pasquali
5. https://www.slideshare.net/WebAble/women-internet-in-bangladesh-by-maya-web-able. Accessed 27 June 2018
6. Schetter, C.D., Tanner, L.: Anxiety, depression and stress in pregnancy: implications for mothers, children, research, and practice. Curr. Opin. Psychiatry **25**(2), 141–148 (2012)
7. Jo, S., Park, H.-A.: Development and evaluation of a smartphone application for managing gestational diabetes mellitus. Healthc. Inform. Res. **22**(1), 11–21 (2016)
8. Lakshmi, B.N., Indumathi, T.S, Ravi, N.: A novel health monitoring approach for pregnant women. In: 2015 International Conference on Emerging Research in Electronics, Computer Science and Technology (ICERECT), Mandya, pp. 324–328 (2015)
9. Lupton, D.: The use and value of digital media for information about pregnancy and early motherhood: a focus group study. BMC Pregnancy Childbirth **16**, 171 (2016)
10. Ayele, T.A., et al.: Prevalence and associated factors of antenatal depression among women attending antenatal care service at Gondar University Hospital, Northwest Ethiopia. PLoS ONE **11**(5), e0155125 (2016). Ed. Klaus Ebmeier
11. https://drive.google.com/file/d/1p9B7V64SHLU5qlNJhwoH3DIhn-ezWzK5/view. Accessed 10 June 2018
12. Bush, J., et al.: Impact of a mobile health application on user engagement and pregnancy outcomes among Wyoming Medicaid members. Telemedicine J. e-Health **23**(11), 891–898 (2017)
13. Olander, E.K., Smith, D.M., Darwin, Z., et al.: Health behaviour and pregnancy: a time for change. J. Reprod. Infant Psychol. **36**(1), 1–3 (2018)

Efficient Software Implementation of Homomorphic Encryption for Addition and Multiplication Operations

Yongwoo Oh[✉], Taeyun Kim, and Hyoungshick kim

Sungkyunkwan University, 2066, Seobu-ro, Jangan-Gu,
Suwon-Si, Gyeonggi-Do 16419, Republic of Korea
{dyddn7997,taeyun1010,hyoung}@skku.edu

Abstract. Fully homomorphic encryption enables any type of calculation on encrypted data. There are several crypto libraries that provide such fully homomorphic encryption. However, since most libraries only support single level binary circuit operations, it is required for developers to efficiently implement basic arithmetic algorithms such as addition, subtraction, multiplication, and division for their own applications. In this paper, we propose fast *binary* addition and multiplication algorithms to support various bit-wise operations. To show the feasibility of the proposed algorithms, we implemented the proposed algorithms for 16, 32, 48, and 64 bits integers using the TFHE library. Our experiment results demonstrate that the proposed addition operation decreases the running time by 11 to 12%, and our multiplication implementation is about 3 to 4 times faster than the non-threaded method for 16, 32, 48 and 64 bits integers.

Keywords: Homomorphic encryption · Binary operation ·
Concurrent calculation

1 Introduction

Homomorphic encryption is an encryption scheme that allows certain calculations to be performed on encrypted data. Since fully homomorphic encryption allows any calculation on encrypted data, it can be applied to various applications protecting user's confidential data.

Homomorphic encryption can be used in cloud computing, biometrics, medical data [1]. Also, homomorphic encryption can be used for blind auctions requiring fair processing among clients [8].

There are several fully homomorphic encryption libraries. However, they only support bit-level circuit operations (binary AND, OR, NAND, NOR, NOT, XOR and multiplexer circuit operations) on encrypted data. Therefore, for developers, it is needed to implement high-level algorithms with those basic circuits. The performance of such implementation can be greatly varied with the developer's

© Springer Nature Switzerland AG 2019
S. Lee et al. (Eds.): IMCOM 2019, AISC 935, pp. 761–768, 2019.
https://doi.org/10.1007/978-3-030-19063-7_61

programming skills and experience because homomorphic encryption operations are typically too slow. TFHE (https://github.com/tfhe/tfhe), one of the fastest homomorphic crypto libraries, takes on average 13 ms to complete one normal circuit operation and 26 ms to complete a multiplexer circuit operation. These results show that homomorphic encryption operations are significantly slower than normal circuit operations.

There were several previous studies that aimed to improve the speed of homomorphic cryptosystem. For example, Brakerski et al. [2] suggested fully homomorphic encryption without bootstrapping, which is called BGV cryptosystem. Ilaria et al. [4] proposed a technique to accelerate bootstrapping that is a performance bottleneck in homomorphic encryption operations.

However, only a few studies (e.g., [5]) have tried to reduce the execution times of arithmetic operations for homomorphic encryption in application level. In this paper, we propose two implementation methods to improve the performance of the multiple bit addition and multiplication operations. Our key contributions are summarized as follows:

- We propose new homomorphic encryption algorithms to fully support various bit-wise addition and multiplication operations. For addition, we calculate concurrently on carry-out bit and sum bit using threads. For multiplication, we hierarchically group operand bits by two and calculating each group concurrently using threads in an iterative manner.
- We evaluate the performance of proposed addition and multiplication algorithms compared with non-threaded versions to show the efficiency of the proposed algorithms.

For evaluation, TFHE library was used in our implementation. However, we claim that our addition and multiplication algorithms can also be implemented by using any homomorphic encryption library with only slight change in a variable form and circuit operation functions.

2   Background

2.1   Homomorphic Cryptosystem

Homomorphic encryption is an encryption scheme that enables calculations on encrypted data [6]. There are two types of homomorphic encryption scheme: partially homomorphic encryption scheme and fully homomorphic encryption scheme. Fully homomorphic encryption scheme supports all kinds of calculation on encrypted data but partially homomorphic one does not. Fully homomorphic encryption scheme uses two keys: a private key to encrypt and decrypt data, and an evaluation key to calculate on encrypted data.

2.2   Efficient Implementation of Arithmetic Operations for Homomorphic Encryption

There have been several attempts to speed up arithmetic operations on homomorphically encrypted data. For example, Seo et al. [7] implemented an arithmetic adder for multiple variables using HElib (https://github.com/shaih/HElib). Their implementation took 130 s for 6 variables to be added and took 195 s for 32 variables' addition under 32-bit environment. In particular, their implementation is efficient when a sequence of addition operations should only be performed on encrypted data. In their method, however, when multiplication and addition operations are performed in a mixed manner, their performance was significantly degraded because this paper was targeting for improving multiple addition-only operations. Yao et al. [3] also implemented arithmetic operations (i.e., addition, subtraction, multiplication and division) on encrypted data using HElib. However, the performance of their operations is not sufficient for real-world applications. In this paper, we present new addition and multiplication implementation techniques using multiple threads to speed up the performance of those arithmetic operations. To the best of our knowledge, this is the first implementation of arithmetic operations using concurrent programming for homomorphic encryption.

2.3   Adder Circuit

When implementing addition and multiplication operations on encrypted data using homomorphic encryption library, we used two adder circuits: a half adder and a full adder. In big-endian environment, a half adder runs on least significant bit, and a full adder runs on the remaining bits.

In a half adder circuit, two result bits, a carryout bit and a sum bit, are calculated as follows:

- Sum $= a \oplus b$
- CarryOut $= a \wedge b$

Here, $\oplus$ and $\wedge$ represent bit-wise XOR and AND operations, respectively.

In a full adder circuit that uses a multiplexer, two result bits, a carryout bit and a sum bit are calculated as follows:

- Sum $= a \oplus b \oplus CarryIn$
- CarryOut $= MUX_{a \oplus b}(a, CarryIn)$

where $MUX_a(b, c)$ is equal to b if $a = 0$; otherwise, $MUX_a(b, c)$ is equal to c if $a = 1$.

3   Methodology

3.1   Addition

Let two encrypted integers be $A = A_1 \cdots A_n$, $B = B_1 \cdots B_n$, and sum of A and B be $S = S_1 \cdots S_n$ where A_i, B_i, and $S_i \in \{0, 1\}$.

764   Y. Oh et al.

First, we calculate the sum of two encrypted integers from a least significant bit (LSB) using a half adder. In this step, we can see that XOR and AND
operations are not dependent to each other (see Fig. 1). Therefore, we can concurrently calculate those operations to avoid the time delay when those operations
are sequentially executed.

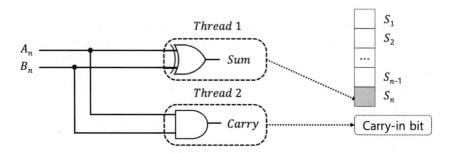

Fig. 1. Half adder in n-bit additions for homomorphic encryption. Here, A_n and B_n
are LSB of two encrypted integers (A and B). XOR and AND operations can be
independently performed.

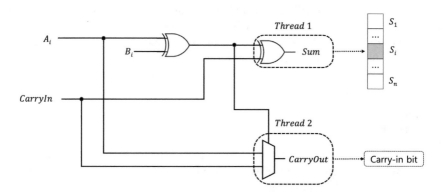

Fig. 2. Full adder in n-bit addition for homomorphic encryption. Here, A_i and B_i are
ith bits of two encrypted integers (A and B). XOR and multiplexer operations can be
independently performed.

After obtaining a carry bit from the LSB calculation, we should successively
calculate the sum of the next bits and a carryout bits. We first calculate $A_i \oplus B_i$.
Next, we use $A_i \oplus B_i$ and $CarryIn$ bit which is calculated as $CarryOut$ in
previous step to calculate a sum bit and a $CarryOut$ bit. As shown in Fig. 2, we
can calculate these two bits concurrently. Similar to the half adder, we repeat
this process sequentially to calculate all carryout and sum bits.

3.2   Multiplication

For a n-bit multiplication operation, we sequentially perform shift and addition operations n times. Because each addition operation takes $O(n)$ time, n addition operations take $O(n^2)$ time. Therefore, we aim to perform addition operations in a parallel manner in order to avoid the time delay caused by the sequential execution of those operations.

Suppose there are two encrypted integers $A = A_1 \cdots A_n$, $B = B_1 \cdots B_n$ for multiplications where A_i and $B_i \in \{0, 1\}$. If we truncate the overflow part of multiplication, we can use the following equation of $A * B = M_1 + M_2 + \cdots + M_n$ where $M_n = A_n \wedge (B_1 B_2 \cdots B_n)$, $M_{n-1} = A_{n-1} \wedge (B_2 \cdots B_n 0)$, $M_{n-2} = A_{n-2} \wedge (B_3 \cdots B_n 00)$, $\cdots$, $M_1 = A_1 \wedge (B_n 0 \cdots 0)$.

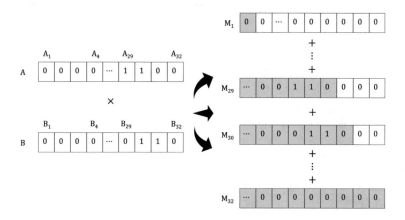

Fig. 3. First step for 32 bit variable multiplication.

In the proposed multiplication algorithm, the first step is to calculate $M_1, M_2, \cdots, M_n$. Interestingly, at this step, all $M_1, M_2, \cdots, M_n$ can be concurrently calculated because each M_i is independent from the other M_j where $i \neq j$. Figure 3 shows an example of multiplication of two 32-bit integers A and B. When calculating M_i, only grey area is to be calculated from A and B, and white area is filled with a constant bit 0. Therefore, all M_i can be concurrently calculated using threads.

Given $M_1, M_2, \cdots, M_n$ for multiplication, we group them by two terms (e.g., M_i and M_j) and calculate them independently. Because every M_k has k valid upper bits, we have to calculate $min(i-1, j-1)$ bits to calculate $M_i + M_j$. Thus, if we match and group by (M_1, M_n), (M_2, M_{n-1}), $\cdots$, $(M_{\frac{n}{2}}, M_{\frac{n}{2}+1})$ to calculate each group concurrently, only $\frac{n}{2} - 1$ bit operation time is needed for performing all operations. Since we grouped terms by two, the number of $\frac{n}{2}$ calculations is needed. Let these be $M1_1, M1_2, \cdots, M1_{\frac{n}{2}}$. These results sequentially have $n - 1$, $n - 2$, $\cdots$, $\frac{n}{2}$ number of zeros in the lower bits. To obtain $M2_1 \cdots M2_{\frac{n}{4}}$, there are $\frac{3n}{4} - 1$ bit operations as every group can be calculated concurrently.

If we repeat this process recursively until the last one is reached, we can obtain $\frac{n}{2} - 1$, $\frac{3n}{4} - 1$, $\frac{7n}{8} - 1$, $\frac{15n}{16} - 1$, $\cdots$ bit operation time for each level. Because the number of operations at each level is between $\frac{n}{2} - 1$ and n, the maximum level is $O(\log n)$. Thus, the time complexity of the entire procedure would be $O(n \log n)$. Figure 4 shows the process of 32-bit multiplication operation. We can see that calculations are performed hierarchically from 16 groups until only one remains.

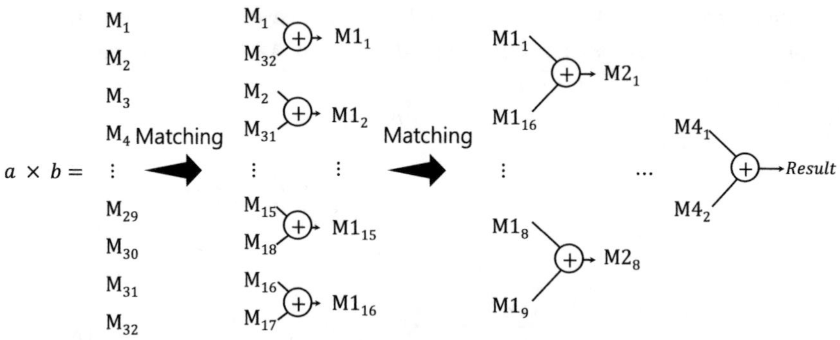

Fig. 4. Example of calculation of M_1 to M_{32} in two 32-bit integers. We group by two at each level and add each other to merge the results iteratively.

If the operand size is not a power of 2, we can pad the remaining M_i with encrypted zeros to increase its size to the next power of 2, and then apply the above procedure.

4   Evaluation

To show the feasibility of the proposed implementation techniques, we implemented the proposed addition and multiplication algorithms with varying bit sizes (16, 32, 48 and 64 bits). In our experiments, we used i5-7600K CPU, 16 GB memory and Samsung EVO 850 pro 250 GB SSD running Ubuntu 16.04 64-bit operating system. To measure the performance of execution time, we used the `perf` program that is widely used in performance measurement. For comparison, we also implemented *non-threaded* addition and multiplication operations with the same condition. To avoid bias, we performed both addition and multiplication operations of our implementations 100 times, respectively, for each condition. The experiment results are shown in Figs. 5 and 6.

Figure 5 shows the performance of binary addition operations. Our proposed implementation method always produced better results for all bit operations than non-threaded implementation. Overall, the proposed method reduced the execution time by about 11 to 12%. Interestingly, the performance of the proposed method is relatively stable compared with the non-threaded implementation for 48 and 64 bit operations.

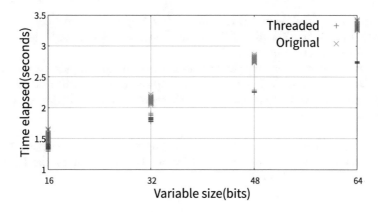

Fig. 5. Performance of addition operations.

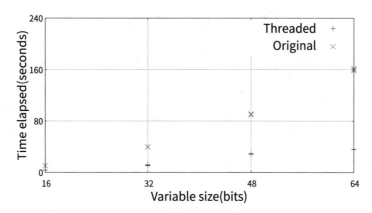

Fig. 6. Performance of multiplication operations.

Figure 6 shows performance of binary multiplication operations. Because the proposed implementation technique significantly increases concurrency during the calculation, we achieved the performance improvement of 2.95, 3.93, 3.19 and 4.48 times for 16, 32, 48 and 64 bit operations, respectively. Again, in the case of multiplication operations, our proposed implementation method always produced better results for all bit operations than non-threaded implementation.

5   Conclusion and Future Work

We proposed new software implementation techniques to improve the performance of addition and multiplication operations on homomorphically encrypted data. To efficiently implement addition and multiplication operations, we suggest the use of threads to increase concurrency in performing operations. To show the feasibility of the proposed algorithms, we implemented addition and multiplication operations with varying bit sizes (16, 32, 48 and 64 bit). According to our experiments, the execution times of those operations can be significantly decreased by about 12 and 448%, respectively, compared with non-threaded implementations for 64 bit operations.

As part of future work, we plan to consider other techniques for rearranging to improve parallelism for basic arithmetic operations. We will also extend the proposed algorithms to other homomorphic cryptography libraries to generalize our results.

Acknowledgements. This work was supported by the ICT R&D programs (No.2017-0-00545) and the National Research Foundation (NRF) funded by the Ministry of Science and ICT (2017H1D8A2031628).

References

1. Archer, D., Chen, L., Cheon, J.H., Gilad-Bachrach, R., Hallman, R.A., Huang, Z., Jiang, X., Kumaresan, R., Malin, B.A., Sofia, H., Song, Y., Wang, S.: Applications of homomorphic encryption. Technical report (2017)
2. Brakerski, Z., Gentry, C., Vaikuntanathan, V.: Fully Homomorphic Encryption without Bootstrapping. Technical report (2011). https://eprint.iacr.org/2011/277
3. Chen, Y., Gong, G.: Integer arithmetic over ciphertext and homomorphic data aggregation. In: IEEE Conference on Communications and Network Security (2015)
4. Chillotti, I., Gama, N., Georgieva, M., Izabachène, M.: Faster Fully Homomorphic Encryption: Bootstrapping in less than 0.1 Seconds. Technical report (2016). https://eprint.iacr.org/2016/870
5. Chillotti, I., Gama, N., Georgieva, M., Izabachène, M.: Improving TFHE: faster packed homomorphic operations and efficient circuit bootstrapping. Techical report (2017). https://eprint.iacr.org/2017/430
6. Gentry, C., Sahai, A., Waters, B.: Homomorphic encryption from learning with errors: Conceptually-simpler, asymptotically-faster, attribute-based. In: Canetti, R., Garay, J.A. (eds.) Advances in Cryptology - CRYPTO 2013. Springer, Heidelberg (2013)
7. Seo, K., Kim, P., Lee, Y.: Implementation and performance enhancement of arithmetic adder for fully homomorphic encrypted data. J. Korea Inst. Inf. Secur. Cryptol. **27**, 413–426 (2017)
8. Suzuki, K., Yokoo, M.: Secure generalized vickrey auction using homomorphic encryption. In: International Conference on Financial Cryptography, pp. 239–249. Springer (2003)

An Approach to Defense Dictionary Attack with Message Digest Using Image Salt

Sun-young Park[1]([⊠]) and Keecheon Kim[2]([⊠])

[1] Department of IT Convergence Information Security, Konkuk University,
Seoul, Korea
dksldl@konkuk.ac.kr
[2] Department of Computer Science and Engineering, Konkuk University,
Seoul, Korea
kckim@konkuk.ac.kr

Abstract. Hash algorithms have been widely used for cryptography. It has been impossible to decrypt the ciphertexts generated through hash algorithms, as an operation that damages the original text is performed. However, various methods of attack occurred over time after the algorithm was developed. The vulnerability of SHA1 (an old hash algorithm) has been revealed, and there has been a great deal of data available for dictionary attacks. Although the industry has been gradually refraining from using SHA1, it remains in use in some existing systems for various reasons. For example, when problems resulting from service interruption or mass update are critical, updating the encryption algorithm can be a burden. In this study, we aim to increase the complexity of ciphertexts by postprocessing hash ciphertext. For that, image salting techniques are used using two-dimensional array masking. This will allow the use of hash ciphertexts with increased complexity in some devices that are forced to use old hash algorithms for various reasons.

Keywords: Hash algorithm · Image processing · Rainbow table ·
Message digest · Two-dimensional array · Image salting

1 Introduction

With the widespread use of the Internet in recent times, network communication methods have developed rapidly to broadband, and the development of networks has heightened the importance of various encryption methods. In particular, the secure hash algorithm (SHA) [1] has been applied to various systems as a secure encryption technique certified by the National Institute of Standards and Technology (NIST.gov).

Hash algorithms, which are the key to hash encryption, are used to change a text message of arbitrary length to a fixed-length message digest. It was considered impossible to replace message directs generated through hash algorithms with the original text, but various attack techniques have been developed. Thus, difficult algorithms have become standards.

However, as a part of the recent emergence of industry 4.0, some services have paid attention to introducing personal health devices. In the process of mounting various

© Springer Nature Switzerland AG 2019
S. Lee et al. (Eds.): IMCOM 2019, AISC 935, pp. 769–777, 2019.
https://doi.org/10.1007/978-3-030-19063-7_62

devices and sensors in various fields, it is sometimes impossible to apply a strong cipher to all sections due to compatibility issues with the existing services or due to the characteristics of devices. In particular, passwords created using hash cryptographs cannot be decrypted again, thereby requiring resetting and initialization, which is not possible in the existing services. As a result, it is not easy to apply the new standards to all sections.

When checking the integrity of the FHIR3 code system of HL7 [2], SHA1 and SHA256 are recommended, and the difference depending on environment is mentioned because of the reason above. The recently proposed algorithm also has the following issue: inferring the original texts by the attack technique using a rainbow table [3]. Therefore, we propose a method to incapacitate the attack technique using the rainbow table while maintaining the interoperability between heterogeneous systems and being applied to the old type OSs (limit hardware) as an additional processing of hash ciphertexts.

2   Background

2.1   Situation with Hash Algorithm Standards

The Secure Hash Algorithms are a family of cryptographic hash functions published by the National Institute of Standards and Technology (NIST) as a U.S. Federal Information Processing Standard (FIPS). A retronym applied to the original version of the 160-bit hash function published in 1993 under the name "SHA" It was withdrawn shortly after publication due to an undisclosed "significant flaw" and replaced by the slightly revised version SHA-1.

At 2005, Cryptographic weaknesses [4] were discovered in SHA-1, and the standard was no longer approved for most cryptographic uses after 2010. SHA-2 Family, which is currently designated as a standard, supports later versions after IE 6, which is the old software application. SHA-3, a hash function formerly called Keccak, chosen in 2012 after a public competition among non-NSA designers. It supports the same hash lengths as SHA-2, and its internal structure differs significantly from the rest of the SHA family.

2.2   Obfuscation

Obfuscation in software development refers to creating sources or machine codes that are difficult for humans to understand. Unlike encryption, obfuscation is characterized by the difficulty proving the performance through accurate calculation.

The book [5], which describes the effects of obfuscation, has a technique called chaff that distributes silver powder as a way to disrupt the radar. It also states that obfuscation can be achieved by spraying data that is not associated with original values, which can be achieved by using Pulse-Echo Method. Track-me-not [6] also uses this technique.

The postprocessing method proposed in the present study is close to obfuscation. It sends the mixed information of the original and other information, and allows for the reorganization of the original information. It does not cause much damage to the original information, but applies slight changes to render it difficult for attackers to stage traditional automated attacks.

2.3 Dictionary Attack

Many online services use a hash function to encrypt the password, but using hash function only cannot prevent attacks. The fastest and most effective way to attack a hash is the dictionary attack using a rainbow table [3]. Using dictionary attack, an attacker can get original text of seized message digest, by making as many pre-computed message digests as possible and then compare it to the seized message.

We can see the danger of using only a simple hash algorithm through this attack scenario. When providing personalized service through login in JAVA API 1.5 and Tomcat 6.0 web server environment, password is stored using SHA-1 when login authentication is performed, and password policy requires 8 alphanumeric characters. It is possible to infer that the password is stored as SHA-1 through the length of the hashed password, and change it to the original password using dictionary attack using the rainbow table. In the case of the 8-digit encryption password, the number of 220 billion cases can be generated, which is 6.6 TB. Considering the recent low-cost hardware capacity, the rainbow table which can convert the hash file to the original text is cheap can do.

Passwords changed in the original text will cause secondary damage to users accessing other services using the same account. An incident similar to this scenario was used in Linkedin's 2012 password attack, which relied solely on the SHA-1 cryptographic scheme to recover the hash, resulting in damage and financial loss. Companies take time to apply various methods to prevent this situation, but it takes more time due to the lack of budget and manpower. The salt technique is used to remove the regularity of the standardized encryption algorithm. However, it is also dependent on the existing encryption algorithm, so it may involve the previously mentioned vulnerability when using a salt value having regularity. In addition, in the case of the SHA2 family, there is the same possibility of decrypting ciphertexts when constructing the rainbow table consisting of the original text and hash ciphertexts.

3 Related Works

3.1 Use Cases of Hash Algorithm Techniques

Hash techniques are used for many services that require data reliability. The HL7 FHIR3 Code System [2] and IHE's Document Digital Signature [7], one of the medical standards, require that the hash values for code and documents be separately stored and checked to verify the authenticity of the code and documents used have. These standards enable the use of SHA1 and SHA256 together. Hashes are easy to tag because they are easy to construct unique values. Therefore, it is proposed [8] to combine QR codes that are used most recently. Furthermore, SDN, which is advantageous for cost saving, also uses the hash algorithm in application program services between controls, and enables performance improvement by using the advantages of hash itself in FlowTable in the Internet of things environment [9]. A hash algorithm is also applied partially in a study proposing fast authentication, and message integrity processing in wireless connection between personal medical devices and smart mobile devices in emergencies [10]. Thus, hash is being used as a device for reliability in various areas.

3.2 Anti-phishing Techniques Using Personalized Images

There are a variety of attacks that threaten Web sites. In The security threat analysis of password authentication systems [11], there are attacks like continuous authentication attempt, phishing, and key logging.

There are various methods for preventing attacks, such as Captcha, keyboard hacking prevention program, virtual keyboard, blocking of overseas IP address, anti-phishing/countermeasures technology, and encryption communication phishing sites. There is also a technique based on user's perception like [12]. Using a browser plug-in, user can turn on balloon, change the address window to green, or show the image they previously specified.

Among the proposed methods, personalized images are used in various sites in recent years. According to [11] analysis, security seal can protect phishing effectively with ease of implementation. In Korea, the above method is used to identify users' sites such as banks, insurance, and securities sites. Phishing attacks occur either by visiting links to fake portal sites or through DNS attacks. There have been many cases of studying various means to prevent phishing. In addition, anti-phishing techniques for personalized images have been patented [13].

Personalized image services require a server that stores images securely. The service includes a technique for displaying an image as a moving layer type in order to provide convenience to the user. This allows the user to recognize whether the service is right or phishing. As individual users can have different image values, this technique can be applied to various personalization services besides phishing prevention. In order to effectively use the contents proposed in this paper, it is more effective to use different images for each individual rather than using one image.

4 Proposed Processing Method of Image Salting

4.1 Description and Application of Technique

Post-processing is performed in the form of masking and creating a two-dimensional array of values of the same size as the result length.

In Fig. 1, The message digest comes from a hash function. Combine this with RAW data extracted from the image file. The process of mixing the data is called the image-salting. Image salting has a different structure from the conventional salt. It is not just a form of adding or subtracting text. Think of adding salt to real food. Salt is scattered rather than concentrated in one place. The process of image-picking is the same as the process of scattering salt. The only difference is that this is data. The message digest collapses messages of various lengths into messages of constant length.

The length of the result is already known because it can be a variety of processing may also propose using a two-dimensional array in this paper and previously proposed [14]. Hashes are used for various services. Also, it is necessary to check whether image salting is available when reflecting on services using hash. If image salting is simply used to store a password as a hash, the process is shown in Fig. 3.

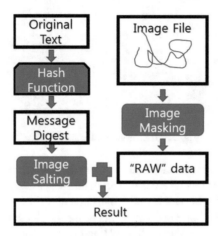

Fig. 1. The process of image-salting.

The process shown in Fig. 2. is an example of combining image salting if you already have a service that has a personalized image [14] feature to prevent phishing. The process is divided into two categories: "registration" and "login". In the registration process, the user selects a specific image to be used by the image server. It combines this into the image salting process when storing the password as a message digest, as well as using it for anti-phishing services. When logging in, a comparison is made with a message digest created by combining a specific image and a password identified through an image server. A similar process is used when a hash algorithm is used as a substitute for a password.

The other point here is to have additional access to the image server to take advantage of image salting. If salt is used in this process, there is a risk that the original text will be restored by dictionary attack and brute-force attack depending on the state of salt. And if you have not applied salt before, you are exposed to dictionary attacks. Image Salt can perform an additional salt operation using the specified image in the case of the data that was used in the past. You do not need to touch an existing hash because it is doing additional work. Also, if the data is stolen and the image is not stolen, the dictionary attack cannot restore the hash to the original text.

4.2 Experiments and Analysis

There are two points of experiments. First, it is to check whether the image salt technique has a speed that does not interfere with the use of a hash service.

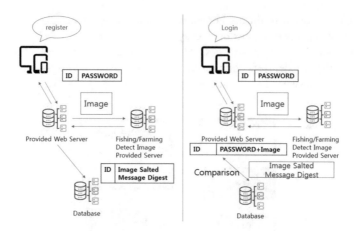

Fig. 2. The process of image-salting.

Second, the other is confirmation that the image salt technique can defend the dictionary attack. The experimental environment in this paper was performed in Windows 7 Professional K OS 64-bit Service Pack 1 environment, PC environment with Intel Core i5-3470 processor (CPU 3.20 Ghz) and 4 GB RAM. Implementation development was performed on Oracle Java Version 1.8.0_121 and Eclipse Neon.2 Release (4.6.2).

When you create a service in Java, you generally use the Security API provided by itself. However, the recently released SHA-3 does not provide a default Security API. In the experiment, it is necessary to confirm the execution time of the existing image and the image-drawing technique. Therefore, I used the bountyCastle API (MIT license) [15] that supports SHA-1, SHA-2 Family, and SHA-3. We used an online web site [16] that consisted of a rainbow table to check whether the value of image salting could defend dictionary attack. The site has a Rainbow table of MD5 and SHA-1.

As of October 2018, it has a total of 12,627,723,871 original texts and message digests, and is usually limited to upper and lower case letters and numbers. When the attacker performs the dictionary attack, it executes using the rainbow table generated in this way. Apart from that, attackers can also generate a rainbow table by analyzing the message digest to determine which hash algorithm was used. In this paper, we study the obfuscation concept that can save time before changing the unsafe algorithm. For this reason, we have mainly conducted experiments using SHA-1, which is not currently recommended but is still widely used.

Previously proposed [14] The image salting process using ImageIO is not so slow as to hinder the service. But it is slower than the basic algorithm. Therefore, if the environment is changed, the performance may be deteriorated. So we did an additional task to improve performance. ImageIO API runs on Java version 1.7 or later. Therefore, it can be judged that it is not suitable for the machining technique which can be used in the target specification. Since InputStream API is not an image-only API, it looks like an image file and has a disadvantage in that it takes a separate process to take pixel values. However, InputStream API is faster than ImageIO API and does not need to consider the Java version, so there is no problem in using Image Salting. In addition to the basic slow image processing, there was additional time required for image processing for masking.

The images consisted of dot graphics and so on, rather than the everyday extractable images that I thought at first, were suitable for image salting. In subsequent tests, we modified the method of image processing to find the average of the distribution of pixel values in the image. And we tried to use it for image salting based on the position of some pixels corresponding to the average value.

Figure 3 shows the results of additional studies based on the above. We can confirm that image salting can be used for service without difference in speed compared with existing hash algorithm. And I confirmed that I can defend the dictionary attack by using the image salt technique.

Processing Time(ms)

0.70	
0.60	
0.50	
0.40	
0.30	
0.20	
0.10	
0.00	

SHA1 Avg   SHA1 mod   SHA2 256 Avg   SHA2 512 Avg   SHA3 256 Avg   SHA3 512 Avg

Fig. 3. SHA processing time and image salting processing time.

Figures 4 and 5 shows as a result of searching 'imcom', which is used to measure execution time, the value of hash using image salt is not retrieved as same as previously proposed [14]. However, in the previous proposal, it was confirmed that imageIO has a slow speed, but in this study, the same effect can be achieved by processing image salting with InputStream.

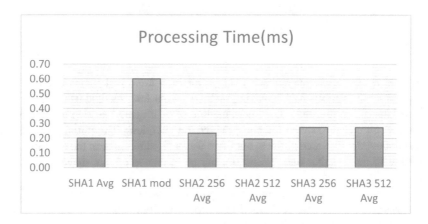

Fig. 4. Before applying image salting.

Fig. 5. After applying image salting.

5 Conclusion

To defend against dictionary attack using a rainbow table that reduces time significantly among the various hash attacking methods, a masking method using images was designed instead of the static Salt, and a method of overlaying masking to the message digest was proposed to avoid affecting the existing services. The initial proposal was able to defend the dictionary attack, but there was a big difference in the execution time compared to the existing one. Therefore, the performance of the image salting technique is improved and it is easy to use it for existing services.

In future, we will apply various hash algorithms and their results based on this and check whether they can be performed in various environments. And we will make it easy to apply the image salt technique based on various images. We will also find examples to demonstrate the stability of the image-salt technique.

Acknowledgment. This work was supported by Institute for Information & communications Technology Promotion(IITP) grant funded by the Korea government(MSIT) (2014-0-00547, Development of Core Technology for Autonomous Network Control and Management) and Next-Generation Information Computing Development Program through the National Research Foundation of Korea (NRF) funded by the Ministry of Science and ICT (No. NRF-2017M3C4A7083678).

References

1. NIST: Secure Hash Standard (SHS) FIPS PUB 180-4 (2015)
2. HL7: HL7 FHIR v3 Code System IntegrityCheckAlgorithm (2016)
3. Oechslin, P.: Making a faster cryptanalytic time-memory trade-off. In: Boneh, D., (eds.) Advances in Cryptology - CRYPTO 2003. CRYPTO 2003. LNCS, vol. 2729. Springer, Heidelberg (2003)
4. Wang, X., Yin, Y.L., Yu, H.: Finding collisions in the full SHA-1. In: Shoup, V., (eds.) Advances in Cryptology–LNCS, vol. 3621. Springer, Heidelberg (2005)
5. Brunton, F., Nissenbaum, H.: Obfuscation: A User's Guide for Privacy and Protest. The MIT Press (2015)

6. Ganter, V., Strube, M.: Finding hedges by chasing weasels: hedge detection using Wikipedia tags and shallow linguistic features. In: Proceedings of the ACL-IJCNLP 2009 Conference Short Papers (ACLShort 2009), Association for Computational Linguistics, Stroudsburg, PA, USA, pp. 173–176 (2009)
7. IHE: IHE IT Infrastructure Technical Framework Supplement Document Digital Signature (DSG) (2015)
8. Ahmed, H.A., Jang, J.W.: Document certificate authentication system using digitally signed QR code tag. In: IMCOM 2018 (2018). Article No. 65
9. Ren, W., Sun, Y.: A hash-based distributed storage strategy of flowtables in SDN-IoT networks. In: GLOBECOM (2017)
10. Wang, C., Zheng, W.: Identity-based fast authentication scheme for smart mobile devices in body area networks. Wirel. Commun. Mob. Comput. **2018**, 7 (2018). Hindawi
11. Noh, H.K., Choi, C.K., Park, M.S., Kim, S.J.: Korea Information Processing Society, vol. 3, no. 12, pp. 463–478 (2014)
12. Kim, J.H., Maeng, Y.J., Nyang, D.H., Lee, K.H.: Cognitive approach to anti-phishing and anti-pharming. J. Korea Inst. Inf. Secur. Cryptol. **19**(1), 113–124 (2009)
13. Ryong, C.D., Yhdatabase CO., LTD.: Personalization Image providing Method for Preventing Phisiing and Pharming and Image Providing System for Same, 101363336 B1 (2014)
14. Park, S., Kim, K.: A study on the processing and reinforcement of message digest through two-dimensional array masking. In: International Conference on Information Networking (ICOIN) (2018)
15. Tau Ceti Co-operative Ltd.: The Legion of the Bouncy Castle Home Page. https://www. bouncycastle.org. Accessed 20 Oct 2018
16. Oelke, M.: Hash Toolkit. http://hashtoolkit.com. Accessed 20 Oct 2018

Real-Time Drone Formation Control for Group Display

Hyohoon Ahn, Duc-Tai Le, Dung Tien Nguyen,
and Hyunseung Choo$^{(\boxtimes)}$

Sungkyunkwan University, Suwon, South Korea
{hyohoon, ldtai, ntdung, choo}@skku.com

Abstract. Advances in electronics and sensor technology have widened the scopes of networked drones to include applications as diverse as surveillance, video recording, operations, entertainment/advertising, signal emission, transportation, and delivery. These applications and services require video recording for their operations. Drones come in various sizes. Large drones are used singly in missions while small ones are used in formations or swarms. The small drones are proving to be useful in civilian applications. These are effective with multiple drones. Consideration of small drones for the applications such as group flight, entertainment, and signal emission lead to deployment of networked drones. To develop group display applications, a real-time drone formation control for group display is proposed. Drones form group displays for an entertainment and displaying application. Simulation shows that drone formation can display messages effectively.

Keywords: Drone · Display · Formation control

1 Introduction

Unmanned Aerial Vehicles, also known as drones, are an emerging technology that can be used for military, public, and civil applications. Military use of drones is more than 25 years old primarily consisting of surveillance, reconnaissance, and strike. Public use is by police, public safety, and transportation management [1].

Drones come in various sizes. Large drones may be used singly in missions while small ones may be used in formations or swarms. The small ones are proving to be useful in civilian applications. Advances in electronics and sensor technology have widened the scope of networked drone applications to include applications as diverse as surveillance, video recording, operations, entertainment/advertising, signal emission, transportation, and delivery.

Many of drone applications are effective with multiple drones. Consideration of small vehicles for the applications such as group flight, entertainment, and signal emission, such as display, leads to deployment of networked drones. Group flight system is more than just a sum of multiple single drones that drones in a group act as a part of system and communicate each other for a mission. By allowing group flight, multiple drones are able to form formation figures and show performances [2]. One of the most famous group flight application is light show performance. For light show,

© Springer Nature Switzerland AG 2019
S. Lee et al. (Eds.): IMCOM 2019, AISC 935, pp. 778–785, 2019.
https://doi.org/10.1007/978-3-030-19063-7_63

drones carries a light bulb and act as a single pixel. However, performing with single light pixel drones requires large number of drones to form figures.

To solve the pixel and number of drone problem, the paper proposes a real-time drone formation control for group display. Drones carry multi pixel display form large display to show text messages and images for alarming application. They are also able to form figures such as traffic sign for warning in traffic accident situations.

The rest of the paper is organized as follows. Section 2 introduces related work. Section 3 proposes our system. Section 4 shows the experimental results. Section 5 presents the conclusion and the future work.

2 Related Work

2.1 Drone Application

Drone has characteristics which can fly, carry sensors and additional devices, and cooperate with other drones. These characteristics enable various applications and these applications can form a software platform. Table 1 shows possible drone applications categories. Presented applications are surveillance, photo/video, operations, entertainment/advertising, signal emission, transportation, and delivery. These applications have common functions and conditions. Most of these applications work with multiple drones [3–5].

Table 1. Drone applications categories.

Applications	Description of application	Functions	Number of vehicles
Surveillance	Conducting short or long range surveillance, image capture, and analytics	Video recording	Single/Multiple
Photo/Video	Using photo and video applications without analytics	Video recording	Single
Operations	Facilitating intensive or difficult tasks	Single/Multiple	Single/Multiple
Entertainment/Advertising	Leveraging drones to perform entertainment or advertise	Forming figures	Single/Multiple
Signal emission	Providing multimedia by emitting signal, video, or sound	Displaying	Single/Multiple

2.2   Group Flight

Multi-drone applications use small drones with onboard processing, vision, GPS navigation, and wireless communications. Communication, sensing, and formation control are important for group flight to efficiently collect information and respond to unknown environments. The drones should fly in one formation, when the drones fly and work together, the formation of the drones must be reconfigured safely according to their mission and surrounding environment. When the drones form a formation and when reconfiguring another formation due to a change in mission, it is necessary to be able to set the destination and route to ensure safety so that there is no collision between the drones [6].

Collision and obstacle avoidance and formation reconfiguration should be ensured for stability in group flight. Collision and avoidance in the drone formation mean collision between the drones forming the formation, not the collision with the external obstacle. Collision avoidance requires cooperative algorithms. Collaborative algorithms control multiple drones that form a formation and set the location and path so that they do not collide. The drones must report each other's sensed value or GPS position via the sensor for collision avoidance.

3   Proposed Group Display

Drone displaying is an attractive application for entertainment, advertisement, and signal emission. The application uses single or multiple drones for the performance. One of the most popular way of using drone for performance is light show. Single drone has limitation of cover area of movement and size of the performance. Drone light show uses multiple drone formation to form figures or text to display. However, for drone light show, the number of drones required is too large that it is difficult and expensive to be performed. Also, large stage is required for many drones to perform. The thesis proposes group display by formation control with drones carrying a multi-pixel display.

Table 2 shows comparison between drone light show and group display. Each drone carrying light bulb performs as a single pixel in current light show system. Group display shows that each drone carries a multi-pixel display that user can show figures and text with a smaller number of drones. A public display such as LED display is a combination of small LED modules. Once the size of the display is decided, it is not able to be changed unless the display is reassemble. Reassembling huge display for different size costs too much. Using the characteristics of drone that it can fly where it wants to go and cooperate with other drones provides various options of the size of the display. By forming group display with drones carrying a display allows users to change size and shape of display without limitation.

Table 2. Comparison between drone light show and group display.

	Drone light show	Group display
Carrying object	Light bulb	Display
Number of pixels	Single pixel	Multiple pixels

The proposed system is consisting of a ground control station and drones. The ground control station controls drone operation by sending message which contains position of each drone and text that display shows. It also works as client application which allow users to input text and formation of the drone. Each drone is carrying LED dot-matrix display.

Possible application scenarios of group display are (a) traffic sign application which goes traffic accident site and guides oncoming vehicles to avoid secondary accident, (b) messaging board application at public place. The operation of the traffic sign application is as follows. When a traffic accident occurs on the road, the accident point and the situation are transmitted to the control station by mobile application. As shown in Fig. 1, the Ground Control Center sends the destination information and mission to the drones. The drones that received the mission from the ground control station move to the target point. The task of the drone is to form a screen of LED signboards through the formation, to form a traffic signal shape by forming a large traffic signal, and to prevent the secondary accident by informing the entering vehicles.

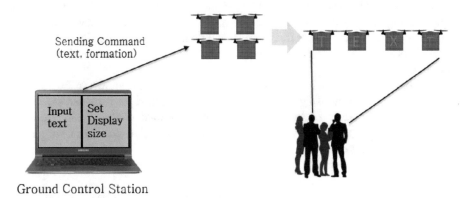

Fig. 1. Group display

There are two methods for traffic signs using drone formation. The first is to form a word or a sentence in the air by making each drone that expresses a letter by using attached LED display to a drone as shown in (a) of Fig. 2. When it is difficult to grasp the accident situation from a distance due to dark environment or obstacles, it is possible to prevent a secondary accident by notifying the driver with the message expressed by LED in the air. The second method is to give the driver a notice with the drone forming a traffic signal as shown in (b) of Fig. 2. If the drones form large traffic signals in the air, vehicles in the distance can identify and contrast the signals. It can help drivers identifying if it is difficult to identify with normal traffic signals at night or in fog.

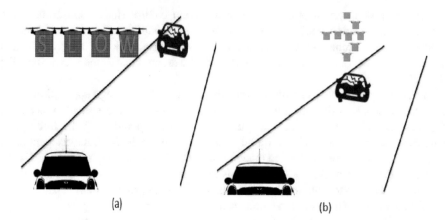

(a)   (b)

Fig. 2. Drone traffic sign

The proposed method is a system that can move freely in the air and perform traffic sign by using the characteristic of the drone which can cooperate with several parties. It is expected that drones that are free to move and do not need to be fixedly installed in a specific location for operation will help prevent secondary accidents after a traffic accident.

The operation of the public messaging board application is shown in Fig. 3. User can send commands by using client application. They can input texts and select formation to set different resolution of the display. Ground control station receives commands from users and delivers to the drone. Drones receive messages from ground control station and perform as the commands.

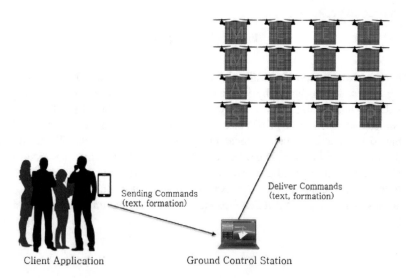

Fig. 3. Public messaging board

Public display can be used as entertainment and advertisement. Current public display system uses mounted display system which has fixed resolution and facing one direction. Using the characteristics of drone group display that can move and form different resolutions by order, public display system can be more effective in using for entertainment and advertisement.

4   Experiment

Testing drone algorithms requires expensive hardware and difficulty in configuring the experiment environment. Errors that occur in a real environment are difficult to reproduce and can damage the drone during the experiment. Figure 4 shows a drone simulator. It was developed to reduce field test time, isolate problems for testing, and reduce the impact of real drones. The simulator developed for high-level tasks such as path planning and multi-drone control provides an environment in which the simulator can equally apply the code to be applied to the actual drone. Theoretically, all components used in the simulation environment can be run on real platforms.

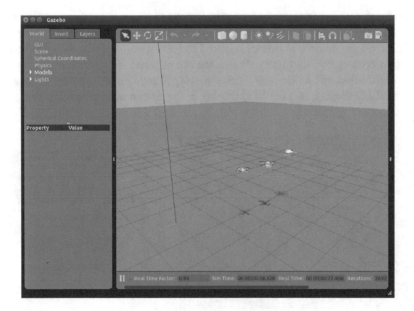

Fig. 4. Drone simulator

In the simulation, the results show construction of the drone formation of the proposed system using the drone simulator. Figure 5 shows that the drones with the LED display form large display resolution and represent letters. Four drones fly in a row to achieve 1 × 4 resolution. If each drone has a LED board module, the message can be conveyed by expressing four letters. Figure 5 shows that the drones form a traffic signal through the formation and deliver the message.

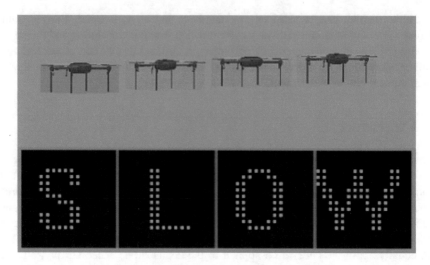

Fig. 5. Drone formation control for group display

In Fig. 6, the eight drones form an arrow to show a bypass sign. Simulation results show that the drones can form traffic signals through the formation. There is no restriction on the number of drones required to form a traffic signal through the drone formation control, and that many drones are not required for general signal representation. In addition, the drones can display messages with a small number of drones by using the LED display, and the LED display in the air can maximize the visual effects.

Fig. 6. Drone formation control for traffic sign

5   Conclusion

Advances in technology have widened the scopes of networked drones to include applications as diverse as surveillance, video recording, operations, entertainment/advertising, signal emission, transportation, and delivery. Many of drone applications are effective with multiple drones. By allowing group flight, multiple drones are able to form formation figures and show performances such as light show. However, performing with single light pixel drones requires large number of drones to form figures. To solve the group display problem, we proposed a real-time formation control framework for group display. Drone formation control for drones that carry multi pixel display provides an entertainment and displaying application with less number of drones. Experimental results show that drone formation can form a group display to show messages effectively.

Acknowledgment. This research was supported in part by Korean government, under G-ITRC support program (IITP-2018-2015-0-00742) supervised by the IITP and Priority Research Centers Program (NRF-2010-0020210), respectively.

References

1. Gupta, L., Jain, R., Vaszkun, G.: Survey of important issues in UAV communication networks. IEEE Commun. Surv. Tutor. **18**(2), 1123–1152 (2016)
2. Yanmaz, E., Yahyanejad, S., Rinner, B., Hellwagner, H., Bettstetter, C.: Drone networks: communications, coordination, and sensing. Ad Hoc Netw. **68**, 1–15 (2018)
3. Hayat, S., Yanmaz, E., Brown, T.X., Bettstetter, C.: Multi-objective UAV path planning for search and rescue. In: IEEE International Conference on Robotics and Automation (ICRA), Singapore, pp. 5569–5574 (2017)
4. Idries, A., Mohamed, N., Jawhar, I., Mohamed, F., Al-Jaroodi, J.: Challenges of developing UAV applications: a project management view. In: International Conference on Industrial Engineering and Operations Management (IEOM), Dubai, pp. 1–10 (2015)
5. Kim, S.J., Jeong, Y., Park, S., Ryu, K., Oh, G.: A survey of drone use for entertainment and AVR (augmented and virtual reality). In: Augmented Reality and Virtual Reality, pp. 339–352 (2018)
6. Ryan, A., Zennaro, M., Howell, A., Sengupta, R., Hedrick, J.K.: An overview of emerging results in cooperative UAV control. In: 43rd IEEE Conference on Decision and Control (CDC), Nassau, pp. 602–607 (2004)

Data Mining and Learning

Summary of Part 5: Data Mining and Learning

Hyunseung Choo, Syed Muhammad Raza, and Tien-Dung Nguyen

With the wake of the digital revolution in the last couple of decades, we are currently experiencing the rise of big data, which will continue with unprecedented rate due to the arrival of the Internet of Things (IoT). This raises numerous challenges starting from the storage of data in resource efficient manner to applying learning algorithms on relevant and correct data set for better consumer services. Advanced database research and techniques are required to not only efficiently store and manage the data but also to retrieve it in simplistic and time effective manner. Data mining plays a vital role in identifying correlation and desired patterns in raw data. Supervised learning algorithms utilize the features of correlated data to train the different variants of neural networks, which are then used in different classification and prediction solutions. The papers collected in this part address the aforementioned challenges revolving around management of data and present forecasting and classification-based services which are developed on top of this data.

The paper, entitled "Sequence Searching and Visualizing over 3D Random Plot of Whole Genome Using Skip List," by Da-Young Lee, Hae-Sung Tak and Hwan-Gue Cho proposes a genome skip list that is modified by reflecting the characteristics of genome data in the skip list data structure, where linear genomic information sequence can be converted into a hierarchical structure. The paper, entitled "A Novel Symbolic Aggregate Approximation for Time Series," by Yufen Yu, deals with the time series data mining applications in which the classical symbolic approach cannot distinguish different time series with similar average values but different trends. The paper, entitled "Making Join Views Updatable on Relational Database Systems in Theory and in Practice," by Yoshifumi Masunaga, Yugo Nagata and Tatsuo Ishii investigate the view updatability in commercial and open source relational database systems alongside the ISO/IEC activities for prescribing view updatability through SQL. The paper, entitled "LP-HD: An Efficient Hybrid Model for Topic Detection in Social Network," by Qingmin Liu, Xiaofeng Gao, and Guihai Chen proposes a link prediction-based hybrid detection model (LP-HD) which combines text feature extraction and topology network for social topic detection. The paper, entitled "Connecting Heterogeneous Electronic Health Record Systems using Tangle," by Emil Saweros and Yeong-Tae Song solves a core problem of how to distribute transparent, updated medical data to a network of caregivers securely. The paper, entitled "Ontology-Based Recommender System for Sport Events," by Quang Nguyen, Luan N. T. Huynh, Tuyen P. Le, and Tae Choong Chung proposes an implementation of a hybrid system which provides recommendations based on smart content-based filtering and social-network-based user profiles for sport events. The paper, entitled "Enhancing Airlines Delay Prediction by Implementing Classification Based Deep Learning Algorithms," by Md. Nazmus Saadat and Md. Moniruzzaman focus on predicting airlines flight delays by analyzing flight data, especially, for the domestic Airlines those moves around the United States of America. The paper, entitled "Mining Regular High Utility Sequential Patterns in Static and

Dynamic Databases," by Sabrina Zaman Ishita, Chowdhury Farhan Ahmed, Carson K. Leung, and Calvin H. S. Hoi proposes a new algorithm for mining regular high utility sequential patterns from static databases. The paper, entitled "Mining Weighted Frequent Patterns from Uncertain Data Stream," by Jesan Ahammed Ovil, Chowdhury Farhan Ahmed, Carson K. Leung, and Adam G. M. Pazdor proposes a novel tree-based approach called WFPMUDS (Weighted Frequent Patterns mining from Uncertain Data Streams)-growth, which is capable of capturing recent behavior of uncertain data streams and only produces significant (weighted) patterns. The paper, entitled "Distributed Secure Data Mining with Updating Database using Fully Homomorphic Encryption," by Yuri Yamamoto and Masato Oguchi proposes the implementation of a master/worker distributed system using the FUP algorithm, which generates candidate item sets efficiently while updating the database. The paper, entitled "LSTM-based Recommendation Approach for Interaction Records," by Yan Zhou and Taketoshi Ushiama proposes an LSTM-based recommendation approach for interaction records. The paper, entitled "MMOU-AR: Multimodal Obtrusive and Unobtrusive Activity Recognition through Supervised Ontology-based Reasoning," by Muhammad Asif Razzaq and Sungyoung Lee aims to propose and provide supervised recognition of Activities of Daily Livings (ADLs) by observing unobtrusive sensor events using statistical reasoning, and investigates their semantic correlations by defining semantic constraints with the support of ontological reasoning. The paper, entitled "The Comparative Analysis of Single-Objective and Multi-Objective Evolutionary Feature Selection Methods," by Syed Imran Ali and Sungyoung Lee presents a comparative analysis of single-objective and multi-objective evolutionary feature selection methods over interpretable models. The paper, entitled "Water Sound Recognition Based on Support Vector Machine," by Tingting Hang, Jun Feng, Xiaodong Li, and Le Yan proposes a sound signal processing method to identify the water sound and link it with stream-flow measurements. The paper, entitled "Measuring Term Relevancy based on Actual and Predicted Co-occurrence," by Yuya Koyama, Takayuki Yumoto, Teijiro Isokawa, and Naotake Kamiura proposes a new measure of term relevancy: a ratio of actual and predicted values of co-occurrence to construct a model predicting co-occurrence for each query as piecewise approximation lines. The paper, entitled "Small Watershed Stream-flow Forecasting Based on LSTM," by Le Yan, Jun Feng, and Tingting Hang investigates the Stream-flow forecasting problem using a time series model based on Long Short-Term Memory (LSTM) that considers past stream-flow data, past weather data and weather forecasts data of the hydrological stations. The paper, entitled "Novel approach for multi-valued truth discovery on conflicting data sources," by Jun Feng, Ju Chen and Jiamin Lu solved the problem of discovering the truth from conflicting data sources by a novel approach for multi-valued truth discovery using Bayesian analysis. The paper, entitled "Information Extraction from Clinical Practice Guidelines: A Step Towards Guidelines Adherence," by Musarrat Hussain and Sungyoung Lee presents a mechanism for extracting meaningful information from Clinical Practice Guidelines, by transforming it into a structured format and training machine learning models. The paper, entitled "A Breast Disease Pre-Diagnosis Using Rule-based Classification," by Suriana Ismail, Roslan Ismail and Tengku Elisa Najiha Tengku Sifizul proposes a Rule Based System to help people to prevent and detect breast cancer early. The paper, entitled "Discovering Correlation in Frequent

Subgraphs," by Fariha Moomtaheen Upoma, Salsabil Ahmed Khan, Chowdhury Farhan Ahmed, Tahira Alam, Sabit Anwar Zahin, and Carson K. Leung presents two measures that help us discover such correlation among frequent subgraphs, based on the observation that elements in graphs exhibit the tendency to occur both connected and disconnected. The paper, entitled "Classifying License Plate Numerals Using CNN," by Tomoya Suzuki and Ryuya Uda proposes a new method to read numbers on a license plate with poor picture quality using CNN. The paper, entitled "Variational Deep Semantic Text Hashing with Pairwise labels," by Richeng Xuan, Junho Shim, and Sang-goo Lee proposes a supervised semantic text hashing method that utilizes pairwise label information. The paper, entitled "SisterNetwork: Enhancing Robustness of Multi-label Classification with Semantically Segmented Images," by Holim Lim, Jeeseung Han and Sang-goo Lee introduces SisterNetwork, a deep learning model to tackle the multi-label classification task for fashion attribute tagging, consisting of two different CNNs to leverage both the original image and the semantic segmentation information. The paper, entitled "A Searching for Strongly Egalitarian and Sex-Equal Stable Matchings with Ties," by Le Hong Trang, Hoang Huu Viet, Tran Van Hoai, and Tran Xuan Hao investigates the stable marriage problem with ties (SMT) problem, and proposes an algorithm based on bidirectional searching is presented for trying to find strongly egalitarian and sex-equal stable matchings. The paper, entitled "Deep Learning Drone Flying Height Prediction for Efficient Fine Dust Concentration Measurement," by Ji Hyun Yoon, Yunjie Li, Moon Suk Lee, Minho Jo proposes a algorithm to save flying distance of drones sensing fine dust information, using deep learning methods.

Sequence Searching and Visualizing over 3D Random Plot of Whole Genome Using Skip List

Da-Young Lee, Hae-Sung Tak, and Hwan-Gue Cho[✉]

Department of Electrical and Computer Engineering, Pusan National University, Busan, South Korea
{schematique,tok33,hgcho}@pusan.ac.kr

Abstract. With the development of next generation sequencing (NGS) technology, genomic research now requires analysis at the entire genome level. Because of easy access to very large amounts of data, it is desirable to look at all the data rather than examine individual bases. At this time, data visualization of the entire genome level can be very useful. However, most visualization tools simply visualize the resulting files derived from external analysis systems. In this study, it was possible to intuitively present the entire sequence to a researcher by converting the data for the entire genome into a 3-dimensional plot. In addition, by compressing the information in 3D space with run length encoding and storing it in a skip list, it is possible to perform fast comparison and search sequences with low complexity by layering base information. As a result, compared to alignment-based sequence comparisons, we obtained improved search results, and we could examine sequences from various angles using layered information.

Keywords: Non-alignment sequence search · Whole genome ·
Efficiency analysis · Approximation string search ·
Sequence preprocessing

1 Motivation

The genomic sequence information is string data composed of repetitions of four letters: adenine (A), guanine (G), thymine (T), and cytosine (C). We used diverse computing technologies to analyze genomic data due to the size and intuitive nature of the data. In particular, a comparison with the previously decoded sequence and search operation occur most frequently in order to compare features and functions in the sequence. The use of a sequence alignment tool based on a dynamic programming technique is appropriate in situations where the object to be analyzed and purpose are specified. This is because high spatial and time complexity are required, but the most accurate result can be obtained by comparing the strings themselves.

© Springer Nature Switzerland AG 2019
S. Lee et al. (Eds.): IMCOM 2019, AISC 935, pp. 791–804, 2019.
https://doi.org/10.1007/978-3-030-19063-7_64

However, if the characteristics and functions of the genome are not well known, the scope of the comparison is not specified and data must be analyzed over a wide unit at the level of the entire genome. At this time, visualization of entire genome units can provide useful information for identifying a wide range of patterns, such as the density of repeated sequences or internal duplication.

In this study, we propose a genome skip list that is modified by reflecting the characteristics of genome data in the skip list data structure [8]. By converting each base information into vector information in 3-D space, compressing it to a rate of 2^n and storing it in the skip list, a linear genomic information sequence can be converted into a hierarchical structure. The advantage of using run length encoding for compression is that the original data can be completely restored while reducing the storage space. In addition, the layered data structure reduces the search space during a sequence search. The precision of the visualization results can be verified at each compression ratio.

2   Related Work

With the development of NGS technology, genomic research now requires analysis at the entire genome level. The goal is to determine the distribution of specific sequences on a chromosome and their relative similarity. Since sequencing at the entire genome level requires analyzing vast amounts of data, research should begin by examining the entire genome rather than scrutinizing individual bases in detail. At this time, visualization of the entire genome is as a very useful analysis tool, and research on genome visualization is active.

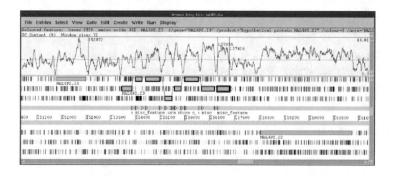

Fig. 1. 'Artemis' visualization tool [10]. (top) A G+C plot and (bottom) forward/reverse DNA sequence visualization results of a spliced gene from plasmodium falciparum chromosome 4.

Figure 1 is an analyzed sequence using visualization software called 'Artemis' [10]. 'Artemis' is a specialized software for visualizing and annotating results of an entire genome sequence derived from an external analysis tool. GC contents, protein properties, coding sequence, forward and reverse DNA, and other aspects can be visualized using different layers. Figure 1 shows a G+C plot and forward/reverse DNA for some sequences extracted from chromosome 4 of plasmodium falciparum. 'Easyfig' [12] is also a window-based sequencing information visualization software using the Python package Tkinter. This can be used to visualize sequence comparisons of BLAST using various colors and transformations. The advantage of displaying these analytical results using visualization tools is that they are interactive. By reviewing the visualization results in 'Artemis', information derived from an analysis tool like EMBL/GenBank can be structurally verified, rather than comparing only numerical data. 'Artemis' also allows you to examine visualization details so that you can check the individual base units as desired and, if necessary, one can check the results compressed in entire units at a glance.

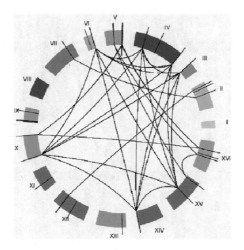

Fig. 2. 'ChromoWheel' [3] was used to visualize the relationship between the lysine degradation genes of saccharomyces cerevisiae. If they have the same E.C. number, the relationship between genes is shown using a connecting line.

'ChromoWheel' [3] and 'Circos' [6] are representative tools that show the relationship of genome sequence in a circular track form. Figure 2 shows a visualization of an information sequence using 'ChromoWheel' [3]. The relationship between lysine degradation genes in the DNA sequence of saccharomyces cerevisiae was visualized. When genes have the same E.C. number, the relationship between several genes is indicated by connecting related genes with lines.

This visualization tool aids genomic research by allowing a heavier sequence analysis program to process data externally. The analysis results are only transmitted to the user by using a visualization module, which is easy to access and use. However, it can only show information that was already analyzed.

On the other hand, sequence analysis using the visualization result itself was performed in some previous studies. Graphical alignment [9] is a typical example, where the 'four line' method is used to assign different values to A, G, T, and C bases. Based on this value, the DNA sequence is sequentially read, and the x value is increased, while y is varied depending on the base content. This analysis does not end with visualization, rather alignment is performed using the results. The process of visualizing two different sequences is shown in Fig. 3. This is one way of calculating the similarity between two sequences using the fact that differences in the y values are found to arise from different bases. However, by increasing the value on the x-axis in proportion to the sequence length, visualizing a large sequence such as a person or a gorilla can render a wide range of chromosome units unobservable.

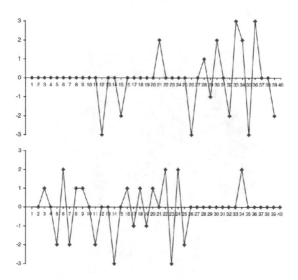

Fig. 3. Graphical alignment [9] process. The x-axis represents the nucleotide sequence index, and the y-axis represents the corresponding value for each base. Alignment is performed using the 'four line' method and with the difference in y values for two sequences (above and below).

In this study, we propose a model that can be used to search the sequence while preserving the intuitive expression of large-scale sequence information, which is an advantage of visualization. Genome studies necessarily require sequence comparisons and searches, and the most commonly used is alignment-based 'BLAST' [5]. However, the use of dynamic programming for large-scale data analysis makes the method very complex, thus it is not suitable for implementation on an individual computer. When the sequence is searched based on the compressed information in the visualization process, it becomes possible to work with much lower complexity. For this purpose, we developed a system that enables visualization and sequence comparison by storing vector information in 3D space as a hierarchical structure in a skip list.

3   Sequence Searching and Visualizing Model

3.1   Base Coordinate Generation

In order to convert the genomic sequence composed of four bases into information in 3-D space, we used a model from the literature [2] in which two bases are combined and assigned to three-dimensional vectors. When a two-dimensional vector is assigned to each base, loss of information occurs when a base has repeating antiparallel vectors. On the other hand, the combination of two bases can reduce information loss and provide intuitive understanding of the contents of the entire sequence through progression of the walk plot in 3-D space. Table 1 shows the 3-D vector assignment method for each 2-mer, and one can see that the Z-axis is used when bases with antiparallel vectors are combined. For example, if the DNA sequence is "AAGCTG-GTA", the result of the vector computed by sequential reading of the two bases is $(+2, 0, 0), (+1, +1, 0), (0, 0, +2), (-1, -1, 0), (-1, +1, 0), (0, +2, 0), (-1, +1, 0),$ $(0, 0, -2)$. The Figs. 4a and b below visualize the contig sequence of sable fish and pacific bluefin tuna, 3 and 10 kbyte in size, respectively. The sequence of Sablefish is relatively shorter, and plot also more monotonous.

Table 1. 3-D vector allocation method for the 2-mer base

2mer	Vector	2mer	Vector
AA	$(+2, 0, 0)$	$\{AG, GA\}$	$(+1, +1, 0)$
CC	$(0, -2, 0)$	$\{AT, TA\}$	$(0, 0, -2)$
GG	$(0, +2, 0)$	$\{CG, GC\}$	$(0, 0, +2)$
TT	$(-2, 0, 0)$	$\{CT, TC\}$	$(-1, -1, 0)$
$\{AC, CA\}$	$(+1, -1, 0)$	$\{GT, TG\}$	$(-1, +1, 0)$

3.2   Skip List Data Structure

In the aforementioned method, when the vector for the 2-mer of the entire sequence is calculated, information actually stored as a single character is transformed into 3-D integers, which is much larger than the size of the data. If you convert a 200 MB sequence of information, such as a person or a gorilla, to graphic information in 1:1 format, the worst-case file size is on the order of GB. Therefore, we convert the 10 cases that can occur in the 2-mer to a number ranging from 0 to 9 and compress it using run length encoding when the same 2-mer is repeated. If the genomic sequence is "AGAGAGAGTTTTT", then it compresses the information into $AG \times 7$, $GT \times 1$, and $TT \times 4$ according to the following equation:

$$< p_i, k_i > \tag{1}$$

P: pattern number
K: the number of pattern

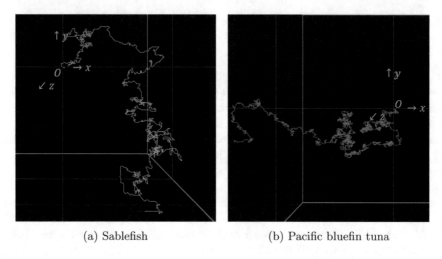

| (a) Sablefish | (b) Pacific bluefin tuna |

Fig. 4. The result of visualizing the contig sequence of marine organisms as a 3D random walk plot model [2]. The contig sequence of Sablefish (figure a) is 3 kbyte and the contig sequence of Pacific bluefin tuna (figure b) is 10 KB. The visualization results of Sablefish, which are relatively shorter, are monotonous.

Also, compressed and 1:1 information are required for fast searching and visualization at various levels. In this study, we set the compression magnification to 2^c, where $c = 3$. Therefore, at the $1 : 2^c$ level, a vector sum of the vectors corresponding to 2^c bp is stored as defined in the following equation:

$$V_i^l = \sum_{t=2^{cl} \times (i)}^{2^{cl} \times (i+1)} vec(< p_t, k_t >) \tag{2}$$

Figure 5 shows the structure of the existing skip list model. The probabilistically selected nodes in the linked list have not only the neighbor nodes but also higher level pointers. In the structure where the level is up to 4, most nodes have only a connection of level 1, that is, a connection pointer to the adjacent node, with a high probability. On the other hand, some probabilistically selected nodes have high-level pointers at the same time, such as nodes where data 500 is stored.

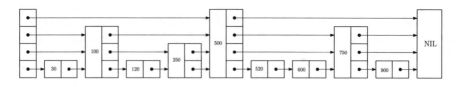

Fig. 5. The structure of the skip list model [8] with up to 4 levels.

This data structures are more efficient than other data structures such as trees, because they are easier to implement and reduce the number of pointers needed. In this paper, we modified the skip list to store vector information. The vector information of the calculated base is input sequentially, and it does not need to be deleted, and the element of the specific position is regularly searched. Since the base information is already reduced in the run length encoding process, it is not necessary to select a node probabilistically, and it is only necessary to link a base vector of a specific order with a higher level.

Figure 6 shows how such vector information is stored in a layered form in the skip list [8] data structure. Each individual list element stores 1:1 information compressed by Run Length Encoding (RLE), and vector information compressed by a factor 2^c is stored directly on top of it. A vector compressed by $(2^c)^2$ information is stored on top of that.

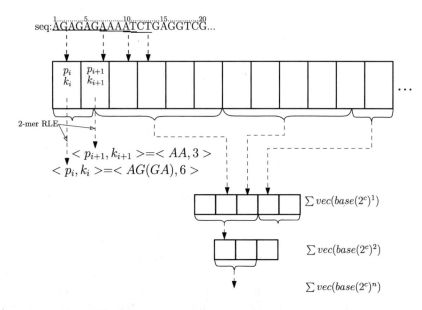

Fig. 6. Hierarchical storage structure of data for the entire genome sequence using run length encoding. Stored vector information compressed by 2^c at higher levels.

3.3   Sequence Comparison and Searching

In this paper, we use a model that compares vectors based on length(L_{AB}) and angle(θ_{AB}) [7]. The vector compression information for the query sequence and the top-level lesson of the reference sequence are obtained, and the length and angle between the two vectors are compared first. Figures 7, 8 shows these two parameters.

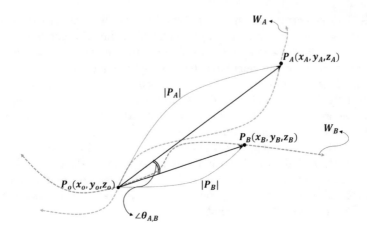

Fig. 7. Comparison parameters $\{\theta_{AB}\}$.

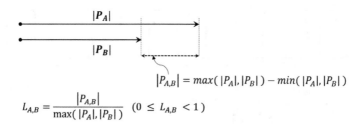

$$|P_{A,B}| = max(|P_A|, |P_B|) - min(|P_A|, |P_B|)$$

$$L_{A,B} = \frac{|P_{A,B}|}{max(|P_A|, |P_B|)} \quad (0 \leq L_{A,B} < 1)$$

Fig. 8. Comparison parameters $\{L_{AB}\}$.

If the lengths of the two vectors are the same, then L_{AB} is 0. If the difference between the two vectors is large, then L_{AB} is close to 1. The sequences are considered similar if the angle between the two vectors and the ratio between the length difference of the two vectors and the long vector are within a range defined by a predetermined reference value. In this way, when comparing two vectors at the highest level and comparing them with each other when the similarity is high, the exact similarity is calculated. However, if the degree of similarity at the upper level is extremely low, the comparison vector and search time can be reduced by skipping the vector without comparing.

4   Experiments

The data used in the experiments were organism data collected from NCBI [1] and the repeat sequences extracted from Retrotector [11]. A module was implemented in C++ that computes the base information as a vector, converts it to run-length encoding, compresses it, and stores it in the skip list. The module that reads compressed information and visualizes it according to the layering level was implemented using Javascript and Web-GL in Windows.

4.1 Biological Data Set Preparation

For the visualization of the compression stage, the target sequence used Human, Gorilla, Flatfish genome sequence provided by NCBI [1] and Tetraodon nigroviridis genome sequence provided by Ensembl [4]. For sequence searching experiment, the reference sequence used the entire Human genome sequence and Flatfish genome sequence provided by NCBI [1]. The query sequence used a partial sequence extracted from the corresponding sequence and a repeat sequence. The data specification is the same as that shown in Tables 2 and 3 below.

Table 2. Human genome sequence of chromosome unit and repeat sequence specification

Chr.	Size (MB)	Generate time (s)	Repeat type	Size (KB)	Generate time(s)
H1	246	33.327	HERV-K102	10	0.004
H5	179	25.281	HERV-K121	10	0.003
H6	169	23.676	HERV-K107	10	0.004
H7	157	21.926	HERV-K109	10	0.004
H8	144	20.260	HERV-K108	10	0.003
H11	133	18.641	HERV-K118	10	0.003
H22	50	5.581	HERV-K101	10	0.003

In the case of humans, the size of the chromosome unit file is 130 MB to 240 MB, while that of the flatfish is 20 MB to 25 MB. Therefore, it took a proportional amount of time to generate vector information for the entire genome. The HERV, LTR, and Gypsy sequences were queried. For Human, it took about 0.004 s to generate vector information and about 0.01 s for the Flatfish.

Table 3. Flatfish genome sequence of chromosome unit and repeat sequence specification

Chr.	Size (MB)	Generate time (s)	Repeat type	Size (KB)	Generate time (s)
F2	20	2.668	LTR	2	0
F8	21	2.823	Gypsy	11	0.015
F9	21	2.825	ERV	13	0.016
F10	25	3.322	Gypsy	13	0.016

4.2   Whole Genome Sequence Visualization in 3D Plot

Figure 9 shows the results of visualizing the first chromosome of a person and a gorilla at different levels. The red walk plot is the human sequence and the green is the gorilla. The walk plot is similar in the second quadrant, so that the similarity of the two sequences is visually confirmed. The layering is performed by exponentiation of 2^c, where c is currently set to 3. The layering level increases from left to right; the first picture is level 1, where a visualization result is matched 1:8 with the actual base. The second figure shows visualization at level 3, which is compressed at a ratio of $1:2^9$ compared to the actual base. The third picture shows visualization at level 4, which is compressed at a ratio of $1:2^12$ compared to the actual base. The fourth picture shows visualization at level 5, which is compressed at a ratio of $1:2^15$ compared to the actual base. The vector information is compressed as the layering level increases, the shape of the visualized walk plot is simple, and processing speed is fast. On the other hand, the information of the vector coincides with the actual base information when the level is lower, and the shape of the walk plot is highly precise and complicated.

Figure 10 shows the results of visualizing first chromosomes of Flatfish and Green spotted puffer at different levels. It visualizes levels 1, 3, 4, and 5 in order from Figure 10a to d. Flatfish and Green spotted puffer are about 10 times shorter in length than humans, and they are similar in size to 19 MB and 21 MB, respectively. When the layering level is 1, the direction of the walk plot can be examined in detail. However, when the level is 5, the information is extensively implied by excessive compression. Because the length of the sequences is about a factor 10 smaller than that of humans, it is best to confirm the degree of simplification at various levels visually.

Large-scale sequences can reduce the processing time because there is little loss of information that can be identified by the user even if they are compressed and visualized. In addition, since it can freely access various layering level, more flexible and efficient access to visualization information becomes possible.

4.3   Sequence Searching Using 3D Random Plot Model

Table 4 shows the results when the query sequence in each reference sequence is searched. H1 corresponds to Human chromosome number 1 and was used as a reference, while HERV-K102, an LTR sequence extracted from the sequence, was searched. 'Comp. Ratio' refers to the compression ratio at the time of a successful search, and the compression ratio is a multiple of 2^3. 'Difference (%)' refers to the ratio of the distance between the found location to the location of the actual query, and the location divided by the total reference length.

When HERV-K102 was searched in H1, the sequence search was successful when information was compressed by a factor 8. The similarity is 24% and the search time was approximately 38 s. Because we searched for information with a factor 8 compression, it was possible to locate a point that required a significant search time and nearly coincides with the location of the actual

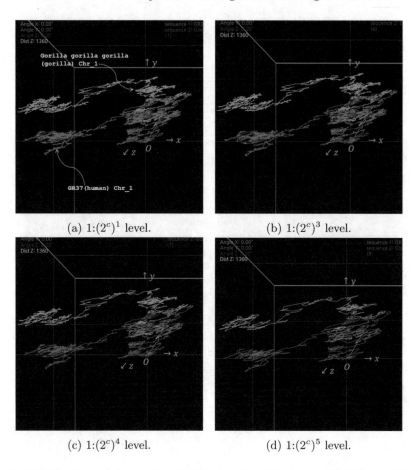

(a) 1:$(2^c)^1$ level.

(b) 1:$(2^c)^3$ level.

(c) 1:$(2^c)^4$ level.

(d) 1:$(2^c)^5$ level.

Fig. 9. Visualization of chromosome 1 sequence for a Human and Gorilla at different levels. In figure a, this is visualized to the extent that it approximately matches the contents of the actual sequence. The shape of the walk plot is highly precise and complex. In figures b, c and d, base information is compressed to $1/2^9$, $1/2^{12}$ and $1/2^{15}$ and visualized, respectively. Because the length of the chromosomes in humans and gorillas is long and the visualization results are complex, compression is still high, even at $1/2^{12}$.

sequence. On the other hand, when searching for query sequence HERV-K107 in H6, only information with a factor 64 compression can be searched successfully; the similarity is 78% and the search time was about 7.5 s. This is because both the query sequence and the reference sequence have less noise. During coordinate compression, other characters other than A, G, T, and C are regarded as noise and are assigned a 0 vector. The 0 vector is not required to visualize the walk plot, but it should not be overlooked because it is required to calculate the position of 1:1 bp.

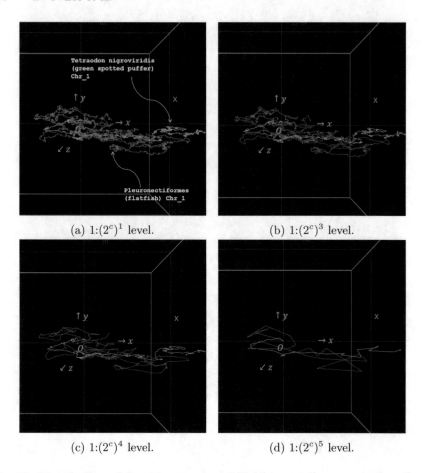

(a) $1{:}(2^c)^1$ level.

(b) $1{:}(2^c)^3$ level.

(c) $1{:}(2^c)^4$ level.

(d) $1{:}(2^c)^5$ level.

Fig. 10. Visualization of first chromosome of Flatfish and Green spotted puffer at different levels. In figure a, this is visualized to the extent that it approximately matches the contents of the actual sequence. The shape of the walk plot is highly precise and complex. In figures b, c and d, base information is compressed to $1/2^9$, $1/2^{12}$ and $1/2^{15}$ and visualized, respectively. Since the sequence length for a Flatfish is about 10 times shorter than that for a human, best to confirm the degree of simplification at various levels visually.

Therefore, more noise in the reference sequence and query sequence required consideration of more compression information, which increases the search time. In case of the Flatfish, F2 used chromosome 2 of the Flatfish as a reference, and the LTR sequence extracted from the corresponding sequence was used as a query. The sequence search could be successfully performed using only a factor 64 compression. The search time required was about 0.09 s, and the error rate between the actual query sequence position and searched position was only 0.61%.

Table 4. Search results for Human and Flatfish query sequences in the reference sequence

Ref. Seq	Que. Seq	Comp. Ratio	Similarity (%)	Difference (%)	Search time (s)
H1	HERV-K102	8	24	0	38.148
H5	HERV-K121	8	30	0.19	30.271
H6	HERV-K107	64	38	0	7.537
H7	HERV-K109	64	26	0.18	7.271
H8	HERV-K108	64	78	0	6.268
H11	HERV-K118	8	50	16.76	46.612
H22	HERV-K101	64	48	5.0	0.646
F2	LTR	64	27	0.61	0.093
F8	Gypsy	8	12	0.84	2.605
F9	ERV	64	13	0	1.872
F10	Gypsy	8	14	8.4	8.315

5   Conclusion and Future Work

Analysis of genomic data is based on a comparison between well-known sequences. In this case, if the comparison target is specified and the space to be searched is relatively limited, one can obtain the most accurate alignment based on dynamic programming. However, if the search space is extended over a wide range of genome units, researchers will need access tools that can quickly scan large volumes of data. At this time, alignment requires considerable analysis time due to its high complexity, but a visualization-based heuristic approach can provide researchers with many intuitive insights. In this study, data from the entire genome was calculated as a vector in 3-dimensional space, converted into visualization information, compressed using run length encoding, and stored in a skip list. It is shown that genome data can be visualized by layering, and layered information can be processed faster compared the use of linear string data in similarity comparison and sequence search.

Future studies will focus on improving the delay involved in the presence of large amounts of noise in the query and reference sequences. The 0 vector generated during compression is not directly used for visualization. However, it is stored because the information is necessary for location calculation during search, thus degrading performance. To solve this problem, we want to apply our robust operation to noisy data by separately storing the information required for searching and visualization.

Acknowledgment. This research was supported by Basic Science Research Program through the National Research Foundation of Korea (NRF) funded by the Ministry of Education (NRF-2017R1D1A1A02018504).

References

1. Coordinators, N.R.: Database resources of the national center for biotechnology information. Nucleic Acids Res. **44**(Database issue), D7 (2016)
2. Da-Young, L., Kyung-Rim, K., Taeyong, K., Hwan-Gue, C.: Comparison-specialized visualization model for whole genome sequences. J. WSCG **24**(2), 43–52 (2016)
3. Ekdahl, S., Sonnhammer, E.L.: ChromoWheel: a new spin on eukaryotic chromosome visualization. Bioinformatics **20**(4), 576–577 (2004)
4. Hubbard, T., Barker, D., Birney, E., Cameron, G., Chen, Y., Clark, L., Cox, T., Cuff, J., Curwen, V., Down, T., et al.: The ensembl genome database project. Nucleic Acids Res. **30**(1), 38–41 (2002)
5. Johnson, M., Zaretskaya, I., Raytselis, Y., Merezhuk, Y., McGinnis, S., Madden, T.L.: NCBI BLAST: a better web interface. Nucleic Acids Res. **36**(suppl-2), W5–W9 (2008)
6. Krzywinski, M.I., Schein, J.E., Birol, I., Connors, J., Gascoyne, R., Horsman, D., Jones, S.J., Marra, M.A.: Circos: an information aesthetic for comparative genomics. Genome Res. **19**(9), 1639–1645 (2009)
7. Lee, D.Y., Tak, H.S., Kim, H.H., Cho, H.G.: Alignment-free sequence searching over whole genomes using 3D random plot of query DNA sequences. Informatica **42**(3) (2018)
8. Pugh, W.: Skip lists: a probabilistic alternative to balanced trees. Commun. ACM **33**(6), 668–676 (1990)
9. Randic, M., Zupan, J., Plavsic, D., et al.: A novel unexpected use of a graphical representation of DNA: graphical alignment of DNA sequences. Chem. Phys. Lett. **431**, 375–379 (2006)
10. Rutherford, K., Parkhill, J., Crook, J., Horsnell, T., Rice, P., Rajandream, M.A., Barrell, B.: Artemis: sequence visualization and annotation. Bioinformatics **16**(10), 944–945 (2000)
11. Sperber, G., Lövgren, A., Eriksson, N.E., Benachenhou, F., Blomberg, J.: Retrotector online, a rational tool for analysis of retroviral elements in small and medium size vertebrate genomic sequences. BMC Bioinformatics **10**(6), S4 (2009)
12. Sullivan, M.J., Petty, N.K., Beatson, S.A.: Easyfig: a genome comparison visualizer. Bioinformatics **27**(7), 1009–1010 (2011)

A Novel Symbolic Aggregate Approximation for Time Series

Yufeng Yu[1]([⊠]), Yuelong Zhu[1], Dingsheng Wan[1], Huan Liu[2],
and Qun Zhao[1]

[1] Hohai University, 1 Xikang Road, Nanjing, Jiangsu, China
hhuheiyun@126.com
[2] Arizona State University, Tempe, USA
huanliu@asu.edu

Abstract. Symbolic Aggregate approximation (SAX) is a classical symbolic approach in many time series data mining applications. However, SAX only reflects the segment mean value feature and misses important information in a segment, namely the trend of the value change in the segment. Such a miss may cause a wrong classification in some cases, since the SAX representation cannot distinguish different time series with similar average values but different trends. In this paper, we present Trend Feature Symbolic Aggregate approximation (TFSAX) to solve this problem. First, we utilize Piecewise Aggregate Approximation (PAA) approach to reduce dimensionality and discretize the mean value of each segment by SAX. Second, extract trend feature in each segment by using trend distance factor and trend shape factor. Then, design multi-resolution symbolic mapping rules to discretize trend information into symbols. We also propose a modified distance measure by integrating the SAX distance with a weighted trend distance. We show that our distance measure has a tighter lower bound to the Euclidean distance than that of the original SAX. The experimental results on diverse time series data sets demonstrate that our proposed representation significantly outperforms the original SAX representation and an improved SAX representation for classification.

Keywords: Time series · Trend feature · Symbolic aggregate approximation · Lower bound · Distance measure

1 Introduction

Time series is a sequence of data changing with time order, which is increasingly important and has attracted an increasing interest due to its wide applications in many domains, such as nature science, engineering technology and social economics. As a consequence, it is said that the time series mining is considered as one of the ten challenging problems in data mining [1]. However, there are a number of challenges in time series data mining, such as high dimensionality, high volumes, high feature correlation and large amount of noises. Moreover, most time series data exist as collections of consecutive values varying continuously in time, which makes many data mining methods ineffective and fragile. As a prerequisite, the consecutive time series

© Springer Nature Switzerland AG 2019
S. Lee et al. (Eds.): IMCOM 2019, AISC 935, pp. 805–822, 2019.
https://doi.org/10.1007/978-3-030-19063-7_65

data need to be taken a way of dimensionality reduction and then formed new representations for time series data mining algorithms.

Symbolic Aggregate approximation (SAX) [2] is a classical symbolic approach for time series data mining, the basic concept of which is to convert the numerical form of a time series into a sequence of discrete symbols according to designated mapping rules. SAX can reduce dimensionality/numerosity of data and has a lower bound to the Euclidean distance, that is, the error between the distance in the SAX representation and the Euclidean distance in the original data is bounded [3]. Therefore, the SAX representation speeds up the data mining process of time series data while maintaining the quality of the mining results. The SAX has been widely used for applications in various domains such as mobile data management [4], financial investment [5] and shape discovery [6].

SAX is based on Piecewise Aggregate Approximation (PAA) [7], so, it has a major limitation inherit from PAA on dimensionality reduction. That is, the symbols in the SAX representation are mapped from the mean values of segments, which may miss other important feature information such as trend and extreme feature of the segments. Furthermore, different segments with similar average values may be mapped to the same symbols, which make the SAX distance between them is 0. For instance, two time series (a) and (b) in Fig. 1 have different trend and extreme feature but their SAX representations are the same as 'feacdb', which may cause misclassifications when using distance- based classifiers.

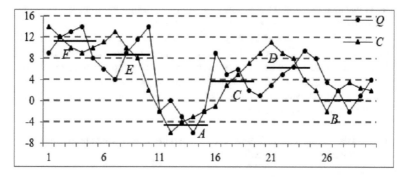

Fig. 1. The same SAX symbolic representation in the same condition

Q and C in Fig. 1 have the same SAX symbolic representation 'FEACDB' in the same condition where the length of time series is 30, the number of segments is 6 and the size of symbols is 6. However, they have different time series. Time series 1 in Q, Time series 2 in C.

The Extend SAX (ESAX) representation overcomes some limitations by tripling the dimensions of the original SAX [8], which adds two new values in each segment based on SAX, i.e., max value and min value, respectively. Therefore, the ESAX representations of time series Q and C in Fig. 1 are 'efffecaaaacffdbcbb' and 'effcefbaafcaadfbbc' respectively. Though ESAX can more approximately represent the time series, it still has the problem in terms of ignoring the important trend feature.

We present a novel symbolic aggregate approximation representation and distance measure for time series, which can not only reflect the segment mean value feature, but also capture trend feature with good resolution, further support time series data mining tasks.

Our work can be simply summarized as follows:

- We extract the trend feature from time series, according to trends distance factor and trend shape factor. Then we design multi-resolution discretization method to transform the trend feature into symbols.
- We propose a novel symbolic dimensionality reduction approach, and call it as Trend Feature Symbolic Aggregate approximation (*TFSAX*) and put forward distance function *TDIST* based on *TFSAX*.
- We demonstrate the effectiveness of the proposed approach by experiments on different datasets. The comprehensive experiments have been conducted in comparison with the SAX and the ESAX representations, experiments validate the utility of our proposed approach.

2 Background

SAX [2] is a symbolic aggregation approximation representation method based on piecewise aggregation approximation PAA [7]. For instance, to convert a time series sequence C of length n, $C = \{C_1, C_2, \dots, C_n\}$, into w symbols, the SAX works as follows:

Step 1: Normalization. Transform raw time series C into normalized time series C' with mean of 0 and standard deviation of 1.

Step 2: Dimensionality reduction. The time series is divided into w equal-sized segments by Piecewise Aggregate Approximation (*PAA*) [7]. That is, $\bar{C} = \{\bar{C}_1, \bar{C}_2 \cdots \bar{C}_w\}$, the ith element of $\bar{C}_i$ is the average of the ith segment and is calculated by the following equation:

$$\bar{c}_i = \frac{w}{n} \sum_{j=\frac{n}{w}(i-1)+1}^{\frac{n}{w}i} c_j \qquad (1)$$

Where c_j is one point of time series C, j is from the starting point to the ending point for each segment.

Step 3: Discretization. According to *SAX* breakpoints search table, choose alphabet cardinality, discretize $\bar{C}$ into symbols and obtain SAX representation $\hat{C}$.

Breakpoints are a sorted list of numbers $B = \beta_1, \beta_2, \dots, \beta_a$ such that the area under a $N(0,1)$ Gaussian curve from β_i to $\beta_{i+1} = \alpha$, where $\beta_1 = -\infty$, $\beta_\alpha = +\infty$. Comparing the segment mean value of $\bar{C}$ with breakpoints, if the segment mean value is smaller than the smallest breakpoint β_1, the segment is mapped to symbol 'a'; if the segment mean value is larger than the smallest breakpoint β_1 and smaller than β_2, the segment is mapped to symbol 'b'; and so forth. These symbols for approximately representing a time series are called a 'word'. Figure 2 illustrates a sample time series converted into the SAX word representation.

Table 1. A lookup table for breakpoints

β_i	a				
	3	4	5	6	7
β_1	−0.43	−0.67	−0.84	−0.97	−1.07
β_2	0.43	0	−0.25	−0.43	−0.57
β_3	−	0.67	0.25	0	−0.18
β_4	−	−	0.84	0.43	0.18
β_5	−	−	−	0.97	0.57
β_6	−	−	−	−	1.07

For the utilization of the SAX in classic data mining tasks, the distance measure was proposed. Give two raw time series $Q = \{q_1, q_2, \ldots, q_n\}$ and $C = \{c_1, c_2, \ldots, c_n\}$ with the same length n, $\widehat{Q}$ and $\widehat{C}$ are their SAX representations respectively with the word size w. In order to measure the similarity based on SAX representation, the SAX distance MINDIST is defined as follows:

$$MINDIST(\widehat{Q}, \widehat{C}) = \sqrt{\frac{n}{w}} \sqrt{\sum_{i=1}^{w} (dist(\widehat{q}i, \widehat{c}i)^2} \qquad (2)$$

where $dist()$ function can be implemented using the lookup table as illustrated in Table 1, and is calculated by the following equation:

$$dist(\widehat{q}i, \widehat{c}i) = \begin{cases} 0 & |\widehat{q}i - \widehat{c}i| \leq 1 \\ \beta max(\widehat{q}i, \widehat{c}i) - 1 - \beta min(\widehat{q}i, \widehat{c}i) & otherwise \end{cases} \qquad (3)$$

Since SAX extracts the mean value information, and misses the trend information, Consequently, MINDIST only measures the similarity of mean value of segments through dimensionality reduction; it can't evaluate the trend similarity.

3   Related Work

SAX is the first proposed symbolic approach which allows dimensionality reduction and supports lower bounding distance measure. Though SAX has a generality of the original presentation and works well in many problems, it can also lead to the loss of trend feature information within the original sequence. There have been some improvements of the SAX representation recently.

Some methods improve the SAX by adaptively choosing the segments. The method in [9] uses the discretization of Adaptive Piecewise Constant Approximation (APCA) [10] to replace the PAA [7] in the SAX. The method in [11] makes use of an adaptive symbolic representation with the adaptive vector of 'breakpoints'. While the two methods above reduce the reconstruction error on some data sets, they still use the average values as the basis for approximation (the latter method uses the same distance measure as the SAX) and do not consider the differences of value changes between segments.

Some methods improve the SAX by enriching the representation of each segment. The method in [8] uses three symbols, instead of a single symbol, to represent a segment in time series. This method triples the dimensions of the SAX and the high dimensionality increases the computational complexity. The method in [12] utilizes a symbolic representation based on the summary of statistics of segments. The method considers the symbols as a vector, including the discretized mean and variance values as two elements. However, it is may be inappropriate to transform the variances to symbols using the same breakpoints for the transformation of the mean values to symbols.

Trend estimation of time series is an important research direction. Many methods have been proposed to represent and measure trends [13–22]. Sun [19] proposed Symbolic Aggregate appro-Ximation based on Trend Distance (SAX-TD), which captures the trends of time series in numerical form by approximating the measure of trends using the difference between the average and the starting point of segment, and the difference between the average and the ending point of segment. Though SAX-TD can better represent time series with different trend characteristics, it is difficult to reflect the trend of the segmentation when the starting point and end point of the sequence are equal to the mean. Yin [22] proposed an improved SAX method called Trend Feature Symbolic Approximation (TFSA), which uses the trend symbols to represent the subsequences after segmentation and allows the subsequences can visually display these trend features. Though TFSA represents biomedical series using trend symbols to improve the classification accuracy of symbolic methods, it also triples the dimensions of the SAX and the high dimensionality increases the computational complexity.

4 Tfsax-Improved Sax Based on Trend Feature

In this section, we employ the sequence mean feature and trend shape feature to represent the time series. It first constructions trends feature triangular shape using trend distance factor and trend shape factor as its right-angle edge, and then, adopt tangent function value of the angle θ which calculate as the ratio of trends distance factor and trend shape factor to quantitatively measure the different trends, at last, incorporate the trend feature variations into the original representation.

4.1 Trend Feature Extraction

4.1.1 Trend Distance Factor

Trends are important characteristics of time series, and they are crucial for the analysis of similarity and classification of time series [24]. However, there is no common definition and a measurement of trend in time series. Therefore, how to define the measure of the distance factor between trend series becomes a difficult problem need to be solved while mining a time series by using trend characters.

Although there are no common definition of trend and a measurement of trend distance in time series, the starting and the ending points are important in segment trend estimation. For example, a trend is up when the value of the ending point is significantly larger than the value of the starting point, while the trend is down when the value of the ending point is significantly smaller than the value of the starting point. It is difficult to qualitatively define a trend, such as the definitions of 'significant up' and 'significant down', 'significant down' and 'slight down'. However, if the trend information of a segment is not utilized, the representations of a time series containing many segments are rough.

Definition 1 Trend distance: Given time series segments q and c with the same length, defined $\Delta q(t_s)$ and $\Delta q(t_e)$ to represent the first order difference between the starting point, the end points and the sequence mean of the segment q separately. The trend distance factor td of them is defined as follows:

$$td = \Delta q(t_s) - \Delta q(t_e) \qquad (4)$$

More obviously, it can see that sequences (a) and (b) in Fig. 2 have the same trend(both rising), but which can be distinguished by the trend distance factor $\Delta p(t_s) - \Delta p(t_e)$.

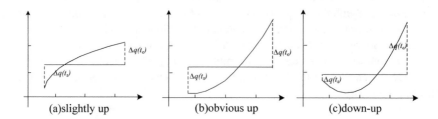

Fig. 2. Several typical segments with the same trend

4.1.2   Trend Shape Factor

For a time series sequence, the start and end points of the sequence can determined the overall trend, but cannot fully indicate the trend shape features in the sequence. For example, there may be local up-trend and in the overall stationary time series, similarly to an overall up-trend time series, which may contain local down-trend and stationary sequence. So, it is necessary to use the feature point such as extreme point or trend point to measure the long time series sequence.

The trend point is the turning point of the time series sequence from one shape to another, which is the point of change that affects the local trend of the pattern.

Definition 2 Trend point: Given a time series X of length N, $X = \{x_1, x_2, \dots, x_n\} \in R^N$ and a data point x_i $(i < m)$, if x_i satisfied (1) $(x_i - x_{i-1})(x_{i+1} - x_i) < 0$; (2) $(x_i - x_{i-1})(x_{i+1} - x_i) = 0$ and $(x_i - x_{i-1}) \neq (x_{i+1} - x_i)$ $(2 \leq i \leq n)$, then x_i is called a up (down, stationary)-trend point.

Correspondingly, the trend shape factor of the time series can be transformed into find the trend point numbers K of the time series. The larger the K takes, the more complex the trend be expressed. Particularly, the value of K is 1 for overall up-, down or stationary trend sequence.

Figure 3 lists several typical sequences which have same (or similar) overall trend but different trend shape features. These sequences can be divided into different trend shape, overall up shape, up after down shape (down-up), down after up shape and compound shape which may contains several combinations of up-down-up shapes.

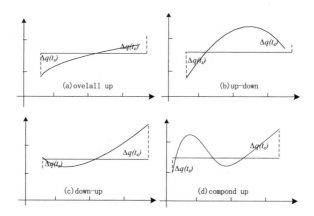

Fig. 3. Several typical patterns with the same (or similar) overall trend but different trend shape

4.1.3 Trend Feature Representation

As we reviewed above, only used trend distance factor or the trend shape factor cannot completely represent the feature of the trend. Therefore, this paper proposes a new approach to represent the overall feature, it first constructions trends feature triangle using trend distance factor $(\Delta p(t_e) - \Delta p(t_s))$ and trend shape factor (K) as its right-angle edge, and then, adopt tangent function value of the angle θ which calculate as the ratio of trend distance factor and trend shape factor to quantitatively measure the different trends, that is:

$$tan\theta = \frac{\Delta p(t_e) - \Delta p(t_s)}{N} \tag{5}$$

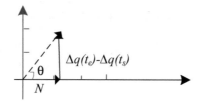

Fig. 4. The construction of trend feature triangle

The construction of trend feature triangle is shown in Fig. 4.

The angle space corresponding to the trend feature triangle is (−90°, 90°), which can be divided into certain numbers of no overlapping intervals, each interval corresponds to a symbol. The trends of segment can be catalogued into three kinds based on the value the angle θ taken. The θ takes a positive value, it indicates that the sequence has the up-trend, moreover, the higher value the θ takes, the faster the trend rising (slow rising, fast rising and sharp rising); so as to the stationary trend and the down-trend. Based on the extent of trend change from judging whether it is high or low and whether it is slow or sharp, design multi-resolution angle break-point intervals search table to map trends to symbols, which is shown as Table 4.

Definition 3 Angle Breakpoints (AB): Angle breakpoints are sorted list of angles, denoted by $B = \theta_1, \theta_2 \ldots \theta_\alpha$. Suppose each interval from θ_i to θ_{i+1} has the equal probability of $1/\alpha$, specifically, $\theta_1 = -\infty$, $\theta_\alpha = +\infty$.

Table 2. Angle breakpoint interval SearchTable

θ_i	a				
	2	3	4	5	6
θ_1	0	−5°	−30°	−30°	−30°
θ_2		5°	0°	−5°	−5°
θ_3			30°	5°	0°
θ_4				30°	5°
θ_5					30°

The corresponding relationship between trend symbol and angle is determined by search table shown as Table 2, for each slope value in the segment, use symbol to represent trend feature according to Table 4. Comparing the angle reflected by slope with a series of angle breakpoints, if the angle of trend segment is smaller than the smallest angle breakpoint θ_1, this trend segment is mapped to symbol 'A'; if the angle of trend segment is larger than the smallest angle breakpoint θ_1 and smaller than θ_2, this trend segment is mapped to symbol 'B'; and so on.

4.1.4   Trend Feature Distance

Distance measurement is an important problem in data mining. So it is necessary to define the trend feature distance after the trend feature representation.

Definition 4 Trend feature distance (*tfdist*): The distance between trend characters can be described by a two-dimensional matrix, the element *tfdist*[i][j] corresponding to the ith row, the jth column can be calculated as follow:

$$tfdist[i][j] = \begin{cases} 0 & |i - j| \leq 1 \\ tan(\theta_{max(i,j)-1} - \theta_{min(i,j)}) & otherwise \end{cases} \quad (6)$$

By checking the corresponding row and column of two symbols in Table 3, the distance between trend symbols can be obtained. For example, *tfdist (A,B) = 0, tfdist (A, C) = tan25°, tfdist (A,E) = tan60°*. When trend cardinality α is arbitrary, the cell'in *tfdist ()* search table can be calculated by Eq. (6).

Table 3. tfdist () search table ($\alpha = 5$)

Trend character	Trend character				
	A	B	C	D	E
A	0	0	tan25°	tan35°	tan60°
B	0	0	0	tan5°	tan35°
C	tan25°	0	0	0	tan25°
D	tan35°	tan5°	0	0	0
E	tan60°	tan35°	tan25°	0	0

4.2   TFSAX Representation

4.2.1   Symbolic Representation

Trend Features based SAX (TFSAX) is a symbolic representation of the mean and trend features of the time series, which not only inherited the advantages of the classic SAX, but also overcomes the shortcomings of the traditional symbolic method that only uses the mean values to describe the original time series, so that it can better represent a different form of time series. Because the continuity of time series data, the ending point of a segment is the starting point of the following segment. Therefore, we only need to add a trend feature to each of the sequence segments to indicate the trend of the segment.

Formally, given two time series Q and C with the length of n, the representations with $w(w < <n)$ words of them are:

$$Q : \Delta q(1)\hat{q}1\Delta q(2)\hat{q}1 \ldots \Delta q(w)\hat{q}_w$$
$$C : \Delta c(1)\hat{C}1\Delta c(2)\hat{C}2 \ldots \Delta c(w)\hat{C}_w$$

where $\hat{C}1\hat{C}2 \ldots \hat{C}w$ are the symbolic representations by the SAX, $\Delta c(1)\Delta c(2) \ldots \Delta c(w)$ are the trend feature representations, and so as to $\hat{q}1\hat{q}2 \ldots \hat{q}w$ and $\Delta q(1)\Delta q(2) \ldots \Delta q(w)$. Compared to the original SAX, our representation adds w dimensions for trend feature.

Based on the two time series shown in Fig. 1, the differences between the several methods under the same conditions are specifically analyzed. The results of the SAX method are "*FEACDB*" and the results of the ESAX are "*EFFFECAAAACFFDBCBB*" and "*EFFCEFBAAFCAADFBBC*". The results of my method are "$F_d E_f A_a C_b D_e B_c$" and "$F_c E_a A_e\ C_f D_b B_d$". For the same number of segments, the dimension of ESAX is three times that of SAX, and the dimension of this method is approximately twice that of SAX. However, the method of my method is much higher than SAX and ESAX in improving the accuracy of distance calculation.

4.2.2   Distance Measures

After the time series are symbolized, it is generally necessary to define distance metrics between the sequences as a metric for subsequent mining. However, for the hydrological time series data mining, the SAX method ignores the trend characteristics of the sequence and discrete the partial precision of the sequence when the time series is discretized. On the other hand, because the trend distance can reflect the trend information characteristics of the sequence segment, and can indicate the trend difference of the sequence segment. Therefore, this section combines the SAX symbol distance and the trend feature symbol distance, and proposes a time series based on the trend feature. The approximate approximation method is used to solve the distance measurement after time series data symbolization. Therefore, we define the distance between two time series based on the trend distance as follows:

$$TDIST(Q,C) = \sqrt{\frac{n}{w}} \sqrt{\sum_{i=1}^{w} \left(dist(\hat{q}_i, \hat{c}_i)^2 + \frac{w}{n} (tfdist(q_i, c_i))^2 \right)} \qquad (7)$$

where $dist(\hat{q}_i, \hat{c}_i)$ is the symbolic distance of the segments q_i and c_i based on SAX representation and $tfdist(q_i, c_i)$ is the trend symbolic distance of the segments q_i and c_i based on TFSAX representation. Note that $\hat{q}_i$ and $\hat{c}_i$ are the new representations of time series Q and C with the same length n, w is the number of segments (or words), and $\hat{q}_i$ and $\hat{c}_i$ are the symbolic representations of segments of segments q_i and c_i, respectively.

From Eq. (6), we see that the influence of the trend distance on the overall distance is weighted by the ratio of dimensionality reduction w/n. w/n is larger when there are more divided segments and each segment is shorter. w/n is smaller when there are fewer divided segments and each segment is longer. This is because in a short segment, the trend is likely to be linear and can be largely captured by two points and hence the weight for the trend distance is high. When the segment is long, the trend is complex, two points are unlikely to capture the trend and hence the weight of the trend distance is low.

4.2.3   Lower Bound

One of the most important characteristics of the SAX is that it provides a lower bounding distance measure. Lower bound is very useful for controlling errors and speeding up the computation. Below, we will show that our proposed distance also lower bounds the Euclidean distance.

The quality of a lower bounding distance is usually measured by the tightness of lower bounding (*TLB*).

$$TLB = \frac{Lower\,Bounding\,Distance(P,Q)}{Euclidean\,Distance(P,Q)} \qquad (8)$$

The value of *TLB* is in the range [0, 1]. The larger the TLB value, the better the quality. Recall the distance measure in Eq. (7), we can obtain that TLB(TDIST) ≥ TLB (MINIDIST), which means the TFSAX distance has a tighter lower bound than the original SAX distance. In conclusion, our improved TFSAX not only holds the lower bounding property of the original SAX, but also achieves a tighter lower bound.

According to [3, 5], the authors have proved that *SAX* distance lower bounds the *PAA* distance and the *PAA* distance lower bounds the Euclidean distance, that is:

$$\sqrt{\sum_{i=1}^{n} (q_i - c_i)^2} \geq \sqrt{\frac{n}{w}} \sqrt{\sum_{i=1}^{w} (\bar{q}_i - \bar{c}_i)^2} \qquad (9)$$

$$n(\bar{p} - \bar{q})^2 \geq n(dist(\hat{p}, \hat{q}))^2 \qquad (10)$$

For proving the TDIST also lower bounds the Euclidean distance, we repeat some of the proofs here. Let $\bar{Q}$ and $\bar{C}$ be the means of time series Q and C respectively. We first consider only the single-frame case (i.e. $w = 1$), Eq. (9) can be rewritten as follows:

$$\sum_{i=1}^{n} (q_i - c_i)^2 \geq n(\bar{Q} - \bar{C})^2 \qquad (11)$$

Recall that $\bar{Q}$ is the mean of the time series, so p_i can be represented in terms of $q_i = \bar{Q} - \Delta q_i$. The same applies to each point c_i in C. Thus, Eq. (11) can be rewritten as follows:

$$\sum_{i=1}^{n} ((\bar{Q} - \Delta q_i) - (\bar{C} - \Delta c_i))^2 \geq n(\bar{Q} - \bar{C})^2 \qquad (12)$$

Rearranging the left-hand side, and then expand by the distributive law as follows:

$$\sum_{i=1}^{n} (\bar{Q} - \bar{C})^2 + \sum_{i=1}^{n} (\Delta q_i - \Delta c_i)^2 - 2 \sum_{i=1}^{n} (\bar{Q} - \bar{C})(\Delta q_i - \Delta c_i) \geq n(\bar{Q} - \bar{C})^2 \qquad (13)$$

Note that Q and C are independent to n, and it was also proved that Δp_i and Δq_i are satisfied:

$$\sum_{i=1}^{n} (\Delta qi - \Delta ci) = \sum_{i=1}^{n} ((\bar{Q} - \Delta qi) - (\bar{C} - \Delta ci)) = (n\bar{q} - \sum_{i=1}^{n} \Delta qi) - (n\bar{c} - \sum_{i=1}^{n} \Delta ci) = 0 \qquad (14)$$

Therefore, after substituting 0 into the third term on the left-hand side, Eq. (14) becomes:

$$n(\bar{Q} - \bar{C})^2 + \sum_{i=1}^{n} (\Delta qi - \Delta ci)^2 \geq n(\bar{Q} - \bar{C})^2 \qquad (15)$$

Obviously, Eq. (15) holds true, that is, *PAA* distance lower bounds the Euclidean distance.

(14) Recall the definition in Eq. (5), $(\Delta p(t_e) - \Delta p(t_s)) = N \tan \theta$, we can obtain an inequality as follows ($i = 1$ is the starting point and $i = n$ is the ending point):

$$\sum_{i=1}^{n} (\Delta qi - \Delta ci)^2 = (\Delta q1 - \Delta c1)^2 + \ldots + (\Delta qn - \Delta cn)^2 \geq (\Delta q1 - \Delta c1)^2 + (\Delta qn - \Delta cn)^2$$

$$= (\Delta q1 - \Delta c1)^2 + ((\Delta q1 - \Delta c1) + (Nq \tan \theta q - Nc \tan \theta c))^2$$

$$\geq 2(\Delta q1 - \Delta c1)^2 + (Nq \tan \theta q - Nc \tan \theta c)^2$$

$$(16)$$

Moreover, according to tangent trigonometric formula, it holds that:

$$\tan \theta q - \tan \theta c = (1 + \tan \theta q \tan \theta c) \tan(\theta q - \theta c) \tag{17}$$

Without loss of generality, let $N_q \geq N_c$(obviously, both N_q and N_c satisfy $N_q \geq 1$), we can get:

$$(Nq \tan \theta q - Nc \tan \theta c)^2 \geq Nc^2 (\tan \theta q - \tan \theta c)^2 \geq (\tan(\theta q - \theta c))^2 \geq (\textit{tfdist}(Q, C))^2 \tag{18}$$

Particularly, Q and C are divided into one segment for the single-frame case, moreover, the trend features of Q or C can divide into no more than two different forms. Thus, the trend feature distance between Q and C is 0 according to Definition 4. Then, Combined with Eqs. (15) and (18), we can get:

$$n(\bar{Q} - \bar{C})^2 + \sum_{i=1}^{n} (\Delta qi - \Delta ci)^2 \geq n(\bar{Q} - \bar{C})^2 + (\textit{tfdist}(Q, C))^2 \geq n(dist(\hat{Q}, \hat{C}))^2 \tag{19}$$

Combining the above in Eqs. (10) and (19), we can get:

$$\sum_{i=1}^{n} (qi - ci)^2 \geq n(dist(\hat{q}i, \hat{c}i))^2 + (\textit{tfdist}(q, c))^2 \tag{20}$$

That is, the *TDIST* distance lower bounds the Euclidean distance on the single-frame case (i.e. $w = 1$).

N frames can be obtained by applying the single-frame proof on every frame, that is:

$$\sqrt{\sum_{i=1}^{n} (qi - ci)^2} \geq \sqrt{\frac{n}{w}} \sqrt{\sum_{i=1}^{w} (dist(\hat{q}, \hat{c})^2 + \frac{w}{n} (\textit{tfdist}(q, c))^2)} \tag{21}$$

The right-hand side of the above inequality is *TDIST(P,Q)* and the left-hand side is the Euclidean distance between P and Q. Therefore, the *TDIST* distance lower bounds the Euclidean distance.

5 Experimental Validation

In this section, we will introduce the results of our experiment. First we present the data set used, the comparison methods and parameter settings. Then, we evaluate the performance of the proposed method in terms of false alarm rate, dimension reduction, and efficiency.

5.1 Data Sets and Parameter Settings

Figure 1 illustrates the edge computing structure differs from the conventional cloud computing structure in that edge devices are not only data producers, but also data consumers. In addition to data collection and transmission, edge computing nodes can process and analyze data.

We performed the experiments on 5 diverse time series data sets, which are provided by the UCR Time Series repository [24]. Some summary statistics of the data sets are given in Table 4. Each data set is divided into a training set and a testing set. The data sets contain classes ranging from 2 to 5, are of size from dozens to thousands, and have the lengths of time series varying from 96 to 470. In addition, the types of the data set are also diverse, including synthetic and real (recorded from some processes).

Table 4. The description of experimental data set

NO	Data set name	#Classes	Size of training set	Size of testing set	Length of time series	Type
1	ECG200	2	100	100	96	Real
2	Two_Pattern	4	1000	4000	128	Synthetic
3	Beef	5	30	30	470	Real
4	Coffee	2	28	28	286	Real
5	CBF	3	30	900	128	Synthetic

The main purpose of the TFSAX method is to improve the accuracy of symbolic representation and approximate distance metric. Therefore, we conduct our experiment on the classification task, of which the accuracy is determined by the distance measure.

To compare the classification accuracy of the original SAX, Extended SAX (ESAX), SAX-TD and our TFSAX, we conduct the experiments using the 1 Nearest Neighbor (1-NN) classifiers. The main advantage is that the underlying distance metric is critical to the performance of 1-NN classifier; hence, the accuracy of the 1-NN classifier directly reflects the effectiveness of a distance measure. Furthermore, the 1-NN classifier is parameter free, allowing direct comparisons of different measures. To obtain the best accuracy for each method, we use the testing data to search for the best parameters w and α. For a given time series of length n, w and α are picked using the following criteria (to make the comparison fair, the criteria are the same as those in [3]): For w, we search for the value from 2 up to n/2, and double the value of w each time.

For α, we search for the value from 3 up to 10.

If two sets of parameter settings produce the same classification error rate, we choose the smaller ones.

5.2   Results

The experimental results of false positive rate (fpr), number of segments (w), number of symbols (α), and dimensionality reduction ratio (ratio) for SAX, ESAX, SAX-TD and TFSAX are shown in Table 5, where entries with the lowest fpr and ratio are highlighted.

Table 5. The overall experiment results

Data set	SAX				ESAX				SAX-TD				TFSAX			
	w	α	ratio	fpr	w	α	ratio	fpr	w	α	ratio	fpr	w	α	ratio	fpr
ECG200	32	10	0.33	0.12	32	10	1	0.1	16	5	0.34	**0.09**	8	5	**0.27**	**0.09**
Two_Pattern	32	10	0.25	0.17	64	10	1.5	0.129	16	10	0.26	0.071	8	10	**0.18**	**0.05**
Beef	128	10	0.28	0.56	32	9	**0.2**	0.52	64	9	0.27	0.2	32	9	**0.2**	**0.14**
Coffee	128	10	0.45	0.496	4	5	**0.04**	0.179	8	3	0.06	**0**	8	3	0.18	0.12
CBF	32	10	0.25	0.104	64	10	1.5	0.138	4	10	**0.07**	0.11	4	10	**0.07**	**0.08**
Average	–	–	0.312	0.33	–	–	0.848	0.213	–	–	0.2	0.1	–	–	**0.16**	**0.09**

5.2.1   False Positive Rate

False positive rate (*fpr*) refers to the error classification caused by the inconsistency between the label of the classification and the label of the original sequence during the classification process. It can be expressed as the ratio of the number of incorrectly classified samples to the number of original samples in the test data set. It is a commonly used measure of the classification and directly reflects the merits of the classification and can be formalized as follow:

$$FPR = (R - (R \wedge C)/C) \times 100\% \qquad (22)$$

Where R represents the detection result of the algorithm and C represents the actual result in the original sequence. It can be seen from the definition that the smaller the fpr is, the better the algorithm is.

The experiment first adopts SAX, ESAX, SAX-TD and TFSAX to symbolize the original time series to obtain the corresponding training set and test set. And then, for each time series of the test data set in Table 4, a 1-NN classifier is used to mine its nearest sequence in the corresponding training set using the above four distance metrics. If the query result of the test object is consistent with the original label of the sequence, it means that the sequence is classified correctly; otherwise it indicates misclassification. Finally, the average value of the classification results of all sequences in the training set is used as the experimental result of the algorithm in that data set.

The fpr of the four algorithms are shown in Table 5, from which we can clearly find that TFSAX has the lowest error in the most of the data sets (4/5) because it using both mean feature and trend morphological feature to represent the sequence; As a contrast,

SAX-TD gets the second lower error (2/5). Moreover, TFSAX significantly has achieved better result than other methods in terms of the average classification error of the test data set.

5.2.2 Dimensionality Reduction

Since one major advantage of the SAX representation is its dimensionality or numerosity reduction, we shall compare the dimensionality reduction of our method with the SAX and the ESAX. The dimensionality reduction ratios are calculated using the w when the three methods achieve their smallest classification error rates on each data set. We will measure the dimensionality reduction ratios as follows:

$$\text{Ratio} = (\text{Number of the reduced data points}/\text{Number of the original data points}) \times 100\% \tag{23}$$

According to the definition of the symbolized representation, the reduction ratios of SAX, ESAX, SAX-TD, and TFSAX are w/n, $3w/n$, $(2w + 1)/n$, and $2w/n$, respectively. The dimensionality reduction results of TFSAX competitive with SAX, ESAX and SAX-TD on the above five data sets are shown in Fig. 5.

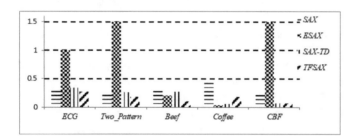

Fig. 5. Comparison of the Dimensionality reduction ratio of different methods

From Fig. 5 we can see that the dimensionality reduction effect of the TSFSA method and the dimensionality reduction effect of the SAX-TD method are basically similar, slightly better than the *SAX* method, while the dimension reduction effect of the *ESAX* is the worst. For each segment, though *TFSAX* increases the trend distance factor and trend shape factor to describe the trend feature information of time series, it also reduces the number of segments needed to achieve the best similarity detection effect. From Table 5 we can see that the average dimension reduction ratio of *TFSAX* in the five experimental data sets is 0.16, which is basically equivalent to 0.2 of SAX-TD.

5.2.3 Lower Bound

In previous section, it has been theoretically proved that distance metric of TFSAX not only satisfies the definition of distance, but also has more compact lower boundary than SAX. This section is validated by experimentation.

The Beef dataset which containing 30 training sets and 30 test sets was used as the test data to verify the algorithm. The experiment first uses the Euclidean distance, SAX distance, SAX-TD distance and TFSAX distance to calculate the distance between any test data set and training data set (each distance measure needs to calculate 900 pairs of distances), and then calculate lower bound of SAX distance, SAX-TD distance, and TFSAX distance under condition of $w = 32$, $a = [3,4,5,6,7,8,9,10]$ and $a = 8$, $w = [2,4,8,16,32,64]$ respectively. The average of 900 experimental results for each method was used as the final lower bound. The experimental results are shown in Fig. 6(a) and (b) respectively.

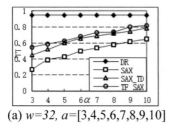 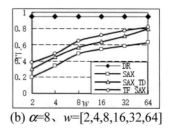

(a) $w=32$, $a=[3,4,5,6,7,8,9,10]$ (b) $\alpha=8$、$w=[2,4,8,16,32,64]$

Fig. 6. Comparison of the lower bound of four methods under different parameters

From Fig. 6 we can see that the lower bound of TFSAX distance metric of the algorithm is infinitely close to the straight line with $y = 1$ on the vertical axis(Euclidean distance) as the number of segments and the number of symbols increase. Obviously, it not only meets the requirements of the lower bound, but also is slightly better than SAX-TD, and is always better than the original SAX. For example, the lower bound of TFSAX at $\alpha = 3$ in 2.8(a) is almost close to that of SAX at $\alpha = 7$, and the lower bound of TSASX at $w = 8$ at 2.8(b) is almost similar to that of SAX at $w = 64$.

6   Conclusion

We have proposed a novel symbolic aggregate approximation representation and distance measure for time series. We firstly extract the trend feature such as trends distance factor and trend shape factor from time series to constructions trends feature triangle to quantitatively measure the different trends. And then, we design multi-resolution discretization method which uses the sequence's trend feature and mean feature to transform the trend feature into symbols. Moreover, we modify the original SAX distance measure by integrating a weighted trend distance. The new distance measure can not only keeps the important property that lower bounds the Euclidean distance, but also is tighter than that of the original SAX. According to the experimental results on diverse data sets, our improved measure decreases the classification error rate significantly and needs a smaller number of words and alphabetic symbols for achieving the best classification accuracy than the SAX and SAX-TD do. Our improved method has slightly better capability of dimensionality reduction and has similar efficiency as the SAX and SAX-TD.

For the future work, we intend to extend the method to other data mining tasks such as clustering, anomaly detection and motif discovery. The proposed method can be applied and improved in the trend feature representation method based on unequal length division.

Acknowledgements. This work has been partially supported by the National Key Research and Development Program of China (No. 2018YFC1508100), the Fundamental Research Funds for the Central Universities (No. 2018B45614) and the CSC Scholarship.

References

1. Yang, Q., Wu, X.: 10 challenging problems in data mining research. Int. J. Inf. Technol. Decis. Making **5**(4), 597–604 (2006)
2. Lin, J., Keogh, E., Lonardi, S., Chiu, B.: A symbolic representation of time series, with implications for streaming algorithms. In: Proceedings of the ACM SIGMOD Workshop on Research Issues in Data Mining and Knowledge Discovery, pp. 2–11 (2003)
3. Lin, J., Keogh, E., Wei, L., et al.: Experiencing SAX: a novel symbolic representation of time series. Data Min. Knowl. Disc. **15**(2), 107–144 (2007)
4. Tayebi, H., Krishnaswamy, S., Waluyo, A., et al.: RA-SAX: resource-aware symbolic aggregate approximation for mobile ECG analysis. In: The IEEE International Conference on Mobile Data Management, pp. 289–290 (2011)
5. Canelas, A., Neves, R., Horta, N.: RA-SAX: A new SAX-GA methodology applied to investment strategies optimization. In: Proceedings of the ACM International Conference on Genetic and Evolutionary Computation Conference, pp. 1055–1062 (2012)
6. Rakthanmanon, T., Keogh, E.: Fast shapelets: a scalable algorithm for discovering time series shapelets. In: Proceedings of the SIAM Conference on Data Mining (2013)
7. Keogh, E., Chakrabarti, K., Pazzani, M., et al.: Dimensionality reduction for fast similarity search in large time series databases. Knowl. Inf. Syst. **3**(3), 263–286 (2000)
8. Lkhagva, B., Suzuki, Y., Kawagoe, K.: New time series data representation ESAX for financial applications. In: The Workshops on the IEEE International Conference on Data Engineering, pp. 17–22 (2006)
9. Hugueney, B.: Adaptive segmentation-based symbolic representations of time series for better modeling and lower bounding distance measures. In: The European Conference on Principles and Practice of Knowledge Discovery in Databases, pp. 545–552 (2006)
10. Keogh, E., Chakrabarti, K., Pazzani, M., Mehrotra, S.: Locally adaptive dimensionality reduction for indexing large time series databases. In: Proceedings of the ACM SIGMOD International Conference on Management of Data, pp. 151–162 (2001)
11. Pham, N., Le, Q., Dang, T.: Two novel adaptive symbolic representations for similarity search in time series databases. In: The IEEE International Asia-Pacific Web Conference, pp. 181–187 (2010)
12. Zhong, Q.: The symbolic algorithm for time series data based on statistic feature. Chin. J. Comput. **31**(10), 1857–1864 (2008)
13. Ljubič, P., Todorovski, L., Lavrač, N., Bullas, J.: Time-series analysis of UK traffic accident data. In: Proceedings of the International Multi-conference Information Society, pp. 131–134 (2002)
14. Yu, G., Peng, H., Zheng, Q.: Pattern distance of time series based on segmentation by important points. In: Proceedings of the IEEE International Conference on Machine Learning and Cybernetics, vol. 3, pp. 1563–1567 (2005)

15. Kontaki, M., Papadopoulos, A., Manolopoulos, Y.: Continuous trend-based clustering in data streams. In: Proceedings of the International Conference on Data Warehousing and Knowledge Discovery, pp. 251–262 (2008)
16. Kontaki, M., Papadopoulos, A., Manolopoulos, Y.: Continuous trend-based classification of streaming time series. In: Advances in Databases and Information Systems, pp. 294–308 (2005)
17. Li, H., Guo, C.: Symbolic aggregate approximation based on shape features. Pattern Recognit. Artif. Intell. **24**(5), 665–672 (2011)
18. Li, G., Zhang, C., Yang, L.: TSX: a novel symbolic representation for financial time series. In: Pacific Rim International Conference on Trends in Artificial Intelligence, pp. 262–273. Springer (2012)
19. Sun, Y., Li, J., Liu, J., et al.: An improvement of symbolic aggregate approximation distance measure for time series. Neurocomputing **138**(11), 189–198 (2014)
20. Yamana, H., Yamana, H.: An improved symbolic aggregate approximation distance measure based on its statistical features. In: International Conference on Information Integration and Web-Based Applications and Services, pp. 72–80. ACM (2016)
21. Li, H., Ye, L.: Similarity measure based on numerical symbolic and shape feature for time series. Control Decis. (2017)
22. Yin, H., Yang, S., Zhu, X., et al.: Symbolic representation based on trend features for biomedical data classification. Technol. Health Care Official J. Eur. Soc. Eng. Med. **23**(s2), 501–510 (2015)
23. Fu, T.: A review on time series data mining. Eng. Appl. Artif. Intell. **24**(1), 164–181 (2011)
24. Keogh, E., Zhu, Q., Hu, B., Hao, Y., et al.: The UCR time series classification/clustering (2011). http://www.cs.ucr.edu/~eamonn/time_series_data/⟩

Making Join Views Updatable on Relational Database Systems in Theory and in Practice

Yoshifumi Masunaga$^{1(\boxtimes)}$, Yugo Nagata^{2}, and Tatsuo Ishii^{2}

^{1} Ochanomizu University (Emeritus), Bunkyo-ku, Tokyo 112-8610, Japan
masunaga.yoshifumi@ocha.ac.jp
^{2} SRA OSS Inc. Japan, Toshima-ku, Tokyo 171-0022, Japan
{nagata,ishii}@sraoss.co.jp

Abstract. Views are a useful tool in relational database application development. Among various types of views, join views are particularly useful alongside selection and projection views. However, unlike selection and projection views, the updatability of join views has been limited in both theory and practice. In this paper, we first show that the updatability of join views, whose underlying base tables are bags in general, equals solving a non-linear simultaneous equation using the merge method for consistent labeling problems in artificial intelligence. This result could provide a more sophisticated way of implementing a view support function on a relational database system based on pro forma guessing of update intention approach, also known as the intention-based approach. Next, we report on the state of the art of view updatability in commercial and open source relational database systems alongside the ISO/IEC activities for prescribing view updatability through SQL, the international standard database language for relational databases. We show that updates to join views are poorly supported in both current relational database systems and the current version of the SQL standard. Finally, we discuss how join views can be made updatable on relational database systems.

Keywords: View · Join view · View update problem · SQL-92 · SQL:1999 · INSTEAD OF trigger · Pro forma guessing of update intention approach · Intention-based approach · PostgreSQL · Non-linear simultaneous equation · Consistent labeling problem

1 Introduction

Almost twenty years ago, Eisenberg and Melton, who contributed to SQL:1999 language standard, mentioned in their report [1] that "A long-standing demand of application writers is the ability to update broader classes of views. Many environments use views heavily as a security mechanism and/or as a simplifier of an applications view of the database. However, if most views are not updatable, then those applications often have to 'escape' from the view mechanism and rely on directly updating the underlying base tables; this is a most unsatisfactory situation". It seems that nothing has changed since that time.

Among various types of views, join views are probably one of the most useful views in relational database application development alongside selection and projection

© Springer Nature Switzerland AG 2019
S. Lee et al. (Eds.): IMCOM 2019, AISC 935, pp. 823–840, 2019.
https://doi.org/10.1007/978-3-030-19063-7_66

views. However, most join views are not updatable, while most selection and projection views are updatable. This is because there is potential ambiguity when an update operation to a join view is translated into update operations against underlying base tables. This problem of finding the necessary and sufficient condition for an update operation to a view to be un-ambiguously translated to base tables is known as the "view update problem" and has been intensively investigated both in theory and in practice.

In theory, much attention has been payed to the view update problem since views were first introduced by Codd [2]. Much research has been done so far, including syntactic [3], semantic [4], interactive [5], and pro forma guessing of update intention [6] approaches. Among these, the pro forma guessing of update intention approach, also known as the intention-based approach, seems promising because it makes join views updatable to the maximum. To test the feasibility of the intention-based approach, a prototype was reported using PostgreSQL 10, an open source relational database system [7].

In practice, SQL, the international standard language for relational databases, which is enacted by ISO/IEC standard groups, first introduced views in SQL-92 with a condition for a view to be updatable. However, as the condition was too restrictive, the updatability condition was revised in SQL:1999 so that certain types of join views and union views became updatable.

Looking at the current state of view support in commercial or open source relational database systems, and inferring from the documents of each system, Oracle and MySQL cope with the view updatability condition set by SQL:1999 to some extent. However, other systems such as DB2, SQL Server, and PostgreSQL remain at the SQL-92 standard level. To compensate for this implementation delay, these vendors recommend that users use INSTEAD OF triggers. But, we would like to stress that this is a dangerous way to achieve view updatability, because abuse of INSTEAD OF triggers can immediately make a database inconsistent.

In this paper, we discuss how to make join views updatable both in theory and in practice.

2 Join View Updatability in Theory

2.1 Theoretical Framework of View Updatability on Relational Database Systems

The view update problem has been intensely studied since views were first introduced by Codd, the inventor of the relational model of data [2].

A view and the updatability of a view is defined by Dayal and Bernstein as follows [3]:

Definition 1 (View Updatability). A view is defined as a mapping function from a database state to a view state. That is, suppose that S_τ represents a database state at time τ and V is the view definition. Then $V(S_\tau)$ is the view state at that time. Suppose that an update operation u is issued to V. Then, u is *translatable* if and only if there exists a translation T of u for an update operation to S_τ so that it has no side effects, is unique, and does not cause any extraneous update.

That is, u is *translatable* if and only if the commutative diagram in Fig. 1 holds. The uniqueness condition is imposed to avoid translation ambiguity, and the no-extraneous-update condition is imposed to indicate that the translation should be minimal in the sense that it should not update the base relations unless u requests otherwise.

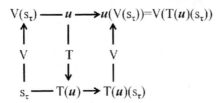

Fig. 1. Commutative diagram representing the translatability of view update u at time τ.

It is interesting to quote a sentence from an article written by Dayal and Bernstein [3] that "Most relational database systems provide a facility for supporting user views. Permitting this level of abstraction has the danger, however, that update operations issued by a user within the context of his view may not translate correctly into equivalent updates on the underlying database. It is the purpose of this paper to formalize the notion of correct translatability, and to derive constraints on view definitions that ensure the existence of correct update mappings. In summary, our theorems show that there are very few situations in which view updates are possible–even fewer, in fact, than intuition might suggest".

However, as we pointed out before, if we adopt the intention-based approach, this situation changes considerably. The difference between the intention-based approach and the syntactic approach comes from the fact that the former treats view updatability at the database instance level very truly, while the latter does it at the database schema level. It is this difference that makes join views updatable to the maximum in the intention-based approach. A related discussion will be given in Sect. 2.3.

2.2 The Intention-Based Approach and the Updatability of Join Views

Natural join views cannot be deleted under traditional approaches. That is, suppose that an update operation is a delete operation d to a view V, which is a natural join of base relations R and S. Then, d is not translatable because translation ambiguity may happen under some combination of the database state, the view definition V, and the delete operation d to V. To avoid such situations, traditional approaches do not accept the delete operation in a natural join view.

However, the idea behind the intention-based approach [6] is the opposite of the idea behind the traditional approach. That is, to break down the wall of tradition, we carefully re-examined what occurs in these cases and found that, in certain cases, the users' view update intention can be estimated by checking the "extension" of each view update translation candidate and can be calculated using temporarily materialized views or temporary tables. This method of view updating was named the pro forma guessing of update intention approach, also known as the intention-based approach. We now show a simple example of how the intention-based approach works.

Example 1 (Natural join views are deletable under the intention-based approach). Let R(A, B) and S(B, C) be base relations whose instances at time τ are shown in Fig. 2, and V = R * S be the natural join view definition. Suppose that a delete operation, *d* is issued:

> *d*: DELETE FROM V
>   WHERE A='a' AND B='b' AND (C='c' OR C='c'')

Then, the following three translation alternatives are obtained by semantic analysis of the delete request to a join view [4]:

> $T_1(d)$: DELETE FROM R
>   WHERE A='a' AND B='b'
> $T_2(d)$: DELETE FROM S
>   WHERE B='b' AND (C='c' OR C='c'')
> $T_3(d)$: Execute both T1(d) and T2(d)

Delete operations to natural join views are not accepted in the traditional approach as mentioned before. However, if we execute $T_1(d)$ temporarily, the temporarily materialized view of R * S, which we call the "extension" of the translation candidate $T_1(d)$, is R * S − {(a, b, c), (a, b, c')}, which is exactly the desired update result of deletion *d*, where − denotes the set difference operation. If we execute $T_2(d)$ temporarily, the temporarily materialized view is R * S − {(a, b, c), (a, b, c'), (a', b, c), (a', b, c')}, which has a side effect, i.e. {(a', b, c), (a', b, c')}is deleted unnecessarily. If we execute $T_3(d)$ temporarily, the same result as that of the $T_2(d)$ case occurs. Although we have three translation alternatives for the realization of delete operation *d*, the commutative diagram of Fig. 1 holds only for $T_1(d)$. That is, the commutativity does not hold for both $T_2(d)$ and $T_3(d)$ in this case. We interpret this phenomenon as follows: although there is no way to determine the exact intention of the delete operation to the natural join view without asking the view updater through human interaction, we estimate that a certain event happened in the real world at the time when the delete operation was issued that corresponds to $T_1(d)$. As two other translation alternatives, $T_2(d)$ and $T_3(d)$, cannot satisfy the commutativity in Fig. 1 in this case, the natural join view R * S is deletable with respect to delete operation *d* under the intention-based approach.

R	
A	B
a	b
a'	b
a	b'

S	
B	C
b	c
b	c'
b'	c

R*S		
A	B	C
a	b	c
a	b	c'
a'	b	c
a'	b	c'
a	b'	c

Fig. 2. Instances of base relations R and S, and the extension of R*S at time τ.

It is not always true that every natural join view is deletable under the intention-based approach, due to translation ambiguity. However, as certain cases exist in which natural join views are deletable, we argue that natural join views could be deleted under the intention-based approach. This discussion also holds for insert and rewrite operations for join views. Now the following theorem holds, where the term update means either delete or insert or rewrite.

Theorem 1. A join view is *updatable* under the intention-based approach if and only if the commutative diagram of Fig. 1 holds under the current database state, the view definition, and the update operation.

Proof. Almost clear by the above discussion. Q.E.D.

2.3 The Essential Difference Between Intention-Based and Traditional Approaches

In Sect. 2.1, we mention that the difference between the intention-based and syntactic approaches comes from the fact that the former treats view updatability at the database instance level, while the latter does it at the database schema level. We stress this difference by pointing out the two main features of the intention-based approach:

First, a relation has two features, i.e. a schema and an instance. That is, when we say that R is a relation, we should be careful to either refer to it as a relation schema or as a relation instance. The schema implies the structure of R, while the instance implies its content. The essential difference between schema and instance is stated as follows: "A property P holds for relation schema R if and only if P holds for every instance R of schema R". For example, a functional dependency $A \rightarrow B$ holds in R if and only if $A \rightarrow B$ holds for every instance R of R. The traditional approaches discuss the view updatability at the schema level. Therefore, if there exists a pair of an instance and an update operation that does not satisfy the commutativity of Fig. 1, then an update operation issued to V is *not translatable*. In contrast, the intention-based approach discusses the updatability of views at the instance level. Therefore, a join view is updatable as was shown in Example 1. This is the reason why the updatability of the traditional approaches is limited, while join view updatability of intention-based approach achieves the maximum.

Second, it should be emphasized again that the intention-based approach is ultimately faithful to Definition 1.

Because of these characteristics, Theorem 1 holds.

2.4 Feasibility Test of the Intention-Based Approach

The feasibility of the intention-based approach was tested by implementing it on PostgreSQL, an open source relational database system [7, 8]. Figure 3 illustrates an overview of the prototype architecture. The left part of this figure shows how PostgreSQL processes queries. A query issued by a client is parsed to produce a query tree.

The query tree is processed by the rule system to produce a rewritten query tree. The rewritten query tree is fed to the planner, i.e., optimizer, to produce a cost-minimum executable plan. The executable plan is performed by the executor, and the execution result is returned to the client. Note that automatically updatable views of PostgreSQL are supported by the rule system.

To implement the intention-based approach, two approaches were examined: a rule-based approach and a trigger-based approach. The former implements the view support mechanism in the rule system context while the latter implements it in the executor context. By examining the pros and cons of the two approaches, we adopted the trigger-based approach. We developed a prototype as an extension of the executor of PostgreSQL 10, the latest version of PostgreSQL [9, 10], as depicted on the right-hand side of Fig. 3. The BEFORE statement-level triggers and the AFTER statement-level triggers with transition tables were used to implement the intention-based view updatability decision algorithm [6]. We have confirmed that the intention-based approach works as it was intended.

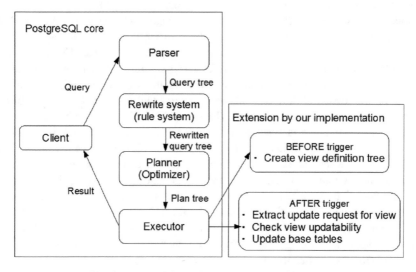

Fig. 3. An overview of the prototype architecture.

However, one of the problems left as a future work was to find a sophisticated method of deciding updatability and finding possible translation alternatives of an update operation to a join view rather than the brute-force method we used to realize the prototype. We will investigate this in the next section.

3 Join View Updatability as the Solvability of a Non-linear Simultaneous Equation

3.1 Formalization of the Updatability of Join Views as a Non-linear Simultaneous Equation

As SQL is designed based on bag semantics, the updatability of views in the intention-based approach should be expanded so that it can accommodate not only sets, i.e., relations in the relational model of data, but also bags, i.e., SQL tables with duplicate rows.

The definition of a bag is given below [8, 11]:

Definition 2 (Bags). Let $R(A_1, A_2, \ldots, A_n)$ be a relation schema. Then, a bag $R(A_1, A_2, \ldots, A_n)$ of R is defined as follows:

$R = \{t_i(k_i) \mid t_i \in \mathrm{dom}(R) \wedge k_i \geq 1\}$, where $\mathrm{dom}(R) = \mathrm{dom}(A_1) \times \mathrm{dom}(A_2) \times \ldots \times \mathrm{dom}(A_n)$, and k_i is the multiplicity of t_i.

Clearly, a bag $R = \{t_i(k_i) \mid i \in \{1, 2, \ldots, m\}\}$ is a relation if and only if $(\forall i \in \{1, 2, \ldots, m\})(k_i = 1)$ holds.

Now, we will show that the updatability of join views under bag semantics is represented as a non-linear simultaneous equation. This will be explained by taking a cross join, i.e., a Cartesian product, as an example.

Example 2. Let $R(A)$ and $S(B)$ be bags and their instances are $\{1(2), 2\}$ and $\{1, 2(2)\}$, respectively. Let V be a view defined as follows:

CREATE VIEW V(A, B)
AS SELECT R.A, S.B
FROM R CROSS JOIN S

Suppose that a delete operation d_1 is issued to V:

d_1 : DELETE FROM V
WHERE A=2

The intention-based approach works in the following manner:

(a) Create a temporary table V_T of V and insert the instance of V in it.

CREATE TEMPORARY TABLE V_T(A INT, B INT)
INSERT INTO V_T(A, B) SELECT A, B FROM V

As a result, $V\_T = \{(1, 1)(2), (1, 2)(4), (2, 1)(1), (2, 2)(2)\}$.

(b) Modify d_1 to d_1^m, and execute it.

d_1^m : DELETE FROM V_T
WHERE A=2

As a result, we know that d_I^m intends to delete $U = \{(2,1)(1),(2,2)(2)\}$ from V_T. Let $W = V\_T -_b U = \{(1, 1)(2), (1, 2)(4)\}$, where $-_b$ represents a "monus" operation in bag algebra.

(c) Corresponding to the multiplicity of rows 1, 2 of R and 1, 2 of S, respectively, let's introduce variables x_1, x_2, x_3, x_4 so that $R^{dI} = \{1(x_1), 2(x_2)\}$ and $S^{dI} = \{1(x_3), 2(x_4)\}$, where R^{dI} and S^{dI} represent the underlying base tables which could realize the intended delete operation d_I to V. As x_1, x_2, x_3, x_4 are the multiplicity of rows after the delete operation d_I is executed, they should satisfy the following conditions:

$$0 \le x_1 \le 2 \, 0 \le x_2 \le 1, \, 0 \le x_3 \le 1, \, 0 \le x_4 \le 2 \tag{1}$$

Note that this restriction insists that the delete operation to a view must also be translated to delete operations to underlaying base tables. We believe that this restriction looks quite reasonable and acceptable. We call it the deletion operation translation principle. Likewise, the insertion operation translation principle is defined. Rewrite is regarded as a sequence of delete and insert.

(d) Calculate $R^{d1} \times S^{d1} = \{(1,1)(x_1 \times x_3),(1,2)(x_1 \times x_4),(2,1)(x_2 \times x_3),(2,2)(x_2 \times x_4)\}$.

(e) If d_I is translatable, then the following equations must hold.

$$x_1 \times x_3 = 2 \tag{2}$$

$$x_1 \times x_4 = 4 \tag{3}$$

$$x_2 \times x_3 = 0 \tag{4}$$

$$x_2 \times x_4 = 0 \tag{5}$$

Now, the translatability problem of a delete operation to a cross join view is reduced to the problem of solving a non-linear simultaneous equation which consists of (1) to (5) in this case.

3.2   Solving Join View Update Translatability Using the Merge Method for the Consistent Labeling Problem

There are many ways to solve a non-linear simultaneous equation. But, in general this is not an easy task because it is similar to the consistent labeling problem (CLP) in artificial intelligence, which is an NP-complete decision problem (where NP stands for nondeterministic polynomial time) [12]. In this paper, we try to solve our problem using the merge method for CLP [13], which can find "all" solutions. The finding-all feature is essentially necessary for our purposes because if there is more than one solution, then there is translation ambiguity, which is not acceptable (Definition 1). Equally, if there is no solution, then the update operation is not translatable. A single solution is needed for view update translatability.

The merge method works in the following manner. We will simulate how it works using the same example used previously.

Example 3. Suppose that R, S, V, and d_1 are the same as given in Example 2. Examine whether d_1 is translatable or not by the merge method.

The first step is to construct a constraint network that represents the non-linear simultaneous equation shown in the previous section. A CLP is then defined as a quadruple (U, L, T, R), where U, L, T, and R represent a set of units, a set of labels, a set of restrictions among labels, and a set of local solutions, respectively. In our case, each component of the quadruple is defined as follows:

$$U = \{x_1, x_2, x_3, x_4\}$$
$$L = \{0, 1, 2\}$$
$$T = \{t_1, t_2, t_3, t_4\},$$

where t_1, t_2, t_3, t_4 represent (x_1, x_3), (x_1, x_4), (x_2, x_3), (x_2, x_4), respectively, corresponding to Eqs. (2) to (5).

$R = \{R_1, R_2, R_3, R_4\}$, where R_1, R_2, R_3, R_4 represent $\{(1, 2), (2, 1)\}$, $\{(2, 2)\}$, $\{(0, 0), (0, 1), (1, 0)\}$, $\{(0, 0), (0, 1), (0, 2), (1, 0)\}$, respectively, each of which corresponds to the set of possible local solutions for each Eq. (2) to (5). The number of elements of R_i is called its size, and (t_i, R_i) is called a constraint pair.

The second step is to introduce a constraint network which is equivalent to Example 2 using the information given by the quadruple shown above. The resultant network is depicted in Fig. 4. Note that the network is undirected.

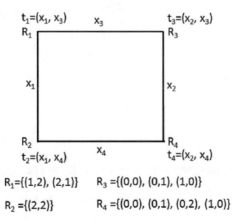

Fig. 4. Constraint network equivalent to Example 2.

The third step is to merge the vertices of the network. Two nodes are merged if they are adjacent. In this example, we pick up the (t_1, R_1)–(t_2, R_2) pair. The merge result is shown in Fig. 5(a). The merge process is continued; let merge $(t_{1,2}, R_{1,2})$–(t_3, R_3) pair so that the merge result $(t_{1,2,3}, R_{1,2,3})$ is obtained as shown in Fig. 5(b). The merge process is continued so that $(t_{1,2,3}, R_{1,2,3})$–(t_4, R_4) pair is merged into a single node $(t_{1,2,3,4}, R_{1,2,3,4})$ as shown in Fig. 5(c), which is the final solution.

Next, we should check the contents of $R_{1,2,3,4}$. In this case, it consists of a single element which is (2, 0, 1, 2). Therefore, we can conclude that the delete operation d_1 to V is translatable. That is, the cross-join view V is deletable with respect to the current database state, the view definition V, and the delete operation d_1.

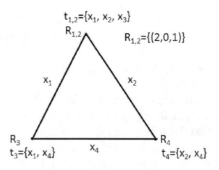

(a) After Merging Nodes (t_1, R_1) and (t_2, R_2) of Fig. 4.

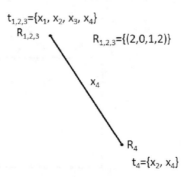

(b) After Merging Nodes $(t_{1,2}, R_{1,2})$ and (t_3, R_3) of Fig. 5 (a).

$t_{1,2,3,4}=\{x_1, x_2, x_3, x_4\}$
$R_{1,2,3,4}$ •
                    $R_{1,2,3,4}=\{(2,0,1,2)\}$

(c) After Merging Nodes $(t_{1,2,3}, R_{1,2,3})$ and (t_4, R_4) of Fig. 5 (b).

Fig. 5. Merge process of constraint network equivalent to Example 2.

Now we can calculate the translation of the d_1 delete operation to the underlaying base tables R(A) and S(B) in the following way:

(a) (2, 0, 1, 2) means that $x_1 = 2$, $x_2 = 0$, $x_3 = 1$, $x_4 = 2$.
(b) The initial values of x_1, x_2, x_3, x_4 are $x_1 = 2$, $x_2 = 1$, $x_3 = 1$, $x_4 = 2$.
(c) By comparing the initial values of x_1, x_2, x_3, x_4 and the values in the final solution, we find that the value of x_2 changed from 1 to 0. This means that the delete operation d_1 to V is translated to the delete operation $T(d_1)$ to the underlying base table R(A), where $T(d_1)$ is defined as follows:

$$T(d_1): \text{DELETE FROM R}$$
$$\text{WHERE A=2}$$

An outline of the program for realizing the merge method is listed below, where V_i represents a constraint pair, and $\min(k_{i,j})$ is the function to select a $V_i - V_j$ pair whose estimated size is minimum. The greedy method could be applied to calculate $\min(k_{i,j})$.

```
begin
    while (two or more vertices remain)
        merge (Vi, Vj) for min(ki,j)
    compute and output all solutions
end
```

To incorporate this function into extension (Fig. 3), of course, preprocessing is necessary to construct a constraint network which is equivalent to the update operation issued.

4 Join View Updatability in Practice

4.1 SQL Standards for View Definitions and Their Updatability

SQL is undoubtedly the most accepted and implemented interface language for relational database systems. The first version of the SQL standards appeared in 1986, and it has been revised frequently since then, including SQL-89, SQL-92, SQL:1999, SQL:2006, SQL:2008, and SQL:2016. Views were first introduced in SQL-92 [14] as defined below:

```
CREATE VIEW view ([ column [, ...]])
    AS SELECT expression [ AS colname] [, ...]
    FROM table [ WHERE condition]
    [ WITH [ CASCADE | LOCAL] CHECK OPTION]
```

The SELECT statement that appeared in this definition is called a "query specification". As a view is not a stored table but merely a definition, SQL treats a query specification itself as if it were a view [14]. To better understand this, we show a selection view defined on a single base table.

Example 4. Suppose that EMP(eno, ename, dept, sal) is a base table, where eno, ename, dept, and sal represent employee number, employee name, department number, and salary, respectively. Then a POOR_EMP view is defined as follows:

> CREATE VIEW POOR_EMP(eno, ename, dept, sal)
>   AS SELECT *
>   FROM EMP WHERE sal < 20

POOR_EMP is updatable because delete, insert, or rewrite operations to it can be translated to that for EMP without any ambiguity, as shown in the next example.

Example 5. Suppose that the manager of department K55 found his or her poor employees, and decided to double their salary:

> UPDATE POOR_EMP SET sal=sal×2
>   WHERE dept='K55'

Without any ambiguity, this update can be translated to the update to base table EMP as follows:

> UPDATE EMP SET sal=sal×2
>   WHERE sal < 20 AND dept='K55'

However, in general, update operations are not translatable. The next shows such an example.

Example 6. Suppose that there are two base tables SUPPLY(sno, pno) and DEMAND (pno, dno), where sno, pno, and dno represent supplier number, part number, and demander number, respectively. Suppose that a natural join view SUPPLY_DEMAND (sno, pno, dno) is defined as follows:

> CREATE VIEW SUPPLY_DEMAND(sno, pno, dno)
>   AS SELECT X.*, Y.dno
>   FROM SUPPLY X, DEMAND Y
>   WHERE X.pnp=Y.pno

Suppose that instances of the two base tables and the extension of the view are shown in Fig. 6, and that the following delete operation is issued to SUPPLY_DEMAND.

> ***d***: DELETE FROM SUPPLY_DEMAND
>   WHERE sno=S1 AND pno=P2 AND dno=D2

Then, **d** could be realized by taking any one of the next three translation candidates to base table(s):

$$T_1(\boldsymbol{d})\text{: DELETE FROM SUPPLY}$$
$$\text{WHERE sno=S1 AND pno=P2}$$
$$T_2(\boldsymbol{d})\text{: DELETE FROM DEMAND}$$
$$\text{WHERE pno=P2 AND dno=D2}$$
$$T_3(\boldsymbol{d})\text{: Execute both } T_1(\boldsymbol{d}) \text{ and } T_2(\boldsymbol{d})$$

However, arbitrary choice is not permitted because the meanings of deletion are different among the three alternatives. That is, $T_1(\boldsymbol{d})$ should be chosen if supplier numbered S1 stopped to supply part P2. But, no one can know what happened in the real world. Therefore, **d** is not translatable.

SUPPLY

sno	pno
S1	P1
S1	P2
S2	P1

DEMAND

pno	dno
P1	D1
P2	D2

SUPPLY_DEMAND

sno	pno	dno
S1	P1	D1
S1	P2	D2
S2	P1	D1

Fig. 6. Tables SUPPLY and DEMAND, and the extension of SUPPLY_DEMAND.

Note here that any query to any view is always translatable by query modification [15].

4.2 Updatable Views in SQL Standards

As mentioned earlier, SQL first introduced views and defined their updatability in SQL-92, and then revised it in SQL:1999.

SQL-92. According to the SQL-92 standard, a query specification is *updatable* if and only if all the following rules are satisfied [14].

Definition 3 (View Updatability in SQL-92).

1. Do not use DISTINCT as a quantifier.
2. The value expressions in the SELECT list are all column references, and none of them appears more than once.
3. The FROM clause has only one table reference, and it identifies either a base table or an updatable derived table.
4. The base table ultimately identified by the table reference isn't referenced (directly or indirectly) in any FROM clause of a sub-query in the WHERE clause (other than as a qualifier to identify a column of the table, of course).
5. Never use a GROUP BY clause or a HAVING clause.

If any of these rules are violated, then the query specification is read-only, that is, we cannot execute DELETE, INSERT, or UPDATE operations on it (note that UPDATE means rewrite in SQL). It is obvious that, by this definition, views defined using the following operators are not updatable in SQL-92: UNION [ALL], EXCEPT [ALL], INTERSECT [ALL], and join operations such as CROSS JOIN, NATURAL JOIN, Condition JOIN, Column Name JOIN, INNER JOIN, or OUTER JOIN. In other words, only a simple view such as a selection view or a projection view on the columns including a key is updatable under the SQL-92 standard. The view POOR_EMP shown in Example 4 is updatable under this definition.

SQL:1999. To enrich the view updatability enacted in SQL-92, SQL:1999 revised the updatability of views in the following way: It introduced several new terms that correspond to SQL-92's "*updatable*". These terms include *potentially updatable, updatable, simply updatable,* and *insertable-into.* Using these terms, SQL:1999 enacted the updatability of views as follows [16]:

Definition 4 (View Updatability in SQL:1999).

1. A <query specification> QS is *potentially updatable* if and only if the following conditions hold:
   (a) DISTINCT is not specified.
   (b) Of those <derived column>s in the <select list> that are column references, no column reference appears more than once in the <select list> .
   (c) The <table expression> immediately contained in QS does not simply contain a <group by clause> or a <having clause> .
2. A < query specification > QS is *insertable-into* if and only if every simply underlying table of QS is insertable-into.
3. If a <query specification> QS is *potentially updatable*, then Case:
   (a) If the <from clause> of the <table expression> specifies exactly one <table reference> TR, then a column of QS is said to be an updatable column if it has a counterpart in TR that is updatable.
   (b) Otherwise, a column of QS is said to be an updatable column if it has a counterpart in some column of some simply underlying table UT of QS such that QS is one-to-one with respect to UT.
4. A <query specification> is *updatable* if it is *potentially updatable* and it has at least one updatable column.
5. A <query specification> QS is *simply updatable* if it is updatable, the <from clause> immediately contained in the <table expression> immediately contained in QS contains exactly one <table reference> , and every result column of QS is updatable.

Eisenberg and Melton mentioned in [1] that "SQL:1999 has significantly increased the range of views that can be updated directly, using only the facilities provided in the standard. It depends heavily on functional dependencies for determining what additional views are updatable, and how to make changes to the underlying base table data to effect those updates".

Of course, by this revision, certain types of join views became updatable [17]. But it seems that the class of updatable join views in SQL:1999 is quite limited. We now show the theorem that is considered to be the basis for their join view support.

Theorem 2. Suppose that relations S(A, B) and T(B, C) are given, where a functional dependency B $\rightarrow$ C holds on T. Then, because of the existence of the functional dependency, there is a one-to-one correspondence between the set of tuples in S to the set of tuples of the natural join V = S(A, B) * T(B, C).

Proof. Let f be a mapping from V = S(A, B) * T(B, C) to S(A, B) such that ($\forall t$ = (a, b, c) $\in$ V) (f(t)= (a, b) $\in$ S). In general, f is a many to one mapping because one tuple (a, b) $\in$ S is joinable with several tuples of T having b as B value. But such a tuple does not exist in T except (b, c) because B $\rightarrow$ C holds. Q.E.D.

That is, the case "3. b)" of Definition 4 holds for V = S(A, B) * T(B, C), where B $\rightarrow$ C holds for T. We will explain it more faithfully in SQL terms:

Example 7. Let S(A, B) and T(B, C) be base relations, where B $\rightarrow$ C holds for T. Define a natural join view V as follows:

```
CREATE VIEW V(A, B, C)
AS SELECT X.A, X.B, Y.C
FROM S X, T Y  WHERE X.B=Y.B
```

V is *insertable-into* and *updatable*. For example, an insert operation to V is translatable:

```
INSERT INTO V VALUES (1,1,1)
```

This is translated to insert operations to underlying relations S and T as follows:

```
INSERT INTO S VALUES (1,1)
INSERT INTO T VALUES (1,1)
```

A delete operation is translated in the following way:

```
DELETE FROM V
WHERE A=1 AND C=1
```

This is translated to the delete operations to S and T:

```
DELETE FROM S
WHERE A=1
DELETE FROM T
WHERE C=1
```

An update operation is accepted in the following way:

$$\text{UPDATE V SET } C=C\times 2$$
$$\text{WHERE } A=1$$

This is translated to the update operation to T:

$$\text{UPDATE T SET } C=C\times 2$$
$$\text{WHERE B IN (SELECT B FROM S}$$
$$\text{WHERE } A=1)$$

All the above translations are possible by Theorem 2.

Note here that the functional dependencies or keys are definable only on relations but not on bags because they are the concepts introduced in the relational model of data which is based on set semantics. Therefore, the updatability that utilizes them does not work for SQL tables and views with duplicate rows. However, the intention-based approach covers both the relation and bag cases because it is formalized under bag semantics.

4.3 Updatability of Join Views on Commercial or Open Source Relational Database Systems

Support of View Updatability Standardized in SQL:1999 or SQL-92—The State of the Art. As mentioned in Sect. 1, at the time of writing this paper and inferring from the documents of each commercial or open source relational database system, Oracle and MySQL seem to cope with the view updatability enhancement standardized by SQL:1999, but other systems such as DB2, SQL Server, and PostgreSQL, seem to only support the view updatability standardized in SQL-92.

Use of INSTEAD OF Triggers. To compensate for the implementation delay, the vendors for DB2, SQL Server, and PostgreSQL, and even Oracle, seem to recommend that users use INSTEAD OF triggers. We can find a tremendous number of Web pages that make statements like "INSTEAD OF triggers - All views are updatable!" or "INSTEAD-OF trigger has the ability to update normally non-updateable views," or "you can create an INSTEAD OF trigger on any view to make it updatable". But, these statements are incorrect and confusing.

As INSTEAD OF triggers are powerful in that they can overwrite any code, we must be careful when using them. Easy use of an INSTEAD OF trigger can cause immediate and serious problems to the database immediately. We must write "correct" code for an INSTEAD OF trigger to avoid violating database consistency. However, asking a user to write correct code is a hard task. After all, this requirement is equivalent to asking him or her to understand and solve the view updatability problem, a problem that has puzzled experts for the last few decades.

4.4   A Way to Make Join Views Updatable in Relational Database Systems

The updatability of join views has been a common concern in relational database theory and practice. However, as reported in this paper, this situation has been changing over time.

Now, there is an opportunity to support join views in commercial or open source relational database systems. There are four alternatives:

(1) Realize the join view update capability suggested by the SQL:1999 standard.
(2) Realize the pro forma guessing of update intention approach, i.e. the intention-based approach.
(3) Realize both (1) and (2) mixed.
(4) Do nothing.

It is possible that this decision depends on a performance versus cost ratio issue. In other words, it may depend on the presence of the use cases and how seriously the database vendors want to make join views updatable to support relational database application development.

5   Conclusions

We examined the updatability of join views from both a theoretical and a practical point of view. In theory, the join view updatability in the traditional approaches, such as the syntactic, semantic, and interactive approaches, are heavily restrictive. However, the pro forma guessing of update intention approach, also known as the intention-based approach, could make join views updatable to the maximum. The difference depends on whether updatability is decided based on the database schema or the database instance. To find a sophisticated way for deciding updatability and finding possible translation alternatives of an update operation to a join view, we revealed that the updatability of a join view is equivalent to solving a non-linear simultaneous equation. A concrete example of this equivalence was shown in detail. We reported on the state of the art of join view updatability as supported by the current commercial and open source relational database systems. We also reported on the ISO/IEC activities for prescribing view updatability through SQL languages. Unfortunately, updates to join views are still poorly supported in practice. To make join views updatable, we propose an alternative way to achieve this goal, but this may deeply depend on the performance versus cost ratio issue.

Acknowledgements. This work was partially supported by JSPS KAKENHI Grant Number 16K00152.

References

1. Eisenberg, A., Melton, J.: SQL:1999, formerly known as SQL3. ACM SIGMOD Rec. **28**(1), 131–138 (1999)
2. Codd, E.F.: Recent investigations in a relational database system. Inf. Process. **74**, 1017–1021 (1974)
3. Dayal, D., Bernstein, P.: On the updatability of relational views. In: Proceedings of 4th VLDB, pp. 368–377 (1978)
4. Masunaga, Y.: A relational database view update translation mechanism. In: Proceedings of 10th VLDB, pp. 309–320 (1984)
5. Sheth, A, Larson, J., Watkins, E.: TAILOR, a tool for updating views. In: LNCS, vol. 303, pp. 190–213. Springer (1988)
6. Masunaga, Y.: An intention-based approach to the updatability of views in relational databases. In: Proceedings of ACM IMCOM 2017, pp. 5–8 (2017)
7. Nagata, Y., Masunaga, Y.: Extending view updatability by a novel theory. In: PGCon 2017, The PostgreSQL Conference, Ottawa, Canada (2017)
8. Masunaga, Y., Nagata, Y., Ishii, T.: Extending the view updatability of relational databases from set semantics to bag semantics and its implementation on PostgreSQL. In: Proceedings of ACM IMCOM 2018, pp. 8–3 (2018)
9. PostgreSQL 10 Documentation, Chapter 50. Overview of PostgreSQL Internals, 50.1. The Path of a Query
10. PostgreSQL 10 Documentation, Appendix E. Release Notes, E.1. Release 10
11. Griffin, T., Libkin, L.: Incremental maintenance of views with duplicates. In: Proceedings of ACM SIGMOD, pp. 328–339 (1995)
12. Tsang, E.: Foundations of Constraint Satisfaction. Herstellung und Verlag: BoD (1996)
13. Nishihara, S.: Consistent labeling problems with applications. Inf. Process. **31**(4), 500–507 (1990). (in Japanese)
14. ISO/IEC 9075:1992 Information technology – Database languages – SQL (1992)
15. Stonebraker, M.: Implementation of integrity constraints and views by query modification. In: Proceedings of ACM SIGMOD, pp. 65–78 (1975)
16. ISO/IEC 9075-2:1999 Information technology – Database languages – SQL – Part 2: Foundation (SQL/Foundation) (1999)
17. Melton, J., Simon, A.: SQL:1999 Understanding Relational Language Components, Morgan Kaufmann (2002)

LP-HD: An Efficient Hybrid Model for Topic Detection in Social Network

Qingmin Liu, Xiaofeng Gao[(✉)], and Guihai Chen

Shanghai Key Laboratory of Scalable Computing and Systems,
Department of Computer Science and Engineering, Shanghai Jiao Tong University,
Shanghai, China
gao-xf@cs.sjtu.edu.cn

Abstract. Social topic detection and analysis has always been one of the hot topics in data mining. In recent years, with the rapid development of Internet technology, the coverage and information volume of social networking data has increased dramatically. The detection of social topics arouses great discussions and inspires extensive researches. The most common method of topic detection is to cluster texts based on textual features or combining textual features and background information. Each cluster obtained represents a topic. This paper proposes a link prediction-based hybrid detection model (LP-HD) which combines text feature extraction and topology network. For topic detection, the model first constructs a topology graph of all features based on corpus. Then it extracts the feature information of each original text to obtain the feature sequence of the text. The information in these feature sequences is added to the topology graph. Every time after a fixed period of time, LP-HD will take a snapshot of the current topology graph and predict the fluctuation range of each edge in the next time window. Combining the current snapshot and the previous snapshot, the actual fluctuation of each edge can be obtained. In combination with the previous forecast, those active edges can be captured. In the end, these active edges define a new hot topic. At the same time, the topics that are being quickly forgotten can be captured through those once active and now inactive edges. In addition, we also analyze the good adaptability of the LP-HD model. Finally, the experiment proves that LP-HD has good performance on both long text and short text in topic detection.

Keywords: Text feature · Topology graph · Link prediction

1   Introduction

Nowadays, with dramatic development of information technology, online social media has become an important channel for information dissemination. Mining new topics in massive social data has become one of the hottest issues in data mining field.

© Springer Nature Switzerland AG 2019
S. Lee et al. (Eds.): IMCOM 2019, AISC 935, pp. 841–857, 2019.
https://doi.org/10.1007/978-3-030-19063-7_67

To detect new topics from the initial data, the definition of this new topic can be varied. For example, a new topic can represent hot events [1–5] involved in many textual data, or it can represent anomalies in some multimedia data [6–8], and it can also refer to dramatic changes in the relationship networks [9]. There are two difficulties in this issue. The first one is that, in order to give full play to the advantages of immediacy, the topic detection algorithm is efficient enough to detect hot topics in real time. Secondly, the coverage and accuracy of topic detection must meet certain requirements. It should not only ensure that the detected event has a high probability of being a real topic, but also ensure that as many hot topics as possible have been detected. Researchers have proposed various methods to solve these problems.

Many researchers try to detect new topics by extracting text features and then clustering the text. This is a relatively mainstream approach. There are two main areas of improvement for such algorithms. The first aspect is an improvement in text feature extraction methods, such as CDLDA [?]. Another aspect is the improvement of clustering algorithms. For example, for one of the most commonly used clustering algorithms, K-means, many researchers propose many variants to optimize it. Various common variants of K-means are summarized in Sect. 3. At the same time, there are some researchers who are not satisfied with the similarity of relying only on text features when considering the criteria of clustering. They suggest that some background information can also provide some useful help for clustering. For example, Wang et al. [11] propose that time stamps and geographical location information can be used to assist in event detection. DBSCAN [?] uses space-based density of text for event detection. These algorithms also achieve very good results. In addition, time-series-based detection algorithms are also a very important branch. An algorithm based on Improved Soft Frequent Pattern Mining (SFPM) [?] achieves very good performance in dynamic keyword detection.

The algorithms based on feature extraction have good scalability, while the algorithms based on time series have better immediacy. We comprehensively use the advantages of these two ideas and propose a new topic detection algorithm link prediction-based hybrid detection model (LP-HD).

In the LP-HD model, we divide the entire event detection process into three steps. In the first step, we first use feature extraction to transform each original information into a set of features it contains. Then build a topology network for all features. Based on the time series, we add the relationships between the features contained in each piece of information to this topology network. In the second step, we will take a snapshot of this topology map after each time window has passed. Then the abnormal topology of this network is then detected against the current topology. Once abnormal fluctuations have been captured, we can think of this as a new hot spot. For those inactive feature pairs, we will record the number of rounds they are not active. If they are still inactive after a certain period of time, then the topic that they were active before has begun to decline rapidly and the influence is rapidly declining.

We also conduct a series of experiments to evaluate the performance of our algorithm. We not only use short text like microblog data, but also use long text like web articles. We also compared LP-HD with other algorithms. Experiments have manifested that our algorithm has a good performance in terms of immediacy and accuracy.

There are three main contributions of this paper:

1 We propose an event detection algorithm based on feature network and time series. This algorithm has good scalability and good performance in event detection.
2 Based on the Sorensen index (SO), we proposed a new method for measuring the similarity of nodes in a weighted topology graph.
3 We can not only detect the emergence of new hot topics, but also detect once-hot topics that have rapidly declined.

The rest of this thesis is organized as follows.

In the second chapter, we will introduce the related research. In Sect. 3, we will give several detailed definitions. In Sect. 4, we will elaborate our detection model LP-HD. Section 5 presents our experiments and conclusion will be shown in Sect. 6.

2   Related Work

The various applications based on network provide people with a new platform to express their ideas and record their daily lives. In this situation, the social network becomes a non-negligible force to witness a dissemination of a social event. Thanks to its instantaneity and high degree of liberty, we can capture topics much more faster than traditional media. As a result, topic detection based on social networks turns into an important and hot issue in data mining field.

2.1   Text Feature Based Model

Until now, most part of works in topic detection is focused on text data. Spontaneously, the most essential task is mining social topics or novel events from these texts. A lot of researches have been carried out rapidly. There are two main phases. The first phase is analysing natural languages, named natural language processing (NLP). The second phase is using these information given by NLP to mine events.

Many researchers devoted themselves into the mining field. In this field, the most frequent used method is clustering, such as the method proposed by Sina Samangooei [12]. The similar messages will be clustered together. Through this technology, each cluster represents a new event. Another topic detection model ReDites [16] is an absolutely automatic system. Basabi Chakraborty also proposed a method to conduct clustering [10]. They calculated the term frequency–inverse document frequency (tf-idf) of each word in the message. In this way, we can calculate the similarity between messages and complete clustering. SED-RHOCC [6] is another high-order co-clustering and robust algorithm for clustering. Besides, construct a topic graph and using Markov decision processes [8] is another novel idea for clustering. The elements for clustering is transformed from messages to topics.

2.2   Mixed Information Based Model

Even though people record event through contents, that does not mean all information is in texts. Some other information also indicates the different aspects of a social event, such as user information, time stamps, and geography information. Taking full advantage of them can contribute to the improvement of detection model.

Wang [11] proposed a method that combines time stamps, geography information, and the features of contents and vision together to detection events. Traag [13] used Bayesian location inference framework to detect event and achieved good performance.

Not only the information of a message can be combined, but also the features acquired by different model can be combined as well. SSMC [8] take full advantages of several models to obtain different features as well as labels.

Since these additional information has great value, some studies focused on the detection based on these values. For example, geography information is one of the most important features, Liu and his team proposed a model Graph-TSS that aimed to not only detect the details of a social event but also acquire the location of an event [18].

2.3   Time Series Based Model

One of the advantages of topic detection based on social networks is instantaneity because it is usually faster than traditional media.

Hyper local events, which can range from an exhibit, a concert, to a protest, to a fire or even an explosion, can be detected by using time series based model [15]. We can easily find abnormal candidate events and judge if it is a new event.

Some studies combined time series and clustering [17]. The method took advantages of text features and time series information. The process of clustering is carried out with the information of already existing clusters and additional clusters in time sequence.

Most topic detection works until now focused on messages, especially plain texts. Time series can also be used for detecting social events with relative information of participants. HQSII [19] is proposed not only for messages but also for all generalized social networks, for example, the personnel changes in a company.

In many previous methods, each piece of data has to be saved for later clustering or classification. In our model, thanks to using the topology map, we need to store less information. At the same time, the time sequence-based approach allows us to react faster to topics.

3   Definitions

Definition 1 (Relevancy). *There are relationships between every pair of features. We call this kind of relationship relevancy $r_{i,j}$. It describes the closeness of them in a period of time. More frequently the two features appear together, larger the relevancy will be. On the contrary, if the relevancy is small between a pair of features, that means there is no event which has relation with both of them.*

The set of relevancies of the whole graph is represented as a matrix R:

$$R = \{r_{i,j}\}_{n_f \times n_f}$$

Definition 2 (Fluctuation). *There is no doubt that the relevancy between any two features can change. The change of two features after an interval time is named fluctuation:*

$$fl_{i,j} = r_{i,j}^{t_{k+1}} - r_{i,j}^{t_k}$$

The set of fluctuation of the whole graph is represented as a matrix FL:

$$FL = \{fl_{i,j}\}_{n_f \times n_f}$$

Definition 3 (Normal fluctuation). *Even though there is not any new events, the messages in two time windows cannot be the same. That means fluctuation between features always exists whether an event takes place or not. This kind of fluctuation will be in a reasonable range and has no meaning for topic detection.*

Definition 4 (Abnormal fluctuation). *When a social event happens, the fluctuation will be indeed drastic. Some relevancy in topological graph will suddenly raise up because we will catch a large number of messages which have relation with the happened event. In other words, the frequency of co-occurrence will be very high. This kind of fluctuation is called abnormal fluctuation that we are interested in.*

Definition 5 (Connection degree). *In the topological graph, every point has more or less connection with others. The set of relevancy between i^{th} point and other points is named the connection degree of ith point Γ_i.*

$$\Gamma_i = \{r_{i,1}, r_{i,2}, ..., r_{i,n_f}\}$$

Definition 6 (Joint connection degree). *Considering two points together, we can see them as an entirety. This entirety has a connection degree as well, named joint connection degree of the ith point and j^{th} point $J_{i,j}$.*

$$J_{i,j} = \{\max\{r_{i,1}, r_{j,1}\}, \max\{r_{i,2}, r_{j,2}\}, ..., \max\{r_{i,n_f}, r_{j,n_f}\}\}$$

Definition 7 (Similarity degree). *Since every point has its connection degree and these connection degrees are different, some connection degrees are more or less similar compared with others. We call this measurement similarity degree $S_{i,j}$. Similarity degree indicates the similarity between the connection degrees of* *two points. In other words, it indicates the similarity between two points in the topological graph.*

$$S_{i,j} = \{\min\{r_{i,1}, r_{j,1}\}, \min\{r_{i,2}, r_{j,2}\}, ..., \min\{r_{i,n_f}, r_{j,n_f}\}\}$$

Definition 8 (Fluctuation range). *As it is mentioned above, fluctuation is inevitable. In each time window, the relevancies can change. However, they should float in a reasonable range if there is no special events. This range is called fluctuation range $flr_{i,j}$, indicating the reasonable range for fluctuation in next time window between i^{th} point and j^{th} point. It is predicted by the current situation. Therefore, it is a function of joint connection degree and similarity degree.*

$$flr_{i,j} = f(J_{i,j}, S_{i,j})$$

4   Methodology

In this section, we explain the LP-HD model, including its framework and details of each part.

4.1   Framework

The initial data is in the form of messages. These social network texts are different from traditional articles because of their short length as well as some irregular words and slang in them. In this case, we use regularization and stop word list to solve this.

Specially, for those long texts, we can use keyword extraction algorithms such as TF-IDF (term frequency–inverse document frequency) algorithm to solve the problem brought by their long length.

Generally, since the messages are very short, each one describes no more than one thing. Therefore, we believe that if several features appeared in a same message, it means that these features describe a same social topic and they all have relationship with this topic. Based on this idea, we propose two versions of model.

Whenever we get a piece of information, we add features extracted from the information to our topology in a weighted or non-weighted way. After a time interval, we compare the topology with that of the previous time window to obtain a change. At the same time, we will also predict fluctuation based on the relationship data between the features of the last time window for catching hot topics.

The ranges of fluctuations of relevancies between two features $flr_{i,j}$ depend on their most recent relevancies $r_{i,j}$, their joint connection degree $J_{i,j}$, and their similarity degree $S_{i,j}$. Their fluctuation ranges have positive relationship with their own strength. Because different features have different degrees of hot spots in social topics. Simultaneously, whether the strength between two features will be enhanced depends on their similarity, which is, their respective connection degree with other features. If they are similar, there is a great possibility that they will appear in the same event. The similarity also depends on their current joint connection degree. With this idea, the prediction of fluctuation range can be completed.

In each time window, there will be some abnormal floating feature relationships, and there will also be some feature relationships within a normal fluctuation range. This information is not useless. On the contrary, combined with the range of relevancy fluctuations, these data can help us detect those declining topics that was once hot topics.

The whole framework is shown in Fig. 1. As we can see in this figure, features in initial data are extracted and added to topology. Then we conduct fluctuation prediction. At last the hot event and declining event can be captured.

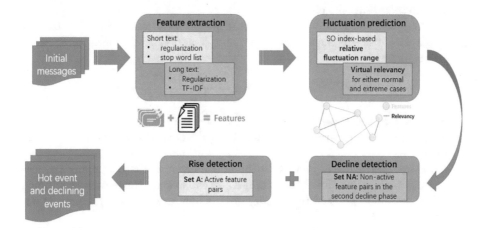

Fig. 1. The framework of model

Last but not least, what has to be emphasized is that LP-HD model has excellent scalability. It is shown in three aspects. First of all, we can use different keyword extraction algorithms to satisfy different requirements. Secondly, the size of the time window can also be adjusted manually. Thirdly, since the corpus of the model is set manually, both the quantity and content of the corpus can be adjusted as needed to improve accuracy and reduce calculating amount.

4.2   Feature Extraction

As mentioned earlier, each text contains more or less features. These features are based on the topics described in the text. Therefore, if we want to get the topic content of this text, we only need to obtain these features, which means, we only need to extract the keywords of each text. The features that these keywords belong to are the characteristics of the text. One of the most critical issues is to choose a suitable range of features. If a feature contains a range that is too large, the keywords that belong to that class will increase accordingly. This can lead to ambiguous divisions, inaccurate descriptions of features, and insensitivity in response to changes in the relationships between features. On the contrary, if the features are too accurate, although an event can be clearly described, due to the large feature population, the time complexity in extracting the features becomes very high, thereby reducing the detection efficiency. At the same time, the increase in the number of features will directly increase the topology network. This will not only lead to a drastic increase in the amount of calculations in the following work, but also require more space to store the data.

Once we have selected the appropriate features, the next thing we need to do is extract the keywords. Many scholars have done in-depth and extensive research on keyword extraction. Examples include Hu [20], Biswas [21], Zimniewicz [22], and their teams. However, in this paper, we hope to use a simple and effective method to maintain efficiency.

For long texts such as WeChat articles, we prefer to use TF-IDF to extract keywords. Because this algorithm is very simple and fast. The value of ith word to j^{th} document is:

$$tf\text{-}idf_{i,j} = \frac{n_{i,j}}{\sum_k n_{k,j}} \times \log \frac{|D|}{1 + |\{d \in D : i \in d\}|}$$

Here, $n_{i,j}$ is the times that i^{th} word appears in j^{th} document. $\sum_k n_{k,j}$ is the total times of appearance of all words in j^{th} document. D is the set of all documents.

After extracting the keywords by the above method and combining the feature corpus we have previously determined, we can extract the features contained in each text. What we need to consider next is how to add these features to our topological graph.

As mentioned earlier, these features are handled in two ways. The first is not considering the differences in the characteristics of the same text. If two features appear in the same text, the relevancy between them will increase by a fixed amount. This method is suitable for short texts. Because short texts often contain only one or two sentences, the relationship between words and topics is very close.

The second is to take into account their differences. That is, if multiple features appear in the same text, the relevancy between them will be different. This method is suitable for long texts. Because long text may describe one thing from multiple views. Even an article may involve other events beside the target event. However, in terms of language habits, the same paragraph basically describes the same event. Therefore, the closer the two sentence that two features correspondingly belong to, the larger the relevancy between the feature pair will be. When they belong to the same sentence, they are the most relevant. We use an exponential function to describe this relationship approximately.

$$\Delta r_{i,j} = e^{-(k_i - k_j)}$$

Here k_i and k_j represents the order of the sentences that i^{th} feature and j^{th} feature belong to respectively.

4.3   Fluctuation Prediction

In each time window, the relevancies between those features will change. If we want to capture those abnormal fluctuations, we must set a reasonable range of fluctuations in advance. If the changes in features are within this reasonable range, they are considered normal. Once their relationship changes beyond this reasonable range, they will be caught. Thus a new event described by these features was detected.

In our model, the floating algorithm is based on the link prediction algorithm. The traditional link prediction algorithm refers to the possibility of predicting the connection of two points that are not yet connected, based on the information of the nodes in the graph and the structural network information. We apply this idea to a floating forecast. The possibility of creating a connection between two points is analogous to the range of variation in the relevancy between two features.

In our algorithm, we mainly use the similarity of nodes in the map to predict the link. In traditional link prediction, there are many ways to assess similarity. Among these various indicators we choose Sorensen index (SO) [23] as the prototype of our prediction model. The similarity of SO is defined as follows:

$$S(i,j) = \frac{2|\Gamma(i) \bigcap \Gamma(j)|}{k(i) + k(j)}$$

Here, For node i, $\Gamma(i)$ and $k(i)$ represent the neighbor node set and the degree, respectively.

We chose SO because it is more in line with our requirements for volatility forecasts. Through SO index, we can get the relative fluctuation range that:

$$flr_{relative}(i,j) = \frac{\sum_{k=1}^{n_f} \min\{r_{i,k}, r_{j,k}\}}{1 + \sum_{k=1}^{n_f} \max\{r_{i,k}, r_{j,k}\}} = \frac{\sum_{s \in S_{i,j}} s}{1 + \sum_{t \in J_{i,j}} t} \tag{1}$$

In Eq. (1), We added a reconciliation parameter 1 to the denominator to prevent the appearance of a certain feature pair that the relevancy between them and any other features are zero.

In our model, we consider the joint connection degree of the two features. That is, consider the two features as a whole and consider the degree of connection between them and all other features. This is the basis for measuring their range of volatility. If the feature group has a large degree of joint connection degree, these two features are very active. Then in the normal fluctuations, their scope should also be greater. For example, if a feature f_i appears on average 10,000 times in each time window, it is normal that it appears more 200 times in the next time window. However, if another feature f_j occurs on average only 100 times, and in the most recent test, it appears more 200 times. This phenomenon is generally not normal. That is, the greater the joint connection degree between pairs of features, the larger their range of fluctuations.

However, in some extreme cases, the fluctuation range of the relevancy is not proportional to the absolute value of the intensity. When the intensity is very small, the absolute value of the fluctuation may be small even with each measurement, but the relative fluctuation of the relative intensity itself will be relatively large. If we follow a completely proportional forecasting model, these tiny fluctuations will be caught and be judged as abnormal fluctuations. In fact, there is probably no related event. For example, a feature pair has a relevancy of 10 for the last test and this test the relevancy becomes 20. Even though the relative fluctuation reaches 100%, there is a high probability that this is still a reasonable fluctuation. Therefore, we have defined a new concept: virtual relevancy r_v. This is not a true intensity, but a virtual strength. Based on it, the absolute range of the next fluctuation can be predicted more reasonably. When the true relevancy is very small, the virtual relevancy becomes very large to satisfy the large relative fluctuations that may occur. When the true relevancy is large enough, the virtual relevancy is infinitely close to the true relevancy. The definition of virtual relevancy is shown in Eq. (2).

$$r_v(i,j) = r(i,j) + \frac{a}{r(i,j)} \tag{2}$$

Here a is a parameter that determined by the actual situation. The higher the threshold for the special case, the greater a will be.

Combining these two points above, we can get the predicted value of the fluctuation range. The fluctuation range is proportional to the relative fluctuation value $flr_{relative}$ and at the same time it is proportional to the virtual intensity r_v. Besides, we added a parameter λ to better fit in actual situations. Therefore, the definition of the fluctuation range is shown in Eq. (3).

$$flr(i,j) = \lambda \times flr_{relative}(i,j) \times r_v(i,j) \tag{3}$$

4.4　Rise Detection

After each time window passes, we will record a snapshot of the topology. In this snapshot, we can get the current intensity information of any two nodes. This information is saved to an adjacency matrix R. At the same time, using this information, we can get the data needed for link prediction, for example, the joint connectivity of any two nodes, their similarity, and their virtual strength. The intensity prediction between all feature pairs will be saved for the next time window. At the same time, in order to capture those abnormal fluctuations in this time window, we need to compare the amount of fluctuation obtained this time with the previous forecast value.

If the fluctuation between two nodes is greater than the predicted range, this feature pair is determined to be active. Therefore, we can obtain a set of abnormal edges in topological graph A, which may or may not be empty sets. If A is an empty set, then no new hotspot events have occurred in the past period of time. If A contains several elements, it indicates that a new hotspot event has occurred in the past period of time. Besides, this new event can be described by the features whose edges are elements in A. At the same time, these abnormally fluctuating edges are marked as active. On the contrary, those edges that are within the normal fluctuation range are marked as inactive. Inactive edges have no effect on event detection. However, they provide information for event declining detection.

4.5 Decline Detection

As we mentioned before, in each snapshot, there are some edges that are active, while others are marked as inactive. These inactive feature sets can be used to detect the decline of a spot event.

In general, the decline of an event is divided into three phases: In the first phase, due to the wide range of events and the large scale of audiences, gradually, a small number of people would no longer care about this incident. In the second phase, the incident begins to decline significantly. Like snowballing, the number of people who do not pay attention to it quickly increases. This also leads to a rapid reduction of information related to the incident. In the third phase, the decline of the incident tended to saturate and the rate of decline slowed again. Because the audience is very few at this time. The incident gradually faded out of sight. Therefore, we use a sigmoid-similar function to describe the fading of this heat over time. The decline function is shown in Fig. 2.

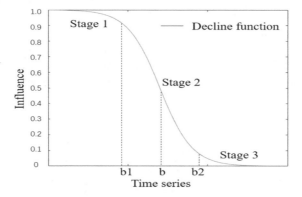

Fig. 2. Decline function

To better characterize this detection process, we define a new variable $l_{ina}(i, j)$. This variable is called inactive level. Whenever a feature pair (f_i, f_j) is inactive in a new time window, $l_{ina}(i, j)$ will increase by one. When the feature pair (f_i, f_j) is marked as active in a new time window, $l_{ina}(i, j)$ is reset to zero. Therefore, in combination with the sigmod function, we can obtain the decline function by Eq. (4):

$$f(l_{ina}(i, j)) = \frac{1}{1 + e^{-l_{ina}(i,j) + \frac{b_1 + b_2}{2}}} \tag{4}$$

We can clearly see from Fig. 2 that if a feature pair was once active, once it is no longer active, it can be seen as entering the first phase of decline. At this stage, the decline is not obvious, and there is a greater possibility of renewed concern. After continuing for b_1 time window, if this feature pair is still continuously inactive, then the related events will enter the second stage of decline. From the prediction aspect, this is the period in which the incident has rapidly declined. To sum up, we detected the incident in the second phase of an incident's decline, which is the most reliable and most valuable.

As a result, for each snapshot, we need to record the level of inactivity of each feature pair. If a feature pair is inactive in the previous snapshot and is also inactive in this snapshot, then its inactive level is incremented by one. If a feature pair is active in the last snapshot and becomes inactive in this snapshot, it is marked as inactive and the inactivity level is reset to one. After each snapshot, we catch these edges which are in level b_1 as a set NA to represent a declining topic.

5   Simulation and Evaluation

5.1   Short Text Topic Detection

In this section, we simulate LP-HD model on sort text. The entire dataset is real data collected from Weibo. Its coverage is from November 2011 to December 2012. It contains more than 1 million users and 20 million pieces of Weibo. The density of Weibo each day ρ is different. We chose different periods which correspond to different ρ for topic detection. What's more, we chose different size of time window ΔT. The settings of parameters are shown in Table 1:

Table 1. Experiment setting

Parameter	Value
Data density (ρ)	$0-25K, 25-50K, 50-75K, \geqslant 75K$ (per day)
Time window size (ΔT)	8 h, 1 day, 1 week, 1 month
Number of features n_f	18
Data topic	Absolutely random

Results of Different Time Window Size ΔT. We firstly set different time window sizes and compare the performance of them. We choose the data in a period of time, about three months. We use two other algorithms as a reference. These two algorithms are based on clustering with text features. In consequence, with the change of ΔT, there is no effects on their detection results. The standardized result is shown in Fig. 3(a).

The result shows that different time window size has an important impact on the final results of topic detection. If ΔT is too small, there will appear lots

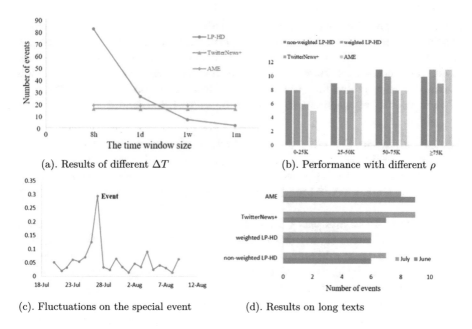

(a). Results of different ΔT

(b). Performance with different ρ

(c). Fluctuations on the special event

(d). Results on long texts

Fig. 3. An example to construct a 2-CDS with 17 nodes

of false events. However, if ΔT is too large, many events will be ignored. This is consistent with our previous theoretical analysis. For different types of target events and data, we should select a time window of the appropriate size.

Performance with Different Data Density. Now that we have determined the appropriate time window size, we will use data sets from time periods which have different data densities ρ in the next experiment. Prior to this, we design two models that correspond to two methods for measuring feature relationships in texts. As we mentioned in the previous section, the relationships between the features in a text can be added equally to the topology graph or added in a weighted way. For these two ideas, we divided our model into two types, weighted LP-HD and non-weighted LP-HD.

In this experiment, we select four different density time periods from the original data set. To ensure that these lengths of time are approximately equal, each time period is approximately 40 days. The experimental results are shown in Fig. 3(b).

The experiments show that the results are relatively stable at various data densities. This may be because the data is distributed more uniformly. When the density increases, only the amount of data increases, and the scope of the topics of texts does not significantly increase. In addition, it can be found that non-weight models are more sensitive to events than weighted models. Because after considering the weight, the value of the relevancy increase brought by the two features in the same text is always less than or equal to that of the non-weight model. Therefore, the topological graph will also be relatively small and therefore relatively insensitive to events. In our experimental dataset, we can see through two reference algorithms that the weighted model detection results are closer to the reference values. However, this does not mean that non-weighted models are inferior to weighted models in any case. Besides, the additional computation cost for calculating weights also needs to be taken into account.

Performance on Specific Events. In this experiment, we manually filter out some microblogs related to a specific event and add some noise data. This artificial dataset was used to test the ability of LP-HD to detect single events. At the same time, we could more intuitively see the fluctuations of the topology graph.

The Olympic Games was held in London from July 27th to August 12th, 2012. We select more than 1,000 pieces of microblog related to it, and at the same time, add noise microblogs on many other topics from July 20 to August 10. For better analysis, we make corresponding adjustments to the corpus. We observe the average relative fluctuations after each snapshot. The experimental results are shown in Fig. 3(c).

In Fig. 3(c), we can observe that the fluctuation started to increase significantly on July 26th, and LP-HD captured abnormal fluctuations, which means it detects a hot topic. On July 27th, the opening day of the Olympic Games, the fluctuations were extremely dramatic. This represents the occurrence of an significant incident. Then fluctuations return to normal again. This does not mean that the topic has ended, but that the hot topic has not changed much compared to the previous day. In the next few days this topic gradually began to decline.

Decline Detection. We also carried out experiments to evaluate the model's ability to detect declining events. We neither have a baseline as a reference, nor find a model or algorithm on this problem to compare. Therefore, we listed the results of the decline detection in Table 2.

Table 2. Decline detection results

Time interval of data	Number of decline events
5/1–5/10	3
5/11–5/20	2
5/21–5/31	4
6/1–6/10	4
6/11–6/20	2

5.2   Long Text Based Topic Detection

We also experimented with long texts. The long text data that we used are the articles collected from the news websites. It was published by Sogou Labs. The dataset has a total of 711 MB. The articles were published from June 2012 to July 2012. We divided the data set into two parts, June and July. Similar to the experiments on short texts, LP-HD did topic detection with them. Accordingly, we compare the test results with that of two reference algorithms. The results are shown in Fig. 3(d).

From Fig. 3(d) we can observe that LP-HD has good performance in long texts. However, the number of events detected by LP-HD is less than the other two algorithms. This is due in part to the fact that the long text datasets we used in this experiment were basically collected on news websites. Therefore, each article itself is an event. This model is not suitable for situations where multiple topics are concurrent. We can only increase the Precision by adjusting the size of the time window. However, the shrinking of the time window may cause a decrease in the Recall. On the other hand, in long texts, good keyword extraction algorithms have significantly improved the final results. Therefore, both reference algorithms have achieved better performance. As a result, in the case where the dataset is consisted by long text, we can consider to sacrifice part of the calculation count to improve the accuracy of keyword extraction.

6   Conclusion

In this paper, we introduce a new model LP-HD that detects hot topics from social network data. The model synthesizes the two concepts of text features and time series. We introduce the relevant background knowledge and defined the problem. On this basis, we describe in detail the three steps of the model: feature extraction, link prediction, and hybrid detection. Compared with the previous methods, LP-HD model has the characteristics of small amount of information stored, fast response to topics, and good adaptability to topics. What is more, it can detect not only hot topics but also declining topics. Finally, we design a series of experiments to verify the good performance of LP-HD.

Acknowledgements. This work was supported by the National Key R&D Program of China [2018YFB1004703]; the National Natural Science Foundation of China [61872238, 61672353]; the Shanghai Science and Technology Fund [17510740200]; the Huawei Innovation Research Program [HO2018085286]; and the State Key Laboratory of Air Traffic Management System and Technology [SKLATM20180X].

References

1. Endo, Y., Toda, H., Koike, Y.: What's hot in the theme - query dependent emerging topic extraction from social streams. In: International Conference on World Wide Web (WWW), Florence, Italy, pp. 31–32 (2015)
2. Khare, P., Torres-Tramón, P., Heravi, B.R.: What just happened? A framework for social event detection and contextualisation. In: Hawaii International Conference on System Sciences (HICSS), Hawaii, USA, pp. 1565–1574 (2015)
3. Wold, H.M., Vikre, L., Gulla, J.A., Özgöbek, Ö., Su, X.: Twitter topic modeling for breaking news detection. In: International Conference on Web Information Systems and Technologies (WEBIST), Rome, Italy, pp. 211–218 (2016)
4. Hashimoto, T., Kuboyama, T., Okamoto, H., Shin, K.: Topic extraction on twitter considering author's role based on bipartite networks. In: International Conference on Discovery Science (DS), vol. 20, pp. 467–476 (2016)
5. Rafea, A., Mostafa, N.A.: Topic extraction in social media. In: International Conference on Collaboration Technologies and Systems (CTS), pp. 239–247 (2017)
6. Bao, B.-K., Min, W., Lu, K., Xu, C.: Social event detection with robust high-order co-clustering. In: The ACM International Conference on Multimedia Retrieval (ICMR), London, Banda Aceh, Indonesia, pp. 135–142 (2013)
7. Wang, Y., Sundaram, H., Xie, L.: Social event detection with interaction graph modeling. In: ACM Multimedia, pp. 865–868 (2012)
8. Yang, Z., Li, Q., Lu, Z., Ma, Y., Gong, Z., Pan, H.: Semi-supervised multimodal clustering algorithm integrating label signals for social event detection. In: IEEE International Conference on Multimedia Big Data (BigMM), Beijing, China, pp. 32–39 (2015)
9. Zhang, X., Zheng, N., Xu, J., Xu, M.: Topic detection in group chat based on implicit reply. In: Pacific Rim International Conference on Artificial Intelligence (PRICAI), Phuket, Thailand, pp. 62–75 (2016)
10. Chakraborty, B., Hashimoto, T.: Topic extraction from messages in social computing services-determining the number of topic clusters. In: IEEE Fourth International Conference on Semantic Computing (ICSC), Pittsburgh, PA, USA, pp. 94–98 (2010)
11. Wang, Y., Xie, L., Sundaram, H.: Social event detection with clustering and filtering. In: Working Notes Mediaeval Workshop (MediaEval) (2011)
12. Samangooei, S., Hare, J., Dupplaw, D., Niranjan, M., Gibbins, N., Lewis, P.: Social event detection via sparse multi-modal feature selection and incremental density based clustering. In: Working Notes Mediaeval Workshop (MediaEval) (2013)
13. Traag, V.A., Browet, A., Calabrese, F., Morlot, F.: Social event detection in massive mobile phone data using probabilistic location inference. In: IEEE International Conference on Privacy, Security, Risk, and Trust, and IEEE International Conference on Social Computing (SocialCom/PASSAT), Boston, MA, USA, pp. 625–628 (2011)
14. Chen, Q., Guo, X., Bai, H.: Semantic-based topic detection using Markov decision processes. Neurocomputing **242**, 40–50 (2017)
15. Xie, K., Xia, C., Grinberg, N., Schwartz, R., Naaman, M.: Robust detection of hyper-local events from geotagged social media data. In: International Workshop on Multimedia Data Mining (MDMKDD), Chicago, IL, USA, pp. 1–9 (2013)

16. Osborne, M., Moran, S., McCreadie, R., von Lünen, A., Sykora, M.D., Cano, E., Ireson, N., Macdonald, C., Ounis, I., He, Y., Jackson, T., Ciravegna, F., O'Brien, A.: Real-time detection, tracking, and monitoring of automatically discovered events in social media. In: Association for Computational Linguistics (ACL), Baltimore, MD, USA, pp. 37–42 (2014)

17. Comito, C., Pizzuti, C., Procopio, N.: Online clustering for topic detection in social data streams. In: IEEE International Conference on Tools with Artificial Intelligence (ICTAI), San Jose, CA, USA, pp. 362–369 (2016)

18. Liu, Y., Zhou, B., Chen, F., Cheung, D.W.: Graph topic scan statistic for spatial event detection. In: ACM International Conference on Information and Knowledge Management (CIKM), pp. 489-498, Indianapolis, IN, USA (2016)

19. Hu, W., Wang, H., Qiu, Z., Nie, C., Yan, L., Du, B.: An event detection method for social networks based on hybrid link prediction and quantum swarm intelligent. World Wide Web (WWW) **20**, 775–795 (2017)

20. Hu, J., Li, S., Yao, Y., Yu, L., Yang, G., Hu, J.: Patent keyword extraction algorithm based on distributed representation for patent classification. Entropy **20**, 104 (2018)

21. Biswas, S.K., Bordoloi, M., Shreya, J.: A graph based keyword extraction model using collective node weight. Expert. Syst. Appl. **97**, 51–59 (2018)

22. Zimniewicz, M., Kurowski, K., Weglarz, J.: Scheduling aspects in keyword extraction problem. In: International Transactions in Operational Research (ITOR), vol. 25, pp. 507–522 (2018)

23. Sørensen, T.: A method of establishing groups of equal amplitude in plant sociology based on similarity of species and its application to analyses of the vegetation on Danish commons. Biol. Skr., 1–34 (1948)

Connecting Heterogeneous Electronic Health Record Systems Using Tangle

Emil Saweros$^{(\boxtimes)}$ and Yeong-Tae Song$^{(\boxtimes)}$

Computer and Information Sciences, Towson University, Towson, MD, USA
esawer1@students.towson.edu, ysong@towson.edu

Abstract. Currently, Electronic Medical Records (EHRs) are centralized and arduous to move around. Getting crucial information in front of a doctor shouldn't be as difficult as it is. We want to solve a core problem of how to distribute transparent, updated medical data to a network of caregivers securely. Due to the current centralized nature of data ownership, there would be no guarantee to deliver untampered up-to-date information unless distributed ledgers or shared data structure is used. The number of patients is continually increasing, and it requires a system that would be able to keep track of a decentralized record of each patient, but at the same time, it should be able to scale to the proportions necessary for handling the increasing network throughput. Our approach from this study is to provide a new enhanced secure and trustable network platform between care providers and patients by sharing data using Tangle, where data is structured using a Directed Acyclic Graph (DAG) to store its EHRs ledger, with the primary motivation of being scalable.

Keywords: IOTA · Distributed Transaction Ledger (DTL) ·
Information Communication Technology (ICT) · Tangle ·
Electronic Health Records (EHR) · Interoperability · Connecting EHR

1 Introduction

EHRs contains highly sensitive private information for diagnosis and treatment in health care, which needs to be frequently distributed and shared among peers such as care providers, insurance companies, pharmacies, researchers, patients families, and among others [1]. There is a need for a secure data sharing infrastructure to handle health data sharing between institutions. At the same time, Continuity of care becomes increasingly important for patients as they age, develop multiple morbidities and complex problems, or grow socially or psychologically vulnerable [2].

The rapid growth of the Information Communication Technology (ICT) opens new opportunities for the development of core activities in the Healthcare sector with the possibilities to improve various aspects of healthcare to have better access to health records, and with the aim to deliver improved way on sharing patient's medical records among caregivers. We are proposing a new interoperable data platform for existing EHR systems to store, maintain, and share Electronic Medical Records which contain sensitive private information in healthcare, which will change the way EHRs are sharing in medical community.

© Springer Nature Switzerland AG 2019
S. Lee et al. (Eds.): IMCOM 2019, AISC 935, pp. 858–869, 2019.
https://doi.org/10.1007/978-3-030-19063-7_68

Blockchains are immutable digital ledger systems implemented in a distributed fashion (i.e., without a central repository) and usually without a central authority [3]. In 2008, the blockchain idea was combined innovatively with several other technologies and computing concepts to enable the creation of modern cryptocurrencies: electronic money protected through cryptographic mechanisms instead of a central repository.

The rise and success of blockchain during the last few years proved that blockchain technology has real-world value. However, this technology also has some drawbacks that prevent it from being used as a generic platform for cryptocurrencies across the globe [4]. In this paper, we discuss an innovative approach by implementing a cryptocurrency solution to connect different EHRs, care providers as we are focusing on the tangle general features, and discussing problems that arise when one attempts to get rid of the blockchain and maintain a distributed ledger to ensure continuity of care.

The paper is organized as follows: in the next section, we discuss the current challenges with the existing EHRs system. In Sect. 3 we provide the basic concepts needed to understand Blockchain background and how Tangle is considered the next evolutionary step of DAG-based architecture of cryptocurrencies. Section 4 presents the related work, while Sect. 5 discusses our approach and briefly discuss the tip selection mechanisms to handle transactions. In the last Sect. 6, we conclude our findings.

2 Challenges

One of the major challenges in connecting heterogeneous EHRs is the development of interoperability as there is a need for hospitals and health systems to exchange medical information with one another. The effective integration of health data and the interoperability between healthcare systems still remains as a challenging task [5]. Moreover, there are other several challenges related to privacy and security. There is an urgency to develop a new version of EHR systems with user-centric access control and privacy preservation. The integration of behavioral health and primary care has gained increasing attention to lower costs, improve quality of care and ensure continuity of care [6].

2.1 Reliability and Validity

The diversity in the data element, study setting, population, health condition, and EHR system studied within this body of literature made drawing specific conclusions regarding EHR data quality challenging [7]. Our approach with this new network platform using Tangle is to ensure the quality of data from specific EHR components and essential data attributes for quality measurement such as granularity, timeliness, and comparability. Besides, it can resolve the factors associated with inadequate or variability in data quality that need to be better understood and effective interventions developed.

2.2   Ownership vs. Control vs. Privacy

Tangle can help patients to become more connected to their data and more involved in their health care processes, which can improve their quality of care. Dr. Josh Landy, a physician, and co-founder of Fig. 1, a text-messaging app for healthcare professionals, he says; Instead of ownership, healthcare professionals and patients should discuss electronic patient data in terms of "stewardship," [8]. As of today, unfortunately for patients, they have little control over it, usually the health care provides who has the authority to change records and patient cannot go in and accept a diagnosis even if it is not accurate. Also, in some cases, patients do not have access to their medical records. However, just because the provider has more control than the patient doesn't mean patients are entirely vulnerable. Depending on the institution in control of that data, most will respect patient privacy. Many healthcare providers would agree that it's beneficial for them to be transparent with patients about their data and that patients have the right to voice their opinions.

2.3   Data Integrity

With the current EHR environment professionals are profoundly dissatisfied with the flow of information as there is a lack of knowledge with document workflow, health information management, and clinical documentation requirements. Data integrity is defined as:

- The extent to which health care data are complete, accurate, consistent and timely.
- A security principle that keeps information from being modified or otherwise corrupted either maliciously or accidentally; also called data quality.

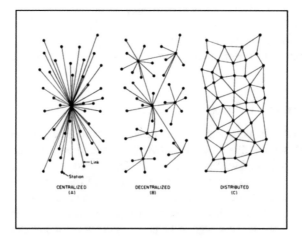

Fig. 1. Centralized and distributed networks (Baran [10])

3 Background (Blockchain vs. Tangle)

3.1 Blockchain

A blockchain is a distributed database that acts as a peer-to-peer ledger system. It is often discussed in the context of Bitcoin, a decentralized digital currency. In electronic currency world, blockchain has been very popular for bitcoin transaction settlement as a public ledger [9].

The centralized network is vulnerable as a destruction of a single central node destroys communication between the end stations. For example; type (B) in Fig. (1) shows the hierarchical structure of a set of stars connected in the form of a larger star with an additional link forming a loop. Such network is called a "decentralized" network, because complete reliance upon a single point is not always required [10].

A blockchain is like a place to store any data semi-publicly in a linear container space (the block). In general, blockchain has some drawbacks, and as we will discuss in the following section, it is not suited for application to some scenarios. Blockchain designed for keeping a financial ledger, the blockchain paradigm can be extended to provide a generalized framework for implementing decentralized compute resources [11, 12]. Each compute resource can be thought of as a singleton state-machine that can transition between states via cryptographically-secured transactions. When generating a new state-machine, the nodes encode logic which defines valid state transitions and upload it onto the blockchain. From there on, the blocks journal a series of valid transactions that, when incrementally executed with the state from the previous block, morph the state-machine into its current state. The Proof of Work consensus algorithm and its underlying peer-to-peer protocol secure the state-machines' state and transitioning logic from tampering, and also share this information with all nodes participating in the system. Therefore nodes can query the state- machines [11] (Fig. 2).

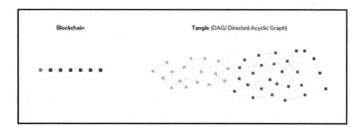

Fig. 2. Blockchain vs. Tangle

3.2 Blockchain Drawbacks

Heterogeneous with the Existing Cryptocurrency Technology. It is considered one of the main blockchain issues as at least two or more need to participate in the system - one issues transactions and the others approve transactions -, which creates conflicts that make all elements spend resources on conflict resolution. Currently, all blockchain consensus protocols (e.g., Bitcoin, Ethereum, Ripple, Tendermint) have a challenging limitation as every fully participating node in the network must process every transaction.

Storage and Transaction Fees. With Electronic health, we have an infinite amount of accumulating data from different participants. The main point of decentralized data is that each participated node has enough needed information to process data without copying the whole actual database. However, with an increasing number of transactions, the size of the database is expanding like the universe. Accordingly, transaction fees are high and paying a fee is larger than the amount of value being transferred or used. Furthermore, it is not easy to get rid of fees in the blockchain infrastructure since they serve as an incentive for the creators of blocks [4].

Scalability. The blockchain is not well suited for high performance (millisecond) transactions involving just one participant with no business network involved, or for replicated database replacement. Additionally, it is not useful as a transaction-processing replacement and is unsuitable for low-value, high-volume transactions For example, PayPal manages 193 transactions per second, and visa manages 1667 transactions per second, Ethereum does only 20 transactions per second while bitcoin manages a large 7 transactions per second! So, we need a solution that these numbers can be improved and work on scalability. DAG resolves the main scalability problems with blockchain in the cryptocurrencies, which are:

- The time is taken to put a transaction in the block.
- The time is taken to reach a consensus.

3.3   Tangle (IOTA)

Currently IOTA foundation is developing a computer simulation of the Tangle network. IOTA differentiates itself from other cryptocurrencies by being based on a non-blockchain data structure with a highly scalable approach to transaction confirmation [13]. IOTA is a distributed ledger which aims to offer a solution to the issues of scalability and high fees which have afflicted blockchain technology. Created in 2015, it gets its name from its long-term objective to power microtrans- actions between IoT devices [4]. IOTA's main difference from existing distributed ledger technology is that it is based on a directed acyclic graph consensus structure called the 'Tangle' rather than a blockchain. Instead of requiring special participants 'miners' to perform computational proof-of-work and validate blocks of transactions in exchange for newly minted tokens, network participants themselves perform consensus by validating two previous transactions each time they wish to make a transaction [4].

The tangle naturally succeeds the blockchain as its next evolutionary step, and offers features that are required to establish a machine- to-machine micropayment system [4]. The Tangle was adapted as cornerstone technology of the IOTA network. IOTA is a scalable, permissionless distributed ledger designed for the Internet of Things industry. DAG based architecture of cryptocurrencies allows for bypassing drawbacks of standard blockchain architecture, like scalability (growing problem of micropayments) and division of users into two groups (issuers and approvers/miners of transactions) [14].

As we mentioned, what makes blockchain hard to implement is the scalability issues. Constant growth of transaction fees is major obstacle in establishing true micropayments. One of the solutions to this problem could be an increase of block size or decrease of time between blocks. However such move brings need for more storage space and that might threaten the very idea of trustless peer-to-peer cryptocurrency [15]. Another characteristic of standard blockchain not suited for IOT industry is the division of users into two groups: those who issue transactions, and those who approve transactions (miners). Both groups have fundamentally different goals and roles.

4 Related Work

Blockchain scalability issues are intensively studied and there are variety of promising solutions like the lightning networks [16], proof of stake protocols [17] but still there are some failure as any system that relies on miners to confirm transactions becomes too expensive and not worth it. Speed also becomes another issue as the network increases in size as more transactions compete for the limited block spaces.

Below are some of the current related work that are based on the Blockchain and Tangle technology, although the proposed solutions of the architecture for secure data exchange does not solve the dilemmas related to data accessibility versus privacy.

- MedRec: A record management system focusing on EMRs using smart contract [11, 18]. The MedRec prototype provides a proof-of-concept system, demonstrating how principles of decentralization and blockchain architectures could contribute to secure, interoperable EHR systems. MedRec is using Ethereum smart contracts to orchestrate a content-access system across separate storage and provider sites, the MedRec authentication log governs medical record access while providing patients with comprehensive record review, care auditability and data sharing.
- Use of Blockchain in Healthcare and Research Workshop by National Institute of Standards and Technology (NIST) [19].
- Electronic Health Records using Blockchain Technology [20].
- The recent paper regarding Equilibria in the Tangle [21]: Paper proved the existence of ("almost symmetric") Nash equilibria for a game in the tangle where a part of players tries to optimise their attachment strategies and they we numerically determined where these equilibria are located.
- Extracting Tangle Properties in Continuous Time via Large-Scale Simulations [14]: Paper demonstrated that the growth of the cumulative weight follows two stages: adoption phase (which appears to be exponential), followed by a linear phase (with slope λ) as an independent finding of the tip selection algorithm; be it URTS or MCMC with some small α.
- Local modifiers in the Tangle [22]: The idea of local modifiers; how the nodes of the network can interact with the ledger in different ways, depending on various kinds of information locally available to them.
- Integrating the Healthcare Enterprise (IHE): IHE is an initiative approach by healthcare professionals and industry to improve the way computer systems in healthcare share information. IHE promotes the coordinated use of established standards such as DICOM and HL7 to address specific clinical needs in support of optimal patient care.

- Volkswagen: German car manufacturer Volkswagen recently announced its collaboration with top IoT-based blockchain project, IOTA, as the two companies prepare to present a new proof of concept utilizing IOTA's tangle network. The concept aims to solve a core problem of how to distribute transparent, incorruptible software updates securely 'over the air' (OTA) to a network of autonomous vehicles. Due to the current centralized nature of data ownership, there would be no guarantee that information delivered around each vehicle is honest and untampered with unless a distributed ledger or some kind was used.
- Fujitsu: Fujitsu announced during the Hannover Mess, one of the largest industrial technology fairs in the world, a software called INTELLIEDGE using IOTA's Tangle network to monitor the industrial environment, through the application it will be possible to monitor a production line of a factory. It is the first application in the world using a DLT.

5 Our Approach

As we move into the future, there appears to be an increasing need for a standardized message block for all digital communication networks. As data rates increase, the velocity of propagation over long links becomes an increasingly important consideration (Fig. 3).

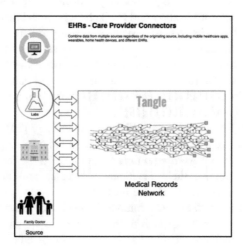

Fig. 3. EHRs – Care provider connector

5.1 Data Source

The new network platform will combine data from multiple sources regardless of the originating source, including mobile healthcare apps, wearables, home health devices, and different EHRs. Importantly, connecting multiple EHRs to provide the same up-to-date information about conditions, treatments, tests, and prescriptions is a solution to improve EHRs interoperability (Fig. 4).

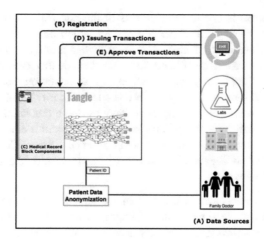

Fig. 4. Connecting EHRs data structure

5.2   Registration

Patient identification is fundamental to the interoperability of health data and the resulting promise of improved care [23]. To join the network and fully benefit from the new tangle network, both patients and care providers can register to ensure Electronic Health Records (EHRs) are safe and confidential records are secured. Managing patient identities across the different healthcare is a challenge as it requires a flexible system, open, interoperable architecture that is scalable. We do not require healthcare organizations to rip out and replace their legacy systems that may have inconsistent, incomplete, or fragmented data. Instead, we enable these legacy systems to share data, despite differences that exist across systems or even within a single vendor's architecture. Registration will help to combine clean data and give the ability to accurately and quickly match and link patient information across records by adding new data elements and or retrieving a healthcare identifier. Also, registration can lower the chances of medical errors, prevent duplicate tests, and may improve the overall quality of care.

5.3   Medical Record Transaction Components

According to Fundamental Skills for Patient Care in Pharmacy Practice Book Chapter 2 [24] and [25], The medical record can be dissected into five primary components, including the medical history (often known as the history and physical, or H&P), laboratory and diagnostic test results, the problem list, clinical notes, and treatment notes. Figure (5) illustrates the transaction components of medical records.

The medical record transaction-based state-machine generalization of the block is informally referred to as smart contracts. Ethereum is the first to attempt a full implementation of this idea. Also, the idea was int builds was introduced by [11], MedRec implemented the smart contract concept saying "This property can enable advanced functionality (multi-party arbitration, bidding, reputation, etc.) to be coded into our proposed system, adapting to comply with differences in regulation and changes in stakeholders needs".

The smart contract is one of the elements of the transaction components. It prioritizes usability by offering a designated contract which aggregates references to all user's patient-provider relationships. Also, it provides a single point of reference to check for any updates to medical history and handles identity confirmation via public key cryptography. Then it employs a DNS-like implementation that maps an existing and widely accepted form of ID (e.g., name, or social security number) to the person's network address. Smart Contact handles security as it is a crucial element to take into consideration especially after the patients become more comfortable with the software. It ensures accurate and better medical coding practices. Smart Contract carries four different security token depends on how participants would like to handle sharing information in the network.

- Private: Some patients would like to manage their healthcare records in the level where they can choose to hide some of their records even from their doctors as it is sensitive to them unless it is absolutely necessary. With a healthcare record solution built on tangle makes this type of granular control possible.
- Public: Includes some healthcare records that may be completely public.
- Protected: Healthcare records can be accessed and inherited automatically within the participated health database.
- Friendly Token: Patients can volunteer their data to researchers, medical studies, pharmacies, hospitals, and for specific conditions and the patient may benefit from the study financially or by getting some discounts.

5.4 Issuing a Transaction

To issue a transaction user (care provider and/or patients) must directly approve N of other transactions where N is an integer and does not necessarily equal 2. If transaction X directly approves Y, it is denoted $X \rightarrow Y$, which means there is an edge between X and Y. In general, the weight of the transaction is proportional to the amount of work that the issuing node invested into it, however, in our approach, all the transactions have the same constant weight (equal to 1).

Issuing a transaction steps:

- The node chooses two other transactions to approve according to an algorithm (as discussed in "E"). In general, these two transactions may coincide.
- The node checks if the two transactions are not conflicting, in the case where there are conflicting transactions, the node needs to decide which transactions will become orphaned (orphaned = are not indirectly approved by incoming transactions anymore).
- For a node to issue a valid transaction, the node must solve a cryptographic puzzle similar to those in the Bitcoin blockchain. This is achieved by finding a nonce such that the hash of that nonce concatenated with some data from the approved transaction has a particular form. In the case of the Bitcoin protocol, the hash must have at least a predefined number of leading zeros.

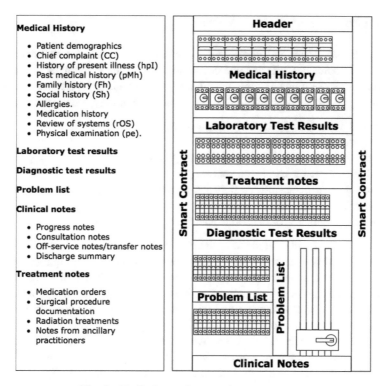

Fig. 5. Medical record transaction components.

5.5 How to Choose Transaction to Approve? Conflict and Approving Transaction

Tangle is Asynchronous, which means it can tolerate conflicting transactions that popped up asynchronously. It believes that any incorrect transaction would be auto-matically orphaned, or erased, as Tangle keeps on growing. Based on formalism and algorithms proposed in the Whitepaper [4] and [14] there are two tips selection mechanism to considered:

- URTS (uniform random tip selection) algorithm - tips are chosen from the list of available tips randomly (uniform distribution).
- MCMC (Markov Chain Monte Carlo) algorithm - random walk of particles towards the tips. Particles are released in the tangle, the first k particles at distinct tips determine which transactions are approve.

6   Conclusion and Future Work

In this paper, we design and implement a new platform to connect healthcare system for health data collection, sharing and collaboration between individuals and healthcare providers. The system can also be extended to accommodate the usage of health data for research purposes. By adopting the new Tangle technology, we are able to provide:

- Improved Time-Data processing to transfer data,
- Data aggregation for research purposes,
- Data-driven guidelines for care plan management,
- Decision support for early intervention,
- Alignment of patients information across time, by disease stage so as to enable comparative effectiveness research, and
- Trusted environment for decision making.

Further research will be done to optimize the default tip selection strategy in a way that minimizes this cost imposed by the selfish approach. Through implementing research methods and techniques from the cross-reactive fields of measure theory, game theory, and graph theory, progress towards resolving the tangle-related open problems has been well underway and will continue to be under investigation.

Also, it is important to observe that the iota network is asynchronous and nodes do not necessarily see the same set of transactions. It should also be noted that the tangle may contain conflicting transactions. Data reduction procedure for IOTA interferograms is another further research, to confirm the accuracy and stability.

References

1. Dubovitskaya, A., Xu, Z., Ryu, S., Schumacher, M., Wang, F.: Secure and trustable electronic medical records sharing using blockchain (2017)
2. Freeman, G., Hughes, J.: Continuity of care and the patient experience. King's Fund, p. 24 (2010)
3. Yaga, D., Mell, P., Roby, N., Scarfone, K.: Blockchain technology overview (NISTIR-8202), p. 59 (2018)
4. Weiner, J.: The Tangle. New Yorker **81**(8), 43–51 (2005)
5. Liang, X., Zhao, J., Shetty, S., Liu, J., Li, D.: Integrating blockchain for data sharing and collaboration in mobile healthcare applications. In: IEEE International Symposium on Personal, Indoor and Mobile Radio Communications, PIMRC, vol. 2017, pp. 1–5, October 2018
6. Milbank Memorial Fund: The Impact of Primary Care Practice Transformation on Cost, Quality, and Utilization | A Systematic Review of Research Published in 2016. Patient-Centered Prim. Care Collab. Robert Graham Cent., July 2017
7. Kitty, J.P.W., Chan, S., Fowles, J.B.: Electronic health records and the reliability and validity of quality measures: a review of the literature (2010)
8. Landy, D.J.: Figure 1. https://figure1.com/about/
9. Tak, P., Liu, S.: Medical record system using blockchain. Big Data Tokenization **9977**, 254–261 (2016)

10. Baran, P.: On distributed communication: introduction to distributed communication network (1964)
11. Ekblaw, A., Azaria, A., Halamka, J.D., Lippman, A., Original, I., Vieira, T.: A case study for blockchain in healthcare: "MedRec" prototype for electronic health records and medical research data MedRec: using blockchain for medical data access and permission management. IEEE Technol. Soc. Mag., 1–13 (2016)
12. Wood, G.: Ethereum: a secure decentralised generalised transaction ledger. Ethereum Proj. Yellow Pap., pp. 1–32 (2014)
13. Tennant, L.: Improving the anonymity of the IOTA cryptocurrency, pp. 1–20 (2017)
14. Kusmierz, B., Staupe, P., Gal, A.: Extracting Tangle properties in continuous time via large-scale simulations (2018)
15. Kusmierz, B.: The first glance at the simulation of the Tangle: discrete model, pp. 1–10 (2017)
16. Poon, J., Dryja, T.: The Bitcoin lightning network: scalable off-chain instant payments. Technical report, p. 59 (2016)
17. BitFury Group: Proof of stake versus proof of work, vol. 2015, pp. 1–26 (2015)
18. Ekblaw, A., Azaria, A.: MedRec: medical data management on the blockchain our motivation approach from fragmented access to comprehensive access, pp. 1–16 (2017)
19. National Institute of Standards and Technology (NIST) Headquarters: Use of Blockchain in Healthcare and Research Workshop. https://oncprojecttracking.healthit.gov/wiki/display/TechLabI/Use+of+Blockchain+in+Healthcare+and+Research+Workshop. Accessed 27 Sept 2016
20. da Conceição, A.F., da Silva, F.S.C., Rocha, V., Locoro, A., Barguil, J.M.: Electronic health records using blockchain technology (2018)
21. Popov, S., Saa, O., Finardi, P.: Equilibria in the Tangle, pp. 1–30 (2017)
22. Popov, S.: Local modifiers in the Tangle, pp. 1–10 (2018)
23. Fernandes, L., O'Connor, M.: Future of Patient Identification (2006)
24. Lauster, C.D., Srivastava, S.B.: Fundamental Skills for Patient Care in Pharmacy Practice (2013)
25. quizlet.com: Components of a Medical Record. https://quizlet.com/25245535/components-of-a-medical-record-flash-cards/

Ontology-Based Recommender System for Sport Events

Quang Nguyen, Luan N. T. Huynh, Tuyen P. Le,
and TaeChoong Chung$^{(\boxtimes)}$

Kyung Hee University, Seoul, Gyeonggi-do 17104, Republic of Korea
{quangnd, luanhnt, tuyenple, tcchung}@khu.ac.kr

Abstract. Sport event participation has changed much recently with effective support of technology. Advanced developments in recommender systems and World Wide Web bring chances such as efficiently distantly booking to alone travelers which allows them to enjoy sport events without being dependent on expensive tourist agencies as in the past. However, currently, popular information sources for such a recommender system are isolated and mainly relied on Web 2.0 formats which are difficult to stored and processed, especially among different platforms and communities. To utilize huge resources of Web 2.0 as well as apply cutting-edge features of Web 3.0 and under-developing Web 4.0, the authors propose an implementation of a hybrid system which collects data from different sources in the Internet (Mashup), apply machine learning to process raw information (Natural Language Processing and Unsupervised Clustering), add semantics to the processed data and make it compatible to latest web generation (Ontology), and provide recommendations based on smart content-based filtering and social-network-based user profiles for sport events. Empirical results show promising applications of such a framework to the market portion of alone travelers and also set an example as a demonstration for the authors' expectation toward Web 4.0 applications in the future.

Keywords: Ontology · Mashup · Natural language processing · Content-based filtering · Unsupervised clustering

1 Introduction

With the explosive growth of Internet and other online content services, recommender systems are very popular and have grabbed much attention from society. With a huge amount of customers' information and item database, recommender systems play in a quest to be able to personalize recommendations as well as improve user experience timely according to users' and items' content. Based on the approach of recommender systems, they can provide similar recommendations to users' inputs (e.g. the implementation of content-based filtering systems) or propose novel and serendipitous recommendations (e.g. the implementation of collaborative filtering or co-occurrence filtering systems). Some recent papers also suggest an integration of multiple recommendation classes to create hybrid recommender systems (e.g. TechLens+ with various combinations of content-based and collaborative filtering [1]). Depending on the

© Springer Nature Switzerland AG 2019
S. Lee et al. (Eds.): IMCOM 2019, AISC 935, pp. 870–885, 2019.
https://doi.org/10.1007/978-3-030-19063-7_69

purpose of the systems, different approaches can be selected to address different intended targets of recommender systems.

Targeted on sport events, recommendations for alone travelers is a special domain of recommender systems. Accommodations and eating places are common demands from this group of users. These needs were addressed in many previous works such as in [2, 3], and [4] serving for accommodation and restaurant recommendations in general. However, the narrow domain for tourism linked with sport events and alone travelers has not been developed much in recent years. With the huge number of sport fans in the world who are willing to follow their favorite team's matches or watch their favorite players by naked eyes, this domain is an open area for applications that support sport fans not only for basic requirements such as hotels and restaurants but also for a more advanced need in connecting to each other for socialization.

With the characteristics of limited historical data from alone travelers, recommender systems must analyze information from alternative sources to build users' profiles. Some approaches to this problem are to utilize the aid of social networks [5–7], or to apply knowledge learned from other domains to create cross-domain recommender systems [8–10]. These implementations, however, have a limitation in data collection since information is located sparsely in the Internet and in multiple isolated sources with different platforms and formats due to historically technical problems in Web 2.0 generation. For example, large social networks such as Facebook and Twitter or services providers such as Google Places have billions of active users [11, 12] but still remain on their own APIs to connect with third-party applications. Serious limitation in connecting and sharing knowledge across platforms and communities is the main problem of Web 2.0 in terms of data unification.

Proposed in 2006 by Markoff [13], the latest generation of WWW - Web 3.0 is the solution to the problem. Web 3.0 or Semantic Web is defined to be a standardization in platform, language, and databases with the introduction of ontology to be the central conceptualization supporting for the tasks of parsing, querying, reasoning, and reusing knowledge. Ontology is proved to bring benefits to recommender systems by improving the system performance in terms of data structure, extending context for the recommendations based on knowledge available in other linked data sources, or using the semantic relations to enhance the accuracy of the suggestions [14–17]. Building ontology-based recommender systems is the obvious trend for knowledge-based systems, especially in the situation that Linked Data were approved as the latest standardization for WWW by W3C.

Aiming to create a framework for cross-domain recommender systems which are beneficial from both the huge resources of database available in Web 2.0 and the advantages of Web 3.0, in this paper, the authors propose a mashup architecture for content-based filtering supporting sport-event scheme with data collected and processed from multiple sources (Facebook, BBC, Google Places) and information stored with semantics according to the definition of Ontology in Web 3.0. During the implementation, some techniques for matching users' profiles utilizing machine learning and content-based recommender system are also presented toward the definition of Web 4.0. The first demonstration of the system had built for the past World Cup event.

The contributions of the authors in this paper are the following:

- A new natural language processing proposal to process social network raw data to build users' profiles.
- An ontology with vocabulary for Sport events created selectively from multiple online sources and users' profiles.
- An ontology-based recommender system with customized query utilized mashup architecture to make knowledge compatible to Web 3.0 and make use of advantages from Web 3.0.

2   Ontology-Based Mashup Recommender System

2.1   System Description

The intended system is created with the intention as a small demonstration for the future of Web. From the definition of Web 3.0 (Linked Data) and Web 4.0 (artificial intelligent Web) [18], the author proposes a prototype in which a system is able to gather all of the information from users smartly and process them to deliver recommendation respecting to users' need. Created the first time in June 2018, it is the application to address to the desire of football fans around the world for accommodation, eating-and-drinking, and connecting-to-each-other needs (Fig. 1).

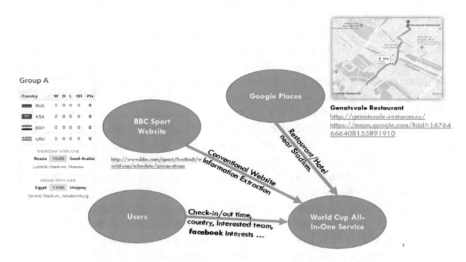

Fig. 1. System description

From simple inputs from users, the application will do the analysis, reasoning, and provides answers to the need of football fans about some-but-not-limited-to following questions while their inputs to the application are simple and at basic level (Username,

Facebook Token Access, Kakao ID, Check-in Date, Check-out Date, Home Country, World Cup Favorite Team).

Q1: Given my log-in data, who are from my home country joining World Cup event?
Q2: Given my Facebook access, who have at least one similar interest with me so that we could gather after the match?
Q3: Given my favorite team, where should I book the hotel to maximize my convenience?
Q4: Given my Facebook access, where is the best place to drink with people like me?
Q5: Given my log-in profile and Facebook access, who is the best soulmate to me?

2.2   System Architecture

Mashup Structure. To realize the idea of the application and make it become the future web demonstration, all of the information is stored in Web 3.0 format under Ontology. It follows the architecture that mashup is utilized to link data from many different database "islands" to create an open and accessible form with semantics of data attached (see Fig. 2). Thereby, the machine can understand the content of the data and can be learned to process them in smart way.

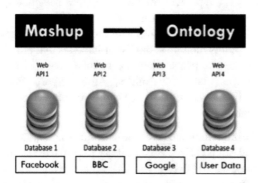

Fig. 2. System mashup structure

Two Levels of System Architecture. For the implementation of data extraction from different database, there are some techniques involved. The first is the extraction data modules: Facebook Graph API, Natural Language Processing (NLP), HTML Parser and Google Places API. The second level are Ontology and Recommender System. Role and functions for each of these modules are described as follows.

Facebook Graph API. This is a low-level HTTP-based API that enables the app to connect with Facebook Server and query for users' information. In Facebook, a social graph is defined to be represented for information. It includes nodes (e.g. individual objects such as a User, a Photo, a Page, or a Comment), edges (e.g. connections between a collection of objects and a single object, such as Photos on a Page or Comments on a Photo), and fields (e.g. data about an object, such as a User's birthday, or a Page's name) [19]. Via Facebook graph API, the proposed ontology-based mashup recommender system will query all of user posts to feed to NLP module (Fig. 3).

Fig. 3. Two levels of system architecture

Natural Language Processing (NLP). This is the module to process the raw strings received from Facebook Graph API module. It applies Machine learning in category predictor to the users' posts by using the statistic algorithm called Term Frequency – Inverse Document Frequency (tf-idf) and Multinomial Bayes Classifier as mentioned in [20]. The outcome of this module is the list of interests that the users have expressed via their historical Facebook posts. They are selected from 13 categories pre-trained with Multinomial Bayes classifier: Politics, Autos, Sport, Electronics, Medicine, Computer Graphics, Windows Stuff, IBM Hardware, Mac Hardware, Religion, Sale, Motorcycles, and Atheism.

HTML Parser. The conventional Website information extraction with the implementation of Beautiful Soup HTML Parser library [21] to extract information in HTML based on the identity of their HTML class. Since the website designs with the old Web architecture (Web 2.0), sophisticated query like the case of ontology, cannot be applied and the semantics of the information is determined based on their markup language. HTML Parser in the proposed application extracts information from BBC website for World Cup with the information of Date, Time, Home Team, Away Team, Stadium, and City. Information will be saved with semantics to Ontology and some pieces will be fed to the Google Places API where places of the matches and restaurants and accommodations for users' convenience are inferred from match data.

Google Places API. This is the HTTP request module to Google places services [22]. Place outcome from this module includes information about the exact location in Google Map and hotel or restaurant website for users' reference need. These pieces of information will be then included in the final ontology with semantics and data and objects properties defined to maintain the relationship with football match data.

Ontology. The complete ontology to express the data semantics and structure is then provided as follows. The box is the class and the arrows represent for the relations (Subclass, Domain, and Range) between classes. The semantics added to the data via the schema and the annotation of the ontology while RDF triples are automatically created from the instances of the classes and the object properties and data properties available from the relationships among classes (Fig. 4).

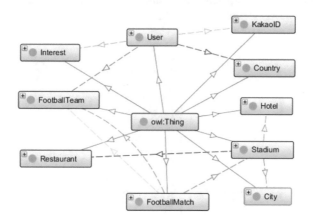

Fig. 4. Ontology for sport system

Recommender System. Based on the information stored on Ontology, a content-based recommendation can be made to the users such as some query to address the 5 questions mentioned in the first part of the section. As inherited from the NLP module, the people matching feature can be done via the implementation of either simple cosine similarity comparison or unsupervised learning clustering techniques applied on the set of users' interests inferred from users' social network posts.

Particularly, each post of a user after processed by NLP module will result in a single category of interest. The sets of multiple posts of a user will form a sequence of interests so that each category will be repeated by a number of times. For examples, a user mentioning Sport in 2 posts, Autos in 3 posts and Politics in 1 post will have the representative string sequence: *Sport Sport Autos Autos Autos Politics*. This string is then converted to tf-idf vector based on the fixed vocabulary of 13 categories mentioned in NLP module. The difference in the repetition times of interests in the final interest sequence stands for the emphasis of importance of each interest category to the characteristic model of according user. The new set of tf-idf vectors among users can be then applied unsupervised clustering methods such as K-Mean, Birch, or Agglomerative to classify people into groups with similar characteristics defined by the string sequence of interests. A custom query will extract the data of group of people when users pull data related to the one member in a clustered group.

2.3   System Implementation

The system is developed with 2 main features for the users: features related to Sign-up and features related to Sign-in. Details of the design are given in Fig. 5.

For the function flowchart, based on the 2 groups of features just mentioned, two flowcharts for the implementation of modules and algorithms are provided in which each of them represents the implementation of ontology creation (see Fig. 6) and the data extraction using SPARQL to query responsible information needed from the users addressing 5 questions as mentioned in Introduction section (see Fig. 7).

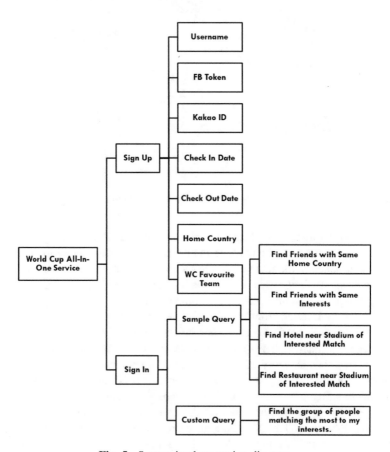

Fig. 5. System implementation diagram

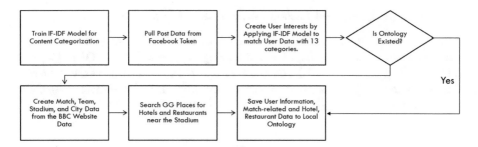

Fig. 6. Sign-up related flowchart

Fig. 7. Sign-in related flowchart

3 Results and Discussion

3.1 Ontology Creation

With data collected from different source, the mashup architecture processes information to group into different classes as shown in Fig. 8. Data from BBC, Google Places, Facebook and User Personal Data are processed and included to the ontology as instances and will be used later to form RDF triples. The check for the completion of the Ontology Creation is then conducted according to the sampled queries to find the answer for question Q1 to Q4 and other more advanced queries according to the querying skill of the users (see Figs. 8 and 9).

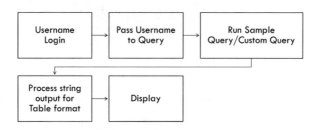

Fig. 8. Classes and properties of created ontology

3.2   SPARQL Query and Recommender System Performance

As tested and confirmed with the 10-users dataset and 120-user dataset, the application runs flawlessly without any mistakes. Addressing to answer question Q1 to Q4, four automatically generated SPARQL queries are generated and matched to expected result. Example of these queries are given as below with particular SPARQL queries for Q1 to Q4 and their corresponding results to the 10-user dataset (see Tables 1, 2, 3, and 4).

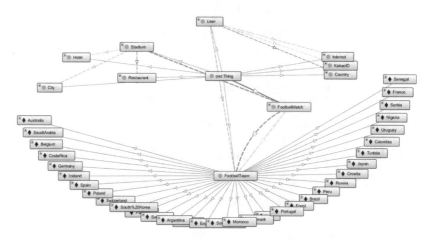

Fig. 9.   Instances examples from created ontology

SPARQL for Question 1. *Given my log-in data, who are from my home country joining World Cup event?*

```
PREFIX WCOnt: <http://Worldcup_Service.fb#>
PREFIX rdf: <http://www.w3.org/1999/02/22-rdf-syntax-ns#>
SELECT DISTINCT ?user ?homecountry ?checkin ?checkout
WHERE {
    ?me WCOnt:hasHomeCountry ?homecountry.
    ?user WCOnt:hasHomeCountry ?homecountry.
    ?user WCOnt:hasCheckInTime ?checkin.
    ?user WCOnt:hasCheckOutTime ?checkout.
    FILTER (?me = WCOnt:Quang)
}
```

Table 1. Result for question 1 query

Username	Home Country	Check-in Date	Check-out Date
Quang	Vietnam	2018-06-14	2018-07-12
Luan	Vietnam	2018-06-13	2018-07-13
Trinh	Vietnam	2018-06-11	2018-07-12

SPARQL for Question 2. *Given my Facebook access, who have at least one similar interest with me so that we could gather after the match?*

```
PREFIX WCOnt: <http://Worldcup_Service.fb#>
PREFIX rdf: <http://www.w3.org/1999/02/22-rdf-syntax-ns#>
SELECT DISTINCT ?user ?homecountry ?kakao ?checkin ?checkout
WHERE {
    ?me WCOnt:hasInterest ?Interest.
    ?user WCOnt:hasInterest ?Interest.
    ?user WCOnt:hasCheckInTime ?checkin.
    ?user WCOnt:hasCheckOutTime ?checkout.
    ?user WCOnt:hasHomeCountry ?homecountry.
    ?user WCOnt:hasKakaoID ?kakao.
    FILTER (?me = WCOnt:Quang)
}
```

Table 2. Result for question 2 query

Username	Home Country	Kakao ID	Check-in	Check-Out
Quang	Vietnam	Quang1234	2018-06-14	2018-07-12
Luan	Vietnam	Luan1234	2018-06-13	2018-07-13
Trinh	Vietnam	Trinh	2018-06-11	2018-07-12
Mushasi	Japan	Mushasi123	2018-06-11	2018-07-12
Cooku	Laos	Cookcuu	2018-06-11	2018-07-12
Uddin	Bangladesh	Uddin12	2018-06-10	2018-07-12

SPARQL for Question 3. *Given my favorite team, where should I book the hotel to maximize my convenience?*

> PREFIX WCOnt: <http://Worldcup_Service.fb#>
> PREFIX rdf: <http://www.w3.org/1999/02/22-rdf-syntax-ns#>
> SELECT DISTINCT ?match ?stadium ?hotel ?map ?website
> WHERE {
>   {?me WCOnt:hasFavouriteTeam ?favouriteteam.
>   ?match WCOnt:hasHomeTeam ?favouriteteam.
>   ?match WCOnt:inStadium ?stadium.
>   ?stadium WCOnt:hasNearestHotel ?hotel.
>   ?hotel WCOnt:hasMapLink ?map;
>   WCOnt:hasWebsite ?website.
>   FILTER (?me = WCOnt:Quang)}
>   UNION
>   {?me WCOnt:hasFavouriteTeam ?favouriteteam.
>   ?match WCOnt:hasAwayTeam ?favouriteteam.
>   ?match WCOnt:inStadium ?stadium.
>   ?stadium WCOnt:hasNearestHotel ?hotel.
>   ?hotel WCOnt:hasMapLink ?map;
>   WCOnt:hasWebsite ?website.
>   FILTER (?me = WCOnt:Quang)}
> }

Table 3. Result for Q3 query

Match	Stadium	Hotel	Map	Website
Spain-Morocco	Kaliningrad	Маяк	https://maps.google.com/?cid=4986911589553952252	https://radiomayak.ru/
Iran-Spain	Kazan Arena	Kazan-Ok	https://maps.google.com/?cid=15196949714704492421	http://kazan-ok.ru/

SPARQL for Question 4. *Given my Facebook access, where is the best place to drink with people like me?*

```
PREFIX WCOnt: <http://Worldcup_Service.fb#>
PREFIX rdf: <http://www.w3.org/1999/02/22-rdf-syntax-ns#>
SELECT DISTINCT ?match ?stadium ?restaurant ?map ?website
WHERE {
    {?me WCOnt:hasFavouriteTeam ?favouriteteam.
    ?match WCOnt:hasHomeTeam ?favouriteteam.
    ?match WCOnt:inStadium ?stadium.
    ?stadium WCOnt:hasNearestRestaurant ?restaurant.
    ?restaurant WCOnt:hasMapLink ?map; WCOnt:hasWebsite ?website.
    FILTER (?me = WCOnt:Quang)}
    UNION
    {?me WCOnt:hasFavouriteTeam ?favouriteteam.
    ?match WCOnt:hasAwayTeam ?favouriteteam.
    ?match WCOnt:inStadium ?stadium.
    ?stadium WCOnt:hasNearestRestaurant ?restaurant.
    ?restaurant WCOnt:hasMapLink ?map; WCOnt:hasWebsite ?website.
    FILTER (?me = WCOnt:Quang)}
}
```

Table 4. Result for question 4 query

Match	Stadium	Restaurant	Map	Website
Spain-Morocco	Kaliningrad	Bavarian restaurant Tsetler/Zötler Bier	https://maps.google.com/?cid=18065314014416804431	http://www.zoetler.ru/leninskij/
Iran-Spain	Kazan Arena	Pashmir	https://maps.google.com/?cid=16355205506115982194	http://www.pashmir.ru/

For the advanced custom query for best match people with logged-in account (Question 5), as mentioned in the architecture, some additional processing steps are needed to be processed before making the query. For the convenience in demonstration, this feature is separated and tested with ground truth to compare the true performance of the applied recommender system.

Tables 5 and 6 summarize the performance of cosine similarity and some unsupervised learning clustering methods. As can be seen from the Table 5, almost the clustering techniques performed well and identify the cluster correctly according to the ground truth table. One exceptional case is Ward Linkage from Agglomerative Clustering techniques. It shows confusion in categorizing User9 to group of {User3, User4, User5}. It comes from the difference in the goal of ward linkage to other linkages although they are all Agglomerative Clustering techniques. To be more specific, ward linkage diminishes the variance of processed clusters while average linkage targets to

Table 5. Ground truth table for dataset with 10 users

User	Interest sequence	Expected category
User0	*Sport Sport Autos Autos Autos Politics*	*0*
User1	*Sport Sport Sport Autos Autos Politics*	*0*
User2	*Sport Sport Autos Autos Autos Politics*	*0*
User3	Electronics Medicine Medicine Medicine Religion	1
User4	Religion Religion Religion Religion Religion	1
User5	Religion Politics Medicine Medicine Medicine	1
User6	*Sale Sale Sale Computer_Graphics Computer_Graphics*	*2*
User7	*Sale Sale Computer_Graphics Computer_Graphics*	*2*
User8	*Sale Sale Sale Sale Mac_Hardware Computer_Graphics*	*2*
User9	*Mac_Hardware Mac_Hardware Mac_Hardware IBM_Hardware Computer_Graphics*	*2*

the mean of the distances of each observation of the clusters and complete linkage uses the maximum distances between all observations of the clusters. Difference in goal leads to different performance in the given dataset of users.

For the cosine similarity tables, the group of users is identified correctly according to the ground truth table. Threshold in experiments is set at 0.05 which eliminates the wrong clustering of User5 to the group of {User0, User1, User2} and enables the correct arrangement of User9 joining group of {User6, User7, User8} in the situation that interest sequence of User9 lacks the dominant feature "Sale" as appeared in those of User6, User7, and User8 (Table 7).

Additional test was performed with a broader dataset of 120 users for the two unsupervised clustering getting the best performance from the 10-user dataset: K-Mean and Birch. The new datasets are created by either creating new samples or repeatedly randomly sampling from the original 10-user dataset. The accuracy percentage is calculated against the truth-table values built for the 120 users in these two datasets. The results for both datasets are summarized in Table 8.

As predicted, with the samples similar to the 10-user dataset, the performance of the tested methods on 120-user randomly repeated dataset is the same as the previous test with 10-user dataset. For the new dataset with larger and much more unpredicted content, performance of both methods degrades. The reason for this degradation is from the limitation in building the Interest Sequence to feed to the clustering methods. Although the accuracy is still kept at acceptable level (approximately 80%), due to the dataset creation constraint, category repetition method to form users' interest sequence needs to be tested under more dynamic situations.

Table 6. Ground truth and clustering result comparison

Username	Ground truth	KMean	Birch	Agglomerative clustering		
				Ward linkage	Average linkage	Complete linkage
User0	*0*	*2*	*0*	*1*	*2*	*2*
User1	*0*	*2*	*0*	*1*	*2*	*2*
User2	*0*	*2*	*0*	*1*	*2*	*2*
User3	1	0	1	0	1	1
User4	1	0	1	0	1	1
User5	1	0	1	0	1	1
User6	*2*	*1*	*0*	*2*	*0*	*0*
User7	*2*	*1*	*0*	*2*	*0*	*0*
User8	*2*	*1*	*0*	*2*	*0*	*0*
User9	*2*	*1*	*0*	**0**	*0*	*0*

Table 7. Cosine similarity for 10-user dataset

Cosine similarity	User0	User1	User2	User3	User4	User5	User6	User7	User8	User9
User0	*1*	*0. 898*	*1*	0	0	0.033	0	0	0	0
User1	*0.898*	*1*	*0.898*	0	0	0.033	0	0	0	0
User2	*1*	*0. 898*	*1*	0	0	0.033	0	0	0	0
User3	0	0	0	*1*	*0.142*	*0.79*	0	0	0	0
User4	0	0	0	*0.142*	*1*	*0.146*	0	0	0	0
User5	0.033	0.033	0.033	*0.79*	*0.146*	*1*	0	0	0	0
User6	0	0	0	0	0	0	*1*	*0.968*	*0.817*	*0.074*
User7	0	0	0	0	0	0	*0.968*	*1*	*0.693*	*0.098*
User8	0	0	0	0	0	0	*0.817*	*0.693*	*1*	*0.17*
User9	0	0	0	0	0	0	*0.074*	*0.098*	*0.17*	*1*

Table 8. Accuracy percentage for 120-user datasets.

Accuracy percentage	120-user randomly repeated dataset	120-user new dataset
K-Mean	100%	78%
Birch	100%	82%

4   Conclusion

The idea of future World Wide Web in smarter direction evolved from the current Web 2.0, Web 3.0, and Artificial Intelligence has been illustrated via the implementation of this ontology-based recommender system framework for sport events. The system utilizes the mashup architecture available between Web 2.0 and Web 3.0, collects data from different sources, and reformat them with the semantics added to create an ontology and utilize reasoning feature to benefit recommending process. Natural Language Processing in other domain (e.g. social network) and Custom Query with the implementation of unsupervised clustering techniques are integrated as an example of Artificial Intelligence application addressing the limited data for a single sport tourism domain. Although the system follows the structure of Linked Data, the linking part of the created ontology to available databases has not been finished. The author leaves this task as the future implementation for completing the idea of creating a broader system able to connect to the other online Linked Data source and satisfying the requirements for 5-star Linked Data from W3C [23].

Acknowledgement. The authors are grateful to the Basic Science Research Program through the National Research Foundation of Korea (NRF-2017R1D1A1B04036354).

References

1. Torres, R., McNee, S.M., Abel, M., Konstan, J.A., Riedl, J.: Enhancing digital libraries with TechLens+. In: Proceedings of the 4th ACM/IEEE-CS Joint Conference on Digital Libraries, Tuscon, AZ, USA (2004)
2. Martinez, L., Rodriguez, R.M., Espinilla, M.: REJA: a georeferenced hybrid recommender system for restaurants. In: IEEE/WIC/ACM International Conference on Web Intelligence and Intelligent Agent Technology, Milan, Italy (2009)
3. Lerttripinyo, T., Jatukannyaprateep, P., Prompoon, N., Pattanothai, C.: Accommodation recommendation system from user reviews based on feature-based weighted non-negative matrix factorization method. In: 2015 12th International Joint Conference on Computer Science and Software Engineering (JCSSE), Songkhla, Thailand (2015)
4. Habib, M.A., Rakib, M.A., Hasan, M.A.: Location, time, and preference aware restaurant recommendation method. In: 2016 19th International Conference on Computer and Information Technology (ICCIT), Dhaka, Bangladesh (2016)
5. Liu, H., Maes, P.: InterestMap: harvesting social network profiles for recommendations. In: Workshop: Beyond Personalization (IUI 2005), San Diego, California, USA (2005)
6. Debnath, S., Ganguly, N., Mitra, P.: Feature weighting in content based recommendation system using social network analysis. In: 17th International Conference on World Wide Web (WWW 2008), Beijing, China (2008)
7. Bonhard, P., Sasse, M.A.: 'Knowing me, knowing you' - using profiles and social networking to improve recommender systems. BT Technol. J. **24**(3), 84–98 (2016)
8. Chung, R., Sundaram, D., Srinivasan, A.: Integrated personal recommender systems. In: Ninth International Conference on Electronic Commerce (ICEC 2007), Minneapolis, MN, USA (2007)
9. Berkovsky, S., Kuflik, T., Ricci, F.: Cross-representation mediation of user models. User Model. User-Adapt. Interact. **19**(1–2), 35–63 (2009)
10. Zhu, F., Wang, Y., Chen, C., Liu, G., Orgun, M., Wu, J.: A deep framework for cross-domain and cross-system recommendations. In: Twenty-Seventh International Joint Conference on Artificial Intelligence (IJCAI-18), Stockholm, Sweden (2018)
11. Statista: Facebook - Statistics & Facts, Statista (2017). https://www.statista.com/topics/751/facebook/
12. Statista: Twitter - Statistics & Facts, Statista. https://www.statista.com/topics/737/twitter/

13. W3C: Linked Data (2018). https://www.w3.org/wiki/LinkedData
14. Bouza, A., Reif, G., Bernstein, A., Gall, H.: SemTree: ontology-based decision tree algorithm for recommender systems. In: Seventh International Semantic Web Conference (ISWC2008), Karlsruhe, Germany (2008)
15. Werner, D., Cruz, C., Nicolle, C.: Ontology-based recommender system of economic articles. In: Eighth International Conference on Web Information Systems and Technologies (WEBIST 2013), Porto, Portugal (2013)
16. Lémdani, R., Polaillon, G., Bennacer, N., Bourda, Y.: A semantic similarity measure for recommender systems. In: Seventh International Conference on Semantic Systems (I-Semantics 2011), Graz, Austria (2011)
17. Wang, Y., Stash, N., Aroyo, L., Hollink, L., Schreiber, G.: Using semantic relations for content-based recommender systems in cultural heritage. In: 2009 International Conference on Ontology Patterns (WOP 2009), Washington DC, USA (2009)
18. Choudhury, N.: World Wide: Web and its journey from Web 1.0 to Web 4.0. (IJCSIT) Int. J. Comput. Sci. Inf. Technol. **5**(6), 8096–8100 (2014)
19. F.D. Team, Facebook for Developers, Facebook (2018). https://developers.facebook.com/docs/graph-api/overview/
20. Joshi, P.: Natural Language Processing. In: Artificial Intelligence with Python, pp. 248–273. Packt Publishing Ltd., Birmingham (2017)
21. Richardson, L.: Beautiful Soup Documentation (2018). https://www.crummy.com/software/BeautifulSoup/bs4/doc/
22. Google: Places API, Google. https://developers.google.com/places/web-service/intro
23. Berners-Lee, T.: Linked Data, W3C (2009). https://www.w3.org/DesignIssues/LinkedData.html

Enhancing Airlines Delay Prediction by Implementing Classification Based Deep Learning Algorithms

Md. Nazmus Saadat[1]($\boxtimes$) and Md. Moniruzzaman[2]

[1] Universiti Kuala Lumpur,
1016, Jalan Sultan Ismail, 50250 Kuala Lumpur, Malaysia
Mdnazmus@unikl.edu.my
[2] Universiti of Malaya,
Jalan Universiti, 50603 Kuala Lumpur, Wilayah Persekutuan, Malaysia
monir@siswa.um.edu.my

Abstract. Technology is evolving in a rapid pace with its numerous discoveries, and nowadays, the rate is more than ever before. Data Analytics has become a knowledge and a tool which significantly contributing relentlessly to majority of the discoveries since this can fetch insights to reduce man-machine interactions. Prediction is an integral part of data analytics that provides meaningful information from historical data to support decisions. Machine learning and deep learning is the core of any predictive analytics where both have their own strengths and weakness. Aviation industry around the world are facing severe problems by the flight delays caused by several factors. In order to achieve its target to provide a hassle-free journey, Aviation Industry is continuously researching to reduce flight delays. This research will focus mainly to predict airlines flight delays by analyzing flight data, especially, for the domestic Airlines those moves around the United States of America. Data science methodology has been implemented in order to fetch the end prediction. In order to transform the high dimension data into a low dimension Principal component analysis is used. Deep learning algorithms, widely popular and state-of-the-art prediction technology, are implemented in the prediction modeling phase. By empirical observation, the research can come to a conclusion that by following the data science methodology better performance could be unlocked to help the aviation industry.

Keywords: Big Data deep learning algorithms · Aviation delay

1 Introduction

Over the past decades, Big Data analytics has been transforming almost every business enormously, airlines, however, is not out of this wave. Travel by air has become the most important than all other means of transportation as everyday millions of people are continuously moving from one place to another by using this mode of transportation. Among numerous problems faced by this mode of transportation, flight delay

S. Lee et al. (Eds.): IMCOM 2019, AISC 935, pp. 886–896, 2019.
https://doi.org/10.1007/978-3-030-19063-7_70

is the most influencing for the traveler using this mode of transportation. Which significantly affects the customer experience as well as the whole business.

According to statistics in the year 2014, around 391 million trips of airlines recorded, in other words, around 1 million trips per day which will be tripled by the year 2015 to 2035 [1]. It is estimated that per flight will create half a terabyte of data [2]. Moreover, if we consider the whole data generates starting from customer booking, departure to arrival, and above all social media data which is a prime factor for the airlines business will be such amount for a particular airlines that it would be immensely hard to manage all factors and keep business going forward with just normal data analytics. In any way this volume, velocity, veracity, variety of data certainly termed as Big Data; therefore, it is of utmost important for any airlines in the world to go for Big Data analytics.

Big data is a word which comes into the play not a long time back. Generally, any big amount of data that cannot be processed by using conventional computing is termed as a Big Data. Every year the amount of data is increasing faster than any other previous years increasing rate. In a recent statistic, it has shown that around one million flights move from one place to another. Which means that a lot of flight data is generating continuously. It is also estimated that per flight generates half a terabyte of data, in volume this is such Big Data or even part of this could not be possible to process in conventional computing.

As the data is unstructured which means that the number of types of data is unknown. Let's look at some of those myriad data types. First, a lot of infographic data is available and shared by users in social media. Secondly, interactive content such as, text data, moving image, audio, video, games etc., even each of these can be categorize and have thousands of distinctive types and properties alone. Some contents that evoke strong positive emotions is the key to mine social media data which is very challenging task to do. Video and image recognition categorize, and analysis is also tremendously industrious job to perform that cannot be done by normal computing.

Since the data generates by social media is partly or totally cannot be judged by its myriad uncertainty that can be categorized by veracity. This property requires big data techniques to handle with. This huge amount of data cannot be visualized by any conventional tools it could be termed as visualization problem. No such visualization tools have been developed to visualize this amount of big data in order to get a first sense of data.

Above all, getting values or insights is the key objective for any data analytical work, and without performing all the data science methods on this amount of big data, we cannot expect any values from the data which can drive us to a better decision making [1]. Many ways Big Data analytics is helping the transformation of airlines business in which some are direct factors and others are indirect factor.

Numerous previous experiments are relying on modelling and simulation techniques. Scenarios that requires to be analyzed are reformed by computing simulations [2]. In order to analyze future scenarios this simulation-based approach is important when we are required to find interactions among the components. The shortcoming of this simulation-based analysis are normally the lower speed of the simulation and inappropriate assumed model. There have been numerous experiments conducted to out lift the speed of simulation [3, 4], however it is still hard to select an appropriate level of abstractions for the quality of the analyses.

By this time, a hierarchical structure of human perception widely used which is termed as deep learning. Some areas of classification and regression such as image and speech recognition machine translation, accuracy is improved by using deep learning algorithms where conventional machine learning algorithms shows lower accuracy [5]. Besides nowadays it is widely used in classifying delay prediction in several industry such as Airlines delay [6]. To be more exact, data analytics of traffic delays in airlines industry it is widely applying which is a direct impact of improved deep learning algorithms.

Deep Recurrent Neural Network (RNN) is used in the experiment carried out by Lawson et al. [8] for analyzing the air traffic information. They used four different methods of RNN some of those are long and some are short-term memory architecture for predicting delays. From the data science perspective their work was eminent as they focused more on machine learning models. They used several data sources which added an advantage and made the whole experiment more meaningful.

According to the experiment by Venkatesh [10], around 18 billion dollar of loss to the traveler cause by flight delays in each year. They have performed a significant literature review on papers related to flight and aviation delays. In this experiment, they used data from Kaggle.com that contains thirty features. In their study they used Random selection to reduce the sample selection bias. In order to predict delay this experiment used mainly two machine learning algorithms namely, Deep Belief Networks (DBN) and Artificial Neural Network (ANN). Both the models are trained by the train dataset. ANN is an algorithm largely depends on the number of nodes and level of depth of the hidden layers. Here ANN are trained by adopting resilient back propagation learning algorithm. The positive side of back propagation is that it trains the neural network faster than the conventional back propagation. Overall, ANN outperforms DBN (deep belief network) with an accuracy of 76.76% which we believe could be improved by selecting features, quality of data, and of course dimensionality reduction.

This literature review observed that although many researches have been done in the soaring field of flight delays, different experiment come up with dissimilar story and outcomes. Which clearly shows the path that the scope of research in this field is vast. Deep learning algorithm can come up with better result, if this can be implemented by following data science methodology.

2 Experimental and Computational Details

This research is to support the primary aim to study and experiment on the flight data in order to efficiently perform analytics, interrelation, and the prediction of flight delays in the USA from 2015 to 2017. The research needs to analyze the intrinsic delay factors. The proposed methodology used in this research is quantitative. The whole experiment will follow one of the widely followed pipeline called CRISP-DM. The process or methodology of CRISP-DM can be explained in these six major steps,

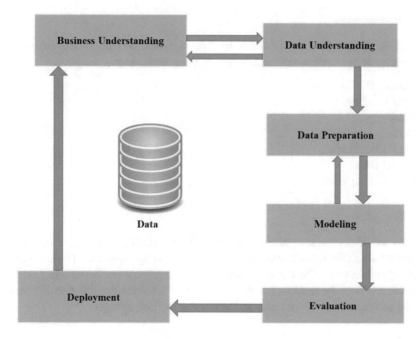

Fig. 1. CRISP DM

2.1 Dataset

The dataset is taken from the Bureau of Transportation Statistics (https://www.transtats.bts.gov/) access date Mar 7, 2018. Flight delay or on-time data is available from the year 1987. However, the experiment is more focused to learn insights of the recent patterns, so we have downloaded the data from the year 2015 to 2017, which is three years in total. The data is in quantitative form, however, there are some features found categorical, which is required to transform into quantitative values.

2.2 Data Volume

The whole data simply cannot download by one click; therefore, we need to download one month at a time, and the file format is comma separated value. After downloading we found that every month consist of around 450 K record which is around 200 MB in size. So, on year consists of 200 * 12 = 2400 MB, in other words, around 2.34 GB of data. So, the three year data will be approximately 7 GB of data that, certainly, cannot be process by conventional processing in a single node. However, the data is divided into two parts where the first part consist of one-year data and the second part will consist of 3 years of data. Primarily, the model will run on 2017 years data which will be 2.3 GB. Then the model will be trained and test overall with three years' data.

2.3   Instruments

Several instruments are used in this research for data collection and analysis, and different tools are used in different stages of the methodology,

Data Collection.
Data Loading.
Sampling.
Data Preprocessing.
Data Visualization.
Model Data.
Result interpretation.

2.4   Sampling

In order to reduce the size of the data we will implement random sampling from the beginning. As of the pseudo-code (Fig. 1) in the step-4 we have done random sampling by using python code. The details of random sampling could be illustrated by the below diagram, if 3% of the random sampling is needed to reduce the size then the scenario will be as below, and every iteration the number will be changed.

2.5   Data Pre-processing

Undoubtedly, the most important part of this research project is data pre-processing. In the first experiment, around 10% data is randomly sampled whereas in the second experiment around 3.33% is sampled in order to keep the size of our desired data fixed to 470 MB. In each experiment around 550 K record is sampled. The below Table 1 is statistical data before and after reduction.

Table 1. Initial data and reduced statistics.

	Number of year	Initial data size	Number of record	Sampled percentage	Sample size	Sample record	Remark
Experiment	3	7 GB	12 * 3 * 410 K	3.33%	450 MB	550 K	Values may vary over time

Firstly, the whole data checked for null value and found that there are approximately 9k missing values under Departure delay, Arrival delay and Actual Elapsed time. The statistics is as below (Table 2),

Table 2. Null value aggregation

Features	Missing values
DEP_TIME	7525
DEP_DELAY	7527
DEP_DELAY_NEW	7527
DEP_DEL15	7527
DEP_DELAY_GROUP	7527
ARR_TIME	8027
ARR_DELAY	9161
……...	………

Therefore, it is required to go deep down and explore more on the dataset to find out the reason. Surprisingly, those have the missing value are either CANCELLED or DIVERTED. In order to prove this, we have sum up the records where flights are either cancelled or diverted, and the result is quite amazing.

$$\text{Number of}(\text{ACTUAL\_ELAPSED\_TIME}) = \text{Number of Flights}(\text{CANCELLED} + \text{DIVERTED})$$

In this experiment, some data are transformed, and it is divided into two parts. In the first part, we were required to get our target variable. However, no such category is defined in the dataset. We have observed that the DELAY is defined by the number of minutes which are integer values ranging from 0 to 1934 min. Therefore, a new variable (delay_response) is created based on the integer value of DEP_DELAY_NEW following a rule that if this value is 15 or more then delay response will be 1.

So the resulting feature (delay_response) is now consist of 0 and 1, which is our target variable. In the second part, selected categorical variables are required to transform into quantitative values. This task can be efficiently and explicitly done by creating dummy variable using python. Pandas get_dummies() function is the tool for this job. After creating dummy variables with Prefix-C the data will be similar as below, where each and every category value will be 1 if the record have the occurrence whereas 0 if no occurrence is observed. It is shown in the below Table 3,

Table 3. Dummy variable list for carrier name

C_AA	C_AS	C_B6	C_DL	C_EV	....
0	0	0	0	0	....
0	0	0	0	0	....
0	0	0	0	1	....
0	0	0	1	0	....
0	1	0	0	0	....

2.6   Visualization

Visualization unlocks the ability to visualize the huge amount of data in a single frame. It is totally impossible to visualize the nature of 550K record; however, python visualization tools can unlock freedom to visualize. Let's visualize some of the features and how they effect on flight delays.

Each record consists of the date, and based on the date, every record has been categorized into four quarter in each year. The significance of quarter is very important since each quarter contains different weather and circumstances. Detailed information about quarter is as below (Table 4),

Table 4. Quarter details code

Quarter details code	Description
1	Quarter1: January 1–March 31
2	Quarter2: April 1–June 30
3	Quarter3: July 1–September 30
4	Quarter4: October 1–December 31

If we plot quarters and the percentage of delay occurrence, the graph shows that there are less delays on the fourth quarter of the year (Fig. 2).

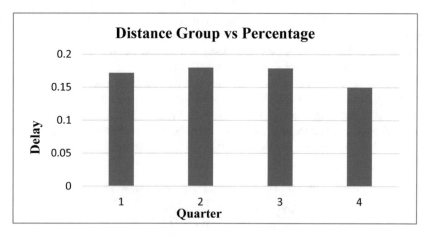

Fig. 2. Effect of distance group on delay occurrence.

Similar visualization techniques are applied for other features to find out its impact to the whole delay count.

2.7   Deep Learning Model

Before deep learning model implementing, dataset need to divide into train and test sets. In order to implement it python train_test_split package needs to be import from sklearn.cross_validation library. Train set is divided randomly 80%, while the test set contains 20% of the whole data. Train set is used to train the deep learning model. Where test set is used to check the accuracy by using confusion matrix performance measures.

The basic of deep learning is to learn data representations through increasing abstraction levels. Hierarchical learning process like this is enormously powerful as it allows a system to learn compound data representation directly from the original dataset.

There are many deep learning algorithms are widely used in predictive learning such as,

- Recursive Neural Network (RNN),
- Deep Neural Network (DNN),
- Convolutional neural network (CNN),
- Deep belief network (DBN) and more.

2.8   Artificial Neural Network

Artificial Neural networks are able to represent complex models that form non-linear hypotheses. In a neural network, we have an input layer, hidden layer(s), and an output layer. Each layer consists of one or more units; units take inputs and produce outputs. For the neural network that we will use for this problem, we will use the logistic function as the activation function in each unit. The input layer represents the features of the input: for example, in our problem, we will have 400 units in the input layer simply because we have 400 features for each input in our problem. Hidden layers enable us to add complexity and represent complicated non-linear models. Finally, the output layer represents the output of the network: for example, in our problem, we will have 10 units in the output layer; each unit represents a label (class) in our problem; and the label of the output unit that has the highest value (i.e. highest probability) will be considered the output label for our input.

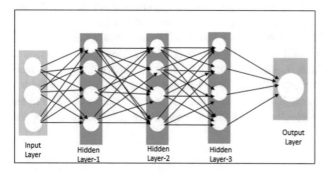

Fig. 3. A typical ANN

Although Deep Learning is a subdivision of Machine Learning extracts more abstract attributes from a large set of training data. Deep Neural network functions mostly similar to how a brain functions. In the above Fig. 3 when there are a lot of hidden layers are included in the network then it could be termed as a Deep Neural Network. Even though this formulation has been successfully used in many applications, the training process is slow and cumbersome. RNN (Recurrent Neural Network) is an NN model intended to perceive structures in streams of data. Unlike feedforward NN that does computations unidirectional from input to output, an RNN computes the present state's output depending on the yields of the previous states. Because of this "memory-like" property, despite learning problems related to vulnerable gradients, RNN applications have grown popularity in many fields involving streaming data.

CNN (Convolutional Neural Network) is a multilayer NN model, inspired by the neurobiology of visual cortex, that consists of convolutional layers followed by fully connected layers. In between these two types of layers, there may exist subsampling steps. They get the better of deep neural networks, which have difficulty in scaling well with multidimensional locally correlated input data. Therefore, the main application of the CNN has been in data sets, where the number of nodes and parameters required to be trained is relatively large. However, in case of large data sets, even this can be intimidating and can be solved using sporadically connected networks.

DA (Deep Autocoder) architecture is obtained by stacking a number of auto encoders that are data driven NN models designed to reduce data dimension by automatically projecting incoming depictions to a lesser dimensional space than that of the input. In an auto encoder, equal units are used in the input/output layers and less units in the hidden layers. Nonlinear transformations are embodied in the hidden layer units to encode the given input into smaller dimensions. Despite that it requires a pertaining stage and suffers from vanishing error, this architecture is popular for its data compression capability and have many variants.

3   Results and Discussion

The main essence of this study is to reduce the number of data dimension before feeding it to the deep learning network. The primary dataset was filtered first from more than 100 feature to one third of it. Some features are found quantitative; however, some are categorical features. These categorical features are quantified by creating dummy variables as a result 808 features were formed at the end.

This study was carried out by using all those features and then it is reduced to 10, 20, 30, 40, 50, 100 features by adopting Principal component analysis (PCA). Interestingly, we have found that with the reduced number of features the accuracy does not change much. Conventional computing hardware has used in order to carry out the whole experiment. By setting an initial number of features of 807 and batch size of 10 below are the results with different neurons in the hidden layers and epoch Size (Table 5).

Table 5. Performance comparison of Deep Artificial Neural Network (DANN) Algorithm with different settings.

Reduced feature features (PCA)	Hidden layer	Number of neurons in the hidden Layer	Epoch	Test set delay prediction accuracy
10	2	5	150	82.10%
10	6	5	100	82.10%
15	2	8	150	82.10%
25	2	13	150	82.08%
50	2	25	150	82.10%
50	3	50	100	82.10%
50	6	25	150	81.80%
100	2	50	100	80.82%

By using 10 features, keeping the number of hidden layer 2, number of neurons 5 and epoch is 150 we find that the test data accuracy is 82.10%. Same result had observed for 15 features. While increasing feature we have observed that at 25 feature the accuracy recorded was 82.08% which is 0.02% less than previous rate. Interestingly, at 50 features the accuracy increase to its previous highest which is 82.10%. Therefore, several experiments have been carried out with the same setup with different number of neurons and hidden layers. Surprisingly, there was no clear differences in accuracy rate. But when the number of hidden layer increase then the accuracy was 81.80%. So, we can conclude that Number of increased hidden layer does not ensure with higher accuracy.

Comparison to similar experimental result: The most significant part of our experiment is that this has enormously improve the previous experiment carried out by Varsha [10] in 2017 where the accuracy recorded was 76.76% using Deep Belief Network (DBN) with sigmoid activation function. However, our experimental result clearly shows 5.34% more accurate while predicting airlines delays which was around 82.10%. The reason behind this may be due to the quality of data, feature availability, and algorithm settings. Besides our experiment successfully implements dimensional reduction which perhaps another reason for the increased accuracy.

In short, principle component analysis (PCA), plays a crucial role in data representation even with very few features, and deep learning algorithm perfectly fits to that data in order to predict with higher accuracy than their other machine learning counterpart.

4   Conclusion

This research is carried out by employing quantitative research method. The outcomes are supported by analysis of the factors associated with flight delays. Significant effort has been dedicated in the research methodology and design which confirms that the features affects enormously to the number of flights delay every day. From the data collection to result interpretation, the crucial task has been made easy by implementing CRISP-DM.

In order to classify whether a flight will be delayed or not, we have used mainstream classification machine learning and deep neural networks. For the machine learning algorithm, Decision Tree is used while for deep neural network as the name stands Deep Artificial Neural network (DANN) has used in this research project. The accuracy of DANN was slightly higher than the Decision Tree, however, even a tiny difference in accuracy is of tremendous valuable since the dataset is enormous and number of flights per day is numerous. In the final part of the experiment we used deep neural network which found to be predicting with the highest accuracy. However, interestingly, this experiment shows improved accuracy in predicting airlines delay than other similar research done recently.

Hadoop ecosystem could perform great in order to turn the Big Data into tiny data making sure that this represents the actual essence of the data. However, this is not the scope of this research project although a random sampling done this job quite well. Implementing the whole process on the cloud platform, such as AWS (Amazon Web Service) Data pipeline or GCP (Google Cloud Platform) would be a better option since it would significantly reduce the testing and implementing cost.

Once the whole testing has done with following proposed system, methodology and tools, a customized application could be developed which can alert an airlines traveller beforehand, which could be a major step towards disturbance free journey with utmost satisfaction. This will surely save multi-billion dollar of business loss.

References

1. Nanji, A.: The incredible amount of data generated online every minute [Infographic] (2017). https://www.marketingprofs.com/charts/2017/32531/the-incredible-amount-of-data-generated-online-every-minute-infographic
2. Droummond, M.: 4 ways airlines can use Big Data to make customers love them (2013). https://w3.accelya.com/blog/4-ways-airlines-can-use-big-data-to-make-customers-love-them
3. Kim, Y.J., Pinon-Fischer, O.J., Mavris, D.N.: Parallel simulation of agent-based model for air traffic network. In: AIAA Modeling and Simulation Technologies Conference, p. 2799 (2015)
4. Wieland, F.: Parallel simulation for aviation applications. In: Proceedings of the 30th Conference on Winter Simulation, pp. 1191–1198. IEEE Computer Society Press (1998)
5. Najafabadi, M.M., Villanustre, F., Khoshgoftaar, T.M., Seliya, N., Wald, R., Muharemagic, E.: Deep learning applications and challenges in Big Data analytics. J. Big Data 2(1), 1–21 (2015)
6. Lv, Y., Duan, Y., Kang, W., Li, Z., Wang, F.-Y.: Traffic flow prediction with big data: a deep learning approach. IEEE Trans. Intell. Transp. Syst. 16(2), 865–873 (2015)
7. United States Department of Transportation: Bureau of transportation statistics (2018). https://www.transtats.bts.gov/DL_SelectFields.asp?Table_ID=236&DB_Short_Name=On-Time
8. Lawson, D., et al.: Predicting flight delays. Technical report, Computer Science Department, CS 229, Stanford University, Stanford (2012)
9. Oza, S., Sharma, S., Sangoi, H., Raut, R., Kotak, V.C.: Flight delay prediction system using weighted multiple linear regression. Int. J. Eng. Comput. Sci. 4(4), 11668–11676 (2015). ISSN 2319-7242
10. Venkatesh, V., et al.: Iterative machine and deep learning approach for aviation delay prediction. In: 4th IEEE Uttar Pradesh, pp. 562–567 (2017)

Mining Regular High Utility Sequential Patterns in Static and Dynamic Databases

Sabrina Zaman Ishita[1], Chowdhury Farhan Ahmed[1], Carson K. Leung[2]($\boxtimes$) ⓘ, and Calvin H. S. Hoi[2]

[1] University of Dhaka, Dhaka, Bangladesh
farhan@du.ac.bd
[2] University of Manitoba, Winnipeg, MB, Canada
kleung@cs.umanitoba.ca

Abstract. Regular pattern mining has been emerged as one of the important sub-domains of data mining with its numerous applications. Although patterns that occur at a regular interval throughout the whole database can lead to interesting knowledge, examining the utility values of these patterns can unveil more interesting useful information. In a sequence database, the task of mining regular high utility patterns can be more challenging. In this paper, we first propose a new algorithm for mining regular high utility sequential patterns from static databases. As handling of the incremental nature of big data brings useful results in many applications in the recent era of big data, we then extend our algorithm to mine regular high utility sequential patterns from dynamic databases. Evaluation results on several real-life datasets show the effectiveness of our two algorithms.

Keywords: Data mining · Regular pattern mining ·
High utility sequential pattern · Incremental mining ·
Information management · Information processing management

1 Introduction

Frequent pattern mining is one of the most common fields of data mining. Mining patterns from sequential database holds more significance because maintaining the order of events in a pattern is more challenging and provides more interesting results. Treating all the items equally in the database may not be desirable in several applications. More interesting patterns can be mined when considering the value or utility [9] of each item in an event of a sequence is regarded as *high utility sequential pattern (HUSP) mining.*

At the end of a database scan, a pattern may have high utility value greater than or equal to the threshold. However, it does not necessarily mean that such a pattern has appeared in a regular manner throughout the whole database. A pattern appearing at a regular user defined interval in the entire database is said to be a *regular pattern.* Hence, regularity can be of great interest to many

© Springer Nature Switzerland AG 2019
S. Lee et al. (Eds.): IMCOM 2019, AISC 935, pp. 897–916, 2019.
https://doi.org/10.1007/978-3-030-19063-7_71

data mining researchers. Applications of regular high utility sequential pattern mining include the following:

- High utility items sold at a regular manner in a retail market,
- web page sequences—with high importance—visited in regular interval, and
- genes occurring at a regular interval in DNA sequences.

The common nature of most real-life databases is to grow incrementally with time. Scanning the whole database every time when a new increment is added is undoubtedly cumbersome. Hence, efficient mining from incremental database can save a lot of the runtime and memory. If incremental mining can be applied in the field of regular high utility sequential patterns, the scope of applications of such patterns will definitely increase.

Existing algorithms in the field of high utility sequential patterns do not consider the regularity nature of the patterns. As a result, patterns mined by such approaches may appear irregularly throughout the database. Moreover, most of the regular pattern mining approaches do not consider the utility factor of the patterns. Hence, some patterns may appear at a regular interval in the database but they may not contribute sufficiently to the total utility of the database. Considering the dynamic nature of the database more efficient mining can be done in this field. These motivated us to propose methods to mine regular high utility sequential patterns from both static and dynamic databases.

Here, an interesting field of frequent pattern mining has been addressed where several fields such as sequential, high utility, regularity and incremental mining collide. Our contributions of the current paper are:

1. Development of a new algorithm called *RHusp* for mining regular high utility sequential patterns from static databases.
2. Thorough testing on several real life datasets to demonstrate the competence of the algorithm for the purpose of practical use.
3. Proposal of a novel algorithm called *RIncHusp* for mining regular high utility sequential patterns from incremental databases continuously over time.
4. Remarkable improvement on the results of incremental mining as compared to static mining.

The remainder of the current paper is organized as follows: Sect. 2 talks about the preliminary concepts related to this work. Section 3 discusses some of the state-of-the-art mining techniques which directly influence the current study. In Sect. 4, the proposed algorithms are developed, and an example is illustrated. Experimental results of the proposed algorithms and comparative study are given in Sect. 5. Finally, conclusions are drawn in Sect. 6.

2   Background

Let I be a set of items $\{I_1, I_2, ..., I_m\}$. A traditional transaction database is a set of transactions, where each transaction is a subset of I. A sequential database SD consists of a set of transactions or sequences $\{S_1, S_2, ..., S_n\}$, where each sequence S_i is a set of events $\{e_1, e_2, ..., e_p\}$. Each of the events is a subset

Table 1. Sequence database *SD*

SID	Sequences
S_1	$<(a,3);[(b,3)(c,2)];(c,1);(d,3)>$
S_2	$<(c,3);[(d,2)(e,4)(f,4)]>$
S_3	$<(a,5);(c,2);[(b,3)(d,1)(e,2)];(d,2)>$
S_4	$<(b,2);[(c,2)(e,3)]>$
S_5	$<(b,4);(c,1);[(d,2)(e,1)]>$
S_6	$<(e,3);(f,5);[(b,2)(c,1)];(d,3)>$

Table 2. External utility table of *SD*

Item	ExtUtility
a	4
b	3
c	7
d	2
e	5
f	1

of *I*. The length of a sequence is the total number of items in the sequence. The order in which the events occur in a sequence is important. For example, $<(abc)a(de)>$ is a sequence. For concision, brackets are omitted if an event contains one item only. Here, (abc) occurs before (a), and (a) occurs before (de). This order must be preserved. The order of items in an event is not important, usually it is taken in alphabetical order. Given a *minsup* threshold, mining subsequences appearing *minsup* times or more in the database is regarded as **sequential pattern mining** [6,10,14].

Example 1. If $\alpha =<(abc)b>$ and $\beta =<(abce)(be)(de)c>$ where a,b,c,d and e are items, then α is a subsequence of β.

In this paper, the terms *sequence* and *pattern* will be used interchangeably.

In high utility pattern mining, every item is associated with an external and an internal utility. The internal utility value comes along with the database, and the external utility value is given separately. For a sequential database, every item i in every event e of a sequence S contains an internal utility value $intU(i,e,S)$. Every distinct item has corresponding external utility value $extU(i)$.

Example 2. A sample sequential database along with the internal utility of all items is given in Table 1. Here, in sequence S_2, there are two events: $<c>$ and $<(def)>$, where $<c>$ occurs before $<(def)>$. The length of S_2 is 4. Each item appears as an $(item, intU(item))$-pair, where $intU(item)$ is the internal utility of the item in that sequence. The internal utility of an item i in an event e in the sequence S is denoted as $intU(i, e, S)$. With this notation, $intU(d, 2, S_2) = 2$ and $intU(b, 3, S_3) = 3$. The corresponding external utility table is given in Table 2, where the external utility of each distinct item $extU(item)$ is shown.

The utility of an item i in an event e of a sequence S is $utility(i, e, S)$, which is the product of its internal utility and external utility. The utility of a pattern P in a sequence S is $utility(P, S)$, which can be derived by summing the utilities of all the items of the pattern appearing in that sequence. The utility of a pattern P in a sequence database SD of n sequences is defined as:

$$utility(P) = \sum_{sid=1}^{n} utility(P, S_{sid}) \tag{1}$$

The utility of an event in a sequence is the summation of utilities of all the items in the event. Then, the utility of a sequence in a sequence database is the summation of all the items in all the events in the sequence. Similarly, the utility of the whole database is the summation of utilities of all the sequences. Given a minimum utility threshold `minUtility`, mining subsequences having utility more than or equal to $minUtility$ is regarded as **mining high utility sequential patterns**.

Example 3. With Tables 1 and 2, the utility $utility(d, 2, S_2)$ of item d in event 2 of sequence S_2 can be computed as:

$$utility(d, 2, S_2) = intU(d, 2, S_2) \times extU(d) = 2 \times 2 = 4.$$

Similarly, $utility\ (b, 3, S_3)$ is 9.

Then, the utility $utility(<bd>, S_5)$ of pattern $<bd>$ in sequence S_5 can be computed as:

$$utility(<bd>, S_5) = (12 + 4) = 16.$$

As a pattern may appear in multiple ways in a sequence, we take the maximum utility. In S_1, pattern $<ac>$ has two orderings with utility values as $(12 + 14) = 26$ and $(12 + 7) = 19$, we take $utility(<ac>, S_2) = 26$.

In the sequence database SD given in Table 1, its utility $utility(SD) = 258$.

A pattern may have utility value greater than or equal to the threshold, but whether the pattern appears regularly in the database or not is not provided in that information. The occurrences of a regular pattern are evenly distributed throughout the whole database. The period of a pattern is the interval between two consecutive sequences where the pattern has appeared. If S_a and S_b are two sequences where a sequential pattern P has appeared consecutively, the period of P defined as $period(P) = S_b - S_a$, where S_a and S_b represent the sequence numbers. If P occurs in m sequences and the set of sequences where P occurs is $\{S_{i1}, S_{i2}, ..., S_{im}\}$, then the maximum period of P is defined as:

$$maxPeriod(P) = \max(S_{i(j+1)} - S_{ij})\ \text{for}\ j \in [1, (m-1)] \tag{2}$$

We have considered that there has been a null sequence with no item at the beginning of the sequence database. Now, according to the above definition, *maxPeriod*(<bc>) in *SD* of Table 1 is 3. The *maxPeriod* of a pattern will serve as the regularity *reg(P)* of a pattern *P*. *P* will be regular if the *reg(P)* value of the pattern does not exceed the maximum regularity threshold *maxReg*. If *maxReg* = 3, then <bc> is a regular pattern.

3   Related Works

A number of works have been done in the field of sequential pattern mining. Of them, GSP [15], SPADE [22], PrefixSpan [13], SPAM [4] are some common ones. All these algorithms follow the antimonotone or downward closure property that if a pattern is found infrequent, none of its super patterns will be frequent. GSP and SPADE maintains the candidate generate and test approach where as SPAM uses bitmap representation for mining sequential patterns. PrefixSpan uses a prefix-projected pattern growth method to recursively project corresponding postfix subsequences into projected databases.

To produce more meaningful and interesting patterns, high utility pattern mining attracts a lot of researchers' interests. It was first introduced by Yao et al. [19]. After that, several algorithms like two-phase [20], IHUP [2], UP-Growth [18] were proposed to mine high utility itemsets. Applying this work in sequence database is more challenging. In this area, works like UL and US [1], USpan [21], HuspExt [3] have been done to produce high utility sequential patterns. *Sequence weighted utilization (SWU)* is used in these works to maintain the downward closure property of patterns. In the later works, instead of the *SWU* upper bound, a tighter upper bound is used to prune the search space more efficiently. It was also applied in our work while mining regular high utility sequential patterns.

Frequent patterns or high utility patterns may appear frequently in the database, but patterns appearing in a regular manner in the database provide more interesting information. To mine regular patterns from transactional databases, RP-tree [16] was proposed. Here, the concept of mining regular patterns was introduced, it mines regular patterns with the help of a tree based data structure called Regular Pattern tree (RP-tree). As a relevant domain, periodic frequent patterns hold interesting meaning too. In this field, mining periodic frequent patters from transactional database [17] and mining periodic high utility sequential patterns [7] have been done.

The common nature of most of the real-life databases is to increment dynamically. It is evidently unsuitable to scan the whole database from scratch every time new increment is added to the database, thus efficient mining is needed. To mine patterns from incremental databases, many algorithms have been developed. To mine sequential patterns from incremental databases several works [5,11,12] have been done. For instance, IncSpan [5] buffers semi-frequent sequences along with frequent sequences. As a result, the whole database needs not to be scanned when new increment occurs.

To the best of our knowledge, there has been no work in the field to mine regular high utility sequential patterns. We are proposing efficient approaches where regular high utility sequential patterns will be mined in both static and dynamic databases. Incremental nature of the database will be handled efficiently to avoid running the process from scratch.

4   Our Regular High Utility Sequential Pattern Mining Algorithms

In this section, we describe our approaches for mining regular high utility sequential patterns from static and dynamic incremental databases. First, we describe the method for mining regular high utility sequential patterns, then provide an incremental scenario and discuss how to handle an incremental database in an efficient manner afterwards.

Definition 1 (Regular High Utility Sequential Pattern). *A pattern P must satisfy the following conditions for being **regular high utility sequential pattern**:*

$$utility(P) \geq minUtility \tag{3}$$

and

$$reg(P) \leq maxReg \tag{4}$$

Here, the utility of the pattern is calculated as in Eq. 1. The thresholds minUtility and maxReg are user defined. If a pattern satisfies both of the conditions, then it is regarded as a regular high utility sequential pattern.

How the regularity value of a pattern is calculated is described later.

The regularity threshold maintains the antimonotone property which means, if a pattern is irregular then none of its super patterns will be regular. So, there is no need for further candidate generation with this pattern as prefix. However, the minimum utility threshold does not maintain the antimonotone property. Even if a pattern does not satisfy the *minUtility* threshold, its super patterns may satisfy the *minUtility* threshold later after adding more weighted items to the pattern. So, for candidate generation, we need to take an upper bound on the utility of the pattern.

Definition 2 (Candidate for Regular High Utility Sequential Pattern). *Instead of taking the whole sequence utility, we will take the summation of the utility of the pattern utility(P) and the remaining utility (remUtility) of the sequence containing the pattern P. Hence, the pattern must satisfy the following conditions for being a potential candidate:*

$$utility(P) + remUtility(P) \geq minUtility \tag{5}$$

and

$$reg(P) \leq maxReg \tag{6}$$

Along with the utility threshold, a pattern or a candidate will also be checked for the regularity threshold. When a candidate is generated, the information of which sequences contain the candidate pattern is also kept. From that information, the regularity of the pattern $reg(P)$ can be calculated as follows:

1. If S_{cur} is the first sequence containing P
2. $\qquad S_{last} = S_{cur}$
3. $\qquad reg(P) = S_{cur}$
4. Else $Period_P = S_{cur} - S_{last}$
5. $\qquad S_{last} = S_{cur}$
6. $\qquad$ If $Period_P > reg(P)$
7. $\qquad\qquad reg(P) = Period_P$

At the end of SD, for each pattern, $Period_P$ will be calculated using S_{cur} as the last sequence in SD and $reg(P)$ will be updated accordingly. First, the length-1 regular high utility sequential pattern candidates will be generated using the above condition.

Example 4. For the database in Table 1, taking *minUtility* as 20% and *maxReg* as 3, the set of length-1 candidates will be $\{<a>, <b>, <c>, <d>, <e>\}$.

Here, $<f>$ is pruned out as it does not satisfy the condition in Definition 2. For each length-1 candidate, projected database will be created. As an item may occur multiple times in a sequence, that information will also be kept for producing patterns with different utilities so that we can get the maximum utility of a pattern. From each of the length-1 candidates, length-2 candidates will be generated from the projected database by concatenating items. Two types of concatenation will take place:

1. *s-concatenation*, where item will be added at the end of the pattern as a different event.
2. *i-concatenation*, where item will be added in the same event of the last item in the pattern

Regardless of the concatenation type, if the new concatenated pattern satisfies the above condition for being a candidate, then it will serve as a candidate pattern for producing length-3 candidates. This way length-$(k+1)$ candidates will be generated from length-k candidates until no candidates are generated. In each of the step, every candidate will be checked for the condition of regular high utility sequential pattern and if the candidate meets the condition then it will be added to the final list of regular high utility sequential patterns.

Example 5. For the database in Table 1, the regular high utility sequential patterns found are $<abd>, <ac>, <ac(be)>, <acd>, <bc>, <bd>, <c>, <cd>, <c(de)>, <ce>$ and $<e>$. For convenience, the maximum pattern length is kept to 4.

4.1 Algorithm *RHusp* for mining static databases

Our algorithm to mine regular high utility sequential pattern from a static database is given in Algorithm 1. It takes the following as inputs:

1. a prefix pattern P,
2. P-projected sequence database $SD_{|P}$,
3. the minimum utility threshold *minUtility*, and
4. the maximum regularity threshold *maxReg*.

Algorithm 1. $RHusp(P, SD_{|P}, minUtility, maxReg)$

Input: (1) pattern P, (2) P-projected sequence database $SD_{|P}$, (3) minimum utility threshold *minUtility*, and (4) maximum regularity threshold *maxReg*

Output: Regular High utility sequences

```
 1: Scan SD_|P to find candidate items for i-concatenation and s-concatenation
 2: Put i-concatenation candidate items in List_i
 3: Put s-concatenation candidate items in List_s
 4: for each item i in List_i and List_s do
 5:     P' ← (P, i)
 6:     if reg(P) ≤ maxReg then
 7:         if utility(P') ≥ minUtility then
 8:             output P'
 9:         end if
10:         if utility(P') + remUtility(P') ≥ minUtility then
11:             make P'-projected database SD_|P'
12:             call RHusp(P', SD_|P', minUtility, maxReg)
13:         end if
14:     end if
15: end for
16: return
```

It outputs the regular high utility sequential patterns in a recursive manner.

First, the algorithm is called with the parameters (1) *null*, (2) the initial complete sequence database SD, (3) minUtility, and (4) *maxReg*. In each of the iterations, the projected database is scanned to find candidate items to concatenate with P by *i-concatenation* or *s-concatenation*. Along with that, the regularity of new pattern P' is also calculated according to the method described above. If the pattern is regular then it is checked for the utility threshold as described in lines 7 to 9. Then it is checked for candidate condition, if satisfies then P'-projected database is created and same procedure will be run again recursively until no candidates are generated. At the end, all the regular high utility sequential patterns will be achieved.

4.2   Incremental Scenario

We have found the set of regular high utility sequential patterns. Now, two consecutive increments occurred to the initial database which is shown in Table 3. The above procedure will run again on the new whole database. Instead, if we can avoid scanning the database from scratch, it will save a lot of our time. For that we will take following measures to handle the incremental nature of the database efficiently.

Table 3. Sequence database SD'

Sequence ID	Sequences
S_1	$<(a,3); [(b,3)(c,2)]; (c,1); (d,3)>$
S_2	$<(c,3); [(d,2)(e,4)(f,4)]>$
S_3	$<(a,5); (c,2); [(b,3)(d,1)(e,2)]; (d,2)>$
S_4	$<(b,2); [(c,2)(e,3)]>$
S_5	$<(b,4); (c,1); [(d,2)(e,1)]>$
S_6	$<(e,3); (f,5); [(b,2)(c,1)]; (d,3)>$
S_7	$<(a,2); (c,2); [(d,2)(e,3)]; (d,1)>$
S_8	$<(b,3); [(c,1)(e,2)]>$
S_9	$<(b,3); [(c,1)(d,6)]; (a,1); (e,1)>$
S_{10}	$<(e,2); [(d,2)(f,2)]>$

Along with buffering regular high utility sequences HS from initial database, we will buffer regular semi-high utility sequences SHS as well. If a sequence does not satisfy the high utility condition then it is checked for the semi-high utility condition as follows:

Definition 3 (Regular Semi-High Utility Sequences). *Conditions for semi-high utility sequences are*

$$utility(P) \geq minUtility \times \mu \tag{7}$$

and

$$reg(P) \leq maxReg \tag{8}$$

where

- *P is a pattern not fulfilling the high utility condition, and*
- *μ is a buffer ratio.*

The utility of P is compared with a fraction of minUtility which is derived from multiplying minUtility by a buffer ratio μ where $0 < \mu \leq 1$. The regularity of P must be less than or equal to maxReg as before. If it satisfies, the sequence is listed in SHS as a regular semi-high utility sequence. Otherwise, it is pruned out.

Now, from these sets of high utility and semi-high utility sequences, we will make an extended trie which will be updated dynamically each time a new increment is added to the database. In this data structure, every node has an item associated with it, along with that, the information of whether the item is s-concatenated or i-concatenated, its utility, regularity and the last sequence where it appeared is saved in that node. The root is a node with null value in all the node parameters. Every node except the leaf ones can be extended for s-concatenation and i-concatenation. Solid lines represent s-concatenations, and dashed lines represent i-concatenations. The path from root to every other node represents a sequence and the information regarding that sequence can be found in the last node of that path. Only the nodes where a pattern ends will have utility, last sequence and regularity values, others will have null in those fields. Because a pattern $<abc>$ having utility 'x' does not mean that pattern $<a>$ also has utility 'x'. So, in the path from root through a-b-c, only $<c>$ will contain the pattern information. A sample extended trie for the initial database in Table 1 is shown in Fig. 1.

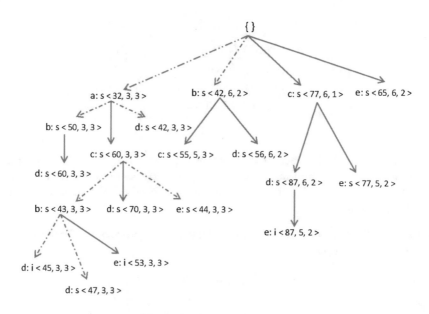

Fig. 1. Extended trie for SD

Here, every node has an item name, a character 's' or 'i' to indicate whether the item is an s-extension or i-extension, then a tuple consisting of three values: first the utility of the pattern ending at that node from the root, the last sequence number where the pattern appeared and the regularity of the pattern. For example, the first branch of the root is a:s $<32, 3, 3>$ which means pattern $<a>$ has utility 32, it has last appeared in S_3 and $reg(<a>) = 3$. The dashed line depicts that it is a semi-high utility sequence. Similarly, in the branch $<c>$, it has a branch $<d>$ as s-concatenation and $<d>$ has a branch $<e>$ as i-concatenation which has the information $e : i<87,5,2>$. It means the pattern $<c(de)>$ has utility 87, it has last appeared in S_5 and $reg(<c(de)>) = 2$.

After creating the trie, for each of the increment to the database, this trie will be dynamically updated. For each of the new sequences, the sequence will be scanned and the node items of the trie will be checked for matching with the items in the sequence. The utility will be updated accordingly. The last sequence number field of the trie will also be updated with the scanned sequence number. Regularity value will be checked for update as well.

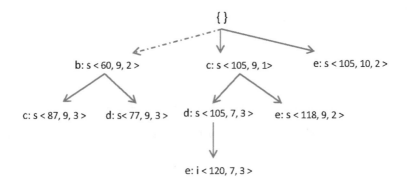

Fig. 2. Extended trie for SD'

Then the utility and regularity of each pattern in the HS and SHS lists will be updated with the help of the extended dynamic trie. Then all the patterns will be checked for the conditions discussed above to decide whether it will go to the new regular high utility sequence list HS' or new regular semi-high utility sequence list SHS'. These two lists will be served as HS and SHS for the next increment in the database. This way all the increments will be handled efficiently without scanning the database from scratch.

After two consecutive increments, the updated trie for database SD' in Table 3 is shown in Fig. 2. Here, we can see the branch $<a>$ of the root is not shown. As $<a>$ appeared to be an irregular pattern, none of its child patterns can be regular. Information in other nodes has been updated according to the database increments.

4.3 Algorithm *RIncHusp* for mining dynamic incremental databases

The algorithm for incremental mining of regular high utility sequential patterns or sequences is given in Algorithm 2. Here, the main method will call the *RHusp* algorithm. Along with making the list of regular high utility sequences HS, it will make the list of regular semi-high utility sequences SHS also. Then it will call the Procedure:*RIncHusp* and send HS and SHS as parameters.

Algorithm 2. *RIncHusp*: Regular High Utility Sequential Pattern mining in Incremental databases

---

Input: initial sequence database SD_{ini}, *minUtility*, *maxReg* and the buffer ratio μ
Output: Regular High Utility Sequences *HS*.
Method:
Begin

1. Let *HS* be the set of regular High utility Sequences and *SHS* be the set of regular Semi-High utility Sequences.
2. $HS \longleftarrow \{\}$, $SHS \longleftarrow \{\}$
3. *HS, SHS* = Call modified RHusp(SD_{ini}, *minUtility*, *maxReg*, μ)
4. **for** each new increment ΔSD in SD_{ini} **do**
5.   HS, SHS = RIncHusp(HS, SHS, ΔSD, minUtility, maxReg, μ)
6.   output *HS*
7. **end for**

End

Procedure: *RIncHusp(HS, SHS, ΔSD, minUtility, maxReg, μ)*

1. Let HS' and SHS' be the set of new regular High utility and Semi-High utility Sequences respectively.
2. $HS' \longleftarrow \{\}$, $SHS' \longleftarrow \{\}$
3. **for** each pattern *P* in *HS* and *SHS* **do**
4.   update *reg(P)*, *utility(P)*
5.   **if** *reg(P)* $\leq$ *maxReg* **then**
6.   **if** *utility(P)* $\geq$ *minUtility* **then**
7.   insert(HS', P)
8.   **else if** *utility(P)* $\geq$ *minUtility* $* \mu$ **then**
9.   insert(SHS', P)
10.   **end if**
11.   **end if**
12. **end for**
13. return HS', SHS'

In that procedure, two new lists *HS'* and *SHS'* will be created for storing new regular high utility and semi-high utility sequences. For each of the pattern in *HS* and *SHS*, the utility and regularity of the pattern will be updated. Then it will be checked for conditions as described in lines 5–11. If the pattern satisfies the maximum regularity threshold, then the utility will be checked to see if it satisfies the high utility condition. If not, then it is checked to see whether it satisfies the semi-high utility condition. According to that, the pattern will go to *HS'* or *SHS'* respectively. Patterns not satisfying both the conditions will be pruned out. The procedure will return the *HS'* and *SHS'* lists which will be used as *HS* and *SHS* for the next increment. At any time point, the set of regular high utility sequential patterns can be found in *HS* list.

5   Experimental Results

In this section, the performance study of both of our approaches over various datasets will be presented. First the performance of *RHusp* will be studied and the results will be compared with USpan [21], a high utility sequence mining algorithm which does not consider the regularity nature of the patterns. Then the performance of *RIncHusp* will be studied and compared with *RHusp* where *RIncHusp* handles the incremental nature of the database efficiently, but *RHusp* does not. The performance of *RIncHusp* is not compared with USpan since the comparison between *RHusp* and USpan is already done beforehand. Both the algorithms were implemented in Java in Windows environment (Windows 10), on a core-i5 intel processor which operates at 3.2 GHz with 8 GB of memory. The datasets were collected from SPMF open-source data mining library [8].

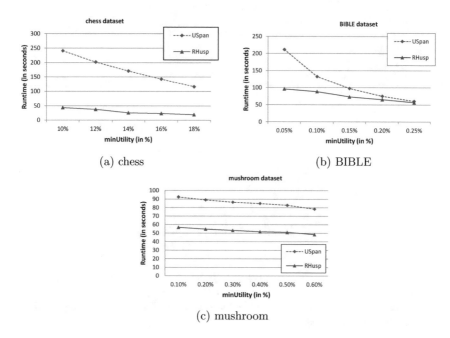

Fig. 3. Runtime of *RHusp* and USpan for varying *minUtility* thresholds in different datasets

5.1   Performance Analysis of *RHusp*

First, the runtime performance of *RHusp* was observed for different *minUtility* thresholds and compared with the result of USpan. The graphical representations of the performances over different datasets are given in Fig. 3. Here, convenient values for *maxReg* were taken. From the figure, it is evident that *RHusp* consumes a lot less amount of time than USpan. This is because *RHusp* considers both the regularity measure and utility of the patterns whereas USpan considers the utility only. Hence, a lot of patterns which appear irregularly in the database are pruned out by *RHusp* unlike USpan. So, in a relatively short amount of time, *RHusp* produces results.

The comparative performances of *RHusp* and USpan with respect to number of patterns generated for varying *minUtility* thresholds in different datasets are shown in Fig. 4. Here, we can see that patterns generated by *RHusp* are less in number. This is due to the regularity nature of the patterns here whereas USpan generates both regular and irregular patterns. Comparative performance analysis of memory usage by *RHusp* and USpan is given in Fig. 5 where we can see *RHusp* performs better in terms of memory also.

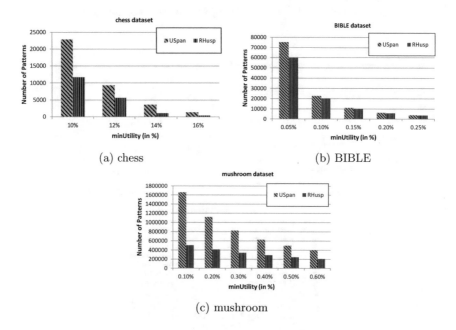

(a) chess   (b) BIBLE

(c) mushroom

Fig. 4. Number of patterns generated by *RHusp* and USpan for varying *minUtility* thresholds in different datasets

Maximum regularity *maxReg* plays an important role in our study. By increasing the value of it, patterns with longer period interval will be buffered. As a result, more patterns will be generated. If we take the value of *maxReg* as 100% then it is same as USpan. Because then the absolute value of *maxReg* will be same as the total number of sequences. Both the regular and irregular high utility sequential patterns will be generated. A suitable *maxReg* value can generate a useful amount of regular high utility sequential patterns. The performances of *RHusp* in foodmart(*minUtility* = 0.01%) and mushroom(*minUtility* = 0.5%) datasets with respect to the number of patterns for different *maxReg* thresholds are given in Figs. 6 and 7 respectively.

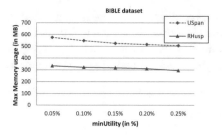

Fig. 5. Memory usage of *RHusp* and USpan for varying *minUtility* thresholds in BIBLE dataset

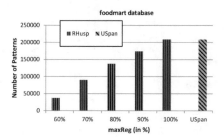

Fig. 6. Number of patterns generated by *RHusp* and USpan for varying *maxReg* thresholds in foodmart dataset

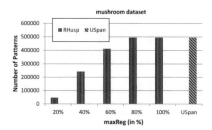

Fig. 7. Number of patterns generated by *RHusp* and USpan for varying *maxReg* thresholds in mushroom dataset

5.2   Performance Analysis of *RIncHusp*

The performance of *RIncHusp* was observed under different performance metrics. The runtime performance of *RIncHusp* was observed for different *minUtility* thresholds in different datasets. It was compared with the performance of *RHusp* which does not handle the incremental nature of the database. The runtime of both the algorithms was calculated as follows: first, for both of the algorithms, an initial part of the database was taken then algorithms were run on it. *RIncHusp* uses the modified *RHusp* to produce regular high utility and regular semi-high utility sequences whereas *RHusp* produces only regular high utility sequences. Then two consecutive increments were added to the database. *RIncHusp* works on the *HS* and *SHS* lists and scan only the incremented portion of the database for each of the increments. *RHusp* scans the whole database from scratch for each increment. For runtime calculation, for both *RIncHusp* and *RHusp*, consumed time in each of the three times (initial, first increment, second increment) were summed.

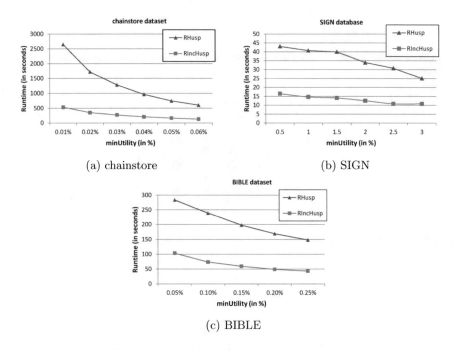

(a) chainstore   (b) SIGN

(c) BIBLE

Fig. 8. Runtime of *RIncHusp* and *RHusp* for varying *minUtility* thresholds in different datasets

The comparative runtime performance of *RIncHusp* and *RHusp* in different datasets for varying *minUtility* thresholds are shown in Fig. 8. Here, convenient values for *maxReg* regularity threshold were taken. From the figure, we can see that *RIncHusp* consumes less amount of time than *RHusp* in all the datasets. And the time difference becomes larger with the decrease of *minUtility* threshold. *RIncHusp* gives way better performance because it effectively handles the increments to database by buffering regular semi-high utility sequences and maintaining dynamic extended trie.

At this point, it is a common assumption that *RIncHusp* cannot generate the complete set of regular high utility sequences, so the number of patterns generated by *RHusp* will be much higher than the number of patterns generated by *RIncHusp*. However, in our experimental results, we have found a significant number of patterns generated by *RIncHusp* as shown in Fig. 9. The difference between the number of patterns by *RIncHusp* and *RHusp* is reasonable considering the extremely fast nature of *RIncHusp*.

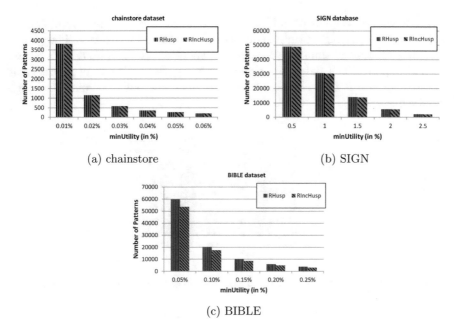

(a) chainstore

(b) SIGN

(c) BIBLE

Fig. 9. Number of Patterns generated by *RIncHusp* and *RHusp* for varying *minUtility* thresholds in different datasets

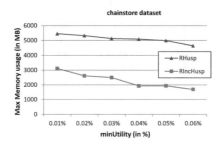

Fig. 10. Memory usage of *RIncHusp* and *RHusp* for varying *minUtility* thresholds in chainstore dataset

Figure 10 shows the performance analysis of *RIncHusp* and *RHusp* with respect to memory usage for different *minUtility* thresholds in chainstore dataset. Memory usage calculation was done same way the runtime calculation was done before. Memory consumed by *RIncHusp* is much smaller than the memory consumed by *RHusp*. Buffering regular semi-high utility sequences plays a vital role to produce a good amount of patterns in a very less amount of time. So, varying the value of μ will result in differences among the number of patterns generated. By lowering the value of the buffer ratio which means lowering the value of *minUtility* (taking smaller percentage of *minUtility* by multiplying it with μ), more patterns can be buffered. On the other hand, if we take 100% as the value of μ, it means no regular semi-high utility sequences will be buffered, thus no contribution to the high utility sequences in the later increments. The graphical representation of number of patterns generated by *RIncHusp* and *RHusp* for varying μ in SIGN dataset is given in Fig. 11. Here, *RHusp* produces same amount of patterns in all the cases since it does not buffer regular semi-high utility sequences.

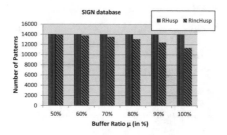

Fig. 11. Number of patterns generated by *RIncHusp* and *RHusp* for varying buffer ratio μ

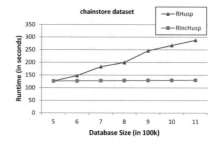

Fig. 12. Scalability test of *RIncHusp* and *RHusp* in chainstore dataset (*minUtility*=0.06%)

A scalability test was run on the chainstore dataset which contains more than one million sequences. We have run both *RIncHusp* and *RHusp* on an initial amount of the database, then six consecutive increments were added to the database and both the algorithms were run on them. *minUtility* threshold was set to 0.06% and *maxReg* threshold was set to 20%. Figure 12 shows the graphical representation of the runtime of *RIncHusp* and *RHusp* for varying database size (in 100k) where we can see *RIncHusp* performs way efficiently than *RHusp* with respect to scalability too. *RIncHusp* consumes significantly less amount of time than *RHusp* since it avoids scanning the whole database when new increments are added.

As a summary of the discussions above, taking less time and memory than USpan, *RHusp* produces regular high utility sequential patterns which are more interesting than those generated by USpan. *RIncHusp* mines regular high utility sequential patterns from incremental databases in a very optimized effective way which we have seen in the comparative performance study of *RIncHusp* and *RHusp* where *RIncHusp* gives way better results than *RHusp* in terms of runtime, memory, scalability.

6   Conclusions

In this paper, we proposed a new algorithm *RHusp* for mining regular high utility sequential patterns from static databases. We also proposed another algorithm—called *RIncHusp*—to mine regular high utility sequential patterns from incremental databases. Our approach runs efficiently by working on the incremented portion of the database, rather than working on the whole database. It also maintains an extended dynamic trie which comes handy while updating pattern's information.

This work is highly applicable in the field where databases are getting constantly new updates. The regularity and utility information of patterns carry interesting information. Areas of application include mining market transactions, weather forecast, improving health-care and health insurance and many others. It can also be used in fraud detection by assigning high utility values to previously found fraud patterns.

As ongoing and future work, we are extending this work to include more research problems to be solved for efficient solutions. We are also extending this work to incremental mining on closed regular high utility sequential patterns, as well as sliding window based mining over data streams.

Acknowledgments. This project is partially supported by NSERC (Canada) and University of Manitoba. In addition, the project is also supported by High-Profile ICT Scholar Fellowship (2017–2018) funded by the Information and Communication Technology (ICT) Division, Ministry of Posts, Telecommunications and Information Technology, Government of the People's Republic of Bangladesh.

References

1. Ahmed, C.F., Tanbeer, S.K., Jeong, B.S.: A novel approach for mining high-utility sequential patterns in sequence databases. ETRI J. **32**(5), 676–686 (2010)
2. Ahmed, C.F., Tanbeer, S.K., Jeong, B.S., Lee, Y.K.: Efficient tree structures for high utility pattern mining in incremental databases. IEEE TKDE **21**(12), 1708–1721 (2009)
3. Alkan, O.K., Karagoz, P.: CRoM and HuspExt: improving efficiency of high utility sequential pattern extraction. IEEE TKDE **27**(10), 2645–2657 (2015)
4. Ayres, J., Flannick, J., Gehrke, J., Yiu, T.: Sequential PAttern mining using a bitmap representation. In: ACM KDD 2002, pp. 429–435 (2002)

5. Cheng, H., Yan, X., Han, J.: IncSpan: incremental mining of sequential patterns in large database. In: ACM KDD 2004, pp. 527–532 (2004)
6. Choi, P., Hwang, B.: Dynamic weighted sequential pattern mining for USN system. In: ACM IMCOM 2017, pp. 19:1–19:6 (2017)
7. Dinh, D.T., Le, B., Fournier-Viger, P., Huynh, V.N.: An efficient algorithm for mining periodic high-utility sequential patterns. Appl. Intell. **48**(12), 4694–4714 (2018)
8. Fournier-Viger, P., Lin, J.C.W., Gomariz, A., Gueniche, T., Soltani, A., Deng, Z., Lam, H.T.: The SPMF open-source data mining library version 2. In: PKDD 2016, Part III. pp. 36–40 (2016)
9. Gunawan, D., Mambo, M.: Set-valued data anonymization maintaining data utility and data property. In: IMCOM 2018, pp. 88:1–88:8 (2018)
10. Hsu, K.: Effectively mining time-constrained sequential patterns of smartphone application usage. In: ACM IMCOM 2017, pp. 39:1–39:8 (2017)
11. Lin, J.C.W., Hong, T.P., Gan, W., Chen, H.Y., Li, S.T.: Incrementally updating the discovered sequential patterns based on pre-large concept. Intell. Data Anal. **19**(5), 1071–1089 (2015)
12. Nguyen, S.N., Sun, X., Orlowska, M.E.: Improvements of IncSpan: incremental mining of sequential patterns in large database. In: PAKDD 2005, pp. 442–451 (2005)
13. Pei, J., Han, J., Mortazavi-Asl, B., Wang, J., Pinto, H., Chen, Q., Dayal, U., Hsu, M.C.: Mining sequential patterns by pattern-growth: the PrefixSpan approach. IEEE TKDE **16**(11), 1424–1440 (2004)
14. Rahman, M.M., Ahmed, C.F., Leung, C.K., Pazdor, A.G.M.: Frequent sequence mining with weight constraints in uncertain databases. In: IMCOM 2018, pp. 48:1–48:8 (2018)
15. Srikant, R., Agrawal, R.: Mining sequential patterns: generalizations and performance improvements. In: EDBT 1996, pp. 3–17 (1996)
16. Tanbeer, S.K., Ahmed, C.F., Jeong, B.S., Lee, Y.K.: Mining regular patterns in transactional databases. IEICE Trans. Inf. Syst. **E91.D**(11), 2568–2577 (2008)
17. Tanbeer, S.K., Ahmed, C.F., Jeong, B.S., Lee, Y.K.: Discovering periodic-frequent patterns in transactional databases. In: PAKDD 2009, pp. 242–253 (2009)
18. Tseng, V.S., Wu, C.W., Shie, B.E., Yu, P.S.: UP-Growth: an efficient algorithm for high utility itemset mining. In: ACM KDD 2010, pp. 253–262 (2010)
19. Yao, H., Hamilton, H.J., Butz, C.J.: A foundational approach to mining itemset utilities from databases. In: SIAM SDM 2004, pp. 482–486 (2004)
20. Yeh, J.S., Li, Y.C., Chang, C.C.: Two-phase algorithms for a novel utility-frequent mining model. In: PAKDD 2007, pp. 433–444 (2007)
21. Yin, J., Zheng, Z., Cao, L.: USpan: An efficient algorithm for mining high utility sequential patterns. In: ACM KDD 2012, pp. 660–668 (2012)
22. Zaki, M.J.: SPADE: an efficient algorithm for mining frequent sequences. Mach. Learn. **42**(1–2), 31–60 (2001)

Mining Weighted Frequent Patterns from Uncertain Data Streams

Jesan Ahammed Ovi[1,2], Chowdhury Farhan Ahmed[1], Carson K. Leung[3](✉) ⓘ, and Adam G. M. Pazdor[3]

[1] University of Dhaka, Dhaka, Bangladesh
farhan@du.ac.bd
[2] East West University, Dhaka, Bangladesh
jesan_ovi@yahoo.com
[3] University of Manitoba, Winnipeg, MB, Canada
kleung@cs.umanitoba.ca

Abstract. In recent years, data mining has become one of the most demanding areas of computer science. For instance, it helps to discover frequent patterns by applying intelligence tools, techniques and methodologies from various kinds of databases. However, with the rapid growth of modern technology, high volumes of data with different characteristics are generated by many applications. Situation has become much more challenging and sophisticated when datasets are uncertain in nature and flowing at high velocity. Many applications also demand real-time analysis of data depending on current characteristics. Several researches have been made to mitigate the challenges regarding uncertain data streams. However, as the datasets are streaming in nature, frequent patterns of those datasets may be huge in size, and thus may require further mining to find the interesting patterns. Interestingness of patterns can be measured by associating weight with each item. In this paper, we propose a novel tree-based approach called WFPMUDS (Weighted Frequent Patterns mining from Uncertain Data Streams)-growth, which is capable of capturing recent behavior of uncertain data streams and only produces significant (weighted) patterns.

Keywords: Data mining · Weighted pattern · Uncertain data · Data stream · Information management · Information processing management

1 Introduction

Today's world is inundated with data as modern applications like to produce huge amount of data. However, at the same time, these applications need to deal with those enormous data to gain knowledge for various purposes, including the following:

© Springer Nature Switzerland AG 2019
S. Lee et al. (Eds.): IMCOM 2019, AISC 935, pp. 917–936, 2019.
https://doi.org/10.1007/978-3-030-19063-7_72

- making decision,
- predicting future situation,
- setting business policy
- etc.

As people has no time to look at this data, some sorts of systems, methodologies, procedures, etc., are in demand that automatically analyze, classify, discover patterns, demonstrate, characterize those data in order to mine useful knowledge. Patterns discovery is one of the most important and demanding aspect of knowledge mining process, which basically performs the following:

- find frequent patterns from different types of datasets,
- show statistical relevancy between various items,
- mine interesting patterns,
- etc.

The discovered patterns help users set future decision making policy.

These demands attract the attention of computer scientists and put challenges to handle different kinds of data. Focusing on the demand, over the year, researchers developed many efficient procedures, tools, techniques, etc. Examples include the following:

- sequential pattern mining [1–5],
- time series mining [6],
- utility patterns mining [7,8],
- etc.

These procedures, tools, techniques, etc. discover various kind of patterns from different types of dataset.

Most of these algorithms, developed by the researchers, use support constrain to prune the item from datasets and only consider the present and absent of item in datasets. However, at the same time, users demand from only interesting patterns that actually matter for them regardless of all frequent patterns. Researchers also mitigated their demand by implementing different procedures like weighted [9] pattern mining methods that only mine significant patterns. In this case, weighted expected support is proposed to prune unnecessary item. Situation has become more challenging and sophisticated when applications produce huge amount of data like network sensor data and demand for real time analysis depending on current behaviours of data that means researchers have to deal with stream databases. DStree [10] was established to handle this kind of datasets. For doing so two important concept those are batch size and window size are introduced. Unlike traditional methods whose count the frequency of each item from start to end of dataset, DStree captured batchwise information of its each node. When number of batch overflows the window size it deletes oldest batch and makes room for new batch by shifting rest of the batches.

Complexities are increased when items in datasets are no more certain like medical database, sensor network, etc., which means each item in transaction is associated with some probability.

Example 1. a: p_1, b: p_2, c: p_3, d: p_4—where a, b, c, d are item and p_1, p_2, p_3, p_4 are their corresponding probability—is an example of uncertain datasets.

In this situation, it is not enough to just calculate frequency of a particular item and using support constrain to decide whether that item is frequent or not as this time item is associated with probability and a particular item may not frequent itself but may be frequent when occurred with another items. Researchers developed several tree structures—and the corresponding algorithms—to handle uncertain nature of datasets. These include the following:

- UF-tree [11],
- CUF-tree [12],
- PUF-tree [13],
- etc.

An important constraint, item cap (I^{CAP}) [13], was introduced with PUF-tree. It is one of the most important concept to prune uncertain item in order to mine frequent patterns. However, some applications like network sensor data analysis in where datasets are uncertain as well as stream in nature, UF-tree, PUF-tree, CUF-tree were not designed to handle the streaming nature of data.

To resolve the problem, USPF-growth [14] was implemented. USPF-growth uses a window based technique like DS-tree to capture the streaming nature of data and a constrain called U^{Cap} to prune uncertain items from dataset. However, this procedure generates all frequent patterns without considering their importance to the applications. Our target is to solve this problem by generating only significant patterns from uncertain data stream. Let us consider the following situation.

Example 2. Suppose that scientists are interested in analyzing the different particles of atmosphere using network sensor data to study the environment of forest to explore which kinds of plant or animal can survive in a particular area or looking for some precious particle like Argon in some region or before exploring some caves and mountains they need to sure whether some particular lethal gas items or their combination are present or not in the air. Network sensor data looks like the following:

- Oxygen: 21%,
- Nitrogen: 71%,
- Argon: 0.08%,
- etc.

Gas items that are looking for may very rare in atmosphere like Argon. In this situation if we use USPF-growth, which just generates all frequent patterns, misses those significant patterns. However, our proposed approach is specially developed to find only those significant items by adding weight to those items.

In this paper, we propose a novel tree-based algorithm called WFPMUDS for weighted frequent pattern mining from uncertain data streams. To elaborate, our *key contributions of the current paper* include the following:

1. Development of an efficient algorithms that perfectly capture the uncertain data streams and mine only significant (weighted) patterns that matter to the applications.
2. Exploitation of the advantages of LMAX (local maximum weight) [15] weight to reduce the number of false patterns during mining.
3. Significant improvement in time and memory requirement to capture the uncertain data stream and mine weighted patterns while applying in various data sets.

As our prior knowledge, we are the first to propose an algorithm of mining weighted uncertain patterns from stream database.

The remainder of this paper is organized as follows: Sect. 2 discusses related works. We then describe our proposed methods in Sect. 3. Experimental results are shown in Sect. 4. Finally, conclusions are drawn in Sect. 5.

2   Related Works

As uncertain databases are one of the most interesting as well as significant sector in data mining, many researchers have been attracted to this field. As a result, several methods have ben developed to mine uncertain datasets to find frequent patterns. Examples of these *uncertain data mining* methods include the following:

- UF-tree [11],
- CUF-tree [12], and
- PUF-tree [13].

Although these methods are very effective on general uncertain datasets, they are not applicable to uncertain data streams. Besides, some methods to deal with data streams have also been proposed. An example of these *data stream mining* methods is as follows:

- DStree [10]

DStree efficiently handles the recent behavior of data streams, but it generates all frequent patterns regardless of their important as well as uncertainty in data was not considered.

In addition, several methods were proposed to mine uncertain frequent patterns from streams. They include the following:

- UF-streaming [16], which is an approximate algorithm; and
- SUF-growth [17], which is an exact algorithm.

As an approximate algorithm, UF-streaming may produce both false positives and false negatives. As an exact algorithm, SUF-growth resolves the problem of the generation of false positives and false negatives but at a potential price of constructing bushy trees.

All those methods we have discussed so far mine just frequent patterns without considering their importance to applications. Methodologies that consider application specific importance [15,18] cannot handle uncertain stream datasets.

Ahmed et al. [19] proposed a tree—and its corresponding mining approach—which is capable of capturing the data streams as well as generates interesting (weighted) patterns. However, this tree works for certain data, which are the striking difference with us.

Shajib et al. [14] discovered a tree structure and mining methods that are more similar to our task. Although their proposed tree is able to capture the data stream where the items are uncertain, it generates all frequent patterns regardless the importance of patterns. As a result, many unnecessary patterns may be generated while some interesting or important patterns may be excluded because of their low frequency.

In contrast, our method is proposed with an intention to solve above limitations using a novel tree structure along with a mining procedure that helps to capture data streams with uncertain data and find only interesting (weighted) patterns by considering the weighted of each item depending on the applications and users need.

3   Our Proposed Approach

Before driving to the original content it is important to explain some necessary terms that will increase the understandability of later contents.

Definition 1 (Transaction Maximum Probability (TMP)). *TMP of a transaction T_i indicates the highest probability of items X_j (where $X_j \in T_i$), i.e.,*

$$TMP(T_i) = MAX_{1 \leq j \leq n} P(X_j) \tag{1}$$

where

- *n is the length of transaction T_i,*
- *X_j is the item of the transaction T_i, and*
- *$P(X_j)$ denotes the probability of X_j in transaction T_i.*

Example 3. Consider a transaction T_1: (a:0.56, b:0.45, c:0.78, f:0.10). Then, the highest probability of all item is the c's probability. So, $TMP(T_1) = P(c) = 0.78$.

Definition 2 (Local Maximum Probability (LMP)). *LMP of a particular item X_j in some transaction T_i is the product of the probability of item X_j and the $TMP(T_i)$, i.e.,*

$$LMP(X_j) = P_i(X_j) \times TMP(T_i) \tag{2}$$

where $P_i(X_j)$ is the probability of X_j in transaction T_i.

Example 4. Consider the transaction T:(a: 0.5, b: 0.56, c: 0.08). Then, $TMP(T) = P(b) = 0.56$, which is highest probability among all item of T. So,

$$LMP(c) = P(c) \times TMP(T) \tag{3}$$

$$= 0.08 \times 0.56 \tag{4}$$

$$= 0.0448. \tag{5}$$

Definition 3 (Total Maximum Probability (TAMP)). *TAMP of a particular item X is the summation of all LMP of that item in a particular batch, i.e.,*

$$TAMP(X) = \sum_i LMP_i(X) \tag{6}$$

for $1 \leq i \leq |batch|$.

Example 5. Let a particular batch B_1 contain two transaction T_1 and T_2 where

- T_1: (a: 0.3, b: 0.5, f: 0.25), and
- T_2: (a: 0.5, c: 0.2, b: 0.7).

Then, we calculate the $TAMP(b)$ as follows:

$$TAMP(b) = LMP_1(b) + LMP_2(b) \tag{7}$$

$$= (P_1(b) \times TMP(T_1)) + (P_2(b) \times TMP(T_2)) \tag{8}$$

$$= (0.5 \times 0.5) + (0.7 \times 0.7) \tag{9}$$

$$= 0.74 \tag{10}$$

3.1 Overview of Our WFPMUDS-Growth Approach

Like most traditional data-mining procedures, our WFPMUDS-growth approach consists of two major phases:

1. Tree construction, and
2. Mining interesting frequent patterns.

There are several sub steps in first phase, include the following:

- insertion,
- partitioning,
- widow shifting,
- etc.

We will discuss these two major phases in the following sections.

3.2   First Phase: Tree Construction

As we are dealing with uncertain data stream, we cannot just calculate some measurement—such as expSup (expected support) as in PUF-tree—to sort the items in the transaction in order to insert into the tree. Moreover, as we also consider the significance importance (weight) of items, it will be not effective to consider some canonical ordering of items like DStree and US-Tree. We took the advantages of the following two weights to help us in miming operation:

- GMAXW, which is the maximum weight of all items; and
- LMAXW, which is the local maximum weight.

Transactions are sorted according to their weights from lower weight to upper to ensure the higher weighted item are at the bottom of the tree, which helps us in pruning step. Thus, we build our header table for our **Weighted Uncertain Data Stream (WUDS)-tree**. Then, following steps are performed to construct our tree structure.

1. We insert each transaction into tree batch wise. In each node, item id and TAMP of that item is saved. Each transaction is inserted in separate branches of the tree if there is no common node found otherwise they are just merged.
2. When new batch comes, new partition is created to each node to hold the information of new batch.
3. When window size is overflowed by the batch number, the oldest batch is deleted from tree and each batch is shifted by one position to make place for new batch.

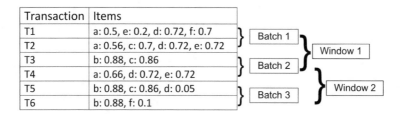

Fig. 1. Uncertain stream database

In first database scan, we read one by one transaction and sorted the items according to their weight. TMP for each transaction is also calculated at that time and inserted into the tree.

Example 6. For demonstration purpose, we used transaction in Fig. 1 where batch size as well as window size is 2. Weight for each item is a: 0.6, b: 0.5, c: 0.4, d: 0.3, e: 0.2, f: 0.1.

When transaction T_1 come, items are sorted according to their weight that is T_1: (f: 0.7, e: 0.2, d: 0.72, a: 0.5) and the $\text{TMP}(T_1) = 0.72$ so each item is inserted into the first partition of the node with their id and LMP. For instance, item f is inserted with

– its id, which is f; and
– value $(0.72 \times 0.7) = 0.504$, which is the LMP($f$).

Similarly, e, d and a are inserted one after another. When T_2 comes, it is started with item e. So, we create a separate branch like FP-tree. Figure 2(a) basically shows how the tree is built after inserting transactions T_1 and T_2. Thus, Batch 1 is inserted into the tree. When Batch 2 arrived, new partition of each node is created, which is shown in Fig. 2(b). Transactions in Batch 2 (i.e., T_3 and T_4) are inserted in the second partition as the same way like Batch 1. Figure 2(c) reflects the changes after insertion of Batch 2.

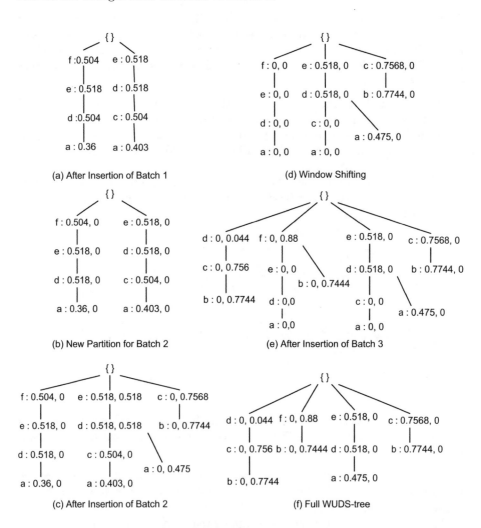

Fig. 2. WUDS-tree construction

As window size is 2, when new batch come at this moment, there is no partition left for this batch. At this point, window shifting is taken place. Batch 1 is deleted from each node, Batch 2 is shifted from partition 2 to partition 1, and Batch 3 is inserted in partition 2. Figure 2(d) shows the effect of window shifting of the WUDS-tree. Now, transaction T_5 and T_6 from Batch 3 are inserted as the same way previous batches did in 2nd partition.

Figure 2(e) reflects the change after T_5 insertion. When all the transactions of Batch 3 are inserted, our final WUDS-tree will look like Fig. 2(e). Then, when the mining request comes at this time, we need to remove all the empty node like the tree in Fig. 2(f).

3.3   Second Phase: Mining Phase

Mining process of WUDS-tree is like FP-growth approach. Like FP-tree mining process is started by the lower item of header table, projected database for the item is built. By removing infrequent item from projected database, conditional tree is built to find the frequent patterns. Recursively mine the conditional tree for longer patterns is performed. Thus, WFPMUDS-growth mines the WUDS-tree to generate the frequent patterns like FP-growth. However, as WFPMUDS is proposed for weighted patterns mining, mining process is not so straightforward like FP-growth. Complexity arises as the weighted patterns do not follow downward closure property. Intelligence is applied here to determine whether a particular item is pruned or consider for longer patterns. The trick is that instead of using WexpSup, we calculate Maximum Weighted Expected support (MWES) which is defined as,

$$MWES(X_i) = TAMP(X_i) \times GMAXW \qquad (11)$$

TAMP of a particular item tells us the highest probability of that item to occur with any other item in the database. So, MWES gives use the highest weighted probability of an item. If MWES of an item does not cross the minimum expected weighted threshold that means this item has no chance to frequent by occurring other items in the database and we can safely prune it from further consideration. If MWES of an item crosses the threshold limit, we consider that for a candidate for frequent patterns, then actual WexpSup (weighted expected support) is calculated for that item. If WexpSup crosses the limit of threshold, then it is considered as frequent patterns.

Example 7. Consider the WUDS-Tree in Fig. 2(f) for describing our mining procedure. Let mining request is come at Window 2 and minimum threshold is 0.2, at this point value of each partition of a particular node is summed. Now, MWES (maximum weighted expected support) for each item is calculated to prune the infrequent items. Here, we found that

$$MWES(f) = TAMP(f) \times GMAXW \qquad (12)$$
$$= 0.088 \times 0.6 \qquad (13)$$
$$= 0.0528 \qquad (14)$$
$$< \text{ minimum weighted expected threshold of } 0.2 \qquad (15)$$

Table 1. *Wsupport* calculation of candidate patterns in Window 2

Item	MWES	WexpSup	Result
a	0.285	0.285	Frequent
b	1.3939	1.1616	Frequent
c	0.907	0.605	Frequent
d	0.337	0.1685	Infrequent but candidate
e	0.337	0.1685	Infrequent but candidate
f	0.0528	Need not calculate	Pruned

Fig. 3. *a* projected database

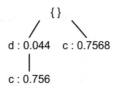

Fig. 4. *b* projected database

Hence, it is pruned from the tree. After that, actual weight support is calculated to determine whether a pattern is actually frequent or not by multiplying TAMP value with actual weight of that pattern. Table 1 shows the details calculation. Now, we will start for looking 2-length frequent patterns by picking lowest entry from our header table that is item 'a' and build a-projected database by taking those branches in where 'a' is appeared. Figure 3 shows the a-projected database. Now, we calculate the maximum and actual weighted support of pattern 'ea' and 'da'. Table 2 shows the details calculation, where actual weighted support of ea is less than threshold. Hence, it is not frequent but MWES crossed the threshold, and thus need further consideration. On the other hand, 'da' is frequent. Now, we will start to find 3-length pattern by taking 'd' projected database, which consists with just one node e: 0.475. MWES of pattern 'eda' = $0.475 \times 0.3 = 0.1425 <$ threshold that means no chance to be frequent in future.

Now, we continue our mining task this time with item 'b'. First, b-projected database is built by taking b's branch. Here when we take b's branch it is enough

Table 2. 2-length frequent patterns

Pattern	Max WexpSup	Actual weight	Actual WexpSup
ea	$0.6 \times 0.474 = 0.285$	0.4	$0.4 \times 0.475 = 0.19$
da	$0.6 \times 0.518 = 0.31$	0.45	$0.45 \times 0.518 = 0.233$

Table 3. Frequent items

Item	Max weighted expSup	Acutal weighted expSup	Result
db	$0.044 \times 0.5 = 0.022$	Need not to calculate	Failed
cb	$(0.756 + 0.7568) \times 0.5 = 0.7564$	$1.51 \times (0.5 + 0.4)/2 = 0.681$	Frequent

to take the lowest TAMP value. For instance, when we take the branch d: 0.044, c: 0.756, b: 0.774, we just take the lowest TAMP value between b and d or c. Because TAMP value indicates the highest probability of a particular item to occur with another item so it is quit enough to chose its own TAMP value when it is smaller, then the TAMP value of item whose projected database is being considered. Figure 4 and Table 3 reflect the b-projected database and 2-length frequent pattern generated from b. In order to calculate maximum weighted expected support we use the weight of b as it to reduce generating false patterns by using LMAX. Here conditional tree of b consists of just one node c: 1.5128 so 3-length patterns cannot be generated. Rest of the tree will be mined in same procedure.

3.4 Analysis of WFPMUDS-Growth Algorithms

Algorithm 1 is the basic procedure of our approach. Uncertain database, weight table, window size and batch size are taken as input at the beginning of the procedure. This procedure generates all the weighted patterns as output. First, using weight table, header table is built in where items are sorted according to their weight.

Afterwards, using the outer loop we actually start to insert each batch into the tree. First if condition in the outer loop check whether the window is full or not. If window is overload it simple delete the oldest batch and shift rest of the batch. After that, inner loop basically inserts each transaction from the batch into the tree by calculating TAM value for that transaction. By this time, if any mining request is come from user, minimum weighted threshold is taken as input. Then, algorithm starts mining using the threshold from the current state of the tree, which is done by the last if condition. At the beginning of mining process, GMAX is calculated from the weight table and mining is started from the bottom of the table. For a particular item-set first, whether the WexpSup of the item-set cross the threshold is checked.

After CandidateTesting procedure is called to generate the frequent patterns, prefix tree is built for that item-set to recursively mine frequent patterns from

Input: (1) Uncertain database, (2) weight table, (3) window size, (4) batch size
Result: List of weighted frequent patterns
initialization;
Create the Header table HT and store the items according to weight descending order.;
foreach *Batch B_i* **do**
    if *Window of tree nodes are already full* **then**
        Delete the oldest batch from nodes;
        Shift each Batch by one position;
    end
    foreach *Transaction $T_j \in B_i$* **do**
        Sort the items in T_j according to HT;
        Calculate TMP_J;
        Insert T_j into Tree;
    end
    if *Any Mining Request From User* **then**
        Input: δ from user
        Calculate GMAXW;
        foreach *item I from Bottom of HT* **do**
            if *$TAMP(I) \times GMAXW \geq \delta$* **then**
                Call CandidateTesting(I,TAMP(I)) ;
                Create Prefix Tree PT_I and Header Table HT_I for item I ;
                Call MiningTree(PT_I , HT_I, I, $Weight_I$) ;
            end
        end
    end
end

Algorithm 1. WFPMUDS-Growth approach algorithms

prefix tree. For each items in the header table, algorithms generate prefix tree and recursively find frequent patterns from prefix tree. CandidateTesting procedure basically finds the actual frequent patterns by calculating actual Weighted Expected support and comparing with the threshold provided by the users. The procedure takes item-set and TAMP for that item-set as input and determines whether the item-set is frequent or not.

Loop in the procedure calculates the total weight of all items belong to the item-set and find the actual weight of the item-set by simply calculating the average weight that is by dividing the total weight by the length of item-set. Then, check whether the actual weighted expected support is more than the threshold or not.

Another procedure of our algorithm is MiningTree which is basically a recursive procedure that is responsible for generating patterns using a particular item, its Prefix tree, header table and LMAX which are given as parameters of the procedure. First, the procedure calculates Maximum WexpSup of the item by multiplying its TAMP value and LMAX value and tests whether the resulted Maximum WexpSup crosses the threshold limit or not. If the item failed, the

Input: Itemset I, Total Maximum Probability of I ($TAMP_I$)
W_I is the actual weight of I ;
Set $W_I = 0$;
foreach *Item $x_i \in I$* **do**
  | $W_I = W_I +$ Weight(x_i) ;
end
$W_I = W_I \div Length(I)$;
if $W_I * TAMP_I > \Sigma$ **then**
  | Add I to the weighted frequent List ;
end

Algorithm 2. Algorithm of CandidateTesting Procedure

test that item is dropped from header table. Otherwise, the procedure creates prefix tree, calls CandidateTesting procedure, and mines the prefix tree. Thus, the procedure recursively produces prefix for particular items and mine frequent patterns until there exist any item in header table. Algorithm 1 provides the basic pseudo code of MiningTree.

4   Experimental Results

To analyze performance of proposed WFPMUDS-growth algorithm, we have conducted experiments on a PC with an Intel Core i3-2100 CPU at 3.10 GHz and 4 GB RAM running MS Windows 10 operating system. Our algorithms are implemented in Java. We analyze performance of our proposed algorithm with respect to runtime and memory requirements for various window size, batch size, GMAX and LMAX. For comparison purpose, we used USPF-growth algorithm as this is the most closely related version of our method.

4.1   Datasets

To evaluate the performance of our proposed work, we used three real life and one synthetic data sets. All the datasets that we used, certain in nature and no weight is associated with each item, are collected from UCI machine learning repository and FIMI repository. For this reason we have included weighted and uncertainty with each item. Some methods that deal with uncertain data, generate probabilistic value randomly for the items [13]. However, we follow normal probabilistic distribution to generate uncertain value depending on the demand of application. Analyzing real life situation patterns whose frequency is average is more demanding rather than extremely frequent patterns or those patterns that are just cross the frequent boundary. That is why we follow normal distribution to assign probability of items by giving highest probability to those items which frequency is average and gradually assign lower probability to less frequent items. For experimental purpose, weights of the items are also generated randomly following normal distribution.

4.2   Effect of Window Size Variation

To perform the runtime analysis of our methods, we conducted several experiment on dataset T1014D100K, Kosarak, Chess and Mushroom. As our proposed
method is sliding window based that captures the most recent data and executes mining to current window, we analyzed runtime and memory requirement
on various window size. For this reason, we vary number of transactions in a
batch and the number of batch in a window to evaluate the corresponding runtime. As the method is about stream data, datasets that contain relatively huge
transaction will be more convenient. That is why we chose T1014D100K and
kosarak to perform runtime analysis while varying batch size and window size.

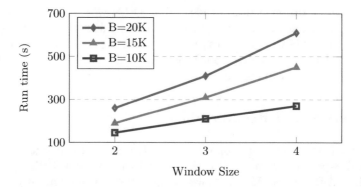

Fig. 5. Effect of window size on runtime (T1014D100k dataset)

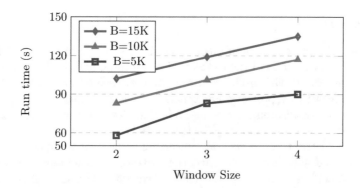

Fig. 6. Effect of window size on runtime (Kosarak dataset)

As mention earlier, T1014D100K is relatively larger dataset containing
100,000 transaction with 870 distinct item and average length of transaction

is 10.1. Figure 5 shows the effect of different window size containing different numbers of transaction in a batch on runtime. The X-axis reflects the window size and Y-axis shows runtime in seconds. We performed analysis for window sizes 2, 3 and 4 where batch size was 10K, 15K and 20K. The mining operation is performed at each window by considering minimum threshold 1.2%. With increase of window size and batch size, runtime also increased. As number of transaction is also increased with batch and window size, which required more time to construct tree, update tree and mine patterns from tree. We also analyze runtime of our algorithm by changing window size and batch size using another dataset Kosarak, which is also good for measuring performance of algorithms that deals with sliding window for its larger data items. Figure 6 reflects the effect of window sizes alone with number on transaction in a batch on runtime. Minimum threshold for this experiment was 0.5%.

4.3   Effect of LMAX and GMAX on Runtime

In order to analyze the runtime of our proposed algorithm another important experiment is made using the concept of LMAX and GMAX. In that experiment we tried to show how LMAX reduce the runtime by eliminating false patterns.

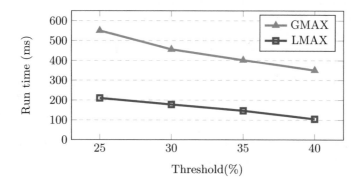

Fig. 7. Effect of LMAX and GMAX on runtime on Chess dataset

Figure 7 described the effect of LMAX and GMAX on runtime using Chess dataset. To perform this experiment we set window size 1 and entire transaction as single batch so that the effect of LMAX and GMAX can be clearly shown. As we recorded the time for entire dataset actually the change of time is easily observed. X-axis of the graph represents the Threshold and Y-axis shows the corresponding runtime. Basically, effect of LMAX depends on the distribution of weight. If the variation of weights is not significant, effect of employing LMAX instead of GMAX is also not be significant. However, in real life variation of weights is significant as different applications are looking only those patterns that have real importance to them. Similar type of analysis is also performed on Mushroom dataset. This time, we set window size 1 and batch size to 8124. Result of this analysis is shown in Fig. 8.

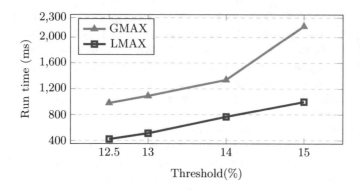

Fig. 8. Effect of LMAX and GMAX on runtime on Mushroom dataset

4.4   Comparison with Existing Algorithms

As the existing algorithms are not suitable for mining weighted frequent patterns from uncertain data streams, for comparison purpose, we choose the closest algorithm, USFP-growth that can only handle uncertain data stream and mine all frequent patterns without judging the importance of patterns to the applications. We keep the necessary element as much similar as possible for fair comparison. We generate probabilistic value of items for each approach using normal distribution. As USFP-growth does not consider weight of items, we set weight of each item to 1 and for our approach we again use normal distribution to generate weight for items. To reflect the difference between two approaches, both dense dataset like chess and sparse dataset like T1014D100K is used. In order to show the runtime difference clearly we set window size to 1 and batch size is equal to the number of transaction of corresponding dataset for ensuring the runtime difference for larger dataset. It is obvious that our approach will take less time to compute frequent patterns as we only generate the interesting patterns.

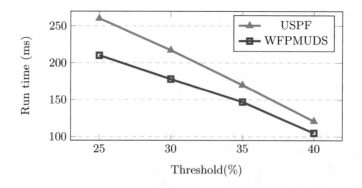

Fig. 9. Runtime comparison on Chess dataset

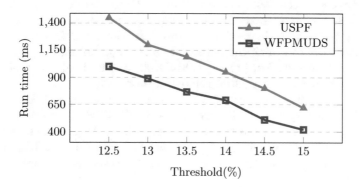

Fig. 10. Runtime comparison on Mushroom dataset

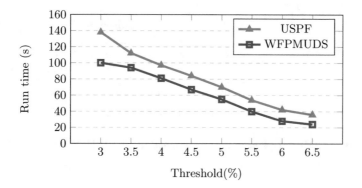

Fig. 11. Runtime comparison on T1014D100K dataset

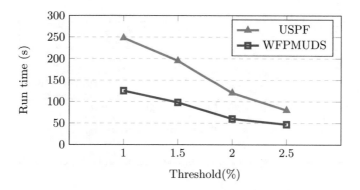

Fig. 12. Runtime comparison on Kosarak dataset

Figure 9 shows the runtime comparison on Chess dataset. X-axis represents the threshold (%) and Y-axis shows runtime. As the dataset is dense, more patterns are generated at the decrease of threshold value and USFP-growth start to take more time to mine the dataset. Similar types of analysis are also performed on datasets like Mushroom, T1014D100k and Kosarak and their runtime comparison is shown in Figs. 10, 11 and 12 respectively.

Another obvious advantage of our algorithm comparing to USFP-growth is it takes less memory than USFP-growth as WFPMUDS-growth only generates those patterns that has significant to the applications. We perform several experiment on memory consumption using T1014D100K and Kosarak dataset. We set different batch size to analyze amount of memory taken by both methods.

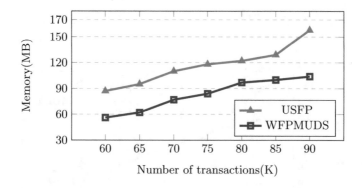

Fig. 13. Memory consumption comparison on T1014D100K dataset

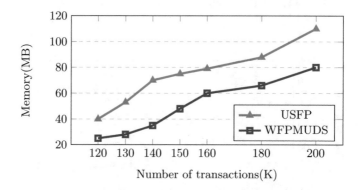

Fig. 14. Memory consumption comparison on Kosarak dataset

Figure 13 shows the memory consumption on T1014D100K dataset. We set Window size to 1 and change the number of transactions in a batch. Thus, we calculate amount of memory required by each algorithms. X-axis represents the number of transaction in a batch and Y-axis denotes amount of memory used by the methods. Figure 14 shows same time of comparison but on Kosarak dataset. In both datasets, WFPMUDS-growth took less memory than USFP-growth. In real life application, memory requirement will be decreased based on the patterns they will be looking for.

5   Conclusions

In this research work, we proposed a novel tree structure, which successfully and efficiently captures uncertain data stream, and a mining algorithm that extract only significant patterns from most recent window. Our main contribution is to mine only important patterns rather than all frequent data without considering their necessity, as recent world applications are hungered only for relevant patterns. Focusing on this demand, we have developed a novel approach called WFPMUDS-growth, which solves the problem of finding significant patterns from uncertain data stream. As our algorithm generates only important patterns, it is less time consuming and also uses less memory resources. As ongoing and future work, we are exploring solutions that aim to generate interesting patterns by not only considering weight of items but also using some other constraints.

Acknowledgments. This project is partially supported by NSERC (Canada) and University of Manitoba.

References

1. Srikant, R., Agrawal, R.: Mining sequential patterns: generalizations and performance improvements. In: EDBT 1996, pp. 1–17 (1996)
2. Pei, J., Han, J., Mortazavi-Asl, B., Wang, J., Pinto, H., Chen, Q., Dayal, U., Hsu, M.: Mining sequential patterns by pattern-growth: the PrefixSpan approach. IEEE TKDE **16**(11), 1424–1440 (2004)
3. Choi, P., Hwang, B.: Dynamic weighted sequential pattern mining for USN system. In: ACM IMCOM 2017, pp. 19:1–19:6 (2017)
4. Hsu, K.: Effectively mining time-constrained sequential patterns of smartphone application usage. In: ACM IMCOM 2017, pp. 39:1–39:8 (2017)
5. Rahman, M.M., Ahmed, C.F., Leung, C.K., Pazdor, A.G.M.: Frequent sequence mining with weight constraints in uncertain databases. In: IMCOM 2018, pp. 48:1–48:8 (2018)
6. Rasheed, F., Alshalalfa, M., Alhajj, R.: Efficient periodicity mining in time series databases using suffix trees. IEEE TKDE **23**(1), 79–94 (2011)
7. Li, Y., Yeh, J., Chang, C.: Isolated items discarding strategy for discovering high utility itemsets. DKE **64**(1), 198–217 (2008)
8. Gunawan, D., Mambo, M.: Set-valued data anonymization maintaining data utility and data property. In: IMCOM 2018, pp. 88:1–88:8 (2018)
9. Yun, U.: Efficient mining of weighted interesting patterns with a strong weight and/or support affinity. Inf. Sci. **177**(17), 3477–3499 (2007)

10. Leung, C., Khan, Q.I.: DSTree: a tree structure for the mining of frequent sets from data streams. In: IEEE ICDM 2006, pp. 928–932 (2006)
11. Leung, C., Mateo, M.F., Brajczuk, D.A.: A tree-based approach for frequent pattern mining from uncertain data. In: PAKDD 2008, pp. 653–661 (2008)
12. Goodman, S.H.: Cup tree. US Patent D393,783, 28 April 1998
13. Leung, C., Tanbeer, S.: PUF-tree: a compact tree structure for frequent pattern mining of uncertain data. In: PAKDD 2013, Part I, pp. 13–25 (2013)
14. Shajib, M., Samiullah, M., Ahmed, C., Leung, C., Pazdor, A.: An efficient approach for mining frequent patterns over uncertain data streams. In: IEEE ICTAI 2016, pp. 980–984 (2016)
15. Ahmed, C., Tanbeer, S., Jeong, B., Lee, Y.: Handling dynamic weights in weighted frequent pattern mining. IEICE Trans. Inf. Syst. **91**(11), 2578–2588 (2008)
16. Leung, C., Hao, B.: Mining of frequent itemsets from streams of uncertain data. In: IEEE ICDE 2009, pp. 1663–1670 (2009)
17. Outten, F.W., Djaman, O., Storz, G.: A suf operon requirement for Fe-S cluster assembly during iron starvation in Escherichia coli. Mol. Microbiol. **52**(3), 861–872 (2004)
18. Ahmed, C., Tanbeer, S., Jeong, B., Lee, Y., Choi, H.: Single-pass incremental and interactive mining for weighted frequent patterns. Expert Syst. Appl. **39**(9), 7976–7994 (2012)
19. Ahmed, C., Tanbeer, S., Jeong, B.: Efficient mining of weighted frequent patterns over data streams. In: IEEE HPCC 2009, pp. 400–406 (2009)

Distributed Secure Data Mining with Updating Database Using Fully Homomorphic Encryption

Yuri Yamamoto$^{(\boxtimes)}$ and Masato Oguchi

Ochanomizu University, 2-1-1 Otsuka, Tokyo 1120012, Japan
yuri@ogl.is.ocha.ac.jp, oguchi@is.ocha.ac.jp

Abstract. Uploading commercial data to third-party cloud services is popular in general. To further promote the active utilization of big data, outsourcing data mining systems that can execute statistical calculations using the uploaded data have been proposed. In this case, personal and sensitive data are required to be encrypted for privacy protection. In previous research, data protection using fully homomorphic encryption (FHE) was proposed for a client/server secret data mining system using the Apriori algorithm. However, this system requires much time because of the computational complexity of FHE calculations. Additionally, although frequent database updates occurred in the practical use of the system, the Apriori algorithm needs recalculation of the whole database at each update. In this study, to solve these two problems, we proposed the implementation of a master/worker distributed system using the FUP algorithm, which generates candidate item sets efficiently while updating the database. We improved execution time of the secure data mining system and made it suitable for practical use.

Keywords: Secure data mining system ·
Fully homomorphic encryption · FUP algorithm · Distributed system

1 Introduction

Data mining is the process of extracting interesting and useful information or characteristic patterns from enormous amounts of data [1, 2]. For example, market basket analysis of consumer purchase histories is a famous example of the commercial utilization of data mining. To accumulate large amounts of data and calculate the statistics, data mining requires large capacity storage and high-performance computers in the computational environment. We think that an outsourcing data mining system, in which organizations can send their data to third-party cloud services and query them and then obtain the statistics from the data, will be widespread in the future. However, since personal customer data owned by companies is confidential and proper security management is required, it is thus desirable to encrypt the data to be concealed from anyone.

© Springer Nature Switzerland AG 2019
S. Lee et al. (Eds.): IMCOM 2019, AISC 935, pp. 937–949, 2019.
https://doi.org/10.1007/978-3-030-19063-7_73

The use of public key encryption is the most common method, but it is not possible for a third party to perform calculations on the data unless they decrypt using a secret key. Additionally, k-anonymization and perturbation are often used as secure data mining techniques, but they protect either the input data or the output data, and there is a possibility of an ambiguous result. For these reasons, the techniques are not suitable for analyzing data safely and obtaining accurate results [3]. Therefore, in recent years, there has been an increase in studies of secure outsourcing systems that utilize fully homomorphic encryption (FHE), which enables polynomial arithmetic of encrypted data.

Previous studies [3,4] proposed a secure data mining protocol using FHE that supports both addition and multiplication of each encrypted data, and the authors make an improvement for accelerating the system by cashing and packing the data as vectors. The authors adapt the Apriori algorithm [5] to calculate the correlation of the items in the database.

However, the calculation time on the server side tends to be excessive because of the large computational complexity of the FHE. Although frequent updates of database occur in the practical use of the system, the Apriori algorithm needs recalculation over the whole database at each update, which takes much useless time.

In this research, we applied the FUP algorithm [6] to improve the optimization of the recalculation while database updating, and adopted a master/worker distributed system for the calculation of the server side to improve the runtime.

The contributions of this paper are as follows: First, we consider the problem of database updating to the outsourcing data mining system for practical use. Recalculation of over the whole database using FHE calculation of each updating is not efficient, so it is not suitable for practical use to adapt a native Apriori algorithm for the system especially using heavy FHE. Second, we adapt the FUP algorithm to reduce candidate item sets of association rule mining. It is important to reduce FHE calculations of the candidate item sets, because the server side should complete calculations without knowing values in the database. Finally, we implement a master/worker distributed system to improve runtime to make the system more suitable for practical use, execute experiments and measure environmental system time, which consists of general Linux machines for cloud computing platforms. The final product results in highly versatile acceleration for practical use.

2   Fully Homomorphic Encryption

FHE is a cipher that supports both the addition and multiplication operations of ciphertexts such as expressions (1) and (2). This property enables the polynomial calculation of ciphertext similar to plaintext. Although FHE provides the function of public key cryptography, it is possible to obtain the encrypted results of the operation on the plaintext from the operation on ciphertext without using a secret key.

$$Encrypt(m) \oplus Encrypt(n) = Encrypt(m + n) \tag{1}$$

$$Encrypt(m) \otimes Encrypt(n) = Encrypt(m \times n) \tag{2}$$

Public key cryptography was first devised in the latter half of the 1970s, and Gentry [7] proposed the method of the FHE algorithm in 2009. First, it appeared to be difficult to use in applications because of its long runtime, but various improvements have been made to the FHE algorithm to make it adequate for simple calculations. However, there is difficulty with the large computational complexity of FHE calculations. The size of the FHE ciphertext tends to be large because it is necessary to add noise to maintain the decryption difficulty of the calculations on the server side; it is also a means of removing the noise at the time of client decryption calculation.

3 Apriori Algorithm

The Apriori algorithm was proposed by Agrawal et al. [8], [5] as an efficient method of extracting association rules of items in a database; it is a well-known algorithm used for frequent data mining, such as association analysis. In the case of the purchase history, the frequency of the item sets, which are the subsets of all items is calculated from a database in which customers purchase transactions are recorded in a binary expression. The frequency is the ratio of the number of item set purchase transactions in all transactions, which is expressed as a support value. The Apriori algorithm derives frequent item sets by repeating the calculation of the support value and pruning while increasing the length of the item sets, as shown in Fig. 1.

We formulate the relationship between the item sets and support values of that subset. Suppose that the item is represented by I_k and the item set $X + Y$ is the item set that adds X to $Y = I_1 I_2 ... I_k$. Additionally, there are x item sets that contain X, among which c pairs is the number of other item sets generated before calculating the item set $X + Y$. We also assume that the relative frequency of item I_j is $f(I_j)$ and the number of transactions is $DB.size$, while the support value $\bar{s}$ of item set $X + Y$ is formula (3).

$$\bar{s} = f(I_1) \times f(I_2) \times ... \times (x - c) / DB.size \tag{3}$$

If the support value of X is small, then the support value of $X + Y$ is small. As a result, item sets that contain infrequent item sets are not frequent. Therefore, Apriori is an efficient mining of association rules algorithm because it stops searching of item sets whose support values do not exceed a minimum support.

4 FUP Algorithm

The Fast UPdate (FUP) algorithm was proposed by Cheung et al., and it is an algorithm that efficiently scans a database when using Apriori calculation while

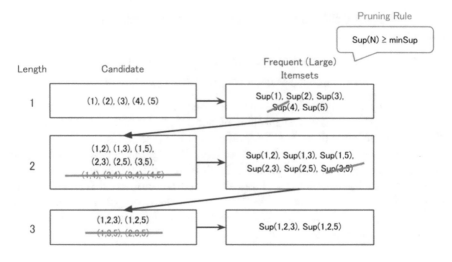

Fig. 1. Apriori algorithm

updating the database [6,9]. Apriori calculates the support value of static data, but when transactions are added to the database, depending on the bias of the transactions, the frequent item sets change at each stage. Then, recalculation of the support value is required. The FUP algorithm reduces the recalculation cost by reusing the frequent item sets of the pre-update database and the pre-update threshold.

We represent the databases before the update and the increasing part as DB, db, and the number of each transactions as $DB.size$, $db.size$, respectively, while i-length candidate item sets are of C_i, and the frequent item sets (large item sets) are denoted by L_i. The support value of item set X is denoted by $Support(X)$. Although the term "support value" differs in meaning in various papers, in this paper, it is distinguished from the support count, which is the number of occurrences. The support value is defined as the ratio of the number of occurrences to the total number of transactions. Additionally, the threshold of the support value (the minimum support value) is expressed as $minS$. The procedure of the FUP algorithm is as follows.

1. $i = 1$
    (a) On the updated database UD, let the sets X following Eq. (4) be L_1' in the pre-update frequent item sets $X \subseteq C_1$.
    (b) Remove the pre-update frequent item sets L_1 from the pre-update candidate item sets C_1, and let C_1' be the item sets that follow the Eq. (5) among X such that $X \subseteq C_1$ on db.
    (c) On the updated database UD, X such that $X \subseteq C_1'$ and following Eq. (4) is added to L_1'; then, let L_1' be the 1-length updated frequent item sets.

2. $i > 1$

    (a) Remove item sets X following $X \subseteq L_{i-1} - L'_{i-1}$ from the pre-update i-length frequent item set L_i

    (b) On the updated database UD, let item sets X following Eq. (4) such that $X \subseteq L_i$ be L'_i.

    (c) On the updates of database db, let the item sets from removing i-length pre-update frequent item sets L_i from i-length item sets, which is generated from L'_{i-1} by the Apriori algorithm, be C_i; then, let X such that $X \subseteq C_i$ following Eq. (5) be C'_i

    (d) On the updated database UD, add sets X such that $X \subseteq C'_i$ following Eq. (4) X to L'_i, and let L'_i be the updated i-length frequent item sets.

$$Support(X)_{DB \cup db} \geq minS \times (DB.size + db.size)/DB.size \qquad (4)$$

$$Support(X)_{db} \geq minS \times db.size/DB.size \qquad (5)$$

5 Previous Research

In this section, an overview of the secure mining of association by FHE as implemented by previous research is provided [3].

5.1 Privacy Preserving Protocol for Counting Candidates

Liu et al. designed a Privacy Preserving Protocol for Counting Candidates (P3CC) as a secure frequent pattern mining over FHE systems [10]. The system is designed as a server/client system that performs Apriori [5] calculation for basket data expressed in binary of 0 or 1 to indicate whether there was a purchase for each transaction. The client who owns the data encrypts each binary data one by one and transmits it to the server. The server that receives the encrypted data performs the Apriori calculation using FHE without knowing the contents of the data. Since the server calculates support values as the result in the encrypted form, the client can safely outsource the data and calculations. Although FHE has the functions of addition and multiplication, a comparison of each encrypted data is extremely difficult. Therefore, to execute comparison with the minimum support values of the Apriori algorithm, the client decodes the results and executes the calculations. As a result, the communication between the server and the client occurs at the maximum for the number of items.

5.2 Polynomial CRT Packing and Caching

Imabayashi et al. [3,4] proposed an efficient frequent pattern mining algorithm over FHE as an improved method of P3CC. To reduce the time and space complexities of the calculations, they adapted polynomial CRT packing, which encrypts multiple transactions as a vector for each itemset and performs FHE calculations over each of the vectors. It utilizes the Chinese Remainder Theorem (CRT) to multiple polynomials as the framework of the FHE instead of the integer-based DGHV scheme in P3CC.

Additionally, the authors proposed optimization by caching of the encrypted support values that were calculated at each stage. Since the Apriori algorithm performs specific repetition of calculations, caching is an efficient technique for reducing the number of FHE calculations [3,4].

Although these methods greatly improved the computation time of the P3CC method, computational complexity of the FHE calculation of each encrypted data is still excessive, and thus, we are working to further reduce the computational complexity of the secure outsourcing data mining system for practical use. In particular, there is room for improvement in the runtime on the server side, where there are many calculations of encrypted data.

6   Proposed Method

We propose a distributed secret FUP calculation system using FHE and perform the following procedure while updating database. The calculation formulas used in the algorithm are those presented in Sect. 4. An overview of the proposed system is shown in Fig. 2.

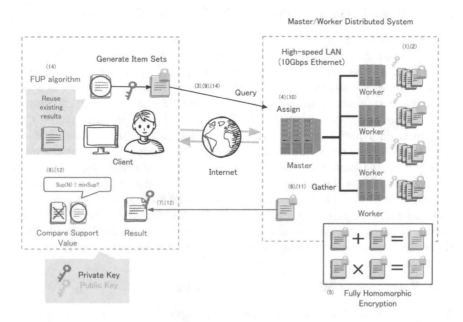

Fig. 2. Overview of the proposed system

(1) The client encrypts data and sends it to the master with the public key.
(2) The master transfers the received data to each worker and the public key.
(3) The client requests the master to calculate support values of candidate item sets and filter frequent item sets before the update. It calculates the support values of nonfrequent itemsets on updates of the database which is small.
(4) The master allocates tasks to each worker on updates of the database (Sect. 4(1)(b)).
(5) Each worker executes the FHE calculation of the item sets.
(6) The master collects the results on the updates of the database and sends them to the client.
(7) The client obtains the results by decoding the received data.
(8) The client compares the results and minimum support value.
(9) The client sends the item sets that exceed the minimum support value and frequent itemsets before the update to the master.
(10) The master allocates tasks about the received item sets to each worker on the entire of database (Sect. 4(1)(a)(c)).
(11) The master collects the results and sends them to the client.
(12) The client obtains the results by decoding the received data and then compares the results and minimum support value.
(13) The client generates an extended set of the received item sets from the results and existing results.
(14) The client sends the item sets that exceed the minimum support value to the master.
(15) Steps (3)–(14) are also repeated in the case of Sect. 4(2) until no item set exceeds the threshold or item set is the longest, and the client obtains the large item sets of the database.

6.1 Division Method

As a method of dividing the tasks of an application, we consider the following three approaches: (1) divide the item sets, (2) divide the highly independent calculations such as the Σ calculation, and (3) divide the transactions. As described in Sect. 5.2, the runtime of the addition and multiplication of the encrypted data on the server side is the problem. The parts of the FHE calculation on the server side are as follows: (a) multiplication of the packing vectors among multiple items to calculate the correlation, (b) calculation of the total value of frequent occurrences in the packing vector, and (c) total the values for each packing vector added by each item set. To divide each of the three tasks (a), (b) and (c), we adopted method (1) to divide the item sets. Additionally, we propose a master/worker distributed system to preserve the environment on the client side.

6.2 Implementation

These protocols were implemented in C++. The FHE calculations of this system are implemented with HElib [11], which is a homomorphic encryption calculation public library on GitHub. We also use Open MPI [12], one library implementation of Message Passing Interface (MPI) [13], which is a standard control and communication system for machines in distribution approach. To use the MPI library efficiently, we use Boost MPI [14] from the C++ extension library.

7 Experimental Results and Discussion

7.1 Experimental Environment

The program of the proposed system was executed on 4 homogeneous machines with an Intel® Xeon® Processor E5-2643 v3 3.4 GHz, 6 cores, 12 threads, memory capacity 512 GB, RAID 0 SSD 480 GB, HDD 2 TB. One machine has the master function and simultaneously has one slot with the worker function, while the other machines each have 2 slots with the worker functions. We conducted experiments to compare the execution time for different numbers of workers by operating with workers in up to seven slots. Similar to previous studies, the generated data by the generator developed by IBM Almaden Quest research group was used as the frequent pattern mining data sets.

7.2 Evaluation of the Execution Time of the Distributed Secret Apriori Calculation

We operated a program for the secret Apriori calculation on the distributed system using data of 50 items and 19,800 transactions. A graph of the execution time for each number of workers is shown in Fig. 3.

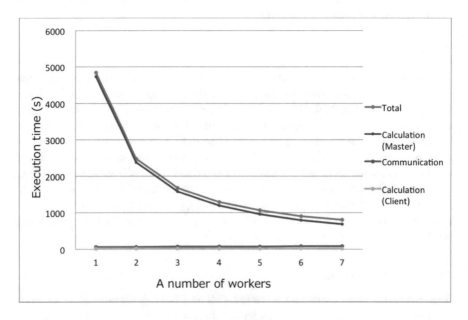

Fig. 3. Execution time of the secret Apriori calculation by the number of workers(s)

As the number of workers increased, the total execution time was reduced, as shown in Fig. 3. In particular, the distribution effect is remarkable in the calculation time on the master side. In addition, as the number of workers increases, the waiting time, including the decoding time of the results by the client and communication time, does not change much. Thus, the overhead in the mutual communication with the client is considered to be small.

7.3 Evaluation of the Execution Time of the Distributed Secret FUP Calculation

We operated a program for a secret FUP calculation on the distributed system using the data of 50 items. A total of 19,470 transactions were used as pre-update data, and 19,800 transactions were used as post-update data. In addition, we operated a program for secret Apriori recalculation using the data of 19,800 transactions to compare. A graph of the execution time for each number of workers is shown in Fig. 4. A graph of comparison of the calculation is shown in Fig. 5.

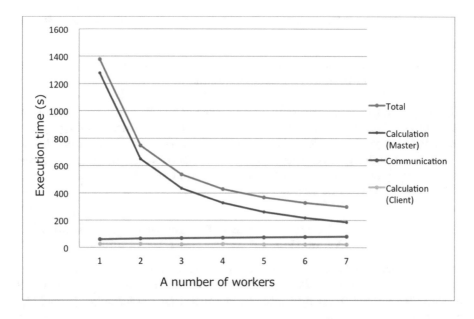

Fig. 4. Execution time of the secret FUP calculation by the number of workers(s)

Figure 4 shows that the total execution time is reduced as the number of workers increased and the distribution effect is remarkable in the calculation time on the master side and similar to the secret Apriori calculation. In the recalculation using the FUP algorithm and the recalculation starting from the beginning by the Apriori algorithm, the calculation time of FUP was approximately 3 to 4 times faster than the Apriori recalculation, as shown in Fig. 5.

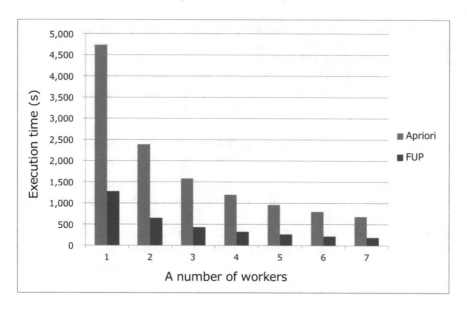

Fig. 5. Comparison of Apriori and FUP calculation time by the number of workers(s)

FUP is the algorithm that reduces the number of candidate item sets and shortens the calculation time by recycling the results of the pre-update database in the recalculation on the updated database. Cheung et al. [6] argues that the FUP algorithm can be expected to be approximately 3 to 7 times faster than the Apriori algorithm. Additionally, in the secret Apriori data mining system using FHE, it was a problem that the calculation time for the encrypted data of each candidate item set was excessive. Therefore, it is considered that the calculation time has been greatly reduced by applying the FUP algorithm to reduce the item set to be calculated.

7.4 Evaluation of the Distribution Efficiency of the Distributed Secret FUP Calculation

As the evaluation of decentralization, we calculate the efficiency from the formula (6) [15] and compare the relationships of the degree of parallelism and decentralization efficiency by Amdahl's law based on the formula (7), and the results are shown in Fig. 6.

$$\text{Efficiency} = \text{Sequential Execution Time (s)}$$
$$/ \text{Parallel Execution Time (s)} \tag{6}$$

$$\text{Efficiency} \leqq \frac{1}{(1 - pctPar) + \frac{pctPar}{p}} \tag{7}$$

Where pctPar is the percentage of the execution time that will be run in parallel, and p is the number of cores on which to run the parallel application.

Dashed lines in Fig. 6 show the efficiency calculated with various percentages of parallel execution time from formula (7).

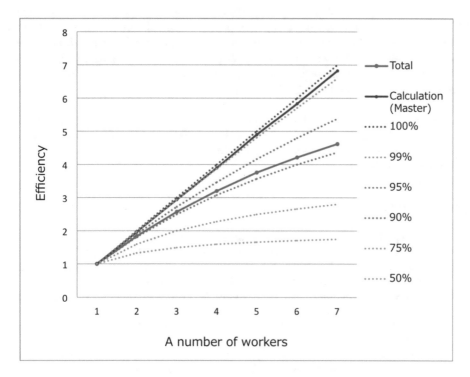

Fig. 6. Efficiency of the experiment using FUP algorithm and efficiency of each parallelism based on Amdahl's law

Figure 6 shows that the runtime of the FHE calculation on the master side is divided at almost 99% to 100% parallelization. However, the total execution time, including the communication time, the calculation time on the client side for encryption and decryption of the data, and the time for file I/O is approximately 90% parallelization. Therefore, we assume from the formula (7) that the maximum efficiency is approximately 10 when increasing the number of workers.

8   Conclusion

We designed a master/worker distributed system for secret data mining of FUP algorithm using FHE for practical use. The calculation task is divided by item sets. As a result, it is possible to reduce the updating calculation time by approximately 3 to 4 times compared with recalculation by the Apriori algorithm. The calculation time on the master side decreased according to the number of workers because of the distribution. In addition, we compared the result with the ideal

efficiency calculated by Amdahl's law. The distributed secret FUP calculation
is expected to have a maximum efficiency of approximately 10 when increasing
the number of workers.

In the future, we will design a distributed system that performs different calculations by workers for the pre-update database and the post-update database
of the secret FUP calculations. Additionally, we will consider depth-first search
strategies instead of breadth-first search strategies for the Apriori algorithm.
We aim to further improve the efficiency of the secret mining of association rules
using FHE while updating the database for practical use.

Acknowledgment. The author would like to thank members of the Yamana Laboratory of Waseda University and the Yamaguchi Laboratory of Kogakuin University for
their valuable advice. This work was partly supported by JST CREST Grant Number
JPMJCR1503, Japan.

References

1. Zhao, Q., Bhowmick, S.S.: Association rule mining: a survey. Technical report,
   CAIS, Nanyang Technological University, Singapore, (2003116):1 (2003)
2. Chen, M.-S., Han, J., Philip, S.Y.: Data mining: an overview from a database
   perspective. IEEE Trans. Knowl. Data Eng. **8**(6), 866–883 (1996)
3. Imabayashi, H., Ishimaki, Y., Umayabara, A., Sato, H., Yamana, H.: Secure frequent pattern mining by fully homomorphic encryption with ciphertext packing.
   In: International Workshop on Data Privacy Management, pp. 181–195. Springer
   (2016)
4. Imabayashi, H., Ishimaki, Y., Umayabara, A., Yamana, H.: Fast and space-efficient
   secure frequent pattern mining by FHE. In: 2016 IEEE International Conference
   on Big Data (Big Data), pp. 3983–3985, December 2016
5. Agrawal, R., Srikant, R., et al.: Fast algorithms for mining association rules.
   In: Proceedings of the 20th International Conference on Very Large Data Bases,
   VLDB, vol. 1215, pp. 487–499 (1994)
6. Cheung, D.W., Han, J., Ng, V.T., Wong, C.Y.: Maintenance of discovered association rules in large databases: an incremental updating technique. In: Proceedings of
   the Twelfth International Conference on Data Engineering, pp. 106–114, February
   1996
7. Gentry, C., et al.: Fully homomorphic encryption using ideal lattices. In: STOC,
   vol. 9, pp. 169–178 (2009)
8. Rakesh, A., Imieliński, T., Swami, A.: Mining association rules between sets of
   items in large databases. In: ACM SIGMOD Record, vol. 22, pp. 207–216. ACM
   (1993)
9. Cheung, D.W.-L., Ng, V.T.Y., Tam, B.W.: Maintenance of discovered knowledge:
   a case in multi-level association rules. In: KDD, vol. 96, pp. 307–310 (1996)
10. Liu, J., Li, J., Xu, S., Fung, B.C.M.: Secure outsourced frequent pattern mining by
    fully homomorphic encryption. In: International Conference on Big Data Analytics
    and Knowledge Discovery, pp. 70–81. Springer (2015)
11. Shoup, V., Halevi, S.: HElib. http://shaih.github.io/HElib/index.html. Accessed
    Jan 2017
12. Open MPI. https://www.open-mpi.org/. Accessed Jan 2017

13. Pacheco, P.S.: Parallel Programming with MPI. Morgan Kaufmann, San Francisco (1997)
14. Boost. http://www.boost.org/. Accessed Jan 2017
15. Breshears, C.: The Art of Concurrency: A Thread Monkey's Guide to Writing Parallel Applications. O'Reilly Media Inc., Sebastopol (2009)
16. Halevi, S., Shoup, V.: Algorithms in HElib. Cryptology ePrint Archive, Report 2014/106 (2014). http://eprint.iacr.org/2014/106
17. Ishimaki, Y., Imabayashi, H., Yamana, H.: Private substring search on homomorphically encrypted data. In: IEEE International Conference on Smart Computing (SMARTCOMP), pp. 1–6. IEEE (2017)

LSTM-Based Recommendation Approach for Interaction Records

Yan Zhou$^{(\boxtimes)}$ and Taketoshi Ushiama

Kyushu University, 4-9-1 Shiobaru, Minami-ku, Fukuoka, Japan
y.zhou.254@s.kyushu-u.ac.jp,
ushiama@design.kyushu-u.ac.jp

Abstract. Interactive platforms such as Spotify and Steam currently play an increasingly important role on the Internet. Users continuously use the content on these platforms. Therefore, the most important data in interactive platforms are interaction records, which contain an enormous amount of information regarding user interests at any given time. However, previous recommendation approaches have been unable to process such records satisfactorily. Therefore, we propose an LSTM-based recommendation approach for interaction records. In our approach, we used a recurrent neural network (RNN) based on LSTM to make recommendations by learning user interests and their changing trend. We propose a pretreatment called serial filling at equal ratio to apply LSTM. Further, we used a dimensionality reduction technique based on matrix factorization to improve the system efficiency. Finally, we evaluated our approach using Steam datasets. As indicated by the results, our approach performs better than other conventional approaches in three aspects: Accuracy, efficiency, and diversity.

Keywords: Recommender system · Interaction records · LSTM

1 Introduction

Over the last few years, interactive platforms have been playing an increasingly important role in the field of Internet service. Spotify, which is a digital music service, and Steam, which is the most popular digital distribution platform for video games, are examples of interactive platforms. Users continuously use the content on these platforms and leave behind their interaction records. An example of an interaction record would be the records of usage, which consists of tracks heard by a user and the usage time. A time series is a series of values obtained by observing continuously or at regular intervals [1]. Therefore, an interaction record can be regarded as a time series.

Recommender systems provide suggestions to help users find the desired information from a huge catalogue. Collaborative filtering is one of the most widely used recommendation algorithms. Its high performance has been verified by many studies. Conventional collaborative filtering is based on static ratings, i.e., a rating matrix [2]. However, for time series data such as interaction records, user interests are dynamic and change with time [3]. Conventional collaborative filtering algorithms usually do not consider such changes. Therefore, it is difficult to select the items that match user interests at the time of recommendation. Many approaches have been proposed to solve

© Springer Nature Switzerland AG 2019
S. Lee et al. (Eds.): IMCOM 2019, AISC 935, pp. 950–962, 2019.
https://doi.org/10.1007/978-3-030-19063-7_74

this problem. However, it is difficult to use the features and latent information from a time series appropriately for conventional recommendation approaches.

In this study, we mainly considered the following four issues regarding interaction records:

- Discovering user interests and their changes from interaction records. This is the most important issue addressed in this study.
- Dealing with implicit feedback. Interaction records when not linked to ratings of items directly is called implicit feedback. We need a method to infer the ratings from interaction records [4].
- Balancing long-term and short-term interests. Every interest has its own lifecycle, which can generally be divided into long-term and short-term interests [5]. It is important to balance them while predicting interests for a user.
- Handling the suboptimal interaction records. Most interactive platforms do not record all events and their exact time. We need to pretreat the records before analyzing them.

We proposed a recommendation model with a RNN based on LSTM for interaction records. RNNs have shown high performance on learning time series. In many RNN units, LSTM proved to be one of the best units because it can learn both long-term and short-term patterns from the time series by using independent memories [6, 7]. Although some RNN units are similar to LSTM [8], they have few differences among them regarding performance [9]. Therefore, we simply chose the basic LSTM to create our model. To use LSTM as a recommendation model, we solved two key issues: normalizing training sets for the best training performance and reducing the system cost for recommendation efficiency.

The remainder of this paper is organized as follows: A brief description of previous related works is presented in Sect. 2. Sections 3 and 4 detail the recommendation approach for interaction records. Descriptions of the datasets, experiments, and discussion are provided in Sect. 5. Finally, the conclusions of this paper are provided in Sect. 6.

2 Related Works

Most traditional recommender algorithms, i.e., memory-based and model-based collaborative filtering, are based on static data. Therefore, researchers previously focused on temporal information in time series data. For example, Koren [10] improved the SVD++ algorithm by considering temporal information in datasets. Ding et al. [3] argued that user decision is mainly affected by recent interests and increased the weight of recent ratings in their recommender algorithms. Gordea et al. [11] proposed a method to set a penalty threshold for rating data. If the time difference between the rating and recommending is greater than this threshold, the rating is no longer considered in the recommender system. Baltrunas et al. [12] proposed a temporal factor-based collaborative filtering model. They trained the model using only the rating data that appeared during the same period. Other studies developed dynamic recommendations based on temporal factors. Zimdars et al. [13] transformed rating data into orderly sets for making recommendations. Lu et al. [14] believed that the features of both users and items change over time and proposed a spatio-temporal model for collaborative filtering recommendations. This model is based on matrix factorization that uses a spatio-temporal filtering approach to estimate user and item factors.

Further studies on time series focused on online, dynamic, and real-time recommendations. Lommatzsch et al. [15] focused on the context information in real time. They proposed an approach combining several types of recommendation algorithms. By analyzing their performance for different contexts, the system can select the relatively best recommendation algorithm according to the context condition in real time. Chen et al. [16] developed a recommender engine from content sources, topic interest models for users, and social voting, by analyzing information streams. Subbian et al. [17] used a hash table to improve recommendation efficiency of item-based collaborative filtering to handle streaming data. Most online and real-time recommendation algorithms ignored the vast majority of historical data to make up-to-date recommendations and reduce the time consumption. Chang et al. [18] and Kille et al. [19] proposed their approaches to combine online and offline recommendations for balancing information loss and recommendation efficiency.

Other researches considered temporal factors as context. Gorgoglione et al. [20] proved that the recommendations using context are easier to obtain trust from users. Panniello et al. [21] proposed a recommendation method considering various time contexts.

Session-based recommendation is similar to our approach based on interaction records. However, the difference between them is that session-based recommendations [22, 23] are usually deployed online and are based on short-term interactions. In contrast, our approach applies relatively long-term interaction records and can be deployed offline.

Interaction records, which are the data that we used, are typical examples of implicit feedback. In case of collaborative filtering, researchers must dig out useful information from such implicit feedback, rather than using it directly. Many studies [24, 25] have proposed collaborative filtering approaches for implicit feedback. However, for recommendations with deep learning, neural networks have been proven to handle such feedback successfully [22, 25, 26]. Therefore, our main task was to set up appropriate datasets to train the neural network model.

3   Data Preparation

3.1   Sliding Window with Interaction Records

Generally, interaction records $\{e_1, e_2, e_3 \cdots\}$ are composed of observed interactions $e = (u, i, t_{start}, t_{end})$, w here $u \in \mathbf{U}$ denotes a user, $i \in \mathbf{I}$ denotes an item, and t_{start} and t_{end} represent the start and end time of the interaction, respectively. Figure 1 shows an example of the interaction records of a user. The top arrow indicates the timeline. Every short line behind the items $\{i_1, i_2, i_3, \cdots\}$ represents the interaction time between the user and items. The blank part on the timeline is blank time; in most cases, blank time occupies most of the timeline.

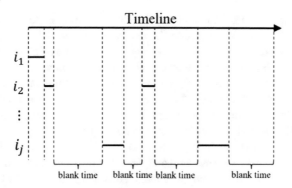

Fig. 1. Example of interaction records

We can use a sliding window to transform such interaction records into time series that can be used to train a neural network model [27]. As shown in Fig. 2, a time window was created based on a given size l and moved at the given slide rate d, where $0 < d \leq l$. On following this process, several copies of the time window still remained. In each copy $t \in \{0, 1, 2, \cdots\}$, we can add interaction time for each item and store it into a vector $\mathbf{r}_u^t$. Consequently, the interaction records $\{(u, i, t_{start}, t_{end}) | u \in \mathbf{U}, i \in \mathbf{I}\}$ are transformed into time series vectors $\{\mathbf{r}_u^t | u \in \mathbf{U}, t \in \{0, 1, 2, \cdots\}\}$.

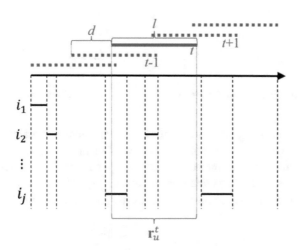

Fig. 2. Sliding window on the interaction records

3.2 Recomposing Interaction Records Datasets

As shown in Fig. 1, blank time occupies most of the timeline for most users. As Fig. 2 indicates, there exists only a short interaction time during most of the windows, which makes the time series vectors sparse, and makes it harder for the neural network model to learn patterns from datasets. Furthermore, the length of blank time in each window is different, which means that the interaction time during any specific period is susceptible to realistic conditions such as whether the user was busy or not during that particular period. This interference information has significant negative effect on the learning performance. Besides, the sliding window technique can only be used on ideal inter-action records, in which the most significant feature is that the data can indicate when the interactions occurred and their exact duration. However, only a few platforms are able to track user behaviors every second. Even if tracking is possible, fetching such records is a costly process for recommender systems. Therefore, how to obtain valuable time series vectors within an acceptable amount of system consumption is a major challenge as well. Figure 3 shows an example of the interaction records, which are the same as the records in Fig. 1, and were gathered only three times. In such interaction records, the order of interactions between two records is knowable, whereas the order of interactions in one record is unknowable. To solve the aforementioned problems, we proposed an approach to transform such suboptimal interaction records datasets into time series vectors called *serial filling at equal ratio*.

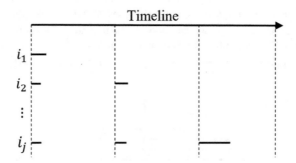

Fig. 3. Example of the interaction records datasets, similar as in Fig. 1

Each record can be regarded as a vector $\mathbf{h}_u^n$, which contains interaction time for each item, where $n \in \{0, 1, 2, \cdots\}$ denotes the number of records. To generate the time series vectors $\mathbf{r}_u^t$, at first, we determined the expected length of time series T, which is: $t \in \{0, 1, \cdots, T - 1\}$. During initialization, we set every time series vector $\mathbf{r}_u^t$ to null, set $n = 0, t = 0$, and calculated $norm_u$, which is the target L^1-norm of vector $\mathbf{r}_u^t$, by Eq. (1).

$$norm_u = \frac{\sum_{n \in \{0,1,\cdots\}} \left\| \mathbf{h}_u^n \right\|_1}{T} \tag{1}$$

Further, for every user $u \in \mathbf{U}$, we repeated the following process until all vectors $\mathbf{h}_u^n$ were processed.

1. Calculate the value of *ratio*, which will then be subtracted from $\mathbf{h}_u^n$ and added to $\mathbf{r}_u^t$ as per Eq. (2).

$$ratio = \frac{norm_u - \left\| \mathbf{r}_u^t \right\|_1}{\left\| \mathbf{h}_u^n \right\|_1} \tag{2}$$

2. Judge *ratio*:
   a. If *ratio* > 1, add $\mathbf{h}_u^n$ to $\mathbf{r}_u^t$ and add 1 to n.
   b. If *ratio* $= 1$, add $\mathbf{h}_u^n$ to $\mathbf{r}_u^t$ and add 1 to both n and t.
   c. If *ratio* < 1, add *ratio* $\cdot \mathbf{h}_u^n$ to $\mathbf{r}_u^t$ and add 1 to t.

After repeating the above for all the vectors, all the interaction data is transformed and stored in time series vectors $\mathbf{r}_u^t$. For the given moment, $\mathbf{r}_u^t$ means the interaction time of user u with each of the items in the tth $norm_u$ interaction time. The pseudocode of *serial filling at equal ratio* is shown below.

Algorithm: *Serial Filling at Equal Ratio*

Input:$\{\mathbf{h}_u^n | n \in \{0,1,2,\cdots\}\}, T$
Output:$\{\mathbf{r}_u^t | t \in \{0,1,\cdots,T-1\}\}$
1: $\mathbf{c} \leftarrow \mathbf{0}$
2: $norm_u = \sum_{n \in \{0,1,2,\cdots\}} \|\mathbf{h}_u^n\|_1 / T$
3: $t \leftarrow 0$
4: $\boldsymbol{foreach}\ n \in \{0,1,2,\cdots\}\ \boldsymbol{do}$
5: $\quad\boldsymbol{while}\ \|\mathbf{c}\|_1 + \|\mathbf{h}_u^n\|_1 > norm_u\ \boldsymbol{do}$
6: $\quad\quad ratio \leftarrow (norm_u - \|\mathbf{c}\|_1)/\|\mathbf{h}_u^n\|_1$
7: $\quad\quad \mathbf{r}_u^t \leftarrow \mathbf{c} + \mathbf{h}_u^n \cdot ratio$
8: $\quad\quad \mathbf{c} \leftarrow \mathbf{0}$
9: $\quad\quad t \leftarrow t + 1$
10: $\quad\quad \mathbf{h}_u^n \leftarrow \mathbf{h}_u^n \cdot (1 - ratio)$
11: $\quad\boldsymbol{end\ while}$
12: $\quad \mathbf{c} \leftarrow \mathbf{c} + \mathbf{h}_u^n$
13: $\boldsymbol{end\ for}$

Interaction time is a type of implicit feedback as argued by us previously. Although neural network showed better performance on learning implicit feedback directly than traditional models [22, 23], we still decided to feed the data that was close to the actual ratings of users into the model. We considered two key points to generate the final training data: item bias and normalization. Firstly, we led the item bias into time series vectors. For each element $r_{u,i}^t$ in time series vector $\mathbf{r}_u^t$, the new element was calculated by Eq. (3).

$$r_{u,i}^{\prime t} = \log\left((e-1)\frac{|\mathbf{U}|}{\sum_{v \in \mathbf{U}} \mathbf{r}_{v,i}^t} r_{u,i}^t + 1\right) \tag{3}$$

The new time series vectors $\mathbf{r}''_u$ were composed of $r''_{u,i}$. Further, we normalized $\mathbf{r}''_u$ by Eq. (4).

$$\gamma^t_u = \frac{\mathbf{r}''^t_u}{\left\|\mathbf{r}''^t_u\right\|_1} \tag{4}$$

4 Recommendations with LSTM

4.1 Recommendation Model

The neural network used in our method is shown in Fig. 4. The hidden layer is composed of the dynamic recurrent neural network with LSTM units to learn the short-term and long-term patterns from the datasets. The input and output layers are full connection layers for mapping the dimension of input/output data to the hidden layer. We normalized the datasets in advance, and hence we set a Softmax layer after the output layer to normalize the output also.

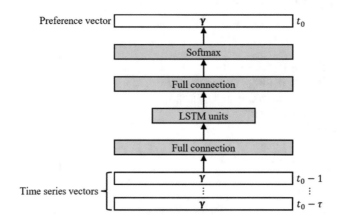

Fig. 4. Concept model of our neural network

As a recommendation model, the expected function of the network is to predict the next preference vector $\gamma^{t_0}_u$ for inputs $X = \left\{\gamma^t_u | t \in \{t_0 - \tau, \cdots, t_0 - 1\}\right\}$, where τ denotes the set value of the time steps. Figure 5 shows the training steps. Inputs and outputs are denoted by X and Y. The initial and final parameters of LSTM are denoted by ρ and ρ'.

For a user, we were able to generate $(T - \tau)$ training sets; we used $(T - \tau) \times |U|$ training sets in total. While recommending, we inputted the recent time series vectors $\left\{\gamma^t_u | t \in \{T - \tau, \cdots, T - 1\}\right\}$ of the target user into the trained model and the final output γ^T_u was regarded as the prediction vector. Finally, we recommended the top-N items according to the prediction vector after excluding the known items of the target user.

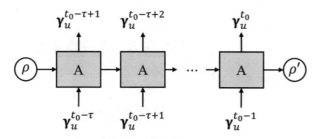

Fig. 5. The training steps

4.2 Dimensionality Reduction

LSTM shows good performance in learning the time series. However, since LSTM has a relatively complicated structure, it costs more than other recommendation models and the deep learning model. It is essential for recommender systems to finish within a tolerable period as they have high number of users. High system efficiency is an essential feature especially for recommender systems based on the time series, because in such systems, it is necessary to use several times the data than that for the systems based on static ratings.

Matrix factorization is a technique often used in the field of recommendation. Matrix factorization can approximate an $m \times n$ user-item rating matrix $\mathbf{R}$ by the product of an $m \times k$ user matrix $\mathbf{P}$ and a $k \times n$ item matrix $\mathbf{Q}$ [28]. In our model, the dimension of the training data was determined by the number of items $|\mathbf{I}|$, which easily exceed hundreds of thousands. The dimension of the training data was considered to be the biggest limitation to the efficiency of the system. To reduce the dimension, we need a mapping matrix with $k \times |\mathbf{I}|$ dimensions. Thus, a $|\mathbf{I}|$-dimensional vector can be mapped to a k-dimensional vector by multiplying the mapping matrix. Therefore, the item matrix $\mathbf{Q}$ through matrix factorization was exactly what we required. However, to use matrix factorization, we need a static user-item rating matrix $\mathbf{R}$, whereas our data consisted of a dynamic time series. In static data, user interests and item features remain the same over time, while in a time series, user interests and item features are considered to change over time. Fortunately, we did not use it to generate recommendations by matrix factorization more than an item matrix $\mathbf{Q}$. Therefore, if the hypothesis that the rate of change of item features is much slower than that of user interests was established, we could obtain $\mathbf{R}$ by Eq. (5) where $\{u_1, \cdots, u_v\} = \mathbf{U}$.

$$\mathbf{R} = \left[\left(\sum_{t \in \{0,1,\cdots,T-1\}} \gamma_{u_1}^t \right)^T, \cdots, \left(\sum_{t \in \{0,1,\cdots,T-1\}} \gamma_{u_v}^t \right)^T \right]^T \quad (5)$$

In the new model shown in Fig. 6, training data would be mapped before the input and after the output layer. Thus, one of the dimensions of the input and output layers would be reduced to k as well.

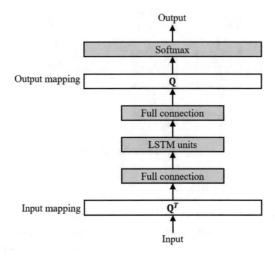

Fig. 6. Dimensionality reduction by mapping matrix Q

5 Evaluation

5.1 Datasets

We used the Steam datasets to test our approach. There are no existing datasets appropriate for evaluating our method; therefore, we gathered the datasets directly using the Steam API, which can return a list of playtimes during the past two weeks, in minutes, for a given user ID. The final datasets included 604,109 interactions of 10,243 users, and 14,649 games in 26 weeks (from November 27, 2016 to May 27, 2017). The dataset would be divided into 13 subsets of interaction records.

5.2 Experiments

To simulate an actual recommendation process, we set the subset of the newest records as the test set and the remaining 12 subsets as the training sets. In this experiment, we set the length of time series $T = 12$ in the serial filling and the aimed dimension $k = 50$ in the dimensionality reduction. Finally, we recommended the top-50 ($N = 50$) items to each user. To evaluate each part of our approach (IR), we set two control experiments. In one of the experiments, we did not apply the serial filling (noSF), and in the other, we did not apply dimensionality reduction (noDR).We compared our approach with collaborative filtering for implicit feedback (IF) [24] and time decay (TD) [3] for the purpose of baselining.

We used the recall rate as evaluation criteria, which is denoted by Eq. (6), to evaluate recommendation accuracy, and used the training and execution times to evaluate system efficiency. Finally, we checked the distribution of recommendation times of the top 300 items to evaluate the diversity of recommendations by each approach.

$$Recall = \frac{\sum_{u \in \mathbf{U}} |\mathbf{Rec}_u \cap (\mathbf{New}_u - \mathbf{Known}_u)|}{\sum_{u \in \mathbf{U}} |\mathbf{New}_u - \mathbf{Known}_u|} \tag{6}$$

We implemented the neural networks on TensorFlow 1.8 using Python 3.6 and deployed the model on a personal computer with GeForce GTX 1080 GPU. Network parameters were set as the following: dynamic learning rate from 0.001 to 0.00001, time steps = 8, batch size = 100, hidden units = 256, and global steps = 200.

5.3   Discussion

Table 1 lists the recall rates in different approaches. It can be seen that IR had a higher recall rate than IRnoSF and the baseline approaches, which proved the validity of IR and serial filling as well. Table 2 lists the time consumed in IR and IRnoDR. According to this table, IR and IRnoDR had similar execution times while IRnoDR spent more time on training than IR. Considering that the recall rate of IR was about the same as that of IRnoDR as seen in Table 1, dimensionality reduction by matrix factorization is a good method to improve the system efficiency. Figure 7 shows the distribution of recommendation times of the top 300 items by IR and IF. It is clear that the items recommended by IR are more widely distributed than that by IF, which means that our approach has better diversity than baseline collaborative filtering.

Table 1. Recall rates in different approaches

IR	**0.1194**
IRnoSF	0.0897
IRnoDR	**0.1202**
IF	0.0608
TD	0.0559

Table 2. Time consumption in IR and IRnoDR

Training time	IR	**54 min 14 s**
	IRnoDR	118 min 37 s
Execution time	IR	3 min 2 s
	IRnoDR	2 min 42 s

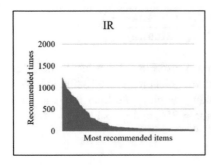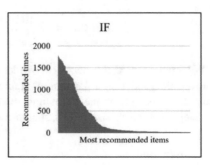

Fig. 7. Distribution of recommendation times of top 300 items

6 Conclusion

In this study, we proposed a recommendation model based on LSTM for interaction records. According to our experiments, our model showed satisfactory performances in three dimensions: accuracy, efficiency, and diversity. In future work, we will apply evaluations to different datasets to confirm the universality of our model. Furthermore, in this work, we treated the interaction time directly as a rating. In fact, ratings are likely to be complicated and influenced by different items and users. Therefore, optimizing the rating vectors should also be an important part of our future work. We also plan to examine other methods and models to deal with the time series.

References

1. Falk, M., et al.: A first course on time series analysis: examples with SAS. GNU Free Documentation Licence (2012)
2. Ricci, F., Rokach, L., Shapira, P.: Kantor: recommender systems handbook. Springer (2010)
3. Ding, Y., Li, X.: Time weight collaborative filtering. In: Proceedings of the 14th ACM International Conference on Information and Knowledge Management, CIKM 2005, pp. 485–492. ACM (2005)
4. Oard, D., Kim, J.: Implicit feedback for recommender systems. In: Proceedings of the AAAI Workshop on Recommender Systems, pp. 80–83. AAAI (1998)
5. Xiang, L., et al.: Temporal recommendation on graph via long- and short-term preference fusion. In: Proceedings of the 16th International Conference on Knowledge Discovery and Data Mining, SIGKDD 2010, pp. 723–732. ACM (2010)
6. Hochreiter, S., Schmidhuber, J.: Long short-term memory. Neural Comput. **9**, 1735–1780 (1997)
7. Sak, H., Senior, A., Beaufays, F.: Long short-term memory recurrent neural network architectures for large scale acoustic modeling. In: Proceedings of the 15th Annual Conference of the International Speech Communication Association, INTERSPEECH 2014, pp. 338–342. ISCA (2014)
8. Greff, K., Srivastava, R., Koutnik, J., Steunebrink, B., Schmidhuber, J.: LSTM: a search space odyssey. IEEE Trans. Neural Netw. Learn. Syst. **28**, 2222–2232 (2017)

9. Gers, F., Schraudolph, N., Schmidhuber, J.: Learning precise timing with LSTM recurrent networks. Mach. Learn. Res. **3**, 115–143 (2002)

10. Koren, Y.: Collaborative filtering with temporal dynamics. In: Proceedings of the 15th ACM SIGKDD International Conference on Knowledge Discovery and Data Mining, KDD 2009, pp. 447–456. ACM (2009)

11. Gordea, S., Zanker, M.: Time filtering for better recommendations with small and sparse rating matrices. In: Proceedings of the 8th International Conference on Web Information Systems Engineering, WISE 2007, pp. 171–183. Springer (2007)

12. Baltrunas, L., Amatriain, X.: Towards time-dependant recommendation based on implicit feedback. In: Workshop on Context-Aware Recommender Systems, CARS 2009, pp. 1–5 (2009)

13. Zimdars, A., Chickering, D., Meek, C.: Using temporal data for making recommendations. In: Proceedings of the 17th Conference on Uncertainty in Artificial Intelligence, UAI 2001, pp. 580–588. ACM (2001)

14. Lu, Z., Agarwal, D., Dhillon, I.: A spatio-temporal approach to collaborative filtering. In: Proceedings of the Third ACM Conference on Recommender Systems, RecSys 2009, pp. 13–20. ACM (2009)

15. Lommatzsch, A., Albayrak, S.: Real-time recommendations for user-item streams. In: Proceedings of the 30th Annual ACM Symposium on Applied Computing, SAC 2015, pp. 1039–1046. ACM (2015)

16. Chen, J., Nairn, R., Nelson, L., Bernstein, M., Chi, E.: Short and tweet: experiments on recommending content from information streams. In: Proceedings of the SIGCHI Conference on Human Factors in Computing Systems, CHI 2010, pp. 1185–1194. ACM (2010)

17. Subbian, K., Aggarwal, C., Hegde, K.: Recommendations for streaming data. In: Proceedings of the 25th ACM International on Conference on Information and Knowledge Management, CIKM 2016, pp. 2185–2190. ACM (2016)

18. Chang, S., et al.: Streaming recommender systems. In: Proceedings of the 26th International Conference on World Wide Web, WWW 2017, pp. 381–389. ACM (2017)

19. Kille, B., et al.: Stream-based recommendations: online and offline evaluation as a service. In: Proceedings of the 6th International Conference on Experimental IR Meets, CLEF 2015, pp. 497–517. Springer (2015)

20. Gorgoglione, M., Panniello, U., Tuzhilin, A.: The effect of context-aware recommendations on customer purchasing behavior and trust. In: Proceedings of the 5th Conference on Recommender Systems, RecSys 2011, pp. 85–92. ACM (2011)

21. Panniello, U., Thuzhilin, A., Gorgoglione, M., Palmisano, C., Pedone, A.: Experimental comparison of pre- vs. post-filtering approaches in context-aware recommender systems. In: Proceedings of the 3rd Conference on Recommender Systems, RecSys 2009, pp. 265–268. ACM (2009)

22. Jannach, D., Ludewig, M.: When recurrent neural networks meet the neighborhood for session-based recommendation. In: Proceedings of the Eleventh ACM Conference on Recommender Systems, RecSys 2017, pp. 306–310. ACM (2017)

23. Quadrana, M., Karatzoglou, A., Hidasi, B., Cremonesi, P.: Personalizing session-based recommendations with hierarchical recurrent neural networks. In: Proceedings of the Eleventh ACM Conference on Recommender Systems, RecSys 2017, pp. 130–137. ACM (2017)

24. Hu, Y., Koren, Y., Volinsky, C.: Collaborative filtering for implicit feedback datasets. In: Proceedings of the 8th International Conference on Data Mining, ICDM 2008, pp. 263–272. IEEE (2008)

25. Takács, G., Pilászy, I., Tikk, D.: applications of the conjugate gradient method for implicit feedback collaborative filtering. In: Proceedings of the 5th ACM Conference on Recommender Systems, RecSys 2011, pp. 297–300. ACM (2011)
26. Covington, P., Adams, J., Sargin, E.: Deep neural networks for YouTube recommendations. In: Proceedings of the 10th ACM Conference on Recommender Systems, RecSys 2016, pp. 191–198. ACM (2016)
27. Hopfgartner, F., Kille, B., Heintz, T., Turrin, R.: Real-time recommendation of streamed data. In: Proceedings of the 9th ACM Conference on Recommender Systems, RecSys 2015, pp. 361–362. ACM (2015)
28. Koren, Y., Bell, R., Volinsky, C.: Matrix factorization techniques for recommender systems. Computer 42–49 (2009)

MMOU-AR: Multimodal Obtrusive and Unobtrusive Activity Recognition Through Supervised Ontology-Based Reasoning

Muhammad Asif Razzaq$^{(\boxtimes)}$ and Sungyoung Lee

Ubiquitous Computing Lab, Department of Computer Engineering,
Kyung Hee University, Seocheon-dong, Giheung-gu,
Yongin-si, Gyeonggi-do 446-701, South Korea
{asif.razzaq,sylee}@oslab.khu.ac.kr

Abstract. The aging population, prevalence of chronic diseases, and outbreaks of infectious diseases are some of the major healthcare challenges. To address these unmet healthcare challenges, monitoring and Activity Recognition (AR) are considered as a subtask in pervasive computing and context-aware systems. Innumerable interdisciplinary applications exist, underpinning the obtrusive sensory data using the revolutionary digital technologies for the acquisition, transformation, and fusion of recognized activities. However, little importance is given by the research community to make the use of non-wearables i.e. unobtrusive sensing technologies. The physical state of human pervasively in daily living for AR can be seamlessly presented by acquiring health-related information by using unobtrusive sensing technologies to enable long-term health monitoring without violating an individual's privacy. This paper aims to propose and provide supervised recognition of Activities of Daily Livings (ADLs) by observing unobtrusive sensor events using statistical reasoning. Furthermore, it also investigates their semantic correlations by defining semantic constraints with the support of ontological reasoning. Extensive experiments were performed with real-world dataset shared by the University of Jaén Ambient Intelligence (UJAmI) Smart Lab in order to recognize the human activities in the smart environment. The evaluations show that the accuracy of the supervised method (87%) is comparable to the one, state of the art semantic approach (91%).

Keywords: Multi-sensor data fusion · Activity Recognition ·
Sampling · Classification · Reasoning

1 Introduction

Over the past few decades, a rapid rise in the advancement of pervasive computing in healthcare has been observed. These advancements include the gathering of Activities of Daily Livings (ADLs) in response to the prevailing challenges

© Springer Nature Switzerland AG 2019
S. Lee et al. (Eds.): IMCOM 2019, AISC 935, pp. 963–974, 2019.
https://doi.org/10.1007/978-3-030-19063-7_75

linked with global healthcare systems. The challenges most often relate to the global issues of an ageing population suffering from physical or mental health, especially related to the chronic diseases [9]. A wide variety of applications are underpinned with state of the art Machine Learning (ML) algorithms for recognizing ADLs in smart environments using obtrusive and unobtrusive sensors. The wearable devices, also called as obtrusive devices, are most commonly engaged by the users for Activity Recognition (AR), however, such devices may not be practically applicable for long-term use because of their maintenance cost, battery life, and discomfort caused by continuously wearing them. This may also lead to the noisy and imprecise state, causing an erroneous classification and recognition. This study explores how to recognize activities based on the available sheer amount of discrete and continuous multimodal data produced by obtrusive, as well as, unobtrusive devices. Above-mentioned factors affect the performance of ML-based AR from multimodal sensory data sources thus the appropriate solution is required, which can lift the performance of the ADLs in monitoring applications [7]. The selected aggregation strategy for data-level fusion determines the way in which multimodal data reach the fusion node [15]. At the fusion node, different ADLs can be best be recognized by the selecting the appropriate fusion strategy with variable window lengths [12]. Because of the promising features of unobtrusive non-wearable sensing devices to recognize human pervasive activities using smart-home applications [6], this study gives a brief overview and usage of human identification technologies categorized namely as, object-based, footstep-based, body shape-based and gait-based identification technologies. Among all, the first type of unobtrusive human identification uses a signal pattern of interaction with an object. The second type of identification strategy uses footstep's pressure, their patterns, sounds, and vibration to identify the ADLs around the home. The third category, body shape-based human identification captures individual's information based on their body shape, height, and width using an ultrasound technology.

Table 1. Unobtrusive sensing technologies applied as non-wearable sensor [5,8]

	Object-based technologies	Footstep-based identification	Body shape-based identification	Gait-based identification
Sensing technologies	• Pressure sensor • RFID • Accelerometer	• Sensor switching • Microphone • Pressure sensor • Electromechanical film • Accelerometer • Piezoelectric • Transducers • Photo-interrupter	• Ultrasonic	• Passive infrared • RF transceiver • Electric potential sensor • Wi-Fi transceiver
Features	• Object use pattern • Object use acceleration	• Walking pattern • Footstep sound • Footstep induced vibration • Centre of pressure trajectory • Geometric and holistic information	• Height • Width • Area • Perimeter • Radius	• Body heat emission • Disruption of RF & Wi-Fi signals • Body electric charge changes

Lastly, gait-based technologies use passive infrared (PIR) detector and Wi-Fi PIR to observe the human body heat emission to recognize the individual and ADLs. Some of the further details and features for the aforementioned technologies are mentioned in Table 1.

This study involves AR, ontology modeling and reasoning [2,14] based on Human Activity Recognition (HAR) dataset shared by the University of Jaén's Ambient Intelligence (UJAmI) Smart Lab [16]. The UJAmI Smart Lab measures approximately 25 square meters divided into five regions: entrance, kitchen, workplace, living room and a bedroom with an integrated bathroom. The need for considering this dataset is to tailor a framework for ADLs recognition and perform the research using sensors, as most of the technologies nowadays are underscored by the elderly people, their health and importance of their occupancy state. So for them, associated activities might affect their functionality of daily life. So ADLs recognition from multimodal sensors for each segment of a daily routine i.e. morning, afternoon and evening in a controlled environment is the primary motivation behind this study. Following are the key objectives being undertaken: (1) proposing, designing and implementing a solution for HAR from the UJAmI event-based dataset by using conventional ML-based method, by preserving temporal states; (2) designing an effective and practical algorithm to train, test and evaluate the ML-based method; (3) using the event-based dataset, such as data from binary sensors to interpret semantic rules by a domain expert, required to build the ontological model, employed further to infer sub-activities; (4) and finally a detailed discussion is performed over the results from a classical supervised ML-based method and ontological reasoning.

The rest of this paper proceeds with an introduction to the proposed model in the Sect. 2, an overview to describe the UJAmI Smart Lab dataset in terms of its structure and format. The experimental evaluations and the results prediction from test data and ontological model are presented in Sect. 3. Finally, a detailed conclusion is drawn with possible improvement, as a future work for this paper in Sect. 4.

2   Methodology

This paper deals with the analysis and recognition of a set of 24 different ADLs, performed by a single male inhabitant under several daily routines for over 10 days in the UJAmI Smart Lab. More information about the collected dataset can be found at the website [16]. The proposed layered ADLs recognition framework is described in Fig. 1. The main processing layers are listed as (a) *Data Sensing Layer*; (b) *Context Acquisition Layer*, which extracts the features, trains the model and classify ADLs; (c) *Context Fusion Layer*, which underscores the patterns and perform data aggregation; (d) the *Semantic Layer*, managed by the ontology with underlying semantic rules by facilitating reasoning using the SPARQL queries; (e) and finally *Application Layer* for disseminating the obtained context or activity.

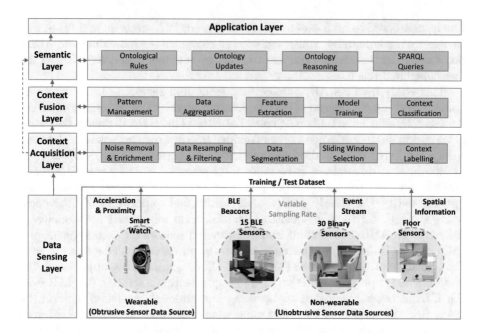

Fig. 1. Proposed framework for HAR using UJAmI Smart Lab dataset

2.1   Dataset and Data Sources Description

The multimodal dataset comprises data collected from four data sources over a period of 10 days. This dataset also preserved different sensor functionalities such as variable signal type from continuous to discrete; sampling frequency spanning from high to low; different protocols for involved magnetic, motion-based, as well as pressure sensors. However, the information related to the underlying wearable and non-wearable sensors technologies and the dataset collected is described in the following subsections:

Unobtrusive Binary Sensor Data. The dataset shared by UJAmI Smart Lab consists of an event stream from a set of 30 binary sensors (BinSens), which comprises of binary values along with the *Timestamps*. These BinSens works on the principles of Z-Wave protocol deployed in an unobtrusive wireless magnetic sensor environment. For example 'Medication box' in use means magnets are detached and such an event is considered as 'open'. When it is set to put back, sensor returns its value 'close'. The inhabitant movement within the sensor range is monitored using the wireless 'PIR sensors'. It works with the ZigBee protocol having maximum IR range of 7 m and sample rate of 5 Hz. The binary values, in this case, are represented by 'Movement' or 'No movement' for 'kitchen', 'bathroom', 'sofa', and 'bedroom' objects. Additionally, the unobtrusive motion sensor sofa, chair, and bed are also equipped with the 'textile layer sensors' to detect the inhabitant's pressure by transmitting 'Present' or 'No present' binary values using the Z-Wave protocol. The BinSens sometimes fire an event rapidly and sometimes may produce a challenging stream, which can last from few seconds to a few minutes or maybe for a few hours, such as pressure sensor stream.

Unobtrusive Spatial Data. An unobtrusive spatial data generated by the suite of capacitive sensors beneath the floor, called SensFloor dataset. It consists of 40 modules, compartmentalized in a 4 × 10 matrix. Each module has eight sensor fields, which are associated with an individual ID. Moreover, SensFloor collects capacitance's changed data with a variable sample rate.

Proximity Data. Another unobtrusive data source provides the proximity data as a set of 15 Bluetooth Low Energy (BLE) beacons at 0.25 Hz sample rate RSSI, which is collected through an android application installed on a smart-watch. These BLE beacons are generated for the objects like a medicine box, fridge, TV controller etc.

Obtrusive Acceleration Data. The ambulatory movements and motion intensities are reflected in an acceleration data stream, which is gathered by the Android application installed on smart-watch worn by the inhabitant. The accelerometer data is collected in tri-orthogonal (x, y, z-axis) directions at a sampling frequency of 50 Hz.

Table 2. Activities recorded in the UJAmI Smart Lab dataset.

ID	Activity name	ID	Activity name	ID	Activity name
Act01	Take medication	Act09	Watch TV	Act17	Brush teeth
Act02	Prepare breakfast	Act10	Enter the SmartLab	Act18	Use the toilet
Act03	Prepare lunch	Act11	Play a videogame	Act19	Wash dishes
Act04	Prepare dinner	Act12	Relax on the sofa	Act20	Put washing into the washing machine
Act05	Breakfast	Act13	Leave the SmarLab	Act21	Work at the table
Act06	Lunch	Act14	Visit in the SmartLab	Act22	Dressing
Act07	Dinner	Act15	Put waste in the bin	Act23	Go to the bed
Act08	Eat a snack	Act16	Wash hands	Act24	Wake up

2.2 Recognising ADLs, Ontology Modelling and Reasoning

The UJAmI Smart Lab dataset covers a maximum of 24 activities in the smart-home environment as mentioned in Table 2. It contains 43,320 training and 27,783 test samples for several routines collected over the period of 10 days. The first step in the proposed framework is to synchronize the pre-segmented dataset from different obtrusive and unobtrusive data sources. The process of synchronization was attained based on the *Timestamps* by taking care of missing values affected by different sample rates. Filling gaps is a challenging task since some data sources generate continuous streams, as well as, discrete data. In the end, generation of a single file for training dataset, over which modeling techniques can be applied to get the classification results, which needs to be applied on an unknown test data ADLs classification. So, to handle such challenges, domain expert analyzes the data tuple patterns critically in order to create SWRL rules for HAR. The Semantic Layer in the proposed framework as mentioned in Fig. 1 presents the ontological operations performed to describe the ADLs recognized from obtrusive and unobtrusive multimodal sensors. The aim behind the development of the ontology is to provide a basic model that not only allows the representation of ADLs but also supports the reasoning as and when it is queried for HAR. The outcome ensures the comprehensiveness of the ontology as shown in Fig. 2 so that it is enough to be able to represent all recorded 24 ADLs in the UJAmI dataset but should also be as concrete to facilitate the use of semantic reasoner and interpretation through SPARQL queries [11]. The ontology defines 24 major ADLs as parent classes. It is enforced with the use of SWRL rules, which provide additional constraints other than object properties and data properties. Some of the excerpt of the SWRL rules used in the modeling of ontology for 24 ADLs are described in the Table 3. The main aim of this work is to recognize activities based on the available dataset, analyze the correlation between activity type and objects involved, evaluate the classification results and address the challenges associated with the UJAmI Smart Lab dataset. Some of the key tasks as performed in [13] to automatically clean and pre-process the dataset are discussed briefly in subsequent sections.

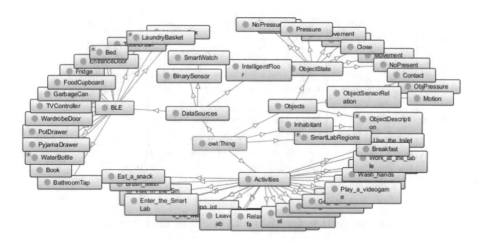

Fig. 2. Excerpt of the developed ontology for UJAmI dataset

Table 3. SWRL/SQWRL definitions: AR based on obtrusive and unobtrusive sensors.

Rule	Activity of daily livings	SWRL/SQWRL obtrusive/Unobtrusive semantic rules
1	Take Medicine	Inhabitant(?Inhab) ∧ hasLocation(?Inhab, ?Kitchen) ∧ hasObject(?Inhab, ?Waterbottle) ∧ hasState(?Waterbottle, ?Open) ∧ hasObject(?Inhab, ?MedicineBox) ∧ hasState(?MedicineBox, ?Open) ⟹ **sqwrl:select(?Inhab, ?Act)**
2	Prepare breakfast	Inhabitant(?Inhab) ∧ hasLocation(?Inhab, ?Kitchen) ∧ makes(?Inhab, ?Product1) ∧ hasType(?Product1, ?Tea) ∧ hasObject(?Inhab, ?Kettle) ∧ hasState(?Kettle, ?Open) ∧ makes(?Inhab, ?Product2) ∧ hasType(?Product2, ?MilkChocolate) ∧ hasObject(?Inhab, ?Microwave) ∧ hasState(?Microwave, ?Open) ∧ sqwrl:makeSet(?opt1, ?Product1) ∧ sqwrl:makeSet(?opt2, ?Product2) ∧ sqwrl:union(?opt3, ?opt1, ?opt2) ⟹ **sqwrl:select(?Inhab, ?opt3)**
3	Dinner	Inhabitant(?Inhab) ∧ hasLocation(?Inhab, ?DiningRoom) ∧ hasActivity(?Inhab, ?Sitting) ∧ hasActivity(?Inhab, ?Eating) ∧ hasObject(?Inhab, ?Pots) ∧ hasState(?Pots, ?Open) ∧ hasObject(?Inhab, ?Dishwasher) ∧ hasState(?Dishwasher, ?Open) ⟹ **sqwrl:select(?Inhab, ?Act)**
4	Watch TV	Inhabitant(?Inhab) ∧ hasLocation(?Inhab, LivingRoom) ∧ hasActivity(?Inhab, ?Sitting) ∧ hasObject(?Inhab, ?MotionSenorSofa) ∧ hasState(?MotionSensorSofa, ?Movement) ∧ hasObject(?Inhab, ?TVRemoteControl) ∧ hasState(?TVRemoteControl, ?Present) ∧ hasObject(?Inhab, ?TV) ∧ hasState(?TV, ?Open) ⟹ **sqwrl:select(?Inhab, ?Act)**
5	Dressing	Inhabitant(?Inhab) ∧ hasLocation(?Inhab, ?Bedroom) ∧ puts(?Inhab, ?Clothes) ∧ hasObject(?Inhab, ?LaundryBasket) ∧ hasState(?LaundryBasket, ?Present) ∧ hasObject(?Inhab, ?Closet) ∧ hasState(?Closet, ?Open) ⟹ **sqwrl:select(?Inhab, ?Act)**

2.3   Activity Recognition Methods

As shown in Fig. 1, the AR framework is a sequence of *Context Acquisition Layer*, which performs data alignment, and pre-processing; *Context Fusion Layer* responsible for applying ML techniques, and model training; and finally *Semantic layer* responsible for ontology manipulation and reasoning tasks.

Context Acquisition Layer

Data Alignment and Mapping: In this layer, the data has to be prepared from the UJAmI Smart Lab dataset corpus in such a way so that it becomes suitable for training and classification evaluation processes. For this, each of the sensor data was reordered and matched into a set of 1-s window slot based on *Timestamps*.

In order to generate uniform *Timestamps*, the data was resampled, segmented and mapped based on the basis of *DateBegin* and *DateEnd* over 1-s time window. It was applied to each instance belonging to the data sources for Spatial, Proximity and Acceleration data. The better performance [3] can be achieved using a sliding window segmentation technique [1] with the step size of 1-s to keep the maximum number of instances [18].

Data Preprocessing: Another important step for preparing data is to resample, filter, remove noise and handle missing data before performing classification and reasoning tasks. In our case, the obtrusive 50 Hz sampled accelerometer data was re-sampled by applying the commonly used time-domain statistical feature, such as a mean filter for each of x, y, and z tri-orthogonal values over a duration of 1-s. The Floor data, which was generated at a variable rate, was also re-sampled within the duration of a 1-s window by taking the mean for floor capacitances. In the case of instances having missing data fields values, which were filled by calculating the average, by taking the preceding 50 samples. All the data preprocessing tasks were accomplished using dedicated software written in Python 3.6 [10].

Context Fusion Layer

Data Modeling: To validate our approach we used 43,320 training data instances (Activity, Acceleration, Proximity, and Floor), aligned based on the *Timestamps*. The feature vector was created using the Waikato Environment for Knowledge Analysis (WEKA) library [17]. As the training dataset contains strings, as well as, numeric tokens, so we have used "StringToWordVector" filter in order to perform tokenization and indexing process. Later on, these feature vectors from the training dataset were used to train an algorithm *FilteredClassifier* to obtain the classification model.

Classification: The results obtained during the modeling phase were analyzed and we opted to choose the best classification model based on the evaluation metrics. We used 10-fold cross-validation technique on the dataset to produce a stable model [4] and predict the correct sequence of activities. We also evaluated our model by using the training dataset for 6 out of 7 days (leave one-day-cross-validation technique) to validate the evaluation metrics on the test data. To keep the vocabulary consistent both the training data and test data were processed using *FilteredClassifier* by ensuring their compatibility. Precision, Recall and F-Measure for the training model can be seen in the Fig. 3. The main observation in Fig. 3 for the activities 'Act08' and 'Act14' ('Eat a snack' and 'Visit in the SmartLab' respectively), showed lower recall leading to overall declined accuracy.

Semantic Layer

Reasoning. The final data prepared to train the classifier was understood by the ontology expert in order to interpret the semantic patterns and rules creation.

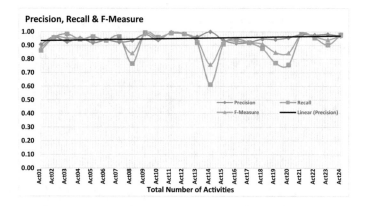

Fig. 3. Evaluation for the training dataset.

These semantic rules were designed and created manually using the training as well as test dataset patterns. We used Description Logic (DL) models to capture the semantic patterns in the dataset, which were converted to OWL2 ontology axioms and SWRL rules. The designed and developed ontology was used to perform reasoning over the unknown test dataset in addition to SPARQL queries for interpreting the axioms. The performance results are shown in Fig. 4, which proves that this method has produced accurate results in terms of predicting correct sequences of activities along with the ML method.

3   Results and Discussion

The proposed solution for this study and its development by using conventional ML method along with ontology-based reasoning is evaluated using performance metrics such as precision, accuracy, and f-measure. The performance evaluation results proved that ontology-based reasoning technique has lead to better accuracy. To improve the overall accuracy, an ontology-based reasoning can mitigate the ADLs recognition veracity. The overall accuracy of 87% was achieved using supervised ML-based classification approach with a mean absolute error of 0.0136. However, the accuracy of the same data when modeled and evaluated using ontology-based reasoning was slightly increased to 91%. The precision, recall, and f-measure for the dataset is mentioned in Fig. 4. The analysis illustrated that the activities 'Act11', 'Act12', 'Act14', 'Act19', and 'Act21' named originally as 'Play a videogame', 'Relax on the sofa', 'Visit in the SmartLab', 'Wash dishes', and 'Work at the table' respectively were mostly misclassified by ML supervised classification method but their accuracy was amplified using ontology-based reasoning.

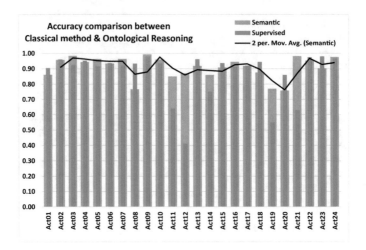

Fig. 4. Prediction vs Actual activities in the test dataset.

4   Conclusions and Future Work

In this paper, we have utilized the Human Activity Recognition (HAR) dataset for recognizing Activities of Daily Livings (ADLs). This dataset was generated by using obtrusive and unobtrusive devices in the smart environment and shared by the University of Jaén Ambient Intelligence (UJAmI) smart lab. The dataset includes data from four data sources, which include binary sensors data, proximity data, spatial data, and acceleration data. In this work, we presented a framework which preprocesses the data, trains a model using conventional Machine Learning (ML) based classifier i.e. *FilteredClassifier* and generates OWL2 axioms required to create an expert-driven ontology. In the ML-based method, during the preprocessing and training phase, several patterns were observed, required to design the basis of ontology. These axiomatic patterns were also converted to generate SWRL rules, which provided additional reasoning support to the ontology in order to infer the ADLs from the test data instances. The obtained results suggest and facilitate comparisons between actual ADLs in test data and predicted ADLs from the supervised ML-based method and ontological reasoning. The presented work discovers some meaningful insights regarding ADLs representation in the form of Ontology and SWRL rules. Further analysis also suggest that it still seems to be having some limitations on data aggregation challenges while processing ADLs using the ML-based supervised learning method. The overall obtained final accuracy of 87% for ML-based classification and 91% for ontology-based semantic reasoning also provide intuition that there seems to be the existence of some limitations while preprocessing the data for missing values and exploiting the correlations between a different set of features. These limitations may have a negative impact on classification and reasoning performance. So our future plan is to investigate these challenges by suggesting a lightweight probabilistic ML-based scheme and resolve the axiomatic conflicts while semantic rule generation.

Acknowledgment. This research was supported by an Institute for Information & Communications Technology Promotion (IITP) grant funded by the Korean government (MSIT) (No. 2017-0-00655). This work was also supported by the MSIT (Ministry of Science and ICT), Korea, under the ITRC (Information Technology Research Center) support program (IITP-2017-0-01629) supervised by the IITP (Institute for Information & communications Technology Promotion) and NRF-2016K1A3A7A03951968.

References

1. Bulling, A., Blanke, U., Schiele, B.: A tutorial on human activity recognition using body-worn inertial sensors. ACM Comput. Surv. (CSUR) **46**(3), 33 (2014)
2. Chen, L., Nugent, C.D., Okeyo, G.: An ontology-based hybrid approach to activity modeling for smart homes. IEEE Trans. Hum.-Mach. Syst. **44**(1), 92–105 (2014)
3. Dernbach, S., Das, B., Krishnan, N.C., Thomas, B.L., Cook, D.J.: Simple and complex activity recognition through smart phones. In: 2012 8th International Conference on Intelligent Environments (IE), pp. 214–221. IEEE (2012)
4. van der Gaag, M., Hoffman, T., Remijsen, M., Hijman, R., de Haan, L., van Meijel, B., van Harten, P.N., Valmaggia, L., De Hert, M., Cuijpers, A., et al.: The five-factor model of the positive and negative syndrome scale ii: a ten-fold cross-validation of a revised model. Schizophr. Res. **85**(1–3), 280–287 (2006)
5. Ghosh, A., Chakraborty, D., Prasad, D., Saha, M., Saha, S.: Can we recognize multiple human group activities using ultrasonic sensors? In: 2018 10th International Conference on Communication Systems & Networks (COMSNETS), pp. 557–560. IEEE (2018)
6. Gochoo, M., Tan, T.H., Liu, S.H., Jean, F.R., Alnajjar, F., Huang, S.C.: Unobtrusive activity recognition of elderly people living alone using anonymous binary sensors and DCNN. IEEE J. Biomed. Health Inf. **23**(2), 693–702 (2018)
7. Gravina, R., Alinia, P., Ghasemzadeh, H., Fortino, G.: Multi-sensor fusion in body sensor networks: state-of-the-art and research challenges. Inf. Fusion **35**, 68–80 (2017)
8. Mokhtari, G., Bashi, N., Zhang, Q., Nourbakhsh, G.: Non-wearable human identification sensors for smart home environment: a review. Sens. Rev. **38**(3), 391–404 (2018)
9. World Health Organization: WHO Expert Committee on Biological Standardization: sixty-eighth report. World Health Organization (2018)
10. Python: Python (2018). https://www.python.org/
11. Razzaq, M.A., Amin, M.B., Lee, S.: An ontology-based hybrid approach for accurate context reasoning. In: 2017 19th Asia-Pacific Network Operations and Management Symposium (APNOMS), pp 403–406 (2017). https://doi.org/10.1109/APNOMS.2017.8094159
12. Razzaq, M.A., Villalonga, C., Lee, S., Akhtar, U., Ali, M., Kim, E.S., Khattak, A.M., Seung, H., Hur, T., Bang, J., et al.: mlCAF: multi-level cross-domain semantic context fusioning for behavior identification. Sensors **17**(10), 2433 (2017)
13. Razzaq, M.A., Cleland, I., Nugent, C., Lee, S.: Multimodal sensor data fusion for activity recognition using filtered classifier, vol. 2, no. 19 (2018). https://doi.org/10.3390/proceedings2191262. http://www.mdpi.com/2504-3900/2/19/1262

14. Riboni, D., Sztyler, T., Civitarese, G., Stuckenschmidt, H.: Unsupervised recognition of interleaved activities of daily living through ontological and probabilistic reasoning. In: Proceedings of the 2016 ACM International Joint Conference on Pervasive and Ubiquitous Computing, pp. 1–12. ACM (2016)
15. Tschumitschew, K., Klawonn, F.: Effects of drift and noise on the optimal sliding window size for data stream regression models. Commun. Stat.-Theory Methods **46**(10), 5109–5132 (2017)
16. UJAmI: UJAmI (2018). http://ceatic.ujaen.es/ujami/sites/default/files/2018-07/UCAmI%20Cup.zip
17. Weka: Weka (2018). https://www.cs.waikato.ac.nz/ml/weka/
18. Yang, J., Nguyen, M.N., San, P.P., Li, X., Krishnaswamy, S.: Deep convolutional neural networks on multichannel time series for human activity recognition. In: IJCAI, vol. 15, pp. 3995–4001 (2015)

The Comparative Analysis of Single-Objective and Multi-objective Evolutionary Feature Selection Methods

Syed Imran Ali and Sungyoung Lee$^{(\boxtimes)}$

Department of Computer Science and Engineering,
Kyung Hee University, Yongin, South Korea
{imran.ali,sylee}@oslab.khu.ac.kr

Abstract. This research presents a comparative analysis of single-objective and multi-objective evolutionary feature selection methods over interpretable models. The question taken in this research is to investigate the role of aforementioned techniques for feature selection on classification model's interpretability as well as accuracy. Since, feature selection is a non-deterministic polynomial-time hardness (NP-hard) problem therefore exhaustively searching for all the possible feature sets is not computationally feasible. Evolutionary algorithms provide a very powerful searching mechanism that is utilized for candidate feature generation in a reasonable time frame. Single-objective (SO) algorithms are generally geared towards finding a subset of candidate feature set which achieves highest evaluation score e.g. classification accuracy. On the other hand, multi-objective (MO) methods are relatively more comprehensive than their counterparts. MO feature selection algorithms can simultaneously optimize two or more objectives such as classification accuracy of a final feature set along with the cardinality of the feature set. In this research, we have selected two representative feature selection algorithms from both the groups. Decision tree and a rule-based classifiers are used for the performance evaluation in terms of interpretability i.e. model size, and predictive accuracy. This research is undertaken to investigate application of SO and MO feature selection methods on small to medium sized classification datasets. The experimental results on 3 interpretable classification models indicate that the relative differences between the two set of models on small datasets may not be much pronounced, yet for the medium-sized datasets MO models provide a promising alternative. Although multi-objective techniques resulted in smaller feature subsets in general, but overall the difference between both the single-objective and the multi-objective feature subset selection techniques, investigated in this study, is not statistical significant.

Keywords: Single-objective feature selection ·
Multi-objective feature selection · Model complexity · Decision tree ·
Rule-based classifier

© Springer Nature Switzerland AG 2019
S. Lee et al. (Eds.): IMCOM 2019, AISC 935, pp. 975–985, 2019.
https://doi.org/10.1007/978-3-030-19063-7_76

1 Introduction

Knowledge discovery in data (KDD) is a wellness methodology for extracting knowledge from data through a series of processes. Feature selection is an important process in knowledge discovery from data [1]. Most of the data classification algorithms are designed to deal with datasets which are relatively clean, complete, and fully accessible for processing [2]. Data classification usually requires two tasks. First a model is constructed from the training data. In this case training data is also called labeled data i.e. class label of each training instance is known beforehand. Later same model is evaluated on an unlabeled test data where predictive accuracy is the main criterion for evaluation of the model [3]. Predictive accuracy of a model is calculated as number of right matches between actual and predicted class labels over all the instances in the test dataset.

$$\text{Predictive accuracy} = (TP + TN)/N \tag{1}$$

Where TP, TN, N are True Positive, True Negative, and N number of instances in a test set.

$$F = 2^*[(\text{Precision} * \text{Recall})/(\text{Precision} + \text{Recall})] \tag{2}$$

A classification model can be either a white-box model or a black-box model. In this regard, the inferred model can be an equation for a decision boundary, i.e. a line or a hyper-plane, which separates different objects into disjoint classes. On the other hand, a white-box model is to interpret due to its transparent underlying logic. Such types of models are also called comprehensible models. Decision tree and rule based classifier are two most popular types of comprehensible classifiers. In case of decision tree, typically depth of the tree is used as a metric for the complexity of the model. Hence, smaller trees are preferred based on the notion of Occam's razor which states that entities should not be multiplied unnecessarily. In this research four feature subset selection methods are used to quantify their effect on the decision tree in terms of classification accuracy and complexity of the model. Figure 1 depicts a decision tree model. Where features are represented by intermediate nodes and leaf nodes accommodate the class labels. Decision tree models are very popular among the domain experts e.g. medical physicians [7]. These models not only depict the nature of relationship among different features but also provides a very intuitive way of comprehending the holistic picture of the domain. Moreover, these models are also very efficient in terms of test case execution. For example, for a given test case as depicted in Fig. 1, first the root node will be evaluated and based on this test half of the tree would be skipped for further processing.

As shown in the example, only one value of the test case i.e. "Hardworking = Yes" would be enough for reaching to a conclusion in the given case.

Rule based classifiers are another important type of comprehensible classification model. These classifiers construct a decision list which contains a number of production rules of the form: IF <conditions> THEN <class>. These rules are normally equivalent to the decision tree's path from the root to different leaf nodes. But

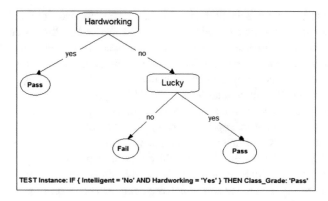

Fig. 1. A generic decision tree model

production rules have no graphical representation. Since in a given rule different conditions can be placed arbitrarily, therefore feature importance, preserved in tree structure, is lost in the case of production rules. Figure 2 depicts a decision list based model.

```
IF <Hardworking=Yes> THEN <Class: Pass>
IF <Hardworking=No> AND <Lucky=No> THEN <Class: Fail>
IF <Hardworking=No> AND <Lucky=Yes> THEN <Class: Pass>
IF <Intelligent=Yes> AND <Lucky=Yes> THEN <Class: Pass>
IF <Intelligent=No> AND <Lucky=No> THEN <Class: Fail>
IF <Lucky=Yes> THEN <Class: Pass>
Default: <Class: Fail>
- - - - - - - - - - - - - - - - - - - - - - - - - - - - - - - - - - - - - - -
Test Instance: IF <Intelligent=No> AND <Hardworking=Yes>
THEN <Class: Pass>
```

Fig. 2. A generic rule based model

When the data have high dimensionality then the standard and widely used classification algorithm not only take a long of time in processing the data but more often than not the resultant model is sub-optimal in terms of predicative accuracy and complex model size [3].

Over the last decade evolutionary feature selection algorithms have shown promising results in the area of feature subset selection [4–6]. Since, evolutionary algorithms are search based global optimization techniques, these methods can provide a powerful mechanism to generate a set of candidate feature subsets and iteratively optimize the feature-set. Another appealing aspect of evolutionary feature selection framework is its ability to handle multiple candidate subsets simultaneously and a parallel manner. Hence, computational time may also be decreased a parallel implementation of the evolutionary algorithms.

Evolutionary feature selection algorithms maybe divided into two categories. Single-objective (SO) feature selection algorithms generated multiple solutions in each generation. For the subsequent generations those solutions which yield higher predictive accuracy determine the search direction of the entire population. Hence, these algorithms are relatively simple in implementation and their searching mechanism is focused on generating such feature subsets which yield higher accuracies. Multi-objective (MO) feature selection algorithms are more comprehensive. These algorithms may employ some of the same operators as that of SO algorithms but their evaluation is based on both predictive accuracy of a candidate feature set as well as cardinality of the set i.e. number of features. Hence, both objectives are simultaneously optimized. For this characteristic of MO feature selection algorithms these methods are more promising in finding feature subsets which are compact and in turn may produce more interpretable classification models.

A lot of research has only focused on the predictive aspect of the classification models [7]. A classification model which is transparent and can be directly interpreted by user is called a comprehensible model such as decision trees and rule-based models. On the contrary a model which is black box (i.e. a set of equations) is called an incomprehensible model e.g. artificial neural networks and support vector machines. Along with predictive accuracy, comprehensibility of the model is also an important characteristic of a classification algorithm [7] for certain domains. Comprehensibility of the model is valuable to such domains as medicine, customer attrition, credit scoring, and bioinformatics etc. Several factors contribute to the importance of comprehensible classification models. User's trust can be achieved if the computer-induced model is understandable and logic of the model is transparent to the user. Another important factor emphasizing the utility of comprehensibility is when an unexpected model is produced. In such cases user's trust can be achieved if model provides good explanations for its decision making process. For example, it can indicate important symptoms which are strongest predictors for a certain disease [7].

2   Evolutionary Algorithms and Data Classifiers

2.1   Evolutionary Feature Selection Algorithms

Particle Swarm Optimization- Nearest Neighbor classifier (PSO-NN) a single-objective feature selection method based on geometric particle swarm optimization algorithm [8]. It uses a geometric framework for utilizing crossover operator. A nearest neighbor classifier is used in a wrapper based feature selection setting. This algorithm creates a feature search space in which a final feature subset is selected through a population of candidate solution sets.

Genetic Algorithm-Correlation based Feature Selection (GA-CSF) is a single-objective filter based feature selection algorithm. It uses a well-known genetic algorithm for feature candidate set generation [9]. Furthermore, worth of feature is considered in terms of feature's correlation with the class and its redundancy in the selected feature set [10].

Non-dominated Sorting II (NS2) is a multi-objective feature selection algorithm based on non-dominated sorting framework. It employs the underlying genetic algorithm framework for solution evolution. A nearest neighbor classifier is used for evaluating the worth of a candidate solution [11].

ENORA is an elitist Pareto-based multi-objective evolutionary algorithm that uses a $(\mu + \lambda)$ survival solution for feature subset selection [12].

2.2 Classification Algorithms

J48 is an alternative implementation of C4.5 which is a comprehensible classification algorithm that has a tree based structure in which leaf nodes represent classes and intermediate nodes are the test conditions. The target is to construct a tree that best fits the training data. Gain ratio is used as a heuristic to select attributes which provide best separation between instances of different classes [13].

JRIP is an implementation of RIPPER algorithm in Weka. JRIP is another popular comprehensible classifier which is based on sequential learning. Once a rule is constructed, all the instances covered and correctly predicted by the rule are removed from the dataset. Stopping criteria are based on description length of the instances and rule set [14].

CART stands for Classification and Regression Trees. Its working is similar to ID3 classifier. CART uses Gini impurity as a heuristic measure. CART can produce both classification and regression trees depending on the class variable [15].

3 Experimentation and Discussion

In our experimental setup we have selected 5 small to medium sized publically benchmark datasets [13]. All the datasets are discretized. Table 1 provides more details of the selected datasets. Weka [14] machine learning tool is used for the experimentation regarding both classification algorithms and feature selection algorithms. Reported results are averaged over 20 runs of each algorithm. All the algorithms are experimented with their default parameters.

Table 1. Datasets

Dataset	No. of features	No. of instances	No. of classes
Ionosphere	35	351	2
Chronic kidney disease	25	400	2
Autos	26	205	6
Sonar	61	208	2
Audiology	70	226	24

As illustrated in Fig. 3, both groups of the feature selection algorithms are fed the same datasets. And their evaluation is performed on 3 comprehensible classification models.

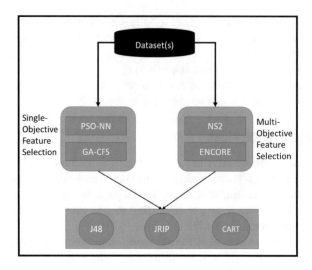

Fig. 3. Algorithms in the experimental setup

3.1 Single-Objective Feature Selection

Table 2 provides detailed results regarding the benchmark with which we will be comparing both SO and MO feature selection algorithms. Any algorithm which out-performs the benchmark results would be deemed useful whereas for the second stage of experimentation we will comparatively analysis of the inter-group results. As shown in Table 2, F-measure is used as an indicator of the performance of a classifier in terms of classification accuracy. F-measure is a harmonic mean of precision and recall measures. For the classification algorithms which provide a decision tree model, such as C4.5 and CART, we have recorded the number of leaves which indicates the spread of the model horizontally and size of the model which takes into account the number of intermediate nodes as well.

Note for Tables 2, 3, 4, 5 and 6. Due to space limitation column headings are replaced by letters. Therefore, A = F-measure, B = number of leaves, C = model size, D = number of rules and E = number of conditions.

Table 2. Experimentation on full dataset

Dataset	J48			JRIP			CART		
	A	B	C	A	D	E	A	B	C
1	0.909	7	13	0.909	5	6	0.9	4	1
2	0.96	14	20	0.951	4	11	0.97	6	11
3	0.749	55	68	0.66	13	26	0.721	23	45
4	0.716	24	47	0.684	3	5	0.685	3	5
5	0.758	32	54	0.694	19	32	0.699	23	45
Average	0.818	26.4	40.4	0.7796	8.8	16	0.795	11.8	21.4

Table 3. Experimentation on PSO-NN

Dataset	J48			JRIP			CART		
	A	B	C	A	D	E	A	B	C
1	0.909	11	21	0.909	5	6	0.9	8	15
2	0.963	18	24	0.973	3	8	0.97	6	11
3	0.736	54	67	0.749	15	33	0.763	27	53
4	0.769	17	33	0.735	6	17	0.719	27	53
5	0.759	27	45	0.724	18	31	0.729	27	53
Average	0.8272	25.4	38	0.818	9.4	19	0.8162	19	37

Table 4. Experimentation on GA-CSF

Dataset	J48			JRIP			CART		
	A	B	C	A	D	E	A	B	C
1	0.9	6	11	0.9	5	6	0.903	6	11
2	0.96	18	24	0.968	4	12	0.97	6	11
3	0.433	11	21	0.225	3	8	0.427	13	25
4	0.677	6	11	0.721	3	7	0.963	10	19
5	0.742	30	49	0.739	14	25	0.708	26	51
Average	0.7424	14.2	23.2	0.7106	5.8	11.6	0.7942	12.2	23.4

Table 5. Experimentation on NS2

Dataset	J48			JRIP			CART		
	A	B	C	A	D	E	A	B	C
1	0.9	5	9	0.906	6	10	0.9	5	9
2	0.794	4	7	0.97	3	9	0.794	4	7
3	0.471	22	23	0.379	13	20	0.476	9	17
4	0.747	14	27	0.729	3	5	0.744	19	37
5	0.779	28	47	0.695	14	27	0.736	13	25
Average	0.7382	14.6	22.6	0.7358	7.8	14.2	0.73	10	19

Table 6. Experimentation on ENORA

Dataset	J48			JRIP			CART		
	A	B	C	A	D	E	A	B	C
1	0.893	9	17	0.9	4	4	0.9	11	21
2	0.963	18	24	0.794	2	3	0.97	6	11
3	0.657	41	57	0.449	9	17	0.594	21	41
4	0.764	20	39	0.692	5	13	0.729	15	29
5	0.747	31	52	0.735	16	29	0.728	23	45
Average	0.8048	23.8	37.8	0.714	7.2	13.2	0.7842	15.2	29.4

On the benchmarked datasets J48 performs better in terms of classification accuracy than the other two classification models. Although CART produces a much smaller decision model which may be deemed favorable for the interpretability criterion. Moreover, since the difference between classification results of both J48 and CART is not statistically significant and given the importance model's interpretability for the domain expert, CART is also a promising alternative to J48 for such domains where domain experts are heavily involved in the decision making. JRIP produced the smallest model i.e. on average 8.8 rules but couldn't achieve comparable accuracy as that of the other two models. Please note that a decision tree model can be converted into a ruleset where number of leaves correspond to the number of rules in the model.

Table 3 details experimental results of PSO-NN feature selection. Comparing the benchmarked results of Table 1, PSO-NN based SO feature selection achieved higher accuracies on all the classification models while model has also reduced. Hence, PSO-NN had a favorable effect on small-medium sized datasets. It is interesting to note that, JRIP ruleset is slightly increased which may not have a significant effect on the overall model size whereas JRIP has comparable accuracy as that of CART. Hence, the rule-based model fared well in terms of model's performance and interpretability.

GA-CFS results are shown in Table 4. Comparing with the benchmarked results, the reduced dataset couldn't improve results of J48 rather some of the important indicators have turned negative, such as, F-measure has decreased. Although the resultant model size has also decreased, this may indicate that some of the important features are not taken into account by J48 while constructing the classification model. Similarly, JRIP has demonstrated the same effects. It is interesting to note that CART has resisted any detrimental effects of the reduced dataset. It has preserved both the predictive accuracy and the model size as that of Table 1. In terms of accuracy, CART has achieved the highest accuracy while maintaining a smaller model size.

In terms of intra-group comparison, PSO-NN achieved on average higher accuracy on decision tree models. The difference in the rule-based classifier is also quite notable. Although GA-CFS produced models with smaller sizes but at the expanse of predictive accuracy. Since, predictive accuracy is of a primary concern therefore we may conclude that PSO-NN fared well in SO feature selection comparison.

3.2 Multi-objective Feature Selection

In this part of the experimentation we have compared MO feature selection with the benchmarked results of Table 1 as well as intra-group results. In this case, average results over the datasets may not provide the important details therefore results would be discussed in terms of small and medium sized datasets. MO algorithms couldn't achieve impressive results on small datasets i.e. ionosphere, chronic kidney disease and autos datasets, whereas, on medium-sized datasets, sonar and audiology results are promising. Table 5 provides details of NS2 results.

As shown in Table 6, ENORA yields relatively higher accuracy as compared to NS2 on decision tree models at the expanse of model size whereas for the rule-based classifier the overall performance is comparable.

3.3   Discussion on Single-Objective and Multi-objective Evolutionary Feature Selection

So far we have discussed the intra-group results of both single-objective and multi-objective feature selection approaches. This section compares both the groups in terms of model's predictive accuracy and complexity.

It is important to note that the single-objective feature selection approaches, discussed in this study, specifically targeted predictive accuracy of the model. In this regard, the candidate feature subsets which score higher on the nearest-neighbor classifier are preferred. It would be interesting to explore the effects of different classifiers on the wrapper based evolutionary feature selection approaches. Unlike the more sophisticated classifiers such as decision trees, the nearest neighbor classifier performs no internal feature selection i.e. embedded feature selection, for constructing the classification model. Therefore, NN plays no role in the cardinality of the selected feature subset.

Table 7 summarizes the results of feature selection algorithms across the datasets used in this study. The value in brackets represent the total number of features selected by an algorithm for the specified dataset. It is interesting to note that PSO-NN achieved highest accuracy among the comparative feature selection approaches with comparable feature subset size. All the other approaches achieved quite similar classification results. Furthermore, as it would have been expected that the multi-objective approaches generated relatively compact feature subsets as compared to the single-objective methods. Although the difference among these two groups is not statistically significant. It is interesting to note that the difference among SO and MO approaches is more pronounced on small-sized datasets. For the medium-sized datasets there is some variation which is difficult to generalize and hence requires further inquiry.

Table 7. Comparison between single-objective-multi-objective evolutionary feature selection

Dataset	PSO-NN	CFS-GA	NS2	ENORA
1	0.90 (14)	0.90 (7)	0.90 (4)	0.89 (6)
2	0.96 (9)	0.96 (12)	0.85 (4)	0.90 (7)
3	0.74 (11)	0.36 (8)	0.50 (2)	0.44 (13)
4	0.74 (26)	0.78 (29)	0.74 (16)	0.72 (28)
5	0.73 (33)	0.72 (24)	0.73 (33)	0.73 (36)
Average	0.82 (18.6)	0.74 (16)	0.74 (11.8)	0.74 (18)

4   Conclusion

In this study we have investigated the role of single-objective feature selection and multi-objective feature selection on the classification model's performance and comprehensibility. Since SO models actively tend to optimize accuracy of the selected feature set whereas MO models also account for the size of the model as well. Through a series of experiments on benchmarked datasets it is observed that although both SO and MO models do enhance the overall quality of the resultant model but the difference among SO and MO models are not much pronounced. In this case given the simplicity of SO models, it is preferred to utilize SO methods for efficient model construction in knowledge discovery process. Furthermore, MO models do tend to find reasonably good trade-off between accuracy and the complexity of the model. It is also observed that for the medium-sized datasets MO models tend to perform comparable or better therefore for the future studies this direction may also be explored. Moreover, MO models tend to take relatively longer processing time. In this regard, the role of instance selection may also be explored for reducing the running time of the MO algorithms.

Acknowledgement. This research was supported by an Institute for Information & Communications Technology Promotion (IITP) grant funded by the Korean government (MSIT) (No. 2017-0-00655). This work was supported by the MSIT (Ministry of Science and ICT), Korea, under the ITRC (Information Technology Research Center) support program (IITP-2017-0-01629) supervised by the IITP (Institute for Information & communications Technology Promotion) and NRF-2016K1A3A7A03951968.

References

1. Buczak, A.L., Guven, E.: A survey of data mining and machine learning methods for cyber security intrusion detection. IEEE Commun. Surv. Tutor. **18**(2), 1153–1176 (2016)
2. Wang, H., Jin, Y., Jansen, J.O.: Data-driven surrogate-assisted multiobjective evolutionary optimization of a trauma system. IEEE Trans. Evol. Comput. **20**(6), 939–952 (2016)
3. Ali, S.I., Kang, B.H., Lee, S.: Application of feature subset selection methods on classifiers comprehensibility for bio-medical datasets. In: International Conference on Ubiquitous Computing and Ambient Intelligence, pp. 38–43. Springer, Cham (2016)
4. Jiang, S., Chin, K.S., Wang, L., Qu, G., Tsui, K.L.: Modified genetic algorithm-based feature selection combined with pre-trained deep neural network for demand forecasting in outpatient department. Expert Syst. Appl. **82**, 216–230 (2017)
5. Zhang, Y., Gong, D.W., Cheng, J.: Multi-objective particle swarm optimization approach for cost-based feature selection in classification. IEEE/ACM Trans. Comput. Biol. Bioinform. (TCBB) **14**(1), 64–75 (2017)
6. Jain, I., Jain, V.K., Jain, R.: Correlation feature selection based improved-binary particle swarm optimization for gene selection and cancer classification. Appl. Soft Comput. **62**, 203–215 (2018)
7. Freitas, A.A.: Comprehensible classification models: a position paper. ACM SIGKDD Explor. Newsl. **15**(1), 1–10 (2014)
8. García-Nieto, J., Alba, E., Jourdan, L., Talbi, E.: Sensitivity and specificity based multiobjective approach for feature selection: application to cancer diagnosis. Inf. Process. Lett. **109**(16), 887–896 (2009)
9. Witten, I.H., Frank, E., Hall, M.A., Pal, C.J.: Data Mining: Practical Machine Learning Tools and Techniques. Morgan Kaufmann, Burlington (2016)
10. Hall, M.A.: Correlation-based feature subset selection for machine learning, Hamilton, New Zealand (1998)
11. Deb, K., Pratap, A., Agarwal, S., Meyarivan, T.A.M.T.: A fast and elitist multiobjective genetic algorithm: NSGA-II. IEEE Trans. Evol. Comput. **6**(2), 182–197 (2002)

12. Jiménez, F., Sánchez, G., García, J.M., Sciavicco, G., Miralles, L.: Multi-objective evolutionary feature selection for online sales forecasting (2016)
13. Quinlan, J.R.: C4.5: Programs for Machine Learning. Elsevier, Amsterdam (2014)
14. Cohen, W.W.: Fast effective rule induction. In Machine Learning Proceedings 1995, pp. 115–123 (1995)
15. Brieman, L., Friedman, J., Olshen, R., Stone, C.: Classification and regression trees. Wadsworth, Google Scholar, Belmont (1984)

Water Sound Recognition Based on Support Vector Machine

Tingting Hang[1,2(✉)], Jun Feng[1], Xiaodong Li[1], and Le Yan[1]

[1] College of Computer and Information, Hohai University, Nanjing, China
[2] Department of Electrical and Information Engineering,
Hohai University Wentian College, Maanshan, China
httsf@hhu.edu.cn

Abstract. Water flow monitoring is of importance for water dynamic controlling in river basins. Currently, most of the water monitoring in the basin is manual-based, there is often some disordered data, which does not conform to hydrological models input standards. In order to solve the above problems, we propose a sound signal processing method to identify the water sound and link it with stream-flow measurements in this paper. Firstly, the water sound features were extracted. On this basis, Support Vector Machine (SVM) and two other classifiers are used to build the classification models, and the model tested using trained and not trained data show the best agreement to identifying the water sound recognition. The best performance of the classification is given by further optimizing the kernel function and penalty factor. The experimental result shows that the model's recognition rate is 98.22% with SVM, and the result of recognition is superior to other classification models.

Keywords: Water sound signal · Feature extraction · Classification · Support Vector Machine · Recognition rate

1 Introduction

Our living environment is full of sounds. Auditory sensations can extract useful information from them to know things going on around us. Moreover, the sound information is more effective as a visual sensation does not work well in dark [1]. With the progress of electronic technology, automatic sound recognition system has been applied in many fields and mobile devices have been equipped with a microphone that can capture the sound signal. This signal can be processed using sound fingerprints techniques to find a match with the sound fingerprints database. This might facilitate an increase in the accuracy of recognition.

Water flow monitor/measurement in the basin is vital for dynamic water control. However, because most of the water flow observations are collected manually, there are often some data disorders, which do not conform to hydrological model standards. Therefore, new technology is urgently needed to record these stream-flow data in real time with reliability.

This paper aims to explore the applications of different approaches, i.e. Support Vector Machine (SVM), k-nearest neighbor (KNN) and Convolutional neural network

© Springer Nature Switzerland AG 2019
S. Lee et al. (Eds.): IMCOM 2019, AISC 935, pp. 986–995, 2019.
https://doi.org/10.1007/978-3-030-19063-7_77

(CNN), classifying the stream-flow with the sound signals. SVM is a machine learning method which can minimize the structural risk. It reasonably addresses problems of the demand for large training samples of the neural network and the flaw of over-fitting. KNN is a non-parametric algorithm used in classification and regression. The KNN classification is prior to determining the most similar classification inside each of the generated clusters. This paper also introduces CNN into the processing of sound signals. Classifications of the heat maps are obtained after feature extraction, and the effect of water sound classification is further analyzed.

2   Related Work

There have been many explorations on sound recognition, and many methods are used in pre-processing of sound, extracting relevant features, and obtaining the classifications or signal recognition [2]. Wang et al. [3] proposed a heart sound recognition method that relied on Mel-frequency cepstral coefficients (MFCC) and an SVM classifier. Chen et al. [4] proposed a deep neural network (DNN) method for recognizing S1 and S2 heart sounds. Sehili et al. [5] achieved daily sound recognition with Gaussian Mixture Models (GMM) and an SVM classifier. Chen et al. [6] proposed to detect activities from the single microphone positioned near the washing basin, also using MFCCs but a more sophisticated Hidden Markov Models (HMM) classifier. Hayashi et al. [7] introduced a new sound recognizing method for daily human activities with a Deep Neural Network (DNN) classifier. Hinton et al. [8] in the other way used DNNs for sound modeling in speech recognition. Chen et al. [9] proposed a novel SVM classifier to identify three kinds of helicopters and one kind of cruise missile. Toyoda et al. [10] proposed a multilayered perceptron NN system for environmental sound recognition. For researches related to water sound, the KNN, SVM and KNN-SVM classifiers are designed to classify the two kinds of underwater targets [11]. Guyot et al. [12] proposed an audio signal processing algorithm to detect water sounds based on a physical model of air bubble sounds, aiming to monitor the daily activities of older adults. However, these studies have not been applied to monitor the stream-flow in the basin.

The main contribution of the paper is to propose an SVM classifier to identify the water sound and link it with stream-flow measurements, the proposed classification model is simple but effective, which fits the problem of classifying stream-flow data in real time with reliability. The outline of the following paper is as follows: Sect. 3 introduces the methods in feature extraction and classification. The experiments and results are introduced in Sect. 4. Discussion and conclusion are presented in the end.

3   The Proposed Method

The experiments consist of two steps: water sound feature extraction, and water sound classification. In the first step, signal preprocessing prepares the sound signal for feature extraction and original features are extracted which are represented by a feature vector in a much simpler and condensed form. In the second step, based on a set of training data containing observations with defined classes, the relations between extracted features and classes are built with the classifier, which can later assign unknown observations to one of the classes.

3.1   Feature Extraction

The water sound series is recorded by 1/16000 s sound amplitude. However, the time series cannot be directly analyzed for classification. A feature extraction stage of the sound wave is necessary for future classification. The first process is to determine a value or vector that can be used as an object or an individual identity [13]. The detailed steps of feature extraction are listed as follows:

(1) The first step is to apply a pre-emphasis filter on the signal to amplify the high frequencies. This filtering process is done for reducing noise during sound cap- ture. The pre-emphasis filter can be applied to a signal x using the first order filter in the following equation, where typical values for the filter coefficient (α) is 0.95 or 0.97.

$$y(t) = x(t) - \alpha x(t - 1) \tag{1}$$

(2) Apply the continuous overlapping segmentation to the framing of water sound to maintain smooth and continuity between two frames. According to the cycle characteristics of water sound signals, the length of each frame ranges from 20 ms to 40 ms and each frame has N sound data samples. The overlapping part between frames is called frameshift, which is within the range of 1/2 and 1/3 of the frame length.

(3) After slicing the signal into frames, we apply a window function to each frame, aiming at reducing the truncation effect of each frame signal to reduce the spectrum energy leakage. If we define the window as w (n), with n ranges from 0 to N − 1, and N is the number of samples in each frame, the result of windowing is one signal:

$$y(n) = x(n)w(n), 0 \leq n \leq N - 1 \tag{2}$$

Where y(n) is the signal result of the convolution between the input signal and the window function and x(n) represents the signal to be convolved by the window function. Nowadays, the commonly used window function is the Hamming window with its formulation as:

$$w[n] = 0.54 - 0.46 \cos\left(\frac{2\pi n}{N - 1}\right) \tag{3}$$

(4) With the above steps finished, we can find that the sound wave is composed of many different frequencies. In order to extract those frequencies and make the data efficiently processed, we decompose the sound waves into components. We use N-point Fourier Transform to separate the bass part, midrange part and the high part where N is typically 256 or 512, and then we compute the power spectrum using the following equation:

$$P = \frac{\left|FFT(X_i)^2\right|}{N} \tag{4}$$

(5) The energy in each frequency band (from low to high) is extracted in this step. For each sound clip, we create a fingerprint which looks like a heat map that represents the energy in the sound clip in each frequency along the time axis. The sound fingerprints are finally used to train tensors with the abstracted representations of the sound. Figure 1 shows the feature extraction steps, and we can get a sound fingerprint. The sound fingerprint can be used to represent the relationship between sound frequency and energy, which also can be used to represent the feature of the water sound. The unit of the horizontal axis is time (s), the unit of the vertical axis is frequency (Hz), and the color in the figure represents the energy of in the frequency band.

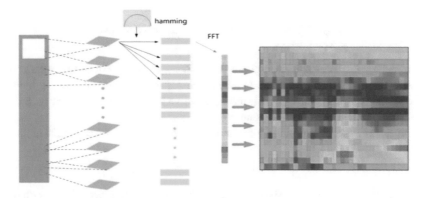

Fig. 1. Water sound feature extraction.

3.2 Classification

The k-nearest neighbor (KNN) [14], Gaussian mixture models (GMM) [5], hidden Markov models (HMM) [6], and support vector machines (SVM) are the commonly used classifiers. Although all these methods continue to be used today, modifications and hybrid classification method have been proposed over the years. Moreover, deep learning methods, such as deep neural networks (DNN) [7, 8], have gained more attention in various pattern recognition problems in recent years. In this paper, three commonly classification methods are selected for comparison, i.e. SVM, KNN and CNN.

SVM. SVM is a new and very promising technique for classification and regression, which could solve various data mining problems via optimized methods. It partly overcomes some traditional problems (e.g. curse of dimensionality). For the two-type classification problem, it is to find a classification plane to separate the two categories, which is formally expressed as:

$$w^T x + b = 0 \tag{5}$$

Any sample point (x_i, y_i) can be substituted into Eq. (6):

$$\begin{cases} w^T x_i + b < 0, & y_i = -1 \\ w^T x_i + b > 0, & y_i = +1 \end{cases} \tag{6}$$

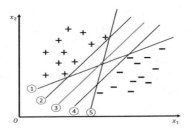

Fig. 2. Schematic diagram of the classification plane.

From Fig. 2, all classification planes from No. 1 to No. 5 give reasonable results, but only no. 3 classification plane is better, the rest of classification planes have the worse disturbance resistance, which means a worse generalization capability. SVM is to find the optimal classification plane.

The Eq. (5) can be classified into the general form of the classification model.

$$f(x) = w^T x + b \tag{7}$$

Our goal is to find the suitable w and b from the training set which contains N sample points for determining the classification plane. The distance between the sample points and classification plane should be as "far" as possible. By introducing the concept of function interval and set interval, we can reform the question into a solution of the optimization problem.

$$\min_{w,b} \frac{1}{2} \|w\|^2 \tag{8}$$

$$s.t. \quad y_i\left(w^T x_i + b\right) \geq 1 \ i = 1, 2, \cdots, N \tag{9}$$

The final solution structure is expressed as follows:

$$w^* = \sum_{i=1}^{N} \alpha_i y_i x_i \tag{10}$$

$$b^* = \frac{\max\limits_{i:y_i=-1} w^{*T} x_i + \min\limits_{i:y_i=+1} w^{*T} x_i}{2} \tag{11}$$

Where α_i is the Lagrange multiplier.

In addition to the advantages of finding the maximum interval classification plane, SVM has its advantages of high anti-perturbation. The kernel trick is introduced to improve the classification model in all kinds of practical situations, especially for the problem of linear inseparability [15]. Table 1 shows several common kernel functions.

The method for SVM is that the eigenvector nonlinearity of low-dimension water sounds is firstly mapped into eigenvector space with high dimensions, to transfer the impartibility of low dimensional into the detachable problem with high dimensions. The key to solving the nonlinearly separable problem is to construct optimal margin hyperplane. The structure of the optimal margin hyper-plane is finally transferred to calculations of the best value and polarization. In SVM, its property is decided by kernel parameter and penalty factors. The intrinsic parameter gamma and penalty factor C have assigned an initial value of the kernel function after the kernel function is selected. The parameter value can be adjusted until the test precision is satisfactory.

Table 1. Several common kernel functions.

Name	Expression	Parameter
Linear nuclear	$K\left(x_i, x_j\right) = x_i^T x_j$	
Polynomial kernel	$K\left(x_i, x_j\right) = \left(x_i^T x_j\right)^d$	$d \geq 1$ is the number of polynomials
Gaussian kernel	$K\left(x_i, x_j\right) = \exp\left(-\frac{\|x_i - x_j\|^2}{2\sigma^2}\right)$	$\sigma > 0$ is the bandwidth of the Gaussian kernel
RBF kernel	$K\left(x_i, x_j\right) = \exp\left(-\beta\|x_i - x_j\|^2\right)$	β is the width of the kernel
Sigmoid kernel	$K\left(x_i, x_j\right) = \tan h\left(\beta x_i^T x_j + \theta\right)$	tanh is the hyperbolic tangent function, $\beta > 0, \theta < 0$

KNN. KNN is a non-parametric algorithm. After the data without tags is input to the model, each feature in the new data is compared with the features corresponding to the data in the sample set. The classification tags of the most similar features in the sample set are extracted as a result. In general, we only select the top k most similar data in the sample data set. This is the source of k in the k nearest neighbour algorithm. Usually, k is an integer no greater than 20. Finally, the category with the most occurrences among the k most similar data is selected as a new classification of the new data. The general steps are as follows:

(1) Split a group of data into two parts, training data and testing data respectively. Also, each data has a data value X and with a label value Y.
(2) Use the training data to train a model by KNN algorithm.
(3) Predict the Y values of the testing data by driving the model with the X value of test data.
(4) Comparisons between the results of predict and Y value of test data.

CNN. Neural networks, such as the convolutional neural network (CNN), have gained significant attention in various pattern recognition problems in recent years, e.g. the image processing. In this section, we are going to apply CNN models to classify the stream-flow. CNN treats the heat map representations as images.

Based on the output of water sound feature extraction, tensor representations of water sound waves are labelled with the stream-flow. The labelled samples are further imported into the neural network models. For each sample, the neural network will try to learn (regress on) the corresponding stream-flow. CNN requires a certain number of samples to train the model so that the model has the associative memory and predictive ability. The training steps are shown as follows:

(1) Network initialization. Determine the number of network input layer, the hidden layer and output layer respectively. The hidden layer includes the convolution layer, Activation layer, Pooling layer, Full connection layer and Normalized layer.
(2) Output calculation of hidden layer. In the hidden layer output, L is the number of hidden layer nodes, F is the active function of the hidden layer.
(3) Output calculation of the output layer. According to the hidden layer output H, the connection weight value w and bias b, we can calculate the predicted output O.
(4) Error evaluation. Based on model prediction output O and expectation output Y, calculate network prediction error e.
(5) Weight update. Update model connection weight w according to the network prediction error e.
(6) Bias update. Update model node bias b according to the network prediction error e.
(7) Determine whether the iteration of the method is over, and if not, return step (2).

4   Result and Analysis

In the experiment, several typical water sound signals are collected using mobile devices and classified into multi-classification by the label. According to cycle characteristics of water sound signals, each frame length is 10 ms–15 ms, the frequency of water sound signal keeps at 16000 Hz. However, during the collecting process, much noise exists in water sounds and will distort the water sounds signal. Thus, the noise of water sound needs to be eliminated before the water sound feature extraction.

4.1   Feature Extraction

In the experiment, the frequency accuracy of the water sound signal is required to be higher and the requirement for the side valve is lower, the Hamming window is selected. In the section, we also take the filter coefficient $\alpha = 0.95$, the frameshift $M = 0.5$, the window length $N = 256$ to get characteristic parameters of water sound signal. According to the feature extraction steps, we can get a sound fingerprint that represents the characteristic in the sound clips. Figure 3 shows the sound fingerprints expressing the different stream-flow. It can be seen from Figure a that there is much low-frequency energy in the sound clips, which is the small stream-flow of water sound. Figure b shows that there is much high-level energy in the sound clips, which is the large stream-flow of water sound.

To build the feature database of water sound signals, 20 features are extracted from the above sound fingerprints. Table 2 shows the features extracted from the water sound recognition database relying on the sound fingerprints. The different rows in Table 2 represent different training samples with each row representing a set of feature values for a training sample. The values represent the magnitude of the stream-flow. If the characteristic of the test data is similar to the training samples, it is recognized as the same stream-flow. In the experiment, we divided the stream-flow into multiple categories. According to the range of stream-flow, and they are identified by different tag values.

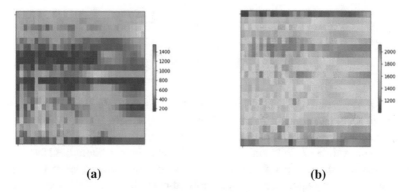

(a)   (b)

Fig. 3. The sound fingerprints expressing the different stream-flow, where (a) is sound fingerprints for small stream-flow and (b) represents sound fingerprints for big stream-flow.

Table 2. The sample of feature extraction of water sound recognition database.

F1	F2	F3	F4	F5	F6	F7	F8	F9	F10
17	26	82	67	36	62	49	104	133	132
23	31	71	40	64	73	81	85	120	120
49	30	72	63	56	86	62	90	105	105
F11	F12	F13	F14	F15	F16	F17	F18	F19	F20
131	127	150	159	163	180	189	192	154	138
117	110	113	140	132	222	220	195	125	83
101	93	108	132	164	194	178	158	143	83

4.2   Classification

Recognition Result and Analysis. Through verification, 3561 collected water sound signals in the experiment are classified between training sample and testing sample based on the ratio of 7 to 3. 2492 samples of water sound signals are used for training and 1069 samples of water sound signals are used for testing, which is proven better for evaluating the classification recognition effect.

Comparison of Different Classifications. In the experiment, in order to further verify the effectiveness of the SVM algorithm, different classification algorithms are used (i.e. SVM, KNN and CNN) for classification on the same basis of the extracted features. In the SVM experiment, through the cross-validation method, different kernel functions are used under the same data conditions, and the kernel function with the smallest error is the best kernel function. Through the simulation experiments, the RBF kernel function is selected. Moreover, the penalty factor C is taken as 1000 and the intrinsic parameter of kernel function gamma as 0.00028. In the CNN experiment, we take the hidden layer active function F = Relu, the output layer active function F = Softmax with the Iteration training times as 1000. In the KNN experiment, we take K = 5. The result of classification is shown in Table 3.

Table 3.   The result of different method.

Method	Time (s)	Accuracy (%)
SVM	193	98.22
KNN	10.27	97.75
CNN	1576	70.29

We can see from the Table 3 that the SVM network classifier is better in the recognition accuracy compared with CNN and KNN network classifier when the training samples are a certain quantity. KNN network classifier has dramatically reduced training time of CNN and SVM network classifier. Although KNN's training time is relatively short, it is not suitable for real-time applications since KNN has no training process. The experimental result shows that the SVM model's recognition rate is 98.22%, and the result of recognition is superior to other classification models.

5   Conclusion

Considering the fact that the majority of the water monitoring method is based on manual records, the corresponding process often suffers from large human efforts and poor performance. The authors propose a sound signal processing method to identify the water sound which can be used to monitor the stream-flow in the basin. Serval machine learning models including SVM, KNN, CNN are employed to classify the signal sources based on the extracted features. Results show that SVM attains the best recognition rate is 98.22%, Which is for real-time application. The framework provides valuable information for Water flow monitor in the basin in an innovative way.

Acknowledgment. This work was supported by the National Key R&D Program of China (Grant No. 2018YFC0407901), partially supported by the Outstanding Young Talents Support Program of Anhui Provincial, partially supported by the National Natural Science Foundation of China under the Grant No. 61370091 and No. 61602151.

References

1. Iwaki, M., Nakayama, S.: A sound recognition method with incrementally taking acoustic stream into consideration. In: 2016 IEEE 5th Global Conference on Consumer Electronics, pp. 1–2 (2016)
2. Pires, I.M., et al.: Recognition of activities of daily living based on environmental analyses using audio fingerprinting techniques: a systematic review. Sensors **18**(1), 1–23 (2018)
3. Wang, Y., et al.: Research on heart acoustic recognition based on support vector machine. In: 2017 Chinese Automation Congress (CAC), pp. 62–65 (2017)
4. Chen, T.E., Yang, S.I., Ho, L.T., et al.: S1 and S2 heart sound recognition using deep neural networks. IEEE Trans. Biomed. Eng. **64**, 372–380 (2016)
5. Sehili, M.A., Istrate, D., Dorizzi, B., et al.: Daily sound recognition using a combination of GMM and SVM for home automation. In: Signal Processing Conference, pp. 1673–1677 (2012)
6. Chen, J., Kam, A.H., Zhang, J., Liu, N., Shue, L.: Bathroom activity monitoring based on sound. In: Gellersen, Hans -W., Want, R., Schmidt, A. (eds.) Pervasive 2005. LNCS, vol. 3468, pp. 47–61. Springer, Heidelberg (2005). https://doi.org/10.1007/11428572_4
7. Hayashi, T., et al.: Daily activity recognition based on DNN using environmental acoustic and acceleration signals. In: Signal Processing Conference, pp. 2351–2355 (2015)
8. Hinton, G., Deng, L., Yu, D., et al.: Deep neural networks for acoustic modeling in speech recognition: the shared views of four research groups. IEEE Signal Process. Mag. **29**(6), 82–97 (2012)
9. Chen, H.H., Zhong, F.P., Xue-Zhong, X.U., et al.: Acoustic recognition of low-altitude flight targets by SVM. Syst. Eng. Electron. **27**(1), 46–48 (2005)
10. Toyoda, Y., Huang, J., Ding, S., et al.: Environmental sound recognition by multilayered neural networks. In: International Conference on Computer and Information Technology. IEEE Computer Society, pp. 123–127 (2004)
11. Huang, J., Zhu, G.P.: Research on the K-D Tree KNN-SVM classifier in underwater acoustic target recognition. J. Ocean Technol. **37**(1), 15–22 (2018)
12. Guyot, P., Pinquier, J., Andre-Obrecht, R.: Water acoustic recognition based on physical models. In: IEEE International Conference on Acoustics, Speech, and Signal Processing, vol. 32, no. 3, pp. 793–797 (2013)
13. Anggraeni, D., et al.: The implementation of speech recognition using Mel-Frequency Cepstrum Coefficients (MFCC) and Support Vector Machine (SVM) method based on Python to Control Robot Arm. In: IOP Conference Series: Materials Science and Engineering, vol. 288, pp. 1–10 (2018)
14. Ibarz, A., Bauer, G., Casas, R., et al.: Design and evaluation of a acoustic based water flow measurement system. In: European Conference on Smart Sensing and Context, pp. 41–54 (2008)
15. Feng, G.: Parameter optimizing for support vector machines classification. Comput. Eng. Appl. **47**(3), 123–126 (2011)

Measuring Term Relevancy Based on Actual and Predicted Co-occurrence

Yuya Koyama$^{(\boxtimes)}$, Takayuki Yumoto$^{(\boxtimes)}$, Teijiro Isokawa, and Naotake Kamiura

University of Hyogo, 2167, Shosha, Himeji, Hyogo, Japan
eil7v0ll@steng.u-hyogo.ac.jp,
{yumoto,isokawa,kamiura}@eng.u-hyogo.ac.jp

Abstract. A measure of term relevancy is important in various applications such as Web search. Although the co-occurrence probability of terms in a database is a simple way to express term relevancy, it suffers from each term having a different co-occurrence tendency. In this paper, we propose a new measure of term relevancy: a ratio of actual and predicted values of co-occurrence (RAP). We construct a model predicting co-occurrence for each query as piecewise approximation lines.

Keywords: Term relevancy · Stochastic model · Web search

1 Introduction

Opportunities for Web search have increased due to the popularization of PCs and smartphones. However, we often obtain unreliable information through web search. For example, when users who do not know about a certain disease (disease X) search the Web and find a Web page stating "Disease X is an infectious disease", they may believe the information regardless of whether it is true. This problem occurs because they cannot judge the relationship between "disease X" and "infectious disease". In this way, for users who do not know much about topics they want to know about, it is important whether the terms on the Web page are related to the query. The term relevancy is also used in various systems such as document search, classification and query recommendation, and it is an important concept.

In this paper, we propose a new measure of term relevancy between query q and term n appearing in search results. For terms co-occurring with query q, we investigate the relationship between co-occurrence and frequency not considering co-occurrence and construct a model to predict co-occurrence from frequency of a term. We express term relevancy using the distance between actual and predicted values of co-occurrence.

© Springer Nature Switzerland AG 2019
S. Lee et al. (Eds.): IMCOM 2019, AISC 935, pp. 996–1005, 2019.
https://doi.org/10.1007/978-3-030-19063-7_78

2   Related Work

There are many researches to support users to judge the credibility of Web pages [1, 2]. Yamamoto and Tanaka developed a system providing multiple measures for judging credibility such as accuracy and objectivity [2]. In this system, the typicality of contents was used as a score representing objectivity. We proposed a method to express typicality on the basis of word co-occurrence in a database [3]. Co-occurrence of terms or words is often used to express the strength of the relation between terms. However, the tendency of term co-occurrence varies depending on each term, so it is difficult to decide a threshold to judge whether the relation between terms is strong.

Term relevancy measures and the related measures have been developed from various aspects. Church and Hanks proposed word association norms using mutual information [4]. In their work, the co-occurrence frequency in a sliding window is used for calculating the mutual information. However, its optimal window size is different for each case, so it needs to find the optimal values. Cilibrasi and Vitanyi focused on the use of Web search engines. They proposed word similarity based on the numbers of results returned by Google's search [5]. However, since many of Web search engines only provide approximate numbers of the results, it is difficult to accurately calculate the similarity. Since Mikolov et al. proposed Word2Vec [6], word embedding has been used for various applications [7, 8]. In word embedding, a word is represented as a vector learned by its surrounding words, and semantic similarity between words can be calculated as cosine similarity of the vectors. However, the similarity between words in antonymous relationship is often high because the surrounding words of the antonymous words are often similar.

3   Term Relevancy Estimation

3.1   Analysis of Conventional Approach

First, we analyze conditional probability,

$$P(q|n) = \frac{|D_{q \wedge n}|}{|D_n|} \tag{1}$$

which is a simple approach for measuring term relevancy. $|D_n|$ is the number of documents containing term n and $|D_{q \wedge n}|$ is the number of documents containing terms q and n in a database. The logarithm of Eq. (1) can be transformed into Eq. (2).

$$\log_{10} P(q|n) = \log_{10} \frac{|D_{q \wedge n}|}{|D_n|} = \log_{10}|D_{q \wedge n}| - \log_{10}|D_n| \tag{2}$$

If $\log_{10}|D_n|$ is rather larger than $\log_{10}|D_{q \wedge n}|$ on the right-hand side in Eq. (2), $\log_{10}|D_n|$ can be dominant to decide the value of $\log_{10} P(q|n)$.

To compare the effects of $\log_{10}|D_{q\wedge n}|$ and $\log_{10}|D_n|$, we show their coefficient correlations with $\log_{10} P(q|n)$ in Fig. 1. The used dataset is described in Sect. 4.1. Figure 1 shows negative correlations between $\log_{10} P(q|n)$ and $\log_{10}|D_n|$. $\log_{10}|D_{q\wedge n}|$ also have mainly negative correlations with $\log_{10} P(q|n)$. This is surprising since we had expected $\log_{10} P(q|n)$ to positively correlate with $\log_{10}|D_{q\wedge n}|$ from the form of Eq. (2). These results prove that $\log_{10}|D_n|$ is more dominant in Eq. (2) than $\log_{10}|D_{q\wedge n}|$. This means that $\log_{10} P(q|n)$ tends to be higher for frequent terms than infrequent terms. To measure term relevancy more accurately, we need to ease this tendency. In our approach, we compare actual co-occurrence probability with the probability expected from the value of $|D_n|$.

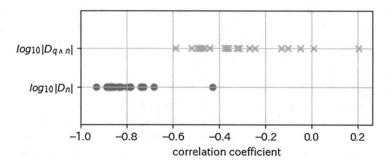

Fig. 1. Correlation coefficients with $\log_{10}P(q|n)$.

3.2 Relevancy Based on Predicted Co-occurrence

3.2.1 Prediction of Term Co-occurrence

To predict term co-occurrence for each query, we use piecewise linear approximation. We split $\log_{10}|D_n|$ into multiple intervals $[i, i+1)(i = 0, 1, 2, \ldots)$. For each interval, we calculate the mean of $\log_{10} P(q|n)$ and plot it at the center of each interval. For example, the mean in $[3.0, 4.0)$ is plotted at $\log_{10}|D_n| = 3.5$. By connecting these points and a point where $\log_{10}|D_n|$ is the maximum to each adjacent point with straight lines, we can obtain piecewise approximation lines. The predicted value of $\log_{10} P(q|n)$ at $\log_{10}|D_n|$ denotes $\log_{10}\widehat{P}(q|n)$.

To compare the effects of $\log_{10}|D_{q\wedge n}|$ and $\log_{10}|D_n|$, we show their coefficient correlations with $\log_{10} P(q|n)$ in Fig. 1. The used dataset is described in Sect. 4.1. Figure 1 shows negative correlations between $\log_{10} P(q|n)$ and $\log_{10}|D_n|$. $\log_{10}|D_{q\wedge n}|$ also have mainly negative correlations with $\log_{10} P(q|n)$. This is surprising since we had expected $\log_{10} P(q|n)$ to positively correlate with $\log_{10}|D_{q\wedge n}|$ from the form of Eq. (2). These results prove that $\log_{10}|D_n|$ is more dominant in Eq. (2) than $\log_{10}|D_{q\wedge n}|$. This means that $\log_{10} P(q|n)$ tends to be higher for frequent terms than infrequent terms. To measure term relevancy more accurately, we need to ease this tendency. In our approach, we compare actual co-occurrence probability with the probability expected from the value of $|D_n|$.

Since infrequent terms tend to be affected by incidental term co-occurrence, we do not predict $\log_{10} P(q|n)$ if $|D_n| < 10$. In other words, $\log_{10} \widehat{P}(q|n)$ is not defined in an interval $[0.0, 0.5)$; it is defined but not used in an interval $[0.5, 1.0)$. We also do not predict it if $|D_{q \wedge n}| = 0$. Figure 2 shows an example where the query is dementia. Each blue point means actual values of $\log_{10} P(q|n)$ at $\log_{10}|D_n|$, and the line means predicted values, $\log_{10} \widehat{P}(q|n)$.

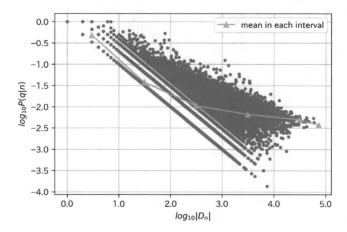

Fig. 2. Piecewise approximation lines when query is "dementia."

Table 1. Queries used for experiments

North America	Madagascar	North Pole	refrigerator
car	computer	air bag	gravity
theory of evolution		atmosphere	atomic power
dementia	diet	sleep	placebo
insomnia	enzyme	Uniqlo (apparel company)	
rugby	ekiben (lunch box sold at train station)		

3.2.2 Ratio of Actual and Predicted Co-occurrence

To express relevancy, we use the difference between actual and predicted values of co-occurrence probability. A larger difference means stronger relevancy between the terms. We use the logarithm of the ratio of the actual and predicted values (RAP) defined in Eq. (3) as relevancy.

$$\log_{10} RAP = \log_{10} \frac{P(q|n)}{\widehat{P}(q|n)} = \log_{10} P(q|n) - \log_{10} \widehat{P}(q|n) \qquad (3)$$

$\log_{10} P(q|n)$ is an actual value of the logarithm of co-occurrence defined in Eq. (2) and $\log_{10} \widehat{P}(q|n)$ is its predicted value defined in Sect. 3.2.1. As $\log_{10} P(q|n)$ is larger than $\log_{10} \widehat{P}(q|n)$, relevancy between the two terms is stronger.

4   Experiment

We compare the proposed method (RAP) with the baseline (Eq. (2)) as the relevancy score. Although we use data written in Japanese for our experiments, neither RAP nor the baseline uses linguistic characteristics. For this reason, we believe the same results would be obtained if we used English data. We use 20 queries about various topics, shown in Table 1. As terms whose relevancy to each query are evaluated, we use terms in the first ten sentences from the Wikipedia article about each query. The total number of terms for all queries is 681.

4.1   Dataset

4.1.1   Noun-Document Database

We created a noun-document database for term co-occurrence. Since we needed to collect pages about various topics for the database, we collected pages bookmarked in Hatena bookmark[1], a popular social bookmarking service in Japan. We randomly collected popular bookmarked pages from content of the collected pages, we extracted nouns by the Japanese morphological analyzer, MeCab [9][2]. This database contains 160,602 URLs, 1,537,808 terms, and 25,788,306 pairs of a term and URL.

4.1.2   Gold Standard of Term Relevancy

To decide the gold standard of relevancy of term pairs, we used crowdsourcing. Ten crowd workers evaluated each pair of terms by selecting one of the following options:

a. I don't know the meaning of one or both of the terms.
b. The terms have almost no relation.
c. The terms have a slight relation.
d. The terms have a moderate relation.
e. The terms have a strong relation.

Eliminating the results selected as option a, we give score 1 to 4 to options b to e and regard their mean scores as the gold standard of term relevance. We refer these scores as evaluators' ratings.

---

[1] Hatena Bookmark, http://b.hatena.ne.jp/.
[2] MeCab: Yet Another Part-of-Speech and Morphological Analyzer, http://taku910.github.io/mecab/.

4.2   Experimental Results and Discussion

We compare the proposed method (RAP) defined in Sect. 3.2 and the baseline mentioned in Sect. 3.1 by a correlation coefficient with the evaluators' ratings. The results are shown in Fig. 3. In this figure, grayed queries denote that RAP was worse than the baseline. In 15 out of 20 queries, RAP obtained higher a correlation coefficient than the baseline. The mean of the correlation coefficient by RAP is 0.592, significantly larger than the mean by the baseline, 0.522 ($p = 0.0032$ by Wilcoxon signed-rank test).

We analyzed both methods from the aspect of relation with $\log_{10}|D_n|$ and evaluators' ratings. We show the results where the query is "north America" as an example where RAP was better than the baseline. Figures 4 and 5 show the results for the baseline and RAP, respectively. In these figures, each point means a term and its $\log_{10}|D_n|$ value can be distinguished by its color and shape. In Fig. 4, the points whose $\log_{10}|D_n|$ is in $[1.0, 2.0)$ or $[2.0, 3.0)$ are distributed in the upper half of the graph, and the points whose $\log_{10}|D_n|$ is in $[3.0, 4.0)$ or $[4.0, 5.0)$ are distributed in the lower half. This means that $\log_{10} P(q|n)$ is greatly affected by $\log_{10}|D_n|$. On the other hand, this tendency is not observed in Fig. 5. Therefore, RAP can suppress the effect of $\log_{10}|D_n|$.

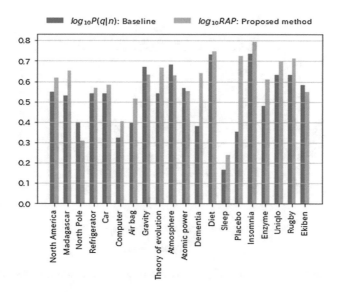

Fig. 3. Correlation coefficient of evaluators' ratings with baseline and proposed method for each query.

However, the results by RAP are worse than those by the baseline in some cases. For one such case, we show the results where the query is "North Pole" in Figs. 6 and 7. From Fig. 6, we can find that $\log_{10} P(q|n)$ is greatly affected by $\log_{10}|D_n|$, which is

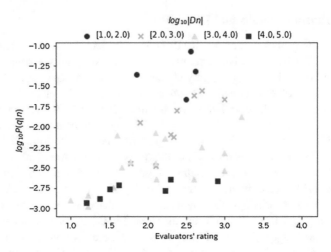

Fig. 4. Relations among $\log_{10}P(q|n)$, $\log_{10}|D_n|$, and evaluators' ratings when query is "North America."

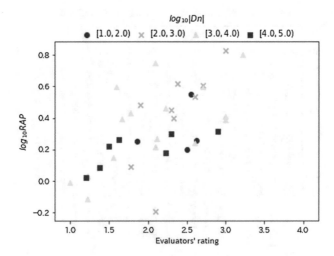

Fig. 5. Relations among RAP, $\log_{10}|D_n|$ and evaluators' ratings when a query is "North America."

similar to Fig. 4. On the other hand, there are points that have high evaluators' ratings but a low RAP value in Fig. 7. To find terms corresponding to these points, we plot terms instead of the points (Fig. 8). Comparing terms whose evaluators' ratings are from 3.0 to 3.5, although common terms such as "Earth" and "polar region" have high RAP values, other terms such as "compass" and "direction" have low RAP values.

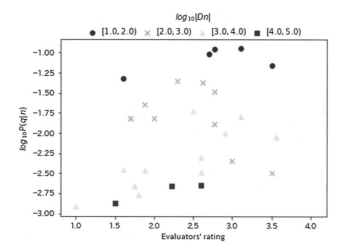

Fig. 6. Relations among $\log_{10}P(q|n), \log_{10}|D_n|$, and evaluators' ratings when query is "North Pole."

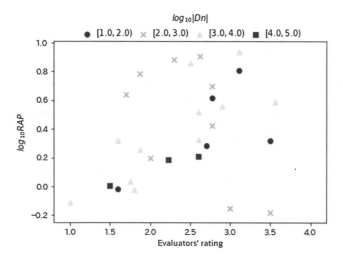

Fig. 7. Relations among RAP, $\log_{10}|D_n|$, and evaluators' ratings when a query is "North Pole."

In other words, "compass" and "direction" have high evaluators' ratings but low RAP values. In these terms, low RAP values are caused by low $\log_{10}P(q|n)$. These terms are often used in a context different from a query term. Therefore, considering the context of each sentence may possibly improve our method.

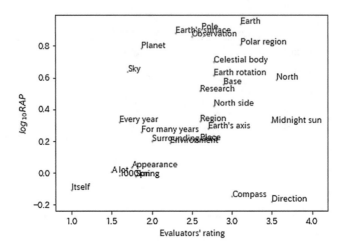

Fig. 8. Relations among RAP, $\log_{10}|D_n|$, and evaluators' ratings when a query is "North Pole."

5   Conclusions

We proposed a measure of relevancy between a query and co-occurring term using the ratio of actual and predicted values of co-occurrence, RAP. We used piecewise approximation lines as a prediction model of co-occurrence for each query. We express term relevancy using the distance between actual and predicted values of co-occurrence. For the experiments, we collected the gold standard of term relevancy by crowdsourcing. In the experiments, we proved RAP was significantly better than the baseline without predicted values. For future work, we will apply RAP to measure the typicality and judge the credibility of Web pages.

Acknowledgements. This work was partially supported by JSPS KAKENHI Grant Number JP17K00429.

References

1. Akamine, S., Kawahara, D., Kato, Y., Nakagawa, T., Inui, K., Kurohashi, S., Kidawara, Y.: WISDOM: a web information credibility analysis systematic. In: Proceedings of the ACL-IJCNLP 2009 Software Demonstrations, pp. 1–4 (2009)
2. Yamamoto, Y., Tanaka, K.: Enhancing credibility judgment of web search results. In: Proceedings of the SIGCHI Conference on Human Factors in Computing Systems, pp. 1235–1244. ACM (2011)
3. Yumoto, T., Yamanaka, T., Nii, M., Kamiura, N.: Finding rare information from the web using social bookmarks and word co-occurrence. Int. J. Biomed. Soft Comput. Hum. Sci. **22**(1), 9–18 (2017)
4. Church, K.W., Hanks, P.: Word association norms, mutual information, and lexicography. Comput. Linguist. **16**(1), 22–29 (1990)

5. Cilibrasi, R.L., Vitanyi, P.M.: The Google similarity distance. IEEE Trans. Knowl. Data Eng. **19**(3), 370–383 (2007)
6. Mikolov, T., Chen, K., Corrado, G., Dean, J.: Efficient estimation of word representations in vector space. ICLR Workshop (2013)
7. Kim, Y.: Convolutional neural networks for sentence classification. arXiv preprint arXiv: 1408.5882 (2014)
8. Poria, S., Cambria, E., Gelbukh, A.: Aspect extraction for opinion mining with a deep convolutional neural network. Knowl. Based Syst. **108**, 42–49 (2016)
9. Kudo, T., Yamamoto, K., Matsumoto, Y.: Applying conditional random fields to japanese morphological analysis. In 2004 Conference on Empirical Methods in Natural Language Processing (EMNLP2004), pp. 230–237 (2004)

Small Watershed Stream-Flow Forecasting Based on LSTM

Le Yan[1(✉)], Jun Feng[1], and Tingting Hang[1,2]

[1] College of Computer and Information, Hohai University, Nanjing, China
{yanle,fengjun,httsf}@hhu.edu.cn
[2] Department of Electrical and Information Engineering,
Hohai University Wentian College, Maanshan, China

Abstract. In recent years, researchers in the field of pattern recognition utilize Artificial Intelligence (AI) based methods to solve the problem of Stream-flow forecasting. Forecasting hydrological time series is of great importance yet very challenging as it is affected by many complex correlations factors in small watershed. In this paper, we forecast the flow values over the next 6 h, using a time series model based on Long Short-Term Memory (LSTM) that considers past stream-flow data, past weather data and weather forecasts data of the hydrological stations. We evaluate our model by root mean square error ($RMSE$), median absolute error ($MedAE$) and coefficient of determination (R^2) in our study case. Compared with the SVM prediction model, the LSTM prediction model has better prediction accuracy, especially the flood peak flow forecast has better performance. The model has achieved satisfactory results in river flow predicting and provided a new method for flood forecasting in small watershed.

Keywords: Machine-learning model · Stream-flow forecasting · LSTM · Small watershed

1 Introduction

Small watershed area have complex and diverse hydrogeological features, boundary conditions and human activities. In addition, there are nonlinear interactions between these factors. Therefore, these make stream-flow forecasting in small watersheds challenging. The conventional black box time series models such as Auto Regressive (AR), Moving Average (MA), Auto Regressive Moving Average (ARMA), Auto Regressive Integrated Moving Average (ARIMA), Auto Regressive Integrated Moving Average with exogenous input (ARIMAX), Linear Regression (LR), and Multiple Linear Regression (MLR) have been applied for stream-flow forecasting [1]. Such models are unable to handle non-stationary and non-linearity involved in hydrological processes. As a result, many researchers have focused on using data-driven techniques for stream-flow forecasting [2]. Generally speaking, traditional data-driven models have exhibited significant progress in forecasting non-linear Hydrological applications and in capturing the noise complexity in data sets.

Recently, with the one of machine learning methods, deep learning reporting huge success by Hinton [3], the deep NN has received increasing attention both academically

S. Lee et al. (Eds.): IMCOM 2019, AISC 935, pp. 1006–1014, 2019.
https://doi.org/10.1007/978-3-030-19063-7_79

and practically [4], and more and more researchers have been trying to apply deep NN methods to stream-flow forecasting [5–7]. On the other hand, stimulated by the success of LSTM on machine translation, a few studies have explored the power of Long Short-Term Memory (LSTM) on time series prediction and obtained promising results [8–10]. So we create LSTM models using hydrological time series data to predict small watershed flow.

This main contribution of the paper is to propose an LSTM network for stream-flow prediction, which has the ability to capture long-term pre-information from hydrological data. To the best of our knowledge, this is the first LSTM architecture for stream-flow prediction in the small watershed. And it is reasonable to help discover and analyze inherent patterns between some hydrological data features and river flow, especially for small watershed whose flood mechanism is too complex to build a physical hydrology model.

The objectives of this research are to investigate the performance of LSTM on forecasting hydrologic time series data collected by a small watershed in China and to compare with the performance of traditional methods such as SVM. The remainder of this paper is organized as follows: In Sect. 2, we describe the related work including data-driven model and LSTM model. In Sect. 3, we provide a general description of the study area, data sets and evaluate the simulation results. We draw conclusions in Sect. 4.

2 Related Work

Considering the relevance to the proposed method, we describe the data-driven model and context algorithms in detail in this section.

2.1 Data-Driven Model

With the development of artificial intelligence technologies, researchers from the machine learning community have proposed some methods to build data-driven model, including, Artificial Neural Networks (ANN), Support Vector Machine (SVM), Fuzzy logic, Evolutionary computation (EC), Wavelet-Artificial Intelligence (W-AI) model, deep learning methods and so on [11]. Kisi [12], using different ANN algorithms make a short-term stream-flow forecast. Four different ANN algorithms are conjugate gradient, back propagation, Levenberg–Marquardt (LM) and cascade correlation, and compared with the basic ANN model, it has better forecast accuracy. Mehr [16] proposed a method based on eight different continuous station forecasting schemes, using feedforward-back propagation (FFBP) neural network algorithm as a cruel search tool to find the best stream-flow forecasting scheme.

Lately, He et al. [13] studied three different methods of data-driven SVM, ANN, and adaptive neuro-fuzzy inference system (ANFIS) for forecasting flow of watershed exportation section in a semiarid environment. A detailed evaluation of the overall performance shows that the SVM model performed superior to ANN and ANFIS in forecasting flow of watershed exportation section for the semiarid region. Adamowski [14] developed a short-term stream-flow forecasting model based on wavelet and evaluated their accuracy compared with ANN models. The wavelet-based models indicate great accuracy for short-term lead times stream-flow forecasting. Wu et al. [18] constructed a hierarchical Bayesian network for stream-flow forecasting in small watersheds. They establish the entity and connection of Bayesian network to represent the variables and physical process of Xin'anjiang model, which is a famous physical model. The model embeds hydrological expert knowledge appropriately and has high rationality and robustness.

Because of the great potential of discovering effective features from hydrological data, some researchers utilizing depth learning technique for stream-flow forecasting. Bai et al. [17] propose a multi-scale depth feature learning (MDFL) method based on a hybrid model is proposed for daily stream-flow forecasting of reservoirs. It is shown that the combination of a deep frame with multi-scale and mixed observation has a good exploration in reservoir inflow prediction. Zhuang et al. [4] establish a new spatio-temporal convolution neural network (ST-CNN) to make full use of the spatio-temporal information to forecasting long-term river flow. It also shows that four hidden layers (two convolutional layers and two pooling layers) can produce comparable results. There are some differences between the above research methods, we focus on short-term stream-flow forecasting in small watershed. On the other hand, inspired by the high potentials of LSTM on machine translation, a few researchers have explored the power of LSTM on time series forecasting and achieved promising results [9, 15]. We utilize the LSTM network to create a short-term flow forecasting model in small watershed whose flood mechanism is too complex to build a base on the physical hydrology model.

2.2   LSTM

Long short-term memory (LSTM) units are units of a recurrent neural network (RNN). An RNN composed of LSTM units is often called an LSTM network. The LSTM replaced the ordinary neuron in the hidden layer with a memory cell and three gates: an input gate, a forget gate and an output gate. The input gate decides how much new input information is added to the cell state, the forget gate decides whether the current cell state is abandon, and the output gate decides whether the current cell value is output.

A schematic of unit of a LSTM is shown in Fig. 1.

Input gate:

$$i_t = \sigma(u_i x_t + w_i h_{t-1} + b_i) \tag{1}$$

Forget gate:

$$f_t = \sigma(u_f x_t + w_f h_{t-1} + b_f) \tag{2}$$

Output gate:

$$o_t = \sigma(u_o x_t + w_o h_{t-1} + b_o) \tag{3}$$

Cell state:

$$c_t = f_t * c_{t-1} + i_t * \tilde{c}_t \tag{4}$$

$$\tilde{c}_t = tanh(u_c x_t + w_c h_{t-1} + b_c) \tag{5}$$

$$h_t = o_t * tanh(c_t) \tag{6}$$

Where i_t, f_t, o_t and c_t are the input, forget, output gates and the LSTM cell state vectors at time t. Similarly, h_{t-1}, h_t denote the output gate of the cell state vector at time t-1, t, respectively. x_i denotes the input of the current cell. σ denotes the Sigmoid activation function. u_i, u_f and u_o denote the matrix of weights from the input, forget, and output gates to the hidden layer, respectively. Similarly, w_i, w_f and w_o denote the weights matrix from the input, forget, and output gate to the input layer, respectively. b_i, b_f, b_o denote the input, forget, and output gate bias vectors. The initial values are $c_0 = 0$ and $h_0 = 0$ and the subscript t indexes the time step.

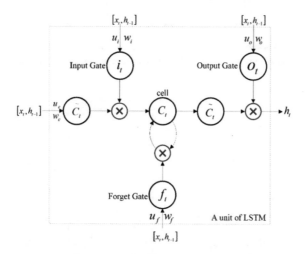

Fig. 1. An LSTM memory unit.

3   Experiments

3.1   Case Study Area

Tunxi (TX) catchment is a small watershed area of 2696.76 km^2 located in the southeast of China. TX catchment physical feature of a place is higher in the West and lower in the east, and the maximum, minimum and average altitude above sea-level are 1398 m, 116 m and 380 m, respectively. The mean annual precipitation is 1600 mm, and the annual distribution of precipitation is uneven. The precipitation is more possibilities lead to flood disasters, especially in April-June. Figure 2 is an overview of TX catchment, in which we can see 11 rainfall stations, 1 evaporation station and 1 river gauging station. In Fig. 2, the rainfall stations are denoted by green dots and river gauging station are denoted by the red triangle. In particular, the TX station is not only a river gauging station but also a rainfall station and evaporation station. This study focuses on forecasting river flow at the TX station.

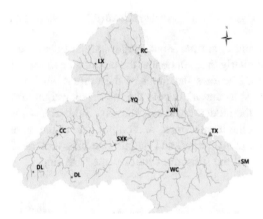

Fig. 2. The map of various types of stations in Tunxi. Note that river gauging Station TX is denoted by the red triangle, others rainfall stations are denoted by green dots.

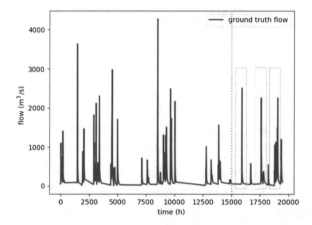

Fig. 3. The data set of the forecast object. The green dash line separates the data into two sets: the training and testing sets. Three floods data in the black dotted box are used to demonstrate the performance of the forecast model.

3.2   Datasets and Model Training

We collected about 19000 data of TX from 1991 to 1999 in flood period as our original dataset. As shown in Fig. 3, we take 15,000 of the data as the training set, and the rest as the test set, and show the forecast results of the three floods using the LSTM model and SVM model.

As the primary aim of this paper, to anticipate the occurrence of flood events and provide early warning, different data-driven models were developed. The purposes of the developed models are to predict the one to six hours river flow in the future. The input data of the model is the past hydrology, past weather and weather forecast data of the stations in the study area. The output of the model is the predicted river flow.

In this study, the kernel function of the SVM model is radial basis function. For LSTM, the ADAM method was selected as the optimization algorithm. The SVM model was implemented using Matlab, R2014b. The LSTM model was programmed using TensorFlow. The dimension of hidden output, the learning rate, weight decay and batch size of LSTM model are set as 128, 0.00225, 10-6 and 100, respectively.

3.3   Model Performance Criteria

The performance of the developed models was evaluated by three criteria, root mean square error (*RMSE*), median absolute error (*MedAE*) and coefficient of determination (R^2).

RMSE is a commonly used evaluated of the differences between values of river flow predicted by a model and the values observed. *RMSE* is always non-negative. In general, a lower *RMSE* is better than a higher one.

The calculation of *RMSE* is as follows:

$$RMSE = \sqrt{\frac{1}{n_{samples}} \sum_{i=1}^{n_{samples}} |y_i^{obs} - y_i^{pre}|} \tag{7}$$

The median absolute error (*MedAE*) is a robust measure of the variability of deviation of observed values of river flow with predict values of river flow. Similarly, a lower *MedAE* is better than a higher one.

The calculation of *MedAE* is as follows:

$$MedAE = median \left(\left| y_1^{obs} - y_1^{pre} \right|, \cdots, \left| y_n^{obs} - y_n^{pre} \right| \right) \tag{8}$$

The coefficient of determination denoted R^2, it is a statistic used in the models whose main purpose is either the prediction of river flow, on the basis of other related hydrological information. It provides a measure of how well-observed outcomes are

replicated by the predicted model, based on the proportion of total variation of outcomes explained by the predicted model. In general, a higher R^2 is better than a lower
one. The calculation of R^2 is as follows:

$$R^2 = 1 - \frac{\sum_{i=1}^{n_{samples}} \left(y_i^{obs} - y_i^{pre}\right)^2}{\sum_{i=1}^{n_{samples}} \left(y_i^{obs} - \bar{y}_i^{pre}\right)^2} \tag{9}$$

Where y_i^{obs} is the i-th observed values of river flow, y_i^{pre} is the i-th predict values of
river flow, $\bar{y}^{pre}$ is the mean of predict values of river flow, $n_{samples}$ is number of test
samples.

3.4   Results

We show the comparison of *RMSE*, *MedAE* and R^2 model between LSTM and SVM in
Table 1. We could find that as the predicted step size increases, the performance of all
models become worse. Compared with the SVM model, the LSTM model has higher
RMSE and R^2 values, it indicates the LSTM model has better prediction accuracy.
Compared with the SVM model, the LSTM model has lower *MedAE* values, it indi-
cates the LSTM model has better robustness.

Table 1. Performance comparisons between SVM model and LSTM model.

Time steps	SVM model			LSTM model		
	RMSE	MedAE	R^2	RMSE	MedAE	R^2
T+2	82.008	18.666	0.911	37.965	1.575	0.981
T+4	95.489	18.784	0.880	57.648	1.961	0.956
T+6	104.043	12.918	0.857	80.480	3.750	0.915
Average	93.847	16.789	0.883	**58.698**	**2.429**	**0.951**

    To further illustrate the performance of AVM and LSTM model in a more intuitive
way, predicted river flow for 3 rainfall events in the testing sets were drawn in Fig. 4.
We compare the 2 h, 4 h and 6 h river flow prediction results of SVM and LSTM with
the ground-truth values, respectively, in every rainfall event. Especially, we find wrong
predictions are easy to occur during flood peak period for SVM methods. Instead, the
LSTM model has a preferable performance in forecasting the time of flood peak
occurrence and the values of flow peak. We could conclude the LSTM model are
suitable to predict floods of small rivers.

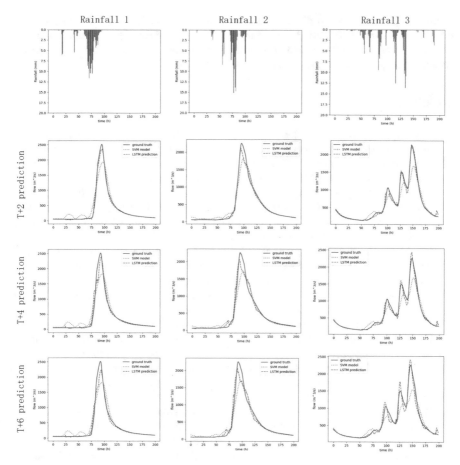

Fig. 4. Comparison with the ground truth river flow and predicted river flow computed by SVM and LSTM. Where each column represents a rainfall event, Lines 2, 3 and 4 represent the comparison of the SVM and LSTM with the predicted step size of 2 h, 4 h and 6 h, respectively.

4   Conclusion

In this study, an LSTM model is proposed for predicting river flow in a small watershed, which is capable to effectively focus on hidden flood factors from previous hydrological sequences. Experimental results on Tunxi dataset show that the proposed method is superior to the comparison method and the proposed LSTM model is effective. The newly proposed model provides a valuable tool for predicting river flow. It can provide a new method to predict flood in small Watershed places with complex hydrogeological characteristics. In the future, we will try to combine the LSTM model with other methods, such as principal component analysis, wavelet transform and attention mechanism. In addition, the proposed method is applied to other hydrological exploration, such as medium-term stream-flow forecasting and stream-flow forecasting in a large watershed.

Acknowledgments. This work was supported by the National Key R&D Program of China (Grant No. 2016YFC0402710), partially supported by the National Natural Science Foundation of China under the Grant No. 61370091 and No. 61602151, partially supported by the Outstanding Young Talents Support Program of Anhui Provincial.

References

1. Yaseen, Z.M., El-Shafie, A., Jaafar, O., Afan, H.A., Sayl, K.N.: Artificial intelligence based models for stream-flow forecasting: 2000–2015. J. Hydrol. **530**, 829–844 (2015)
2. Valipour, M.: Long-term runoff study using SARIMA and ARIMA models in the United States. Meteorol. Appl. (2015). http://dx.doi.org/10.1002/met.1491
3. Han, S., Coulibaly, P.: Bayesian flood forecasting methods: a review. J. Hydrol (2017)
4. Zhuang, W.Y., Ding, W.: Long-lead prediction of extreme precipitation cluster via a spatiotemporal convolutional neural network. In: Proceedings of the 6th International Workshop on Climate Informatics: CI (2016)
5. Liu, F., Xu, F., Yang, S.: A flood forecasting model based on deep learning algorithm via integrating stacked autoencoders with BP neural network. In: Proceedings of IEEE International Conference on Multimedia Big Data, pp. 58–61 (2017)
6. Krizhevsky, A., Sutskever, I., Hinton, G.E.: ImageNet classification with deep convolutional neural networks. In: NIPS (2012)
7. Lecun, Y., Bengio, Y., Hinton, G.: Deep learning. Nature **521**(7553), 436–444 (2015)
8. Zaytar, M.A., Amrani, C.E.: Sequence to sequence weather forecasting with long short-term memory recurrent neural networks. Int. J. Comput. Appl. **143**, 7–11 (2016)
9. Ma, X., Tao, Z., Wang, Y., Yu, H., Wang, Y.: Long short-term memory neural network for traffic speed prediction using remote microwave sensor data. Trans. Res. Part C: Emerg. Technol. **54**, 187–197 (2015)
10. Xingjian, S.H.I., Chen, Z., Wang, H., Yeung, D.Y., Wong, W.K., Woo, W.C.: Convolutional LSTM network: a machine learning approach for precipitation nowcasting. In: Advances in Neural Information Processing System, pp. 802–810 (2015)
11. Nourani, V., Baghanam, A.H., Adamowski, J., Kisi, O.: Applications of hybrid wavelet–artificial intelligence models in hydrology: a review. J. Hydrol. **514**, 358–377 (2014)
12. Kisi, Ö.: Streamflow forecasting using different artificial neural network algorithms. J. Hydrol. Eng. **12**, 532–539 (2007). http://dx.doi.org/10.1061/(ASCE)1084-0699
13. He, Z., Wen, X., Liu, H., Du, J.: A comparative study of artificial neural network, adaptive neuro fuzzy inference system and support vector machine for forecasting river flow in the semiarid mountain region. J. Hydrol. **509**, 379–386 (2014). https://doi.org/10.1016/j.jhydrol.2013.11.054
14. Adamowski, J.: River flow forecasting using wavelet and cross-wavelet transform models. Hydrol. Process. **22**, 4877–4891 (2008)
15. Xingjian, S.H.I., Chen, Z., Wang, H., Yeung, D.Y., Wong, W.K., Woo, W.C.: Convolutional LSTM network: a machine learning approach for precipitation nowcasting. In: Advances Neural Information Processing System, pp. 802–810 (2015)
16. Mehr, A.D., Kahya, E., Sahin, A., Nazemosadat, M.J.: Successive-station monthly streamflow prediction using different artificial neural network algorithms. Int. J. Environ. Sci. Technol. (2014). http://dx.doi.org/10.1007/s13762-014-0613-0
17. Bai, Y., Chen, Z., Xie, J., Li, C.: Daily reservoir inflow forecasting using multiscale deep feature learning with hybrid models. J. Hydrol. **532**, 193–206 (2016)
18. Wu, Y., Xu, W., Feng, J., Shivakumara, P., Lu, T.: Local and global bayesian network based model for flood prediction. In: Proceedings of ICPR (2018)

Novel Approach for Multi-valued Truth Discovery

Jun Feng, Ju Chen$^{(\boxtimes)}$, and Jiamin Lu

College of Computer and Information, Hohai University, Nanjing, China
{fengjun, jiamin.luu}@hhu.edu.cn, ChenJu_Stu@163.com

Abstract. Although several methods have been developed in the past to discover the truth from conflicting data sources, many of them share the drawback of assigning the same default weight to all data sources in the beginning of truth discovery, without introducing any a priori knowledge concerning the data sources. This weakness limits the applicability of these methods' resulting in them being greatly discounted. In the worst case, when untrusted data sources make up the majority, the process of truth discovery can be completely derailed. To address this issue, a novel approach for multi-valued truth discovery called MTDF is proposed in this paper, which attempts to solve the problem by adding an initial data source weight building component. Besides, for the uniqueness of multi-valued truth discovery problems in terms of the single-valued ones, the value list obtained from data sources for a certain object is separated into atomic values, and a Bayesian analysis is used to calculate the probability of each atomic value being true. As a result, a list of atomic values that have a high probability to be true is obtained. Experiments conducted on three real datasets show that MTDF has a better performance in multi-valued truth discovery tasks than other methods.

Keywords: Multi-valued truth discovery · Truth discovery · Bayesian analysis

1 Introduction

In this era of big data, variety and volume are the two most important characteristics of data. Owing to the rapid development of information technology, many frontier technologies such as knowledge graph construction and question answering (QA) systems can integrate information from a variety of data sources. However, conflicts often occur among the values obtained from different sources concerning the same object due to the lack of background knowledge, outdated values, spelling mistakes, misprints, malicious deception, etc. Hence, in order to obtain high-quality information and maximize resource utilization, researchers began the study of truth discovery. It aims at resolving conflicts among multiple potential values concerning the same object and obtaining the correct answers. This concept was first proposed in 2008 by Yin et al. [1].

Most existing truth discovery methods [1–13] assume that there is only one truth for each object, which is not always the case. For example, one person may have more than one child or a book may be authored by several people. Research on multi-valued

© Springer Nature Switzerland AG 2019
S. Lee et al. (Eds.): IMCOM 2019, AISC 935, pp. 1015–1028, 2019.
https://doi.org/10.1007/978-3-030-19063-7_80

truth discovery has begun recently [14–19] and the current strategies are not yet mature enough. A serious weakness that exists in both multi-valued and single-valued truth discovery methods is that all data sources are assigned the same default weight at initialization, limiting the applicability of these methods since it requires most of the data sources involved to be reliable, and the direction of truth discovery may be far from the truth otherwise. Li et al. [20] list several possible improvements for data fusion and propose that using the sampled trustworthiness as initial data source weight can improve the results for all truth discovery methods. Their findings concur with our point of view, but unfortunately, their study was not an in-depth study and they did not provide concrete methods for initial data source weight building by sampled trust-worthiness. Furthermore, the uniqueness of the multi-valued truth discovery gives rise to more complex problems than those of single-valued truth discovery. For example, false positives and false negatives should differentiate in the former. In this paper, we attempt to provide solutions to the aforementioned problems.

To summarize, our main contributions are as follows:

1. We assign initial weights to data sources using stratified sampling instead of assigning the same default weight to every participating data source. We stratify objects according to the difficulty of discovering their multi-valued truth and use information entropy to reflect the difficulty degree. Experiments using the existing method TruthFinder with the initial source weight assignment component of our solution show that this change can effectively improve the precision of multi-valued truth discovery tasks.
2. In addition to the precision, we propose the use of a true negative rate to com-prehensively measure the quality of data sources in multi-valued truth discovery problems. For the uniqueness of the multi-valued truth discovery problems in terms of the single-valued ones, we separate each value list obtained from different sources for each object into atomic values and apply Bayesian analysis to calculate the probability of each atomic value being true. As a result, a list of atomic values that have high probability of being true is obtained.
3. The experimental results for three real data sets show that our proposed method MTDF generally achieves a better performance in both precision and recall than other existing methods.

2   Related Work

Based on the type of problem that has to be solved, the existing works on truth discovery can be divided into single- and multi-valued truth discovery methods. The former can be further divided into iterative [1–5], optimization-based [6–10], and cluster-based [11–13] methods. However, in this paper, we will mainly focus on multi-valued truth discovery methods.

To the best of our knowledge, there is not a large amount of research on multi-valued truth discovery methods. Fang [14] proposes a full-fledged graph-based model called SmartMTD that incorporates important implications for truth discovery of multi-valued objects. Wang et al. [15] developed a new definition of mutual exclusion to reflect the inter-value implication in the multi-truth discovery context and a finer-grained copy detection method to cope with sources that have large profiles. Furthermore, Wang et al. [16] incorporate two measures, namely the calibration of imbalanced positive/negative claim distributions and the consideration of the implication of values' co-occurrence in the same claims to improve the truth discovery accuracy. Li et al. [17] propose a model called HYBRID that jointly makes two decisions: how many truths there are, and what they are. It considers the conflicts between values as important evidence for ruling out wrong values, while keeping the flexibility of allowing multiple truths. Lin and Chen [18] propose an integrated Bayesian approach to incorporate the domain expertise of data sources and the confidence scores of value sets. Zhao et al. [19] propose a probabilistic graph model called the latent truth model (LTM) that introduces the truth as a latent variable and models the generative error process and two-sided source quality using the Bayesian approach.

3 Methods

This section is structured as follows. Firstly, we present a brief problem formulation. Next, we illustrate the first step in the method – initial data source weight building. Following this, we explain the process of calculating the probability that an atomic value is true. The updating model of quality measures of each data source is then presented. Finally, we present our algorithm.

3.1 Problem Formulation

Mathematical symbols used in this paper are listed in Table 1.

Table 1. Mathematical symbols used in this study.

Name	Description
S	Set of data sources
O	Set of objects
V_j^i	List of atomic values provided by s_i on object o_j
S_o^V	Set of sources that provide value list V on object o
$S_o^{\bar{V}}$	Set of sources that do not provide value list V on object o
S_o^v	Set of sources that provide atomic value v on object o
$S_o^{\bar{v}}$	Set of sources that do not provide atomic value v on object o
S_o	Set of sources that provide values on object o
V(s)	Set of lists provided by source s
O(s)	Set of objects for which source s provide values
V_j	Ground true value list for object o_j
V_j'	List of distinct false atomic values consisting of all sources on o_j
TV_j^i	List of true negatives from source s_i on object o_j

Generally, multi-valued truth discovery problems involve a set of sources $S = \{s_1, s_2, \ldots, s_m\}$ that provide values on a set of objects $O = \{o_1, o_2, \ldots, o_n\}$. Each source s_i provides a list of atomic values V_j^i for object o_j. For example, a certain bookstore may provide a list of authors for a certain book, and each author is considered as an atomic value. However, similar to reality, different data sources may give different list of values for the same object, causing conflict. We focus on resolving conflicts among data sources and providing a list of values with the highest probability of being true.

Unlike with single-valued truth discovery, in the case of multi-valued truth discovery, the value list provided by different data sources cannot simply be categorized into right and wrong cases. For example, some sources may provide a part of a true list with some omissions, while some sources may provide the full part of a true list but include some mistakes. In these cases, we cannot simply use precision (the probability of an atomic value provided by a certain source is true) to distinguish the data sources. In this study, tn represents the number of atomic values that are not provided by a source indeed do not belong to the truth list, fn represents the number of the atomic values that are not provided by a source that indeed belong to the truth list, tp represents the number of atomic values that are provided by a source that indeed belong to the truth list, and fp represents the number of the atomic values that are provided by a source that indeed do not belong to the truth list. To comprehensively reflect the quality of a data source, the Precision score and True negative score are used to evaluate every data source involved. They are defined as follows:

Definition 1 (Precision score). The precision score of a data source s, denoted by Pre(s), can be calculated using $\mathrm{Pre}(s) = \frac{tp}{tp+fp}$. It is the probability of an atomic value provided by s being true.

Definition 2 (True negative score). The true negative score of a data source s, denoted by Tne(s), can be calculated using $\mathrm{Tne}(s) = \frac{tn}{tn+fn}$ is the probability of a value not occurring in s being false.

3.2   Initial Data Source Weight Building

Since the distribution of initial data source weights has a significant influence on the entire truth discovery process, an initial data source weight building component was added in the truth discovery process, reflecting the credibility of data sources as objectively as possible. Furthermore, to improve the efficiency, the trustworthiness of data sources was assumed by exploiting as few samples as possible.

Based on the considerations above, objects were divided into three levels: simple, medium, and difficult according to how difficult it is to discover the truths associated with them. Since we believe that the more the candidate values of a certain object, the more the conflicts, and the harder it is to find true values, information entropy was used to measure the degree of conflict for different objects. The difficulty score of an object o is calculated using the following formula:

$$\text{Diff}(o) = -\sum_{V \in V_o} \frac{|S_o^V|}{|S_o|} \log \frac{|S_o^V|}{|S_o|} \tag{1}$$

Next, the objects were sorted by their difficulty score and two parameters - θ_1, θ_2 were chosen based on the difficulty score distribution. Objects with $\text{Diff}(o) \leq \theta_1$ are simple level objects, objects with $\theta_1 < \text{Diff}(o) \leq \theta_2$ are medium level objects, and objects with $\text{Diff}(o) > \theta_2$ are difficult level objects. Then, a certain number of objects in each level is randomly selected and a Precision score and a True negative score are given to each source based on a comparison of the values they provided and the ground truth. The truths for all objects in the sample are collected via manual judgement. For each data source, a vector representing all the values it provides is created, i.e., $V^i = \left(V_1^i, V_2^i, \ldots, V_j^i, \ldots, V_n^i \right)$, where V_j^i represents the value list given by s_i on o_j. Accordingly, a ground truth vector $T = (V_1, V_2, \ldots, V_n)$ is also established. Each value list - V_j^i, is split into atomic values, scores are assigned to V_j^i in terms of the Precision and True negative rate, which can be represented as follows:

$$\text{Pre}\left(V_j^i \right) = \frac{\left| Inter\left(V_j^i, V_j \right) \right|}{\left| V_j^i \right|} \left(\sum_{v \in Inter\left(V_j^i, V_j \right)} \log \frac{|S_{o_j}|}{|S_o^v|} \right) \tag{2}$$

$$\text{Tne}\left(V_j^i \right) = \frac{\left| V_j' - \left(V_j^i - Inter\left(V_j^i, V_j \right) \right) \right|}{\left| V_j' - \left(V_j^i - Inter\left(V_j^i, V_j \right) \right) \right| + \left| V_j - Inter\left(V_j^i, V_j \right) \right|} \left(\sum_{v \in TV_j^i} \log \frac{|S_o^v|}{|S_{o_j}|} \right) \tag{3}$$

In both formulas, $V_j' = \bigcup_1^{|O|} \left(V_j^i - Inter\left(V_j^i, V_j \right) \right)$, $TV_j^i = V_j' - \left(V_j^i - Inter \left(V_j^i, V_j \right) \right)$. The popularity of true positives and true negatives are taken into consideration while calculating the Precision score and True negative score of the provided value list. Little credit is assigned to the atomic true value that most sources can provide, and this rule is reflected in the formulas above. The following formulas are used to obtain a final score for source s_i:

$$\text{Pre}(s_i) = w_1 \cdot simple_i^{Pre} + w_2 \cdot medium_i^{Pre} + w_3 \cdot difficult_i^{Pre} \tag{4}$$

$$\text{Tne}(s_i) = w_1 \cdot simple_i^{Tne} + w_2 \cdot medium_i^{Tne} + w_3 \cdot difficult_i^{Tne} \tag{5}$$

Simple, medium, and difficult correspond to the average score that s_i gets based on formulas (2) and (3) on these three levels respectively, and w_1, w_2, w_3 represent different weights assigned to the different difficulty levels of objects. The general principle of MTDF in assigning initial weights to data sources is that the data sources are treated as students undergoing examinations where each is given a test that includes a list of questions with different difficulty levels and different corresponding scores.

3.3   Probability of Atomic Value Being True

Consider a dataset obtained from an online bookstore aggregator, www.abebooks.com. Suppose we want to collaborate with the authors of the book named "C++ Coding standards: Rules, Guidelines, and Best Practices" (ISBN:0321113586). Seven bookstores (sources) in the aggregator provided author information for this book, but only two of them, Gunter Koppon and Stratford Books, provided true values (Table 2). The naïve voting method would predict {Herb, Sutter} as the correct list of authors, as this value list is supported by more sources. Such results may not always be reasonable because they do not take the book stores' Precision score and True negative score into account. If we know that the first three book stores have high Precision score and low True negative score, we would assume the information provided by them is more likely to be true, but there are still possible answers they do not provide. Similarly, if we know that Bobs Books tends to provide information that is as complete as possible, as they do not want to miss any possible correct answers and suffer from low Precision score but enjoy high True negative score, we would assume that the atomic values that do not occur in this source are more likely to be false. In order to obtain an author list with high precision and recall, we must consider not only the Precision score of sources providing each atomic value, but also the True negative score of sources that do not provide them.

Table 2. Websites (i.e., sources) that provide author information about a book.

Source	List of authors
TheBookCom	Herb, Sutter
The Book Depository	Herb, Sutter
SWOOP	Herb, Sutter
Gunter Koppon	Herb, Sutter, Andrei, Alexandrescu
Technischer Overseas	Sutter
Stratford Books	Herb, Sutter, Andrei, Alexandrescu
Bobs Books	John Fuller, Herb, Sutter, Andrei, Alexandrescu

In order to obtain a correct and complete value list for each object, each participating data source should be assessed from both the angle of Precision score and True negative score. Each value list is separated into atomic values, and the probability of each atomic value being true is calculated. For an atomic value v, Pr(v) denotes the probability of v being true, $\Pr(\bar{v})$ denotes the probability of v being false and $\varphi(o)$ denotes the possible world. We assume the a priori belief of each value being true is $\alpha(v)$, which is calculated as:

$$\alpha(v) = \frac{|S_o^v|}{|S_o|} \tag{6}$$

Based on this, Bayesian analysis is used to calculate the probability of v being true, i.e., a posteriori probability $\alpha'(v)$, which is represented using formula (7) below:

$$
\begin{aligned}
\alpha'(v) = \Pr(v) = \Pr(v|\varphi(o)) &= \frac{\Pr(\varphi(o)|v)\Pr(v)}{\Pr(\varphi(o)|v)\Pr(v) + \Pr(\varphi(o)|\bar{v})\Pr(\bar{v})} \\
&= \frac{\Pr(\varphi(o)|v)\alpha(v)}{\Pr(\varphi(o)|v)\alpha(v) + \Pr(\varphi(o)|\bar{v})(1 - \alpha(v))} \\
&= 1 / \left(1 + \frac{\Pr(\varphi(o)|\bar{v})}{\Pr(\varphi(o)|v)} \cdot \frac{1 - \alpha(v)}{\alpha(v)}\right)
\end{aligned}
\tag{7}
$$

Then, the two conditional probabilities, $\Pr(\varphi(o)|v)$ and $\Pr(\varphi(o)|\bar{v})$, are represented based on an analysis of the sources that provide this atomic value and that do not provide this atomic value. S_o^v represents the set of sources providing the atomic value v for object o and $S_o^{\bar{v}}$ represents the set of sources do not provide v for o. To calculate the probability of an atomic value being true, both the Precision score of the data sources that provide it and the True negative score of the data sources that do not provide it should be considered. Thus, formulas (8) and (9) are obtained, which correspond to the probability of v being true and false, respectively.

$$\Pr(\varphi(o)|v) = \prod_{s \in S_o^v} Pre(s) \prod_{s \in S_o^{\bar{v}}} (1 - Tne(s)) \tag{8}$$

$$\Pr(\varphi(o)|\bar{v}) = \prod_{s \in S_o^{\bar{v}}} Tne(s) \prod_{s \in S_o^v} (1 - Pre(s)) \tag{9}$$

Based on formulas (7), (8), and (9), a posteriori probability of v being true $\alpha'(v)$, can be obtained. The advantages of MTDF are that the probability of each value being true is calculated separately, and the Precision score of the sources that provide a certain atomic value and the True negative score of sources not provide this value are both considered when calculating the probability of an atomic value being true, and vice versa.

Now that the procedure for calculating the probability of a certain atomic value being true has been presented, the next question is how many atomic values should be presented in a multi-valued truth discovery. As in reality, the requirements for accuracy and recall vary in different areas. For example, for product recommendation systems, in order to cause as little disturbance to the user as possible, it is desirable that the recommended content is of interest to the user; therefore, it is necessary to improve the precision rather than recall. On the contrary, in a fugitive information retrieval system, it is desirable to not miss the fugitive; hence, it is necessary to increase the recall. Based on this, a parameter σ is set to control the number of atomic values presented. Each atomic value that satisfies $Pr(v) > \sigma$ will be regarded as one of the truths. As a result, a list of atomic values $V_o^{true} = \{v | v \in V(o), Pr(v) \geq \sigma\}$ is returned.

3.4   Updating Quality Measures of Data Sources

As more knowledge regarding the probability of each atomic value being true is obtained, and the truth list is obtained as well. The Precision score and True negative score of the sources must be updated. These scores can be calculated as follows:

$$Pre(s_i) = \sum_1^{|O(s_i)|} \left(\frac{\left| Inter\left(V_j^i, V_j \right) \right| \left(\sum_{v \in Inter\left(V_j^i, V_j \right)} log \frac{|S_{o_j}|}{|S_o^v|} \right)}{\left| V_j^i \right| \cdot (|O(s_i)|)} \right) \tag{10}$$

$$Tne(s_i) = \sum_1^{|O(s_i)|} \frac{\left| V_j' - \left(V_j^i - Inter\left(V_j^i, V_j \right) \right) \right| \left(\sum_{v \in TV_j^i} log \frac{|S_o^v|}{|S_{o_j}|} \right)}{\left| V_j' - \left(V_j^i - Inter\left(V_j^i, V_j \right) \right) \right| + \left| V_j - Inter\left(V_j^i, V_j \right) \right|} / (|O(s_i)|) \tag{11}$$

3.5   Algorithm

This section explains the steps that comprise the algorithm. In line 1, the threshold σ is set. When the probability of each atomic value being true exceeds σ, the value will be regarded as one of the truths in the list for each object. In lines 2 to 7, initial weights are assigned to the data sources involved based on a comparison of the ground truth and the value they provide for objects with different difficulty level. In lines 8 to 17, the probability of each atomic value being true is calculated and the measures of the quality of the data source, namely, Precision score and True negative score are updated. Finally, ground true value list for each object is returned.

Algorithm 1 MTDF: Novel multi-valued truth discovery framework

Input: The sampled object sets O'. The real object set O; The data sources set S; The value lists V_o^s for each s in object o;

Output: $\{v \mid v \in V(o), \Pr(v) > \sigma\}$ for all $o \in O$;

1: $\sigma \leftarrow$ default value; // initialization begin
2: for each $s_i \in$ S do
3:      for each $o \in$ O' do
4:          Pre $(s_i) \leftarrow$ Equation (4)
5:          Tne $(s_i) \leftarrow$ Equation (5)
6:      end for
7: end for // initialization end
8:      for each $s_i \in$ S do
9:          for each $o \in$ O do
10:             for each $v \in V_o$ do
11:                 $\alpha(v) \leftarrow$ Equation(6)
12:                 $\alpha'(v) \leftarrow$ Equation (7), (8), (9)
13:             end for
14:             Pre $(s_i) \leftarrow$ Equation (10)
15:             Tne $(s_i) \leftarrow$ Equation (11)
16:         end for
17:     end for
18: **return** lists of truth for objects

---

4 Experiments

A description of the three real-world datasets used in the experiments is provided in Sect. 4.1. The experimental settings are explained in Sect. 4.2. The performance of the proposed method is explained and comparisons are made with other methods in Sect. 4.3, followed by a discussion.

4.1 Dataset Description

Book-Author Dataset: The first dataset is a book-author data set [1, 2]. It contains 1263 books about computer science and engineering published by Addison-Wesley, McGraw-Hill, Morgan Kaufmann, or Prentice Hall. The dataset contains 24,364 listings from 894 book stores (data sources). The ISBN of each book is used to search for it on www.abebooks.com. Each book is listed on a set of online bookstores that provide the authors of the book. The goal of the experiment is to find the correct list of authors for each book. One hundred books were randomly selected the ground truth were obtained from each book's cover.

Parent-Children Dataset: The biography dataset [5] contains 11,099,730 editing records about people's birth/death dates, spouses, and parents/children on Wikipedia. We obtained records about 55,259 users claiming children for 2,579 people. The goal of the experiment is to find the complete list of parents and children list. We randomly pick 200 parents and decide the latest editing records as the gold standard.

Director-Movie Dataset: The director-movie dataset [18] contains 1,134,432 records from 15 major movie websites. The records describe the association between directors and 16,955 movies. The goal of the experiment is to find the complete list of movies that a director has completed. Three hundred movies were randomly selected and label the directors from the movie posters to get the ground truth.

4.2   Experimental Settings

Methods to compare. The proposed approach MTDF was compared with the methods listed below.

Majority Voting: regards a value as true if it is provided by the majority of the participating data sources.

TruthFinder [1]: considers a value as true if its veracity score exceeds 0.5.

Sample-based TruthFinder: a modified version of the TruthFinder in which an initial data source weight building component is added through stratified sampling, as described in Sect. 3.2.

LTM [19]: considers two types of errors under the scenarios of multi-valued truth, i.e., false positive and false negative, and creates a graphical model to determine the source quality and truthfulness of each value provided for an object.

MBM [15]: develops a new definition of mutual exclusion to reflect the inter-value implication in the multi-truth discovery context and a finer-grained copy detection method to cope with sources that have large profiles.

HYBRID [17]: considers the conflicts between values as important evidence for ruling out wrong values, while keeping the flexibility of allowing multiple truths and making two decisions, i.e., how many truths there are and what they are.

Implementations. Whenever applicable, $\sigma = 0.65$. In addition, 2.5% of each level of objects was used for sampling and the initial weight for each source was obtained. All methods were implemented using Java.

4.3   Experimental Results

Table 3 and Figs. 1, 2, 3 and 4 show the performance of the different methods for the three datasets in terms of precision, recall and F-measure. As shown in the table, the proposed algorithm MTDF achieves the best precision on the three datasets, and attains a relatively higher recall among the other methods. Since when we calculate the probability of each atomic value being true, the Precision score of each source that provides the atomic value and the True negative score of each source that does not provide the atomic value are both taken into consideration.

In addition to the comparisons of the precision, recall, and F-measure, the efficiency of the method was also examined. Six small datasets were created by randomly

Table 3. Comparison of precision and recall of different methods for the three datasets. The best, second best, third best are in bold.

Methods	Book-Author		Parent-Child		Director-Movie	
	P	R	P	R	P	R
Majority voting	0.912	0.753	**0.919**	0.774	**0.920**	0.790
TruthFinder	0.921	**0.890**	0.910	0.841	0.916	**0.882**
Sample-based TruthFinder	**0.936**	0.877	**0.924**	0.846	0.922	0.850
LTM	0.885	0.841	0.906	0.855	**0.923**	**0.885**
MBM	0.863	**0.901**	0.882	**0.894**	0.810	0.817
HYBRID	**0.940**	0.878	0.897	**0.856**	0.861	**0.893**
MTDF	**0.943**	**0.902**	**0.915**	**0.883**	**0.941**	0.880

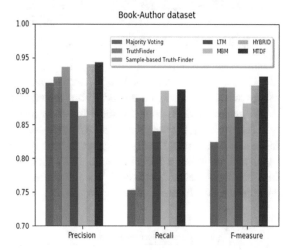

Fig. 1. Comparison of Book-Author dataset

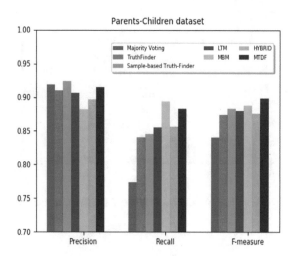

Fig. 2. Comparison of Parents-Children dataset

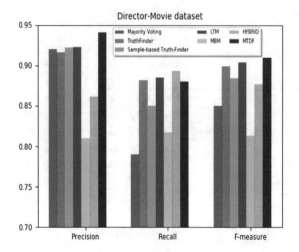

Fig. 3. Comparison of Director-Movie dataset

sampling 2000 to 10000 movies from a set of 111,987 movies and the execution time for each dataset using each method was compared. From the results, which are presented in Fig. 4, LTM generally takes long to finish the truth discovery tasks. This because its procedure is more complex which makes LTM more sensitive to data scale. Majority voting takes the least time mainly because it usually involves a simple counting process. Our method takes a moderate amount of time when compared to other methods while having a better F-measure. To summarize, the results of the experiments demonstrate the effectiveness of our approach.

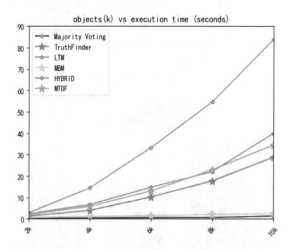

Fig. 4. Execution time of different methods for different numbers of objects

5   Conclusion

We studied the problem of multi-valued truth discovery for an object from conflicting sources. First, initial weights were assigned to data sources via stratified sampling, and information entropy was used to measure the degree of conflict among different candidate values. The value list obtained from different data sources was split to obtain the atomic values, and Bayesian analysis was used to find the probability of each atomic value being true. Then, a parameter was set to control the number of outputs according to real requirements, making our method more flexible. In the process of calculating the probability of each atomic value being true, the precision of each data source that provides the value and the true negative rate of the data sources that do not provide it are considered. The results of the experiments using three real datasets prove that our proposed approach is effective.

Acknowledgements. This paper has been supported in part by The National Key R&D Program of China (Grant No. 2018YFC0407901), and Natural Science Foundation of China (61602151).

References

1. Yin, X., Han, J., Philip, S.Y.: Truth discovery with multiple conflicting information providers on the web. IEEE Trans. Knowl. Data Eng. **20**(6), 796–808 (2008)
2. Dong, X.L., Bertie-Equille, L., Srivastava, D.: Data fusion: resolving conflicts from multiple sources. In: Wang, J., Xiong, H., Ishikawa, Y., Xu, J., Zhou, J. (eds.) 14th International Conference on Web-age Information Management, vol. 7923, pp. 64–76. Springer, Beidaihe (2013)
3. Dong, X.L., Bertie-Equille, L., Srivastava, D.: Truth discovery and copying detection in a dynamic world. Proc. VLDB Endowment **2**(1), 562–573 (2009)
4. Dong, X., Saha, B., Srivastava, D. Less is more: selecting sources wisely for integration. Proc. VLDB Endowment **6**(2), 37–48 (2012)
5. Pasternack, J. Roth, D.: Knowing what to believe (when you already know something). In: Proceedings of the 23rd International Conference on Computational Linguistics (Coling 2010), pp. 877–885. Tsinghua University Press, Beijing (2010)
6. Li, Q., Li, Y., Gao, J., et al.: Resolving conflicts in heterogeneous data by truth discovery and source reliability estimation. In: ACM SIGMOD International Conference on Management of Data, pp. 1187–1198. ACM, UT (2014)
7. Zhang, C., Zhu, L., Xu, C., et al.: LPTD: achieving lightweight and privacy- preserving truth discovery in CIoT. Future Gener. Comput. Syst. (2018). arXiv preprint arXiv:1804.02060 [cs.CR]
8. Yao, L., Su, L., Li, Q., et al.: Online truth discovery on time series data. In: Proceedings of the 2018 SIAM International Conference on Data Mining, pp. 162–170. SIAM, California (2018)
9. Li, Q., Li, Y., Gao, J., et al.: A confidence-aware approach for truth discovery on long-tail data. Proc. VLDB Endowment **8**(4), 425–436 (2014)
10. Yin, X., Tan, W.: Semi-supervised truth discovery. In: 20th International Conference on World Wide Web, pp. 217–226. ACM, Hyderabad (2011)

11. Wang, X., Sheng, Q.Z, Fang, X.S., et al.: Approximate truth discovery via problem scala reduction. In: Proceedings of the 24th ACM International Conference on Information and Knowledge, pp. 503–512. ACM, Melbourne (2015)
12. Zhang, L., Qi, G., Zhang, D., et al.: Latent Dirichlet truth discovery: separating trustworthy and untrustworthy components in data sources. IEEE Access **6**, 1741–1752 (2018)
13. Lamine Ba, M., Horincar, R., Senellart, P., et al.: Truth finding with attribute partitioning. In: Proceedings of the 18th International Workshop on Web and Databases, pp. 27–33. ACM, Melbourne (2015)
14. Fang, X.: Truth discovery from conflicting multi-valued objects. In: 26th International Conference on World Wide Web Companion, pp. 711–715. ACM, Perth (2017)
15. Wang, X., Sheng, Q.Z., Fang, X.S., et al.: An integrated bayesian approach for effective multi-truth discovery. In: Proceedings of 24th ACM International on Conference on Information and Knowledge Management, pp. 493–502. ACM, Melbourne (2015). https://doi.org/10.1145/2806416.2806443
16. Wang, X., Sheng, Q.Z., Yao, L., et al.: Truth discovery via exploiting implications from multi-source data. In: Proceedings of the 25th ACM International on Conference on Information and Knowledge Management, pp. 861–870. ACM, New York (2016)
17. Li, F., Dong, X.L., Langen, A., et al.: Discovering multiple truths with a hybrid model (2017). arXiv preprint arXiv:1705.04915v[cs.DB]
18. Lin, X., Chen, L.: Domain-aware multi-truth discovery from conflicting sources. Proc. VLDB Endowment **11**(5), 635–647 (2018)
19. Zhao, B., Rubinstein, B., Gemmell, J., et al.: A bayesian approach to discovering truth from conflicting sources for data fusion. Proc. VLDB Endowment **5**(6), 550–561 (2012)
20. Li, X., Dong, X.L., Lyons, K., et al.: Truth finding on the deep web: is the problem solved? Proc. VLDB Endowment **6**(2), 97–108 (2012)

Information Extraction from Clinical Practice Guidelines: A Step Towards Guidelines Adherence

Musarrat Hussain and Sungyoung Lee[✉]

Department of Computer Science and Engineering,
Kyung Hee University, Yongin-si, South Korea
{musarrat.hussain, sylee}@oslab.khu.ac.kr

Abstract. Clinical Practice Guidelines (CPGs) are an essential resource for standardization and dissemination of medical knowledge. Adherence to these guidelines at the point of care or by the Clinical Decision Support System (CDSS) can greatly enhance the healthcare quality and reduce practice variations. However, CPG adherence is greatly impeded due to the variety of information held by these lengthy and difficult to parse text documents. In this research, we propose a mechanism for extracting meaningful information from CPGs, by transforming it into a structured format and training machine learning models including Naïve Bayes, Generalized Linear Model, Deep Learning, Decision Tree, Random Forest, and Ensemble Learner on that structured formatted data. Application of our proposed technique with the aforementioned models on Rhinosinusitis and Hypertension guidelines achieved an accuracy of 82.10%, 74.40%, 66.70%, 66.79%, 74.40%, and 83.94% respectively. Our proposed solution is not only able to reduce the processing time of CPGs but is equally beneficial to be used as a preprocessing step for other applications utilizing CPGs.

Keywords: Information extraction · Clinical practice guidelines processing · Guidelines adherence · Text processing

1 Introduction

Clinical Practice Guidelines (CPGs) are "systematically developed statements to assist practitioners and patient decision about appropriate healthcare for specific circumstances" [1]. It has an indispensable role in disseminating medical knowledge, enhancing healthcare quality, reducing cost, and decreasing practice variations. CPGs aims to help healthcare providers and patients to make the best decision about treatment for a particular condition, by picking most suitable strategies in a specific clinical situation [2]. It can either be used by healthcare providers or can be transformed to machine interpretable format to be part of the Clinical Decision Support System (CDSS) to support clinicians at the point of care.

Despite the valuable goal and importance, the adherence rate of CPGs varies between 20% to 100% depending upon clinical scenario and the nature of the CPG [3]. The main hurdle in the adherence of CPGs is the current format (unstructured

© Springer Nature Switzerland AG 2019
S. Lee et al. (Eds.): IMCOM 2019, AISC 935, pp. 1029–1036, 2019.
https://doi.org/10.1007/978-3-030-19063-7_81

document) of the CPG and clinician/healthcare provider unawareness about CPGs. Most of the healthcare providers are unaware about the existence of CPG and they face difficulties in understanding on directing them toward a specific CPG [4].

There are many other obstructs in adherence to CPGs related to clinicians, patients behaviors and CPGs itself [4]. Besides other obstacles, one of the major hurdles in adherence to CPG is the format of the CPGs. Most of the CPGs are published online in medical journals having an unstructured format. CPG contents based on proximity to technical solution can be categorized into two parts: background information and disease specific information. Background information includes anecdotes and thoughts of the author. While more concrete information, relating to causes, consequences, and actions form the disease-specific information. Due to the wide range of variation in the nature of CPGs, it is very important to understand each CPG before transforming it into a machine interpretable format. It requires a lot of time to locate scenario/disease specific information at a limited time during real practice. Therefore, clinicians avoid utilizing CPGs during real practice. Also, it creates difficulties during the conversion of CPGs to computer interpretable format. The adherence rate can be increased by finding a mechanism that can extract relevant information from the CPG and filter out irrelevant information.

The primary goal of this research is to find and extract relevant information also called recommendation statements from CPGs and filter out background information. To achieve this goal, we transform hypertension [5] and Rhinosinusitis [6] CPG text to structured format (word vector) and train machine learning models including Naïve Bayes, Generalized Linear Model, Deep Learning, Decision Tree, Random Forest, and Ensemble Learner. The trained models achieved accuracy of 82.10%, 74.40%, 66.70%, 66.79%, 74.40%, and 83.94% respectively on extracting recommendation statements. This technique has twofold advantages. It can extract disease specific information for a clinician in real time at the point of care. It can also be used as a preprocessing step for CPGs transformation to computer interpretable format, which can increase CPG adherence, improve the healthcare quality and can eventually reduce healthcare cost.

2   Related Work

The history of CPGs started in late 1970 by the National Institute of Health Consensus Development Program. The objective of the program was to improve healthcare quality by identifying and adopting best practices [7].

Yetisgen-Yildiz et al. [8] proposed a text processing pipeline that can identify and extract recommendation sentences from radiology reports using Natural Language Processing (NLP) and supervised text classification techniques. In the pipeline, the task performed in the sequence involves section segmentation, sentence segmentation, and sentence classification. The author used 800 anonymized radiology reports for training the machine learning model which was annotated by radiologist and internal medicine physician independently. The trained model was then used to classify the new input sentence into a recommendation or non-recommendation sentence. However, the nature of radiology reports is completely different from that of CPGs. Radiology report has semi-structured nature while CPGs are completely unstructured documents. These reports are divided into different sections specified for one type of information. The recommendation sentences are written in *Impression Sections* of the report. And also all the reports follow the same structure, therefore, a model can easily be trained on these document.

Khalifa et al. [9] proposed a mechanism that detects cardiovascular risk factors in clinical notes of diabetes patients by using the existing NLP techniques and tools. The risk factors include high blood pressure, high cholesterol levels, obesity, and smoking status. They used existing tools: Apache UIMA Textractor and cTAKES for text preprocessing and the risk factors identification using a regular expression. They considered smoking status risk factor in the study. However, the mechanism is effectively identifying risk factors in the cardiovascular domain. Each domain has different risk factors and a generalized solution is required that can identify and extract the required information from all types of documents.

Priyanta et al. [10] did the comparative analysis of machine learning and rule-based models for sentence subjectivity classification. The rule was generated using opinion patterns. The experiment was performed on classifying Indonesian news sentences into two categories subjective and objective. The machine learning model used for the classification and comparison were Multinomial Support Vector Machine (SVM) and Naïve based classifier (NBC). The experiment was performed on 2550 document, containing 46393 sentences. The experimental result showed that the rule-based classifier outperformed by achieving 80.36% accuracy as compared to SVM 74% and NBC 71%.

Hematialam et al. [11] proposed an automatic technique of finding and extracting recommendation statements in CPGs. The authors used a supervised machine learning model (Naïve Bayes, J48, and Random Forest) that classify CPG sentences into three categories: NC (no condition), CA (condition-action), and CC (condition consequence). The domain expert annotated three type of guidelines (hypertension, chapter 4 of asthma, and rhinosinusitis) and the authors used these guidelines as a training set for training machine learning model. The authors used Part of Speech (POS) tags as a feature to make the model more domain independent. Each action-condition statement has a modifier, and the most used modifiers in the CPGs used by authors in their study were "If", "in", "for", "to", "which", and "when". The candidate statements were found by using regular expressions. The identified candidate statements were transformed/paraphrased to "if condition, then consequences" format for rules generation.

In our previous study [12] we used a semi-automatic technique to extract recommendation statements from CPG documents. We manually analyze the same annotated hypertension guideline [5] as used in this study and extract heuristic patterns that can recognize recommendation statements in the unseen CPGs. The identified patterns were able to extract recommendation statements with 85.54% accuracy. However, the limitation of that work is the manual effort required for pattern extraction. Also, the extracted heuristic patterns depend on the contents of CPGs which may vary in different CGPs format. Therefore, the heuristic patterns may not be able to perform well on all CPGs. To overcome the manual effort, in this research, we used an NLP pipeline that can extract recommendation statements with satisfactory performance.

3   Proposed Methodology

The primary focus of this research is to automatically and efficiently extract recommendation statements from a CPG and filter out background information using state-of-the-art machine learning algorithms. To achieve this goal, we devised an NLP pipeline as depicted in Fig. 1. The proposed pipeline accomplished the aforementioned goal in two major steps, it transforms a CPG into a structured format (Word Vector) and then trained an ensemble learning model that uses the base classifier including Naïve Base, Generalized Liner Model, Random Forest, Deep Learning, and Decision Tree on the generated structured document. The trained model is then used to classify unseen CPG statements to recommendation statement (RS) or non-recommendation statement (NRS). The detail of the process is given in the following sections.

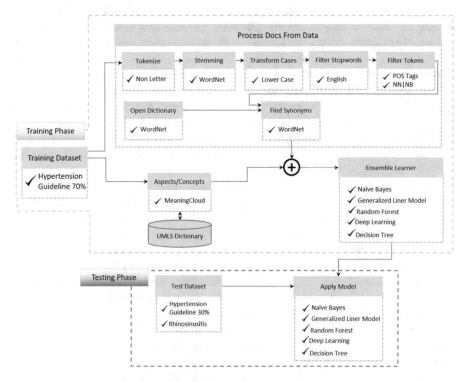

Fig. 1. Recommendation statements extraction pipeline.

3.1   Unstructured to Structured Conversion

In order to transform a given unstructured CPG to a structured format (word vector) multiple steps are designed in sequence.

Initially, the CPG sentences along with label are loaded to working space. We used the Term Frequency-Inverse Document Frequency (TF-IDF) scheme for the creation of word vector. The *Process Docs From Data* consist of others operators ranging from tokenization to synonyms identification. The token operator split the input text into tokens based on word spacing scheme. The words of each token are then transformed to its base format using WordNet stemming followed by *Transform Case* which converts all tokens to its lower case to maintain symmetry. Some of the word tokens despite maximum usage in the document may have limited impact known as stop words removed by *Filter Stopwords*. We applied the Part-of-Speech (POS) using PENN Tree Scheme, employed pattern (NN|NB) to filter the names and verbs used in the input text. As we notice that the CPG recommendation statements mostly consists of disease/medicine name and action on them. We also employed the word expansion mechanism to make the word vector more comprehensive for effective classification. For word expansions, we added synonyms component to the pipeline. We used WordNet dictionary for synonyms identification.

3.2 Aspects/Concepts Extraction

We used MeaningCloud services to find and extract aspects of the input text. We created the local copy of Unified Medical Language System (UMLS) dictionary at MeaningCloud and then used the APIs services for the aspects/concepts extraction based on the created dictionary. The aspects/concepts addition to the data increased the performance of basic classifier as well as ensemble learning base classification as discussed in the result section.

The final outcome of this process is a structured data (word vector) consists of all tokens of interest along with synonyms and their aspects/concepts. This document/ structured data will be then used in the following section for training the machine learning algorithms.

3.3 Ensemble Learner

Ensemble learning combines and applies multiple models on the same instance of data to accurately predict the class label for this instance to reach the final conclusion. The algorithm considered in this study includes Naïve Bayes, Generalized Liner Model, Random Forest, Deep Learning, and Decision Tree. The majority voting technique was used to get the final decision. In this technique, the computed results of each algorithm are analyzed in order to determine the final class recommended by most of the algorithms.

4 Result and Discussion

We performed multiple experiments with different settings on annotated hypertension CPG [5] consists of 78 recommendation statements among total 278 statements. In this study, the CPG statements annotated as CA, CC, or A are considered as RS statements. The CPG was split into 70% and 30% for training and testing part. The training part of the CPG consist of total 195 statements including 58 recommendation statements. While the testing part consists of total 83 statements including 20 recommendation statements. The trained models were also validated on another CPG (Rhinosinusitis [6]) to authenticate the performance (in term of accuracy) of the models.

The experiment that outperformed among others achieved the best accuracy of 79.82% by Ensemble Learner algorithm as shown in Fig. 2. In this experiment, we used TF-IDF for word vector generation, Non Letters for tokenization, WordNet for stemming and English stopwords were filtered out. In the filter tokens component, we observed from multiple experiments that the NN and NB tokens have the maximum contribution in achieving the accurate result. Therefore, we filter out all other tokens. We find the synonyms of the remaining token using WordNet dictionary. However, in this experiment the aspects/concepts for input dataset were not included.

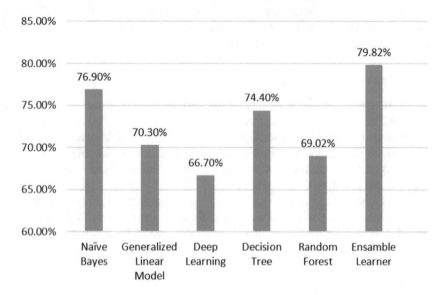

Fig. 2. Results achieved by each algorithm without aspects.

The experiment was repeated with the same setting as earlier but we also find and include aspects because clinical guidelines describe clinical scenarios and normally uses clinical terminology specific to a target disease. To extend the scope of the mechanism to be applicable on any CPG irrespective of the target disease we find the category (Aspect/Concepts) by utilizing UMLS medical dictionary. The final structured data generated is consists of word tokens, their synonyms and their aspect along with occurrence frequency.

We trained and tested machine learning models on aforementioned two CPGs. The models considered for the study includes Naïve Bayes, Generalized Liner Model, Deep Learning, Decision Tree, Random Forest, and Ensemble Learner as shown in Fig. 3. The models achieved 82.10%, 74.40%, 66.70%, 66.70%, 74.40%, and 83.94% accuracies respectively.

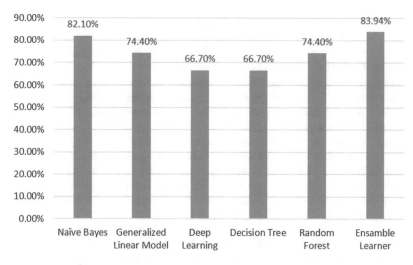

Fig. 3. Results achieved by each algorithm with aspects.

5 Conclusion and Future Work

Clinical practice guidelines have the potential to overcome all deficiencies of healthcare. However, due to the nature and format of guidelines, it has lower adherence rate and faces difficulties in achieving this goal. Some of the deficiencies can be reduced by filtering out irrelevant information from CPG and provide disease-specific information at the point of care. This paper focuses on the transformation of the CPG to structured format (word vector) and uses machine learning models to filter out irrelevant information from CPG. This mechanism can provide two-fold benefits. First, it can be used for filtering out irrelevant information from guidelines. Which will increase the effectiveness of guidelines, improve healthcare quality, help in providing evidence-based practice, and reduce processing time for identifying disease-specific information. Secondly, it can be used as a preprocessing step for other text mining applications.

In the future, we are planning to extend the existing work to further improve accuracy. We are also working on a system that can generate a machine interpretable CPG which will play an imperative role in CPG adherence by integrating it into CDDS system. The current proposed work will serve as an essential part of the conversion process.

Acknowledgement. This research was supported by an Institute for Information & Communications Technology Promotion (IITP) grant funded by the Korean government (MSIT) (No. 2017-0-00655). This work was supported by the Ministry of Science and ICT (MSIT), Korea, under the Information Technology Research Center (ITRC) support program (IITP-2017-0-01629) supervised by the Institute for Information & communications Technology Promotion (IITP) and NRF- 2016K1A3A7A03951968.

References

1. Lohr, K.N., Field, M.J.: Clinical Practice Guidelines: Directions for a New Program. National Academies Press, Washington, D.C. (1990)
2. Wenzina, R., Kaiser, K.: Identifying condition-action sentences using a heuristic-based information extraction method. In: Riaño, D., Lenz, R., Miksch, S., Peleg, M., Reichert, M., ten Teije, A. (eds.) Process Support and Knowledge Representation in Health Care, pp. 26–38. Springer, Cham (2013)
3. Rello, J., Lorente, C., Bodí, M., Diaz, E., Ricart, M., Kollef, M.H.: Why do physicians not follow evidence-based guidelines for preventing ventilator-associated pneumonia?: a survey based on the opinions of an international panel of intensivists. Chest **122**, 656–661 (2002)
4. Kilsdonk, E., Peute, L.W., Riezebos, R.J., Kremer, L.C., Jaspers, M.W.: From an expert-driven paper guideline to a user-centred decision support system: a usability comparison study. Artif. Intell. Med. **59**, 5–13 (2013)
5. James, P.A., Oparil, S., Carter, B.L., Cushman, W.C., Dennison-Himmelfarb, C., Handler, J., Lackland, D.T., LeFevre, M.L., MacKenzie, T.D., Ogedegbe, O.: 2014 evidence-based guideline for the management of high blood pressure in adults: report from the panel members appointed to the Eighth Joint National Committee (JNC 8). JAMA **311**, 507–520 (2014)
6. Chow, A.W., Benninger, M.S., Brook, I., Brozek, J.L., Goldstein, E.J., Hicks, L.A., Pankey, G.A., Seleznick, M., Volturo, G., Wald, E.R.: IDSA clinical practice guideline for acute bacterial rhinosinusitis in children and adults. Clin. Infect. Dis. **54**, e72–e112 (2012)
7. Jacobsen, P.B.: Clinical practice guidelines for the psychosocial care of cancer survivors. Cancer **115**, 4419–4429 (2009)
8. Yetisgen-Yildiz, M., Gunn, M.L., Xia, F., Payne, T.H.: A text processing pipeline to extract recommendations from radiology reports. J. Biomed. Inform. **46**, 354–362 (2013)
9. Khalifa, A., Meystre, S.: Adapting existing natural language processing resources for cardiovascular risk factors identification in clinical notes. J. Biomed. Inform. **58**, S128–S132 (2015)
10. Priyanta, S., Hartati, S., Harjoko, A., Wardoyo, R.: Comparison of sentence subjectivity classification methods in Indonesian News. Int. J. Comput. Sci. Inf. Secur. **14**, 407 (2016)
11. Hematialam, H., Zadrozny, W.: Identifying condition-action statements in medical guidelines using domain-independent features. ArXiv Prepr. arXiv:1706.04206 (2017)
12. Hussain, M., Hussain, J., Sadiq, M., Hassan, A.U., Lee, S.: Recommendation statements identification in clinical practice guidelines using heuristic patterns. In: 2018 19th IEEE/ACIS International Conference on Software Engineering, Artificial Intelligence, Networking and Parallel/Distributed Computing (SNPD), pp. 152–156. IEEE (2018)

A Breast Disease Pre-diagnosis Using Rule-Based Classification

Suriana Ismail[1]($\boxtimes$), Roslan Ismail[1],
and Tengku Elisa Najiha Tengku Sifizul[2]

[1] Universiti Kuala Lumpur-Malaysian Institute of Information Technology,
1010, Jalan Sultan Ismail, 50250 Kuala Lumpur, Malaysia
{suriana, drroslan}@unikl.edu.my
[2] Software Engineering Programme, Universiti Kuala Lumpur-Malaysian
Institute of Information Technology, 1010, Jalan Sultan Ismail,
50250 Kuala Lumpur, Malaysia
tengkuelisats@gmail.com

Abstract. The Manual detection of breast disease cannot detect occurrences of potential high risk problem for a patient at the early stage. Since it is a commonly found disease among women, it is critical to overcome the problem as fast as possible. In this paper the design of the proposed Rule Based System will be presented and the symptoms of the breast cancer disease and possible ways to prevent it will be outlined. The proposed Rule Based System was produced to help people to Prevent and early detection breast cancer, because it is known that this disease does not have medication or cure yet. Android and web application are the two platform used in the designing of the proposed ruled based system. The application to breast cancer pre-diagnosis utilizes the features of each highly potential symptoms, obtained from patient perceived condition, to discriminate benign from malignant breast lumps. This allows an accurate pre-diagnosis without the urgency need for a surgical biopsy.

Keywords: Breast disease · Rule-based · Artificial intelligence

1 Introduction

Breasts cancer is an important issue in all women's life, not just in current life but also was in the past and is in the future. It is a threat for many people females and males. But it affects females more frequency than male. It is well known that female breast cancer incidence is the largest in proportion among other type of cancers in general; where the annual breast cancer achieves the largest proportion among cancers [1–3].

Detecting diseases at an early stage is crucial to overcome problems and treating the effected patient. Breast diseases is the common disease that have been effected to most people especially women not only in Malaysia but also around the world. The National Cancer Registry (NCR) 2006 reported that there were 3,525 female breast cancer cases in Malaysia and this made the most commonly diagnosed cancer in women (29.9% of all new cancers) [1]. Breast cancer was the most common cancer in all ethnic group and age groups of females from the age 15 years old onwards [3].

© Springer Nature Switzerland AG 2019
S. Lee et al. (Eds.): IMCOM 2019, AISC 935, pp. 1037–1044, 2019.
https://doi.org/10.1007/978-3-030-19063-7_82

Breast diseases are divided into two which are benign (non-cancerous lump) and malignant (cancerous lump) [3, 8]. Breast cancer is the third most common cancer worldwide and is the most common cancer in women [2]. Since 1991, breast cancer has been the second leading cause of cancer admissions in Ministry of Health hospitals, and deaths due to breast cancer is in fourth place in terms of cancer deaths accounting for 6–8% of all cancer deaths [2]. Benign breast lumps usually have smooth edges and can be moved slightly when you push against them [1]. Breast cancer is the most common invasive cancer in women, and the second main cause of cancer death in women, after lung cancer [5]. The symptoms that may be caused by breast diseases are a lump or thickening of the breast, and changes to the skin or the nipple.

The reason why the system proposes on breast diseases because there are still lack of awareness of breast diseases among women. Women mostly afraid of regaining some knowledge on breast diseases. This is because they fear of husband leaving them because they have flaws on their body. Then, they are shy and fear of doctors and hospital also one of the reason why the awareness of breast diseases are still lacking. Table 1 shows the female breast age specific cancer incidence per 100,000 populations, by Ethnicity and Sex, Peninsular Malaysia 2006 [3]. The manual encoding of the pathology reports requires special knowledge for each cancer type and the transferal is a complicated and time consuming task where the coders have to read and interpret the content of each report. There is therefore a need of a system capable of automatic information extraction. The system should be able to accurately extract the relevant fields for each type of cancer.

Table 1. Female breast age-specific cancer incidence per 100,000 populations, by ethnicity and sex, Peninsular Malaysia 2006

		Age groups (year)							
		0–9	10–19	20–29	30–39	40–49	50–59	60–69	70+
Female	Malay	0	0.2	3.2	27	73.8	114.7	78.9	43.4
	Chinese	0	0	4.9	26.9	96.6	176.7	143.3	118.8
	Indian	0	0	3.2	16.7	82.3	111.1	138.3	140.5

2 Background Studies

2.1 Preliminary

Particularly in Malaysia, a lot of people nowadays did not aware of how important for us to take care of our own body. Some did not realize they have been effected by diseases. Manual way of health screening still dominating where patients need to go to the general hospital and wait for long queues. Most of the people in Malaysia has already been diagnose in stage 3 or 4 even for their first visit. This is because they are shy and fear of doctors and hospital. In order to reduce this problem, this project propose an integrated web and mobile application for Early Diagnosis System for Breast Diseases. The system has built for three users which are admin, doctors and patients. This is to create an early diagnosis system where the people can check their body condition and if they are already being effected by the disease. The system will suggest the specialist doctor to treat the patients' illness after a pre-diagnosis is completed.

2.2   Techniques for Medical Diagnosis

Rule-Based System

A Rule-based system is a set of "if-then" statements that uses a set of assertions, to which rules on how to act upon those assertions are created. Rule-based systems are used as a way to store and manipulate knowledge to interpret information in a useful way. They are often used in artificial intelligence applications and research. Figure 1 shows and example of rule-based system [7].

The advantages of using rule-based system are all the knowledge is expressed in the same format. There are also a few of downside using rule-based system. The rule-based system requires deep knowledge of the domain as well as a lot of manual work and generating rules for a complex system is quite challenging and time consuming. If the application is too complex, building the rule-based system can take a lot of time and analysis.

If there symptom(s) Dyspnea (mild)

AND shock (mild)

AND weak in heart sound (mild)

AND chest pain (mild-moderate)

AND Gallop in heart sound (mild)

AND paradoxical splitting of 2nd heart sound (mild)

AND Bl.P(Hypertension)

AND pulse (Tachycardia)|

AND Duration of pain (not <20 min and not >20 min)

OR fever (mild)

OR nausea (mild)

OR palpitation (mild)

THEN the Disease is Angina pectoris

Fig. 1. Example of rule-based system

Fuzzy Expert System

A fuzzy expert system is a form of artificial intelligence that uses a collection of membership functions (fuzzy logic) and rules (instead of Boolean logic) to reason about data. Unlike conventional expert systems, which are mainly symbolic reasoning engines, fuzzy expert systems are oriented toward numerical processing. Fuzzy expert systems are oriented towards numerical processing. It takes numbers as input, and then translates the input numbers into linguistic terms like small, medium and large. Then the task of the rules is to map the input linguistic terms onto similar linguistic terms describing the output. Finally, the translation of output linguistic terms into an output numbers is done [6]. The disadvantages of using the fuzzy expert system are there is no systematic approach to fuzzy system designing, they are understandable only when simple and they are suitable for the problems which do not need high accuracy.

Expert System

Expert System is a branch of computer science that transfers the intelligence of human experts into machines and it is now a very prominent cross disciplinary area [4]. It uses the database of expert knowledge to offer some advice or make some decision to user easier. The expert system is very useful to make decision making such as diagnosis, explaining, suggest alternate option to a problem, etc.

2.3   Existing Mobile Based Application

A Mobile Based Expert System for Disease Diagnosis and Medical Advice Provisioning designed a rule based mobile system to diagnose ten common diseases in Nigeria, provides medical advices and prescriptions where applicable [4]. Their proposed system acts as a medical assistant where it provides medical diagnosis to common health conditions through the users' input together with the users' personal health record and then the application will generate medical advices to patients based on the diagnosis result. The system is designed with Android Operating System by using the Android Studio.

Another similar development of mobile application is a Mobile Based Health Care System for Patient Diagnosis using Android OS for diabetic patient. The application takes care of the daily life health care issues by taking various check-up either by patient or its assistance that can be able to take his/her medical check-up by using medical peripherals and upload the report by its mobile phone to server where expert system could suggest precautionary steps or diagnosis along with patient status [7].

3   Methodology

3.1   Data Requirement

Based on the scope of the study, data and required information on generic and specific symptoms are gathered from Medical Officer in Surgical Out Patient Department (SOPD) from Hospital Sultanah Nur Zahirah, Kuala Terengganu, one of local hospital in Malaysia. A series of interviews and symptoms list are extracted and carefully analysis in order to design the algorithm.

3.2   Identification of Set of Rule

Table below depicted the overview of the system flow intended for the study (Table 2).

Table 2. Overview of the system flow

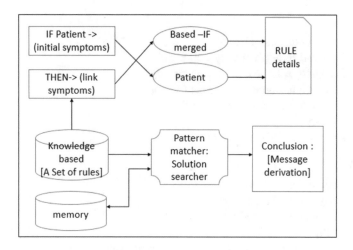

Next, the following excerpt of proposed algorithm are designed.
[Breast Disease] : Rule-based algorithm

Rule 1:

IF The patient has abnormal tissue felt from the rest of the breast tissue
OR IF Presence of lump(s) of varying size(s) around breast
OR IF May or not feel pain
OR IF Clear colored or non-blood-stained nipple discharge
OR IF History of bacterial infection
THEN Benign (non-cancerous lump)

Rule 2:

IF The patient has changes in breast size or shape
OR IF Skin changes (puckering, dimping or 'peau-de-orange' like skin)
OR IF Presence of lumps of varying size(s)
OR IF May or not feel pain
OR IF Lump(s) in the armpit
OR IF Presence of lump(s) around the collarbone
OR IF Single duct nipple discharge especially blood-stained discharge
OR IF Inverted nipple (pulled in)
OR IF Redness on the skin/nipple
THEN Malignant (cancerous lump)

There are three (3) types of risk factor for malignant (cancerous lump)

Rule 3:

IF The patient consume alcohol
OR IF First full term pregnancy >30 years
OR IF Presence of lumps of varying size(s)
OR IF Undergo hormone replacement therapy
OR IF Consume oral contraceptive pill
OR IF Obesity
THEN Low Risk of Breast Cancer

Rule 4:

IF The patient's age above 40 years old
OR IF Early menarche (<12 years old)
OR IF Late menopause (>55 years old)
OR IF Nulliparity
OR IF Benign breast diseases with proliferation with atypia
OR IF Dense breast
THEN Moderate Risk of Breast Cancer

Rule 5:

IF The patient has personal history of invasive breast cancer
OR IF Benign breast disease with atypical hyperplasia
OR IF Ionising radiation from treatment of Breast Cancer Hodgkin's disease, etc.
OR IF Significant first degree family history
THEN High Risk of Breast Cancer

3.3   Implementation

The modules of web application for the system, using C# Language on Microsoft Visual Studio 2012 software. While for the modules of mobile application, using Android Studio Application. Screenshot of the main and possible result report of a patient are as in Fig. 2 below.

3.4   Testing and Evaluation

Functional unit testing is carried using black box testing. All use cases developed for the proposed system are followed closely. Automation testing for the test cases automated is carried out.

Fig. 2. Screenshot of main view and patient report view

4   Result and Discussion

Two main testing are evaluated as to ensure the proposed system execute the intended requirement gathered and passed the designed functional unit test. As depicted in Fig. 3, Use case-functional scope by user ideally passed all use cases at minimum 89%. There are six uses cases involved in the testing. Figure 4 also depicted the function test by both platforms of mobile and web by setting at 92% passes value. The test is executed on all eight functional test by each platform.

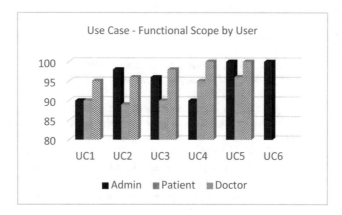

Fig. 3. Use case-functional scope by user

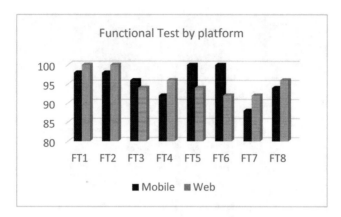

Fig. 4. Functional test by platform

5   Conclusions

In summary, we have performed both an experimental and testing evaluation on the proposed system. The proposed study demonstrates the practical benefits to target users in avoiding late diagnosis while helping the user be more confident into getting the first pre-diagnosis without any hassle. Further evaluation and testing need to be carried out when the identified set of potential patient case by the hospital authority is ready to be tested.

Acknowledgement. This work was supported by Universiti Kuala Lumpur-Malaysian Institute of Information Technology. Thank you to those involved in improving this paper.

References

1. Benign Breast Lumps (2018). Accessed from webmd.com. https://www.webmd.com/breastcancer/benign-breast-lumps#1
2. CPG Secretariat: Management of breast cancer. In: Clinical Practice Guidelines, Malaysia (2002)
3. CPG Secretariat: Management of breast cancer. In: Clinical Practice Guidelines, Malaysia, p. 81 (2010)
4. Isinkaye, F.O., Awosupin, S.O., Soyemi, J.: A mobile based expert system for disease diagnosis and medical advice provisioning. Int. J. Comput. Sci. Inf. Secur. (IJCSIS) **15**, 568–572 (2017)
5. Nordqvist, C.: What you need to know about breast cancer, 27 November 2017. Accessed from medicalnewstoday.com. https://www.medicalnewstoday.com/articles/37136.php
6. Singla, J., Grover, D., Bhandari, A.: Medical expert systems for diagnosis of various diseases. Int. J. Comput. Appl. **93**, 36–43 (2014)
7. Soltan, R., Rashad, M., El-Desouky, B.: Diagnosis of some diseases in medicine via computerized experts system. Int. J. Comput. Sci. Inf. Technol. (IJCSIT) **5**, 79–90 (2013)
8. Mohamad, D.S.: Interview with Dr. Siti Hartinie on Breast Diseases. (T. E. Tengku Sifzizul, Interviewer), April 2018

Discovering Correlation in Frequent Subgraphs

Fariha Moomtaheen Upoma[1], Salsabil Ahmed Khan[1],
Chowdhury Farhan Ahmed[1], Tahira Alam[1], Sabit Anwar Zahin[1],
and Carson K. Leung[2(✉)] ⓘ

[1] University of Dhaka, Dhaka, Bangladesh
farhan@du.ac.bd
[2] University of Manitoba, Winnipeg, MB, Canada
kleung@cs.umanitoba.ca

Abstract. In today's networked world, graph data is becoming increasingly more ubiquitous as the complexities, layers and hierarchies of real life data demand to be represented in a structured manner. The goal of graph mining is to extract frequent and interesting subgraph from large graph databases. It is even more significant to perform correlation analysis within these frequent subgraphs, as such relations may provide us with valuable information. However, unfortunately much work has not been done in this field even though its necessity is enormous. In this paper, we propose two measures that help us discover such correlation among frequent subgraphs. Our measures are based on the observation that elements in graphs exhibit the tendency to occur both connected and disconnected. Evaluation results show the effectiveness and practicality of our measures in real life datasets.

Keywords: Data mining · Graph mining · Correlation discovery ·
Information management · Information processing management

1 Introduction

Pattern mining is one of the most important topics in data mining. The core idea is to extract relevant "nuggets" of knowledge describing parts of a database. Frequent pattern mining is a domain of data mining that imposes immense significance in everyday transactions. It searches for recurring relationships in a given dataset and leads to the discovery of correlations and associations among itemsets in large transactional and relational datasets.

Another essential domain of data mining is that of graph mining [7]. Graph mining is essentially the problem of discovering repetitive subgraphs in large graph databases. Datasets that do not correspond to flat transaction setting, which contain structures, layers, hierarchy and/or geometry may be represented with graph structures. This graph structure may be used to model complicated relationships among data with the help of its diverse attributes, such as the following [12]:

© Springer Nature Switzerland AG 2019
S. Lee et al. (Eds.): IMCOM 2019, AISC 935, pp. 1045–1062, 2019.
https://doi.org/10.1007/978-3-030-19063-7_83

- labels,
- directions,
- weights,
- etc.

Mining such subgraphs also imposes challenges, overcoming which leads to many compelling results.

Graphs can be used to model complicated structures. Examples of these complicated structures include the following:

- images,
- chemical compounds,
- circuits,
- social networks,
- the Web,
- biological networks,
- workflows,
- XML documents, and
- protein structures [8,9].

A large number of graph search algorithms have been evolved in computer vision [13,14], video indexing, text retrieval and chemical informatics. Correlation analysis bears immense significance in discovering correlation among the patterns that are generated from frequent pattern mining. However, this correlation analysis is different in case of transactional database and graph database. This is because graph databases are much more complex than transactional databases. Our *key contributions of this paper* can be summarized as follows:

- We propose two new measures for discovering correlation in frequent subgraph:
    1. ConfTog, which determines the confidence between two elements in a subgraph when they are connected by an edge; and
    2. ConfAll, which determines the confidence between two elements whenever they are present in a subgraph regardless of being connected or disconnected.
- We also propose a maximum difference threshold, which is a variable to be set as per the interest of the end users. This threshold helps us determine whether the elements in a subgraph are truly internally correlated.

The remainder of this paper on discovering correlation in frequent subgraphs is organized as follow:

- Section 2 contains a review of related graph mining methods.
- Section 3 consists of an elaborate description and explanation of our proposed measures and related definitions with intricate examples.
- Section 4 includes experiments and results to validate our proposed measures.
- Section 5 concludes our research as a whole and also introduces some future scopes of work in this field.

2   Related Works

Not a lot of work has gone into this particular area of graph mining. For instance, Fan et al. [10] proposed *graph-pattern association rule* (*GPARs*) for social media marketing. It help discover regularities between entities in social graphs by extending association rules for itemsets and identify potential customers by exploring social influence. They provide a parallel scalable algorithm that guarantees a polynomial speedup over sequential algorithms with the increase of processors.

In order to interpret human behaviour in meeting discussions, Yu et al. [15] proposed a tree based mining method to extract frequent patterns of human interaction based on the captured content of face-to-face meetings. This method represents human interactions by a tree and analyzes it to determine frequent interactions, typical interaction flows, and relationships between different types of interactions. However, such a tree based method can not capture all kinds of triggering relationships between interactions in a meeting such as those about correlated interactions.

A different type of approach is proposed by Fariha et al. [11]. In their approach, a meeting can be modeled as a *weighted directed acyclic graph* (*DAG*), from which weighted frequent interaction patterns can be discovered.

Bringmann and Nijssen [6] provided a more appropriate way to find structural regularities or anomalies in large graph network. It revised the conventional definition of support measure to fit more accurately in single-graph setting.

Tiwari et al. [1] introduced an algorithm called *FP-Growth-Graph*, which uses graph rather than tree to arrange the items for mining frequent itemsets. Their method contains three main parts. At first, it scans the database only once for generating graph for all item. Secondly, it prunes the non-frequent items based on given minimum support threshold and readjusts the frequency of edges, and after that constructs the FP graph. According to their study, the advantage of using graph structure comes in the form of space complexity because graph uses an item as node exactly once rather than two or more times as was done in tree.

Balaji Raja and Balakrishnan [2] introduced two ideas. The first one analyzes the ability to explore the demographic parameters of the underlying entities and their inter relations using the traditional graph theory approach. The second is the improved algorithm based on graph and clustering based mining association rules. This improved algorithm is named *Graph and Clustering Based Association Rule Mining* (*GCBARM*). The GCBARM algorithm scans the database of transaction only once to generate a cluster table and then clusters the transactions into cluster according to their length. The GCBARM algorithm is used to find frequent itemsets and will be extracted directly by scanning the cluster table. Their method reduces memory requirement and time to retrieve the datasets and hence it is scalable for any large size of the database.

AlZoubi [3] proposed an improved approach to mine strong association rules from an association graph, called *graph based association rule mining* (*GBAR*) method, where the association for each frequent itemset is represented by a subgraph, then all sub-graphs are merged to determine association rules with high confidence and eliminate weak rules, the proposed graph based technique is self-motivated since it builds the association graph in a successive manner. These rules achieve the scalability and reduce the time needed to extract them.

Amal Dev [4] proposed a new graph based algorithm for associative rule mining. It can be used to improve decision making in a wide variety of applications such as: market basket analysis, medical diagnosis, bio-medical literature, protein sequences, census data, logistic regression, fraud detection in web, customer relationship management (CRM) of credit card business etc. The proposed system works by constructing a graph. The main advantage is that the system needs only a single database scan, it is also possible to mine rules related to some particular items only.

3 Proposed Method

Here, we present our proposed measures and discuss them in detail with an example scenario.

3.1 Preliminaries

In this section, we will discuss about the measures and thresholds proposed by us and other concepts required to construct our algorithm. We will consider the graph database in Fig. 1. This is a chemical dataset where the each node represents a chemical compound and the bonds between these chemical compounds are represented by the edges between the nodes. We label the edges representing single bond with '1' and those representing double edges with '2'.

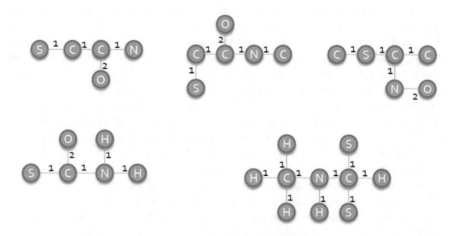

Fig. 1. Labeled graph database

3.2   Elements in a Graph

A graph is made up of a set of vertices and edges. An element in a graph may be a vertex, an edge, two edges and so on. Multiple elements in a graph may form a pattern.

Example 1. Let us consider the following graph. Here, each of the vertices a, b, c and d is an element. Again, the edges a-b, a-a, a-c, a-d, etc., are also elements. The patterns a-a-d, a-c-d, a-c-d-c and so on are also considered as elements in a graph.

3.3   Confidence Together (ConfTog)

The gSpan [5] algorithm generates the frequent subgraph patterns from a graph database. In order to determine the internal correlation between elements in a subgraph, correlation measures can be used. We propose two such measures. One of them is ConfTog. ConfTog is a correlation measure that determines the confidence between two elements in a subgraph when they are directly connected by an edge. It can be shown by the following formula:

$$\begin{aligned}
&\text{ConfTog}(a \to b) \\
&= \frac{\text{support\_count of } a \text{ and } b \text{ directly connected by an edge}}{\text{support\_count of } a}
\end{aligned} \quad (1)$$

Here, a and b are two elements. So, it is seen that *ConfTog* is the ratio of the number of times a and b occur in a graph together, i.e., connected by an edge and the number of times a occurs in a graph. If this pattern occurs multiple times in a subgraph, we only count them once.

Example 2. Let us consider Fig. 1 and calculate the ConfTog of the edge C-C. We can see that C-C occurs in 3 out of 5 graphs. And C occurs in all 5 graphs. Hence,

$$\text{ConfTog}(C \to C) = \frac{3}{5} = 60\% \quad (2)$$

3.4   Confidence All (ConfAll)

We propose another new measure ConfAll, that determines the confidence between two elements in a subgraph whenever they are present, either connected or disconnected. This measure can be also used to determine the internal correlation between elements a subgraph. It can be shown by the following formula:

$$\begin{aligned}
&\text{ConfAll}(a \to b) \\
&= \frac{\text{support\_count of } a \text{ and } b \text{ being present in a graph}}{\text{support\_count of } a}
\end{aligned} \quad (3)$$

Here, a and b are two elements. So, it is seen that *ConfAll* is the ratio of the number of times a and b occur in a graph regardless of being connected or disconnected and the number of times a occurs in a graph. If this pattern occurs multiple times in a subgraph, we only count them once.

Example 3. Revisit Fig. 1 and calculate the ConfAll of the edge C-C. The two elements C and C occur separately in 4 out of 5 graphs. And C occurs in all 5 graphs. Hence,

$$\text{ConfAll}(C \rightarrow C) = \frac{4}{5} = 80\% \tag{4}$$

3.5   Minimum Confidence Threshold

A confidence value is pre-specified against which we check the value of ConfTog and ConfAll of any association rule, this is called *minimum confidence threshold.*

3.6   Maximum Difference Threshold

The differences between the two kinds of confidence measures are checked to stay within a maximum value for any association rule, that value is called the *maximum difference threshold.*

3.7   Problem Definition

The maximum difference threshold is introduced to differentiate between the two kinds of confidence measures. It is an upper bound limit, beyond which we do not consider elements to be strongly bonded. If the difference between ConfTog and ConfAll for an association rule stays within this limit, we say that the elements have *strong affinity.*

Example 4. In the case of the previously mentioned example, if we set the maximum difference threshold to be 20%, for the edge C-C we get the difference value = 80% − 60% = 20%. Although both confidence measures do not satisfy the confidence threshold of 70%, we can say that, in a graph where a C exists and if another C co-exists in that graph the probability of them to be connected together is pretty notable. Hence, they have strong affinity.

3.8   A Long Example

Let us now see an example workout to better understand how our measures perform. This simulation works on the dataset consisting of 5 graphs of Fig. 1 and progresses up to the discovery of 2 edged frequent patterns for simplicity.

Suppose, minimum support threshold = 3/5, minimum confidence threshold = 70% and the maximum difference threshold = 20%.

Based on the minimum support threshold we generate frequent sub-graphs consisting of a single edge. We get the result in Fig. 2. Let us calculate the confidence of the frequent subgraphs using the two measures.

In case of the association rule C → C, we see that the nodes C and C occur in 3 out of 5 graphs connected by an edge. And C occurs in all 5 graphs. So, ConfTog is 60%. However, C and C occur in 4 graphs, regardless of being

Fig. 2. 1-edge candidates

Table 1. Table of confidence measures and their difference for edge C-C

Association rule	ConfTog	ConfAll	Difference
C → C	3/5 = 60%	3/5 = 80%	20%

connected or disconnected. So, ConfAll is 80%. Their difference satisfies the maximum difference threshold. See Table 1.

In case of association rule C → N, we see that C and N occur in all 5 graphs together. And, C is present in all of them. So, both ConfTog and ConfAll is 100%. They satisfy the minimum confidence threshold. Hence we consider this association rule. And their difference meets the maximum difference threshold. So, the two elements can be said to be strongly internally correlated. See Table 2.

Table 2. Table of confidence measures and their difference for edge C-N

Association rule	ConfTog	ConfAll	Difference
C → N	5/5 = 100%	5/5 = 100%	0%
N → C	5/5 = 100%	5/5 = 100%	0%

In case of association rule N → C, we have noticed that N and C occur together in all 5 graphs. And, N is present in all 5 graphs as well. So, both ConfTog and ConfAll is 100%. They satisfy the minimum confidence threshold. So, we can consider this association rule. Again, their difference also satisfies the maximum difference threshold. Thus, the two elements are said to be *strongly internally correlated*. See Table 2.

Similarly, for rest of the association rules, we get Table 3.

We use the GSpan algorithm to extend the 1-edge subgraphs. In case of the edge C-C, by rightmost extension we add the forward edges C, N, S, O and H starting from the node farthest from the root to the root. This is shown in Fig. 3.

Table 3. Table of confidence measures and their difference for edges S-C and C-O

Association rule	ConfTog	ConfAll	Difference
S → C	5/5 = 100%	5/5 = 100%	0%
C → S	5/5 = 100%	5/5 = 100%	0%
C → O	3/5 = 60%	4/5 = 80%	20%
O → C	3/4 = 75%	4/4 = 100%	25%

Here, six new candidates are pruned because their support does not satisfy the minimum support threshold. These are the candidates with the supports written in red. Those that satisfy the minimum support threshold are written in green. Another two candidates are pruned because they have a non-minimum DFS code. Thus we get two new candidates from the edge C-C through rightmost extension. Similarly, considering the rightmost extension of the rest of the edges we get total six more new candidates. There are isomorphs among the newly generated candidates. We prune the redundant graphs based on the minimum DFS code. The final 2-edge subgraphs are shown in Fig. 4. Now, we calculate the confidence of each frequent subgraphs using the two measures.

When we consider the elements C and C-N, we see that they occur together in 3 out of 5 graphs and C is present in all 5 graphs. So, the ConfTog is 60%. Again the two elements are present in 4 graphs. So, the ConfAll is 80%. Their difference is 20% which satisfies the maximum difference threshold. In case of association rule (C-C) → N, we see that the elements C-C and N occur together

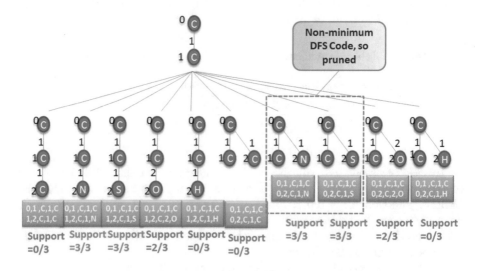

Fig. 3. Rightmost extension for the edge C-C

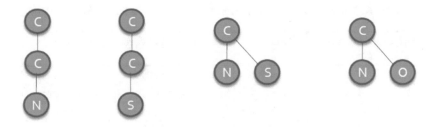

Fig. 4. 2 edge subgraphs

in 3 graphs and C-C occurs in 3 graphs as well. So, ConfTog is 100%. The two elements are present in the 3 graphs, connected or disconnected. So, ConfAll is also 100%. Their difference is 0%, which satisfies the maximum difference threshold. So, this pattern is considered interesting and the elements can be said to be strongly internally correlated (Table 4).

Table 4. Table of confidence measures and their difference for subgraph C-C-N

Association rule	ConfTog	ConfAll	Difference
C → (C-N)	3/5 = 60%	4/5 = 80%	20%
(C-C) → N	3/3 = 100%	3/3 = 100%	0%
N → (C-C)	3/5 = 60%	3/5 = 60%	0%
(C-N) → C	3/5 = 60%	4/5 = 80%	20%

In case of the association rule N → (C-C), we see that N and C-C occur in 3 graphs together and N occurs in 5 graphs. So, ConfTog is 60%. Again the two elements are present in those 3 graphs, regardless of being connected or disconnected. So, the ConfAll is also 60%. Their difference value is zero, which means they have the strongest affinity, whenever vertex N and edge C-C appears in the same graph it is highly probable that they will be in there together. Finally, in case of association rule (C-N) → C, the elements C-N and C occur together in 3 graphs ans C-N occurs in 5. So, ConfTog is 60%. Again C-N and C are present in 4 graphs either connected or disconnected. So, ConfAll is 80%. Their difference is 20% which satisfies the maximum difference threshold.

We show how we can mine interesting patterns from the given dataset. Here, we present Table 5 for all the association rules upto 2 edges and the final pruned result. Out of the 23 association rules, six rules (namely rules 6, 16, 17, 18, 19 and 22) are pruned out as not interesting. We can say that the remaining 17 rules are truly internally correlated.

Table 5. Table of confidence measures and their differences for all edges

Serial no.	Association rule	ConfTog	ConfAll	Difference	Decision
1	C → C	3/5 = 60%	4/5 = 80%	20%	Interesting
2	C → N	5/5 = 100%	5/5 = 100%	0%	Interesting
3	N → C	5/5 = 100%	5/5 = 100%	0%	Interesting
4	S → C	5/5 = 100%	5/5 = 100%	0%	Interesting
5	C → S	5/5 = 100%	5/5 = 100%	0%	Interesting
6	C → O	3/5 = 60%	4/5 = 80%	20%	Interesting
7	O → C	3/4 = 75%	4/4 = 100%	25%	Not interesting
8	C → (C-N)	3/5 = 60%	4/5 = 80%	20%	Interesting
9	(C-C) → N	3/3 = 100%	3/3 = 100%	0%	Interesting
10	N → (C-C)	3/5 = 60%	3/5 = 60%	0%	Interesting
11	(C-N) → C	3/5 = 60%	4/5 = 80%	20%	Interesting
12	C → (C-S)	3/5 = 60%	4/5 = 80%	20%	Interesting
13	(C-C) → S	3/3 = 100%	3/3 = 100%	0%	Interesting
14	S → (C-C)	3/5 = 60%	3/5 = 60%	0%	Interesting
15	(C-S) → C	3/5 = 60%	4/5 = 80%	20%	Interesting
16	S → (C-N)	3/5 = 60%	5/5 = 100%	40%	Not interesting
17	(C-N) → S	3/5 = 60%	5/5 = 100%	40%	Not interesting
18	N → (S-C)	3/5 = 60%	5/5 = 100%	40%	Not interesting
19	(S-C) → N	3/5 = 60%	5/5 = 100%	40%	Not interesting
20	N → (C-O)	3/5 = 60%	3/5 = 60%	0%	Interesting
21	(C-N) → O	3/5 = 60%	4/5 = 80%	20%	Interesting
22	O → (C-N)	3/4 = 75%	4/4 = 100%	25%	Not interesting
23	(C-O) → N	3/3 = 100%	3/3 = 100%	0%	Interesting

4   Experimental Results

Here, we run experiments to validate our proposed measure. The experiments are run on our modified code over several datasets. First, we try to show that there is considerable difference between the two kinds of confidence measure. We do this by generating the number of association rules satisfying a pre-specified confidence threshold and comparing those numbers for both kinds of confidences. The next experiment focuses on determining the difference threshold between those two confidences.

4.1   Experiment 1: Comparison Between ConfTog and ConfAll

We have run our codes on 3 real world datasets to ponder the effectiveness of our proposed measure. The datasets are:

1. Chemical 340,
2. MCF-7 (active), and
3. MOLT-4 (active).

The last two datasets are taken from Yan's website which assembled 11 graph datasets from the PubChem website (http://pubchem.ncbi.nlm.nih.gov).

We run the 1st version of our code by setting the minimum confidence threshold to 60% and record the number of association rules acquired from the frequent patterns generated from the implementation of gSpan algorithm, for both ConfTog and ConfAll that satisfy the minimum confidence threshold. Then, gradually increase the conf threshold and document the data. We plot the confidence thresholds on the x-axis and number of association rules for both ConfTog and ConfAll on the y-axis from the data table for each dataset. Thus, we get two lines for both kinds of confidence.

Figure 5 shows the results for Chemical-340 dataset. We can see from the chart that the difference is not that prominent between the ConfTog and ConfAll lines for this dataset. They also merge at one point where the number for association rules satisfying confidence threshold for both measures becomes the same.

Figure 6 shows that the two lines maintain a moderate distance throughout the whole experiment for MCF-7 dataset. They neither merge nor becomes notably distant from each other.

In Fig. 7, we can see that, the distance between MOLT-4 lines for two measures also shows the similar characteristics as MCF-7 lines. Though their distance is a bit more noticeable and adequate.

The distances between the ConfTog and ConfAll lines are noteworthy. We can safely say that this difference cannot be neglected. If we indulge ourselves to look past the structural requirements and dive into this clearly distinguished area between the two kinds of confidence, we can attain some very interesting knowledge. A varying range of decisions can be made depending on the type of graphs (e.g., social graph, chemical structures, etc.).

4.2   Experiment 2: Distribution of Difference Values

This part of the experiment is the same for both experiments. However, the later parts are different. As a result, they will be discussed separately from this point on.

We mentioned determining a maximum difference threshold beyond which we can say that the bonding is not that strong. In order to have a clear picture about the distinction between elements that have strong affinity and those that do not, we can make a difference-value chart. Minor adjustment is made to the code to create the second version, which outputs the difference values only, between the ConfTog and ConfAll of the association rules formulated from the 1-edged and 2-edged frequent patterns.

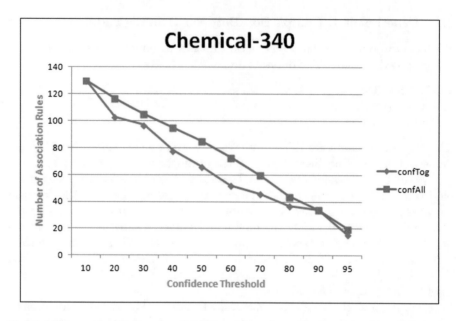

Fig. 5. Minimum confidence threshold vs number of association rules graph for dataset Chemical-340

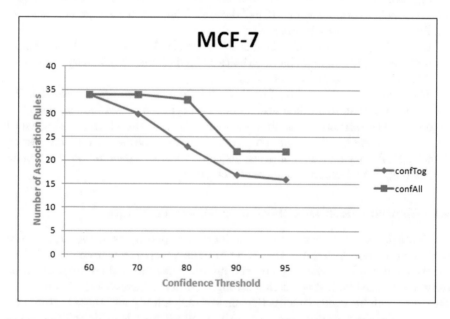

Fig. 6. Minimum confidence threshold vs number of association rules graph for dataset MCF-7

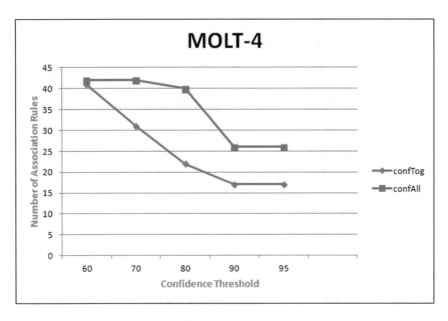

Fig. 7. Minimum confidence threshold vs number of association rules graph for dataset MOLT-4

We construct two types of chart for clear perception. In the 1st chart the horizontal axis shows the values of variable (in this case the difference values) with an interval of 10 and the vertical axis shows the frequencies associated with these values, that is, how many association rules have difference values within the corresponding interval. The second chart is almost similar, it has the same labels for the x- and y-axes. However, it shows the cumulative heights of the bars. Its vertical axis is responsible for depicting the cumulative frequencies associated with the values on the x-axis, that is, how many association rules have difference values in the range from 0 to that particular value.

Figures 8 and 9 show the Chemical-340 charts, where we see that the largest bar is for the 1st interval and the 2nd chart clearly points out that the height does not increase that much after the 1st interval, meaning most of the rules fall within 0–10% difference values. This can fairly mean the elements of this dataset have strong affinity among them. If we set the difference threshold to 10%, it will mean the rules with difference values less than equals 10% are strongly affiliated.

In Figs. 10 and 11, the MCF-7 charts show that setting the difference threshold to 10% gives us roughly 50% association rules. Setting to 20% will result in having approximately 90% rules which can hinder our purpose of distinguishing bonding among elements. So, 10% can be an acceptable threshold.

Finally, in Figs. 12 and 13, we can see that for MOLT-4 there are more than 50% values in the 1st interval. Setting difference threshold to 10% leads to almost half the rules being strongly affiliated.

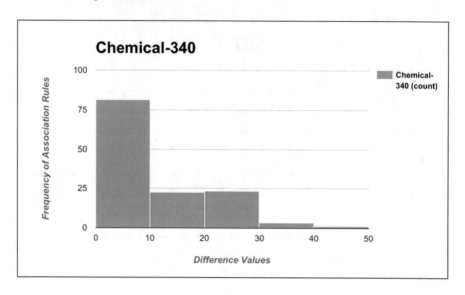

Fig. 8. Difference values vs frequency of data chart for dataset Chemical-340

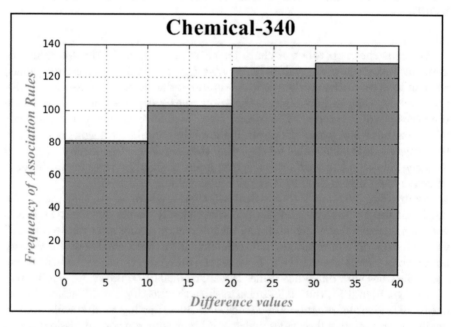

Fig. 9. Difference values vs cumulative frequency of data chart for dataset Chemical-340

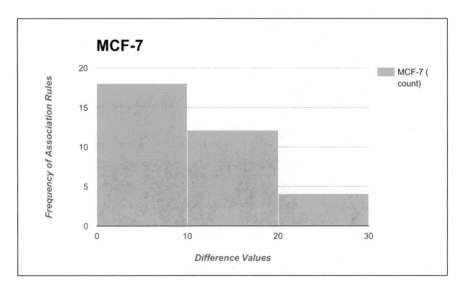

Fig. 10. Difference values vs frequency of data chart for dataset MCF-7

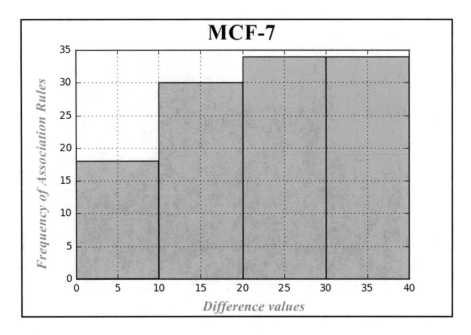

Fig. 11. Difference values vs cumulative frequency of data chart for dataset MCF-7

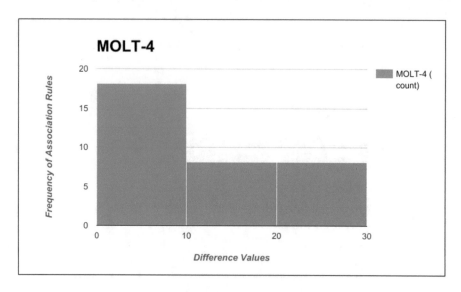

Fig. 12. Difference values vs frequency of data chart for dataset MOLT-4

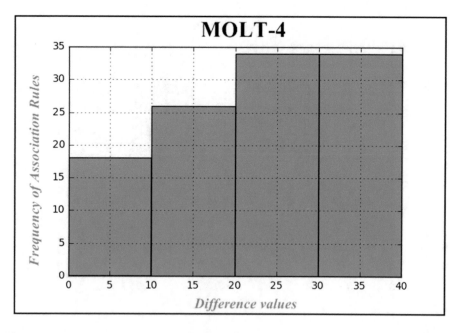

Fig. 13. Difference values vs cumulative frequency of data chart for dataset MOLT-4

In all of the above 5 cases, 10% seems to be popular choice for our proposed maximum difference threshold. However, this choice is rather erratic and different contexts can require this threshold to be set differently according to varying demands. If we want to conclude in a more concrete where we can say that if let, 60% association rules falls within the maximum difference threshold then that dataset is very strong, it has solid affinity among all its elements. In that sense our 1st dataset, Chemical-340 is strong and the rest are not.

5   Conclusions

We have proposed two new measures that help us determine the correlation in frequent subgraphs. We have also explained the significance of our measures with the help of a detailed example workout on a chemical dataset. We have shown how these two measures give us different levels of confidence and helps us discovering interesting association rules in real-life datasets. Based on a maximum difference threshold, the difference between the two measures also help us determine which elements are truly internally correlated. In future, we wish to devise a generalized version of this algorithm so that it can compute measures up to n-edged frequent subgraphs. Our approach can also be extended for applying in dynamic databases.

Acknowledgments. This project is partially supported by NSERC (Canada) and University of Manitoba.

References

1. Tiwari, V., Tiwari, V., Gupta, S., Tiwari, R.: Association rule mining: a graph based approach for mining frequent itemsets. In: ICNIT 2010, pp. 309–313 (2010)
2. Balaji Raja, N., Balakrishnan, G.: An innovative approach to association rule mining using graph and clusters. In: CCSIT 2012, pp. 59–67 (2012)
3. AlZoubi, W.: An improved graph based method for extracting association rules. IJSEA **6**(3), 1–10 (2015)
4. Amal Dev, P., Sobhana, N.V., Joseph, P.: GARM: a simple graph based algorithm for association rule mining. Int. J. Comput. Appl. **76**(16), 1–4 (2013)
5. Yan, X., Han, J.: gSpan: graph-based substructure pattern mining. In: ICDM 2002, pp. 721–724 (2002)
6. Bringmann, B., Nijssen, S.: What is frequent in a single graph? In: PAKDD 2008, pp. 858–863 (2008)
7. Chakrabarti, D., Faloutsos, C.: Graph Mining: Laws, Tools, and Case Studies. Synthesis Lectures on Data Mining and Knowledge Discovery. Morgan & Claypool Publishers, San Rafael (2012)
8. Chittimoori, R.N., Holder, L.B., Cook, D.J.: Applying the subdue substructure discovery system to the chemical toxicity domain. In: AAAI FLAIRS 1999, pp. 90–94 (1999)
9. Dehaspe, L., Toivonen, H., King, D.R.: Finding frequent substructure in chemical compounds. In: KDD 1998, pp. 30–36 (1998)

10. Fan, W., Wang, X., Wu, Y., Xu, J.: Association rules with graph patterns. PVLDB **8**(12), 1502–1513 (2015)
11. Fariha, A., Ahmed, C.F., Leung, C., Abdullah, S.M., Cao, L.: Mining frequent patterns from human interactions in meetings using directed acyclic graphs. In: PAKDD 2013, Part I, pp. 38–49 (2013)
12. Gudes, E., Vanetik, N., Cohen, M., Shimony, E.: Graph and web mining - motivation, applications and algorithms. Presentation, Department of Computer Science Ben-Gurion University, Israel
13. Klviinen, H., Oja, E.: Comparisons of attributed graph matching algorithms for computer vision. In: STEP 1990, pp. 354–368 (1990)
14. Piriyakumar, D.A.L., Murthy, C.S.R., Levi, P.: A new A* based optimal task scheduling in heterogeneous multiprocessor systems applied to computer vision. In: HPCN-Europe 1998, pp. 315–323 (1998)
15. Yu, Z., Yu, Z., Zhou, X., Becker, C., Nakamura, Y.: Tree-based mining for discovering patterns of human interaction in meetings. IEEE TKDE **24**(4), 759–768 (2010)

Classifying License Plate Numerals Using CNN

Tomoya Suzuki$^{(\boxtimes)}$ and Ryuya Uda

School of Computer Science, Tokyo University of Technology,
1404-1 Katakuramachi, Hachioji, Tokyo 192-0982, Japan
c0115188eb@edu.teu.ac.jp, uda@stf.teu.ac.jp

Abstract. Nowadays, security cameras are usually set in various places in Japan. The cameras are effective for criminal investigation. Especially, a license plate on a car which is in images by the cameras can identify the car. However, numbers on the license plate photographed by the cameras sometimes unreadable for humans since the image of the numbers is often poor picture quality, and noise and light decrease the quality much more. Therefore, we propose a new method to read numbers on a license plate with poor picture quality and we evaluated our method by experiments. In this paper, we described the method, experiments, evaluation and plans in future. The main idea is to read the numbers by machine learning on CNN which a lot of images of numbers created by three dimensional rotations and retouching are put in. The retouching processes in this paper are shift, cropping, smoothing, noise assignment, brightness changing and random erasing. A model created by the learning with the created images is saved and used for the classification of numbers on license plates. We think that the method is technically new since we have never heard the method to use three dimensional virtual numbers for the classification of numbers on real license plates. We prepared photos of real license plates and experimentally classified them by decreasing their resolution in stages. As a result, images with only 2 by 4 square pixels resolution were able to be classified with a probability of 99%. On the other hand, the same image with different cropping area was sometimes classified with a quite low probability. We will identify the cause of the problem in future.

Keywords: License plate · Classifying · CNN · Machine learning ·
Deep learning

1 Introduction

Yoshiura et al. pointed out surveillance cameras are required for criminal investigation [1]. Especially, in Japan, surveillance cameras are usually set at supermarkets and convenience stores. Also, they are additionally set at various public spaces such as roads, stations and parks. Local governments have set surveillance cameras on local roads from 2014 in Japan. In Tokyo, total 6,500 cameras was

© Springer Nature Switzerland AG 2019
S. Lee et al. (Eds.): IMCOM 2019, AISC 935, pp. 1063–1075, 2019.
https://doi.org/10.1007/978-3-030-19063-7_84

planned to set on school roads of each of 1,300 public elementary schools. The surveillance cameras have been used for criminal investigation. However, numbers on a license plate are sometimes difficult to be read even when they appear in images by the surveillance cameras. The resolution of images by surveillance cameras is usually set to low to economize on memory or to protect privacy of people. In addition, images deteriorate by blur, noise, haze and non-uniform light. Therefore, we propose a new method to read numbers on a license plate with poor picture quality and we evaluated our method by experiments.

2   Related Works

A license plate recognition system with neural networks is proposed by Wang et al. [2]. Amirgaliyev et al. considered various methods of ALPR (Automatic License Plate Recognition), and described their opinions about the most admirable recognition method [3]. Ashtari et al. showed a method to detect the place of license plates on Italian cars by template matching which was corrected by them and analysis of pixels of target colors [4]. They resulted that the detection rate was 96% at most. Rashid also proposed a method to detect license plates with high percentage without high quality images taken by expensive hardware devices [5]. He described an algorithm of high-speed automatic detection of the license plates and resulted that the detection rate was achieved to 96% with a small data set. The research results of the above four researches are important for the detection of license plates in images, but it is not mentioned to read numbers on the license plates. On the other hand, we focus on to read numbers on numbers on the license plates detected in images.

Haneda et al. mentioned the method to extract numbers from license plates [6]. They resulted that they were able to extract the numbers with high accuracy. Wang et al. proposed an algorithm to recognize letters on license plates of Chinese cars with high accuracy [7]. They told that their algorithm was completely met the requirements for actual automatic letter recognition. Xing et al. proposed a method of letter segmentation and recognition with improved Radon transform by specifying the place of a license plate [8]. In these three researches, numbers which can be read by humans are extracted and read automatically. On the other hand, we focus on to read numbers which cannot be read by humans.

Jingu et al. proposed a method to specify the numbers in images which are taken by surveillance cameras [9]. Their objective is the same as ours, and their results showed that they were able to specify a number to one digit with a percentage of 95%. On the other hand, our goal of the percentage is higher than that of theirs. We use CNN to increase the accuracy of the reading while they did not use it.

Hata et al. proposed a letter recognition method with ensemble learning [10]. Their paper is not for license plates but they classified letters by CNN. Ensemble learning in their paper is supposed to be effective to increase the accuracy of classification for our method. We will consider using ensemble learning while we do not apply ensemble learning to our method in this paper.

3   Proposed Method

We explain the proposed method in this section.

3.1   Creation of Type Designation Numbers Samples

First, we prepared image samples of type designation numbers by taking photos of actual license plates. The definition of the type designation numbers is from one to four digits Arabic numerals which are written from left to right on a license plate of a car in Japan. When the number of digits is less than four, the blank digits are filled by middle dots. The type designation numbers were manually extracted from a photo with the Arabic numerals and the blank digits, and gray-scaled. Next, we prepared 28 pixels width times 53 pixels height white rectangle image with PNG format. One Arabic numeral which is shrunk to be put in the rectangle image became one number image, and total eleven images with PNG format were created. On the other hand, 64 pixels times 64 pixels images are used for learning with CNN (Convolutional Neural Network) which is mentioned later. The 28 times 53 square pixels images are put in 64 times 64 square pixels images. There is no reason that the 28 times 53 square pixels images must be 28 times 53 square pixels, but we decided the size was appropriate by several trials of creating the images since the size was fit when each digit is put in one image. The shrinking rate is the same for all digits, and each digit is deployed with two-pixel margin at top and bottom. The middle dot is deployed at the center of the image. On the other hand, we think the strict deployment is not so important in experiments in this paper since original image of the digits were extracted from snapshots and the images were processed by some processes described later. The created eleven images are called basic images in this paper.

3.2   Creation of Training Samples

We explain how to create training samples for CNN. Before putting images to CNN, some processes are executed to the basic images to increase the accuracy of classification.

Shift. N_{shift} numbers images are created from one basic image by randomly shifting to top, bottom, left and right at most one fourth of the image size. (N_{shift} is a natural number.) The shift means moving images determined pixels to top, bottom, left and right. When shifting, the size of the images is not changed. Protruded pixels are cut off and blank pixels are filled by white. The logic of the shift process is matched to the description that the strict deployment is not so important in experiments in Sect. 3.1.

Cropping. $N_{cropping}$ numbers images are created from one 1.2-time-expanded basic image by cropped to 28 pixels times 53 pixels in random position. ($N_{cropping}$ is a natural number.) The logic of the cropping process is also matched to the description that the strict deployment is not so important in experiments in Sect. 3.1.

Three Dimensional Rotation. Three dimensionally rotated images are created from the 28 times 58 square pixels basic images, shifted images or cropped images. Images are rotated from $-\pi/4$ to $\pi/4$ in each X, Y or Z axis. We decided what number equal parts an image was divided from $-\pi/4$ to $\pi/4$. When an image is divided into N_x, N_y and N_z in each axis, the number of images which is shown in Eq. (1) is created from one image. The number is called N_{3D}. This rotation is executed in order to recognize a license plate which is diagonally photographed.

$$N_{3D} = \left\{ (N_x + 1) * (N_y + 1) * (N_z + 1) \right\} \tag{1}$$

Smoothing. $N_{smoothing}$ numbers images are created from a rotated 28 times 58 square pixels basic image. ($N_{smoothing}$ is a natural number.) Smoothing is a process in which luminance values are set to the average values of pixels around a target pixel. Random size kernels for both width and height are used. Also, the width and the height are equal, and the minimum value is 12 and the maximum value is 30. The kernel size was decided by our eyes when we saw the processed images. The minimum value was decided when we could completely recognize the smoothing process, and the maximum value was decided when we could distinguish numbers. The process is for a countermeasure against optical blur from the limitation of image sensor performance and for the correct classification of resized images by bilinear interpolation.

Noise. N_{noise} numbers images with impulse noise are created from a rotated 28 times 58 square pixels basic image by replacing N_{white} number pixels to black and N_{black} number pixels to white. (N_{white} is a random natural number, N_{black} is a random natural number, and $N_{smoothing}$ is a natural number.) The process is for a countermeasure against noise which comes from compensation of photos at dark places.

Change of Brightness. $N_{brightness}$ numbers images with randomly changed brightness are created from a rotated 28 times 58 square pixels basic image. ($N_{brightness}$ is a natural number.) Also, the same numbers of both lighter and darker images are created. (When the number is odd, the number of darker images is one more than that of lighter images.) The process is for a countermeasure against clipped whites and crushed shadows. Clipped white means that all pixels are turned to white although the pixels originally have different brightness. Crushed shadow also means that all pixels are turned to black although the pixels originally have different brightness.

Random Erasing. $N_{erasing}$ numbers images with Random Erasing Data Augmentation by Zhong et al. [11] are created from a rotated 28 times 58 square pixels basic image. ($N_{erasing}$ is a natural number.) Actually, a grayscale rectangle is deployed in a random place, and the rectangle is created with decided parameters in which pixel value is set to from 0 to 255, mask size is set to from 0.02 to 0.4 times the original image, and mask aspect ratio is set to from 0.3 to 3. The process is for capturing not partial feature but whole feature of a image. Also, it is expected that images can be classified by CNN even when a part of a image is hidden by an obstacle.

Creation of Square Image. New images are created by all processes mentioned above, and they are deployed at the center of 64 times 64 square pixels black image. The all new images are used as training samples. The number of the training samples is shown by the Eq. (2).

$$\left\{11 + 11 * (N_{shift} + N_{cropping})\right\} * N_{3D}$$
$$* (N_{smoothing} + N_{noise} + N_{brightness} + N_{erasing}) + 11 \qquad (2)$$

Shaping of Images for Classification. First, photos for classification were prepared. Note that, we do not manage images in which over a half part of a number is hidden by obstacles or numbers are extremely changed. Also, in our experiments, we have known the correct answer of the type designation numbers since the experiments are not for real situation but for evaluation of our method. We extracted a number by trimming a rectangle in which one of the type designation numbers may be included. The place of the trimming was not strictly decided. The rectangle image was placed at the center of the 64 times 64 square pixels black image after resizing 28 times 53 square pixels. We resized the image by bilinear interpolation. In bilinear interpolation, four pixels around a pixel is used, and the interpolation is good at prevention from deterioration of image quality.

Learning and Test with CNN. We used CNN for learning and test. In experiments in this paper, we used a network called AlexNet by Krizhevsky [12]. There are five convolution layers and three fully-connected layers in this network. First, different labels are put on samples with each Arabic numeral and middle dot. Next, training is executed and the weight a model is saved. Actual images as test samples mentioned in Subsect. 3.2 are classified with the model.

4 Implementation

In this section, we explain implementation of our method described in Sect. 3. Python 3 is used for the implementation except the implementation of three dimensional rotation.

4.1   Creation of Training Samples

Shift. Image is shifted by cv2.warpAffine() function. We reason why we adopted this function was that the function uses affine transformation which can easily shift images. This function requires input images, 2 by 3 transformation matrix and the size of output images as parameters. When the value of movement to (x, y) is defined (t_x, t_y), the transformation matrix M as the shift is expressed as shown in Eq. (3).

$$M = \begin{bmatrix} 1\ 0\ t_x \\ 0\ 1\ t_y \end{bmatrix} \qquad (3)$$

When M is set as NumPy array with the type of numpy.float32, the M is expressed numpy.float32($[[1, 0, t_x], [0, 1, t_y]]$). In our experiments, t_x and t_y are generated by random.randint() function since both of them are randomly decided. A value from zero to one fourth of the image size is inputted as an argument of random.randint() function, and the generated integer by this function is defined as t_x. t_y is also generated as same as t_x. The size of output image is the same as the size of input image. The generated numpy.ndarray type images are saved as PNG images. The process is repeated N_{shift} times.

Cropping. Images which are expanded by cv2.resize() function are cropped to 28 times 53 square pixels. Coordinates of both top and left pixels are randomly decided by the generated values of random.randint() function. The coordinate of the top is decided from zero to the value that is the height of the expanded image minus the height of the original image. The coordinate of the left is decided from zero to the value that is the width of the expanded image minus the width of the original image. The generated images are saved as PNG images. The process is repeated $N_{cropping}$ times.

Three Dimensional Rotation. Three dimensional rotation is only implemented not by Python but by Java. Java 3D which is a Java extension API is used. Box which is a type of primitive types are created in simple universe. The size of Box is 28 times 58 times 0. Also, the background color of the Box is white. Under the situation, the Box is rotated to X, Y and Z directions by rotX(), rotY() and rotZ() methods in Transform3D class. In addition, an angle is required as a parameter. For example, when the Box is rotated to X direction, a loop is executed N_x times and $-\pi/4$ is inputted in specific loops. The specific loops are the first loop and $(-\pi/4 + \pi/2/i)$th loop. (i is an integer from 2 to $N_x + 1$.) The arguments for Y and Z directions are decided as same as that for X direction. Images of all patters are transformed to 28 times 53 square pixel image by renderOffScreenBuffer() method in Canvas3D class, and saved as PNG images. N_{3D} numbers of images are created by the process.

Smoothing. Images are smoothed by cv2.blur() function. We reason why we adopted this function was that smoothing of images with a normalized box type filter is executed. The function requires width and height of kernel as parameters. The width and the height are randomly decided as described in Sect. 3. Actually, numpy.random.randint() function is used for the generation of random numbers. numpy.random.randint() function has two parameters which are the minimum value of the generated values and the value next to the maximum value of the generated values. In our implementation, 12 is set to the minimum value and 31 is set to the value next to the maximum value. The generated random integers are inputted to cv2.blur function as the width and the height of kernel, and the generated images are saved as PNG images. The process is repeated $N_{smoothing}$ times. If cv2.boxFilter() function is used instead of cv2.blur() function, the process can be implemented with the same procedure. We chose cv2.blur() for the implementation since the name of the function is shorter than the other name.

Noise. We originally implemented a function of adding noise. In the function, a random integer from 1 to 28 and a random integer from 1 to 53 are generated N_{white} times each in order to determine coordinates of pixels which are replaced with white. numpy.random.randint() function is used for the generation. The combinations of the random integer from the first to N_{white} number are the coordinates of the pixels. The RGB value of the pixels is replaced to $(255, 255, 255)$. N_{black} numbers of coordinates of pixels which are replaced with black are selected by the same procedure as that for white, and the RGB value of the pixels is replaced to $(0, 0, 0)$. The generated images are saved as PNG images. The process is repeated N_{noise} times.

Change of Brightness. Image and ImageEnhanceImage modules in PIL library are imported for the implementation of change of brightness. Image.fromarray() function is used to transform numpy.ndarray type images to Image type images. Brightness of an image with Image type is changed by Enhance.Brightness() method. The change is controlled by setting a float value to enhance() method as an argument. When the value is set to 1.0, the copy of the inputted image is returned. When the value is set to 0.0, an image with all black pixels is returned. There is no upper limitation of the value. We reason why we adopted this function was that brightness of all pixels of an image is equally changed. numpy.random.randint() function is used for the change since brightness is randomly changed in our experiments. When getting dark, 1 as the minimum value and 10 as the maximum value are set to the arguments of numpy.random.randint() function. When getting bright, 10 as the minimum value and 100000 as the maximum value are set to the arguments of numpy.random.randint() function. The generated integers are casted to float type, and set to enhance() function as an argument after divided by 10. Finally, the images are transformed from Image type to numpy.ndarray type by numpy.asarray() function, and saved as PNG images. The process is repeated $N_{brightness}$ times. The half of the images are turned to dark, and the others are turned to bright as mentioned in Sect. 3.2.

Random Erasing. We originally implemented a function of Random Erasing. All values are based on the recommended values by Zhong et al. First, the mask size is randomly determined from 0.02 times to 0.4 times of the size of the original image by numpy.random.randint() function. In our experiments, the minimum value is 28 times 53 times 0.02, and the maximum values is 28 times 53 times 0.4. Next, the aspect ratio of the mask is randomly determined from 0.3 to 3 by numpy.random.rand() function. The random value can be generated from 0.3 to 3 by adding 0.3 to 3 times the generated value since numpy.random.rand() function generates a value from 0 to 1 when no argument is set. The width and the height of the mask is determined by the size and the aspect ratio after the size and the aspect ratio are determined. The height is determined to the square root value of the division of the size by the aspect value by numpy.sqrt() function after the cast to int type. When the height is bigger than the height (53) of the original image, it is fixed to the height - 1 (53 - 1). The width is also determined to the square root value of the division of the size by the aspect value by numpy.sqrt() function after the cast to int type. When the width is bigger than the width (28) of the original image, it is fixed to the width - 1 (28 - 1). After that, the position of the mask is randomly determined by numpy.random.randint() function. The top value is from 0 to a value which is the distance between the height of the original image and that of the mask. The left value is from 0 to a value which is the distance between the width of the original image and that of the mask. Finally, an integer from 0 to 255 is generated by numpy.random.randint() as a pixel value of the mask, and the value is set to the determined place and the determined size. The images of numpy.ndarray type generated by the procedure are saved as PNG images. The process is repeated $N_{brightness}$ times.

Creation of Square Image. The square image mentioned in Sect. 3 is created with cv2.add() function by placing a rectangle image at the center of a 64 by 64 pixels PNG image which is painted with black. The created images are saved as PNG images again.

Learning and Test with CNN. We implemented CNN with TensorFlow which is software library for machine learning and keras which is one of the neural network library for TensorFlow. The version of TensorFlow is 1.12.0 and that of keras is 2.2.4. We created eleven folders named as 0 to 9 numbers and middle dot and put created PNG image files mentioned above in the folders since the images are classified to eleven with CNN. Note that, the name of the folder for middle dot is 10, and the names of the others are the same as each number. Sample images are trained by fit() function and the parameters of the function are training samples, batch size and the number of epochs. Trained models are saved as h5 type by model.save_weights() function. The h5 type files are loaded by model.load_weights() function. The value of precision of training samples is outputted by predict() function.

5   Evaluation

In this section, we evaluate the results of our experiments.

5.1   Evaluation of Learning

Samples were trained under the condition as follows.

$N_{shift} = 5$
$N_{cropping} = 5$
$N_x = N_y = N_z = 4$
$N_{smoothing} = N_{noise} = N_{brightness} = N_{erasing} = 5$
$N_{black} = N_{white} = 1000$
The number of epoch is 20
batch size is 32

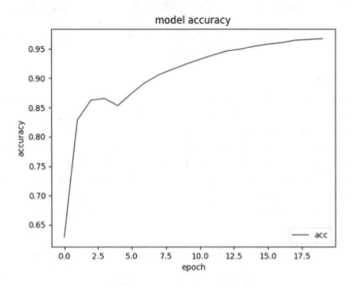

Fig. 1. Example values of accuracy of each epoch in a training.

The numbers above are determined by our trial experiments. When we tried several times, the numbers indicated that training had finished around 10 h and accuracy with the model has converged around 0.97. Therefore, we used the numbers and there is no reason that other numbers can be applied. In future, we will evaluate our method again with other numbers, but the numbers are fixed the values above in this paper. For example, the values of accuracy of each epoch in a training are shown in Fig. 1.

5.2   Evaluation of Difference in Resolution

One image of a number is prepared extracted from a license plate. We evaluated the classification accuracy in stages of decreasing the resolution of the image. An example of the extracted original image which is shaped described in Sect. 3.2 is shown in Fig. 2. Also, examples of the images which are shrinking in stages and shaped are shown in Figs. 3, 4, 5, 6 and 7. The titles of the figures indicate the size of images before shaping. That is, the image in Fig. 7 is constructed with two pixels, but the edge is smooth since it is resized by bilinear interpolation as shown in Sect. 3.2.

Fig. 2. Example of 36 by 58 image.   **Fig. 3.** Example of 18 by 29 image.   **Fig. 4.** Example of 9 by 14 image.

The predicted values for each image by predict() function are shown in Table 1. An integer number between single quotations indicates a label i.e. a number among zero to nine and middle dot. A float value besides the integer number indicates the probability in which the number is classified in the number. All images except 1 by 2 pixel image are correctly classified with percentage of 99% or more. The results show that a robust model against decreasing resolution is created. The 1 by 2 pixel image is thought to be difficult to be classified with high accuracy since only two pixels can express quite few patterns.

Table 1. Prediction results of images with low resolution.

Size	Predicted numeral
36 × 58	'6', 0.99995863
18 × 29	'6', 0.99996996
9 × 14	'6', 0.9999702
4 × 7	'6', 0.9998155
2 × 4	'6', 0.9988733
1 × 2	'1', 0.34142175

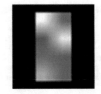

Fig. 5. Example of 4 by 7 image.   **Fig. 6.** Example of 2 by 4 image.   **Fig. 7.** Example of 1 by 2 image.

5.3 Evaluation of Difference in Extraction Range

Some number images extracted from a license plate are prepared. When the images are extracted, margin around the number must be different. Each image is shaped as described in Sect. 3.2 and classified. Examples of different extraction of the same number from the same license plate are shown in Figs. 8 and 9. Image in Fig. 9 is example with wider extracted area than that in Fig. 8.

Fig. 8. Example of a number image with narrow extracted area.   **Fig. 9.** Example of a number image with wide extracted area.

The predicted values of each image by predict() function are shown in Table 2. Number images with narrow extracted area are correctly predicted with a percentage of 78%. On the other hand, number images with wide extracted area are predicted to '0' with a percentage of 50%. We consider that there are few training samples which are correspond to the images. That is, much more images with wide margin around the number must be required. Moreover, aspect ratio had changed in shaping the images so that the shape of a number in images in training samples were different from that in test samples. Furthermore, the total number of training samples might be too few. We think that the accuracy can be increased when the number of images, especially, shifted images and three dimensionally rotated images, are increased. The same consideration can be applied to number images with narrow extracted area. To increase the number of training samples will be our future work.

Table 2. Prediction results of images extracted in different areas.

Range	Predicted numeral
Narrow	'4', 0.78227735
Wide	'0', 0.50750625

6 Summary

In this paper, we evaluated our method of classifying numbers on license plates and considered how to improve the method in future. In the experiments, many number images were created by three dimensional rotation after various image processing, and the created images were trained with CNN. The models which were created with CNN were saved, and used for the classification of numbers on real license plates. The results indicate that our method is robust against low resolution images. On the other hand, some images were classified with quite low accuracy under a specific condition of the extraction process. We think that the solution of the problem is one of the top priority matters.

We will improve our method by considering the problems mentioned in Sect. 5. Also, we think the number of experiments is few, so we will do them again and again for the evaluation. We will also consider the change of the number of each process and values of arguments to evaluate the change of the accuracy as shown in Sect. 3.2.

References

1. Yoshiura, N., Kato, S., Takita, A., Ohta, N., Fujii, Y.: Analysis of questionnaire result on installing security cameras on school routes. IPSJ J. **59**(3), 1106–1118 (2018). (in Japanese)
2. Wang, C., Liu, J.: License plate recognition system. In: 2015 12th International Conference on Fuzzy Systems and Knowledge Discovery (FSKD), pp. 1708–1710 (2015). https://doi.org/10.1109/FSKD.2015.7382203
3. Amirgaliyev, B.Y., Kenshimov, C.A., Kuatov, K.K., Kairanbay, M.Z., Baibatyr, Z.Y., Jantassov, A.K.: License plate verification method for automatic license plate recognition systems. In: 2015 Twelve International Conference on Electronics Computer and Computation (ICECCO), pp. 1–3 (2015). https://doi.org/10.1109/ICECCO.2015.7416892
4. Ashtari, A.H., Nordin, M.J., Fathy, M.: An Iranian license plate recognition system based on color features. IEEE Trans. Intell. Transp. Syst. **15**(4), 1690–1705 (2014). https://doi.org/10.1109/TITS.2014.2304515
5. Rashid, A.E.: A fast algorithm for license plate detection. In: 2013 International Conference on Signal Processing, Image Processing Pattern Recognition, pp. 44–48 (2013). https://doi.org/10.1109/ICSIPR.2013.6497956
6. Haneda, K., Hanaizumi, H.: A study on numbers extraction method in automatic license plates recognition system. In: Proceedings of the 75th National Convention of IPSJ, No. 1, pp. 449–450 (2013). (in Japanese)

7. Wang, N., Zhu, X., Zhang, J.: License plate segmentation and recognition of Chinese vehicle based on BPNN. In: 2016 12th International Conference on Computational Intelligence and Security (CIS), pp. 403–406 (2016). https://doi.org/10.1109/CIS.2016.0098

8. Xing, J., Li, J., Xie, Z., Liao, X., Zeng, W.: Research and implementation of an improved radon transform for license plate recognition. In: 2016 8th International Conference on Intelligent Human-Machine Systems and Cybernetics (IHMSC), vol. 01, pp. 42–45 (2016). https://doi.org/10.1109/IHMSC.2016.52

9. Jingu, A., Ota, N.: Numeral identification of low resolution license plates photographed by security cameras. In: Proceedings of the 73rd National Convention of IPSJ, No. 1, pp. 527–528 (2011). (in Japanese)

10. Hata, Y., Komori, K., Kawana, H., Oeda, S.: Evaluation of authenticity judgment of character recognition by ensemble learning of CNN. In: Proceedings of the 75th National Convention of IPSJ, No. 1, pp. 707–708 (2018). (in Japanese)

11. Zhong, Z., Zheng, L., Kang, G., Li, S., Yang, Y.: Random erasing data augmentation. CoRR abs/1708.04896 (2017)

12. Krizhevsky, A., Sutskever, I., Hinton, E.: ImageNet classification with deep convolutional neural networks. In: Advances in Neural Information Processing Systems 25, pp. 1097–1105 (2012)

Variational Deep Semantic Text Hashing with Pairwise Labels

Richeng Xuan[1], Junho Shim[2(✉)], and Sang-goo Lee[1]

[1] Department of Computer Science and Engineering, Seoul National University,
Seoul, Republic of Korea
{xuanricheng,sglee}@europa.snu.ac.kr
[2] Department of Computer Science, Sookmyung Womens University,
Seoul, Republic of Korea
jshim@sookmyung.ac.kr

Abstract. With the rapid growth of the Web, the amount of textual data has increased explosively over the past few years. Fast similarity searches for text are becoming an essential requirement in many applications. Semantic hashing is one of the most powerful solutions for fast similarity searches. Semantic hashing has been widely deployed to approximate large-scale similarity searches. We can represent original text data using compact binary codes through hashing. Recent advances in neural network architecture have demonstrated the effectiveness and capability of this method to learn better hash functions. Most encode explicit features, such as categorical labels. Due to the special nature of textual data, previous semantic text hashing approaches do not utilize pairwise label information. However, pairwise label information reflects the similarity more intuitively than categorical label data. In this paper, we propose a supervised semantic text hashing method that utilizes pairwise label information. Experimental results on three public datasets show that our method can exploit pairwise label information well enough to outperform previous state-of-the-art hashing approaches.

Keywords: Natural language processing · Semantic hashing ·
Machine learning · Similarity search

1 Introduction

In recent years, the number of web documents has been increasing rapidly. Text data is the most basic type of data on the Internet, and it exists in large amounts. There are many applications based on big data in text form. With the advent of word embedding [13], text data has investigated in a variety of studies, including those focusing on sparse one-hot vectors or high-dimensional data. When text data becomes high-dimensional data, many document embedding [10,15] approaches can be used to convert all of the words in the article into a fixed-size representation to reduce the size of the data. Although the textual data after

© Springer Nature Switzerland AG 2019
S. Lee et al. (Eds.): IMCOM 2019, AISC 935, pp. 1076–1091, 2019.
https://doi.org/10.1007/978-3-030-19063-7_85

this process runs can be represented by a smaller amount of data, the method remains not accurate enough to be used with large-scale text data.

While textual big data with large volumes become more pervasive in many applications, increasing attention is drawn to work on approximate nearest neighbors (ANN) searches of text data with improvements in both computation efficiency and search quality levels. Due to the computational and storage efficiency of compact binary codes, hashing is a good resolution for text ANN searches, as this method can transform high-dimensional data into compact binary codes and generate similar binary codes for similar data items. With a compact size of the hash code, we can determine the nearest neighbors from large amounts of text data in a real-time fashion.

Existing hashing methods can be roughly divided into the data-dependent and data-independent approaches. In data-independent approaches, the well-known locality-sensitive hashing (LSH) [1] approach is useful for cases without a pre-training dataset because the hash function is typically randomly generated, making it independent of any training data. Data-dependent approaches attempt to learn the hashing function by utilizing training data. Recently, ranging from deep learning to hashing [9,11,31] have shown that both feature representation and hash codes can be learned more effectively using deep neural networks.

Existing data-dependent text hashing approaches train hash functions typically from class label data [3,18]. In these approaches, the evaluation metric is the percentage of documents from among 100 retrieved documents that have a label identical to the query document. Although this metric can suitably simulate the finding of identical class data, in practical applications, using pairwise label data is a better match considering the requirements. For example, in a question-and-answer system, when a new question is submitted by a user, the system attempts to find a semantically similar question instead of a question in the same class. Semantic similarity is often marked with a pairwise label in the corpus. A paraphrase identification dataset is a typical pairwise labeled dataset. These datasets assign a tag of 1 if two articles are semantic duplicates, and vice versa. This represents the essentially difference from tagged class label to one article. In many real-world examples, using pairwise information to learn an ANN search is more meaningful.

In this paper, we propose a variational pairwise supervised text hashing (VPSH) which exploits pairwise label information to learn compact binary text hashing functions. The contributions of this paper are as follows. First, to the best of our knowledge, we present the first VAE based pairwise supervised text semantic hashing method that learns pairwise label information to improve the performance. Second, we propose a method to learn the compact binary code for the well-known pairwise labeled corpus. The experimental results show that our method outperforms existing hashing approaches. Finally, we propose a label weight annealing technique to help our method learn the pairwise label information, and demonstrate its efficiency through experimental evaluations.

The rest of this paper is organized as follows. In Sect. 2, we review previous hashing approaches. Section 3 presents our semantic text hashing scheme. The experimental results are shown in Sect. 4. Finally, Sect. 5 concludes this paper.

2   Related Work

The paraphrase identification task is one of the most common tasks used to handle the pairwise labeling of text data. In this task, the label is 1 if one sentence is a paraphrase of another, and the label is 0 otherwise. For the text nearest-neighbor retrieval problem, finding semantically similar text is actually a paraphrase-identification task. A number of deep-learning models have proposed in relation to this problem [6,23]. These works commonly exploit a complex neural network structure, making the model more consistent with the task and increasing the prediction accuracy. However, the model becomes more complex. To simplify the model, a latent variable model for paraphrase identification has been studied [19]. This method uses a convolution-deconvolution autoencoder to infer the sentence representation for semi-supervised learning. With the power of a novel encoder-decoder, this model works well with limited amounts of labeled data. Nevertheless, these works all use two instances of raw text data as the input to calculate the result. The prediction can be accurate even for a large amount of data, though it becomes a time-consuming task with greater amounts of data. It is not practical to execute a real-time ANN search for a large dataset.

Due to the computational and storage efficiency levels of compact binary codes, hashing has been widely used for ANN searches. Traditional data hashing methods such as LSH [4,30] and Spectral hashing [24] have widely been used. When supervised information such as class labels or relative similarity is available, a supervised hashing method is a better choice. Supervised hashing has also been used to exploit information with which to learn the hashing function [14]. With the revival of neural networks, deeper learning models are used for hashing [11,31]. These methods can be applied to general high-dimensional data or image data. However, if directly used in the text keyword vector space, they usually fail fully to capture semantic similarity with regard to the original text. Hence, many text hashing methods have been proposed. Initially, text semantic hashing using an autoencoder was developed in order to learn hash functions [16]. These methods build multiple Boltzmann machines [20] to learn the binary unit capable of modelling the input text word count data. After training, the binary hash code of any document is acquired by simply thresholding the output of the deepest layer. Furthermore, several studies have explored the power of convolutional neural networks (CNNs) for text hashing with the help of word embedding [25,28].

In recent years, the probabilistic generative model has attracted much attention. The variational auto-encoder (VAE) [8] is indeed an appealing framework for generative modeling as it couples variational inference [22] with a deep neural network. VAEs acquire the advantages of both deep learning and probabilistic generative models. VAEs achieve state-of-the-art performance in many problems, especially with image data [26]. In the domain of natural language processing, the neural answer selection model (NASM) [12] is a generic variational inference framework for conditional models of text. This model extracts the semantics between a question and answer pair. It uses the question and answer text data as input to calculate the relationship result instead of a hash code, which however makes this model unsuitable to process large amounts of data. For text hashing, unsupervised and supervised variational deep semantic hashing (VDSH) [3] methods have been developed to preserve each content from a document during the text hashing process. These approaches use a VAE framework similar to that presented here. However, VDSH differs from our hashing scheme in that it utilizes only categorical class label data to improve the accuracy of semantic text hashing. Pairwise label data have not been used in previous hashing methods.

3   Proposed Method

3.1   Text Hashing Using VAEs

Let x denote the input text and z denote the hash code of the given input text. z has a real value or binary code of n. We also refer to n as the number of bits. The encoding process is to infer z from document x, while the decoding process is to reconstruct x from the latent variable z. Intuitively, the latent variable learns from the corpus that captures the key semantic features from x. The hash function is the distribution of the encoding $p(z|x)$. All previous studies attempted to learn this distribution from the corpus. Let x be the bag-of-words representation of a document. x has a length of $|V|$, where V is the vocabulary of words that have appeared in the corpus. TF-IDF [17] schemes are used for weighing

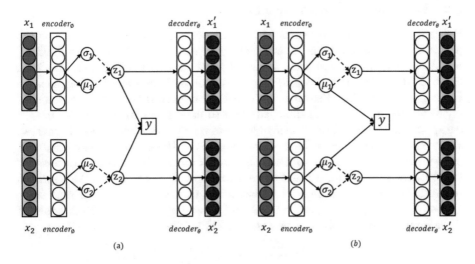

Fig. 1. Architectures of variational pairwise supervised text hashing (VPSH) methods: (a) stochastic method (VPSH-ST), (b) deterministic method (VPSH-DE)

the x representation. Recent advances in VAEs for semantic text hashing [3,18] have demonstrated the effectiveness of the model in supervised text hashing. In this model, the approximation encoding distribution is $q_\phi(z|x)$ and the decoding distribution is $q_\theta(x|z)$, where ϕ and θ are the parameters of the encoder and decoder, respectively. Based on the VAE framework [8], we maximize the variational lower bound instead of the marginal distribution. Finally, the objective function of text hashing is as follows.

$$L_{vae} = E_{q_\phi(z|x)}[logq_\theta(x|z)] - D_{KL}(q_\phi(z|x)||p(z)) \qquad (1)$$

The Kullback-Leibler (KL) divergence D_{KL} covers the posterior distribution $q_\theta(z|x)$ and prior $p(z)$. Here, $p(z)$ in VDSH [3] is a multivariate standard distribution. This model approximates the posterior distribution $q_\theta(z|x)$, which is assumed to be a Gaussian $N(\mu, \sigma^2)$ distribution with a mean of μ and variance of σ^2 as the output of encoder parameterized typically by ϕ in the domain of a neural network. In order to determine hash code z from the encoder outputs, we can sample z from $N(\mu, \sigma^2)$. However, the sampling operation may not induce a gradient, which makes the neural network untrainable. In practice, we can use a reparameterization trick [8] to divert the non-differentiable sampling operation out of the network. This operation is represented by the dashed lines from μ_1 and σ_1 to z_1 in Fig. 1. This trick initially undertakes sampling $\epsilon \sim N(0, 1)$ and then implements the sampling operation of z_1 as follows:

$$z_1 = \mu_1 + \sigma_1^{1/2} * \epsilon \qquad (2)$$

It should be noted that during the inferencing process, we consider μ as a hash code because the hashing function cannot suitably handle randomness.

3.2   Pairwise Supervised Hashing

For supervised learning, previous studies utilize the categorical label information y. For the categorical label information, we can add the minimizing discriminate loss $p(y|z)$ directly and use the label weight W_{label} to control its influence. Supervised hashing is utilized to allow the hash code of a similar document to be similar by learning. Previous works using categorical label information simply allow documents in the same category to have a similar hash code. In text data, pairwise label information should contain the most directly way to marked similarities or relationships between the two texts. Here, we use the more direct pairwise label information to learn the hash function because this data-set is annotated with a similar content or another semantic relationship. We approach semantic text hashing from a joint discriminate pairwise objective perspective.

Let x_1 and x_2 be two input texts and z_1 and z_2 denote the hash codes of these inputs, respectively. In this case, Eq. (1) will be extended as follows:

$$\begin{aligned} L_{vae}(\phi, \theta; x_1, x_2) =& E_{q_\phi(z|x)}[\log q_\theta(x_1|z_1) + \log q_\theta(x_2|z_2)] \\ &- D_{KL}(q_\phi(z_1|x_1)||p(z_1)) \\ &- D_{KL}(q_\phi(z_2|x_2)||p(z_2)) \end{aligned} \qquad (3)$$

The parameters ϕ and θ are shared for processing x_1 and x_2, as shown in Fig. 1. In addition, $p(z_1)$ and $p(z_2)$ in this method are multivariate standard distributions.

For pairwise supervised hashing, the goal here is to address the distribution $p(y|z_1, z_2)$, where y is the label information of text x_1 and x_2. We assume that the label information is binary (y is 0 or 1). In order to make the model balance the variational lower bound and discriminate objective, the total objective is given as follows,

$$L_{total} = L_{vae}(\phi, \theta; x_1, x_2) + W_t \cdot L_{label}(z_1, z_2, y) \tag{4}$$

where W_t is the label weight that controls the supervised influence.

The input of the decoder is a sample of a latent variable distribution, which is parameterized by the encoder output. Therefore, it is possible directly to compare z_1 and z_2 as discriminator objects, as shown in Fig. 1(a). In order for the generated hash code to have similar properties, we can use the similarity of z_1 and z_2 directly on the discriminator. This method is referred to *stochastic* because z_1 and z_2 can be regarded as the sampling result from $N(\mu_1, \sigma_1{}^2)$ and $N(\mu_2, \sigma_2{}^2)$. The loss function of this method is as follows,

$$L_{label} = y - sim(z_1, z_2) \tag{5}$$

where z_1 and z_2 are real value vectors with length n, and $sim()$ is the similarity function of the two inputs. In this work, we use *cosine similarity*, although the similarity function can vary in practice according to the actual requirements. According to the reparameterization trick, as expressed by Eq. (2), L_{label} can be defined as follows:

$$L_{label} = y - sim(\mu_1 + \sigma_1^{1/2} \cdot \epsilon, \mu_2 + \sigma_2^{1/2} \cdot \epsilon) \tag{6}$$

In other words, the *stochastic* method utilizes all outputs of the encoder (μ_1, μ_2, σ_1, σ_2).

In previous supervised semantic text hashing methods, a fully connected softmax layer was often used to map the hash code z and label information y. There are two z variables in our model; hence, if we only concatenate z_1 and z_2 as the input of the softmax layer, z_1 and z_2 will not be similar to each other. During the inferencing of text hashing, the hash code z is μ in the output of the encoder. Accordingly, we can also use the similarity between μ_1 and μ_2 as an indicator. The *deterministic* method is another approach that computes the similarity between μ_1 and μ_2, as shown in Fig. 1(b). Because μ_1 and μ_2 in this case are the values before the sampling operation, the similarity will be more directly reflected in the encoder parameters. The loss function in this method is as follows:

$$L_{label} = y - sim(\mu_1, \mu_2) \tag{7}$$

3.3 Label Weight Annealing

KL cost annealing [2] is a simple approach to deal with the *vanishing latent variable problem*. This problem often leads to straightforward VAEs with the decoder failing to encode meaningful information. This annealing technique increases the weight of the KL divergence from 0 to 1 during the training process. When using the proposed method, the addition of pairwise similarity also causes a similar problem. The similarity becomes easier to converge compared to the other cases during training, and we propose the label weight annealing technique to address this problem. This technique controls the label information weight during training to obtain the most aptly supervised learning.

Similar to the KL cost weight in *KL cost annealing*, W_t in Eq. (4) is the label weight that controls the effects of pairwise supervised influence. The KL cost weight increases from 0 in training step 0. However, our W_t during *Label weight Annealing* is not the same as *KL cost annealing*. It does not necessarily increase from the beginning of training. It may also increase after training for a while:

$$W_t = \begin{cases} c \cdot t, & \text{if } t > t_{start} \\ 0, & \text{otherwise.} \end{cases} \tag{8}$$

Here, t_{start} is the step at which the increase starts during training and c is the ratio of the increase in each step t. The t_{start} value is determined according to the influence of unsupervised learning. For example, if excessive unsupervised learning degrades the performance, we set a smaller value of t_{start}. If the influence of supervised learning is only a supplement to unsupervised learning, we set a larger value of t_{start}. The value of c must take into consideration whether or not supervised learning would destroy unsupervised learning if supervised learning and unsupervised learning are carried out at the same time. If it is destroyed, c should be smaller to minimize such conflicts. Our practice shows that c should not have a large value when the number of bits n is large.

Because the convergence of similarity is completely different from that in the other cases, letting the label information learn from the beginning with a large weight will force the method to depend on the initial variable during training. A somewhat poor initial value will be devastating to the entire training process.

4   Experiment

4.1   Dataset Detail

To demonstrate the effectiveness of our model, we use three standard publicly available datasets in two tasks for training and evaluation. First, we compare the outcome of our model with those of others during a paraphrase identification task. This task aims to determine the paraphrase of two sentences, a problem considered as a touchstone in natural language understanding. The Microsoft Research Paraphrase Corpus (MSRP) [5] dataset is a popular benchmark which contains 4,076 sentence pairs for training and 1,725 for testing. Each text pair with human annotations indicates whether each pair captures a paraphrase/semantic equivalence relationship. The Quora Question Pairs[1] dataset

---

[1] https://data.quora.com/First-Quora-Dataset-Release-Question-Pairs.

contains numerous question pairs, with each annotated with a binary value indicating whether the two questions are a paraphrase of each other. More specifically, this dataset consists of 384,348 sentence pairs for training, and 10,000 pairs for testing. We measure the accuracy performance of our algorithm on these datasets.

We also evaluate our method on a candidate answer ranking task. WikiQA [27] is an open-domain question-answer dataset. The task is to rank the candidate answers based on their relatedness to the question. It is important to note that a question and answer pair in this case is not a similar pair as is typically understood. Questions and correct answers do not often have the same words, and they do not have the same meaning. Owing to negative sample effects, we assume that there is at least one correct answer to a question [29]. After filtering, the corresponding dataset consists of 20,360 question-answer pairs for training, and 2,352 pairs for testing. The performance measures used on WikiQA are the mean average precision (MAP) and the mean reciprocal rank (MRR).

4.2   Baselines and Setup

We evaluate the following baselines for comparisons: Locality Sensitive Hashing (LSH)[2] [4], Spectral Hashing (SpH)[3] [24], and Variational Deep Semantic Hashing (VDSH)[4] [3]. Because previous supervised methods do not utilize pairwise label information, we do not compare the performance with these methods.

The Adam optimizer is widely used in VAE [3,18]. We use the Adam optimizer [7] with a learning rate of 0.001 and use the learning rate exponential decay with a factor of 0.96 for every 10,000 steps. For large datasets, we set the learning rate set to 0.08 in Quora for fast training. We also use the dropout technique [21] with a value of 0.9 to alleviate over-fitting. Dropout has essentially become a standard in deep learning. We found that VAE based text hashing and several other models exploiting the same datasets use dropout at 0.9 in common. Therefore, we also used the same value. We set the starting label weight parameter t_{start} in Eq. (8) to 40 and to 30 for the stochastic method and the deterministic method, respectively.

All experiments are implemented in Tensorflow[5] and conducted on a server with an Intel i7-6850K CPU, a NVIDIA GeForce GTX TITAN X GPU, and 16 GB of main memory.

4.3   Paraphrase Identification Task

In this task, all of the methods convert the training sentences into the binary codes, and the label information of the training dataset is used to learn the optimal threshold of the similarity. In the test, if the similarity of a pair of sentences

---

[2] http://pixelogik.github.io/NearPy/.

[3] https://github.com/wanji/sh.

[4] https://github.com/unsuthee/VariationalDeepSemanticHashing.

[5] https://www.tensoflow.org.

Table 1. Accuracy for the paraphrase identification task in MSRP dataset

Methods	16 bits	32 bits	64 bits	128 bits	256 bits
LSH	0.5551	0.5533	0.5406	0.5386	0.6223
SpH	0.5311	0.5742	0.5879	0.5857	0.5922
VDSH	0.5848	0.6045	0.6244	0.6015	0.5043
VPSH-ST	0.7017	0.6984	0.6600	0.7163	0.6867
VPSH-DE	0.6669	0.6670	0.6875	0.6725	0.6282

Table 2. Accuracy for the paraphrase identification task in Quora dataset

Methods	16 bits	32 bits	64 bits	128 bits	256 bits
LSH	0.6701	0.6696	0.6794	0.7078	0.7072
SpH	0.6672	0.6701	0.6725	0.6817	0.6852
VDSH	0.6684	0.6701	0.6713	0.6684	0.6701
VPSH-ST	0.6684	0.6678	0.6690	0.6701	0.6678
VPSH-DE	0.6696	0.6864	0.6933	0.6713	0.684

exceeds this threshold, we predict this sentence pair as a paraphrase. Therefore, although the baselines are not applied to the training set label information during the construction of the binary code, the label information data is referenced during the process of the judgment of the paraphrase. Tables 1 and 2 shows the results of our model and the baselines for Quora and MSRP.

In the Quora dataset, the proposed VPSH-ST and VPSH-DE methods outperform the baselines with various numbers of bits. Although they all use the label information of the training set, our methods work significantly better than the others in the Quora dataset, indicating that our model can effectively allocate documents with informative and meaningful hashing codes. Comparing our two methods, we find that the stochastic method works better than the deterministic method. It indicates that with a sufficient number of training sets, the stochastic method with μ and σ is better than the deterministic method with only the mean value μ from the latent variable distribution.

In the MSRP dataset, VDSH and VPSH-ST are worse than the traditional LSH and SpH methods with various numbers of bits. It indicates that the VAE based method is not trained completely in this dataset, mainly due to the resulting training information problem. In the absence of a training set, if we set more latent variables, the poor training set cannot make the model training successful. The MSRP dataset has only 4,076 pairs of sentences for training, and the bag-of-words representation scheme as used here ignores the word order. Our VPSH-DE method outperforms the baselines at 32 bits and 64 bits, while with other numbers of bits when using VPSH-DE, the accuracy is worse than that of LSH. Thus, VPSH-DE is easier to train than VDSH or VPSH-ST when training information is scarce.

Comparing the results with the two data sets, we find that our model works better on a large dataset (Quora), as a large amount of training data allows the probability generation model to learn more stable parameters. Compared to the image data, training information in text data is extremely rare. In short, with a large amount of training data, our method is suitable for learning pairwise information.

Table 3. MAP for the candidate answer ranking in WikiQA dataset

Methods	16 bits	32 bits	64 bits	128 bits	256 bits
LSH	0.4643	0.5056	0.5121	0.5231	0.5381
SpH	0.4704	0.4843	0.4749	0.4600	0.4900
VDSH	0.5581	0.5614	0.5624	0.4741	0.4574
VPSH-ST	0.5840	0.6418	0.6358	0.6253	0.6428
VPSH-DE	0.6312	0.6425	0.6228	0.6141	0.5939

Table 4. MRR for the candidate answer ranking in WikiQA dataset

Methods	16 bits	32 bits	64 bits	128 bits	256 bits
LSH	0.4745	0.5122	0.5233	0.5270	0.5472
SpH	0.4759	0.4892	0.4863	0.4663	0.4995
VDSH	0.5684	0.5691	0.5725	0.4790	0.4653
VPSH-ST	0.5854	0.6427	0.6353	0.6273	0.6433
VPSH-DE	0.6327	0.6431	0.6222	0.6147	0.5938

4.4   Candidate Answer Ranking Task

In the WikiQA dataset, candidate answers are ranked according to the hamming distance of two binary codes. In this task, the calculation of the MAP requires the value of the confidence for sorting, and the baseline cannot utilize the training data indirectly in this task. We test two of our stochastic methods VPSH-ST and the deterministic method VPSH-DE described in Sect. 3.2. As mentioned earlier, the questions and answers in the WikiQA dataset may not be semantically similar. Therefore, working on unsupervised learning too much may cause the model to deviate from the correct training. We assign t_{start} a small value in this data so that our method introduces pairwise label information as early as possible.

Tables 3 and 4 show the MAP and MRR outcomes of our methods and the baselines in the WikiQA dataset. First, we observe that our two methods are significantly better than the baseline hashing methods. This indicates that our method can capture the question-answer relationship from a pairwise dataset.

More importantly, the MAP outcome of a state-of-the-art approach in the candidate answer ranking task is known to be 0.7069 [12]. This is not very different from our experiment result. Our hashing method can change the computation time scale and storage space without a significant MAP reduction.

Second, VPSH-ST is better than VPSH-DE with 64 bits, 18 bits and 256 bits, indicating that the VPSH-DE method works well with relatively few bits. On the other hand, the VPSH-ST method becomes better as the number of bits increases. It appears that the difference between the VPSH-ST method and the state-of-the-art approach in the candidate answer ranking task [12] will be smaller when the number of bits is larger.

Finally, VDSH [3] and VPSH-DE do not always improve the performance as the number of bits increases, possibly due to the model over-fitting. It suggests that using a long hash code does not always enhance the outcomes of these methods. These results illustrate that VPSH-ST is more helpful to prevent over-fitting.

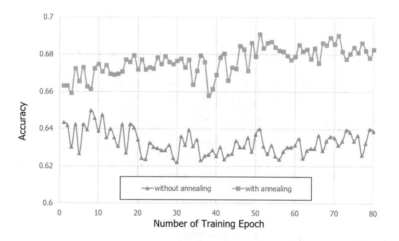

Fig. 2. The worst test accuracy case without label weight annealing in the MSRP dataset for each training epoch

4.5   Effect of Label Weight Annealing

Applying supervised learning in a simple manner is actually worse than expected. Figure 2 shows the worst test accuracy case without label weight annealing in the small dataset (MSRP). Note that the label weight W_t in *without annealing* case is a constant value which we set to be same as the maximum value in *with annealing* case. The degree of similarity depends strongly on weight initialization. As shown in the figure, a slightly bad start and a joint learning pairwise label will not increase the accuracy until the end of the train. In other words, if there is no label weight annealing technique, there is a certain probability that the method will not learn very well.

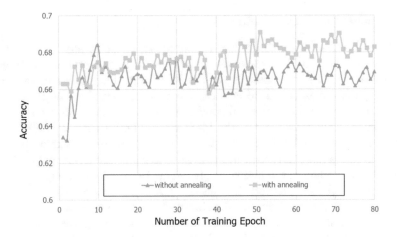

Fig. 3. Accuracy with the MSRP dataset for each training epoch with label weight annealing and without label weight annealing

Figure 3 presents the results of a comparison of the label weight annealing technique with the best cases when using label weight annealing and when not using label weight annealing. With label weight annealing, t_{start} in this experiment is 40, and we find that the accuracy starts to increase at 40 epochs. Without this technique, the method achieves the best accuracy in the first few epochs, while during subsequent training the accuracy of the model gradually decreases. All of these observations indicate that the label weight annealing technique will facilitate more pairwise label information.

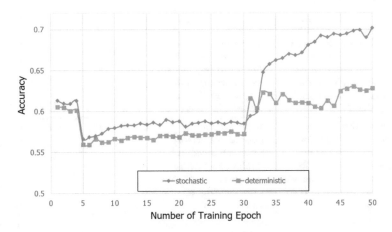

Fig. 4. Accuracy with the Quora dataset for each training epoch with the stochastic method and the deterministic method

4.6   Stochastic vs Deterministic

In order to compare the two proposed methods more intuitively, we present the difference between the two methods when trained on a large dataset (Quora), as shown in Fig. 4. Both methods use the label annealing technique, and t_{start} in this experiment is set to 30 epochs. We observe that there is no difference between the two methods before 30 epochs. When the pairwise label loss is added from 30 epochs, the stochastic method is obviously better than the deterministic method. As mentioned earlier, the stochastic method uses a randomly chosen value in the probability distribution of the latent variable, although this can be problematic when the training set is small. When the training set is large enough, the stochastic method can effectively prevent over-fitting.

In order to determine whether the size of the training data affects the performance of either the stochastic or deterministic method, we performed a number of comparative experiments. Although the stochastic method is better with a large amount of training data, there is not much difference between the stochastic and deterministic methods with a small amount of training data. The deterministic method shows the convergence easier as it can more directly capture the similarity in the encoder. On a large training set, the rapid convergence feature may cause the model to fall into the local optimal.

Table 5. Example of the predicted label on each hashing method in Quora dataset

No.	Sentence 1	Sentence 2	Grand truth	VPSH-ST	VDSH	SpH	LSH
1	are animals capable of feeling/experiencing emotions unknown to humans?	do animals have the same emotions as humans?	1	1	0	0	0
2	is energy in vacuum real? how do we know that this energy that can be borrowed and returned immediately is real if virtual particles didn't exist then?	can energy be borrowed from vacuum -lrb- to be returned immediately -rrb-? if this happens how do we know vacuum energy is real if virtual particles don't exist?	1	1	0	0	0
3	how are the laws regarding self defense weapons enforced in australia and how do they compare to the way they're enforced in the philippines?	how are the laws regarding self defense weapons enforced in australia and how do they compare to the way they're enforced in norway?	0	0	1	1	1
4	should i eat before i workout?	what should i eat before and after a workout?	0	0	1	1	1

4.7   Case Study

Table 5 presents some examples of the input sentences and the predicted label for each hashing method on the Quora dataset.

First, we found that the previous hashing methods will predict False if two sentences, such as examples 1 and 2, have only a few words in common. As TF-IDF [17] weighting schemes do not treat these identical words as keywords, the weights of these words will not be high. In such a case, the previous method would perform incorrectly and assign a label of 0.

Second, we observe that the previous hashing methods will predict True if two sentences have a higher degree of word overlap. In example 3, if philippines in the first sentence is replaced with word norway in the second sentence, the similarity between the two articles is high; therefore, LSH and SpH will assign a value of 1 for this example. In example 4, the second sentence adds a number of words to the first sentence. This type of example can cause the similarity of two sentences to be very high, similar to example 3. In general, the vocabulary is larger than the sentence length, and bag-of-words representation is very sparse representation. When we learn the hash function from this representation, overlap becomes a very powerful feature. Our VPSH method will directly use many false examples as a means of compensation, and it will accurately return an outcome of false in examples 3 and 4.

5   Conclusions

In this paper, we presented a novel semantic text hashing method which exploits the pairwise label information. More specifically, the method can learn label information in a more direct manner. The hashing function can generate more informative hash codes to capture the pairwise label information. Using this binary hash code, we observed the effectiveness of the proposed method through experiments in which we compared its performance with those of other state-of-the-art approaches. We also propose a label information annealing technique to integrate the pairwise label information into the hash code more efficiently.

Acknowledgments. This research was supported in part by the National Research Foundation of Korea (NRF) funded by the Ministry of Science, ICT and Future Planning through the Basic Science Research Program under Grant 2017R1E1A1A03070004. It was also supported in part by the Ministry of Science, ICT and Future Planning through the ITRC (Information Technology Research Center) support program (IITP-2017-2015-0-00378) supervised by the IITP (Institute for Information and communications Technology Promotion).

References

1. Andoni, A., Indyk, P.: Near-optimal hashing algorithms for approximate nearest neighbor in high dimensions. In: 47th Annual IEEE Symposium on Foundations of Computer Science, FOCS 2006, pp. 459–468. IEEE (2006)
2. Bowman, S.R., Vilnis, L., Vinyals, O., Dai, A., Jozefowicz, R., Bengio, S.: Generating sentences from a continuous space. In: Proceedings of The 20th SIGNLL Conference on Computational Natural Language Learning, pp. 10–21 (2016)
3. Chaidaroon, S., Fang, Y.: Variational deep semantic hashing for text documents. In: Proceedings of the 40th International ACM SIGIR Conference on Research and Development in Information Retrieval, pp. 75–84. ACM (2017)
4. Datar, M., Immorlica, N., Indyk, P., Mirrokni, V.S.: Locality-sensitive hashing scheme based on p-stable distributions. In: Proceedings of the Twentieth Annual Symposium on Computational Geometry, pp. 253–262. ACM (2004)
5. Dolan, B., Quirk, C., Brockett, C.: Unsupervised construction of large paraphrase corpora: exploiting massively parallel news sources. In: Proceedings of the 20th international conference on Computational Linguistics, p. 350. Association for Computational Linguistics (2004)
6. Hu, B., Lu, Z., Li, H., Chen, Q.: Convolutional neural network architectures for matching natural language sentences. In: Advances in Neural Information Processing Systems, pp. 2042–2050 (2014)
7. Kingma, D.P., Ba, J.: Adam: a method for stochastic optimization. In: ICLR 2015 (2015)
8. Kingma, D.P., Welling, M.: Auto-encoding variational Bayes. In: ICLR 2014 (2014)
9. Lai, H., Pan, Y., Liu, Y., Yan, S.: Simultaneous feature learning and hash coding with deep neural networks. In: Proceedings of the IEEE Conference on Computer Vision and Pattern Recognition, pp. 3270–3278 (2015)
10. Le, Q., Mikolov, T.: Distributed representations of sentences and documents. In: International Conference on Machine Learning, pp. 1188–1196 (2014)
11. Li, W.-J., Wang, S., Kang, W.-C.: Feature learning based deep supervised hashing with pairwise labels. arXiv preprint arXiv:1511.03855 (2015)
12. Miao, Y., Yu, L., Blunsom, P.: Neural variational inference for text processing. In: International Conference on Machine Learning, pp. 1727–1736 (2016)
13. Mikolov, T., Sutskever, I., Chen, K., Corrado, G.S., Dean, J.: Distributed representations of words and phrases and their compositionality. In: Advances in Neural Information Processing Systems, pp. 3111–3119 (2013)
14. Norouzi, M., Fleet, D.J., Salakhutdinov, R.R.: Hamming distance metric learning. In: Advances in Neural Information Processing Systems, pp. 1061–1069 (2012)
15. Pennington, J., Socher, R., Manning, C.: Glove: global vectors for word representation. In: Proceedings of the 2014 Conference on Empirical Methods in Natural Language Processing (EMNLP), pp. 1532–1543 (2014)
16. Salakhutdinov, R., Hinton, G.: Semantic hashing. Int. J. Approximate Reasoning 50(7), 969–978 (2009)
17. Schütze, H., Manning, C.D., Raghavan, P.: Introduction to Information Retrieval, vol. 39. Cambridge University Press, Cambridge (2008)
18. Shen, D., Su, Q., Chapfuwa, P., Wang, W., Wang, G., Carin, L., Henao, R.: Nash: toward end-to-end neural architecture for generative semantic hashing. arXiv preprint arXiv:1805.05361 (2018)
19. Shen, D., Zhang, Y., Henao, R., Su, Q., Carin, L.: Deconvolutional latent-variable model for text sequence matching. arXiv preprint arXiv:1709.07109 (2017)

20. Smolensky, P.: Information processing in dynamical systems: foundations of harmony theory. Technical report, Colorado Univ at Boulder Dept of Computer Science (1986)
21. Srivastava, N., Hinton, G., Krizhevsky, A., Sutskever, I., Salakhutdinov, R.: Dropout: a simple way to prevent neural networks from overfitting. J. Mach. Learn. Res. **15**(1), 1929–1958 (2014)
22. Wainwright, M.J., Jordan, M.I., et al.: Graphical models, exponential families, and variational inference. In: Foundations and Trends® in Machine Learning , vol. 1, pp. 1–2 (2008). 1–305
23. Wang, Z., Hamza, W., Florian, R.: Bilateral multi-perspective matching for natural language sentences. arXiv preprint arXiv:1702.03814 (2017)
24. Weiss, Y., Torralba, A., Fergus, R.: Spectral hashing. In: Advances in Neural Information Processing Systems, pp. 1753–1760 (2009)
25. Xu, J., Wang, P., Tian, G., Xu, B., Zhao, J., Wang, F., Hao, H.: Convolutional neural networks for text hashing. In: IJCAI 2015, pp. 1369–1375 (2015)
26. Yan, X., Yang, J., Sohn, K., Lee, H.: Attribute2Image: conditional image generation from visual attributes. In: European Conference on Computer Vision, pp. 776–791. Springer (2016)
27. Yang, Y., Yih, W.-t., Meek, C.: WikiQA: a challenge dataset for open-domain question answering. In: Proceedings of the 2015 Conference on Empirical Methods in Natural Language Processing, pp. 2013–2018 (2015)
28. Yang, Z., Hu, Z., Salakhutdinov, R., Berg-Kirkpatrick, T.: Improved variational autoencoders for text modeling using dilated convolutions. In: International Conference on Machine Learning, pp. 3881–3890 (2017)
29. Yin, W., Schütze, H., Xiang, B., Zhou, B.: ABCNN: attention-based convolutional neural network for modeling sentence pairs. arXiv preprint arXiv:1512.05193 (2015)
30. Youn, J., Shim, J., Lee, S.-G.: Efficient data stream clustering with sliding windows based on locality-sensitive hashing. IEEE Access **6**(1), 63757–63776 (2018)
31. Zhu, H., Long, M., Wang, J., Cao, Y.: Deep hashing network for efficient similarity retrieval. In: AAAI 2016, pp. 2415–2421 (2016)

SisterNetwork: Enhancing Robustness of Multi-label Classification with Semantically Segmented Images

Holim Lim$^{(\boxtimes)}$, Jeeseung Han$^{(\boxtimes)}$, and Sang-goo Lee

IntelliSys Corp., Seoul National University, Seoul, South Korea
{ihl7029,jshan,sglee}@europa.snu.ac.kr

Abstract. The rapid growth of the online fashion market has raised the demand for fashion technologies, such as clothing attribute tagging. However, handling fashion image data is challenging since fashion images likely contain irrelevant backgrounds and involve various deformations. In this paper, we introduce SisterNetwork, a deep learning model to tackle the multi-label classification task for fashion attribute tagging. The proposed model consists of two different CNNs to leverage both the original image and the semantic segmentation information. We evaluate our model on the DCSA dataset which contains tagged fashion images, and we achieved the state-of-the-art performance on the multi-label classification task.

Keywords: Multi-label classification · Semantic segmentation · Fashion

1 Introduction

The online fashion market is growing rapidly, and the demand for related technology is increasing. When dealing with fashion, it is important to exploit image data since visual information is the most principal feature in the fashion domain. Among those technologies related for online fashion retail, extracting human-understandable fashion attributes from images is one of the most crucial tasks since it serves as a basic module for most retail applications such as fashion search [1] or fashion recommendation [2].

However, dealing with unconstrained fashion images is a highly challenging task in computer vision for two major reasons. First, "user photos" such as outfit-of-the-day photos and selfies are likely to contain irrelevant information such as background and other objects. Second, fashion images frequently suffer various deformations such as deviation of focus, diverse contrast depending on the position of the light source, pose of the human model and etc. Therefore, it is difficult to extract robust image features that can handle a wide range of the fashion domain.

In this paper, we introduce SisterNetwork, a deep learning model that leverages the augmented information to tackle a robust multilabel classification task for tagging fashion images. The model consists of two different Convolution Neural Network

---

H. Lim and J. Han—These authors contributed equally to this work.

© Springer Nature Switzerland AG 2019
S. Lee et al. (Eds.): IMCOM 2019, AISC 935, pp. 1092–1099, 2019.
https://doi.org/10.1007/978-3-030-19063-7_86

(CNN) that accept original images and semantically segmented images, respectively, followed by "fusing" layer and fully-connected layer for classifying. A semantically segmented image is a preprocessed image from which irrelevant background pixels are removed. To "fuse" two different outcomes of CNNs, we first apply max-pooling respectively to obtain vectors of the same dimension and then fuse them through the fusing layer. In the fusing layer, we adopt two methods: elementwise addition and Hadamard product.

For evaluation, we performed a multi-label classification task for fashion attribute tagging on the DCSA dataset [9], which contains 1,856 "user photo" images with 26 labels. Through the experiment, we found that exploiting semantic segmentation information yield, considerable improvement, compared to the baseline model that only considers original images. As a result, the proposed model achieved the state-of-the-art performance.

2 Related Works

2.1 Learning from Noisy Fashion Data

In the clothing detection task, focusing on a region of interest is necessary in order to minimize noises. For this purpose, there are object detection methods [3–5] that detects the bounding box localization of objects of interest and semantic segmentation methods [6–8] to segment the specific object for guiding fashion-related models.

In addition to, pose-based approaches have been suggested. Chen et al. [9] used the pose estimation method to estimate the skeleton on the upper body and remove the background except the upper body area using the grab-cut based on the ratio set by the heuristic method. Liu et al. [10] proposed an approach called Landmark detection. They used the method to detect the position of a certain point which can be a feature of a fashion item and to use the feature by pooling the surrounding information.

We propose that the semantic segmentation method approach to solving the multi-label classification problem would improve the supervised CNN network.

2.2 Dataset

As a fashion domain dataset, we used the DCSA dataset, which was introduced by Chen et al. [9]. DCSA dataset contains 1856 photos with 26 ground truth clothing attributes such as "long sleeves", "has a collar", and "striped pattern". The labels were collected using Amazon Mechanical Turk. We find that the tagset designed in DCSA is intuitive and clearly identifiable. For these reasons, we choose DCSA as the target task for our studies.

3   Approach

We propose a novel deep model structure, SisterNetwork which simultaneously exploits original images and semantically segmented images. We name this structure "Sister" as original images go through the bigger backbone network and preprocessed images go through the other smaller network. Then two outputs of sister networks are combined and fed into the last multi-label classification layers.

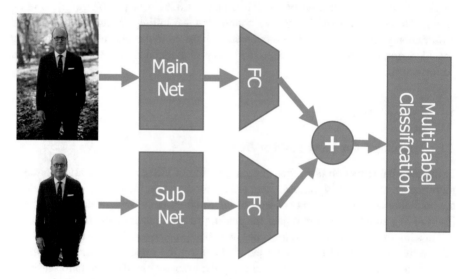

Fig. 1. Overview of Sister-A structure

3.1   Semantic Segmentation

In this paper, our goal is to classify annotated labels on unconstrained images. Hence, we argue that using images preprocessed by semantic segmentation models could be useful. As [9] shows, color labels and bigger pattern like 'solid' or 'graphics' achieve higher accuracy. But labels like 'neckline' which can be inferred from fine-grained features exhibit lower accuracy. So to exploit these local features, we need to use both original images and semantic segmentation images at the same time.

3.2   Sister Structure

The most intuitive approach to encoding two similar images is to use a symmetric (or twin) network in which two identically shaped image-encoding networks encode respective images. However, in our experiments, we find that the network for encoding processed images requires less expressive power, hence employing a completely same network architecture performs worse than a more specialized network. In this paper, we show that our asymmetric architecture is more effective than the counterpart.

4   Experiments and Details

We trained and evaluated our models on DCSA [9] using 16-fold cross-validation. DCSA contains 23 binary labels and 3 multiclass labels. For the binarized labels, we compute scores indicating the likelihood of the visual presence for those labels. And the other multi-class labels, probabilities for each class are computed. We evaluate our models using the unweighted mean of all label accuracies.

4.1   Base Settings and Segmentation Models

We use Resnet [11] pretrained on ImageNet. Input images are feed into Resnet to predict 26 labels. The last average pooling layer is replaced with 7×7 max pooling layer. Also, the last fully connected layer is changed to 2 fully connected layers. The first fully connected layer is followed by ReLU and the number of filters is set to 1024. Also, the model is optimized with Adam at learning rate 10^{-5}. We denote this experimental setting as *the base setting*.

Before the main experiments, we select the best segmentation under the base setting. We choose the model that achieves the highest mean accuracy using images processed by each approach. Specifically, we use Resnet-50 at batch size 64 to select the segmentation model.

Table 1. MACC using segmented features on Resnet50

	V3-Black	V3-White	FCN-White
Title Set	CFPD	CFPD	VOC
MACC	90.43	90.45	**91.06**

We consider FCN-8 s [6] and DeepLab V3 [7] as our candidates. Table 1 shows the results of each model and training data for semantic segmentation. DeepLab V3 models are trained on CFPD [13] and remove backgrounds in the DCSA [9] images. V3-Black and V3-White make backgrounds black or white each. FCN model is trained on VOC data, and only leaves pixels classified as a person. As FCN achieves the best results, we use it for other experiments.

4.2   Baselines and Compared Methods

Baseline. The baseline for each experimental setting is the network that predicts based on the unprocessed images. ResNet variants are used as the encoder network. We trained baseline models under the base setting. Batch size was 64 in the Resnet-50 baseline and 32 for other cases.

Sister-A. These approaches are shown in Fig. 1. Sister-As utilize both processed images and original images. Unprocessed images are fed into the backbone and processed images are fed into the small one which is Resnet-50. And each output goes

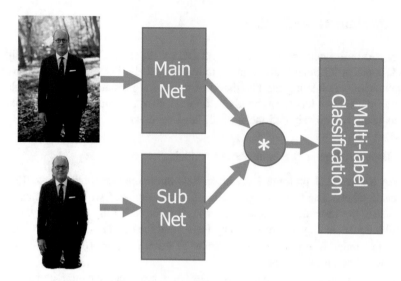

Fig. 2. Overview of Sister-B structure

through one fully-connected layer followed by ReLU. The number of filters is set to 1024. Then they are fused with elementwise addition. The last layer computes scores from this summed vector. In the Resnet-152 case, batch size was 16. And in other cases, batch sizes were 32.

Sister-B. These approaches are shown in Fig. 2. These models are similar to Sister-As. But they do not have a fully connected layer following the backbone neither the small one. Also, the output features are not summed but element-wise multiplied (Hadamard product). Batch sizes were the same as in Sister-A.

Table 2. Comparison between Naïve and Twin models

Backbone	Baseline	Sister-A	Sister-B
Resnet-50	91.40	**91.63**	91.62
Resnet-101	91.66	**91.73**	90.50
Resnet-152	91.65	**91.91**	90.97

Table 2 shows mean accuracies of the approaches. We see that the baselines using Resnet-101 and Resnet-152 show similar results although Resnet-152 has a larger number of parameters. On the other hand, our Sister-A models gain higher accuracies as backbone networks get larger.

We also reevaluate Resnet-101 Sister-A model with a small modification that the small network is set to be Resnet-101. In this case, we can get a mean accuracy of 91.59%. This shows that the asymmetricity of our model is effective.

5   Discussion

5.1   Employing Gated Methods

As sister models present higher accuracies than baselines, we conduct additive experiments using other gate methods. These experiments use Resnet-50 as backbone networks. Detailed structures are shown in Fig. 3. These models are optimized with Adam at learning rate 10^{-5} and batch sizes are 32. We note that the gating approaches exhibit lower accuracies compared to Resnet-50 baseline (Table 3).

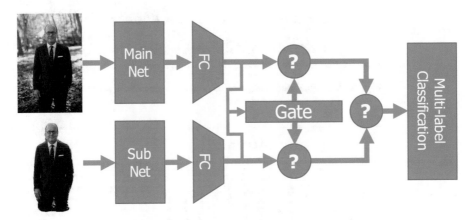

Fig. 3. Structure employing gated methods

Table 3. Gated Twin models with Resnet-50

Model	Single-value gate	Element-wise gate (Add)	Element-wise gate (Concat)
MACC	90.12	90.32	90.43

5.2   Micro and Macro Features

Figure 4 shows 'neckline' and 'category' label accuracies using Resnet-50 baseline, FCN-White, and Resnet-50 Sister-A. The baseline model achieves higher accuracy in the category. And FCN-White achieves higher accuracy in the neckline. We argue that using images processed by semantic segmentation leads the model to learn detail features and using whole images leads the model to learn overall features. Combining these two, our model can gain more accuracies from both micro and macro features (Fig. 5).

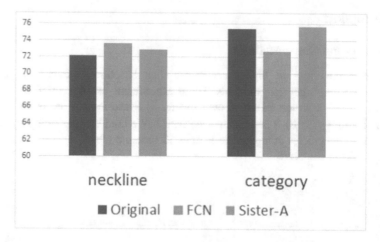

Fig. 4. 'neckline' and 'category' accuracies using Resnet-50

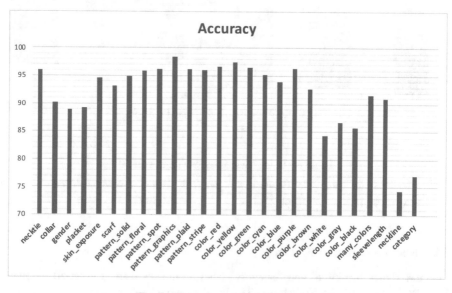

Fig. 5. Sister-A model label accuracy

6   Conclusions

In this paper, we proposed and addressed the multi-label, multiclass classification in the fashion domain. We proposed novel deep learning architecture, namely SisterNetwork which leverages both whole images and preprocessed images. Exploiting semantic segmentation models, we conducted diverse experiments to achieve higher accuracies. A key feature of our model is the use of semantic segmentation information and an asymmetric network architecture.

Albeit our model is quite simple and elegant, there might be a room for developing more elaborate approaches. Although gated methods that we used were not significant yet, we believe further work would be beneficial for the multi-label classification task.

Acknowledgments. This work was supported by the Technology development Program (S2646078) funded by the Ministry of SMEs and Startups (MSS, Korea).

References

1. Di, W., et al.: Style finder: fine-grained clothing style recognition and retrieval. In: Computer Vision and Pattern Recognition Workshops (2013)
2. Zhou, W., et al.: Fashion recommendations using text mining and multiple content attributes (2017)
3. Girshick, R.: Fast R-CNN. In: Proceedings of the IEEE International Conference on Computer Vision (2015)
4. Liu, W., et al.: SSD: single shot multibox detector. In: European Conference on Computer Vision. Springer, Cham (2016)
5. Redmon, J., Farhadi, A.: YOLO9000: better, faster, stronger. arXiv preprint (2017)
6. Long, J., Shelhamer, E., Darrell, T.: Fully convolutional networks for semantic segmentation. In: Proceedings of the IEEE Conference on Computer Vision and Pattern Recognition (2015)
7. Chen, L.-C., et al.: Encoder-decoder with atrous separable convolution for semantic image segmentation. arXiv preprint arXiv:1802.02611 (2018)
8. He, K., et al.: Mask R-CNN. In: IEEE International Conference on Computer Vision (ICCV). IEEE (2017)
9. Chen, H., Gallagher, A., Girod, B.: Describing clothing by semantic attributes. In: European Conference on Computer Vision. Springer, Heidelberg (2012)
10. Liu, Z., et al.: DeepFashion: powering robust clothes recognition and retrieval with rich annotations. In: Proceedings of the IEEE Conference on Computer Vision and Pattern Recognition (2016)
11. He, K., et al.: Deep residual learning for image recognition. In: Proceedings of the IEEE Conference on Computer Vision and Pattern Recognition (2016)
12. Liu, S., et al.: Fashion parsing with weak color-category labels. IEEE Trans. Multimedia **16** (1), 253–265 (2014)

A Searching for Strongly Egalitarian and Sex-Equal Stable Matchings

Le Hong Trang[1(✉)], Hoang Huu Viet[2], Tran Van Hoai[1], and Tran Xuan Hao[2]

[1] Ho Chi Minh City University of Technology, VNU-HCM,
268 Ly Thuong Kiet, Ho Chi Minh City, Vietnam
lhtrang@hcmut.edu.vn
[2] Vinh University, 182 Le Duan, Vinh City, Nghe An Province, Vietnam

Abstract. The stable marriage problem with ties (SMT) is a variant of the stable marriage problem in which people are permitted to express ties in their preference lists. In this paper, an algorithm based on bidirectional searching is presented for trying to find *strongly egalitarian* and *sex-equal* stable matchings. We indicate that the use of two simultaneous searches in the algorithm not only accelerate the finding of solutions but also is appropriate for the strong stability criterion of SMT. The algorithm is implemented and tested for large datasets. Experimental results show that our algorithm is significant.

Keywords: Strongly stable · Egalitarian · Sex-equal · Ties · Bidirectional search

1 Introduction

An instance I of the classical stable marriage problem (SMP) of size n involves n men and n women. Each man ranks n women to give himself a preference list, and similarly each woman ranks n man to also give herself a preference list. A matching M in I is an one-one correspondence between the men and women of I. For a pair of man and woman $(m, w) \in M$, we denote by $M(m)$ and $M(w)$ the partner of m and w in M, respectively, i.e., $w = M(m)$ and $m = M(w)$. A pair (m, w) is said to be blocking pair for M, if m and w are not partners and m (w, respectively) prefers w (m, respectively) to $M(m)$ ($M(w)$, respectively). A matching that admits no blocking pair is said to be stable, otherwise it is unstable. Let us denote by $p_m(w)$ ($p_w(m)$, respectively) the position of w (m, respectively) in m's (w's, respectively) preference list. It was shown by Gale and Shapley that every instance of SMP admits at least a stable matching [1], and the matching can be found in $O(n^2)$. A stable matching found by Gale and Shapley's algorithm is man- or woman-optimal.

© Springer Nature Switzerland AG 2019
S. Lee et al. (Eds.): IMCOM 2019, AISC 935, pp. 1100–1111, 2019.
https://doi.org/10.1007/978-3-030-19063-7_87

For a stable matching M, we define the *man cost*, denoted by $sm(M)$, and the *woman cost*, denoted by $sw(M)$, as follows:

$$sm(M) = \sum_{(m,w) \in M} p_m(w),$$

$$sw(M) = \sum_{(m,w) \in M} p_w(m).$$

We also define the egalitarian and sex-equal costs by:

$$c(M) = sm(M) + sw(M), \tag{1}$$
$$d(M) = |sm(M) - sw(M)|. \tag{2}$$

Let $\mathcal{M}$ be the set of all stable matchings of an instance I of the SMP. For $M \in \mathcal{M}$, M is called to be *egalitarian* (*sex-equal*, respectively) if $c(M)$ ($d(M)$, respectively) is minimum among all stable matchings in $\mathcal{M}$ [14]. The fairness is emphasized for the egalitarian and sex-equal matchings in which we attempt to obtain the balance of preferences between men and women in a stable matching. Several approaches were proposed for finding egalitarian and sex-equal matching such as genetic algorithm [14], ant colony system [15], and approximation algorithm [9].

A generalization of SMP called the stable marriage problem with ties (SMT) arises when people are permitted to express ties in their preference lists. Particularly, each person does not need to rank members of the opposite sex in *strict order*. Some of those involved might be indifference among members of the opposite sex. When once the ties is allowed in the preference lists, stability of a matching can be defined in three possible forms [6]. In particular, a matching M is *weakly stable* if there is no couple (x, y), each of whom strictly prefers the other to his/her partner in M. Also, a matching M is *strongly stable* if there is no couple (x, y) such that x strictly prefers y to his/her partner in M, and y either strictly prefers x to his/her partner in M or is indifferent between them. Finally, a matching M is *super-stable* if there is no couple (x, y), each of whom either strictly prefers the other to his/her partner in M or is indifferent between them. We note that a person p *strictly prefers* a person q in a preference list, p precedes q in the list.

The SMT problem has been received many attentions of researchers (examples can be found in [3,5,6,8]). In order to deal with the problem of large size, a useful approach based on local searching was studied. In 2013, Gelain et al. proposed a local search method to speed up the process of finding solutions [2]. Munera et al. have also addressed the problem using local search approach, based on Adaptive Search [13]. However, these methods only find a stable matching of a given instance.

In this paper, we present a local search algorithm to address SMT for strongly egalitarian or sex-equal stable matchings. Based on the distributive lattice structure of strongly stable matchings in an SMT instance, a search scheme consisting of two simultaneous local searches is performed in both man and woman points of view. In each locally searching step, an adapted breakmarriage operation is used to determine the neighbor set of a strongly stable matching. By the dominance relation which is maintained by the breakmarriage operation, a stop condition is established to terminate the algorithm. The algorithm is implemented and tested for large datasets. Experimental results are given to show the time performance of our algorithm.

The rest of the paper is organized as follows. Section 2 recalls the distributive lattice structure formed by the set of strongly stable matchings, which is the essential structure for establishing the searching scheme in our algorithm. The proposed algorithm is given in Sect. 3. Section 4 is devoted to implementation and testing for large datasets. Finally, some concluding remarks are given in Sect. 5.

2   Preliminaries

The lattice structure of stable matchings plays an important role for solving some problems associated with SMP, such as finding all stable pairs, all stable matchings [4], and also an egalitarian stable matching [7] of a given instance of SMP. Utilizing the dominance properties in the lattice structure, a number of local search algorithms have been proposed for finding optimally stable matchings, i.e. sex-equal and egalitarian, of SMPs [16,17].

Since the preference condition is not symmetric in the definition of strongly stability, the distributive lattice of strongly stable matchings is thus not obviously derived. Fortunately, Manlove in [11] indicated that under a equivalence relation defined on the set of strongly stable matchings for a given SMT instance, the set of equivalence classes forms a distributive lattice under a dominance relation.

Let M and M' be matchings of a given SMT instance I, and q be a person in I. We say that q strictly prefers M to M' (is indifferent between M and M', respectively) if q strictly prefers $p_M(q)$ to $p_{M'}(q)$ (is indifferent between $p_M(q)$ to $p_{M'}(q)$, respectively).

Definition 1 (Equivalence relation [11]). *Given a SMT instance I, let $\mathcal{M}$ be the set of strongly stable matchings in I. An equivalence relation, denoted by $\sim$, is defined on $\mathcal{M}$ as follows: for any $M, M' \in \mathcal{M}$, $M \sim M'$ if and only if each man is indifferent between M and M'.*

A dominance relation is defined on the set of stable matchings in an SMP instance (see [4]). Such a relation on strongly stable matchings can be defined as follows. Let I be an SMT instance and $\mathcal{M}$ be the set of strongly stable matchings in I. Let $M, M' \in \mathcal{M}$. M *dominates* M', written $M \preceq M'$, if each man strictly prefers M to M', or is different between them [11]. We denote by $\mathcal{C}$ the set of equivalence classes of $\mathcal{M}$ under $\sim$, and by $[M]$ the equivalence class containing M, for $M \in \mathcal{M}$. A partial order, denoted by $\preceq^*$, is defined on equivalence classes as follows.

Definition 2 ([11]). *For two equivalence classess* $[M], [M'] \in \mathcal{C}$, $[M] \preceq^* [M']$ *if and only if* $M \preceq M'$.

The distributive lattice structure of strongly stable matchings for an SMT instance is given in the following, which helps us to develop a searching for optimal matchings (described in Sect. 3).

Theorem 1 (Distributive lattice [11]). *Let I be an SMT instance, $\mathcal{M}$ be the set of strongly stable matchings in I. Let $\mathcal{C}$ be the set of equivalence classes of $\mathcal{M}$ under $\sim$, and $\preceq^*$ be the dominance partial order on $\mathcal{C}$. Then $(\mathcal{C}, \preceq^*)$ forms a finite distributive lattice.*

3   Searching Algorithm

Our scheme is to locally search, among neighborhoods of a strongly stable matchings, a better matching with respect to an optimal criterion, i.e. egalitarian or sex-equal. The selection of a better matching is performed due to the dominance relation. The distributive lattice of strongly stable matchings for an SMT instance (as indicated in Theorem 1), is the essential structure for such a search.

Given an SMT instance, the existence of a strongly stable matching can be determined and such a matching can be found in $O(n^4)$ [6,10]. If a strongly matching exists, the search starts at the matching and then iteratively performs the followings: (i) determining a set of neighbors of the matching, (ii) then selecting a better matching among them due to an optimal criterion. The search terminates when meeting a top condition (mentioned in the end of this section). Task (i) can be performed by using a modification of the breakmarriage operation as below.

Breakmarriage for Strongly Stable Matchings

The concept of breakmarriage operation was introduced by McVitie and Wilson [12]. Let M be a matching in a given SMP instance and let (m, w) be a pair in M. The operation breaks the marriage of m and w. Starting with m proposing the woman following w in his preference list, it then performs a sequence of proposals, acceptances, and rejections as given by the Gale and Shapley's Algorithm [1]. The operation terminates if w is engaged to m' she prefers to m or some man is rejected by all women. If the operation terminates with the former case, i.e. all men are engaged, we obtain a new matching, denoted by M'. It was shown in [12] that M' is stable too. Furthermore, M' dominates M.

Unlike in SMP, the acceptance and rejection of a woman w in the sequence of proposals, acceptances, and rejections are different for strongly stability in SMT instances. Namely, by the definition of strongly stability, w can accept a man she prefers to her current partner or the man and her current partner are indifferent. Let (m, w) be a pair in strongly stable matching in an SMT instance. The breakmarriage operation should be modified as follows. Let us consider the situation when w is received a proposal from some man:

(a) If the man is indifferent between his partner and w, w accepts the man only if she strictly prefers the man to m.

(b) Otherwise, w accepts the man if she strictly prefers the man to m or is indifferent between them.

Proposition 1. *Let M be a strongly stable matching in a given SMT instance. If a matching M' is obtained by performing the breakmarriage operation due to (a) and (b), M' is also strongly stable.*

Proof. Regarding the stability, we use the similar argument in the proof of Theorem 3 in [12]. In particular, all pairs in M' that keep the same in M, are still stable. The remaining pairs in M' are generated in which a man gets a better partner and a woman gets a worse partner, or he/she is indifferent between the current partner and new one. In the meanwhile, by the rules (a) and (b), the strongly stability is maintained.

The Algorithm

We now describe the algorithm. Let M be a strongly stable matching in a given SMT instance. Procedure FINDNEXT(M) determines a better matching for the search. It computes the set of neighbors using the modified breakmarriage operation, and then chooses a matching in the set that has minimum egalitarian or sex-equal cost, given by Eqs. (1) and (2). Because our algorithm is local search based, the search can get stuck at a local minimum. We can overcome this issue by randomly choosing the next matching due to a small value of p.

```
 1: procedure FINDNEXT(M)
 2:     neighborSet := ∅
 3:     for m := 1 to n do
 4:         M_child := BREAKING(M, m)
 5:         if (M_child ≠ NULL) then
 6:             neighborSet := neighborSet ∪ M_child
 7:         end if
 8:     end for
 9:     if (small random probability p) then
10:         M_next := a random matching in neighborSet
11:     else
12:         M_next := arg min_{M∈neighborSet}(f(M))
13:     end if
14:     return M_next
15: end procedure
```

The notation m in Procedure FINDNEXT(M) means that we perform breakmarriage with respect to man. If the role of man and woman is swapped, a breakmarriage operation with respect to woman is established in the same manner.

Algorithm 1. Searching algorithm for strongly stable matchings

---

1: **Input:** an SMT instance with preference lists A and B.
2: **Output:** a strongly stable matching.

3: $M_{left} := \text{MANOPTIMAL}(A, B)$
4: $M_{right} := \text{WOMANOPTIMAL}(B, A)$
5: **if** (A strongly stable matching exists) **then**
6:     $M_{best} := \arg\min_{M \in \{M_{left}, M_{right}\}} (f(M))$
7:     $forward := true$
8:     $backward := true$
9:     **loop**
10:        **if** ($forward$) **then** searching w.r.t man point of view
11:           $M_{next} := \text{FINDNEXT}(M_{left})$
12:           **if** ($f(M_{next}) > f(M_{left})$) **then**
13:              $forward := false$
14:              **if** $f(M_{best}) > f(M_{left})$ **then**
15:                 $M_{best} := M_{left}$
16:              **end if**
17:           **end if**
18:           $M_{left} := M_{next}$
19:        **end if**
20:        **if** ($backward$) **then** searching w.r.t woman point of view
21:           $M_{next} := \text{FINDNEXT}(M_{right})$
22:           **if** ($f(M_{next}) > f(M_{right})$) **then**
23:              $forward := false$
24:              **if** $f(M_{best}) > f(M_{right})$ **then**
25:                 $M_{best} := M_{right}$
26:              **end if**
27:           **end if**
28:           $M_{right} := M_{next}$
29:        **end if**
30:        **if** ((not forward) and (not backward)) **then**
31:           **if** ($sm(M_{left}) \leq sm(M_{right})$) **then**
32:              $forward := true$
33:              $backward := true$
34:           **else**
35:              **break**
36:           **end if**
37:        **end if**
38:     **end loop**
39:     **return** M_{best}
40: **else**
41:     There is no strongly stable matching.
42: **end if**

From that observation, we can develop an algorithm consisting of two simultaneously local searches. One carries out with respect to man point of view, and the other is with respect to woman point of view. These two searches not only aims to speed up the algorithm but also needs to improve the accuracy solutions obtained by the algorithm. To this end, it is important to determine appropriately an initial matching as well as the searching direction for each search. The distributive lattice structure as described in Sect. 2 allows again us to solve this.

Our search is given in Algorithm 1. On the lattice, we performs two searches in which one starts at a man-optimal strongly stable matching and the other at a woman-optimal one. The former called *forward*, by the breakmarriage operation with respect to man, computes the set of neighbors and then chooses a better matching due to the optimal cost $f(M)$ (that is the egalitarian or sex-equal cost). In the meantime, the later called *backward* does in the same manner, with respect to woman point of view. Two searches are performed iteratively until a so-call meeting condition is satisfied. The algorithm, thereby, travels on the lattice due to the dominance relation. Because of the dominance, man costs of matchings found during forward searching should be increased, while those during backward searching should be decreased. The meeting condition is defined to be the moment when the man cost of a matching by forward search is greater than that by backward one, i.e., $sm(M_{left}) > sm(M_{right})$. The idea behind of Algorithm 1 was actually introduced in [16], which is called the bi-directional local search. It, however, defers from which in [16] that the breakmarriage operation here is modified to adapt to the strongly stability. Furthermore, some computational aspects of the set of stable marriages in an SMT instance, are also utilized to improve the performance of the algorithm as described in the next section.

4   Implementation

4.1   Simulation Results

We used the method given in [3] to generate SMT instances. Due to the method, we take two parameters: the problem's size n and a probability, say p_t, of ties. Given a 2-tuple $\langle n, p_t \rangle$, an instance of SMT is generated iteratively as follows:

1. A random preference list of size n for each man and woman is produced.
2. We iterate over each person's (men and women's) preference list: for a man m_i and for his choices c_i from his second to his last, a random value p such that $0 \leq p < 1$, is generated; if $p \leq p_t$ then the preference for his c_i^{th} choice is the same as c_{i-1}^{th} choice.

An instance generated as $\langle n, 0.0 \rangle$ will be a classical SMP. Table 1 shows an SMT instance which is randomly generated with $\langle 8, 0.5 \rangle$ (ties are denoted by braces).

The algorithm is implemented in Matlab 2016a and run on the platform OS X, Core i5 2.5 GHz with 8 GB RAM. We run the algorithm for SMT instances of different sizes. For each instance, the algorithm is tested for 10 times obtain the average costs and the probability of ties p_t is varied from 0.0 to 1.0 in steps of 0.01.

We first investigate how the parameter p_t influences the egalitarian and sex-equal costs of an SMT instance. Figure 1 shows the average cost of egalitarian

Table 1. An SMT instance generated with $\langle 8, 0.5 \rangle$.

Men's list	Women's list
1: (5 7) (1 2) 6 (8 4) 3	1: 5 3 7 (6 1 2 8) 4
2: 2 (3 7) 5 4 1 8 6	2: 8 6 (3 5 7 2 1 4)
3: (8 5 1) 4 6 2 3 7	3: (1 5 6) (2 4) 8 7 3
4: (3 7 2) 4 1 (6 8 5)	4: 8 (7 3 2) (4 1 5) 6
5: (7 2) 5 1 3 (6 8 4)	5: (6 4 7 3 8) (1 2) 5
6: (1 6) (7 5) 8 (4 2 3)	6: 2 8 (5 4) (6 3) 7 1
7: (2 5) 7 6 3 4 (8 1)	7: (7 5 2) 1 8 (6 4) 3
8: (3 8 4 5) (7 2 6 1)	8: (7 4 1) 5 2 3 6 8

and sex-equal matchings found by the algorithm for $\langle 100, p_t \rangle$, varying p_t. The trend of egalitarian costs is decreased as p_t increases. This is because of the definition of egalitarian cost, i.e. $sm(M) + sw(M)$ (see Eq. (1)), for a given matching M. When p_t increases, there are more people which are indifferent in preference lists. Therefore, the values of $sm(M)$ and $sw(M)$ should be decreased. For the sex-equal cost, by (2), it is defined to be $|sm(M) - sw(M)|$. The figure shows that the probability of the cost that is closer to zero, is high as higher value of p_t. These observations on the behaviors of egalitarian and sex-equal costs, when varying p_t, thus also indicate the correctness of the algorithm.

In Figs. 2 and 3, we show the results of testing for datasets of large size. We report here the average execution time of the algorithm which runs for five instance sizes $n = 50, 100, 200, 300, 500$, varying p_t in steps of 0.01. The running times for finding egalitarian and sex-equal matchings are quite similar for each instance size. It also seems that the variability of running times is higher as p_t increases. This can be caused by increasing the number of ties in preference lists. Then, the number of strongly stable matchings can also be increased. The searching space should be larger and thus the necessary time for searching might sometimes be increased. On the other hand, a high people number of ties can also help to quickly obtain solutions, since the stop condition of the algorithm can be quickly reached. Finally, we would also like to show here the performance of the algorithm for large datasets. For $n \leq 300$, the running time mostly is not exceeded 5 (s) for both egalitarian and sex-equal costs. For largest size of 500, the times in worse cases mostly are about 30 (s) and 35 (s) for egalitarian and sex-equal costs, respectively. Such execution times indicate the significance of our algorithm.

4.2   Algorithm Acceleration

Procedures MANOPTIMAL(A, B) and WOMANOPTIMAL(B, A) in Algorithm 1 can be given by STRONG given in [6] which is an extension of Gale and Shapley's algorithm.

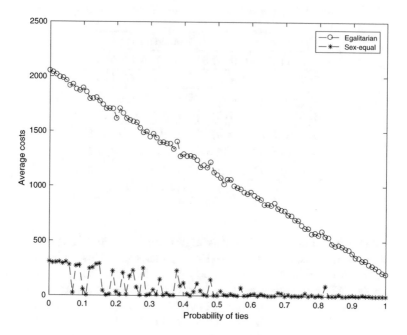

Fig. 1. The average cost of egalitarian and sex-equal matchings for $\langle 100, p_t \rangle$.

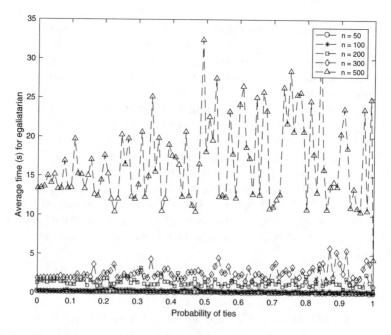

Fig. 2. The average execution time of finding egalitarian matchings for $\langle n, p_t \rangle$.

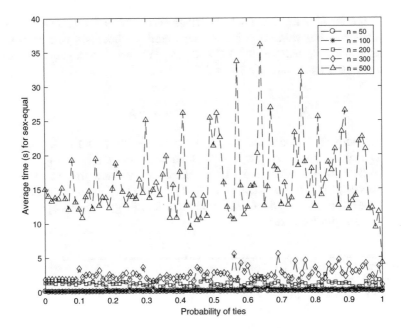

Fig. 3. The average execution time of finding sex-equal matchings for $\langle n, p_t \rangle$.

In Algorithm 1, For each iteration of finding a next matching (Procedure FINDNEXT(M)), the algorithm performs a number of breakmarriage operations (Procedure BREAKING(M, m)). This number should be very large as the size of instances increases. Consequently, the computational cost of the breakmarriage operation is high. It is possible to reduce the cost by using a so-called shortlist. The concept of shortlists for a given SMP instance was given in [7]. In particular, if a woman w accepts a proposal from a man m, then the woman never accepts a proposal from a man following m in her preference list. Then, all men following m should be removed from the list. We also remove w from preference lists of the men. It was known that if w (m, respectively) is absent in the shortlist of a man m (w, respectively), (m, w) is not a stable pair. Thereby, in the sequence of proposals of a breakmarriage operation, before proposing to a person of opposite sex, he/she checks if the person is in his/her shortlist. If the person is absent in the list, the pair of them is not stable. Therefore, we can ignore the stability checking for the pair.

In the case of ties, for a given SMT instance, shortlists with respect to man point of view can be obtained during performing Procedure MANOPTIMAL(A, B) as follows. When a woman w accepts the proposal from a man m:

1. remove all men whose strictly worse preference than m from the preference list of w,
2. remove also w from the preference lists of those men.

Let us denote by X^m and Y^m such shortlists (i.e. obtained with respect to man point of view). From woman point of view, similar shortlists can be obtained in the same manner, denoted by X^w and Y^w. We now define shortlists for the instance, denoted by X and Y,

$$X = X^m \wedge X^w \quad \text{and} \quad Y = Y^m \wedge Y^w,$$

where the operator $\wedge$ is defined by $X(m_i, w_j) = X^m(m_i, w_j)$ if $X^m(m_i, w_j) = X^w(m_i, w_j)$, otherwise $X(m_i, w_j) = 0$ (meaning that w_j is absent from m_i's preference list). It is similar to Y. Using shortlists, we now improve the Procedure $\text{FINDNEXT}(M)$ to $\text{FINDNEXT}(X, Y, M)$, where the breakmarriage takes X and Y as parameters to reduce the number of stability checking operations for pairs of a man and woman.

1: **procedure** $\text{FINDNEXT}(X, Y, M)$
2:     $neighborSet := \emptyset$
3:     **for** $m := 1$ to n **do**
4:         $M_{child} := \text{BREAKING}(X, Y, M, m)$
5:         **if** $(M_{child} \neq NULL)$ **then**
6:             $neighborSet := neighborSet \cup M_{child}$
7:         **end if**
8:     **end for**
9:     **if** (small random probability p) **then**
10:        $M_{next} :=$ a random matching in $neighborSet$
11:    **else**
12:        $M_{next} := \arg\min_{M \in neighborSet}(f(M))$
13:    **end if**
14:    **return** M_{next}
15: **end procedure**

---

5   Conclusion

The paper presented a searching for strongly egalitarian and sex-equal matchings of SMT instances. The algorithm performs two simultaneous local searches on the distributive lattice structure of strongly stable matchings in SMT instances to find optimal solutions. By dominance relation in the lattice, the algorithm terminates when the man costs of two searches "meet" each other. The simulation results obtained on large datasets show that the algorithm can find solutions in a reasonable time. In order to speed up the searching, a use of shortlist concept also was presented, which helps to reduce the number of checking pairs in the breakmarriage operation. It however has not implemented yet in this paper. This is performed in a general locally searching scheme in which we aims to develop for several variants of SMP.

Acknowledgement. This research is funded by the Vietnam National Foundation for Science and Technology Development (NAFOSTED) under **grant number 102.01-2017.09**.

References

1. Gale, D., Shapley, L.S.: College admissions and the stability of marriage. Am. Math. Monthly **69**(1), 9–15 (1962)
2. Gelain, M., Pini, M., Rossi, F., Venable, K., Walsh, T.: Local search approaches in stable matching problems. Algorithms **6**(4), 591–617 (2013)
3. Gent, I.P., Prosser, P.: An empirical study of the stable marriage problem with ties and incomplete lists. In: Proceedings of European Conference on Artificial Intelligence, Lyon, France, pp. 141–145 (2002)
4. Gusfield, D.: Three fast algorithms for four problems in stable marriage. SIAM J. Comput. **16**(1), 111–128 (1987)
5. Irving, R.W.: An efficient algorithm for the "stable roommates" problem. J. Algorithms **6**, 577–595 (1985)
6. Irving, R.W.: Stable marriage and indifference. Discrete Appl. Math. **48**, 261–272 (1994)
7. Irving, R.W., Leather, P., Gusfield, D.: An efficient algorithm for the "optimal" stable marriage. J. ACM **34**(3), 532–543 (1987)
8. Iwama, K., Manlove, D., Miyazaki, S., Morita, Y.: Stable marriage with incomplete lists and ties. In: Proceedings of the 26th International Colloquium on Automata, Languages, and Programming, LNCS, vol. 1644, pp. 443–452. Springer (1999)
9. Iwama, K., Miyazaki, S., Yanagisawa, H.: Approximation algorithms for the sex-equal stable marriage problem. ACM Trans. Algorithms **7**(2) (2010). Article no. 2
10. Manlove, D.: Stable marriage with ties and unacceptable partners. Technical report TR-1999-29, University of Glasgow (1999)
11. Manlove, D.F.: The structure of stable marriage with indifference. Discrete Appl. Math. **122**, 167–181 (2002)
12. McVitie, D.G., Wilson, L.B.: The stable marriage problem. Commun. ACM **14**(7), 486–490 (1971)
13. Munera, D., Diaz, D., Abreu, S., Rossi, F., Saraswat, V., Codognet, P.: Solving hard stable matching problems via local search and cooperative parallelization. In: Proceeding of 29th AAAI Conference on Artificial Intelligence, Austin, Texas, United States, pp. 1212–1218 (2015)
14. Nakamura, M., Onaga, K., Kyan, S., Silva, M.: Genetic algorithm for sex-fair stable marriage problem. In: Proceedings of IEEE International Symposium on Circuits and Systems, Seattle, WA, ISCAS 95, vol. 1, pp. 509–512 (1995)
15. Vien, N.A., Viet, N., Kim, H., Lee, S., Chung, T.: Ant colony based algorithm for stable marriage problem, vol. 1, pp. 457–461. Springer, Dordrecht (2007)
16. Viet, H.H., Trang, L.H., Lee, S., Chung, T.: A bidirectional local search for the stable marriage problem. In: Proceedings of the 2016 International Conference on Advanced Computing and Applications, Can Tho City, Vietnam, pp. 18–24 (2016)
17. Viet, H.H., Trang, L.H., Lee, S., Chung, T.: An empirical local search for the stable marriage problem. In: Proceedings of the 14th Pacific Rim International Conference on Artificial Intelligence: Trends in Artificial Intelligence, Phuket, Thailand, pp. 556–564 (2016)

Deep Learning Drone Flying Height Prediction for Efficient Fine Dust Concentration Measurement

Ji Hyun Yoon[1], Yunjie Li[1], Moon Suk Lee[1], and Minho Jo[2]([✉])

[1] Department of Computer and Information Science, Korea University,
Sejong Metropolitan City, South Korea
jihyun931@gmail.com, liyuanjie@gmail.com,
winnyiee@gmail.com
[2] Department of Computer Convergence Software, Korea University,
Sejong Metropolitan City, South Korea
minhojo@korea.ac.kr

Abstract. Fine dust concentration has been collected so far by fine dust concentration monitoring towers at fixed heights. However fine dust concentration level varies significantly with heights. It is possible for people to get informed of wrong dust concentration information. Drone equipped with a fine dust sensor can fly up and down to sense fine dust concentration. Drone can solve the wrong fine dust concentration information problem. But we face too much energy consumption problem of drone and possibly delayed information because drone should fly from ground up to top. To solve this problem, we propose to cut drone flying height by predicting the height, using deep learning methods, at which maximum fine dust concentration can be sensed. Experimental results show that the proposed methods save 58.28% of flying distance.

Keywords: Fine dust concentration · Drone · Deep learning · RNN · CNN · Prediction of maximum flying height

1 Introduction

Healthcare problem caused by air pollution has been a big issue among people these days. The concentration of fine dust is very harmful for human being's health. Fine dust is one of main causes of air pollution problem. To avoid the threat of fine dust, the Korean government has installed fine dust sensing stations in the country. However, the location of the installed stations didn't consider the height. Unlike other air pollutants, the concentration of fine dust is strongly affected by air flow, temperature, and other weather conditions [1]. Because air flows move fine dust, fine dust has different levels of concentration at different heights in the same place.

As given in Table 1, fine dust shows different concentration levels between high rooftop and ground. This illustrates that the fine dust station installed on the rooftop does not give true fine dust concentration information for real life. Table 2 shows 224 of 264 fine dust sensing stations are installed at over 10 m height above ground in South Korea.

© Springer Nature Switzerland AG 2019
S. Lee et al. (Eds.): IMCOM 2019, AISC 935, pp. 1112–1119, 2019.
https://doi.org/10.1007/978-3-030-19063-7_88

Table 1. Different levels of fine dust concentration by height [2].

	PM_{10} ($\mu g/m^3$)		$PM_{2.5}$ ($\mu g/m^3$)		$PM_{1.0}$ ($\mu g/m^3$)	
Rooftop	40.7	30.9	24.3	18.1	20.0	15.0
Ground	48.9	34.7	29.9	19.7	25.2	16.0
Difference	+8.2	+3.8	+5.6	+1.6	+5.2	+1.0
	(20.2%)	(12.3%)	(23.0%)	(8.9%)	(26.2%)	(6.7%)

* PM_n: Particulate matter n micrometers or less in diameter.

Table 2. Number of fine dust sensing stations by height in South Korea [3, 4].

	5 m	10 m	15 m	20 m
Numbers	9	31	203	21

Because of height limitation of the installed stations, it is impossible to measure the worst (maximum) concentration level of fine dust in real time. To solve this serious problem, we propose to use drones for measuring fine dust. Drone has demonstrated a possibility to measure air quality including fine dust, pressure, wind, relative humidity, ozone, and other gases [5–7]. In Fig. 1, a drone is equipped with a fine dust sensor. Thus, the drone flies up and down in order to measure fine dust concentration at different heights. The drone can measure the level of fine dust concentration at allowed heights. The drone will send the maximum (highest) level of fine dust concentration and its corresponding height (m) as well as level of concentration-height matching data. We can be informed of more realistic and true fine dust concentration information in real time. In addition, if we are informed of both maximum level of concentration and its corresponding height, it will be helpful for the people who live in the city with buildings and apartments.

But we face one practical problem in carrying out fine dust concentration measurement at different heights by drones. A drone cannot fly up to the maximum allowable height every minute. It will consume too much energy of drone and may not give us real time fine dust concentration information. Thus, we need to cut flying height of drone. If we can predict the height point of maximum fine dust concentration level, we can stop drone flying-up at the predicted height. A machine learning approach, reinforcement learning, has proven to control drone in 2017 and 2018 [8, 9]. In our research, we implemented deep learning approaches such as deep neural network (DNN) and recurrent neural network (RNN) to predict height point of maximum fine dust concentration level. We compared predicted height points to actual height points for verification of our proposed work. Experiment results have shown a practical possibility of our proposed work. In experiment the proposed DNN model has shown 80.3% of accuracy and the proposed RNN model, 84.5%, respectively and the DNN has saved 57.31% of flying distance while the RNN, 58.28%, respectively.

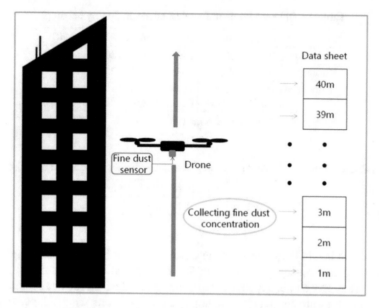

Fig. 1. Measuring fine dust concentration by height.

2 Prediction of Maximum Fine Dust Concentration and Flying Height

To obtain maximum flying heights of drone, fine dust concentration data at different heights were collected. The actual data were collected (measured) by drone, ranging from ground to 40 m in height for 14 h. Collection was made every 10 cm whenever the drone flied up. From the collected actual data, a normal distribution model of fine dust concentration levels at each height increment of 10 cm, was derived. Thus, we obtained a total of 400 different normal distributions. The actual data were used to train the DNN and RNN models. The one-year past (historical) fine dust concentration data were obtained from [10, 11]: average fine dust concentration per hour, average temperature per hour, average wind speed per hour, average ground fine dust concentration per hour, and average humidity per hour were used as the input to generate training and test data sets.

As shown in Fig. 2, a sample data of 800 h was selected in one-year past data. The sample data are within range of 10% centered from the collected actual data. The range of 10% is measured by avg. fine dust concentration/hour, temperature/hour, wind speed/hour, and humidity/hour, respectively. The past data were applied to the actual fine dust concentration data by the following formula:

$$G_m = (Actual_m / Past) * Average_{past}$$

G_m: Normalized data for training and testing

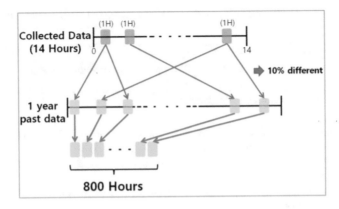

Fig. 2. 800 h sample data generation.

m: flying heights in meters ranging from 1 (ground) to 40 (top).

$Actual_m$: Collected actual data at m height (every 5 min)

$Past$: Past data at the same time as the actual data

$Average_{past}$: One-year averaged data of past for a specific hour.

The normalized data were used as the mean value of the standard normal distribution. Thus, we can generate the realistic fine dust concentration data x by the following Z formula:

$$x = Z \times \sigma + G_m$$

$*\sigma$: Standard deviation

We generated 10,000 data sets for training while 1,000 data sets for testing accuracy from the above standard normal distribution.

Our deep learning models, DNN and RNN will be trained using the above data sets. The proposed DNN and RNN models will predict the height at maximum fine dust concentration level. The proposed DNN model has 6 input nodes and 1 input node given in Table 3 and has 6 hidden layers. The proposed RNN model has 7 sequence lengths and 1 hidden size (Output). No hidden layer was used, and LSTM (Long Short-Term Memory models) was applied for cells.

The height points of maximum fine dust concentration the drone will be predicted by DNN and RNN, respectively, as shown in Fig. 3. In order to predict the height point, input data will be entered in DNN and RNN models, respectively. By predicting the height point in advance, the drone does not need to fly up to the allowable top point, 40 m to find the maximum (highest) fine dust concentration level. Figure 4 shows the

Table 3. Input & output data type for training DNN and RNN models (PM_{10}).

	Data type					
Input	Ground PM_{10} Concentration	Ground Temp.	Avg. PM_{10} per Hour	Avg temperature	Avg wind speed	Avg humidity
Output (Target)	Height of maximum concentration					

* *Ground PM_{10} Concentration*: Measured on the ground by drone in real time (current time).
* *Ground Temperature*: Measured on the ground by drone in real time.
* *Ave. PM_{10} per hour*: Averaged fine dust concentration per hour from historical data which the meteorological administration provides.
* *Ave. Temperature (per hour)*: Averaged temperature per hour from historical data which the meteorological administration provides.
* *Avg. Wind Speed (per hour)*: Averaged wind speed per hour from historical data which the meteorological administration provides.
* *Avg. Humidity (per hour)*: Averaged humidity per hour from historical data which the meteorological administration provides.

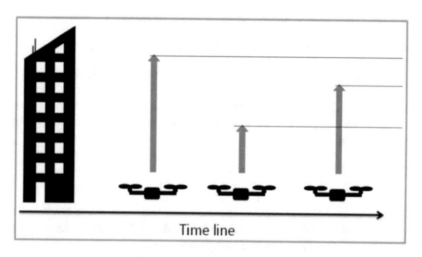

Fig. 3. Target height of drone.

flowchart of the proposed fine dust concentration measurement processes by drone. The detailed algorithm steps are as following:

Step 1. When the system is operated, input data of data sets, which were recalled from database (DB), will be entered in two different deep learning prediction models, DNN, RNN. The input data will be 'Ground fine dust concentration,' 'Ground temperature,' 'Average fine dust concentration,' 'Average temperature,' 'Wind speed,' and 'Humidity'.

Step 2. The shutdown sequence will be checked to stop the system or not.

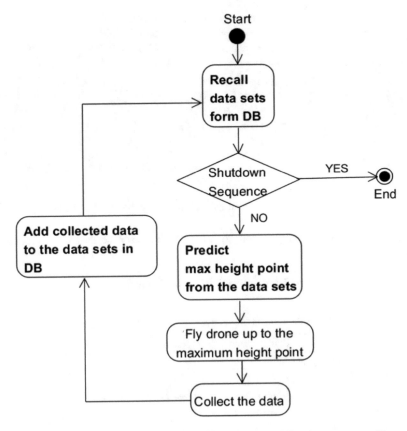

Fig. 4. Flowchart for the prediction of highest point of fine dust concentration.

Step 3. The height point of maximum fine dust concentration level is predicted by DNN and RNN, respectively.

Step 4. Drone receives the maximum height point and will fly up to the point.

Step 5. When drone flies, it will collect fine dust concentration data. The collected data are 'Average fine dust concentration of every 10 cm of height', and 'Average temperature of every 10 cm of height'.

Step 6. The collected data will be added to the data sets in DB.

3   Simulation Results

The actual data and generated data were used for training DNN and RNN models, respectively. Figure 5 shows accuracy of prediction of DNN and RNN by comparing the original data. The accuracy of DNN prediction is 72.67% on average while the one of RNN prediction has 80.61% accuracy on average. The prediction accuracy results mean that the proposed approach can be practically applied. The gap between predicted

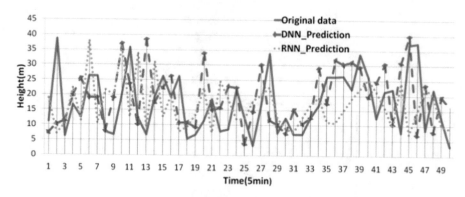

Fig. 5. Comparison of actual and predicted data by DNN and RNN.

heights (10 cm) is too narrow and so we consider error of ±40 cm. Accuracy is calculated by:

$$Predicted\ max\ height / Actual\ max\ height$$

Table 4. Cost-saving and accuracy for 10-hour flight.

	Flying distance (m)	Accuracy (%)	Saving (%)
5-min cycle	4,800 m	–	–
DNN	2,366.4 m	72.67%	50.7%
RNN	1,886.7 m	80.61%	60.69%

In Table 4, according to the predicted height by DNN, the drone flew 2,366.4 m out of 4,800 m and by RNN, it flew 1,886.7 m out of 4,800 m for 10 h of flight. Both used 5-min flying cycle of measurement. DNN max concentration height point prediction shows 50.7% saving of full flying distance, 4,800 m while RNN max concentration height point prediction saves more flying distance by 60.69%.

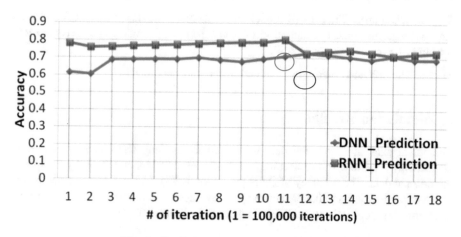

Fig. 6. Prediction accuracy vs. training time.

Figure 6 illustrates the prediction accuracy depending on training time (the number of iterations). The DNN prediction model provides the best accuracy, 72.67% at 12-time training point (12 × 100,000 = 1200,000 iterations) while the RNN prediction model shows the best one, 80.61% at 11-time training point.

4 Conclusion

In this paper we propose the deep learning prediction methods of drone flying height at which maximum fine dust concentration can be sensed. Our experimental results show that the proposed DNN and RNN prediction models can be practically applied to save energy of drone and to provide real time fine dust concentration information. It is necessary for us to obtain more real actual data with drone flight to increase reliability of experiment in the future. We will do more research on optimal measurement cycle time to have the best prediction accuracy in the future.

References

1. Chen, J., Yu, X., Sun, F., Lun, X., Fu, Y., Jia, G., Zhang, Z., Liu, X., Mo, L., Bi, H.: The concentrations and reduction of airborne particulate matter (PM10, PM2.5, PM1) at shelterbelt site in Beijing. In: Atmosphere, vol. 6, pp. 650–676 (2015). https://doi.org/10.3390/atmos6050650
2. Kim, H.T., Ho, S.: Breathing on the Ground, Half of the Fine Dust Stations are over 10 Meters. KBS News. http://mn.kbs.co.kr/news/view.do?ncd=3054574. Accessed 15 Aug 2018
3. Airkorea. Station Information. https://www.airkorea.or.kr/stationInfo. Accessed 15 Aug 2018
4. Kim, H.I.: 20 m is High' Points Out. Measurement of Fine Dust at 19.9 m and 19.5 m. Chosun News (2018). http://news.chosun.com/site/data/html_dir/2018/03/28/2018032800241.html. Accessed 18 Aug 2018
5. Baxter, R.A., Bush, H.: Use of small unmanned aerial vehicles for air quality and meteorological measurements. In: National Ambient Air Monitoring Conference 2014, Atlanta, GA, USA (2014)
6. Villa, T.F., Salimi, F., Morton, K.M., Morawska, L., Gonzalez, F.: Development and validation of a UAV based system for air pollution measurements. Sensors 16(12), 1–22 (2016). https://doi.org/10.3390/s16122202
7. Alvear, O., Zema, N.R., Natalizio, E., Calafate, C.T.: Using UAV-based systems to monitor air pollution in areas with poor accessibility. J. Adv. Transp. 2017, 1–14 (2017). https://doi.org/10.1155/2017/8204353. Article ID 8204353
8. Hwangbo, J., Sa, I., Siegwart, R., Hutter, M.: Control of a quadrotor with reinforcement learning. In: IEEE Robotics and Automation Letters, Pre-print Version (2017). https://doi.org/10.1109/cca.2010.5611206
9. Pham, H.X., La, H.M., Feil-Seifer, D., Nguyen, L.V.: Autonomous UAV navigation using reinforcement learning. In: Pre-Print Version. https://arxiv.org/abs/1801.05086 (2018). https://doi.org/10.1109/icarcv.2016.7838739
10. Airkorea. Confirmed data by sensing station. https://www.airkorea.or.kr/pastSearch. Accessed 13 Aug 2018
11. The Meteorological Administration. Surface weather observation, http://sts.kma.go.kr/jsp/home/contents/statistics/newStatisticsSearch.do?menu=SFC. Accessed 08 Aug 2018

Correction to: Emotion-Aware Educational System: The Lecturers and Students Perspectives in Malaysia

Rasheed M. Nassr, Abdulaziz Hadi Saleh, Hassan Dao,
and Md. Nazmus Saadat

Correction to:
Chapter "Emotion-Aware Educational System:
The Lecturers and Students Perspectives in Malaysia"
in: S. Lee et al. (Eds.): *Proceedings of the 13th International*
Conference on Ubiquitous Information Management
and Communication (IMCOM) 2019, **AISC 935,**
https://doi.org/10.1007/978-3-030-19063-7_49

The original version of this chapter was published without a reference to an earlier chapter. This has now been rectified and the reference has been added.

The updated version of this chapter can be found at
https://doi.org/10.1007/978-3-030-19063-7_49

© Springer Nature Switzerland AG 2019
S. Lee et al. (Eds.): IMCOM 2019, AISC 935, p. C1, 2019.
https://doi.org/10.1007/978-3-030-19063-7_89

Author Index

© Springer Nature Switzerland AG 2019
S. Lee et al. (Eds.): IMCOM 2019, AISC 935, pp. 1121–1123, 2019.
https://doi.org/10.1007/978-3-030-19063-7

Printed in the United States
By Bookmasters